Basic Pro/ENGINEER® with References to PT/Modeler

Louis Gary Lamit
De Anza College

with technical assistance provided by
James Gee, Wilbur Jorgenson, John Shull
De Anza College

PWS Publishing
An Imprint of Brooks/Cole Publishing Company
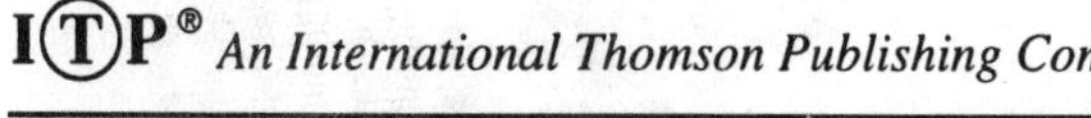
I(T)P® *An International Thomson Publishing Company*

Pacific Grove • Albany • Belmont • Bonn • Boston • Cincinnati • Detroit • Johannesburg • London • Madrid • Melbourne
Mexico City • New York • Paris • Singapore • Tokyo • Toronto • Washington

Project Development Editor: Suzanne Jeans
Marketing Team: Nathan Wilbur, Margaret Parks
Production Editor: Tom Novack
Cover Design: Roger Knox

Cover Photo: Courtesy of CADTRAIN
Cover Printing: Patterson Printing
Printing and Binding: Patterson Printing

Parametric Technology Corporation
128 Technology Drive
Waltham, MA 02154
617-398-5000

Some material in this book was provided by CADTRAIN, Inc. and is based on their CBT product for Pro/ENGINEER®.
We wish to thank CADTRAIN for the use of their product, COAch for Pro/ENGINEER.
CADTRAIN, Inc.
2429 West Coast Highway
Newport Beach, CA 92663
800-631-5757

For more information, contact PWS Publishing at Brooks/Cole Publishing Company:

BROOKS/COLE PUBLISHING COMPANY
511 Forest Lodge Road
Pacific Grove, CA 93950
USA

International Thomson Editores
Seneca 53
Col. Polanco
11560 México D. F., México

International Thomson Publishing Europe
Berkshire House 168-173
High Holborn
London WC1V 7AA
England

International Thomson Publishing GmbH
Königswinterer Strasse 418
53227 Bonn
Germany

Thomas Nelson Australia
102 Dodds Street
South Melbourne, 3205
Victoria, Australia

International Thomson Publishing Asia
60 Albert Street
#15-10 Albert Complex
Singapore 189969

Nelson Canada
1120 Birchmount Road
Scarborough, Ontario
Canada M1K 5G4

International Thomson Publishing Japan
Hirakawacho Kyowa Building, 3F
2-2-1 Hirakawacho
Chiyoda-ku, Tokyo 102
Japan

Printed in the United States of America

10 9 8 7 6 5 4 3 2 1

Library of Congress Cataloging-in-Publication Data
Lamit, Louis Gary
Basic Pro/ENGINEER with references to PT/Modeler / Louis Gary Lamit; with technical assistance provided by James Gee, Wilbur Jorgenson, and John Shull.
p. cm.
Includes index.
ISBN 0-534-95411-1
1. Computer-aided design. 2. PT/Modeler. 3. Pro/ENGINEER.
4. CAD/CAM systems—Computer programs. I. Title.
TA174.L345 1998
620'.0042'0285—dc21

98-28207
CIP

Dedication

To Dennis, for having been such a great friend all these years

About the Author

Louis Gary Lamit is the former head of the drafting department and CAD facility manager, and is currently an instructor, at De Anza College in Cupertino, California, where he teaches computer-aided drafting and design using Pro/ENGINEER.

Mr. Lamit has worked as a drafter, designer, numerical control (NC) programmer, and engineer in the automotive, aircraft, and piping industries. Most of his work experience is in the area of mechanical and piping design. Mr. Lamit started as a drafter in Detroit (as a job shopper) in the automobile industry, doing tooling, dies, jigs and fixture layout, and detailing at Koltanbar Engineering, Tool Engineering, Time Engineering, and Premier Engineering for Chrysler, Ford, AMC, and Fisher Body. Mr. Lamit has worked at Remington Arms and Pratt & Whitney Aircraft as a designer, and at Boeing Aircraft and Kollmorgan Optics as an NC programmer and aircraft engineer.

Since leaving industry, Mr. Lamit has taught at all levels (Melby Junior High School, Warren, Michigan; Carroll County Vocational Technical School, Carrollton, Georgia; Heald Engineering College, San Francisco California; Cogswell Polytechnical College, San Francisco, and Cupertino, California; Mission College, Santa Clara, California; Santa Rosa Junior College, Santa Rosa, California; Northern Kentucky University, Highland Heights, Kentucky; and De Anza College, Cupertino, California).

Mr. Lamit has written a number of textbooks, including ***Industrial Model Building***, with Engineering Model Associates, Inc. (1981), ***Piping Drafting and Design*** (1981), ***Descriptive Geometry*** (1983), and ***Pipe Fitting and Piping Handbook*** (1984) for Prentice Hall; ***Drafting for Electronics***, with Sandra Lloyd (3rd edition, 1998), and ***CADD***, with Vernon Paige (1987), were published by Charles Merrill (Macmillan-Prentice Hall Publishing). ***Technical Drawing and Design*** (1994), ***Principles of Engineering Drawing***, with Kathy Kitto (1994), ***Engineering Graphics and Design***, with Kathy Kitto (1997), and ***Engineering Graphics and Design with Graphical Analysis***, with Kathy Kitto (1997) were published by West Publishing (ITP/Delmar). ***Basic Pro/ENGINEER in 20 Lessons*** (1998) was published by PWS Publishing (ITP).

Mr. Lamit received a B.S. degree from Western Michigan University in 1970 and did Masters' work at Wayne State University and Michigan State University. He has also done graduate work at the University of California at Berkeley and holds an NC programming certificate from Boeing Aircraft.

Contents

Part Two Assemblies

Part Three Generating Drawings

Part Four Advanced Capabilities

Appendixes

Index

Preface

Pro/ENGINEER® (and PT/Modeler™) is one of the most widely used CAD/CAM software programs in the world today. This book introduces you to the basics of the program and enables you to build on these basic commands to expand your knowledge beyond the scope of the book.

The book does not attempt to cover all of Pro/E's features, only to provide an introduction to the software, make you reasonably proficient in its use, and establish a firm basis for exploring and growing with the program as you use it in your career or classroom. For information on new releases of Pro/E, as well as data files and other useful tools pertaining to this book, point your web browser to **www.pws.com/ge/lamit.html**.

The basic premise of this book is that the more parts, assemblies, and drawings you create using Pro/E (or PT/Modeler), the better you learn the software. With this in mind, each lesson introduces a new set of commands (with an incrementally harder set of parts), building on previous lessons. The parts created in the first 13 lessons are used later in the text to create assemblies (two lessons) and generate drawings (five lessons). This procedure allows you to work with actual completed parts, assemblies, and drawings in a short lesson format, instead of building large complex projects where new commands may get lost in the process.

Also new to this edition are references to ***PT/Modeler***. PT/Modeler is a less expensive, basic version of Pro/ENGINEER. Models and drawings created with PT/Modeler are upwardly compatible with Pro/ENGINEER. ***The Pro/ENGINEER® Student Collection from Addison-Wesley*** (see the **Resources** section of the front matter) is a very inexpensive version of PT/Modeler that can be purchased by the student for use in learning Pro/E. References to PT/Modeler will appear in the left-hand column of the text where PT/Modeler commands differ or are not available for the procedure.

The book is divided into one set of ten *Sections* and four *Parts*: **Part One--Creating Parts, Part Two--Assemblies, Part Three--Generating Drawings** and **Part Four--Advanced Capabilities**. Part Four-Advanced Capabilities contains one lesson. Each new edition of the text will add one or more lessons to this part. Each **part** has a variety of individual **lessons**.

Every lesson introduces a new set of commands and concepts that are applied to a *part*, an *assembly*, or a *drawing*, depending on where in the book you are working.

Lessons involve creating a new part, an assembly, or a drawing, using a set of Pro/E commands that walk you through the process step by step. Each lesson starts with a list of objectives and ends with a **lesson project**. The lesson project consists of a part, assembly, or drawing that incorporates the lesson's new material and uses and expands on previously introduced material from other lessons. Projects use planning sheets from Appendix D as tools for establishing the design intent.

The **Appendixes** contain advanced projects, a glossary, reference materials, and planning sheets.

An online tutorial from **CADTRAIN**, called **COAch for Pro/ENGINEER®**, has been referenced throughout the book. It is one of the best ways available for teaching and learning CAD/CAM software. **COAch** is used within lessons and as a reference. A sampler CD is included in the back of the text. This CD contains a variety of modules from various products offered in ***CADTRAIN's COAch for Pro/ENGINEER®***. You will need Pro/ENGINEER software installed to run the tutorial and step-by-step functions of the CD. You can view the tutorial and step-by-step instructions without performing the procedures if you have **Netscape® 3.0** or later. You can try to run PT/Modeler with the CD, but in general you will be forced to create your own reference models and assemblies. The models that come with the CD are Pro/ENGINEER models and they do not have downward-converting mobility.

After working and teaching in drafting, design, and engineering graphics-oriented areas for the last 34 years, I feel that Pro/E is one of the most comprehensive, intuitive, productive, and stimulating programs I have experienced. This book was written so that students and professionals will have an up-to-date Pro/E software manual. The book serves well as a home study guide for those wishing to expand their knowledge of Pro/E, as a training guide, or as a reference for the Pro/E user.

Pro/E runs on a **Windows® 95**, **Windows NT®**, or UNIX platform. UNIX and Windows NT are preferred for both Pro/E and PT/Modeler. PT/Modeler (student collection) *requires* Windows NT to run PT/Render.

Engineering students and professionals with technical training in Pro/E are finding that knowledge of this powerful tool can open up a whole new career path with unlimited possibilities.

If you wish to contact me concerning questions, changes, additions, suggestions, comments, and so on, please send email to one of the following:

llamit@ix.netcom.com (Lamit and Associates)
lglamit@yahoo.com (De Anza College)
http://www2.netcom.com/~llamit/small.html

Acknowledgments

I would like to thank the following people for assistance in preparing this manuscript:

- **James Gee** (checking and corrections)
- **John Shull** (technical assistance)
- **Ken Louie** (checking)
- **Wilbur Jorgenson** (checking and corrections)
- **Sanjiv Sheth** (illustrations)
- **Sunmoon Suhaimi** (parts)
- **Francis Nicholson** (checking)
- **Roenna Del Rosario** (checking)
- **Steven Koko Washington** (systems administration)
- **Harry Kenny** (parts)
- **Steven Toms** (checking)

I would also like to thank Jason Perry, Design Engineer, PPC, for valuable comments during the review of the manuscript. A special mention of Steven Toms for input on the differences between PT/Modeler and Pro/ENGINEER commands must also be made.

I also want to thank the following people and organizations for the support and materials they granted the author:

Parametric Technology Corporation for their support and timely software updates.

- **Dave Pettine** *Parametric Technology Corporation*
- **Larry Fire** *Parametric Technology Corporation*

CADTRAIN, makers of COAch for Pro/ENGINEER®, for their Pro/ENGINEER tutorial software.

- **Dennis Ftajic** CADTRAIN
- **Kathy Bennett** CADTRAIN
- **Ron Gates** CADTRAIN

Resources

A variety of information, books, online products, and job opportunities (**www.pejn.com**) are available. We have listed some of the more useful and important resources. You can also do a search on the web with PTC, Proe, Pro/E, Pro/ENGINEER, and so on as keywords. Resources include web site addresses for a variety of companies, a description of **The Pro/ENGINEER Student Collection (PT/Modeler)**, a list of books on Pro/ENGINEER, links to **InPart Design**, an internet-based company that supplies 3D models of standard parts over the Web, and **CADTRAIN**; the creator of **COAch for Pro/ENGINEER**.

Parametric Technology Corporation

Release 19.0 delivers the most comprehensive and tightly integrated applications available today to leverage the product model throughout the enterprise. With Release 19.0, PTC provides new and powerful functionality, including enhanced information management tools; improved capabilities for migration of legacy data; and dynamic tools to allow interaction with Pro/ENGINEER through a Netscape® browser.

This production-proven, fully mature release offers enhanced capabilities for leveraging both legacy data and Pro/ENGINEER product model information throughout the product development process.

See
www.ptc.com
www.ptc.com/register.htm
www.ptc.com/cs/index.htm

Access the PTC web site at **www.ptc.com**. Register for an account on the system at **www.ptc.com/register.htm**. An email address and a valid software serial number are required for you to register. After registering, you will have your own login and password to enter the customer service site for technical support at **www.ptc.com/cs/index.htm**. Explore the different sections, including the Knowledge Base.

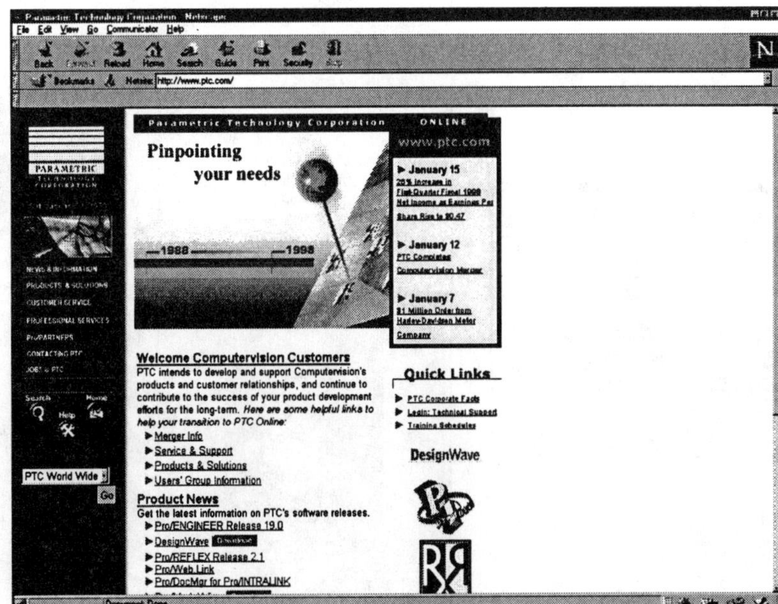

InPart DESIGN

InPart is an innovative leader in creating and publishing CAD content for the mechanical engineering community worldwide. InPart, headquartered in Saratoga, California, is the maker of **DesignSuite™**. DesignSuite is the first comprehensive design standardization solution for every designer's desktop.

See **www.inpart.com**

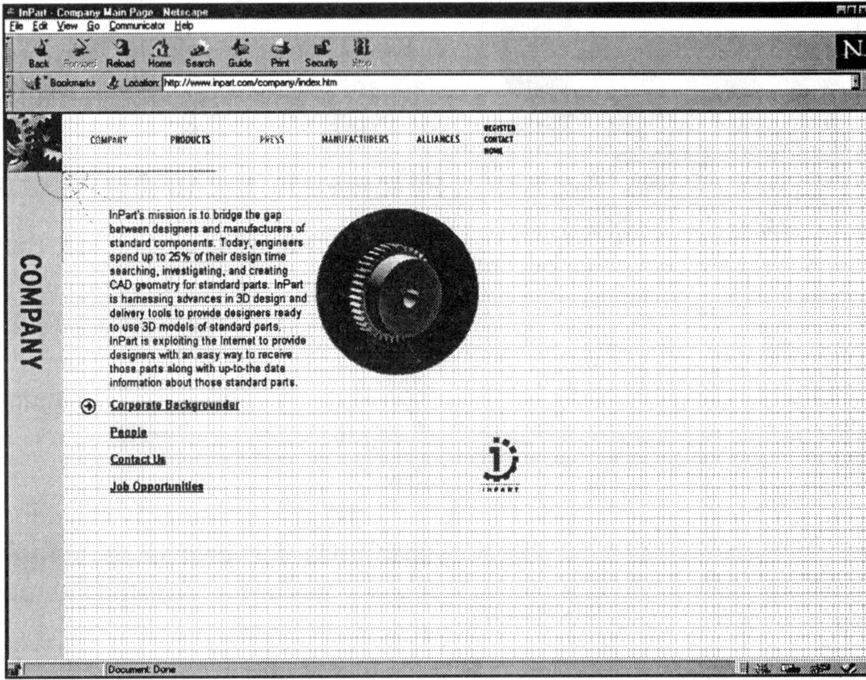

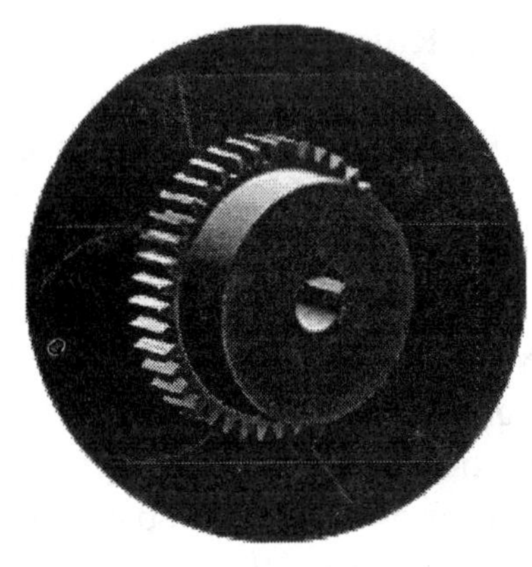

DesignSuite's powerful search engine can locate and retrieve the right standard part for your Pro/E design. Efficient search algorithms give DesignSuite an unprecedented high-performance query capability--results are returned to you within seconds.

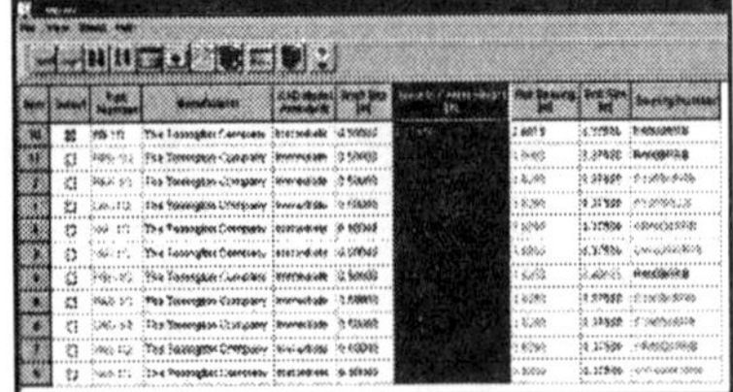

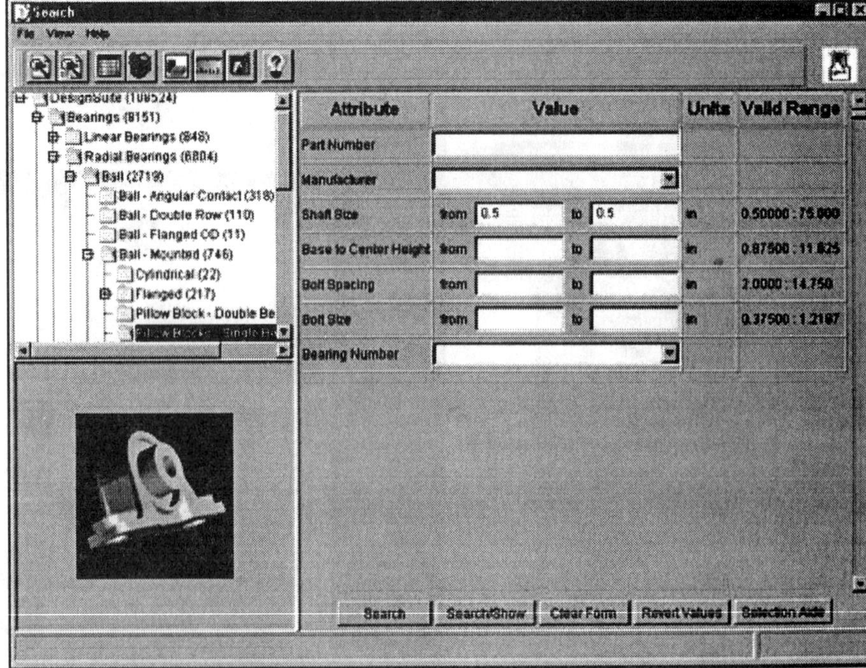

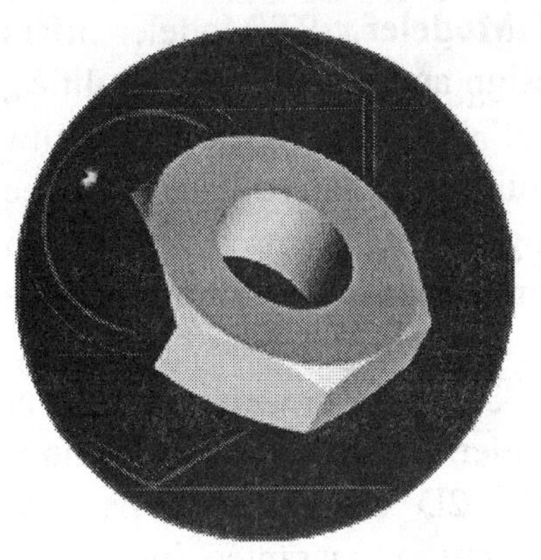

You can:
Search by part hierarchy, part graphics, and/or parametric data.
Search an extensively categorized database of vendor parts.
Search across top manufacturers in each part category.
Search a complete list of part manufacturers.
Search through only the manufacturers you select.
Search by pertinent technical specifications.
Search by a specific part number.
Search by any combination of the above.
Search for the part you need.

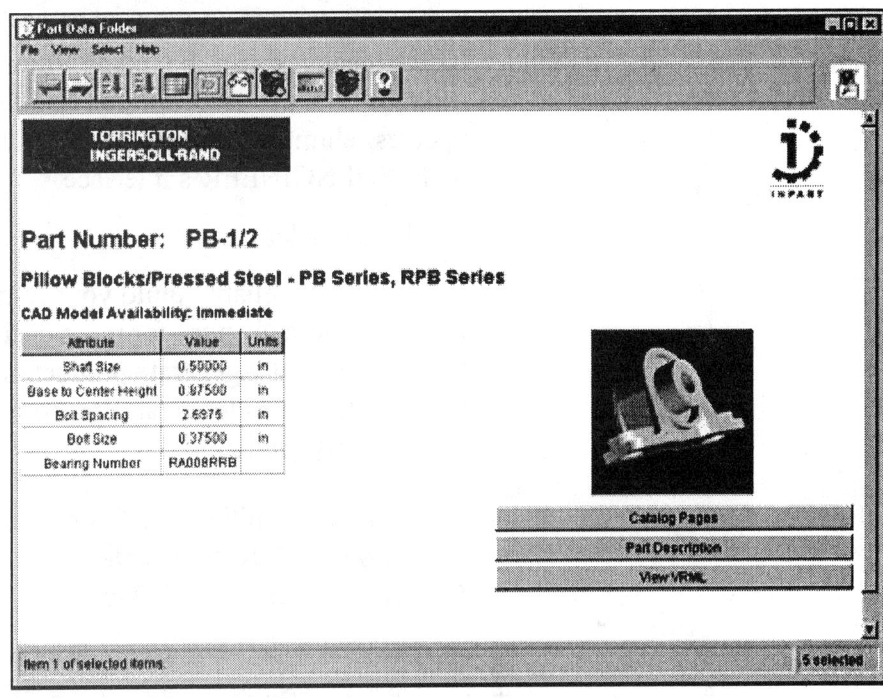

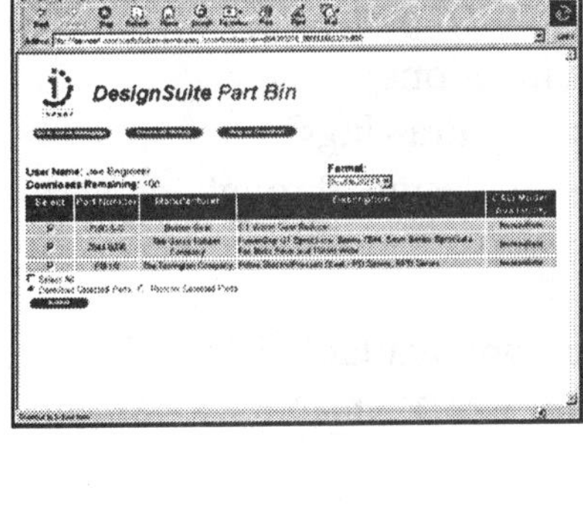

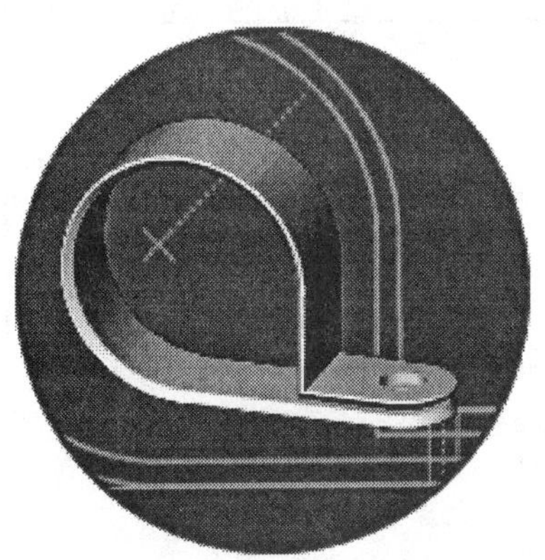

Parts are located by navigating the Part Class Tree and entering specifications in the Query Form. As the leaves in the Part Class Tree are selected, pertinent technical specifications or attributes are presented in the Query Form on the right panel. A toolbar is provided for quick access to commonly used commands.

To locate a part, select a leaf in the Part Class Tree by selecting a specific part category. As the category is selected, related technical attributes are shown in query form. Refine your search by navigating through the hierarchy of part names or by clicking on the associated image. Limit your search by entering attribute ranges. (You can easily toggle between decimal and S.I. units.) Use the Selection Aide to help you select the right part by providing information on product configuration and use. Simply click on [Insert image].

Once you've located the part or parts you need, click on Submit Query. DesignSuite sends a request to the DesignSuite servers and presents you with results.

Alert assurance option guarentees you get email updates when parts change. See **www.inpart.com.**

The Pro/ENGINEER Student Collection (PT/Modeler)

This text can also be used to learn **PT/Modeler**. PT/Modeler offers parametric, feature-based modeling, design and assembly capabilities, production drawings, visualization, and data import tools. All this functionality is contained in a single fully associative database. The user interface is based on common engineering terms coupled with menus derived from the Windows environment, making the interface natural and intuitive.

Starting with familiar 2D sketches, 3D solid models are quickly built using PT/Modeler. Since PT/Modeler is parametric and feature-based, changes can be made from the 2D environment, part, or assembly. Full associativity and integration guarantee immediate updates, eliminating mistakes. PT/Modeler's interface is compatible with Pro/ENGINEER's interface.

PT/Modeler can:

- Easily and reliably build your fully detailed part
- Make rapid changes for more design iterations
- Automatically generate production-ready drawings
- Facilitate top-down and/or bottom-up design through powerful assembly modeling

Some capabilities of PT/Modeler are not available in the Student Collection. **This text references the Pro/ENGINEER *Student Collection* version of PT/Modeler.**

See
www.aw.com/cseng/cad/ptc

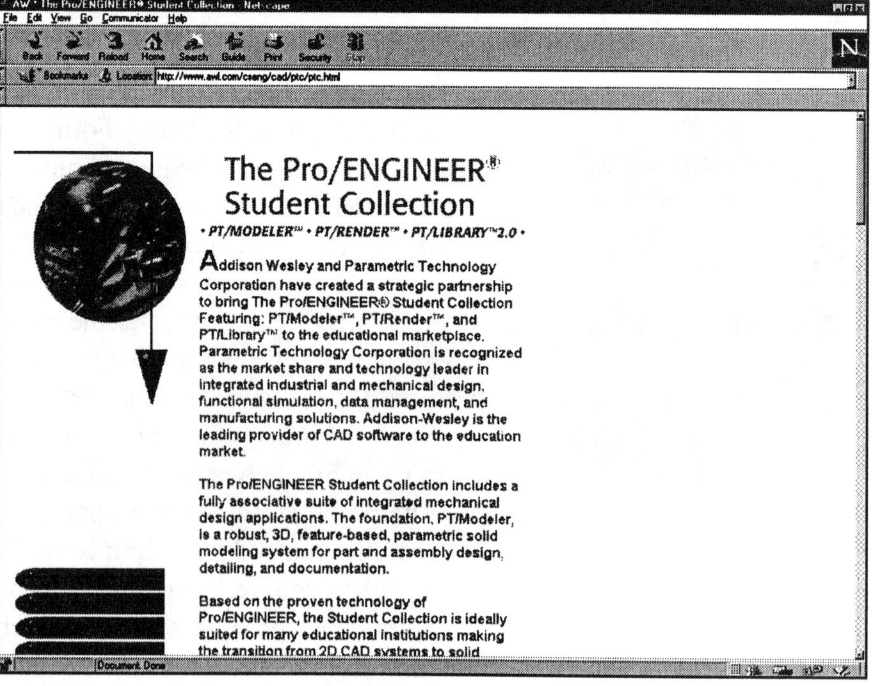

See www.cadtrain.com

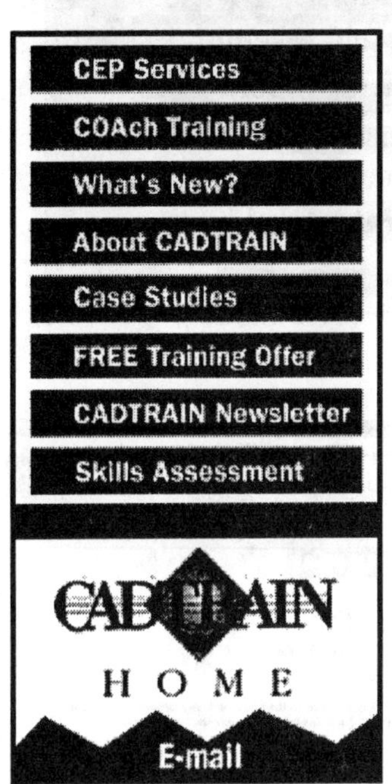

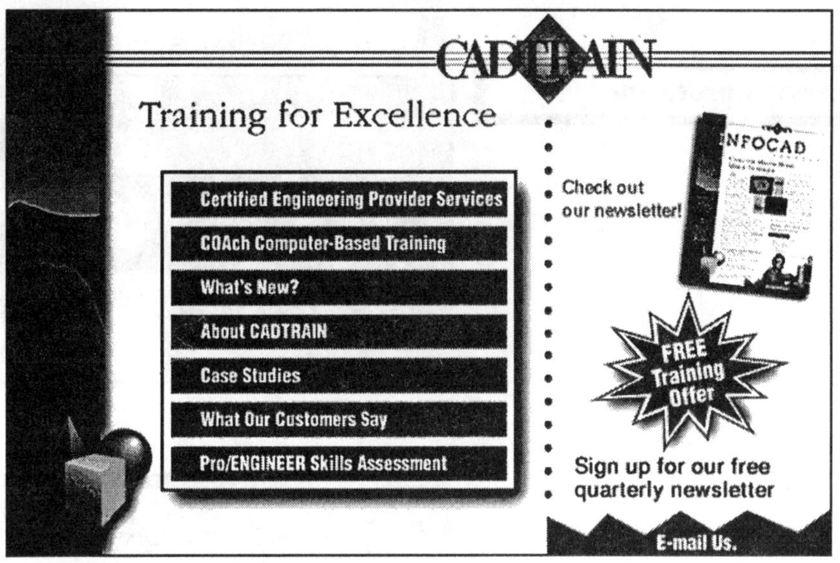

CADTRAIN's COAch for Pro/ENGINEER

COAch is a computer-based training (CBT) product designed to provide a comprehensive and affordable training program for Pro/ENGINEER users in their actual CAD environment. This self-paced onscreen training tool enables engineers, designers, drafters, and NC programmers to customize their training experience by following the learning sequence best suited to their individual needs. Working in an interactive environment, students are introduced to key concepts, are guided through demonstration exercises, test questions, and complete comprehensive modeling projects.

COAch runs interactively with Pro/ENGINEER; not PT/Modeler. If you do not have Pro/ENGINEER, view the COAch CD that accompanies this text, using Netscape, and use the information to learn Pro/E.

NOTE

The CD provided at the back of this text contains a sampler of CADTRAIN's COAch for Pro/ENGINEER and includes the following modules:

- **Modeling**
  - **General Interaction**
  - **What Is a Parametric Model**
  - **Basic Modifications to a Part**
  - **Using Datums**
- **Drawings**
  - **Creating Drawings**
  - **Detailing**
- **Assemblies**
  - **Overview of Assemblies**
  - **Creating an Assembly**
- **Super Users**
  - **Advanced Modeling**
    - **User-Defined Features**
    - **Family Tables**

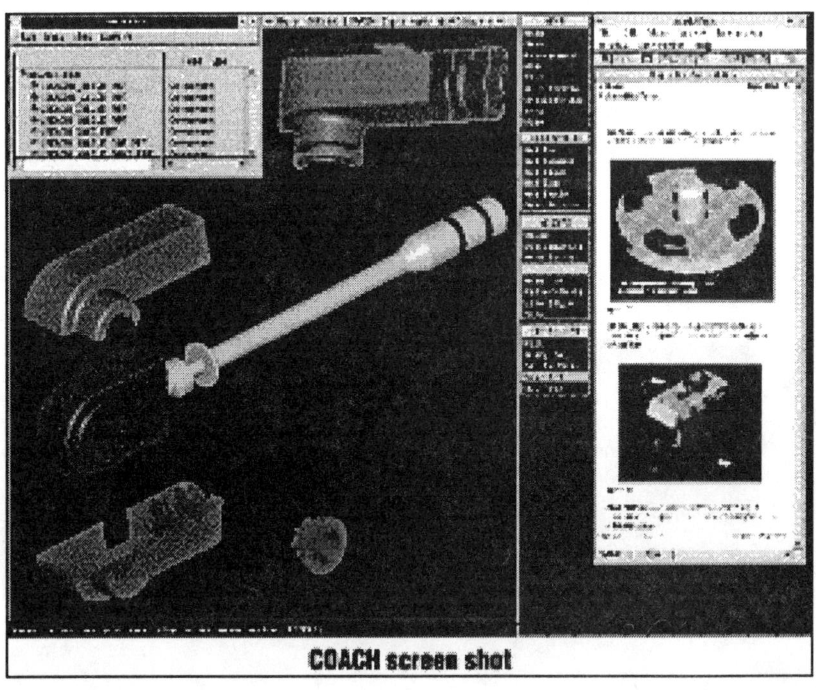

COACH screen shot

See www.proe.com

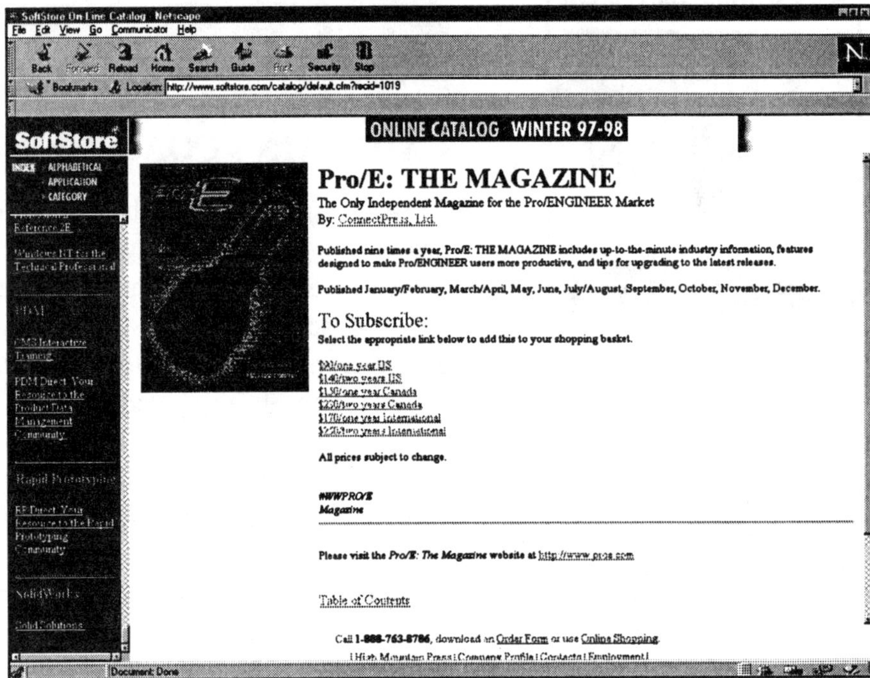

Pro/E: THE MAGAZINE

An independent magazine for the Pro/ENGINEER market by ConnectPress, Ltd. Published nine times a year, Pro/E: THE MAGAZINE includes up-to-the-minute industry information, features designed to make Pro/ENGINEER users more productive, and tips for upgrading to the latest releases.

Books on Pro/ENGINEER

Automating Design in Pro/ENGINEER With Pro/Program: The Professional User's Guide to Programming With Pro/Program
Mark Henault (Editor) et al. / Paperback / Published 1996

Inside Pro/ENGINEER: The Professional User's Guide to Designing With Pro/ENGINEER Book and Disk
James Utz, W. Robert Cox / Paperback / Published 1995

Inside Pro/Surface
Norm Ladouceur, Norman Ladouceur / Paperback / Published 1997

Pro/ENGINEER Exercise Book
Bill Paul / Paperback / Published 1995

Pro/ENGINEER Tips and Techniques
Tim McLellan, Fred Karam / Paperback / Published 1996

Thinking Pro/ENGINEER: Mastering Design Methodology
David Bigelow / Paperback / Published 1995

Inside Pro/ENGINEER: The Professional User's Guide to Pro/ENGINEER
Dennis Steffen / Paperback / Published 1997

Pro/ENGINEER Quick Reference
Robert Monat / Published 1995

Design Modeling Using Pro/ENGINEER
James E. Bolluyt / Paperback

Introduction to Pro/ENGINEER
Jeffrey S. Freeman, Andrew E. Whelan / Paperback / Published 1997

The Pro/ENGINEER Exercise Book: Optional Instructor's Guide
Onword Press Development Team / Paperback / Published 1993

Pro/ENGINEER in Practice
David Bigelow / Paperback / Published 1998

Pro/ENGINEER Tutorials (Release 18)
Roger Toogood / Paperback / Published 1997

Solid Modelling With Pro/ENGINEER
Clarence W. Mayott, Geraldine B. Milano / Hardcover / Published 1993

See www.pejn.com

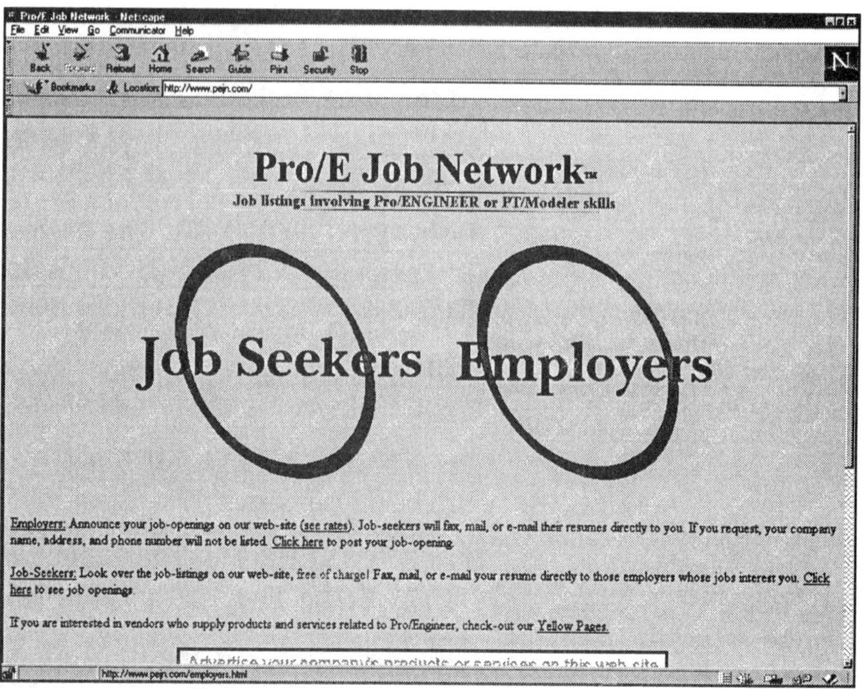

Pro/ENGINEER and PT/Modeler Job Network

A variety of services are available over the Internet:

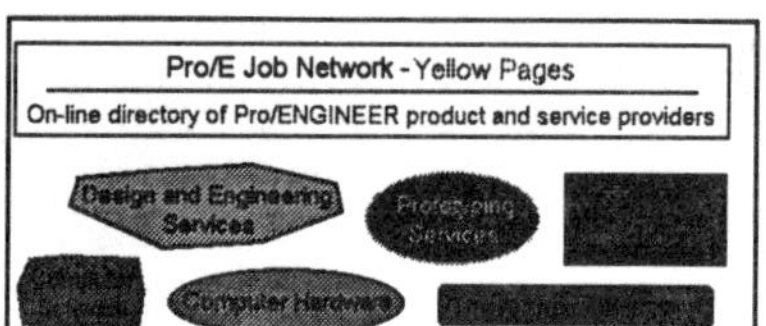

Employers: Announce your job-openings on our web site. Job seekers will fax, mail, or email their resumes directly to you. If you request it, your company name, address, and phone number will not be listed.

Job Seekers: Look over the job listings on our web site, free of charge. Fax, mail, or email your resume directly to those employers whose jobs interest you.

To see a list of jobs, pick on a section of the country where you are interested in working

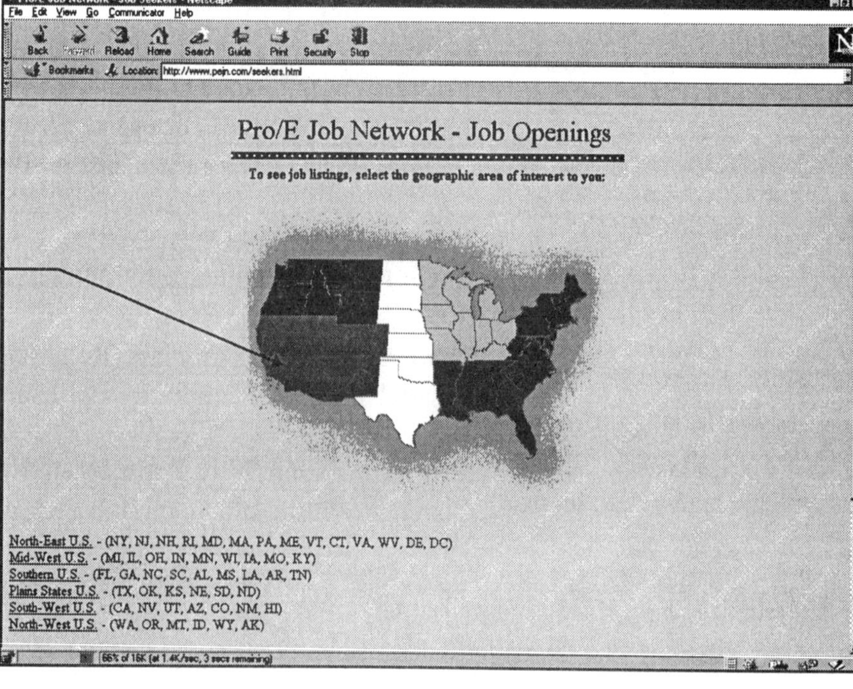

PT/Modeler

See
www.aw.com/cseng/cad/ptc

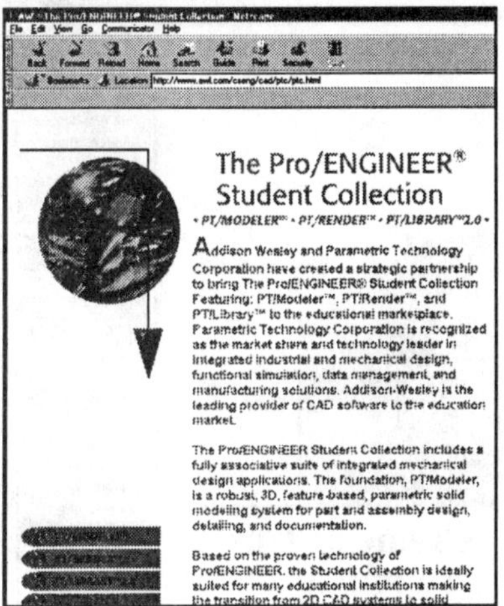

The Pro/ENGINEER Student Collection (PT/Modeler)

This textbook can also be used to learn **PT/Modeler**. PT/Modeler offers parametric, feature-based modeling, design and assembly capabilities, production drawings, visualization, and data import tools. This functionality is contained in a single fully associative database. The user interface is based on common engineering terms coupled with menus derived from the Windows environment, making the interface natural and intuitive. Remember, a few capabilities of PT/Modeler are not available in the Student Collection. **This text references the *Student Collection* version of PT/Modeler.**

The available modes on your system are determined by what licenses (options) were purchased with **Pro/ENGINEER** or **PT/Modeler**. With **The Pro/ENGINEER Student Collection,** only four modes are provided. **PT/Modeler**, when purchased as a full product, has the same modes as the Student Collection, but can also be purchased with other options, such as PT/Sheetmetal and PT/Manufacturing.

Pro/ENGINEER

NOTE

If you are using this text with the purchased version of PT/Modeler (not the Student Collection), you may see slightly different and expanded capabilities but this text is compatible with all three products--***Pro/ENGINEER, PT/Modeler,*** and ***The Pro/ENGINEER Student Collection.***

PT/Modeler
(The Pro/ENGINEER Student Collection)

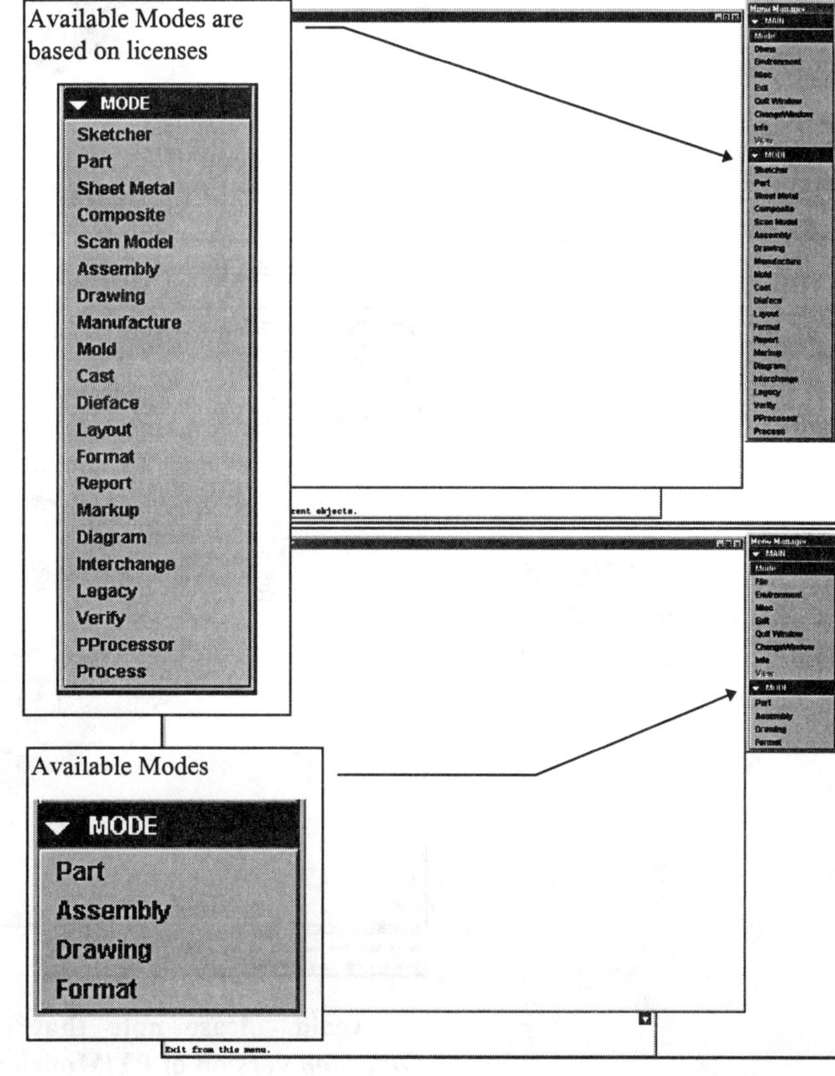

Although PT/Modeler's (and the Student Collection) interface is compatible with Pro/ENGINEER's interface, there are some differences in the command sequence. A few of PT/Modeler's commands are truncated or slightly simplified. As an example:

Feature ⇒ Create ⇒ Cut *becomes* **Feature ⇒ Cut**

The menu structure also reflects these differences, as shown in the next two illustrations.

Pro/ENGINEER

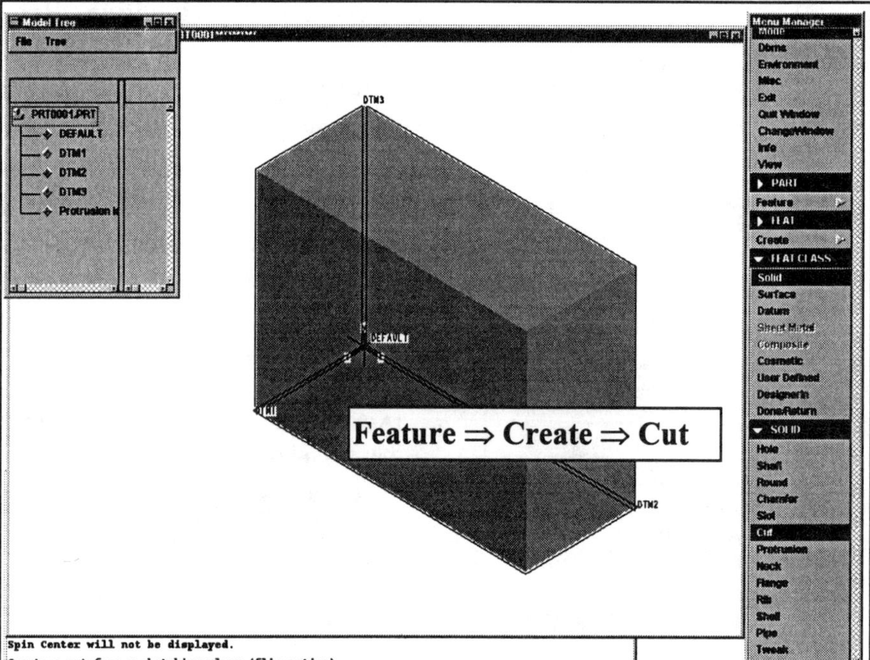

PT/Modeler
(The Pro/ENGINEER Student Collection)

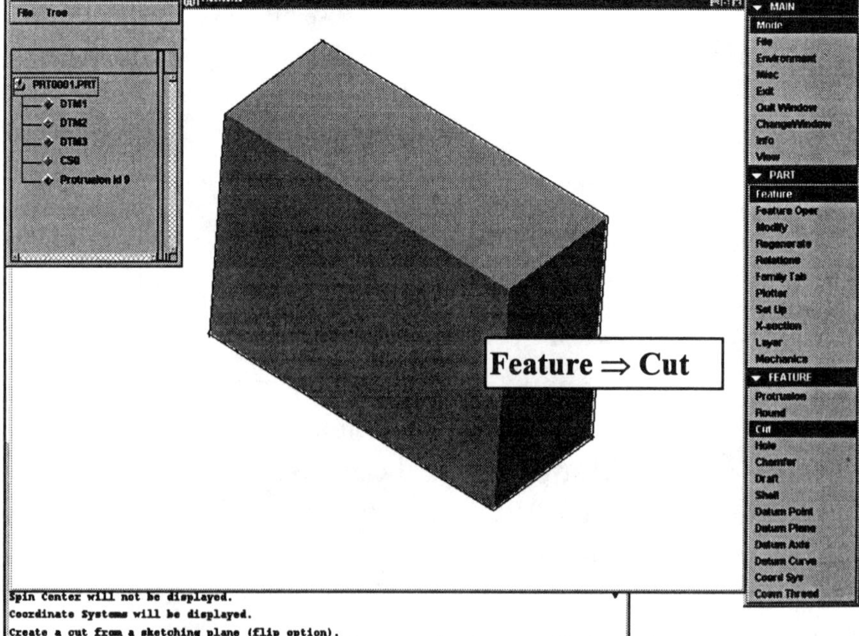

Again, please note that this text references the *Student Collection* version of PT/Modeler, not the purchased PT/Modeler.

Sections

Section 1 Introduction
Section 2 Using the Text
Section 3 Fundamentals
Section 4 Utilities
Section 5 Configuration Files and Mapkeys
Section 6 Environment
Section 7 Setup and Information
Section 8 Layers
Section 9 The Sketcher
Section 10 Interface

The following ten sections cover the basics of Pro/ENGINEER. PT/Modeler differences are *not* noted. Lessons 1-21 provide PT/Modeler commands where they differ from those of Pro/ENGINEER.

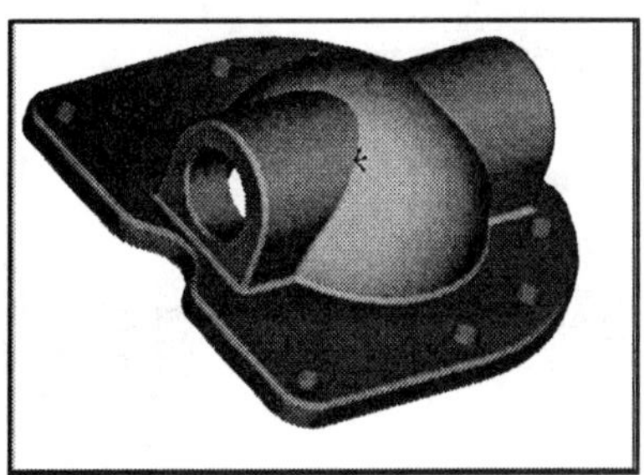

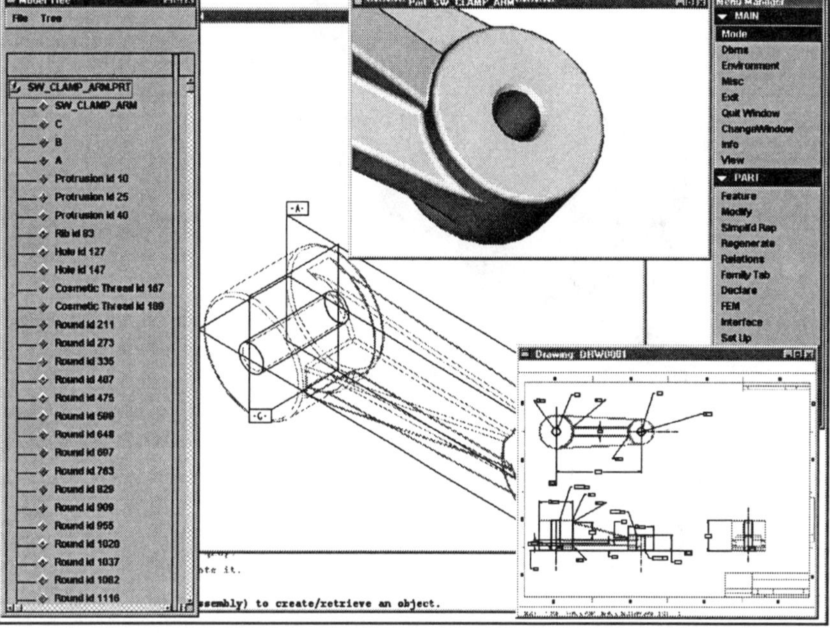

Part Design

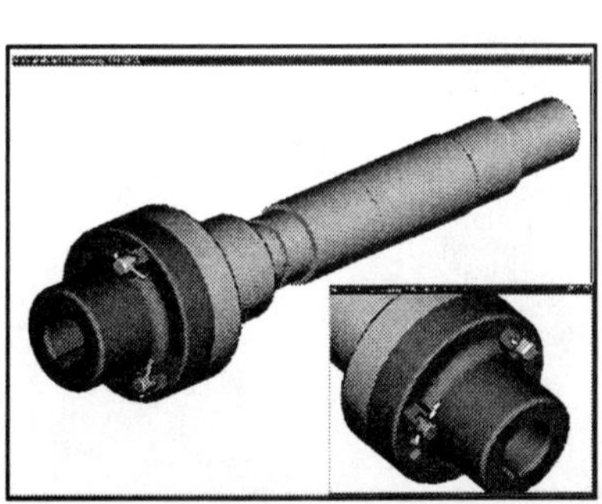

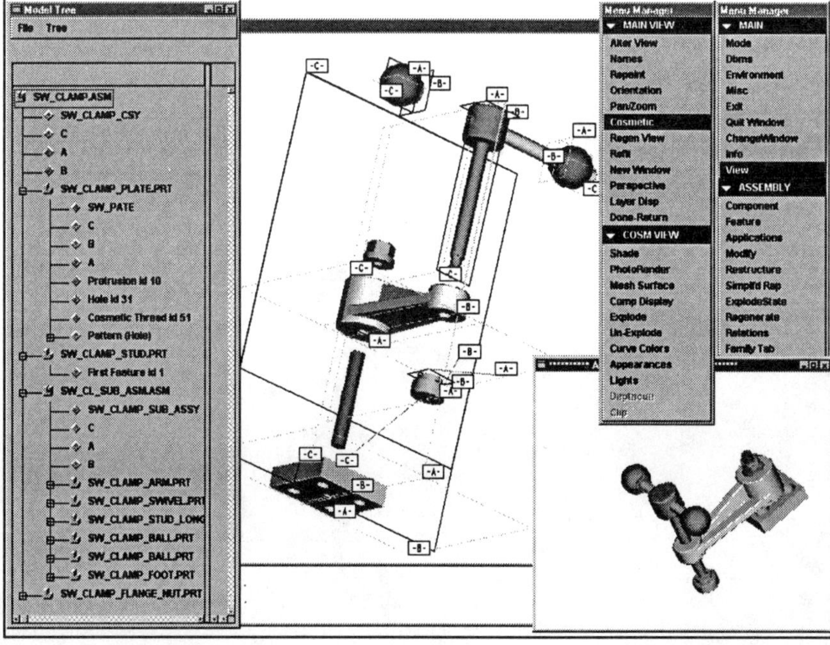

Assembly Design

Assembly, Part, and Drawing

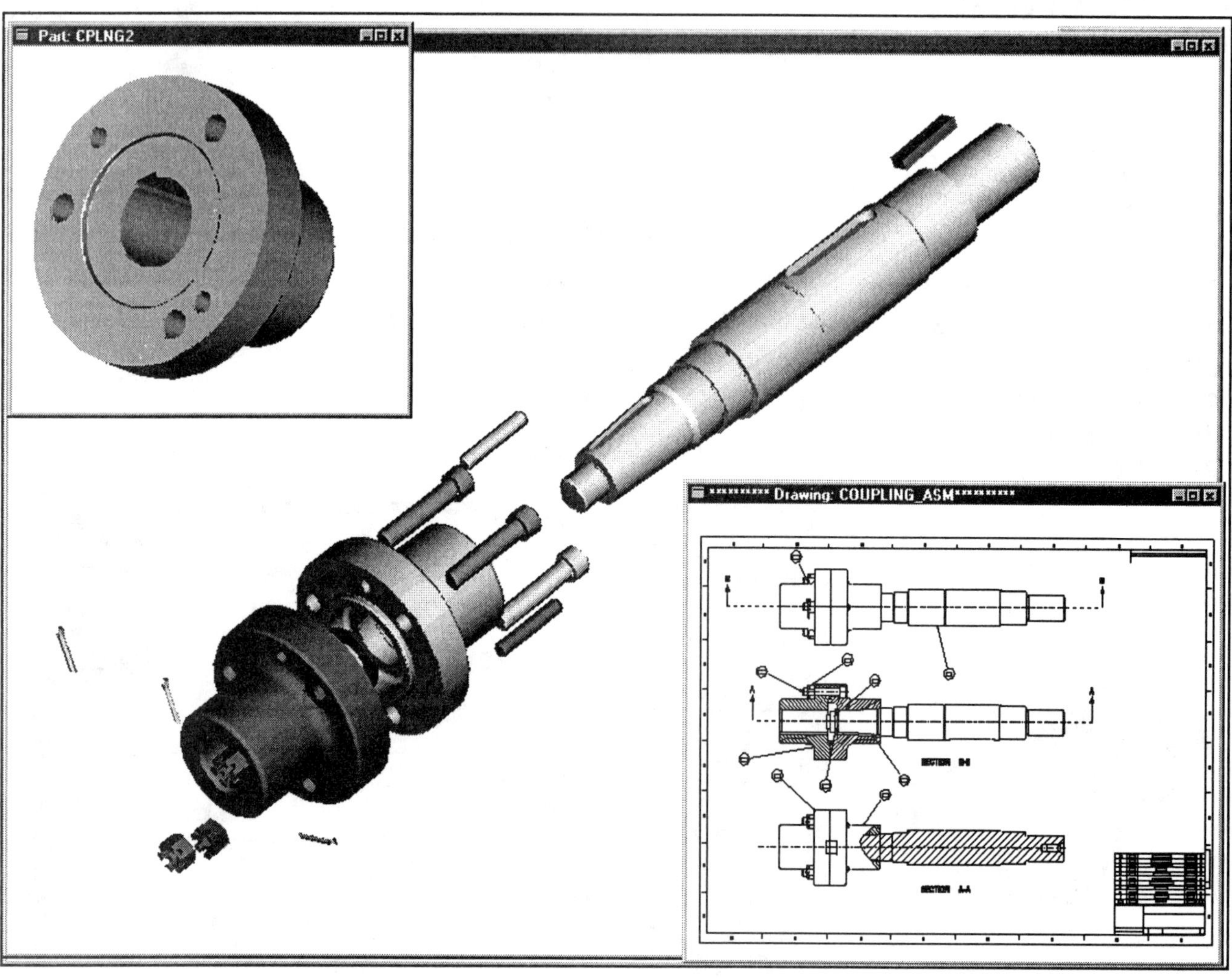

Section 1

Introduction

This work text introduces the basic concepts of parametric design using Pro/ENGINEER (Pro/E) to create and document individual parts, assemblies, and drawings. **Parametric** can be defined as *any set of physical properties whose values determine the characteristics or behavior of an object.* **Parametric design** enables you to generate a variety of information about your design: its mass properties, a drawing, or a base model. To get this information, you must first model your part design (Fig. 1.1).

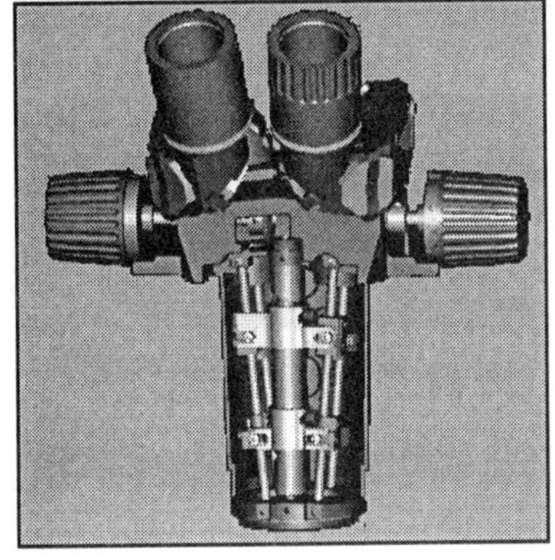

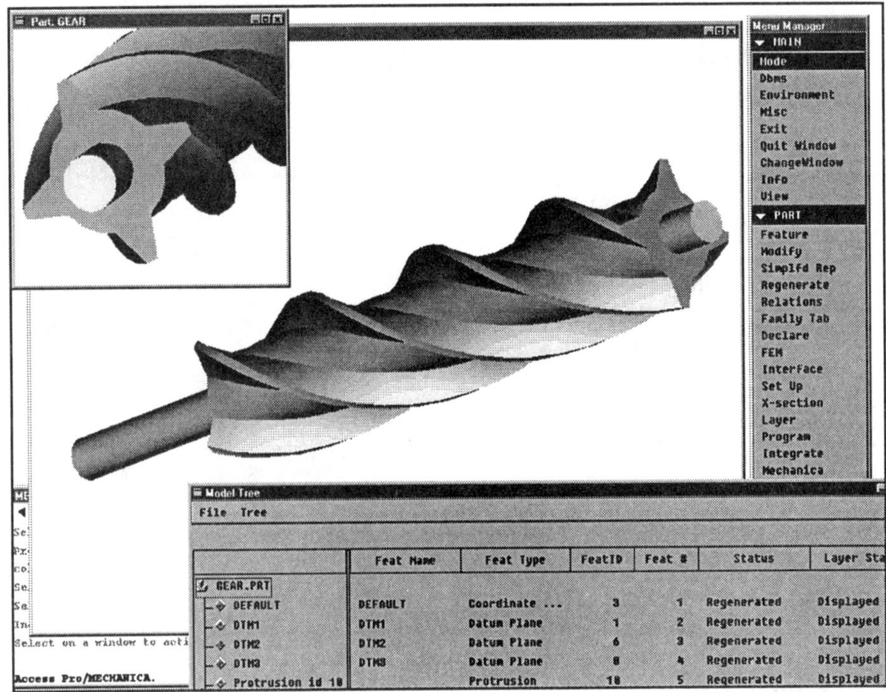

Figure 1.1
Part Design

This section is intended to introduce you to parametric modeling philosophies used in Pro/E, including:

Feature-Based Modeling Parametric design represents solid models as combinations of engineering features (Fig. 1.2).

Creation of Assemblies Just as features are combined into parts, parts may be combined into assemblies, as shown in Figure 1.3.

Capturing Design Intent The ability to incorporate engineering knowledge successfully into the solid model is an essential aspect of parametric modeling (Fig. 1.4).

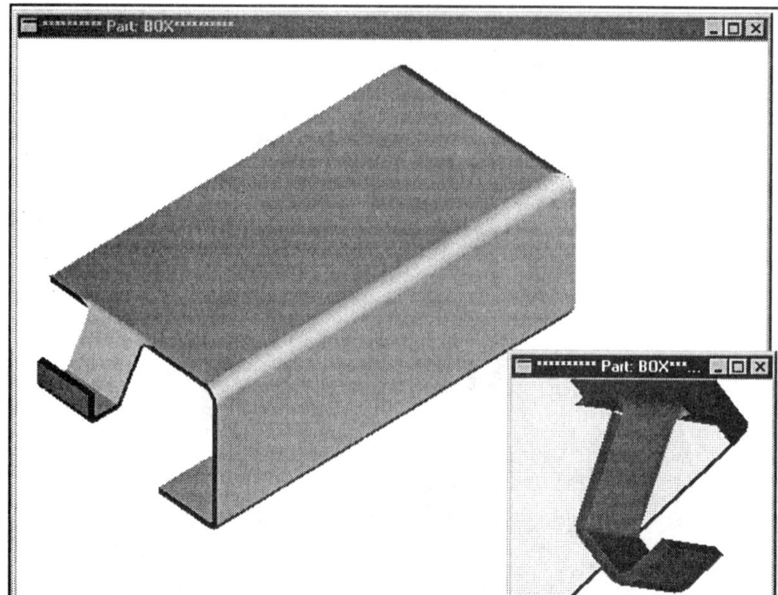

Figure 1.2
Feature-Based Modeling

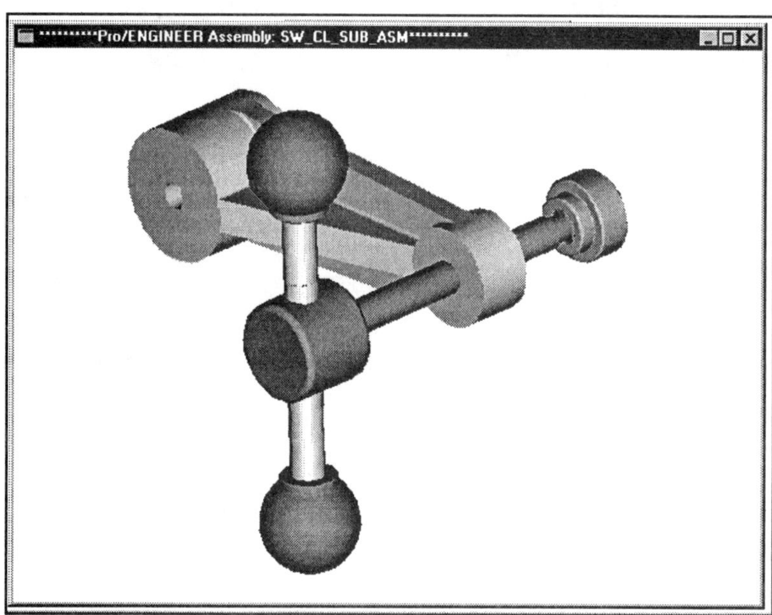

Figure 1.3
Creation of Assemblies

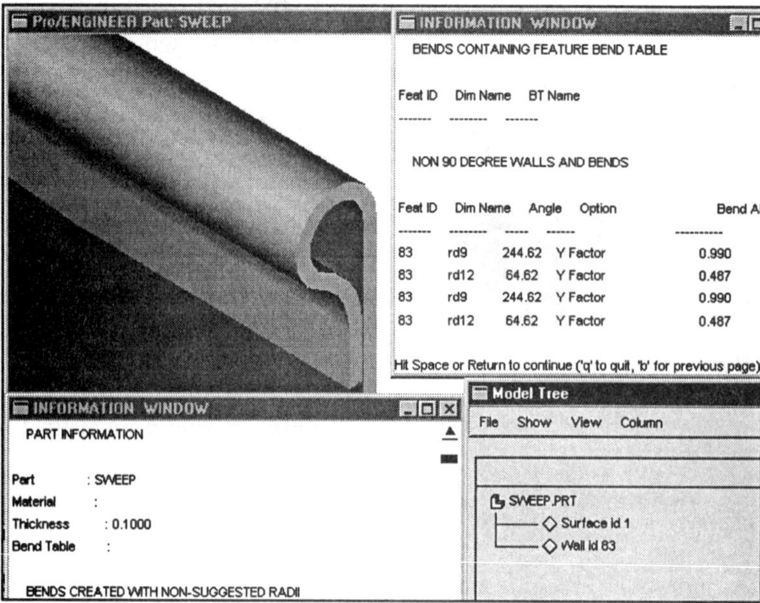

Figure 1.4
Capturing Design Intent

These methodologies are the principal aspects of successful parametric solid modeling. Figure 1.5 illustrates the role of each in the modeling process.

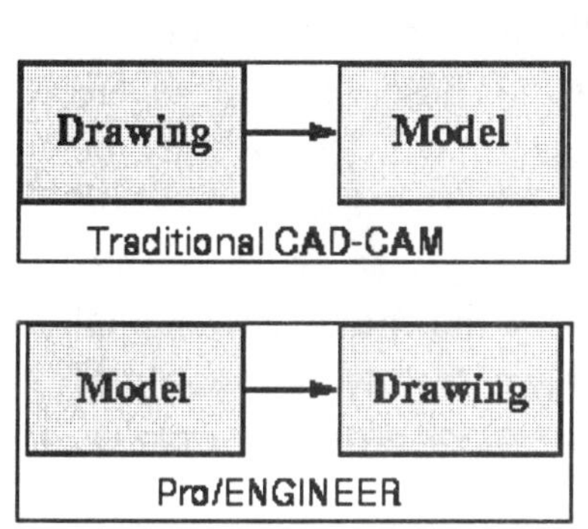

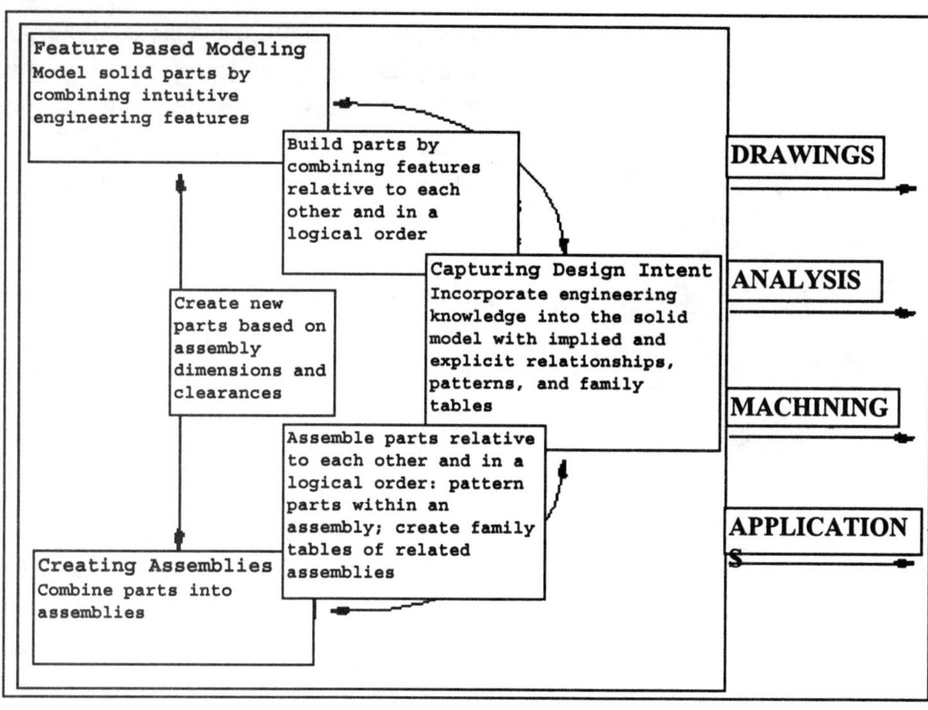

Figure 1.5
Parametric Design Methodology

Modeling vs. Drafting

A primary and essential difference between parametric design and traditional computer-aided drafting packages is that parametric design models are three-dimensional. Increasingly, designs are represented in the form of solid models that capture design intent as well as design geometry. Engineering designs today are constructed as mathematical solid models instead of as 2D drawings. A solid model is one that represents a shape as a 3D object that has volume and mass properties.

There are two main reasons for the move to solid models. First, solid modeling software packages can serve as an easy means of portraying parts for study by cross-functional concurrent-engineering teams. The solid model can be understood even by nontechnical members of the team, such as those from the marketing and sales departments.

Second, the capabilities of solid modelers have been upgraded so that the model can represent not only the geometry of the part being designed, but also the intent of the designer. This facility is most significant when the designer needs to make changes to the part geometry.

The designer must make far fewer changes in later-generation parametric solid models that capture design intent than in previous CAD/CAM modeling software. In parametric design, drawings are produced as views of the 3D model, rather than the other way around.

Parametric design models are not drawn so much as they are *sculpted* from solid volumes of materials.

Parametric Design Overview

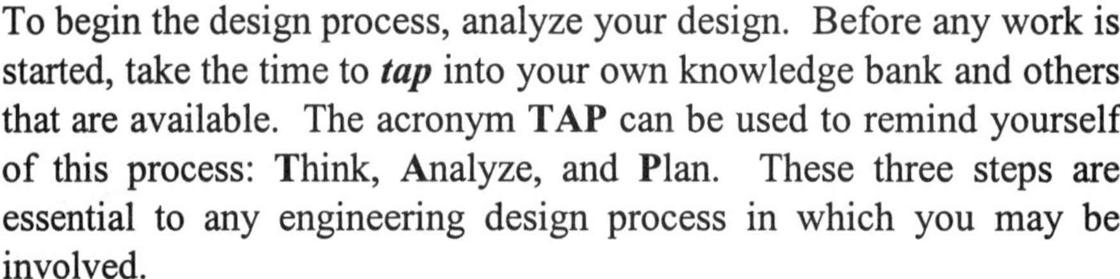

To begin the design process, analyze your design. Before any work is started, take the time to ***tap*** into your own knowledge bank and others that are available. The acronym **TAP** can be used to remind yourself of this process: **T**hink, **A**nalyze, and **P**lan. These three steps are essential to any engineering design process in which you may be involved.

Break down your overall design into its basic components, building blocks, or primary features. Identify the most fundamental feature of the part to sketch as the first, or base, feature. A variety of **base features** can be modeled using *protrusion-extrude, revolve, sweep,* and *blend* commands.

Sketched features (*neck, flange,* and *cut*) and pick-and-place features called **referenced features** (*holes, rounds,* and *chamfers*) are normally required to complete the design. With the **SKETCHER,** you use familiar 2D entities (points, lines, circles, arcs, splines, and conics). There is no need to be concerned with the accuracy of the sketch. Lines can be at differing angles, arcs and circles can have unequal radii, and features can be sketched with no regard for the actual parts' dimensions. In fact, exaggerating the difference between entities that are similar but not exactly the same is actually a far better practice when using the SKETCHER.

The software enables you to apply logical geometric constraints to the sketch. **Constraints** clean up the sketch geometry according to the software assumptions. **Geometry assumptions** and constraints close ends of connected lines, align parallel lines, and snap sketched lines to horizontal and vertical orthogonal orientations. Additional constraints are added by means of **parametric dimensions** to control the size and shape of the sketch.

Feature-Based Modeling

Features are the basic building blocks you use to create a part. Features "understand" their fit and function as though "smarts" were built into the features themselves. For example, a hole, neck, or cut feature "knows" its shape and location and the fact that it has a negative volume. As you modify a feature, the entire part automatically updates after regeneration. The idea behind feature-based modeling is that the designer constructs a part so that it is composed of individual features that describe the way the geometry is supposed to behave if its dimensions change. This happens quite often in industry, as in the case of a design change.

Parametric modeling is the term used to describe the capturing of design operations as they take place, as well as future modifications and editing of the design (Fig. 1.6). The order of the design operations is significant. Suppose a designer specifies that two surfaces are parallel such that surface 2 is parallel to surface 1. Therefore, if surface 1 moves, surface 2 moves along with it to maintain the specified design relationship. Surface 2 is a **child** of surface 1 in this example. Parametric modelers allow the designer to **reorder** the steps in the part's creation.

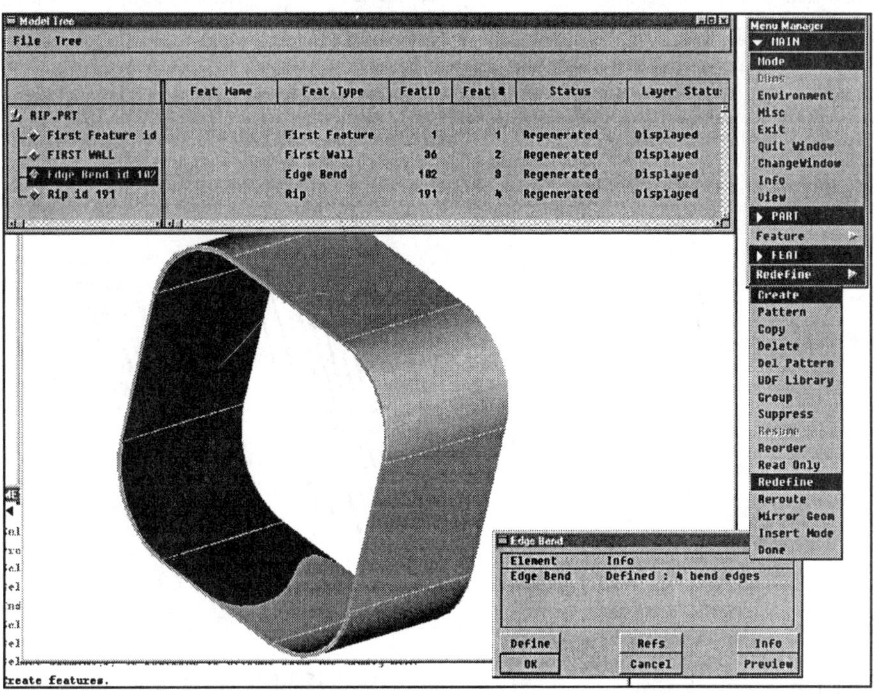

Figure 1.6
Editing a Design

The "chunks" of solid material from which parametric design models are constructed are called **features**. Features generally fall into one of the following categories:

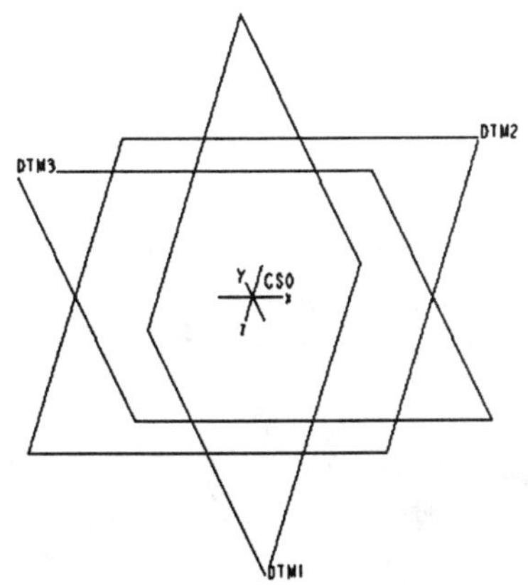

Base Feature The base feature may be either a sketched feature or datum plane(s) referencing the default coordinate system. The base feature is important because all future model geometry will reference this feature directly or indirectly; it becomes the root feature. Changes to the base feature will affect the geometry of the entire model (Fig. 1.7).

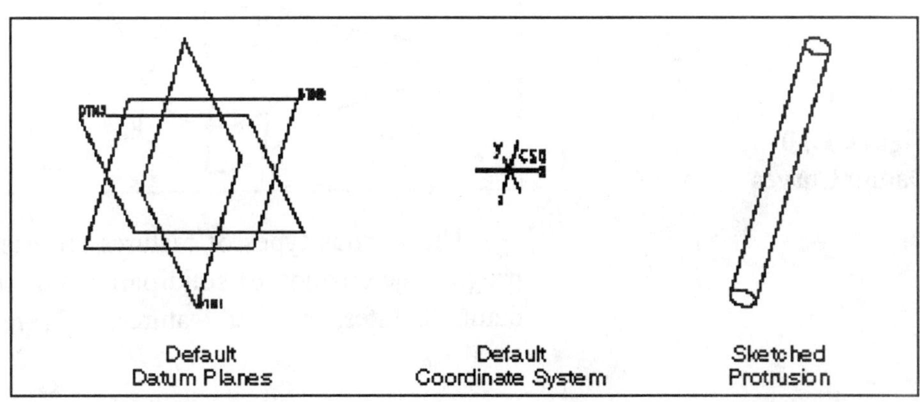

Figure 1.7
Base Features

Sketched Features Sketched features are created by extruding, revolving, blending, or sweeping a sketched cross section. Material may be added or removed by protruding or cutting the feature from the existing model (Fig. 1.8).

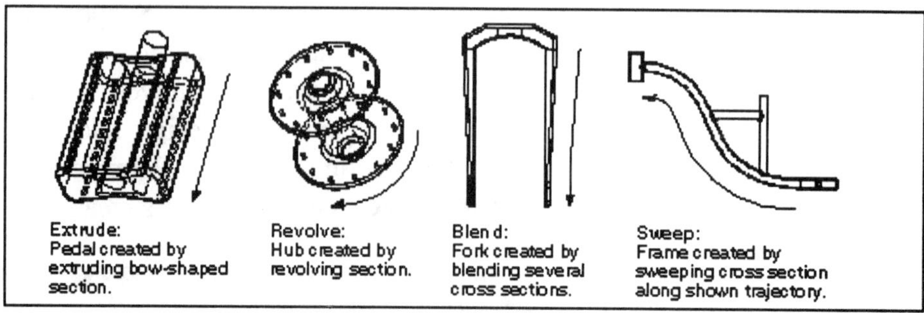

Figure 1.8
Sketched Features

Referenced Features Referenced features reference existing geometry and employ an inherent form; they do not need to be sketched. Some examples of referenced features are rounds, drilled holes, and shells (Fig. 1.9).

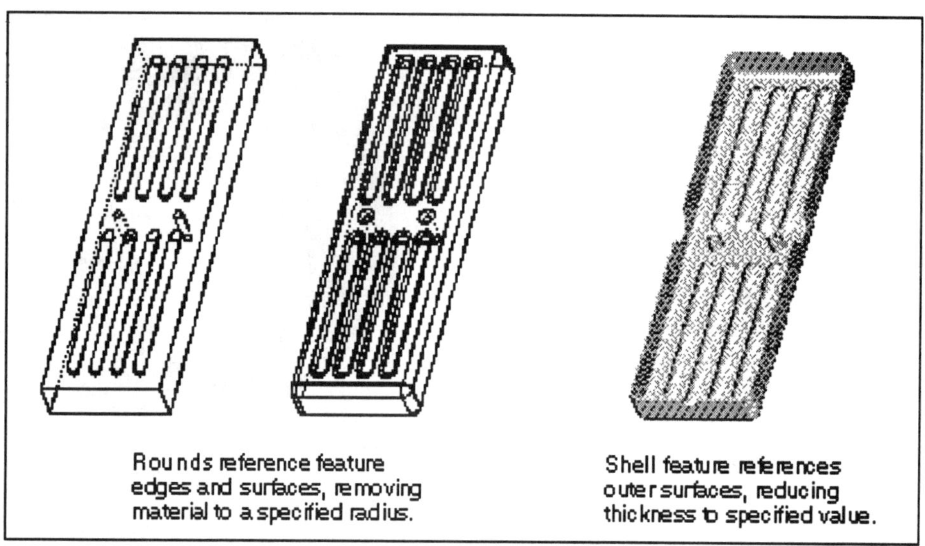

Figure 1.9
Referenced Features

Datum Features Datum features, such as planes, axes, curves, and points, are generally used to provide sketching planes and contour references for sketched and referenced features. Datum features do not have physical volume or mass and may be visually hidden without affecting solid geometry (Fig. 1.10).

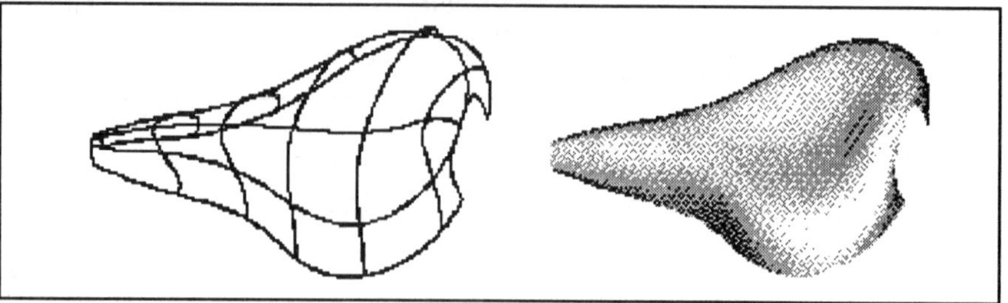

Figure 1.10
Datum Curves

The various types of features are used as building blocks in the progressive creation of solid parts. Figure 1.11 shows base features, datum features, sketched features, and referenced features.

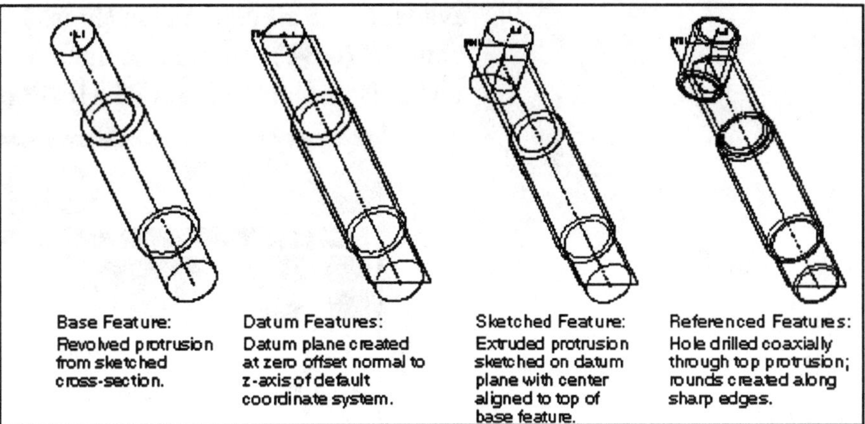

Figure 1.11
Feature Types

Conclusion

As you progress through this text's lessons, you will sometimes be prompted to return to a referenced section to review it. At this point in the process of learning Pro/E, you cannot expect to understand and apply every concept involved in this very complex software. The sections in the text are really references that are used throughout each lesson's step-by-step construction and each lesson's project. You will become more familiar with the concepts and capabilities of Pro/E as you complete each project. Reading the reference sections will help you improve this understanding, but nothing takes the place of actual practice. In reality, the lessons are presented in a building-block format that increases your knowledge incrementally and without unnecessary and complicated discussions. Feel free to return to the various sections to confirm material that you encounter throughout the lessons.

Besides consulting the sections, you should get comfortable with the online documentation available in Pro/E. **Pro/HELP** can be accessed at any time during a working Pro/E session (Fig. 1.12).

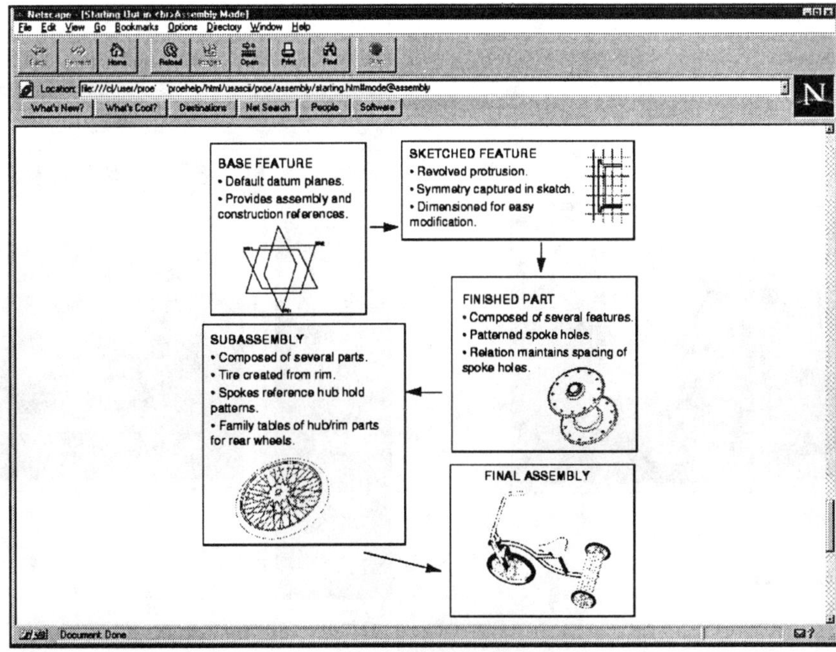

Figure 1.12
Pro/HELP

When working on PT/Modeler (or the Student Collection), help is available for all commands (Figs. 1.13 and 1.14). A computer-based training (CBT) program comes with the Student Collection (Fig. 1.15). Both the CBT and PT/HELP run with Netscape® 3.0 or later.

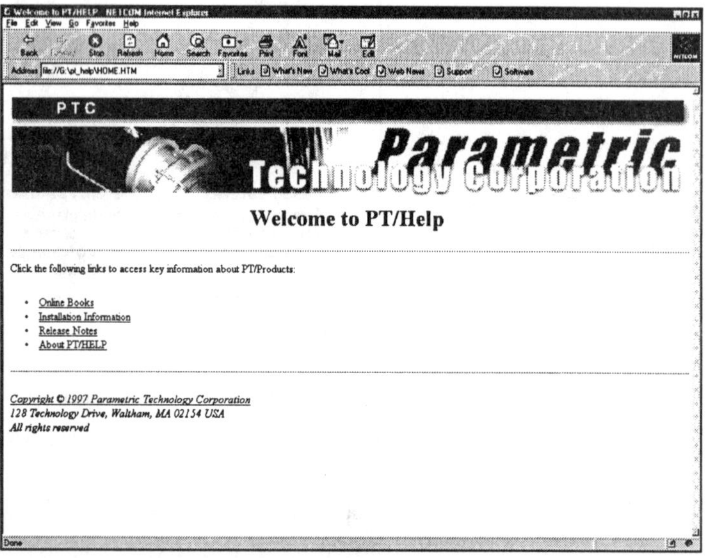

Figure 1.13
PT/HELP

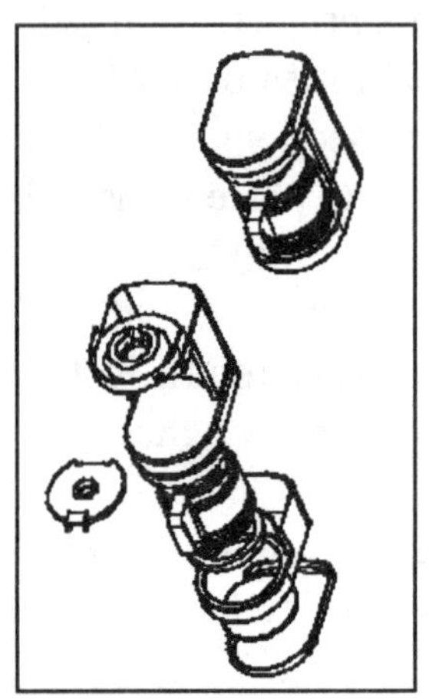

In a similar way you can select Child Refs from the Parent/Child menu and show child references for any selected component. Ref Info from the Child Refs menu lists this information in an information window.

Bill of Materials

As you add components into an assembly, PT/Modeler maintains a quantitative list of the components in the assembly. This is called a "bill of material" (BOM). You can list the BOM by choosing Info, BOM. PT/Modeler displays an information window listing all the parts and their quantities. PT/Modeler simultaneously creates a text file copy of this list on disk with the name *assemblyname.bom.1*.

The following illustration shows a typical BOM.

A Bill of Materials listing for a Coffee Maker

Note:

Figure 1.14
PT/HELP for Assemblies

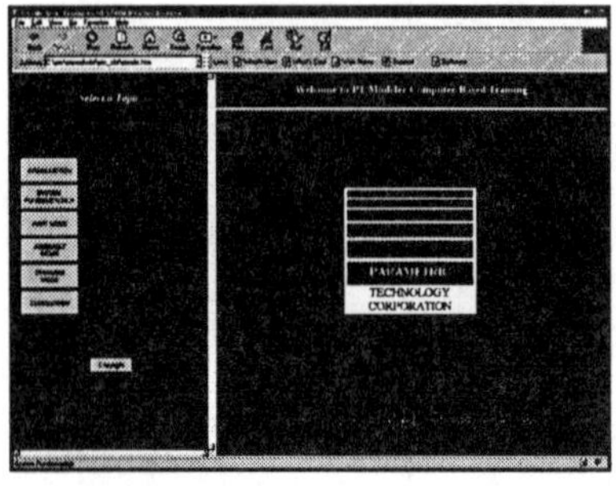

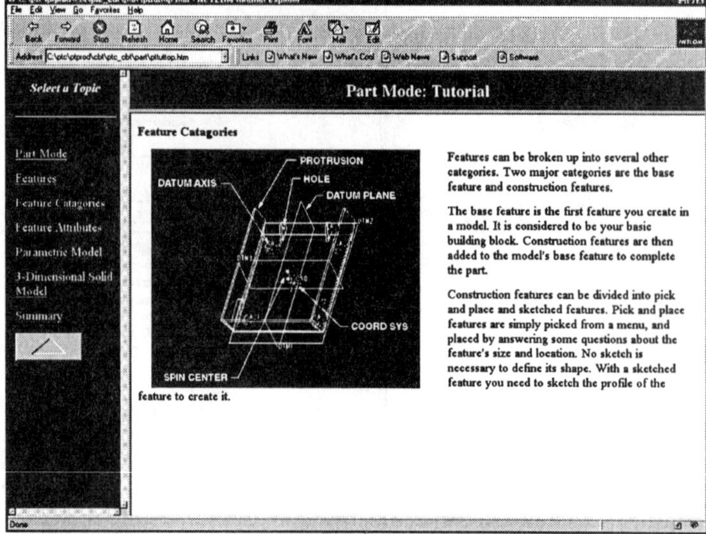

Figure 1.15
CBT

Section 2

Using the Text

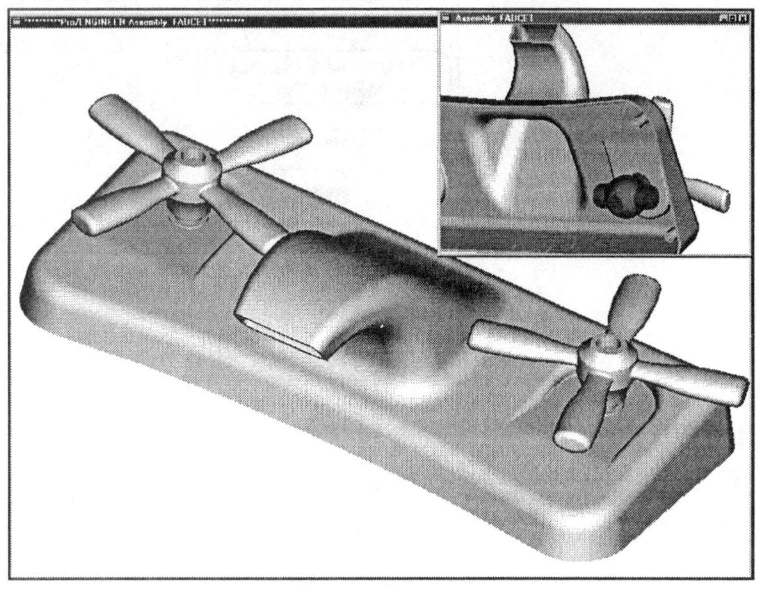

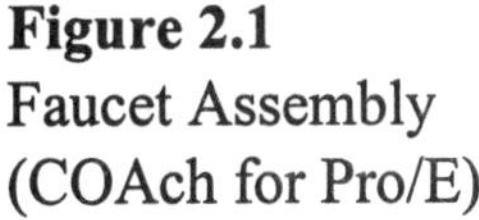

Figure 2.1
Faucet Assembly
(COAch for Pro/E)

Following the **Sections** portion of the book over viewing Pro/E, the main body of the text is divided into four parts: **Part One--Creating Parts**, **Part Two--Assemblies, Part Three--Generating Drawings**, and **Part Four--Advanced Capabilities**. Each **part** has a variety of individual **lessons**. Every lesson introduces a new set of commands and concepts that are applied to a *part*, an *assembly*, or a *drawing*, depending on where in the text you are working.

Lessons involve creating a new part, an assembly, or a drawing using a set of commands that walk you through the process step by step. Each lesson starts with a set of objectives and ends with a new project that requires you to apply the material you have just learned in that lesson (also building on previous lessons).

Each lesson ends with a **lesson project**. The lesson project consists of a part, assembly (Fig. 2.1), or drawing that incorporates the lesson's new material and expands on and uses material introduced in earlier lessons. Student projects use planning sheets from Appendix D as tools for establishing the design intent.

Page S2-3 shows the typical lesson page layout. The left-hand column holds symbols that prompt you, at appropriate places, with a variety of messages, including when to save, ask for help, set up, configure, implement engineering change orders, incorporate hints, and apply notes. Sample menus will also be shown in this column.

Design Intent Planning Sheets

The *design intent* of a feature, a part, or an assembly is thought out and established *before* you rush into modeling. Appendix D provides a variety of sketching formats (Figs. 2.2 and 2.3) for planning a design. These sheets are referred to in the text as **DIPS** (**D**esign **I**ntent **P**lanning **S**heets). Design intent is present throughout this textbook.

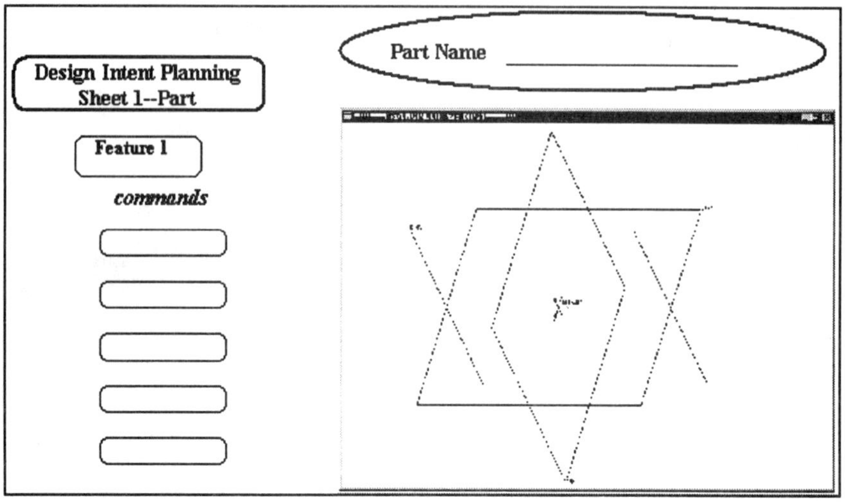

Design Intent Planning
Sheet 1--Part
Feature 1
commands
Part Name

Figure 2.2
Design Intent Planning
Sheet 1--Part

A few minutes of planning and sketching on paper will save countless hours of redoing your design on the computer system. Skipping this step in the design of a feature, part, or assembly is a recipe for disaster. In industry, there are thousands of stories of how a designer or engineer created a graphically correct part or assembly that looked "visually" correct. Upon closer examination, the model or assembly had too many or too few datum planes and a variety of parent-child relationships that were nothing but examples of the designer's incorrect use of Pro/E.

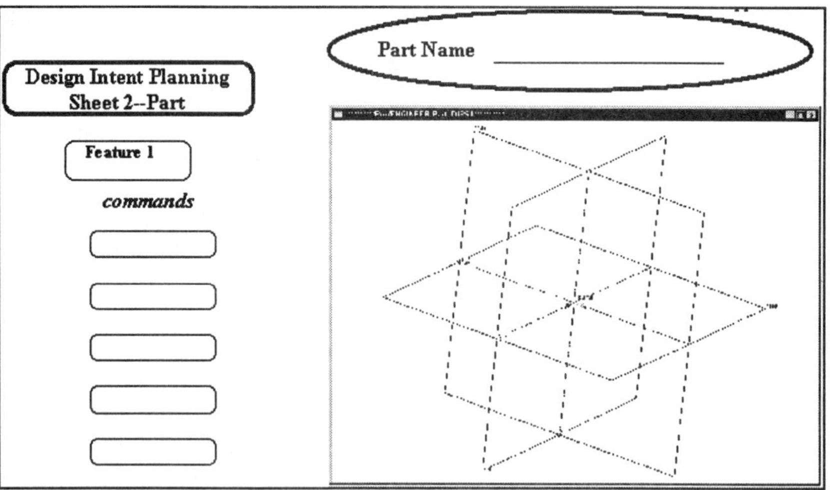

Figure 2.3
Design Intent Planning
Sheet 2--Part

A variety of other problems will occur downstream in the design and manufacturing process, including possible feature failures that result when minor engineering change orders are introduced. Without proper process planning, organization, and a well-defined design intent, the model is useless. In most cases, it would take more time to reorder, modify, redefine, and reroute the model to correct poor design habits than it would to remodel the part.

To get you used to the fact that changes are part of the design process, **engineering change orders** (ECOs) are introduced in lesson step-by-step assignments and in the lesson projects.

Symbols	Meanings
 **Dbms (File**--PT/Modeler) ⇒ **Save** ⇒ **enter** **Purge** ⇒ **enter** ⇒ **Done-Return**	***DBMS*** The **DBMS** symbol will remind you to save your object or design work as you complete a set of tasks or commands. **Save** and **Purge** your file every few minutes.
? Pro/HELP	***Help*** Move the cursor and **highlight** the command that you need online help with, then press the right button on your mouse.
Set Up ⇒ Units ⇒ Millimeter ⇒ Done	***Commands*** Commands are boldface and sometimes enclosed in a box. The ⇒ symbol means to initiate the next command pick. When the command line has ⇒ **enter**, it means to press the Enter key.
CONFIG.PRO **def_layer_datum**	***Config.pro*** Configuration file settings are given to change default settings and help you become comfortable with customizing Pro/ENGINEER. From the **Main** menu, pick **Misc** ⇒ **Edit Config**.
PT/Modeler™	***PT/Modeler*** References to **PT/Modeler** (The Student Collection) commands when they differ from the corresponding commands in Pro/ENGINEER.
E C O	***ECO*** **Engineering change orders** are introduced at various times throughout the lessons. Changes may entail adding *parametric relations*, using *insert mode*, *redefining* the feature, *modifying* dimension values, *rerouting*, or *reordering* features.
COAch™ for Pro/ENGINEER	***COAch*** Refers the user to specific CADTRAIN tutorials.
HINT **DATUM PLANES** will be the first feature on all parts and assemblies.	***Hints*** Hints about commands or procedures are provided to assist you in completing each lesson.
 ***NOTE***	***Notes*** Notes are given to explain an aspect of design intent.
EGD REFERENCE **Engineering Graphics and Design** by L. Lamit and K. Kitto Read Chapter See page	***EGD REFERENCES*** EGD references refer to chapter and page numbers in *Engineering Graphics and Design* or *Fundamentals of Engineering Graphics and Design,* by Lamit and Kitto, West/Delmar Publishing Co., 1997, that can be consulted for more information on a lesson part, project, or concept.

Section 3

Fundamentals

The design of parts and assemblies and the creation of related drawings form the foundation of engineering graphics. With Pro/ENGINEER, many of the previous steps in the design process have been eliminated, streamlined, altered, refined, or expanded. The model you create as a part that must be manufactured forms the basis for all engineering and design functions. The part model contains the geometric data describing the part's features, but also includes nongraphical information embedded in the design itself. The part, its associated assembly, and the graphic documentation (drawings) are parametric. The physical properties described in the part drive (determine the characteristics and behavior of) the assembly and drawing. Any data established in the assembly mode in turn determine that aspect of the part and, subsequently, the drawings of the part and the assembly. In other words, all the information contained in the part, the assembly, and the drawing is interrelated, interconnected, and parametric (Fig. 3.1).

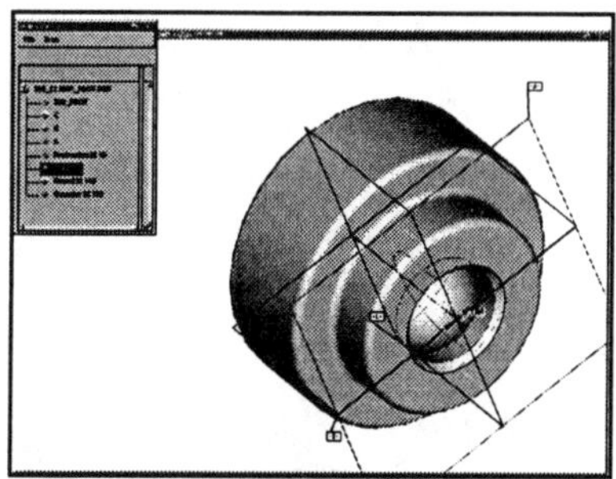

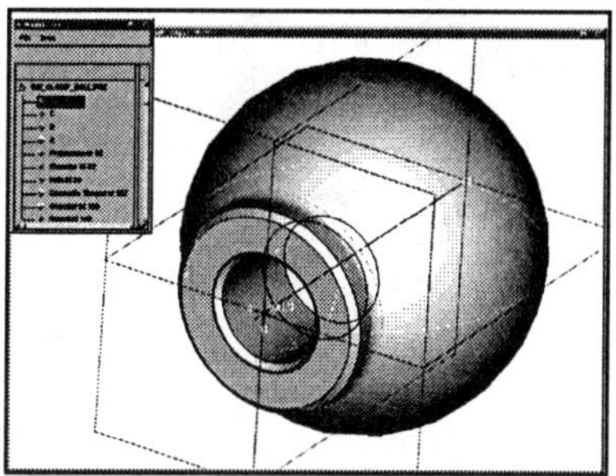

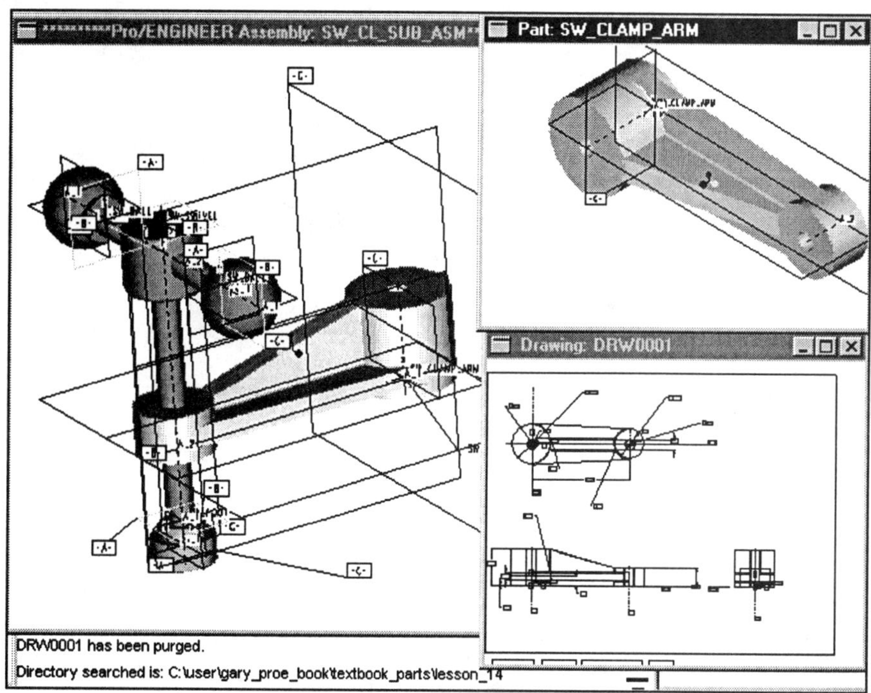

Figure 3.1
Part, Assembly, and Drawing

In most cases, the part will be the first component of this interconnected process. Therefore, in this text, the first set of lessons (1 through 13) covers part design.

Part Design

The part function in Pro/E is used to design components. Parts are started by sketching the basic form to produce the part's **base feature**.

During part design (Fig. 3.2), you can accomplish the following:

* Define the base feature
* Define and redefine construction features to the base feature
* Modify the dimensional values of part features (Fig. 3.3)
* Embed design intent into the model using tolerance specifications and dimensioning schemes
* Create detail drawings of the part
* Create pictorial and shaded views of the component
* Create part families (family tables)
* Perform mass properties analysis and clearance checks
* List part, feature, layer, and other model information
* Measure and calculate model features

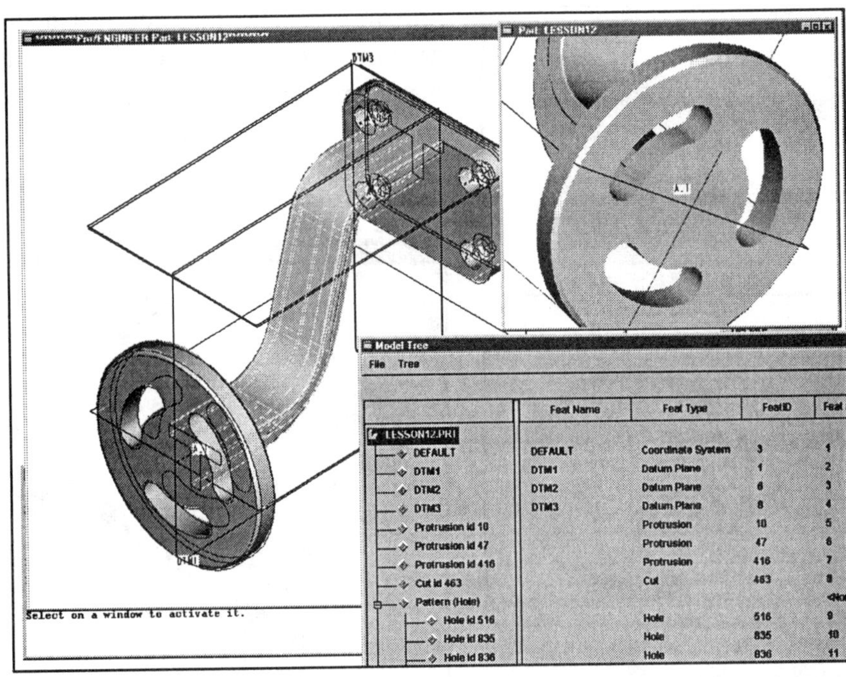

Figure 3.2
Part Design

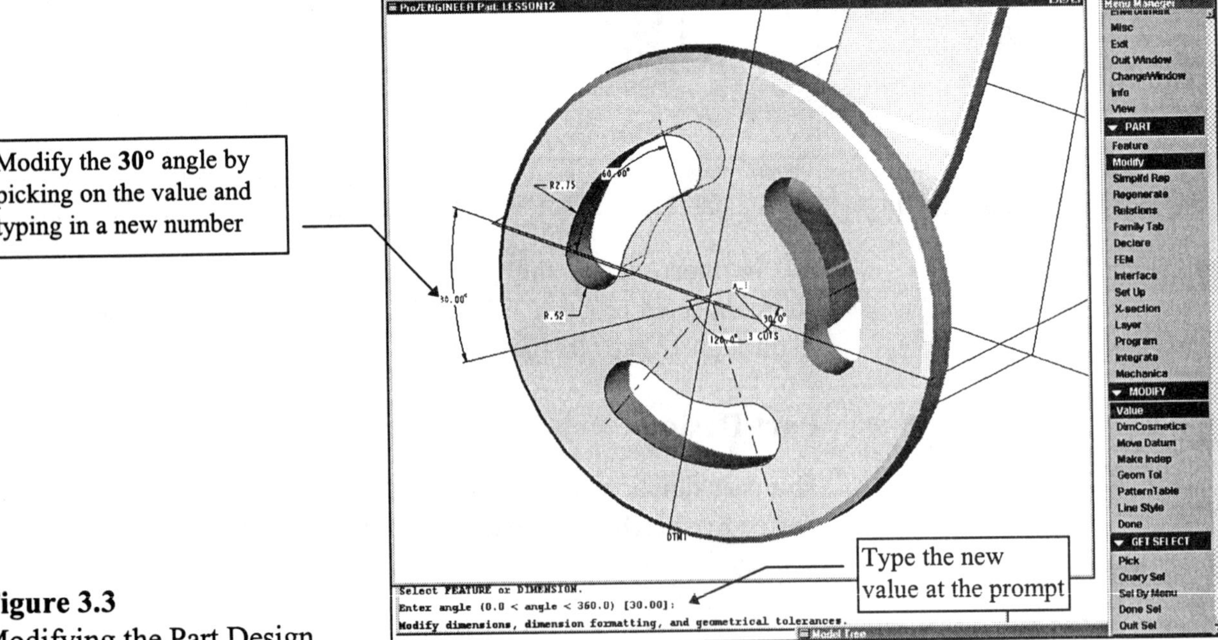

Figure 3.3
Modifying the Part Design

Establishing Part Features

The design of any part requires that the part be *confined*, *restricted*, *constrained*, and *referenced*. In parametric design, the easiest method to establish and control the geometry of your part design is to use three datum planes. Pro/E allows you to use the **primary datum** to start your base feature. By creating the default datum planes (**DTM1**, **DTM2**, and **DTM3** in Pro/E), you can constrain your design in all three directions. In Figure 3.4, three **default datum planes** and a **default coordinate system** have been created. Note that in the **Model Tree** window they have become the first features of the part, meaning that they will be the **parents** of the features that follow.

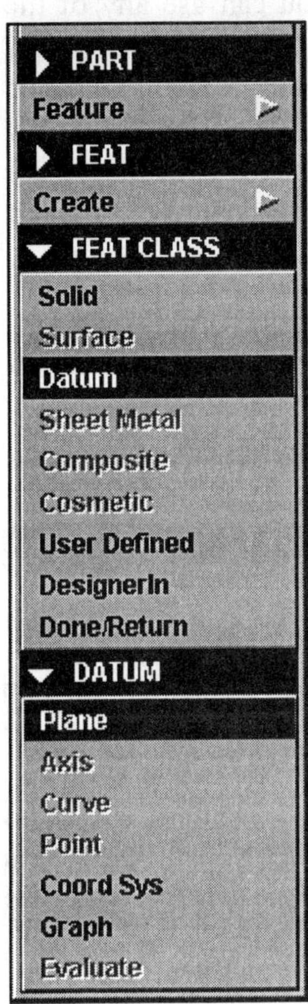

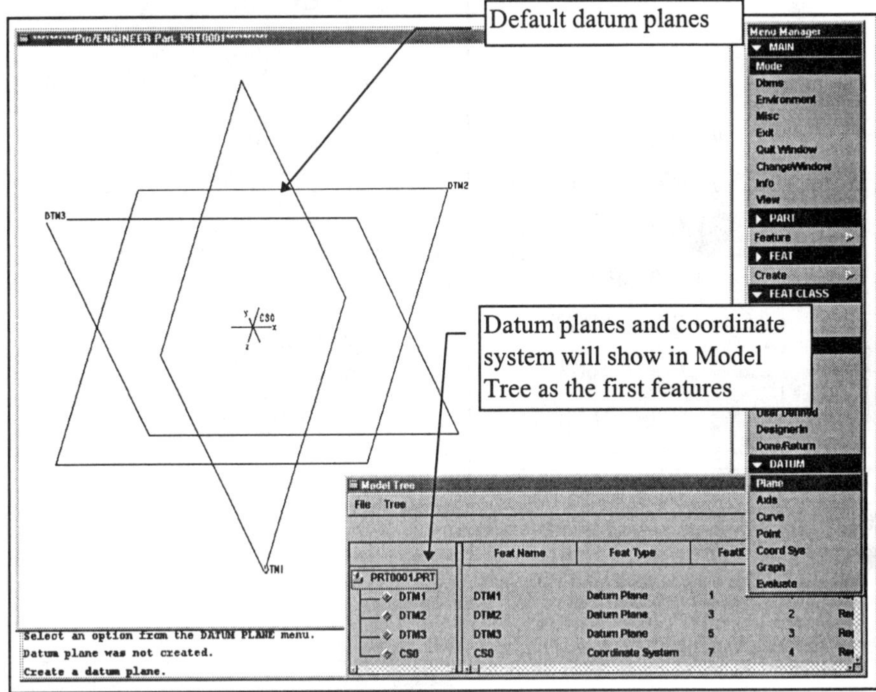

Figure 3.4
Default Datum Planes and Coordinate System

In order to see how datum planes work in the design of a part, try a simple exercise. Take a book or a box and put it on the floor of a room in your house or school. Choose the most important plane (flat side). You have now established **datum A**, the **primary datum** plane (**DTM3** in many of the lessons and projects in this text). Slide the book or box up to a wall near the corner of the room. Choose the longest or second most important plane (flat side). You have now established **datum B**, the **secondary datum** plane (**DTM2** for many of the lessons and projects in this text). Lastly, push the book or box up against the other wall. You have now established **datum C**, the **tertiary datum** plane (**DTM1** for many of the lessons and projects in this text). The book is now constrained by three planar surfaces. If the box were a real workpiece or stock material, you could secure the part to the floor with clamps and machine it into a required design as if it were on a milling table.

Although this exercise and description are simplified and will not work for many parts, they do demonstrate the process of establishing your part in space using datums. In Pro/E, you can use any of the datums as sketching planes or, for that matter, any of the part planes for sketching geometry. Any number of other datums can be introduced into the part as required for feature creation, assembly operations, or manufacturing applications. The bracket in Figure 3.5 has four datum planes. Datum 4 is the fourteenth feature of this part and is the parent of the three "slot" cuts (Fig. 3.6).

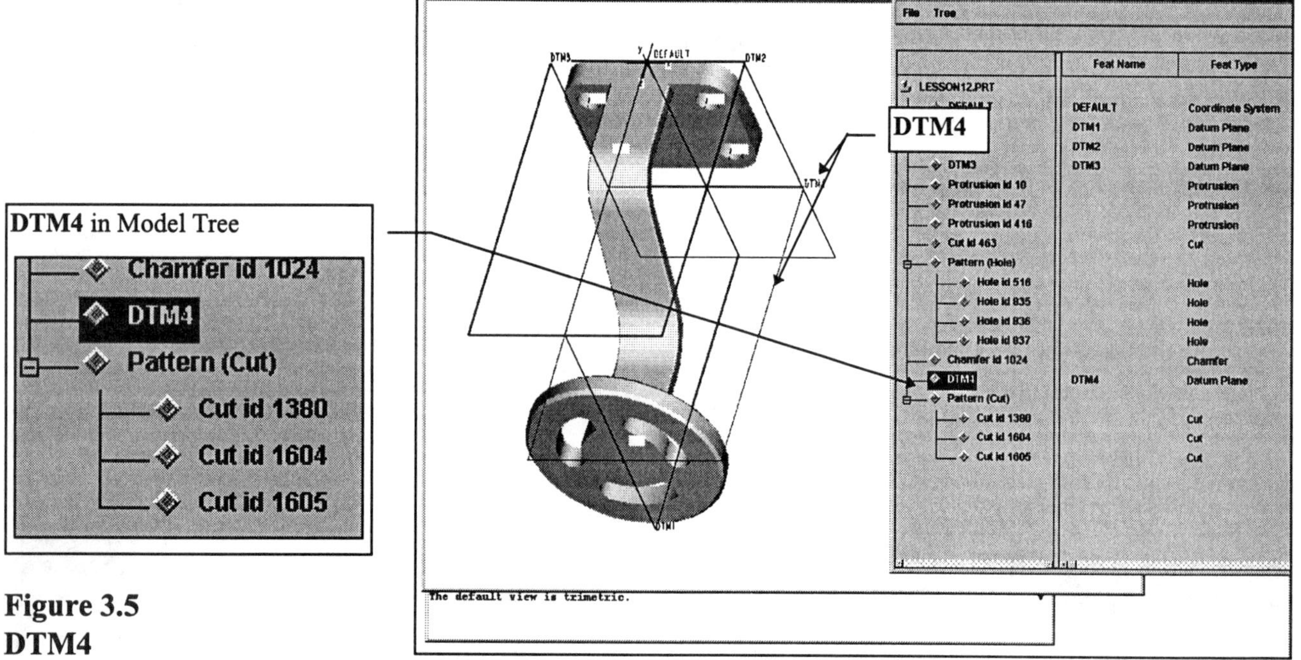

Figure 3.5
DTM4

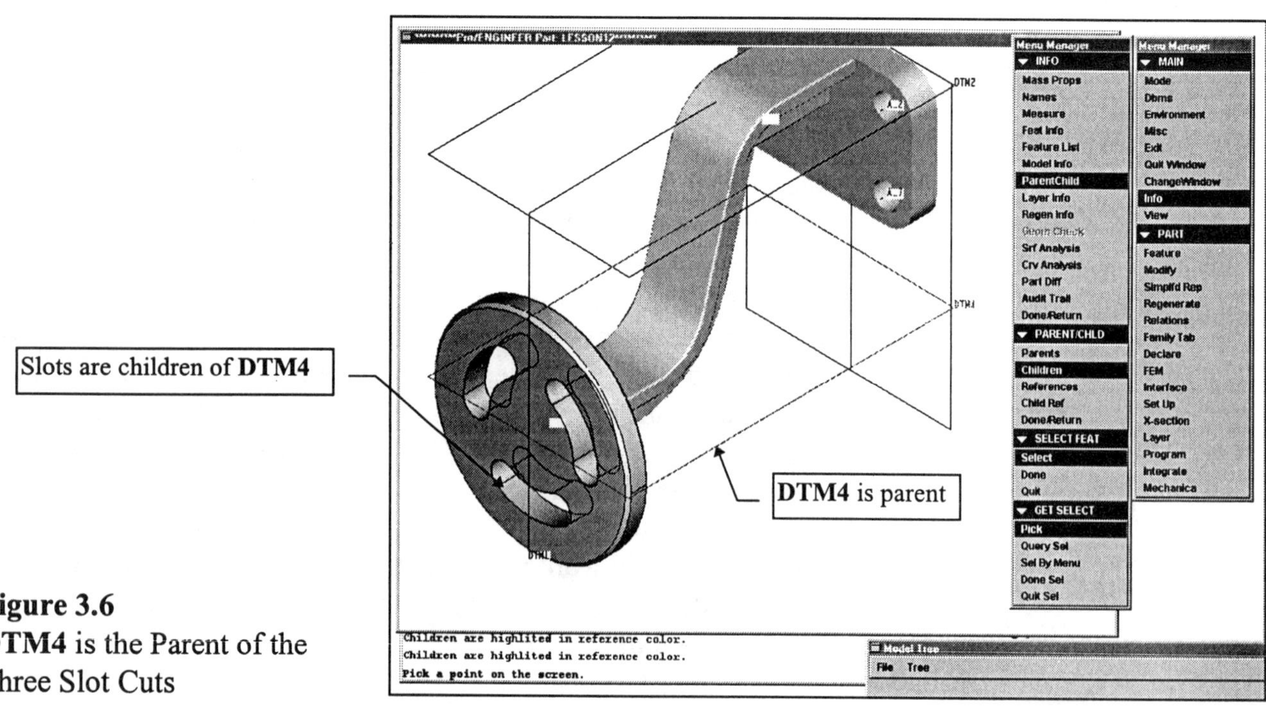

Figure 3.6
DTM4 is the Parent of the Three Slot Cuts

Datum Features

Datum features are planes, axes, and points you use to place geometric features on the active part. We have discussed *default* datums. Datums other than defaults can be created at any time during the design process, as was done in the construction of the bracket shown in Figure 3.6, where datum 4 was introduced and then used to create the slots. There are three (primary) types of datum features: **datum planes**, **datum axes**, and **datum points**. (There are also ***datum curves*** and ***default datum coordinate systems.***) You can display all types of datum features, but they do not define the surfaces or edges of the part or add to its mass properties. In Figure 3.7, a variety of datum planes are used in the creation of the cell phone.

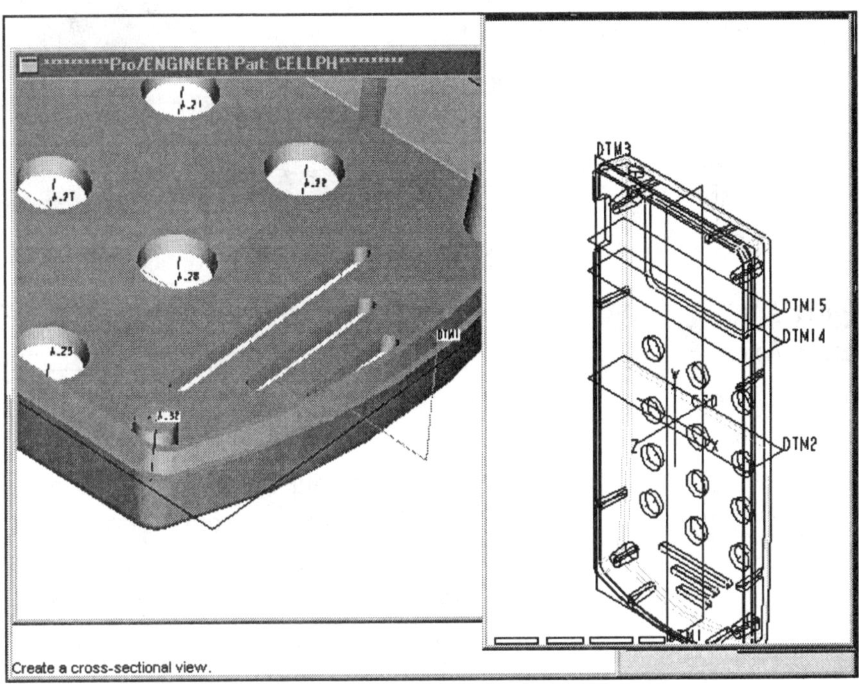

Figure 3.7
Datums in Part Design

Datum planes are infinite planes located in 3D model mode and associated with the object that was active at the time of their creation. To select a datum plane, you can pick on its name or anywhere on the perimeter edge. Datum planes are *parametric*--geometrically associated with the part. Parametric datum planes are associated with and dependent on the edges, surfaces, vertices, and axes of a part. For example, a datum plane placed parallel to a planar face and on the edge of a part moves whenever the edge moves and rotates about the edge if the face moves. As you create parametric datum planes, relationships to the active part are determined by defining combinations of a placement option that link the datum plane to the part.

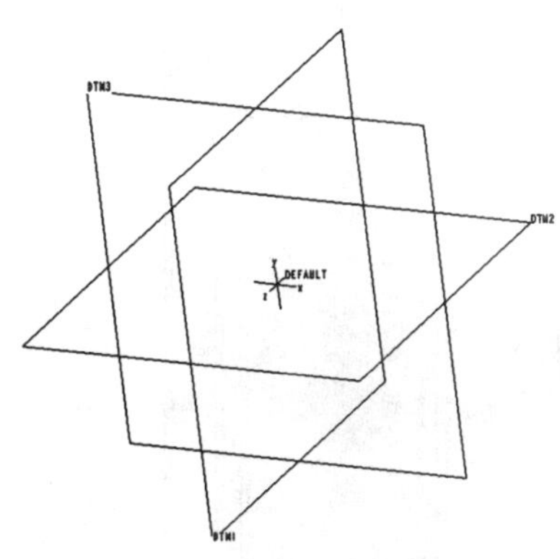

Datum planes are used to create a reference on a part that does not already exist. For example, you can sketch or place features on a datum plane when there is no appropriate planar surface. You can also dimension to a datum plane as if it were an edge.

A datum is created by specifying constraints that locate it with respect to existing geometry. For example, a datum plane might be made to pass through the axis of a hole and parallel to a planar surface. Chosen constraints must locate the datum plane relative to the model without ambiguity. You can also use and create datums in assembly mode.

Besides datum planes, datum axes and datum points can be created to assist in the design process. In Figure 3.8, a datum axis has been created through the cylindrical feature.

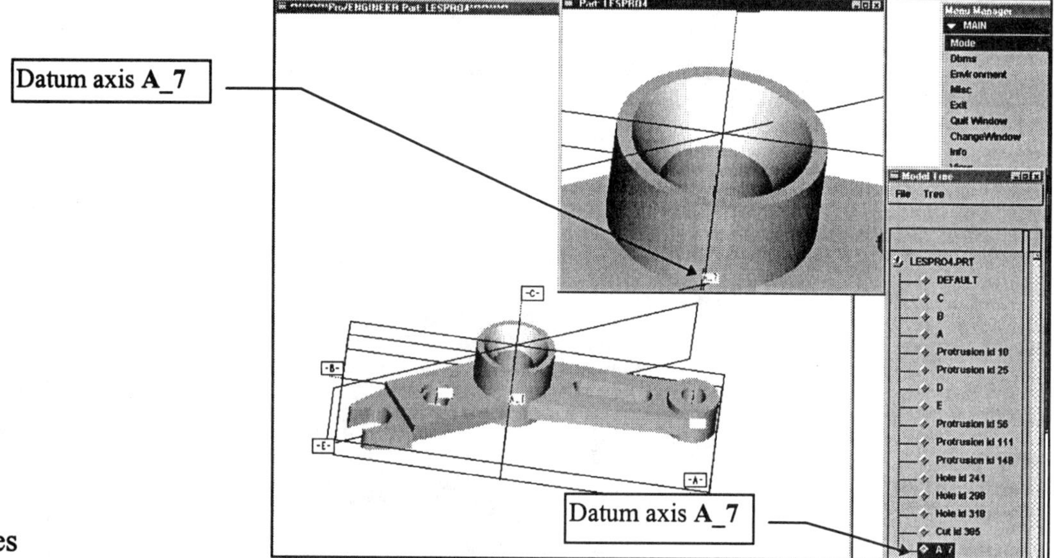

Figure 3.8
Datum Axes

You can also automatically create datum axes through cylindrical features such as holes and solid round features by setting this as a default in your Pro/E configuration file (the holes in the cell phone in Fig. 3.7 all have axes). The part in Figure 3.9 shows **A_1** and **A_2** through the hole. **A_1** was the default axis (of the circular cut) and doesn't show in the Model Tree window. Axis **A_2** was created as a datum axis and therefore is a feature shown in the Model Tree window.

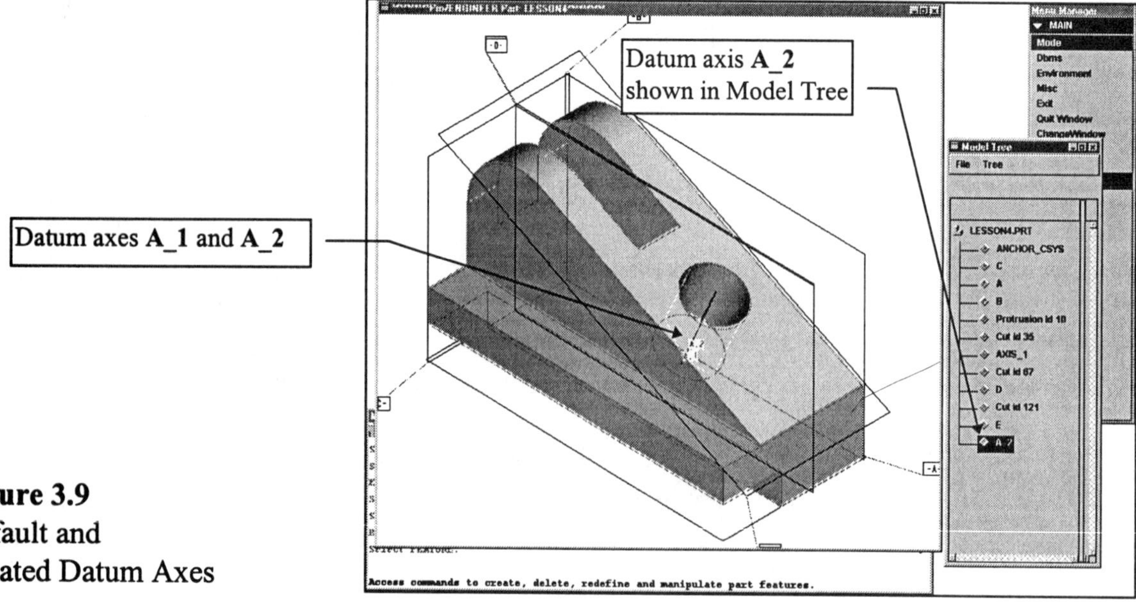

Figure 3.9
Default and
Created Datum Axes

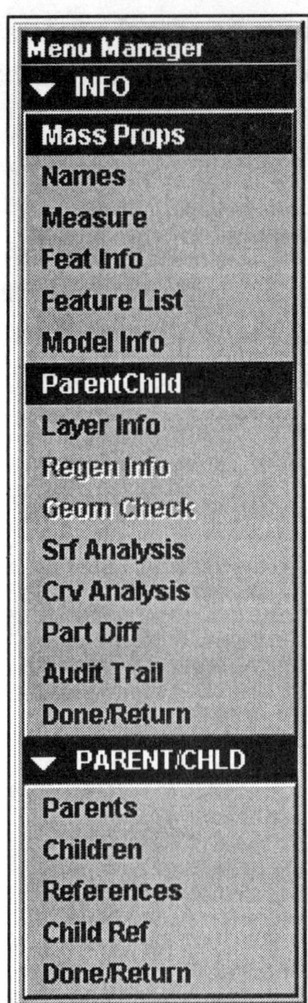

Parent-Child Relationships

Because solid modeling is a cumulative process, certain features must, of necessity, precede others. Those that follow must rely on previously defined features for dimensional and geometric references. The relationships between features and those that reference them are termed **parent-child relationships**. Because children reference parents, features can exist without children, but children cannot exist without their parents. Using **Model Info**, Pro/E lists the models' information, including the parent-child relationships (Fig. 3.10).

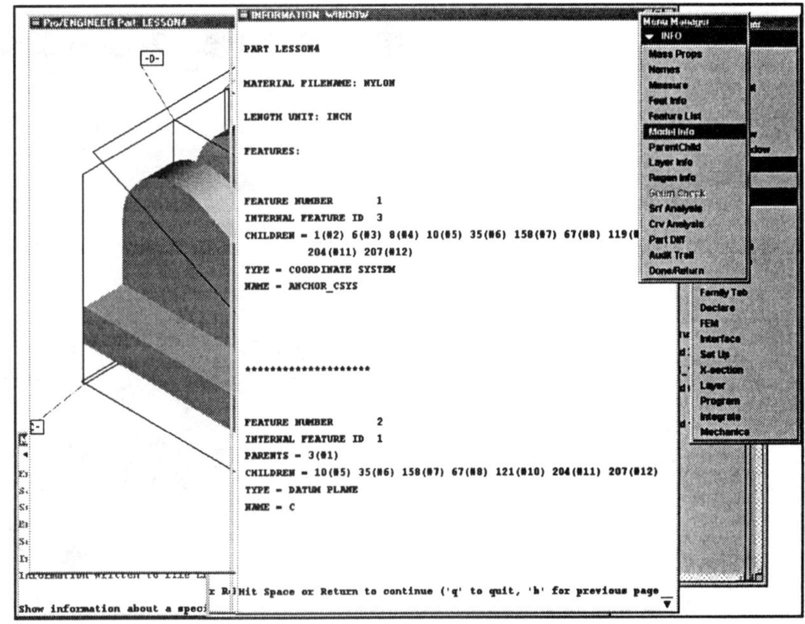

Figure 3.10
Parent-Child Relationships

To get the parent-child information for a feature, choose the following commands:

Info ⇒ **ParentChild** ⇒ **Children** ⇒ (select and pick the feature, as shown in Fig. 3.11)

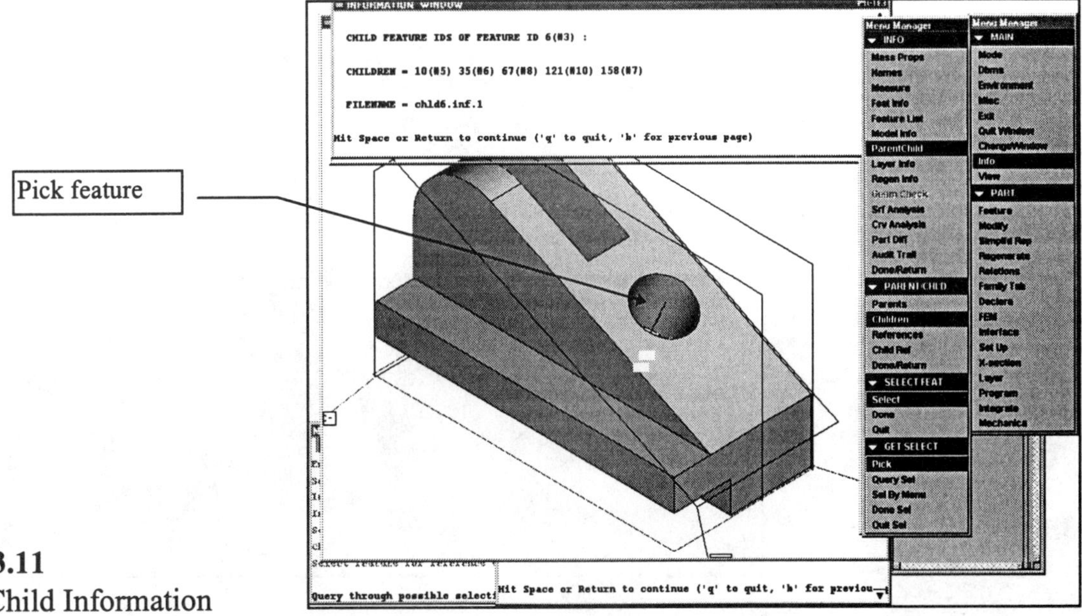

Figure 3.11
Parent-Child Information

The parent-child relationship is one of the most powerful aspects of parametric design. When a parent feature is modified, its children are automatically recreated to reflect the changes in the parent feature's geometry. It is essential to reference feature dimensions so that design modifications are correctly propagated through the model/part. As an example, the modification to the length of a part is automatically propagated through the part and will affect all children of the modified feature. This is shown in Figures 3.12 and 3.13, where the base feature of the part has been modified in width, length, and front cut height.

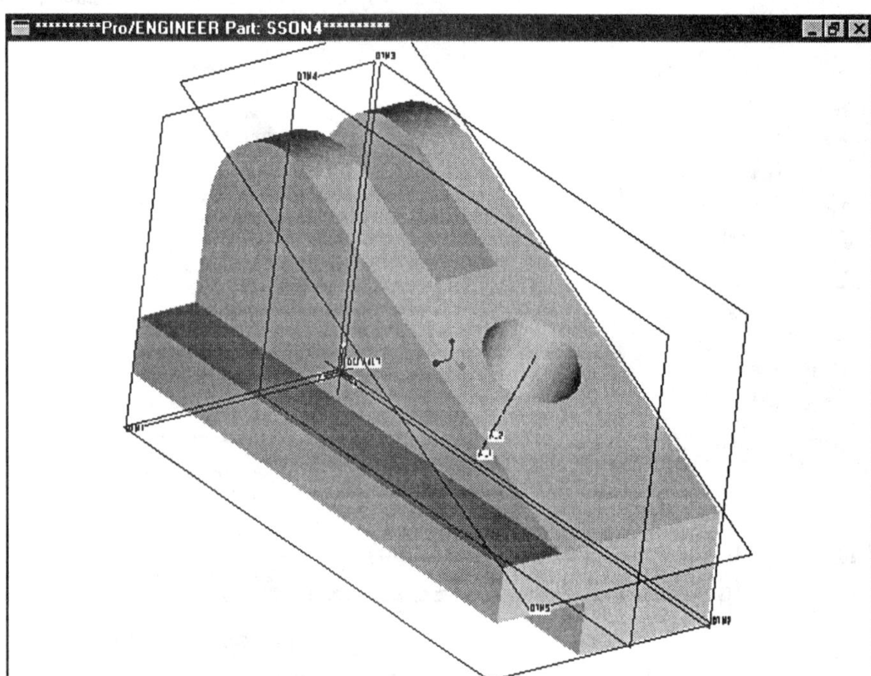

Figure 3.12
Original Design

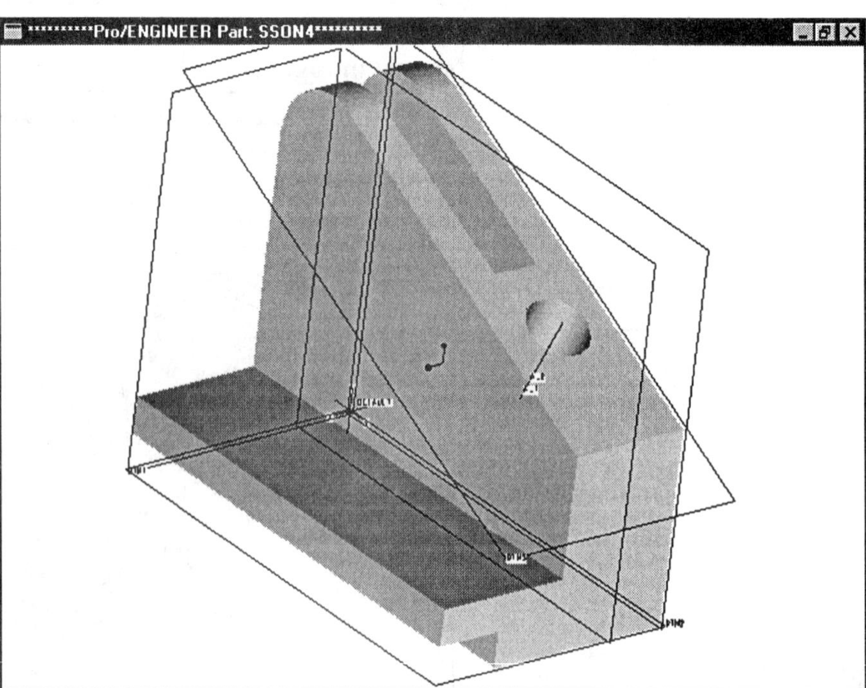

Figure 3.13
Modified Base Feature

Capturing Design Intent

Family of Parts

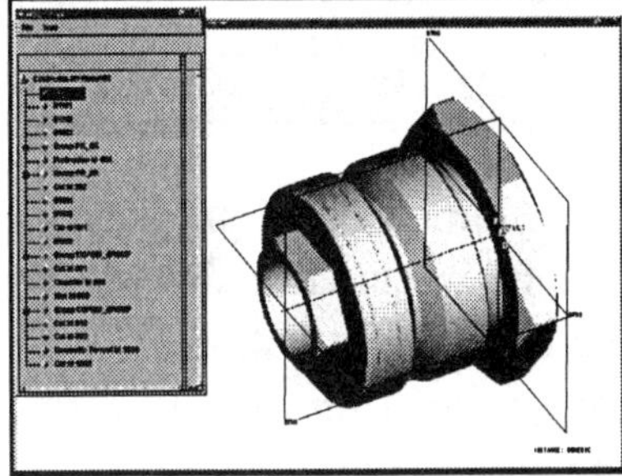

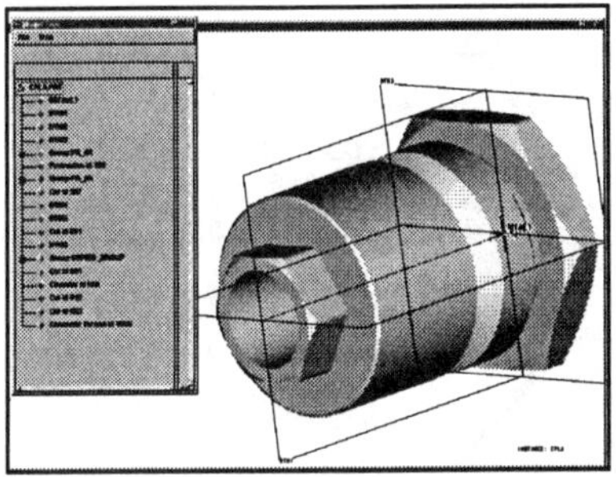

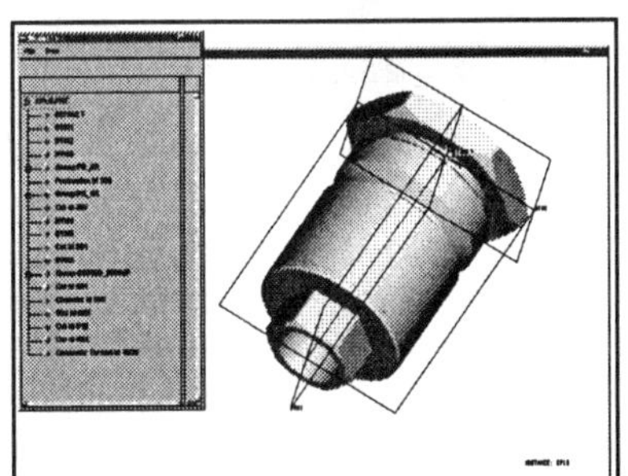

A valuable characteristic of any design tool is its ability to **render** the design and at the same time capture its **intent**. Parametric methods depend on the sequence of operations used to construct the design. The software maintains a *history of changes* the designer makes to specific parameters. The point of capturing this history is to keep track of operations that depend on each other. Whenever Pro/E is told to change a specific dimension, it can update all operations that are referenced to that dimension.

For example, a circle representing a bolt hole may be constructed so that it is always concentric to a circular slot. If the slot moves, so does the bolt circle. Parameters are usually displayed in terms of dimensions or labels and serve as the mechanism by which geometry is changed. The designer can change parameters manually by changing a dimension or can reference them to a variable in an equation (**relation**) that is solved either by the modeling program itself or by external programs such as spreadsheets.

Parametric modeling is particularly useful in modeling whole **families** of similar parts and in rapidly modifying complex 3D designs. It is most effective in working with designs where changes are likely to consist of dimensional changes rather than radically different geometry.

Feature-based modeling is the construction of geometry as a combination of **form features**. The designer specifies features in engineering terms, such as holes, slots, or bosses, rather than geometric terms, such as circles or boxes.

Features can also store nongraphic information. This information can be used in activities such as drafting, numerical control (**NC**), finite-element analysis (**FEA**), and kinematics analysis.

Capturing design intent is based on incorporating engineering knowledge into a model by establishing and preserving certain geometrical relationships. The wall thickness of a pressure vessel, for example, should be proportional to its surface area and should remain so, even as its size changes. Parametric design captures these relationships in several ways:

Implicit Relationships Implicit relationships occur when new model geometry is sketched and dimensioned relative to existing features and parts. An implicit relationship is established, for instance, when the section sketch of a tire (Fig. 3.14) uses rim edges as a reference.

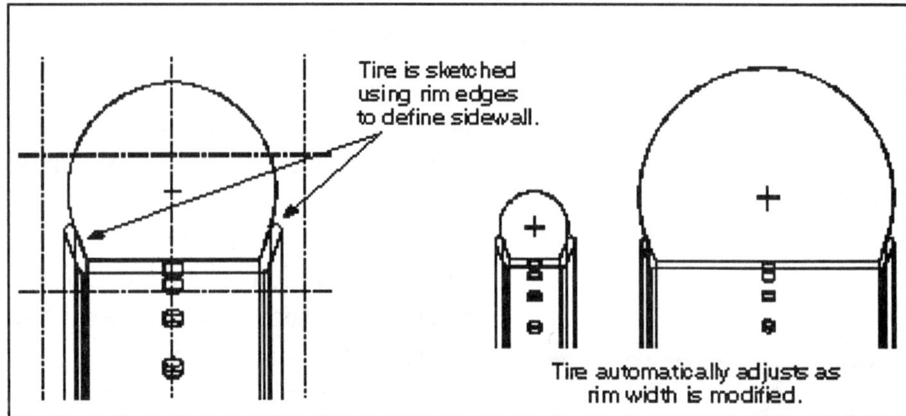

Figure 3.14
Tire and Rim

Patterns Design features often follow a geometrically predictable pattern. Features and parts are patterned in parametric design by referencing either construction dimensions or existing patterns. One example of patterning is a wheel hub with spokes (Fig. 3.15). First, the spoke holes are radially patterned. The spokes can then be strung by referencing this pattern.

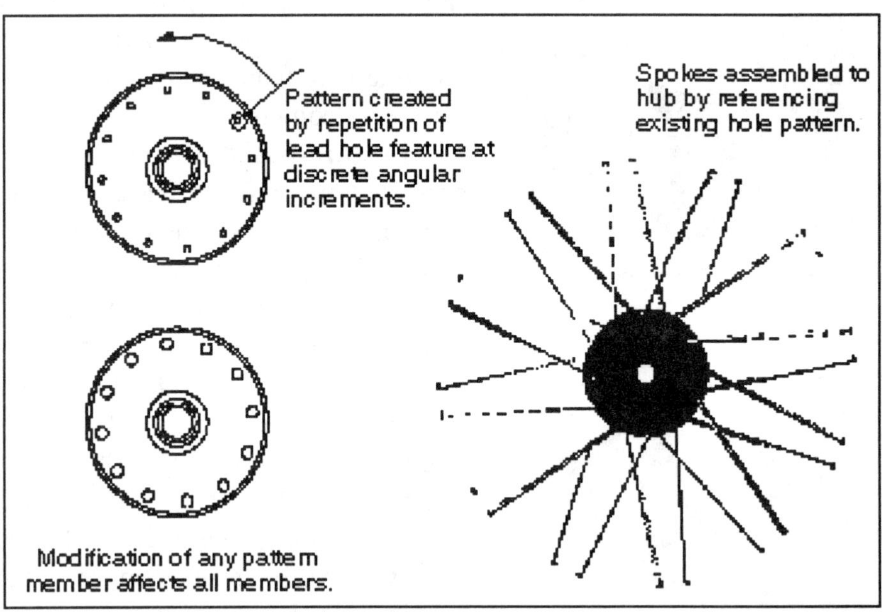

Figure 3.15
Patterns

Modifications to a pattern member affects all members of that pattern. This helps capture design intent by preserving the duplicate geometry of pattern members.

Explicit Relations Whereas implicit relationships are implied by the feature creation method, an explicit relation is mathematically entered by the user. This equation is used to relate part and feature dimensions in the desired manner. An explicit relation might be used, for example, to ensure that any number of spoke holes will be evenly spaced around a wheel hub (Fig. 3.16).

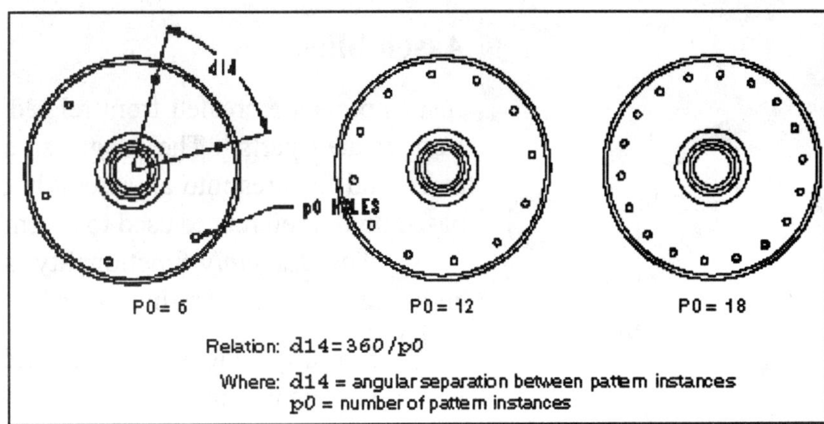

Figure 3.16
Adding Relations

Family Tables Family tables are used to create part families from generic models by tabulating dimensions or the presence of certain features or parts. A family table might be used, for example, to catalog a series of wheel rims with varying width and diameter as shown in Figure 3.17.

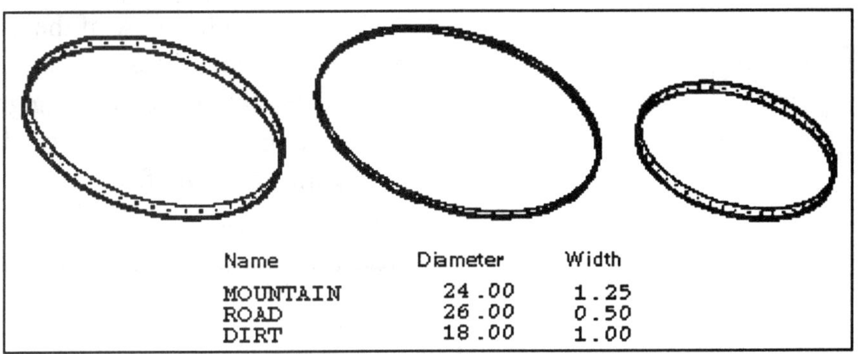

Name	Diameter	Width
MOUNTAIN	24.00	1.25
ROAD	26.00	0.50
DIRT	18.00	1.00

Figure 3.17
Family Table

The modeling task is to incorporate the features and parts of a complex design while properly capturing design intent to provide flexibility in modification. Parametric design modeling is a synthesis of physical and intellectual design (Fig. 3.18).

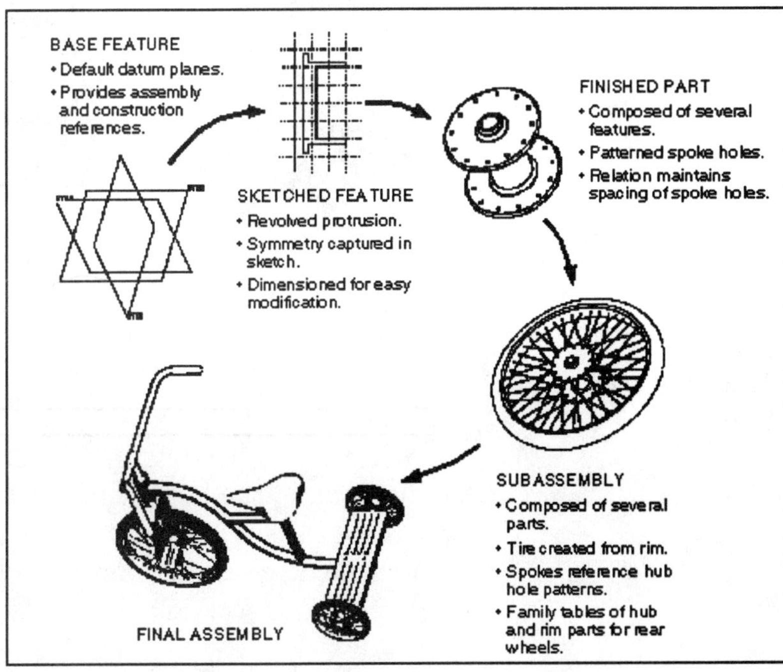

Figure 3.18
Relations

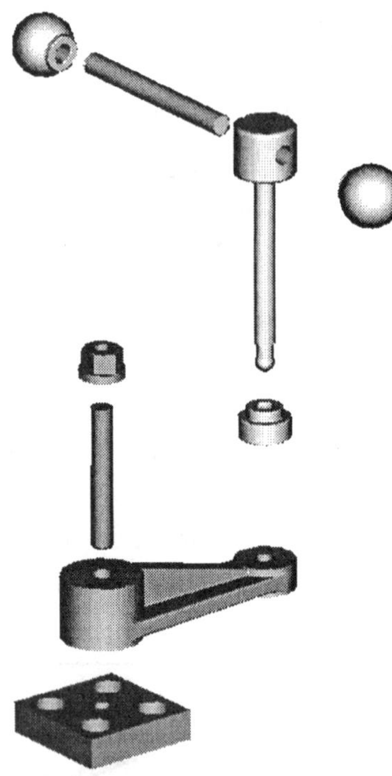

Assemblies

Just as parts are created from related features, **assemblies** are created from related parts. The progressive combination of subassemblies, parts, and features into an assembly creates parent-child relationships based on the references used to assemble each component (Fig. 3.19).

The *Assembly* functionality is used to assemble existing parts and subassemblies. During assembly creation, you can:

* Simplify a view of a large assembly by creating a simplified representation
* Perform automatic or manual placement of component parts
* Create an exploded view of the component parts
* Perform analysis, such as mass properties and clearance checks
* Modify the dimensional values of component parts
* Define assembly relations between component parts
* Create documentation drawings of the assembly
* Create a shaded view of the assembly
* Use the assembly as a subassembly
* Perform automatic interchange of component parts
* Create parts in Assembly mode
* Create Assembly features

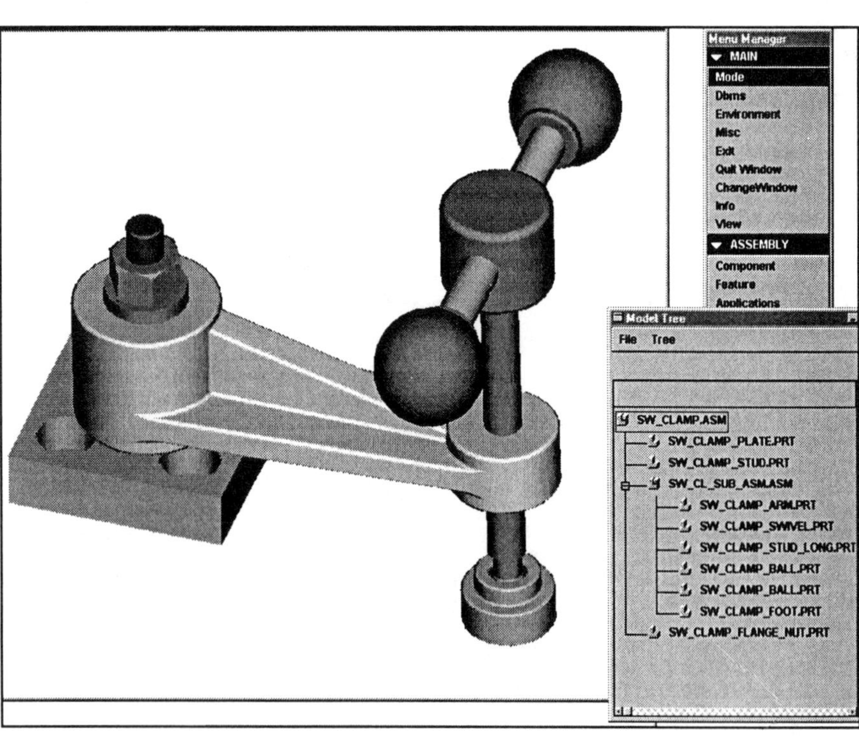

Figure 3.19
Clamp Assembly and Model Tree

Just as features can reference part geometry, parametric design also permits the creation of parts referencing assembly geometry. Assembly mode allows the designer both to fit parts together and to design parts based on how they should fit together.

In Figure 3.19, the assembly of the clamp was shown. Figure 3.20 shows the clamp subassembly. A *Bill of Materials* report is generated for the assembly in Figure 3.21. **Info ⇒ BOM** is chosen from the INFO menu.

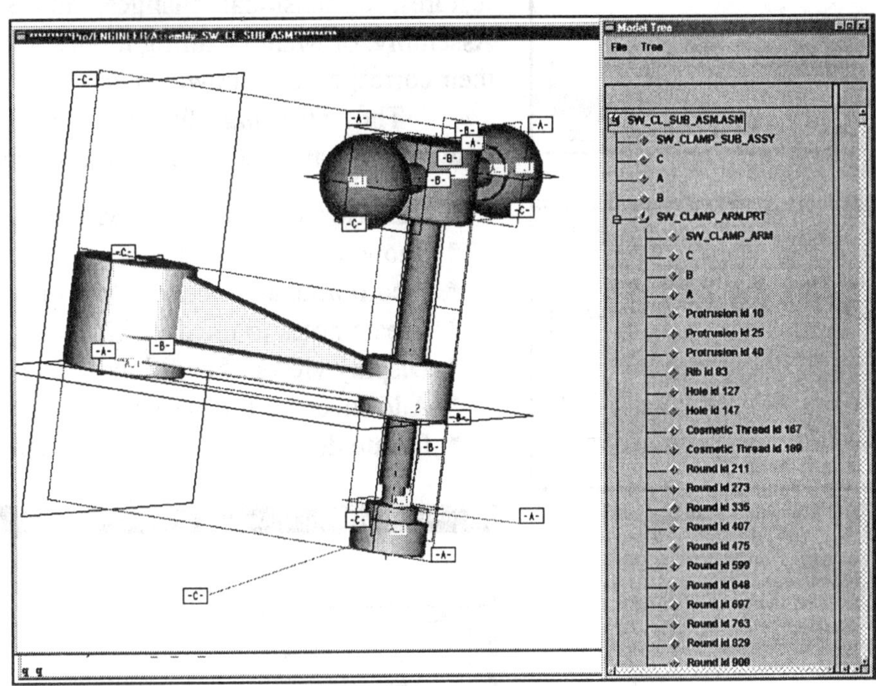

Figure 3.20
Clamp Subassembly

Figure 3.21 Clamp Assembly and BOM

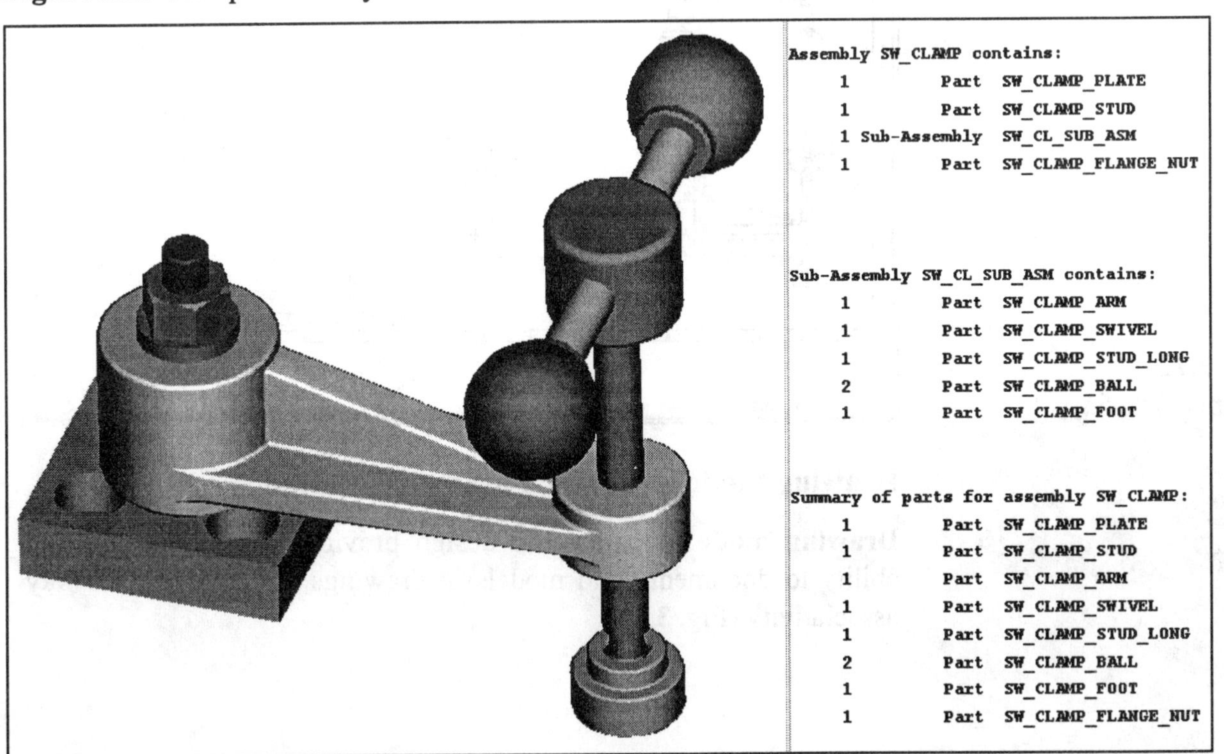

Assembly SW_CLAMP contains:

1	Part	SW_CLAMP_PLATE
1	Part	SW_CLAMP_STUD
1	Sub-Assembly	SW_CL_SUB_ASM
1	Part	SW_CLAMP_FLANGE_NUT

Sub-Assembly SW_CL_SUB_ASM contains:

1	Part	SW_CLAMP_ARM
1	Part	SW_CLAMP_SWIVEL
1	Part	SW_CLAMP_STUD_LONG
2	Part	SW_CLAMP_BALL
1	Part	SW_CLAMP_FOOT

Summary of parts for assembly SW_CLAMP:

1	Part	SW_CLAMP_PLATE
1	Part	SW_CLAMP_STUD
1	Part	SW_CLAMP_ARM
1	Part	SW_CLAMP_SWIVEL
1	Part	SW_CLAMP_STUD_LONG
2	Part	SW_CLAMP_BALL
1	Part	SW_CLAMP_FOOT
1	Part	SW_CLAMP_FLANGE_NUT

Drawings

You can create **drawings** of all parametric design models or get some of them by importing files from other systems. All model views in the drawing are **associative**: if you change a dimensional value in one view, other drawing views update accordingly. Moreover, drawings are associated with their parent models. Any dimensional changes made to a drawing are automatically reflected in the model. Any changes made to the model (e.g., addition of features, deletion of features, dimensional changes, and so on) in Part, Sheet Metal, Assembly, or Manufacturing modes are also automatically reflected in their corresponding drawings.

The *Drawing* functionality is used to create annotated drawings of parts and assemblies. During drawing creation, you can:

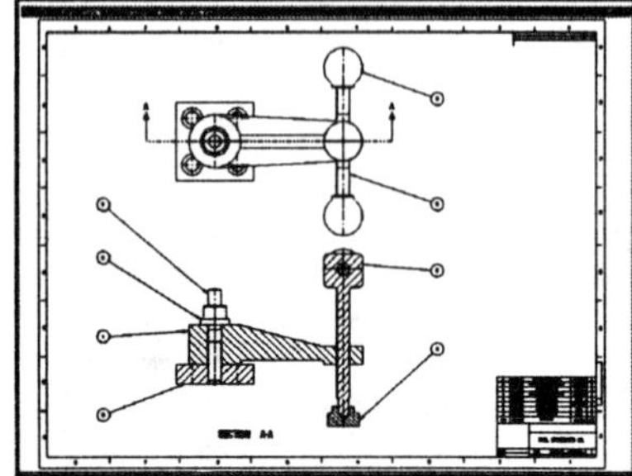

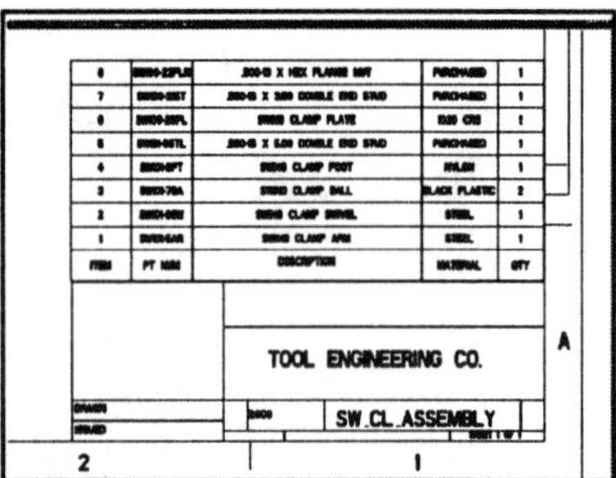

* Add views of the part or assembly
* Show existing dimensions
* Incorporate additional driven or reference dimensions
* Create notes to the drawing
* Display views of additional parts or assemblies
* Add sheets to the drawing
* Create draft entities on the drawing

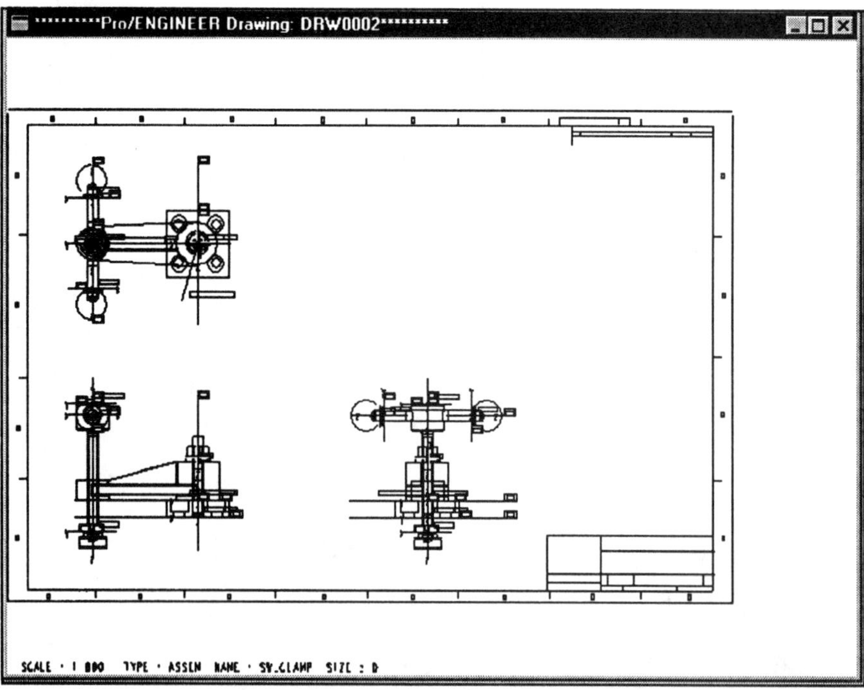

Figure 3.22
Assembly Drawing

Drawing Mode and Basic Parametric Design

Drawing mode in parametric design provides you with the basic ability to document solid models in drawings that share a two-way associativity (Fig. 3.22).

Changes that are made to the model in the Part or the Assembly mode will cause the drawing to automatically update and reflect the changes. Any changes made to the model in Drawing mode will be immediately visible on the model in Part and Assembly modes. The part shown in Figure 3.23 has been detailed in Figure 3.24.

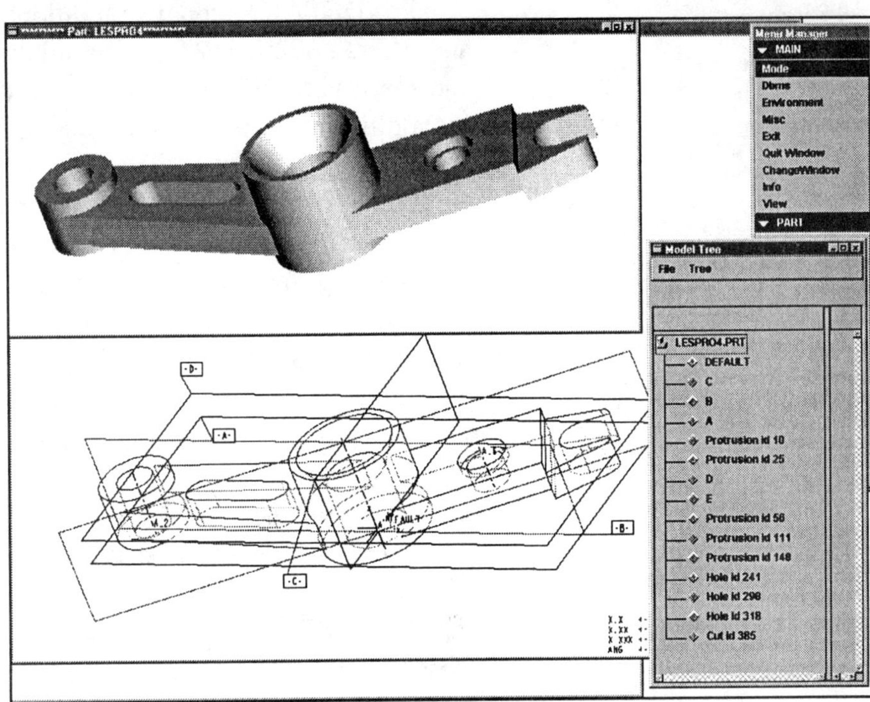

Figure 3.23
Angle Frame Model

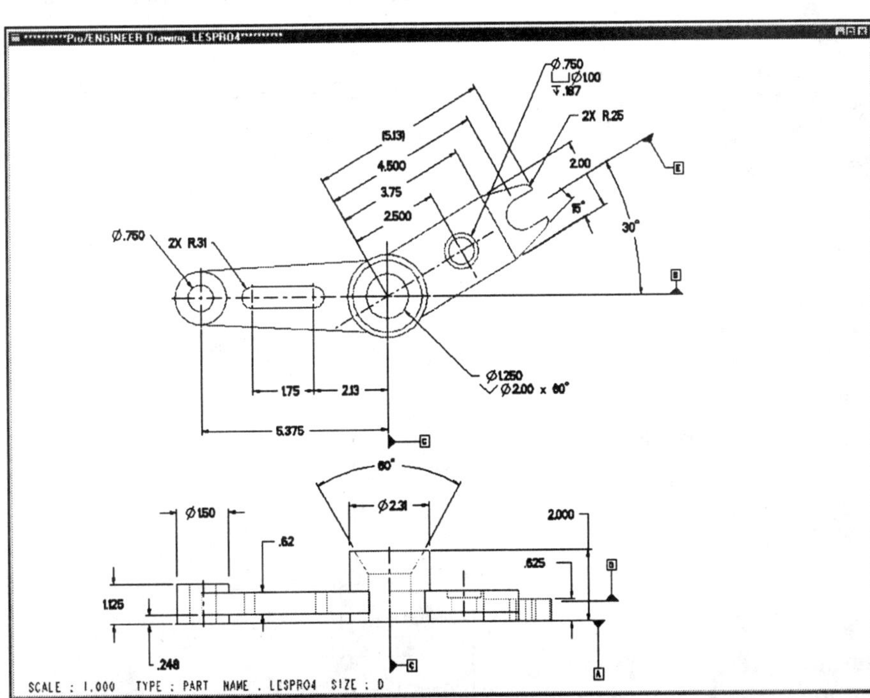

Figure 3.24
Angle Frame Drawing

Basic Pro/E (without the optional module Pro/DETAIL) allows you to create drawing views of one or more models in a number of standard views with dimensions.

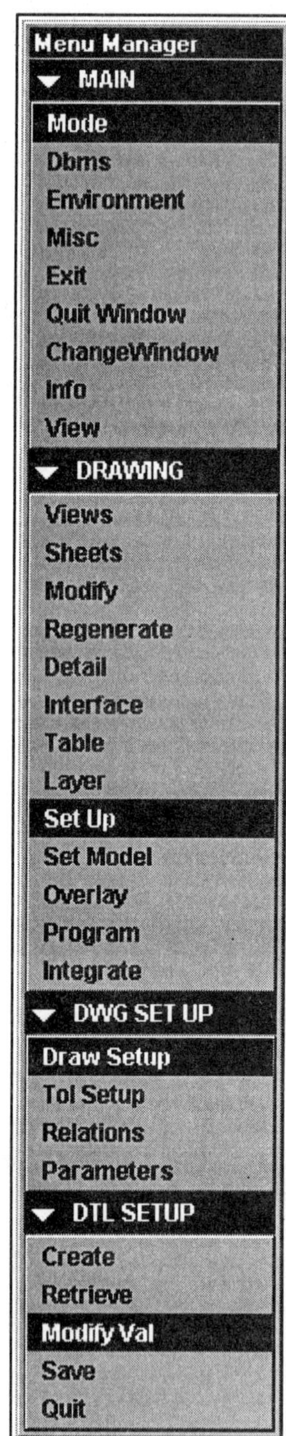

Figure 3.25
Setting **.dtl** File Configuration Defaults with **Modify Val**

You can annotate the drawing with notes, manipulate the dimensions, and use layers to manage the display of different items on the drawing. The optional module **Pro/DETAIL** can be used to extend the drawing capability or as a stand-alone module allowing you to create, view, and annotate models and drawings.

Pro/DETAIL supports additional view types and multi-sheets, and offers commands for manipulating items in the drawing, adding and modifying different kinds of textural and symbolic information. In addition, the abilities to customize engineering drawings with sketched geometry, create custom drawing formats, and make numerous cosmetic changes to the drawing are available.

Drawing parameters are saved with each individual drawing and drawing format. Drawing parameters determine the height of dimension and note text, text orientation, geometric tolerance standards, font properties, drafting standards, and arrow lengths. Parameter values are stored by Pro/E in files with **.dtl** extensions. The drawing parameters can be altered by picking **Set Up** ⇒ **Modify Val** (Fig. 3.25).

When you regenerate a drawing, the drawing and the model that it represents are recreated, not simply redrawn. This means that if any of the model's dimension values were changed while you are in drawing mode, regenerating the drawing causes the model to update these changes. The regenerated drawing displays the updated model and any changes that were made to it.

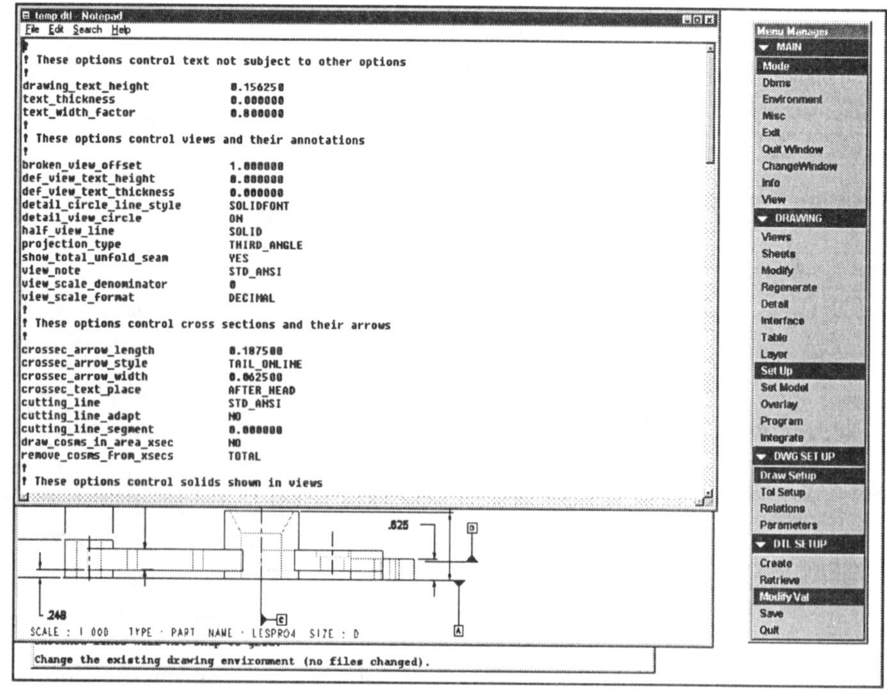

Manufacturing and Parametric Design

Parametric design systems also provide the tools to program and simulate numerical control manufacturing processes (Fig. 3.26). The information created can be quickly updated if the engineering design model changes. NC programs in the form of ASCII cutter location (CL) data files, tool lists, operation reports, and in-process geometry can be generated.

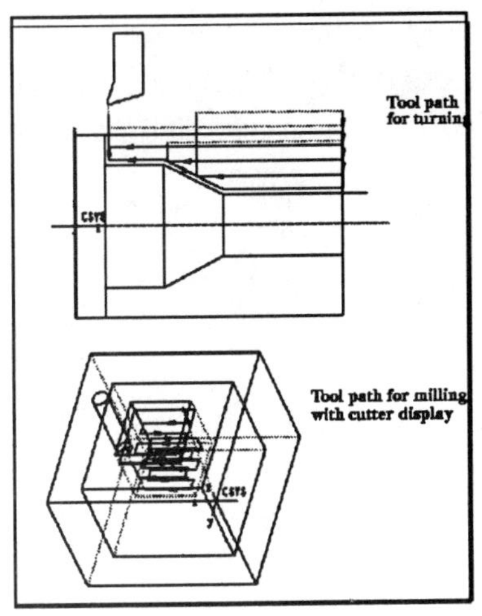

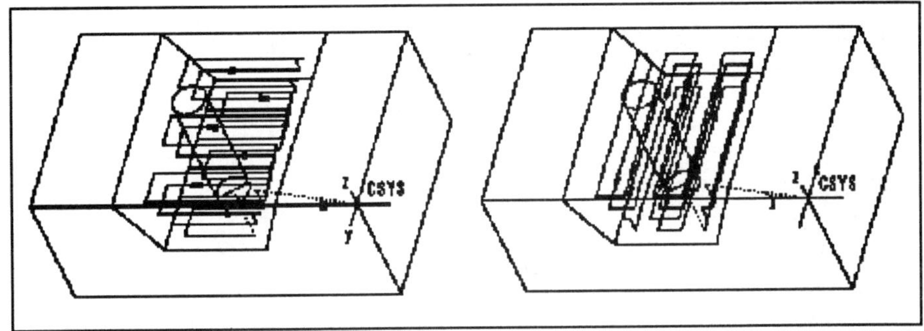

Figure 3.26
Numerical Control Machining Simulation

Pro/MANUFACTURING

Pro/MANUFACTURING software creates the data necessary to drive an NC machine tool to machine a part. It does this by providing the tools to let the manufacturing engineer follow a logical sequence of steps to progress from a design model to an ASCII CL data file that can be post-processed into NC machine data (Fig. 3.27).

Design Model
Workpiece
Manufacturing Model
Machine Tools (Workcells)
Fixture Setups
Set Up Manufacturing Database
Tools
Set Up Operation
Define NC Sequences
Pro/MFG
Create CL Data Files (APT)
Produce In-Process Model
Post-Process
Drive NC Machine Tool

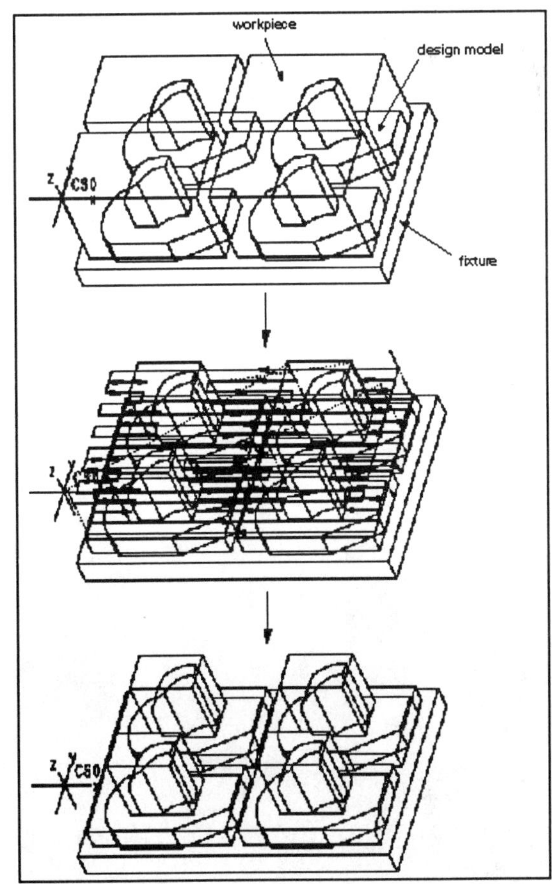

Figure 3.27
Multiple Part Machining Using Pro/MANUFACTURING

Section 4

Utilities

Several utilities are used during the design of parts, assemblies, and drawings. The abilities to access and change directories, edit a trail file, give operating system-level commands, and edit and load new configuration settings are essential to the efficient and accurate creation of project databases.

Operating Modes

The **Mode** option in the **MAIN** menu allows you to enter the major modes of Pro/E operation. When this option is chosen, a submenu appears showing the various operational modes. The modes available on your system are dependent on the modules (options) purchased, licensed, and installed on your workstation. In Figure 4.1, the Mode option has been chosen and the module selection menu appears. Note that the system has a variety of modules available, including **Sheet Metal**, **Cast**, **Composite**, **Diagram**, **Dieface**, **Mold**, and **Markup**.

More than one module can be active in a Pro/E session at one time. You can have a part, the assembly where it is used, and its drawing on the screen simultaneously. In Figure 4.2, part, assembly, and drawing modes have been activated. You can move among the windows and work on the drawing, the part, or the assembly. The part database essentially drives the drawing and the assembly. Since Pro/E is *parametric*, any editing or modifications of the part will be reflected in the assembly and drawing modes after regeneration. The same is true of drawing mode and assembly mode changes.

If you have the Student Collection of PT/Modeler, your available modes will be **Part**, **Assembly**, **Drawing**, and **Format**.

Menu Manager
MAIN
Mode
Dbms
Environment
Misc
Exit
Quit Window
ChangeWindow
Info
View
MODE
Sketcher
Part
Sheet Metal
Composite
Scan Model
Assembly
Drawing
Manufacture
Mold
Cast
Dieface
Layout
Format
Report
Markup
Diagram
Interchange
Legacy
Verify
PProcessor
Process

Figure 4.1
Mode Command

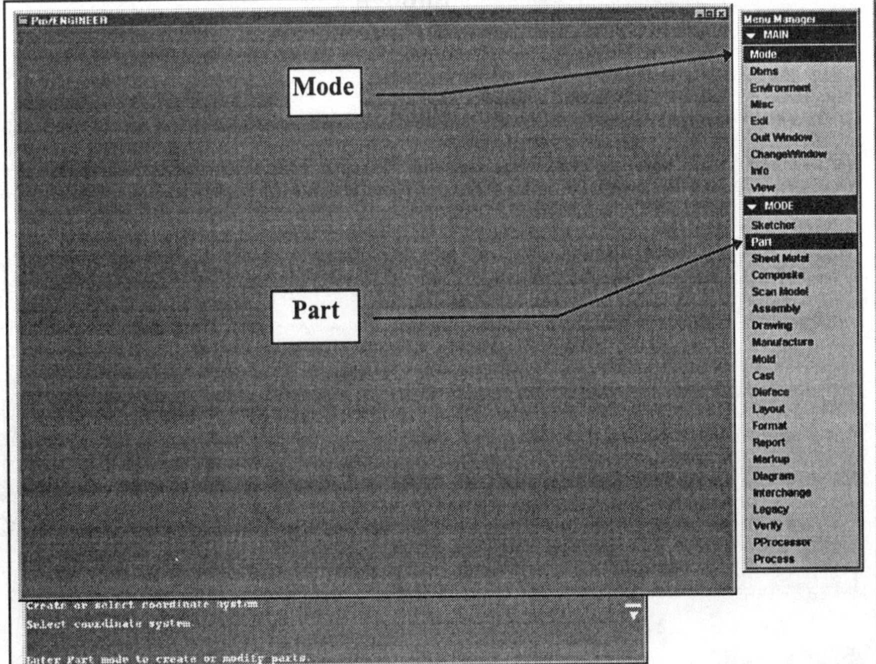

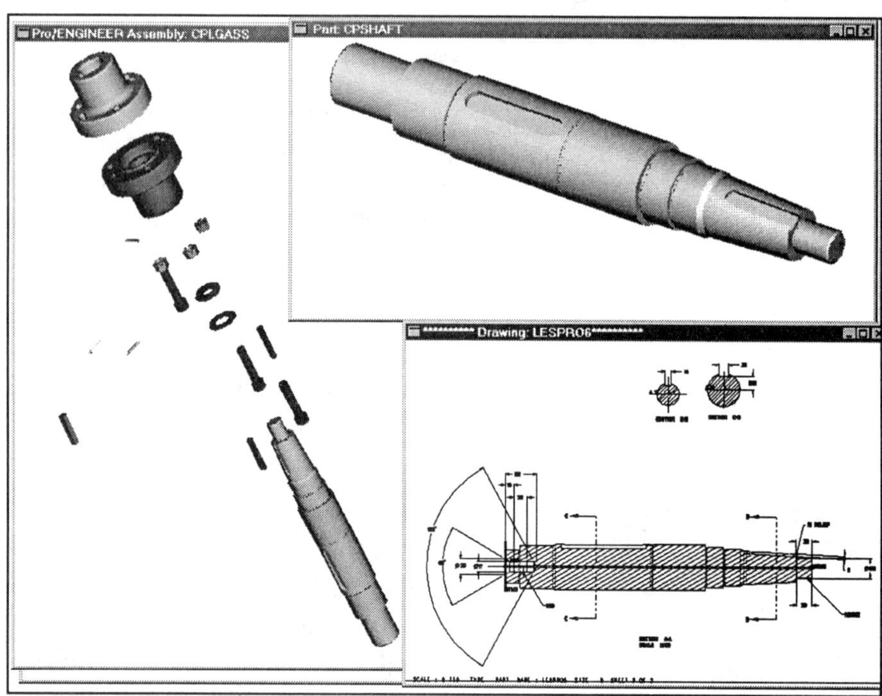

Figure 4.2
Part, Assembly, and Drawing Modes Open in the Same Pro/E Session

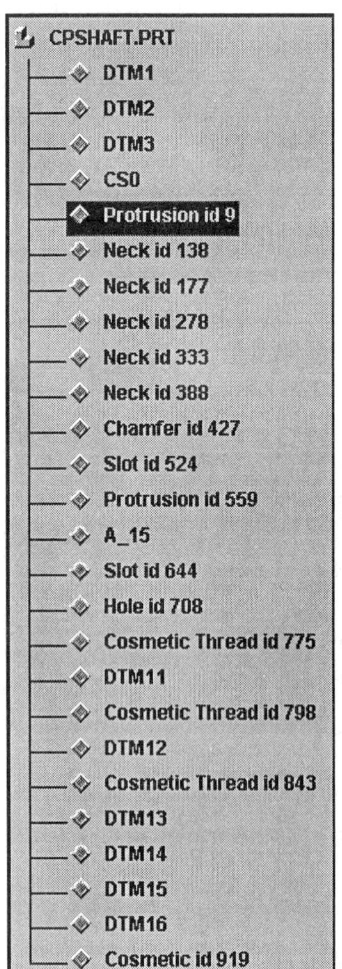

The primary modules covered in this text are:

Part This mode is used to create parts from *section sketches*, modify parts, and add features to parts (Fig. 4.3). It is the module that you will be working in most of the time. This mode is used in Lessons 1 through 13 and Lesson 21.

Assembly This mode is used to assemble components and is covered in Lessons 14 and 15.

Drawing This mode is used to create fully dimensioned drawings of parts and assemblies. The Drawing mode is used in Lessons 16 through 20.

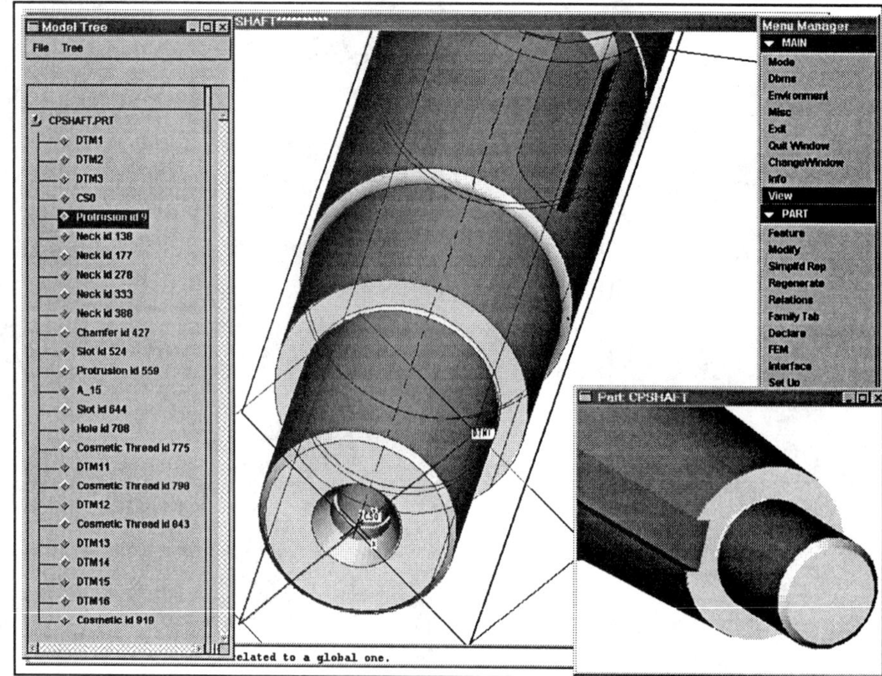

Figure 4.3
Part Mode

File Handling

File Naming

The rules for file naming, such as file name length and the effects of uppercase/lowercase names, are dependent on the operating system. Be aware of these rules before you file "objects" in Pro/E. Generally, all UNIX file names and directory names must be lowercase. DOS file names can be only eight characters long and have only one extension with three characters.

The following is an example of default-naming conventions that are used by Pro/E to aid file management. See Appendix A in *Fundamentals of Pro/E* for other naming conventions.

xxx.sec	Cross section
s2d####.sec	Default sketch
xxx.prt	Part
prt####.prt	Default part
xxx.drw	Drawing
drw####.drw	Default drawing
xxx.frm	Drawing format
xxx.asm	Assembly
asm####.asm	Default assembly
color.map	Shaded view color settings
names.inf	File used for Names INFO listing
rels.inf	Temporary file for relations listing
partname.inf	Temporary file used for Part INFO listing
partname.m_p	Temporary file used for Part Mass Properties listing
assembyname.inf	Temporary file used for Assembly INFO listing
feature.inf	Temporary file used for Feature INFO listing
config.pro	Default configuration options file
trail.txt	Default name for Trail file
xxxx.cfg	Model Tree Settings file

Because Pro/E adds a period (.) between the file name entered and the appropriate extension, do not use a period (.) in any file names you create.

File Management

Within the Pro/E environment, files (objects-where an "object" may be a part, assembly, drawing, layout, sketch, or manufacturing model) are managed through the **DBMS** menu. This menu item is found in the **MAIN** menu (Fig. 4.4). The submenu under DBMS has a variety of options. *You can abort the options' function by pressing the <ESC> key.* The following options are available:

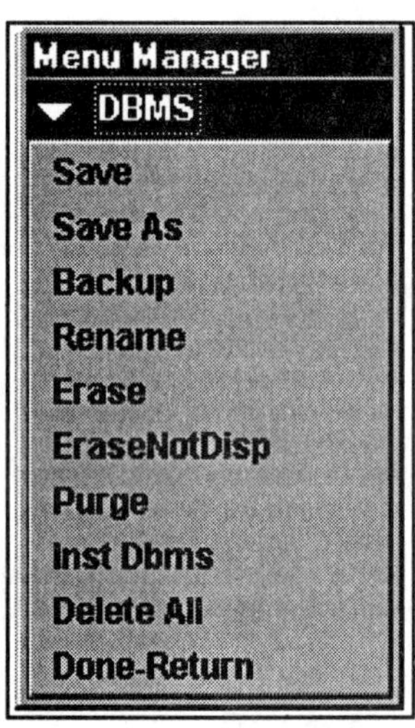

Save Files the object to the disk.
Save As Duplicates the current object and saves it with a new file name.
Backup Saves a backup copy of an object to a specified directory.
Rename Changes the file name of the current object to a new name. *Have the part, drawing, and its assembly active (in memory) before changing names!*
Erase Removes an object from workstation working memory.
EraseNotDisplay Removes all objects from workstation working memory *except* those that are currently displayed and any objects referenced by the displayed objects.
Purge Deletes all but the last object version from disk. *Use this every time you save, and it will eliminate all versions except the one you just saved. If you do not do this, then every time you save the object, you will have created a file reflecting the condition of the object at that time. If you save 27 times, you will have 27 different versions!*

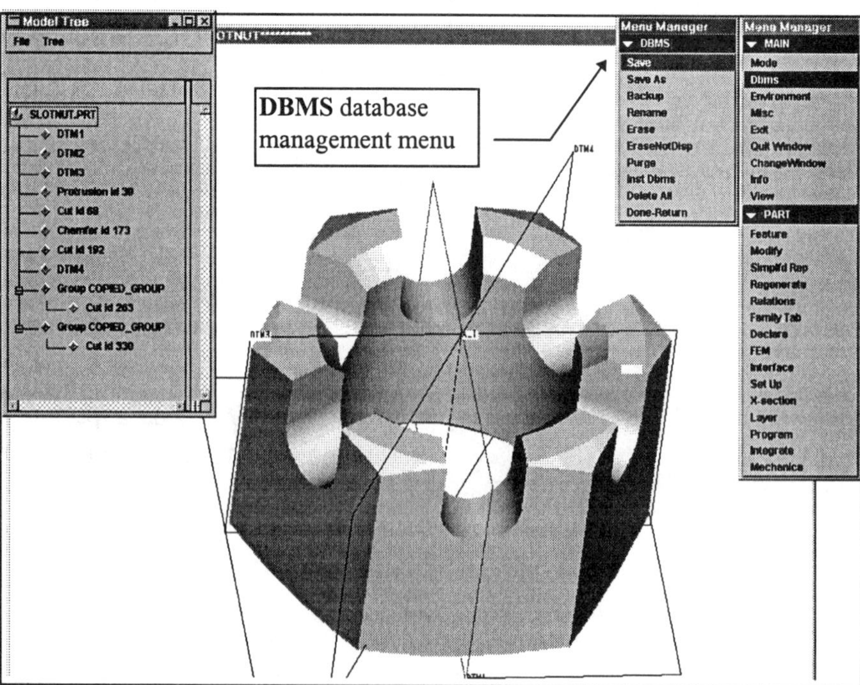

Figure 4.4
DBMS Menu

Inst Dbms Accesses the family table DBMS function.

Delete All Deletes all current and saved object versions from the disk and workstation memory. *Since it deletes all versions of the object, be very careful when using this command.*

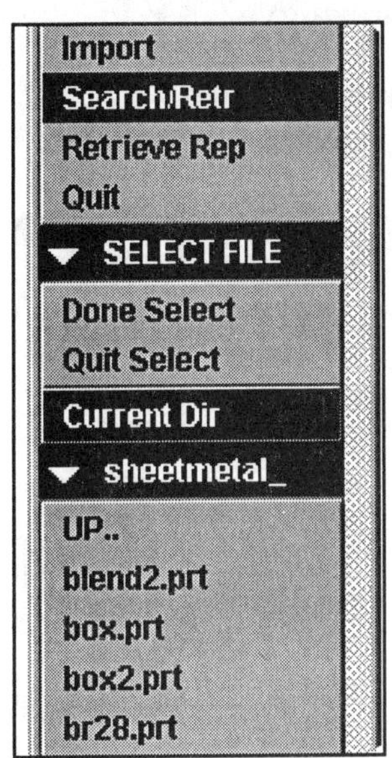

In all Pro/E modes (Sketcher, Part, Assembly, Drawing, and so on), objects can be created, retrieved, imported, and listed through an ENTERSECTION, ENTERPART, ENTERASSY, or ENTERDRAWING menu, depending on what mode you are in. For you to work in these modes, a new object must be created or an existing object retrieved. If different versions of an object exist, entering its name, extension, and version number will retrieve it. As a matter of practice, get used to using **Search/Retr**. This will allow you to navigate the current directory and change to another directory. If you use **Retrieve** and give a wrong object name or if you are in the wrong directory, Pro/E may take a considerable amount of time searching for the object.

Choosing **List** from the menu will display the objects that have been created in that mode and note any objects that have not been stored using **Dbms**. There is a configuration option that will prompt you to file only unsaved objects upon exiting Pro/E.

Choosing **Search/Retr** from the menu (Fig. 4.5) displays a list of all the objects of the desired type and directories the user can pick from instead of typing in the name of the object to be retrieved.

Figure 4.5
Using **Search/Retr**

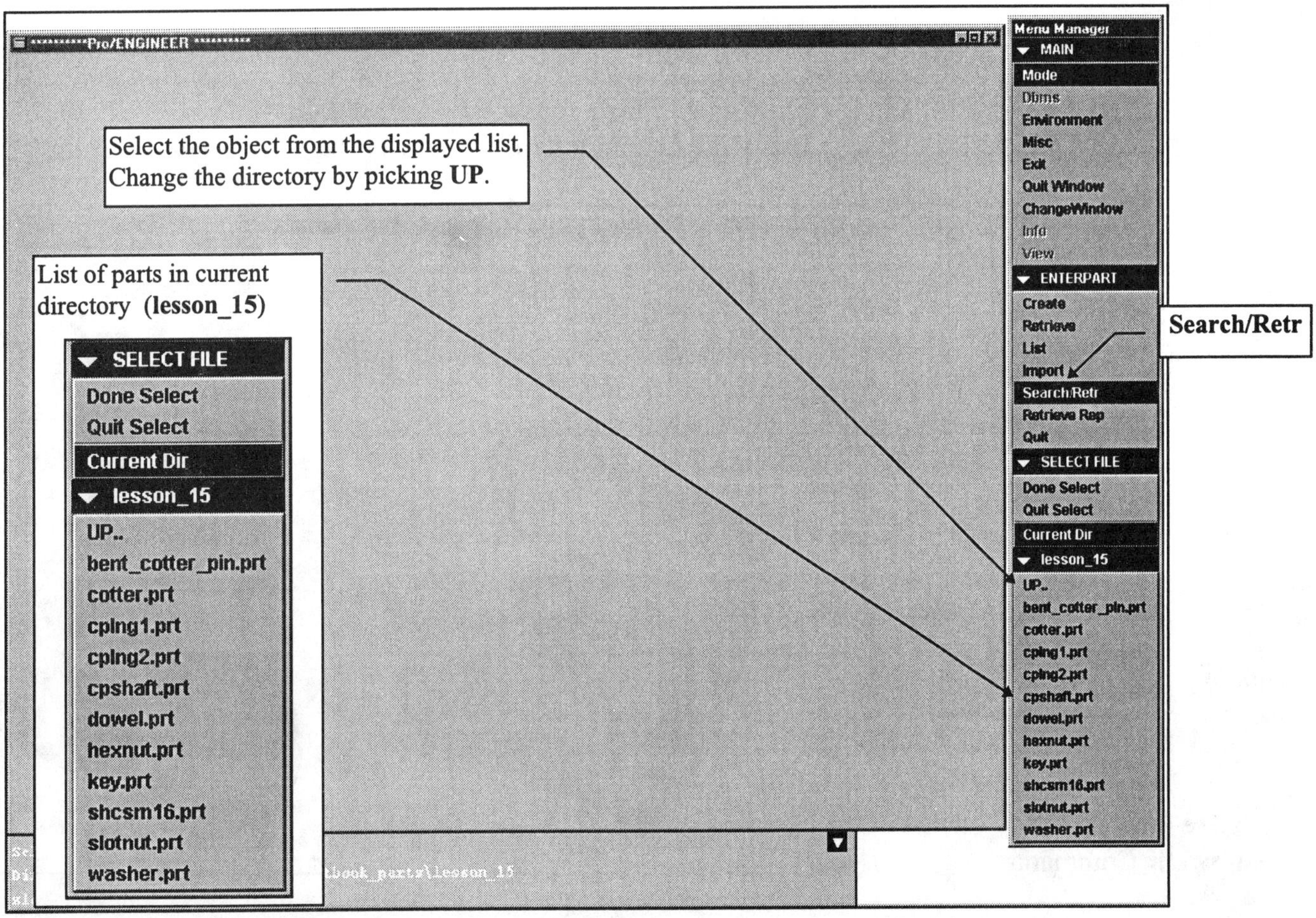

Select the object from the displayed list. Change the directory by picking **UP**.

List of parts in current directory (**lesson_15**)

Search/Retr

Operating System Access

To access the operating system or perform a system command, choose **Misc** (Fig. 4.6) from the MAIN menu. The MISC menu enables you to access various applications, including the following:

List Dir Lists the contents of the current working directory (Fig. 4.6). This is one of the most frequently used system commands.

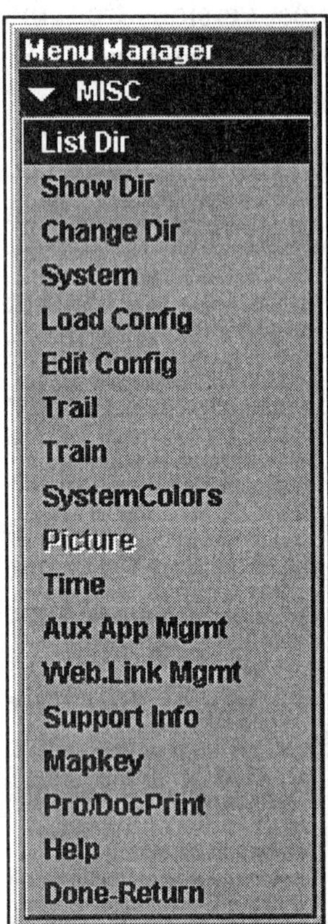

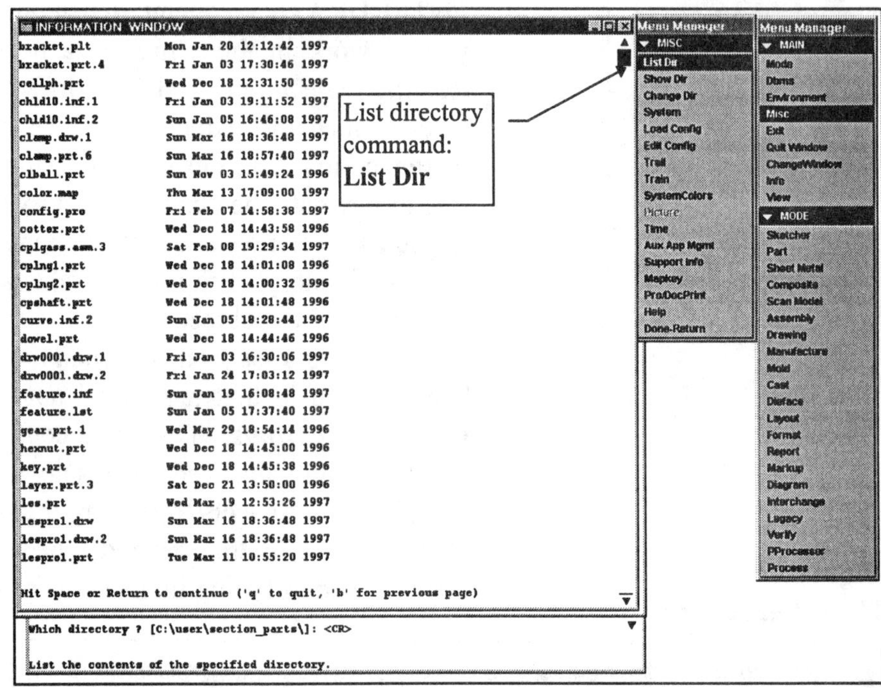

Figure 4.6
List Dir Command

Show Dir Shows the current working directory at the command line at the bottom left of the screen (Fig. 4.7).

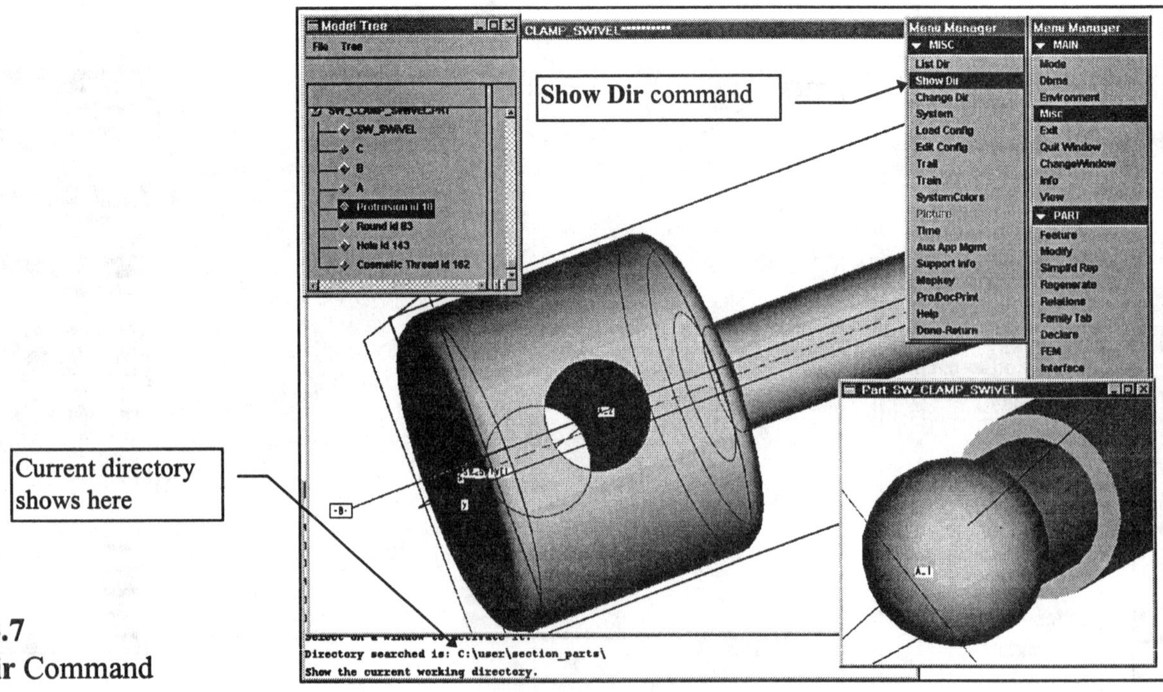

Figure 4.7
Show Dir Command

Change Dir Changes to another working directory (Fig. 4.8). Enter a **?** at the prompt to see current options. By picking **UP**, you go to the next-higher-level directory. In this example, the directories have been named **Lessons**. To make another directory the working directory, pick it from the choices and then choose **HERE**.

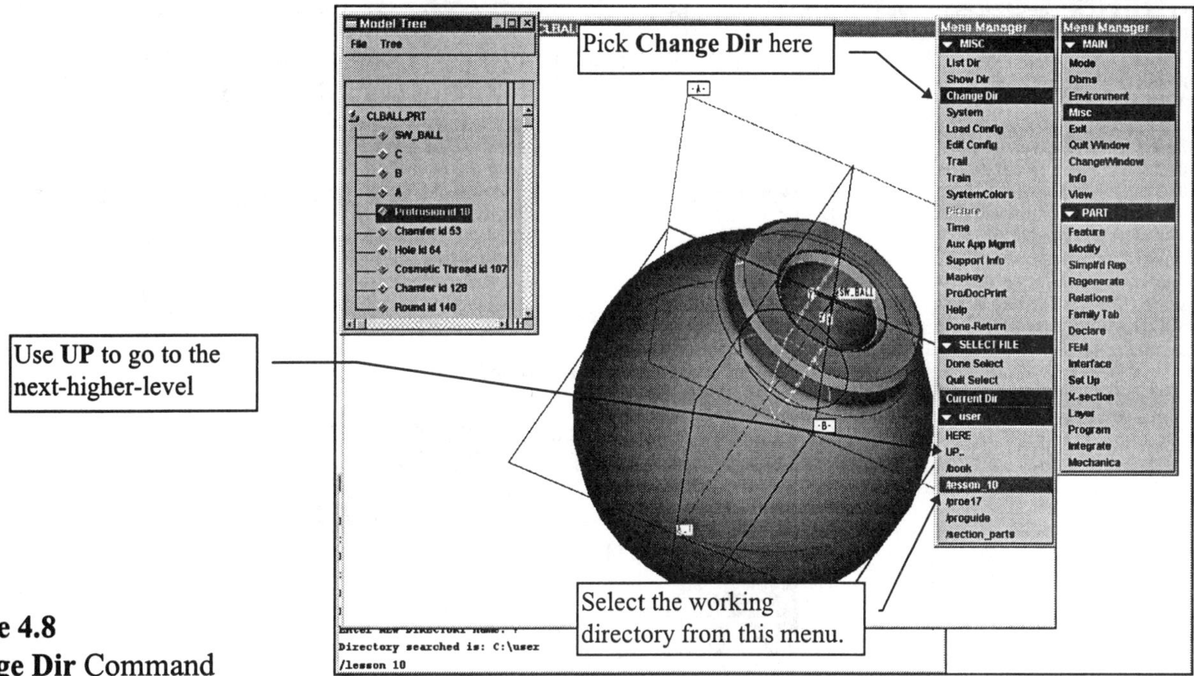

Figure 4.8
Change Dir Command

System Enters the system window without exiting Pro/E (Fig. 4.9). This window allows you access to the system-level commands. UNIX or Windows system commands can be entered here. The type and level of commands here are determined by your system privileges.

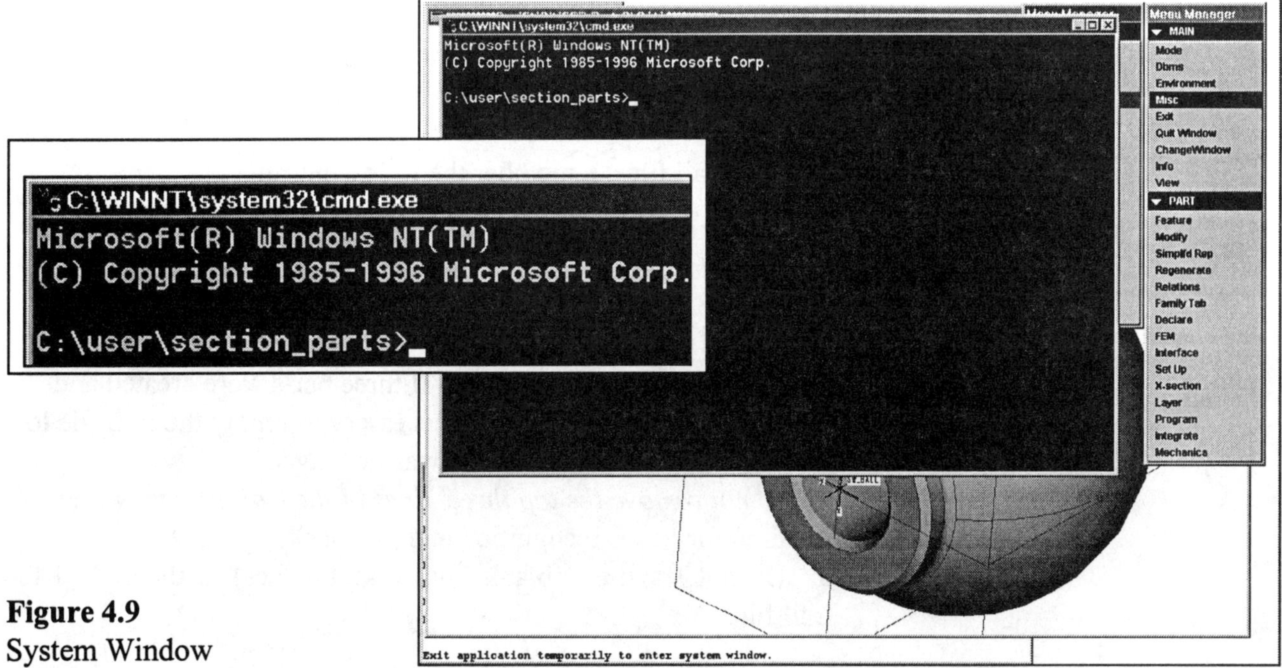

Figure 4.9
System Window

Load Config Loads a configuration file (this is explained in detail in Section 5).

Edit Config Edits a configuration file. If you use Pro/TABLE, this allows you to see all possible configuration options as well as applicable values (see Section 5).

Trail Runs a Pro/E trail file (Fig. 4.10). A **trail file** is a record of all the menu picks, selections, and keyboard entries for a single Pro/E working session. A new trail file is started every time a session of Pro/E is started, and it is saved when the session is ended. The trail file is named **trail.txt.#**. You can see and edit a trail file by opening it with your system editor.

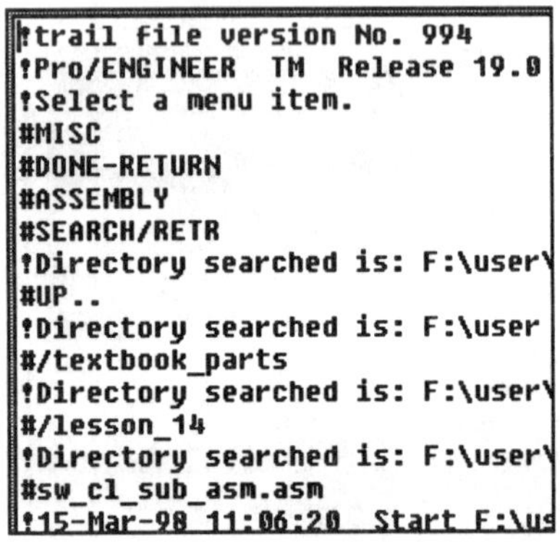

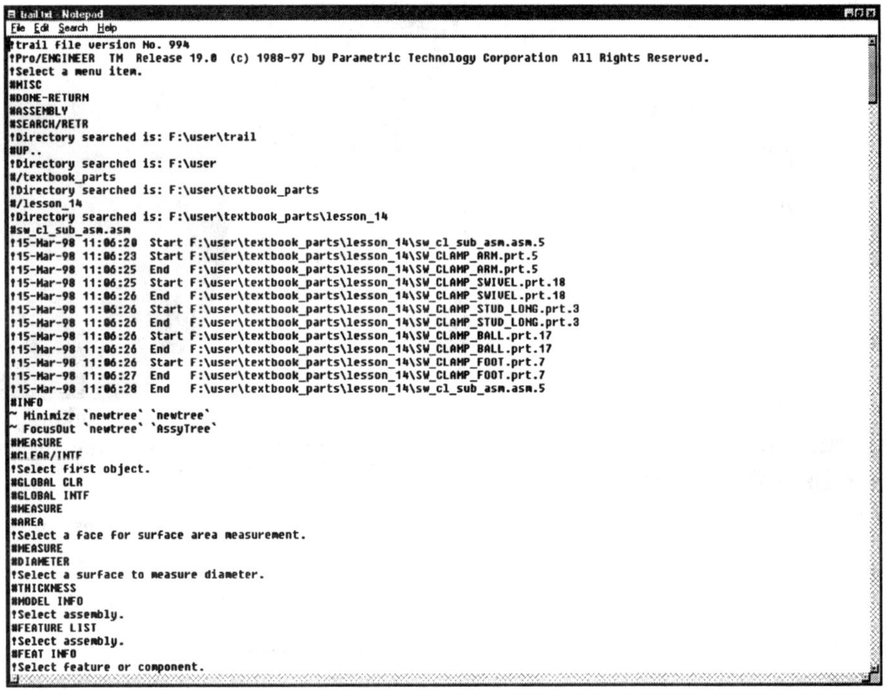

```
!trail file version No. 994
!Pro/ENGINEER  TM  Release 19.0
!Select a menu item.
#MISC
#DONE-RETURN
#ASSEMBLY
#SEARCH/RETR
!Directory searched is: F:\user\
#UP..
!Directory searched is: F:\user
#/textbook_parts
!Directory searched is: F:\user\
#/lesson_14
!Directory searched is: F:\user\
#sw_cl_sub_asm.asm
!15-Mar-98 11:06:20  Start F:\us
```

```
trail.txt - Notepad
File Edit Search Help
!trail file version No. 994
!Pro/ENGINEER  TM  Release 19.0  (c) 1988-97 by Parametric Technology Corporation  All Rights Reserved.
!Select a menu item.
#MISC
#DONE-RETURN
#ASSEMBLY
#SEARCH/RETR
!Directory searched is: F:\user\trail
#UP..
!Directory searched is: F:\user
#/textbook_parts
!Directory searched is: F:\user\textbook_parts
#/lesson_14
!Directory searched is: F:\user\textbook_parts\lesson_14
#sw_cl_sub_asm.asm
!15-Mar-98 11:06:20  Start F:\user\textbook_parts\lesson_14\sw_cl_sub_asm.asm.5
!15-Mar-98 11:06:23  Start F:\user\textbook_parts\lesson_14\SW_CLAMP_ARM.prt.5
!15-Mar-98 11:06:25  End   F:\user\textbook_parts\lesson_14\SW_CLAMP_ARM.prt.5
!15-Mar-98 11:06:25  Start F:\user\textbook_parts\lesson_14\SW_CLAMP_SWIVEL.prt.18
!15-Mar-98 11:06:26  End   F:\user\textbook_parts\lesson_14\SW_CLAMP_SWIVEL.prt.18
!15-Mar-98 11:06:26  Start F:\user\textbook_parts\lesson_14\SW_CLAMP_STUD_LONG.prt.3
!15-Mar-98 11:06:26  End   F:\user\textbook_parts\lesson_14\SW_CLAMP_STUD_LONG.prt.3
!15-Mar-98 11:06:26  Start F:\user\textbook_parts\lesson_14\SW_CLAMP_BALL.prt.17
!15-Mar-98 11:06:26  End   F:\user\textbook_parts\lesson_14\SW_CLAMP_BALL.prt.17
!15-Mar-98 11:06:26  Start F:\user\textbook_parts\lesson_14\SW_CLAMP_FOOT.prt.7
!15-Mar-98 11:06:27  End   F:\user\textbook_parts\lesson_14\SW_CLAMP_FOOT.prt.7
!15-Mar-98 11:06:28  End   F:\user\textbook_parts\lesson_14\sw_cl_sub_asm.asm.5
#INFO
~ Minimize `newtree` `newtree`
~ FocusOut `newtree` `AssyTree`
#MEASURE
#CLEAR/INTF
!Select first object.
#GLOBAL CLR
#GLOBAL INTF
#MEASURE
#AREA
!Select a face for surface area measurement.
#MEASURE
#DIAMETER
!Select a surface to measure diameter.
#THICKNESS
#MODEL INFO
!Select assembly.
#FEATURE LIST
!Select assembly.
#FEAT INFO
!Select feature or component.
```

Figure 4.10
Trail File

The file can be replayed to retrieve information that was lost in a Pro/E session (an object that was not saved to disk). Choose **Misc** ⇒ **Trail** and enter the trail file name to execute the file.

When using trail files, remember the following rules:

1. Copy the trail file to something other than **trail.txt**. Do not keep *trail* as the file name.
2. Edit the copied file as required. As an example, remove commands such as #EXIT, #SYSTEM, #DBMS, and #EXPORT.
3. Replay only what is needed. If three parts were created and two were saved, then it is only necessary to replay the trail file to recreate the remaining part that was not saved.
4. *Do not remove the top three lines of the trail file;* they are required for Pro/E recognition and playback.
5. Do not leave any blank lines (empty lines) at the end of the trail file.

Train Runs a Pro/E training trail file. You can edit and save a trail file as a training file for education or training programs.

System Colors Modifies the colors used by Pro/E for background (Fig. 4.11), geometry, letters, and so on.

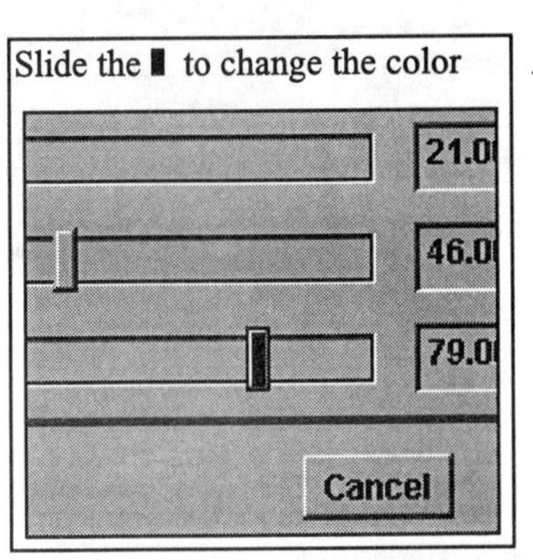

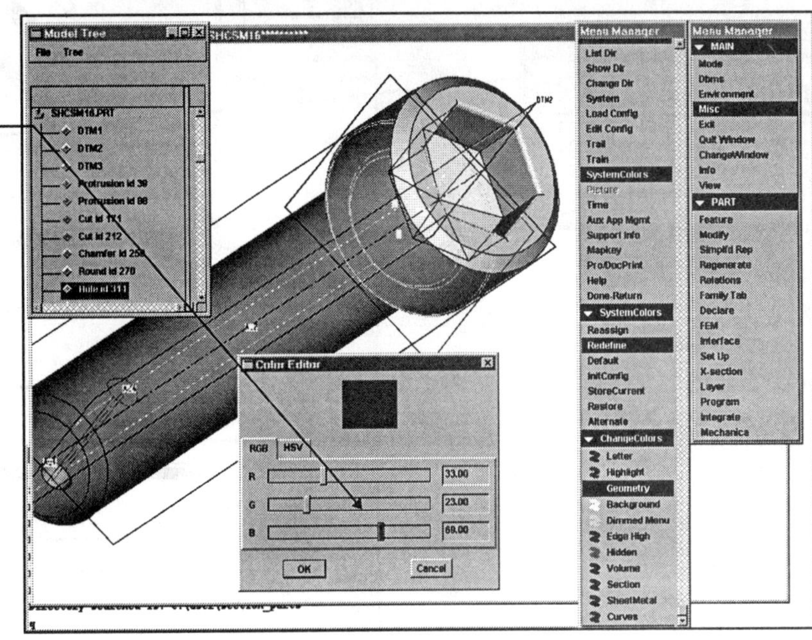

Figure 4.11
System Colors Command

Web.Link Mgmt Manage interaction with Web.Link applications.

Support Info Provides information about your system and current Pro/E products installed on your system, including addresses and phone and FAX numbers for support (Fig. 4.12).

Time Displays the current time at the command line.

Mapkey Creates a macro interactively (Fig. 4.13).

Aux App Mgmt Manages auxiliary applications (Fig. 4.14).

Pro/Doc Print Prints a Pro/Help document (Fig. 4.15).

Help Starts online help system.

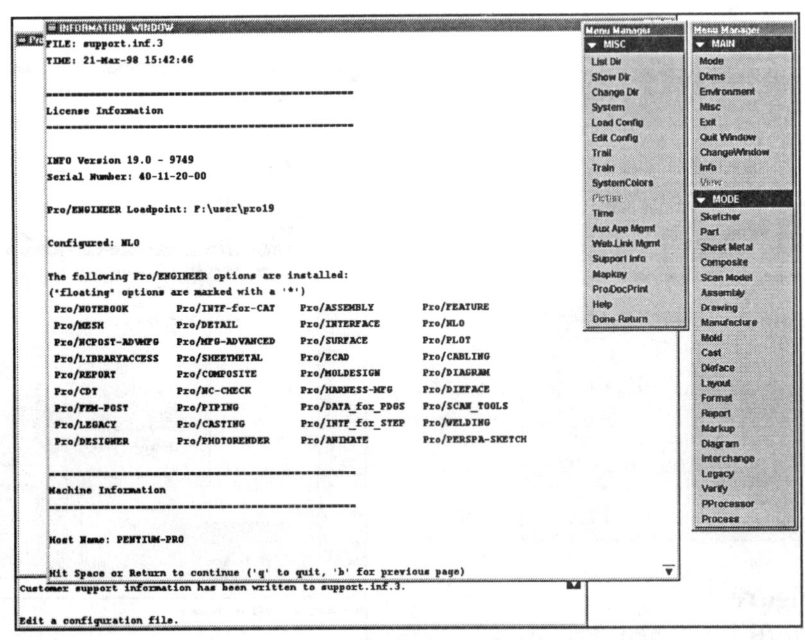

Figure 4.12
Support Info Command

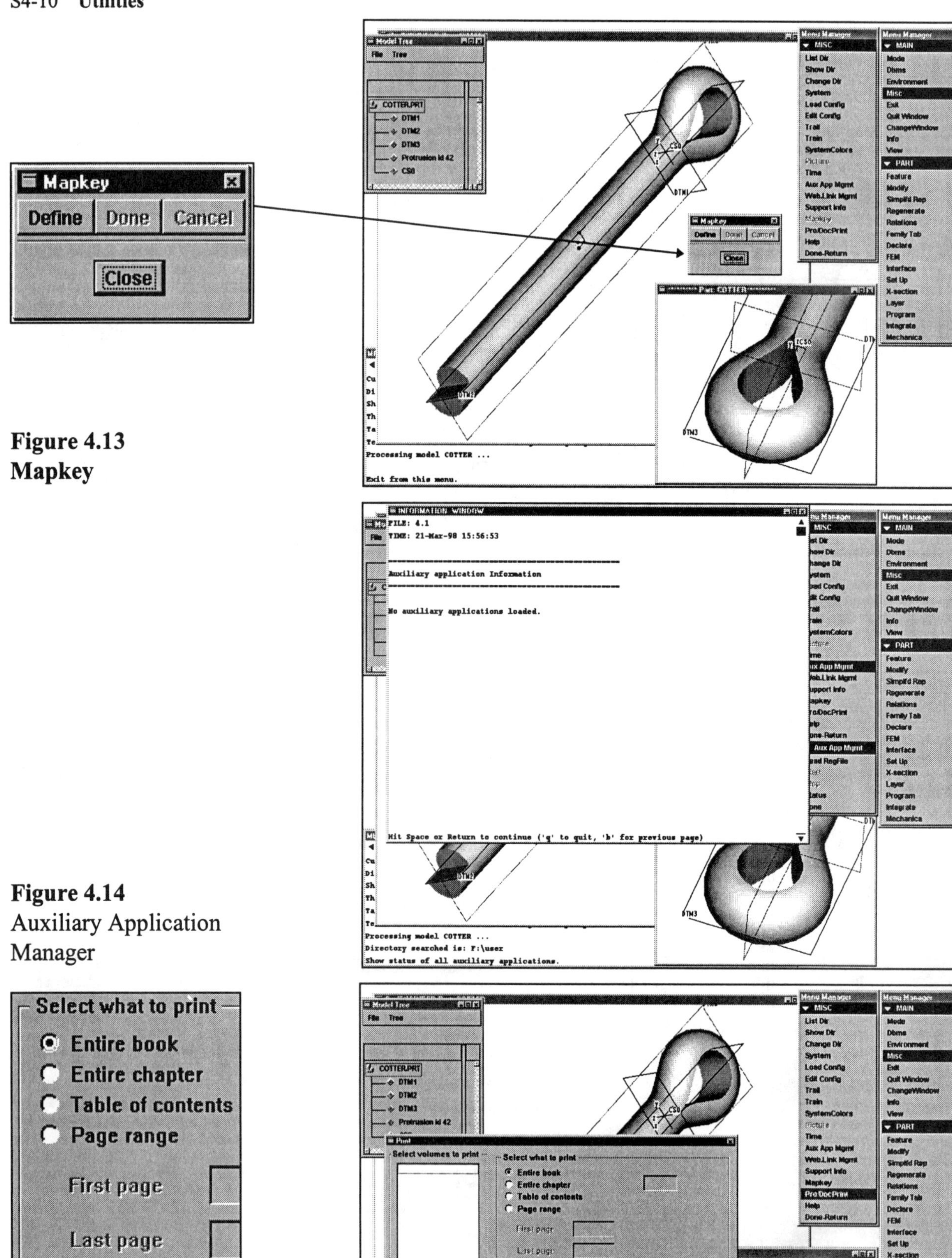

Figure 4.13
Mapkey

Figure 4.14
Auxiliary Application Manager

Figure 4.15
Pro/E Document Printer

Section 5

Configuration Files and Mapkeys

A **configuration file** allows you to configure (set up) a Pro/E working session to your project or company requirements. A configuration file can set school or company standards for storage, formatting, establishing default units for new parts (such as millimeters instead of inches). Configuration files are important in establishing the path to the location of directories that contain your library items.

Each user can have individual configuration files, but establishing a system-wide configuration file is a way to ensure compliance of project standards.

You can control the environment in which Pro/E runs, that is, specify the way your objects are oriented, plotting configurations, environment settings, table editor, system colors, and so on. There are two ways to set the working environment: *editing the configuration file* and *using the environment menu.*

Configuration Files (*config.pro*)

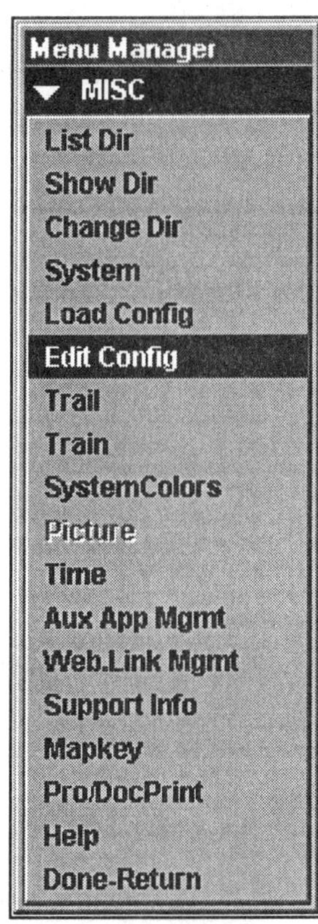

The default configuration file name is *config.pro*. You can enter the working environment settings into this configuration file before starting a new Pro/E session. These settings will then be activated every time you start a new session. If you do not enter a particular setting in the configuration file and there is no master configuration file, Pro/E will use its own default settings. Figure 5.1 shows a typical personalized configuration file.

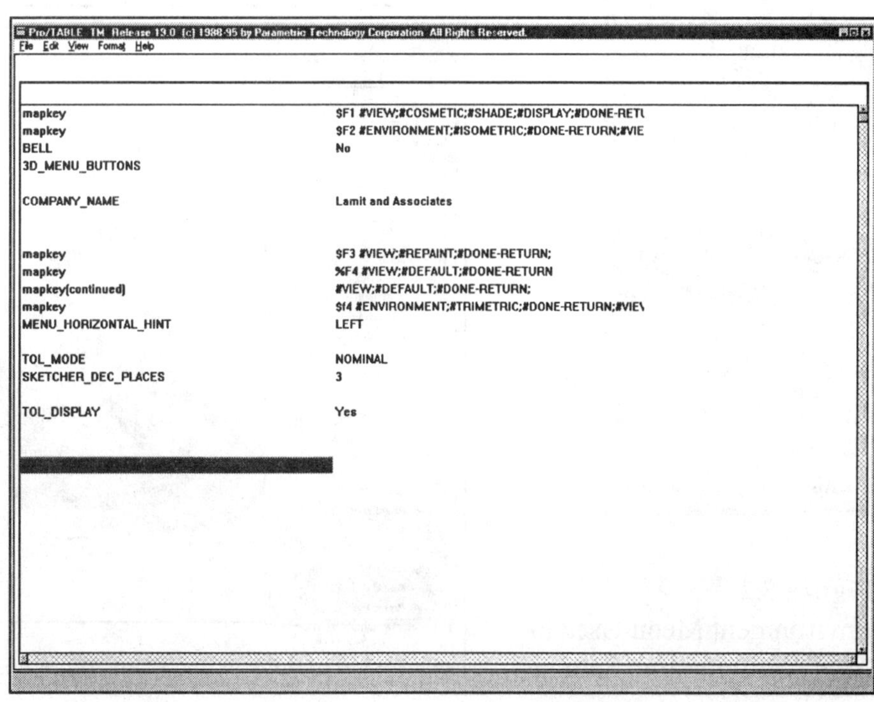

Figure 5.1
Configuration File

Using the Environment Menu

Frequently used working environment settings are included in the ENVIRONMENT menu, which is accessible from every Pro/E mode. This enables you to change the settings easily during the current session. When you enter Pro/E, the values of the environment settings are set to those in your *config.pro* file; otherwise, they are set to the default. In Figure 5.2, the ENVIRONMENT menu is shown. To change a setting, just pick the item; a check (✔) will appear to activate this capability. To remove a setting, pick the item and the check will disappear.

Setting Configuration Files

All the environment options and other global settings can be preset, before starting Pro/E, by entering the relevant options and their settings into the configuration file. *If Pro/TABLE is not available, a system window comes up. You can then edit your config.pro file with your system editor.* Configuration files can also include settings for tolerance display, formats, calculation accuracy, and the number of digits used in Sketcher dimensions. Pro/E assumes default settings for variables that can be specified in the configuration file but are not.

You may add comments to the configuration file by entering an exclamation point (!) at the beginning of a line (Fig. 5.3).

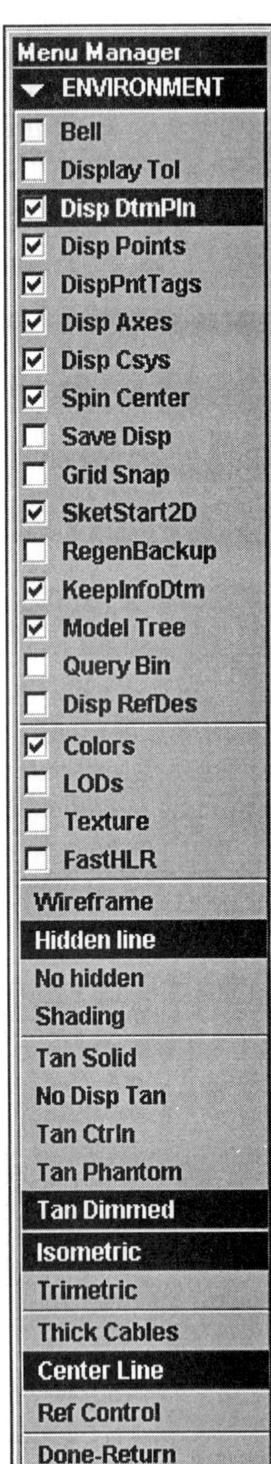

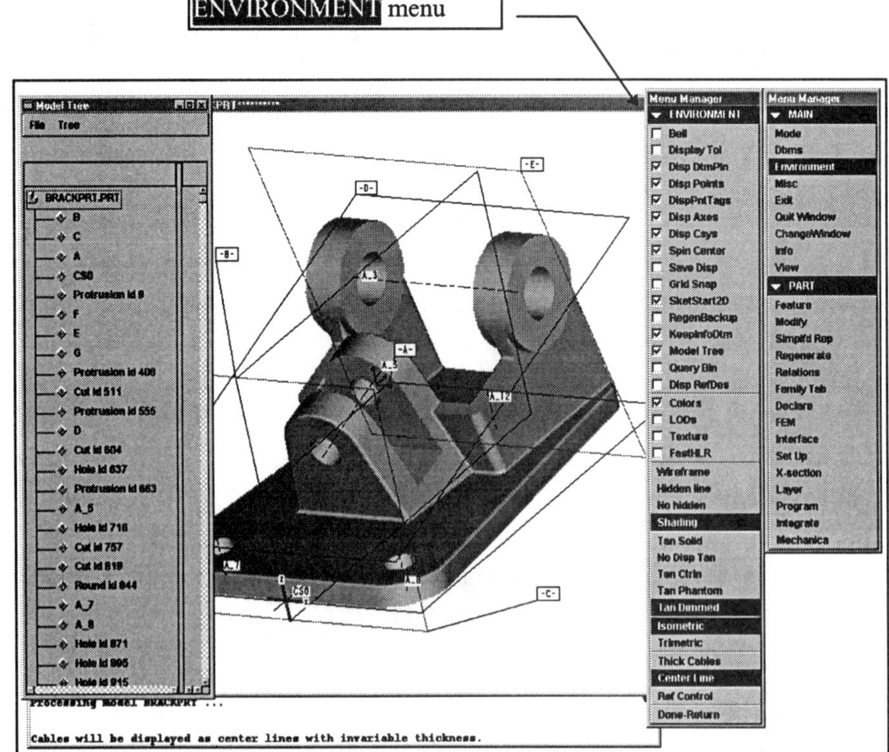

Figure 5.2
Environment Menu Used to Alter Configuration Settings

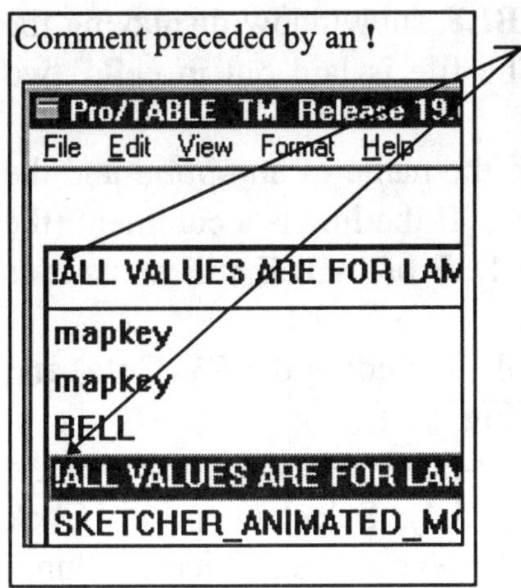

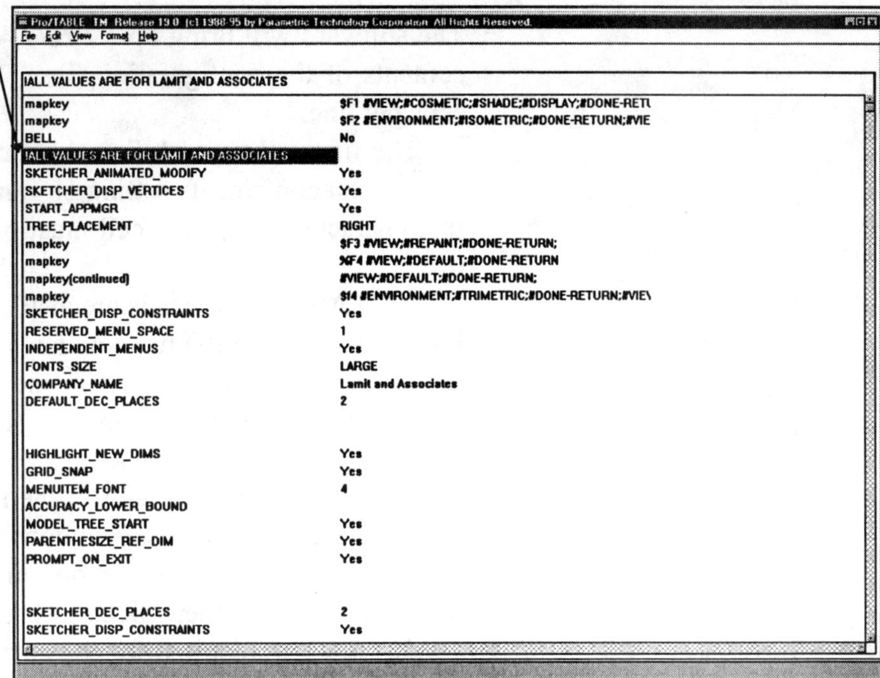

Figure 5.3
Configuration File with Comment

Pro/E can read configuration files from several areas. If a particular option is present in more than one configuration file, the latest option and its setting are the ones used by the software.

At startup, Pro/E first reads in a protected system configuration file called *config.sup*. Next, it reads configuration files from the following directories, in the order in which they are listed:

1. *Loadpoint*/text (*loadpoint* is the directory in which Pro/E is installed): Your system administrator may have put a configuration file in this location to support your organization's standards for formats and libraries. Anyone accessing Pro/E from this loadpoint will use the settings in this file.
2. Login directories: This is the home directory for your login ID. Placing your configuration file here lets you start Pro/E from any directory without having a copy of the file in each directory.
3. Startup directory: This is your current or working directory when you start Pro/E. The local *config.pro* (in your startup directory) is the last to be read; therefore, it will override any conflicting *config.pro* option entries.
4. Lastly, Pro/E uses the default settings. These settings are built into the software and automatically take effect unless a configuration file option specifies otherwise.

Editing a Configuration File

To edit a configuration file during a Pro/E session, choose **Edit Config** from the MISC menu, enter the file name (*config.pro* is the default), and press **enter**.

The software will bring up a **Pro/TABLE** subwindow displaying the contents of the configuration file. The file is laid out in cells, two cells per line.

The first cell on each line contains the name of an option and the second cell contains its setting or value. If the line is a comment (the first character of the first cell must be !), Pro/TABLE still treats it as two cells.

Pro/TABLE commands are available, including the **F3** (Goto) and **F4** (Choose keywords) function keys (Fig. 5.4).

After highlighting a cell in the left-hand column, press **F4**; the **Choose Keyword** window appears (Fig. 5.4). Select an option. Click on **OK** and the *config.pro* setting will appear in the left-hand column (Fig. 5.5). Now select a cell in the right-hand column and press **F4**. The **Choose Keyword** subwindow appears, with a list of possible (nonnumeric) values for the option in the corresponding left cell. Select one of them, then click on **OK** to enter it in the cell.

If the value required for the current option is a number (e.g., for "angular_tol") or a user-defined string, such as "Mapkey", the **Choose Keyword** subwindow does not appear.

Figure 5.4
Choose Keyword Window

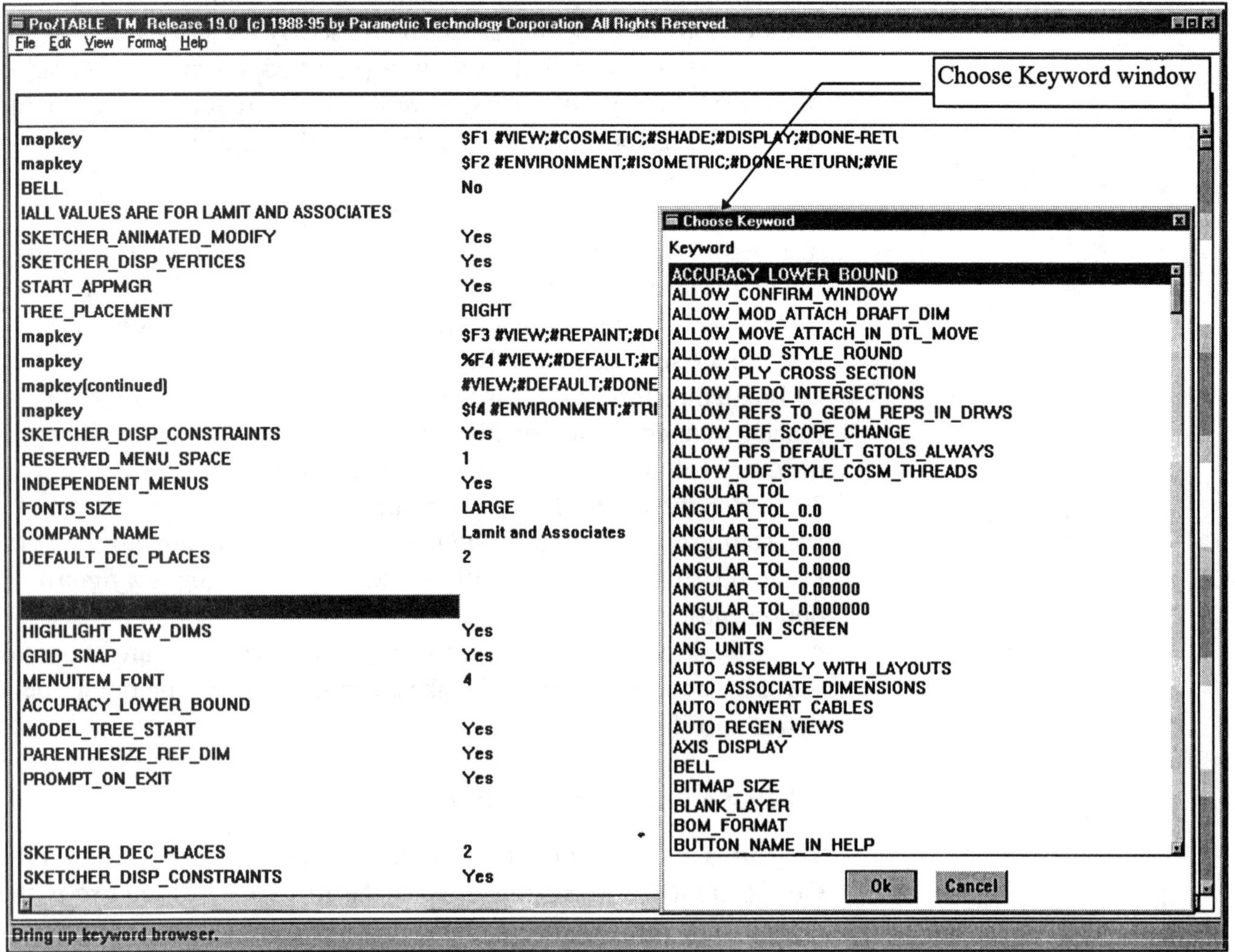

In Figure 5.5, the GRID_SNAP configuration file option was chosen, and the **Choose Keyword** window gives a choice of **Yes** or **No** for the value.

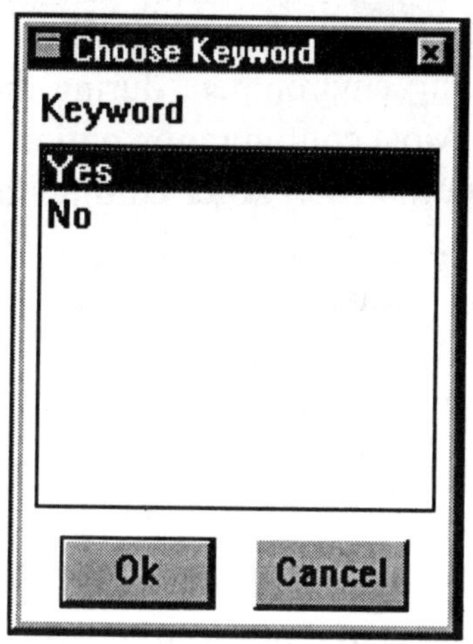

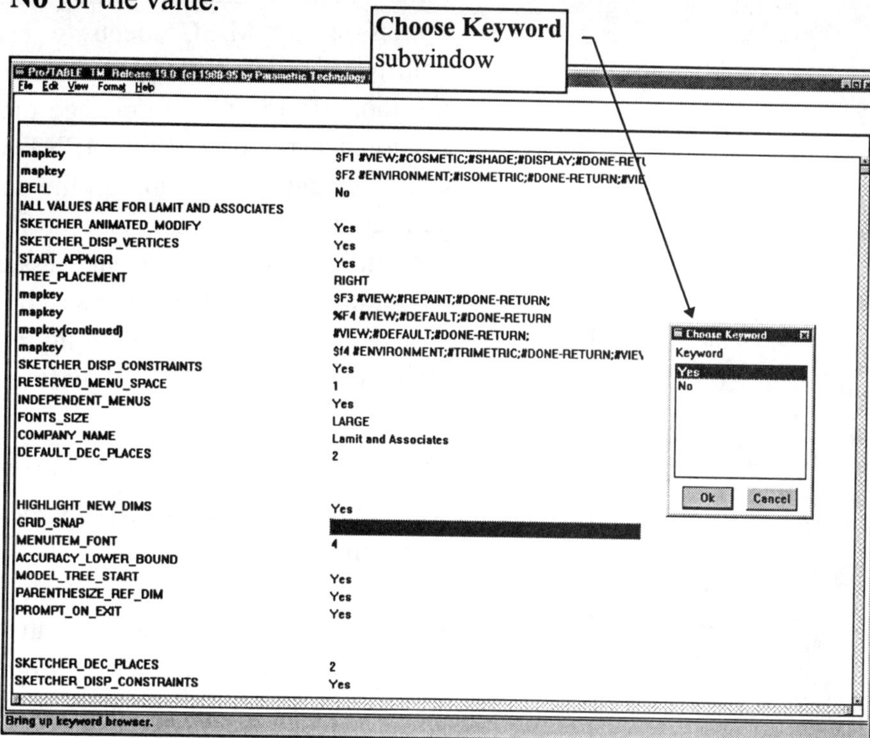

Figure 5.5
Choose Keyword Value Subwindow

A configuration file line cannot be more than 80 characters long. After creating the required configuration options and setting their values, pick **File** from the menu and **Exit** to save the new settings and close Pro/TABLE (Fig. 5.6).

Again, if Pro/TABLE is not available, a system window comes up. You can then edit your *config.pro* file with your system editor.

Exit Alt+F4
saves *config.pro* file and exits Pro/TABLE

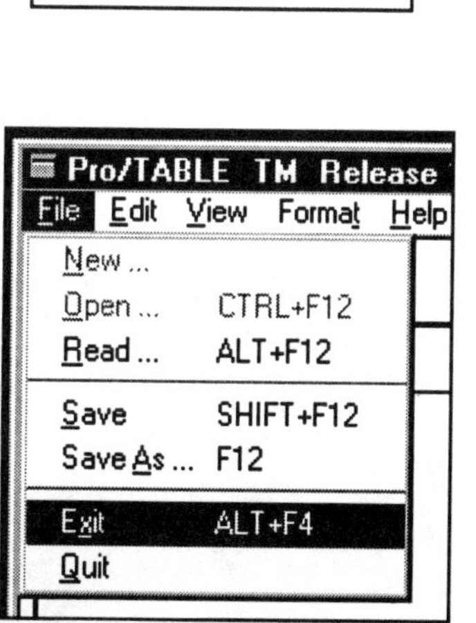

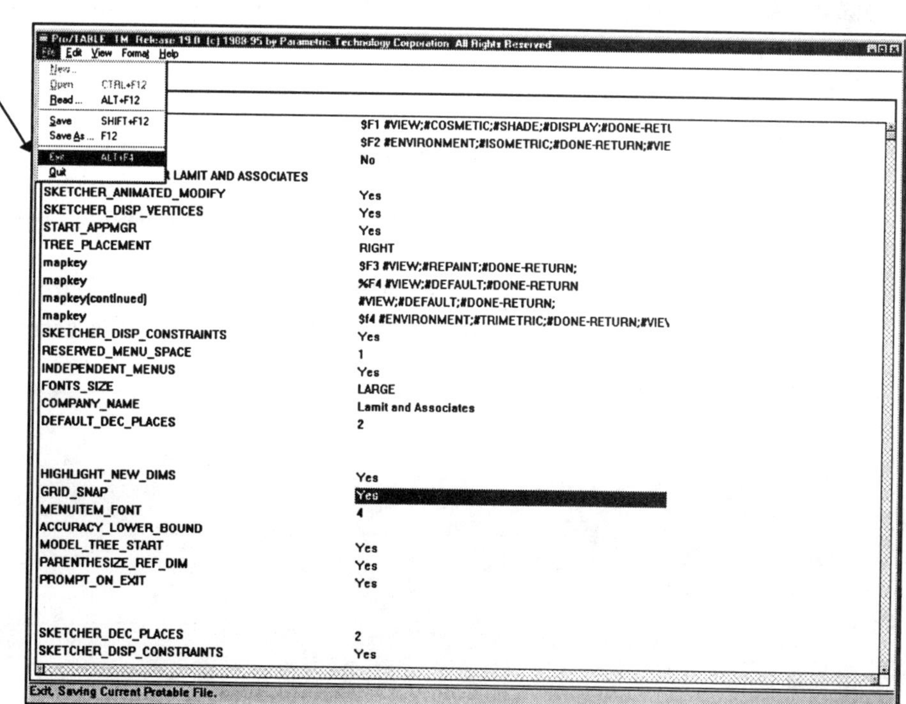

Figure 5.6
Saving and Exiting *config.pro*

Loading a Configuration File

After you have finished editing, load the file, using the **Load Config** option in the MISC menu to enable the new settings. Default configuration file settings are made in the process of starting a Pro/E session. If you want to change the working environment during the session, use the ENVIRONMENT menu. Most configuration options, can be changed only through the configuration file. After editing the configuration file, you must load it. For many configuration options, you must exit the current session and restart Pro/E, from the beginning, so that the new configuration settings are installed properly. If you are working on a part, assembly, or drawing, remember to save your work before exiting and restarting Pro/E.

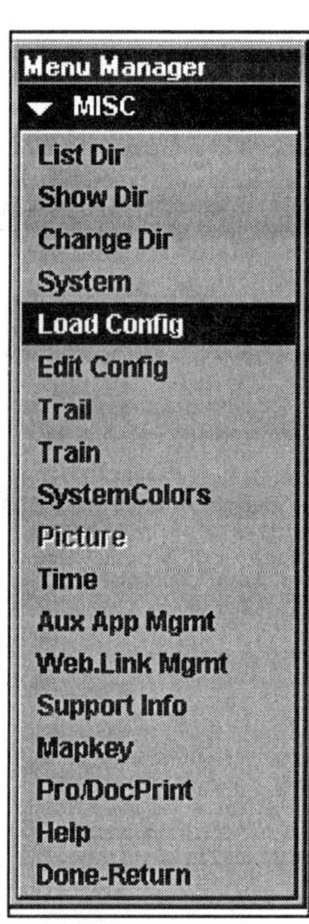

To load a configuration file during a session:

1. Choose **Misc** from the MAIN menu, then **Load Config** from the MISC menu.
2. Enter the full configuration path and file name (*config.pro* being the default). Pro/E updates with the new configuration file settings. If there are errors in the new configuration file, a message is displayed in the startup window.

Mapkeys

You can create keyboard macros, which map frequently used command sequences to certain keyboard keys. The macros are saved in the configuration file. Each macro begins on a new line. Keyboard macros require the following format:

mapkey keyname #command;#command; . . .

As an example, the following shows how to create a macro to map the shading-command sequence to the keyboard character sequence **vs**:

mapkey vs #view;#cosmetic;#shade;#display;#done-return;

Entering **vs** from the keyboard during a current Pro/E session causes the current model (part or assembly) to be shaded.

In Figure 5.7, a macro to save and purge the current model is associated with function key **F5**:

mapkey $F5 #dbms; #save; #purge; #done-return

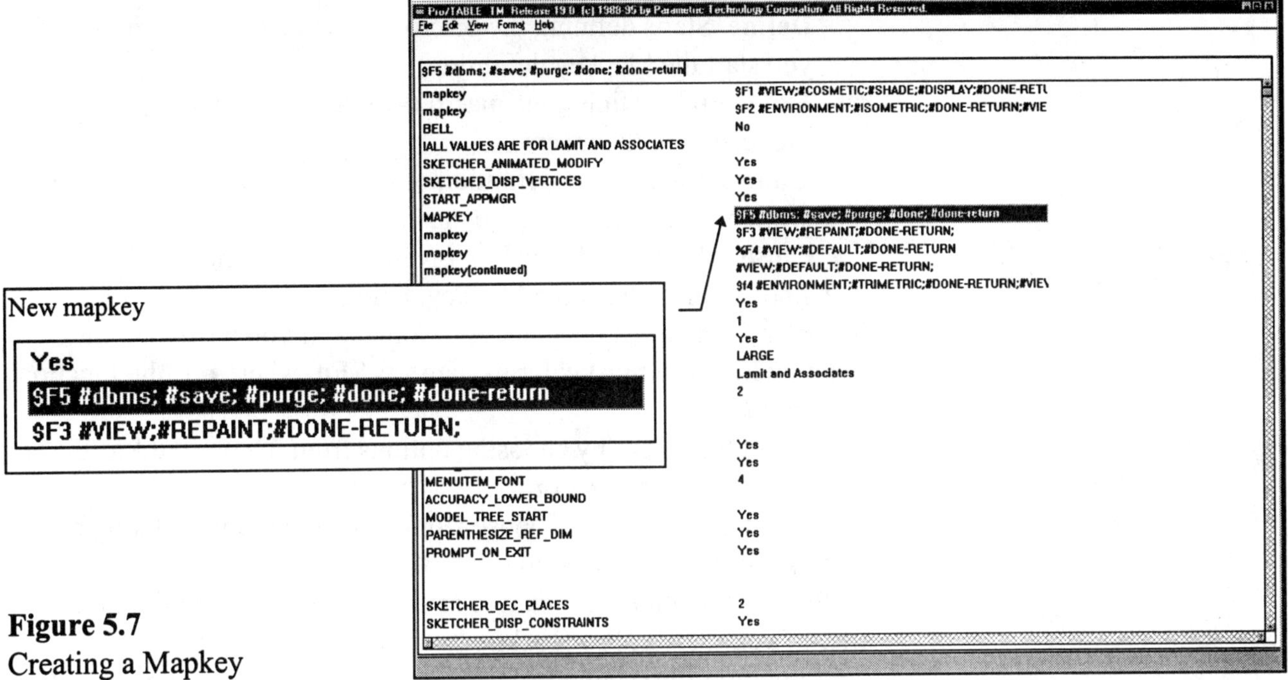

Figure 5.7
Creating a Mapkey

Creating Macros

You can create keyboard macros in two ways:

1. Create the macro interactively during the current session. Pro/E records the keystrokes. Upon completion, Pro/E can append the macro to the configuration file in your current directory.
2. Edit the current configuration file.

When you are creating keyboard macros, you must follow these syntax requirements:

1. The mapped keys can be either keyboard **function keys** or regular keyboard **characters**. You cannot use mixed sets, that is, function key plus keyboard character.
2. The **$** character is reserved for specifying function keys. If you wish to use a function key, you must enter its name as **$Fn,** where **n** is the key number. You are limited by the number of function keys on your keyboard, usually 12. (Pro/E supports up to 20 function keys.)

Creating Macros Interactively

Starting with Pro/E 17, a macro can be created interactively by following these steps (Figs. 5.8 and 5.9):

1. Choose **Mapkey** from the MISC menu. The Mapkey dialog box appears, with the following options:

   Define Starts defining the macro (available only before you start the Mapkey definition).

   Done Ends defining the macro (available only while you are recording the macro).

   Cancel Cancels the definition of the current macro (available while you are recording the macro).

   Close Closes the dialog box (available at all times).
2. Choose **Define** from the Mapkey dialog box.
3. Enter the name of the mapkey. Note that if you wish to use a function key, you must enter its name as **$Fn**, where **n** is the function key number.
4. Record the macro by choosing options from the desired Pro/E menus in the appropriate order.
5. To end the definition, choose **Done** from the Mapkey dialog box. The macro is now defined in session.
6. Pro/E asks you if you want to save the macro to a configuration file. If you reply **y**, Pro/E asks you to enter the name of a configuration file (*config.pro* is the default). Pro/E then appends the macro to the current configuration file. If you reply **n**, the macro is defined only in the current session. Pro/E does *not* append it to a configuration file. When you end the current Pro/E session, the macro disappears.
7. To define another macro, repeat steps 2 through 6.
8. Choose **Close** to end this process.

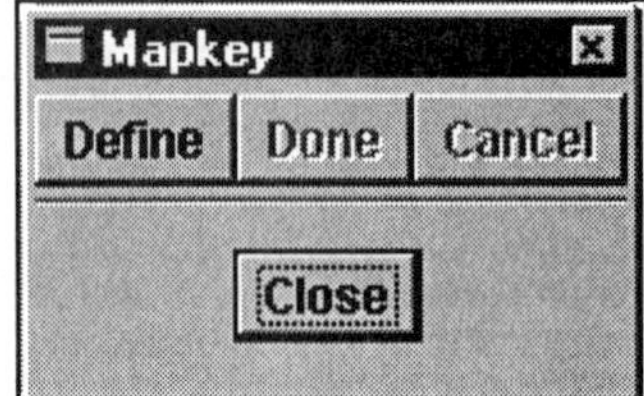

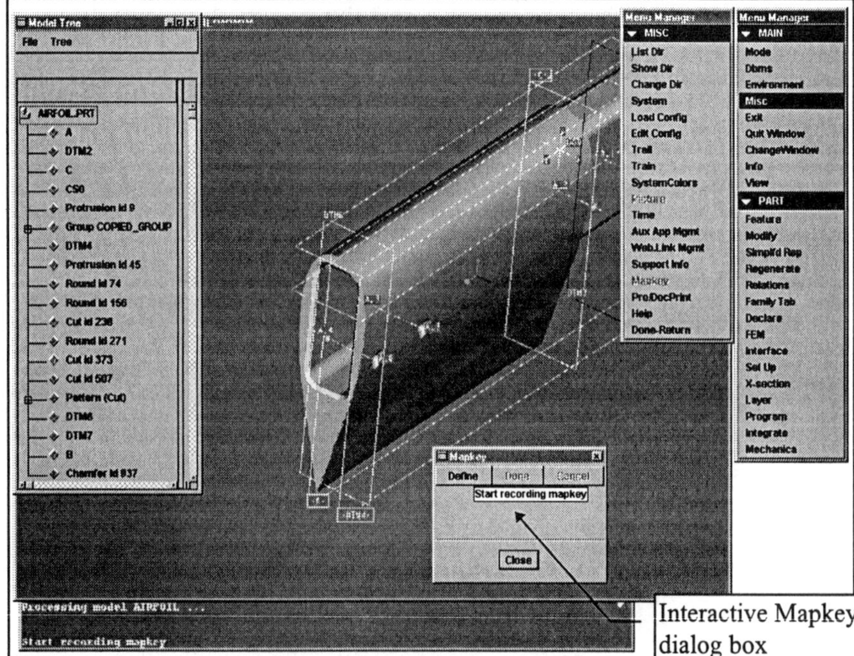

Figure 5.8 Creating a Mapkey Interactively

Choose the command sequence that you wish to become a macro (**Mapkey**)

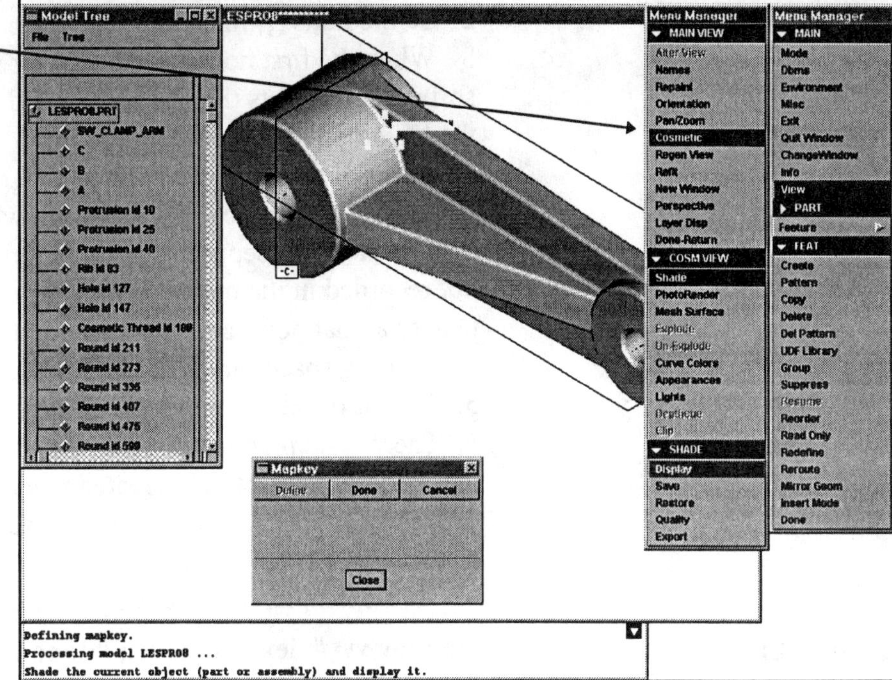

Figure 5.9
Mapkey Creation

The Mapkey dialog box can be left open and will not affect the operation of Pro/E. You can see and edit your interactively created mapkey by picking **Misc** ⇒ **Edit Config** ⇒ **enter**. Figure 5.10 shows the open configuration file and the new **vp** mapkey.

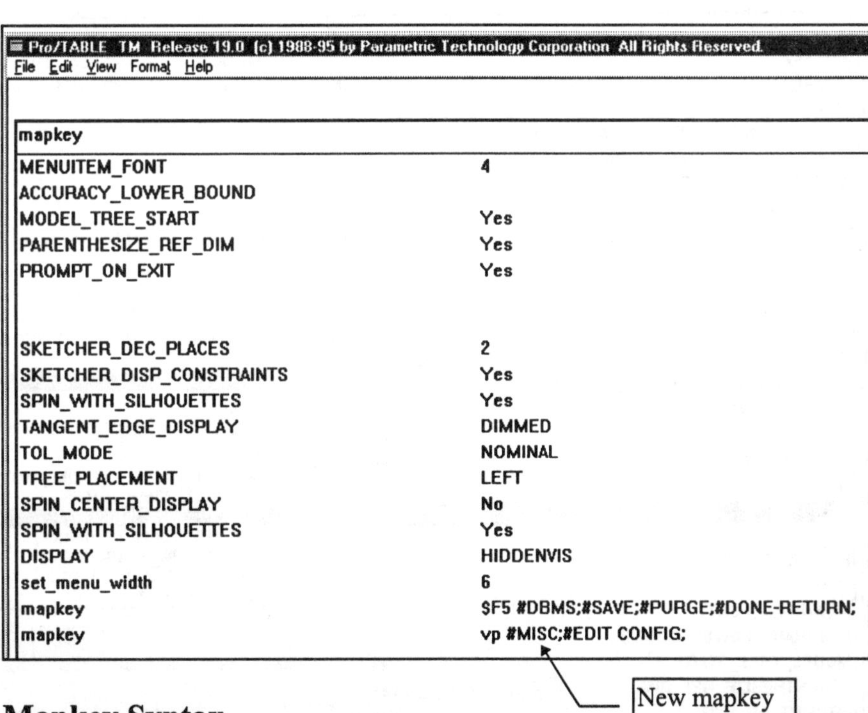
Pro/TABLE TM Release 19.0 (c) 1988-95 by Parametric Technology Corporation All Rights Reserved.

File Edit View Format Help

mapkey	
MENUITEM_FONT	4
ACCURACY_LOWER_BOUND	
MODEL_TREE_START	Yes
PARENTHESIZE_REF_DIM	Yes
PROMPT_ON_EXIT	Yes
SKETCHER_DEC_PLACES	2
SKETCHER_DISP_CONSTRAINTS	Yes
SPIN_WITH_SILHOUETTES	Yes
TANGENT_EDGE_DISPLAY	DIMMED
TOL_MODE	NOMINAL
TREE_PLACEMENT	LEFT
SPIN_CENTER_DISPLAY	No
SPIN_WITH_SILHOUETTES	Yes
DISPLAY	HIDDENVIS
set_menu_width	6
mapkey	$F5 #DBMS;#SAVE;#PURGE;#DONE-RETURN;
mapkey	vp #MISC;#EDIT CONFIG;

New mapkey

Figure 5.10
Configuration File Showing the Interactively Created Mapkey

Mapkey Syntax

A variety of rules must be applied when you are creating macros or editing macros that were originally created *interactively* or by editing the configuration file (*config.pro*):

1. Precede each command with a # sign.
2. Semicolons (;) must be used to separate commands.
3. When the first non-space character in a field is not a # sign, the software interprets the rest of the field as if you had entered it from the keyboard in response to a prompt.

If Pro/E prompts you for a keyboard input while you are creating the macro, you can enter a response, which is then recorded in the macro. If you accept the default by pressing **Enter**, the response is *not* recorded in the macro. When you execute the macro, Pro/E pauses at that point and prompts you for keyboard input.

4. Leading spaces are ignored by the software.
5. Except for keyboard input, entries are not case-sensitive.
6. There is no practical limit to the length of a macro (the maximum length of an interactively created macro is 2048 characters). Use the backslash (\ ) as a continuation character. Do *not* have any characters or spaces after the backslash.

For example, the following macro, **vs**, was created as shown:

**mapkey vs #view; #default; #done-return; #view; **
#cosmetic; #shade; #display; #done-return;

Appendix B has a sample mapkey file that you can create and save in your current directory.

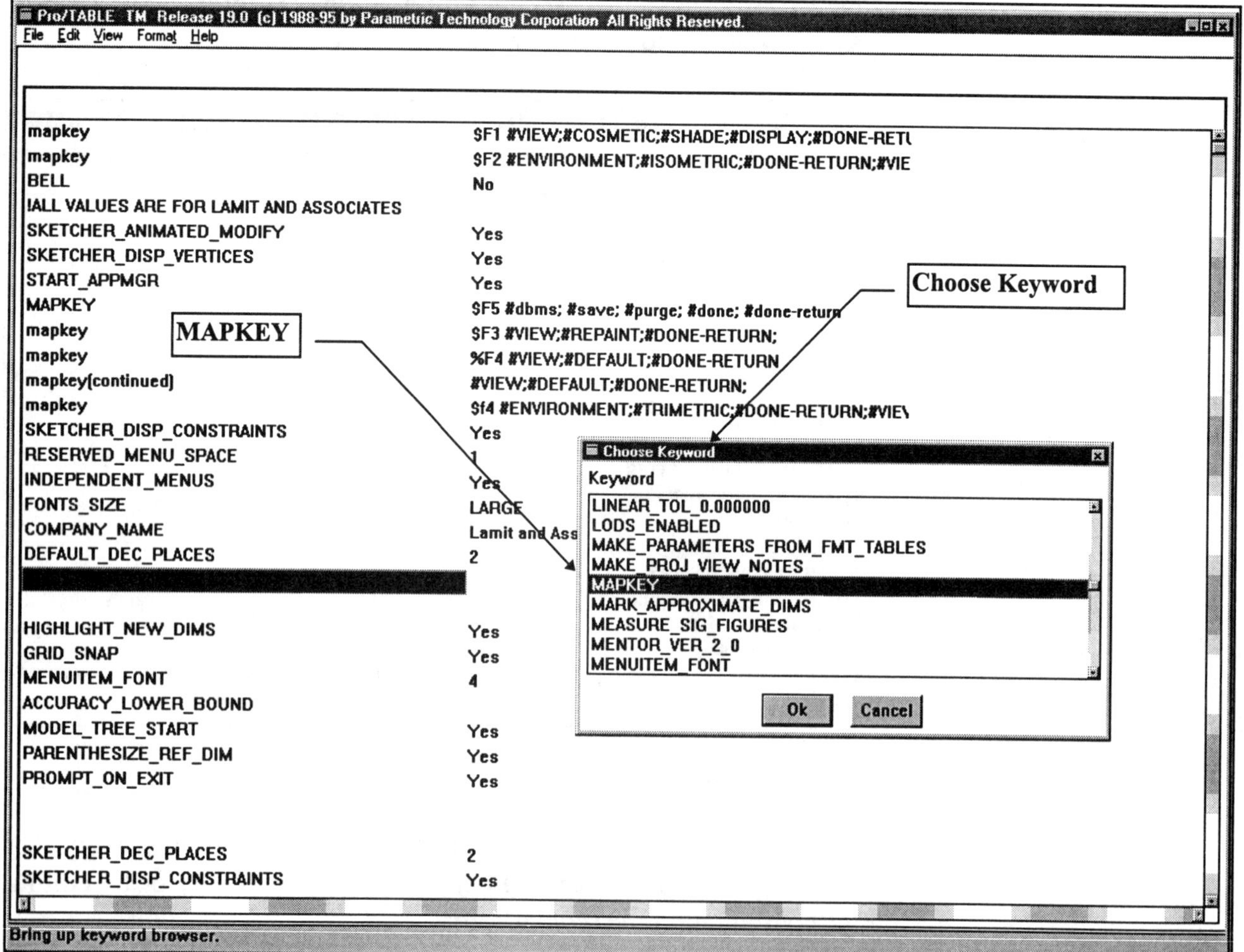

Figure 5.11
Configuration File Showing the **Choose Keyword** Menu Used to Create a Mapkey

Section 6

Environment

Pro/E lets you work in a multiple-window environment where you can have anywhere from two to ten or more windows open at a time, depending on your computer system's capabilities. You start working in the **Main working window** by default.

If the Main window is being used and you start a **Mode** that requires a new window, a new, smaller window is created for you to work in. As an example, when you are working on a part, if you then create or retrieve a different part without first quitting the Main working window, the new part will be created or retrieved in a new (smaller) window. In Figure 6.1, the swivel was retrieved first (main working window) and then the arm was retrieved. You can easily change your working window from one to another window.

The **MAIN** menu contains two window options: **ChangeWindow** and **Quit Window**. You can work in any window by picking **ChangeWindow** from the MAIN menu and picking inside the window in which you wish to work.

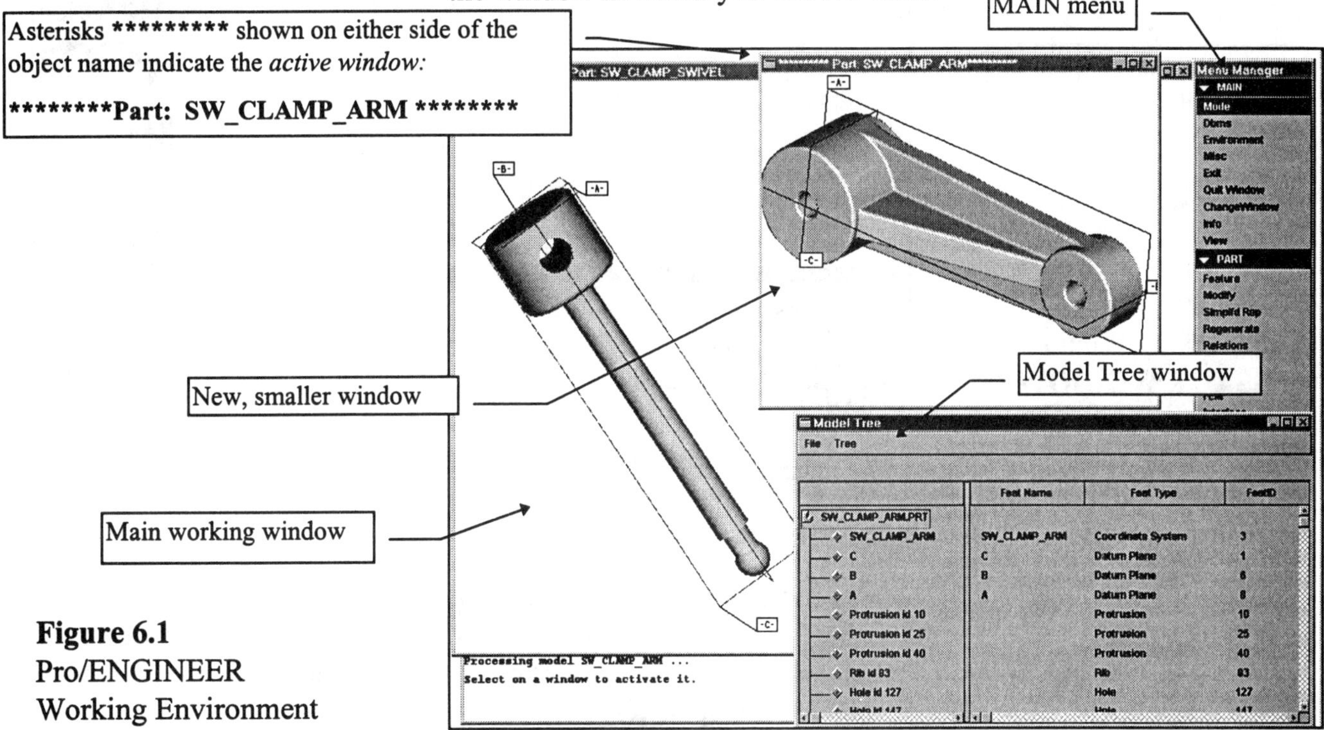

Figure 6.1
Pro/ENGINEER
Working Environment

Quit Window blanks out the currently active window. When multiple windows are open, **Quit Window** is used to clear the active window. The software *does not* automatically assign a remaining window to be active. **ChangeWindow** must be used to activate one of the remaining windows.

The active window will have asterisks on either side of the object's name: ************Part: SW_CLAMP_ARM************

The **Model Tree window** for the model in the selected window reflects the active part model or assembly model (Fig. 6.1, lower right corner).

Screen Layout

When you start a Pro/E session, the workstation screen is divided into three main areas:

Main working window Most of the graphical work is accomplished in this window (**Pro/ENGINEER** in Fig. 6.2).
Pop-up menus Pro/E modes and commands are selected using the mouse in these menus (**MAIN** and **MODE** in Fig. 6.2).
MESSAGE WINDOW Pro/E prompts and operation information are displayed here (**MESSAGE WINDOW** in Fig. 6.2).

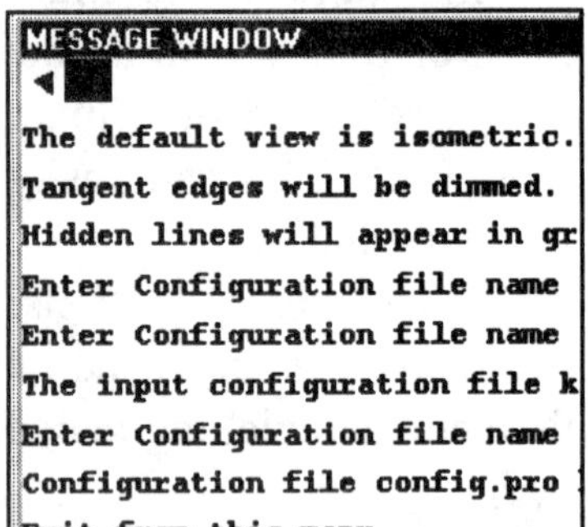

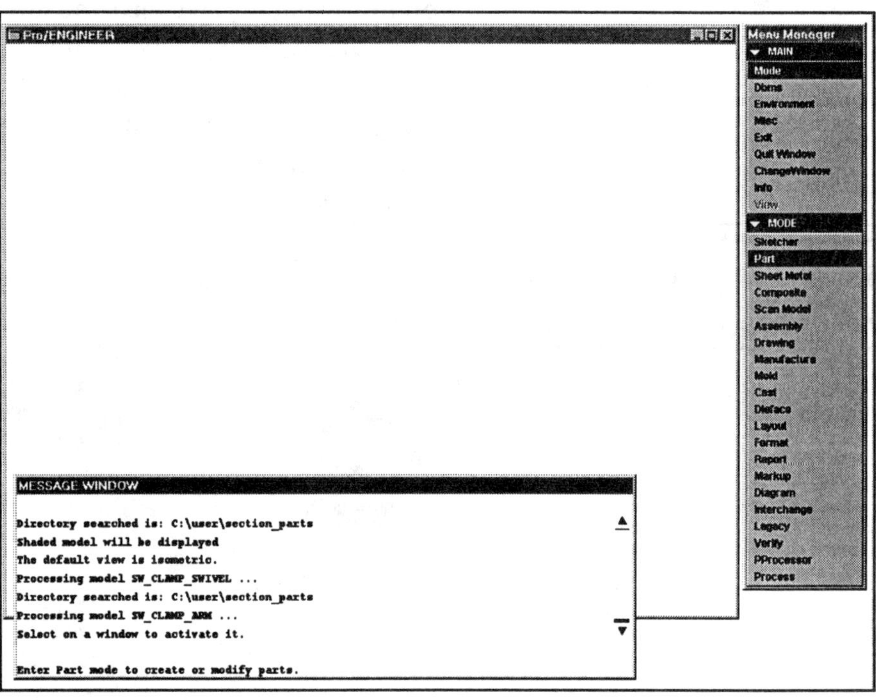

Figure 6.2
Main Working Window, **MESSAGE WINDOW**, **MAIN** Menu, and **MODE** Menu

Menu Boxes and Dialog Boxes

Menu and **dialog boxes** are used for *creating* and *redefining* features in Pro/E. You can create or redefine a feature by selecting information blocks, called options and elements, using the following menus and dialog boxes:

Feature Option Menus During the **Feature** ⇒ **Create** ⇒ **Solid** ⇒ (**Cut or Protrusion**) sequence, you can select solid feature options, such as **Extrude**, **Revolve**, **Sweep**, and **Blend**, and decide whether the feature will be **Solid** or **Thin** (Fig. 6.3). *Options cannot be redefined.*
Feature Element Menus Used to control the process of defining the elements required to create the feature, such as **Direction** and **Depth** (for example, **Depth** from the **SPEC TO** menu, as shown in Fig. 6.4). *Elements can be redefined.*

Dialog Boxes Provide a visual method of displaying and changing the feature **Elements** and their current status. Information regarding the feature and its references can also be accessed. The **Preview** option provides a view of the feature or operation prior to creation, allowing you to either accept the design or redefine the feature's elements. In Figure 6.4, the **CUT: Extrude** dialog box is displayed.

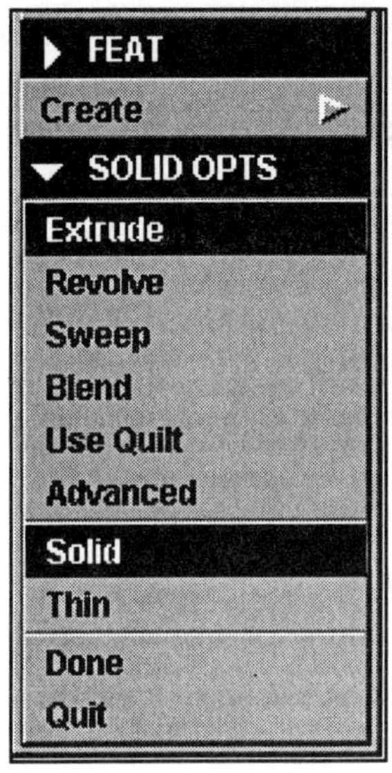

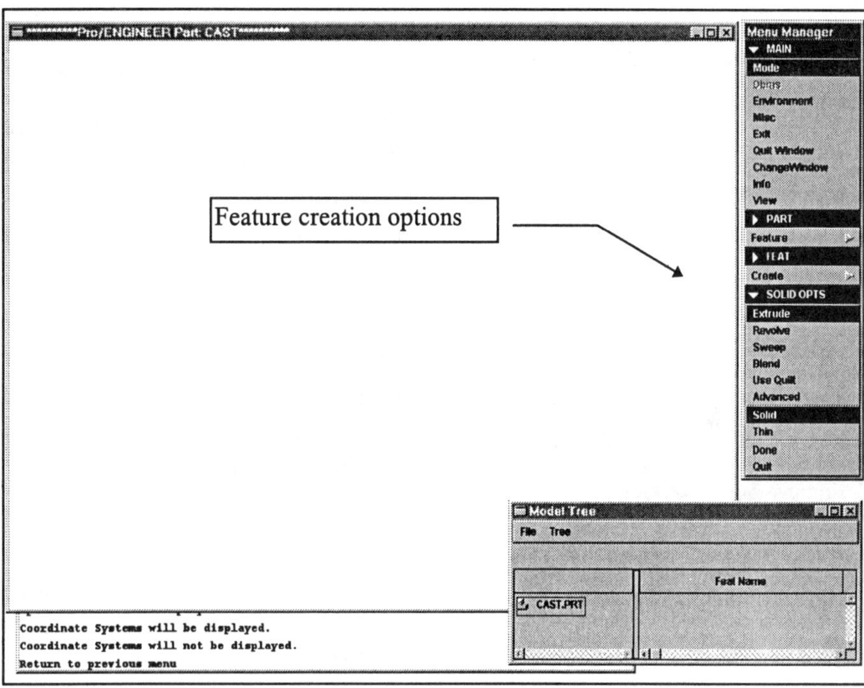

Figure 6.3
Feature Options Menu

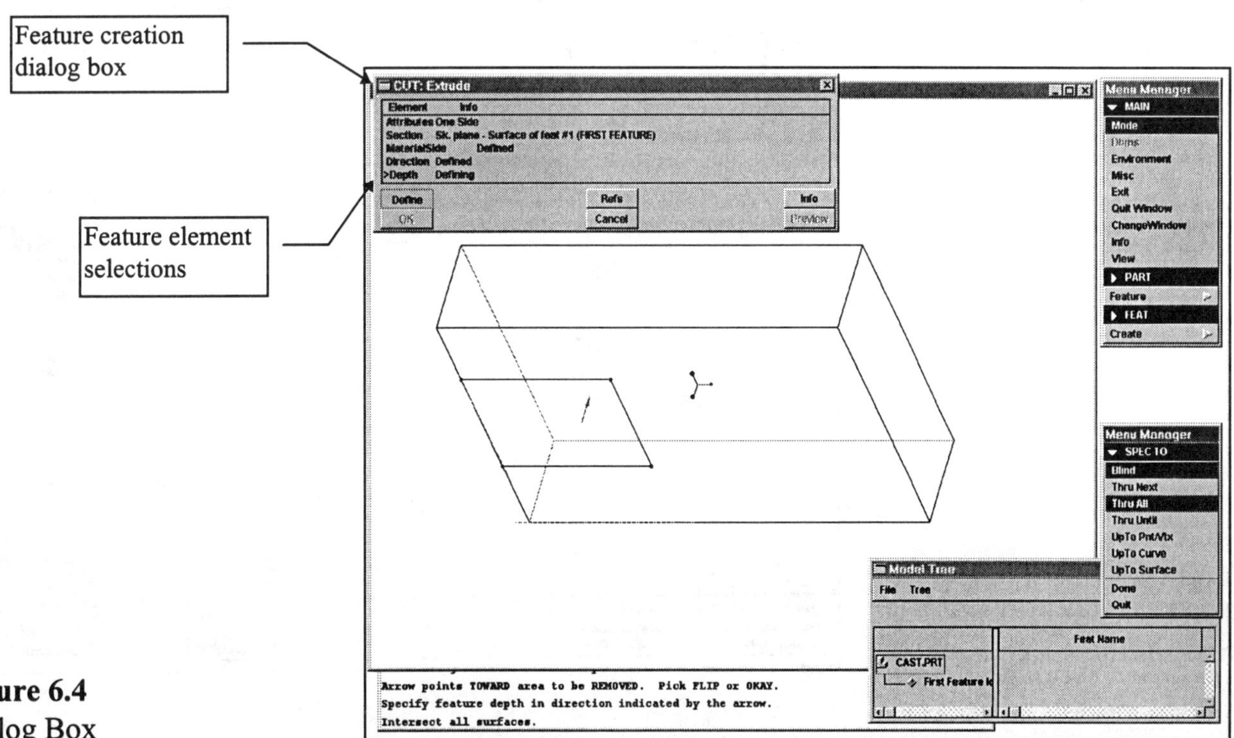

Figure 6.4
Dialog Box

Feature Creation

When creating a feature, values are required for the feature options and elements. The options specified will determine the elements required to complete the feature. As an example, the cut feature in Figure 6.4 was extruded; its elements are **Attributes**, **Section**, **MaterialSide**, **Direction**, and **Depth,** and a sweep cut will have the elements of **Trajectory**, **Section**, and **MaterialSide**.

There are three types of elements:

Required Essential for the feature definition
Optional Not essential for the feature definition
Conditional Required as a result of the selection of another required element

The example in Figure 6.5 shows a feature being added to an existing part. The feature class has been defined as **Solid** and the feature type as a **Shell**. Once you have specified the feature options, Pro/E will display an information dialog box.

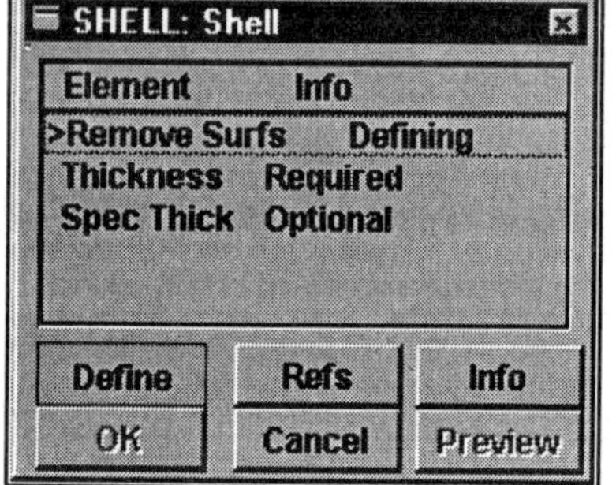

The dialog box title is based on the feature type and option defined (**Shell** in Fig. 6.5). The dialog box contains an **Element** list and action buttons. The Element list (**Remove Surfs**, **Thickness**, and **Spec Thick** in Fig. 6.5) is a tabulated list that contains the **Element** name and **Info** describing the current state of the feature's elements, such as **Defining** (or **Undefined**), **Required**, and **Optional**.

Shows all information (**Elements**) required to create a shell feature:
Remove Surfs
Thickness
Spec Thick

Figure 6.5
Shell Command and the Dialog Box

The dialog box action buttons (Fig. 6.6) enable you to perform a variety of functions:

Define Defines or *redefines* all marked elements for the feature, working from the top to the bottom of the element list. Pro/E displays the menus for the elements currently being processed and displays messages and required prompts in the message window.

Refs Displays the references for the selected elements and allows the user to step through each reference by navigating through the SHOW REFS menu.

Info Displays information on the feature being created.

OK Completes feature creation and closes the dialog box.

Cancel Aborts the current feature definition process and closes the dialog box.

Preview Displays the feature geometry as it will be built with the current definitions. Preview has been used in Figure 6.6, where the shell command was given, and the Preview button allows you to see the feature as it will be created before accepting the design input. At this preview state you could change (define/redefine) the shell thickness or change/add reference surfaces to be removed.

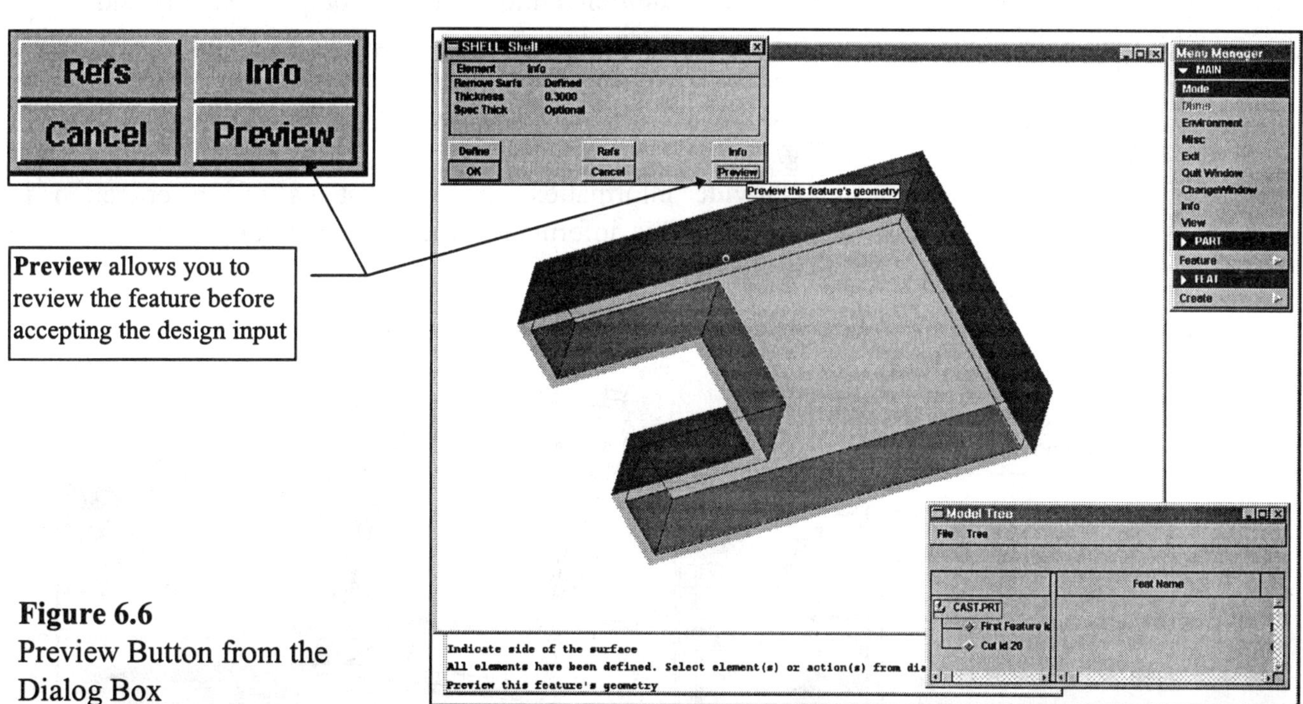

Figure 6.6
Preview Button from the Dialog Box

Choosing Dialog Box Items

Elements listed in the dialog box are selected by moving the cursor over one or more items and clicking with the *left* mouse button. To select one element at a time, move the cursor over one element and click once with the *left* mouse button.

For a range selection, move the cursor to the top or bottom of the range of elements. Click the left mouse button and hold it down, then drag the cursor to the opposite end of the range you wish to select. Another method is to move the cursor to the top or bottom of the range, then click once. To select more than one element, move the cursor to the element on the opposite end of the range and, while pressing the **Shift** key, click once with the *left* mouse button.

Action buttons in the dialog boxes are selected using the *left* mouse button. For example, a user would select the **Define** button after highlighting the desired elements in the tabulated list. Double-clicking on an element in the tabulated list will automatically initiate the **Define** action button.

System Messages

System messages appear in the **MESSAGE WINDOW**, usually at the bottom of the main working window. The messages have the following functions:

* Request additional information to complete a command.
* Provide a *one-line help message* about a menu item being chosen. Help messages appear (normally in yellow) at the bottom of the Message window (**"Place coaxially to an axis"** in Fig. 6.7).
* Provide information about the status of an operation in progress. This information (in white letters) scrolls up in the message window.

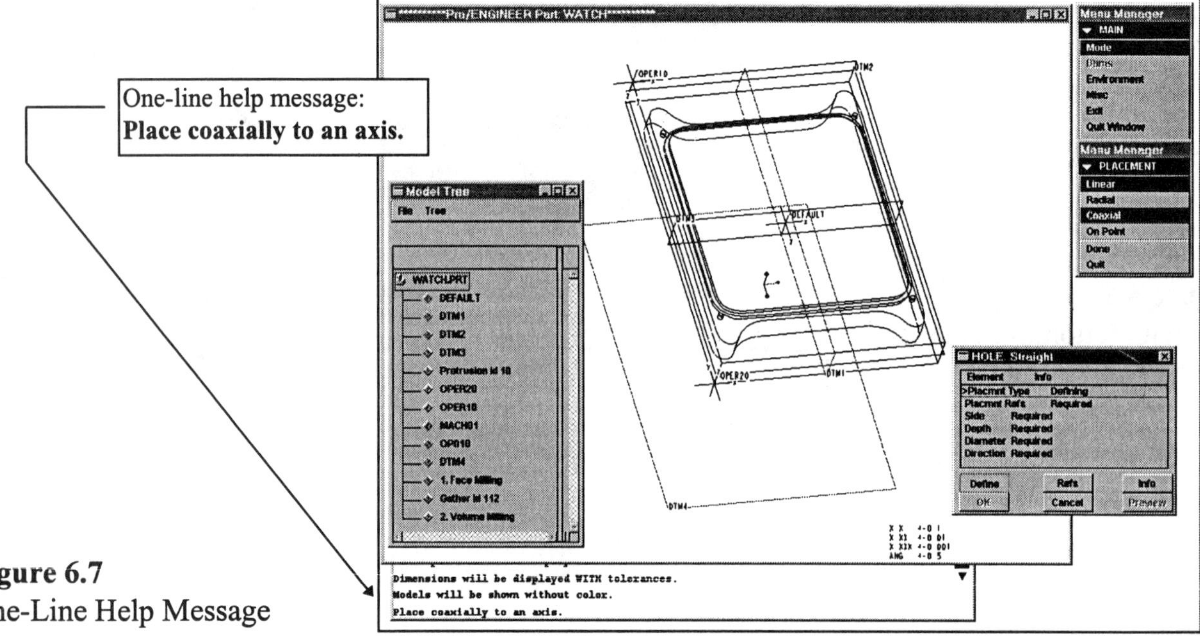

Figure 6.7
One-Line Help Message

When Pro/E is providing information, you can proceed with another function and the message will disappear. If Pro/E is asking for additional information, provide the information by entering it from the keyboard or indicating it with a mouse pick.

A bell will sound to notify you when input is needed (unless the bell has been turned off in the ENVIRONMENT menu manually or as a *config.pro* setting; most users turn the bell off, since it is extremely annoying, especially if there are 30 users in the same room!). When data need to be entered, a prompt ending with a colon will appear in the MESSAGE WINDOW. When data input is required, all other functions are temporarily disabled until data entry is complete.

The MESSAGE WINDOW can be expanded using the workstation's window management functions to "pop" the window in front (Fig. 6.8). The main working window can overlap the MESSAGE WINDOW, showing up to five message lines at startup, though the default is one line.

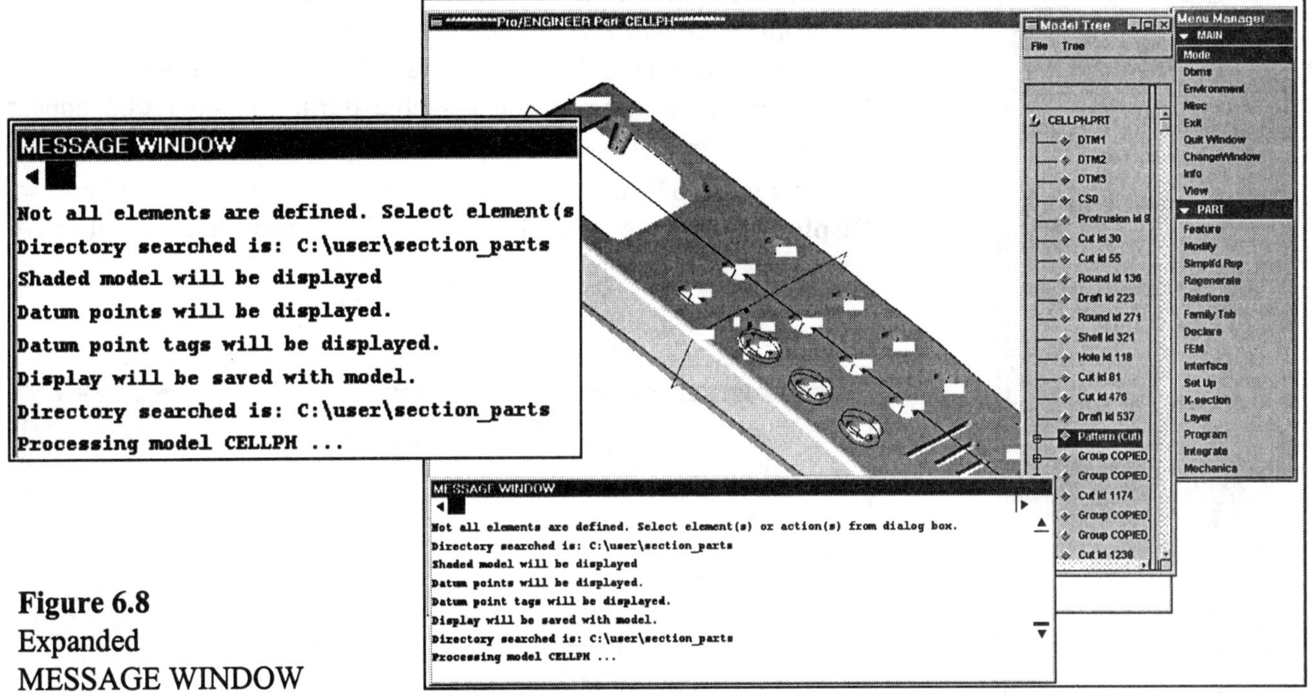

Figure 6.8
Expanded
MESSAGE WINDOW

The MAIN Menu

The **MAIN** menu provides access to the major modes of Pro/E operation and other system-related functions:

Mode Allows you to access major Pro/E modes. It brings up the MODE menu that allows you to choose the desired mode.

Dbms Used to manage Pro/E objects.

Environment Affects the operating modes. It is used interactively to change the working environment (e.g., display mode settings, prompt bell status, object storage settings).

Info This menu item lets you access information functions without having to quit your current Pro/E action.
Misc Gives you access to the workstation's operating system functions without exiting Pro/E, and allows the rerun of journal and training files in the Pro/E environment.
Exit Ends the Pro/E working session.
Quit Window Quits from the current window being used.
ChangeWindow Changes the window being worked in, from one to another.
View Changes the view of an object.

Choosing Menu Items

Menu items are selected using a three-button mouse. To choose a menu item, move the cursor over the item until it is highlighted. Click the *left* mouse button to activate the selection. The menu item is then highlighted in a different color and the function is executed. *Submenu items that are already highlighted are the default options.* To change from the default, choose a different item in that submenu. Menu selections that are not available during an operation appear dimmed and are disabled.

In Figure 6.9, the commands **View** ⇒ **Cosmetic** ⇒ **Shade** ⇒ **Display** have been chosen and the model is shaded in the small active window.

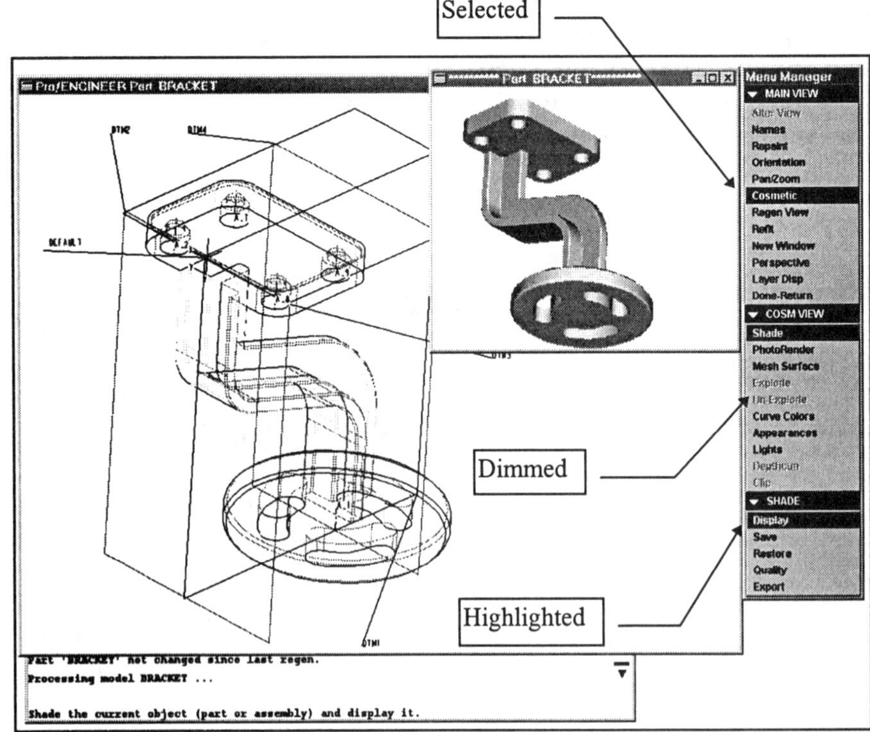

Figure 6.9
Selected, Highlighted, and Dimmed Menu Items

Online Help (OnLine Documentation with Pro/HELP)

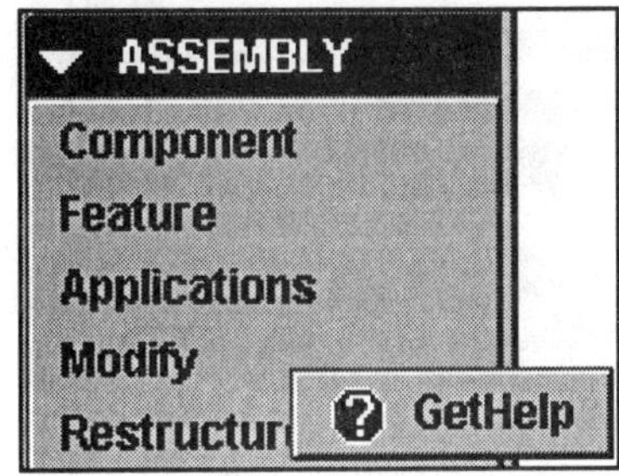

When you scroll the cursor up and down the menu items, a one-line description of the menu item that is highlighted appears in the MESSAGE WINDOW at the very bottom left of the screen.

If more detailed help is needed, then, with the menu item highlighted, click the *right* mouse button to access **online documentation**. In Assembly mode (Fig. 6.10), the item **Modify** has been highlighted. Clicking the *right* mouse button has brought scrolled pages of the **Assembly Modeling User's Guide** to the screen. You can scroll to other pages, or go to a table of contents or index, or do a search by scrolling to the menu items at the very top or bottom of the online documentation.

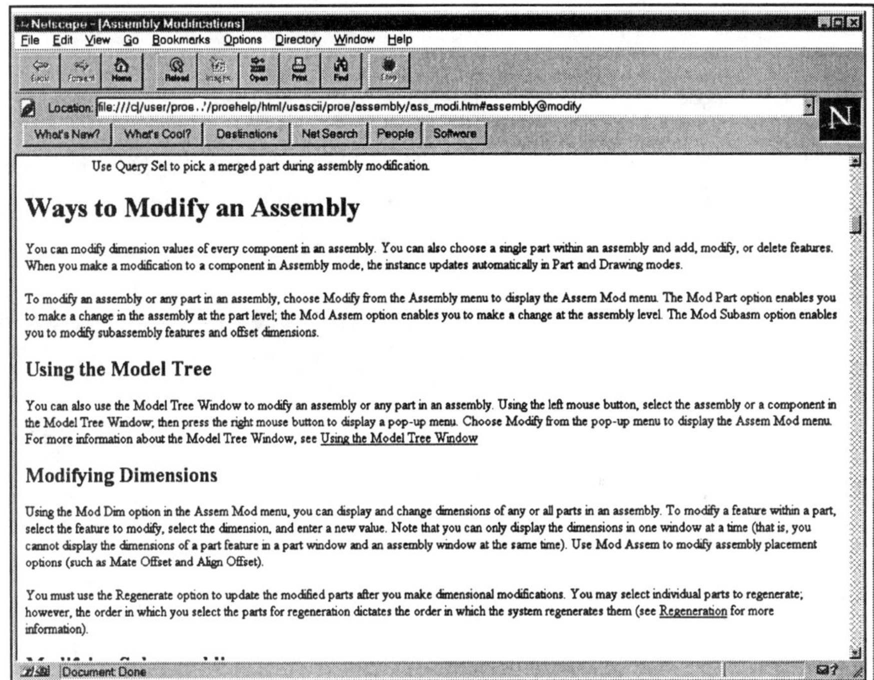

Figure 6.10
Pro/HELP

Pro/HELP is a tool that allows you to view Pro/E user guides and support documentation on your computer screen. When installed and configured on your system, Pro/HELP opens to the appropriate document and page to provide information and help whenever the *right* mouse button is clicked on a menu item.

This gives you access to various user guides and libraries installed on your system. You have the ability to:

Browse User Guide documentation
Establish bookmarks
Search for keywords
Print pages (Fig. 6.11)

The Print dialog box is activated when you choose **Print**, found along the top of the Netscape Menu (Fig. 6.11).

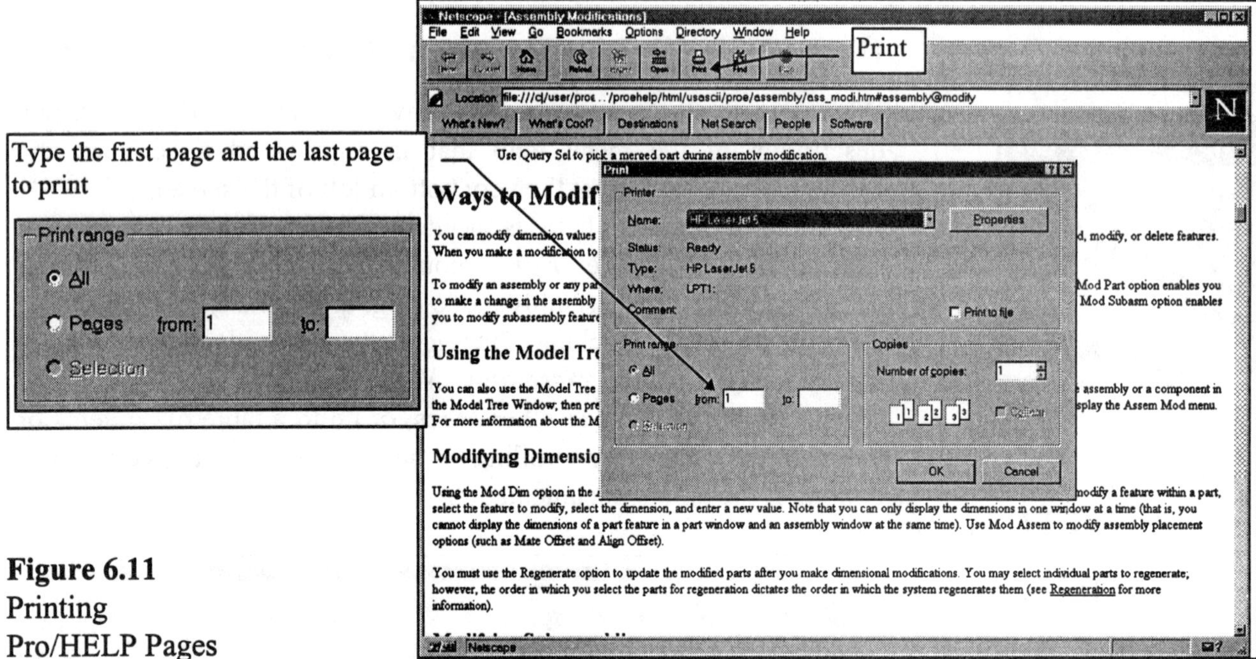

Figure 6.11
Printing
Pro/HELP Pages

Besides providing help, Pro/HELP can access any library that you have installed on your system. In Figure 6.12, the Tooling library has been accessed.

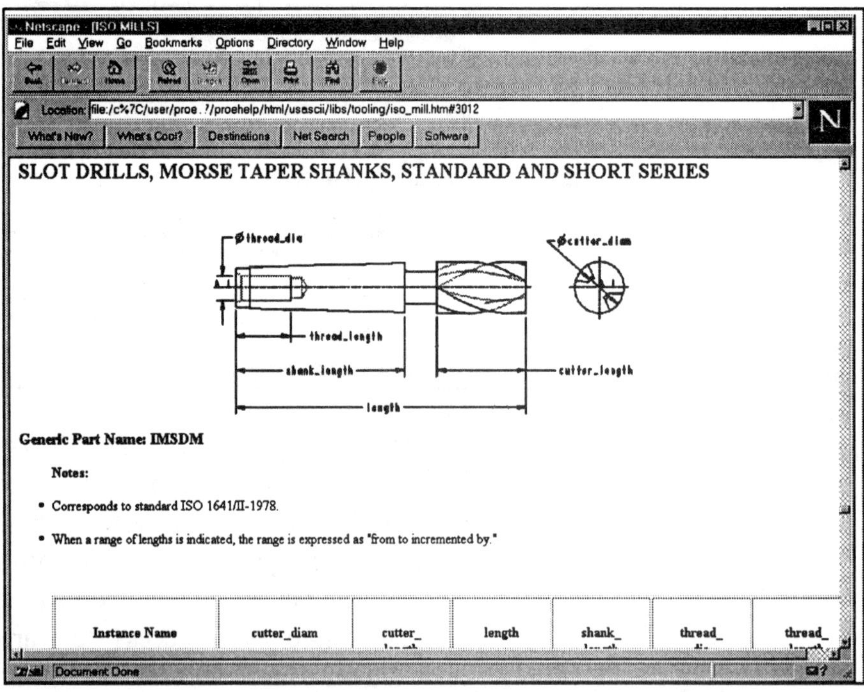

Figure 6.12
Pro/HELP
Tooling Library

Throughout the text, you will be prompted to access help for the menu items that you will be learning in each lesson. There is no way that this text can cover all the commands and capabilities of Pro/E. You must use part of every session looking up information on commands and discovering the wide range of options available in Pro/E. It is the only way to expand your understanding of Pro/E and become a more complete and competent user.

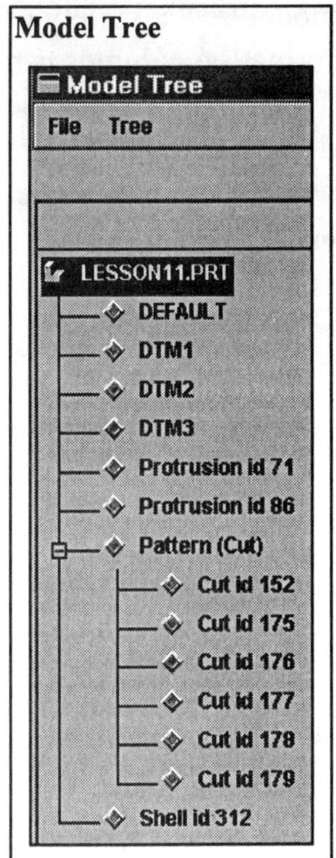

Figure 6.13
Model Tree Window

Model Tree

The **Model Tree** displays your part model or assembly model structure. Each feature (Fig. 6.13) or part (if it's a component of an assembly) is displayed along with information about the object or feature.

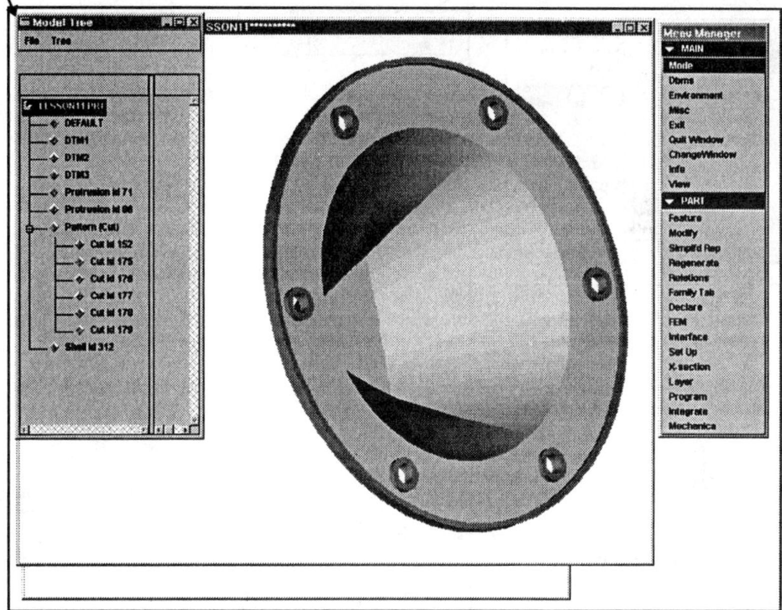

In Figure 6.14, an assembly is shown with its model tree displayed. The assembly and its components are listed in a tree structure.

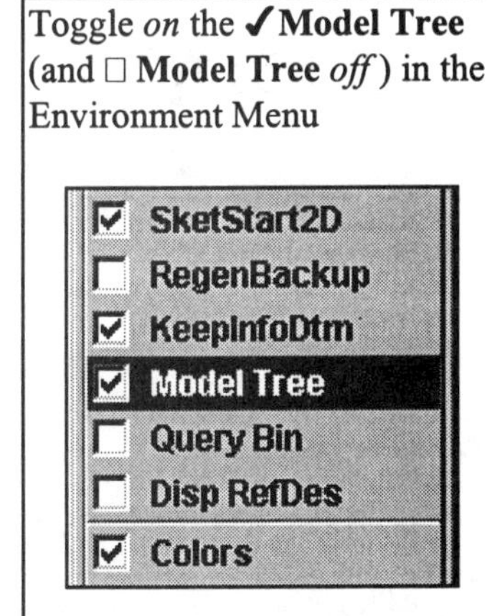
Toggle *on* the ✓**Model Tree** (and □ **Model Tree** *off*) in the Environment Menu

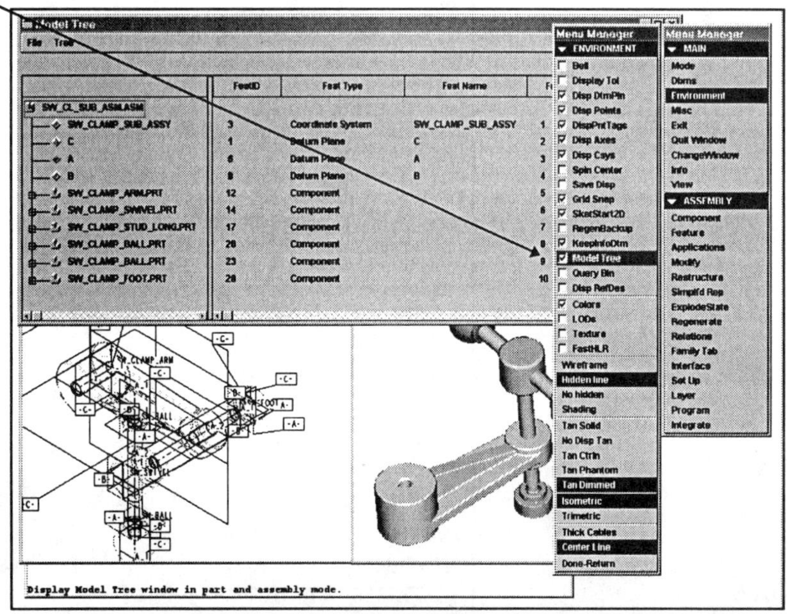

Figure 6.14
Assembly Model Tree

The model tree can be disabled by toggling off (□) the check mark (✓) in the ENVIRONMENT menu.

Besides the feature or object listing in the model tree, a variety of information columns can also be displayed. Information columns can be added or removed and their format altered. The format can also be saved for later use. A model tree, in Part mode, is shown in Figure 6.13. In Figure 6.15, columns for information about the part model have been enabled by choosing **Tree ⇒ Columns ⇒ Add/Remove.**

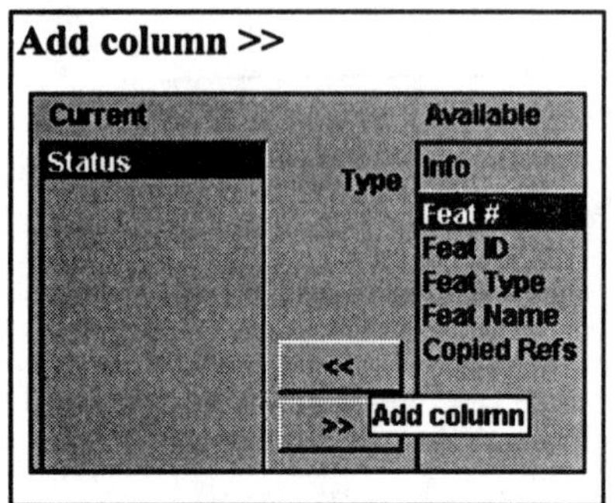

Figure 6.15
Adding Columns to the Model Tree

After the columns have been added, the model tree looks like the screen capture in Figure 6.16.

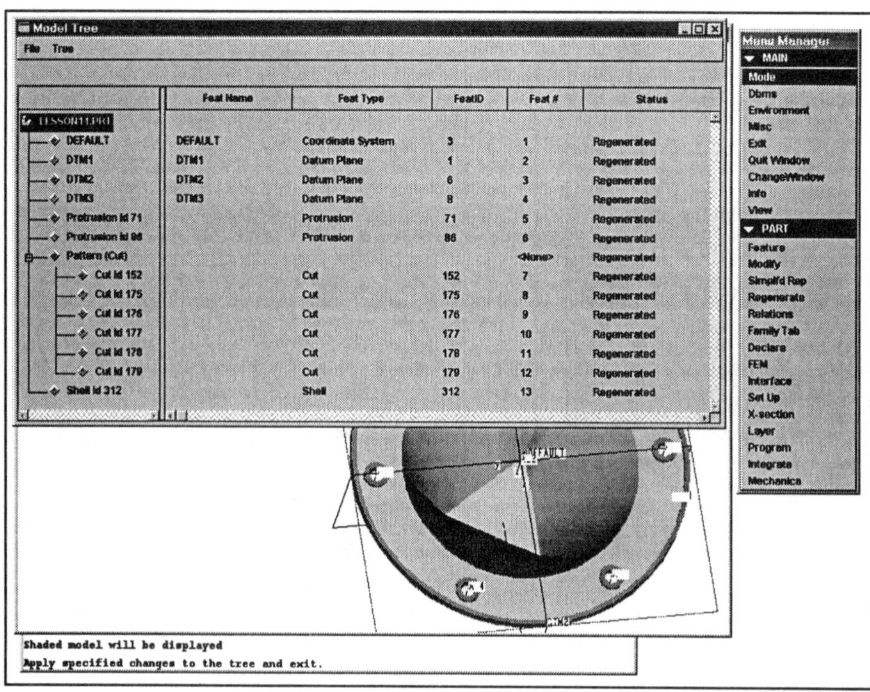

Figure 6.16
Expanded Model Tree

The columns can be altered by picking **Tree ⇒ Columns ⇒ Format** and changing the width of each category (Fig. 6.17).

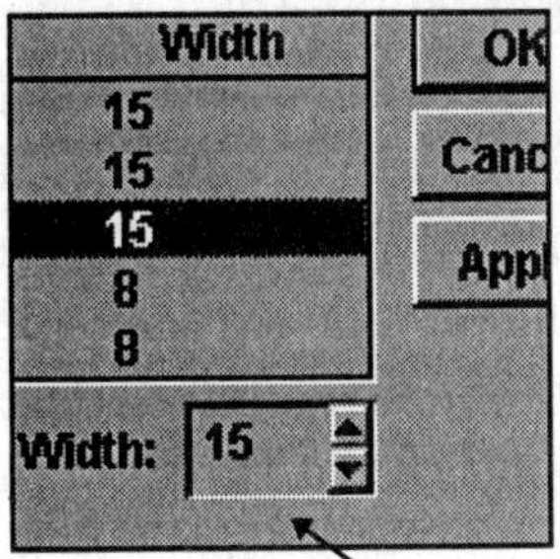

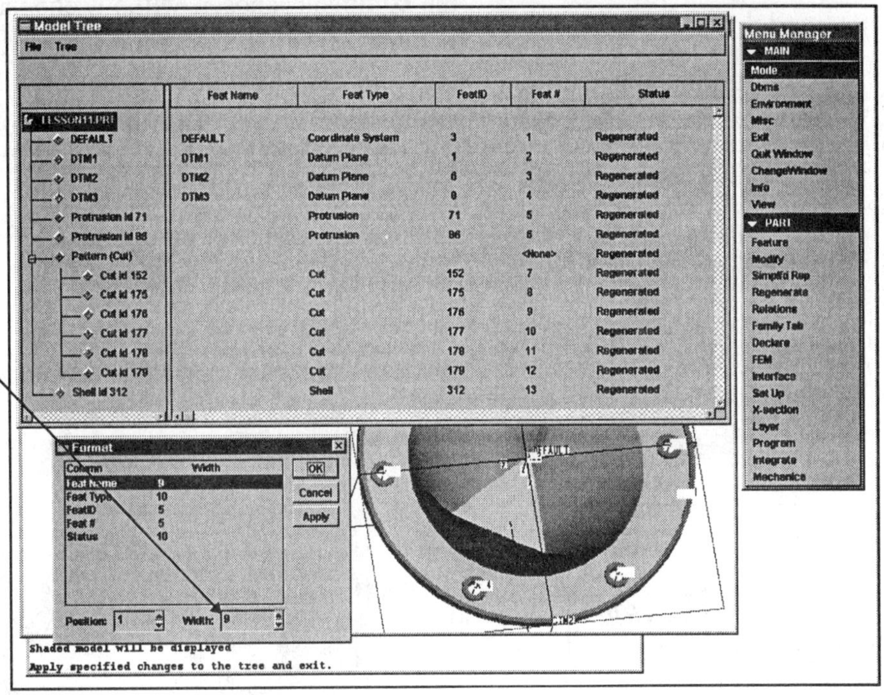

Change the format of the model tree by altering the width of the columns

Figure 6.17
Reformatting the Columns in a Model Tree

Since the model tree window is interactive, you can either select objects directly from it or select **Sel By Menu** from the GET SELECT menu and then pick a component or feature from a menu list.

Figure 6.18 is an example of selecting parts of an assembly from the model tree. The **SW_CLAMP_ARM.PRT** has been selected and is shown highlighted.

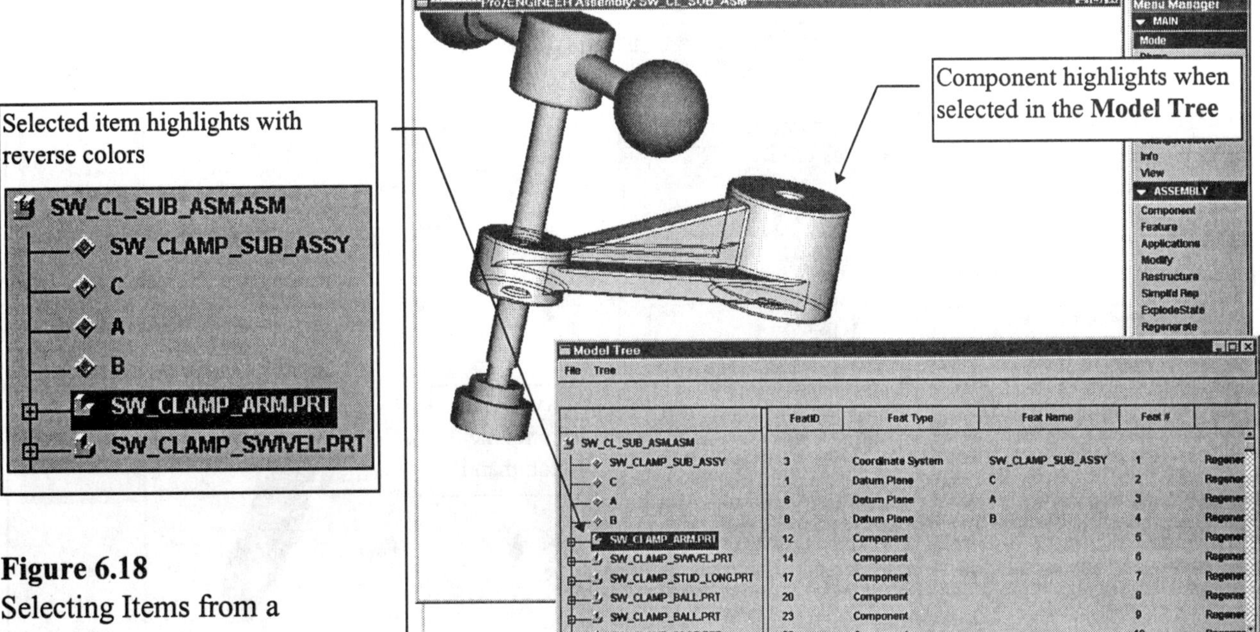

Figure 6.18
Selecting Items from a Model Tree

NOTE

Choose the feature from the model tree

Highlight and select from the cascading choices

The model tree is also used to access a variety of commands that were formerly available only through the menu structure. Commands such as **Modify**, **Redefine**, **Suppress**, **Reroute**, **Info**, and **Delete** can be chosen directly from the model tree. The feature is selected and the command chosen without choosing from the menu structure and without selecting the feature from the working window. In Figure 6.19, the **Shell** feature was selected from the model tree with the *left mouse button*. The **Info** command and the **Model Info** option were then highlighted and selected with the *right mouse button*. In Assembly mode (Fig. 6.20), the model tree can be expanded to include the selected component's features. The model tree is also used to select a variety of commands similar to those for a part, such as **Pattern** and **Feature Create** (Fig. 6.20).

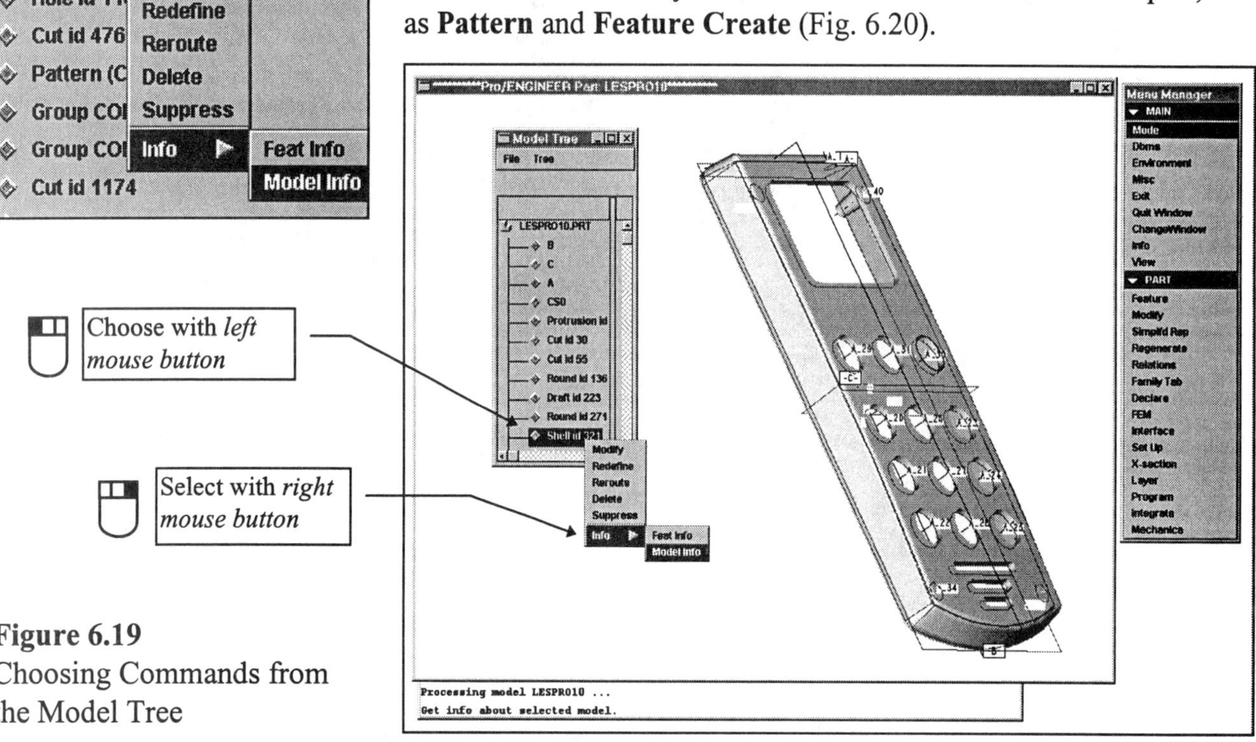

Figure 6.19
Choosing Commands from the Model Tree

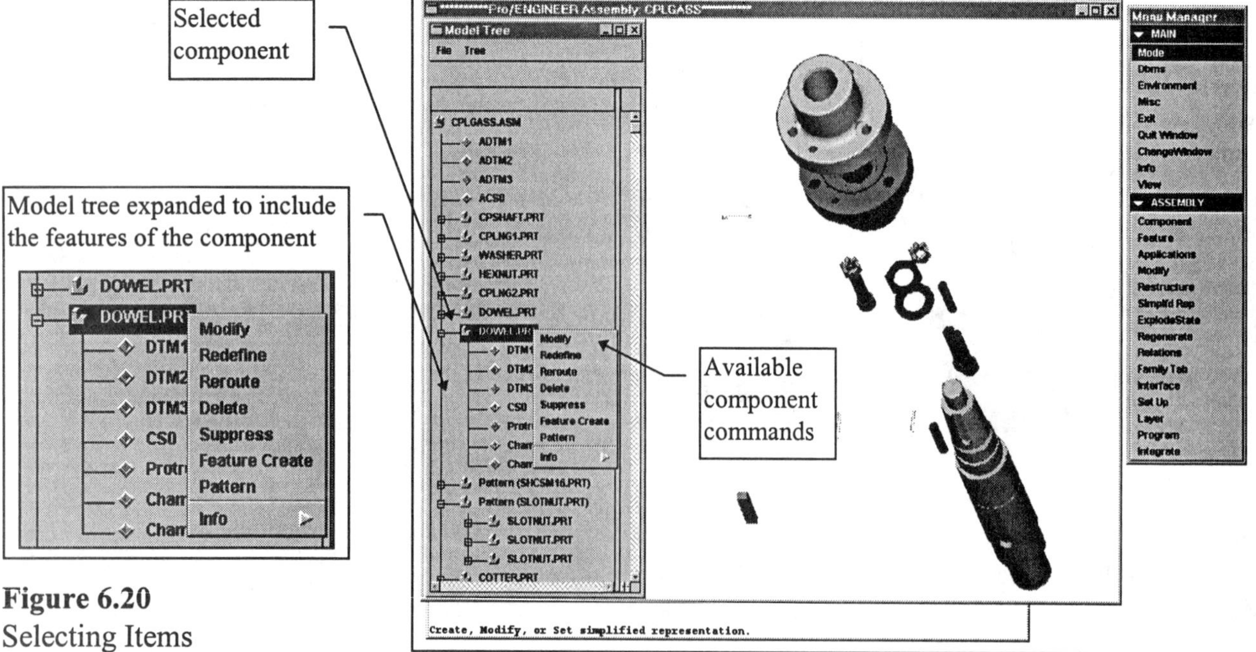

Figure 6.20
Selecting Items

Section 7

Setup and Information

Several capabilities are used during the design of parts, assemblies, and drawings. In this section, **Set Up** and **Info** functionality are covered. Before a project is started, whether it is a part or an assembly, you must set up the working standards for that project. Units must be selected. Dimension tolerance standards are set to ANSI (1982), ASME (1994), ISO, JIS, or DIN. Materials should be selected for the part. Parameters are input. Accuracy is set. Geometric tolerances are established, and other specific design settings are organized and input at this point.

The units, accuracy, and other settings established with the **Set Up** command will of course be reflected when information is requested about the model.

At any time in the design process, **Info** (information) about the part or assembly can be requested. Mass properties (Fig. 7.1), surface analysis (Fig. 7.2), geometric tolerances, measurements, and a variety of information listings are available with the **Info** command.

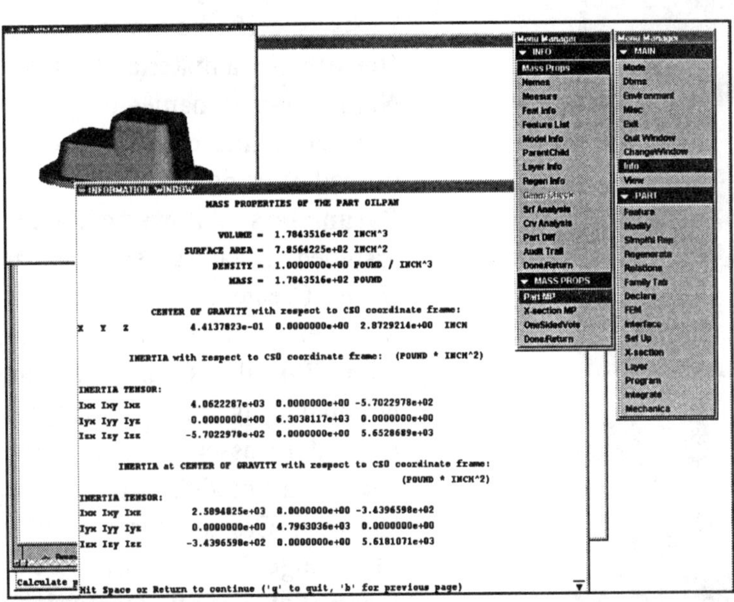

Figure 7.1
Mass Properties

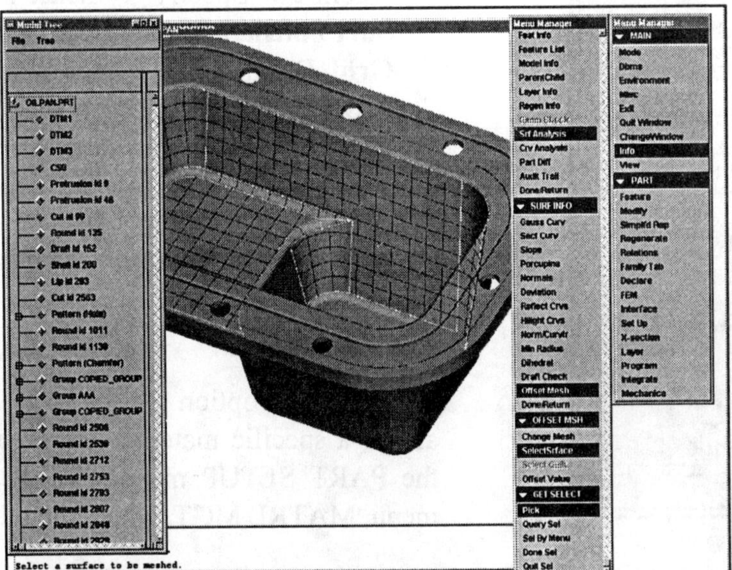

Figure 7.2
Surface Analysis

Setup (Part and Assembly)

PART
Set Up
PART SETUP
Material
Accuracy
Units
Density
Name
Parameters
Notes
Mass Props
Dim Bound
Ref Dim
Shrinkage
Geom Tol
Surf Finish
Grid
Tol Setup
Interchange
Done

In *part mode* and *assembly mode*, the SETUP menu allows you to define various attributes of the model. The most important of these capabilities are discussed in detail in this section. The following is a listing of options:

Material Create and modify material data files.
Accuracy
Relative Modify relative part accuracy (Part mode). This is the computational accuracy of Pro/E geometry calculations and has a default value of **0.0012**. Part accuracy is relative to the size of the part. Increasing the part's accuracy increases regeneration time.
Absolute This functionality improves the matching of parts of different size or different accuracy (e.g. imported parts created on another system). In most cases, use relative accuracy.
Units Specify the dimension units for the part or assembly.
Density Set a material density value to be assigned to the part.
Name Assign names to features, assembly members, and so on. You can name or rename datum planes and coordinate systems instead of using default names.
Parameters Allows editable user-defined parameters to be assigned to features, surfaces, models, etc.
Notes Create, modify, or remove notes associated with the model.
Mass Props Create a file of mass properties.
Dim Bound Change a dimension regeneration value from nominal to its upper, lower, or midpoint tolerance boundary. This is used in assemblies to compute clearance/interference checks and on part models to perform tolerance stack-up studies.
Ref Dim Create reference dimensions for the model.
Shrinkage Modify shrinkage of part dimensions.
Geom Tol Specify geometric tolerances for surfaces and features.
Surf Finish Define surface finish symbols for the part model.
Grid Define a 3D grid for the model.
X-Section Create, modify, and display *assembly* cross sections.
Declare Establish declarations to a layout.
Tol Setup Specify tolerance standards (ANSI, ISO/DIN).
Interchange Shows information about, or removes references to, interchange groups.

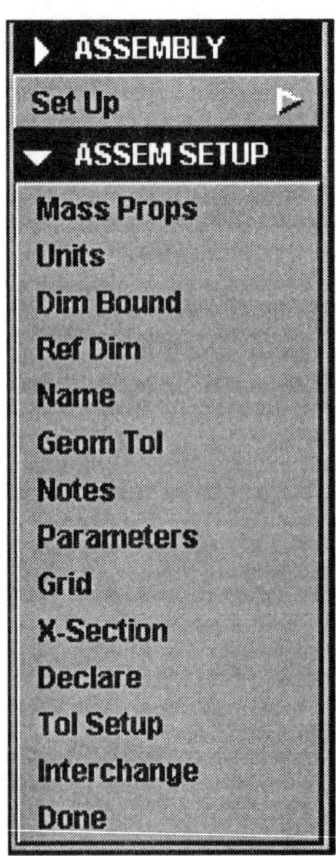

Material

The **Material** option is used to create and modify material data. To assign a specific material to the current part, choose **Material** from the PART SETUP menu. This calls up the material management menu, MATRL MGT, which includes the following options:

Define Define the properties of a new material. The system editor displays a default specification file (Fig. 7.3) that must be edited in order to add the desired values for the material parameters. Different materials can be created by building new material files with the system editor (Fig. 7.4).

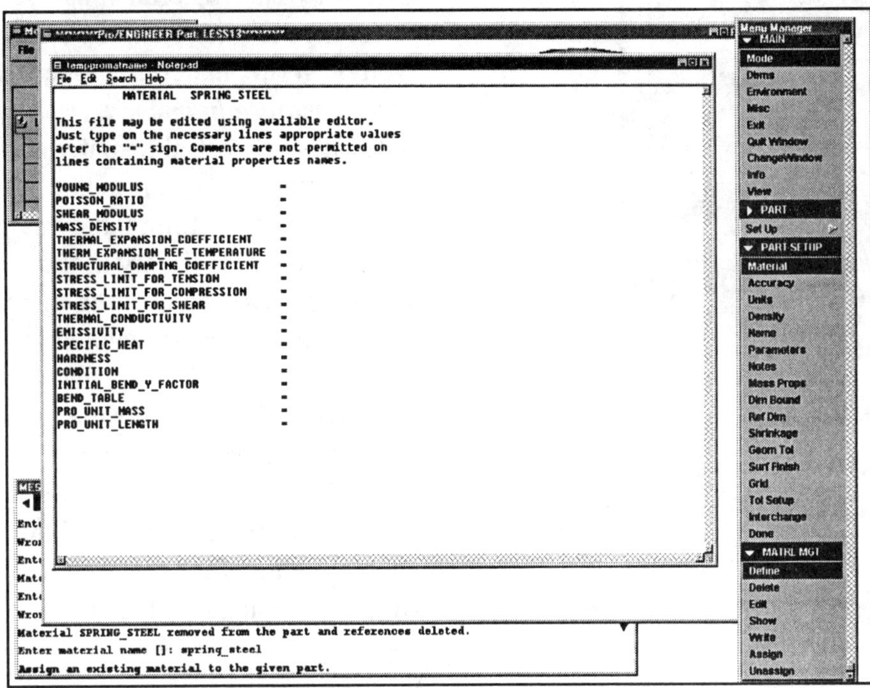

Figure 7.3
Material Defining

Material Steel_1040		
Young_Modulus	=	29000000
Poisson_Ratio	=	0.27
Shear_Modulus	=	11000000
Mass_Density	=	0.00879
Thermal_Expansion_Coefficient	=	6.78
Thermal_Expansion_Ref_Temperature	=	32.0
Structural_Damping_Coefficient	=	0.01
Stress_Limit_For_Tension	=	36000
Stress_Limit_For_ Compression	=	36000
Stress_Limit_For_ Shear	=	36000
Thermal_Conductivity	=	
Emissivity	=	
Specific_Heat	=	
Hardness	=	
Condition	=	
Initial_Bend_Y_Factor	=	
Bend_Table	=	steel_1040

Figure 7.4
Steel Material File

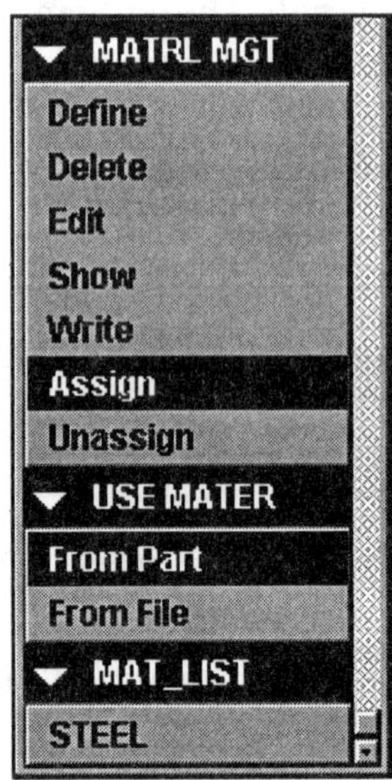

Delete Remove a material from the part's internal database. Pick the material from the MAT_LIST namelist menu.

Edit Edit the material specification file using the system editor. The material must be in the part's internal database. Select the material from the MAT_LIST namelist menu.

Show Displays the material specification file in an information window. The material must be in the part's internal database. Select the material from the MAT_LIST namelist menu.

Write Write material properties from the part to a disk file named *materialname.mat* (*materialname* is the name of the material). Select the material from the MAT_LIST namelist menu, then specify the name under which the material will be stored. After defining the materials required in your design, create a materials library in the appropriate directory.

Assign Assigns an existing material to the part. This material is used in all analysis calculations of the part. From the USE MATER menu, select the material list to choose from:

> **From Part** From the part's internal database, use the material as listed in the MAT_LIST namelist menu (Fig. 7.5).
>
> **From File** Use material that is stored in a disk file. Provide the name of the material to be retrieved.

Unassign Unassign the currently assigned part material.

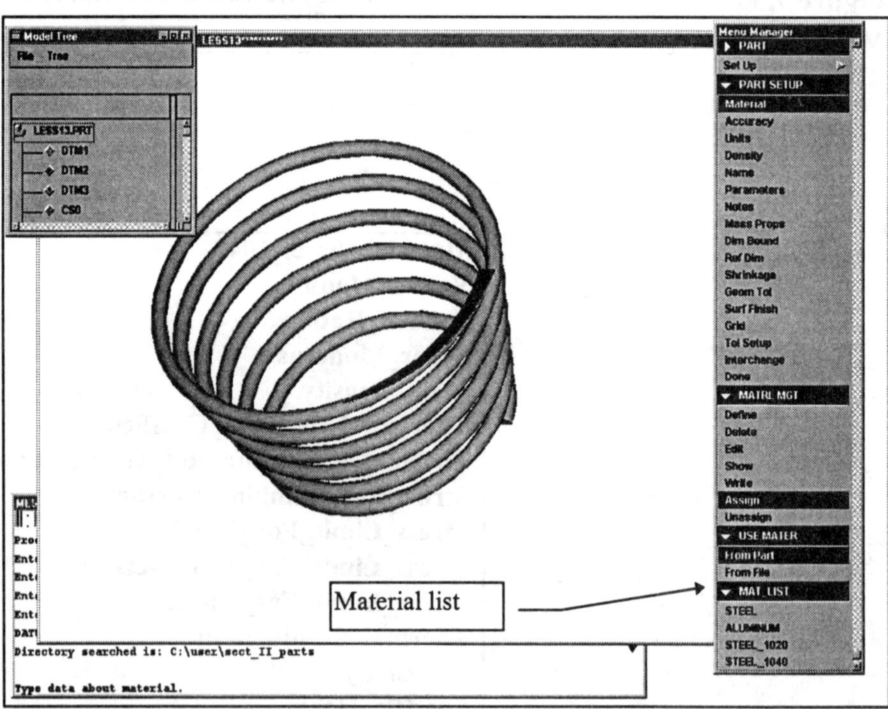

Figure 7.5
Assigning Material to a Part

Changing Material Parameters in a Part

You can specify or change material parameters by choosing **Material ⇒ Edit** to edit the material properties stored in the part's internal database or by choosing **Material ⇒ Assign** to change the material file assigned to the part. After regeneration, the material values are updated.

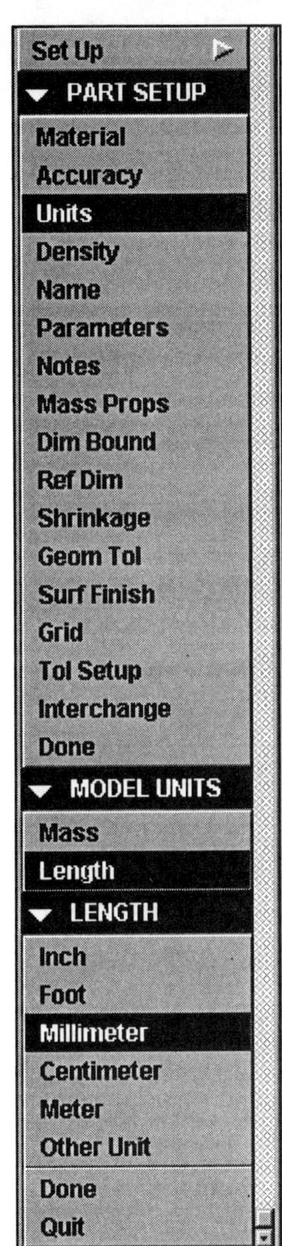

Units

The **Units** option in the PART SETUP menu enables you to specify the dimension units to be associated with the part (Fig. 7.6). The **Units** option in the ASSEM SETUP menu governs the units of assembly features, placement offsets, and explode distances.

Units for a part are established, at startup of Pro/E, by means of the configuration file option pro_unit_length. The default can be set to:

unit_inch	**unit_foot**	**unit_mm**	**unit_cm**	**unit_m**

Rules about the use of units:

1. Since relations are not scaled along with the model, modifying units may invalidate relations. Nonparametric features, such as cosmetic ones, and IGES models are also not scaled with the change of units.
2. A part can have only one set of units.
3. Unless otherwise set by the pro_unit_length configuration file option, parts default to inches and pounds.
4. The internal units for an assembly are those of its base component. If, however, the units of the base component have been changed, the assembly units will not automatically change. You must specify the units of the assembly again.
5. You cannot change the units of an assembly that contains assembly features that intersect a part.
6. User-defined parameters have no units. The appropriate conversion factors in relations must be included.
7. Cross sections do not have units.

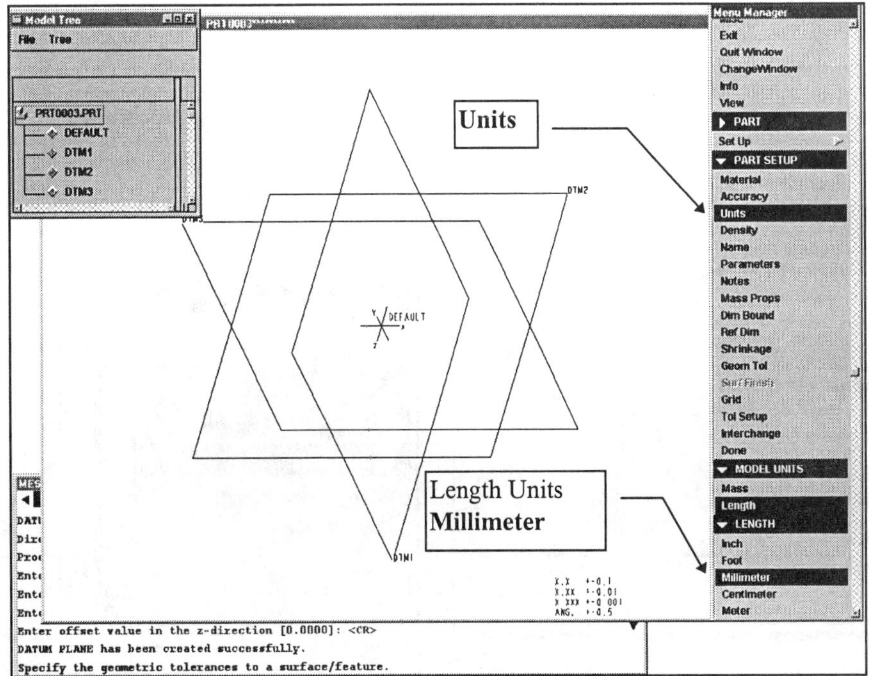

Figure 7.6
Setting Units at the Beginning of Part Creation

Specifying Units

To specify the units of a part:

1. Choose **Set Up** ⇒ **Units**.
2. For length units, choose **Length** and one of these: **Inch, Foot, Millimeter, Centimeter, Meter,** or **Other unit**. Once selected, these become the active, displayed units. For mass units, choose **Mass** and one of these: **Ounce, Pound, Ton, Gram, Kilogram, Tonne, Other Unit**. Chose **Done** from the Length menu.
3. If a model exists when length units are changed, the SCALE menu appears (Fig. 7.7):

   Same Size Tells Pro/E to keep the model the *same physical size;* therefore, the values of the dimensions will change.

   Same Dims Tells Pro/E to keep the dimension values the same; therefore, the *size of the model changes*.

   User Scale Scales the part by a specified amount. Enter the scale factor in terms of a size relative to the current part (e.g., **1.0** will not change the part's current size, **.25** makes the part one quarter (**25%**) of its current size).
4. Choose one of the above options, then choose **Done** from the SCALE menu.

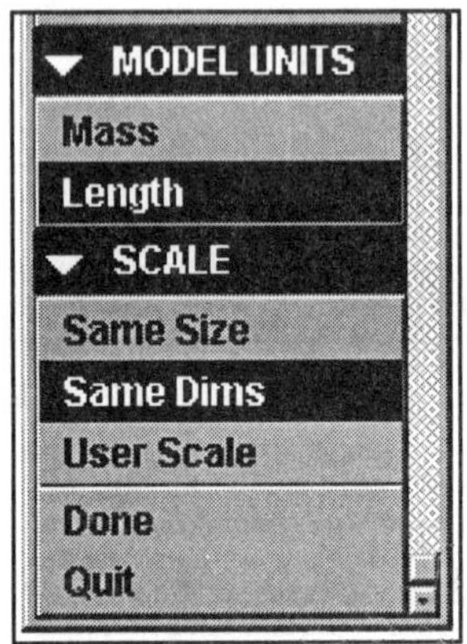

Figure 7.7
Changing Part Units

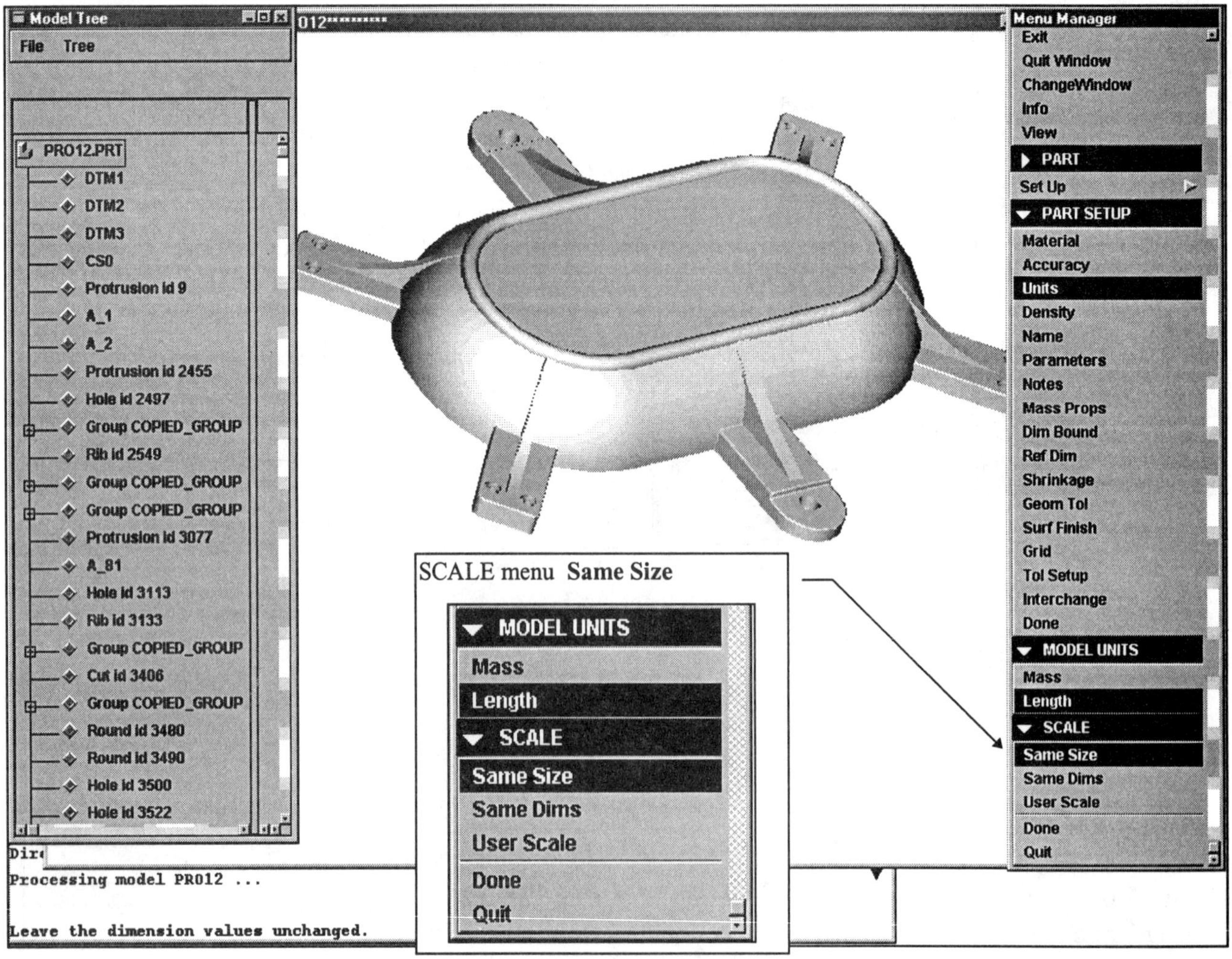

Dimension Tolerance Setup for Parts or Assemblies

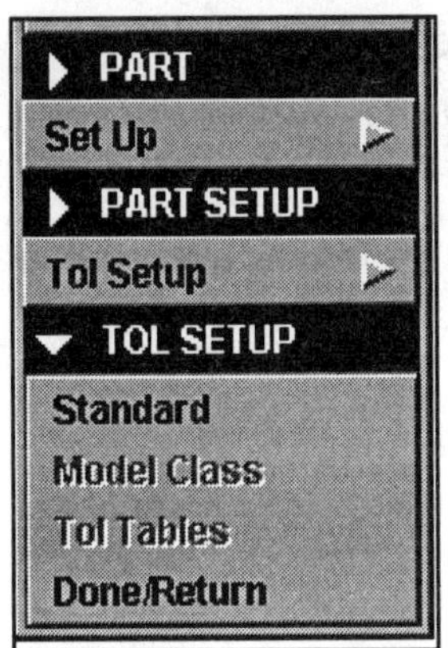

The application of dimension tolerances is governed by either ANSI or ISO/DIN standards. By default, the configuration file option **tolerance_standard** is set to ANSI. ANSI dimension tolerances are assigned according to the number of digits specified. ISO/DIN dimension tolerances are driven by a set of ISO/DIN tolerance tables. To change to ISO/DIN tolerances, change the **tolerance_standard** option to ISO/DIN.

Setting Up Dimension Tolerances

To use the **Tol Setup** command in Part or Assembly mode (Fig. 7.8), do the following:

1. Choose **Set Up** in the PART menu or ASSEMBLY menu.
2. Choose **Tol Setup** in the PART SETUP menu or ASSEM SETUP menu.
3. The TOL SETUP menu is displayed with the following options:

Standard Changes the tolerance standard of the current model from the TOL STANDARD menu.

ANSI Switches the standard from ISO/DIN to ANSI. New tolerances are determined for all dimensions based on the number of digits in the dimensions. Tolerance tables, if any, will be deleted.

ISO/DIN Switches the standard from ANSI to ISO/DIN. The system tolerance tables and any available user-defined tolerance tables are loaded.

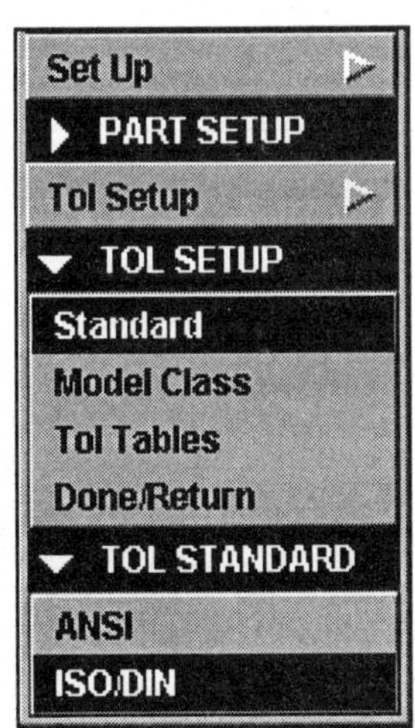

Model Class Changes the tolerance class of the current model using TOL CLASSES menu options.

Tol Tables Specifies an action using the TOL TBL ACT menu options.

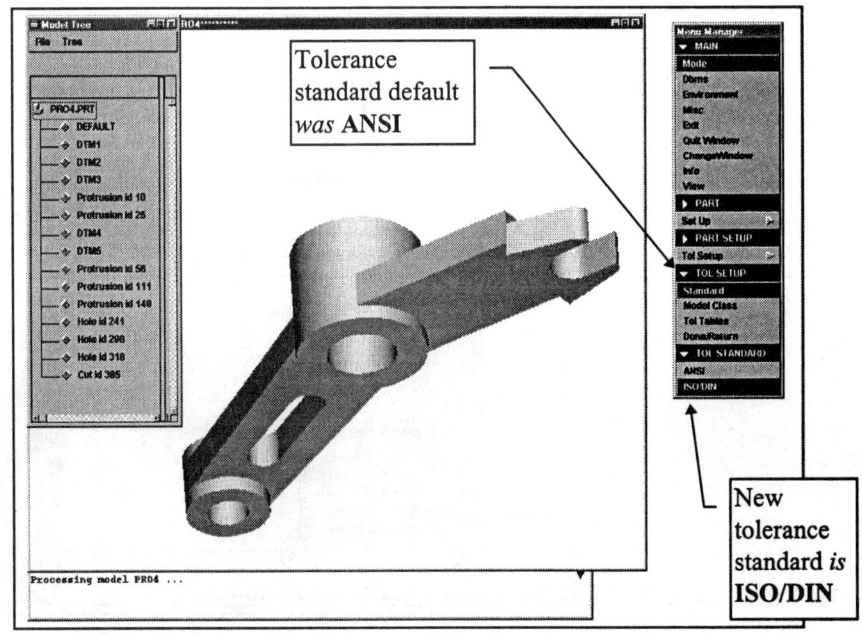

Figure 7.8
Tolerance Selection

Information

Choosing **Info** brings up the INFO menu, which provides options to obtain information about the geometric properties of models, names of models and sections, clearance between parts or surfaces, and so on. Figure 7.9 shows the command option **Model Info** in the INFO menu.

A variety of different types of information and analysis are available, including:

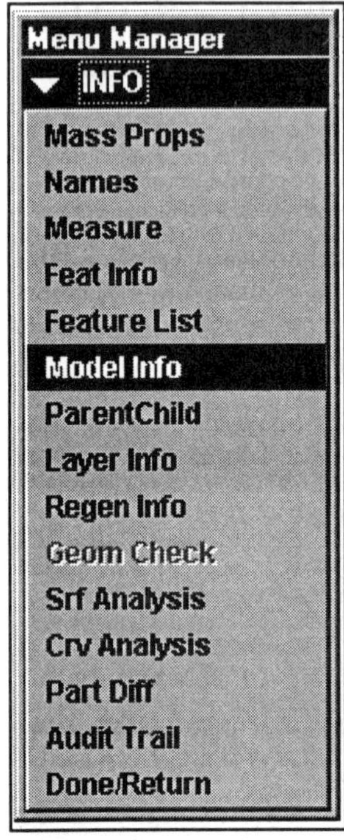

Mass property computation
File name listing
Measure clearance and interference
Measurement
Feature information
Feature list
Model (*part, assembly*) **information**
Parent/Child information
Layer information
Regeneration information
Geometry checking
Surface analysis
Curve analysis
Bill of materials (*Assembly mode*)
Part difference information
Audit trail

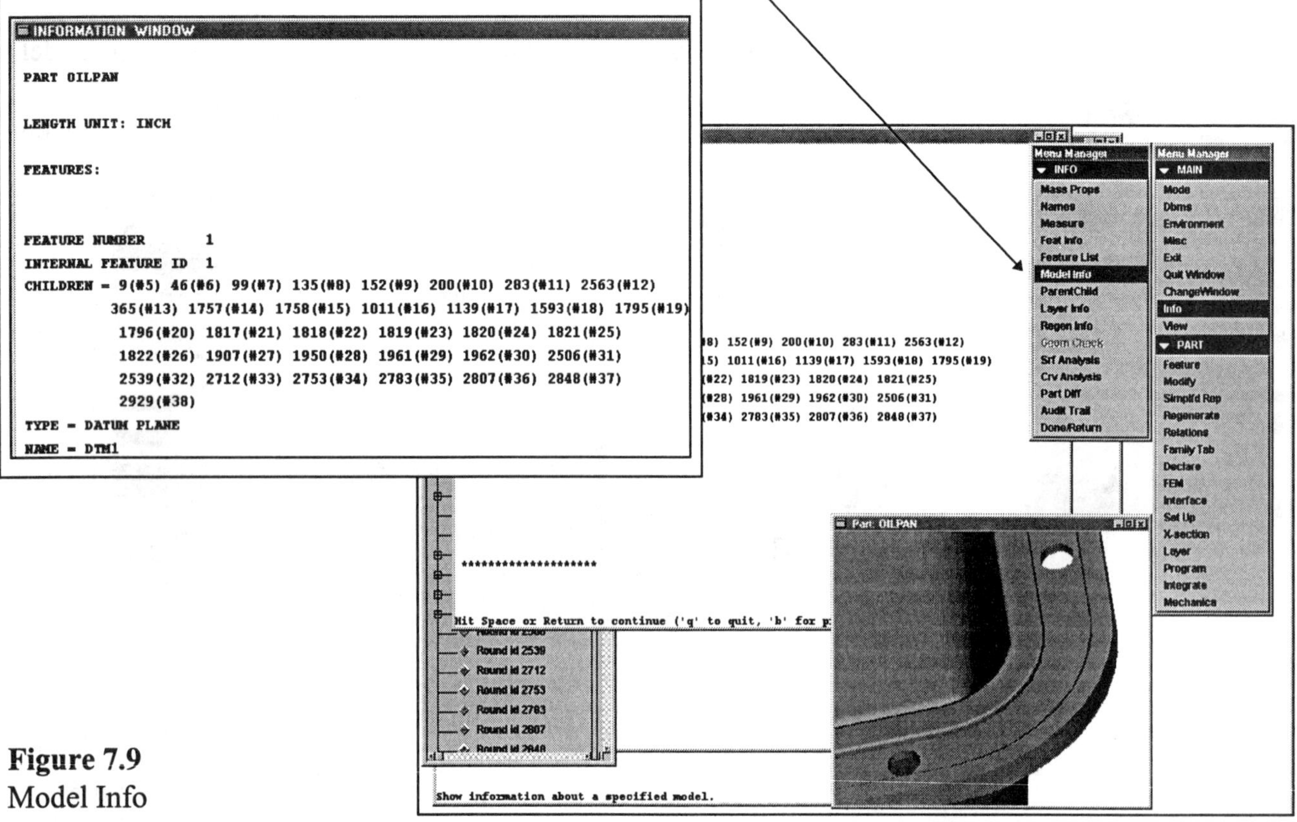

Figure 7.9
Model Info

The Info Menu Options

The **Info** option in the MAIN menu can be used to compute mass properties, measure distances, clearances, and interferences, and obtain model information. The **BOM** (bill of materials) option is available only in Assembly mode. The **Regen Info** option is available in both Part mode and Assembly mode. Mass properties for parts, assemblies, and cross sections can be computed by choosing **Mass Props** from the INFO menu.

Measure

The **Measure** command in the INFO menu is used to analyze model and draft geometry. It is available in all modes. Figure 7.10 shows the **Measure ⇒ Area ⇒ Query Sel ⇒ Accept** commands given with the circular surface of the boss selected and accepted.

MEASURE
Area
AREA
Actual
Projected
Surface
Quilt
Done/Return
GET SELECT
Pick
Query Sel
Sel By Menu
Done Sel
Quit Sel

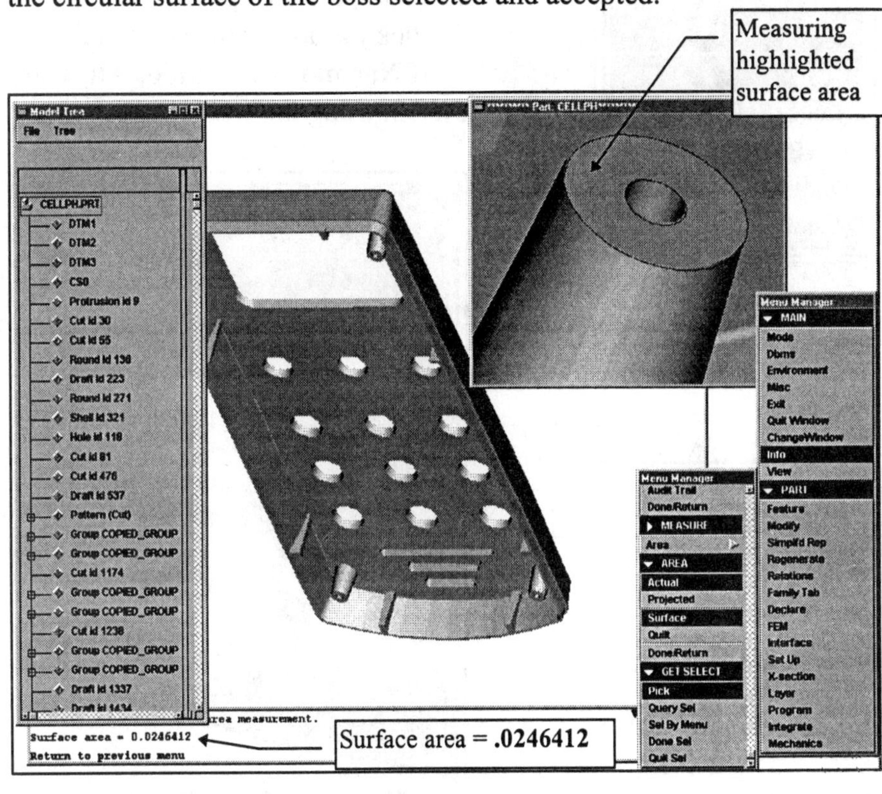

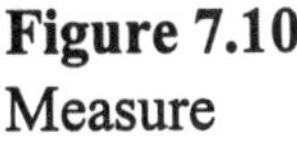

Figure 7.10
Measure

Curve/Edge

Part edges and datum curves are measured using the **Curve/Edge** option from the MEASURE menu. Available **Info** options for edge or curve information include:

Length Measures and displays the length of an edge or a curve segment.
Type Displays the type of the edge or curve.
Normal You have to select or create a coordinate system to measure normals. Displays, in red, the normal vector (second derivative) to the edge or curve at the selected point.

Tangent Displays, in red, the tangent vector (first derivative) to the edge or curve at the selected point. The coordinates of the tangent vector are displayed in the message window. You have to select or create a reference coordinate system.

Curvature Calculates and displays the curvature of the edge or curve at the selected point.

Radius Calculates the radius at the selected point on the curve or edge.

All Displays the combined information provided by Length, Type, Normal, Tangent, and Curvature. In Figure 7.11, the cell phone's front edge has been selected and information is displayed in the MESSAGE WINDOW at the bottom of the screen.

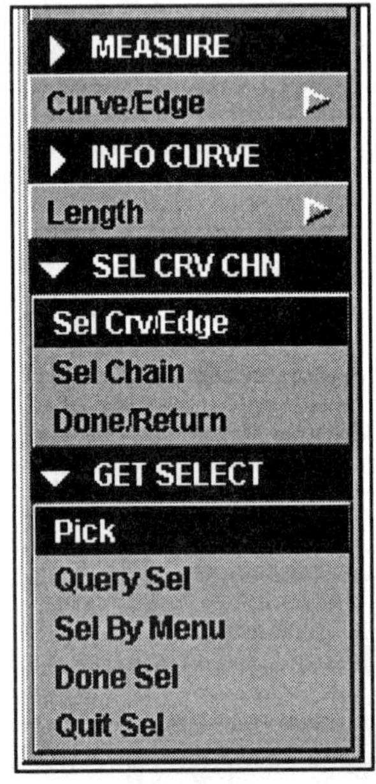

To measure an edge or datum curve:

1. Pick **Curve/Edge** from the MEASURE menu.
2. Pick the appropriate option from the INFO CURVE menu.
3. If **Normal, Tangent,** or **All** is chosen, select or create a coordinate system using the GET COORD S menu options.

Measuring this edge
EDGE LENGTH = 2.84061

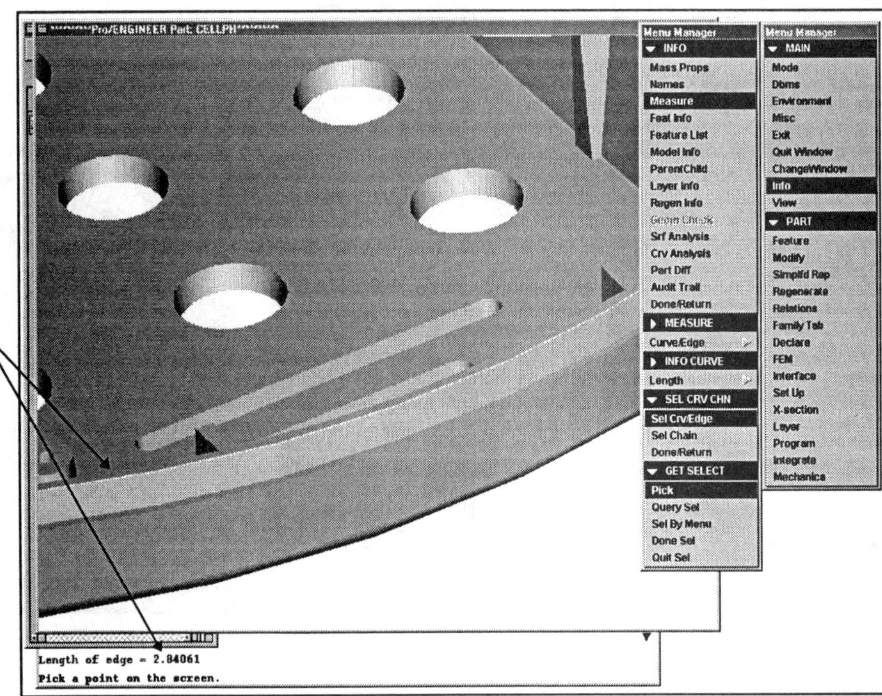

Figure 7.11
Measure Curve/Edge

4. If **Curvature** or **Radius** is chosen, select an option from the POINT OPT menu:

   Select The measurements will be made for the pick point.

   End point The measurements will be made for the nearest endpoint of the edge or curve segment.

5. Select the edge or curve to be measured. Use the **Sel Chain** command when measuring the lengths of several curves/edges lying on the same surface.

Angle

The **Angle** command is used to measure the angle between axes, planar curves, and planar nonlinear edges. The angle measured depends on where you select the edges. To measure angles between planes, do the following:

1. Pick **Angle** from the MEASURE menu.
2. Select **Plane**, select the plane surface to measure from, and then select the second plane surface. The selected geometry is highlighted in blue, and the angle will be displayed in the MESSAGE WINDOW.

When you measure the angle to/from a planar nonlinear edge or planar curve, Pro/E calculates the angle to/from the plane in which the indicated edge or curve lies. When computing the value of an angle between two planes (Fig. 7.12) or between a plane and a line, Pro/E selects the smaller of the two possible angle values. In this example, the angle measurement between the angled surface and the front ledge surface is **25°**.

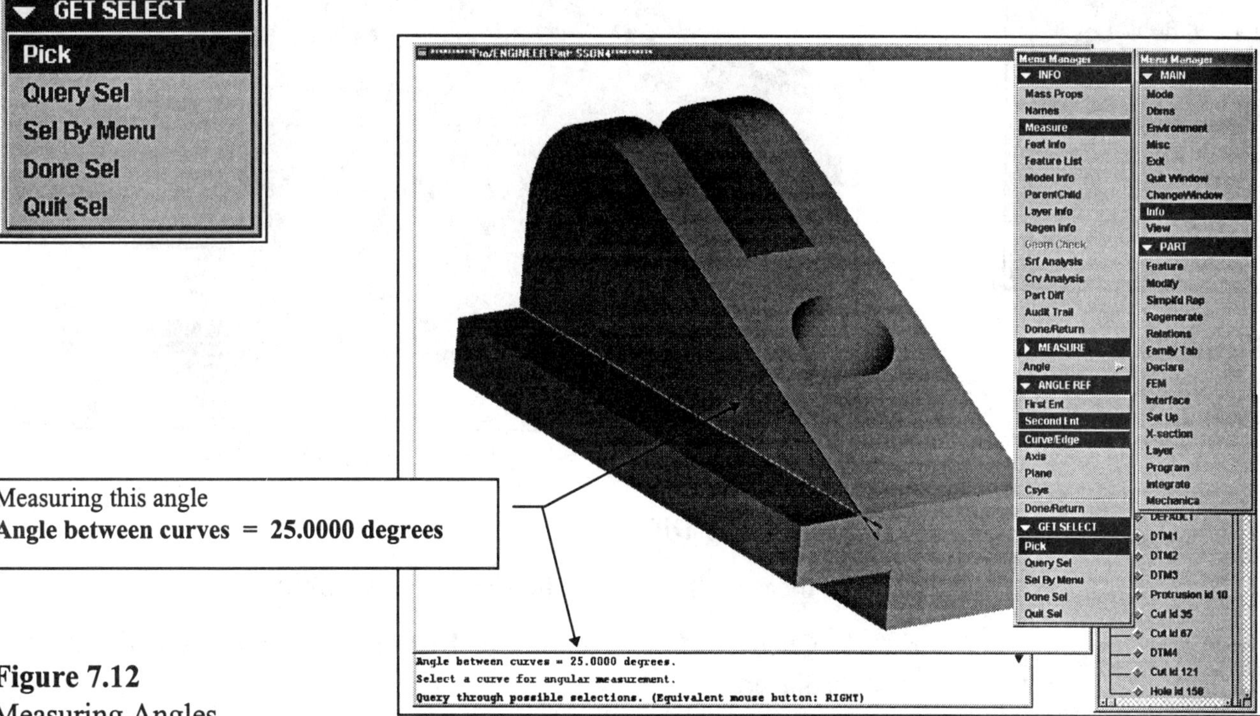

Figure 7.12
Measuring Angles

Distance

Distance is measured with respect to a basis entity. The *basis entity* is the one from which you measure, that is, the first entity selected when you start measuring **Distance**. After the *basis entity* is selected, you can make as many measurements as you like by selecting various entities to measure to. Each distance will be calculated with respect to the first entity, until you restart the measuring process by selecting a new *basis entity*.

When measuring distances, you have to specify the entity type before selecting an entity from which, or to which, to measure. The entity types are:

From Point A point on the part surface, or a datum point.
From Vertex A vertex of the part (Fig. 7.13).
From Plane A planar part surface or datum plane.
From Lin Ent A linear entity, such as an axis or feature edge.
From Csys A coordinate system. When you select **Csys** as the entity type, the GET COORD S menu is displayed. You can create or select a coordinate system to measure **From** or **To**.

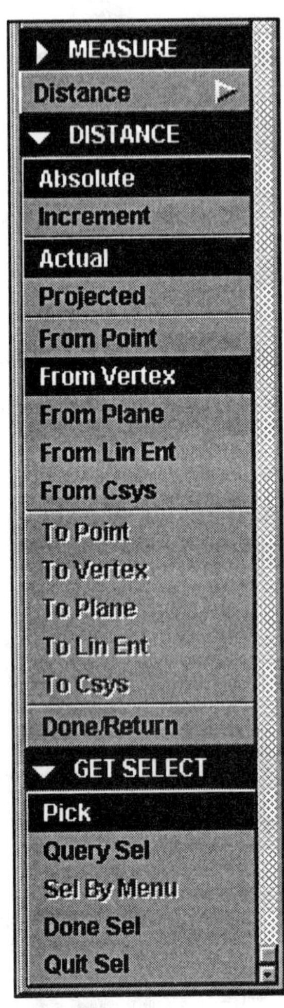

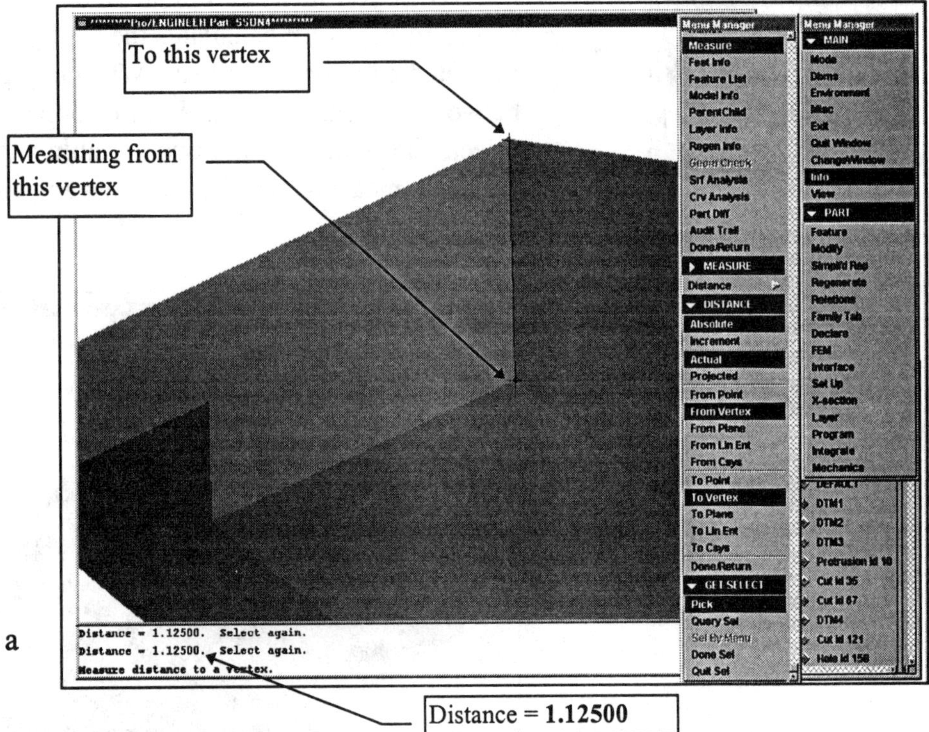

Figure 7.13
Measuring from a Vertex to a Vertex

Measuring Distance

To measure the distance between any two entities:

1. Choose **Info ⇒ Measure ⇒ Distance**. The DISTANCE menu is displayed.
2. Select absolute or incremental coordinates:

Absolute Display the measured distance with respect to the basis coordinate system.
Increment Display both the measured distances and the differences between their coordinates, or incremental coordinates (e.g., **dx, dy, dz**), with respect to the basis coordinate system. Both the **From** and **To** options are inaccessible until you specify or create a coordinate system from which to determine the measurements.

3. Choose the appropriate **From** option to establish the *basis entity.*
   From Point Measure from an arbitrary point.
   From Vertex Measure from a vertex.
   From Plane Measure from a plane.
   From Lin Ent Measure from a linear entity.
   From Csys Measure from a coordinate system. The GET COORD S menu will be displayed.
4. Choose the appropriate **To** option.
   To Point Measure to an arbitrary point.
   To Vertex Measure to a vertex.
   To Plane Measure to a plane.
   To Lin Ent Measure to a linear entity.
   To Csys Measure to a coordinate system. The GET COORD S menu will be displayed.
5. Make as many measurements as you want from the *basis entity.* You can change the type of entity you are measuring to at any time.
6. To restart the measuring process for a new *basis entity*, repeat the process from step 3, choosing a new *basis entity* option (e.g., **From**).
7. To end the measuring process, choose **Done/Return.**

Figure 7.14
Measuring from the Coordinate System

In Figure 7.14, **Info ⇒ Measure ⇒ Distance ⇒ From Csys ⇒ Sel By Menu ⇒ Select** (pick the coordinate system) ⇒ **Done ⇒ To Plane** (use **Query Sel** to select the bottom circular plane of the part) was used to measure and show the **.750000** distance.

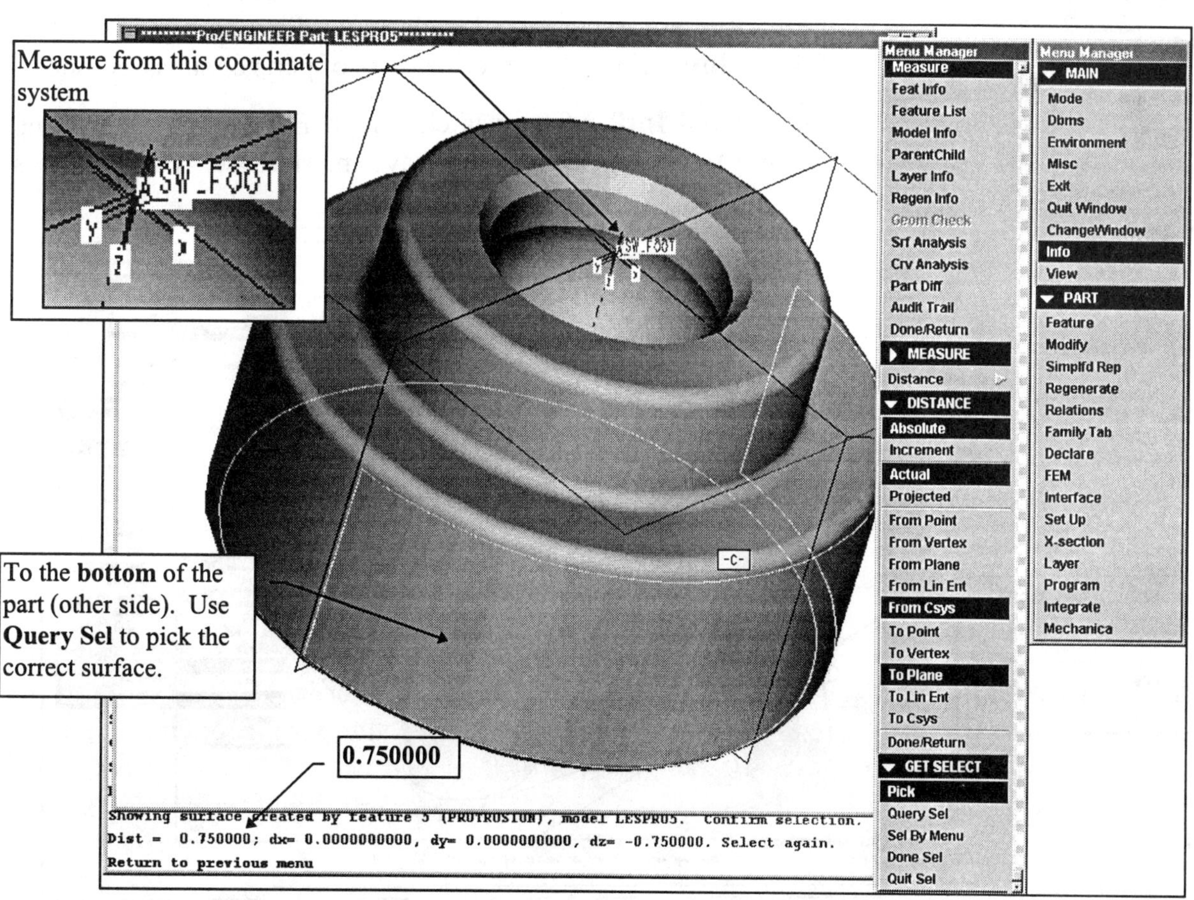

Clearance and Interference Calculations

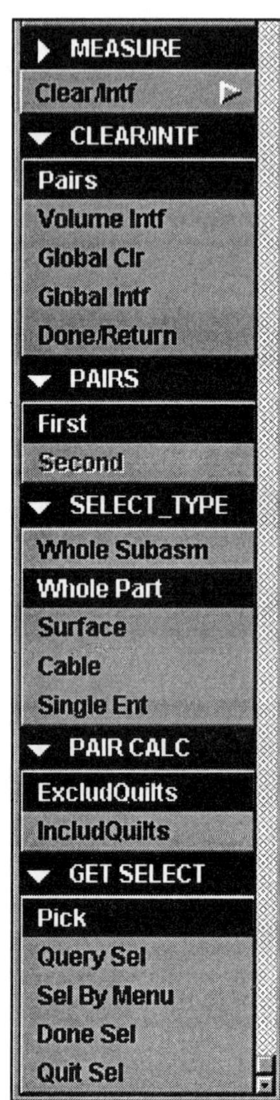

The **Clear/Intf** option calculates and displays either the clearance distance or the interference between pairs of any combination of subassemblies, parts, surfaces, cables, or entities. Figure 7.15 shows the menu structure for measuring clearance and interference.

If the objects selected do not interfere, the minimum clearance is displayed graphically as a red line. A small red circle with crosshairs will display at each end of the red line to identify the location at which the clearance is being measured. The clearance value is displayed in the MESSAGE WINDOW. If there is interference between the two objects, Pro/E highlights the volume of interference and provides the value or highlights the curve or point of intersection, as appropriate for the items selected. The following is the procedure to determine a clearance or interference:

1. Pick **Measure** from the INFO menu.
2. Select **Clear/Intf** from the MEASURE menu.
3. Choose the desired option from the CLEAR/INTF menu:

Pairs Get clearance or volume of interference between pairs of any combination of subassemblies, parts, surfaces, cables, and/or entities.

Vol Intf Used in Pro/ECAD to ensure that keep/in and keep/out areas have not been violated.

Global Clr Find all pairs of parts or subassemblies that have clearances less than a specified clearance distance.

Global Intf Find all interfering pairs of parts or subassemblies.

The **Global Intf** option was chosen in Figure 7.15 (the clamp subassembly). Note that four pairs have interference. Pair 1 of 4 is shown.

Interference value shown in command line

(0.0477 inch^3)

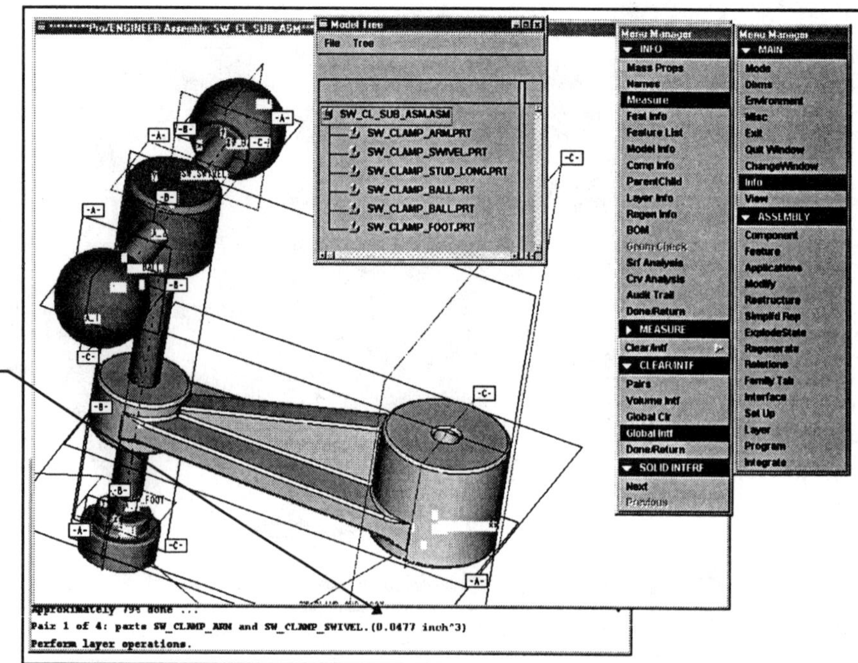

Figure 7.15
Global Interference

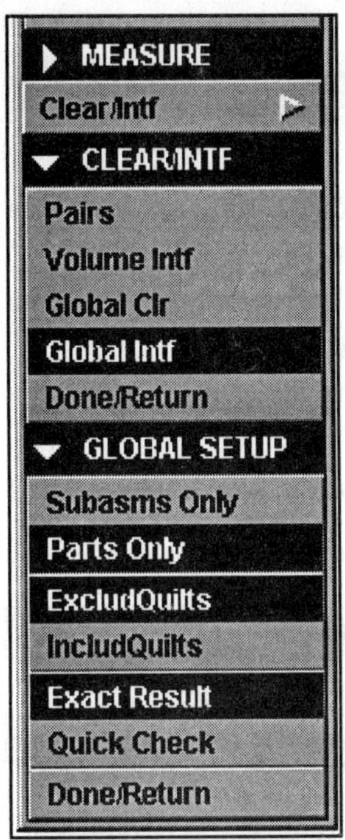

Global Clearance Within an Assembly

This measurement command is used to identify, in an entire assembly, all component parts or subassemblies for which clearances are less than or equal to a clearance distance that is specified in the design (Fig. 7.16):

1. Select **Measure** from the INFO menu.
2. Pick **Clear/Intf** from the MEASURE menu.
3. Choose **Global Clr**. The GLOBAL SETUP menu is displayed. The options will appear in mutually exclusive pairs. Select the desired pair options, then choose **Done/Return.**

Pair 1:
Subasms Only Does a global check for clearances between all subassemblies, but not within individual subassemblies.
Parts Only Performs global checking for clearances between all parts in the assembly.
Pair 2:
ExcludQuilts Computes the clearance between solids.
IncludeQuilts Not available for global clearance processing.

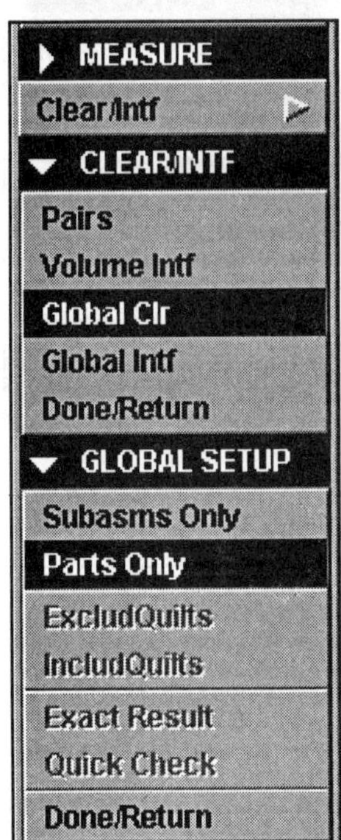

4. When prompted in the MESSAGE WINDOW, enter the clearance distance. Pro/E will determine if any components of the assembly are within the specified clearance distance. All interferences are included.
5. Use **Next** or **Previous** in the GLOBAL CLR menu to step through the display of identified pairs (Fig. 7.16). To exit the process, choose **Done/Return** in the CLEAR/INTF menu.

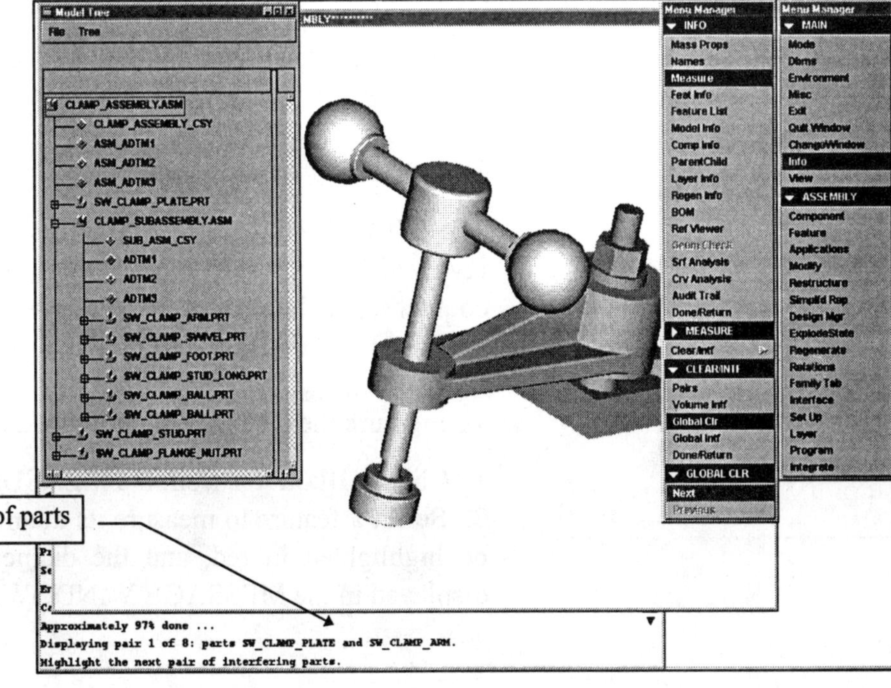

Listing of clearances between pairs of parts

Figure 7.16
Global Clearance

Surface Area

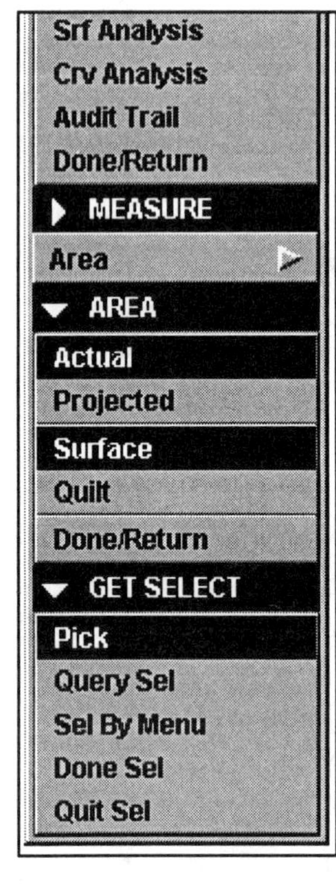

The MEASURE option **Area** measures the area of any surface on the part or the area of a datum surface.

1. Pick **Area** from the MEASURE menu. Choose an option:
   Actual Calculate the area of a surface or quilt.
   Projected Specify a projection direction and calculate the area of a surface or quilt projected in that direction.
   Surface Calculate the surface area of a face.
   Quilt Calculate the surface area of a quilt.
2. Select a surface for area measurement. The surface selected will be highlighted in red (Fig. 7.17).

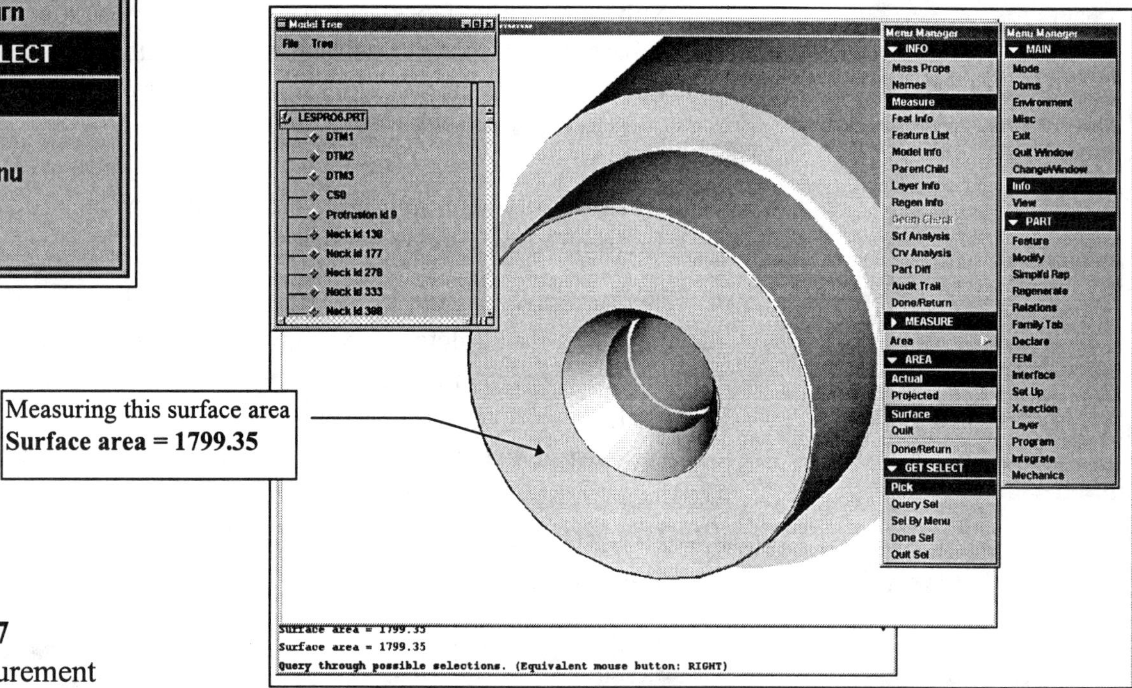

Figure 7.17
Area Measurement

Diameter

The MEASURE option **Diameter** measures the diameter of any revolved surface of a part. In Figure 7.18, the edge of the counterbored hole was chosen and the diameter of **1.00000** is displayed in the MESSAGE WINDOW.

To measure the diameter of a revolved feature:

1. Choose **Diameter** from the MEASURE menu.
2. Select a feature to measure its diameter. The surface selected will be highlighted in red, and the diameter *at the pick point* will be displayed in the MESSAGE WINDOW.

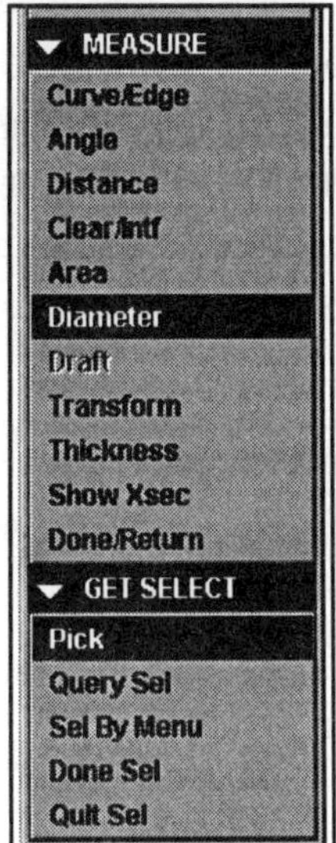

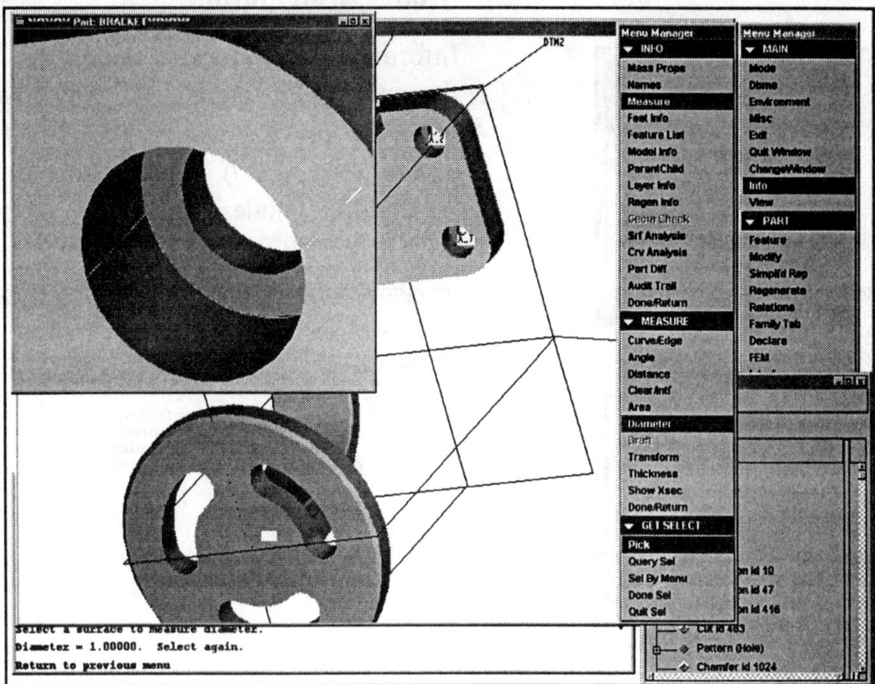

Figure 7.18
Diameter Measurement

Thickness

Thickness measures the thickness of a part to determine if a specified region has a thickness that is greater than or less than a user-specified maximum or minimum value. The region is displayed as in Figure 7.19. If a region's thickness is over the maximum allowable, it will be displayed with a red border. If a region's thickness is below the minimum allowable, it will be displayed with a blue border.

Measuring thickness the of part

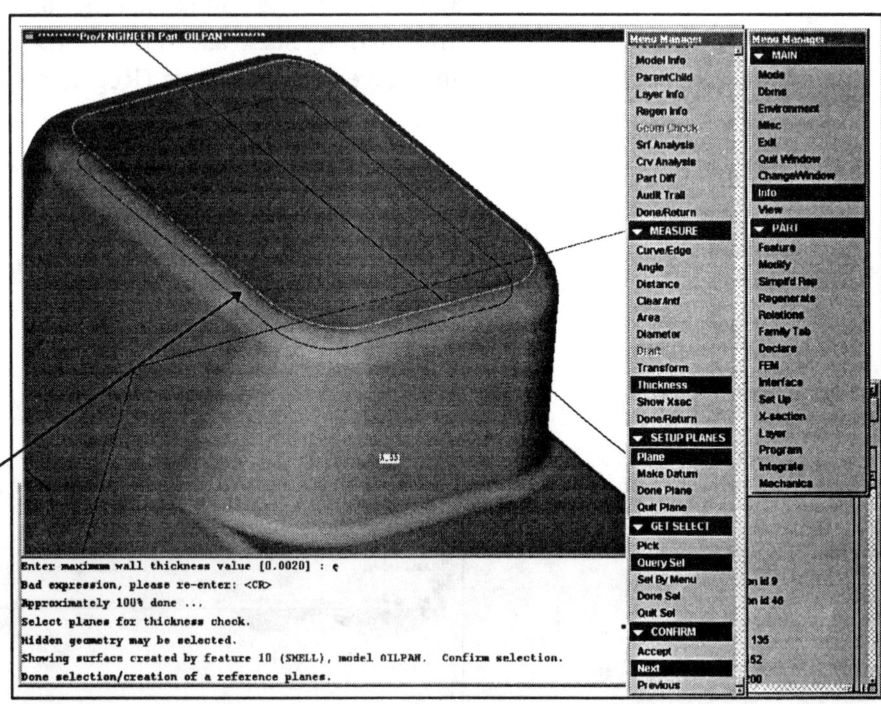

Figure 7.19
Thickness Measurement

Model Information

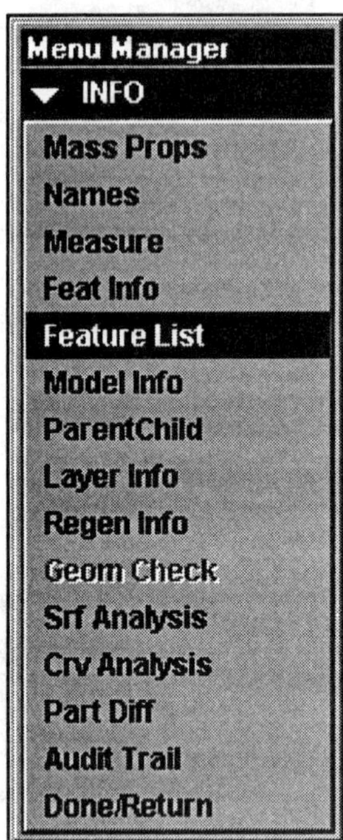

Information about features can be accessed by choosing **Feat Info** from the INFO menu. After choosing **Feat Info**, specify a feature either by picking it with the mouse or by using the **Sel By Menu** option (Fig. 7.20). **Feat Info** can be chosen in both Part and Assembly modes. Feature information is shown in the INFORMATION WINDOW.

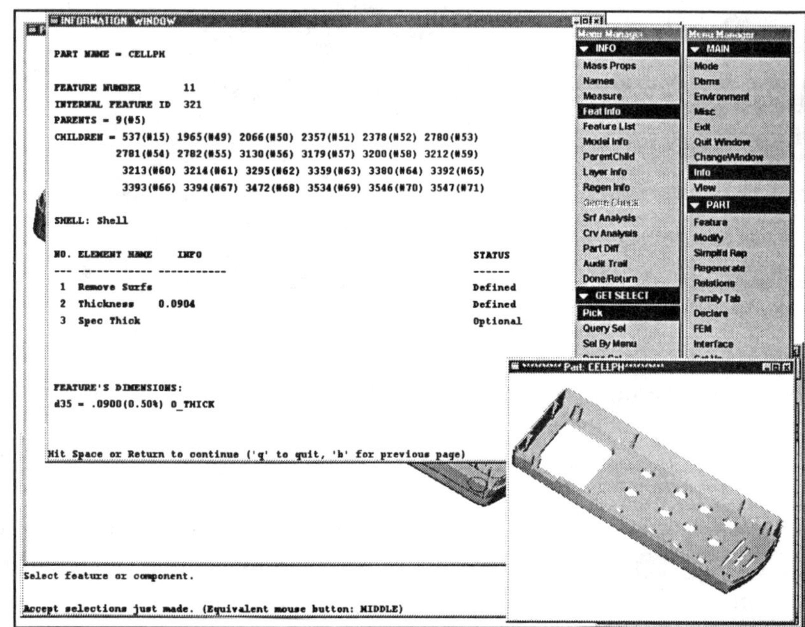

Figure 7.20
Feature Information

Feature List

When you choose **Feature List**, the INFORMATION WINDOW appears with a table listing the features in order and giving information, such as Number, ID, Name, Type, Suppression Order and Regeneration Status (Regenerated, Unregenerated, Failed, and so on). An example is shown in Figure 7.21.

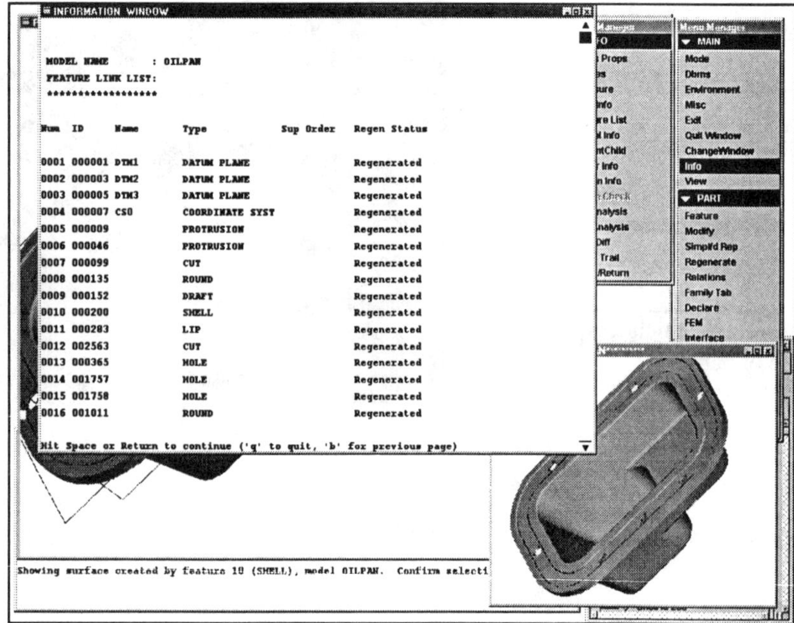

Figure 7.21
Feature List

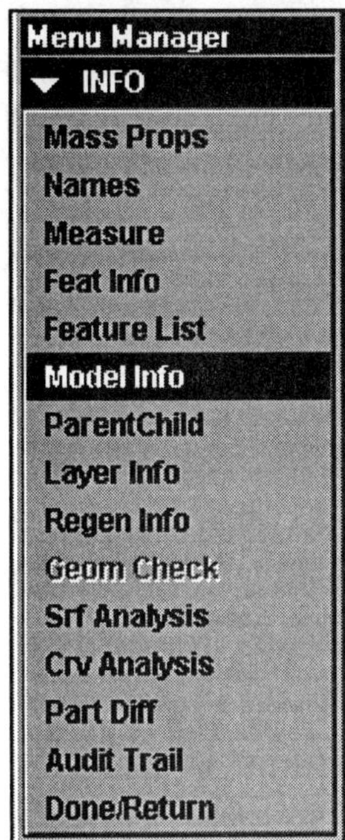

Part Information

Information about every feature of a part can be accessed by choosing **Model Info**. In Part mode, the INFORMATION WINDOW will appear immediately. In Assembly mode (Fig. 7.22), choose **Part** from the MODEL INFO menu and specify a part, either by picking a part with the mouse (**Pick**) or by entering the name of the part (**Name**).

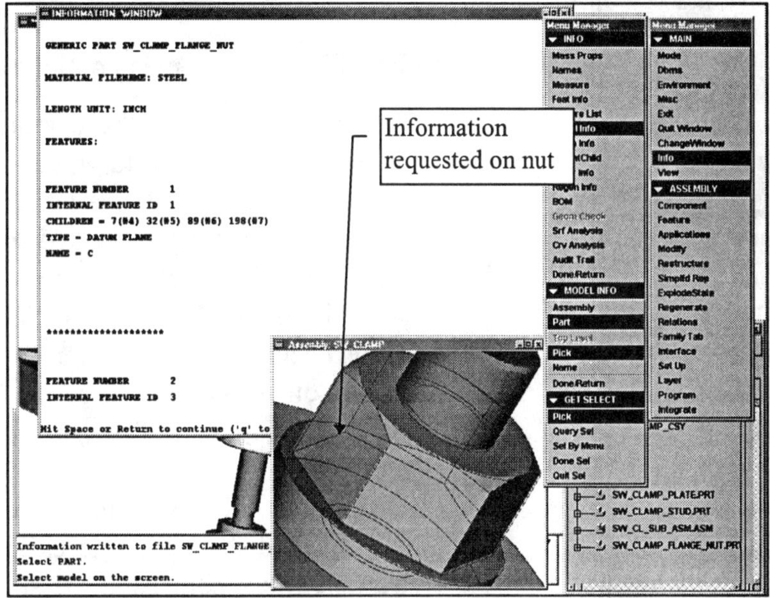

Figure 7.22
Part Model Information

Assembly Information

Assembly information can be accessed by choosing **Model Info** and then **Assembly** from the MODEL INFO menu, then either picking on an assembly or specifying the name of an assembly. An INFORMATION WINDOW displaying the assembly information will then appear. The names of the components in the assembly are displayed in a hierarchical structure to show how it was assembled (Fig. 7.23).

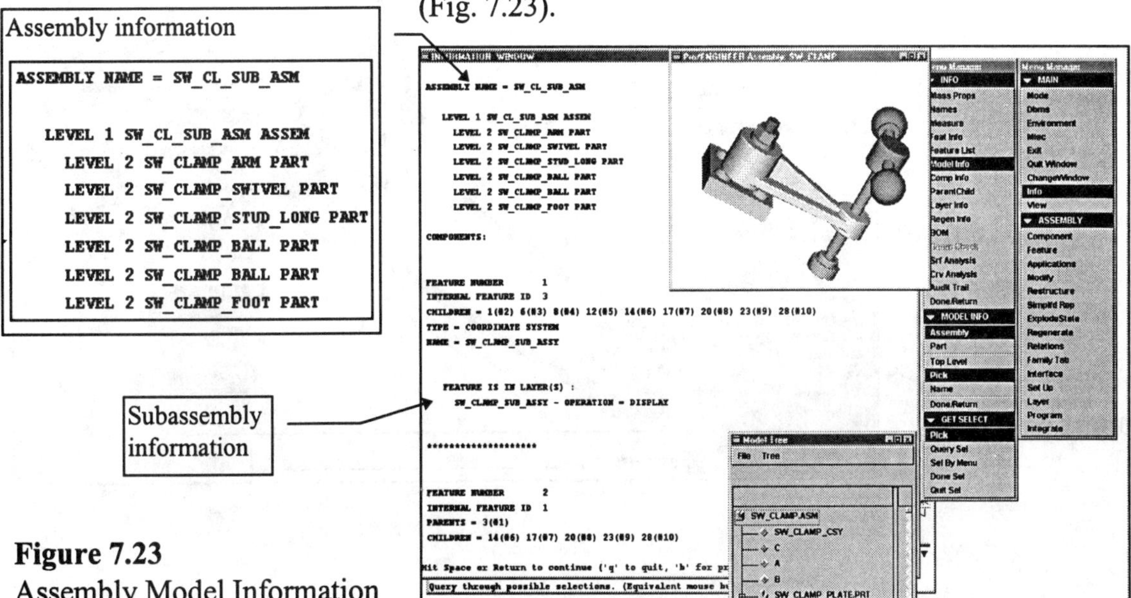

Figure 7.23
Assembly Model Information

ParentChild Information

The **ParentChild** option in the INFO menu is used to highlight the relationships between features. If you select either **Parents** or **Children**, you can then choose either to create an information file or to highlight the appropriate geometry on the screen. If the file option is chosen, the information is written to a file and displayed in the INFORMATION WINDOW.

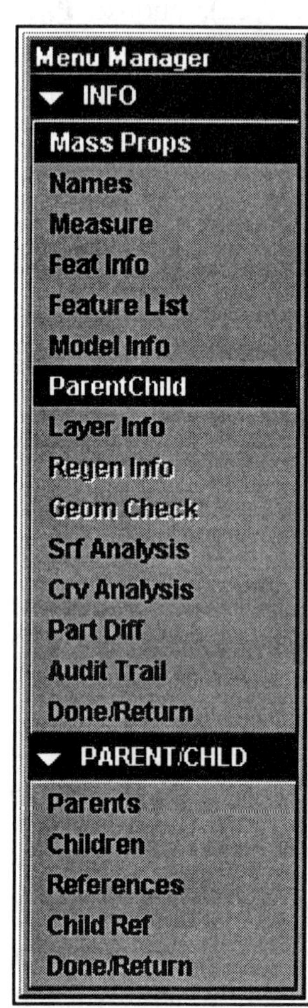

The PARENT/CHLD menu options are as follows:

Parents Shows all the parents for the selected feature. The parent features are highlighted in the Reference color.

Children Shows all the children for the selected feature. The child features are highlighted in the Reference color. In Figure 7.24, the cylindrical protrusion was chosen and its children were listed in the INFORMATION WINDOW.

References All the references used to construct a feature are shown one at a time. These can be axes, datums, surfaces, edges, or other features. They are shown in the Internal Reference Color.

Child Ref Shows all the references for each of the children of the selected feature. The child features are highlighted in the Surface Mesh color, and the surfaces, edges, or points that they reference are highlighted in the Internal Quilt Edge color.

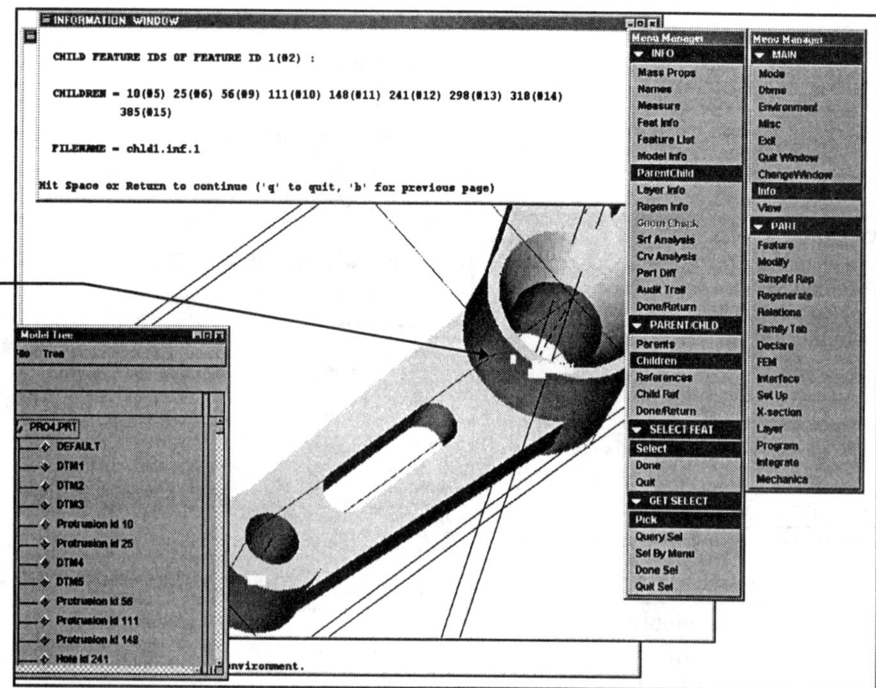

Revolved protrusion was selected

Figure 7.24
Children Information for a Part Feature

Displaying ParentChild Information

ParentChild
Layer Info
Regen Info
BOM
Ref Viewer
Geom Check
Srf Analysis
Crv Analysis
Audit Trail
Done/Return
▼ PARENT/CHLD
Parents
Children
References
Child Ref
Done/Return
▼ FILE/HILITE
File
Highlight

This command is used to show parent/child relationships of a particular feature of a component or if the information is needed for an assembly:

1. Select **ParentChild** from the INFO menu.
2. Choose one of the options from the PARENT/CHLD menu.
3. Chose **Parents** (Fig. 7.25) or **Children** (Fig. 7.24), then select one of the options from the FILE/HILITE menu:
   File A text file with the corresponding feature IDs listed will be written to disk and displayed in the INFORMATION WINDOW. The file name is displayed (Fig. 7.25).
   Highlight The appropriate geometry will be highlighted with the default color codes.
4. Select the feature. In Figure 7.25, the swivel component of the assembly was selected.

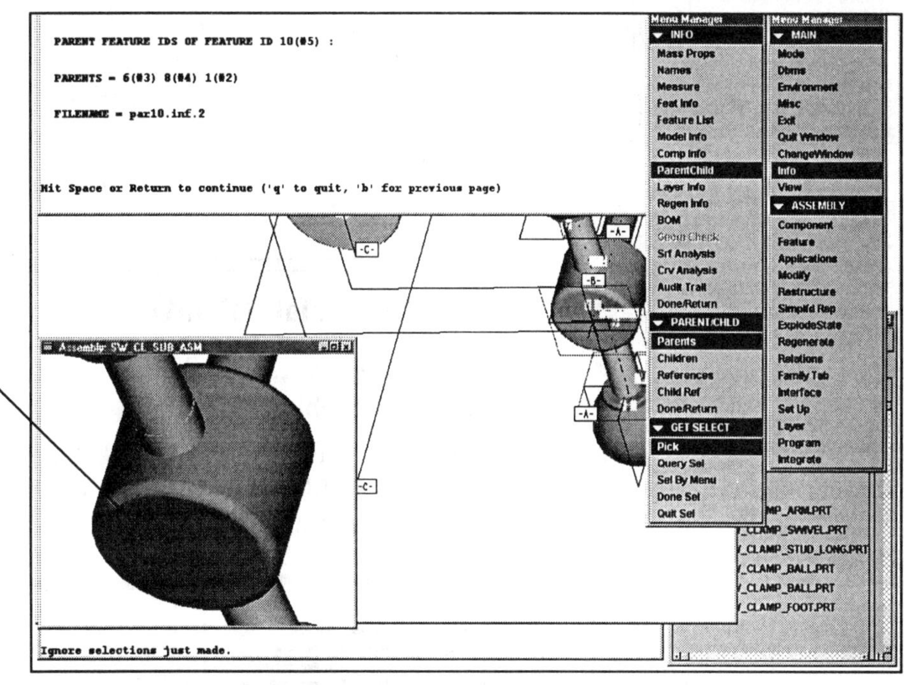

Figure 7.25
Parent Information

5. If **References** is chosen, select the desired feature or component and the SHOW REF menu appears. Choose one of the following:
   Next or **Previous** Select the next or previous reference.
   Info Displays information about the highlighted reference in the INFORMATION WINDOW.
6. If **Child Ref** is chosen, select the desired feature or component and the CHILD REFS menu appears. Choose one of the following:
   Next or **Previous** Selects the next or previous reference.
   Ref Info Displays information about the highlighted reference in the INFORMATION WINDOW.
   Child Info Using the highlighted reference, displays information about the child feature(s) in the INFORMATION WINDOW.

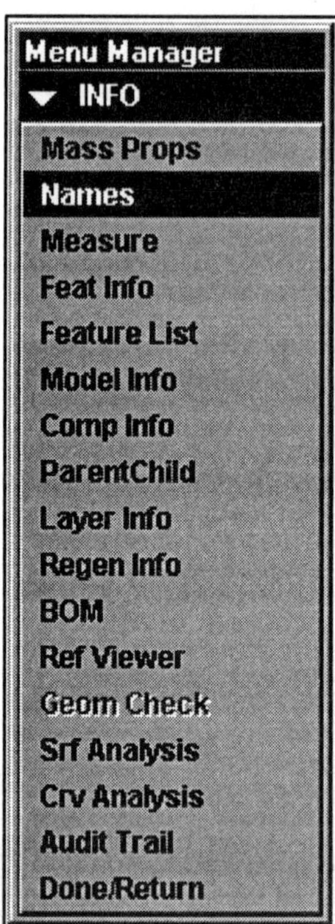

File Name Listing

The INFO menu option **Names** displays a list of all stored files in an INFORMATION WINDOW (Fig. 7.26). The first portion of the listing shows parts, assemblies, drawings, layouts, and sections in memory. The second portion gives a complete listing of all Pro/ENGINEER objects in the working directory.

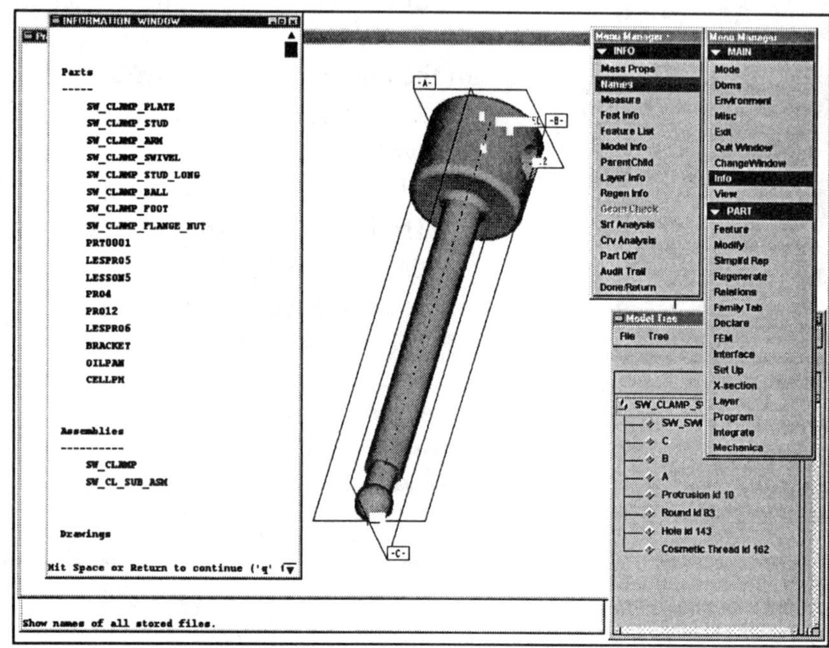

Figure 7.26
Names

Bill of Materials (BOM)

A *bill of materials* is a listing of all parts and part parameters in the current assembly (Fig. 7.27). You can customize the output formats to produce a particular form of presentation and content. BOMs can be created for assemblies in Assembly mode or from assembly drawings in Drawing mode.

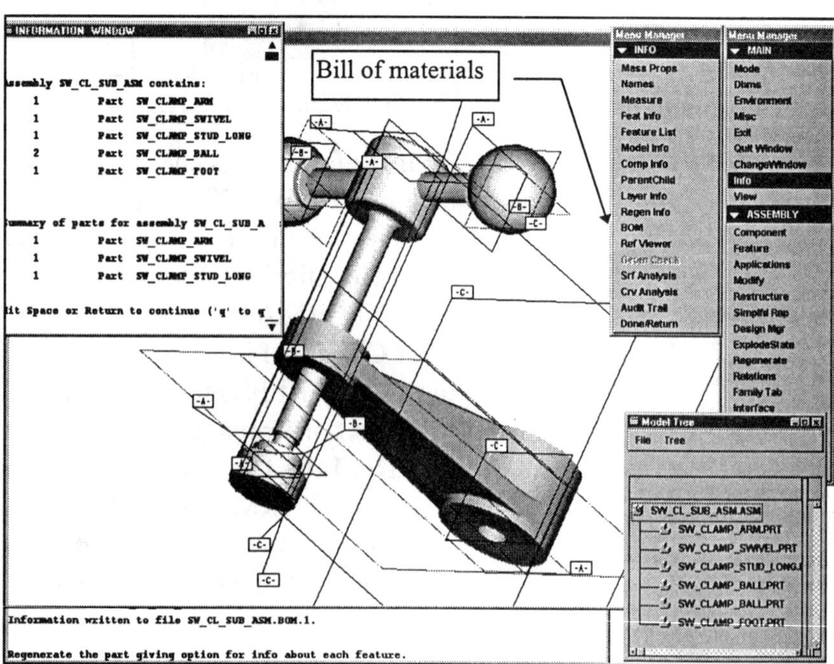

Figure 7.27
BOM

The information that follows shows how to create and format simple text BOMs, which are stored as text files. The optional module Pro/REPORT provides functionality for creating BOM reports: graphical BOMs with complex formatting and indexing.

The source of the bill of materials (BOM) output format can be configured by the configuration file. An example of the configuration file option for a user-defined format is:

bom_format	bomcompany.fmt

The default output format for the BOM is divided into two sections:

Breakdown Lists the name, type, and number of instances of each member and submember.

Summary Lists the total quantity of each part included in the assembly. It amounts to a "shopping list" of all the parts needed to build the assembly from the part level.

The BOM in Figure 7.28 contains the subassembly shown in Figure 7.27. Note that the **Model Tree** also displays information about an assembly.

```
Assembly SW_CLAMP contains:
    1          Part  SW_CLAMP_PLATE
    1          Part  SW_CLAMP_STUD
    1 Sub-Assembly  SW_CL_SUB_ASM
    1          Part  SW_CLAMP_FLANGE_N


Sub-Assembly SW_CL_SUB_ASM contains:
    1          Part  SW_CLAMP_ARM
    1          Part  SW_CLAMP_SWIVEL
    1          Part  SW_CLAMP_STUD_LON
    2          Part  SW_CLAMP_BALL
    1          Part  SW_CLAMP_FOOT


Summary of parts for assembly SW_CLAMP
    1          Part  SW_CLAMP_PLATE
    1          Part  SW_CLAMP_STUD
    1          Part  SW_CLAMP_ARM
    1          Part  SW_CLAMP_SWIVEL
    1          Part  SW_CLAMP_STUD_LON
    2          Part  SW_CLAMP_BALL
    1          Part  SW_CLAMP_FOOT
    1          Part  SW_CLAMP_FLANGE_N
```

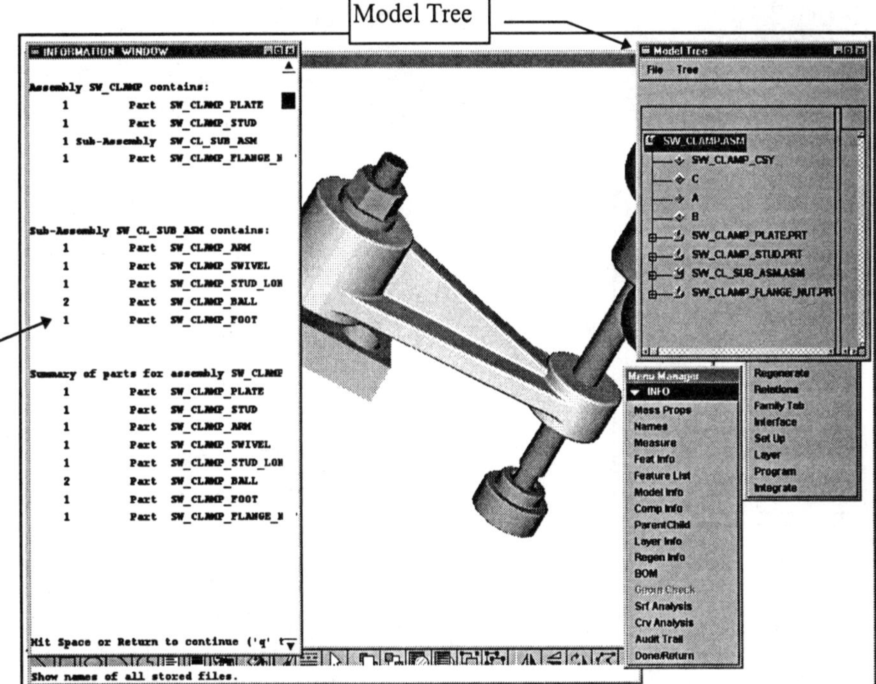

BOM

Figure 7.28
Assembly BOM

A user-defined BOM output format specifies the formats of the breakdown section and the summary section. You can include one or both sections, but you must specify the column titles, row content, and display format for each included section.

The Interface Menu

To import or export information through Pro/E, the INTERFACE menu is available (Fig. 7.29). The options that appear in this menu will change depending on the mode you are working in and the Pro/E modules that are available (e.g., Pro/INTERFACE, Pro/PLOT, Pro/ECAD, Pro/STEP, etc.). Pro/E can import and export data in a variety of formats, including IGES, IGES Groups, STEP, DXF, SET, VDA, CGM, STL, Plotter files, Neutral files, Render, Inventor, 3DPAINT, PATRAN Geom, COSMOS Geom, SUPRTB Geom, CatiaFacets, PDGS, ECAD, CGM, TIFF, PHOTORENDER, CATIA, CATIA IIF, CDRS, and VRML.

Plotting can be accomplished with the **Interface** option from the PART menu, ASSEMBLY menu, DRAWING menu, and other menus. Your system administrator will set the path to the plotting and printing devices on your system. Choose **Interface** ⇒ **Export** ⇒ **Plotter More Plotters** ⇒ (choose plotter) ⇒ **OK** ⇒ **OK**.

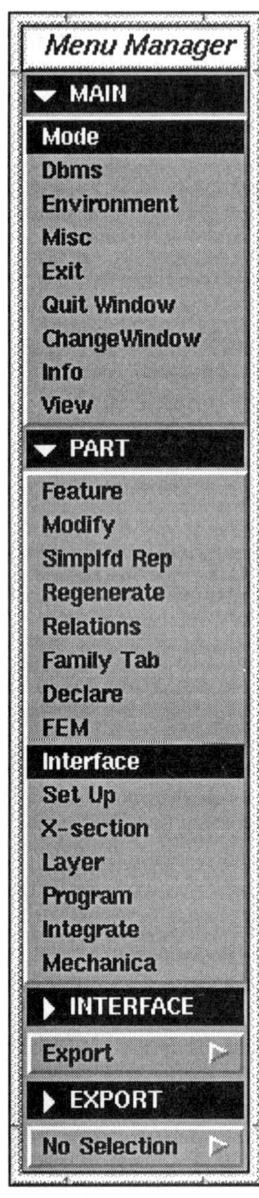

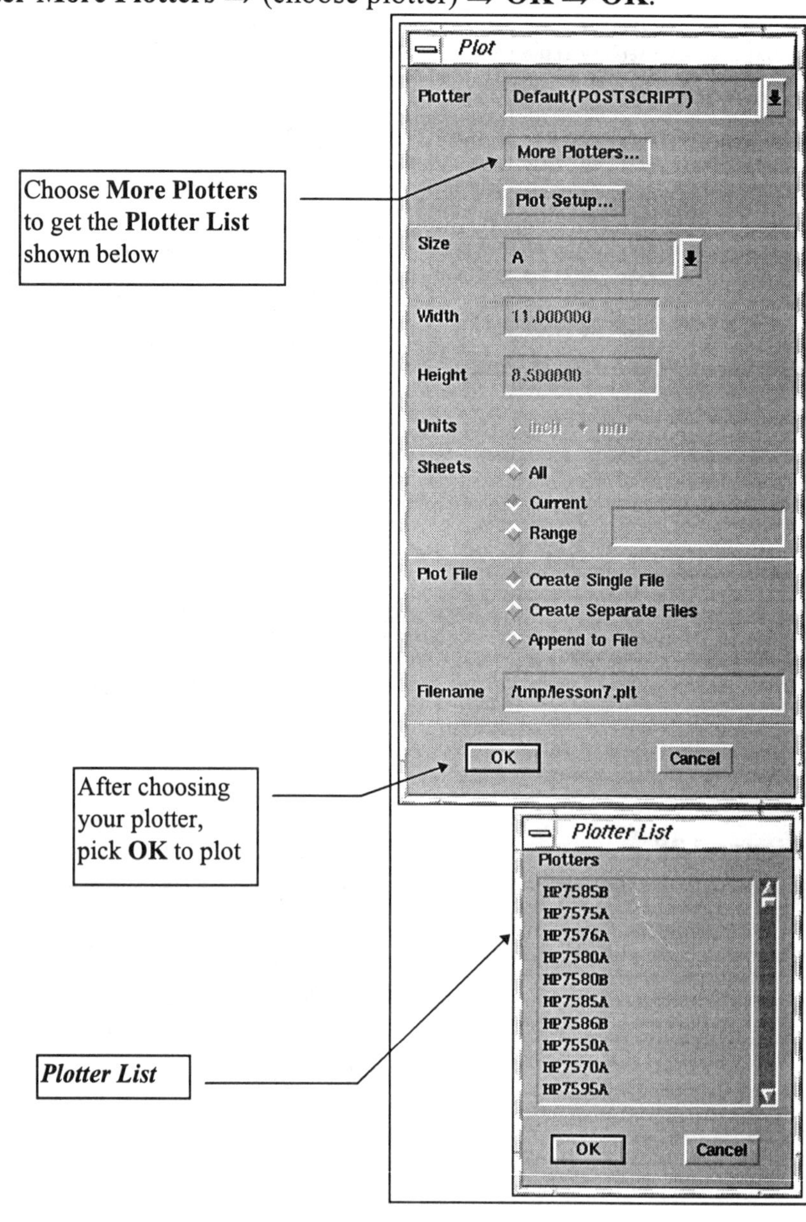

Figure 7.29
Interface

Section 8

Layers

Layers are an essential tool for grouping items and performing operations on them, such as selecting, displaying or blanking, plotting, and suppressing. Any number of layers can be created. User-defined names are available, so layer names can be easily recognized. Most companies have a layering scheme that serves as a *default standard* so that all projects follow the same naming conventions and objects/items are easily located by anyone with access. Layer information, such as display status, is stored with each individual part, assembly, or drawing.

Using Layers

The LAYERS menu (Fig. 8.1) contains five operations that can be performed on layers:

Setup Layer Performs operations on layers such as creating, deleting, and renaming; manipulating layers using the from file options; and adding parameters when exporting layers to another system.
Set Items Adds or removes items from selected layers.
Set Display Changes the display status (display or blank) of selected layers.
SetDefLayer Accesses the default layer editor.
Info Displays layer information.

Figure 8.1
LAYERS Menu

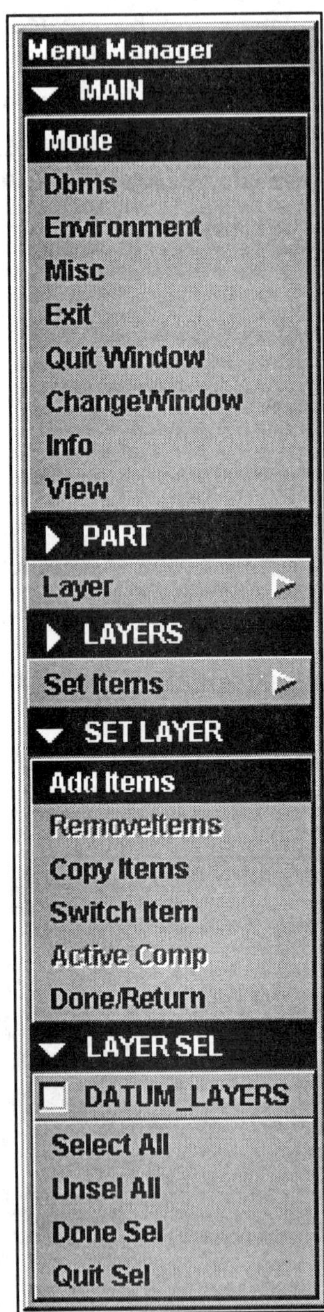

To use layers:

1. Set up the layers to which items will be added.
2. Add items to the specified layers:
   * Some items can be automatically added to layers as they are created, using configuration file options.
   * An item can be added by selecting its type and then picking the items themselves.
   * Features are also added by selecting an option that adds all features of a particular type to the layer.
   * A range of features can be added to a layer.
   * One layer can be added onto another layer.
   * Items can be copied from one layer to an existing layer or to a new layer.
3. Set the display status of the layers:
   * A layer's name can be picked from a Namelist menu.
   * A layer can be retrieved from a file that contains the desired layer status, which automatically sets each layer to the status specified in the file.
   * The current layer file can be edited. All changed layers will reflect the new status once the file is saved.

Layer Names

Layers are identified by name. Layer names can be expressed in numeric or alphanumeric form, with a maximum of 31 characters per name. When layers are displayed using **LAYER SEL**, numeric layer names are sorted first, then alphanumeric layer names. Layer names in alphabetic form are sorted alphabetically.

Setting Up Layers

The SETUP LAYER menu contains the following options:

Create Creates a layer by entering its name.
Delete Deletes selected layers.
Rename Used to rename a selected layer.
Specify Id Used to set or remove an interface layer ID.
From File Allows manipulation of a layer by modifying parameters in the layer file.

Creating a Layer

Before you can place items on a layer, you must first create the layer using the following method (Fig. 8.2):

1. Pick **Layer** from the PART or ASSEMBLY menu (select the active model if you are in Assembly mode).
2. Choose **Setup Layer** from the LAYERS menu.
3. Select **Create** from the SETUP LAYER menu, and enter the layer name. Continue to create layers by entering new names at the prompt. Press **Enter** to quit the creating mode.

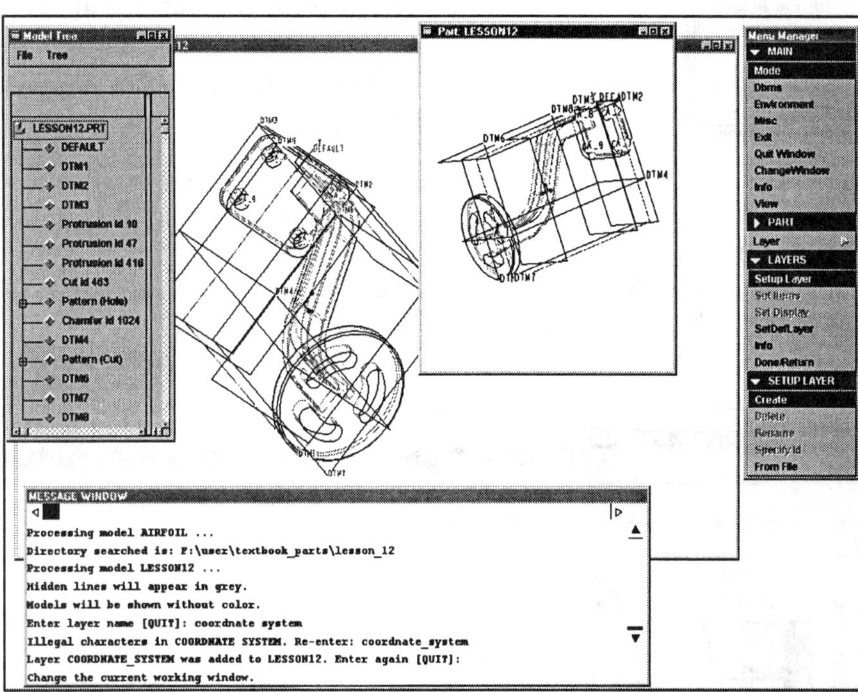

Figure 8.2
Creating Layers

In Figure 8.2, the first attempt to enter a layer name called **coordinate system** was met with a system response of:

```
Enter layer name [QUIT]: coordnate system
Illegal characters in COORDNATE SYSTEM. Re-enter: coordnate_system
Layer COORDNATE_SYSTEM was added to LESSON12. Enter again [QUIT]:
Change the current working window.
```

This message is saying that there *can be no spaces* in the layer name. Therefore, the next attempt included an underline character: **coordinate_system**. This layer has been created for this part but still does not contain items. Another layer, called **datum_plane**, will also be created. The next step is to set the items we want on each layer. We want to put the coordinate system on one layer and the three datum planes on the other.

Adding Items to Layers

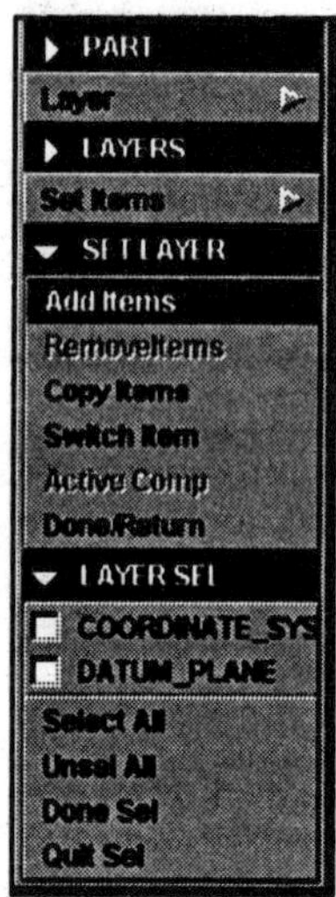

After you have created a layer, you can associate items to it by taking the following steps:

1. Choose **Layer** from the PART or ASSEMBLY menu (select the active model if you are in Assembly mode).
2. Choose **Set Items** from the LAYERS menu.
3. Select **Add Items** from the SET LAYER menu.
4. Check (✓) one or more layers on which you are going to add items and then pick **Done Sel**.
5. The LAYER OBJ menu appears, with a list of possible item types. Select the desired item type.
6. The GET SELECT menu appears. Select the desired objects or features by:

   * Picking them from the screen.
   * Picking them from the model tree by highlighting each item, as shown in Figure 8.3.
   * Selecting them by navigating through the menu structure, also shown in Figure 8.3.

Figure 8.3
Using the Model Tree to **Set Items** on Layers

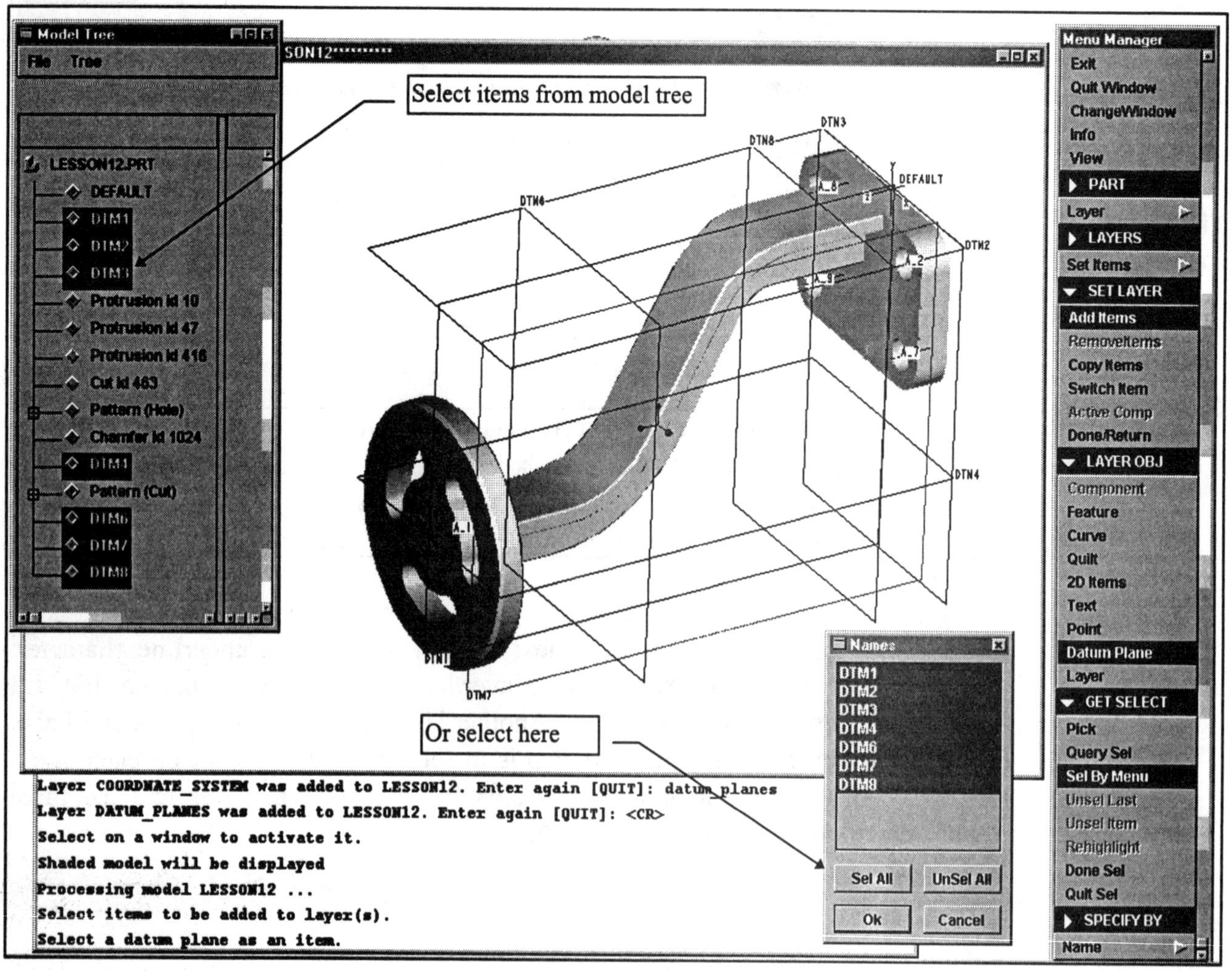

Select items from model tree

Or select here

Layer COORDNATE_SYSTEM was added to LESSON12. Enter again [QUIT]: datum_planes
Layer DATUM_PLANES was added to LESSON12. Enter again [QUIT]: <CR>
Select on a window to activate it.
Shaded model will be displayed
Processing model LESSON12 ...
Select items to be added to layer(s).
Select a datum plane as an item.

After the items have been set, you can verify that they have been placed by picking **Info ⇒ Layer Info ⇒ Disp Status ⇒ Select All ⇒ Done Sel**, as shown in Figure 8.4 for a part.

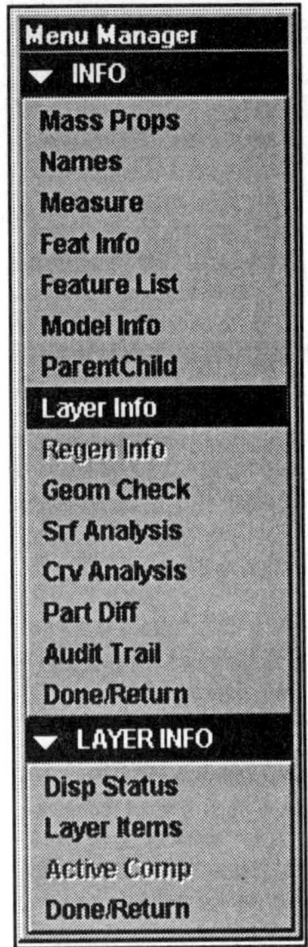

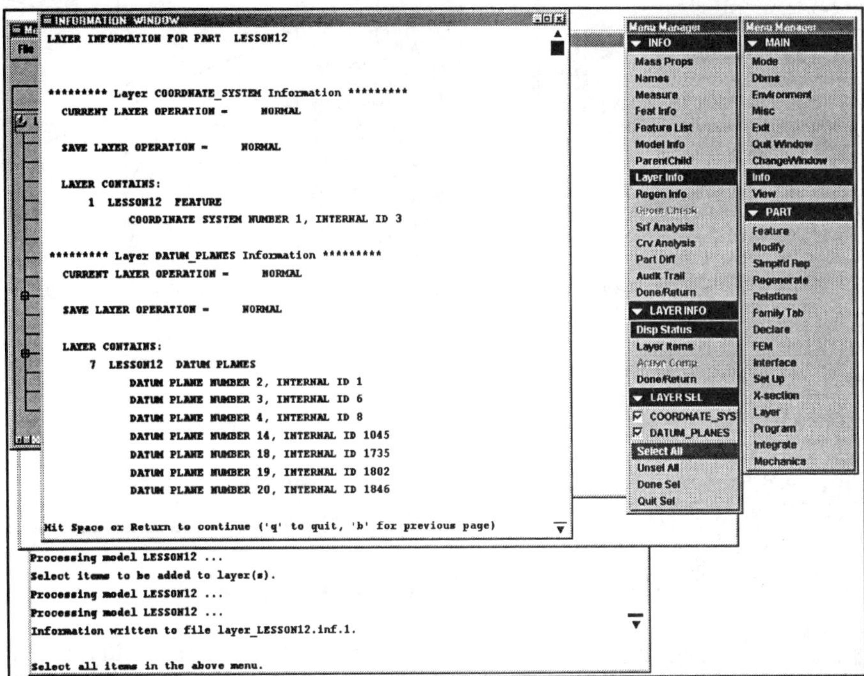

Figure 8.4
Layer Info

Removing Items from Layers

When you remove items from a layer in the active model, you disassociate them from the layer. In Figure 8.5, the **RemoveItems** command has been given and the datum plane layer was chosen. The procedure is similar to the one for adding items. Use the following steps:

1. Pick **Layer** from the PART or ASSEMBLY menu (select the active model if you are in Assembly mode).
2. Choose **Set Items** from the LAYERS menu.
3. Select **RemoveItems** from the SET LAYER menu. The LAYER SEL menu appears.
4. Check the layers from which you wish to remove items, then choose **Done Sel**.
5. The DEL ITEM menu appears, with these choices:

   Specify Allows you to select the items you want to remove from the selected layer(s). If you choose this, the LAYER OBJ and GET SELECT menus appear, just as they appear when you are adding items.

   Remove all Gives you the ability to remove all the items on the selected layer(s). Pro/E prompts you to confirm your choice by entering **y**.

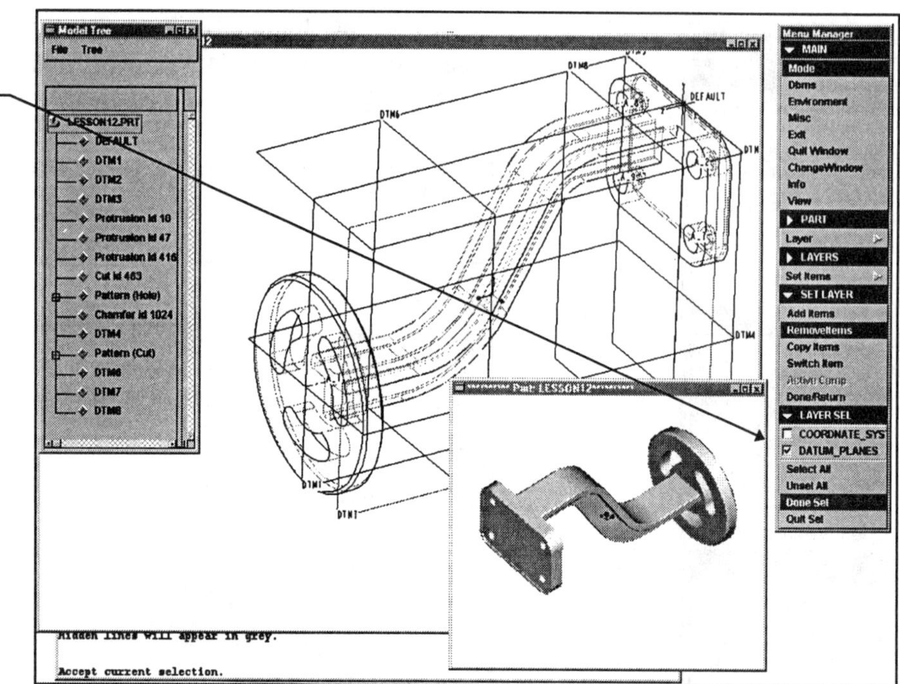

Figure 8.5
Remove Items Menu

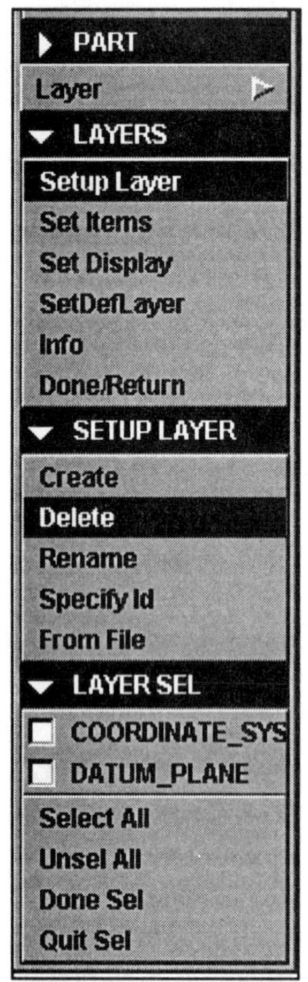

Renaming a Layer

To rename an existing layer, choose the following command sequence:

1. Choose **Layer** from the PART or ASSEMBLY menu (if you are in Assembly mode, select the active model).
2. Pick **Setup Layer** from the LAYERS menu.
3. Choose **Rename** from the SETUP LAYER menu.
4. Select the name of the layer you want to rename from the LAYER SEL menu.
5. Provide a new name for the layer.

Deleting a Layer

To delete an existing layer, choose the following command sequence:

1. Choose **Layer** from the PART or ASSEMBLY menu (select the active model if you are in Assembly mode).
2. Pick **Setup Layer** from the LAYERS menu.
3. Choose **Delete** from the SETUP LAYER menu.
4. Select the names of the layer(s) you want to delete from the LAYER SEL menu. A check mark (✓) to the left of the (names) indicates the one(s) you have selected.
5. Choose **Done Sel** and the layer(s) will be deleted.

System Default Layering

Pro/E automatically places certain types of items on specified layers when they are created, using the configuration file option **def_layer**. Use the following format:

Def_layer type-option layername

where *type-option* determines the item type and *layername* is the name that you assign to the layer. A few valid options are as follows:

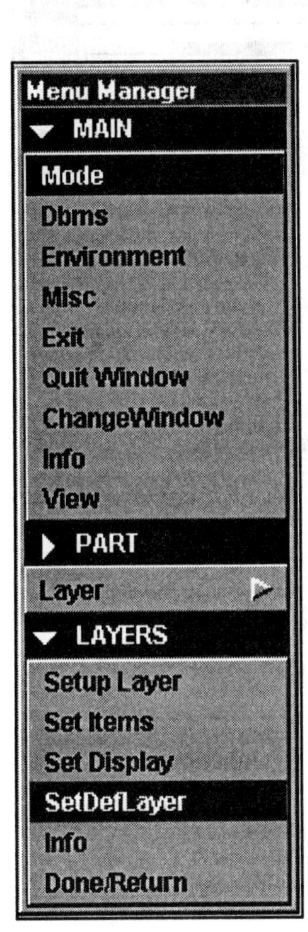

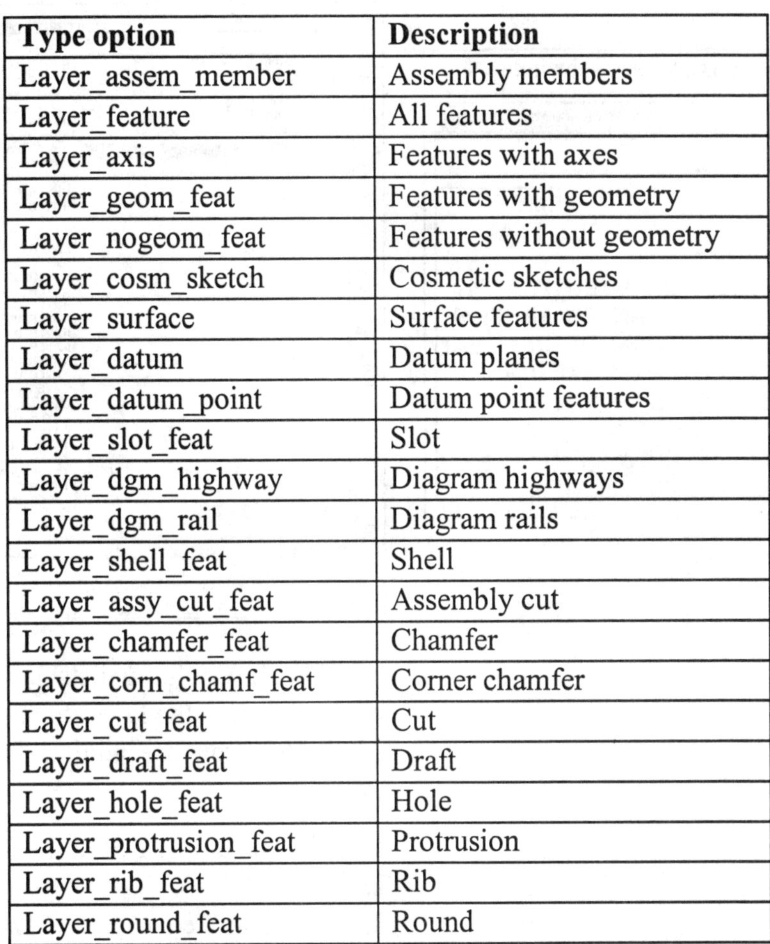

Type option	Description
Layer_assem_member	Assembly members
Layer_feature	All features
Layer_axis	Features with axes
Layer_geom_feat	Features with geometry
Layer_nogeom_feat	Features without geometry
Layer_cosm_sketch	Cosmetic sketches
Layer_surface	Surface features
Layer_datum	Datum planes
Layer_datum_point	Datum point features
Layer_slot_feat	Slot
Layer_dgm_highway	Diagram highways
Layer_dgm_rail	Diagram rails
Layer_shell_feat	Shell
Layer_assy_cut_feat	Assembly cut
Layer_chamfer_feat	Chamfer
Layer_corn_chamf_feat	Corner chamfer
Layer_cut_feat	Cut
Layer_draft_feat	Draft
Layer_hole_feat	Hole
Layer_protrusion_feat	Protrusion
Layer_rib_feat	Rib
Layer_round_feat	Round

When you create an entity of one of these types, Pro/E will automatically add it to the specified default layer. If a feature (e.g., a hole) has an axis (other than a datum axis), that axis can be automatically placed on two default layers, one for features with axes (option **layer_axis**) and one for the particular type of feature (e.g., hole option **layer_hole_feat**).

As an alternative to editing the configuration file, you can use the **SetDefLayer** option to edit the default layer table (Fig. 8.6). Layering options defined or changed here will *not* reflect back to the configuration file and are good only for *new* features created in the current session in *all* models. It is a good practice to set up the default layering *before* you start work on a part, assembly, or drawing.

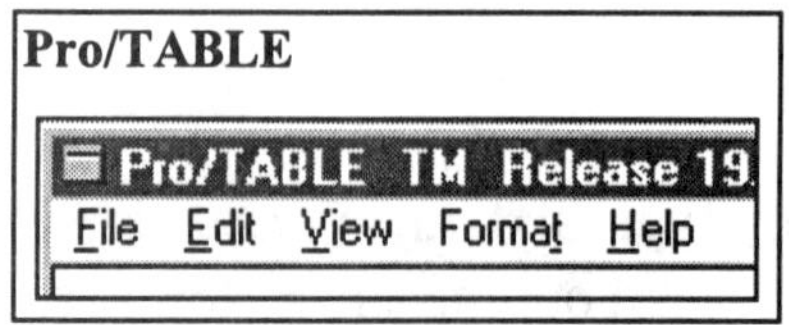

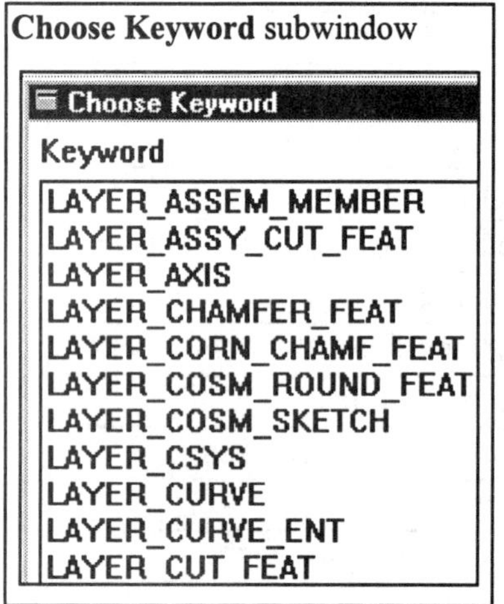

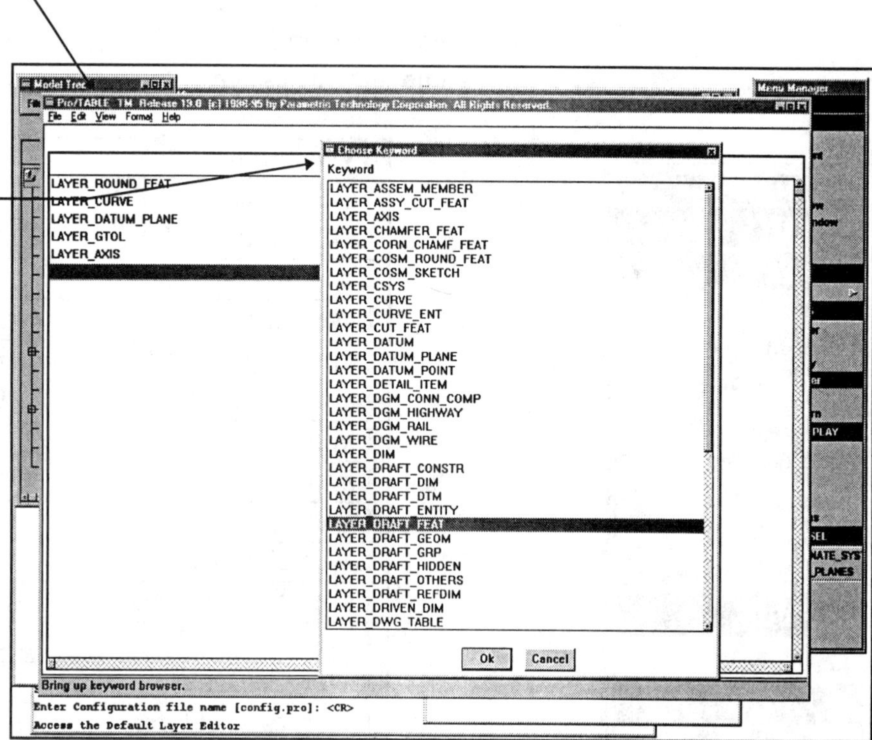

Figure 8.6
Setting Default Layers

The left column lists layer items. The right column lists the layers to which the layer items are assigned.

When you are adding layer items using Pro/TABLE, do the following:

1. Select an empty cell in the left column.
2. Press **F4**.
3. Pick on an item (here the **LAYER_DRAFT_FEAT** keyword is highlighted) and the editor will copy it into the active cell (Fig. 8.6).

Pro/E supplies a number of default layer names:

Layer_dim	**Def_dims**
Layer_corn_chamf_feat	**Def_chamfers**
Layer_protrusion_feat	**Def_protrusions**
Layer_axis	**Def_axis**
Layer_assem_member	**Def_components**
Layer_assy_cut_feat	**Def_features**
Layer_chamfer_feat	**Def_chamfers**

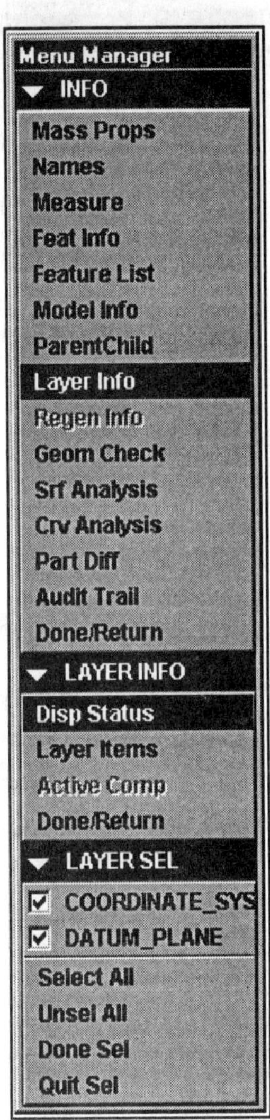

You can enter one of them into the right column by:

1. Selecting an empty cell in the right column.
2. Picking the **Choose Keywords** option in the EDIT menu. This will bring up the CHOOSE KEYWORD NAMELIST menu of default layer names.
3. Picking on a layer name. Choose OK and the editor will copy it into the active cell. Alternatively, you can type a layer name from the keyboard, as was done in the example. When finished, choose **Exit** from the FILE menu.

In Figure 8.7, the **Info** command has been given and the screen shows that a set of datums and a coordinate system have been added in Part mode. You can get information on layers from the **Info ⇒ Layer Info ⇒ Disp Status** command in the MAIN menu (see sidebar illustration on this page) or from the PART menu after choosing **Layer ⇒ Info ⇒ Disp Status** (Fig. 8.7).

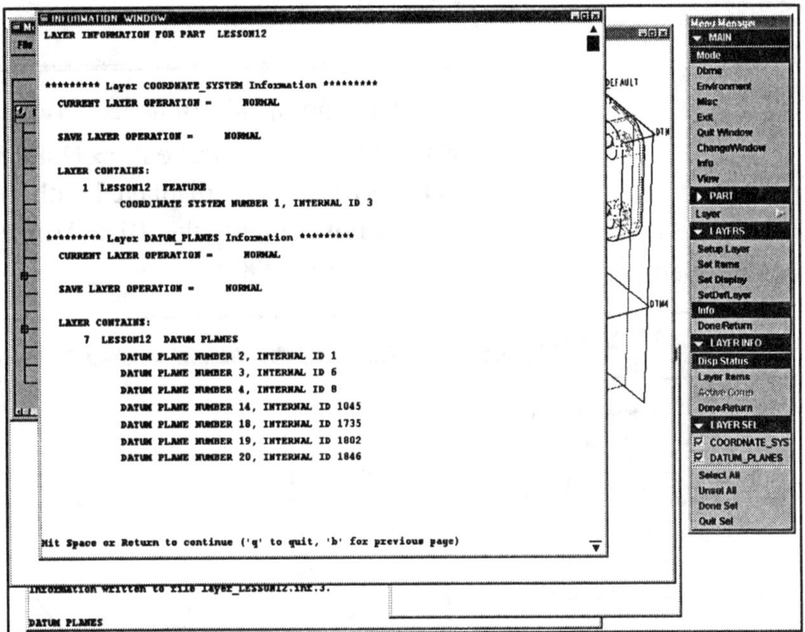

Figure 8.7
Layer Info

Displaying Layers

In the MAIN menu, we can choose to turn on or off a variety of items, including datum planes, coordinate systems, and datum axes. Another way to control the display of the items on the screen is to put groups of items on layers and display or blank them as needed. Later you will learn that the datums can be set as default geometric tolerance features. Datum planes set in this way *cannot be turned off using environment settings;* therefore, when the object is displayed in Part, Assembly, or Drawing mode, the only way to control the basic datum display is to blank its layer. This will become obvious when you see assemblies with 5 to 2000 components, all with basic datums displayed at once! Blanking layers becomes essential when this happens.

In Figure 8.8, the datums have been set as basic and are blanked.

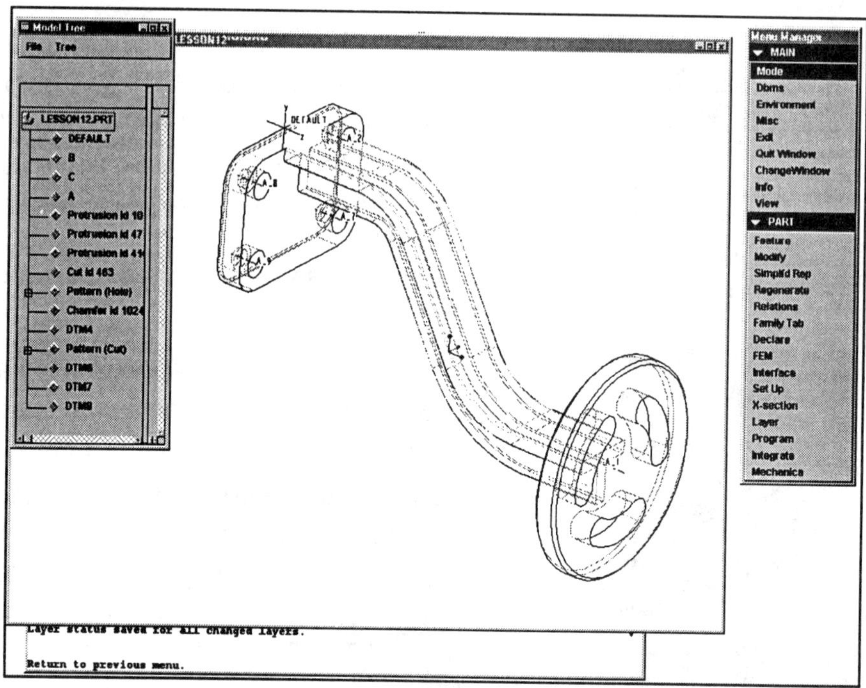

Figure 8.8
Blanked Layer

The command **View** ⇒ **Layer Disp** ⇒ **Blank** (the default) ⇒ ✓**datum_plane** (checked) ⇒ **Done Sel** was given. The results show that the part is now displayed without datum planes. Note that if you pick **Environment**, the **Disp DtmPln** selection is still checked, as shown in Figure 8.9.

Figure 8.9
Environment Settings

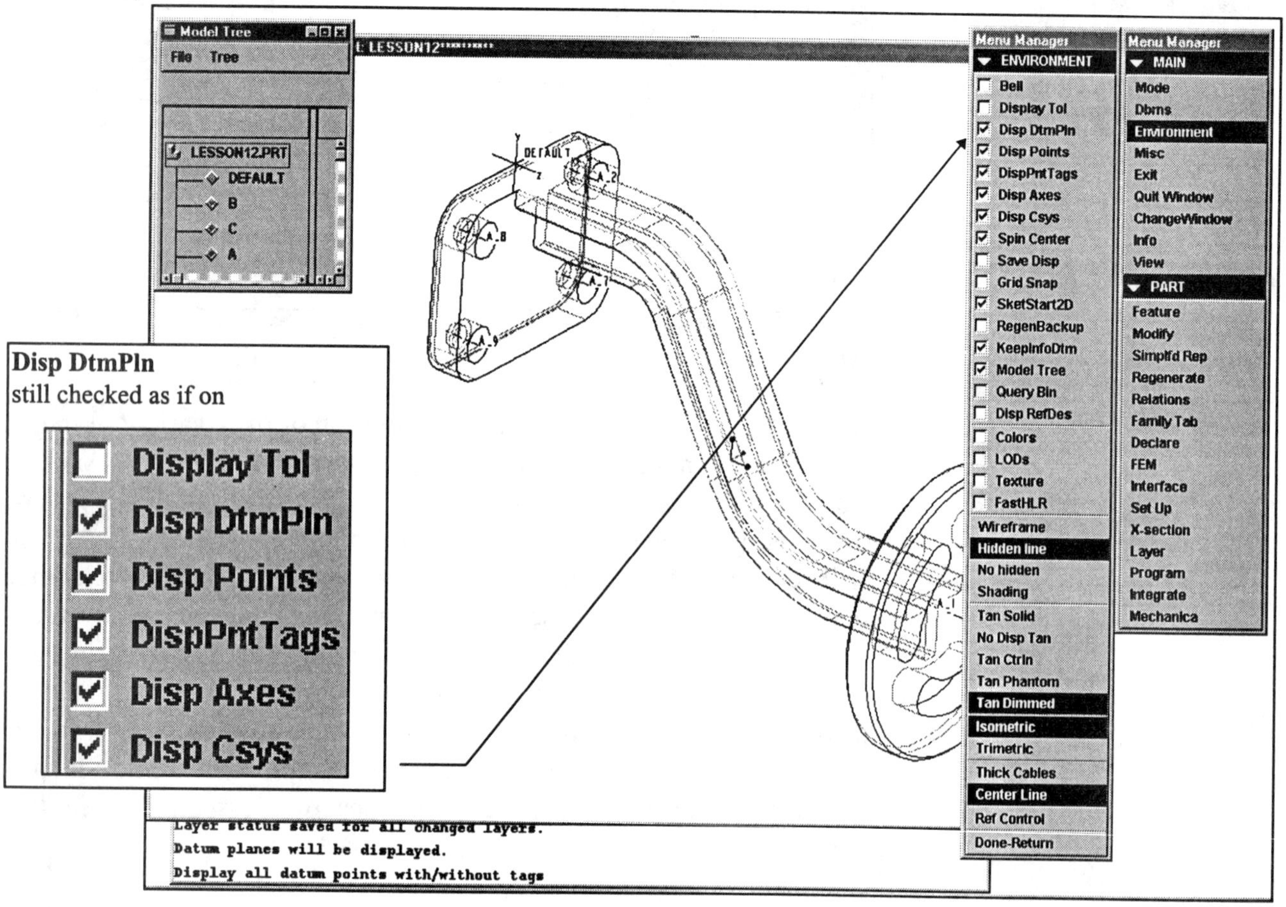

Disp DtmPln
still checked as if on

Layer status saved for all changed layers.
Datum planes will be displayed.
Display all datum points with/without tags

Section 9

The Sketcher

In Pro/ENGINEER, *sketching* is done in the **Sketcher**. Almost all traditional lines, circles, arcs, and their variations can be accomplished on the screen without creating exact and perfectly constructed geometry. Pro/E will assume a variety of conditions, such as tangency, similar sizes for same-type geometry, parallelism, perpendicularity, verticals, horizontals, coincident endpoints, tangent points, and symmetry.

The Sketcher can be entered by picking **Mode ⇒ Sketcher**; you are automatically put into the Sketcher when creating most geometry in Part mode. If you go directly into the Sketcher, you will be creating sections that can be recalled later during a part feature creation. The sections are like sheets of graph paper with sketched geometry on them. They can be saved for later use in any feature creation where a section is called for. The section in Figure 9.1 is an example of a section created in Sketcher mode **(Pro/ENGINEER Section: S2D0001** is shown in the window bar). Note that you must give it a unique name or use the default, **S2D0001**.

A section created in Part mode or Assembly mode can be saved for later use as well as being used to create the geometry needed for the present feature.

Sketcher techniques are used in many areas of parametric design. The purpose of the Sketcher, like that of parametric design, is to enable the quick and simple creation of geometry for your model. The Sketcher requires you to create and dimension this geometry, but during the sketching process you do not have to be concerned with exact dimension sizes or the creation of perfect and accurate geometry.

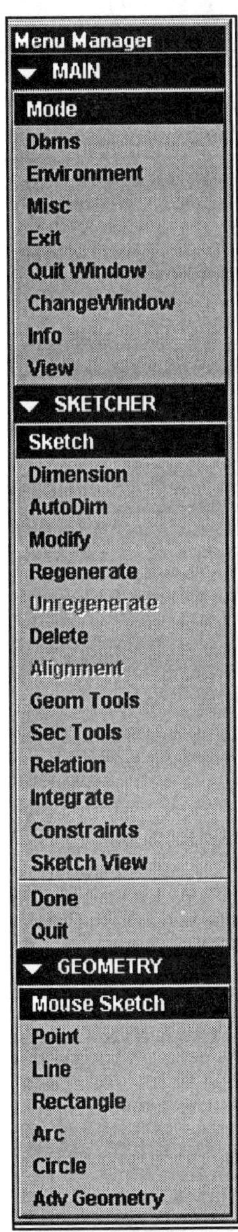

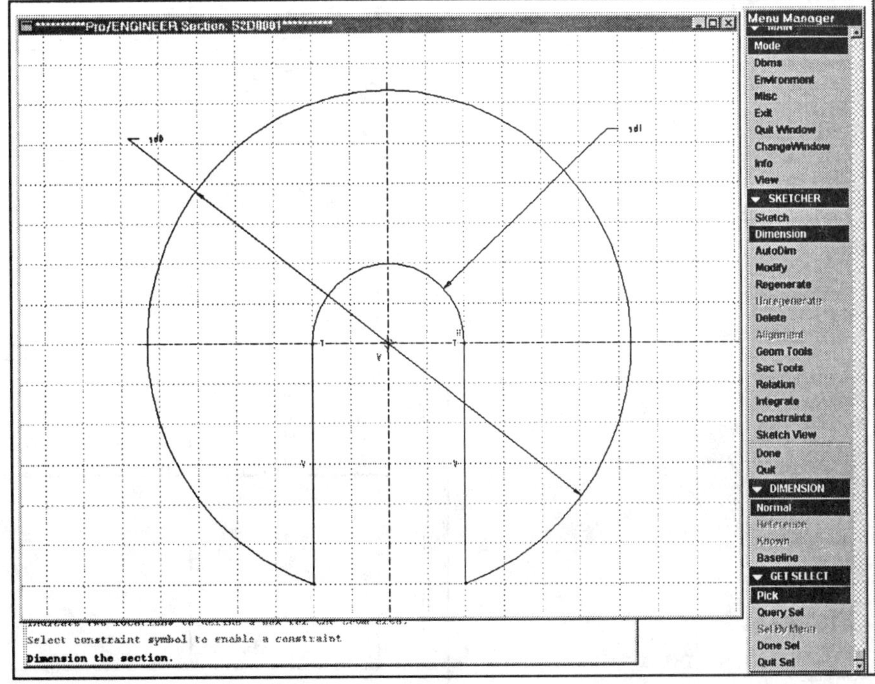

Figure 9.1
Sketcher Mode

Creating sections within the SKETCHER menu is not difficult. There are only a few steps to remember:

1. **Sketch** Sketch the section geometry. Use SKETCHER tools to create the section geometry (Fig. 9.2).
2. **Dimension** Dimension the section. Use a dimensioning scheme that you want to see in a drawing. Dimension to control the characteristics of the section geometry (Fig. 9.3).
3. **Alignment** Align the section geometry to a datum feature or to a part feature (Fig. 9.3).
4. **Regenerate** Regenerate the section. Regeneration solves the section sketch based on your dimensioning scheme (Fig. 9.4).
5. **Relations** Add section relations. Add relations to control the parametric behavior of your section.

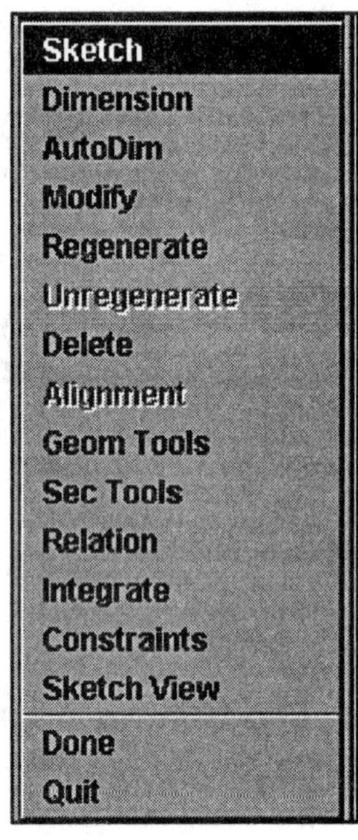

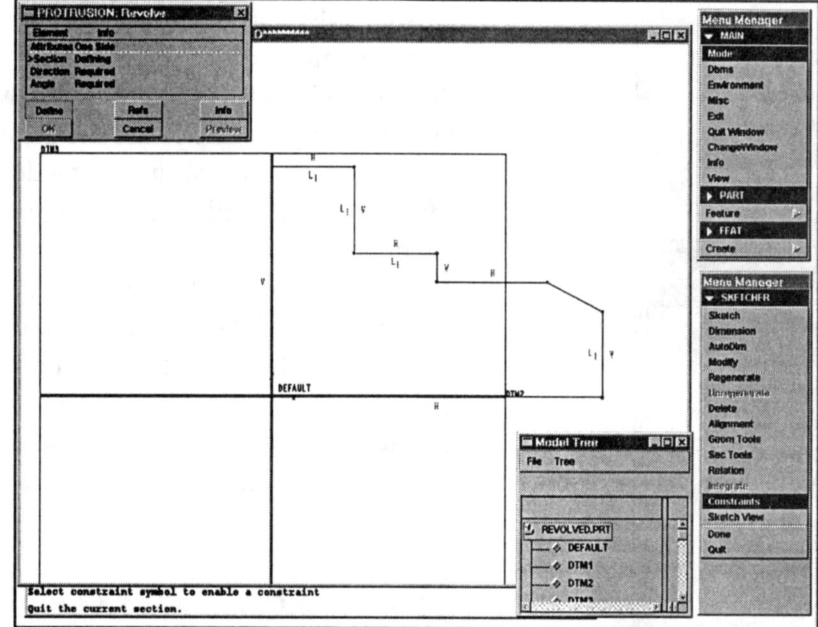

Figure 9.2
Sketch the Features

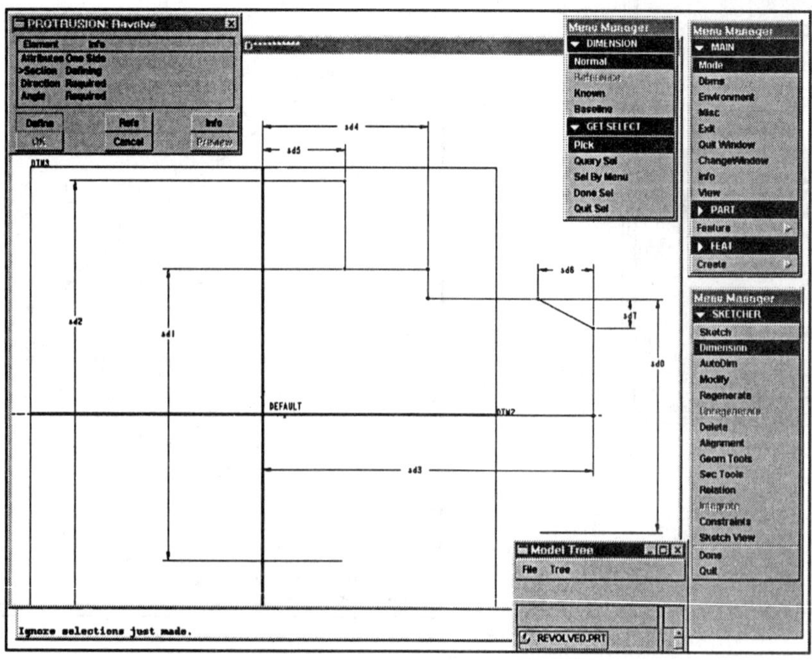

Figure 9.3
Dimension and
Alignment of the Sketch

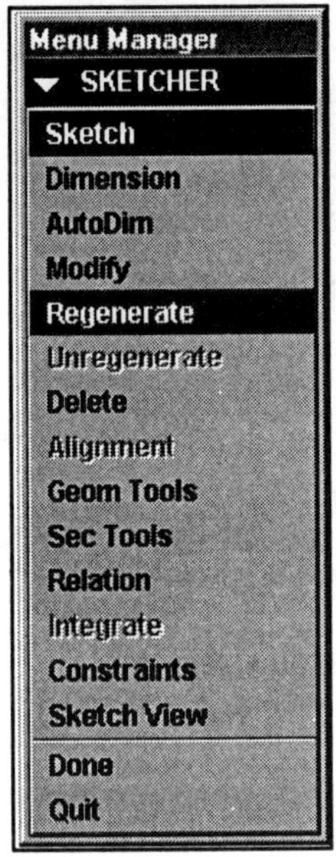

6. **Modify** Change the sketch dimension values to the required design sizes.
7. Repeat step 4.

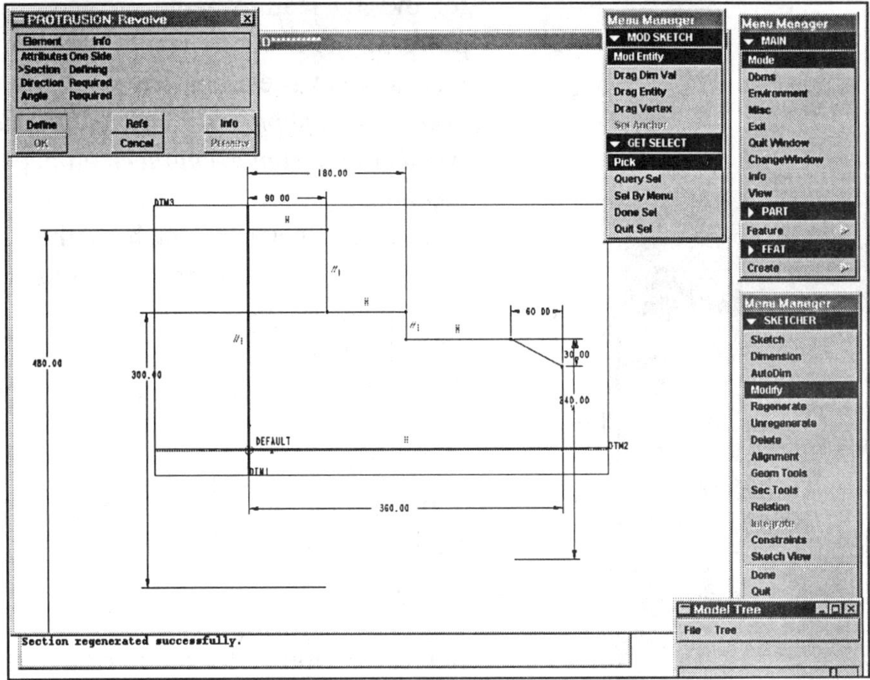

Figure 9.4
Regenerate the Section

After Pro/E regenerates the sketch, the feature can be completed, as shown in Figure 9.5. More features can then be added, using *sketched* features or *pick-and-place* features such as holes and chamfers.

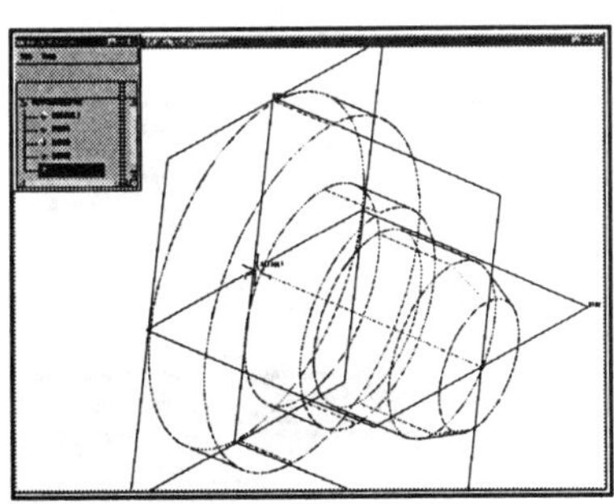

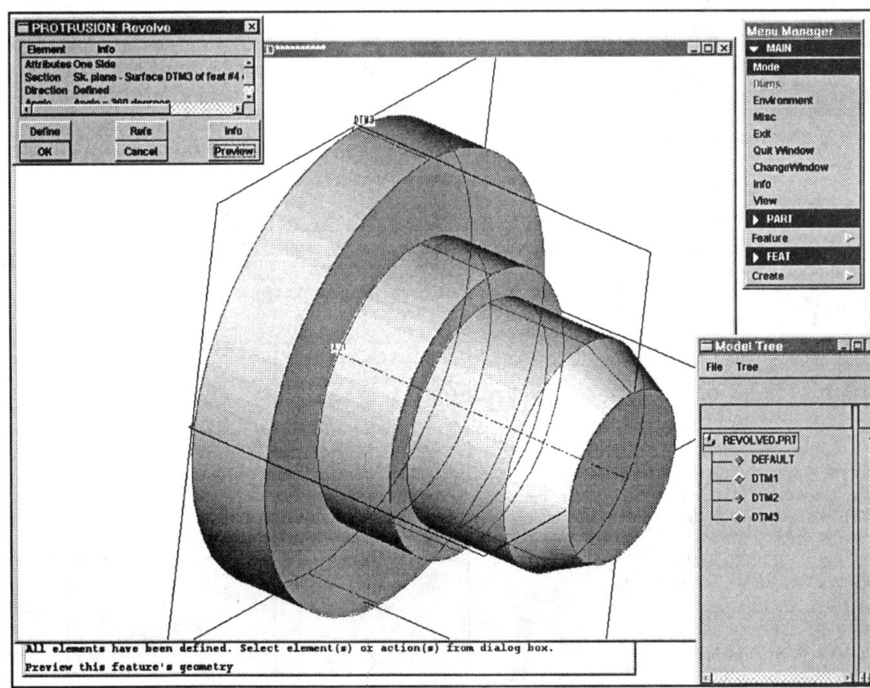

Figure 9.5
Revolved Feature Created from the Sketch

Sketcher and the Mouse

The **SKETCHER** is used to establish 2D sections that are the basis for the 3D feature being created. In order to understand just how powerful the Sketcher is in a parametric design system, you need only look at what sketching has been throughout the ages: *Sketching is a process of simply and efficiently establishing the basic design and intent of a designer-engineer* on paper, and that is now possible with the mouse.

In the Sketcher, much of the section geometry can be created using the three buttons on the mouse.

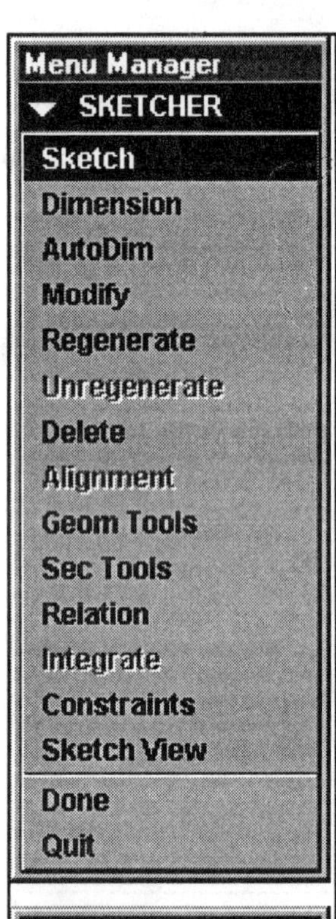

LEFT button Used to create lines. The line command in the Sketcher chains lines together. This button also aborts the creation of circles.

MIDDLE button Used to create circles. The first pick is the center of the circle; the second pick is a location on the diameter of the circle. Also used to end line creation.

RIGHT button Used to create tangent arcs. *There must be an existing line, arc, or spline to reference for tangency.* Place the cursor near the tangent entity, and press the right button to start the arc. Press the right button again to set the endpoint of the arc.

These functions are available when the **Sketch** option is selected from the SKETCHER menu and **Mouse Sketch** (Fig 9.6) is left as the default in the GEOMETRY menu. Also, there are other geometry functions in the GEOMETRY menu.

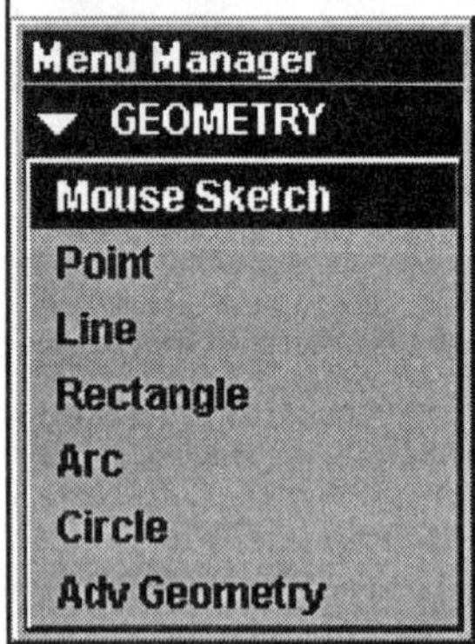

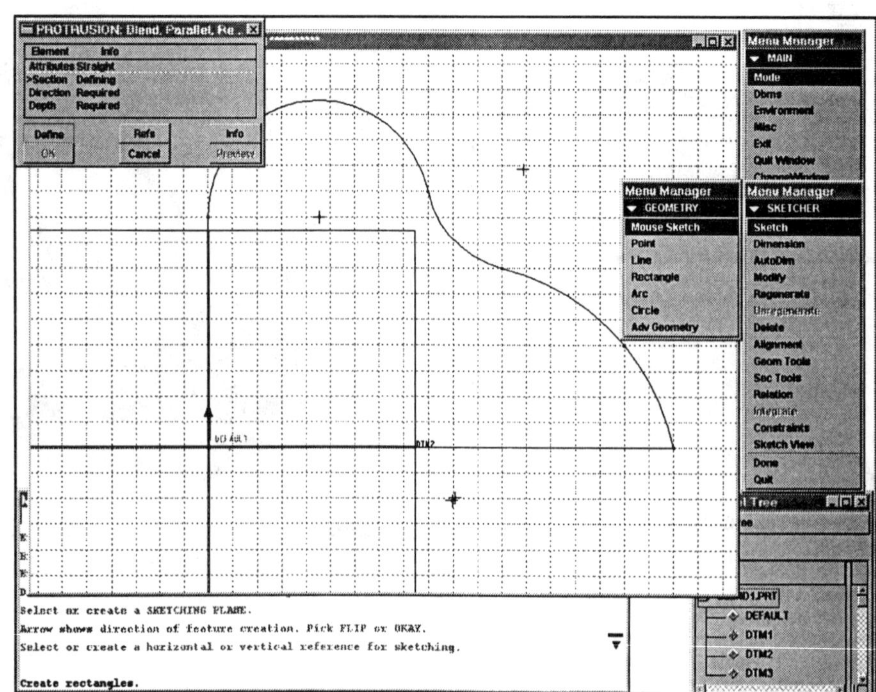

Figure 9.6
Mouse Sketch

Regenerating a Section Sketch

Menu Manager
ENVIRONMENT
Bell
Display Tol
Disp DtmPln
Disp Points
DispPntTags
Disp Axes
Disp Csys
Spin Center
Save Disp
Grid Snap
SketStart2D
RegenBackup
KeepInfoDtm
Model Tree
Query Bin
Disp RefDes
Colors
LODs
Texture
FastHLR
Wireframe
Hidden line
No hidden
Shading
Tan Solid
No Disp Tan
Tan Ctrln
Tan Phantom
Tan Dimmed
Isometric
Trimetric
Thick Cables
Center Line
Ref Control
Done-Return

In the past, designers have sketched on paper, showing lines, arcs, circles, and other geometric forms in rough, simplified outline and internal forms. The sketched shapes are assumed to be what they *sort of* look like. Round shapes approximating a circle are assumed by the reader of the sketch to be circles, curved shapes are assumed to be arcs, and lines drawn straight up or down are assumed to be vertical. Lines drawn left to right are assumed to be horizontal. Lines sketched at an angle are straight lines that are angled. Dimensions roughly sketched on a less-than-perfect drawing of a part are assumed to represent the exact, perfect shape desired by the person sketching. All this seems obvious to most people involved in engineering design. With the introduction of parametric design, we can now sketch on the screen and allow Pro/E to make all the assumptions that were traditionally made by a person creating a sketch or reading a sketch. These assumptions include, but are not limited to, the following: *symmetry*, *tangency*, *parallelism*, *perpendicularity*, *equal angles*, *same-size arcs* and *circles*, and *coincident centers*.

The sketch started in Figure 9.7 was completed without the **Grid Snap** activated. Note that the lines and arcs are not sketched perfectly. After regeneration, Pro/E will straighten the lines and align the features according to a set of assumptions or rules. Almost-vertical lines will be vertical, close-to-horizontal lines will become horizontal, and so on. We suggest that you keep the grid snap on (**✓Grid Snap**) most of the time while sketching the first features of a part; experienced users normally keep it off and trust the Pro/E assumptions to clean up any sketching inconsistencies.

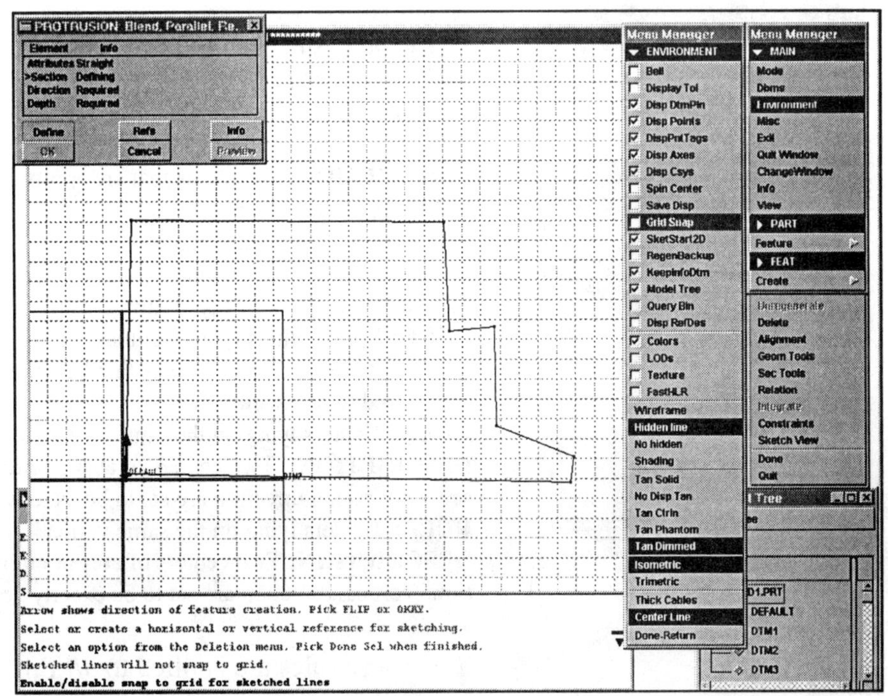

Figure 9.7
Sketching Without **Grid Snap**

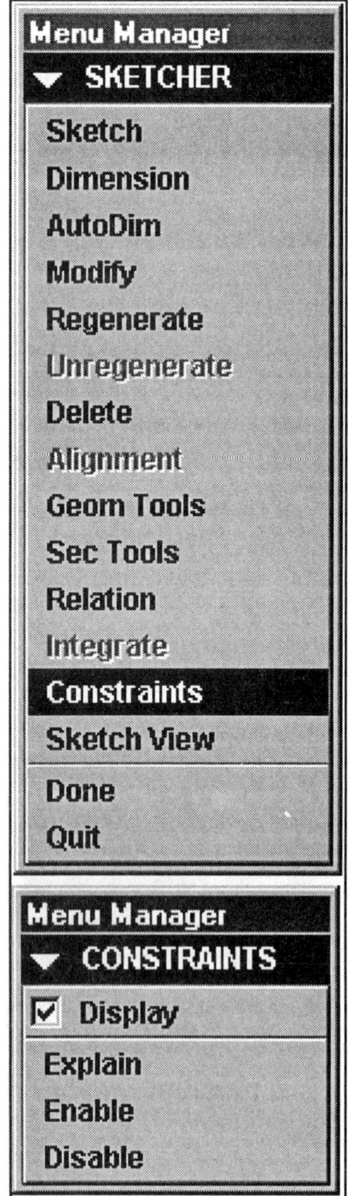

During regeneration, Pro/E checks to make sure that it understands your dimensioning scheme and that you have created a complete and independent set of parameters. Pro/E analyzes your section based on the geometry you have sketched and the dimensions you have created. In the absence of explicit dimensions, implicit information based on the sketch is used. You can quickly understand and control the assumptions made in solving the sketch. This streamlines the process of sketch creation and behavior diagnosis.

Modifications made in sketches are *animated* over a brief time during regeneration. If a sketch fails, it changes shape up to the point of failure, allowing you to view the section at the point of failure with the option of restoring the dimensions to the old values. This provides an understanding of how the sketch fails by showing the point of failure, and by displaying through animation how that point was achieved. Corrective actions can then be taken.

Here is a list of implicit information that Pro/ENGINEER uses to regenerate a section:

RULE: Equal radius/diameter
DESCRIPTION: If two or more arcs or circles are sketched with approximately the same radius, they are assigned the same radius value.
RULE: Symmetry
DESCRIPTION: Entities sketched symmetrically about a centerline are assigned equal values with respect to the centerline.
RULE: Horizontal and vertical lines
DESCRIPTION: Lines that are approximately horizontal or vertical are considered to be exactly horizontal or vertical.
RULE: Parallel and perpendicular lines
DESCRIPTION: Lines that are sketched approximately parallel or perpendicular are considered to be exactly parallel or perpendicular.
RULE: Tangency
DESCRIPTION: Entities sketched approximately tangent to arcs or circles are assumed to be tangent.
RULE: 90°, 180°, 270° arcs
DESCRIPTION: Arcs are considered to be multiples of 90° if they are sketched with approximately horizontal or vertical tangents at the endpoints.
RULE: Collinearity
DESCRIPTION: Segments that are approximately collinear are considered to be exactly collinear.
RULE: Equal segment lengths
DESCRIPTION: Segments of unknown length are assigned a length equal to that of a known segment of approximately the same length.
RULE: Point entities lying on other entities
DESCRIPTION: Point entities that lie approximately on lines, arcs, or circles are considered to be exactly on them.
RULE: Centers lying on the same horizontal
DESCRIPTION: Two centers of arcs or circles that lie approximately along the same horizontal direction are set to be exactly horizontally aligned.
RULE: Centers lying on the same vertical
DESCRIPTION: Two centers of arcs or circles that lie approximately along the same vertical direction are set to be exactly vertically aligned.

These rules are applied to *all* Pro/ENGINEER sketches.

Sketcher Mode and Constraints

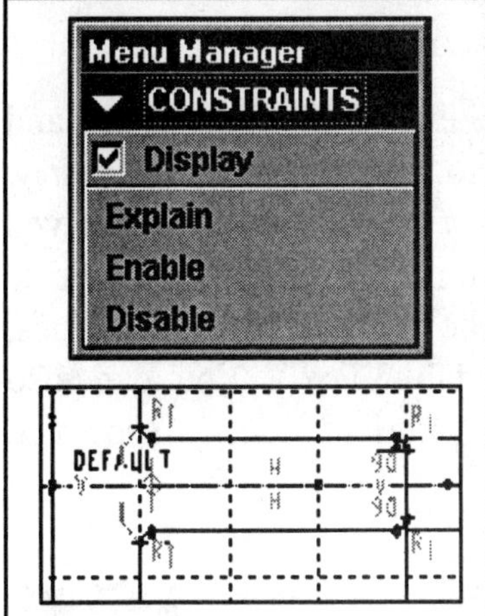

Constraints used in solving a sketch are displayed graphically on the sketch with the aid of *small symbols* that appear next to the entities to which they apply (Fig. 9.8). You can turn off the display of these symbols. You can also click on the symbols to disable or enable the constraints and to obtain a brief explanation. Also, endpoints of section entities are highlighted with the aid of small dot symbols. This graphical display of symbols for constraints replaces the old user interface of a constraints dialog box.

Sketcher Constraint Symbols

Symbol	Meaning
H	Horizontal
V	Vertical
T	Tangent
R1	Equal Radius/Diameter
L1	Equal Length
⊥	Perpendicular
//	Parallel
▬ , ▌	Collinearity, centers lying on same vertical/horizontal
→←	Symmetry

Figure 9.8
Sketch with Constraints Displayed

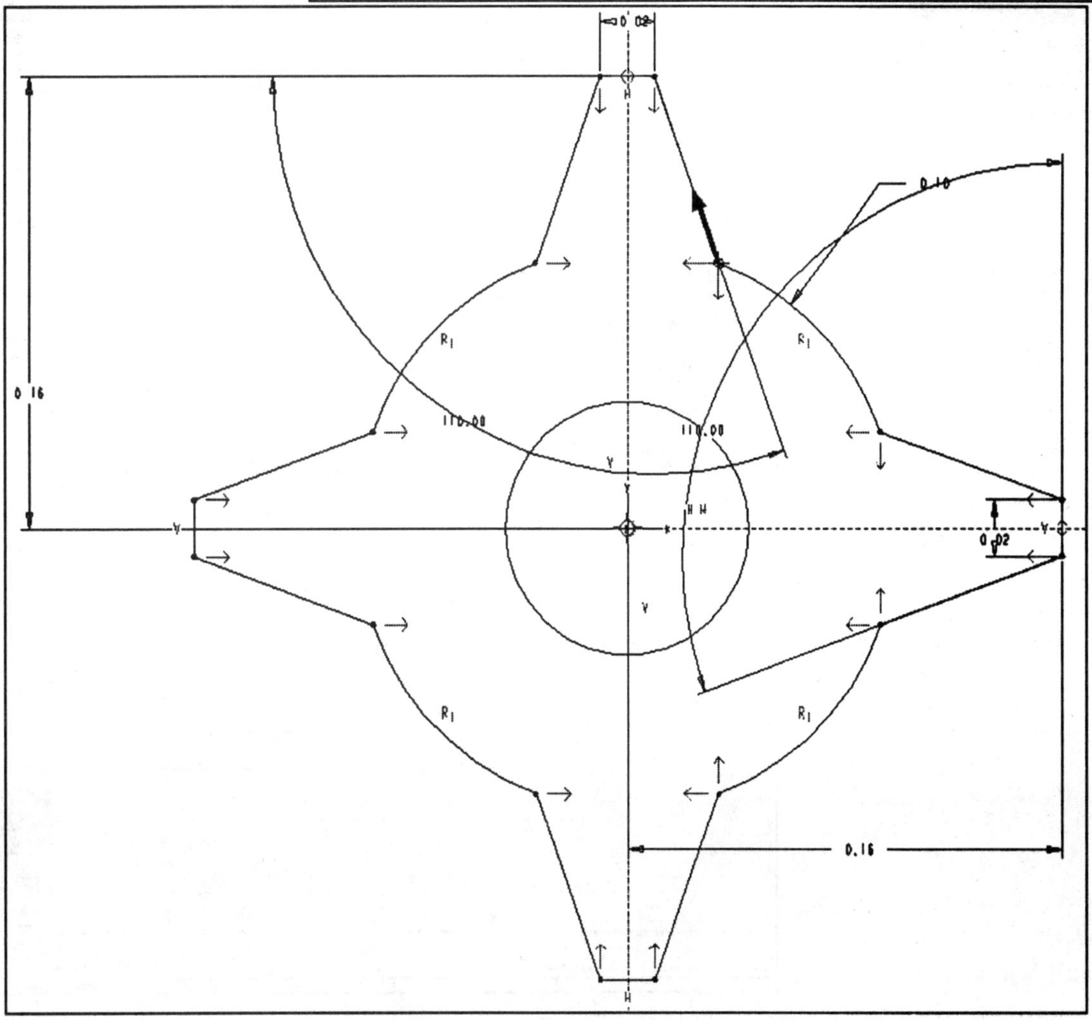

Sketching Lines

You can create two types of **lines** with Pro/E: geometry lines and centerlines. **Geometry lines** are used to create *feature geometry*. **Centerlines** are used to define the *axis of revolution* of a revolved feature, to define a line of symmetry within a section, or to create construction lines. In Figure 9.9, the section contains lines, centerlines, and arcs shown in 2D **Sketch View**. You can also reorient the sketch into a 3D **Default** view orientation, as shown in Figure 9.10.

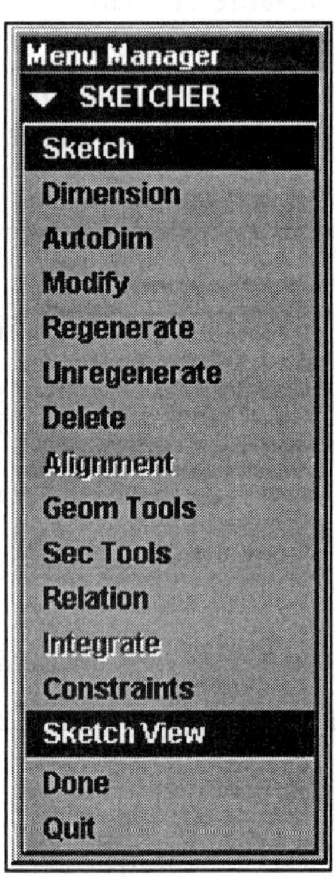

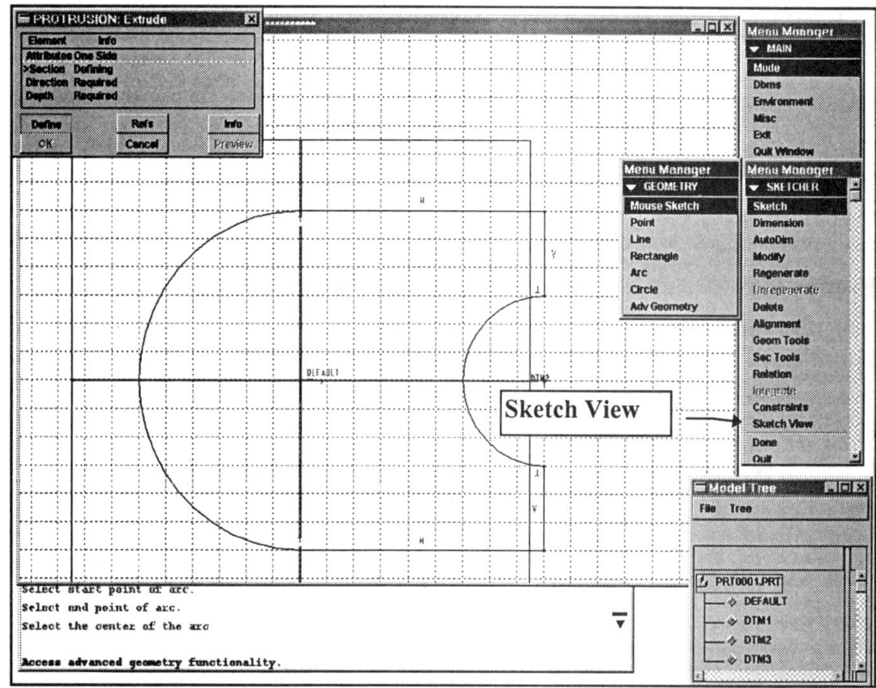

Figure 9.9
2D Sketch View

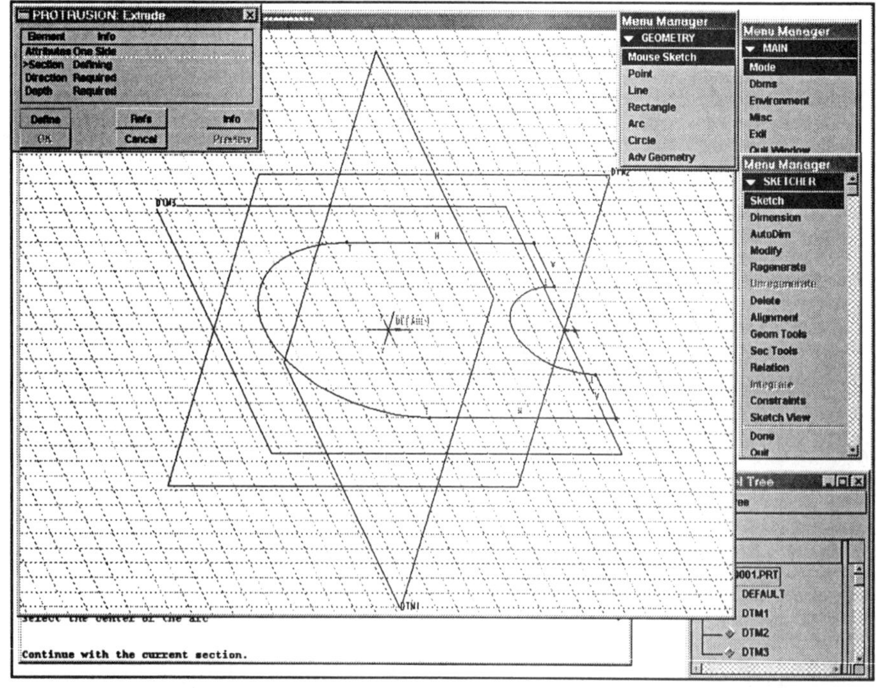

Figure 9.10
3D Default View

To sketch **geometry lines** or **centerlines**, do the following:

1. Choose **Line** from the GEOMETRY menu. The LINE TYPE menu appears.
2. Choose **Geometry** or **Centerline** from the top portion of the menu to indicate the type of line that you want.
3. Choose an option from the bottom portion of the menu to indicate how to create the line:

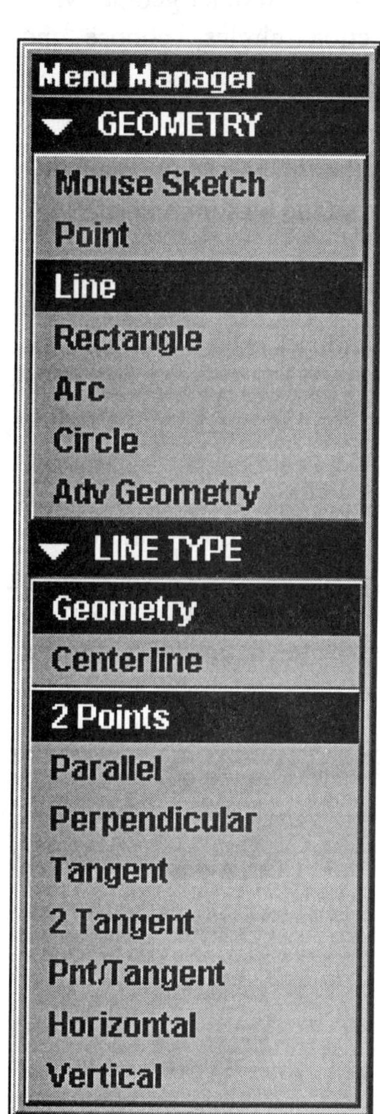

2 Points Create a line by picking the start point and endpoint. Geometry lines created using this command will automatically be chained together.

Parallel Pick an existing line to determine the new line's direction, then pick the start point and endpoint. For a centerline, only a single pick is needed to determine the parallel placement of the line, and the ends of the *centerline* will be chosen to fit model or section outlines.

Perpendicular Pick an existing line to determine the new line's direction, then pick the start point and endpoint. For a centerline, only a single pick is needed to determine the perpendicular placement of the line, and the ends of the *centerline* will be chosen to fit model or section outlines.

Tangent Pick an endpoint of an arc or spline to start the new line and determine its direction, then pick the endpoint of the line. For a centerline, only a single pick is needed to determine the tangent placement of the line, and the ends of the *centerline* will be chosen to fit model or section outlines.

2 Tangent Pick two arcs, splines, or circles to determine the direction of the new line. The line is automatically created between the selected entities, and *splits the entities at the tangency points*. A 2 Tangent line created to construction entities will not split the entity. A 2 Tangent *centerline*, created as a 2 Tangent line defined using two circles, will not split the circles.

Pnt/Tangent Pick a point anywhere in the current section, then pick an arc, spline, or circle to which the line must be tangent.

Horizontal Create a line that is horizontal relative to the orientation of the section. For a geometry line, the endpoint is automatically the starting point of a chained vertical line. For a *centerline*, only a single pick is needed to determine the horizontal location of the line.

Vertical Create a line that is vertical relative to the orientation of the section. For a geometry line, the endpoint is automatically the starting point of a chained horizontal line. For a *centerline*, only a single pick is needed to determine the vertical location of the line.

Circles

A variety of **geometry circles** and **construction circles** can be created in the Sketcher (Fig. 9.11). Geometry circles are used to create feature geometry, whereas construction circles serve as guides and references but do not create feature geometry. Construction circles are displayed in the same color as circles, but with a *phantom line font* rather than the solid font that is used for circles and other geometry.

To sketch geometry and construction circles, choose the following command picks:

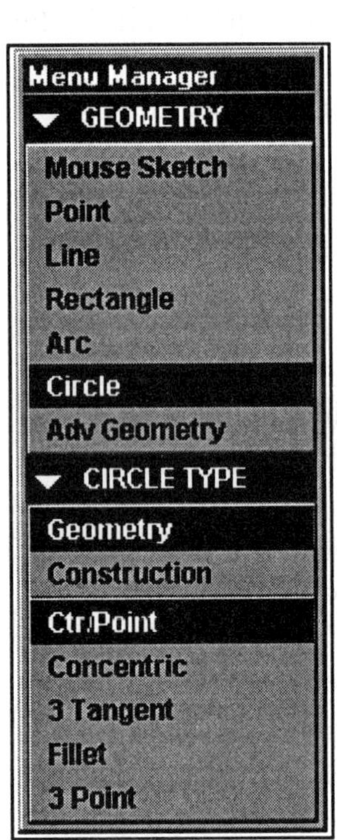

1. Select **Circle** from the GEOMETRY menu.
2. Pick **Geometry** or **Construction** from the top part of the menu.
3. Choose one of the following options from the bottom part of the menu:

   Ctr/Point Basically the same as creating a circle using **Mouse Sketch** and the middle button of the mouse, except that here the left button is used for the picks.
   Concentric Pick an existing circle or arc, then pick on the radius of the new circle.
   3 Tangent Create a circle connecting three existing reference entities. These can be centerlines, construction features, or geometry features.
   Fillet Create a circle tangent to two existing entities.
   3 Point Pick three points you wish to define the circle's circumference.

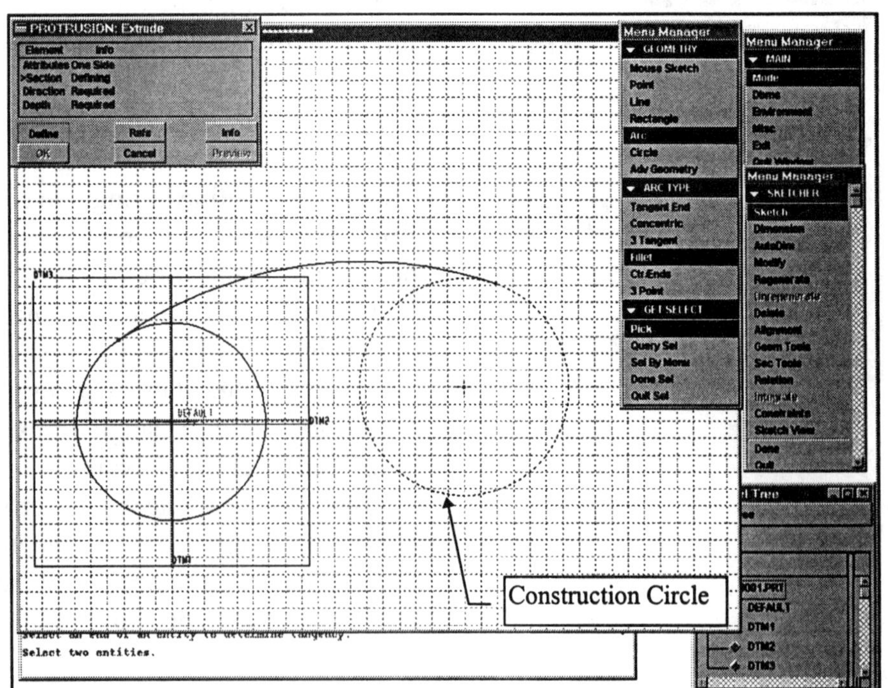

Figure 9.11
Geometry Circles, Arcs, and Construction Circles

Arcs

Arcs are sketched using the menu or the mouse. To sketch arcs:

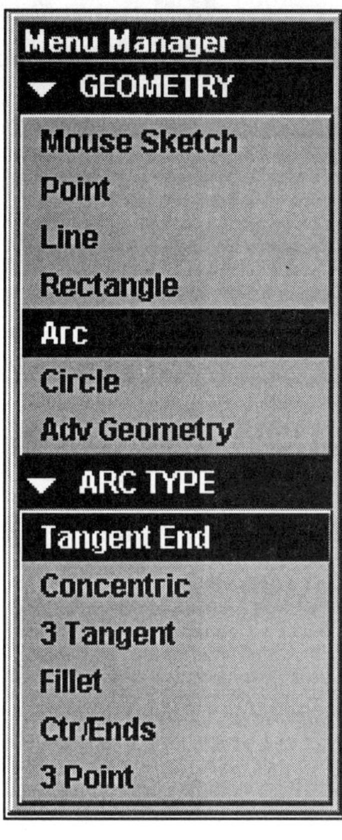

1. Select **Arc** from the GEOMETRY menu. The ARC TYPE menu appears.
2. Pick one of the following options from the ARC TYPE menu:

 Tangent End This is the same as creating an arc using **Mouse Sketch**, except that you must use the left mouse button. Pick an end of an entity to determine tangency, then pick the endpoint of the arc.

 Concentric Pick an existing circle or arc as a reference, then pick the endpoints of the new arc. As you create the arc, a radial line will appear through its center to assist you in aligning the endpoint.

 3 Tangent Select three entities for the new arc to be tangent to, then create the arc in the same direction as the reference picks.

 Fillet Pick two entities to connect by creating a tangent arc (Fig. 9.12).

 Ctr/Ends Pick the center point of the arc, then pick the arc's endpoints.

 3 Points Pick the endpoints of the arc, then pick a point on the arc.

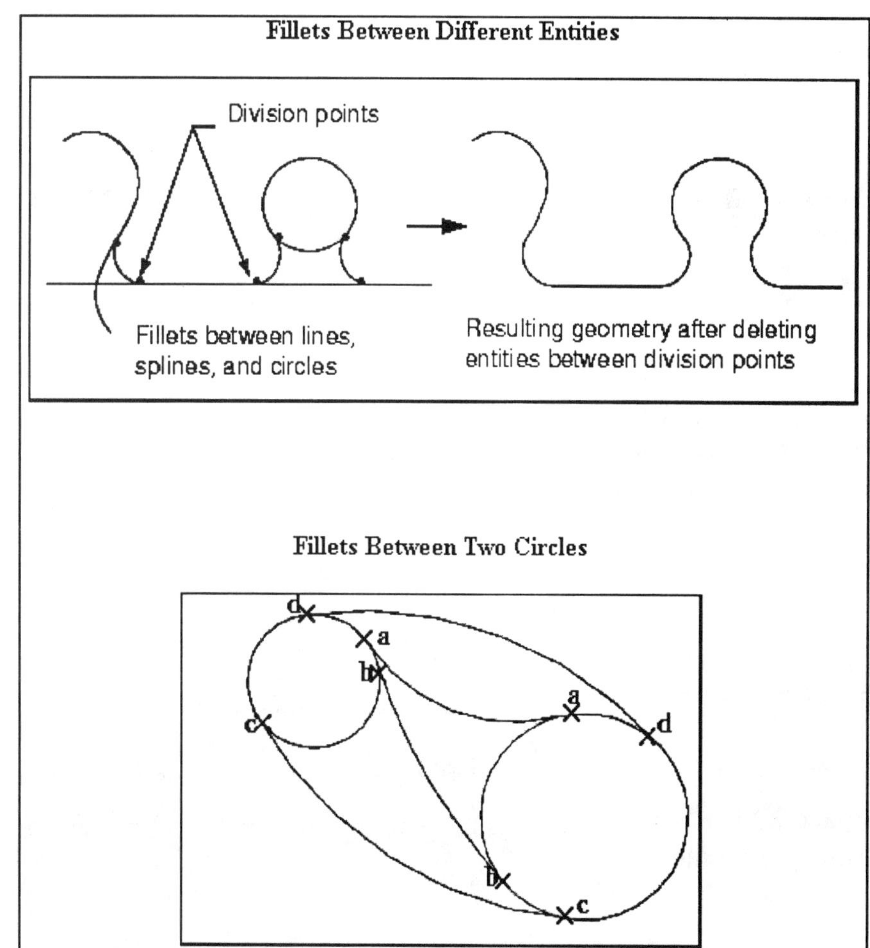

Figure 9.12
Arc Fillets and Fillets Between Circles

Advanced types of geometry, such as conics (Fig. 9.13) and splines (Fig. 9.14), can also be created in the Sketcher.

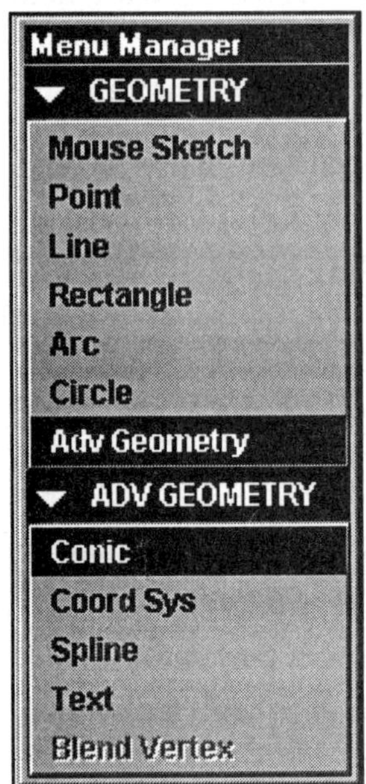

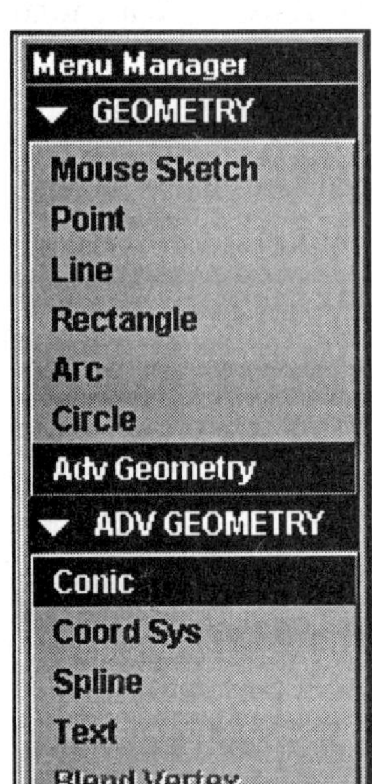

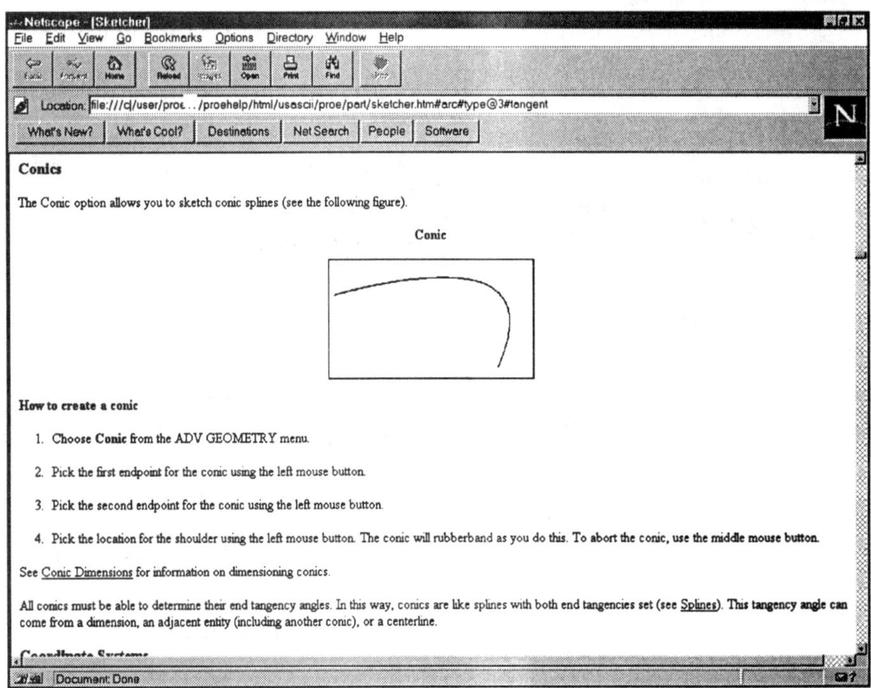

Figure 9.13
Sketching Conics

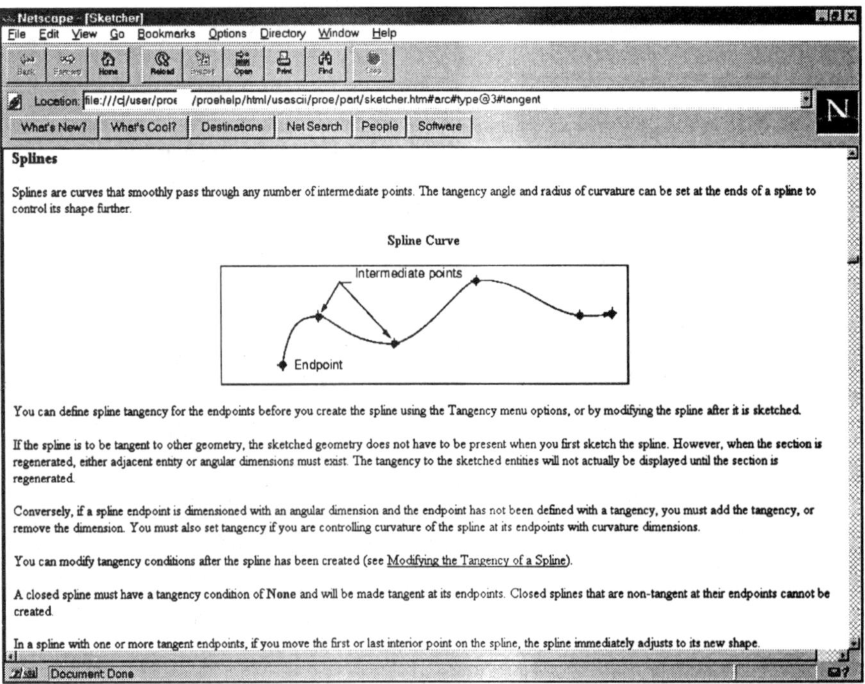

Figure 9.14
Sketching Splines

Dimensioning Sections

To regenerate a sketch successfully, it must be properly dimensioned. The Sketcher provides the ability to dimension a sketch with just a push of a button. If any references are required to locate a section, you are prompted to select the desired references and then complete the dimensioning scheme. There are two steps in dimensioning an entity: pick the entity or entities with the ***left mouse button***, then place the dimension at the desired location using the ***middle mouse button***.

HINT

You can specify critical dimensions to identify key design intent, and the rest of the dimensioning can be completed automatically.

Linear Dimensions

Linear dimensions indicate the length of a line segment or the distance between two entities. The dimension value (Fig. 9.15) is displayed as a symbol until the sketch is successfully regenerated. Only horizontal and vertical dimensions are allowed when you create a dimension between two arc or circle extents (tangency points). The dimension is created to the tangency point closest to the pick point.

Dimension symbol **sd16** is the center-to-center dimension; dimension symbol **sd18** is the tangent-to-tangent dimension.

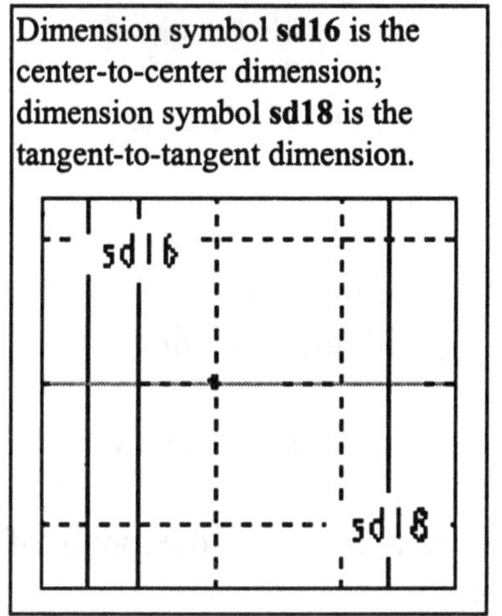

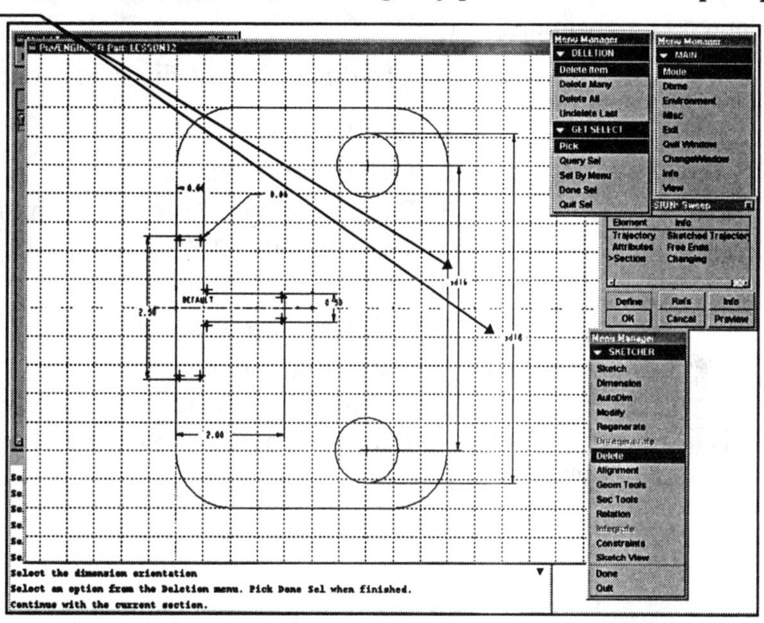

Figure 9.15
Linear Dimensioning

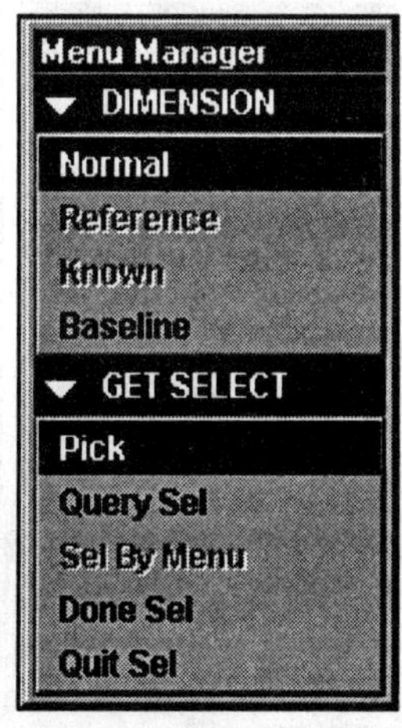

Linear dimensions (Fig. 9.15) let you do the following:

* Dimension the explicit length of a line: pick the line and then place the dimension.
* Dimension the distance between two parallel lines: pick the two lines and then place the dimension.
* Dimension the distance between a point and a line: pick the line, pick the point, and place the dimension.
* Create a dimension between two points (centerpoints and coordinate systems are included, but vertices are excluded): pick the points and location for the dimension. The DIM PNT PNT menu appears:

Horizontal Horizontal distance between the points
Vertical Vertical distance between the points

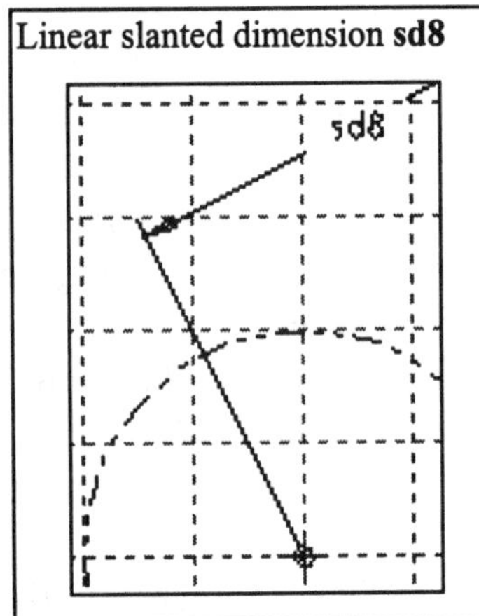

Slanted Shortest distance between the points or, as in Figure 9.16, the shortest distance between a point and a circle

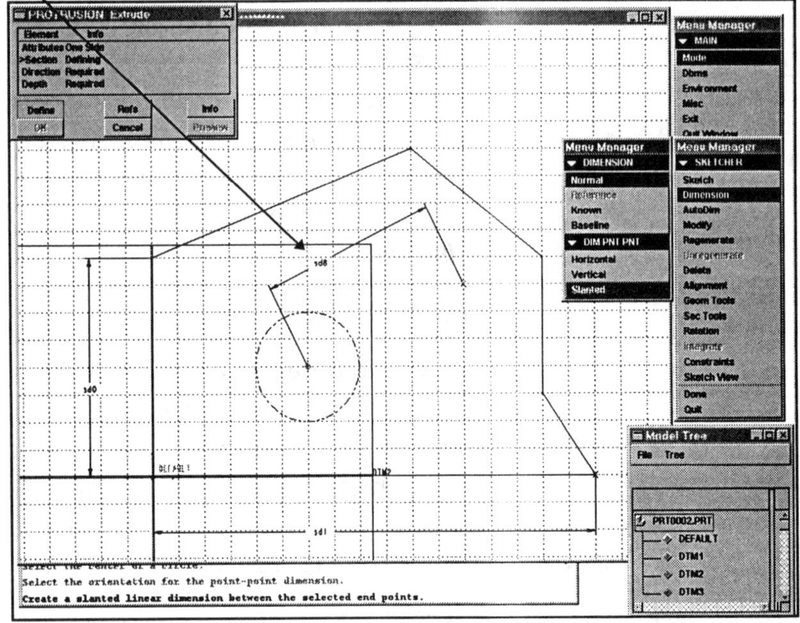

Figure 9.16
Linear Slanted Dimensioning

To dimension the distance between a line and a circle or arc:

1. Pick the line.
2. Pick the arc or circle.
3. Place the dimension (with the middle mouse button).
4. The ARC PNT TYPE menu will appear, with thc following options:

   Center Use to dimension between the arc or circle center and the line (Fig. 9.17).

   Tangent Use to dimension between a line and the point of nearest tangency on the arc or circle.

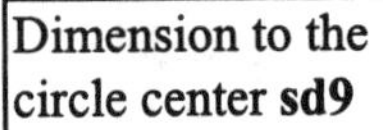

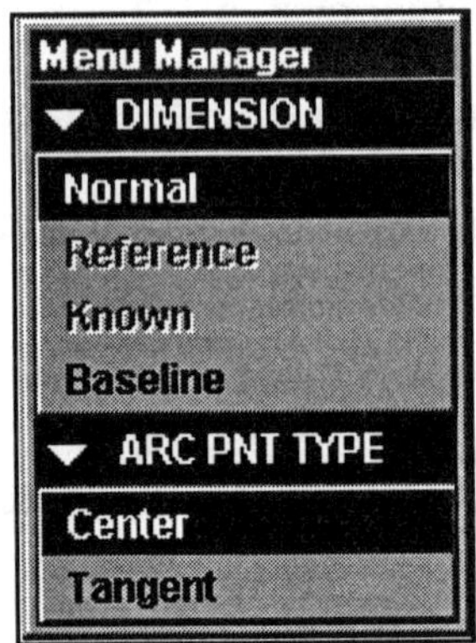

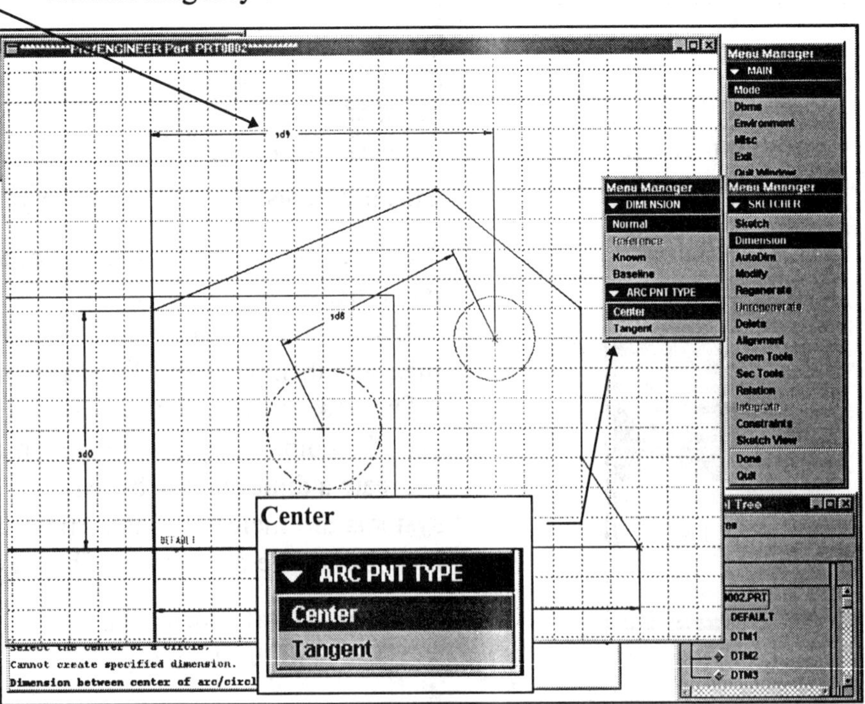

Figure 9.17
Linear Center Dimensioning

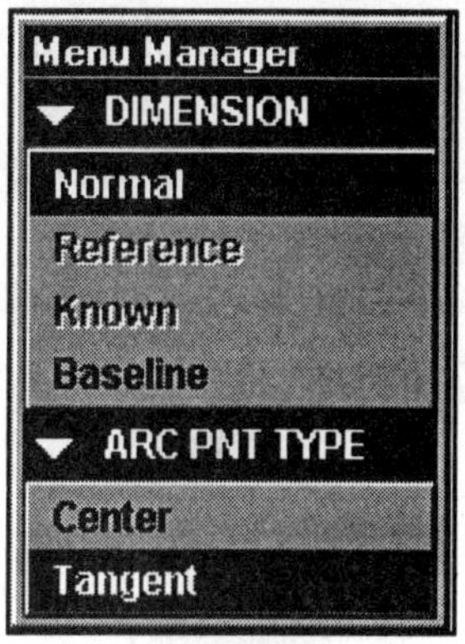

Figure 9.18 shows the dimension between a line and a circle at its tangency. To dimension between tangencies (Fig. 9.19):

1. Pick the first arc or circle.
2. Pick the second arc or circle.
3. Place the dimension.
4. Select **Tangent** from the ARC PNT TYPE menu.
5. Select either **Vert** or **Horiz** for the proper orientation.

Menu Manager
DIMENSION
Normal
Reference
Known
Baseline
ARC PNT TYPE
Center
Tangent

Center-to-center

Tangent-to-line dimension **sd10**

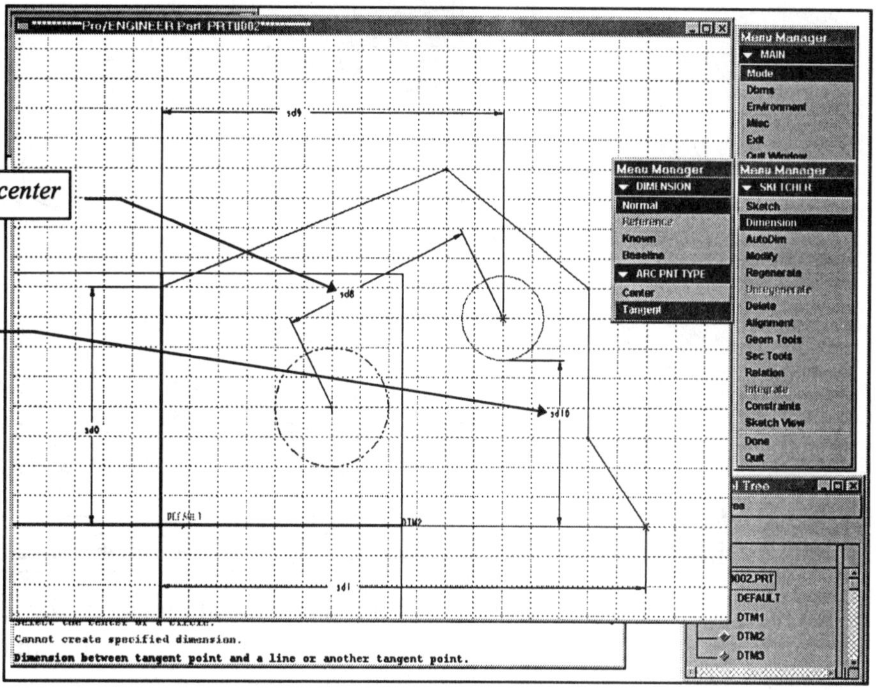

Cannot create specified dimension.
Dimension between tangent point and a line or another tangent point.

Figure 9.18
Tangent Dimensioning

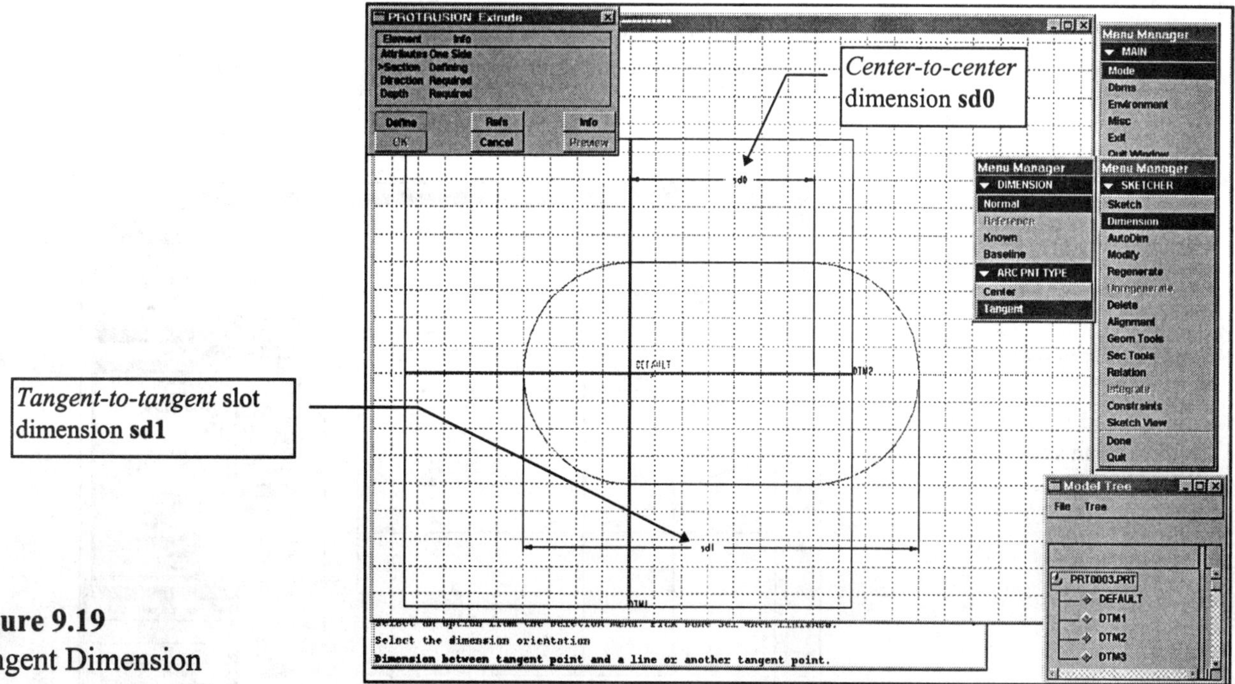

Figure 9.19
Tangent Dimension

Diameter Dimensions

Diameter dimensions measure the diameters of sketched circles and arcs or the diameters for sketching sections about the axis.

To create a diameter dimension for an arc or a circle, pick on the arc or circle twice, then place the dimension (Fig. 9.20).

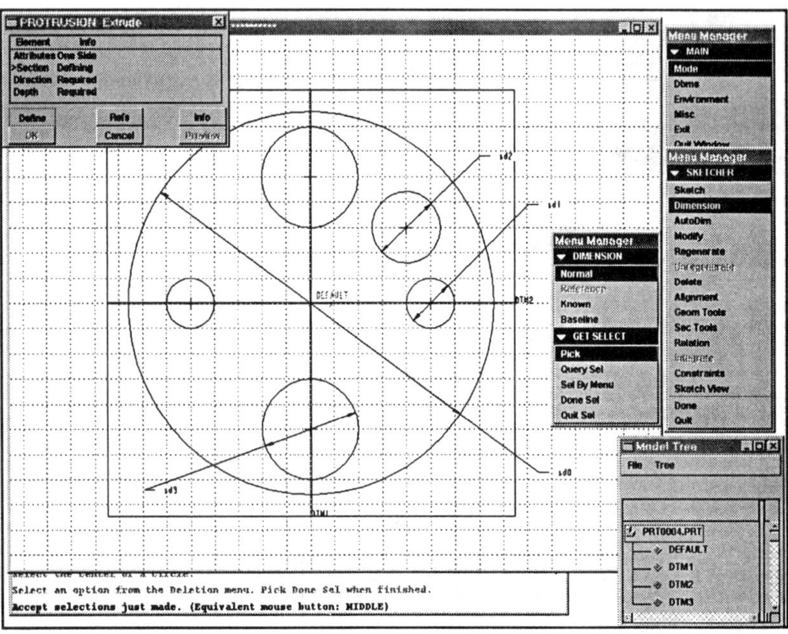

Figure 9.20
Diameter Dimensions

The diameter dimension for a revolved feature will extend beyond the centerline, indicating that it is a diameter dimension and not a radius dimension (Fig. 9.21). To create a diameter dimension for a section that will be revolved:

1. Select the entity to be dimensioned.
2. Pick the centerline that will be the axis of revolution.
3. Pick the entity again.
4. Place the dimension.

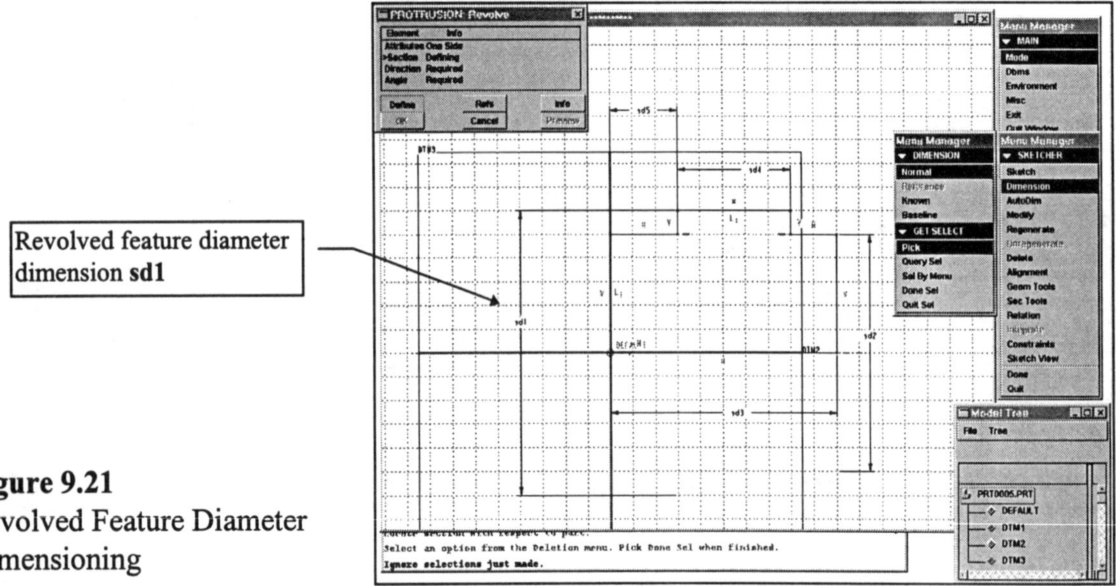

Figure 9.21
Revolved Feature Diameter Dimensioning

Radial Dimensions

Radial dimensions measure the radii of circles and arcs and the radii of circles and arcs created by revolving a sketched section about an axis.

To create a radial dimension for an arc or circle, pick on the arc or circle and then place the dimension. In general, circles are dimensioned as diameters and arcs as radii (Fig. 9.22).

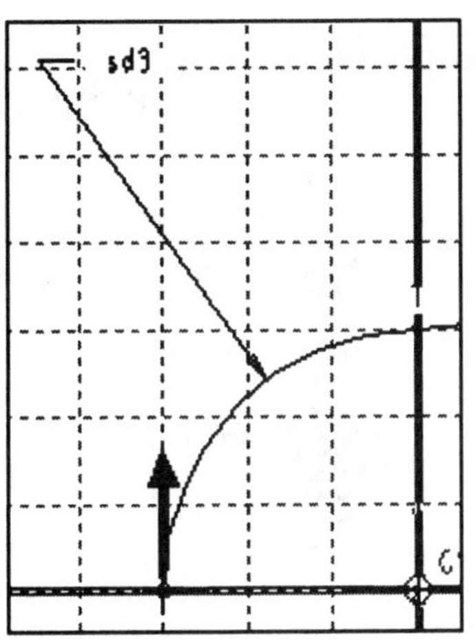

Figure 9.22
Arc Dimensioning

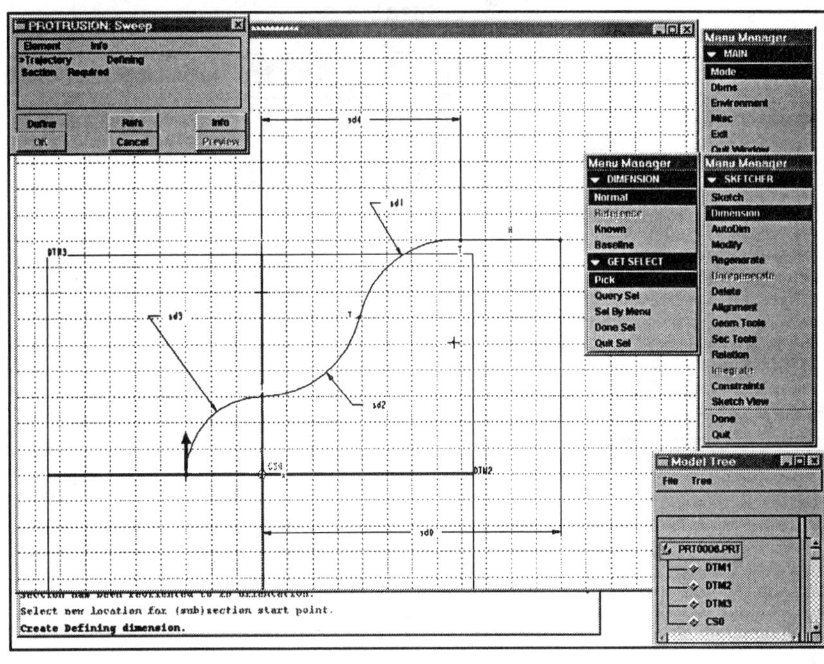

Again, to create a diameter dimension for a section that will be revolved, pick on the entity, pick on the centerline axis, pick on the entity, and then place the dimension. The example in Figure 9.23 was dimensioned in the 3D **Default** view instead of the 2D **Sketch View**.

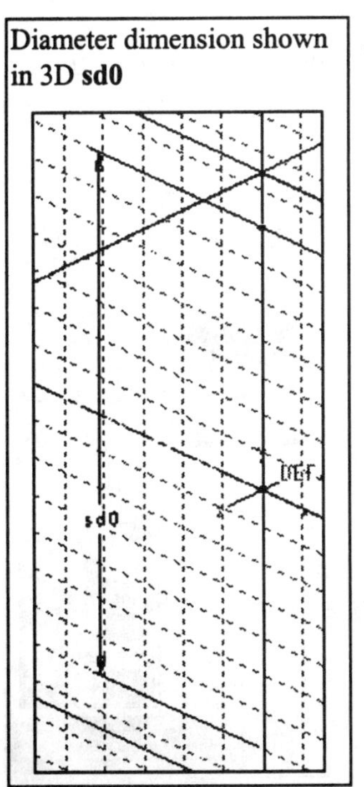

Figure 9.23
3D Default View Diameter Dimensioning

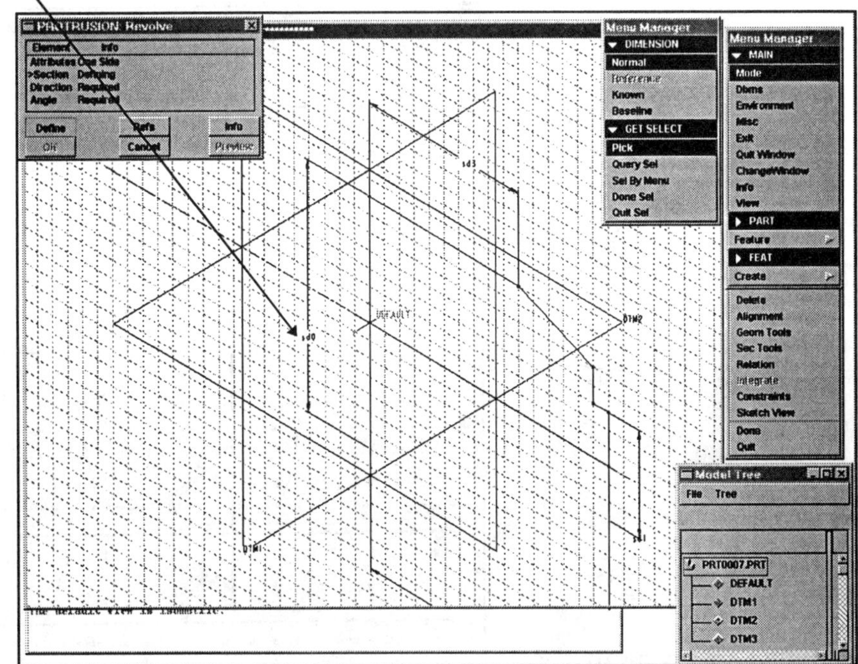

Angular Dimensions

Angular dimensions measure the angle between two lines (Fig. 9.24) or the angle of an arc between its endpoints.

To create an angular dimension between lines, pick the first line, pick the second line, and then place the dimension. Where you place the dimension determines how the angle is measured (either acute or obtuse).

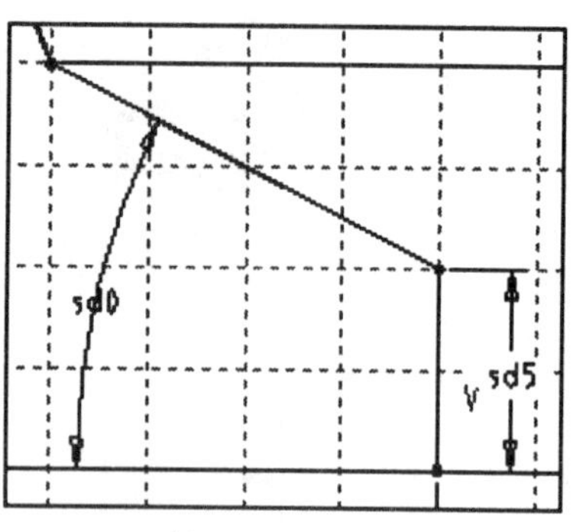

Figure 9.24
Angle Dimensioning

To create an arc angle dimension, pick one endpoint of the arc, pick the other endpoint of the arc, pick on the arc, and then place the dimension (Fig. 9.25).

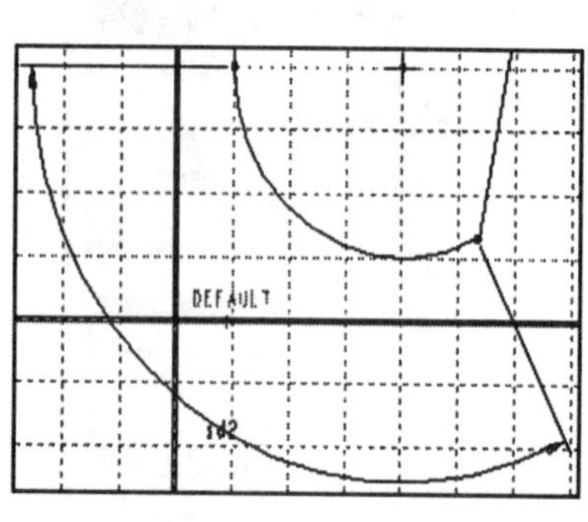

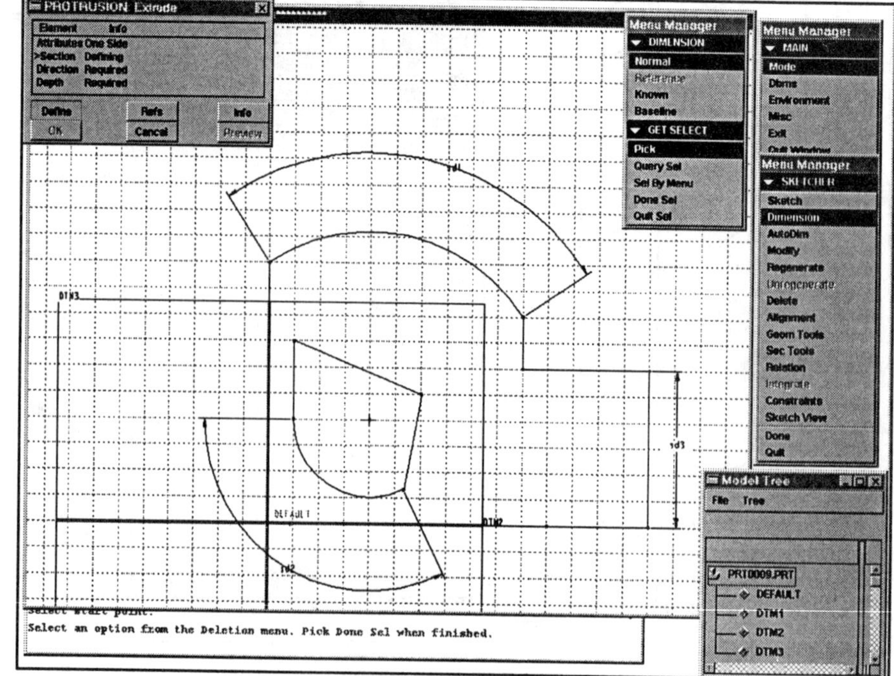

Figure 9.25
Arc Angle Dimensioning

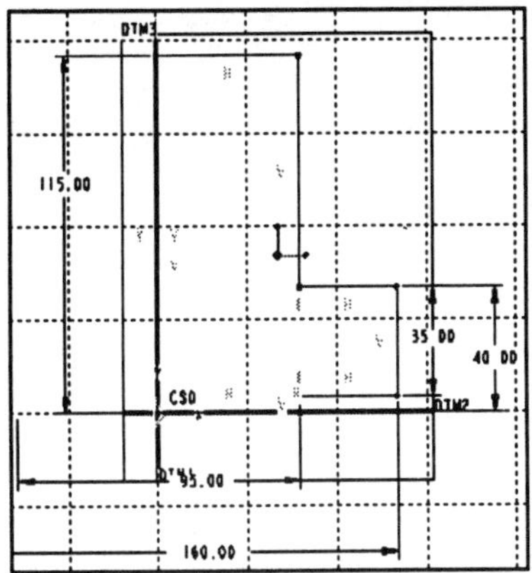

To sum up, the aim of the Sketcher in parametric feature-based design is to create quick and simple geometry for your model. The sketching process enables you to create and dimension the geometry for a feature or set of features based on your design. Remember, during the sketching process you need not concern yourself with creating perfect geometry or exact dimensions. You can modify your dimensions later in the design process (Figs. 9.26 and 9.27).

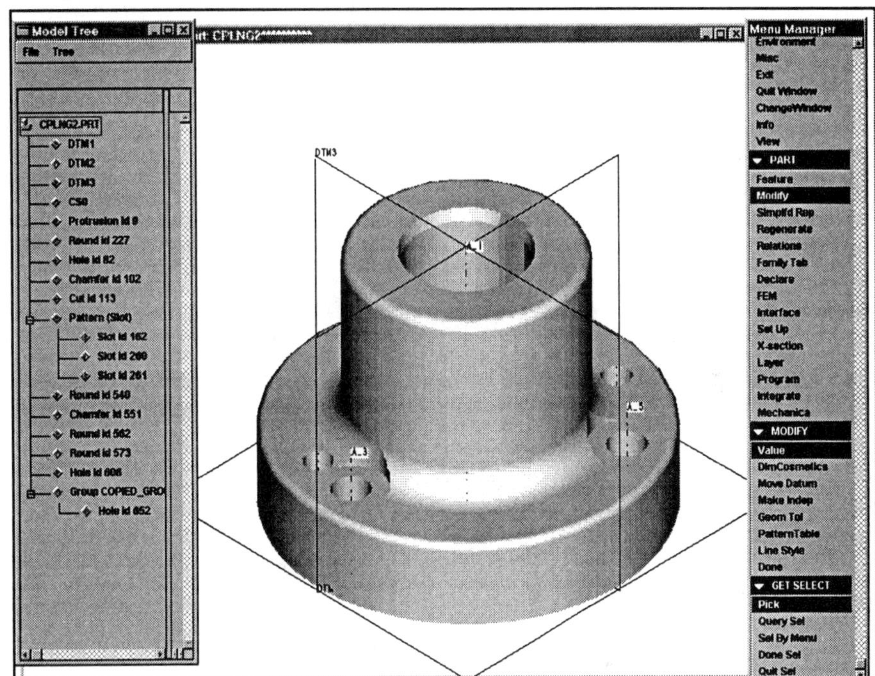

Figure 9.26
Original Part Design

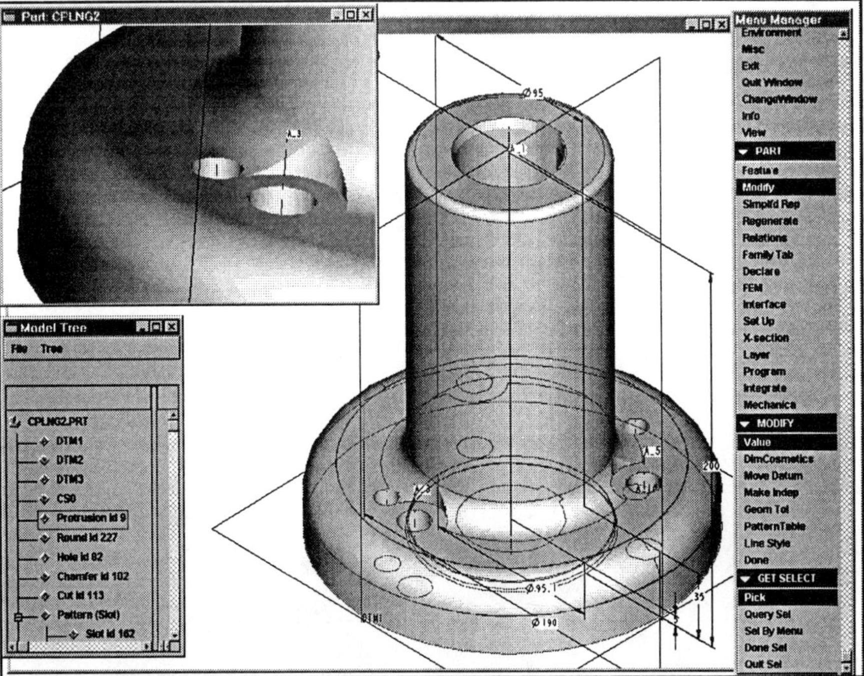

Figure 9.27
Modified Final Design

Section 10

Interface

This section teaches you the basic skills required to begin a **Pro/E** session, and is extracted from CADTRAIN's Fundamentals version of COAch for Pro/ENGINEER. You will learn how to interact with the **Pro/E** User Interface by performing some basic view manipulation operations. You will also learn some basic file management techniques.

Pro/E uses a windows interface on the workstation to display a series of interactive work areas (Fig. 10.1). The **Pro/E** environment is divided into several main areas: the **Main Working Window** (the graphics window), the **Menu Manager**, the **Message Window**, and the **Model Tree**.

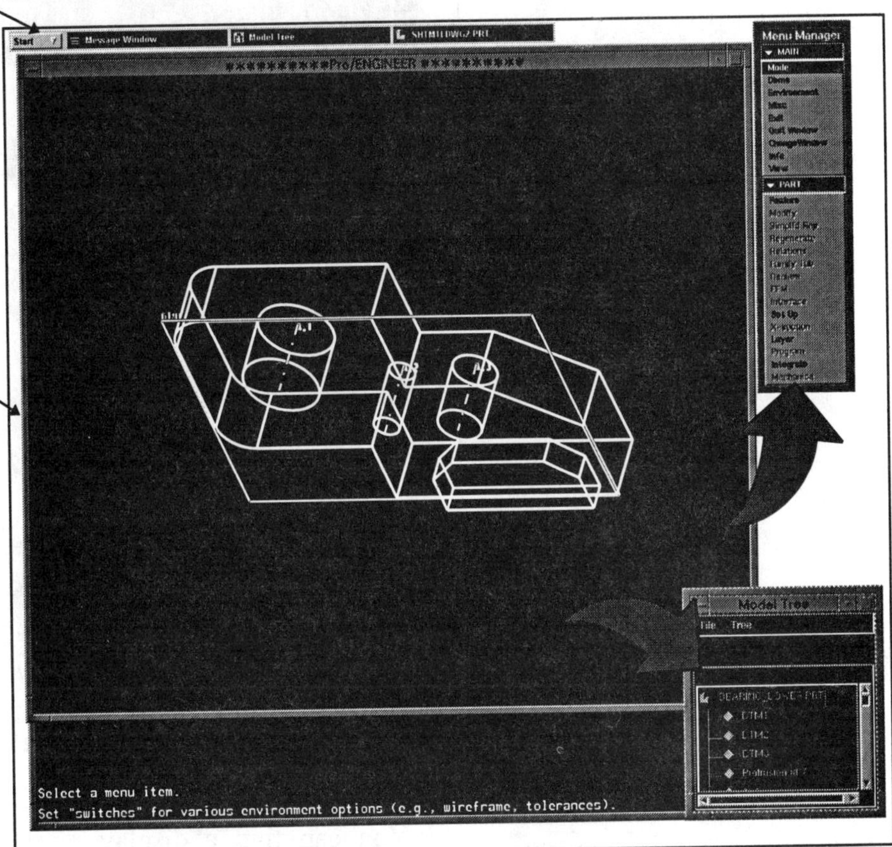

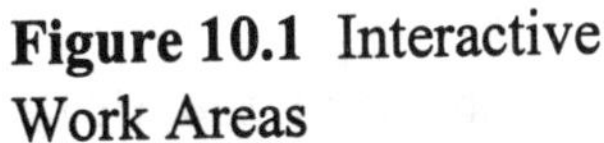
Figure 10.1 Interactive Work Areas

If your system is set up to use the **Application Manager**, you will also find it displayed on the screen.

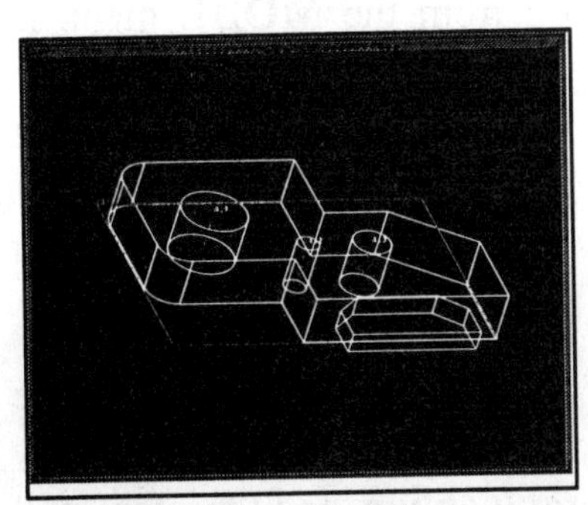

The Main Working Window

The largest area is the Main Window. This is where the object is displayed and manipulated. This area is much like the screen on a television set. You have control over the scale and angle at which you view your object.

The Main Window is also the "parent" window in terms of the window manager. If you iconify the Main Window, all other **Pro/E** windows automatically disappear (until you restore the Main Window).

The Menu Manager

You interact with **Pro/E** via a series of hierarchical menus. All menus are displayed in the **Menu Manager**. The top-level menu is called the **MAIN** menu. When you make a choice from this menu, **Pro/E** displays another menu, which offers you more choices, as shown in Figure 10.2.

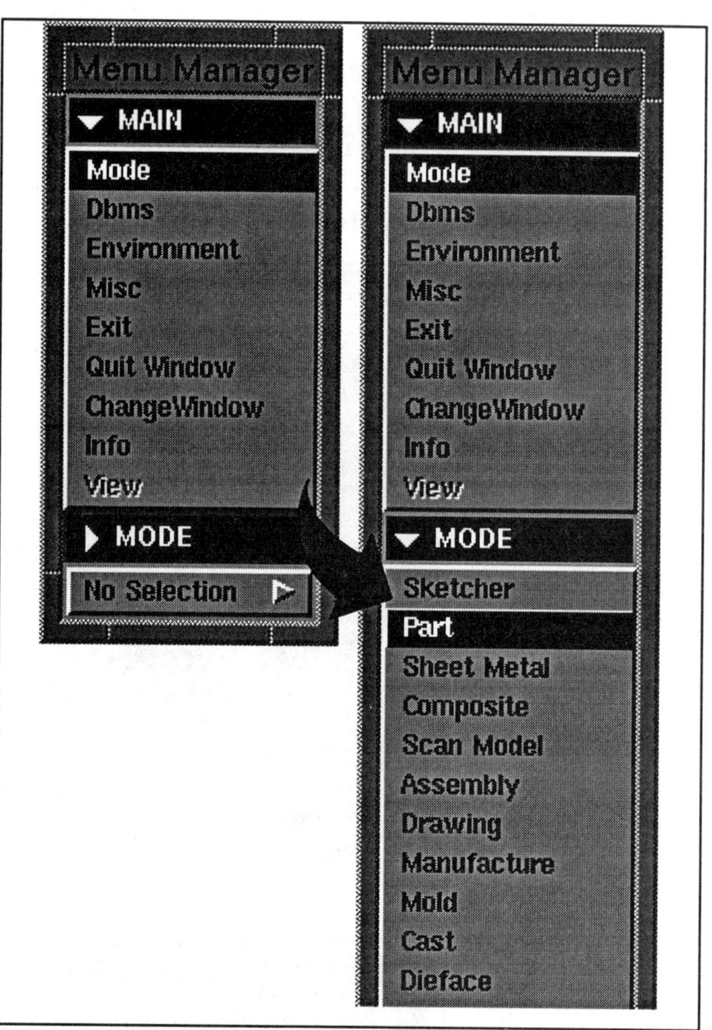

Figure 10.2 Hierarchical Menus

You can find a display of the menu hierarchy in the **Pro/E** Fundamentals Manual.

The most common option to choose from the **MODE** menu is **Part**. This is especially true when you are just starting to learn **Pro/E**. Later on, when you have more experience, you will use other options from this menu.

Understanding Menu Selections

It is also important to understand how **Pro/E** displays options in the **Menu Manager**. When a menu first appears with an option already highlighted (in REVERSE), it is the ***default*** option, and therefore will be automatically selected, as shown in Figure 10.3.

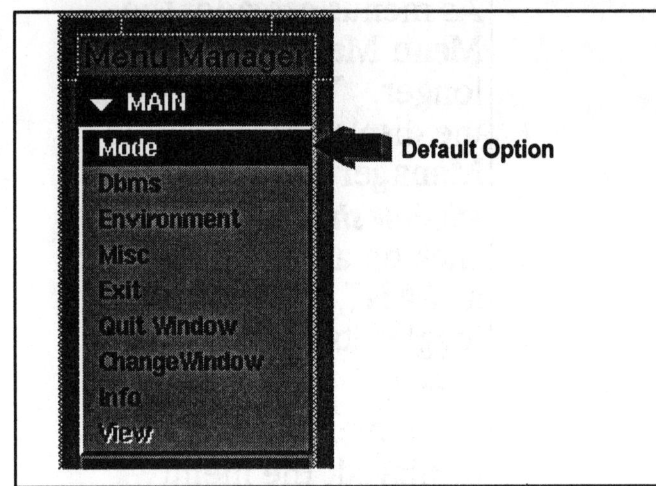

Figure 10.3 Menu Default Options

When an option is **gray**, it is not available (Fig. 10.4). In this example, the **View** option in the **MAIN** menu is not available because there is no open object (i.e. part) and therefore there are no views to modify. As soon as you open an object, you will see that this option becomes available and that the text becomes **black,** as shown in Figure 10.5.

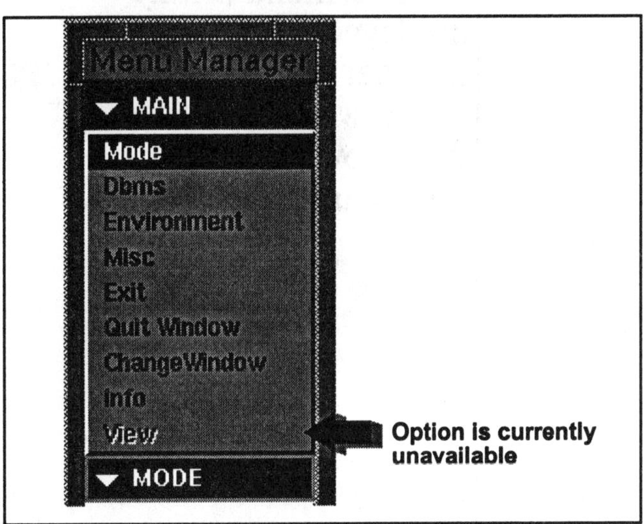

Figure 10.4 Unavailable Menu Options

When a series of menus cascades, Pro/E expects you to pick a menu option from the lowest menu. However, you can select options from any displayed menu.

Some menus have multiple lists of options, separated by horizontal lines. In general, you work your way down through the lists, choosing an option from each one.

In Figure 10.6, you must select one option from each of the first two lists. Here, **Extrude** and **Solid** have been chosen. You must make yet a *third* selection in order to proceed to the next set of menus (which will be displayed in place of the current set of menus).

As menus cascade, the **Menu Manager** gets longer. You can shorten the display of the **Menu Manager** by utilizing ***window shades***. When you click on a menu title, such as **PART**, the system toggles its window shade state.

If the menu is fully displayed, the menu collapses, thus "*raising the window shade.*" When the menu collapses, the arrow next to the menu title turns 90 degrees to point to the right.

When a menu is fully displayed.....
In other words, the "*window shade is lowered,*" the arrow points downward.

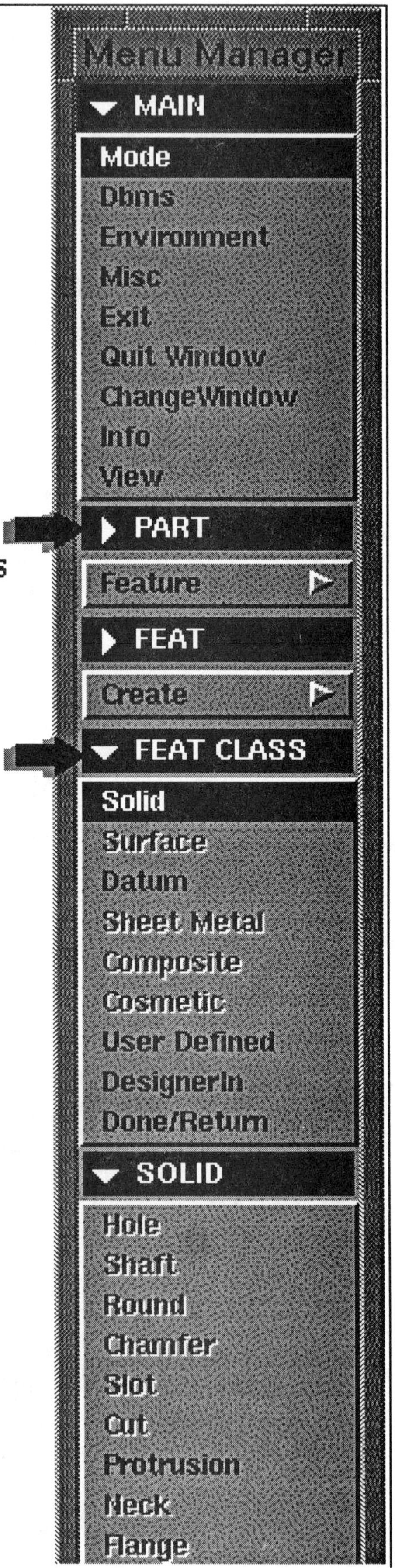

Menu Manager
MAIN
Mode
Dbms
Environment
Misc
Exit
Quit Window
ChangeWindow
Info
View
PART
Feature
FEAT
Create
FEAT CLASS
Solid
Surface
Datum
Sheet Metal
Composite
Cosmetic
User Defined
DesignerIn
Done/Return
SOLID
Hole
Shaft
Round
Chamfer
Slot
Cut
Protrusion
Neck
Flange

Figure 10.5 Menu Options

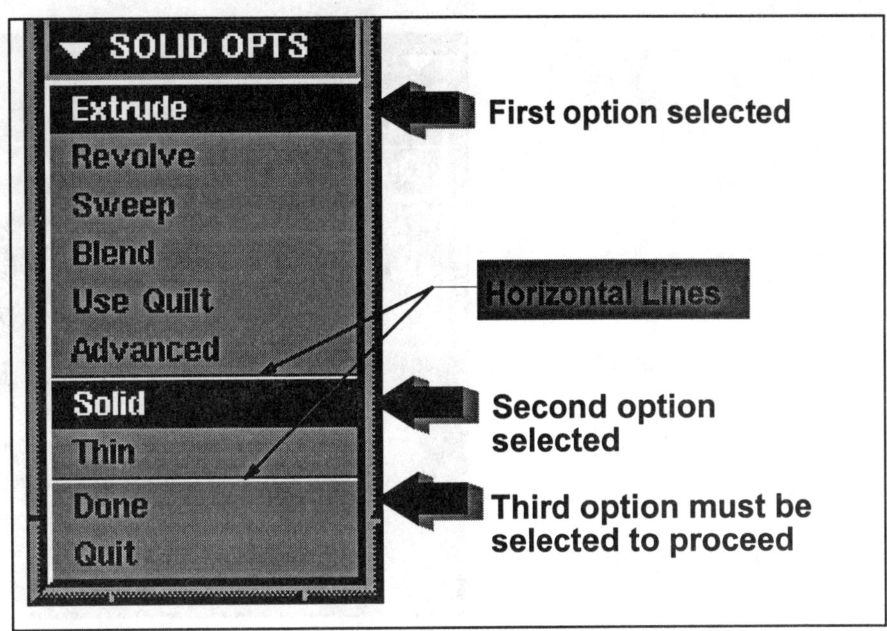

Figure 10.6 Selecting Menu Options

To move on to the next set of menus in **Pro/E**, you usually must select **Done** or **Done Sel**. To terminate an operation, you usually select **Quit** or **Quit Sel**. To go back up a level in the menu structure, you usually select **Done-Return** or **Done/Return**. Keep in mind that these are only general rules, for you will discover that there are times when these rules are broken.

Selection of Items

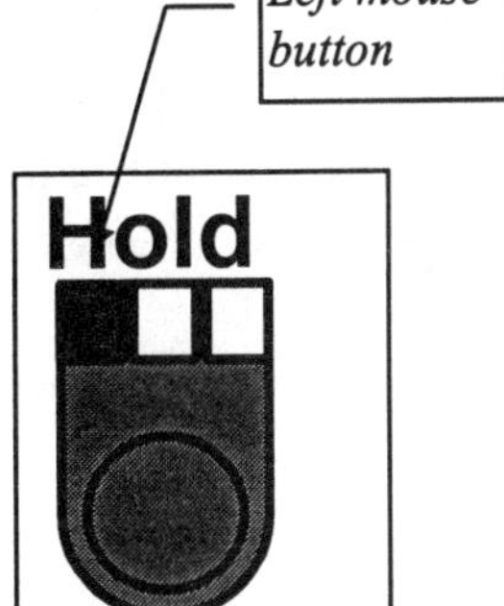

Pro/E makes use of all three buttons on your mouse. Most of the time, however, you will use the left mouse button. The left mouse button is used to choose menu options, as well as to select objects in the Main (graphics) Window. You will learn more about the functionality of the other mouse buttons in another section.

The Message Window

As you move the mouse cursor over the menu options, they become momentarily highlighted in gray. If you press ("click") the left mouse button on an option, it will be selected and highlighted in black.

You can get additional information about **Pro/E** commands via menu items. Placing the mouse cursor on top of a menu option displays a one line help message or hint about the menu option in the **Message Window**. The hint is displayed in YELLOW at the bottom of the Message Window, explaining what that command will accomplish.

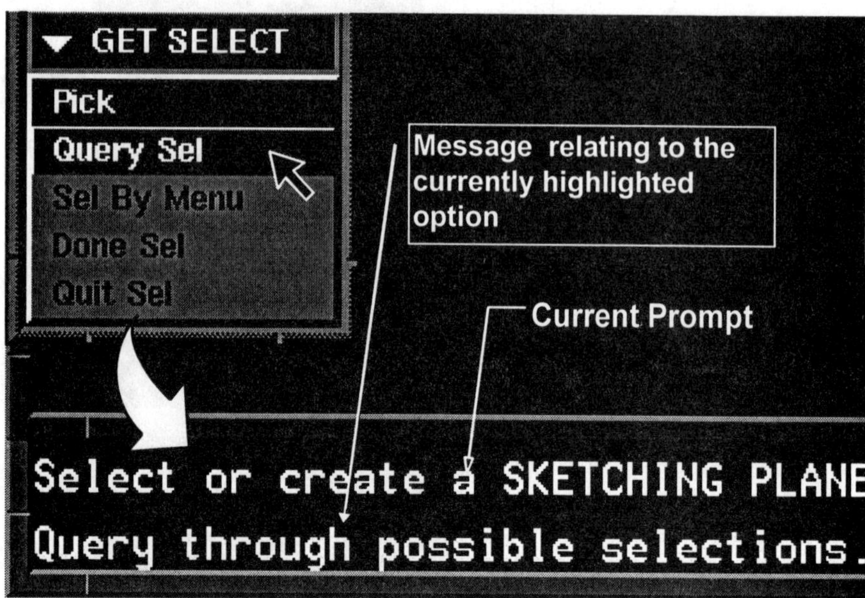

Figure 10.7 Message Window

This capability is also very important when you are looking at a list of object (file) names. Since these names are often longer than the width of the menu, you need a quick way to see the entire name. If you move the mouse cursor over the names on the menu, the full names are displayed in the **Message Window**.

Reading the Message Window is very important. Not only does **Pro/E** display the menu hints as you move the mouse cursor over the menu options; it also displays a prompt that is related to the current operation (Fig. 10.7). The current prompt message is displayed in 

. The previous prompts are scrolled upwards in the Message Window, and can be examined if necessary. This allows you to review the operations you have completed.

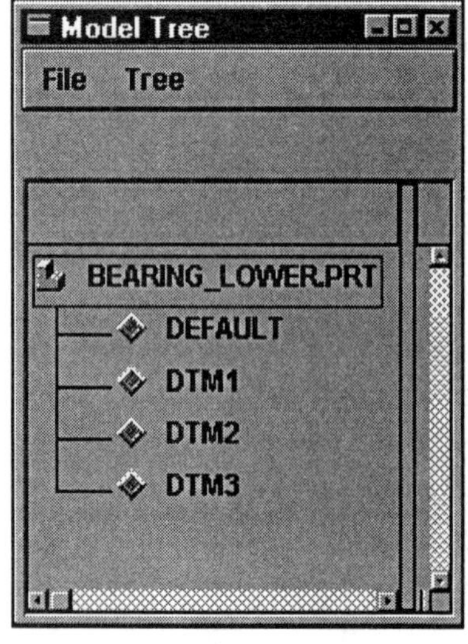

The Model Tree

When you actually retrieve an object, Pro/E displays a window called the **Model Tree** (illustration at left). The **Model Tree** is used to navigate through, and select objects in, a **Pro/E** model. You will learn more about how to use this tree later on. You can turn it on and off using the **Model Tree** option in the **Environment** menu.

Orientation

You can think of the Main Window as being just like a camera. It displays a picture of the current object on your screen, based on the current orientation of the object and on how close the "camera" is to the object (Fig. 10.8).

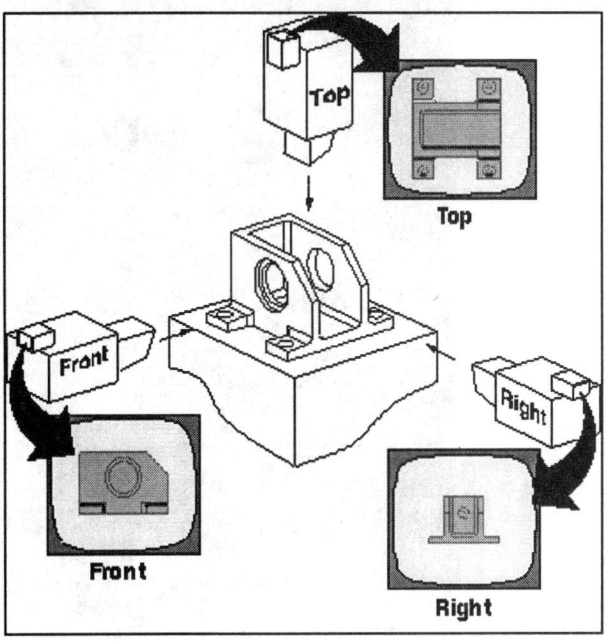

Figure 10.8
View Orientation

The function you use to look at the object from different angles is called **Orientation.** This term is analogous to the photography term of *rotating* the camera to look at the objects in the view finder from a different angle.

The function you use to move closer to, or farther away from, the object is called **Zoom.** This term is analogous to the camera term of *zooming* in or out to make the objects in the picture appear closer or farther away.

You can also move the camera in a way that it does not change the orientation, nor its distance from the object. You can **Pan** the view side to side, or up and down. Keep in mind that you are moving the camera, not the object itself (Fig. 10.9).

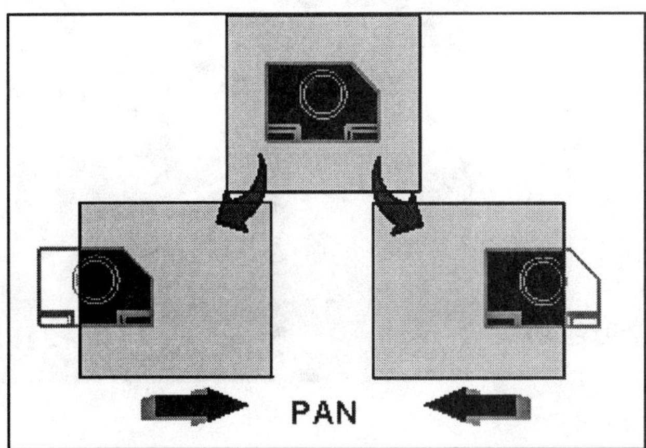

Figure 10.9 Pan

When you look at an object in a 3D view, while it is displayed in **Wireframe** mode, it often "flips" inside out. By default, Pro/E helps you visualize objects by displaying them in **Hidden line** mode (Fig. 10.10). In **Hidden line** mode, edges that are *behind* the object appear dimmer. This setting can be changed in the **Environment** menu.

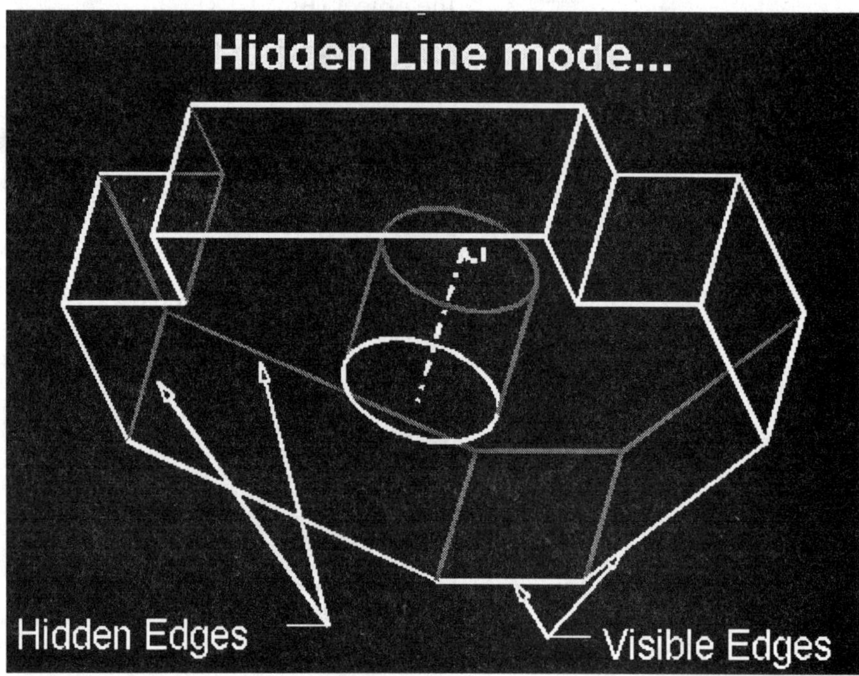

Figure 10.10 Hidden Line Mode

Orienting the View of an Object

You can create customized views of any model, or use the built-in **Default** view [either the **Isometric** or the **Trimetric** (Fig. 10.11) view of the object]. You can also define 3D views of an object by rotating it with specified angles, or by rotating it dynamically (using the **Spin** option).

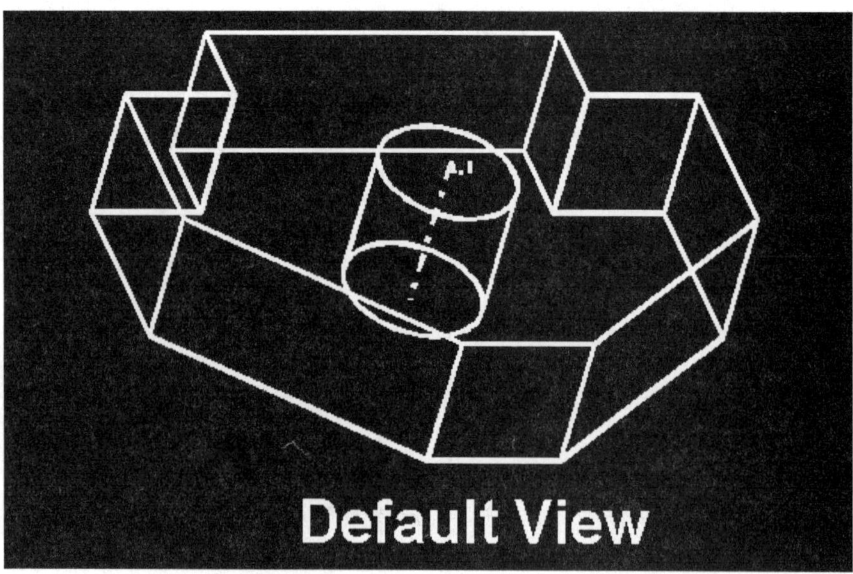

Figure 10.11 Default Trimetric View Displayed in Wireframe

To define an orthographic view, you use the **Orientation** function to specify model faces (or planes) that represent the front, back, top, bottom, right, or left face of the object. *Two* items are required to define any view, such as **Right** and **Front** (Fig. 10.12) or **Top** and **Right**.

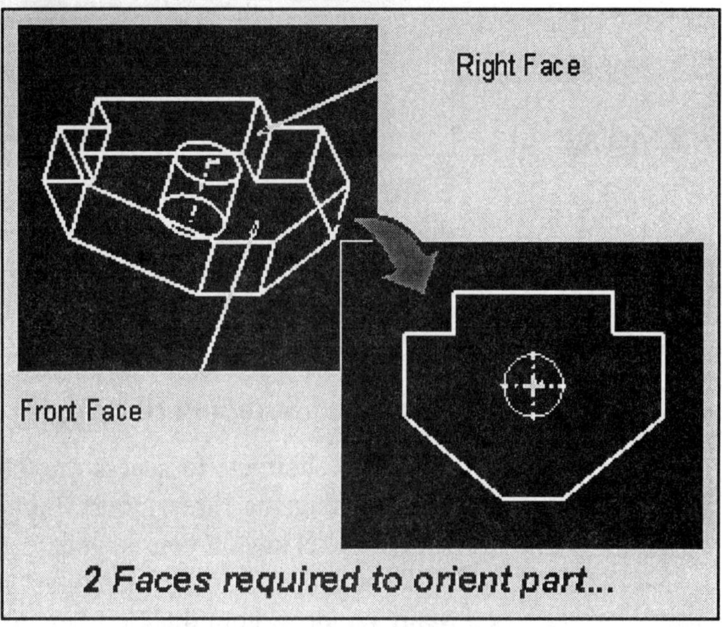

Figure 10.12 Orientating a View

Whenever you have a view orientation that you want to reuse, you can save it with a user-defined name using **View** ⇒ **Names** ⇒ **Save** ⇒ (enter a name) ⇒ **enter**. You can recall them by selecting their name from a list using **View** ⇒ **Names** ⇒ **Retrieve** ⇒ (pick the name).

The Basic Zoom/Spin/Pan Functions

The functions in **Pro/E** that change the view of an object are **Zoom**, **Spin**, and **Pan**. These functions can be accessed from the **View** option in the **MAIN** menu. You can use the viewing functions even while you are in the middle of another operation (e.g., creating a feature):

Zoom In/Zoom Out Increases or decreases the size of the object on the screen. *Zoom does not change the size of the object, only its display size on the screen.*

Spin Moving the cursor spins the object around on the screen, relative to the center of the model.

Pan Moving the cursor drags the object. Basically, you can "pick up" the object with the cursor and drag it to anywhere you want within the Main Window.

There is also a **Rectangle Zoom** mode (Fig. 10.13), in which you define a rectangular box that expands to "fill the screen." This allows you to **Zoom** and **Pan** simultaneously, since it changes the size of the view and its center on the screen.

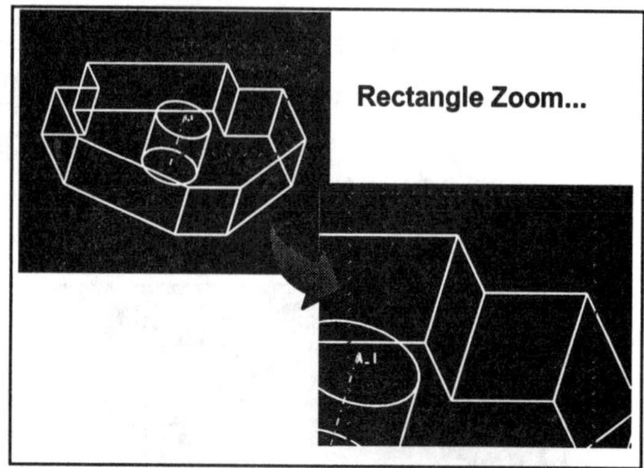

Figure 10.13 Zooming

To perform a **Rectangle Zoom,** you place the cursor at one of the diagonal corners of the zoom box you want to define. **Click** (do not hold) the **left** mouse button firmly. Move the cursor to the other diagonal point (drawing a box), and click the **left** button again--the **Zoom** will immediately take effect.

The Zoom/Spin/Pan Shortcuts

There are shortcuts to access the **Zoom**, **Spin**, and **Pan** functions in **Pro/E**--using the three mouse buttons (Fig. 10.14), in combination with the **Ctrl** key on your keyboard.

To use these mouse mappings, simply move the cursor into the Main Window, hold the **Ctrl** key down, and click and hold down the desired mouse button (for **Zoom**, **Spin**, or **Pan**). Then moving the cursor changes the view of the object, according to the view option. Once you begin a viewing action, it continues until you release the mouse button.

The mouse **Zoom** functionality is slightly different from the menu options. **Zoom In/Zoom Out** works like this: moving the cursor down or to the left zooms in on the part, and moving the cursor up or to the right zooms out. To perform a mouse **Rectangle Zoom,** hold down the **Ctrl** key and place the cursor at one of the diagonal corners of the zoom box you want to define. **Click** (do not hold) the **left** mouse button firmly. Move the cursor to the other diagonal point, and click the **left** button again to finish.

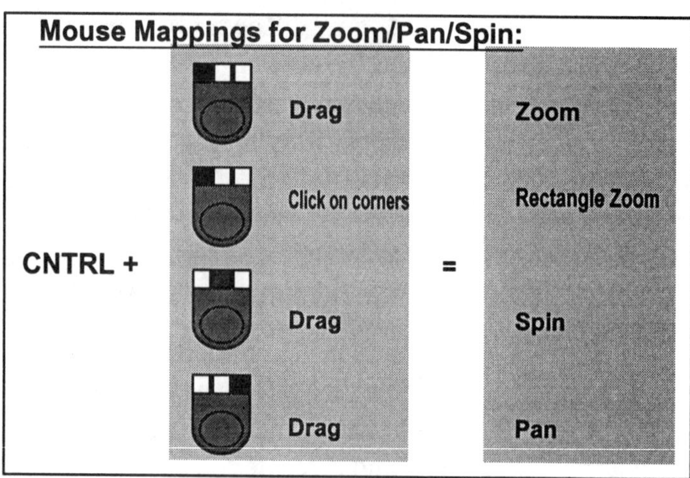

Figure 10.14 Mouse Functions

Pro/E allows you to manipulate the view of the active object. You can control the view display of the object in the active Graphics Window by altering its scale and/or viewing angle using the **Zoom, Spin**, and **Pan** functions.

After opening a object you can manipulate the view. When a part is displayed in the Graphics Window, the Title area of the Graphics Window will read:

******** Pro/ENGINEER Part: INTRO1 ********

The Default view is displayed in the Graphics Window as shown in Figure 10.15. The **Default** view will be either **Trimetric** or **Isometric** depending on the **Environment** setting. The Model Tree window is also displayed when the part model is retrieved or activated (Fig. 10.16). Notice that the part is displayed in **Hidden line** mode.

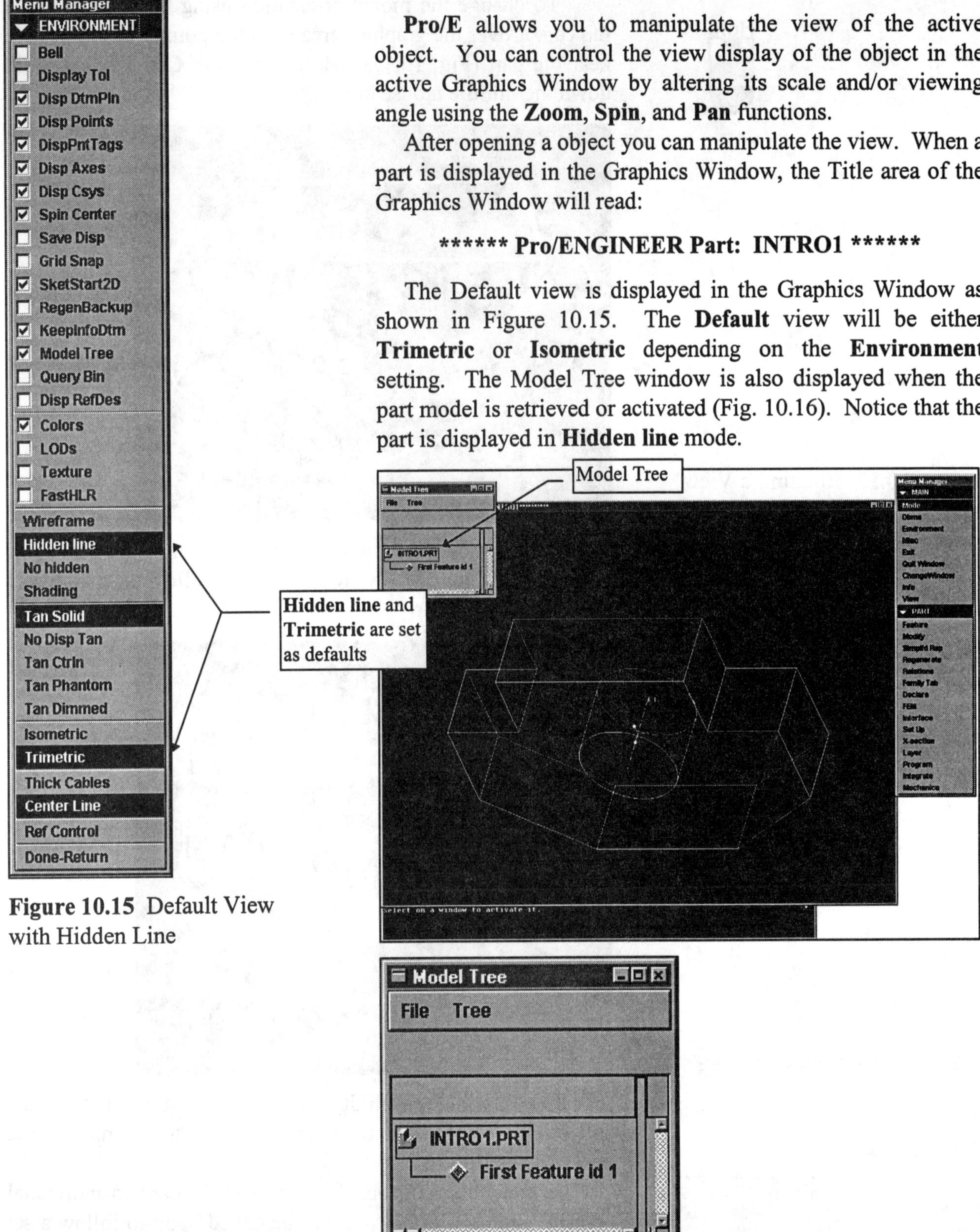

Figure 10.15 Default View with Hidden Line

Figure 10.16 Model Tree

The **Model Tree** can be turned off in the **Environment** menu by choosing **Environment** ⇒ □ **Model Tree** ⇒ **Done-Return**.

To change the model orientation using the mouse, move the cursor over the graphics area until the pointer is somewhere near the part (Fig. 10.17). Hold down the **Ctrl** key and hold down the **middle** mouse button.

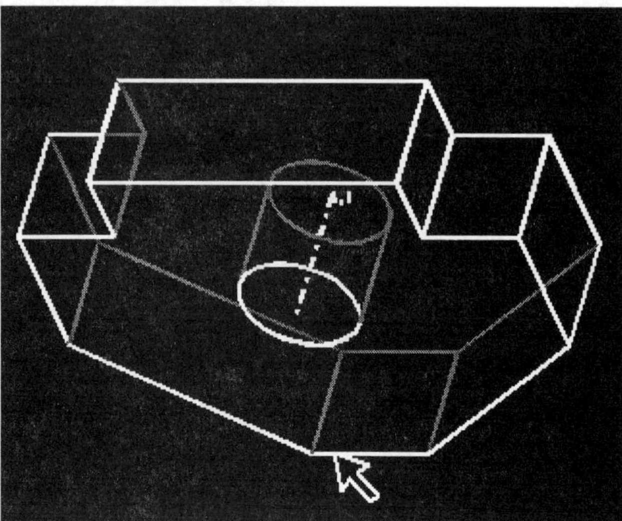

Figure 10.17 Rotating a View using the Mouse

Move the cursor around on the screen until the part spins around as in Figure 10.18. Release the **middle** mouse button to stop spinning the part.

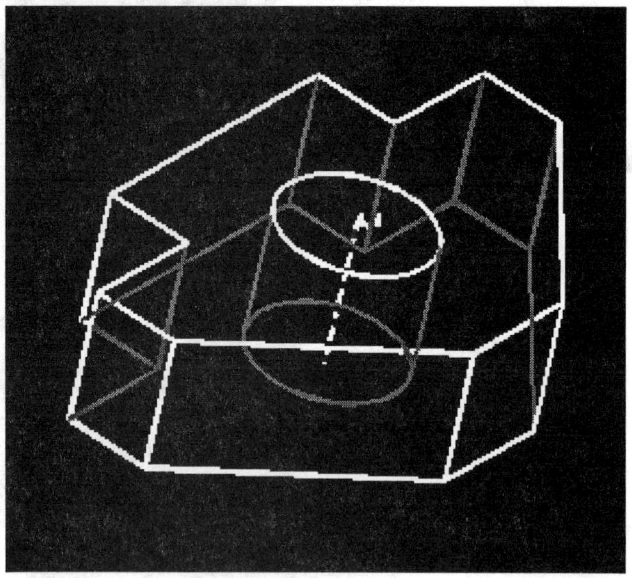

Figure 10.18 Rotated View

Other Interface capabilities are introduced throughout the following lessons as they are needed to complete the assignments.

The remaining portion of this text is devoted to individual lessons. In each **lesson** you will be called upon to follow a set of steps to complete the **lesson part**, which in most cases includes the creation of a variety of sketches. After you complete the lesson part, the material that was introduced in that lesson will be applied to the subsequent **lesson project**.

Part One

Creating Parts

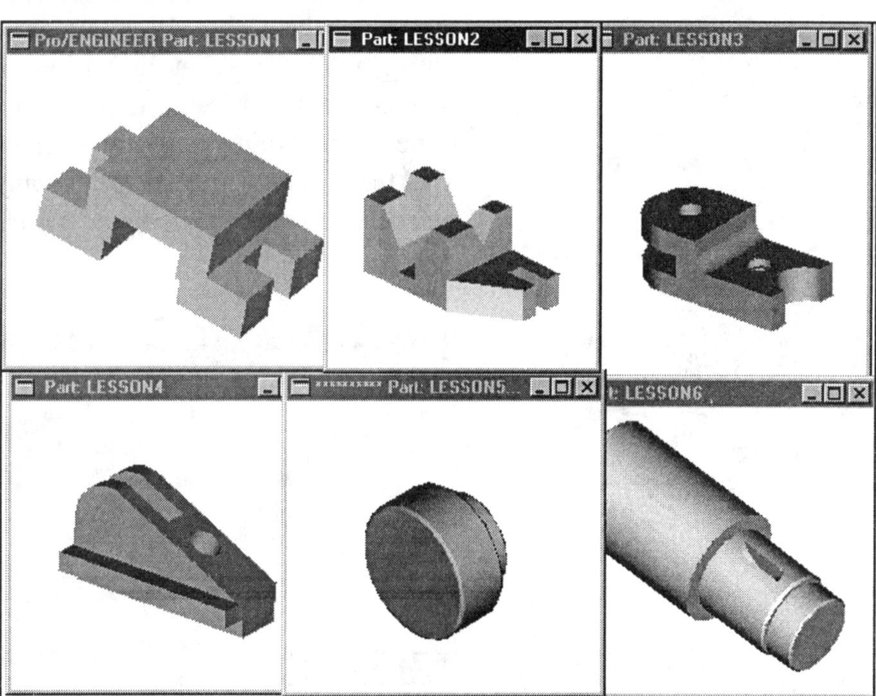

Lesson Parts 1-6

Creating Parts

The *design intent* of a feature, a part, or an assembly (and even a drawing) should be established before any work is done with Pro/ENGINEER. Skipping this step in the design process is a recipe for disaster. In industry, there are thousands of stories of how a designer created a *graphically correct* part or assembly that *"looked"* visually precise.

Upon closer examination, the model or assembly had too many or too few datum planes, parent-child relationships that were nothing but an example of the designers incorrect use of Pro/E, and massive feature failures that resulted when minor ECOs were introduced after the original design was complete. Pro/E is only as good as the person designing with it.

Without proper process planning, organization, and a well-defined design intent, the part model is useless. In most cases, such poor design habits result in the parts being remodeled, since it would take more time to reorder, modify, redefine, and reroute. In fact, most poor designs can't be fixed.

Part Design Philosophy: Design Intent

The **design intent** of a project must be understood before modeling geometry is started. Use the Design Intent Planning Sheets provided in Appendix D to sketch and analyze your part before modeling.

The *dimensioning scheme* will establish the dimensions that are critical for the design: What dimensions on the part might be modified during an *ECO*? What dimensions are required for *manufacturing* the part economically and to the correct *tolerances*? Are there any dimensional *relationships* that must be established and maintained? Will the part be a member of a *family of similar parts*? How does the part relate to other *parts in the assembly*?

Lesson Parts 7-13

Use the following basic guideline to create a typical part:

1. Establish the system environment that you will work in for the project, including environment setting and *config.pro* settings
2. Use setup to establish the material and units.
3. Establish the datum planes and coordinate system.
4. Create a layering scheme, and set the datum planes and the coordinate system on a layer.
5. Rename the datum planes per the part and geometric tolerance requirements.
6. Determine the base feature and protrusion type. Sketch the base feature on the appropriate datum plane.
7. Establish the dimensioning scheme for the feature.
8. Determine what construction features should be used on the part.
9. Build a construction feature using a dimensioning scheme, keeping relationship requirements in mind.
10. Add relations to control the feature where desired, per the design intent.
11. Adjust dimension cosmetics as desired.
12. Create new layers, and establish a layering scheme for dimensions and features.
13. Add reference dimensions required for documentation.
14. Repeat steps 1-14 for each feature of the part until the part is completed.

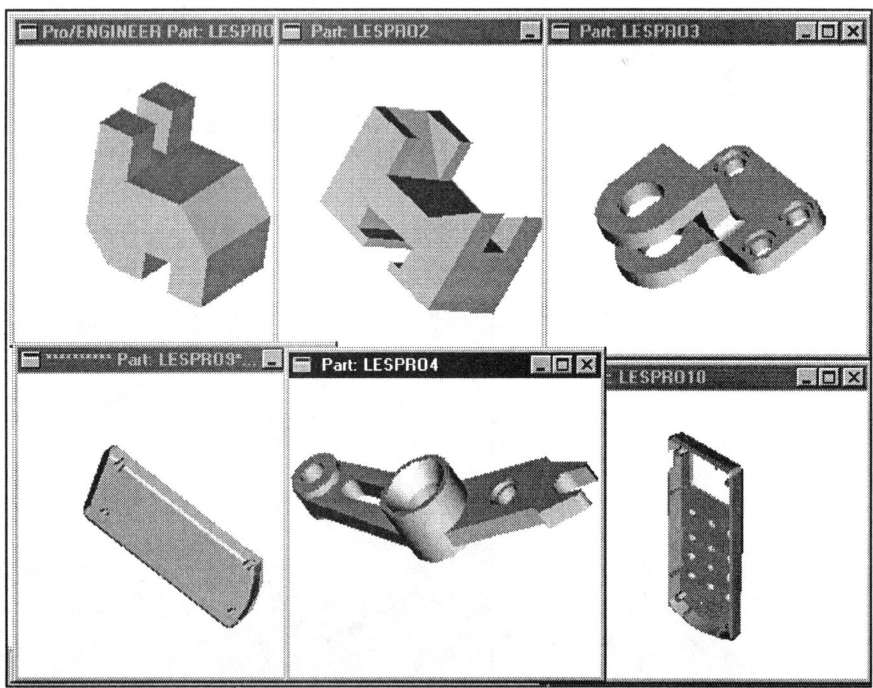

Lesson Project Parts

A wide variety of features can be created with Pro/ENGINEER. The following 13 part creation lessons incorporate most of the following capabilities:

Protrusions (Lesson 1) Part features that add material
Slots and Cuts (Lesson 1) Features that remove material from a part
Holes (Lesson 3) Creates different types of holes - through, counterbored (sketched), blind, etc.
Shafts Creates shafts
Rounds (Lesson 3) Creates many types of rounds
Chamfers (Lesson 6) Creates edge and corner chamfers, which remove flat sections of material to create a beveled surface
Necks (Lesson 5) Creates a neck, which is a special type of revolved slot that creates a groove around a revolved part or feature
Flanges Creates a flange, which is analogous to a neck, except it adds material to the revolved solid
Ribs (Lesson 8) Creates a rib, which creates a thin fin or web that is attached to a part
Tweak features Creates drafts (Lesson 9), local pushes, domes, ears, lips, patches, bends, and free-form features
Shells (Lesson 10) Creates a shell feature, which removes a surface or surfaces from the solid, then hollows out the inside of the solid, leaving a shell of a specified wall thickness
Pipes Creates a pipe, which is a three-dimensional centerline that represents the centerline of a pipe
Cosmetic features (Lesson 6) Creates cosmetic features, sketched, thread, groove, and user-defined

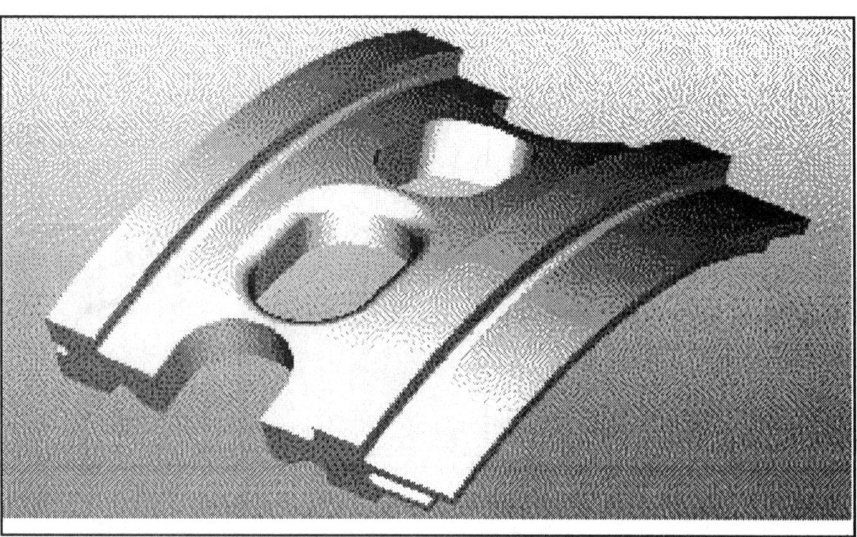

Lesson 1

Protrusions and Cuts

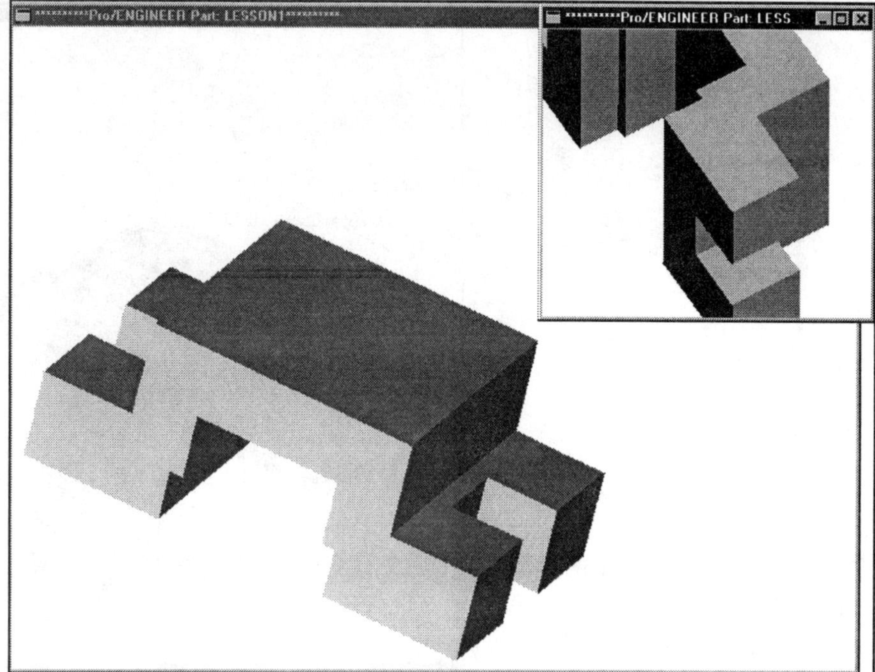

Figure 1.1
Clamp

OBJECTIVES

1. **Create a base feature using an extruded protrusion**
2. **Understand setup and environment settings**
3. **Define and set a material type**
4. **Create and use datums**
5. **Sketch a protrusion and a cut feature in the Sketcher**
6. **Understand the feature dialog box**
7. **Learn how to align sketch geometry**
8. **Shade a part**
9. **Copy a cut feature**
10. **Save and purge a part file**

EGD REFERENCE
Engineering Graphics and Design with Graphical Analysis *or* **Fundamentals of Engineering Graphics and Design**
by L. Lamit and K. Kitto
Read Chapters 5, 10
See pages 111, 192-193, 302, 409

COAch™ for Pro/ENGINEER
If you have **COAch for Pro/ENGINEER** on your system, go to SEARCH and do the Segment shown in Figures 1.4 and 1.6.

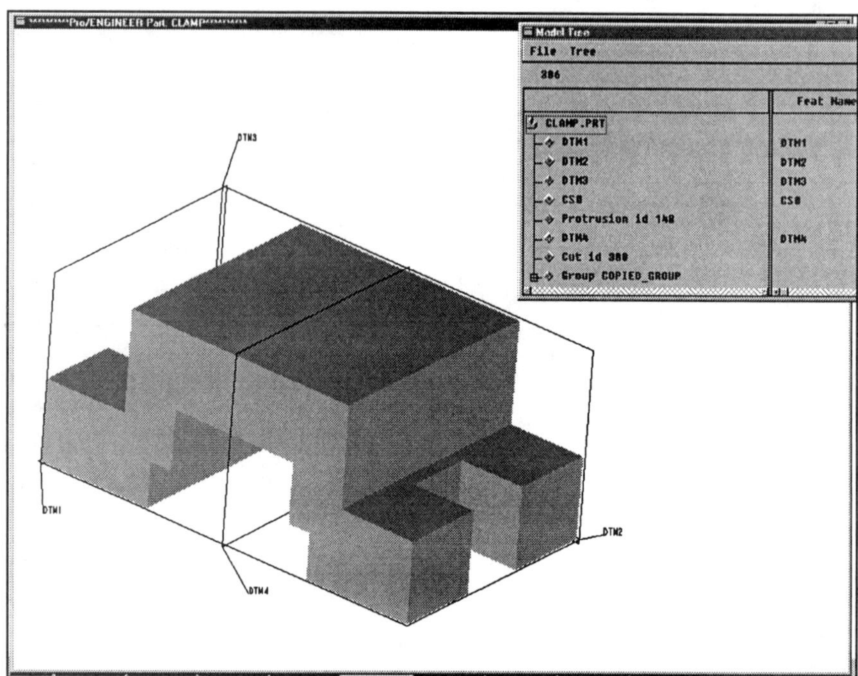

Figure 1.2
Clamp Part Showing the Datum Planes and the Model Tree

PROTRUSIONS AND CUTS

A **protrusion** is a part feature that adds material. You can sketch different geometry by combining a variety of form options and attributes during creation of the protrusion feature. **Cuts** and **slots** are used to remove material from existing solid features. Figures 1.1 and 1.2 show a simple protruded part with identical cuts on both sides.

> ***NOTE***
> **The first few pages of each lesson contain general information (online documentation, references to COAch, etc.) on the commands of features introduced in that lesson. The actual step-by-step sequence comes after this material. PT/Modeler commands, where they differ from Pro/ENGINEER commands, are shown *only* in the step-by-step section of the lesson. PT/Modeler differences are *not* provided in the first few pages introducing information about the lesson's concepts.**

Protrusions

A protrusion (Fig.1.3) is *always the first solid feature created.* This can be the **base feature** or the first feature created after a base feature of datum planes.

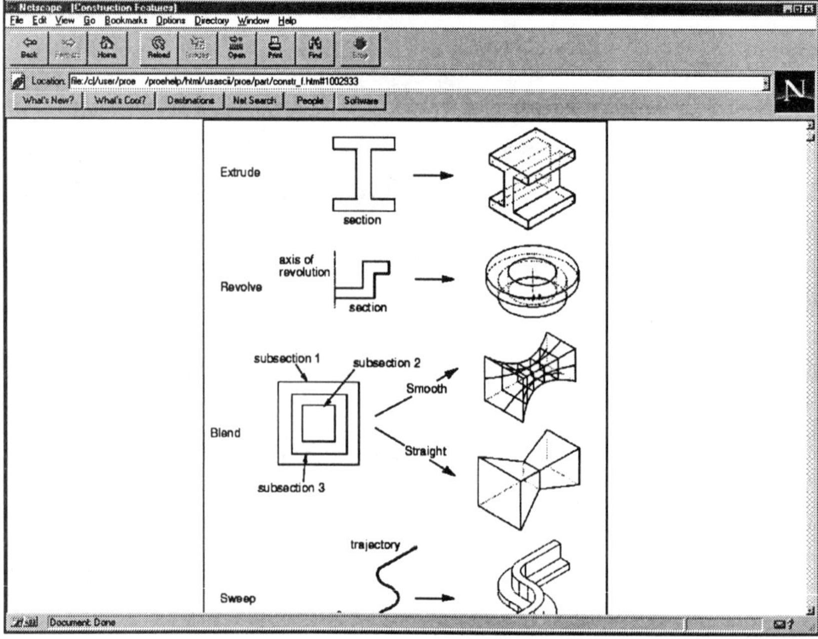

Figure 1.3
Online Documentation Protrusions

FEAT
Create
FEAT CLASS
Solid
Surface
Datum
Sheet Metal
Composite
Cosmetic
User Defined
DesignerIn
Done/Return
SOLID
Hole
Shaft
Round
Chamfer
Slot
Cut
Protrusion
Neck
Flange
Rib
Shell
Pipe
Tweak
Intersect

The first solid feature is always a **Protrusion**

To create an extruded protrusion:

1. Choose **Feature** from the PART menu, then **Create** from the FEAT menu.
2. Choose **Protrusion** from the SOLID menu.
3. Choose **Extrude** ⇒ **Solid** ⇒ **Done** ⇒ from the SOLID OPTS menu.
4. Pro/E displays the PROTRUSION: Extrude dialog box, which lists the elements needed for creating this type of protrusion.
5. Pro/E displays the ATTRIBUTES menu, which lists the following options:

   One Side Creates the feature on one side of the sketching plane.

   Both Sides Creates the feature on both sides of the sketching plane.
6. Choose **One Side** or **Both Sides** ⇒ **Done** from the ATTRIBUTES menu.
7. Select the sketching plane and the sketch orientation reference.
8. **Sketch** the protrusion.
9. Align (**Alignment**) and **Dimension** the section geometry.
10. **Regenerate** the section.
11. **Modify** and **Regenerate** the section sketch.
12. Specify the depth of the protrusion and choose **OK**.

Cuts (and Slots)

To remove material from a part, use one of the following features:

Cut Removes material from a specified side.

Slot Removes material within a closed section (Fig. 1.5).

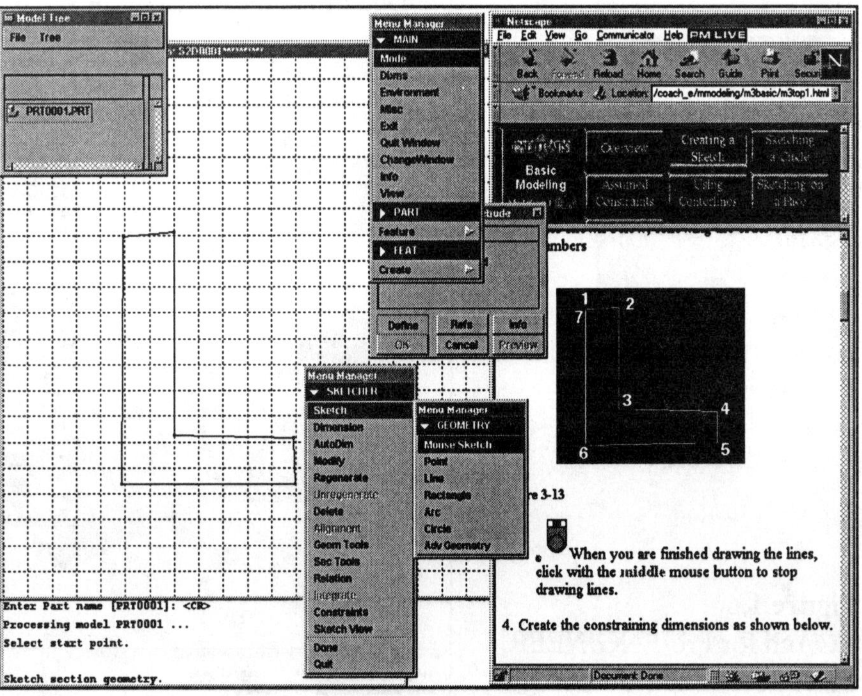

Figure 1.4
COAch for Pro/ENGINEER, Basic Modeling (Creating a Sketch)

Netscape - [Construction Features]

File Edit View Go Bookmarks Options Directory Window Help

Multiple Contour Protrusions

A protrusion with two outside loops

A protrusion with one outside loop and two inside loops

Slots and Cuts

To remove material from a part, use one of the following features:

- *Slot*-Remove material within a closed section.
- *Cut*-Remove material from a specified side. See Specifying the Side for more information.

For information on how to use these options and how to sketch sections for cuts and slots, see Creating Features.

How to create a cut or slot

1. Choose **Feature** from the PART menu, then **Create** from the FEAT menu.
2. Choose **Slot** or Cut from the SOLID menu.

Figure 1.5
Online Documentation for Slots and Cuts

To create a cut (Fig. 1.6) or slot, do the following:

1. Choose **Feature** from the PART menu, then **Create** from the FEAT menu.
2. Choose **Slot** or **Cut** from the SOLID menu.
3. Choose **Extrude** ⇒ **Solid** ⇒ **Done** ⇒ (SOLID OPTS menu).
4. The appropriate dialog box is displayed.
5. Choose **One Side** or **Both Sides** ⇒ **Done** (ATTRIBUTES menu).
6. Select the sketching plane on the part and the part's orientation.
7. Sketch, **Alignment**, **Dimension**, and **Regenerate** the section.
8. **Accept** the cut direction or flip the arrow (**Cut** option only).
9. Determine the depth of the cut or slot.

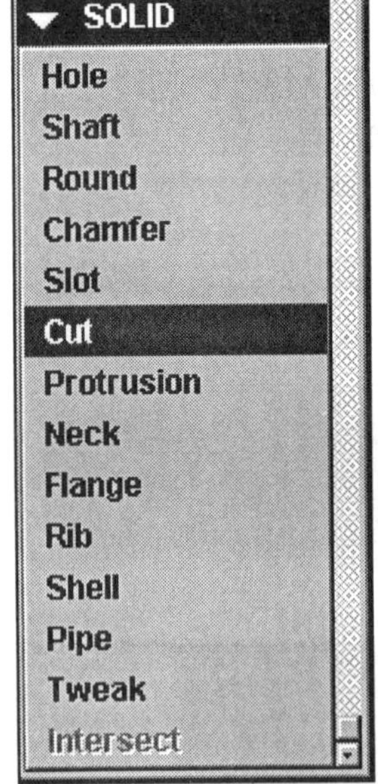

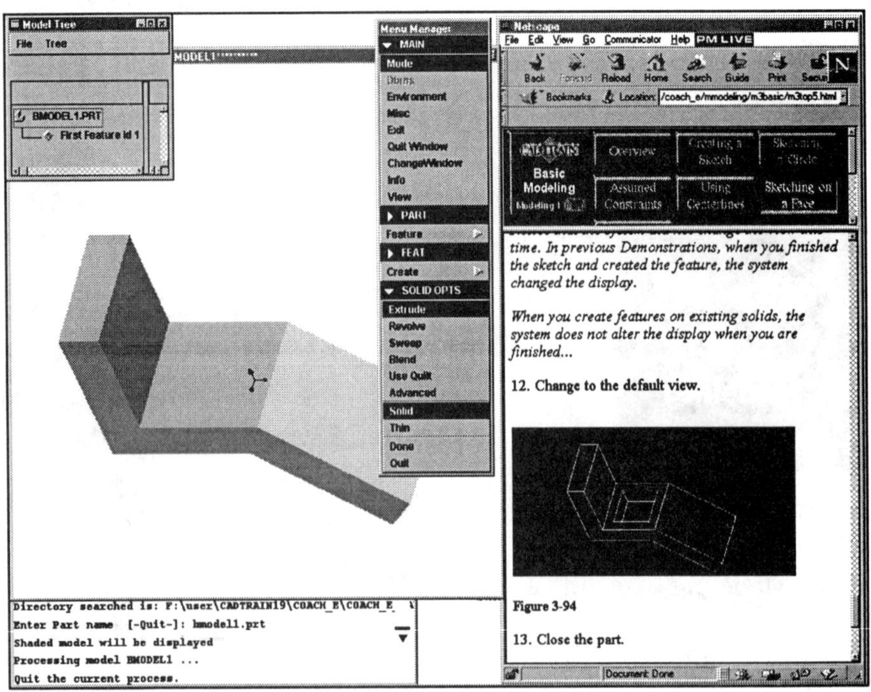

Figure 1.6
COAch for Pro/ENGINEER, Basic Modeling (Sketching on a Face)

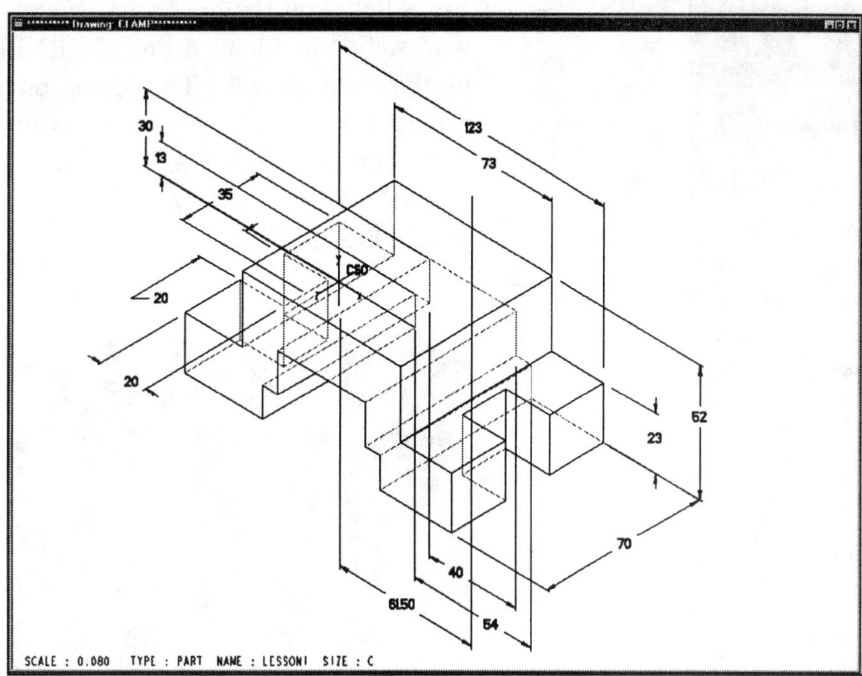

Figure 1.7
Clamp Dimensions

Clamp

The clamp (Fig. 1.7) will be our first ***lesson part***. It is composed of a simple protrusion and a cut. A number of things need to be established before you actually start modeling. These include setting up the *environment*, selecting the *units*, and establishing the *material* for the part.

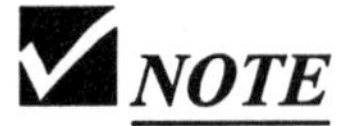

NOTE

PT/Modeler commands, where they differ from Pro/ENGINEER commands, are shown only in the step-by-step section of the lesson.

Before you begin any part using Pro/E, you must plan the design. The **design intent** will depend on a number of things that are out of your control and a number that you can establish. Asking yourself a few questions will clear up the design intent you will follow: Is the part a component of an assembly? If so, what surfaces or features are used to connect one part to another? Will geometric tolerancing be used on the part and assembly? What units are being used in the design, SI or decimal inch? What is the part's material? What is the primary part feature? How should I model the part? And what features are best used for the primary protrusion (the first solid mass)? On what datum plane should I sketch to model the first protrusion? These and many other questions will be answered as you follow the step-by-step lesson part. But you must answer many of the questions on your own when completing the ***lesson project***, which does not come with step-by-step instructions.

HINT

Before you start modeling with Pro/E, copy the **DIPS** from Appendix D to have them available for planning parts, feature sketches, assemblies, and drawings.

Using the appropriate ***Design Intent Planning Sheet*** for each project will increase your chance of having a part, assembly, or drawing with the appropriate design intent and project sequence. For the lesson part step-by-step designs and the lesson projects found in Lessons 1-13, you should uses **DIPS 1**, **2**, **7**, and **8**. Block out some trial feature sequences to establish the parent-child relationships required by your design intent using **DIPS 1** and **2**. Use **DIPS 7** and **8** for your feature geometry sketches and dimensioning scheme.

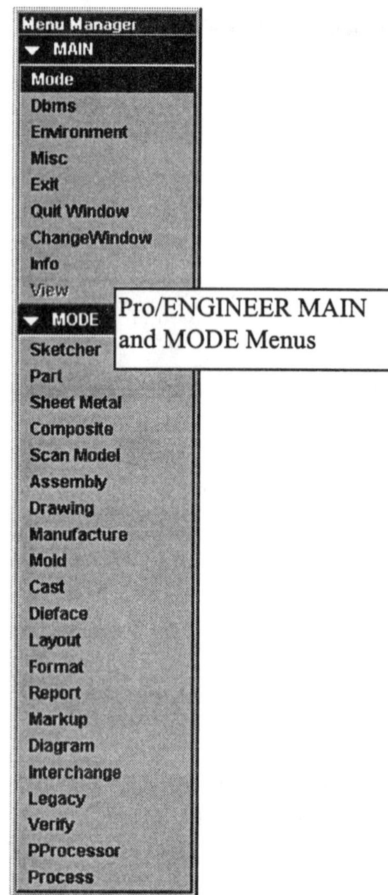

After you have started your UNIX, NT, or Windows 95 workstation and loaded Pro/ENGINEER (or PT/Modeler), you can set up the environment. The screen on your system will look similar to Figure 1.8. The MAIN menu is in the upper right-hand portion of your monitor. The MAIN menu has a number of choices available. We will be concerned with the **Environment** command first, to set up certain Pro/E defaults.

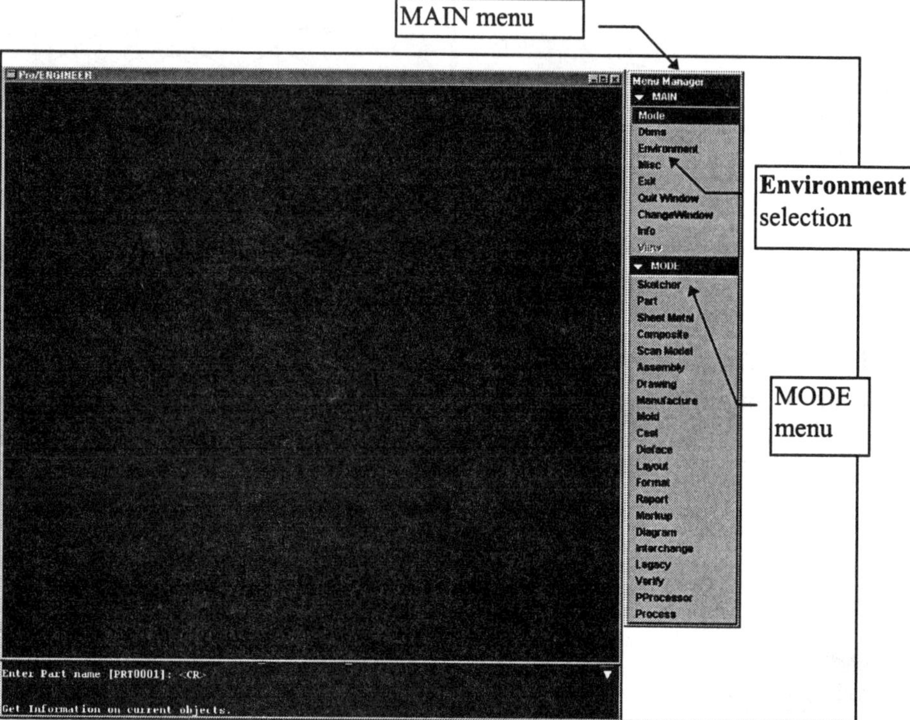

Figure 1.8
Starting a Pro/E Session

Your screen will have a *dark blue background* (by default). For most of this text, an alternate system color will be used for the background color. The text illustrations of screen captures show a white MAIN WINDOW color, not a solid black one as you see in Figure 1.8.

When you are starting a new part, assembly, or drawing, the first thing you should always do is check your environment settings. The following command block will appear at the beginning of each lesson in the text (you must do this without being prompted when starting the lesson projects). The command block will also prompt you to *give the* **Set Up** *command after the part mode, assembly mode, or drawing mode has been activated.* You can change the environment settings before or after **MODE** has been selected and at any time afterward during an active Pro/E session.

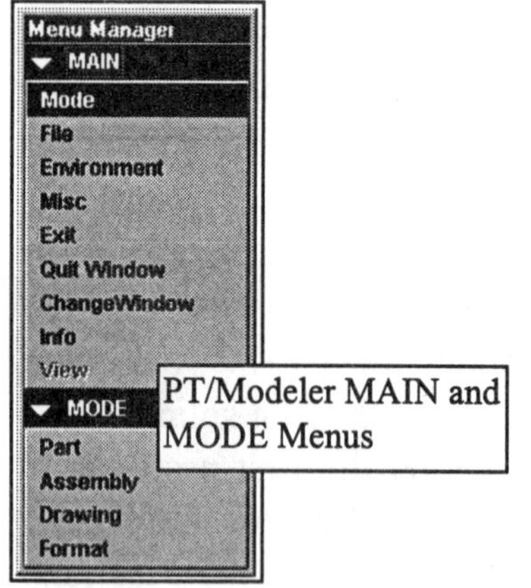

HINT

For setting material, choose **Material** ⇒ **Define** ⇒ (type **Steel** and then press **enter**) ⇒ *remember that the material properties window will appear on the screen and you must fill in the data, choose* **File** ⇒ **Exit** *and then* **Assign** ⇒ (pick **Steel**) ⇒ **Accept** ⇒ **Done**

ENVIRONMENT AND SETUP

Environment ⇒ **✓Grid Snap ✓ Model Tree Hidden Line Tan Dimmed**
Set Up ⇒ **Units** ⇒ **Millimeter** ⇒ **Done** ⇒ **Material** ⇒ **Define** ⇒ (type **Steel** and then press **enter**) ⇒ **Assign** ⇒ (pick **Steel**) ⇒ **Accept** ⇒ **Done**

NOTE

Always use **Set Up** to set your **Units** and **Material** at the beginning of every part.

If material files have not been generated, Pro/E will prompt you to enter the material properties into a file.

The first command you will give in this lesson will be to set the *environment* for the lesson part to be created. Using your mouse, highlight the **Environment** command. Choose it by clicking the left mouse button. This will bring up the ENVIRONMENT menu, as shown in Figure 1.9. To activate an environment setting, pick on the small box □ to the left of the selections. A ✓ means that the choice is activated. To deactivate a choice, pick on a box □ to remove the ✓. To set the model visibility, choose **Hidden line** and **Tan Dimmed**, as shown in Figure 1.9. Also, activate the **Grid Snap** if it is not already on (✓**Grid Snap**). When you are finished with your selections, pick **Done-Return** from the ENVIRONMENT menu.

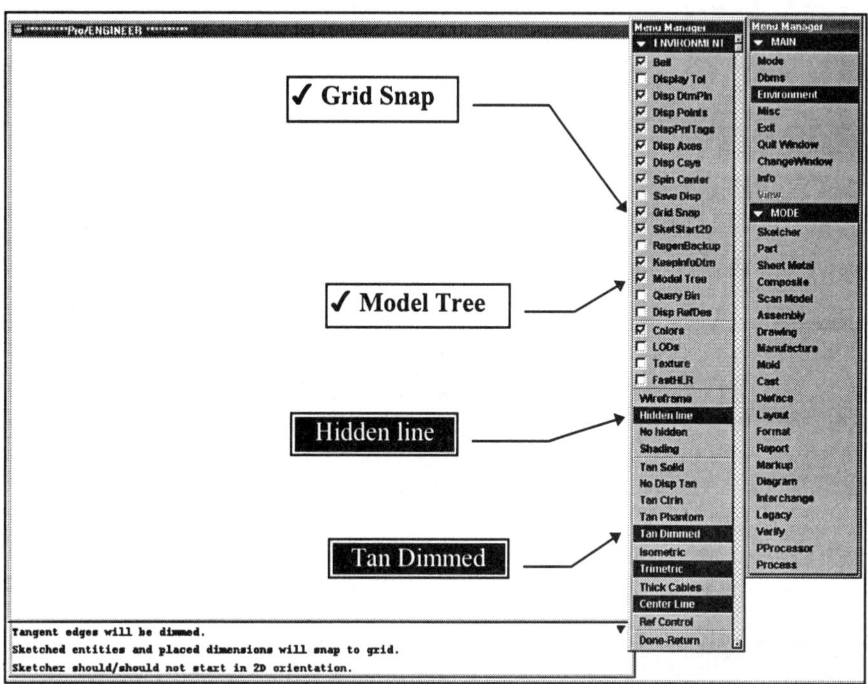

Figure 1.9
Setting the Environment

To start the actual part construction, you must choose a mode from the MODE menu. Since you will be creating a part, the **Part** mode will be selected (Lessons 1-13 and Lesson 21). In Lessons 14 and 15, the **Assembly** mode will be used, and in Lessons 16-20 the **Drawing** mode is accessed. Choose the following commands:

Part ⇒ **Create** ⇒ (type the part name at the command prompt provided in the lower left-hand of the screen) **CLAMP** ⇒ **enter** [press **enter** on your keyboard (Fig 1.10)]

The screen will show the **Model Tree** in the upper left side of the screen. You can move the Model Tree window to another spot, out of the way of the working area, the MAIN WINDOW of the screen. Using the mouse, pick on the upper bar of the Model Tree and hold down the left mouse button while dragging the window to a new screen location. Notice that the part name appears along the top of the MAIN WINDOW, as shown in Figure 1.10. The part name also appears in the Model Tree, which has been moved in Figure 1.11.

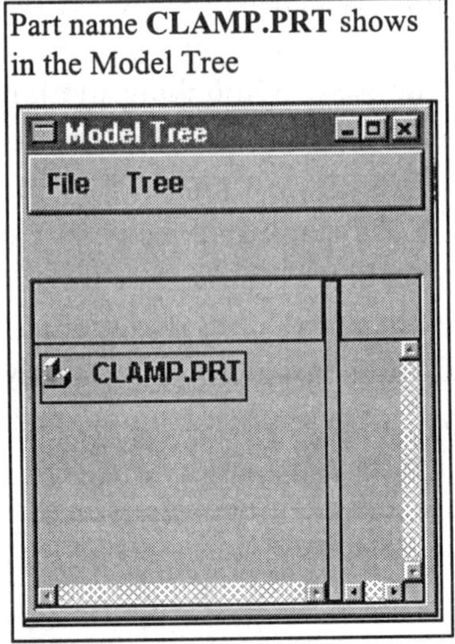

Figure 1.10
Part Name and Model Tree

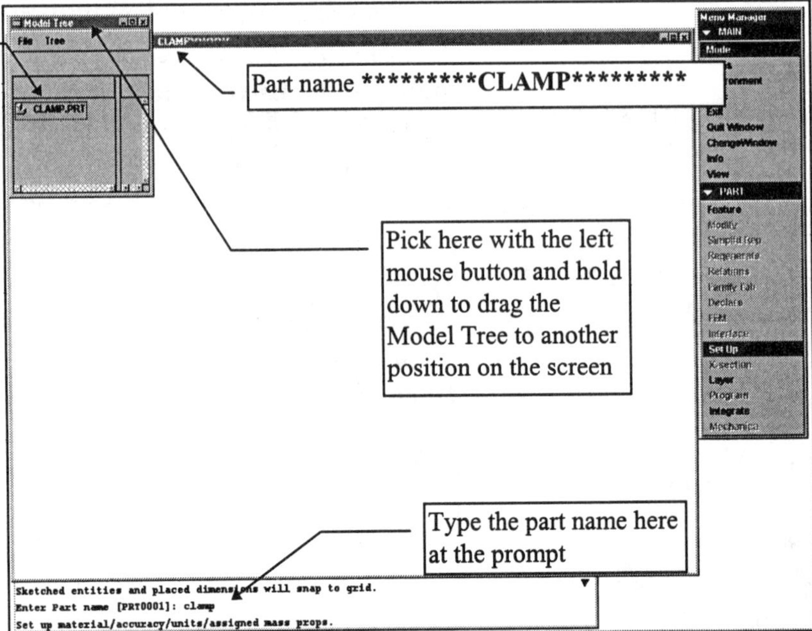

The next step is to set up the part. Normally, the text will prompt you to carry out this step using the **Environment** and **Setup** command box shown earlier in this lesson. It is repeated here to provide a more detailed understanding of the process. Choose the following commands:

Set Up ⇒ **Units** ⇒ **Millimeter** ⇒ **Done** ⇒ **Material** ⇒ **Define** ⇒ (Type **Steel** at the prompt and press **enter**. A material *Table* appears on the screen--enter the values) ⇒ **File** (from material table) ⇒ **Save** ⇒ **File** ⇒ **Exit** ⇒ **Assign** ⇒ (pick **Steel**) ⇒ **Accept** ⇒ **Done**

This will set the part's units as millimeters (Fig. 1.11) and the material as steel.

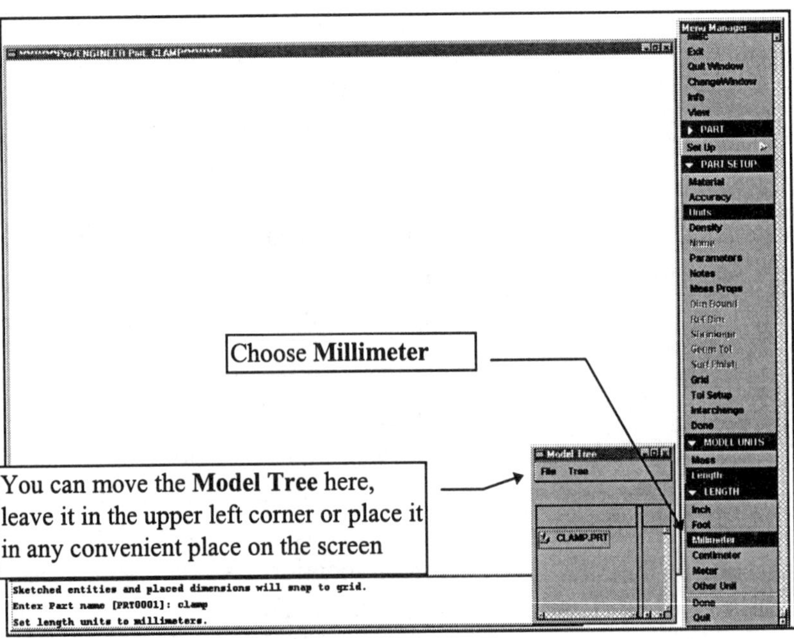

Figure 1.11
Setting Up the Part Units

The **Material** command brings up the Material Table using the default text editor, as shown in Figure 1.12. You can fill in the information if it is available from your instructor or just save and exit the table. **Assign** the material to the part after exiting the table.

NOTE
Your text editor may differ.

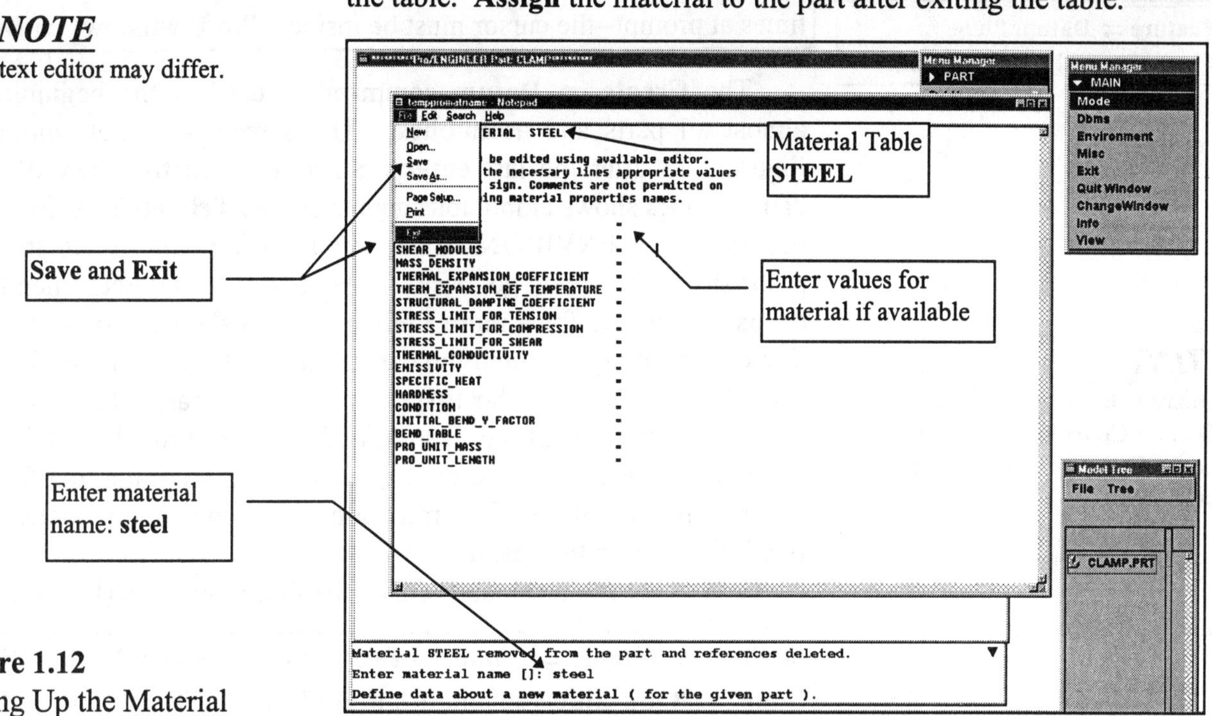

Figure 1.12
Setting Up the Material

At this point, it may seem that there is a lot to do before any modeling takes place. You are building not just a solid model, but a *complete integrated database* that controls all aspects of the design, including the part's material properties, units, geometry, and other important parameters that capture the design intent of the project. The next step is to create three default datum planes and a default coordinate system (Fig. 1.13).

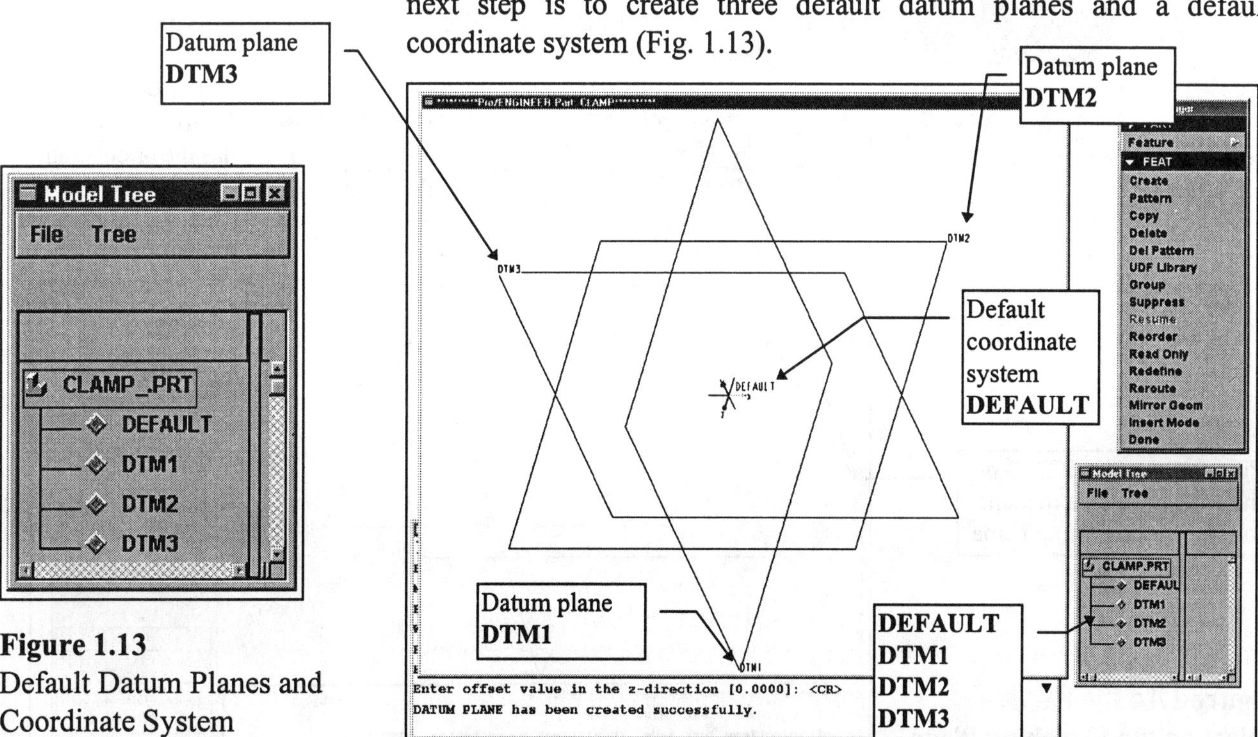

Figure 1.13
Default Datum Planes and Coordinate System

The datum planes will be used to sketch on and orient the part's section geometry. Choose the following commands:

Feature ⇒ Create ⇒ Datum ⇒ Plane ⇒ Offset ⇒ enter (three times at prompt--the cursor must be inside a Pro/E window) ⇒ **Done**

PT/Modeler™

Feature ⇒ Datum Plane ⇒ Coord Sys ⇒ Default ⇒ Done

The **Create ⇒ Datum** command is used at the beginning of almost all parts and assemblies. The screen will look similar to Figure 1.13 after the last **enter**. Since an isometric view of three datum planes shows ambiguous geometry, the **Trimetric** setting is the default in the ENVIRONMENT menu. After you create the first protrusion for a part, go back to the ENVIRONMENT menu and choose **Isometric** for a more lifelike view of the part (or assembly). The coordinate system and the datum planes show in the Model Tree (Fig. 1.14). They are the first features of the part. These features become the *parents* of any feature tied to them through sketching or alignment. It is always good to have datum planes as the first features of a part instead of the first protrusion. This gives the modeler more flexibility later in the design.

HINT

Default Datum Planes and the **Default Coordinate System** will be the first features on all parts *and* assemblies.

The following commands create the first protrusion (Fig. 1.14):

Feature ⇒ Create ⇒ Solid ⇒ Protrusion ⇒ Extrude ⇒ Solid ⇒ Done ⇒ One Side (default from ATTRIBUTES menu) ⇒ **Done** (pick **DTM3** as the sketching plane) ⇒ **Okay** (for selecting the direction of protrusion projection--*red arrow*)

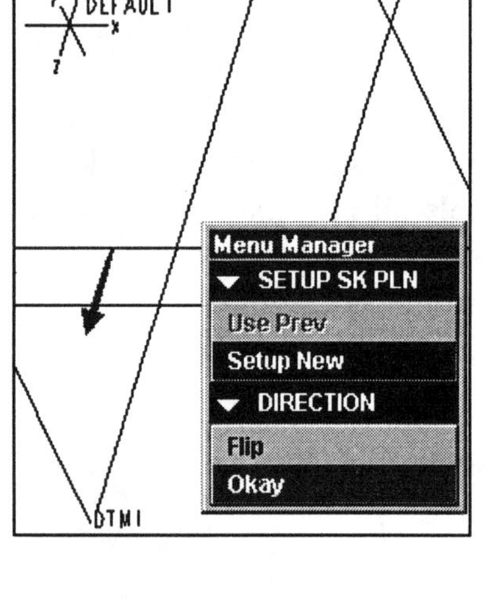

Pick on **DTM3** as shown in Figure 1.14. This is the plane on which you will be sketching the first protrusion's section geometry. The protrusion's feature dialog box appears on the screen at this time. For many of the actions you will be performing on this part, you will use the dialog box buttons.

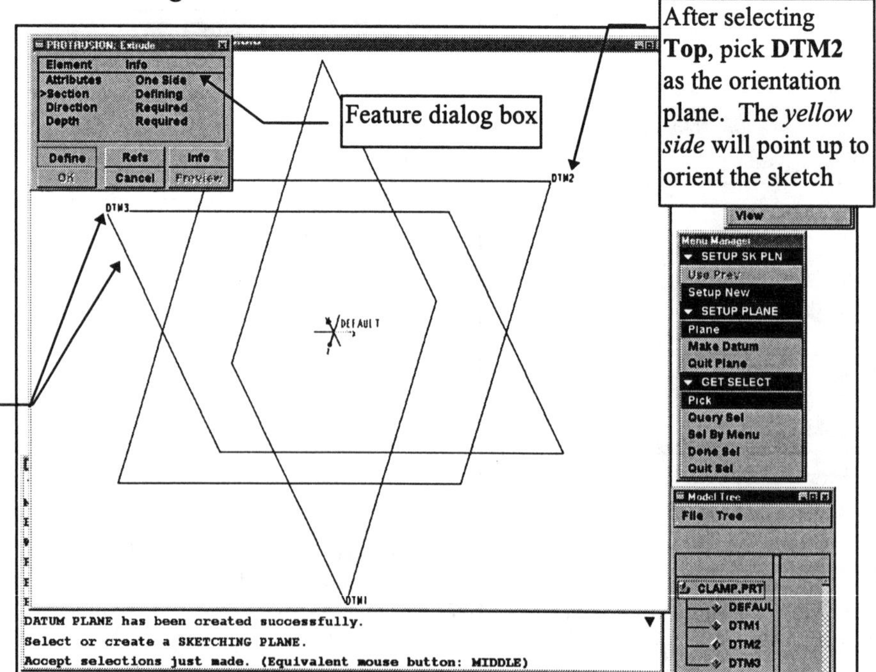

Pick on the name **DTM3** or the edge of the datum plane to select the sketching plane

Figure 1.14
Selecting the Sketching Plane

NOTE

Most sketches will be created on **DTM3** and aligned to **DTM1** and **DTM2**

You must still orient the sketch. Complete the process by choosing **Top** and picking **DTM2**. The sketch is now oriented as shown in Figure 1.15. The sketch is displayed in 2D, with the coordinate system at the middle of the sketch, where **DTM1** and **DTM2** intersect. The **X** coordinate arrow points to the right and the **Y** coordinate arrow points up. The **Z** arrow is coming toward you (out from the screen). The square box you see is the limited display of **DTM3**, which is like a piece of graph paper you will be sketching on when you create the protrusion's geometry. The coordinate system is used for a few commands and is required for Pro/MANUFACTURING. Pro/E is not a coordinate-based software, so you need not enter geometry with **X**, **Y**, and **Z** coordinates as with many other CAD software packages. In reality, you don't need to turn on the coordinate system, but many modelers prefer to see the comforting axes displayed on the screen using the right-hand rule (similar to those in other software they may have used in the past).

HINT

DTM3 (yellow side) is facing you and **DTM2** (*yellow side*) is facing the top of the screen (you are seeing it as an edge)

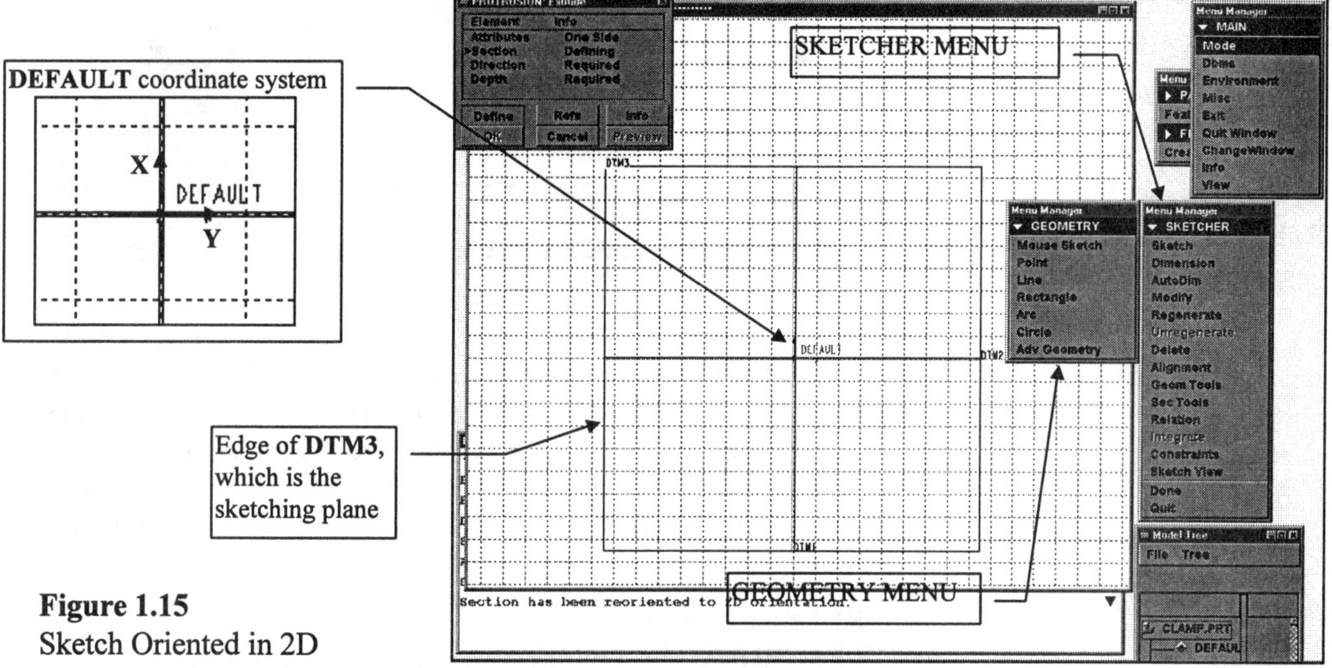

Figure 1.15
Sketch Oriented in 2D

You should reread Section 9, The Sketcher, at this time. It will help you understand some of the Sketcher's capabilities and what is required before a successful feature can be completed. In general, many of the part base protrusion features you will be modeling start with sketching in the first quadrant (Fig. 1.16). To make this more convenient and have a greater sketching area, we need to change the position and size of the sketch, as shown in Figure 1.16. To resize the sketch on the screen, place the mouse cursor in the center of the screen and hold down the control key (**Ctrl**) while simultaneously holding down the left mouse button; moving the mouse will enlarge or shrink (*zoom*) the display of the sketch.

Zooming the graphics in the Main Window with the mouse

Move the mouse up or down until the sizes of the screen and grid are correct. While you press **Ctrl**, the right mouse button *pans* the sketch across the screen and the middle button *rotates* the sketch in 3D. You can sketch in 2D or 3D with Pro/E; for now, let's stick with 2D sketching. If you rotate your sketch and wish to return to the 2D orientation, pick **Sketch View** from the SKETCHER menu.

HINT

If you double-click with the left mouse button (with the **Ctrl** key depressed), you start a zoom box. Move the cursor to the desired end location of the zoom box and pick once with the left mouse button.

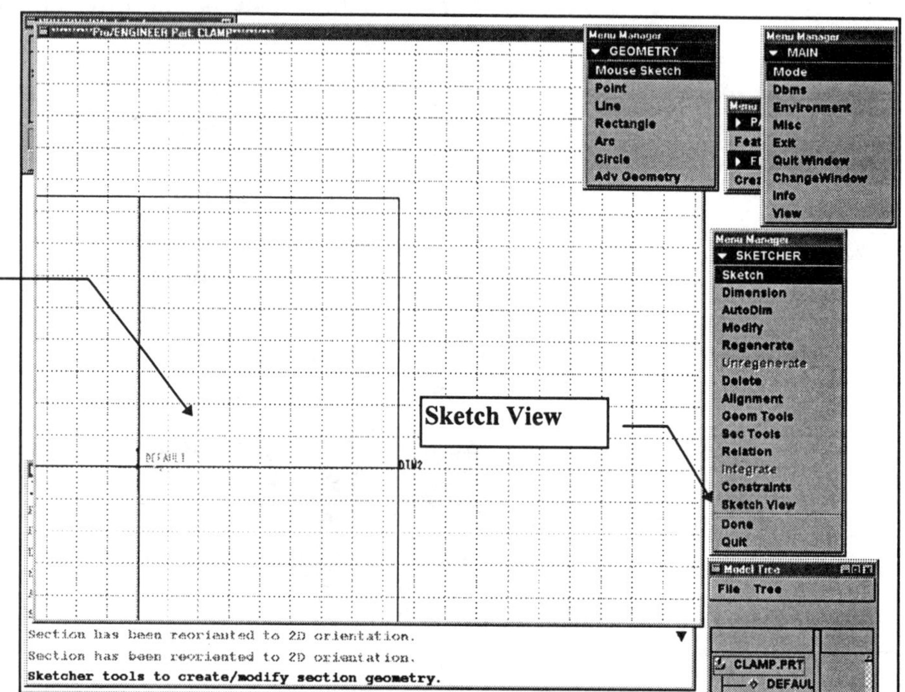

Figure 1.16
2D Sketch

Since you turned on the **Grid Snap** in the ENVIRONMENT menu, you can now sketch by simply picking grid points representing the part's geometry (outline). Since this is a sketch in the true sense of the word, you need only create geometry that *approximates* the shape of the feature; the sketch does not have to be accurate as far as size or dimensions are concerned. Pro/E constrains the geometry according to rules. In Section 9, the rules for the **constraints** are listed and include the following:

? Pro/HELP

Highlight the **Constraints** command and press the right mouse button

RULE: Symmetry
DESCRIPTION: Entities sketched symmetrically about a centerline are assigned equal values with respect to the centerline.
RULE: Horizontal and vertical lines
DESCRIPTION: Lines that are approximately horizontal or vertical are considered to be exactly horizontal or vertical.
RULE: Parallel and perpendicular lines
DESCRIPTION: Lines that are sketched approximately parallel or perpendicular are considered to be exactly parallel or perpendicular.
RULE: Tangency
DESCRIPTION: Entities sketched approximately tangent to arcs or circles are assumed to be exactly tangent.

READ THIS!

Do not draw lines or other entities on top of each other. Pro/E will not regenerate the sketch, since it will think you have two sections, not one. It is better to **Delete ⇒ Delete All** and start the sketch again. Later you will feel more comfortable troubleshooting a failed sketch.

The **Mouse Sketch** default is active in the GEOMETRY menu, so we can begin the sketching process using the mouse buttons without selecting another command. The left button on the mouse creates lines; the middle button, circles; and the right mouse button, arcs that are tangent to the end of a line/arc.

The outline of the part's primary feature is sketched using a set of connected lines. The part's dimensions and general shape are provided in Figure 1.7. Since much of the part can be defined by sketching an outline similar to its front view, you can complete most of the part's geometry with one protrusion. The cuts on the sides will be the second feature created. The part will have its base (bottom horizontal edge) aligned with the edge of **DTM2**, its left edge aligned with **DTM1**, and, since you are sketching on **DTM3**, its back face aligned with that datum. If you are at all confused about this, reread Sections 3 and 9. It is important not to create any unintended constraints while sketching. Therefore, remember to exaggerate the sketch geometry and not to align edges that have no relationship. Pro/E is very smart: If you draw two lines at the same horizontal level, Pro/E thinks they are horizontally aligned even if you later dimension them differently! Figure 1.17 shows how to sketch vertical lines that *are not in line with each other.*

Place the mouse near the center of the coordinate system at the intersection of the **DTM1** and **DTM2** datum planes and click the left button. Continue picking until you have sketched an outline (Fig. 1.17) approximating the primary feature of the clamp part shown in Figures 1.2 and 1.7.

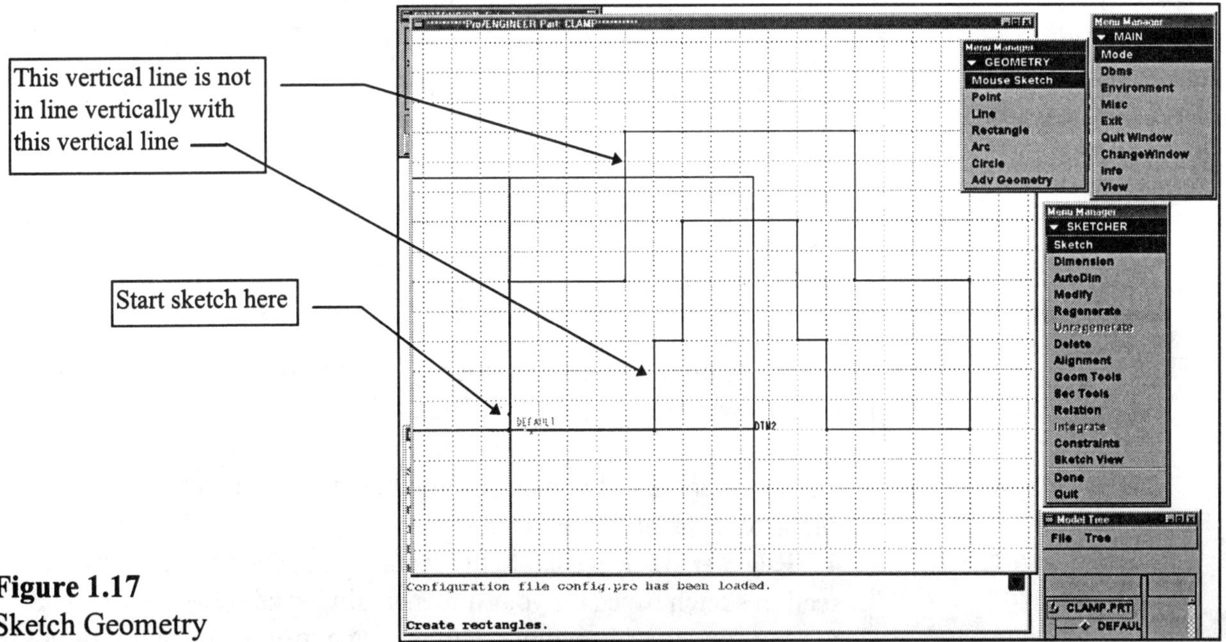

Figure 1.17
Sketch Geometry

After completing the sketch outline (section), select **Regenerate** from the SKETCHER menu. Notice that the endpoints of section entities are highlighted with the aid of small colored dot symbols (•).

Sketcher constraints can be turned on or off (enabled or disabled) while you are sketching. A Sketcher *constraint symbol* appears next to the entity that is controlled by that constraint. An **H** next to a line means horizontal, a **T** means tangent. To disable a constraint, pick **Disable** from the CONSTRAINTS menu and then pick a symbol on the screen. To obtain a brief explanation, pick **Explain** from the CONSTRAINTS menu and then pick a symbol on the screen. If the constraints are not displayed on the sketch, select **Constraints** from the SKETCHER menu and check **✓Display** in the CONSTRAINTS menu (Fig. 1.18). The constraints will appear on the section. They will also be displayed when the sketch is oriented in 3D.

HINT

Choose **View** ⇒ **Default** ⇒ **Done-Return** to orient the sketch in 3D

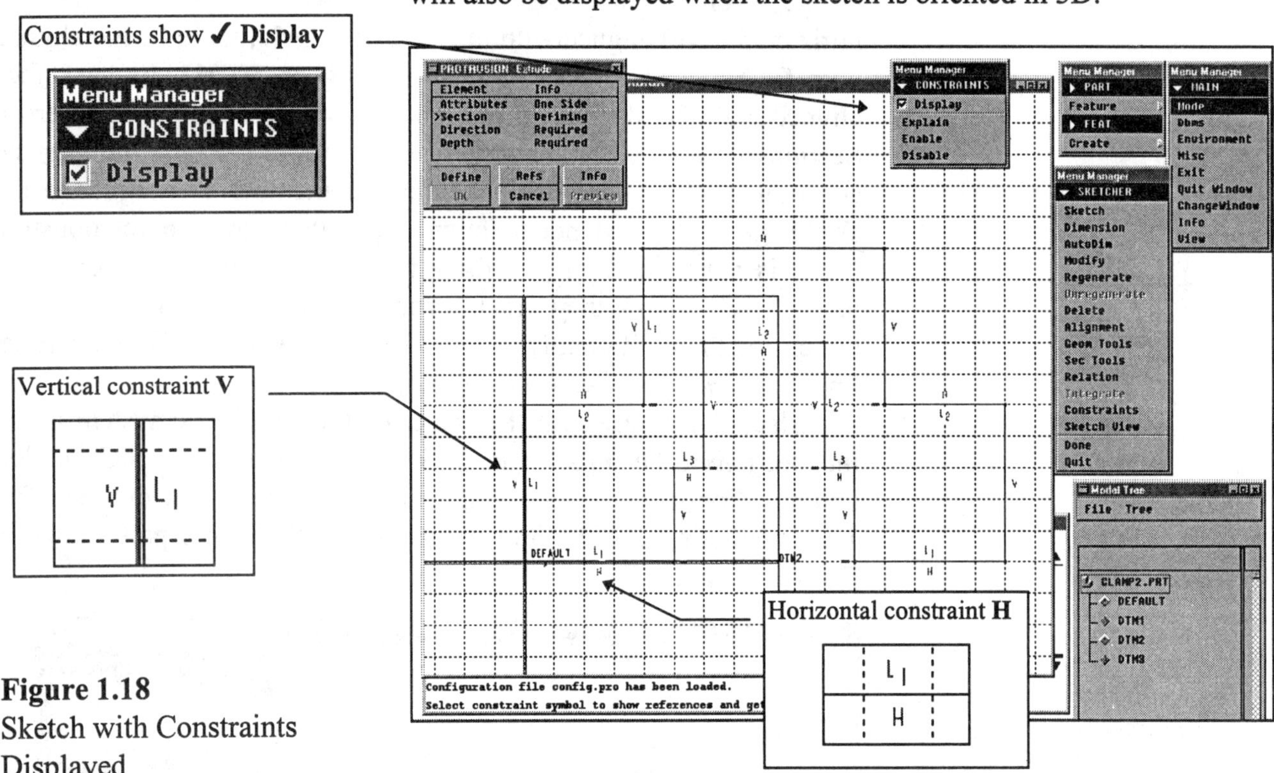

Figure 1.18
Sketch with Constraints Displayed

In general, the following six steps are used when sketching a section:

1. **Sketch** Sketch the section geometry. Use Sketcher tools to create the section geometry.
2. **Alignment** Align the section geometry to a datum feature or to an existing part feature (if applicable).
3. **Dimension** Dimension the section. Use a dimensioning scheme that you want to see in a drawing. Dimension to control the characteristics of the section.
4. **Regenerate** Regenerate the section. Regeneration solves the section sketch based on your dimensioning scheme.
5. ***Relations*** Add section relations, to control the behavior of your section. *This step depends on the feature.*
6. **Modify** Modify the dimension values. Change the dimension values to reflect the design intent. **Regenerate** again after modifying.

HINT

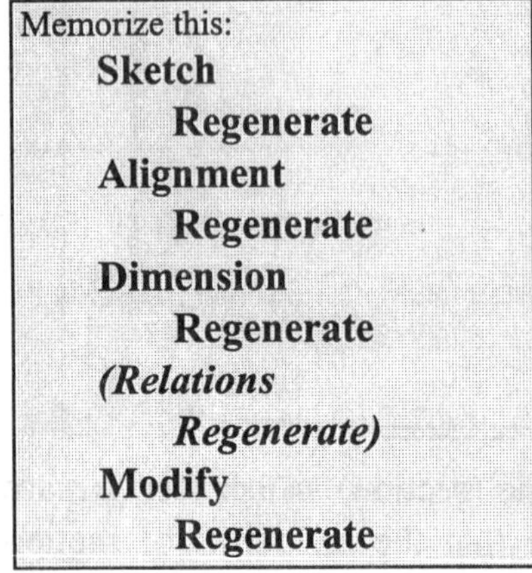

Next, add a centerline to the sketched section:

Sketch ⇒ **Line** ⇒ **Centerline** ⇒ **Vertical** (Pick on a vertical grid line running through the center of the sketch) ⇒ **Regenerate**

NOTE
You can change the size and type of grid in the Sketcher. You will learn how to do this in Lesson 2.

This is where your sketching skills come in handy. Did you sketch the section so it is symmetrical? Did you leave an even number of grid squares on each side of the section's center? It's not mandatory that you do so, but when you are learning Pro/E, it helps to simplify the sketch and pay attention to details, since you do not have the necessary skills, at this time, to get the sketch to regenerate easily.

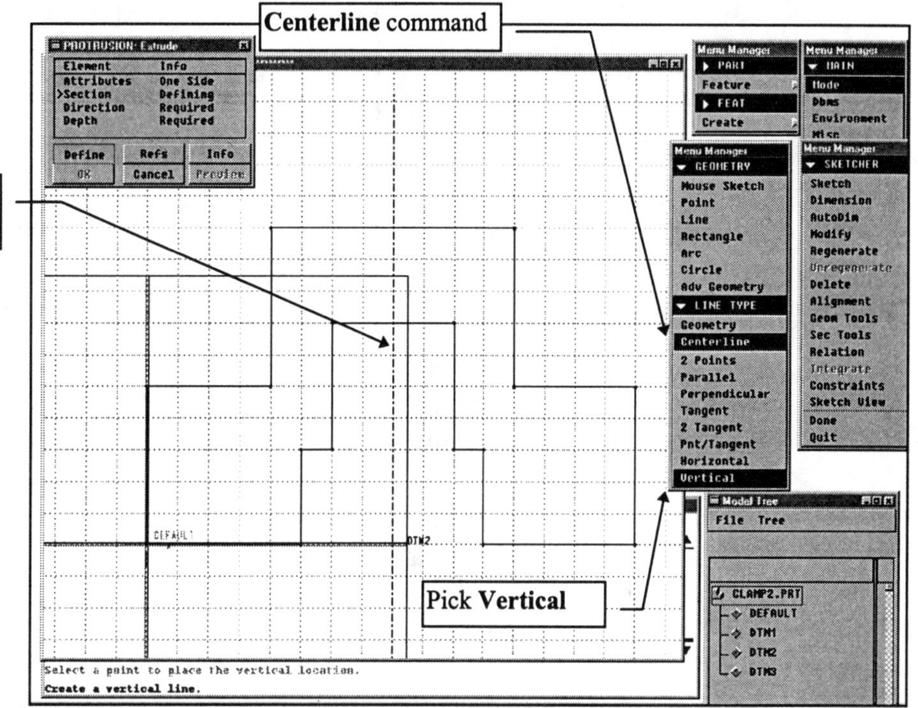

Figure 1.19
Sketching a Centerline

It is now time to *align* the sketch. This simple procedure sometimes confuses the beginner. Even though you have sketched the base and left edge against **DTM2** and **DTM1**, you still need to tell Pro/E what part of the sketch will be aligned with **DTM1** and **DTM2.** Choose **Alignment** from the SKETCHER menu and pick twice with the left mouse button on **DTM2** and the horizontal line of the sketch (you need not move the cursor between the picks), as shown in Figure 1.20. Do the same with **DTM1** and the left edge vertical line of the sketch. The command line will say **ALIGNED**. Remember, since the line and edge of **DTM1** are coincident, you need only click twice with the left mouse button on exactly the same spot. The alignment of the sketch tells Pro/E that the bottom edge of the part will be on **DTM2** and the left side of the part will be against **DTM1**.

In a later lesson, you will find out how to set the datum planes according to **ASME Y14.5M 1994** geometric tolerance standards, as *basic* datums, and rename the datum planes to **A, B,** and **C**.

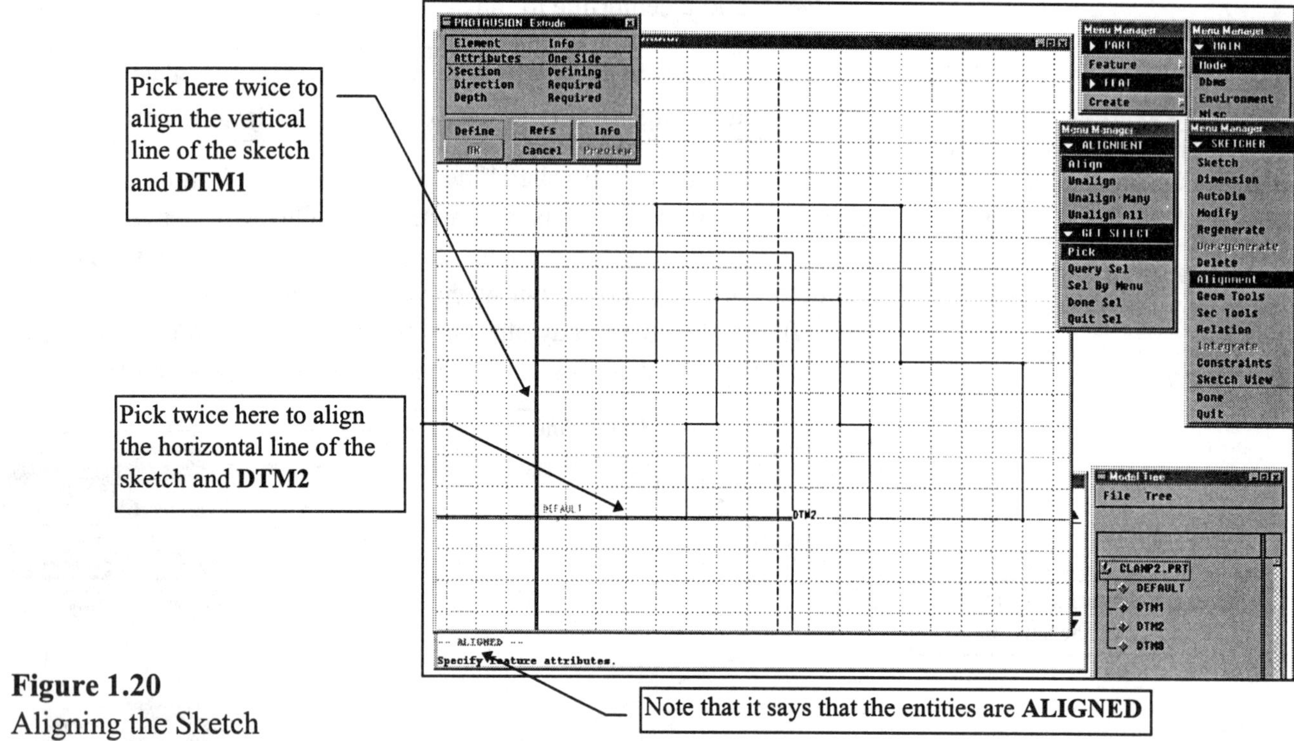

Figure 1.20
Aligning the Sketch

Look at the part again in Figures 1.1, 1.2, and 1.7. Datum **DTM2** could be datum **A**, **DTM3** datum **B**, and **DTM1** datum **C**, if geometric tolerancing was used on the part.

The sketch now needs to be dimensioned according to the design intent of the part. Look at Figures 1.7 and 1.21 to see the dimensioning scheme used to create the part's geometry. *Pro/E can automatically dimension a sketch, but the design intent may not be what you want.*

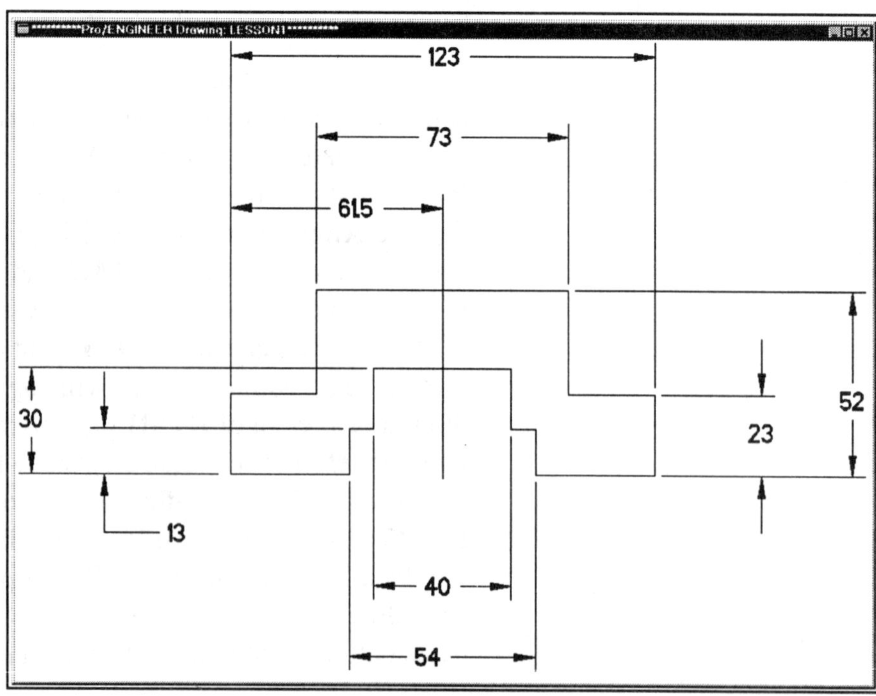

Figure 1.21
Part Dimensions

Place the dimensions as shown in Figure 1.22. Do not be concerned with the perfect positioning of the dimensions, but try to follow the spacing and positioning standards found in the **ASME Geometric Tolerancing and Dimensioning** standards. It will save you time when you create a drawing of the part.

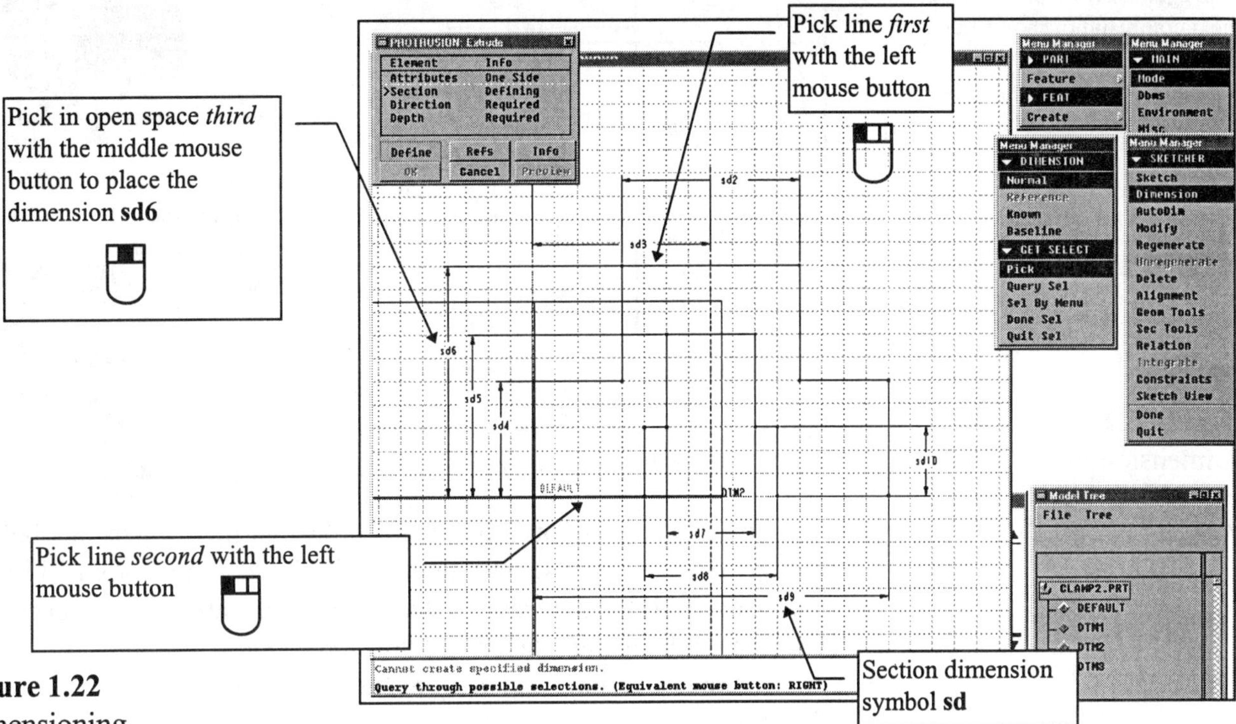

Figure 1.22
Dimensioning

Dimensions placed at this stage of the design process are displayed on the drawing document by simply asking to show all the dimensions. The most important thing is to get enough dimensions on the sketch to regenerate. The dimensions should be ones required to manufacture the part. To dimension between two lines, simply pick the lines with the left mouse button and place the dimension value with the middle button. At this stage of the dimensioning process, the dimensions are displayed as **sd** symbols: **sd1**, **sd2, sd3**, etc.

HINT

If you pick too many lines or choose an incorrect sequence of entities, simply pick the **Dimension** command to restart the dimensioning process.

It is now time to see if the sketch and dimensions will regenerate. Choose **Regenerate** from the SKETCHER menu (Fig. 1.23). The command line says **Section Regenerated Successfully**. The **Modify** command is automatically initiated at the successful regeneration of the section. You can now change/modify the dimensions to the *design sizes*. Note that the regenerated dimensions are based on a grid size of **30** units, here **30** millimeters. All the dimensions are too big. Pick on each dimension individually and type the correct value at the prompt using the correct dimension from Figures 1.7 and 1.21. Make sure you change every dimension correctly. If you changed all but the center cut dimension and left it as a value greater than the part's width, the section will fail at the next regeneration. If this happens, immediately pick **Unregenerate** from the SKETCHER menu and correct your neglected dimension(s).

HINT

If the section fails, always **Unregenerate** *before* going to the next command or modifying dimensions. Modify only real dimensions, not dimensions displayed with **sd** values.

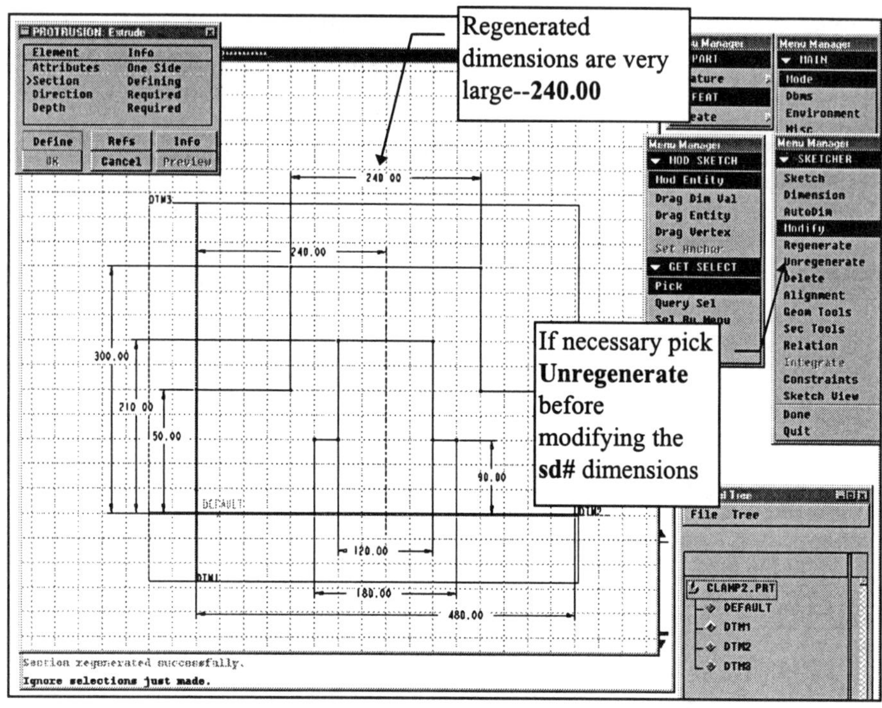

Figure 1.23
Dimensioning

For a failed section, do not try to change the dimensions before unregenerating, because initiating certain Pro/E commands will block the use of **Unregenerate**. The **Unregenerate** choice becomes *dimmed* on the menu when it is unavailable (Fig. 1.24).

Make corrections and **Regenerate** the sketch. The correct sizes for the features are now displayed according to the design intent dimensions, and the section is complete. Turn off the constraint display to see the sketch more clearly (Fig. 1.24).

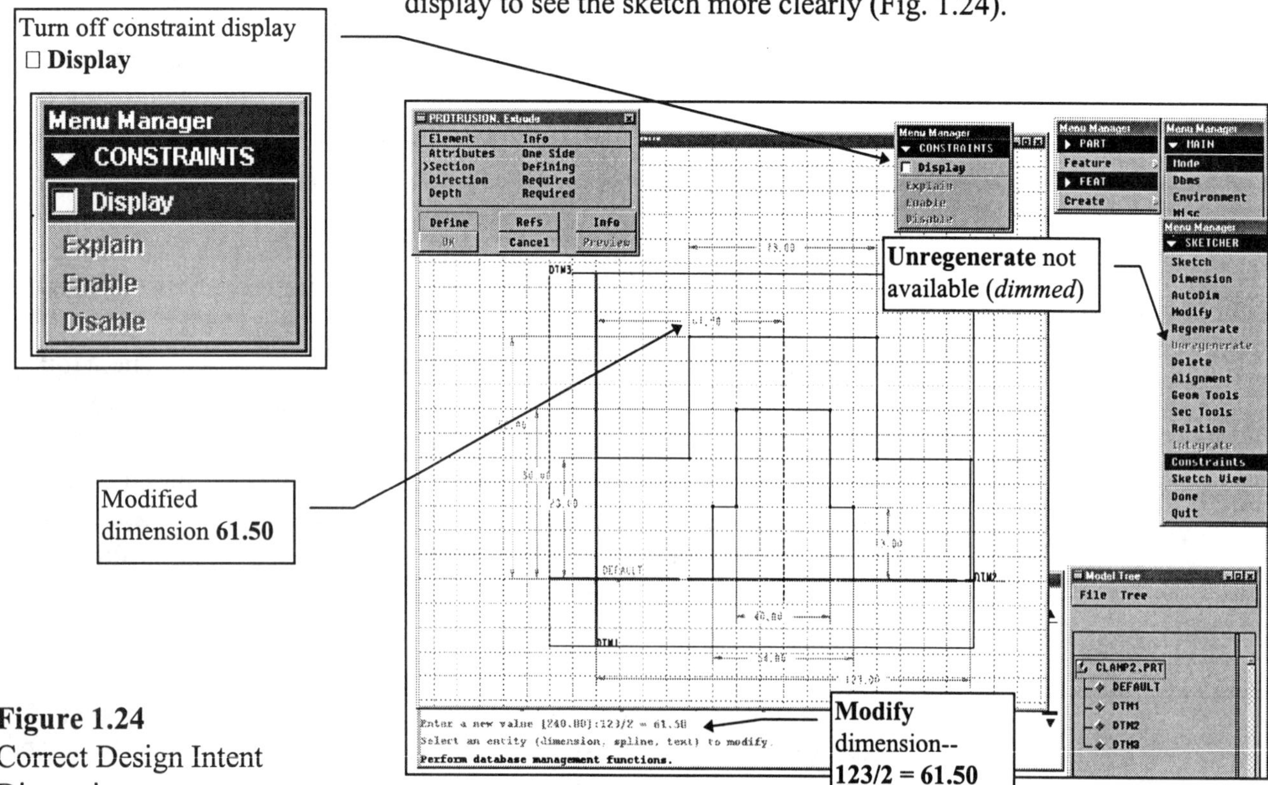

Figure 1.24
Correct Design Intent Dimensions

Complete the protrusion by choosing **Done** (in the Sketcher) ⇒ **View** (from the MAIN menu) ⇒ **Default** ⇒ **Done-Return** (from the ORIENTATION menu) to see the sketch in 3D trimetric. Pro/E now prompts for the depth in the *direction* of protrusion creation and displays an arrow ➔ pointing in the direction previously set when **DTM3** was selected as the sketch plane (Fig. 1.25).

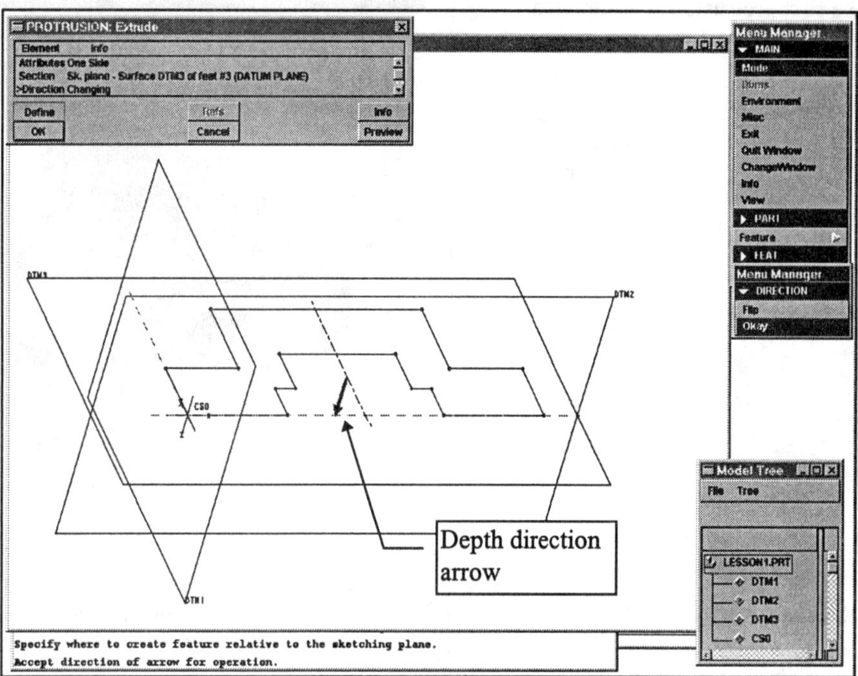

Figure 1.25
3D View of Sketch Showing the Arrow Direction

Choose **Blind** (from the SPEC TO menu) ⇒ **Done**, and at the prompt, type the depth dimension of the part (**70**). Pick **Preview** from the dialog box to see the feature (Fig. 1.26).

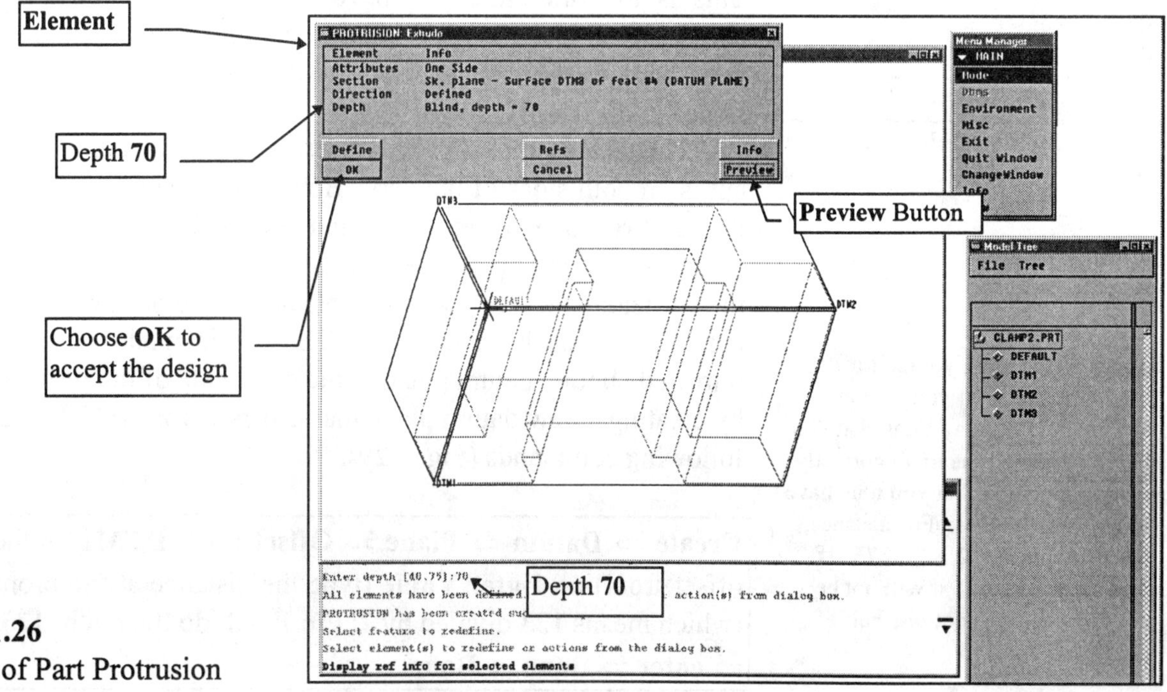

Figure 1.26
Preview of Part Protrusion

HINT

Use **Trimetric** before the first protrusion is created, and use **Isometric** as the view default after the first protrusion is completed.

At this point, you could pick one of the elements from the dialog box and **Define** to change anything completed up to this point. Instead, pick **OK** from the dialog box. To see the model in a shaded state, choose **Environment** and turn off the spin center, datum planes, and datum axis. Make **Isometric** and **Shading** defaults, then pick **Done-Return**. The part will be displayed as in Figure 1.27.

Shaded protrusion

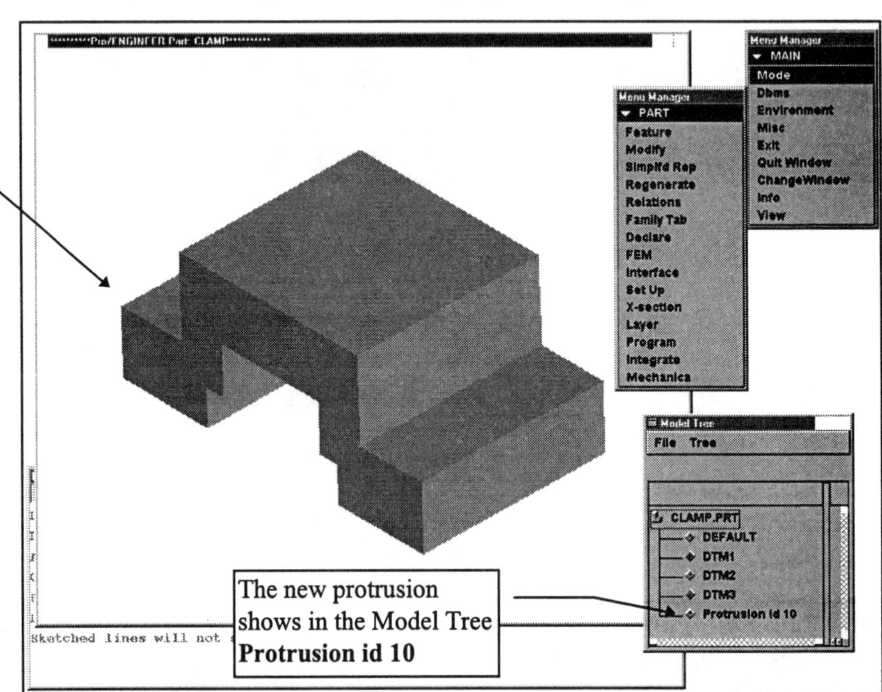

Figure 1.27
Shaded Part in Isometric View

Dbms (**File**--PT/Modeler) ⇒ **Save** ⇒ **enter** **Purge** ⇒ **enter** ⇒ **Done-Return**

PT/Modeler™

PT/Modeler uses **File** in place of **Dbms**

Choose **Dbms** (from the MAIN menu) ⇒ **Save** ⇒ **enter** ⇒ **Purge** ⇒ **enter** ⇒ **Done-Return**. This will save the part at its present stage. Always use **Purge** after the **Save** command to eliminate previous versions of the part that may have been saved. This is the first time you have saved this part, so **Purge** is not necessary. But it is a good habit to get into so that you do not fill up your hard drive with versions of the same part saved at different stages in the design process.

The next feature will be a cut (Fig. 1.28). The **20 X 20** centered cut is on both sides of the part. Before we create this feature, change the **Environment** back to **Hidden line** and turn on the settings for datum planes, datum axes, etc. Leave the view set as isometric. Since the cut feature is identical on both sides of the part, we can mirror the cut after it has been created. The cut is made on one side and mirrored about a datum plane to the other side of the protrusion. Start by creating a new datum plane that is offset from **DTM1**. Choose the following commands (Fig. 1.29):

HINT

Never do the math yourself if Pro/E can complete it for you. Sometimes you may think that you have completed the math correctly without noticing that you may have rounded the values. For instance, **5.255** divided by **2** is **2.6275**. If you round to **2.628** it will not be correct since it is not one half of **5.255**.

Create ⇒ **Datum** ⇒ **Plane** ⇒ **Offset** (pick **DTM1** as the plane to offset from) ⇒ **Enter Value** (type the distance at the prompt) **123/2** (which means **123** divided by **2**; Pro/E will do the math: **123/2 = 61.5**) ⇒ **enter** ⇒ **Done** ⇒ **Done**

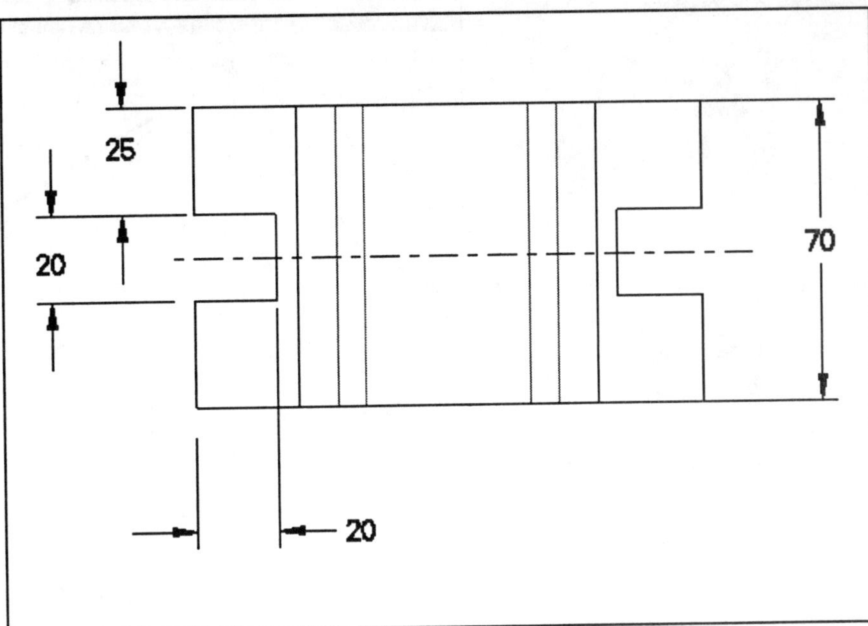

Figure 1.28
Cut Dimensions

The new datum plane will be offset from **DTM1**

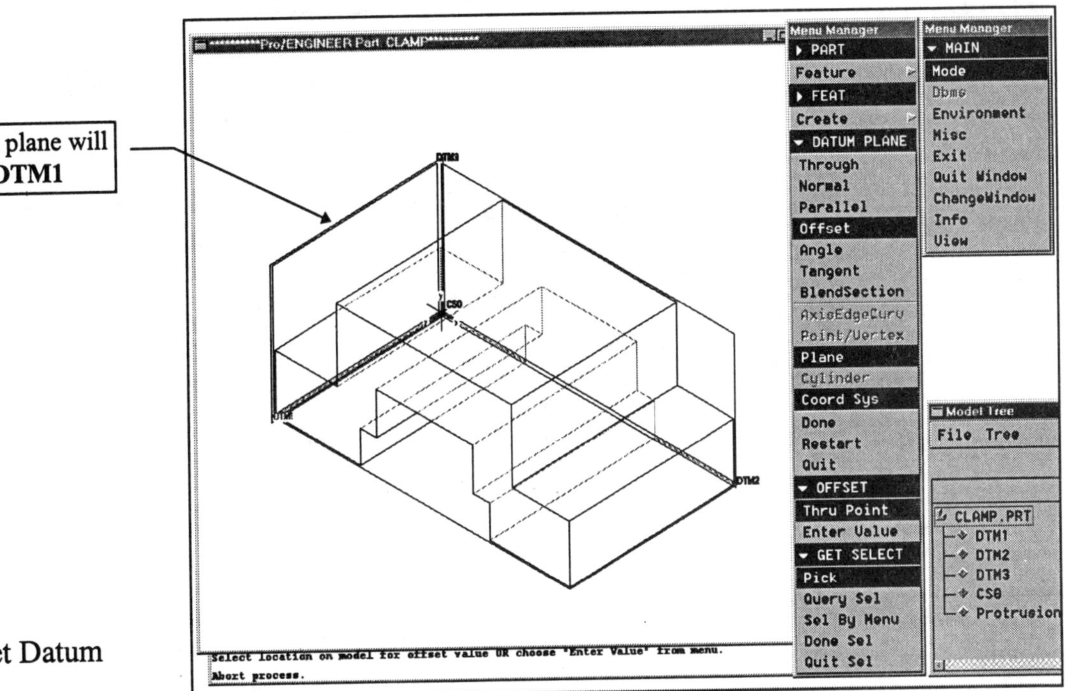

Figure 1.29
Creating an Offset Datum Plane

The datum plane is offset from **DTM1,** which is its *parent feature*. The new datum plane is shown on the screen passing through the center of the part and also appears in the Model Tree (Fig. 1.30). **DTM4** can now be used to construct other features as required by the design.

The cut will be a sketched feature similar to the first protrusion, except that it will remove material. Choose the following commands:

PT/Modeler™
Feature ⇒ Cut ⇒ etc.

Feature ⇒ Create ⇒ Cut ⇒ Extrude (default) ⇒ **Solid** (default) ⇒ **Done ⇒ One side** (default) ⇒ **Done** (pick **DTM2** as the sketching plane) ⇒ **Flip ⇒ Okay ⇒ Top** (pick **DTM3** to orient the sketch)

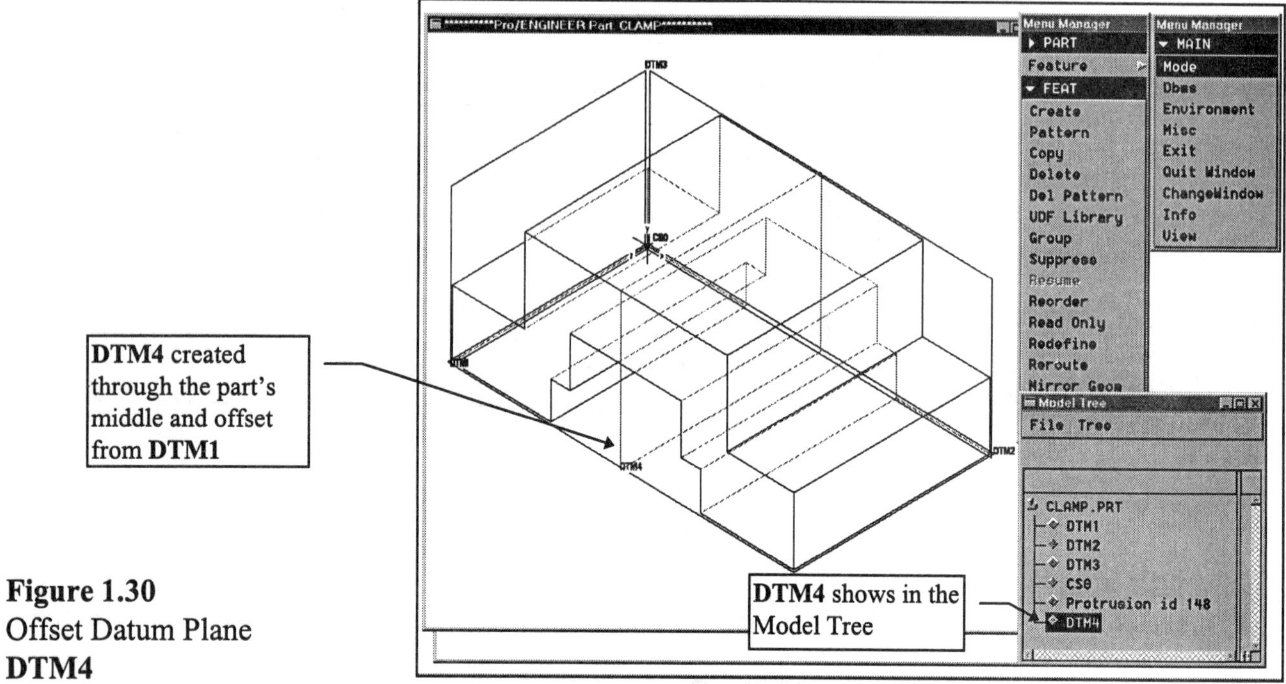

Figure 1.30
Offset Datum Plane
DTM4

In Figure 1.31, you are being prompted for the **Direction of Feature Creation**. Flipping the arrow changes the direction of the feature creation so that it will cut the part protrusion and not *air!*

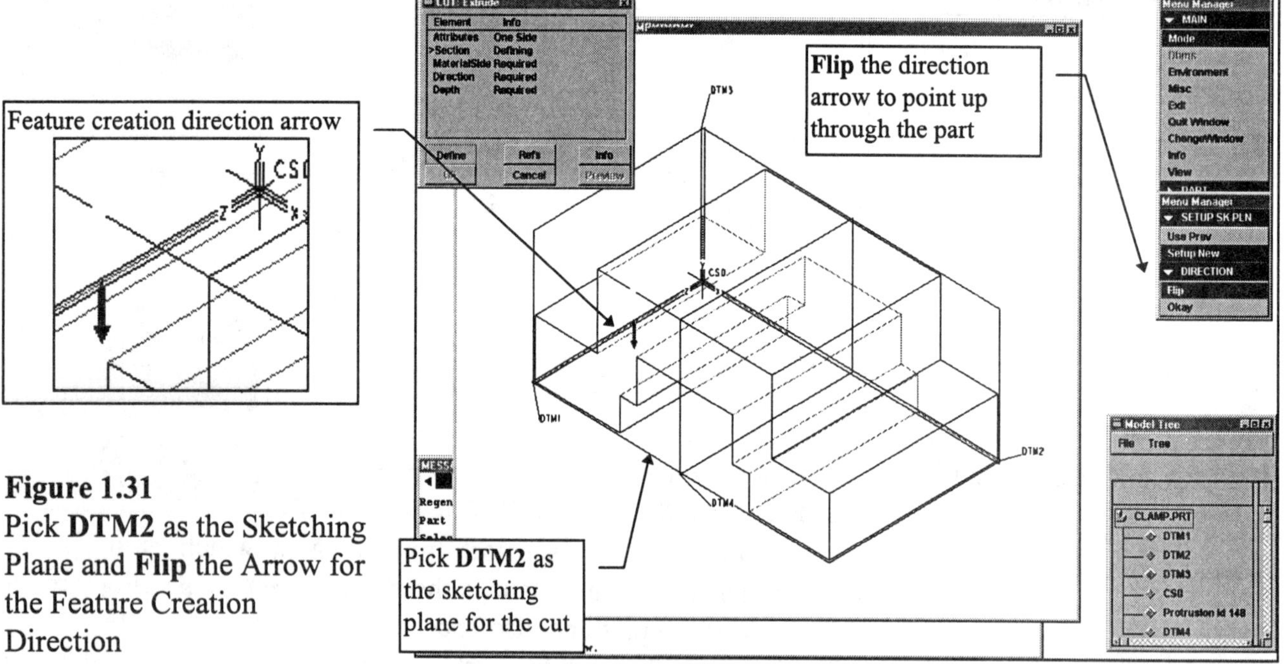

Figure 1.31
Pick **DTM2** as the Sketching Plane and **Flip** the Arrow for the Feature Creation Direction

In Figure 1.32, you are looking at the *bottom of the part*. At first, it is hard to see the orientation of the part. To get visually oriented, note the location of the coordinate system. **DTM1** and **DTM3** show as edges and will be used to dimension the sketch. The sketch is composed of three lines forming an *open section*. The endpoints of the two lines that touch **DTM1** will be aligned with that datum plane. The dimensioning scheme and values will be the same as those shown in Figure 1.28. The edge of the cut is dimensioned from **DTM3**.

Turn the **Grid Snap** off in the ENVIRONMENT menu, and continue the commands (Fig. 1.32):

Environment ⇒ ☐ **Grid Snap** ⇒ **Done-Return** ⇒ **Sketch** ⇒ **Line** ⇒ **Horizontal** (pick on **DTM1** to start the line)

Complete the three continuous lines by picking four endpoints, as shown in Figure 1.32. After the second pick, the **Sketcher** toggles from horizontal to vertical line option automatically.

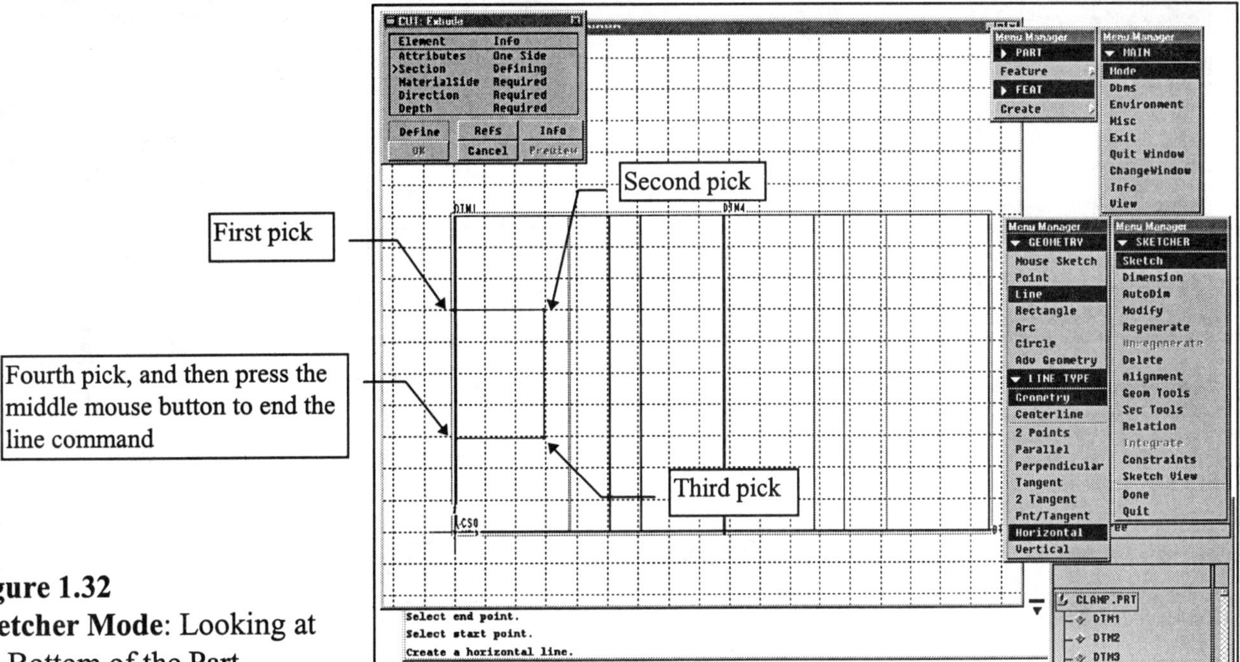

Figure 1.32
Sketcher Mode: Looking at the Bottom of the Part

Next, align the endpoints of the cut where they touch **DTM1** (Fig. 1.33). Three dimensions are needed to fix the feature's position on the part. The view shown in Figure 1.33 was panned and zoomed.

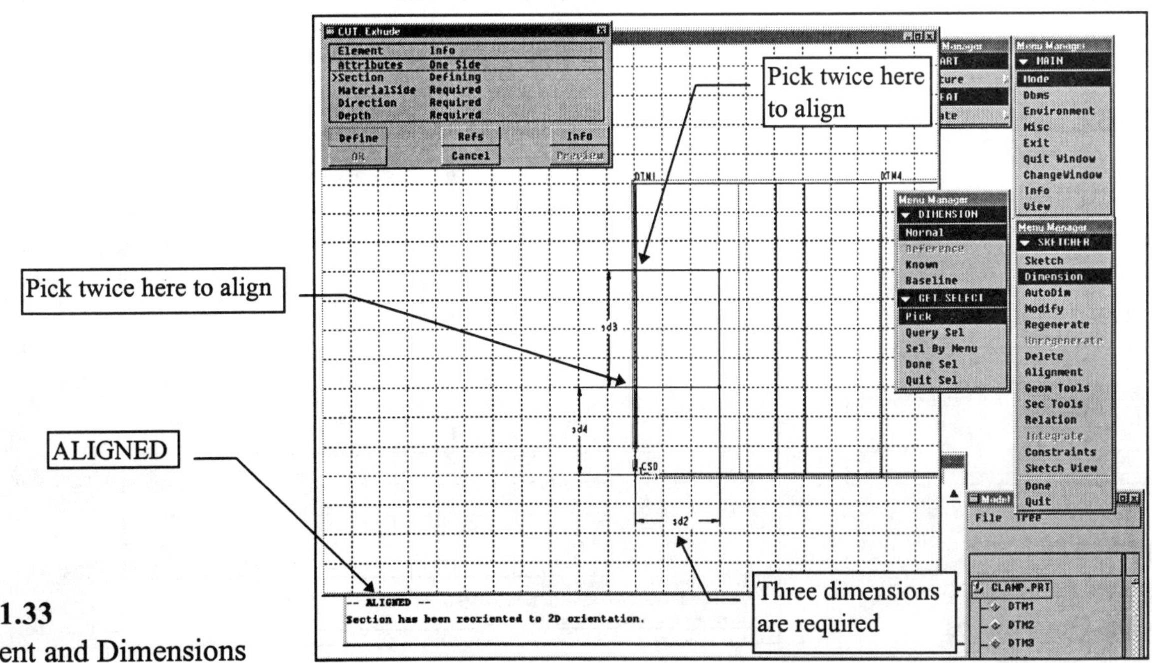

Figure 1.33
Alignment and Dimensions

It is now time to **Regenerate** the sketch to see if you created, aligned, and dimensioned it correctly (Fig. 1.34). After the successful regeneration of the rough sketch, the sketch constraints show onscreen and the MOD SKETCH menu options are available. Pick on each dimension and type the design sizes for the dimensions. When you are finished modifying the values, choose **Regenerate** again.

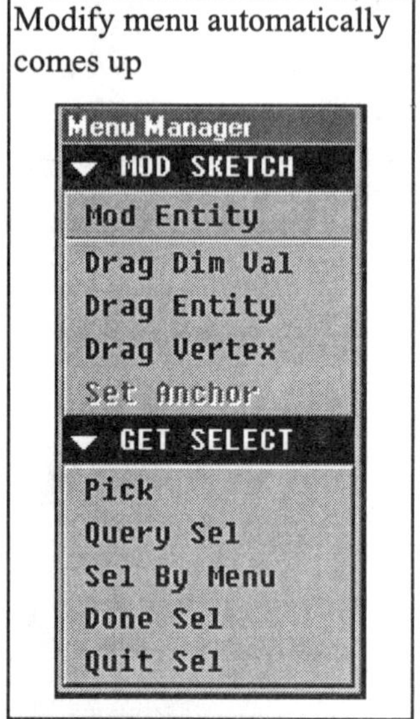

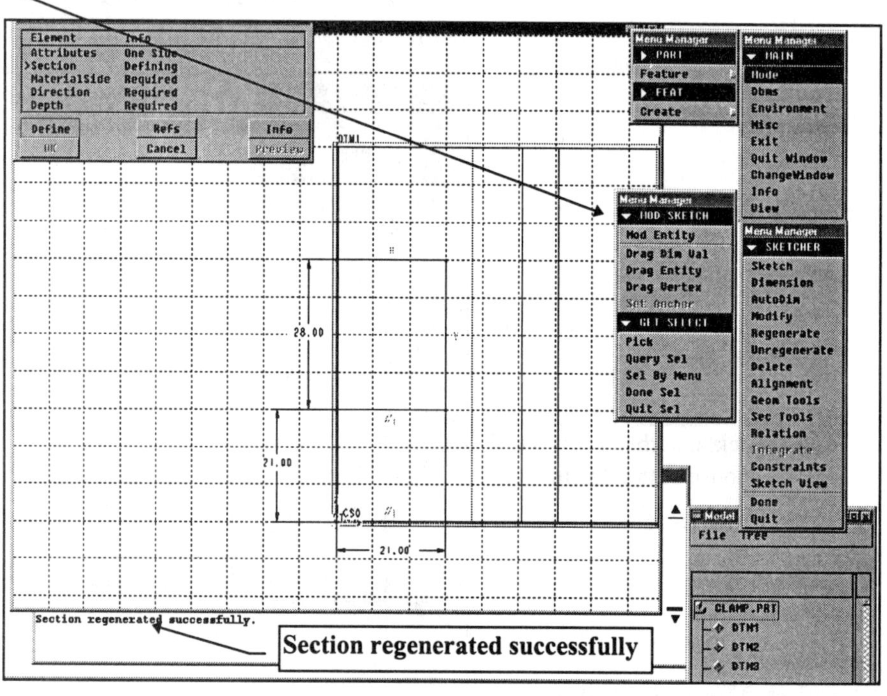

Figure 1.34
Regenerated Sketch with Section Dimensions and Constraints Displayed

The Main Graphic Window will *animate* the sketch while regenerating. The cut's section is now complete (Fig. 1.35). Choose **Done** from the SKETCHER menu.

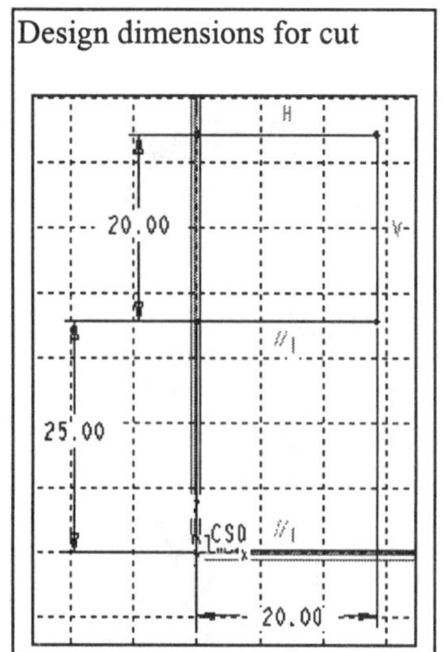

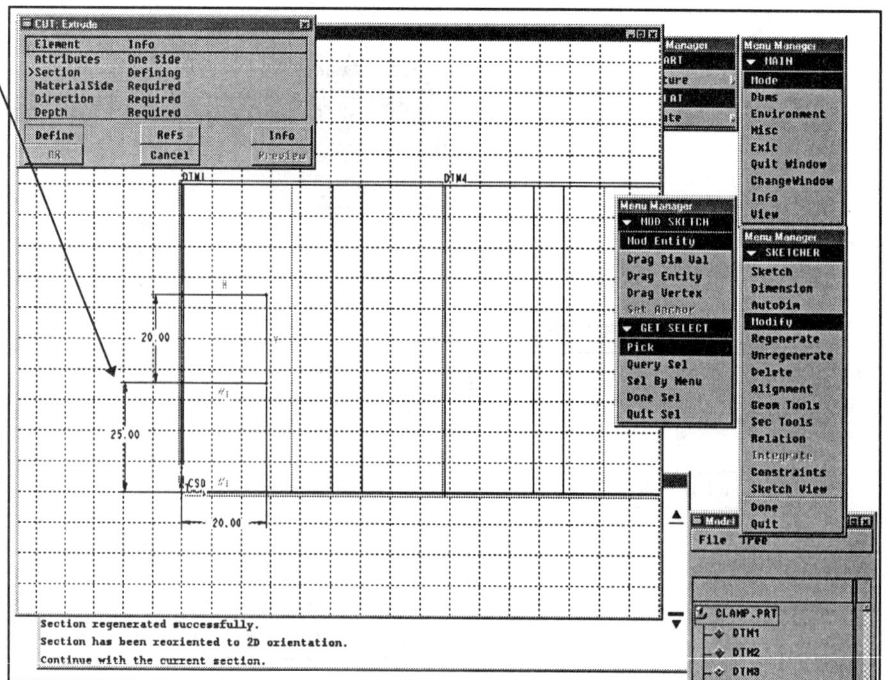

Figure 1.35
Regenerated Sketch with Design Dimensions

The direction of material removal is required at this stage of the cut creation. An arrow is displayed showing the direction in which the material will be removed (Fig. 1.36). Here, the direction arrow points toward the area to be removed for the cut; therefore, pick **Okay** from the DIRECTION menu. If you select the incorrect direction for the arrow material removal direction, the whole part will be cut away and the small **20 X 20** square piece will be all that's left of your part!

HINT

If the removal direction was pointing away from the cut and all that was left after the command was completed was a small square block of material, pick **Direction** and **Define** from the dialog box and redo the command.

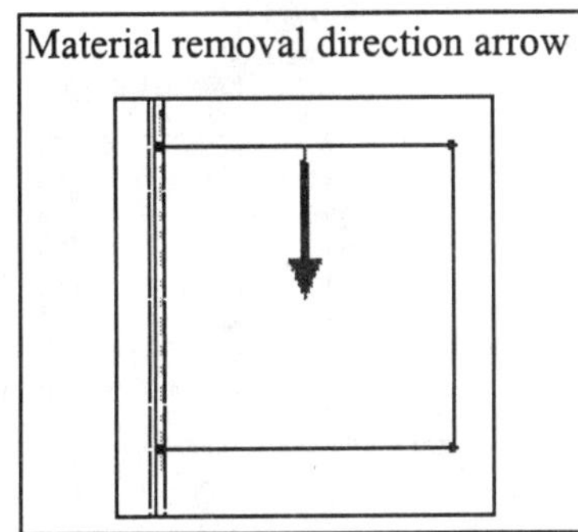

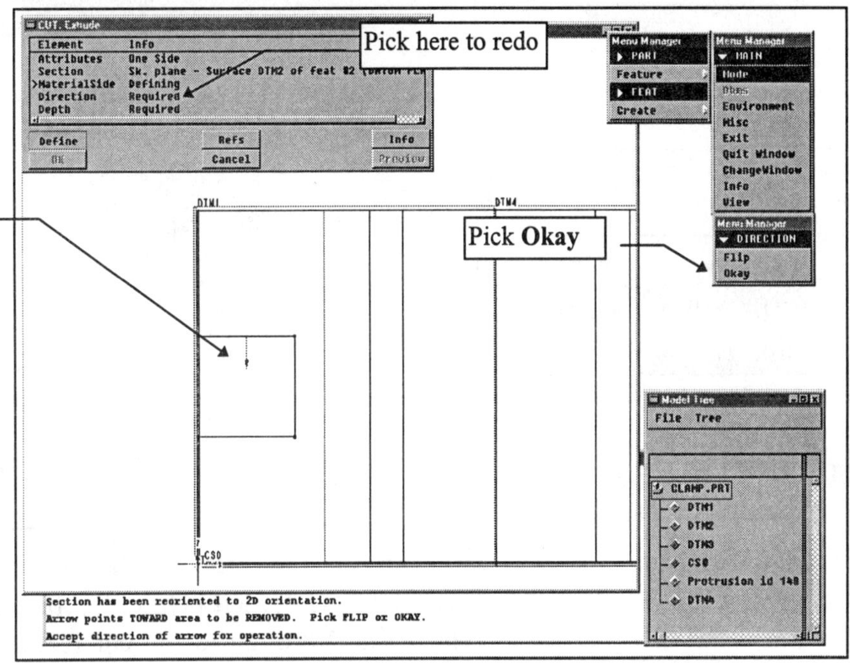

Figure 1.36
Direction of Material Removal

We recommend changing to a 3D view orientation in order to verify the correct cut-depth direction (Fig. 1.37).

Choose **View** ⇒ **Default** ⇒ **Done-Return** ⇒ **Okay** (make sure the material removal direction arrow points *into* the cut area to be removed) ⇒ **Thru All** (from SPEC TO menu)

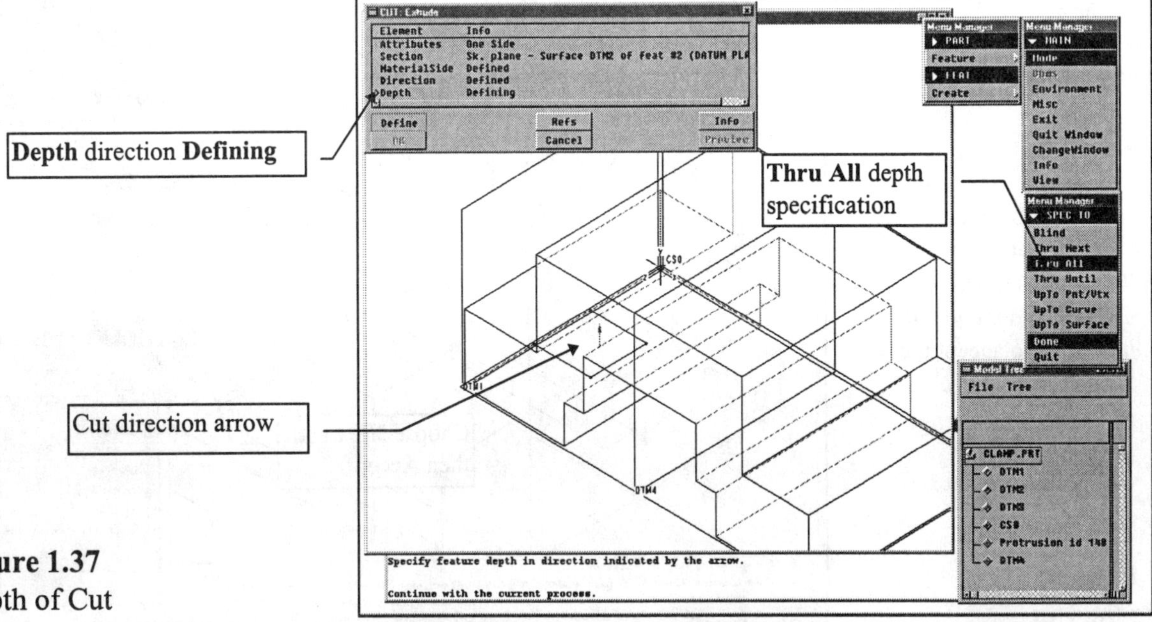

Figure 1.37
Depth of Cut

Choose **Done** from the SPEC TO menu and **Preview** from the dialog box; or, if you don't wish to the see the feature previewed, choose **OK** from the dialog box (Fig. 1.38).

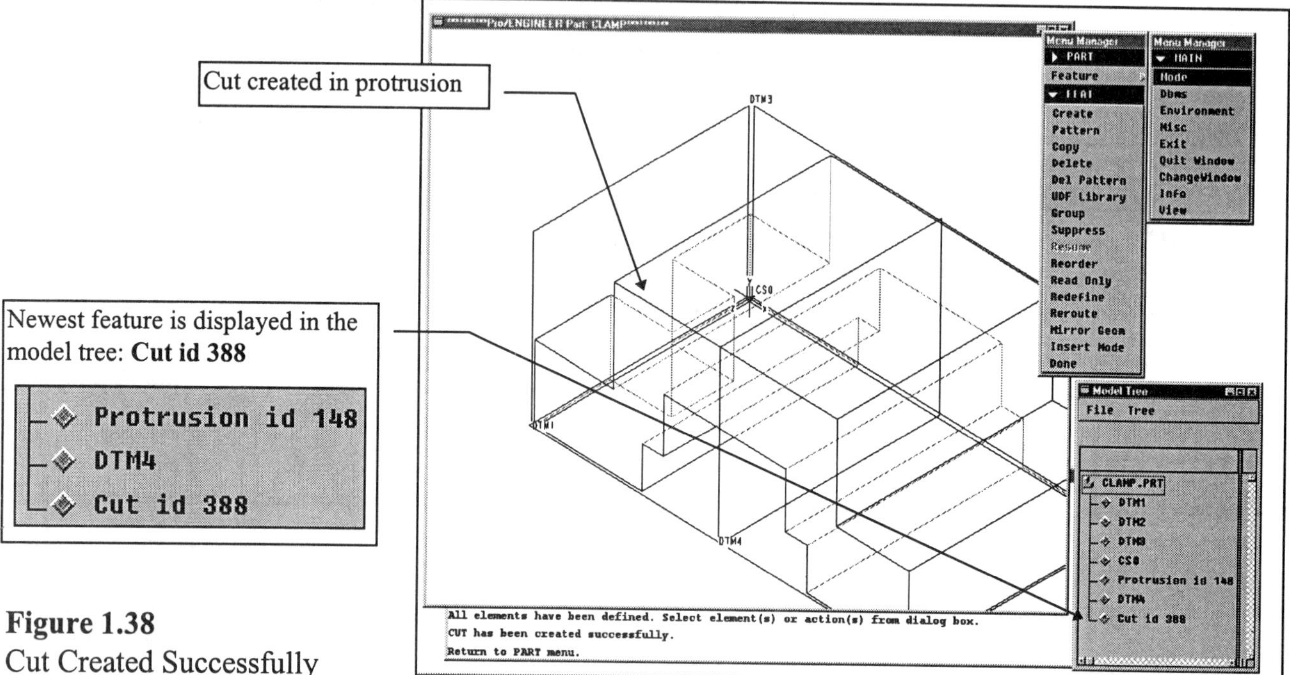

Figure 1.38
Cut Created Successfully

The last step is to copy the cut feature to the right side of the part. The feature is hidden behind the protrusion, so you must use **Query Sel** to filter through to the cut feature. It will be highlighted when selected. From the FEAT menu, choose the following (Fig. 1.39):

Copy ⇒ **Mirror** ⇒ **Independent** ⇒ **Done** ⇒ **Query Sel** (pick on the cut feature) ⇒ **Next** (from the CONFIRM menu) ⇒ **Accept** ⇒ **Done Sel** ⇒ **Done**

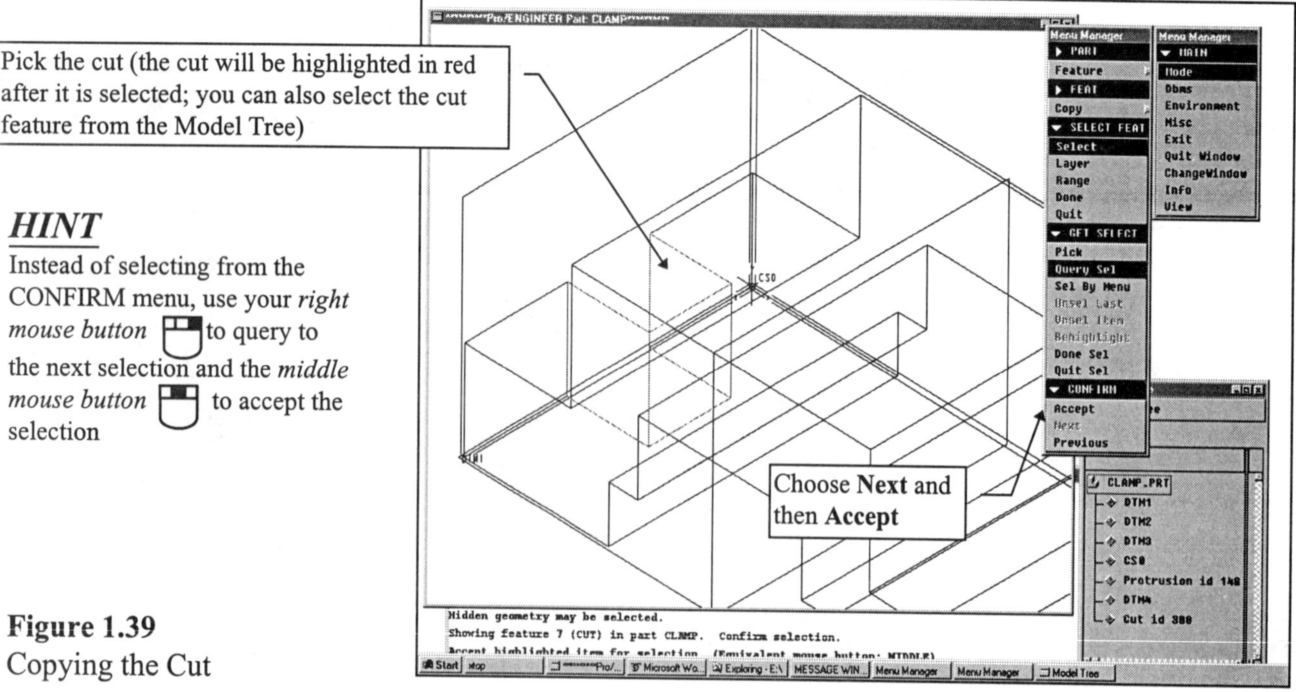

HINT

Instead of selecting from the CONFIRM menu, use your *right mouse button* to query to the next selection and the *middle mouse button* to accept the selection

Figure 1.39
Copying the Cut

Pro/E now prompts you to select the plane or datum to use as the mirroring plane (Fig. 1.40). The cut will be mirrored about **DTM4**.

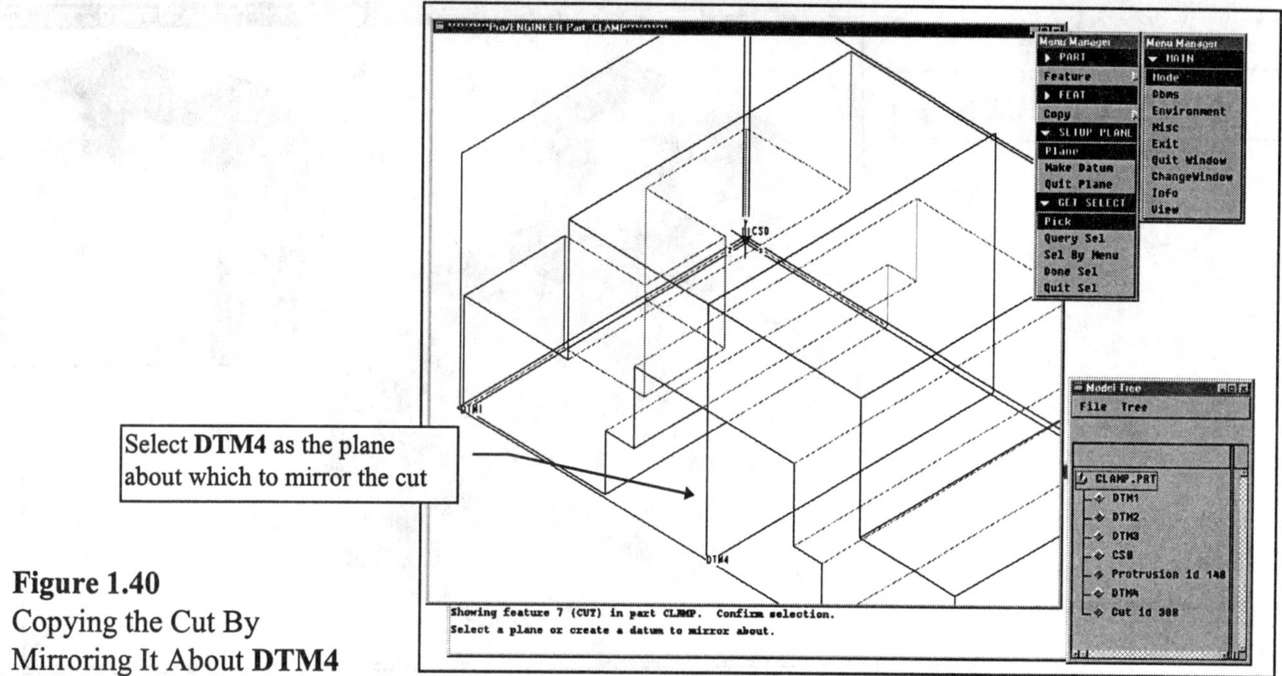

Figure 1.40
Copying the Cut By Mirroring It About **DTM4**

Pick on the edge or name of datum plane **DTM4**. You can use **Query Sel** here if you wish, or use your mouse buttons. Once you **Pick** or **Accept** after **Query Sel**, the cut is mirrored automatically. Pro/E completes the command and displays the part. You have completed your first Pro/E part! To see the protrusion and the cuts clearly, rotate and shade the part as shown in Figure 1.41:

(**File**--PT/Modeler, not **Dbms**)

Environment ⇒ Shading ⇒ Done-Return ⇒ View ⇒ Default ⇒ Done-Return ⇒ Dbms ⇒ Save ⇒ enter ⇒ Purge ⇒ enter

HINT

Hold down the **Ctrl** key and press the right (**Pan**), left (**Zoom**), or middle (**Rotate**) mouse button to orient the part on the screen

Pan **Zoom** **Rotate**

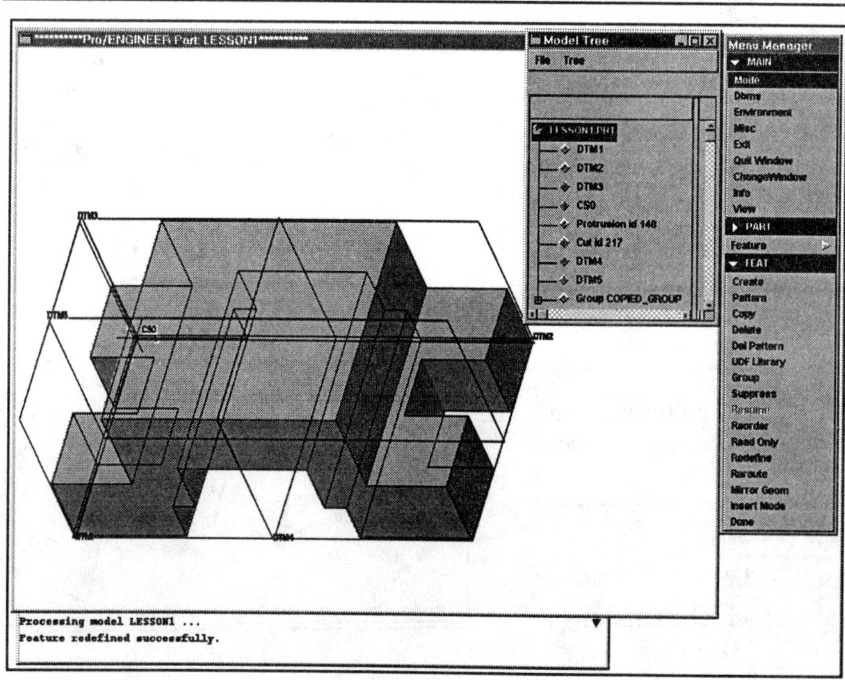

Figure 1.41
Rotated and Shaded Clamp

Lesson 1 Project

Angle Block

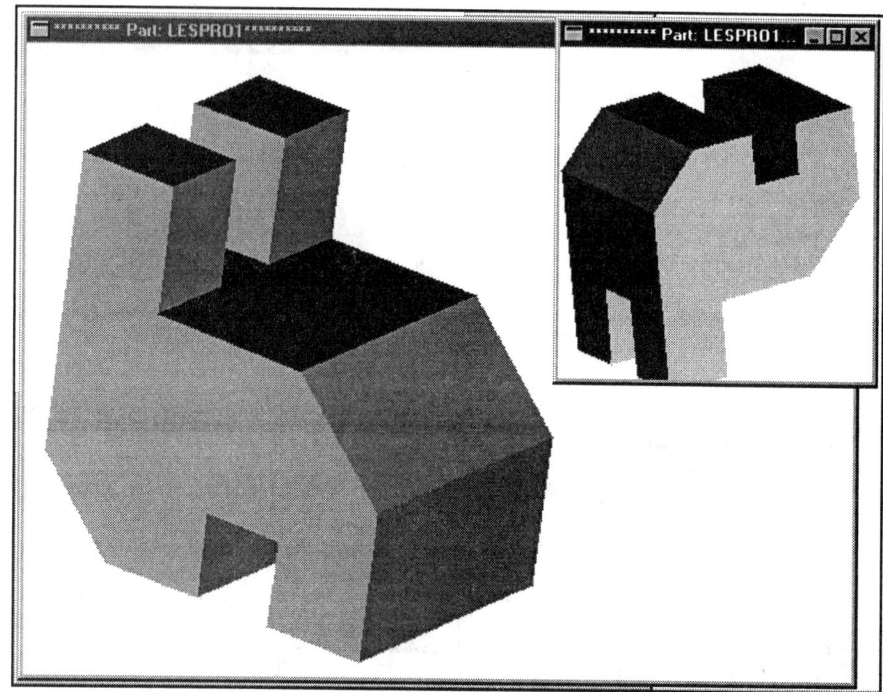

Figure 1.42
Angle Block

? Pro/HELP

Remember to use the help available on Pro/E by highlighting a command and pressing the right mouse button ? GetHelp

Angle Block

The first **lesson project** is a simple block that requires many of the same commands as the **Clamp**. First, create a default coordinate system and the default datum planes. Using the datums, create the part shown in Figures 1.42 through 1.46. Sketch the protrusion on **DTM3**. Sketch the cut on **DTM1** and align it to the upper surface/plane of the protrusion, as shown in Figure 1.43.

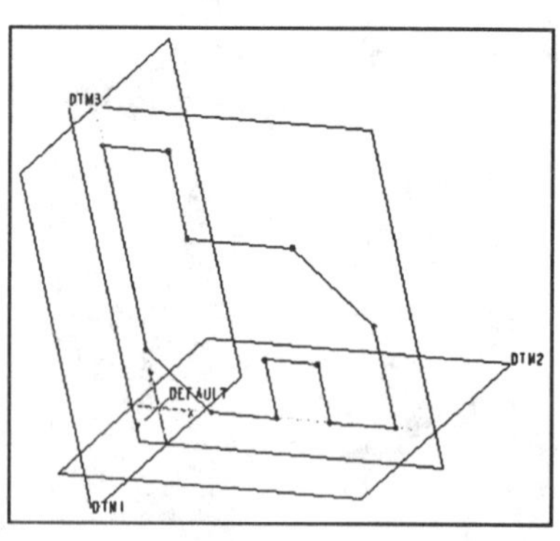

Sketch the first protrusion with this surface/plane as the section outline

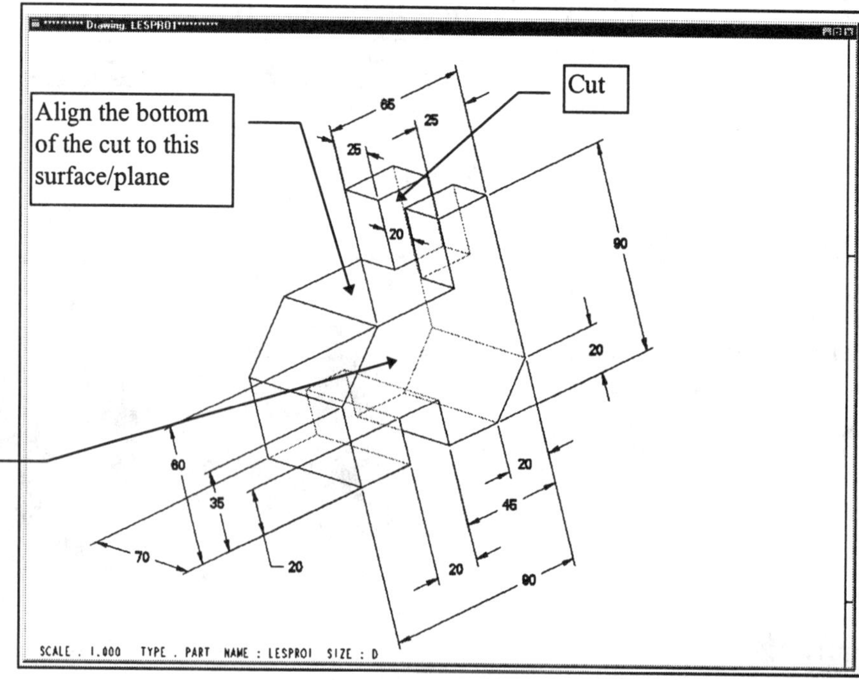

Figure 1.43
Angle Block Dimensions

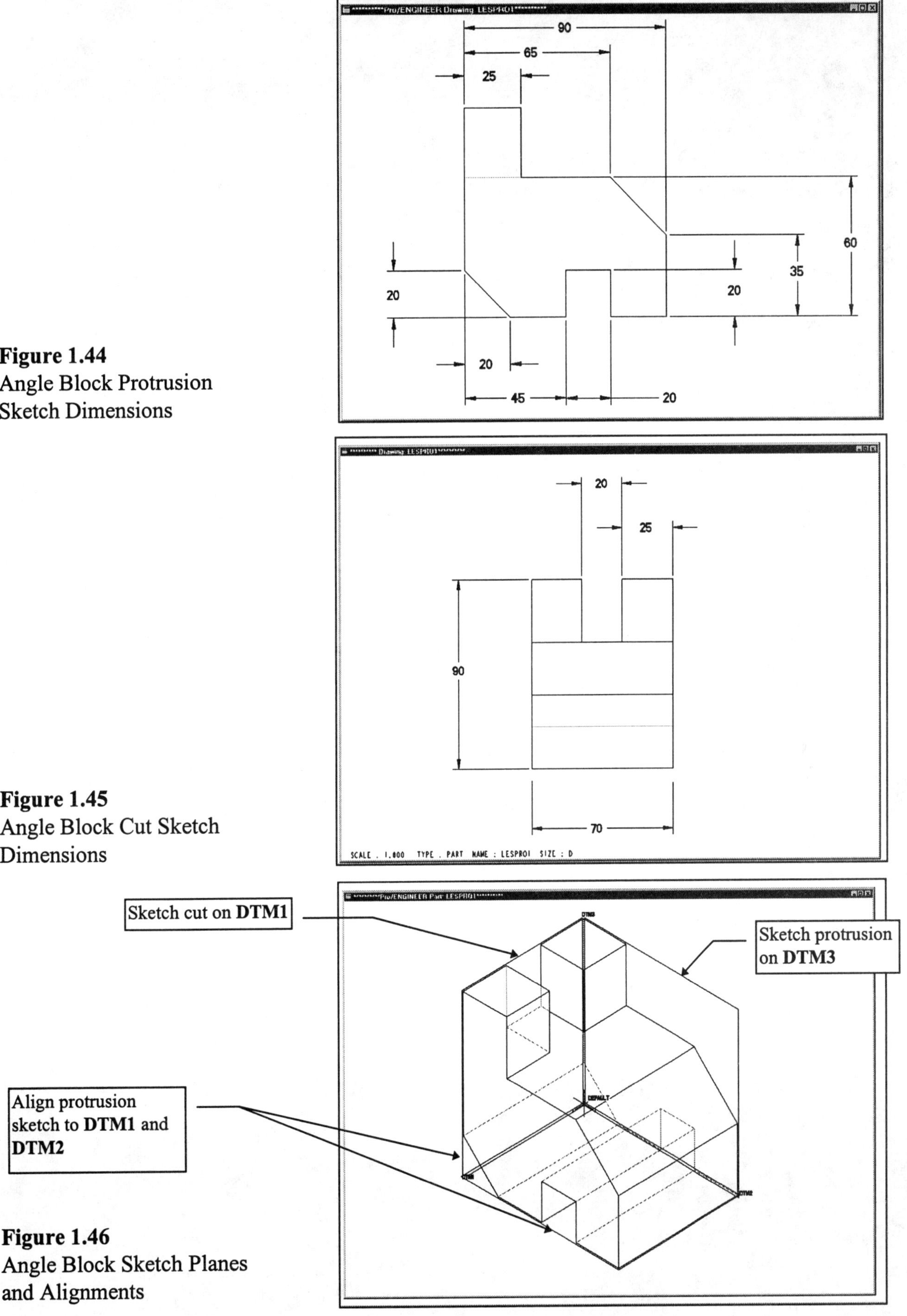

Figure 1.44
Angle Block Protrusion Sketch Dimensions

Figure 1.45
Angle Block Cut Sketch Dimensions

Figure 1.46
Angle Block Sketch Planes and Alignments

Lesson 2

Modify and Redefine

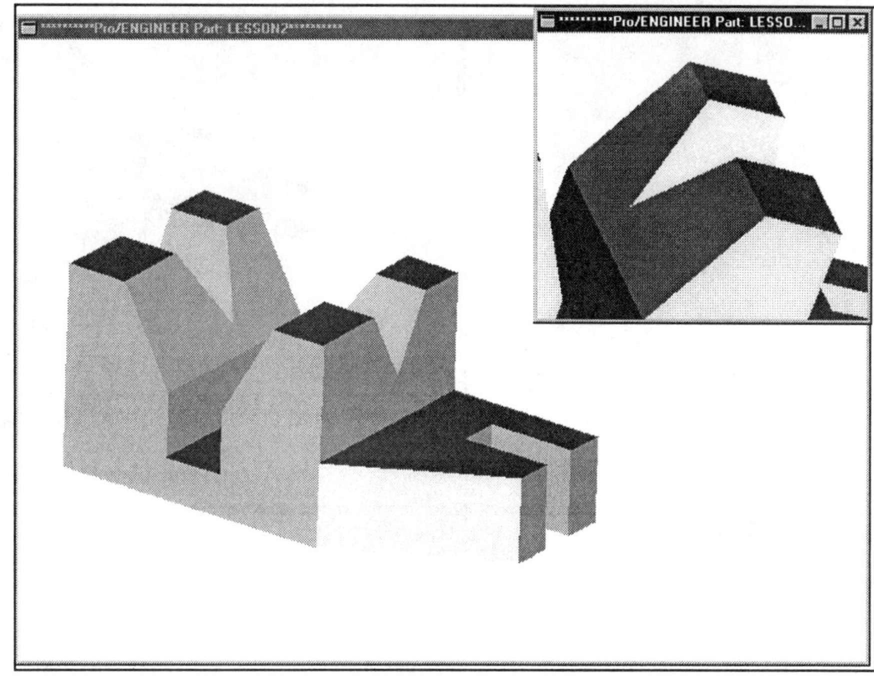

Figure 2.1
Base Angle

OBJECTIVES

1. Change existing features by modifying dimensions

2. Alter the view consideration in dimension display

3. Redefine a part's features

4. Modify the number of decimal places of dimensions

5. Regenerate modified and redefined parts

6. Input a *config.pro* change

7. Use the Info command to get feature and model information

8. Use the Model Tree to modify, redefine, and get information

EGD REFERENCE
Engineering Graphics and Design with Graphical Analysis *or* **Fundamentals of Engineering Graphics and Design**
by L. Lamit and K. Kitto
Read Chapter 10
See pages 298, 399

COAch™ for Pro/ENGINEER

If you have **COAch for Pro/ENGINEER** on your system, go to SEARCH and do the Segment shown in Figures 2.3 and 2.5.

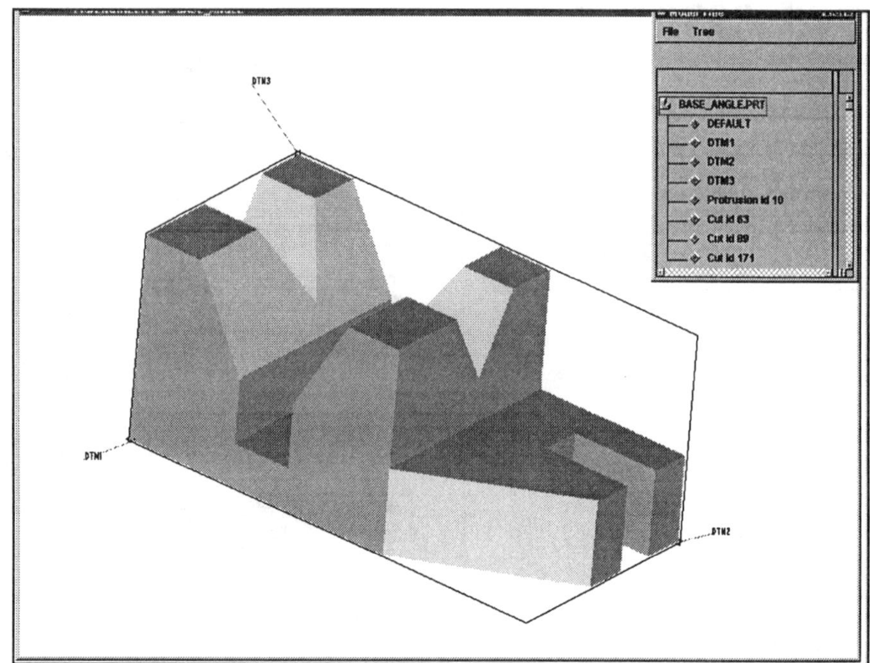

Figure 2.2
Base Angle Showing Datums, and Model Tree

MODIFY AND REDEFINE

Both **Modify** and **Redefine** are important capabilities in the design of parts (Fig. 2.1). With **Modify**, you can change any dimension used in the creation of a feature. With **Redefine**, the feature's attribute, the placement plane, the placement/orientation references, and the size and configuration of the feature can be redone. **Modify** is for simple dimensional changes, and **Redefine** is for more comprehensive changes to the model's features. You will be creating the **BASE ANGLE** shown in Figure 2.2 and then modifying and redefining some of its features. If you have **COAch for Pro/ENGINEER**, complete the appropriate segment (Fig. 2.3).

Figure 2.3
COAch for Pro/E, Basic Modifications (Changing Parameters)

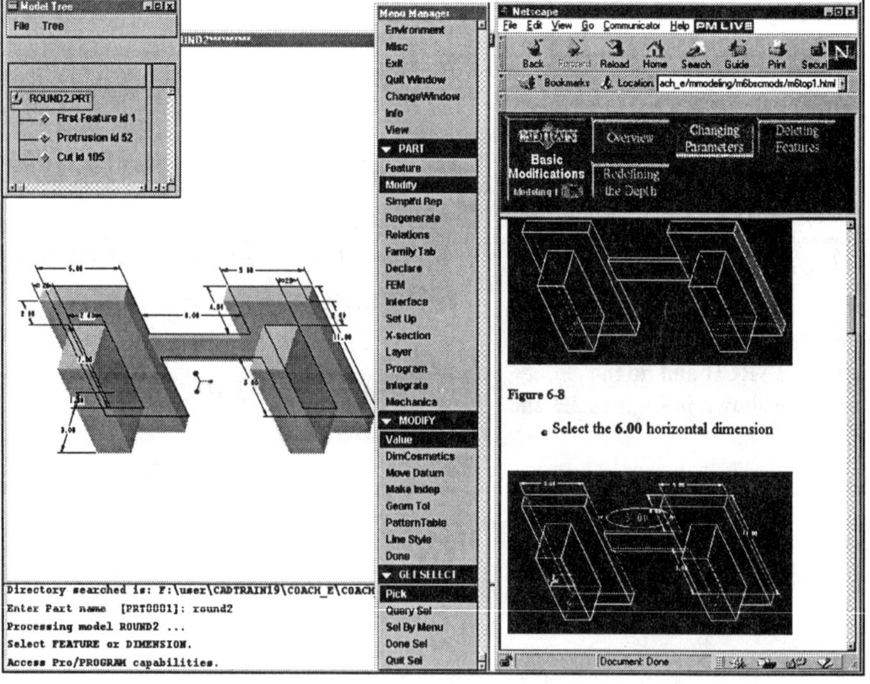

Modify

To modify dimensions, choose **Modify**, then **Value** (default) from the MODIFY menu, and select a feature. Pro/E displays all the dimensions associated with the selected feature. If you pick an edge that is shared by two features while you are using the **Query Sel** option, Pro/E highlights the associated features in turn. Pro/E displays the CONFIRM menu, which, after you pick **Next**, lets you step back and forth through the highlighted features to choose the one you want.

View Orientation When Displaying Dimensions

The view orientation of a part can be adjusted to improve clarity in viewing dimensions. The need to change the view orientation becomes apparent when the dimensions overlap, the dimensions are in planes that are perpendicular to the current view, or Pro/E displays the section dimensions on the original sketching plane. Hold down the **Ctrl** key and press the middle button of the mouse to rotate the model to the desired position. Figure 2.4 shows a portion of the available help for the modify-view consideration.

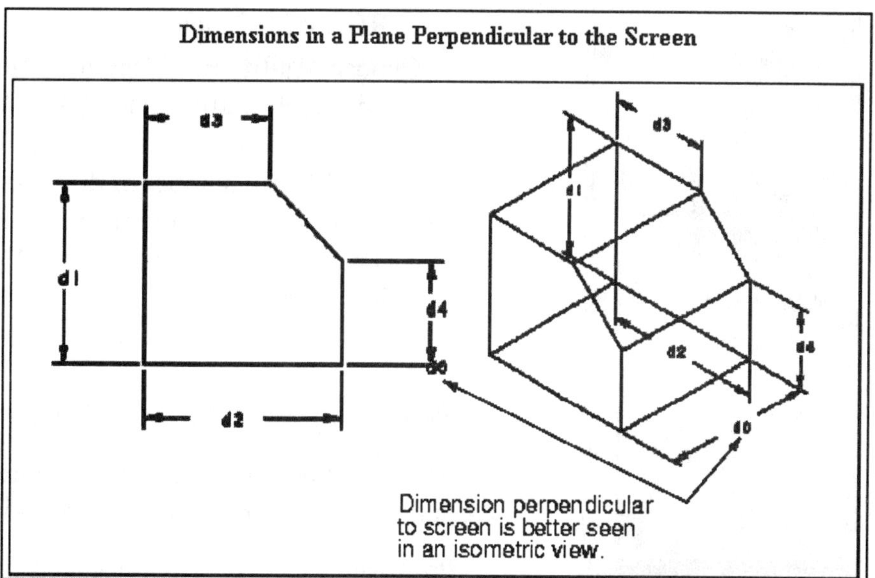

Figure 2.4
Online Documentation,
Modify (View Consideration)

Modifying Dimension Values

When you modify the value of a dimension, you can enter a new number or a **relation**. Pro/E supports the use of negative dimensions. The value entered depends on the displayed sign of the dimension. By default, Pro/E displays all dimensions as positive values, and entering a negative value tells Pro/E to create the section geometry on the opposite side, but the *direction* of a feature creation cannot be changed by entering a negative number. Use the **Redefine** option to redefine the direction of the feature.

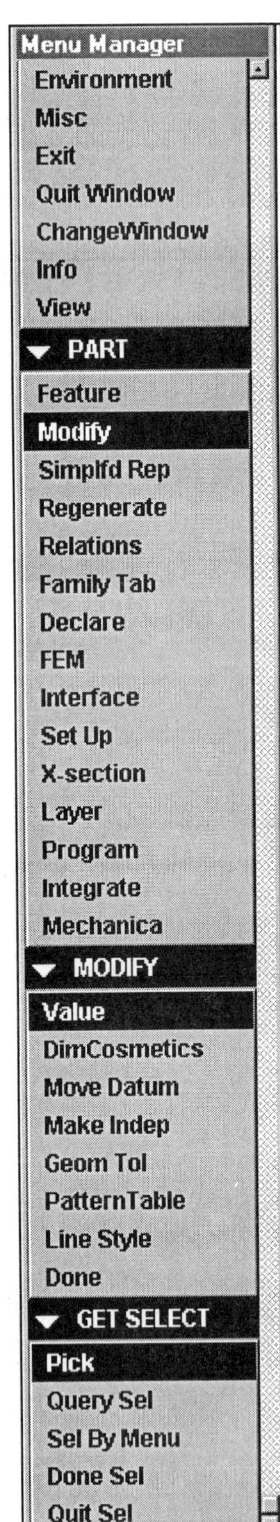

To modify a dimension value:

1. Choose **Modify** from the PART menu.
2. Display the dimensions of a feature by picking on any surface of the desired feature.
3. Pick the dimension to change. The value is highlighted in red, and Pro/E displays a prompt in the MESSAGE WINDOW.
4. Enter a new value, or accept the current value by pressing **Enter**. In many cases, this value can be negative. This new value (displayed in white) replaces the old value.
5. Modify other dimensions as required.
6. When you have completed all the changes, choose **Regenerate** to recalculate the part using the new dimension values.

Modifying the Number of Decimal Places for Dimensions

The default number of decimal places for dimensions is two. To increase the precision of a particular dimension, enter a new value with the desired precision. Modifying the number of decimal places, for a dimension, rounds the value of the dimension.

To decrease the precision of a particular dimension:

1. Turn the tolerances on by choosing **Display Tol** from the ENVIRONMENT menu.
2. Choose **Modify ⇒ DimCosmetics ⇒ Format.**
3. Choose **Nominal** from the DIM FORMAT menu, then pick a dimension. Its tolerance display changes to nominal.
4. To modify the number of decimal places to display for one or more dimensions (including reference dimensions), choose the DIM COSMETIC menu and pick **Num Digits**. Enter the number of significant digits.
5. Select the dimensions whose display is to be changed.

Redefining Features with Elements

You redefine the following features using the Feature Definition dialog box to change the elements with which they were created:

Protrusions	Cuts	Slots
Shells	Rounds	Holes
Some surface features	Drafts	Draft offsets
Some datum curves	Shafts	

To redefine a feature that has elements:

1. Choose the **Redefine** option from the FEAT menu, and pick the feature to be redefined.
2. Pro/E displays the Feature Definition dialog box. Each element and its current value are listed. Select the element to redefine, then select the **DEFINE** button. Pro/E prompts for the information needed to redefine the element.

Redefining Features

The **Redefine** option in the FEAT menu allows you to change the way a feature is created, including section geometry. The types of changes you can make depend on the selected feature.

A cut has a *section;* therefore, it can be redefined. If you have **COAch for Pro/ENGINEER** on your system, go to SEARCH and complete the appropriate Segment shown in Figure 2.5.

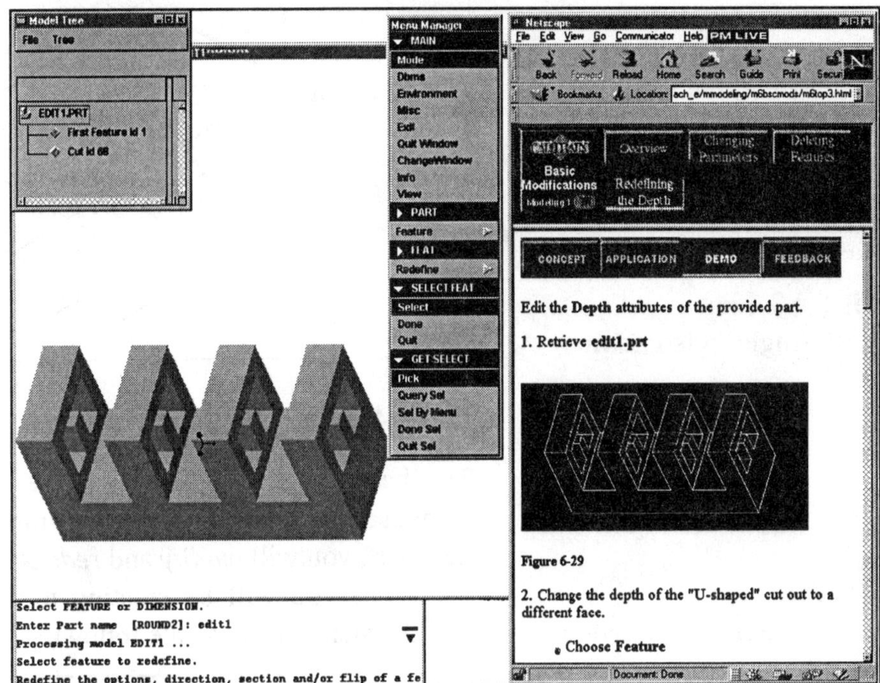

Figure 2.5
COAch for Pro/E, Basic Modifications (Redefining the Depth)

Pro/E recreates the feature using the new feature definitions. When feature sections are redefined, you may need to redimension any child feature whose reference edge or surface was replaced. Redimension the child feature using the options **Redefine** and **Scheme**, or **Reroute**. If you make any changes to the feature that cause the feature creation to abort, you enter the **Resolve** environment.

When using the **Redefine** option for a feature that was created with the options **Copy**, **Mirror**, and **Dependent**, Pro/E issues a warning message stating that the selected feature is a dependent copy of the highlighted feature. If **Continue** is chosen from the WAITING menu, Pro/E will display the REDEFINE menu with the elements **Attributes**, **Direction**, **Section**, and **Depth**. For example, if you choose **Section** after you select the option to redefine, the section of the selected feature can be changed, yet upon completion is still dependent on the referenced feature.

When you apply the redefinition, Pro/E removes the feature geometry and creates temporary geometry for your changes. When you exit from the user interface, Pro/E regenerates the part.

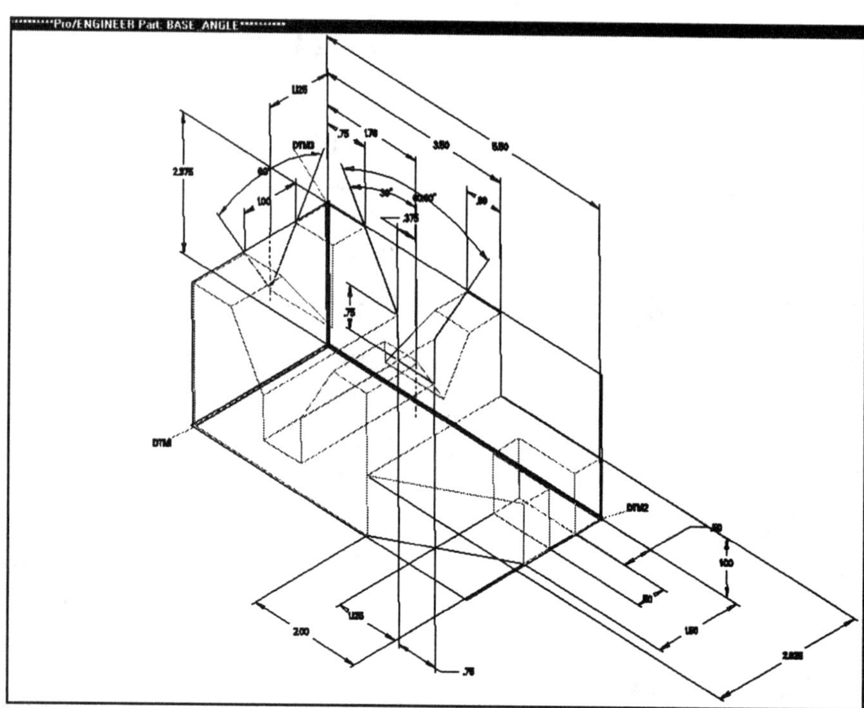

Figure 2.6
Base Angle In Isometric

Base Angle

The base angle (Fig. 2.6) will be our second ***lesson part***. It is composed of one protrusion and three cuts. Along with creating a new part, you will *modify* and *redefine* the part after it is completed.

Since you will be creating a new part, choose **Part** from the MODE menu. Choose the following commands:

NOTE
You must include an underline character when creating file names that have a space. **BASE ANGLE** needs to be typed as **BASE_ANGLE**. No spaces are allowed in object (part, assembly or drawing, etc.) names.

Mode ⇒ **Part** ⇒ **Create** ⇒ (type the part name at the command prompt) **BASE_ANGLE** and press **enter**.

As in Lesson 1, the *environment*, *units*, and *material* for the part need to be established:

ENVIRONMENT AND SETUP

Environment ⇒ **✓Grid Snap ✓ Model Tree**
Hidden line Tan Dimmed

Set Up ⇒ **Units** ⇒ **Length** ⇒ **Inch** ⇒ **Done** ⇒ **Material** ⇒ **Define** ⇒ (type **Aluminum**, then press **enter**) ⇒ (table of material properties, change or add information) ⇒ **File** ⇒ **Save** ⇒ **File** ⇒ **Exit** ⇒ **Assign** ⇒ (pick **Aluminum**) ⇒ **Accept** ⇒ **Done**

Create a default coordinate system and the three default datum planes (Fig. 2.7). The datum planes are used to sketch on and orient the part's section geometry. Choose the following commands:

PT/Modeler™

Feature ⇒ **Datum Plane** ⇒ **Coord Sys** ⇒ **Default** ⇒ **Done**

Feature ⇒ **Create** ⇒ **Datum** ⇒ **Plane** ⇒ **Offset** ⇒ (press **enter** three times at the prompt) ⇒ **Done**

The next command creates the first protrusion for the part shown in Figure 2.7:

PT/Modeler™

Feature ⇒ Protrusion etc.

Feature ⇒ Create ⇒ Protrusion ⇒ Extrude ⇒ Solid ⇒ Done ⇒ One Side ⇒ Done ⇒ (pick **DTM3** as the sketching plane) **⇒ Okay** (for selecting the *direction* of protrusion projection) **⇒ Top ⇒** (pick **DTM2** as the *orientation* plane)

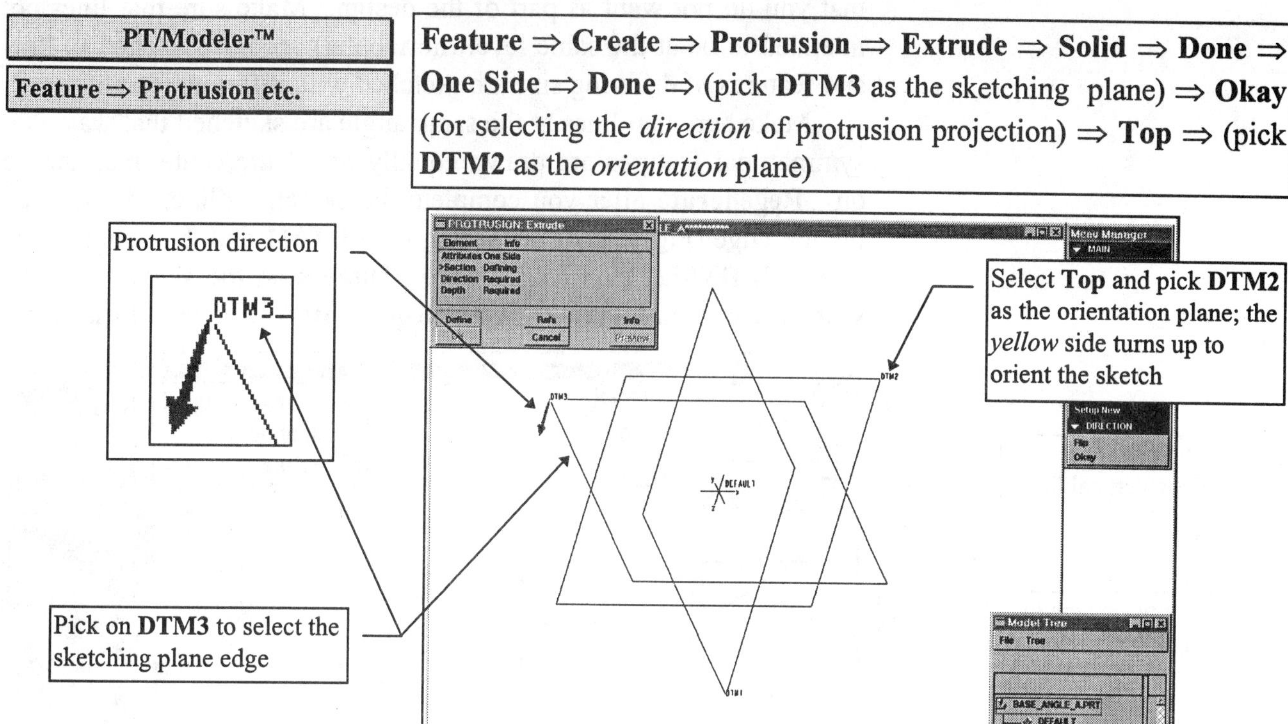

Figure 2.7
Selecting the Sketching Plane

The **Grid Snap** is on and the **Mouse Sketch** default is active in the GEOMETRY menu, so you can begin the sketching process using the mouse buttons. Sketch the outline of the part's primary feature using a set of connected lines, as was done in Lesson 1. The part's dimensions and general shape are provided in Figures 2.1 and 2.6. Most of the part can be defined by sketching an outline similar to its front view. Therefore, you can complete most of the part's geometry with one protrusion. The cuts on the top and side will be created later. The part will have its base aligned with the edge of **DTM2,** its left edge aligned with **DTM1,** and, since you are sketching on **DTM3,** its back face aligned with that datum.

The default number of digits for the sketcher is **2**. You can set the number of digits using the configuration file option **SKETCHER_DEC_PLACES** (a value in the range **0** to **14**).

HINT

default_dec_places sets the default number of decimal places to be displayed in all model modes for nonangular dimensions; it does not affect the *previously displayed* number of digits as modified using NUM DIGITS.

sketcher_dec_places controls the number of digits displayed when you are in the Sketcher.

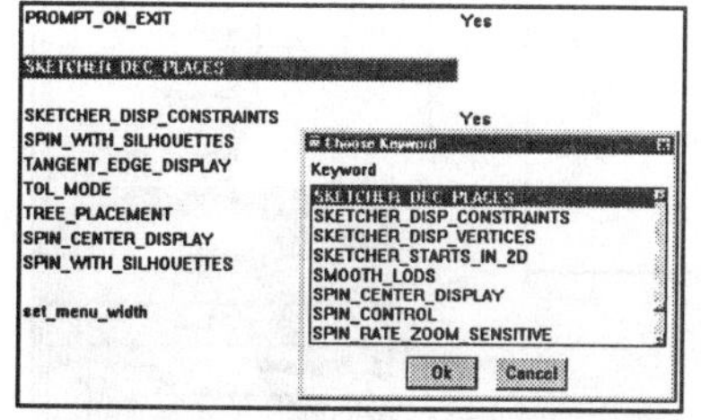

Misc ⇒ Edit Config ⇒ enter ⇒ (select an empty cell) **⇒** press **F4 ⇒** (move down the menu to pick **SKETCHER_DEC_PLACES**) **⇒ Ok ⇒** (select adjacent cell and type **3** as the new default) **⇒ enter ⇒ File ⇒ Save ⇒ Exit** (the table) **⇒ Load Config ⇒ enter ⇒ Done-Return**

In the example that follows, we have left the number of decimal places to be two. Pro/E will round three places at two (**1.125** becomes **1.13**).

Sketch the part's outline as shown in Figure 2.8. Add a vertical centerline down the middle of the slot. Do not create any constraints that you do not want as part of the design. Make sure that lines not intended to be at the same level (horizontal) are not sketched in line, and features of differing size are sketched with different lengths.

Make sure that lines at the same angle are sketched that way, that symmetrical features are drawn equally about the centerline, and so on. **Regenerate** after you complete the sketch outline. Next, align the left edge (Fig. 2.8) of the sketch with **DTM1** and the bottom edge line with **DTM2**, then **Regenerate**. Dimension the sketch with the scheme shown in Figure 2.9, then **Regenerate** again. (See Lesson 1.)

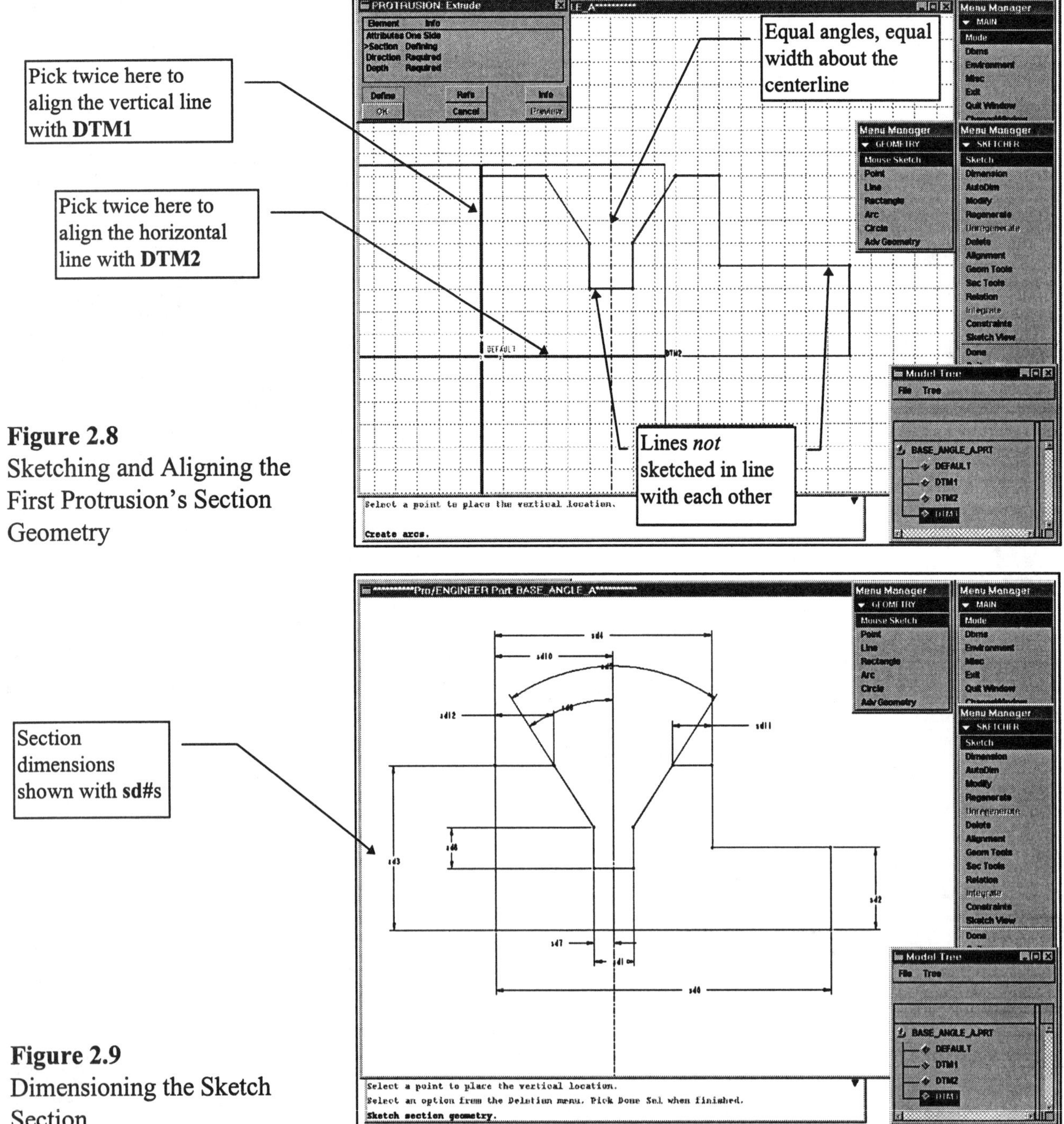

Figure 2.8
Sketching and Aligning the First Protrusion's Section Geometry

Figure 2.9
Dimensioning the Sketch Section

Figure 2.10 shows the successfully regenerated sketch. All the dimension values are larger than the design dimensions at this point in the process. The sketch grid was **30** units and decimal inch was selected at the beginning of the part. Therefore, each grid is **30** inches square, which accounts for the huge size of the sketch.

30-by-**30** unit grid spacing (**30** x **30** inches default)

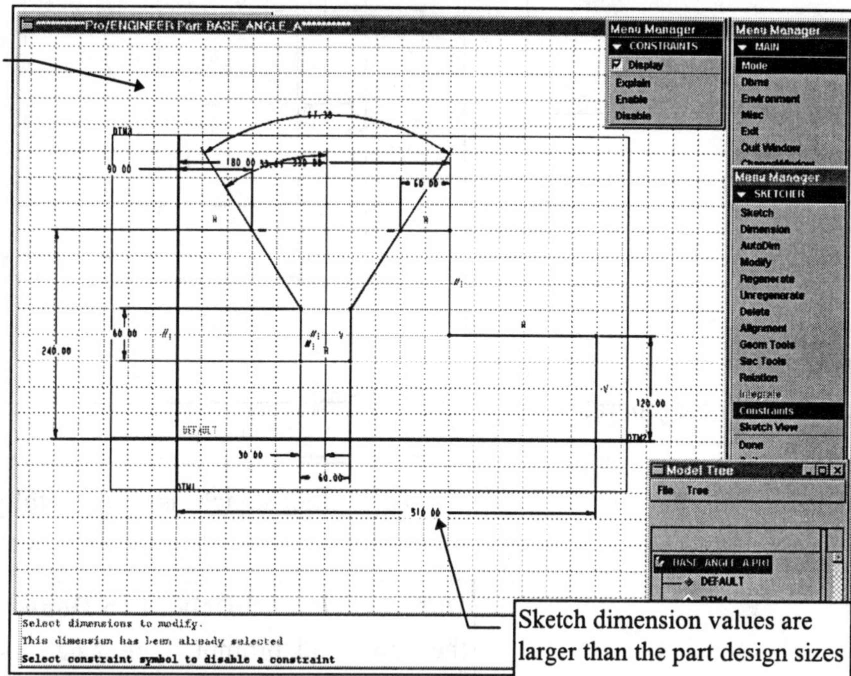

Figure 2.10
Sketch Dimensions:
Section regenerated successfully

Modify the dimensions one at a time using the design values shown in Figure 2.11, then **Regenerate** the sketch as shown in Figure 2.12.

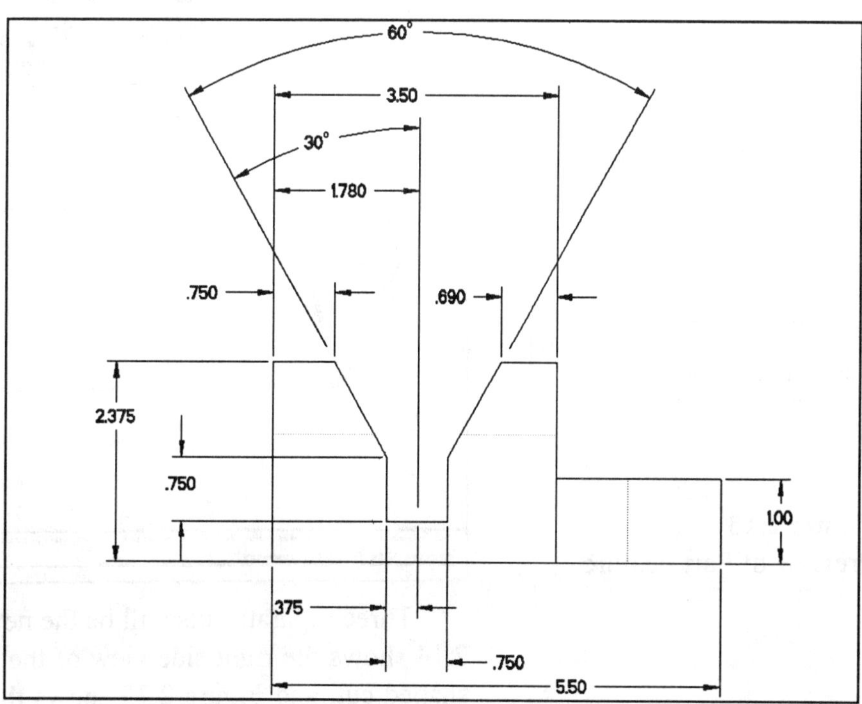

Figure 2.11
Part Dimensions

Sometimes it is difficult to see the regenerated dimensions. To rotate the sketch hold down the **Ctrl** key and the middle mouse button while slowly moving the mouse about the pad.

Figure 2.12
Successfully Regenerated Section

After the sketch is successfully regenerated with the design dimensions, choose the following commands:

Done ⇒ **View** ⇒ **Default** ⇒ **Done-Return** ⇒ **Blind** ⇒ **Done** ⇒ (at the command prompt, type **2.625** as the part's depth) ⇒ **enter** ⇒ **Preview** (Fig. 2.13) ⇒ **OK** ⇒ **Done**

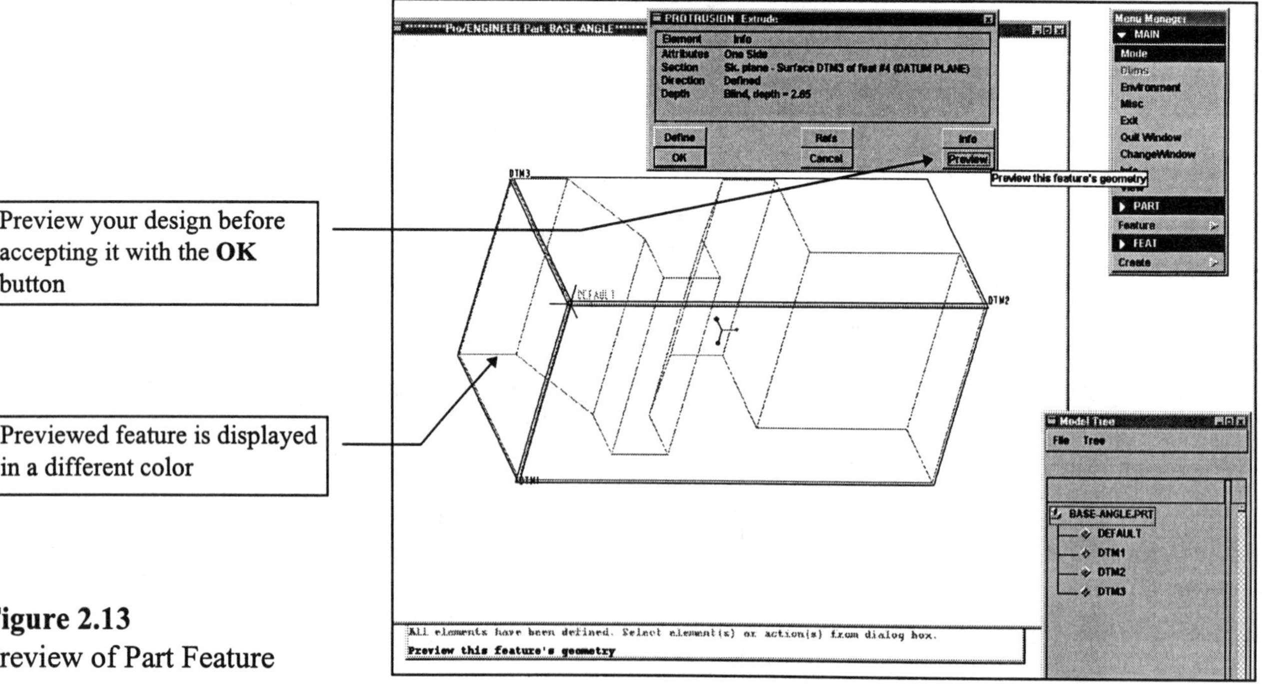

Figure 2.13
Preview of Part Feature

Three separate cuts will be the next features to be created. Figure 2.14 shows the right side view of the part with dimensions for the V-shaped cut, and Figure 2.15 shows the top view with dimensions for two more cuts.

Dbms (File--Pt/Modeler) ⇒ Save ⇒ enter Purge ⇒ enter ⇒ Done-Return

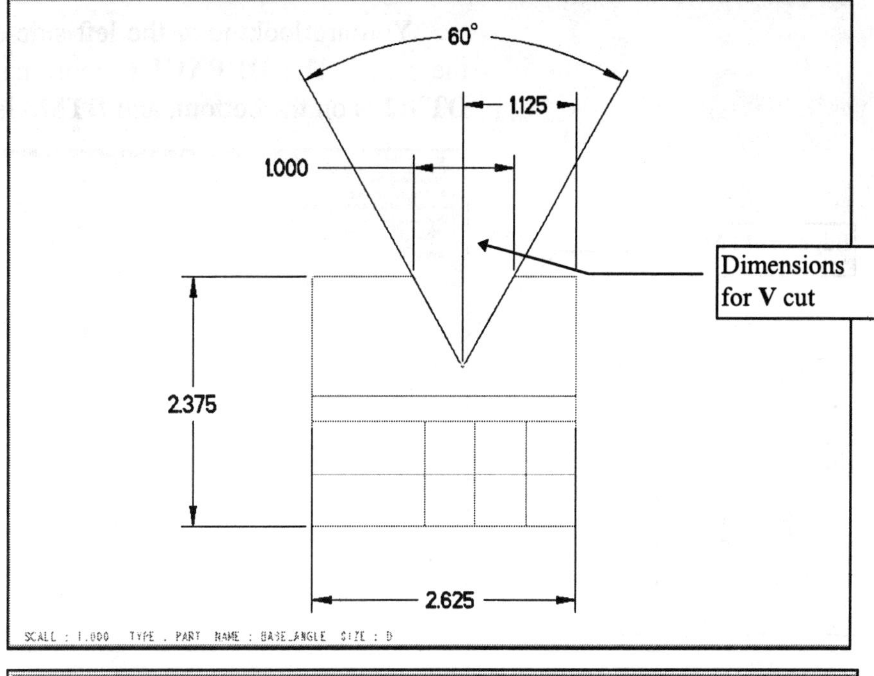

Figure 2.14
Right Side View with Dimensions

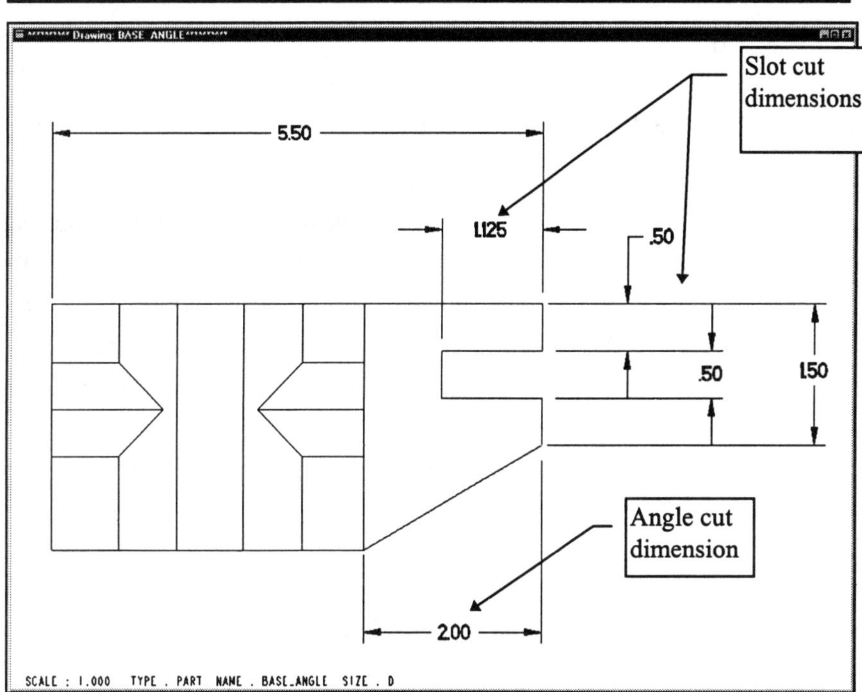

Figure 2.15
Dimensioned Top View

The next feature that will be created is the V-shaped cut. The dimensions for the cut are shown in Figure 2.14. The **V** cut is sketched similar to the first protrusion except that it removes material. Choose the following commands:

PT/Modeler ™
Feature ⇒ Cut etc.

Feature ⇒ Create ⇒ Cut ⇒ Extrude (default) **⇒ Solid** (default) **⇒ Done ⇒ One side** (default) **⇒ Done** (pick **DTM1** as the sketching plane) **Flip ⇒ Okay ⇒ Top** (pick **DTM2** to orient the sketch)

If necessary, use **Query Sel** to pick the datum planes. Pro/E now enters the **Sketcher** automatically and displays the part and datum planes as shown in Figure 2.16.

You are looking at the left side of the part in this orientation of the part. The **DEFAULT** coordinate system is in the lower left, **DTM2** is on the bottom, and **DTM3** is along the left.

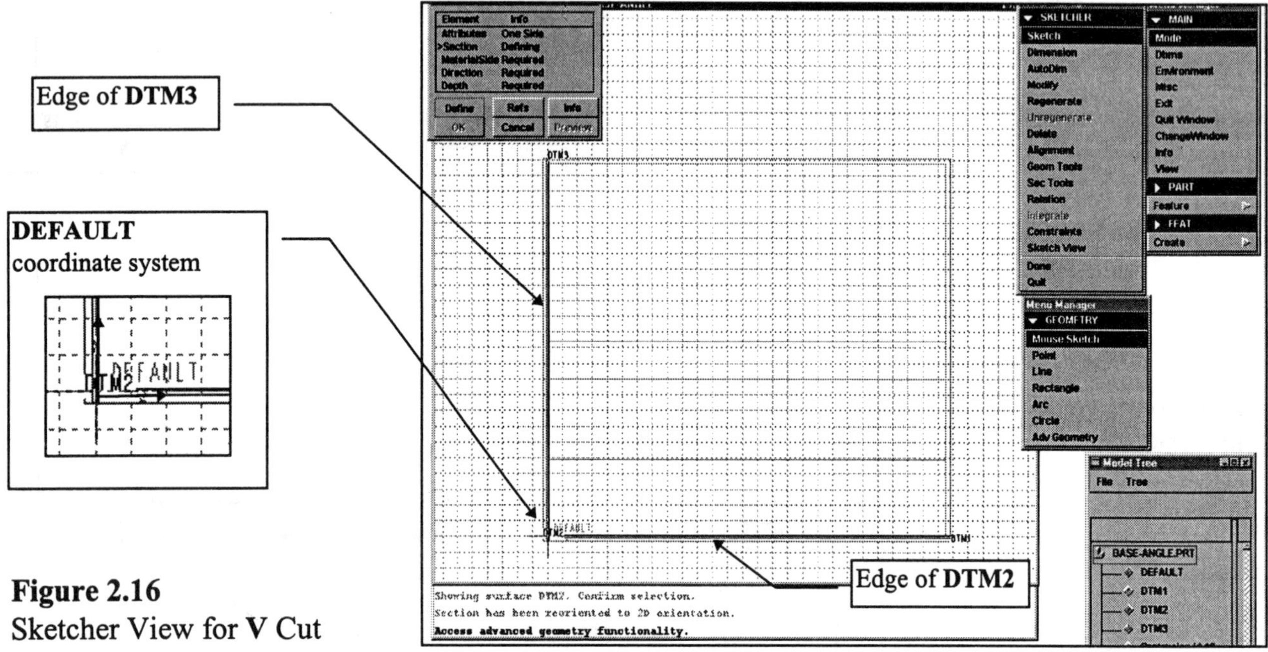

Figure 2.16
Sketcher View for **V** Cut

Sketch the **V** cut symmetrically about a vertical centerline, align the two endpoints to the top of the part, and dimension the **V** shape as shown in Figure 2.17. The centerline establishes the symmetrical **V** angle without your having to give a **30°** angle dimension. If the sketch is not drawn symmetrically, you may need to add the **30°** half-angle dimension, regenerate the sketch, and then delete the **30°** dimension.

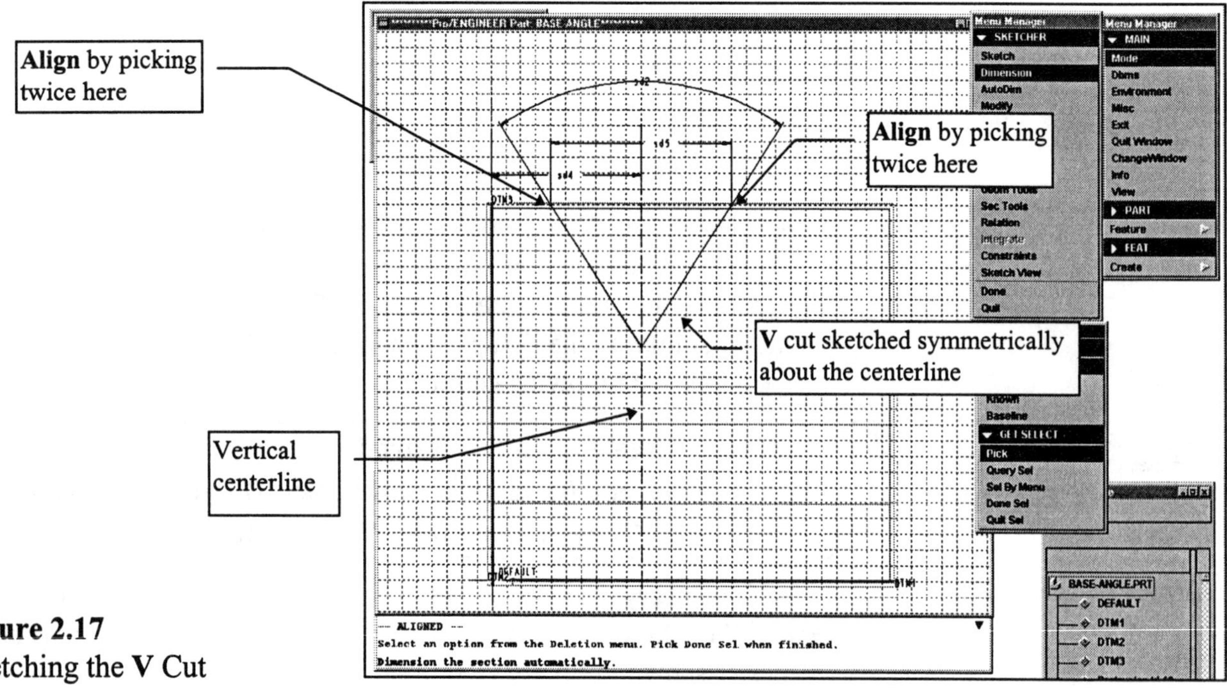

Figure 2.17
Sketching the **V** Cut

Regenerate the sketch as shown in Figure 2.18.

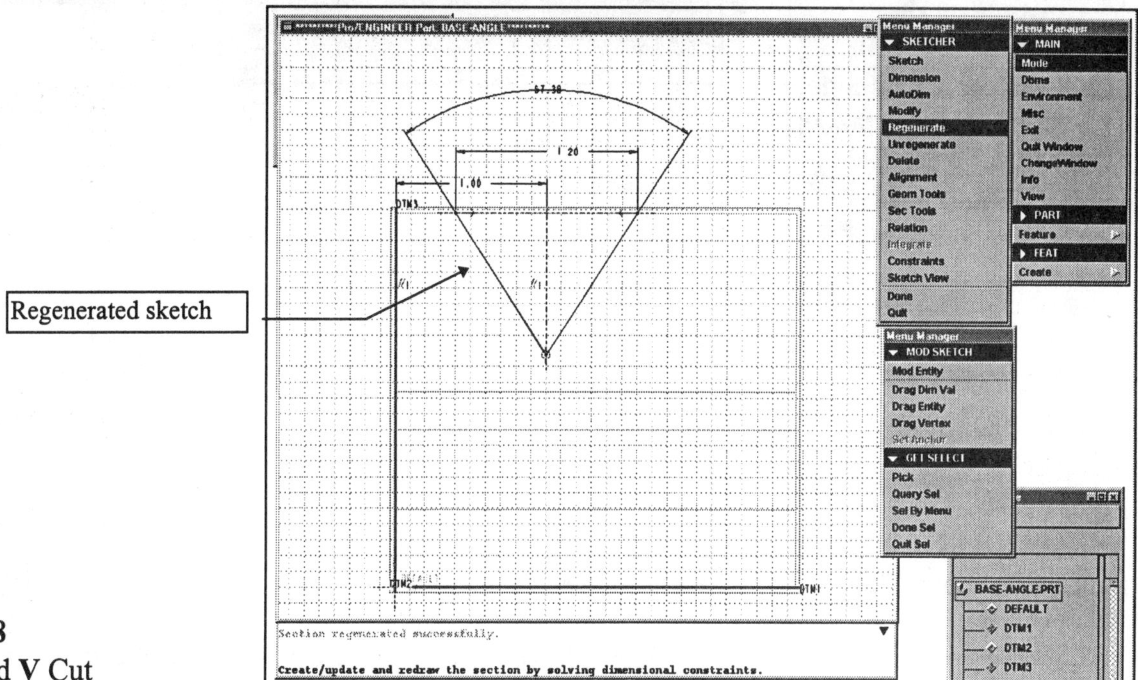

Figure 2.18
Regenerated **V** Cut

Modify the dimensions using the design dimensions shown in Figure 2.14 and **Regenerate** the sketch as shown in Figure 2.19.

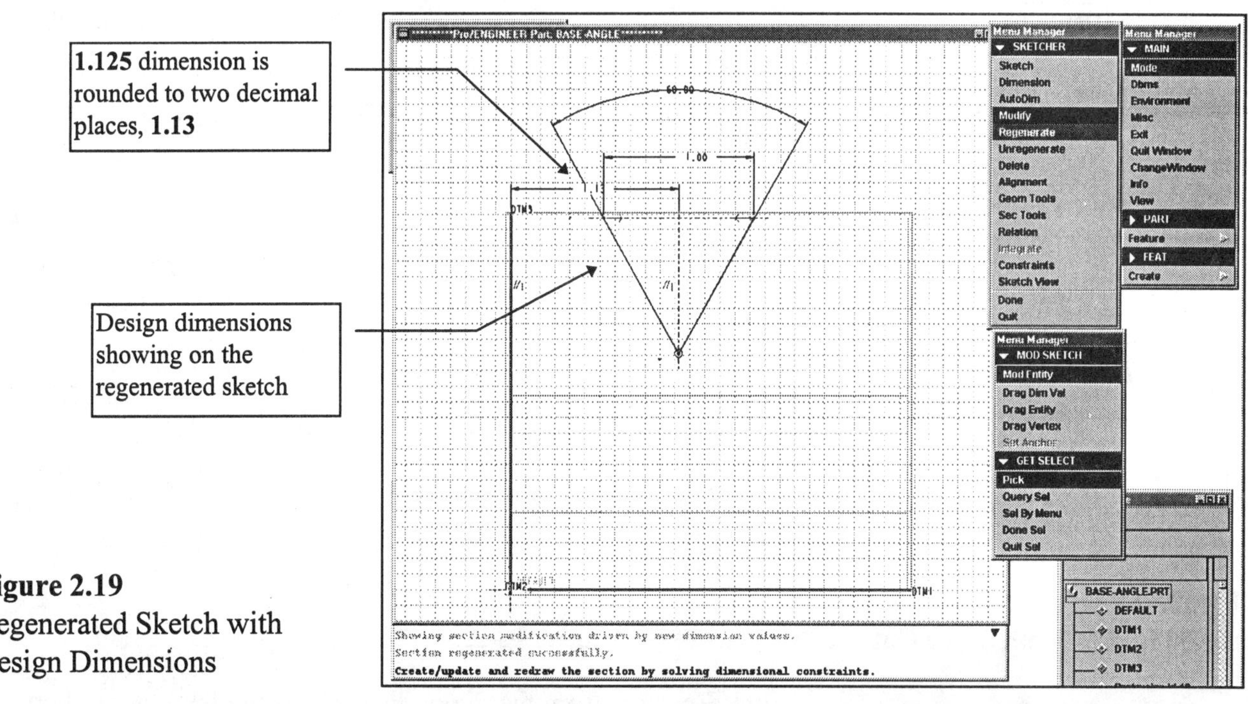

Figure 2.19
Regenerated Sketch with Design Dimensions

After the sketch is regenerated correctly, choose **Done**. The next **Element** that needs defining is the material removal direction. An arrow is displayed (Fig. 2.20), and you are prompted with **Arrow points TOWARD area to be REMOVED.** Pick **Flip** or **Okay.** Choose **Okay**, since the arrow is oriented correctly.

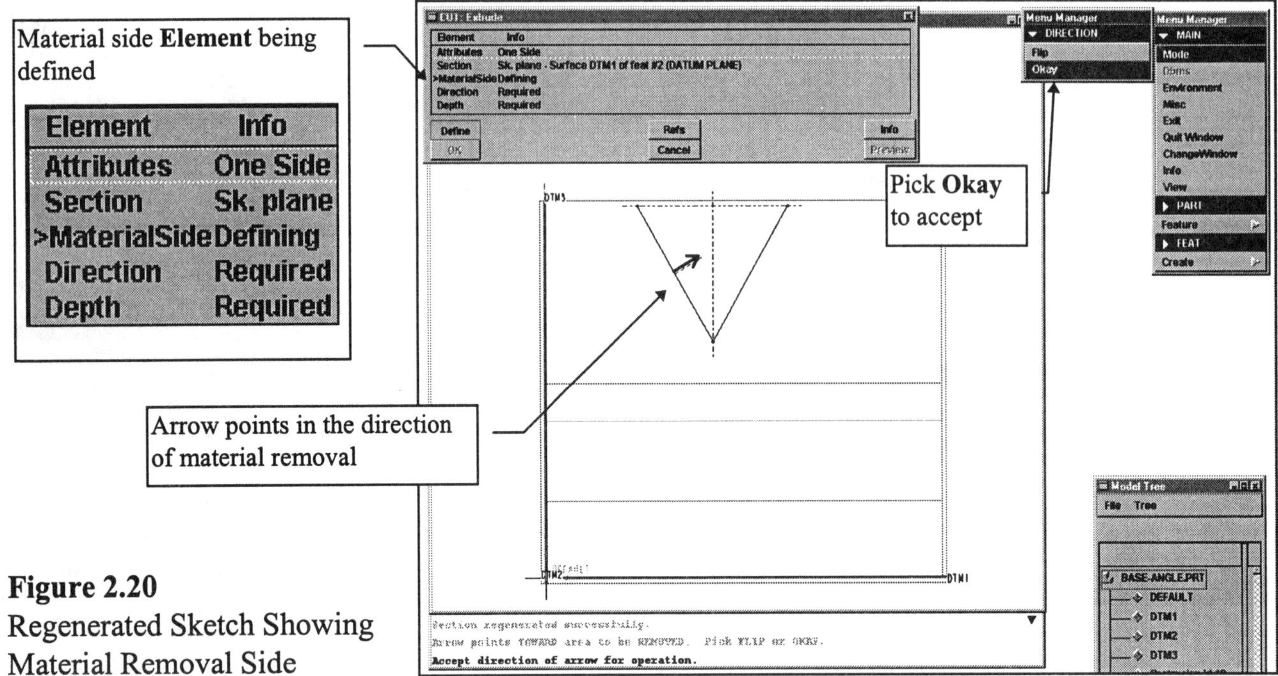

Figure 2.20
Regenerated Sketch Showing Material Removal Side

After picking **Okay**, choose **View** ⇒ **Default** ⇒ **Done-Return** ⇒ **Thru Until** ⇒ **Done** ⇒ (pick the vertical plane shown in Fig. 2.21).

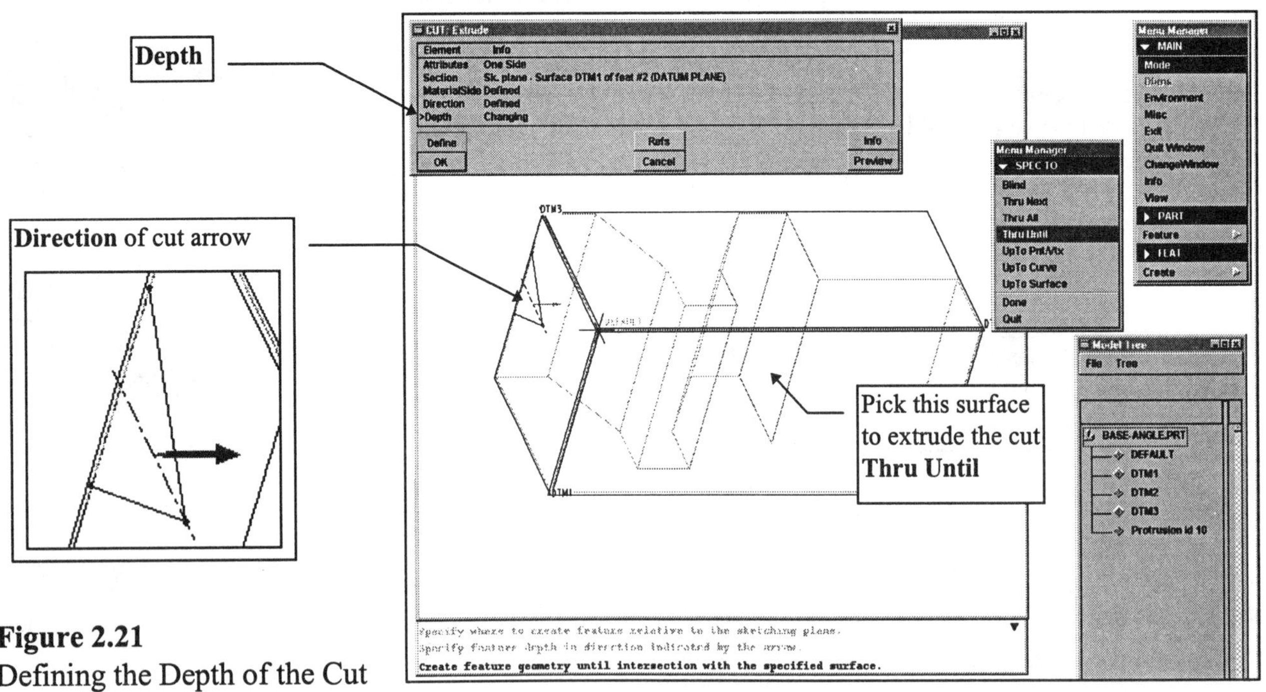

Figure 2.21
Defining the Depth of the Cut

Pick **Preview** from the dialog box to display the cut, as shown in Figure 2.22. Instead of accepting the design by picking the **OK** button, pick the **Depth** element in the dialog box and then pick **Define**. Choose **Thru Next** from the SPEC TO menu (Fig. 2.23), then **Done**. Choose **Preview** from the dialog box.

PT/Modeler™

PT/Modeler does not have **Upto Surface**

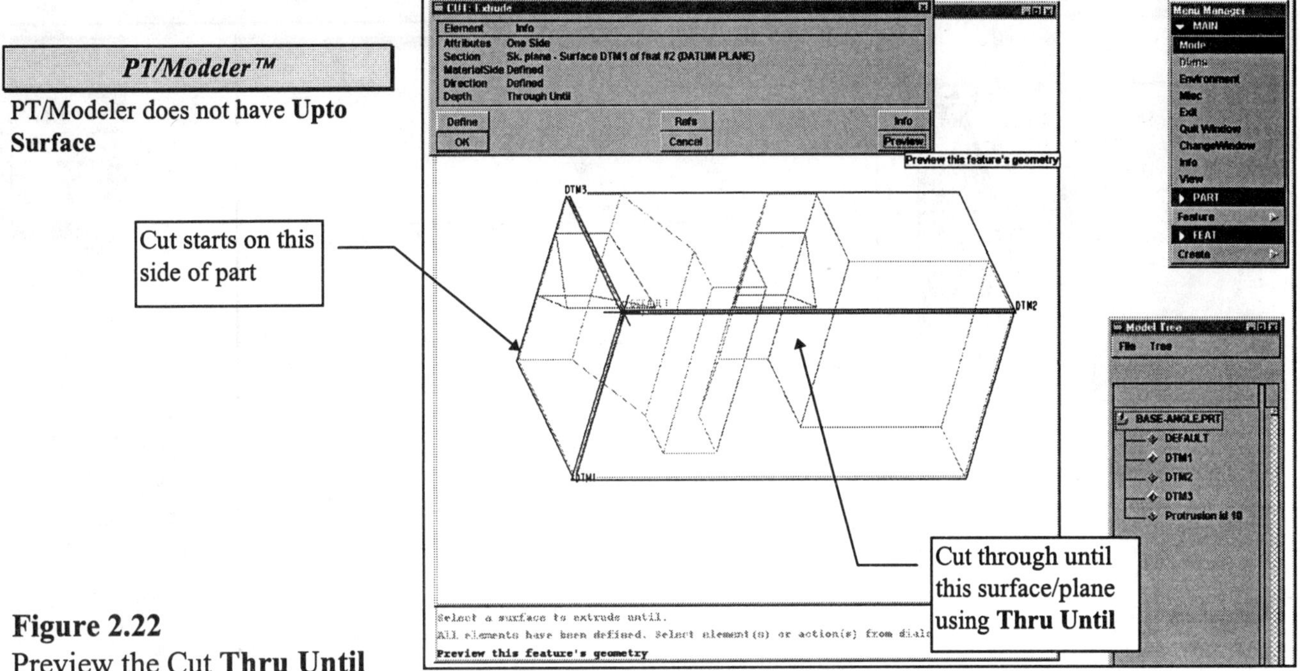

Figure 2.22
Preview the Cut **Thru Until**

The cut now goes up to the next feature it encounters and stops there (Fig. 2.23).

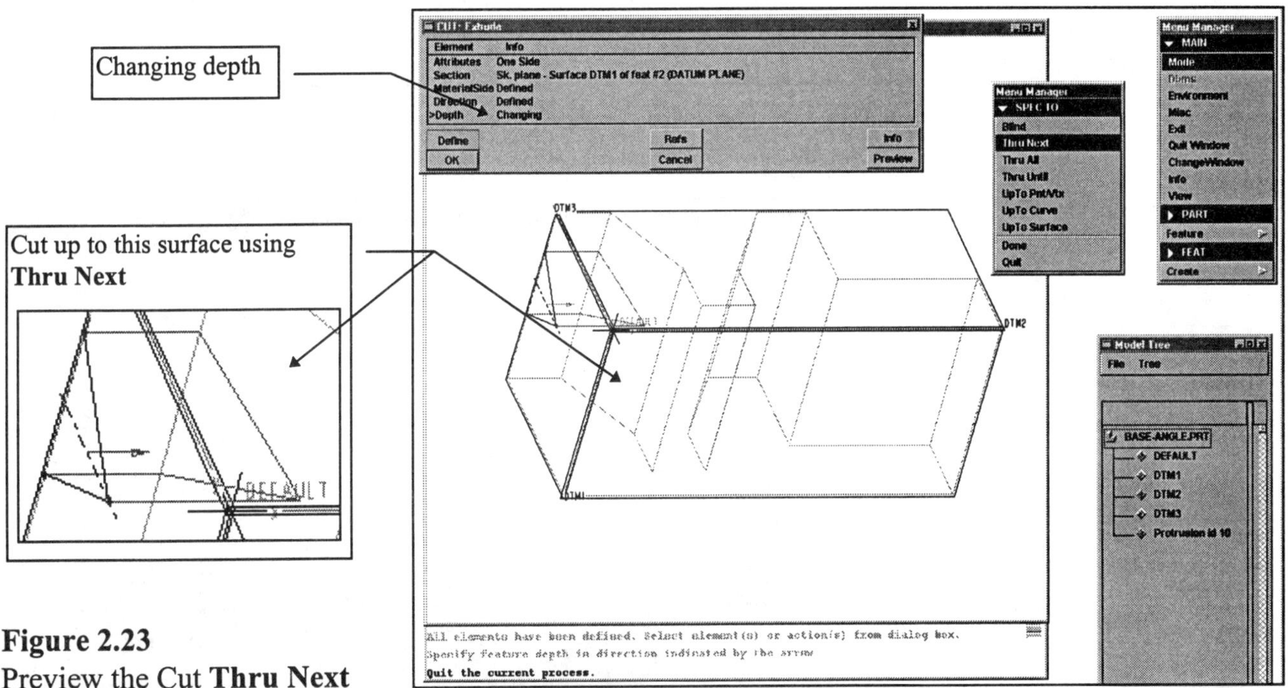

Figure 2.23
Preview the Cut **Thru Next**

Pick **Depth** and **Define** from the dialog box again, select **Upto Surface**, then **Done,** and pick the same surface as in the first example when you selected **Thru Until** ⇒ **Preview** ⇒ **OK** (from the dialog box) ⇒ **Done**. The part should look like Figure 2.24.

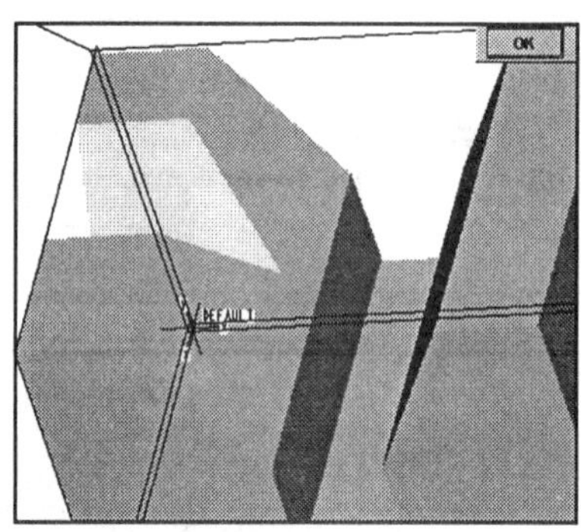

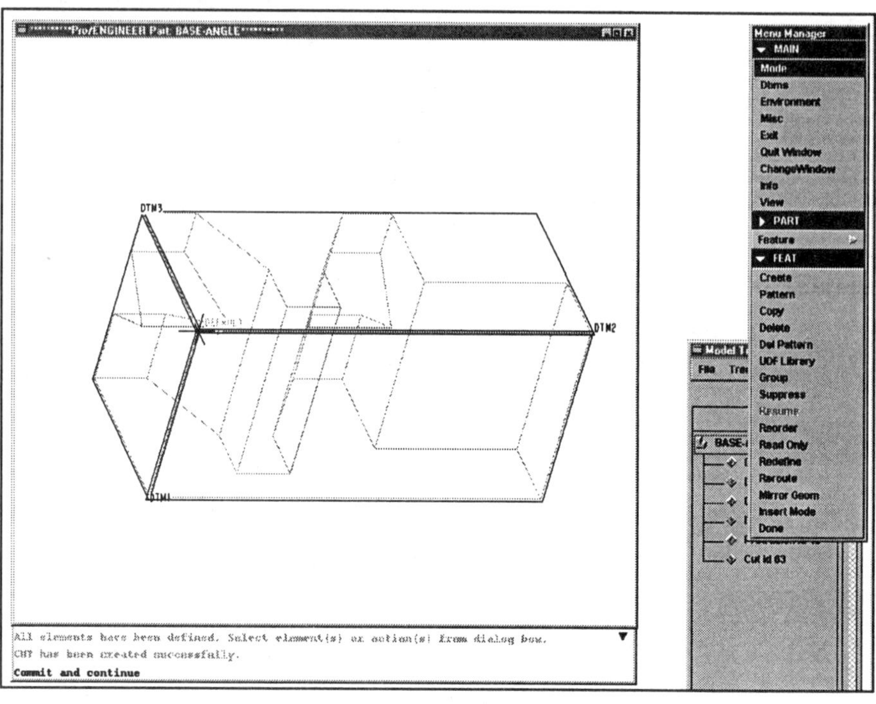

Figure 2.24
Preview the Cut
Upto Surface

Change the part's environment, shading, orientation, and position on the screen to those shown in Figure 2.25. Change the environment by choosing the following commands:

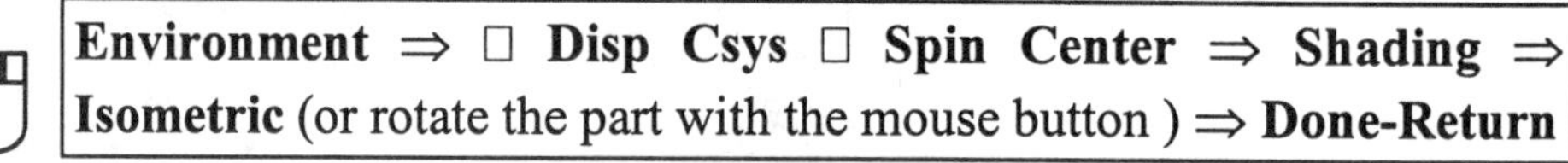

Environment ⇒ □ **Disp Csys** □ **Spin Center** ⇒ **Shading** ⇒ **Isometric** (or rotate the part with the mouse button) ⇒ **Done-Return**

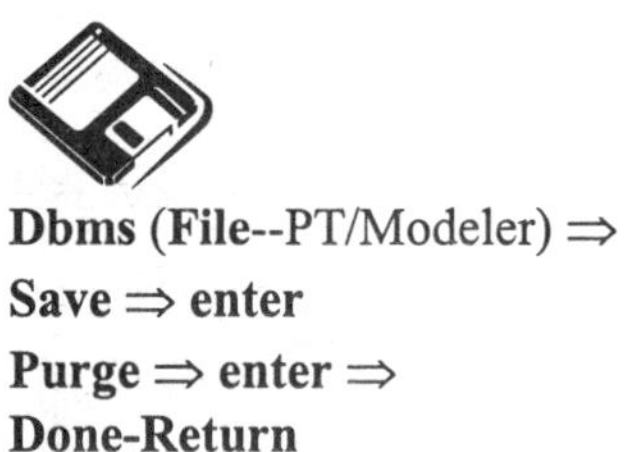

Dbms (File--PT/Modeler) ⇒
Save ⇒ **enter**
Purge ⇒ **enter** ⇒
Done-Return

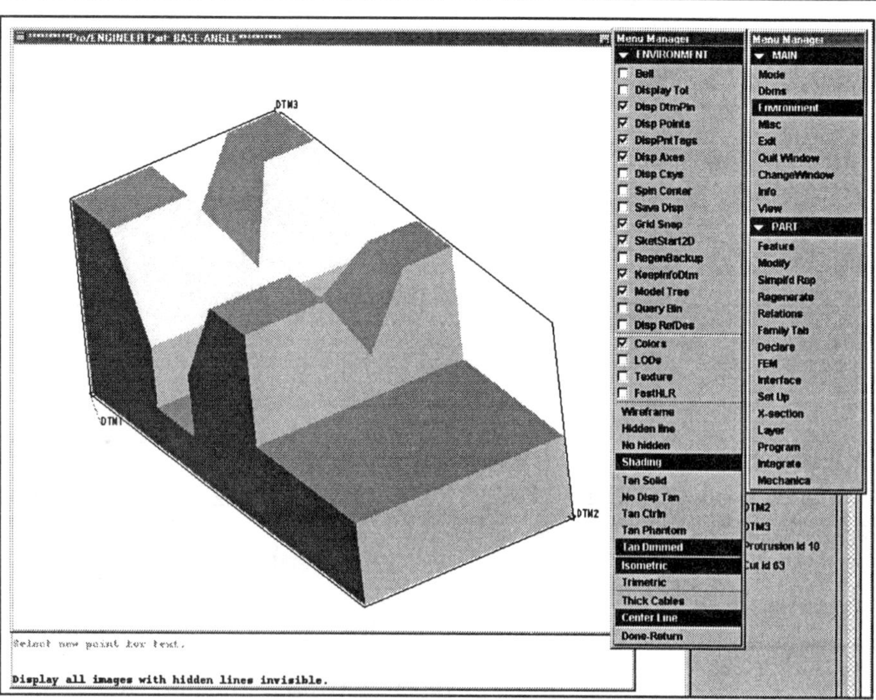

Figure 2.25
Shaded Part

The cut is now complete. Next, the angle cut and the **U**-shaped cut need to be created.

Before creating the next cut, set the environment and view with the following commands:

Environment ⇒ Hidden line ⇒ Isometric ⇒ Done-Return ⇒ View ⇒ Repaint ⇒ Done-Return ⇒ View ⇒ Default ⇒ Done-Return

HINT

You can make only *one open* section cut at a time. You can make *multiple closed* sections as one feature, but this is considered poor design practice.

The next feature will be another cut. Two cuts are still required for the completion of the part. They can be created together using closed sections or separately with one open section at a time. It is better design intent to make the cuts separately, since they do not have any particular relationship except that they cut the same direction and start on the same surface/plane. A *closed section* is a sketched set of entities that start and end at the same position, like a square □ **shape.** An *open section* does not form a closed figure; an example would be a ∩ or ∪ **shape.**

Start the angled cut by the choosing the following commands:

PT/Modeler™

Feature ⇒ Cut etc.

Feature ⇒ Create ⇒ Cut ⇒ Extrude ⇒ Solid ⇒ Done ⇒ One Side ⇒ Done ⇒ (pick **DTM2** as the sketching/placement plane) ⇒ **Flip** (change the direction of the cut to pass through the part, not out into space) ⇒ **Okay ⇒ Top ⇒** (pick **DTM3** as the orientation plane)

In Figure 2.26, the direction of cut is selected.

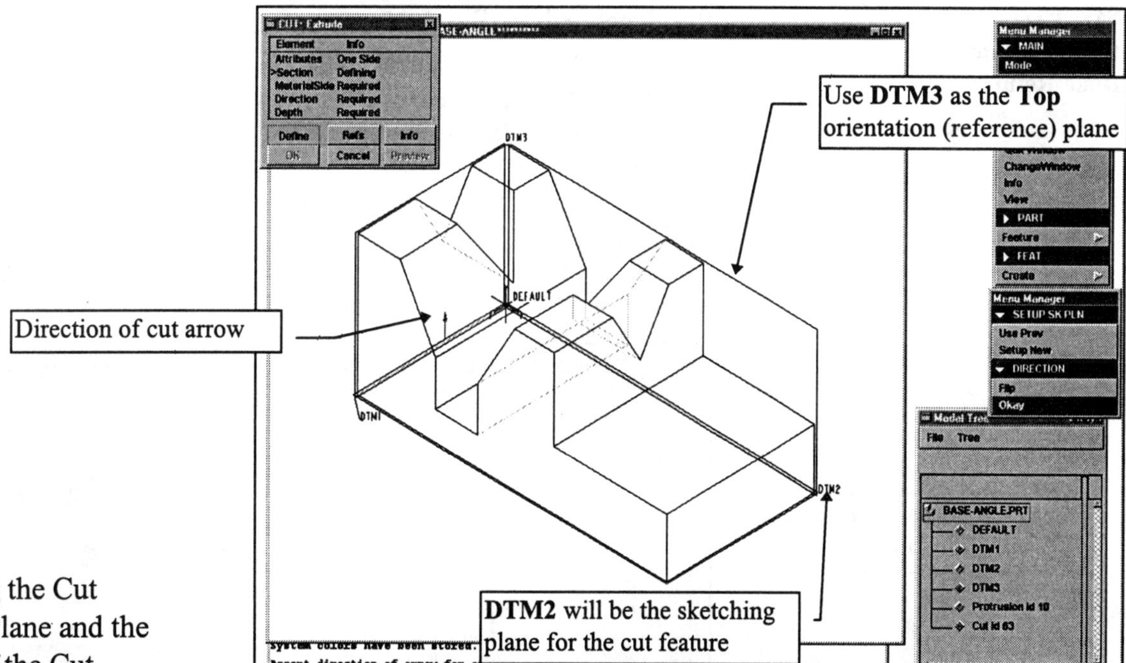

Figure 2.26
Establishing the Cut Placement Plane and the Direction of the Cut

Turn off the **Grid Snap** in the ENVIRONMENT menu: **Environment ⇒ □ Grid Snap ⇒ Done-Return**. Your screen should show the part in the Sketcher looking from the bottom, through the part, as in Figure 2.27. Because you are looking at the bottom, the cut will be sketched on the upper right of the part.

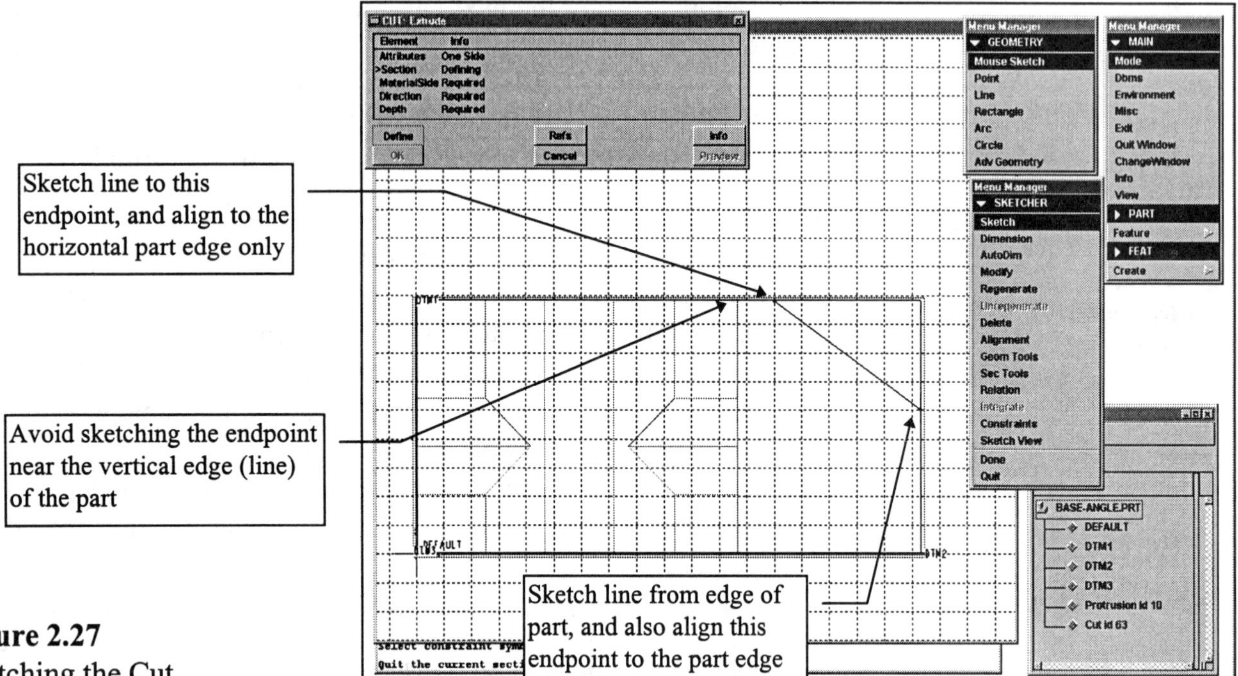

Figure 2.27
Sketching the Cut

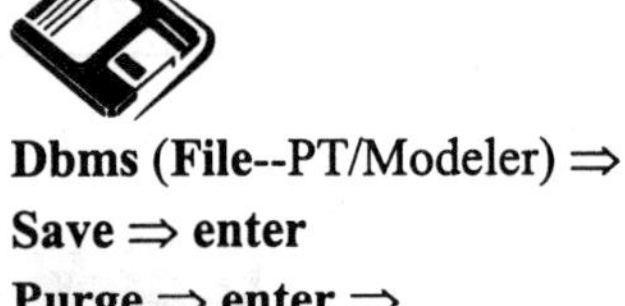

Dbms (**File**--PT/Modeler) ⇒
Save ⇒ **enter**
Purge ⇒ **enter** ⇒
Done-Return

Do not sketch the endpoint of the line near the left vertical edge (Fig. 2.27) of the part feature, in order to avoid an unwanted assumption that that point and the part edge are exactly at the same position. This will allow the cut to be modified later by changing the value of the cut dimension, then **Regenerate.** Align the endpoints of the sketched line (Fig. 2.27), then **Regenerate.**.

Dimension, **Regenerate** (Fig. 2.28), **Modify** the dimensions using the design values from Figure 2.15, and **Regenerate** the cut. After you pick **Done**, the side of material removal must be selected (Fig. 2.29).

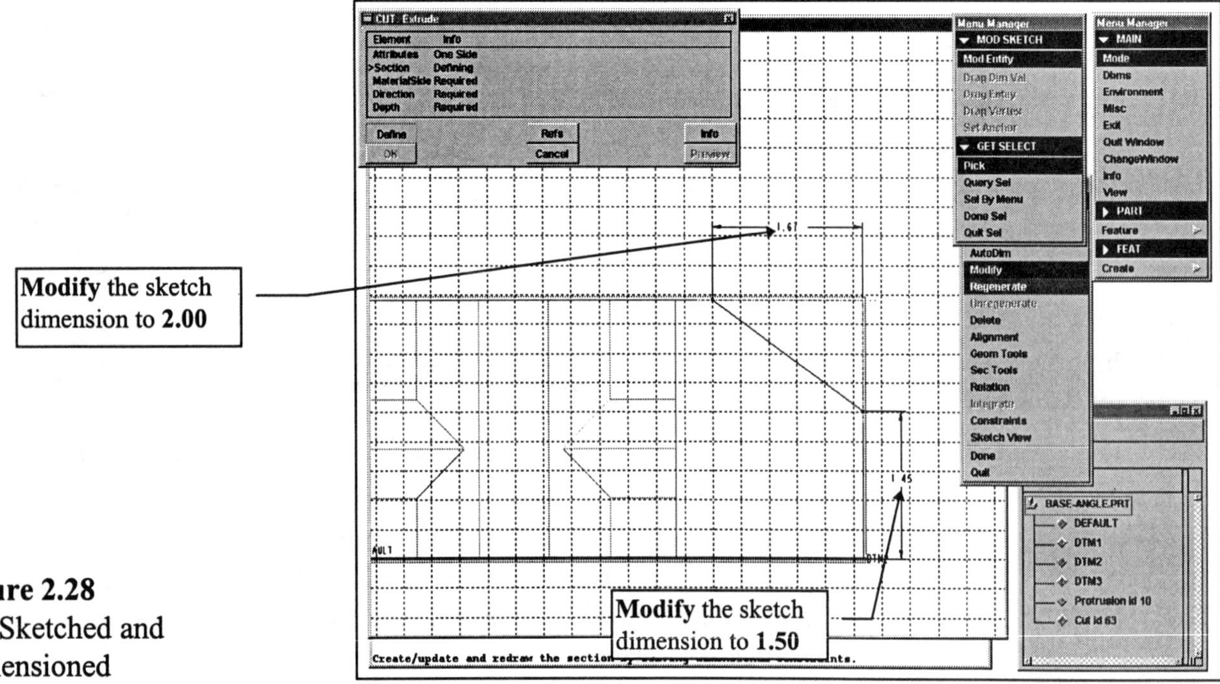

Figure 2.28
Cut Sketched and Dimensioned

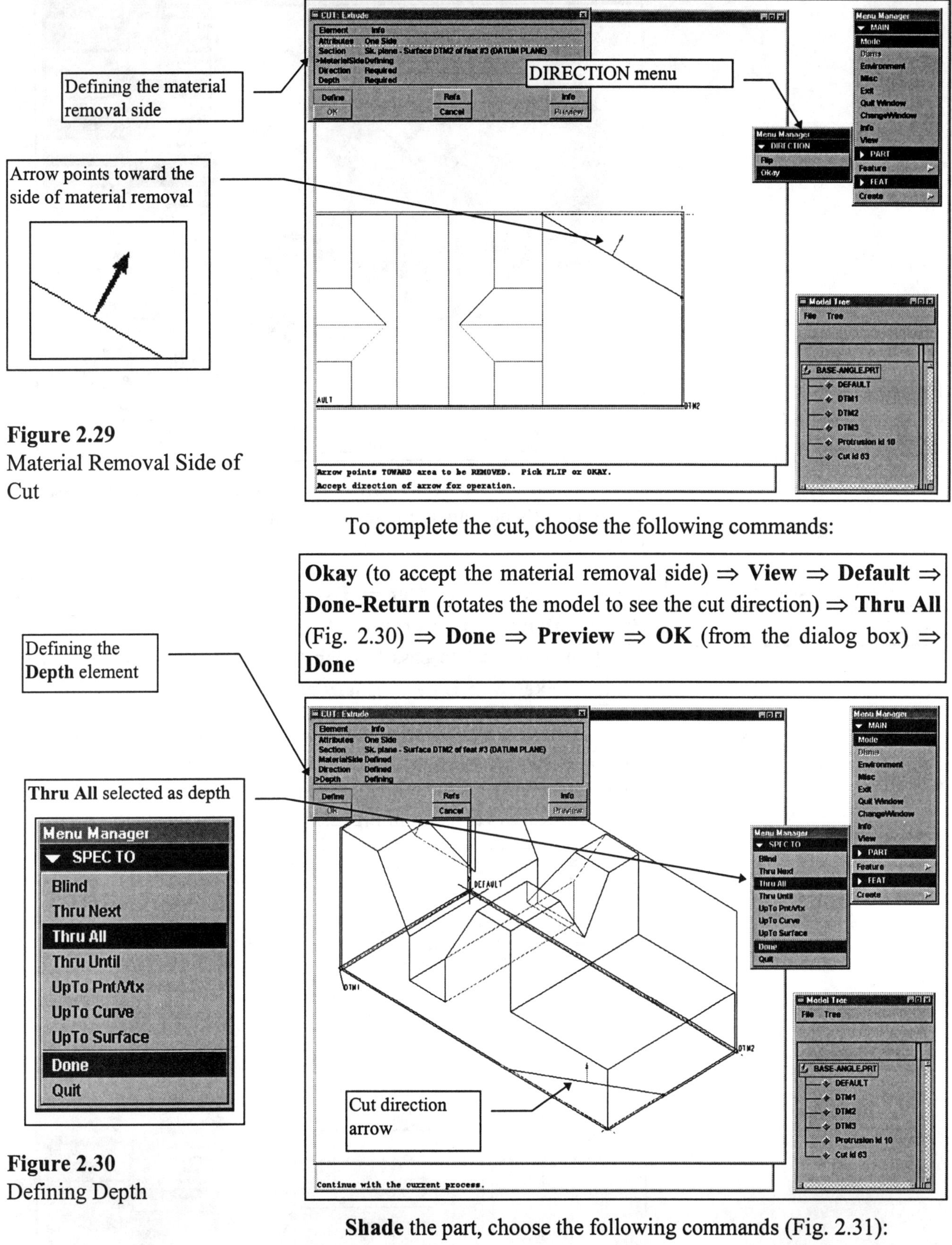

Figure 2.29
Material Removal Side of Cut

To complete the cut, choose the following commands:

Okay (to accept the material removal side) ⇒ **View** ⇒ **Default** ⇒ **Done-Return** (rotates the model to see the cut direction) ⇒ **Thru All** (Fig. 2.30) ⇒ **Done** ⇒ **Preview** ⇒ **OK** (from the dialog box) ⇒ **Done**

Figure 2.30
Defining Depth

Shade the part, choose the following commands (Fig. 2.31):

View ⇒ **Cosmetic** ⇒ **Shade** ⇒ **Display** ⇒ **Done-Return** ⇒ **View** ⇒ **Repaint** ⇒ **Done-Return**

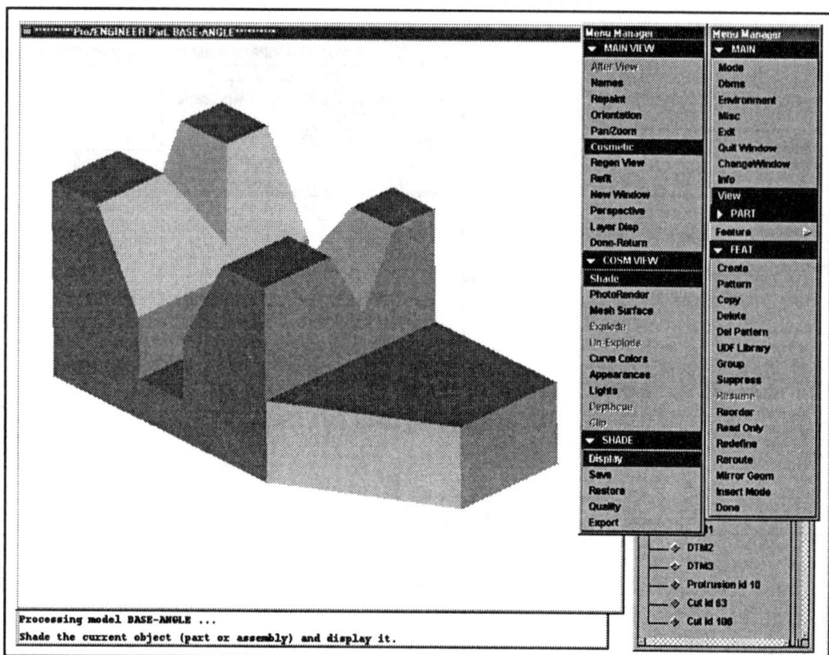

Figure 2.31
Completed Cut Shown as a Shaded Model

The last feature of the part is the **U**-shaped cut. We can use the previous cut's placement plane and orientation. Choose the following commands:

PT/Modeler™

Feature ⇒ Cut etc.

Feature ⇒ Create ⇒ Cut ⇒ Extrude ⇒ Solid ⇒ Done ⇒ One Side ⇒ Done ⇒ Use Prev (sketching plane) **⇒ Flip** (change the direction of the cut to pass through the part, not out into space) **⇒ Okay ⇒ Sketch ⇒ Line ⇒ Horizontal**

Sketch the three lines as shown in Figure 2.32. Align the endpoints of the two horizontal lines with the right side of the part, then **Regenerate. Dimension**, **Regenerate**, **Modify** (the dimensions using the design values from Fig. 2.15), and **Regenerate** the cut.

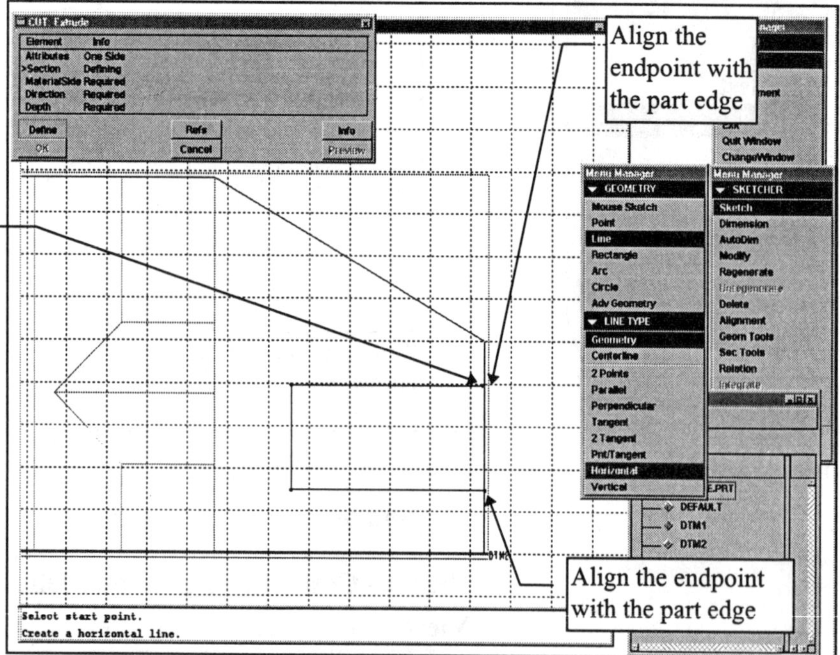

Figure 2.32
U-Shaped Cut

The cut is dimensioned, modified, and regenerated in Figure 2.33.

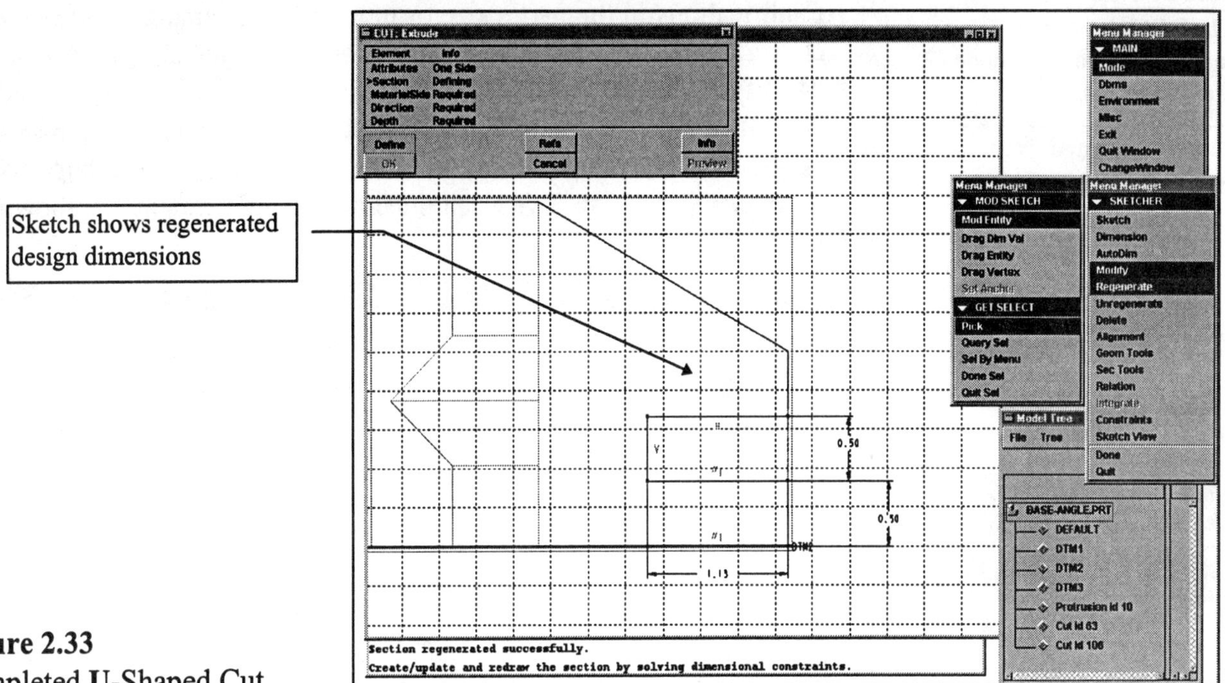

Figure 2.33
Completed **U**-Shaped Cut

Complete the part with the following commands (Fig. 2.34):

Done ⇒ **Okay** (accepts material removal side) ⇒ **Thru All** ⇒ **Done** ⇒ **OK** (from dialog box) ⇒ **View** ⇒ **Default** ⇒ **Done-Return**

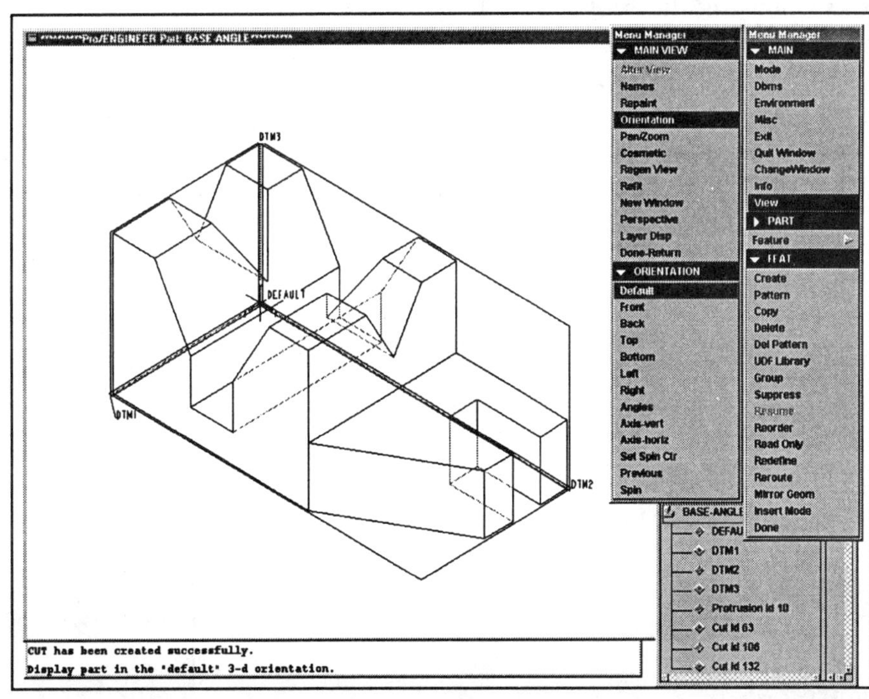

HINT

Save and **Purge** after every new feature is created:
Dbms (**File**--PT/Modeler) ⇒
Save ⇒ **enter**
Purge ⇒ **enter** ⇒
Done-Return

Figure 2.34
Completed **BASE_ANGLE**

Since you will be modifying and redefining the part, it might be a good idea to save this version under another name:

Dbms (File--PT/Modeler) ⇒ **Save As** ⇒ **enter** ⇒ **BASE_ANGLE_A** ⇒ **enter** ⇒ **Done-Return**

NOTE

The **ECO** shown here is provided in the Pro/E **Format** mode. To see the format, choose:

Mode ⇒ **Format** (from the MODE menu) ⇒ **Search/Retr** ⇒ **Format Dir** ⇒ **ecoform.frm**

Very few projects make it through the design, engineering, and manufacturing sequence without changes. The changes can be simple modifications in the part's size or more extreme changes in the part's configuration. When simple dimensional changes are requested, the **Modify** command is used; when configuration changes are needed, the **Redefine** command is called upon. The organization (engineering, manufacturing, and so on) issuing the changes will normally release an **ECO** (engineering change order), as shown in Figure 2.35.

ECO

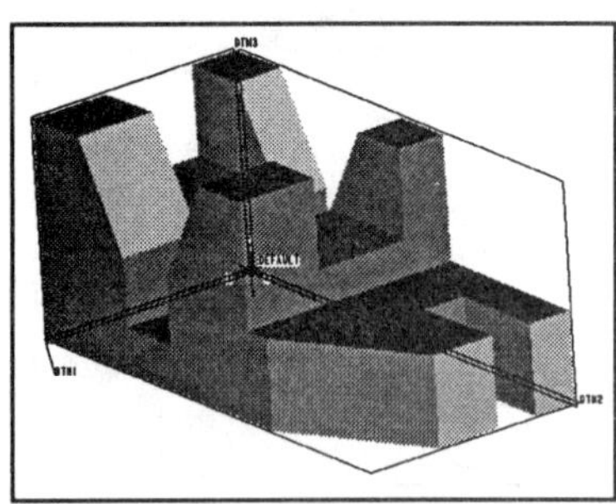

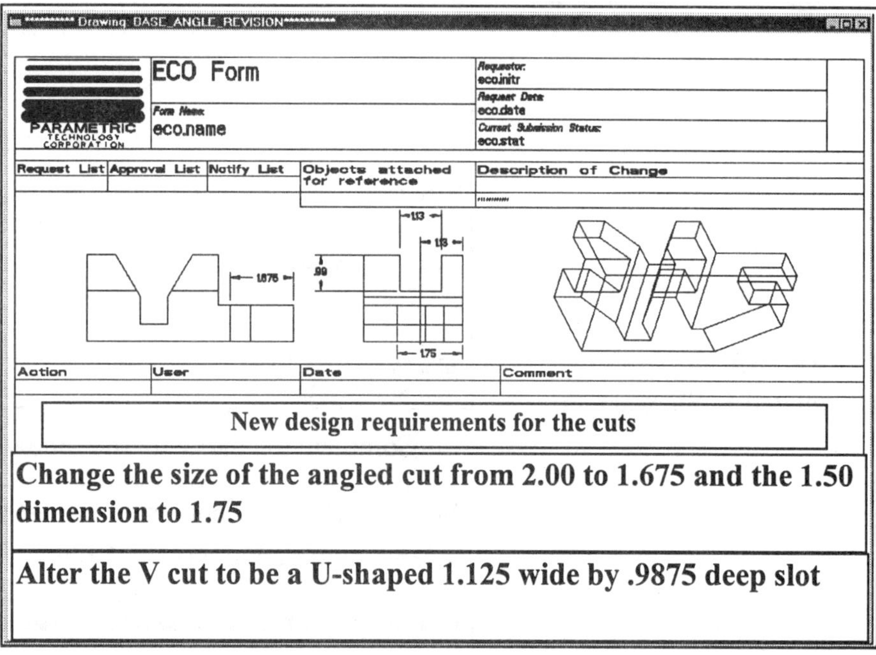

Figure 2.35
ECO Changing the Angled Cut Size and the **V** Cut

You can **Modify** and **Redefine** from the menu structure or from the Model Tree. Use the **Model Tree** for both ECOs (Fig. 2.36).

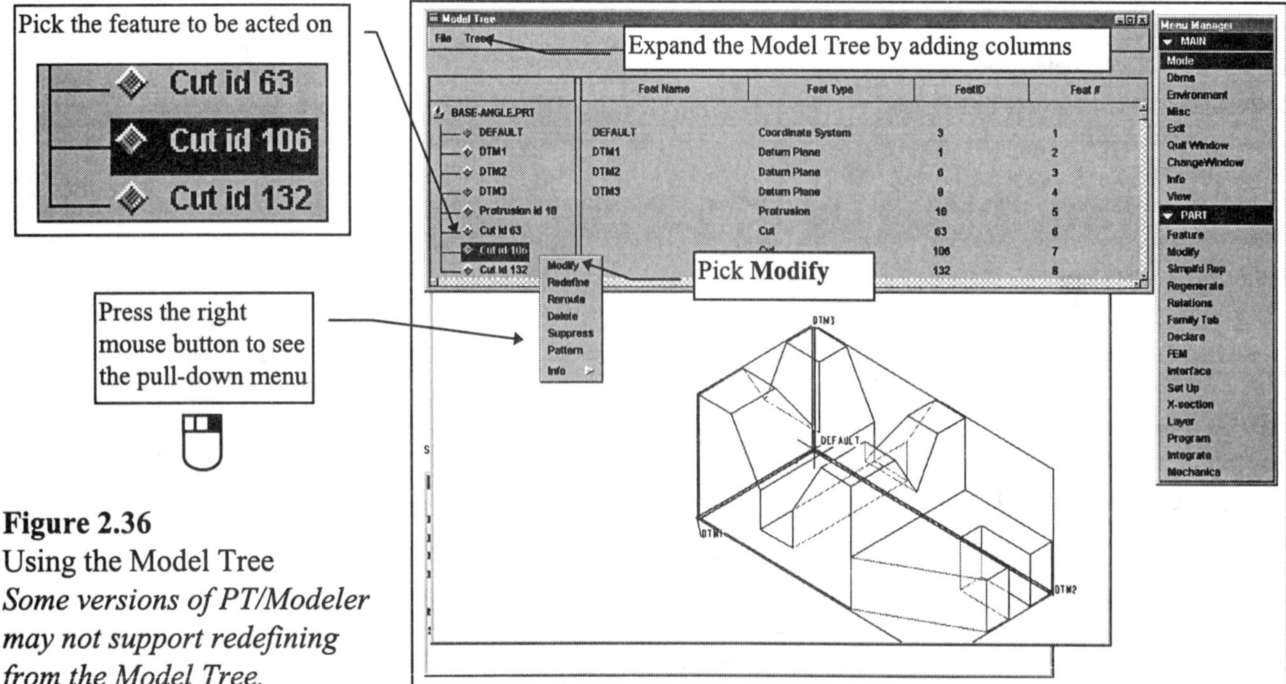

Figure 2.36
Using the Model Tree
Some versions of PT/Modeler may not support redefining from the Model Tree.

After selecting the feature to be modified from the Model Tree, press and hold down the right mouse button to display the pull-down menu (the mouse cursor must be somewhere in the Model Tree for this to work). Move the mouse to highlight the **Modify** command and then release the mouse button. Pick the **2.00** dimension, type the new value in the command line (**1.675**), and press **enter**; do likewise to the **1.50** dimension to change it to **1.75** (Fig. 2.37). **Regenerate** the part to see the changes. **Shade** and rotate the part to see the changes clearly (Fig. 2.38).

PT/Modeler™

Access **Modify** from the PART menu and **Redefine** from the FEATURE OPER menu.

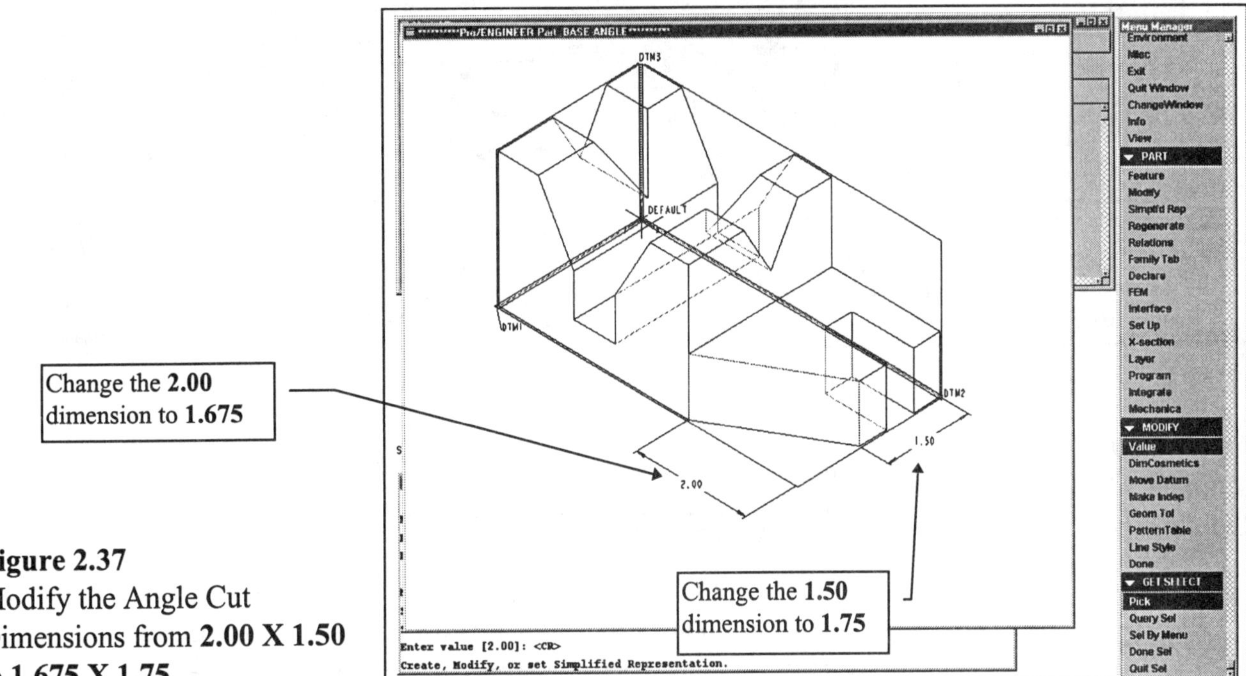

Figure 2.37
Modify the Angle Cut Dimensions from **2.00 X 1.50** to **1.675 X 1.75**

Dbms (File--PT/Modeler) ⇒
Save ⇒ **enter**
Purge ⇒ **enter** ⇒
Done-Return

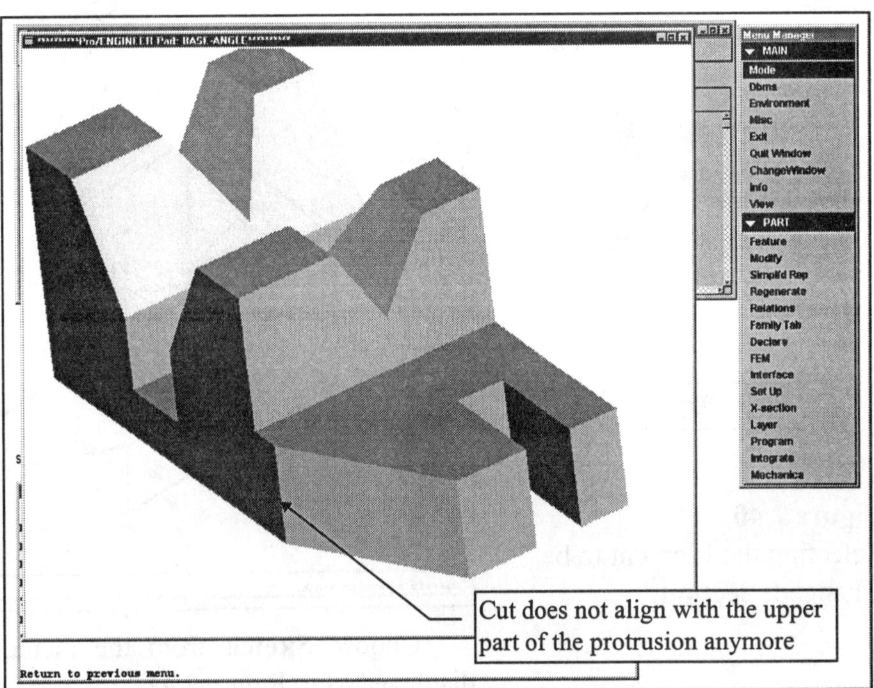

Figure 2.38
Modified Cut

The other requirement from the ECO was to change the shape and the size of the **V** cut. Alter the **V** cut to be a **U**-shaped slot **1.125** wide by **.9875** deep. Again let's use the Model Tree for the feature and the command selection (Fig. 2.39). For the proper pop-up menu to display, the PART menu must appear in the menu structure.

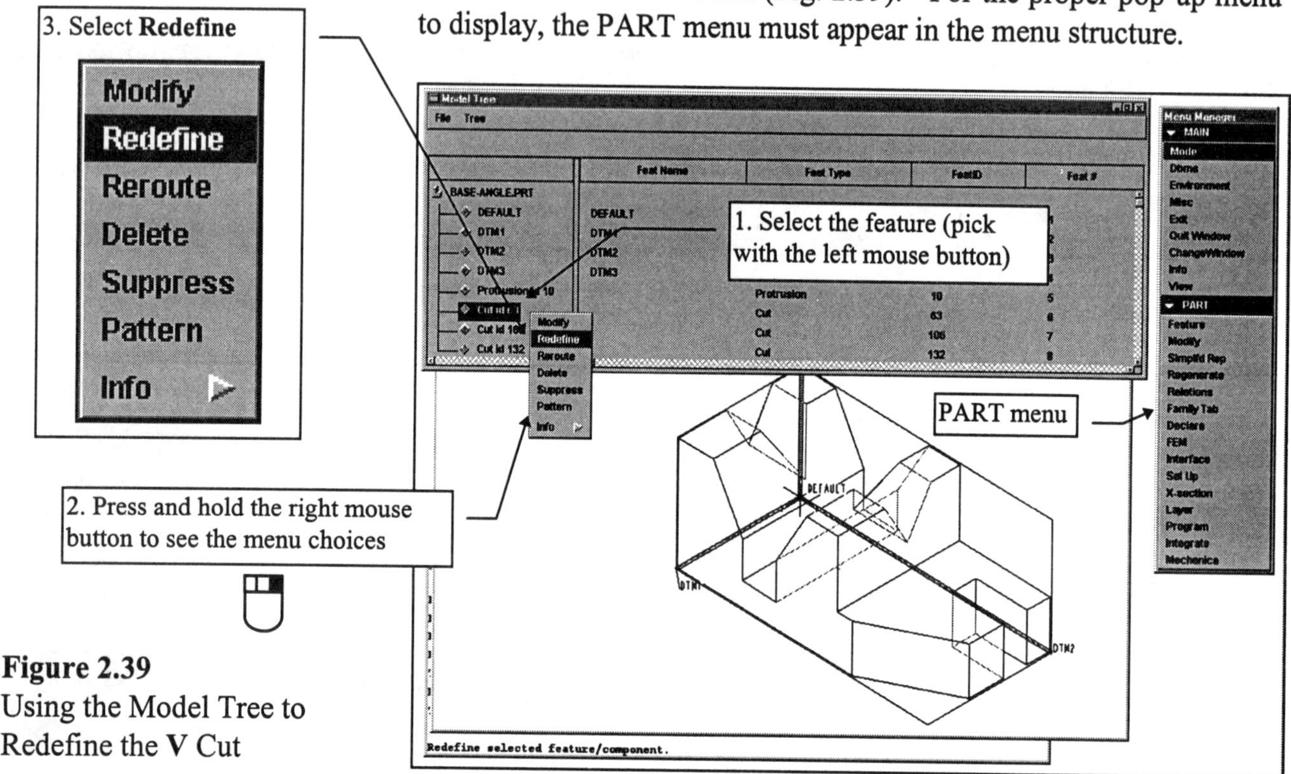

Figure 2.39
Using the Model Tree to Redefine the **V** Cut

From the dialog box, pick **Section** and then choose **Define**, as shown in Figure 2.40. You are now able to redefine the section.

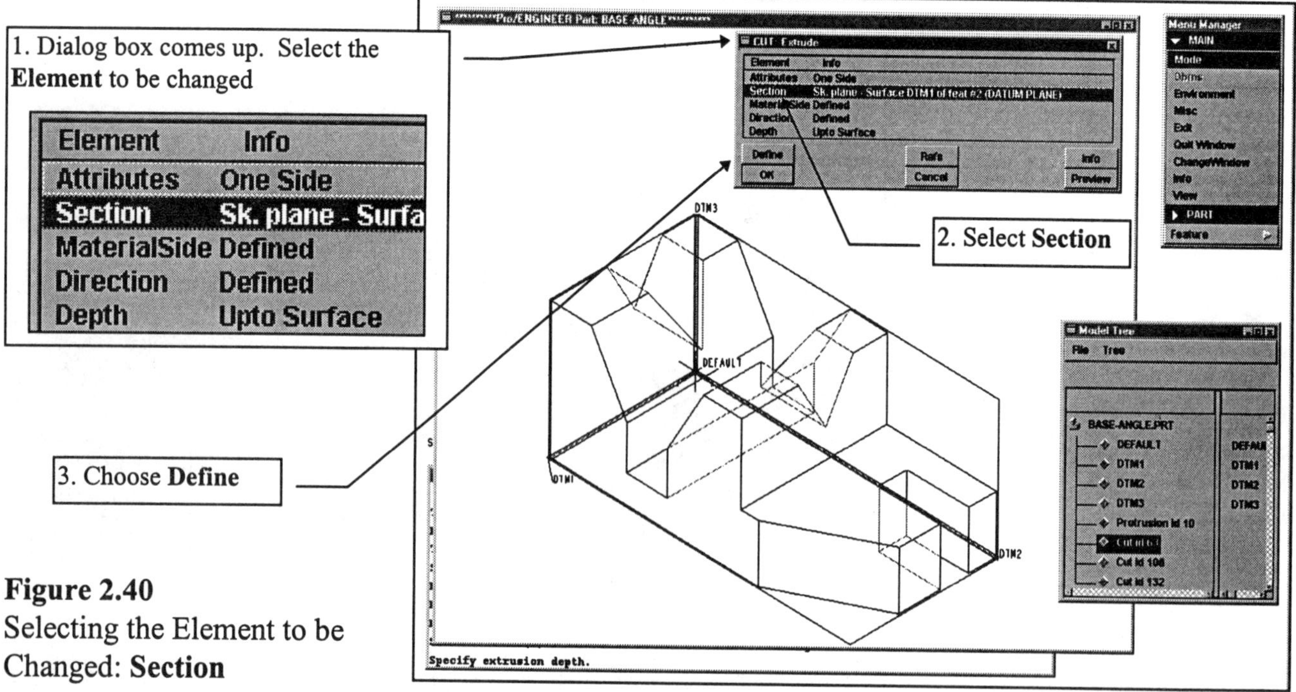

Figure 2.40
Selecting the Element to be Changed: **Section**

Choose **Sketch** from the menu. The section sketch will be displayed as in Figure 2.41.

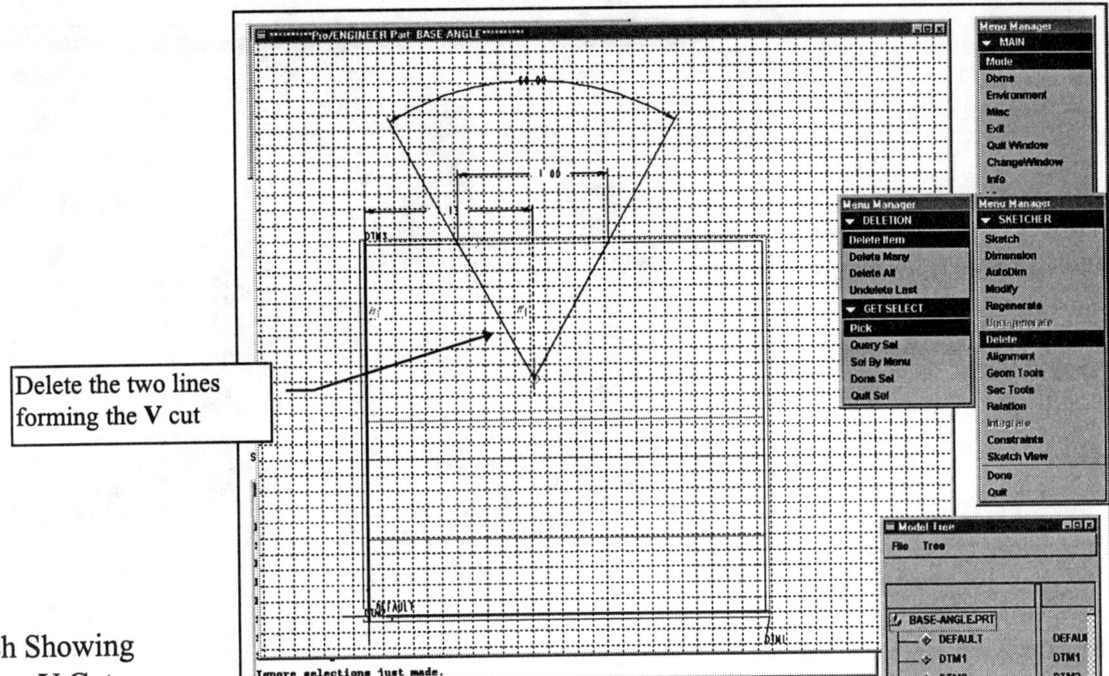

Figure 2.41
Section Sketch Showing Original Design **V** Cut

Use the same centerline for the redefined cut section. Delete the two lines forming the **V** and sketch three lines of the slot as shown in Figure 2.42. Align the vertical sketch lines to the top of the part, then dimension using the **ECO** sizes of **1.125** wide by **.9875** deep (Fig. 2.42). The **.9875** dimension rounds to **.99**. **Regenerate** the sketch, then pick **Done** from the SKETCHER menu and **OK** from the dialog box. **Shade** and spin the part (Fig. 2.43). Save the modified and redefined part.

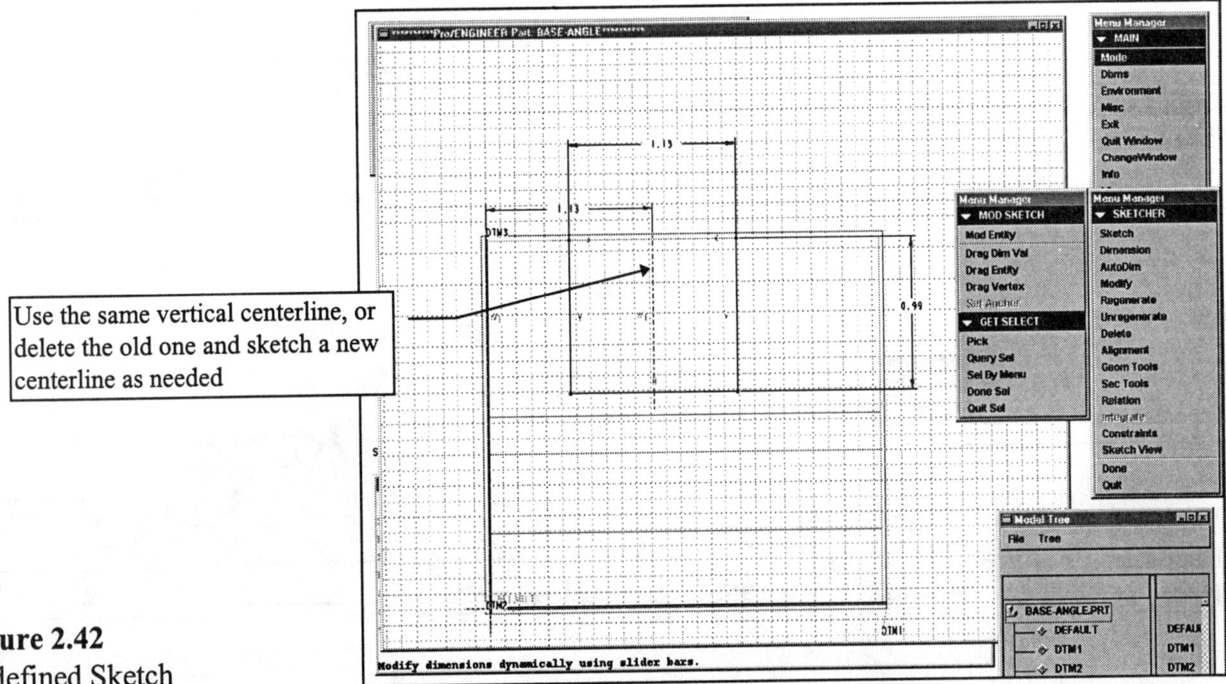

Figure 2.42
Redefined Sketch

Dbms (**File**--PT/Modeler) ⇒
Save ⇒ **enter**
Purge ⇒ **enter** ⇒
Done-Return

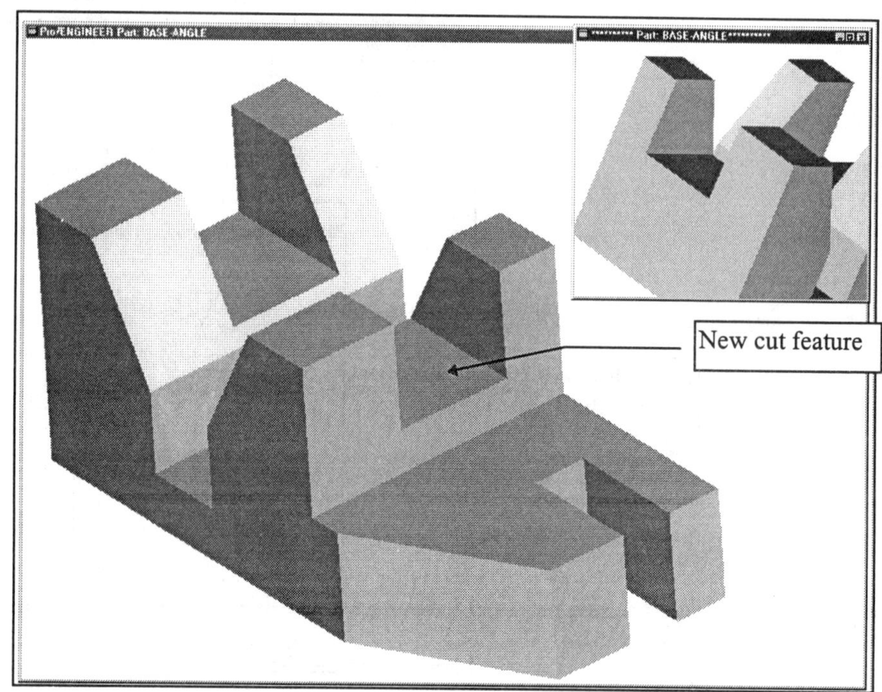

Figure 2.43
ECO Changes

You can also get information regarding your model at any time in the design or redesign process. Using the Model Tree, request information on the part feature just redefined (Fig. 2.44). A variety of information is displayed on the screen, including the **Feature's Dimensions**, **Feature Number**, **Internal Feature ID Number**, and **Parents** (Fig. 2.45).

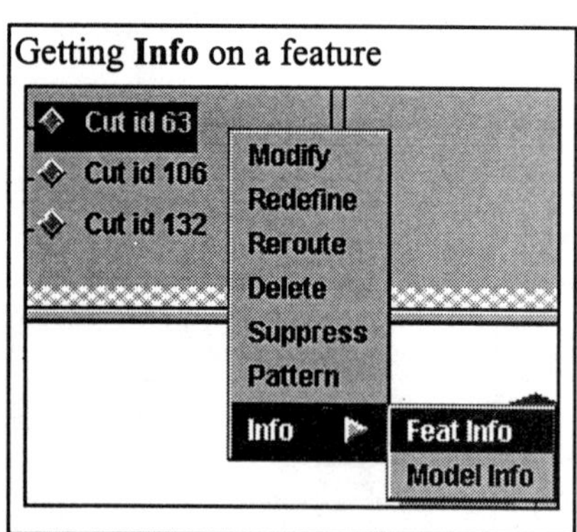

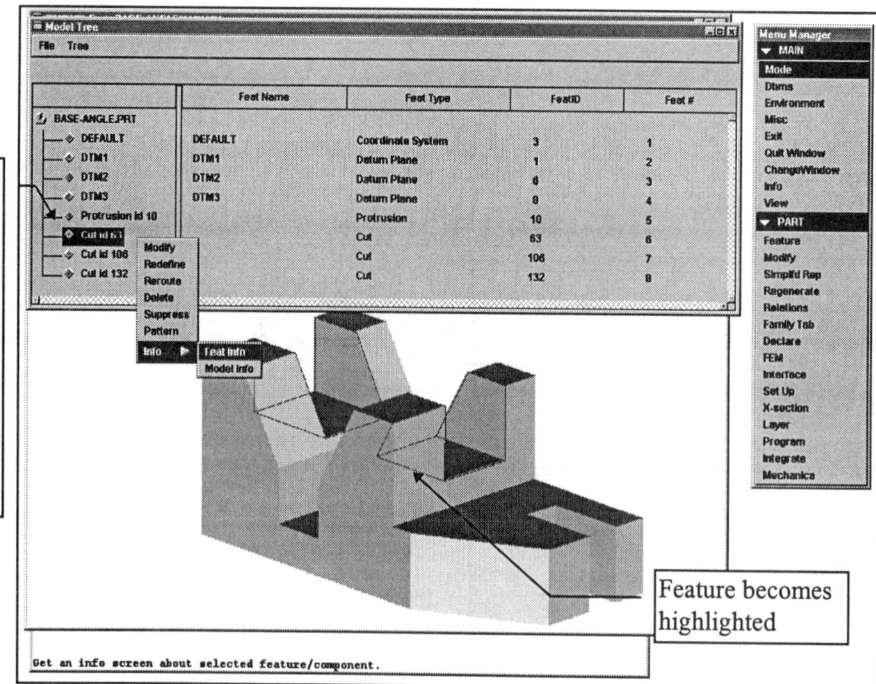

Figure 2.44
Requesting Feature Information

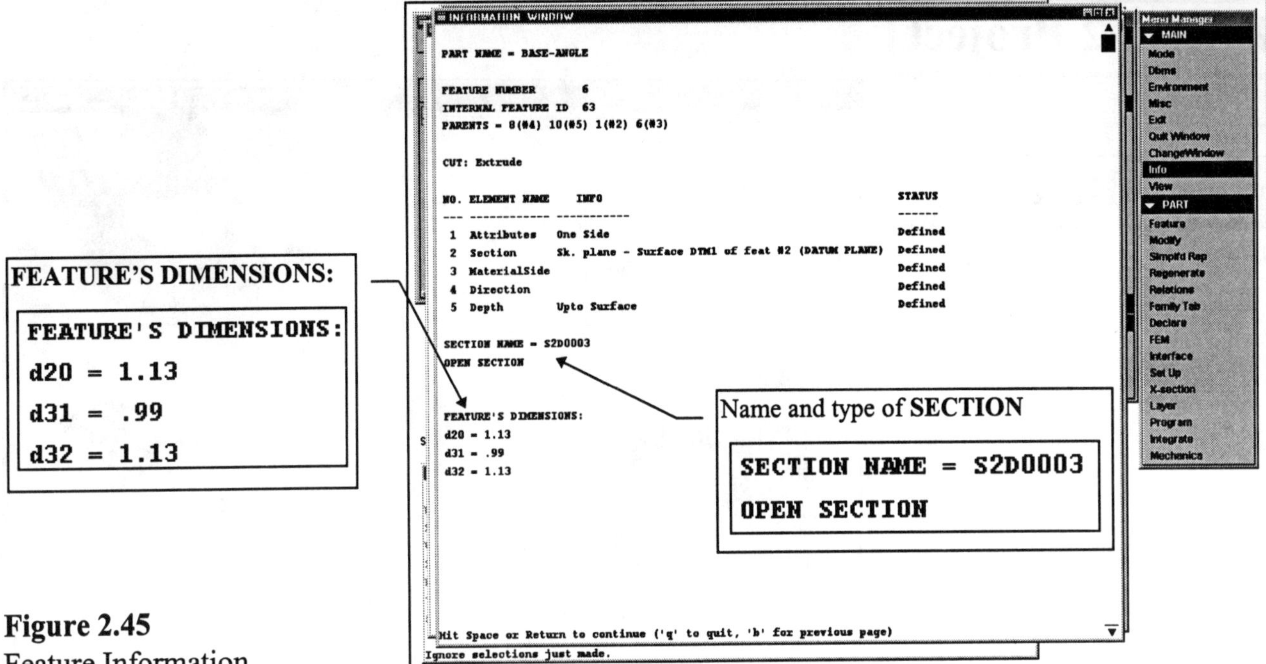

Figure 2.45
Feature Information

Besides feature information, you can extract information on the whole model by choosing the following commands:

Info ⇒ **Model Info** (Fig. 2.46) ⇒ type **q** to quit, **b** for previous page, or press **enter** or the **Spacebar** to continue (mouse cursor must be somewhere inside the INFORMATION WINDOW)

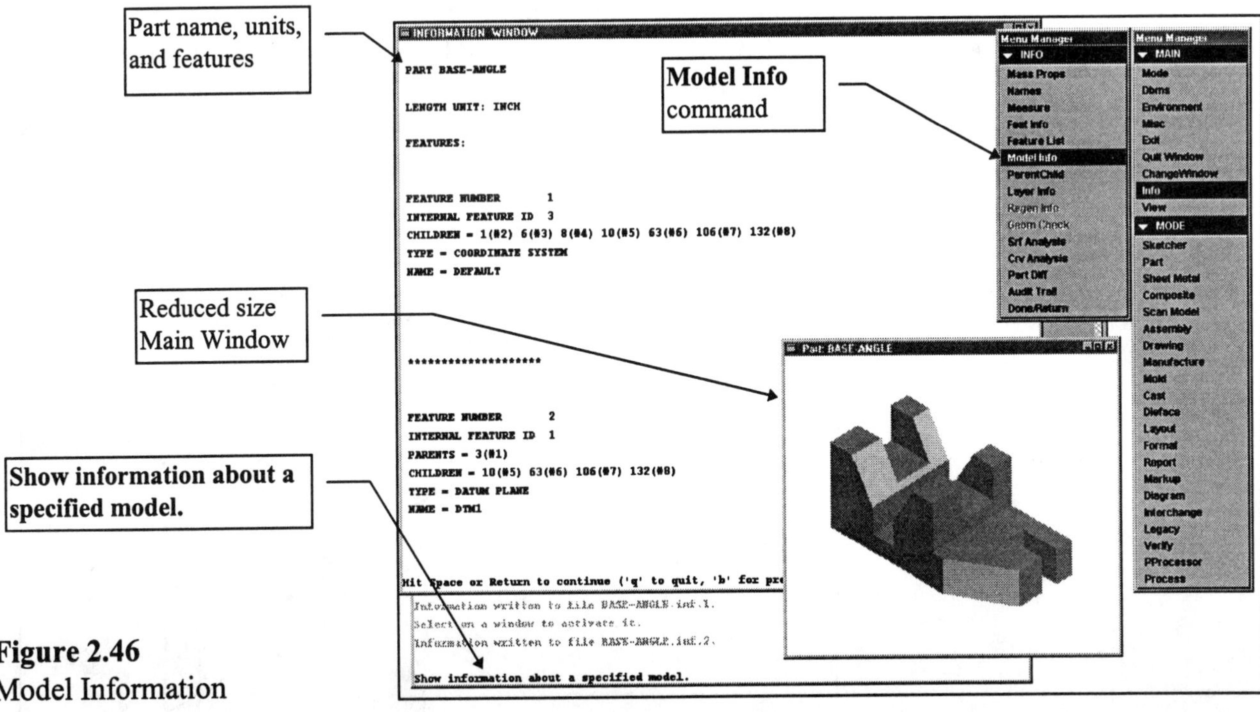

Figure 2.46
Model Information

Lesson 2 Project

T-Block

Figure 2.47
T-Block

T-Block

The second **lesson project** is a block (Figs. 2.47 through 2.51) that is created with types of commands similar to those for the **BASE_ANGLE**. Create datum planes and a default coordinate system. Sketch the protrusion on **DTM3**. After the T-Block is modeled, you will be prompted to modify and redefine a number of its features from an **ECO**.

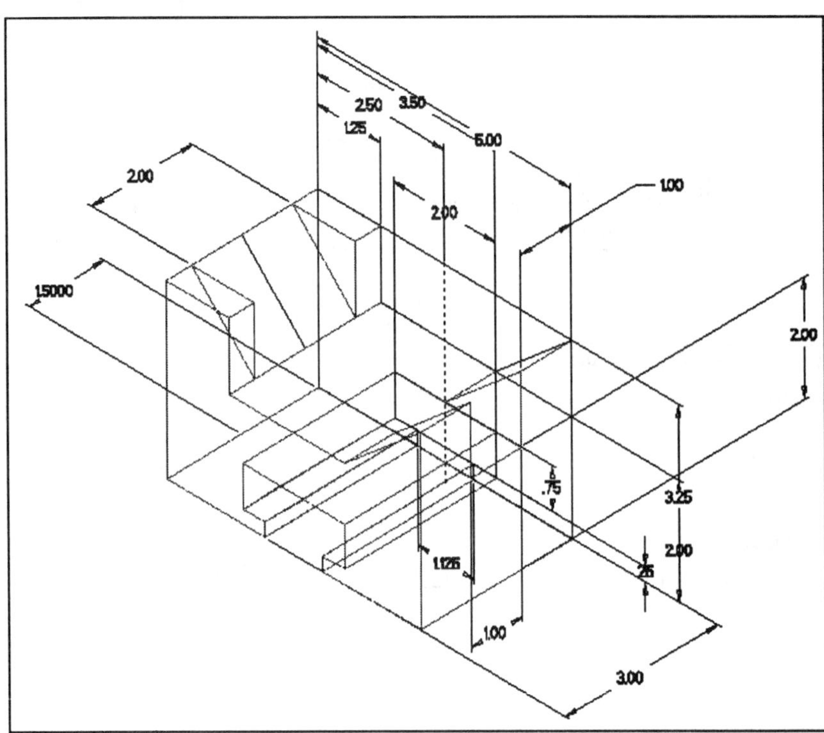

Figure 2.48
T-Block Dimensions

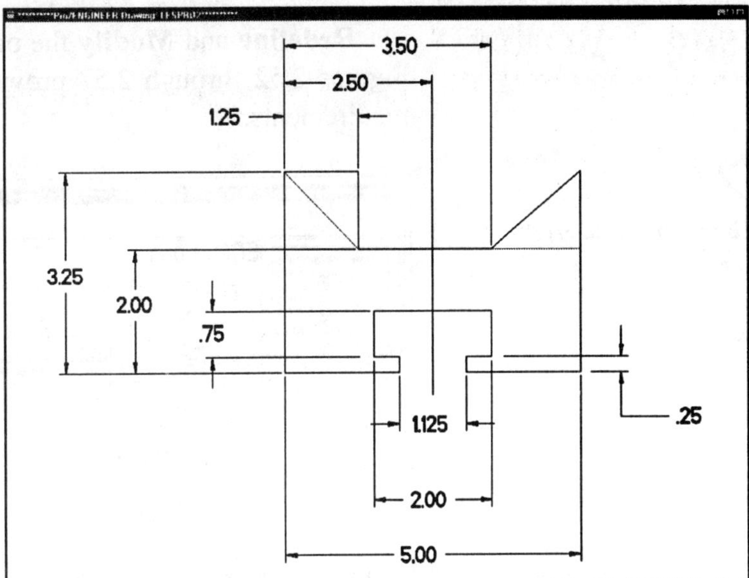

Figure 2.49
Front View

NOTE

The **T**-shaped slot is symmetrical and is located at the center of the part. Later, when completing the ECO, be careful not to assume that this condition still applies!

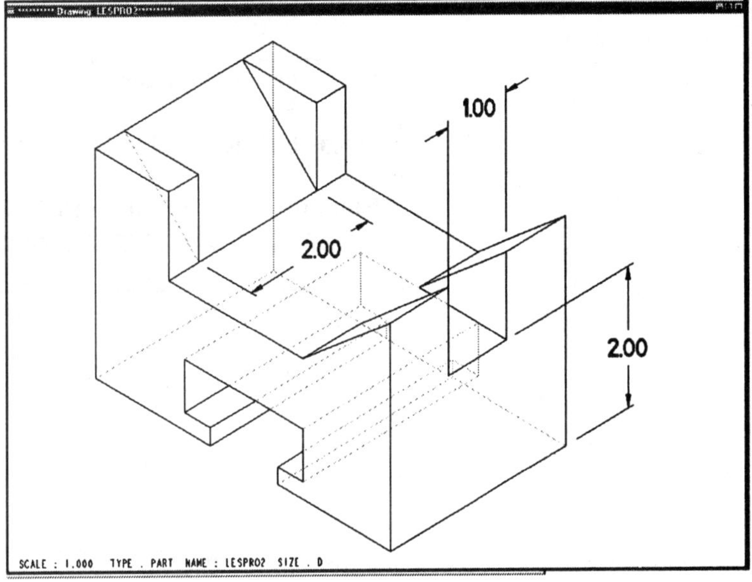

Figure 2.50
Pictorial View

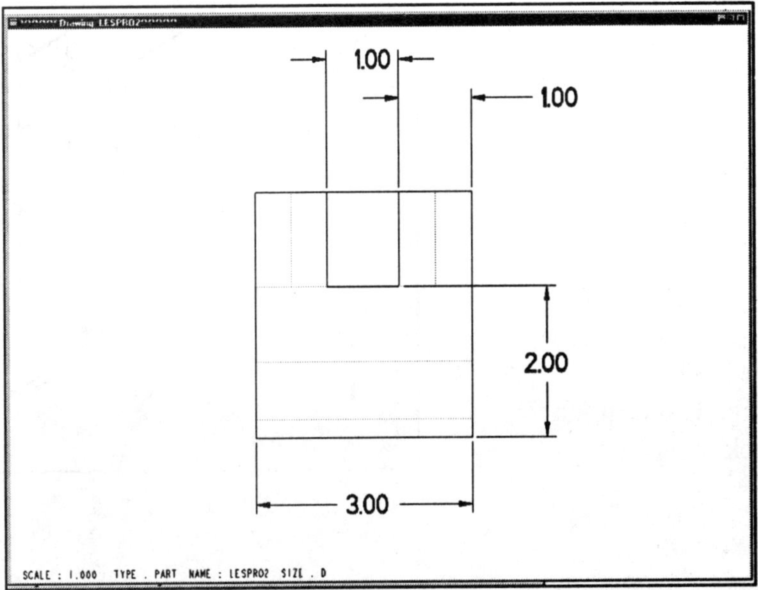

Figure 2.51
Right Side View

Redefine and **Modify** the part after you save it to another name. Figures 2.52 through 2.57 provide the ECO and feature redefinition requirements.

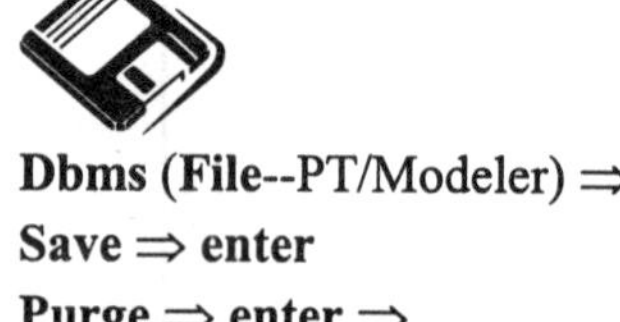

Dbms (File--PT/Modeler) ⇒
Save ⇒ enter
Purge ⇒ enter ⇒
Done-Return

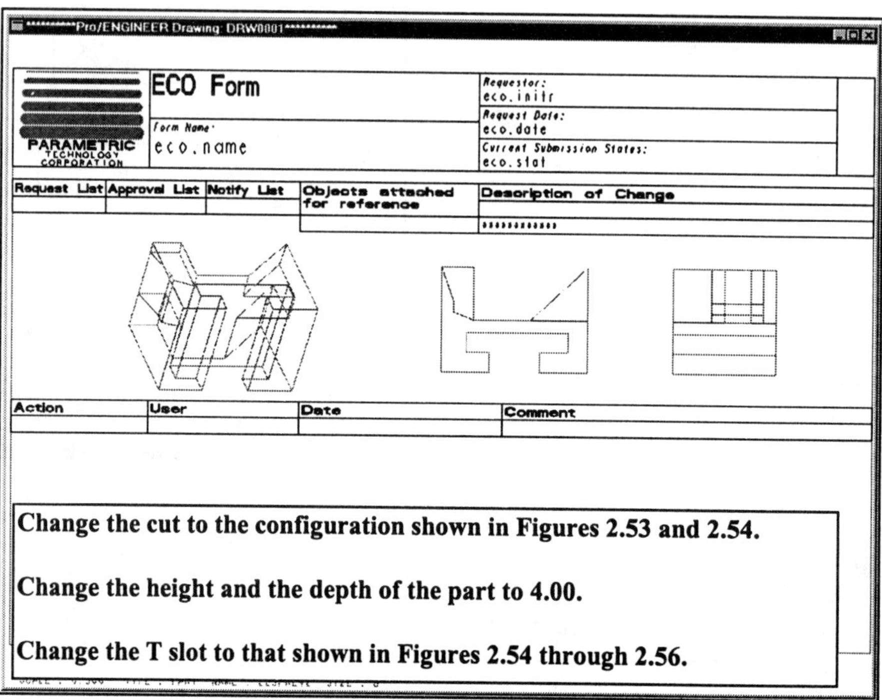

Pro/ENGINEER Drawing: DRW0001

ECO Form

PARAMETRIC TECHNOLOGY CORPORATION

Form Name:
eco.name

Requestor:
eco.initr

Request Date:
eco.date

Current Submission Status:
eco.stat

Request List	Approval List	Notify List	Objects attached for reference	Description of Change

Action	User	Date	Comment

Change the cut to the configuration shown in Figures 2.53 and 2.54.

Change the height and the depth of the part to 4.00.

Change the T slot to that shown in Figures 2.54 through 2.56.

Figure 2.52
ECO

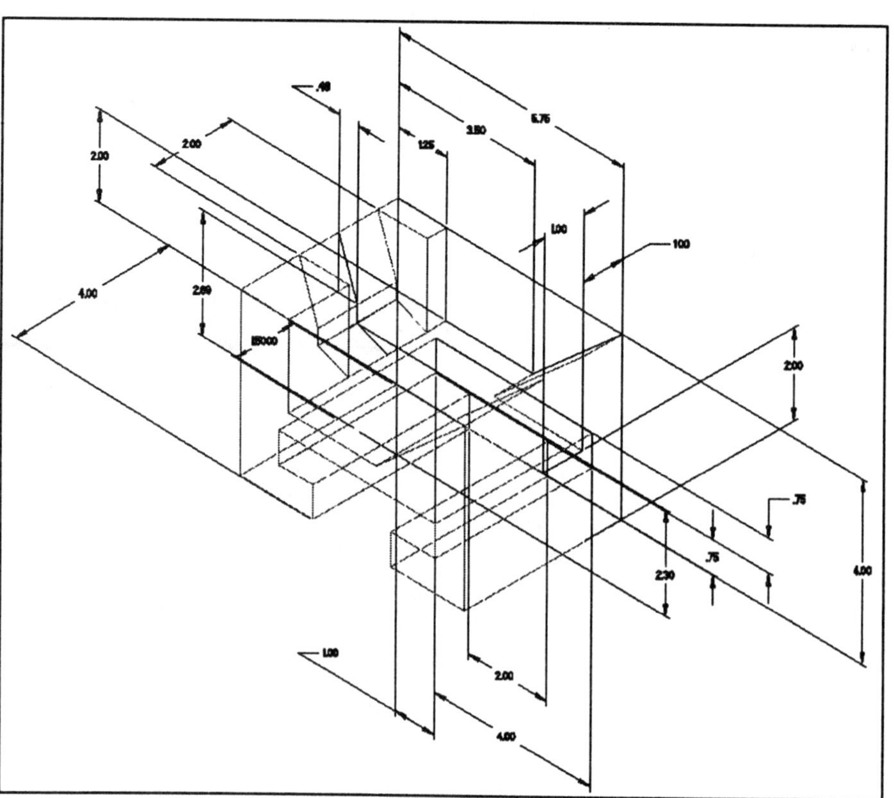

Figure 2.53
Modified Dimensions for T-Block

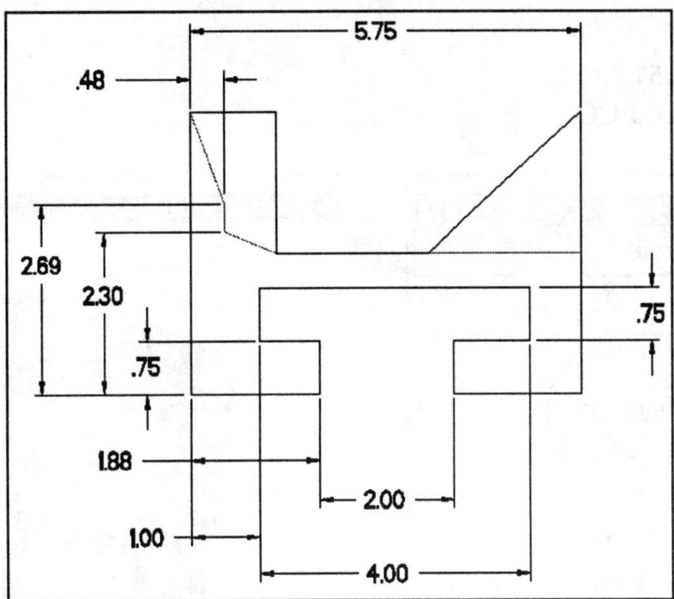

Figure 2.54
Front View

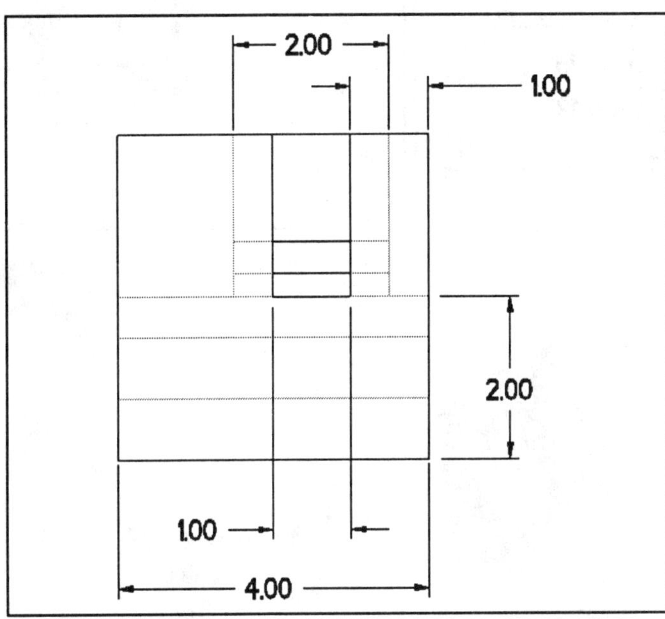

Figure 2.55
Right Side View

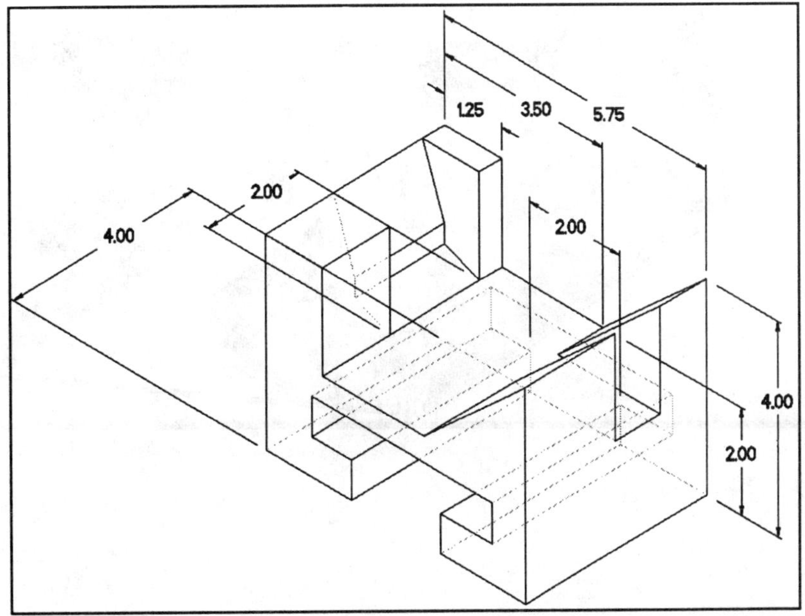

Figure 2.56
Pictorial View

Figure 2.57
Completed ECO

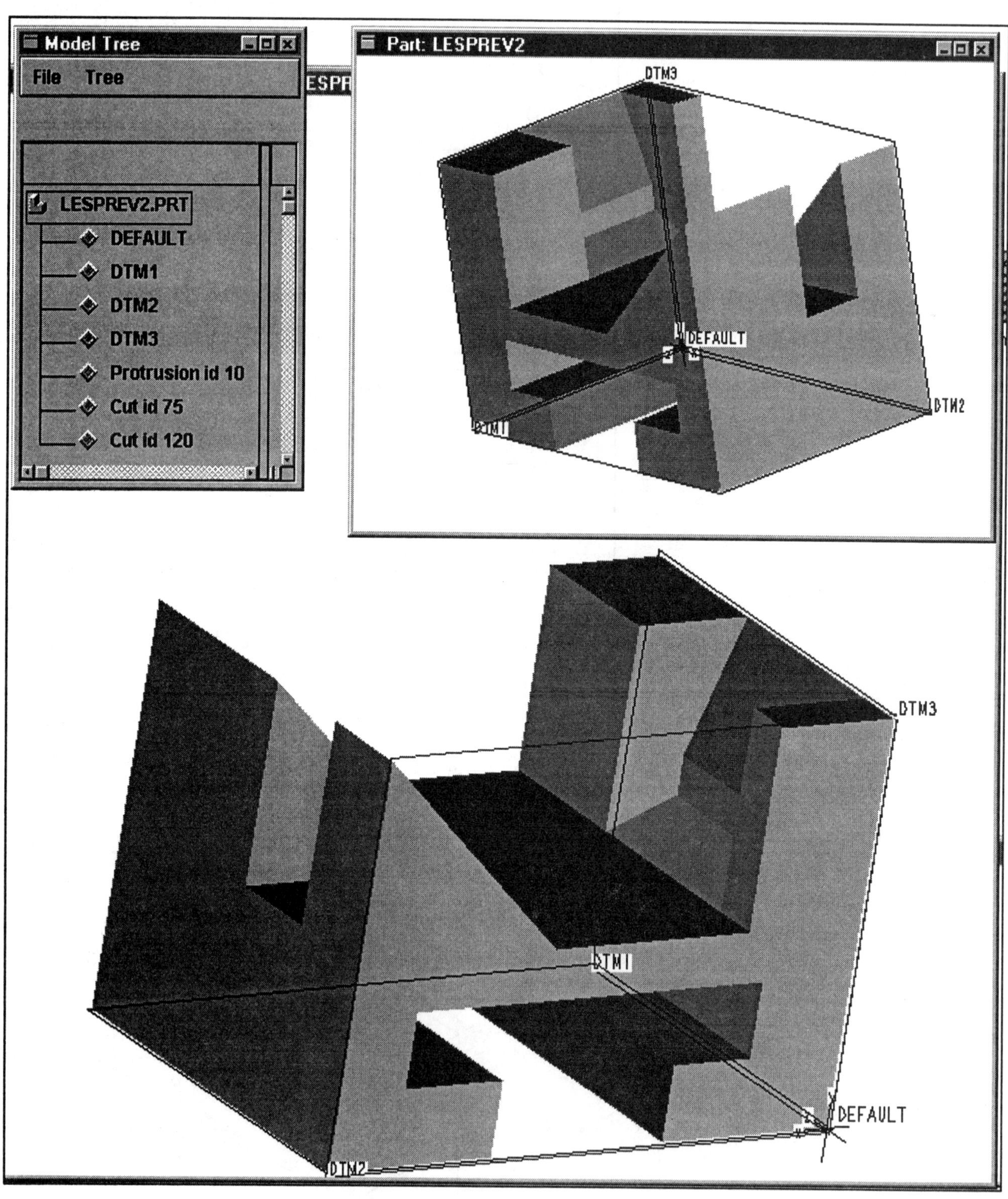

Lesson 3

Holes and Rounds

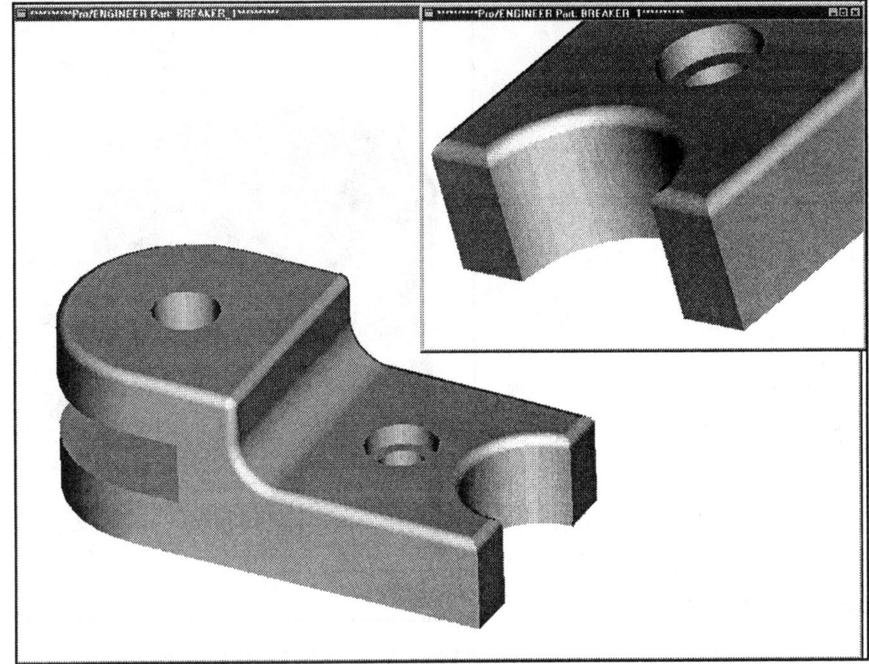

Figure 3.1
Breaker

OBJECTIVES

1. **Create simple rounds along model edges**
2. **Sketch arcs on sections**
3. **Create a straight hole through a part**
4. **Complete a sketched hole**
5. **Understand the difference between sketched and pick-and-place features**
6. **Understand the options for specifying hole depth, including Blind, Thru Next, Thru All, and Thru Until**
7. **Use the Info command to extract feature information**
8. **Understand the types of round creation options**

EGD REFERENCE
Engineering Graphics and Design with Graphical Analysis *or* **Fundamentals of Engineering Graphics and Design**
by L. Lamit and K. Kitto
Read Chapter 10
See pages 286-341, 364, 570

COAch™ for Pro/ENGINEER

If you have **COAch for Pro/ENGINEER** on your system, go to SEARCH and do the Segments shown in Figure 3.4 and Figure 3.6.

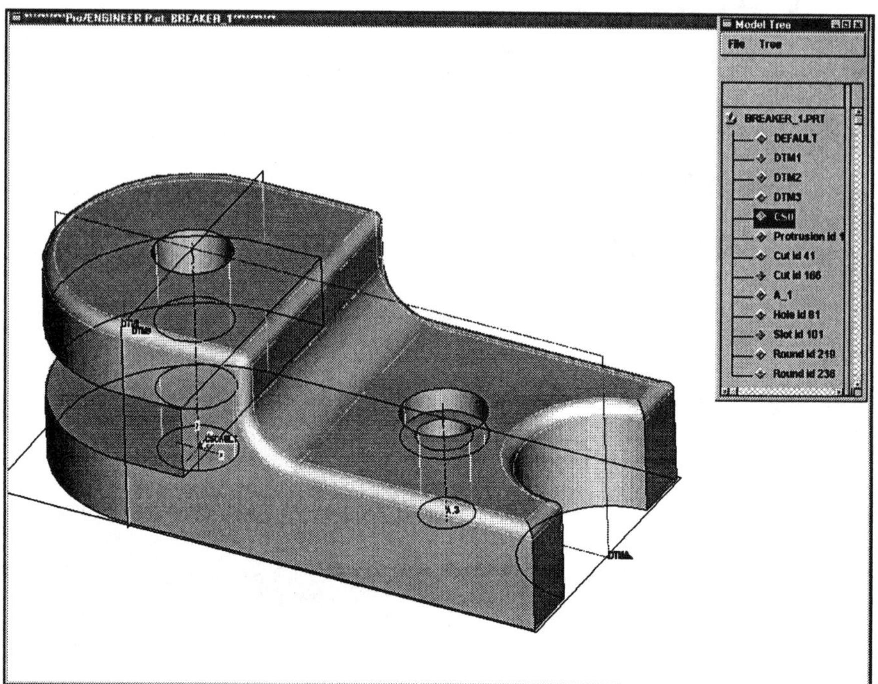

Figure 3.2
Breaker with Axes, Datums, Coordinate System, and Model Tree

HOLES AND ROUNDS

A variety of geometric shapes and constructions are accomplished automatically with Pro/E, including *holes* and *rounds*. These features are called *pick-and-place* features, since they are created automatically from your input and then placed according to prompts by Pro/E. A hole can also be created using **Cut,** but it must be sketched. In general, pick-and-place features are not sketched (except for the **Sketched** option when you are creating a complex hole shape such as a countersink hole or counterbore, as in Figs. 3.1 and 3.2). **Round** creates a fillet, or a round on an edge, that is a smooth transition with a circular profile between two adjacent surfaces.

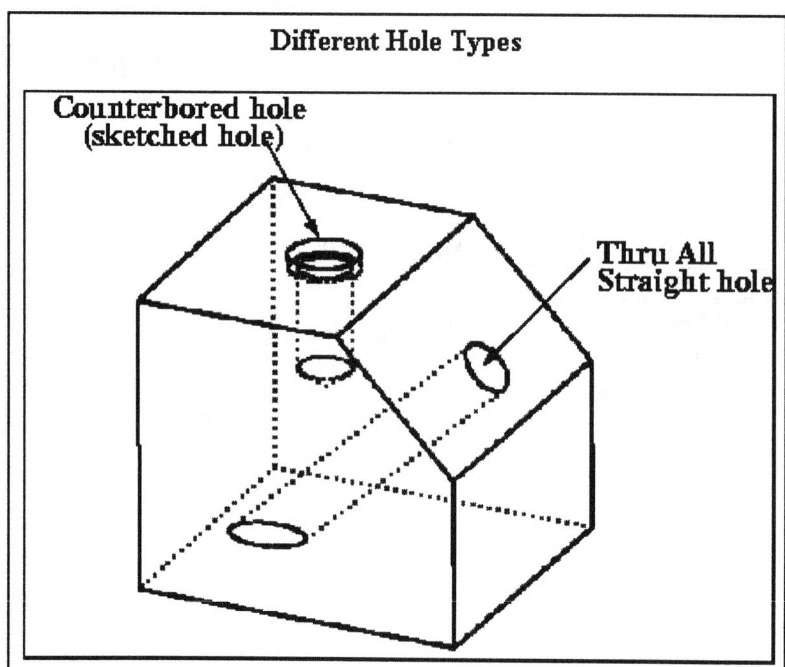

Figure 3.3
Online Documentation, Holes

Holes

The **Hole** option creates many types of holes, including through, counterbore, and blind. Figure 3.3 shows a small sample of online help available for the **Hole** command. All holes are based on two basic types of hole geometry:

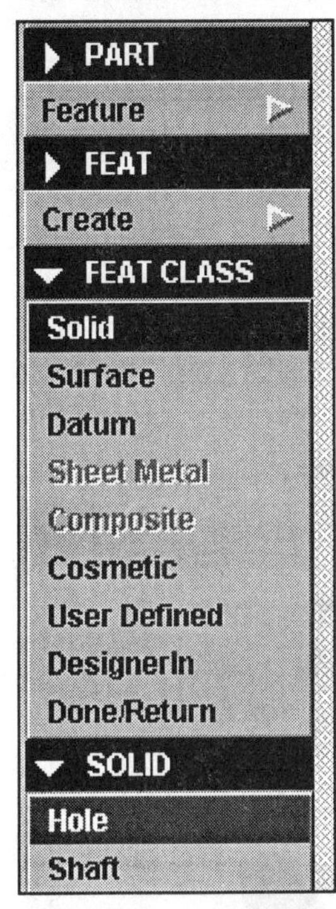

Straight hole An extruded slot with a circular section. This type passes from the placement surface to the specified end surface.
Sketched hole A revolved feature defined by a sketched section. Counterbore and countersink holes, for example, are created as sketched holes.

Straight Holes

All straight holes are created with constant diameter. To create a straight hole:

1. Choose **Hole** from the SOLID menu.
2. Pro/E displays the HOLE OPTS menu. Choose **Straight** ⇒ **Done**.
3. Pro/E displays the Feature Creation dialog box and the PLACEMENT menu, which lists the options **Linear**, **Radial**, **Coaxial**, and **On Point**. Choose one of these options, then **Done**.
4. Select the placement plane.
5. Select the first reference (an edge, axis, planar surface, or datum).
6. Enter the distance from the first reference in the message area.
7. Select the second reference.
8. Enter the distance from the second reference in the message area.
9. Pro/E displays the SIDES menu. Choose **One Side** or **Both Sides**, then **Done**.
10. Select the extent to which the hole will be created, then choose **Done**. The SPEC TO menu options include:

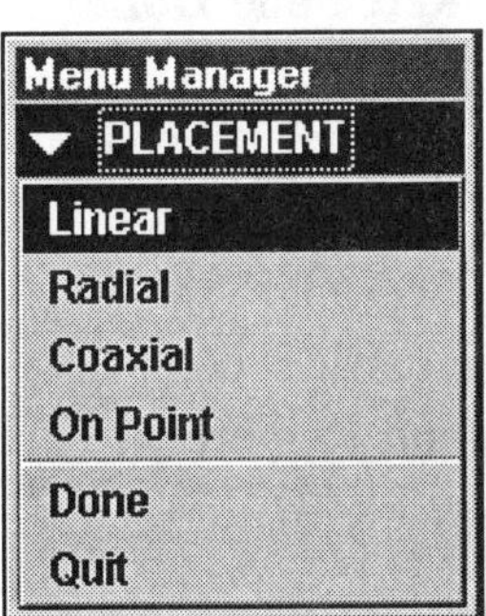

Blind Creates a hole with a flat bottom.
Thru Next Creates a hole that continues until it reaches the next part surface.
Thru All Creates a hole that intersects all the surfaces.
Thru Until Creates a hole that goes through all the surfaces until it reaches intersection with the specified surface.
UpTo Pnt/Vtx Creates a hole with a flat bottom that continues until it reaches the specified point or vertex.
UpTo Curve Creates a hole with a flat bottom that continues until it reaches the specified curve that you draw in a plane parallel to the placement plane.
UpTo Surface Extrudes the hole from material until the bottom of the hole conforms to the selected bounding surface.

11. Enter the depth of the hole at the prompt, if **Blind** was selected.
12. Enter the diameter of the hole at the prompt.
13. Select the **OK** button in the dialog box to create the hole.

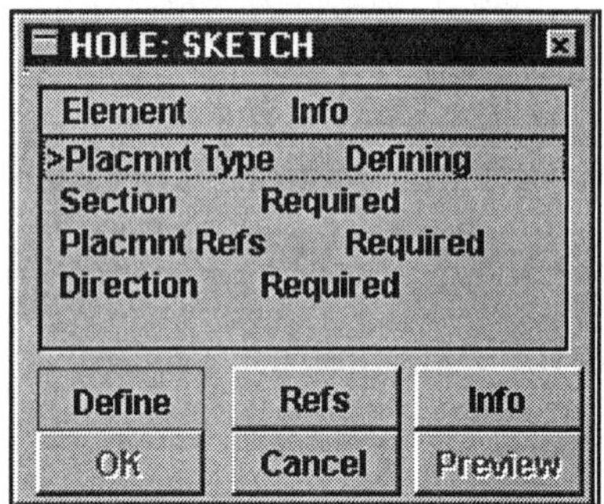

Sketched Holes

A *sketched hole* is created by sketching a section for revolution and then placing the hole on the part. Sketched holes are always blind and one-sided. Sketched holes must have a vertical centerline (*axis of revolution*), with at least one of the entities sketched normal to the axis centerline. Pro/E aligns the normal entity with the placement plane. The remainder of the sketched feature is cut from the part, as with a revolved cut. You can also use the revolved cut command to create holes. If you have **COAch for Pro/ENGINEER** on your system, go to SEARCH and do the Sketched Holes segment as shown in Figure 3.4.

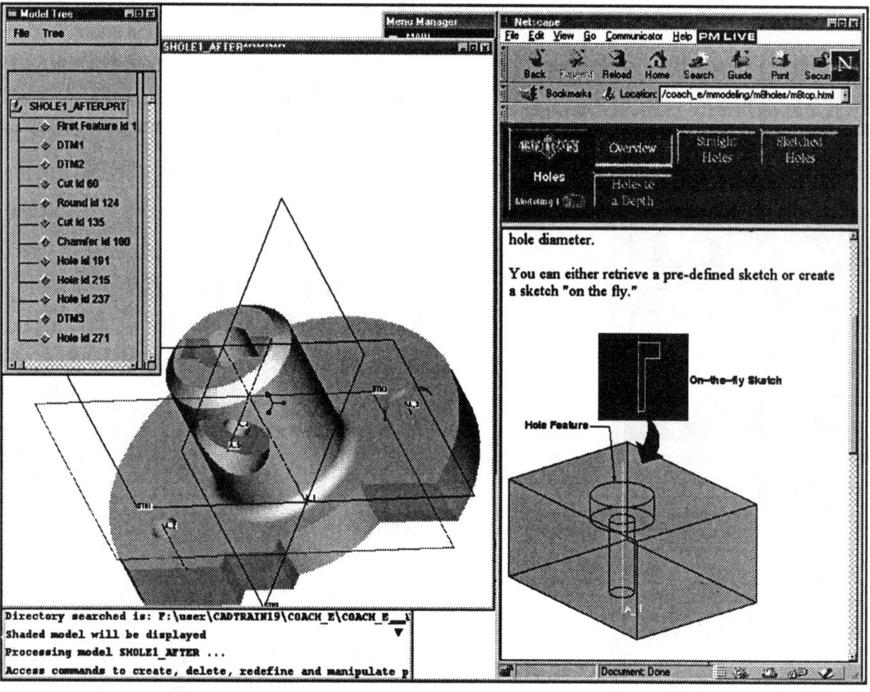

Figure 3.4
COAch for Pro/E, Holes (Overview)

To create a sketched hole:

1. Choose **Hole** from the SOLID menu. Choose **Sketch** and **Done** from the HOLE OPTS menu.
2. Pro/E displays the feature creation dialog box.
3. Choose the dimensioning scheme for the hole using the PLACEMENT menu options. Choose **Done.**
4. Pro/E displays a grid in a subwindow. Sketch a vertical centerline and then the cross section of the hole. **Dimension** and **Regenerate** the section. **Modify** the dimensions. Choose **Done.**
5. Select the placement plane.
6. Select the first reference edge, planar surface, or axis.
7. Enter the distance from the first reference at the prompt.
8. Select the second reference, enter the distance, and choose **OK**.

Rounds

Rounds (Fig. 3.5) are created at selected edges of the part. Tangent arcs are introduced as rounds between two adjacent surfaces of the solid model. There are cases in which rounds should be added early, but in general, wait until later in the design process to add the rounds. Introducing rounds into a complex design early in the project can cause a series of failures later in the design. You might also choose to place all rounds on a layer and suppress that layer to speed up your working session. There are a number of basic rounds to consider, including **Edge Chain, Surface-Surface, Edge-Surface,** and **Edge Pair**.

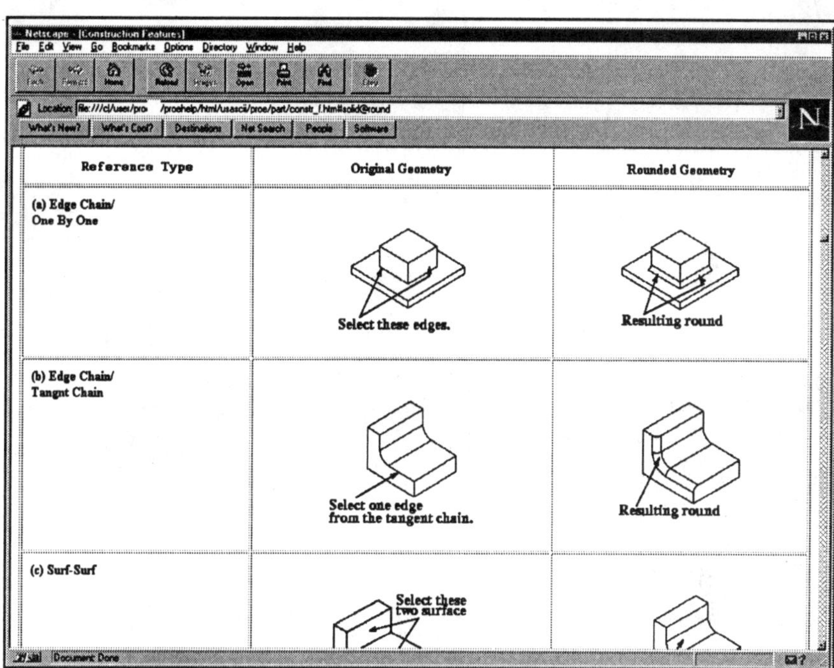

Figure 3.5
Online Documentation, Rounds

Two categories of rounds are available: simple and advanced. Much of the time, you will create *simple* rounds. These rounds smooth the hard edges between two adjacent surfaces. If you have **COAch for Pro/ENGINEER** on your system, go to SEARCH and do the appropriate segment for rounds, shown in Figure 3.6.

Creating a Simple Round

Figures 3.7 (edge pair) and 3.8 (advanced variable) show some of the online documentation available on rounds. You should read your manual, or highlight the **Round** command on the screen, press the right-hand mouse button, and choose ? GetHelp.

The basic steps to create a simple round include:

1. Choose **Round** from the SOLID menu.
2. A dialog box appears, listing elements of the round feature.
3. Choose **Simple** and **Done** from the ROUND TYPE menu.
4. The Attributes element is selected by default. Use the RND SET ATTR menu options to specify the round's attributes.

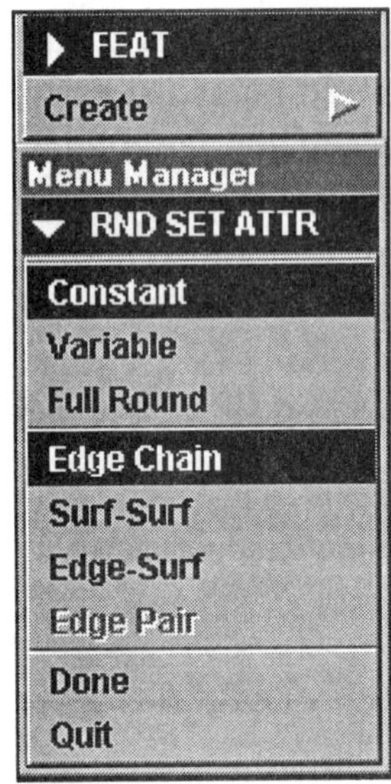

Specify the type of round by selecting one of these options:

Constant Creates a round between two sets of surfaces with a constant radius.
Variable Creates a round between two sets of surfaces with variable radii. Specify radii at the ends of the chain of edges or at the ends of the spine (when the spine is required) and, optionally, at additional points along the edges or along the spine.
Full Round Creates a round by removing a surface; the consumed surface becomes a round.

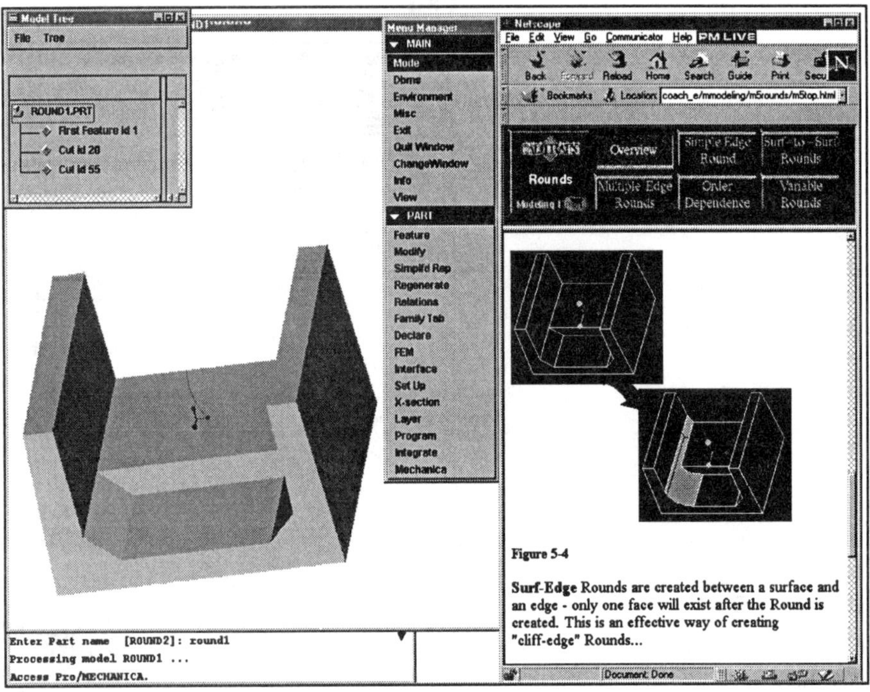

Figure 3.6
COAch for Pro/E, Rounds (Overview)

Select one of the following options to specify the type of references for placing the round:

Edge Chain Places a round by selecting a chain of edges. To select the chain, use options in the CHAIN menu.
Surf-Surf Places a round by selecting two adjacent surfaces.
Edge-Surf Places a round by specifying an edge and a tangent surface.
Edge Pair Places a full round by specifying a pair of edges.

5. Choose **Done** from the RND SET ATTR menu.
6. Pro/E prompts you to select the placement references.
7. For other than a full round, enter the radius for the round.
8. For other than a full round, define the extension boundaries of the round by specifying the **Round Extent** element (optional).
9. If required, define the **Attach Type** element.
10. Choose **OK** from the dialog box.

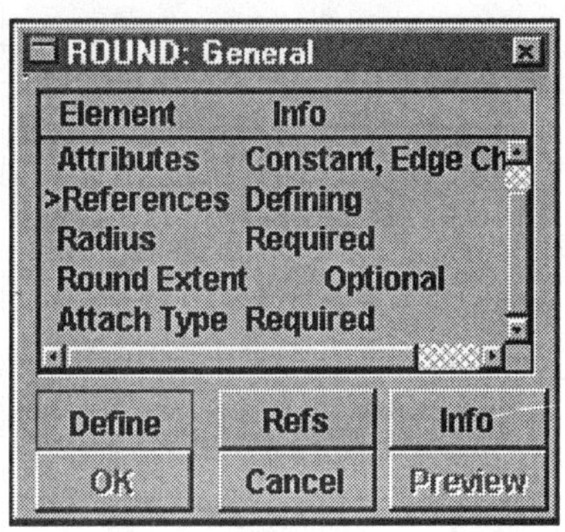

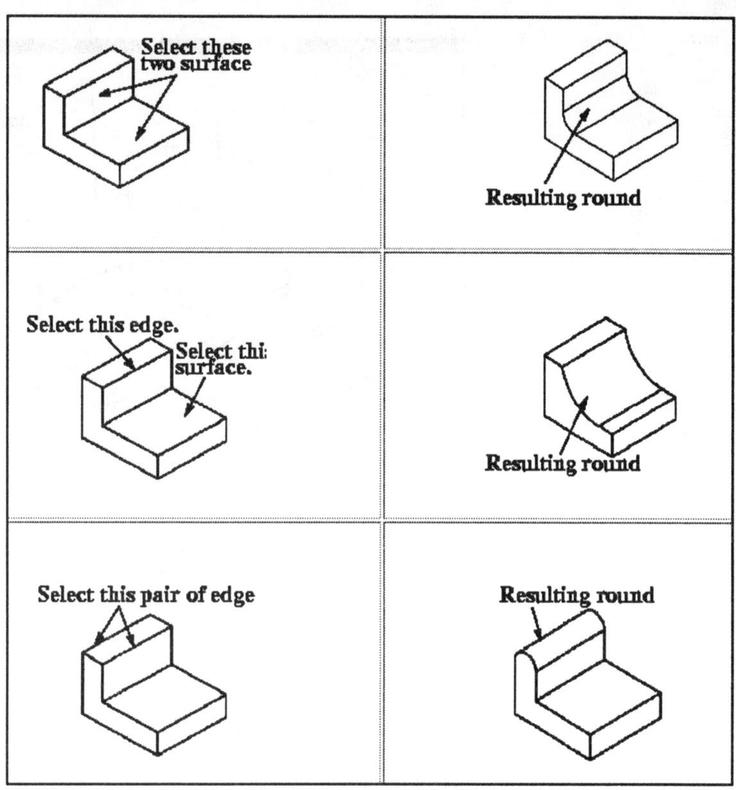

Figure 3.7
Online Documentation, Rounds

Using the Chain Menu Options

When you select reference edges with the **Edge Chain** option, Pro/E displays the CHAIN menu. Note that you can choose more than one option: choose an option, select the references as prompted by Pro/E, and then choose the next option. The CHAIN menu lists the following options:

One By One Define a chain one at a time by selecting individual edges and curves.
Tangnt Chain Define a chain by selecting an edge. All tangent edges are included in the selection.
Surf Chain Define a chain of edges by selecting a surface.
Unselect Unselect references from one of the preceding options.

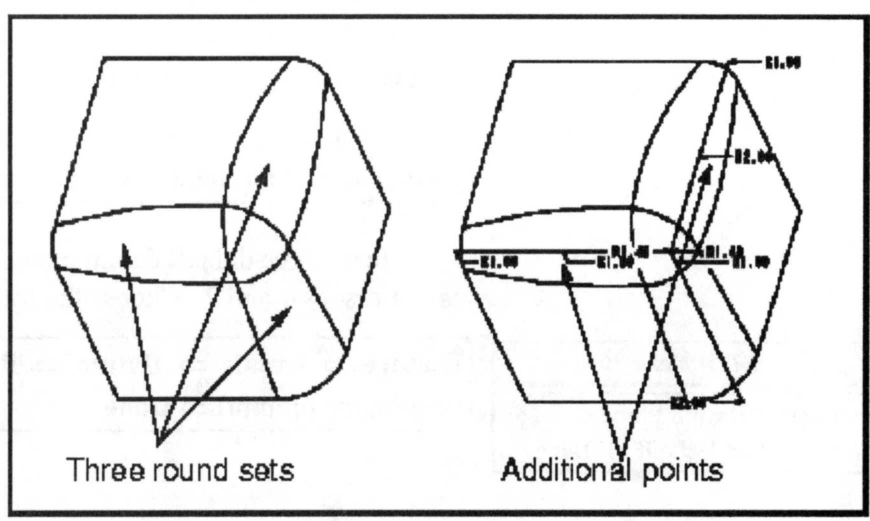

Figure 3.8
Online Documentation, Variable-Radius Rounds

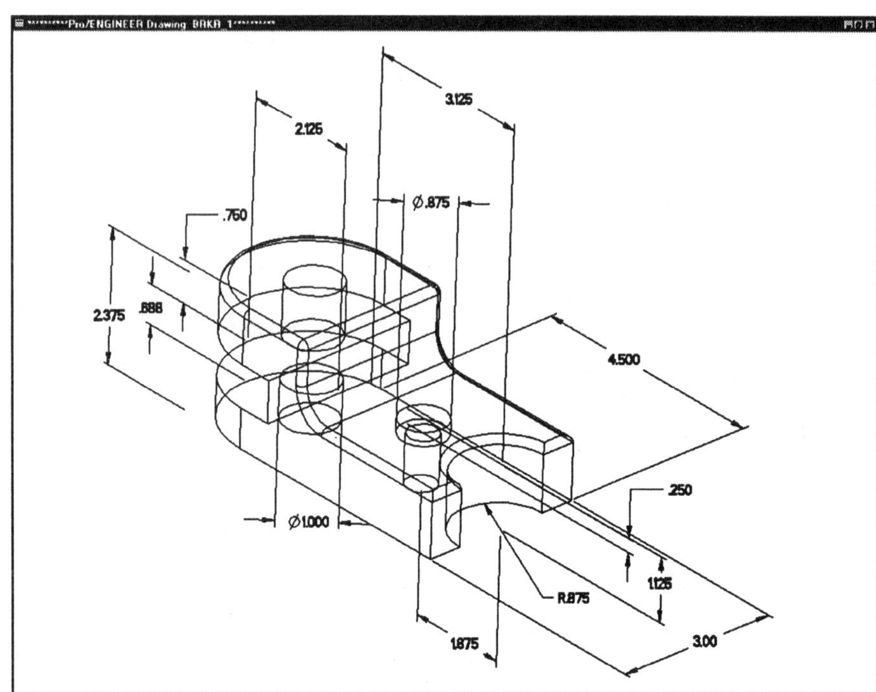

Figure 3.9
Breaker Dimensions

Breaker

The **Breaker** (Fig. 3.9) is the third lesson part. This part introduces two new features, *holes* and *rounds*. Also, in the Sketcher, the **Arc** command will be used to create the rounded end and the half-circle cut of the Breaker. Choose **Part** from the MODE menu. Choose the following commands:

Part ⇒ **Create** (type the part name **BREAKER**) ⇒ **enter**

As always, the *environment*, *units*, and *material* for the part need to be established:

CONFIG.PRO
sketcher_dec_places 3

SETUP AND ENVIRONMENT

Set Up ⇒ **Units** ⇒ **Length** ⇒ **Inch** ⇒ **Done** ⇒ **Material** ⇒ **Define** ⇒ (type **Aluminum**, then press **Enter**) ⇒ (table of material properties, change or add information) **File** ⇒ **Save** ⇒ **File** ⇒ **Exit** ⇒ **Assign** ⇒ (pick **Aluminum**) ⇒ **Accept** ⇒ **Done**

Environment ⇒ ✓**Grid Snap**
Hidden line **Tan Solid** (try this for a different look)

Create three default datum planes and a default coordinate system as in Lessons 1 and 2. Choose the following commands:

PT/Modeler™
Feature ⇒ **Datum Plane** ⇒ **Coord Sys** ⇒ **Default** ⇒ **Done**

Feature ⇒ **Create** ⇒ **Datum** ⇒ **Plane** ⇒ **Offset** ⇒ **enter** (three times at the prompt) ⇒ **Done**

The first protrusion for the Breaker is created using an extruded protrusion and sketching (on **DTM3)** the outline of the part as seen from its top. Choose the following commands to set up the section for the sketch:

Feature ⇒ Create ⇒ Protrusion ⇒ Extrude ⇒ Solid ⇒ Done ⇒ One Side ⇒ Done ⇒ (pick **DTM3** as the sketching plane) **⇒ Okay** (for selecting the *direction* of feature creation) **⇒ Top ⇒** (pick **DTM2** as the *orientation* plane)

Though it is not really necessary for the sketching of this section, sometimes the grid spacing needs to be altered to a different size. Change the size of the grid spacing by choosing the following commands:

PT/Modeler™

Sec Tools ⇒ Grid ⇒ X Spacing (type **15** at prompt) **⇒ Y Spacing** (type **15** at prompt) **⇒ enter ⇒ Sketch**

Sec Tools ⇒ Sec Environ ⇒ Grid ⇒ Params ⇒ X&Y Spacing ⇒ (type **15** at the prompt) **⇒ enter ⇒ Done/Return**

Your screen should look similar to Figure 3.10. The grid will now be twice as dense.

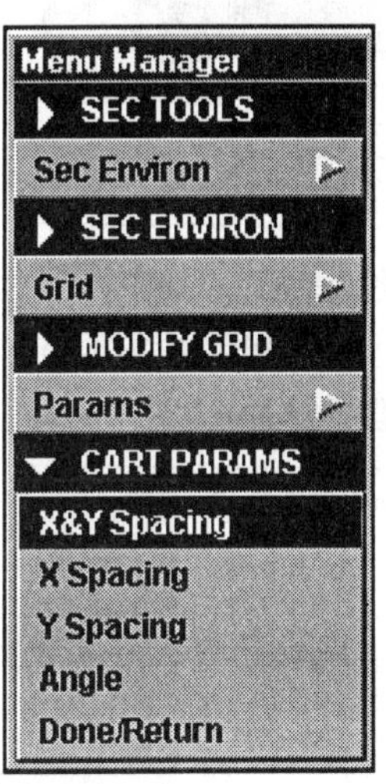

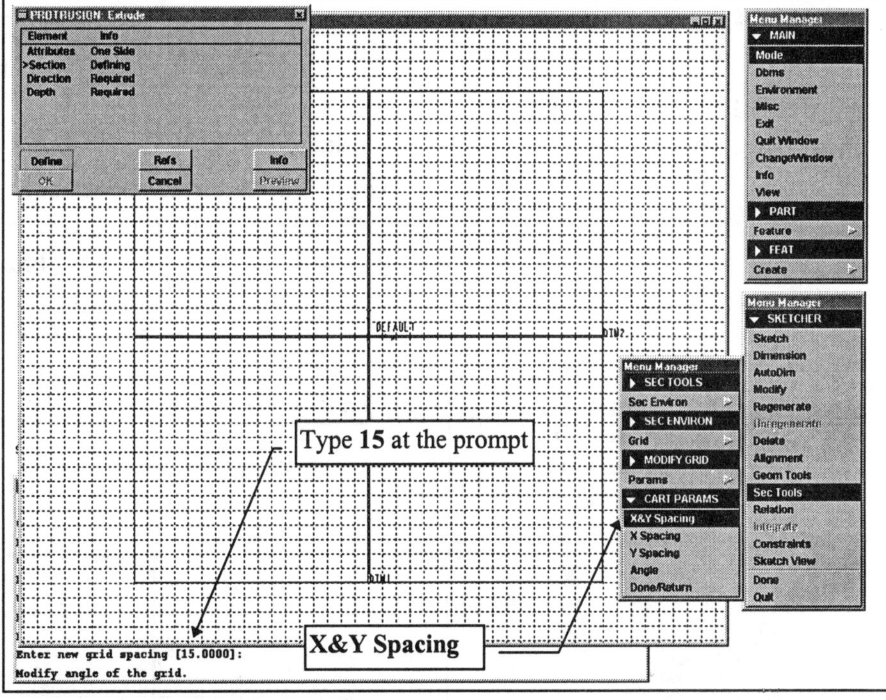

Figure 3.10
Grid Size Changed to **15** Units

You will sketch the section by creating an arc representing the part's curved end. The origin of the arc will be at the intersection of **DTM1** and **DTM2**, as shown in Figure 3.11. The section will be composed of two arcs, four lines, and a horizontal centerline. Align the arcs with **DTM1** and **DTM2**, and the centerline to **DTM2**.

Create the arcs first (Fig. 3.11), then add the lines (Fig. 3.12). Choose the following commands:

Sketch ⇒ **Arc** ⇒ **Ctr/Ends** (pick the intersection of **DTM1** and **DTM2** as the arc's center, then pick the starting and ending points; repeat the process for the smaller arc) ⇒ **Sketch** ⇒ **Mouse Sketch** ⇒ (create the lines of the section geometry) ⇒ **Regenerate**

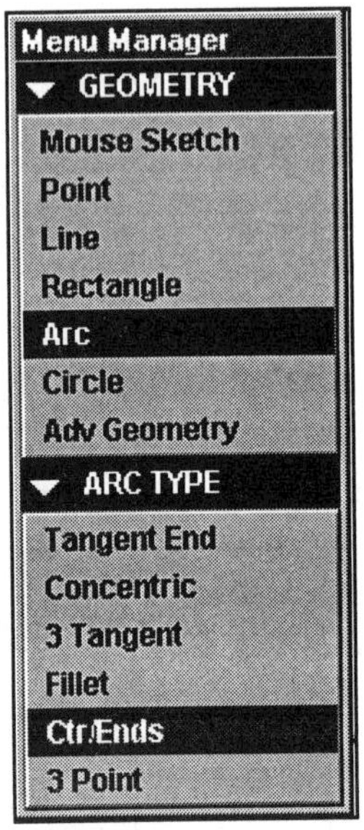

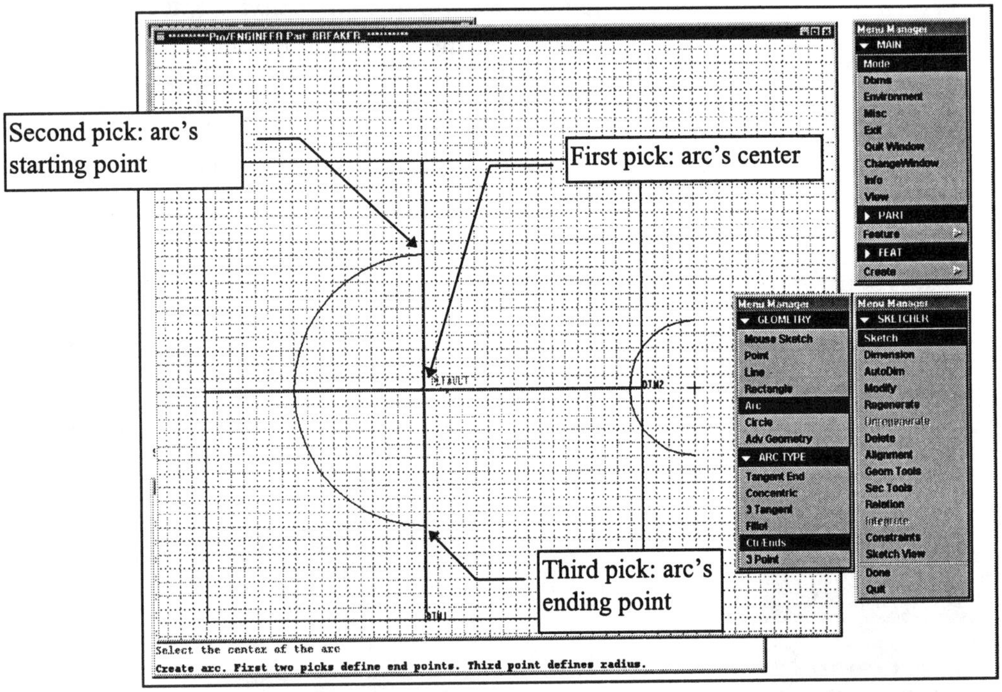

Figure 3.11
Sketched Arcs

Align the centerline by double-picking on the entity and datum plane. To align the arcs, pick on the arc itself (or its center point) and then on the datum to which it will be aligned. **Regenerate** again.

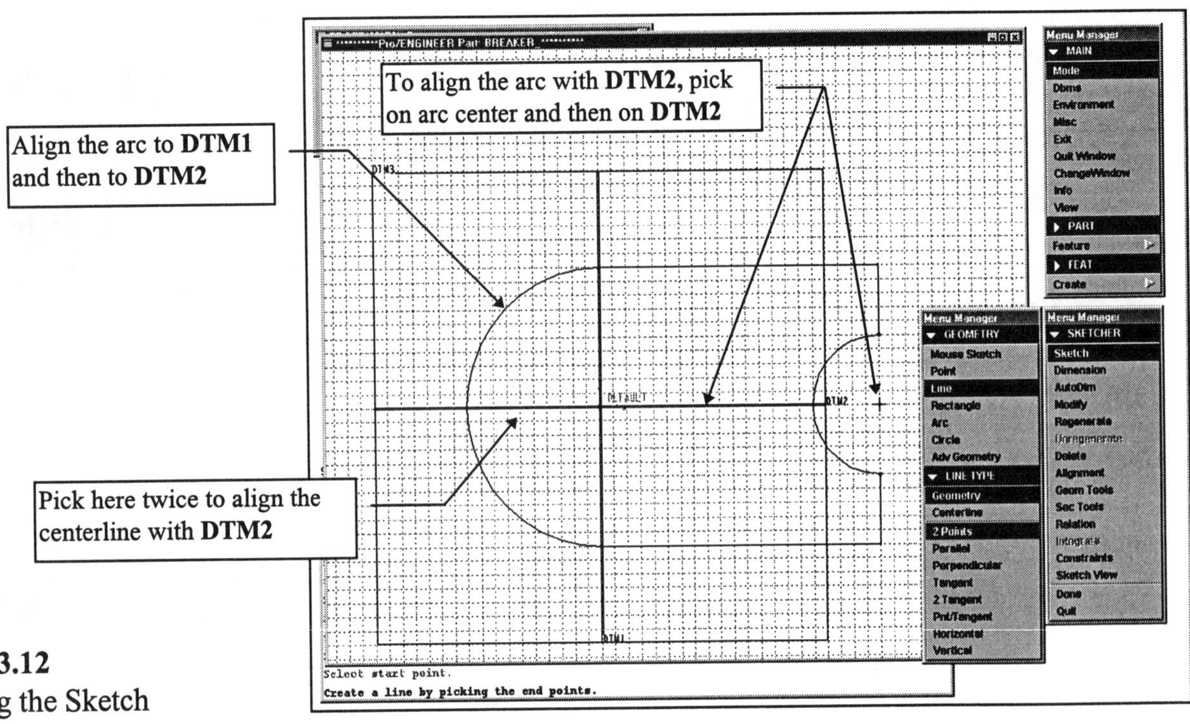

Figure 3.12
Aligning the Sketch

Use the dimensions provided in Figure 3.13 as the dimensioning scheme. **Dimension**, **Regenerate**, **Modify**, and **Regenerate** the sketch as shown in Figure 3.14. Only three dimensions are required for the successful regeneration of the sketch.

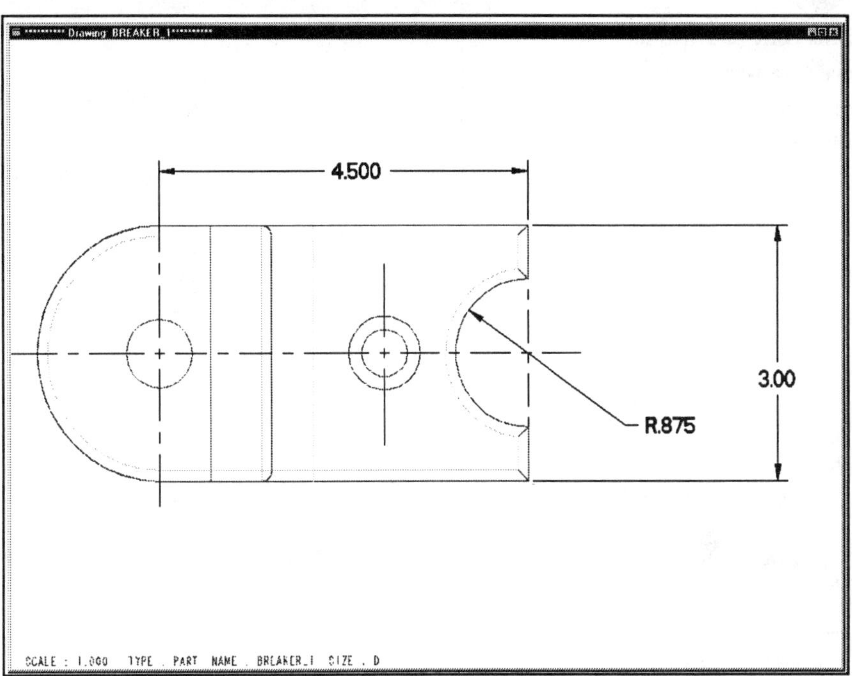

Figure 3.13
Design Intent Dimensions

HINT
You can change the grid size at any point in the sketching process. After the sketch is regenerated, the **15.00** inch grid zooms out of view beyond the **5.00** inch long part model. Change the grid to **.25.**

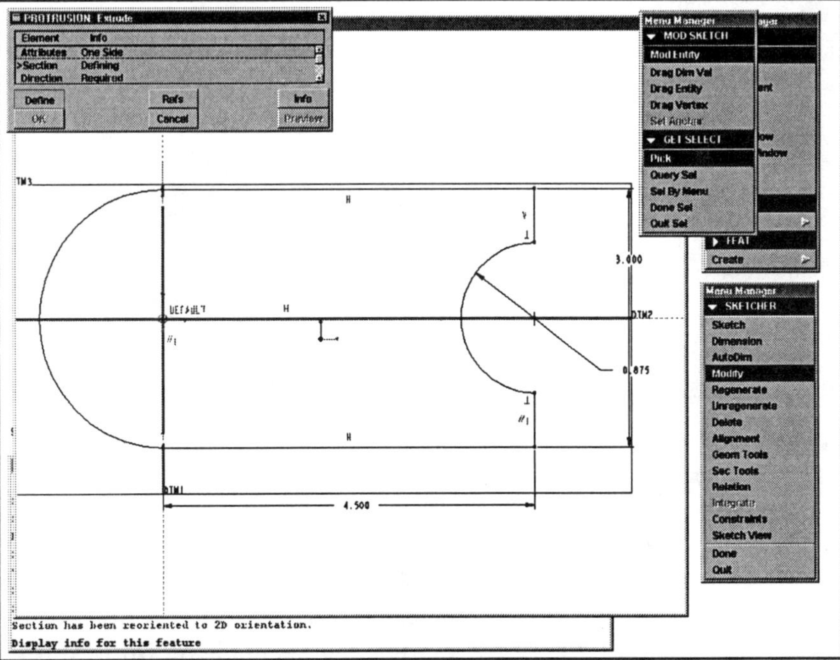

Figure 3.14
Modified and Regenerated Sketch

To complete the feature, choose the following commands:

Regenerate ⇒ **Done** ⇒ **Blind** ⇒ **Done** ⇒ **2.1875** (as the depth of protrusion) ⇒ **enter** ⇒ **OK** (from the dialog box) ⇒ **View** ⇒ **Default** ⇒ **Done-Return** (to view the protrusion as in Fig. 3.15) ⇒ **Done** ⇒ **View** ⇒ **Cosmetic** ⇒ **Shade** ⇒ **Display** (shade the part as in Fig. 3.16) ⇒ **Done-Return**

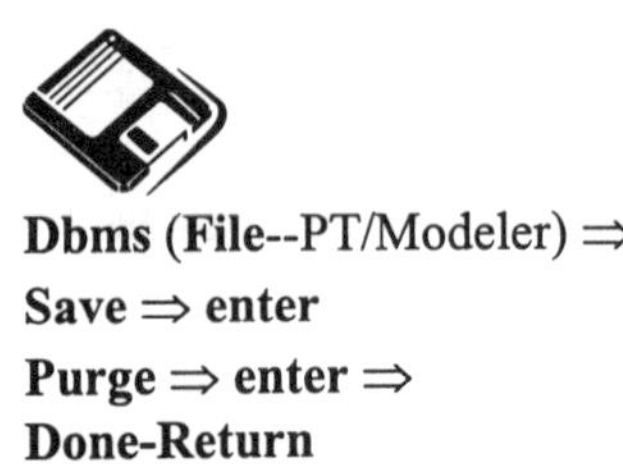

Dbms (**File**--PT/Modeler) ⇒
Save ⇒ **enter**
Purge ⇒ **enter** ⇒
Done-Return

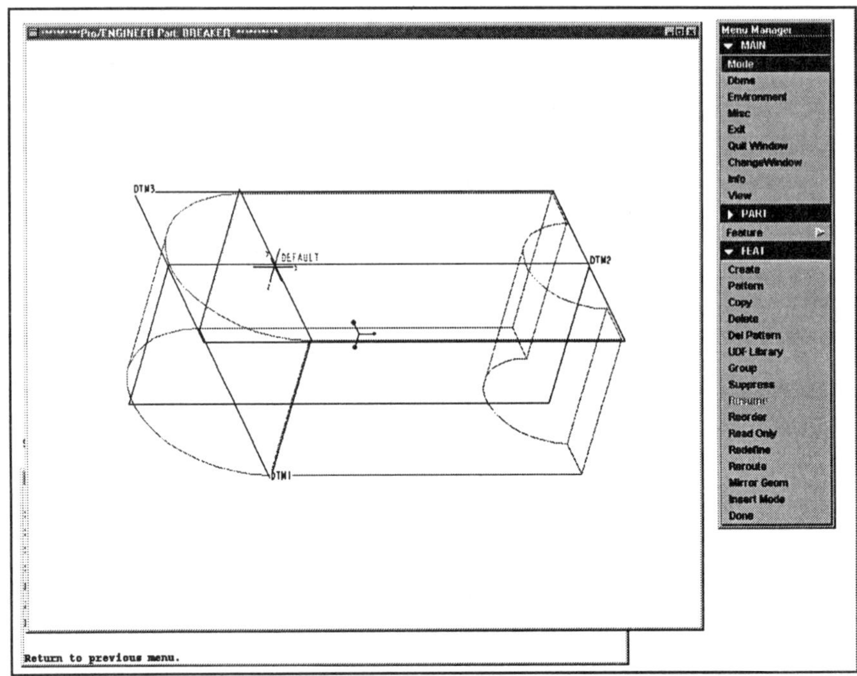

Figure 3.15
Completed Protrusion

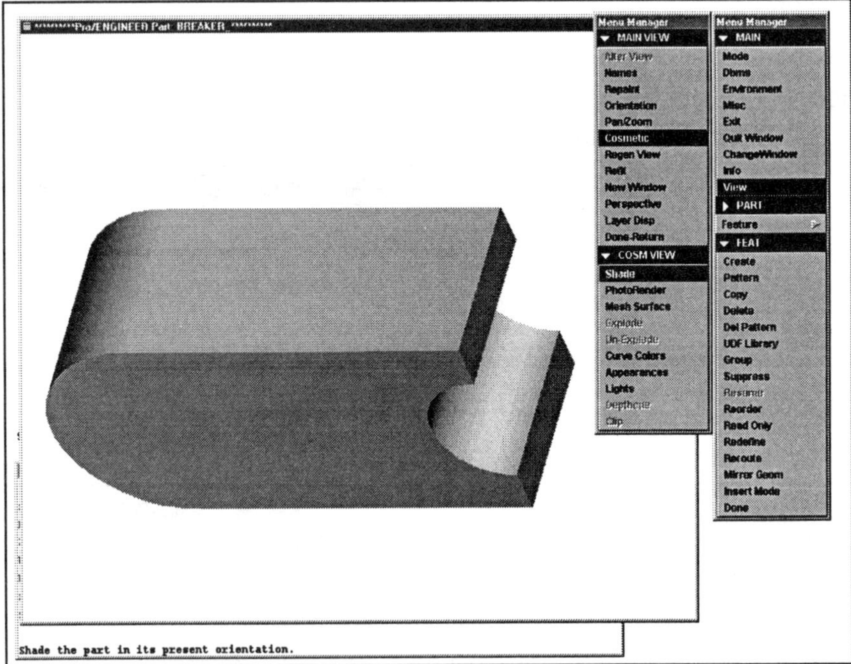

Figure 3.16
Shaded Protrusion

The next features will be the cuts created to remove portions of the protrusion. The cuts will complete the primary features of the part. **DTM2** is used as the sketching plane for both cuts, and the extruded cut is made on **Both Sides**.

In general, leave *holes* and *rounds* as the final features of the part. Most holes are *pick-and-place* features that are added to the model at a similar step, such as when they are drilled, reamed, or bored during actual manufacturing. In most cases, this means after most of the machining has been completed. Rounds are the very last features created. A good many model failures happen when a set of rounds is being created. Leaving them for the final features reduces the effort needed to resolve modeling problems.

The next feature that will be created is the cut on the top of the part. The dimensions for the cut are shown in Figure 3.17. Choose the following commands:

PT/Modeler™
View ⇒ Repaint ⇒ Done-Return ⇒ Feature ⇒ Cut etc.

View ⇒ Repaint ⇒ Done-Return ⇒ Feature ⇒ Create ⇒ Cut ⇒ Extrude ⇒ Solid ⇒ Done ⇒ Both Sides ⇒ Done (pick **DTM2** as the sketching plane) ⇒ **Flip ⇒ Okay ⇒ Top** (pick **DTM3** to orient the sketch)

If necessary, use **Query Sel** to pick the datum planes (Fig. 3.18). Pro/E now enters the **Sketcher** and displays the part and datum planes, as shown in Figure 3.19.

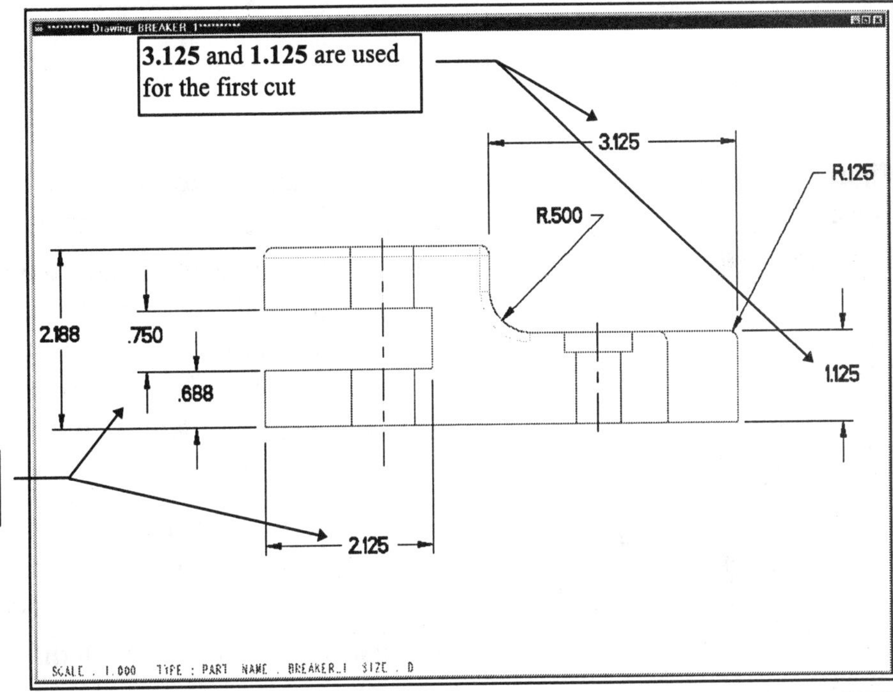

Figure 3.17
Front View of Part

DTM2 is the *sketching plane* and is picked first

DTM3 is the sketch *orientation plane* and is picked second

Figure 3.18
Selecting the Sketching and Orientation Planes

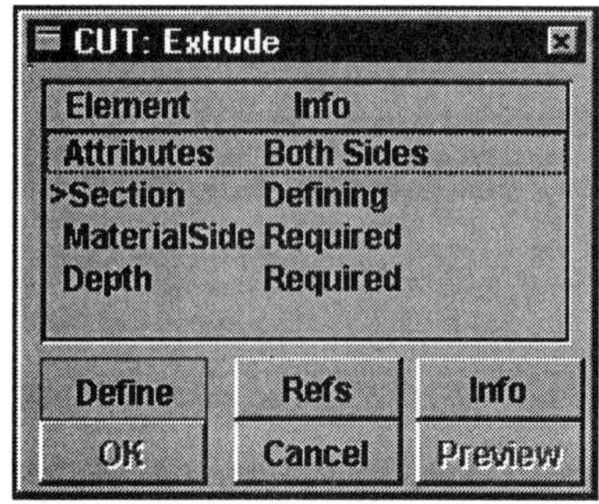

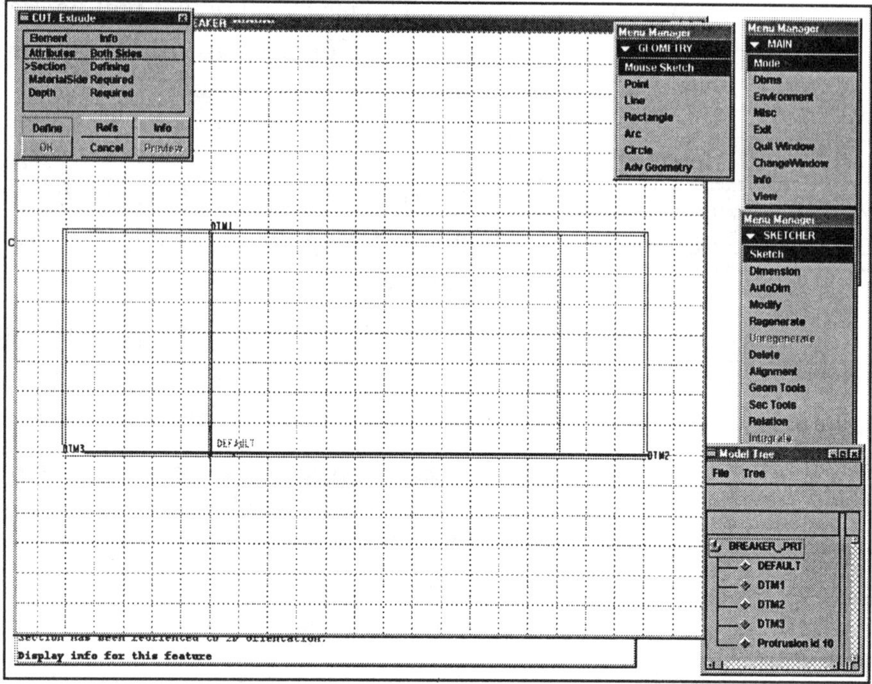

Figure 3.19
Sketcher

Before you start sketching, turn off the □ **Grid Snap**. Sketch the three endpoints of the two lines as shown in Figure 3.20. Align the two endpoints that touch the part's edges and **Regenerate** the sketch. **Dimension** (use the dimensions shown in Figure 3.17), **Regenerate**, **Modify**, and **Regenerate** to complete the sketch (Fig. 3.21). Use the following commands:

Environment ⇒ □ **Grid Snap** (turn off the Grid Snap) ⇒ **Done-Return** ⇒ **Sketch** ⇒ **Line** ⇒ **Vertical** (pick the three endpoints of the lines) ⇒ **Regenerate** ⇒ **Alignment** ⇒ **Regenerate** ⇒ **Dimension** ⇒ **Regenerate** ⇒ **Modify** ⇒ **Regenerate**

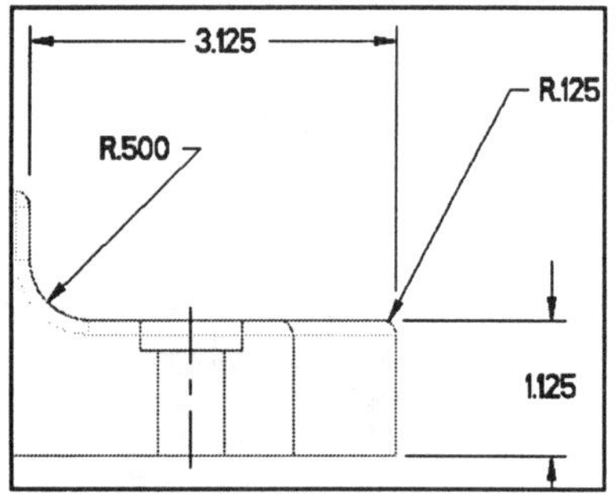

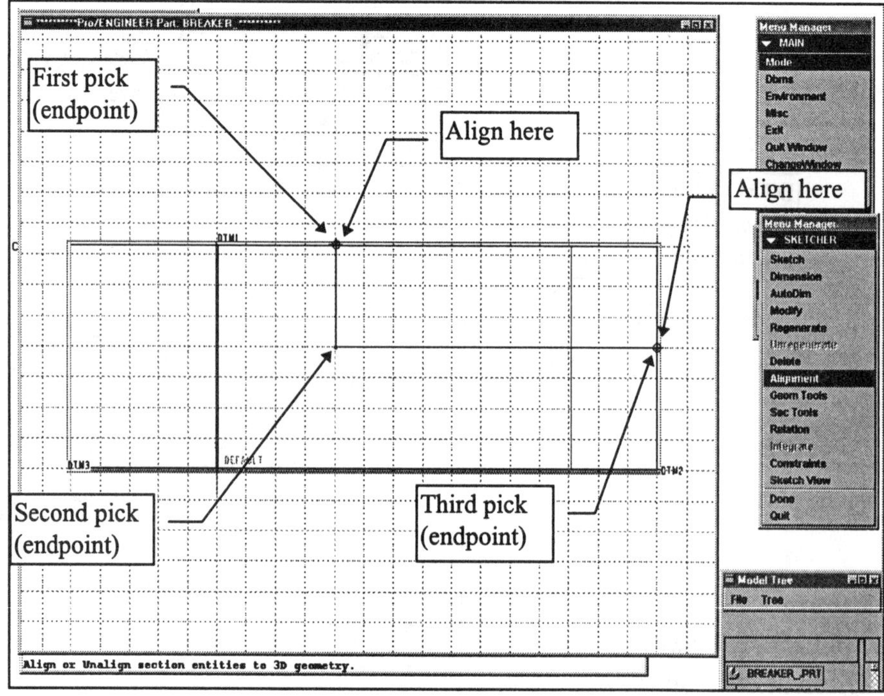

Figure 3.20
Section Sketch of Cut

Dimension 3.125
Dimension 1.125

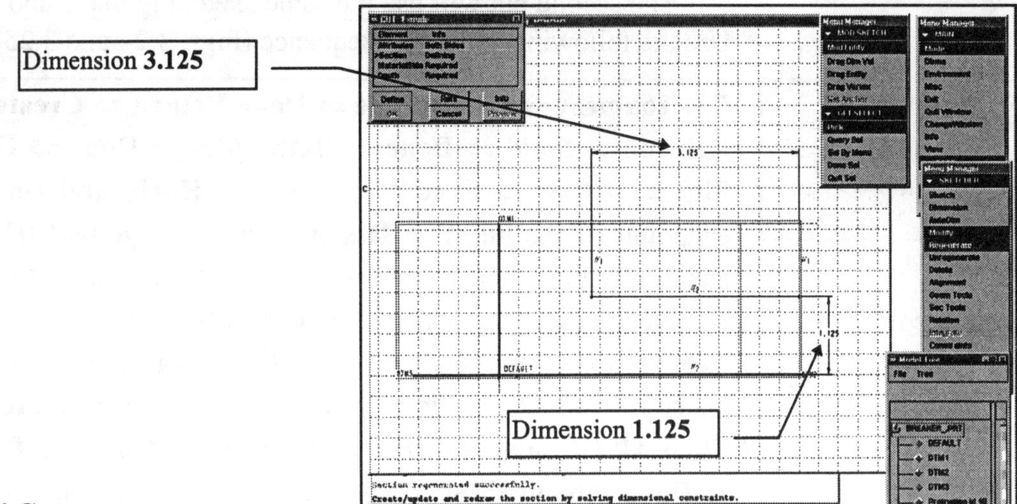

Figure 3.21
Dimensioned Cut

Complete the cut: **Done** (from SKETCHER menu) ⇒ **Okay** (for the material removal direction, as shown in Fig. 3.22) ⇒ **View** ⇒ **Default** ⇒ **Done-Return** ⇒ **Thru All** (first side) ⇒ **Done** ⇒ **Thru All** (second side) ⇒ **Done** ⇒ **Preview** ⇒ **OK** ⇒ **Environment** ⇒ **Shading** ⇒ **Isometric** ⇒ **Done-Return** (Fig. 3.23)

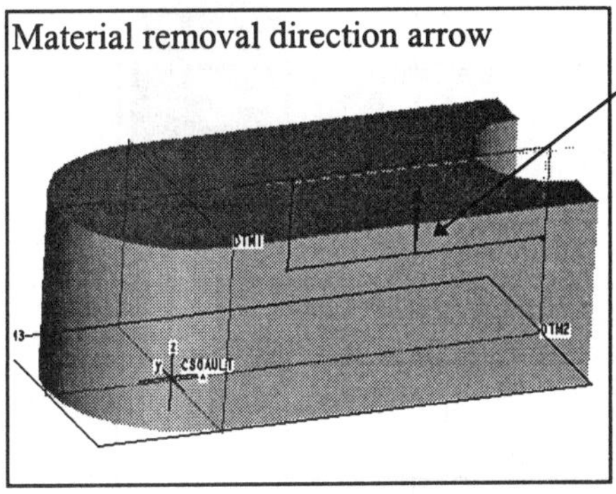
Material removal direction arrow

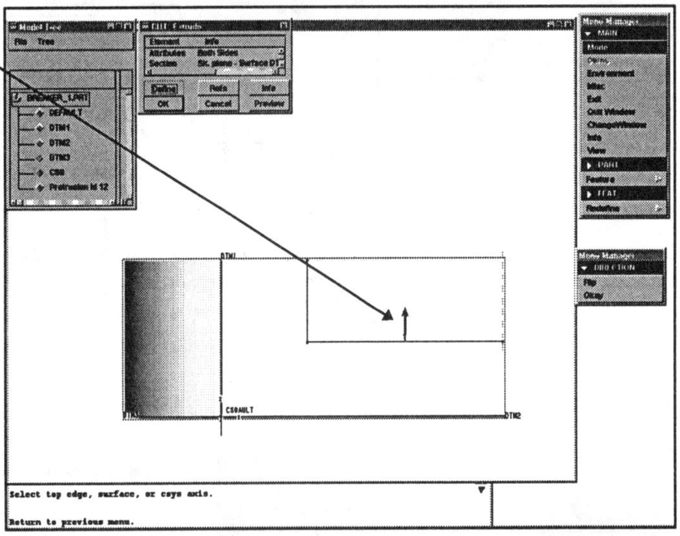

Figure 3.22
Material Removal Direction

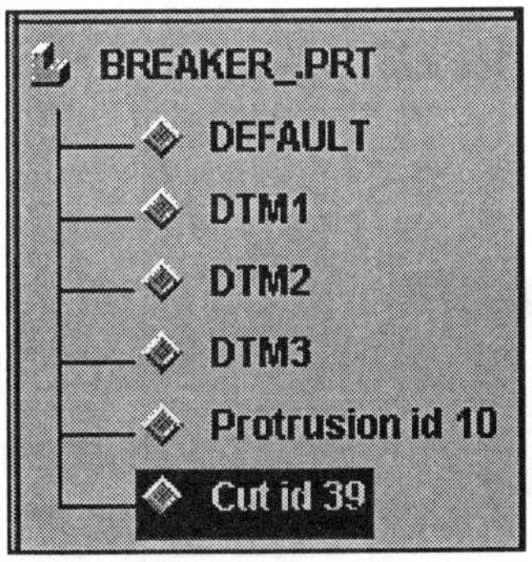
BREAKER_.PRT
DEFAULT
DTM1
DTM2
DTM3
Protrusion id 10
Cut id 39

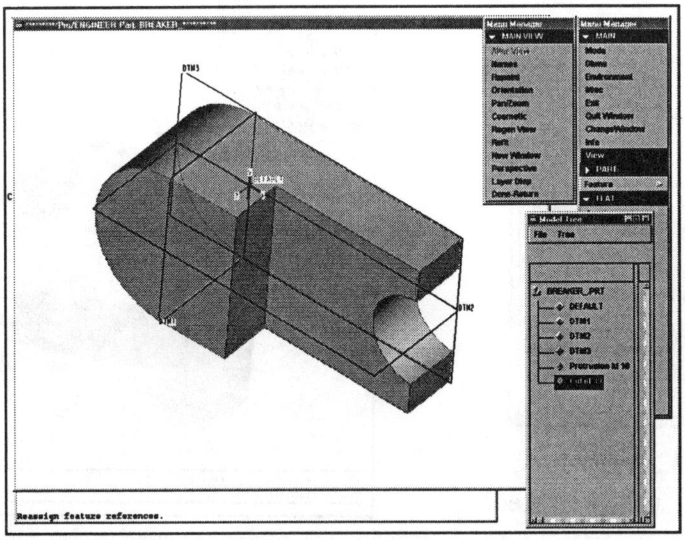

Figure 3.23
Completed Cut

The second cut will use the same sketching plane and orientation. Use the following command sequence (Figs. 3.24 and 3.25):

PT/Modeler™
Environment ⇒ Hidden Line ⇒ Done-Return ⇒ Feature ⇒ Cut etc.

Environment ⇒ Hidden line ⇒ Done-Return ⇒ Create ⇒ Cut ⇒ Extrude ⇒ Solid ⇒ Done ⇒ Both Sides ⇒ Done ⇒ Use Prev ⇒ Flip ⇒ Okay ⇒ Sketch ⇒ Line ⇒ Horizontal (pick the four endpoints of the lines) ⇒ **Regenerate ⇒ Alignment** (align the two endpoints that touch the part's left edge) ⇒ **Regenerate ⇒ Dimension** (see Fig. 3.17) ⇒ **Regenerate ⇒ Modify** (Fig. 3.24) ⇒ **Regenerate ⇒ Done** (from SKETCHER menu) ⇒ **Okay** (for the material removal direction) ⇒ **View ⇒ Default ⇒ Done-Return ⇒ Thru All** (1st side) ⇒ **Done ⇒ Thru All** (2nd side) ⇒ **Done ⇒ OK ⇒ Environment ⇒ Shading ⇒ Ctrl** (rotate the part) ⇒ **Done-Return** (Fig. 3.25)

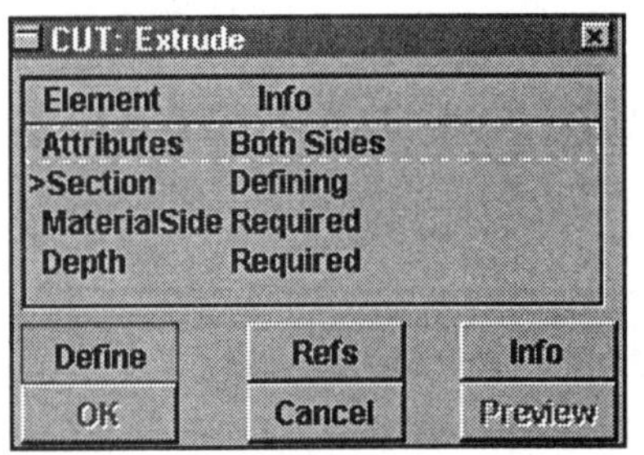

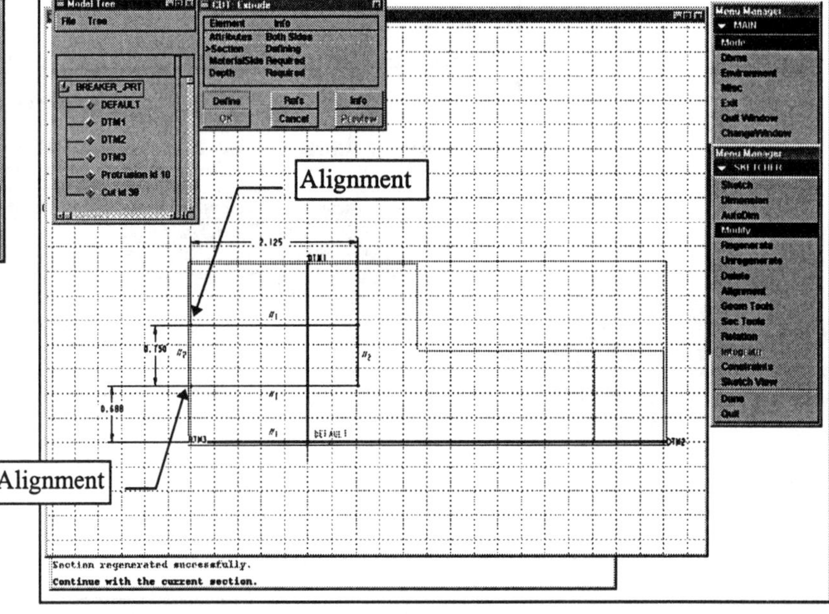

Figure 3.24
Second Cut Section

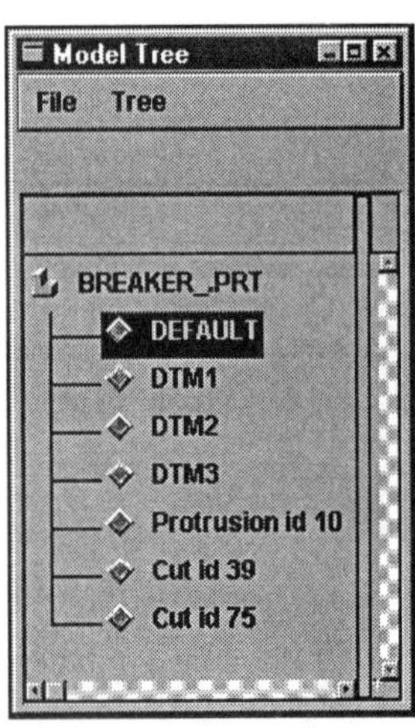

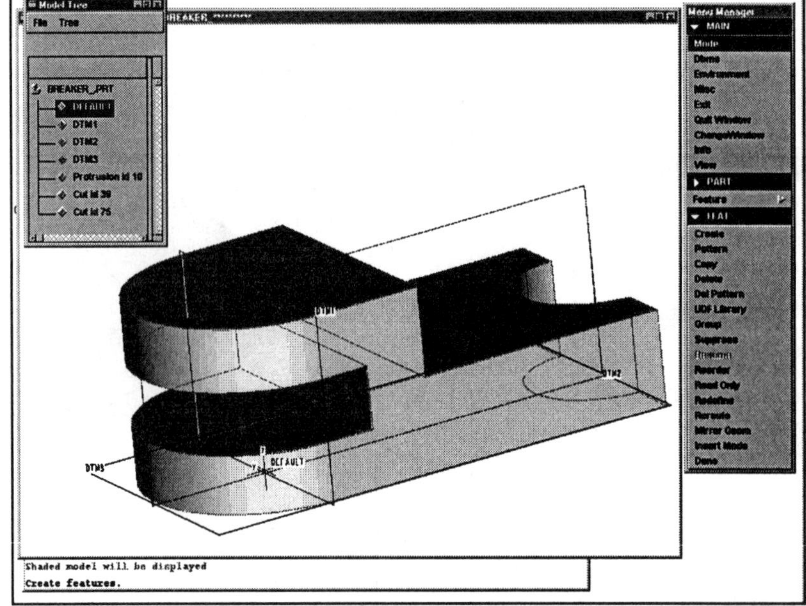

Figure 3.25
Second Cut Completed

The next feature to be created is a hole. This is a *pick-and-place* feature that does not require a sketch. The placement plane is the part's top surface; the dimensioning edges/planes are **DTM1** and **DTM2** (Fig. 3.26). The hole will be at the intersection of the two datum planes; therefore the distance from both will be **0** inches. Choose the following commands:

? Pro/HELP

To get more information about holes, highlight the **Hole** command and press the right mouse button and **? GetHelp**.

Create ⇒ Hole ⇒ Straight ⇒ Done ⇒ Linear ⇒ Done ⇒ (pick the hole's placement plane) ⇒ (select the first edge to dimension from, which will be **DTM2**, and type **0** at the prompt) ⇒ **enter ⇒** (select the second edge to dimension from, which is **DTM1**, and type **0** at the prompt) ⇒ **enter ⇒ One Side ⇒ Done ⇒ Thru All ⇒ Done ⇒** (enter **.8125** at the prompt) ⇒ **enter ⇒ OK** (Fig. 3.27) ⇒ **Done**

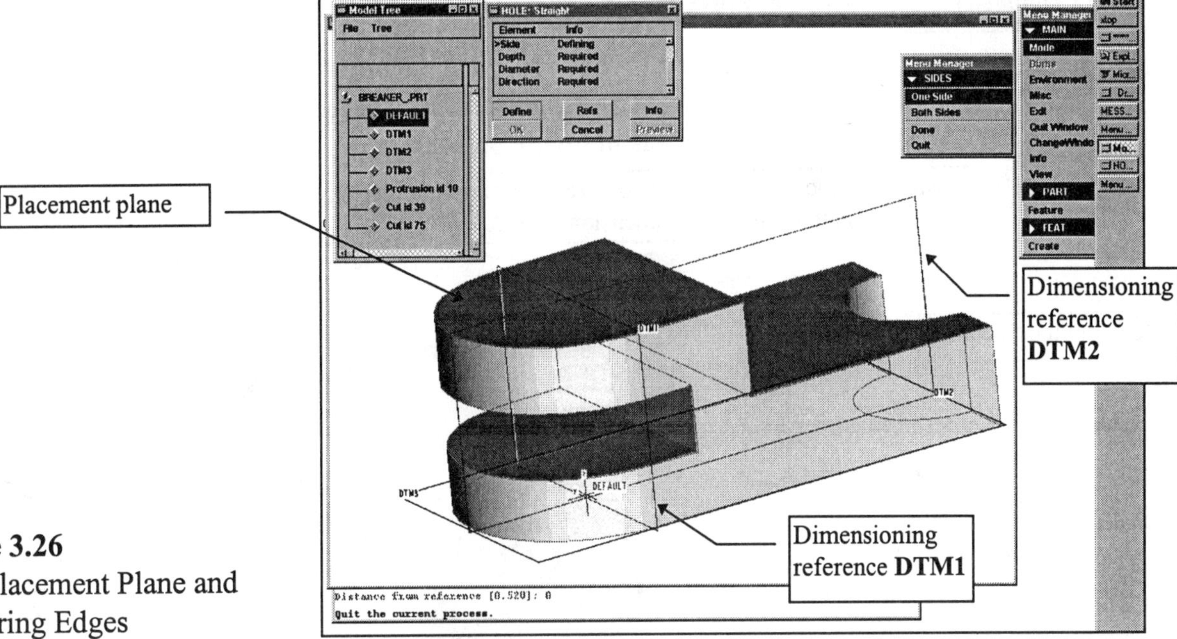

Figure 3.26
Hole Placement Plane and Measuring Edges

Dbms (**File**--PT/Modeler) ⇒
Save ⇒ enter
Purge ⇒ enter ⇒
Done-Return

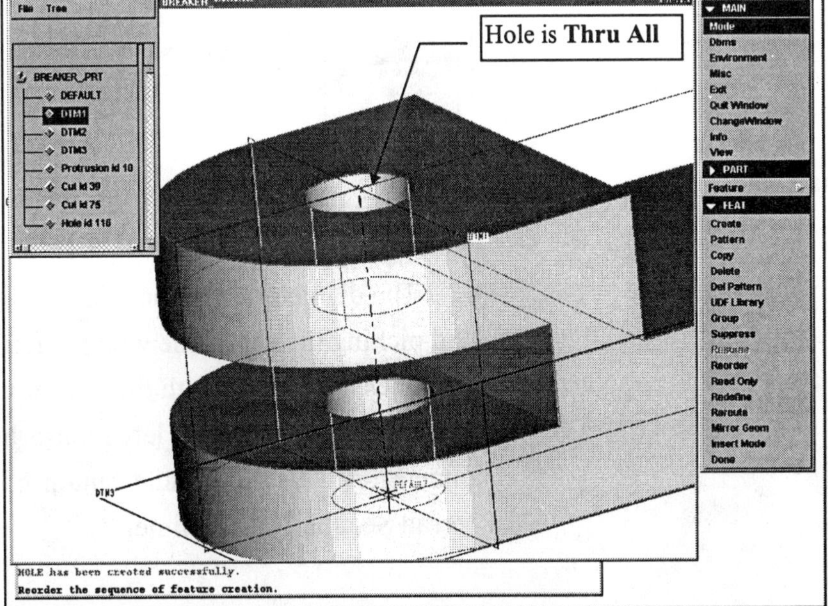

Figure 3.27
Completed Hole

The next hole will be a *sketched hole*. A sketched hole can be almost any configuration. Sketched holes can be used to create counterbore holes, countersunk holes, spotfaced holes, and so on. At this point in the lesson, you will create a counterbore hole. Sketched holes are really nothing more than revolved cuts.

NOTE

All entities of a sketched hole must be on one side of a vertical centerline, and the section must be closed.

Sketched holes are created with a section sketch, just like a cut. ***The section must have a vertical centerline, and all entities must be on one side of that centerline.*** Always start by sketching the vertical centerline first. Next, sketch the entities required to describe the hole's shape (half of the shape). No alignment is necessary, but dimensions are required. ***The section must be closed.*** Use the following commands:

PT/Modeler™
Feature ⇒ Hole etc.

Feature ⇒ Create ⇒ Hole ⇒ Sketch ⇒ Done ⇒ Linear ⇒ Done

A small window appears on the screen at this point. Start with a vertical centerline, create the required *closed* section, and dimension the sketch as in Figure 3.28.

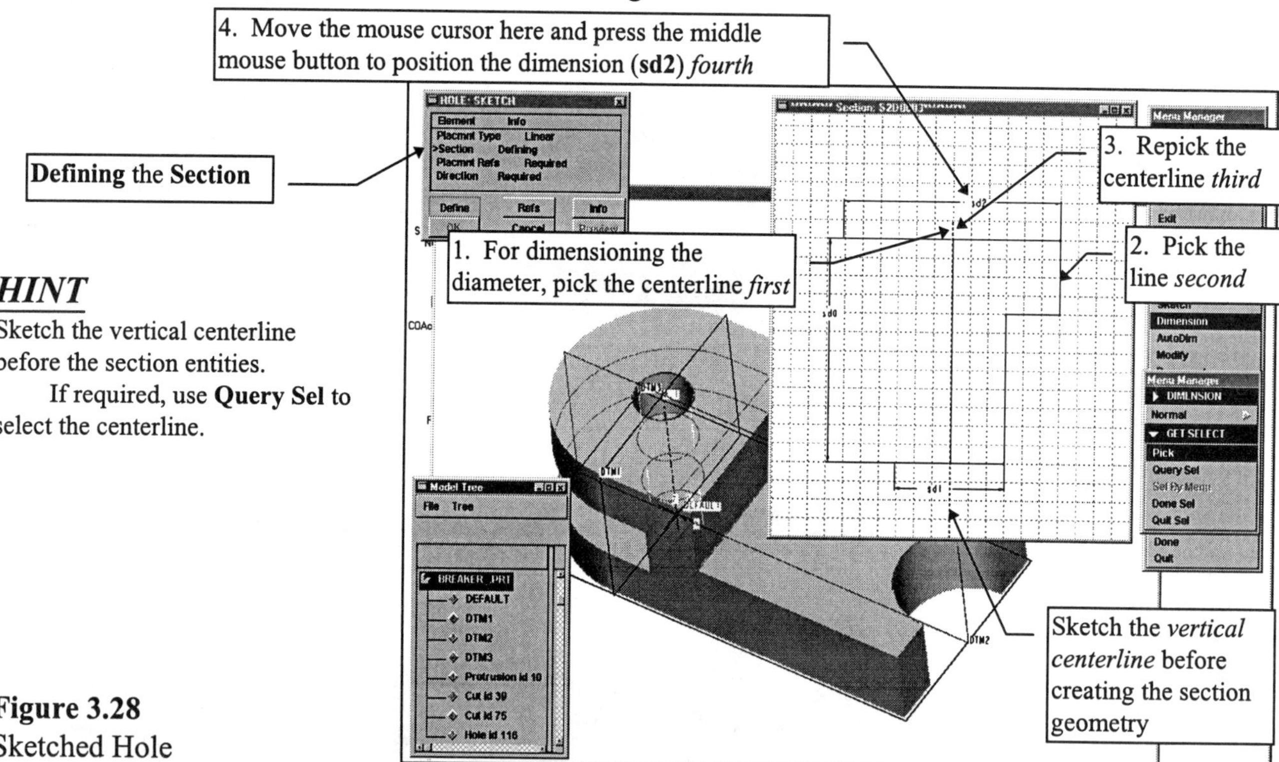

HINT

Sketch the vertical centerline before the section entities.

If required, use **Query Sel** to select the centerline.

Figure 3.28
Sketched Hole

The diameter of the hole is dimensioned by choosing **Dimension** and picking the centerline with the left mouse button, the edge to be dimensioned also with the left mouse button, and the centerline a second time with the left mouse button, and then placing the dimension with the middle button of the mouse. You should reread Section 9 at this time.

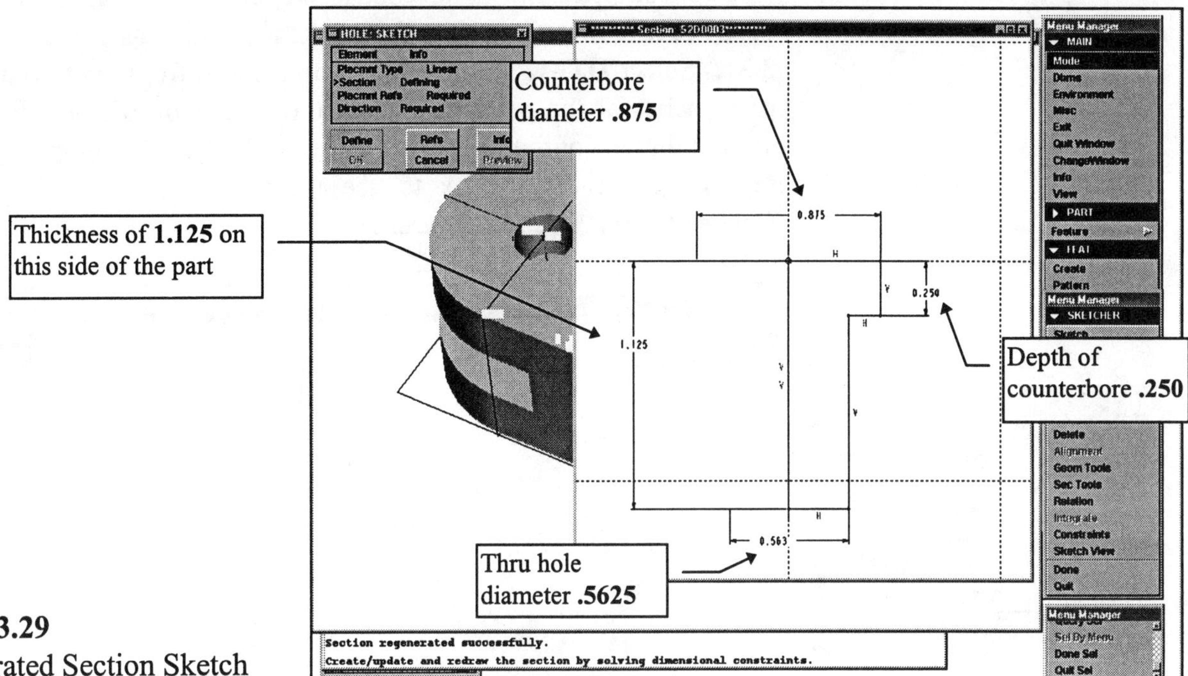

Figure 3.29
Regenerated Section Sketch

Regenerate the sketch and **Modify** the dimensions (Fig. 3.29). The counterbore diameter is **.875** and the thru hole diameter is **.5625** (Fig. 3.30). The depth of the hole is the same as the thickness of the part where the hole is placed, **1.125** (If you want to keep the thru hole *through* the part, regardless of modifications to the part's thickness, you must use a relationship. This will be covered in Lesson 8).

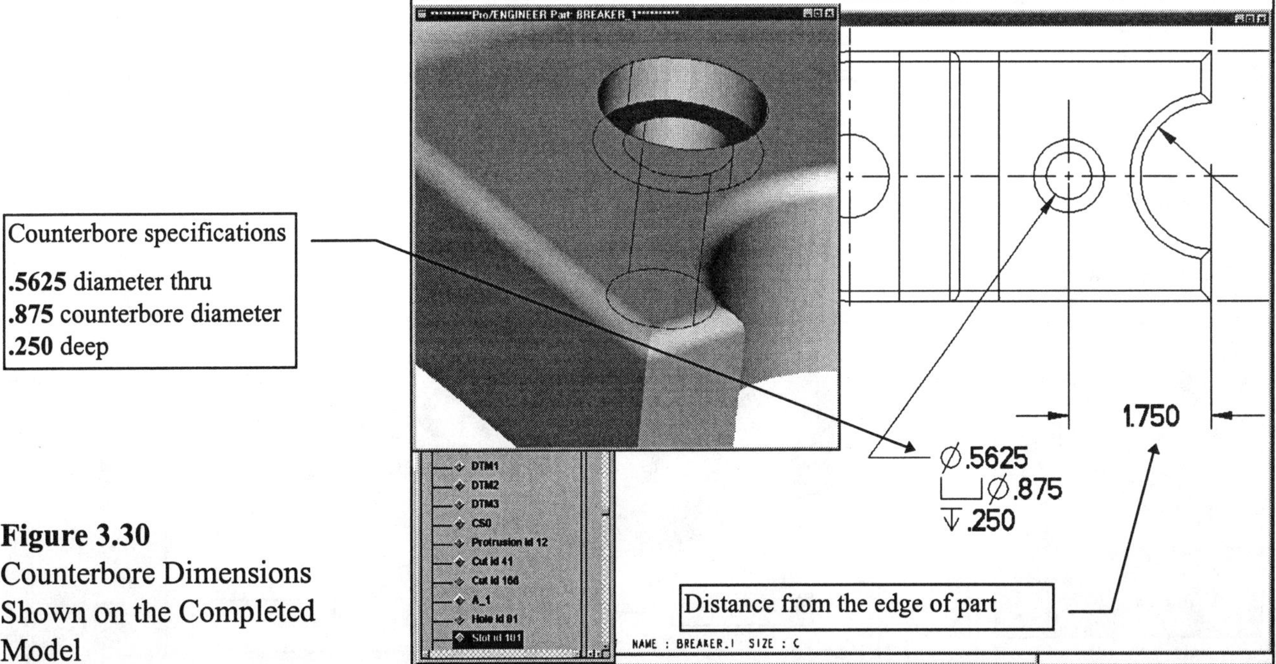

Figure 3.30
Counterbore Dimensions Shown on the Completed Model

To complete the hole, choose **Regenerate ⇒ Done** (from the SKETCHER menu). Select the placement plane as shown in Figure 3.31. Select **DTM2** as the edge to dimension from, and give a value of **0** inches at the prompt. Select the right side of the part for the second dimensioning reference, and type **1.75** at the prompt. Choose **OK** from the dialog box to see the completed hole (Fig. 3.32). Choose **Done.**

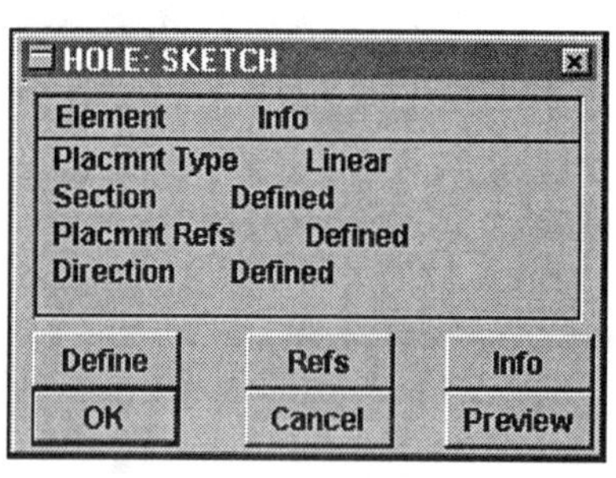

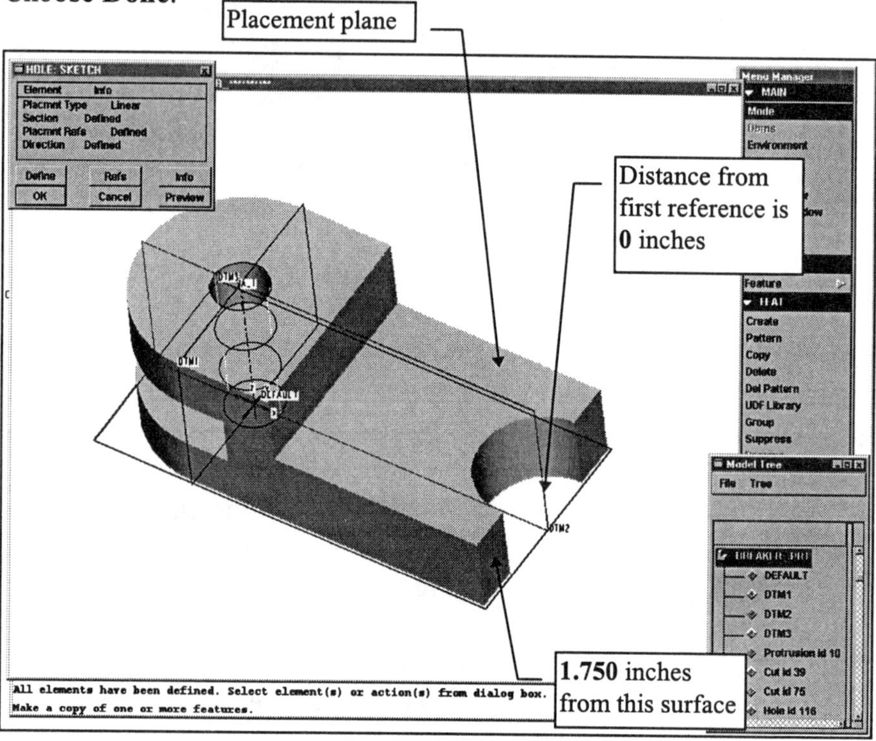

Figure 3.31
Placement Plane and Dimensioning References

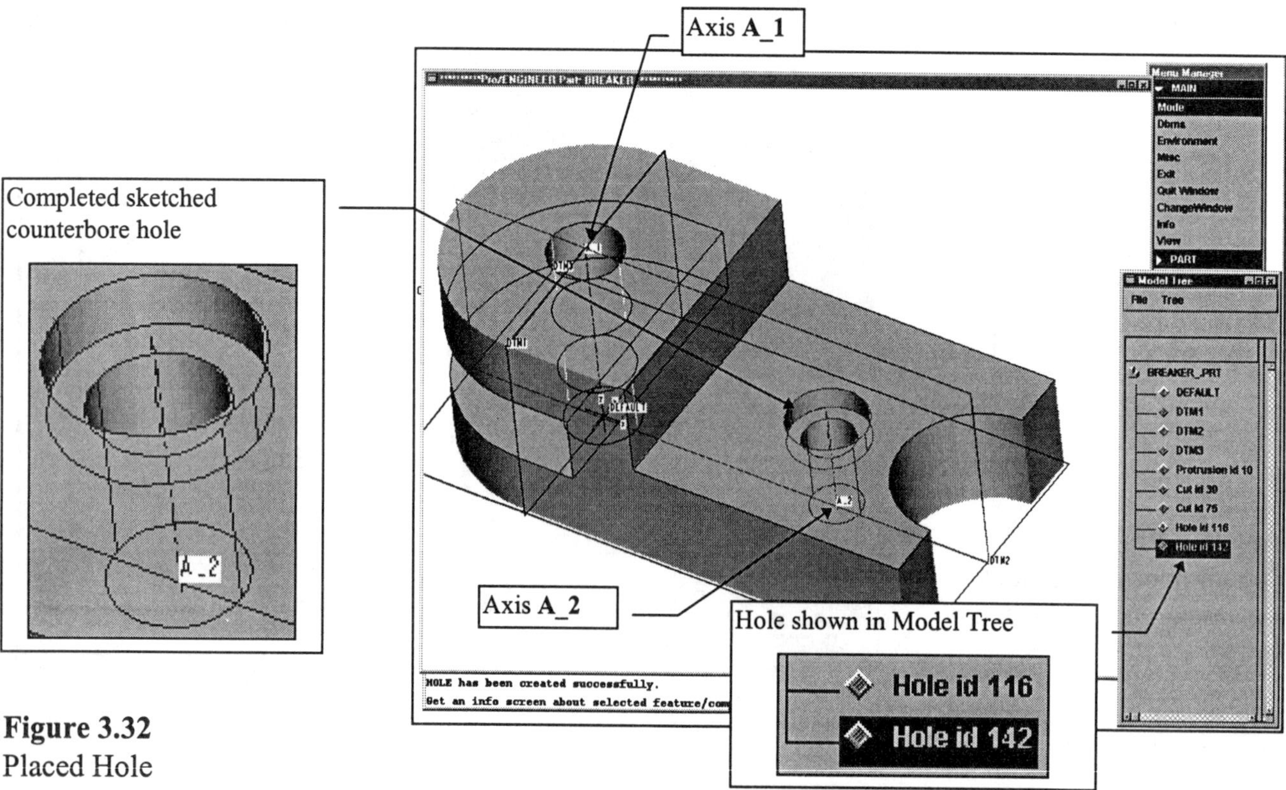

Figure 3.32
Placed Hole

Open a new window and zoom in on the counterbore (Fig. 3.33). Highlight the feature in the Model Tree and press the right mouse button. Select **Info ⇒ Feat Info**. The INFORMATION WINDOW will appear on your screen, as in Figure 3.34.

New window:
View ⇒ New Window ⇒ View ⇒ Pan/Zoom ⇒ Zoom In (pick twice to define the zoom window) **⇒ Done-Return**

Dbms (**File**--PT/Modeler) ⇒
Save ⇒ enter
Purge ⇒ enter ⇒
Done-Return

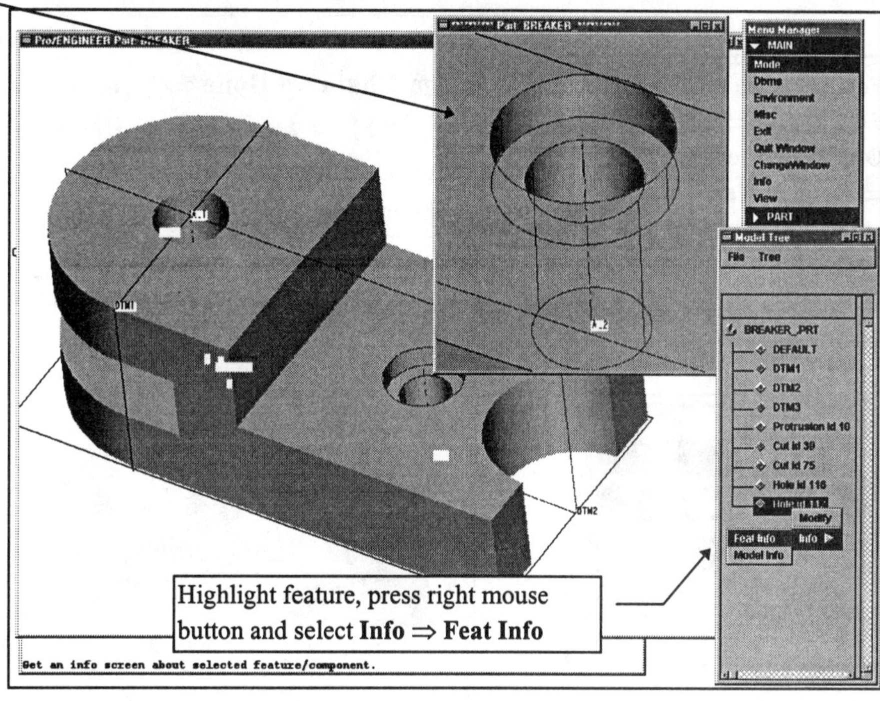

Figure 3.33
New Window and Feature Information Requested from the Model Tree

Information about a feature, part, assembly, etc. can be requested as needed during a Pro/E Session.

To close an INFORMATION WINDOW, type **q** to quit (make sure the mouse cursor is inside the INFORMATION WINDOW). Choose **Quit Window ⇒ Change Window** and pick in the Main Window.

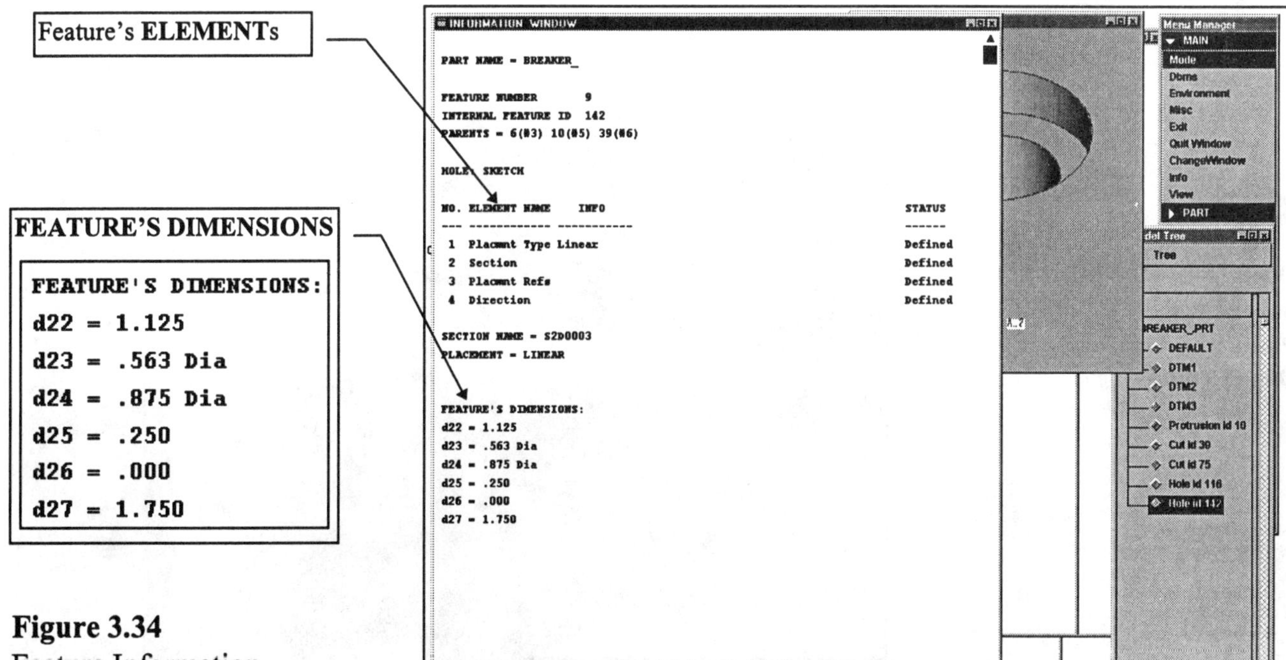

Figure 3.34
Feature Information

To complete the part, a number of rounds need to be created. The first round is an edge round between the vertical and horizontal faces of the first cut. Choose the following commands:

PT/Modeler™
Feature ⇒ Round ⇒ Constant ⇒ Edge Chain ⇒ Done ⇒ One By One ⇒ (pick edges) ⇒ **Done Sel ⇒ Done ⇒ New Value** ⇒ (**.50**) ⇒ **enter ⇒ Preview ⇒ OK**

Feature ⇒ Create ⇒ Round ⇒ Simple ⇒ Done ⇒ Constant ⇒ Edge Chain ⇒ Done ⇒ One By One ⇒ (pick the edge as shown in Fig. 3.35) ⇒ **Done Sel ⇒ Done ⇒ New Value** ⇒ (type the round radius value at the prompt, **.50**) ⇒ **enter ⇒ Preview ⇒ OK** (round will appear as in Fig. 3.36) ⇒ **Done**

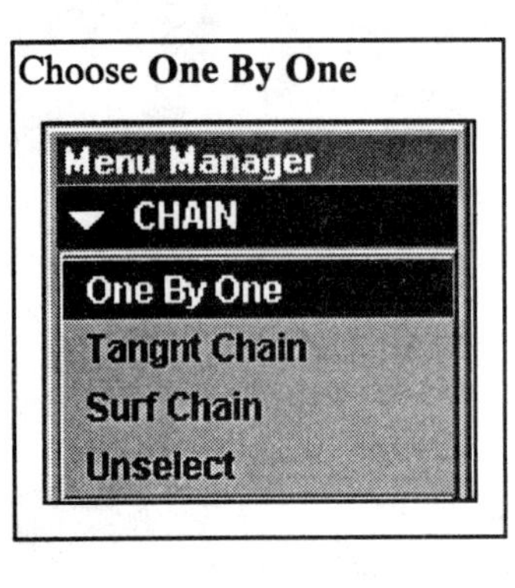

Pick this edge

Coordinate Systems will not be displayed.
Accept selections just made. (Equivalent mouse button: MIDDLE)

Figure 3.35
Edge to Be Rounded

Dbms (**File**--PT/Modeler) ⇒
Save ⇒ enter
Purge ⇒ enter ⇒
Done-Return

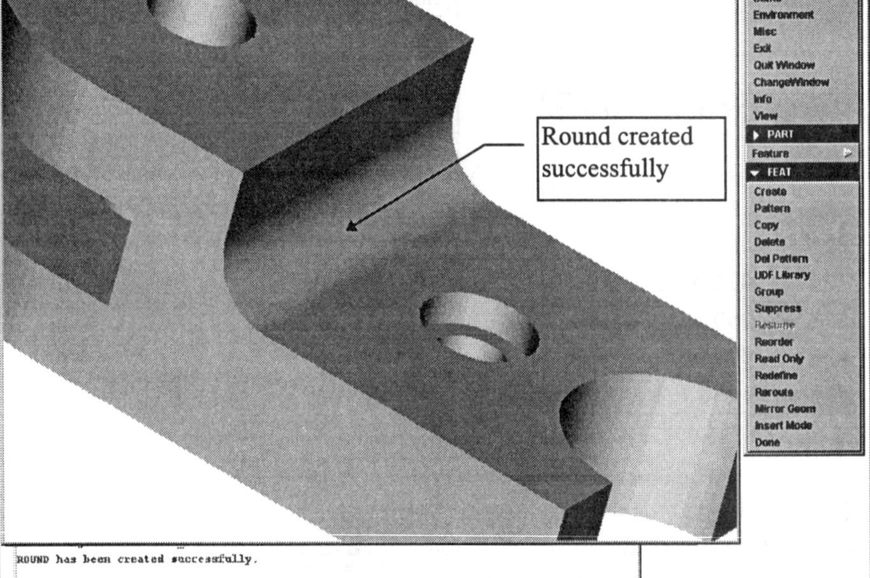

Figure 3.36
Round

Finally, the last round is a set that can be created all at the same time. Choose the following commands:

PT/Modeler™
Feature ⇒ Round ⇒ Constant etc.

Feature ⇒ Create ⇒ Round ⇒ Simple ⇒ Done ⇒ Constant ⇒ Edge Chain ⇒ Done ⇒ Tangnt Chain (pick the upper edge as shown in Fig. 3.37) ⇒ **One By One** ⇒ (pick the remaining edges as shown in Fig. 3.37 in sequence) ⇒ **Done Sel** ⇒ **Done** ⇒ **New Value** ⇒ (type the round radius value at the prompt, **.125**) ⇒ **enter** ⇒ **Preview** ⇒ **OK** (Fig. 3.38)

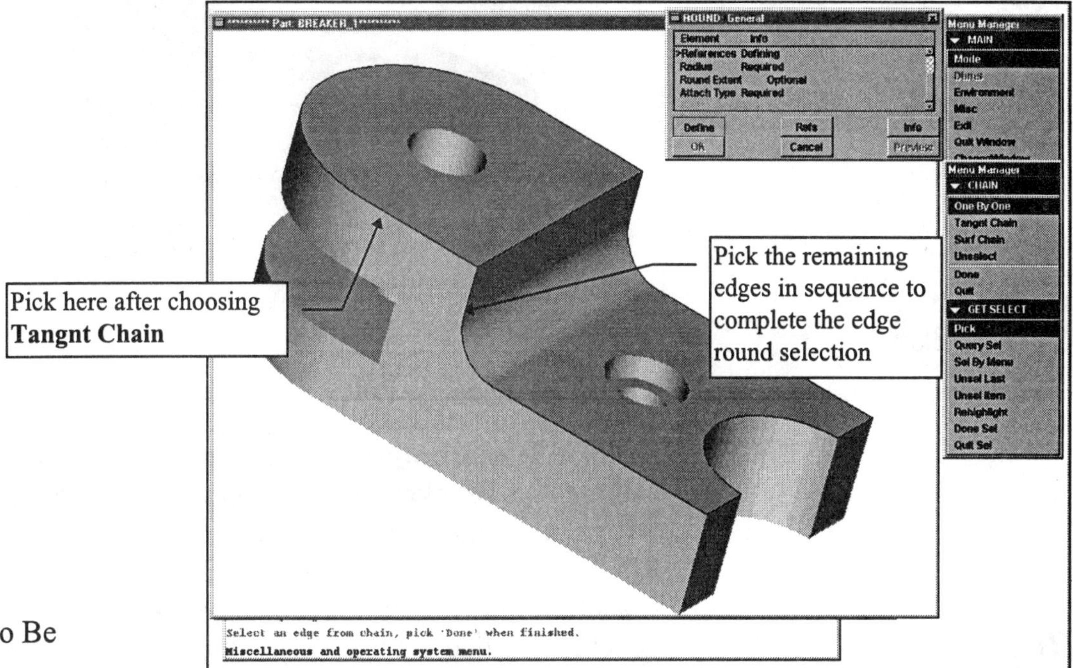

Figure 3.37
Pick the Edges to Be Rounded

Dbms (**File**--PT/Modeler) ⇒ **Save** ⇒ **enter** **Purge** ⇒ **enter** ⇒ **Done-Return**

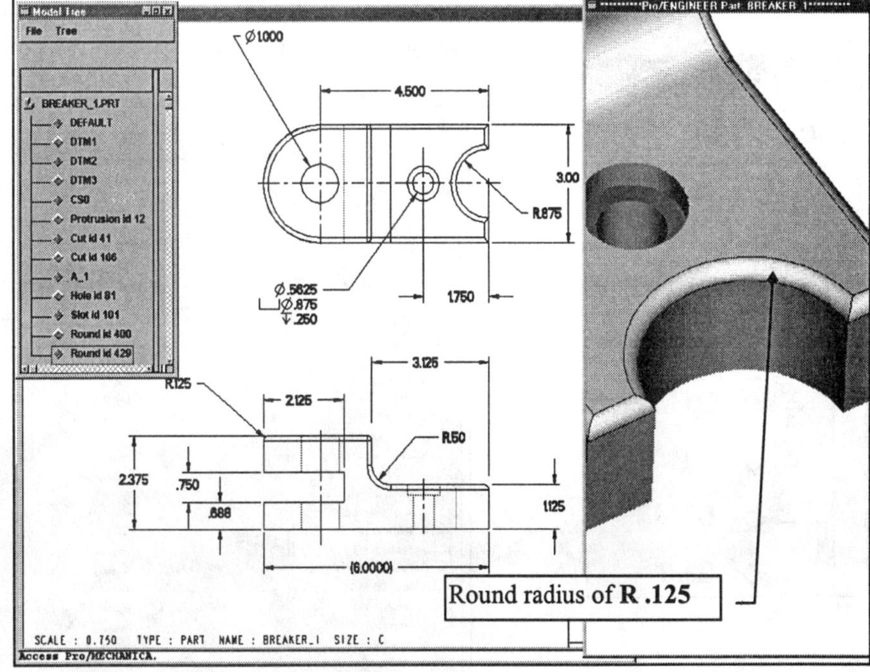

Figure 3.38
Completed Part

Lesson 3 Project

Guide Bracket

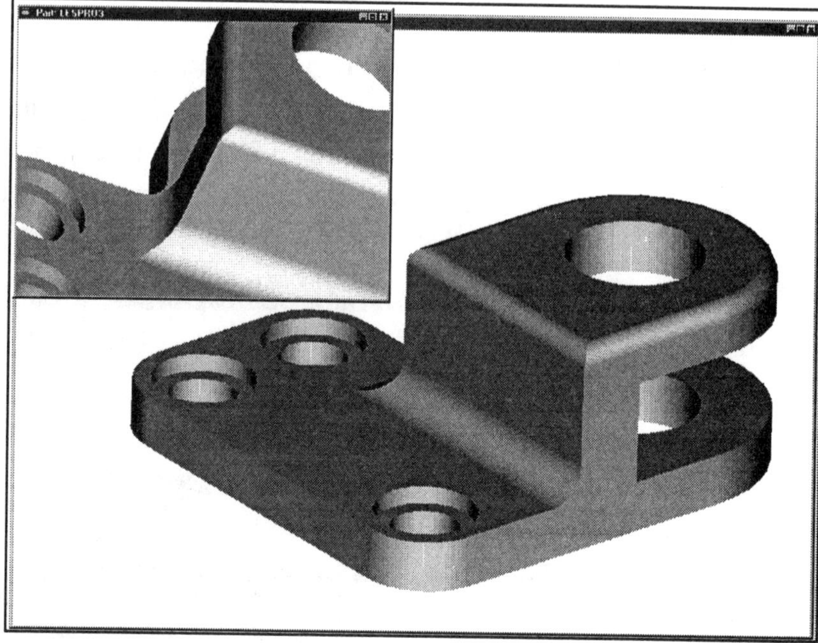

Figure 3.39
Guide Bracket

Guide Bracket

The third **lesson project** is a machined part that requires commands similar to the **Bracket**. Simple rounds and straight and sketched holes are part of the exercise. Create the part shown in Figures 3.39 through 3.45. At this stage in your understanding of Pro/E, you should be able to analyze the part and plan out the steps and features required to model it. You do not have to use the exact design intent shown here in the lesson project. You must use the same dimensions and dimension scheme, but the choice and quantity of datum planes and the sequence of modeling features can be different.

HINT
DATUM PLANES will be the first features on all parts and assemblies.

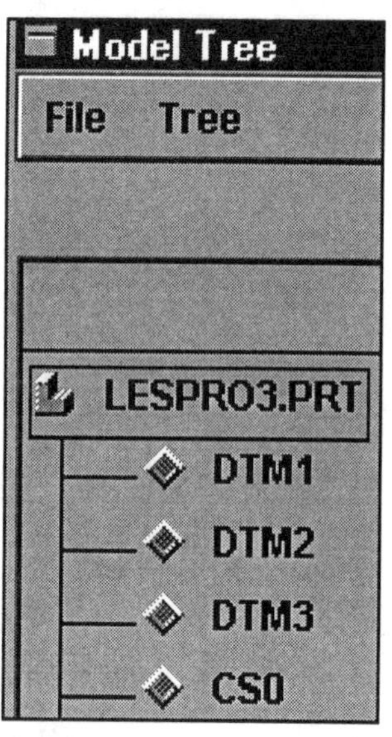

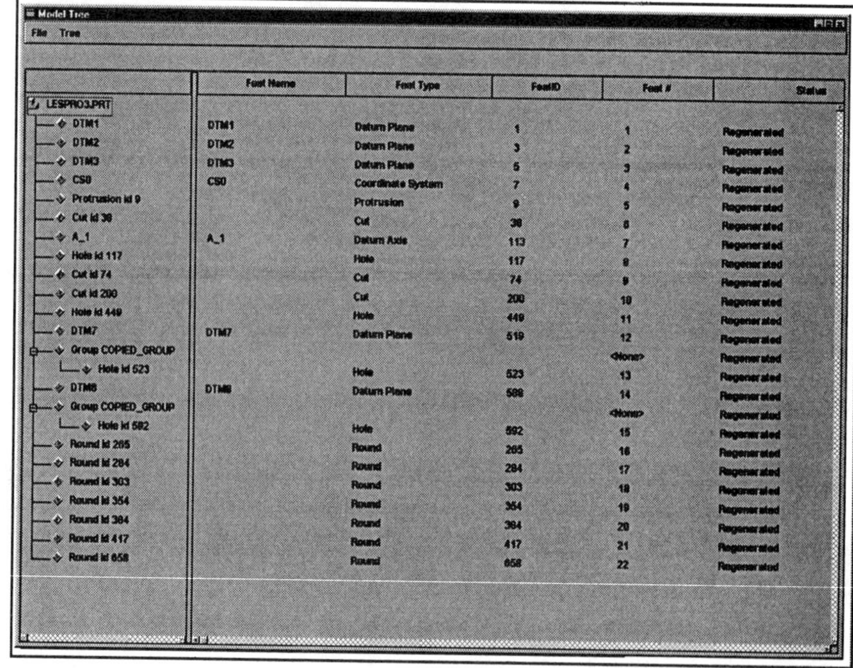

	Feat Name	Feat Type	FeatID	Feat #	Status
LESPRO3.PRT					
DTM1	DTM1	Datum Plane	1	1	Regenerated
DTM2	DTM2	Datum Plane	3	2	Regenerated
DTM3	DTM3	Datum Plane	5	3	Regenerated
CS0	CS0	Coordinate System	7	4	Regenerated
Protrusion id 9		Protrusion	9	5	Regenerated
Cut id 38		Cut	38	6	Regenerated
A_1	A_1	Datum Axis	113	7	Regenerated
Hole id 117		Hole	117	8	Regenerated
Cut id 74		Cut	74	9	Regenerated
Cut id 200		Cut	200	10	Regenerated
Hole id 449		Hole	449	11	Regenerated
DTM7	DTM7	Datum Plane	519	12	Regenerated
Group COPIED_GROUP				<None>	Regenerated
Hole id 523		Hole	523	13	Regenerated
DTM8	DTM8	Datum Plane	589	14	Regenerated
Group COPIED_GROUP				<None>	Regenerated
Hole id 592		Hole	592	15	Regenerated
Round id 265		Round	265	16	Regenerated
Round id 284		Round	284	17	Regenerated
Round id 303		Round	303	18	Regenerated
Round id 354		Round	354	19	Regenerated
Round id 384		Round	384	20	Regenerated
Round id 417		Round	417	21	Regenerated
Round id 658		Round	658	22	Regenerated

Figure 3.40
Guide Bracket Model Tree

NOTE

Use the DIPS to plan out your feature creation sequence and the selection of datum sketching planes.

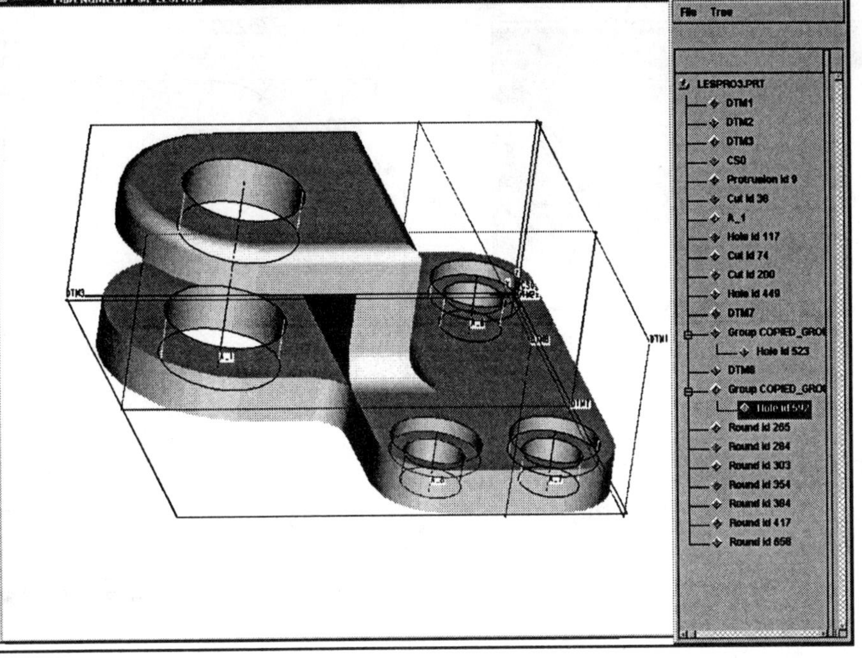

Figure 3.41
Guide Bracket Showing Datums and Coordinate System

Figure 3.42
Guide Bracket Drawing

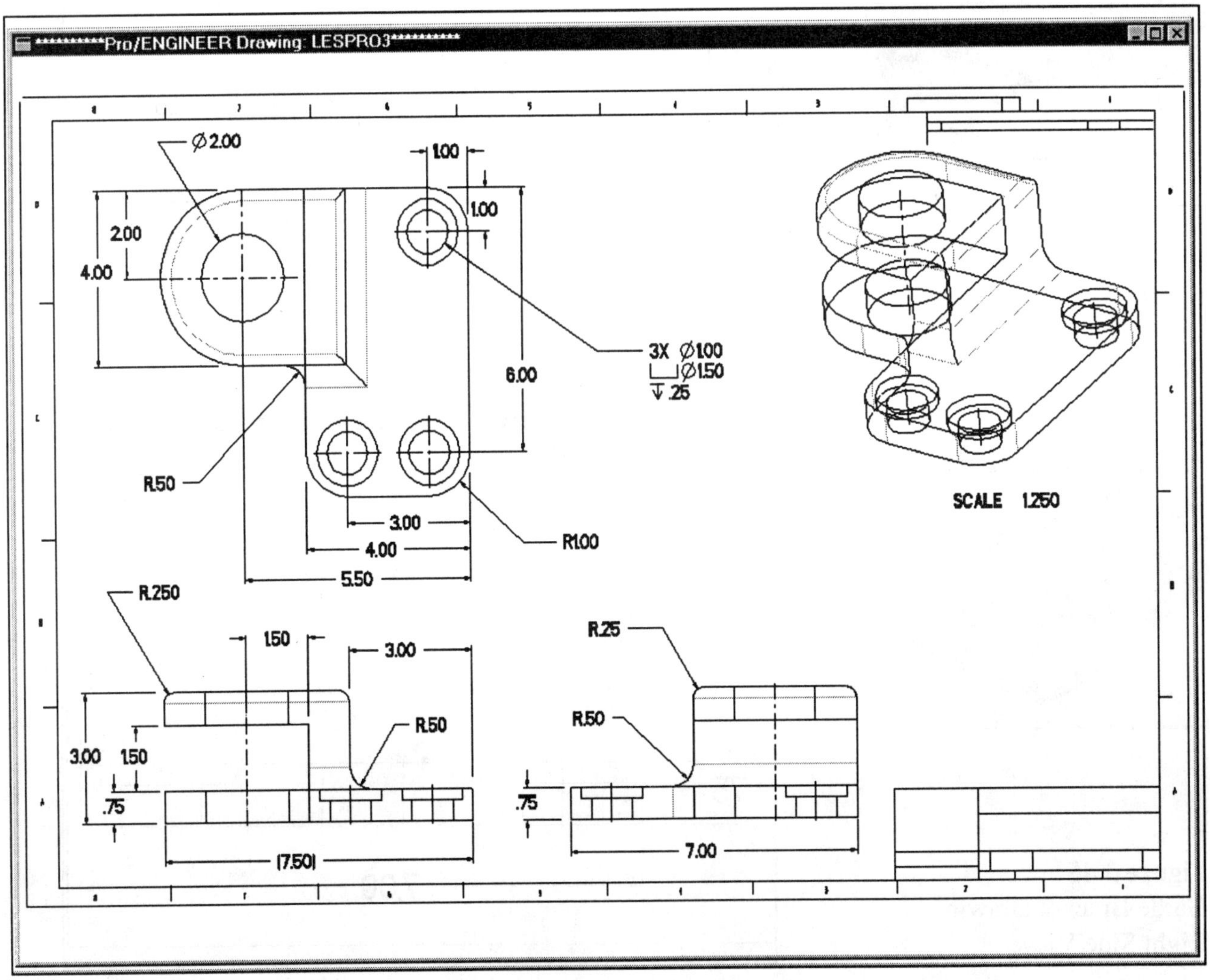

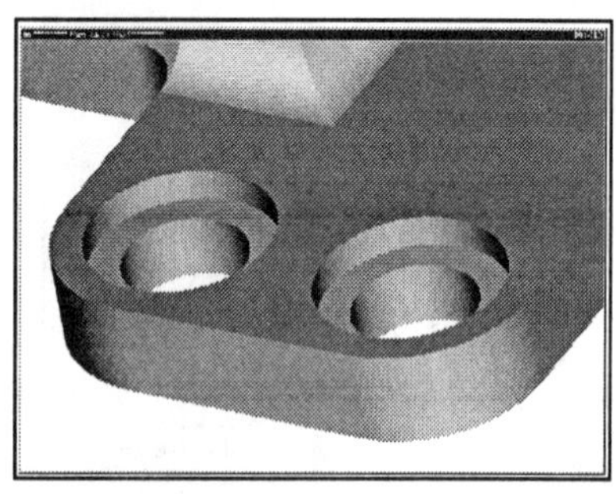

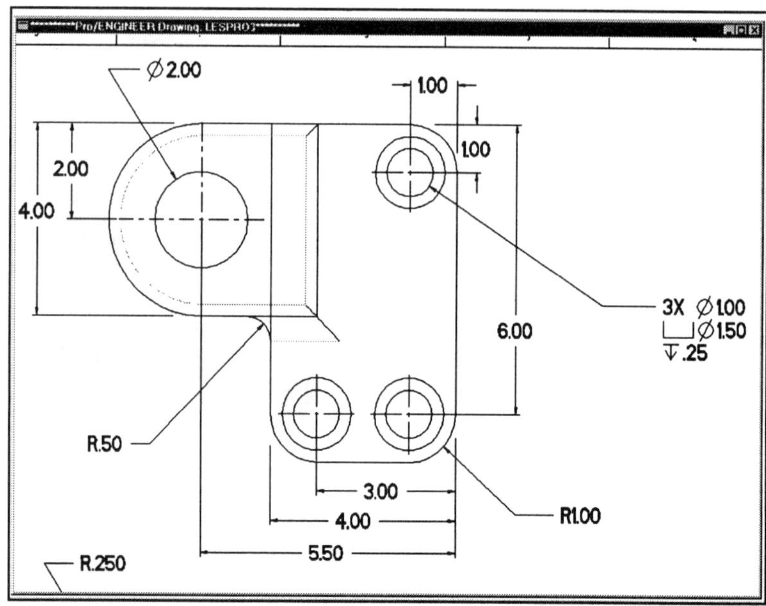

Figure 3.43
Guide Bracket Drawing,
Top View

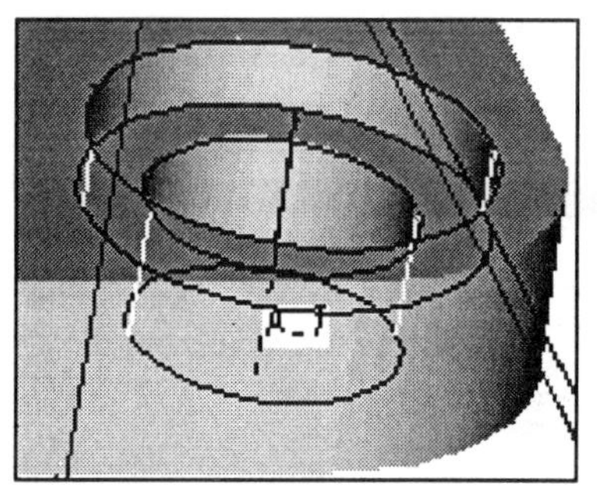

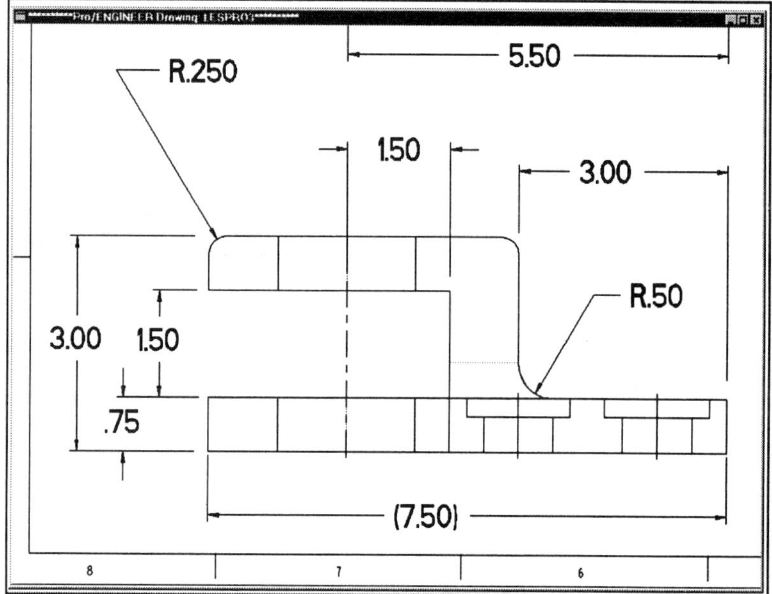

Figure 3.44
Guide Bracket Drawing,
Front View

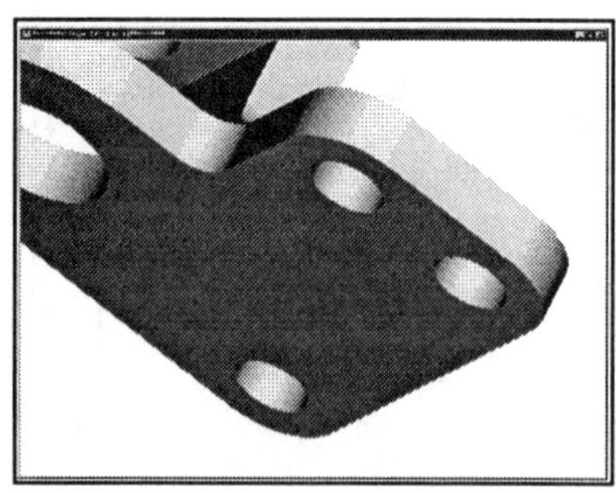

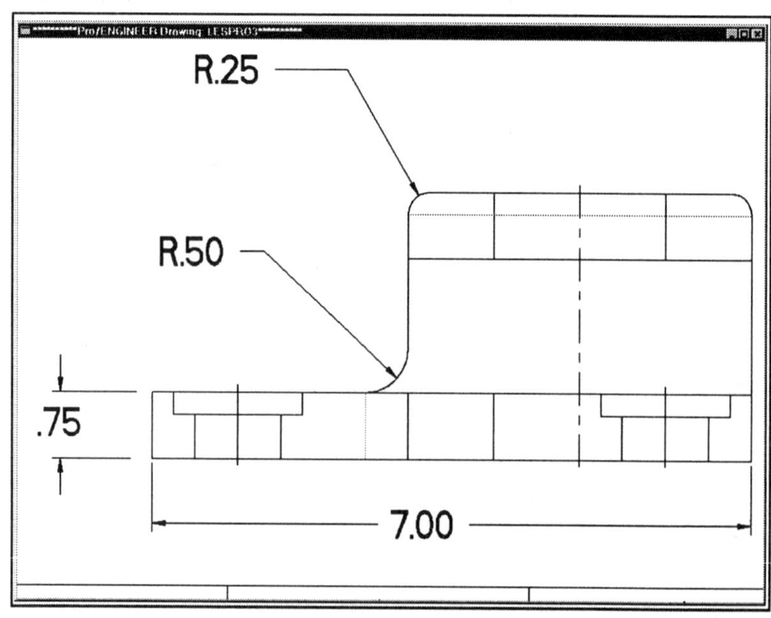

Figure 3.45
Guide Bracket Drawing,
Right Side View

Lesson 4

Datums and Layers

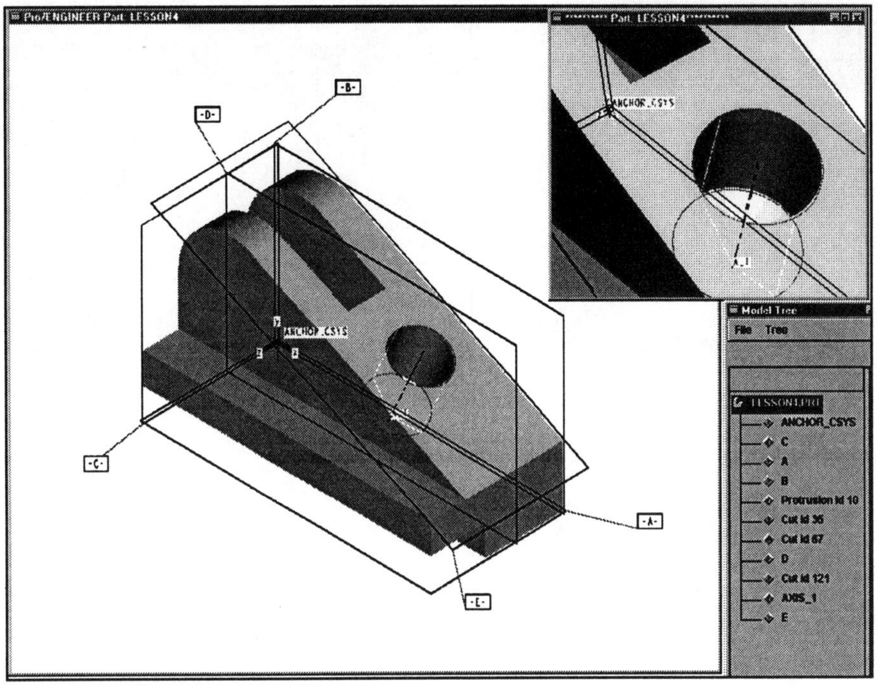

Figure 4.1
Anchor

EGD REFERENCE
Engineering Graphics and Design with Graphical Analysis *or* **Fundamentals of Engineering Graphics and Design**
by L. Lamit and K. Kitto
Read Chapters 10, 25
See pages 388, 549, 929-932

COAch™ for Pro/ENGINEER

If you have **COAch for Pro/ENGINEER** on your system, go to SEARCH and do the Segment shown in Figure 4.3.

OBJECTIVES

1. **Create datums to locate features**
2. **Use layers to organize part features**
3. **Set datum planes for geometric tolerancing**
4. **Rename datums**
5. **Add a simple relation to control a feature**
6. **Use datum planes to establish sections**
7. **Reroute a features references**
8. **Use Info command to get layer information**
9. **Learn how to change the color and shading of models**

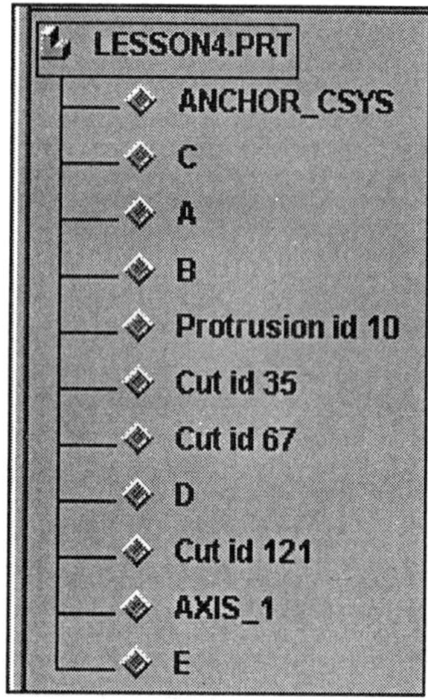

Figure 4.2
Anchor with Datums

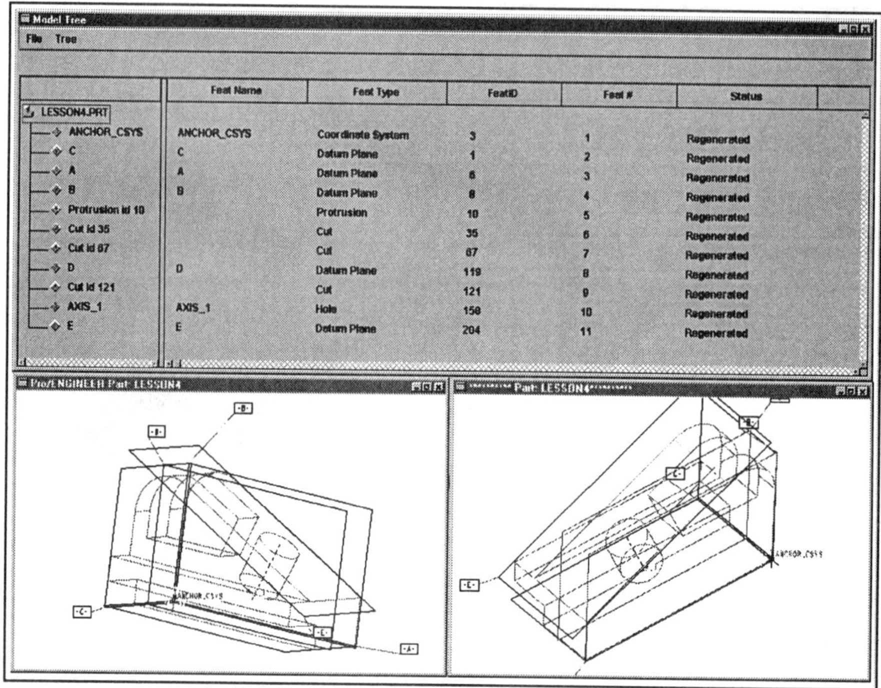

DATUMS AND LAYERS

Datum planes and layers are two of the most useful mechanisms for creating and organizing your design (Figs. 4.1 and 4.2). **Layers** were covered in detail in Section 8 and that section should be reread at this point. Datum features such as *datum planes* and *datum axes* are essential for the creation of all parts, assemblies, and drawings using Pro/E. If you have **COAch for Pro/ENGINEER** on your system, do the appropriate segment, shown in Figure 4.3.

Figure 4.3
COAch for Pro/E, Datums (Overview)

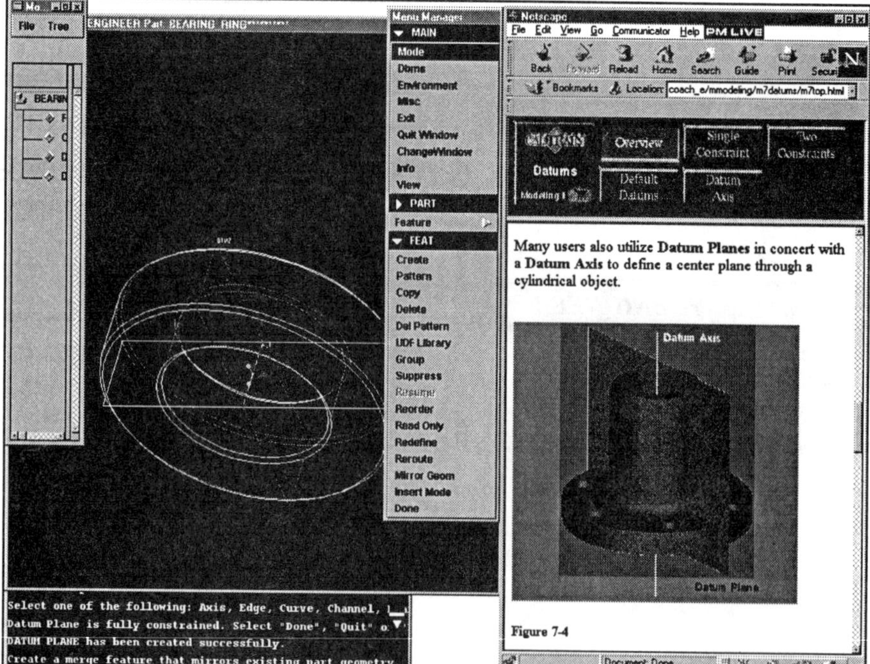

Datum Planes

Datum planes are used to create a reference on a part that does not already exist. For example, you can sketch or place features on a datum plane when there is no appropriate planar surface; you can also dimension to a datum plane as if it were an edge. When you are constructing an assembly, you can use datums with assembly commands. All datums have a red side and a yellow side so that you know on which side you are working.

Datum planes can be used as references, as sketching planes, and as parent features for a variety of nonsketched part features.

In most part designs, the first features created will be three default datums: **DTM1, DTM2,** and **DTM3** (Fig. 4.4). The base construction feature (a protrusion) is created using these datums.

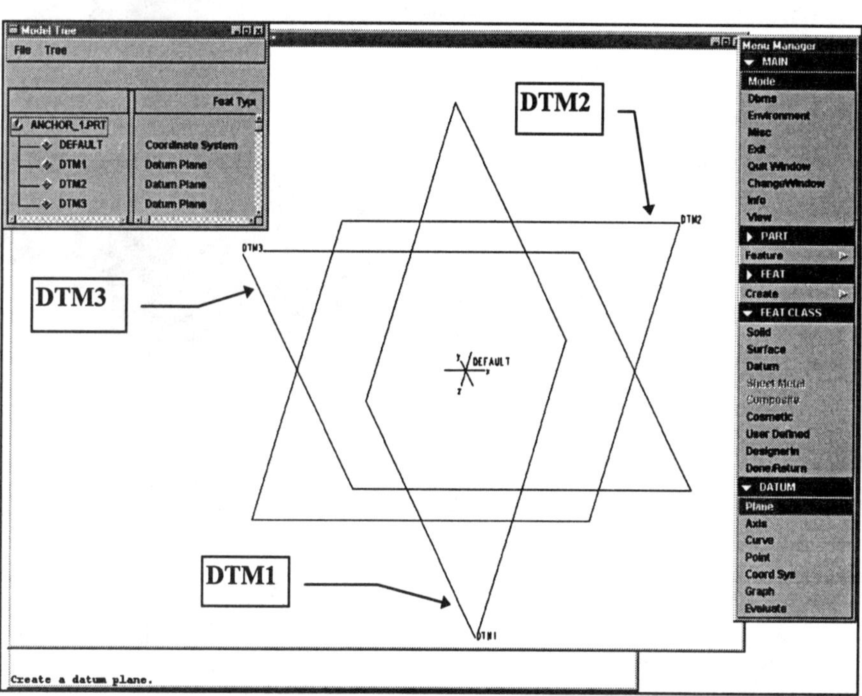

Figure 4.4
Default Datums

A nondefault datum is created by specifying **constraints** that locate it with respect to existing geometry. As an example, a datum plane might be created to pass tangent to a cylinder and parallel to a planar surface. Selected constraints must locate the datum plane relative to the model without ambiguity. Figure 4.5 shows an angled datum used to model the part. The following datum constraints are used alone, since each locates the datum plane completely:

Through/Plane Creates a datum plane coincident with a planar surface.

Offset/Plane Creates a datum plane that is parallel to a plane and offset from the plane by a specified distance.

Offset/Coord Sys Creates a datum plane that is normal to one of the coordinate system axes and offset from the origin of the coordinate system. When this option is selected, you are prompted to select the axis the plane will be normal to, then to enter the offset distance along this axis.

BlendSection Creates a datum plane through the section that was used to make a feature. When sections exist, as for a blend, you will be prompted for the section number.

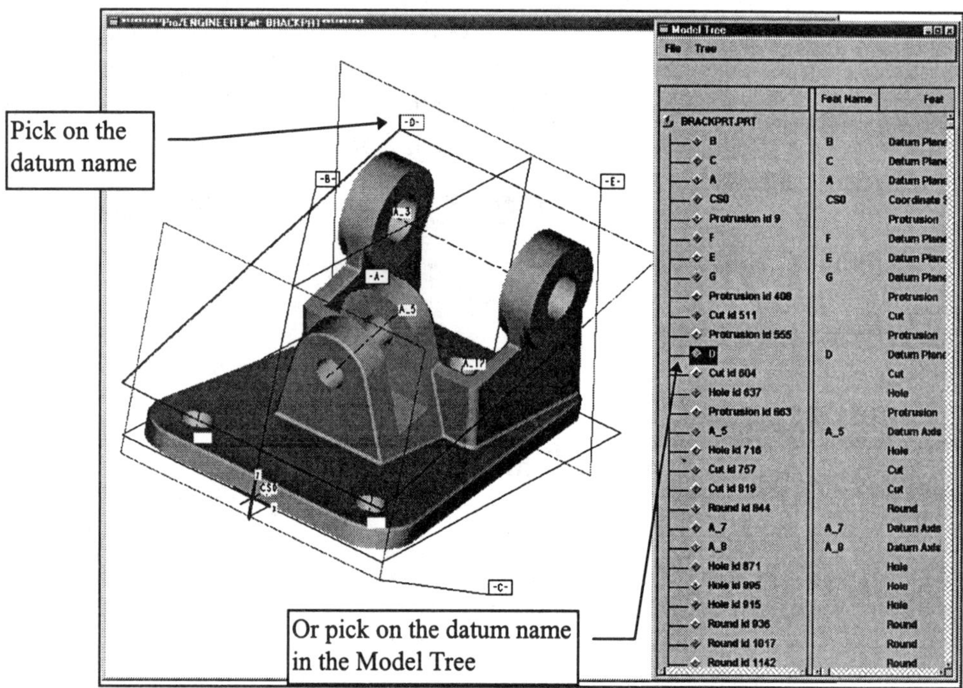

Figure 4.5
Nondefault Datums

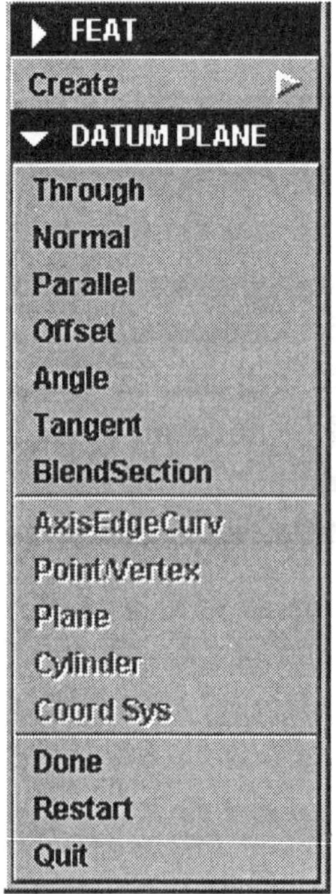

To create a datum plane:

1. Choose **Datum** from the FEAT CLASS menu (or **Make Datum** from the SETUP PLANE menu), then choose **Plane** from the DATUM menu.
2. Choose the desired constraint option from the DATUM PLANE menu. All appropriate geometry options in the lower section of the menu will be selected automatically. To limit the items to select, click on highlighted menu options to unhighlight them.
3. Pick the necessary references.
4. Repeat Steps 2 and 3 until the required constraints have been established. When the maximum number of constraints have been specified, Pro/E notifies you by *dimming out* all options except **Done**, **Restart**, and **Quit**. Although they are actually *infinite planes*, datum planes are displayed scaled to the model size. To select a datum plane, you can pick on its name or select one of its boundaries, or select it by picking on its name in the **Model Tree** (Fig. 4.5).

The size of a displayed datum plane changes with the dimensions of a part. All datum planes, except those made ***on the fly*** (within other commands using **Make Datum**), can be sized to specific geometry using the **Redefine** command. These allow you to make your datum plane as big as the model or as small as an edge or surface on the model. Figure 4.6 shows a page of the help available on datum planes.

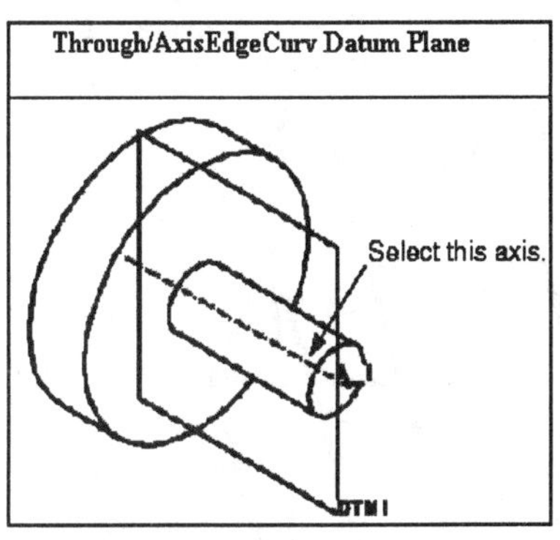

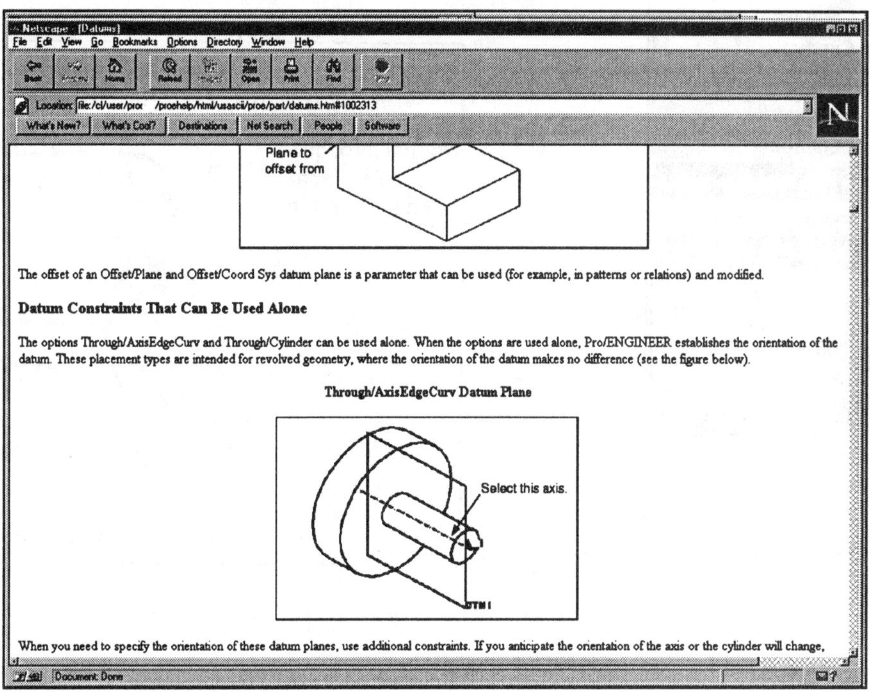

Figure 4.6
Online Documentation, Datums

The options available for sizing the datum plane outline are:

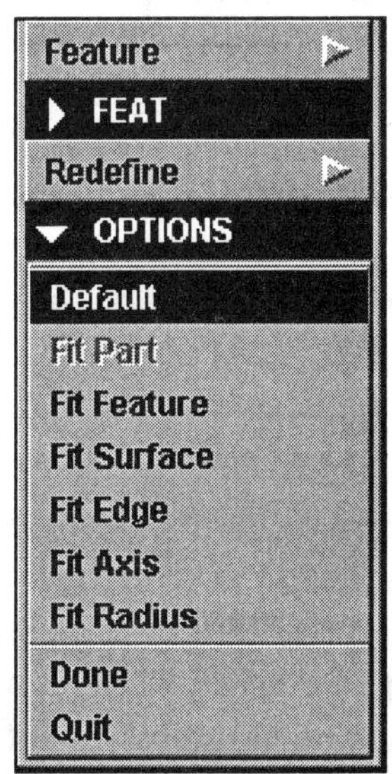

Default The datum plane is sized to the model (part or assembly).
Fit Part Sizes the datum plane to a part in Assembly mode.
Fit Feature Sizes the datum plane to a part or assembly feature.
Fit Surface Sizes the datum plane to any surface.
Fit Edge Sizes the datum plane to fit an edge.
Fit Axis Sizes the datum plane to fit an axis.
Fit Radius Sizes the datum plane to fit a specified radius, centering itself within the constraints of the model.

Datum Axes

Datum axes (Fig. 4.7) can be used as references for feature creation, such as the coaxial placement of a hole. They are particularly useful for making datum planes, for placing items concentrically, and for creating radial patterns. Axes can be used to measure from, place coordinate systems, and place specific features. The angle between a feature and an axis, the distance between an axis and a feature, and so on can be determined using the **Info** command. Axes (appearing as centerlines) are automatically created for:

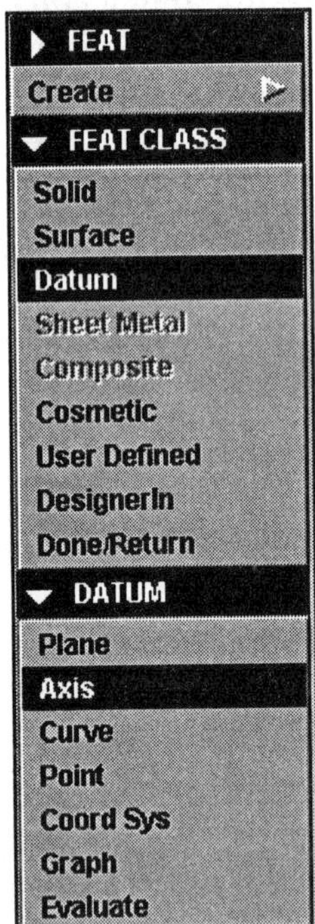

Figure 4.7
Online Documentation
Datum Axes

Revolved features All features whose geometry is revolved, including revolved base features, holes (Fig 4.8), shafts, revolved slots, cuts, and circular protrusions (Fig. 4.9).

Extruded circles An axis is created for every extruded circle in any extruded feature.

Extruded arcs An axis can be created automatically for extruded arcs only when you set the configuration options.

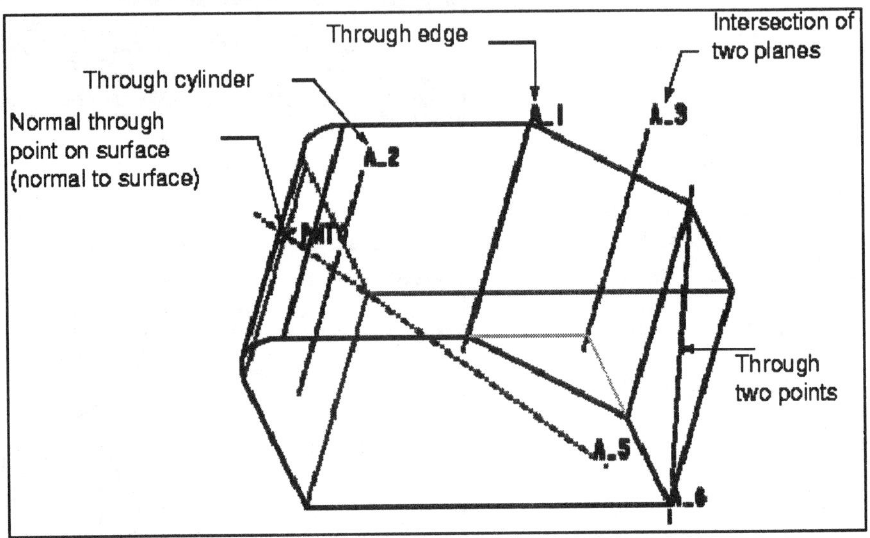

To create a datum axis:

1. Choose **Datum** from the FEAT Class menu and then **Axis** from the DATUM menu.
2. Choose the desired constraint option from the DATUM AXIS menu:

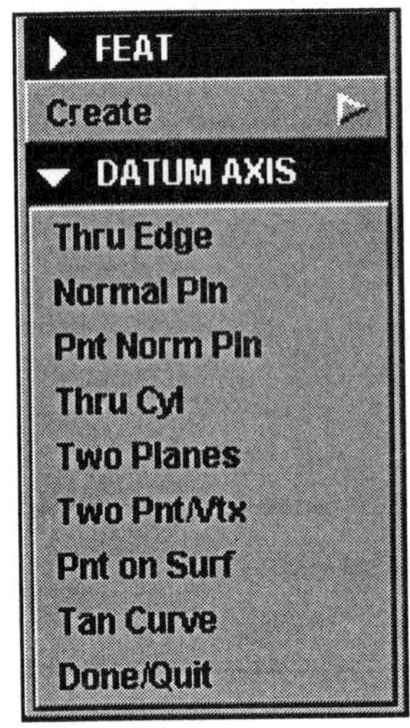

Thru Edge Creates a datum axis through a straight edge. Select the edge.

Norm Pln Creates an axis that is normal to a surface, with linear dimensions locating it on that surface.

Pnt Norm Pln Creates an axis through a datum point and normal to a specified plane.

Thru Cyl Creates an axis through the "imaginary" axis of any surface of revolution (where an axis does not already exist). Select a cylindrical surface or a revolved surface. Note that some features that only appear to be cylindrical, such as a revolved feature round, cannot be selected.

Two Planes Creates a datum axis at the intersection of two planes (datum planes or surfaces). Select two planes; they cannot be parallel, but they need not be shown to intersect on the screen.

Two Pnt/Vtx Create an axis between two datum points or edge vertices. Select datum points or edge vertices.
Pnt on Surf Create an axis through any datum point located on a surface; the point does not need to have been created using **On Surface**. The axis will be normal to the surface at that point.
Tan Curve Create an axis that is tangent to a curve or edge at its endpoint. Select the curve/edge for it to be tangent to, then select an endpoint of the curve/edge.

3. Pick the necessary references for the selected option.

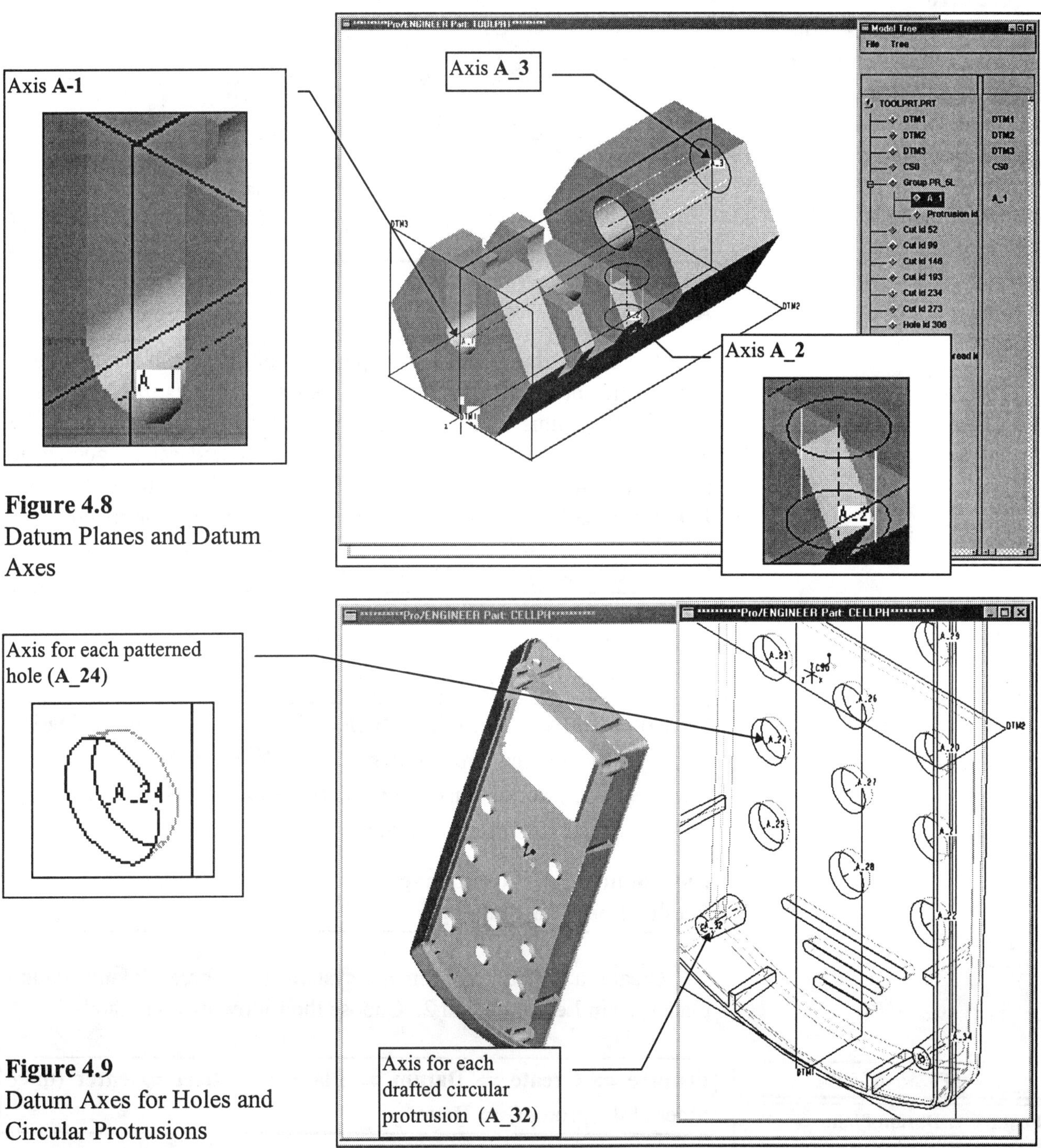

Figure 4.8
Datum Planes and Datum Axes

Figure 4.9
Datum Axes for Holes and Circular Protrusions

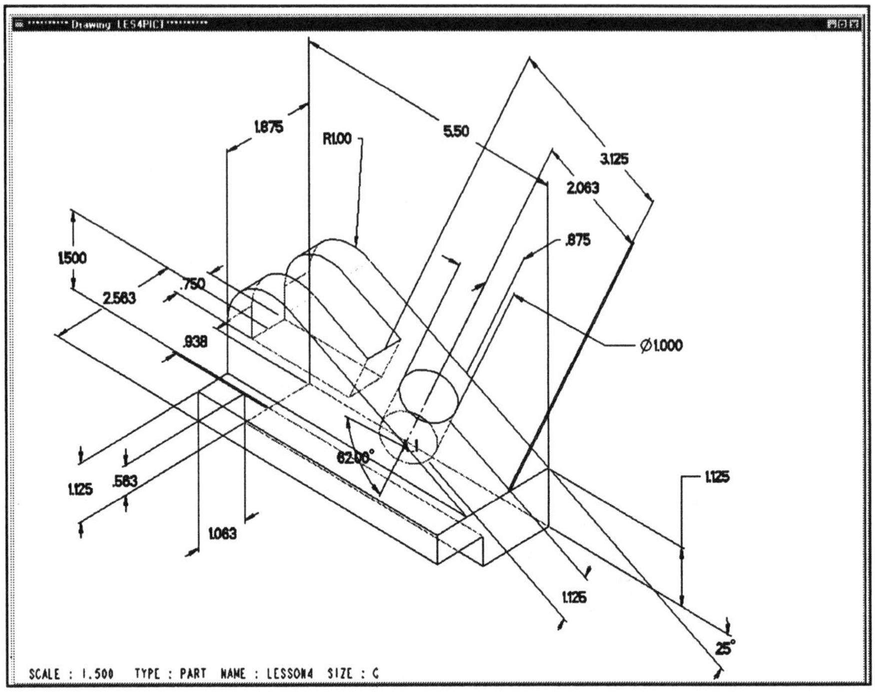

Figure 4.10
Anchor Dimensions (note that many of the dimensions are rounded since this is a part model shown in default view with dimensions displayed using **Modify**, not a dimensioned pictorial drawing)

Anchor

The **Anchor** is the fourth lesson part. Though default datums have been sufficient in previous lessons, Lesson 4 requires the creation of nondefault datums and the assignment of datums to layers. The datum planes will be put on a separate layer and set as geometric tolerance features. Reread Section 8, Layers, at this time. Choose **Part** from the MODE menu and choose the following commands:

Part ⇒ **Create** (type the part name **Anchor**) ⇒ **enter**

Set up the units and the environment:

SETUP AND ENVIRONMENT

Set Up ⇒ **Units** ⇒ **Length** ⇒ **Inch** ⇒ **Done** ⇒ **Material** ⇒ **Define** ⇒ (type **Steel**, then press **enter**) ⇒ (table of material properties, change or add information) ⇒ **File** ⇒ **Save** ⇒ **File** ⇒ **Exit** ⇒ **Assign** ⇒ (pick **Steel**) ⇒ **Accept** ⇒ **Done** **Environment** ⇒ ✓**Grid Snap** **Hidden line Tan Phantom**

Create a default coordinate system and three default datum planes, as in Lessons 1 and 2. Choose the following commands:

PT/Modeler™
Feature ⇒ **Datum Plane** ⇒ **Coord Sys** ⇒ **Default** ⇒ **Done**

Feature ⇒ **Create** ⇒ **Datum** ⇒ **Plane** ⇒ **Offset** ⇒ **enter** (three times at the prompt) ⇒ **Done**

The first protrusion will be sketched on **DTM3,** as in previous lessons. Use Figure 4.11 for the protrusion dimensions, and sketch the outline in Figure 4.12. Only **5.50, R1.00, 1.125,** and **25°** are needed. Choose the following commands:

PT/Modeler™
Feature ⇒ Protrusion etc.

Feature ⇒ Create ⇒ Protrusion ⇒ Extrude ⇒ Solid ⇒ Done ⇒ One Side ⇒ Done ⇒ (pick **DTM3** as the *sketching* plane) **⇒ Okay** (for selecting the *direction* of feature creation) **⇒ Top ⇒** (pick **DTM2** as *orientation* plane) **⇒ Sketch ⇒ Mouse Sketch**

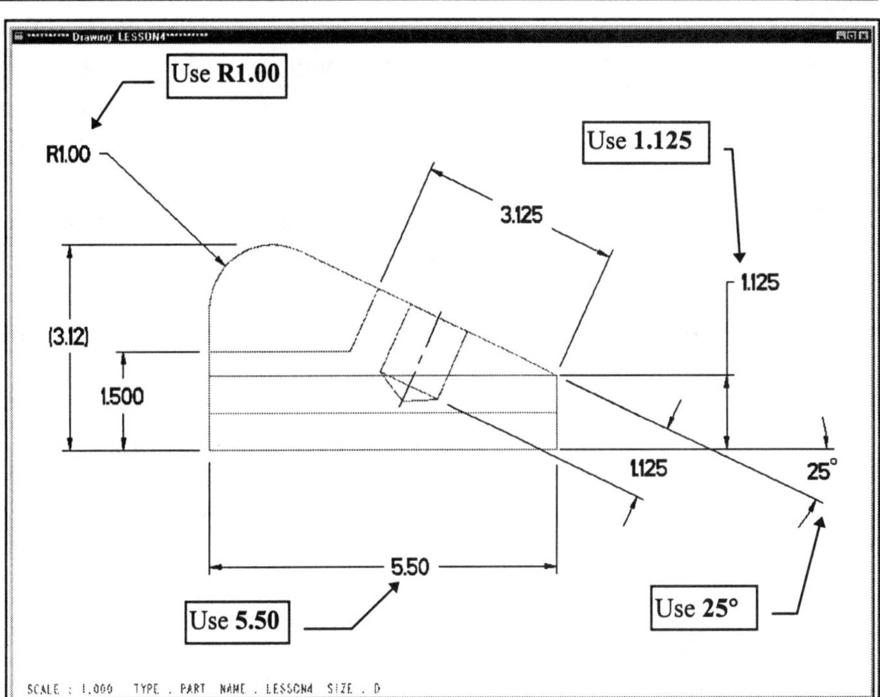

Figure 4.11
Protrusion Dimensions

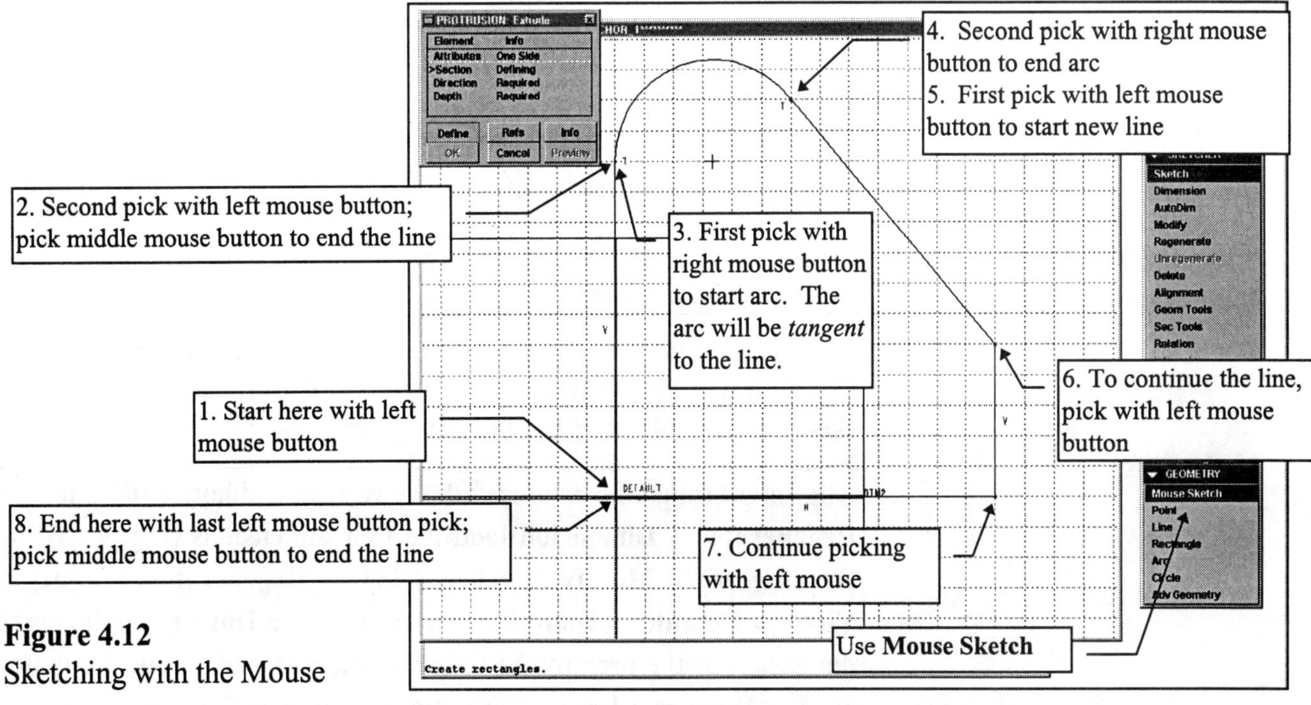

Figure 4.12
Sketching with the Mouse

Sketching with the mouse is a fast and efficient method of creating geometry. The **left mouse button** is used to create a continuous set of *lines*. The **middle mouse button** is used for creating *circles*, and the **right mouse button** creates *tangent arcs* when the first pick is near the end of an existing line or arc.

? Pro/HELP

Use online documentation to understand the use of the mouse buttons better.

Start the sketch (Fig. 4.12) by picking the first endpoint of the vertical line with the left mouse button (1). Pick the second endpoint at a position needed to create a vertical line (2). Use the middle mouse button to end the line. Next, using the right mouse button, pick near the last endpoint created for the line to start an arc (3). The arc will rubberband tangent from the line. Pick with the right mouse button to end the arc (4). Now use the left mouse button to finish the section sketch (5-8). The sketch must be closed and composed of single entities. Do not draw lines or arcs on top of one another or the sketch will fail to regenerate. The sketch does not have to have exactly the same dimensional scale as the physical part. As long as the outline is similar, Pro/E corrects the sketch when the dimensions are modified. After the sketch is complete, align the edges, add dimensions, and regenerate the section sketch (Fig. 4.13).

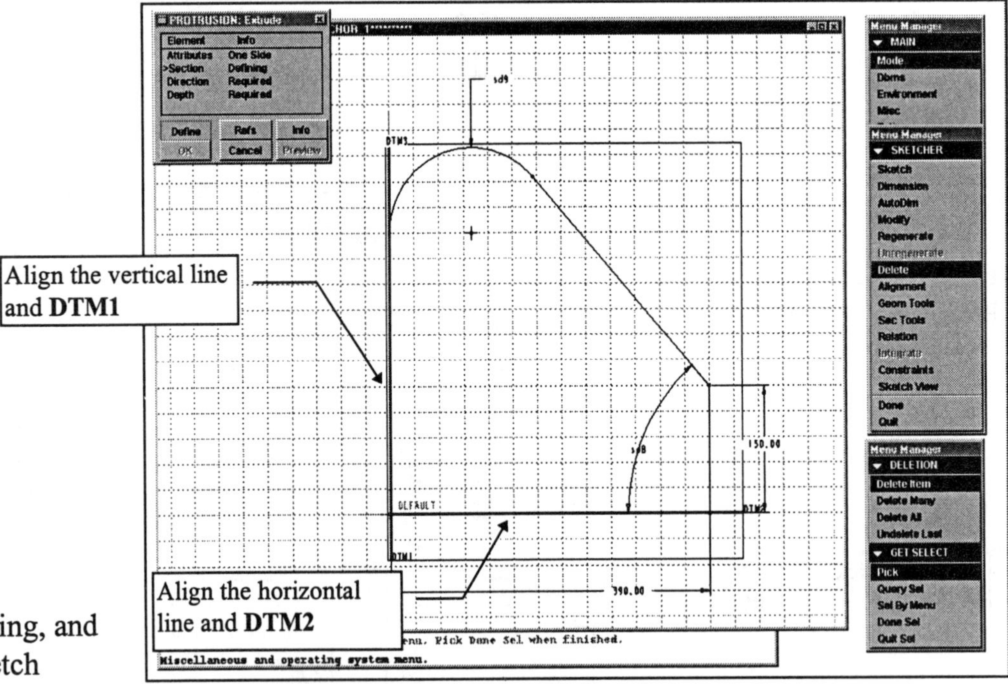

Figure 4.13
Aligning, Dimensioning, and Regenerating the Sketch

After the section has successfully regenerated, modify the values to the design dimensions shown in Figure 4.11 and regenerate the sketch (Fig. 4.14) using the following commands:

Alignment (align the datums with the vertical and horizontal lines) ⇒ **Regenerate** ⇒ **Dimension** [add the four dimensions (Fig. 4.13)] ⇒ **Regenerate** ⇒ **Modify** (see Fig. 4.11) ⇒ **Regenerate** ⇒ **Done** ⇒ **View** ⇒ **Default** ⇒ **Done-Return** ⇒ **Blind** ⇒ **Done** (type the depth dimension at the prompt: **2.5625)** ⇒ **enter** ⇒ **OK** (from the dialog box) ⇒ **Done** ⇒ **View** ⇒ **Cosmetic** ⇒ **Shade** ⇒ **Display**

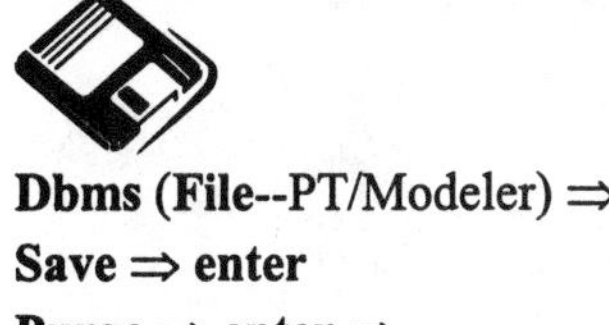

Dbms (File--PT/Modeler) ⇒
Save ⇒ enter
Purge ⇒ enter ⇒
Done-Return

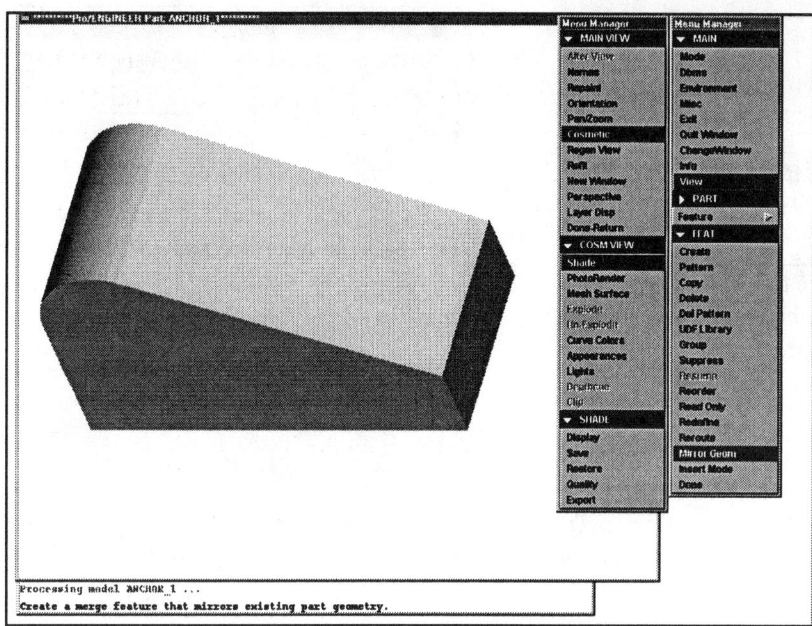

Figure 4.14
Shaded First Protrusion

At this point, let's change the color of the model. In general, try to avoid the colors that are used as defaults. Since feature and entity highlighting is defaulted to red, colors similar to red should be avoided. Datum planes have red and yellow sides; therefore, yellow should be avoided. Shades of blue and green work well. It is really up to you to choose colors that are pleasant to look at when you're gazing at the monitor for hours on end!

View ⇒ Cosmetic ⇒ Appearances ⇒ Define (will activate the **Appearance Editor** on your screen, as shown in Fig. 4.15. Pick on the color box to initiate the **Color Editor**, as shown in Fig. 4.16)

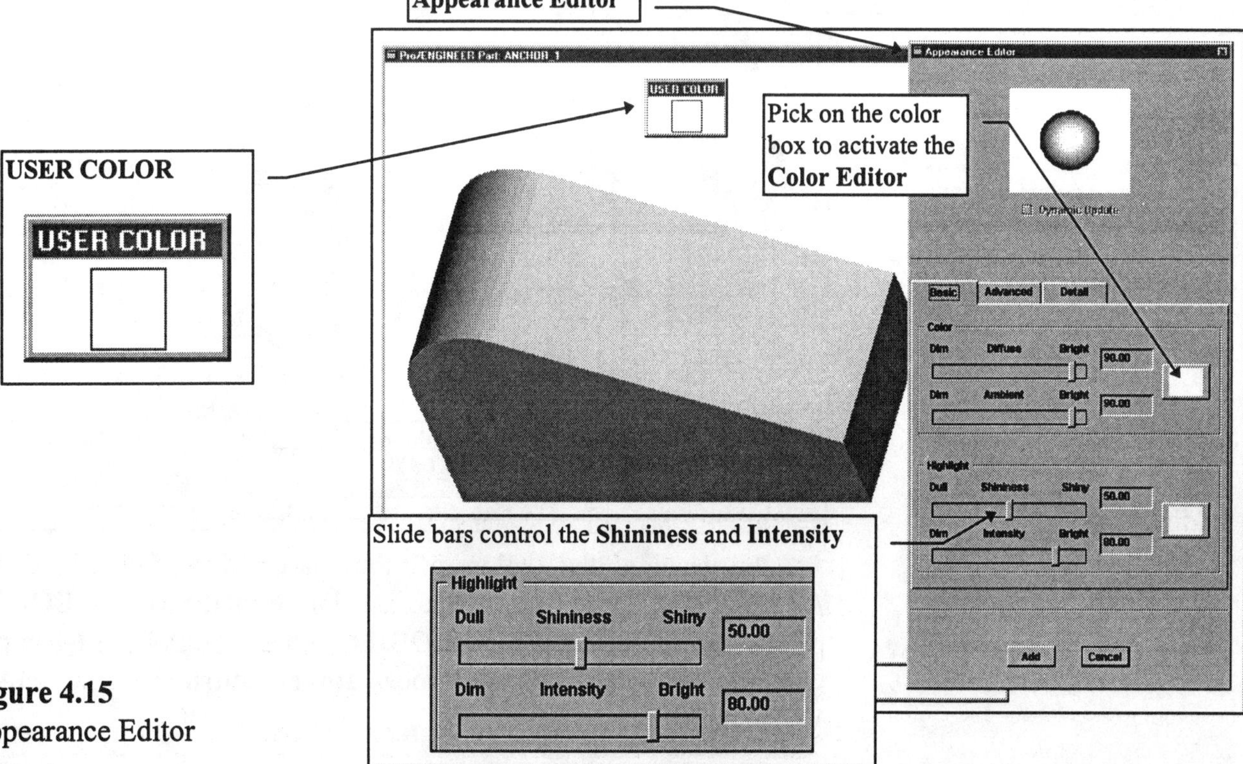

Figure 4.15
Appearance Editor

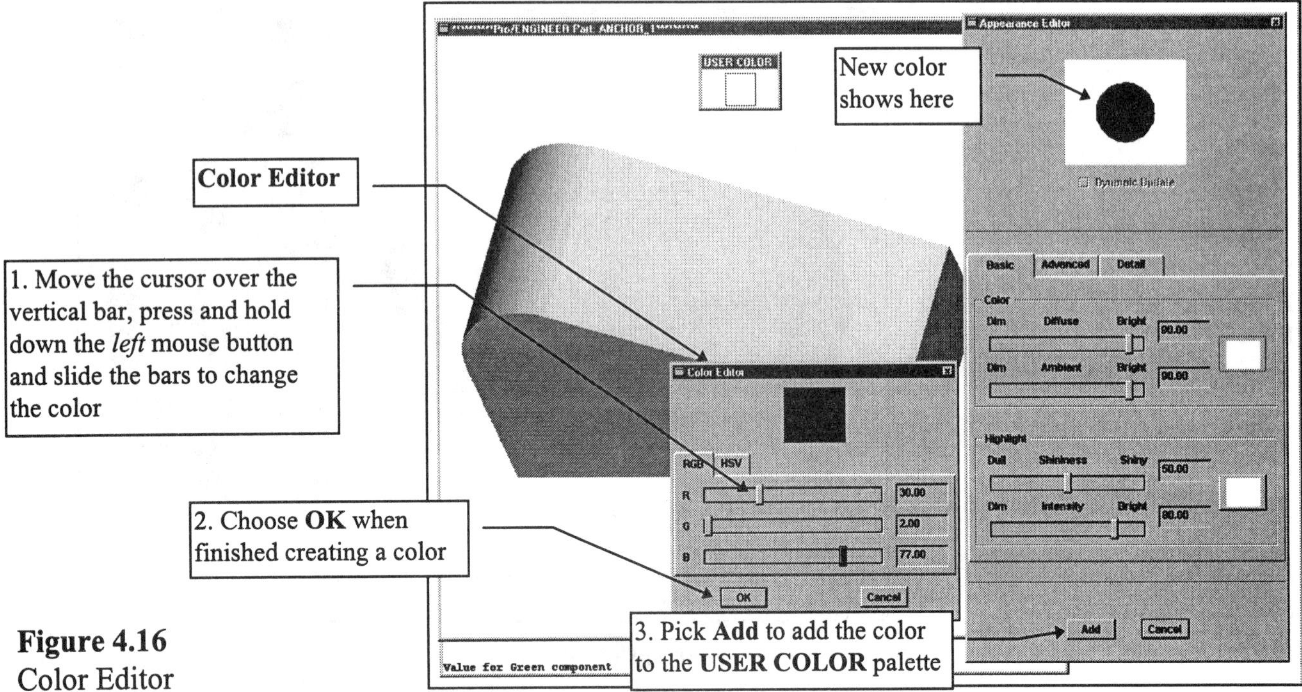

Figure 4.16
Color Editor

Slide the RGB bars to create useful colors. Add a color to the **USER COLOR** palette by choosing **OK** from the **Color Editor** and then **Add** from the **Appearance Editor** (Fig. 4.17). Continue making a few more colors, adding each to the palette. Finally, select the **Cancel** button from the **Appearance Editor.**

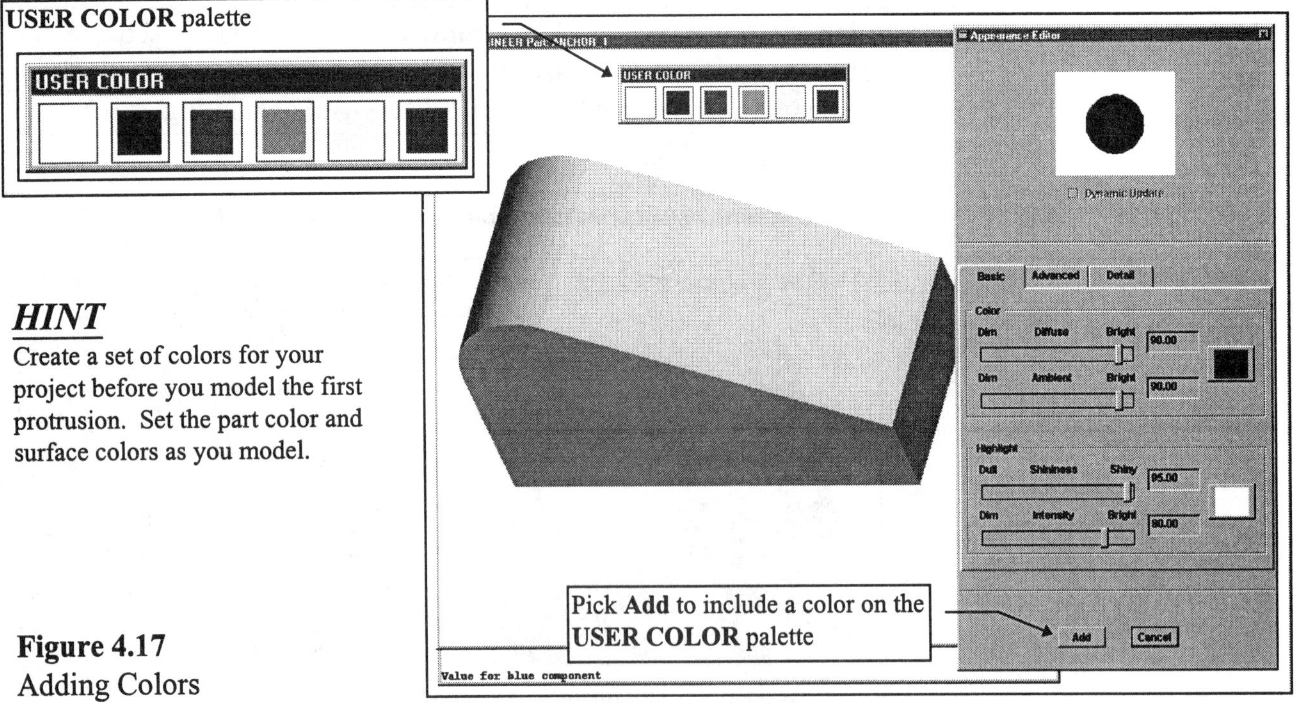

HINT
Create a set of colors for your project before you model the first protrusion. Set the part color and surface colors as you model.

Figure 4.17
Adding Colors

To change the color of the part, pick **Set** (from the APPEARANCES menu). Select a color from the **USER COLOR** palette and then **Part** from the OBJECT menu (Fig. 4.18). The part will now be a new color. Choose **Done-Return** from the MAIN VIEW menu to end the process.

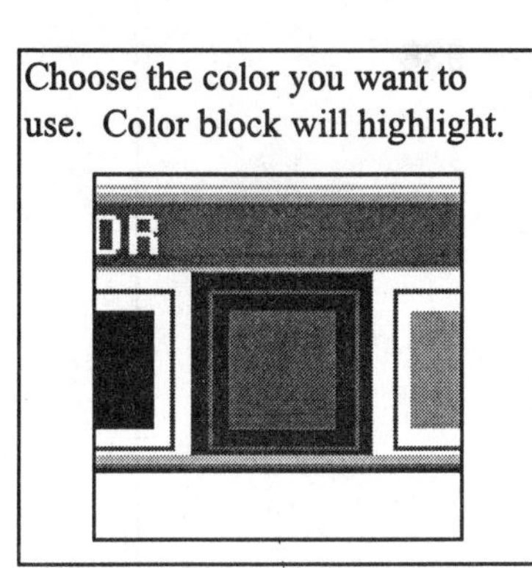

Figure 4.18
Coloring the Model

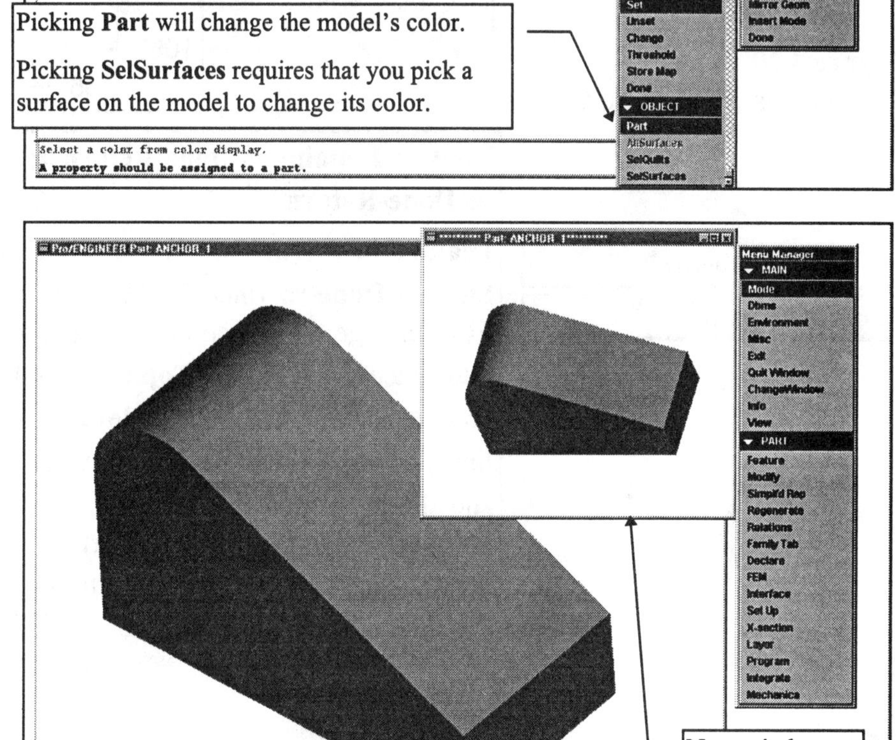

Dbms (**File**--PT/Modeler) ⇒
Save ⇒ **enter**
Purge ⇒ **enter** ⇒
Done-Return

Figure 4.19
Model with New Color

Unfortunately, this text is not in color, so you cannot see the changes to the model (Fig. 4.19). Experimenting with the colors is most students' idea of fun. Enjoy yourself, but don't create too many colors; they really aren't necessary at this stage of the project.

The next two features will be cuts. For both of the cuts, use **DTM1** as the sketching plane and give **Thru All** as the depth. Each cut requires just two lines, two alignments, and two dimensions. Use the first cut's sketching/placement plane and reference/orientation (**Use Prev**) for the second cut. Create the cuts separately, each as an open section. Figure 4.20 shows the dimensions for each cut. Choose the following commands:

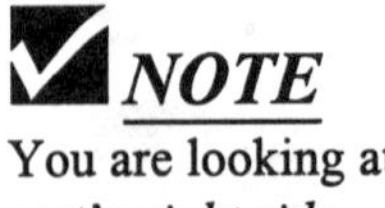

NOTE

You are looking at the part's *right side.*

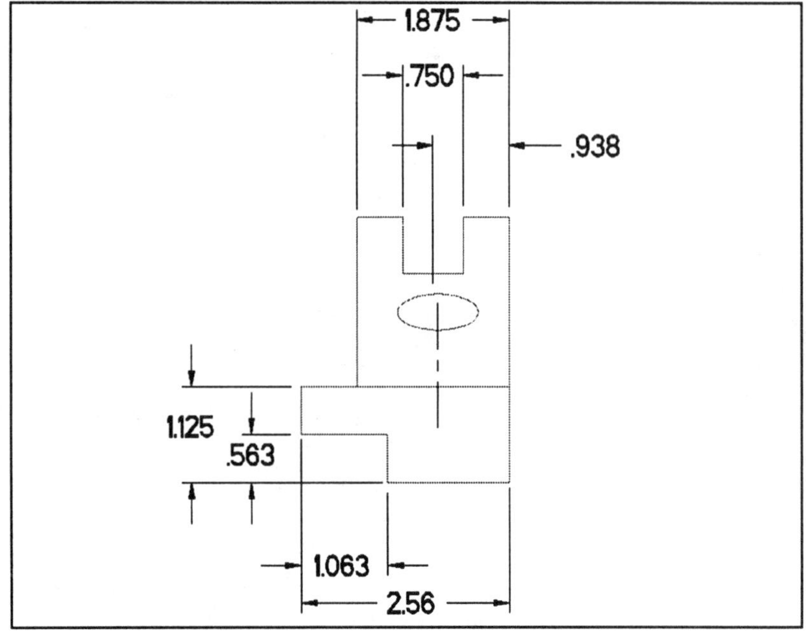

Figure 4.20
Dimensions for the Two Cuts

View ⇒ Repaint ⇒ Done-Return ⇒ Environment ⇒ □ Grid Snap ⇒ Done-Return

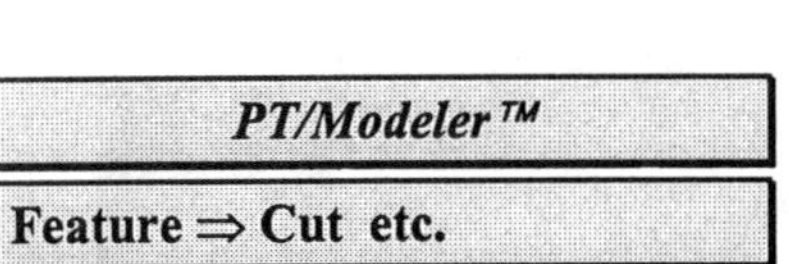

Feature ⇒ Create ⇒ Cut ⇒ Extrude ⇒ Solid ⇒ Done ⇒ One Side ⇒ Done ⇒ (pick **DTM1** as the sketching/placement plane) **⇒ Flip** (change the direction of the cut to pass through the part, not out into space) **⇒ Okay ⇒ Top ⇒** (pick **DTM2** as the orientation plane; you are looking at the left side) **⇒ Line ⇒ Vertical** (sketch the two lines) **⇒ Alignment** (align the vertical line with the top of the part and the horizontal line with the right side edge) **⇒ Regenerate ⇒ Dimension** (add the dimensions) **⇒ Regenerate ⇒ Modify** (change the dimensions to **1.125** and **1.875**) **⇒ Regenerate** (Fig. 4.21)

NOTE

You are looking at the part's *left side.*

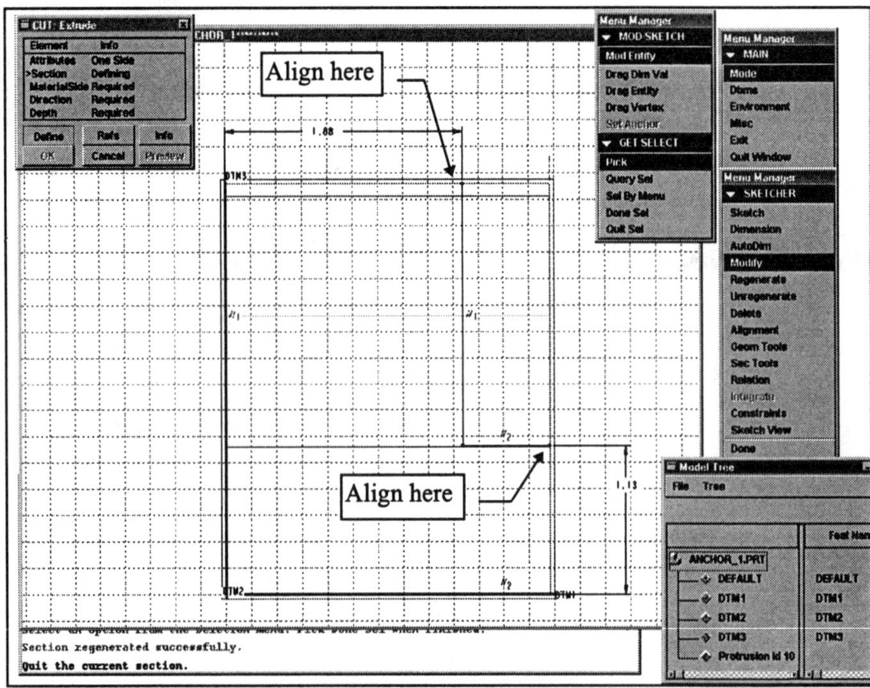

Figure 4.21
Cut Section Dimensions

Complete the cut (Fig. 4.22) using the following commands:

Done ⇒ Okay ⇒ Thru All ⇒ Done ⇒ View ⇒ Default ⇒ OK ⇒ View ⇒ Cosmetic ⇒ Shade ⇒ Display ⇒ Done-Return ⇒ Done

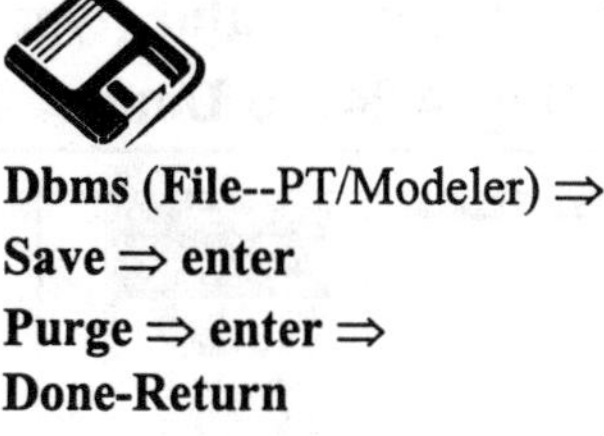

Dbms (File--PT/Modeler) ⇒
Save ⇒ enter
Purge ⇒ enter ⇒
Done-Return

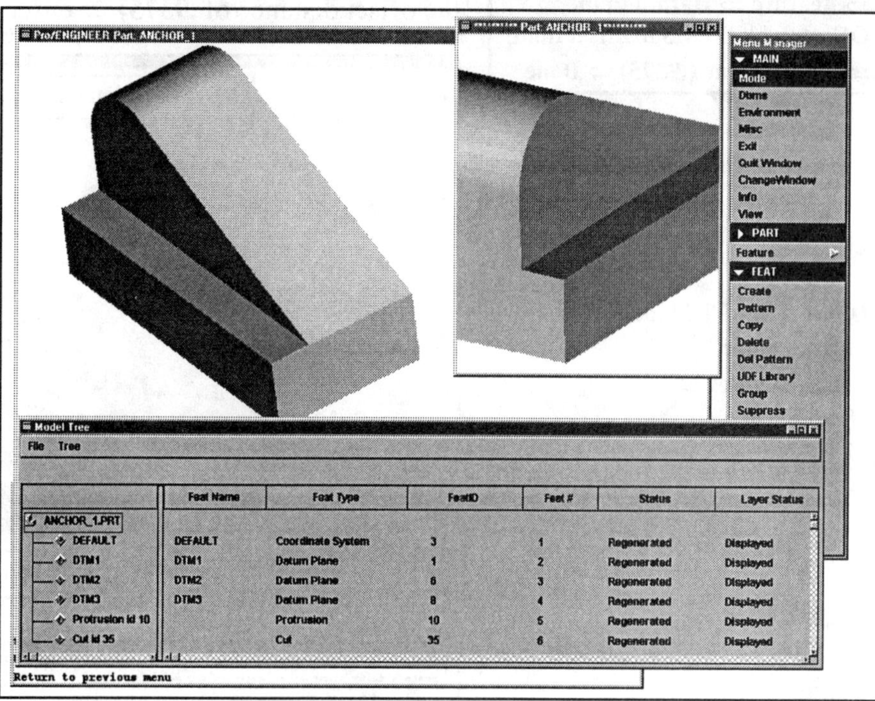

Figure 4.22
Completed Cut

Complete the second cut using the last two command blocks (Fig. 4.23). Everything will be the same except for the dimensions **.563** and **1.063**. You can expedite things by selecting **Use Prev** for the sketching and orientation planes.

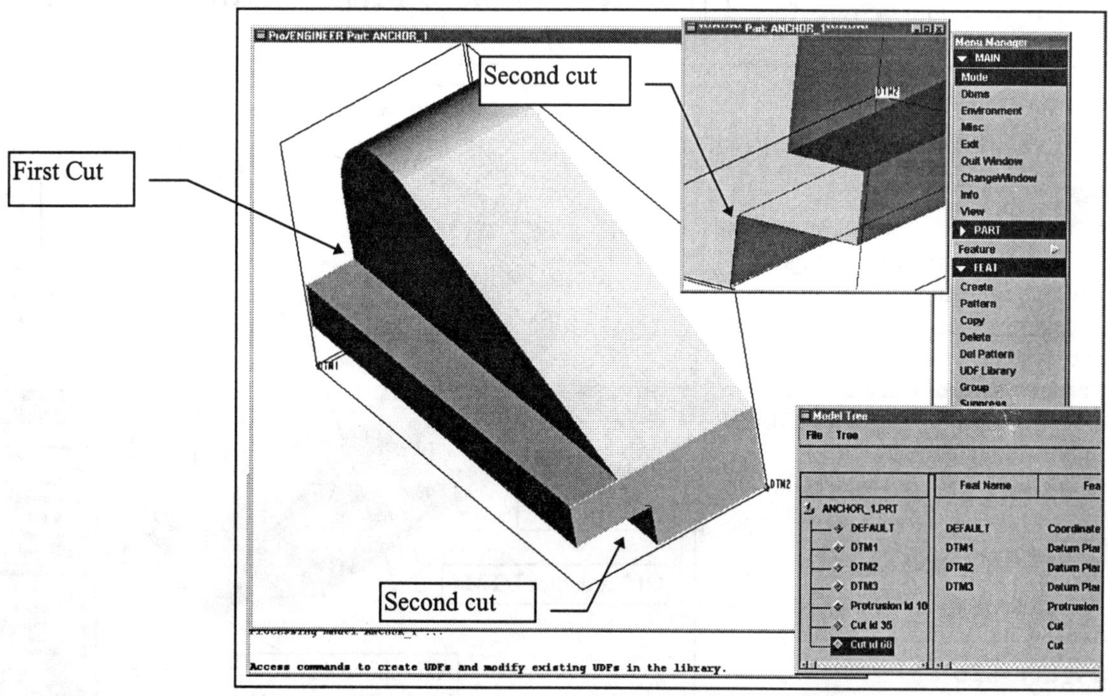

Figure 4.23
Second Cut

On the next feature, we will create a new datum plane to sketch on. A datum axis will also be created at this time. Choose the following commands:

PT/Modeler™

View ⇒ Repaint ⇒ Done-Return ⇒ Feature ⇒ Datum Plane ⇒ Offset (pick **DTM3** to offset from) **⇒ Enter Value (.9375) ⇒ Done**

View ⇒ Repaint ⇒ Done-Return ⇒ Feature ⇒ Create ⇒ Datum ⇒ Plane ⇒ Offset (pick **DTM3** to offset from) **⇒ Enter Value** (type the offset distance of **.9375**) **⇒ enter ⇒ Done** (Fig. 4.24) **⇒ Done**

Dbms (File--PT/Modeler) **⇒ Save ⇒ enter Purge ⇒ enter ⇒ Done-Return**

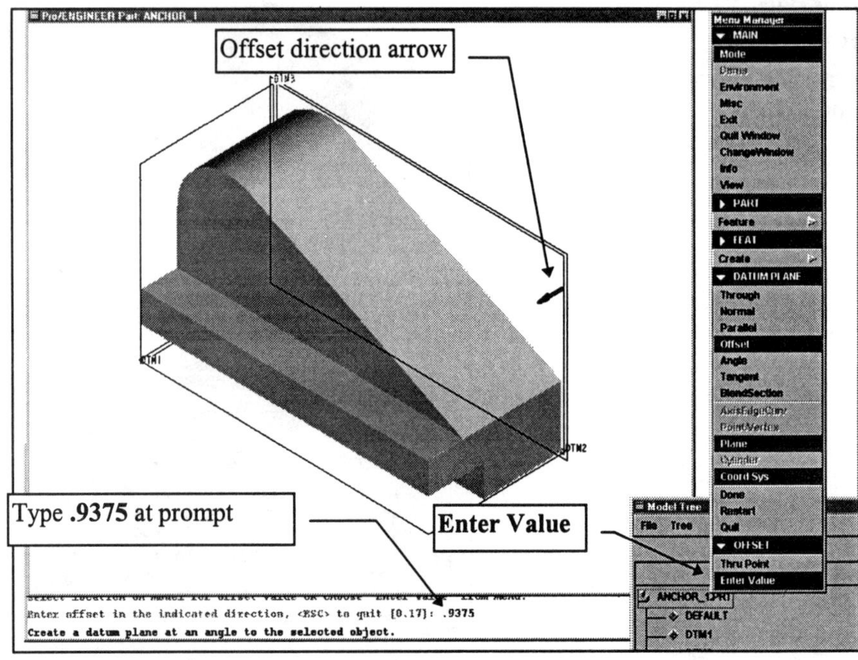

Figure 4.24
Creating an Offset Datum Plane

The new datum plane is **DTM4**. A datum axis will now be created through the curved top of the part, using the following commands:

PT/Modeler™

Feature ⇒ Datum Axis ⇒ Thru Cyl etc.

Feature ⇒ Create ⇒ Datum ⇒ Axis ⇒ Thru Cyl (pick as shown in Fig. 4.25) **⇒ Done**

NOTE
An axis can be inserted through any curved feature. Holes and circular features automatically have axes when they are created. Features created with arcs, fillets, and so on need to have axes added afterward, unless set in the *configuration file* as the default.

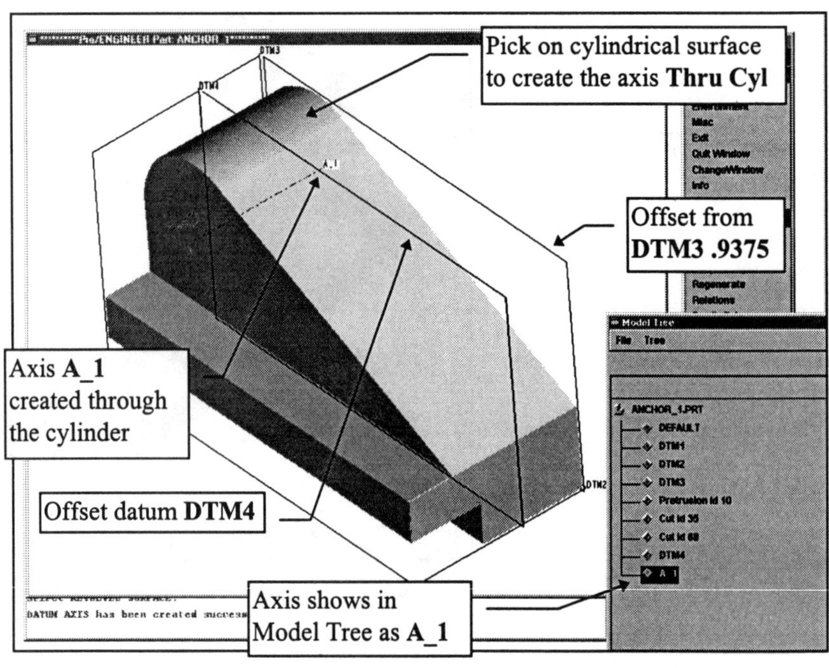

Figure 4.25
Creating a Datum Axis

The cut on the top of the part is created by sketching on **DTM4** and projecting it toward both sides. Use the Model Tree to select the appropriate datum planes. Choose the following commands:

PT/Modeler™

Feature ⇒ Cut etc.

Feature ⇒ Create ⇒ Cut ⇒ Extrude ⇒ Solid ⇒ Done ⇒ Both Sides ⇒ Done ⇒ (pick **DTM4** as the sketching/placement plane) **⇒ Okay ⇒ Top ⇒** (pick **DTM2** as the orientation plane) **⇒ Line ⇒ Perpendicular** (pick the angled edge and then the starting and ending points of the line) **⇒ Line ⇒ Horizontal ⇒** (pick the endpoints of the horizontal line as shown in Fig. 4.26) **⇒ Regenerate ⇒ Alignment** (align the angled line's endpoint with the angled edge of the part and the horizontal line with the left side edge) **⇒ Regenerate ⇒ Dimension** (add the dimensions) **⇒ Regenerate ⇒ Modify** (change the dimensions to **3.125** and **1.50,** as in Fig. 4.11) **⇒ Regenerate** (Fig. 4.26)

HINT

To get the slanted dimension of **3.125**, select the slanted line itself and then the end of the block. Place the dimension properly.

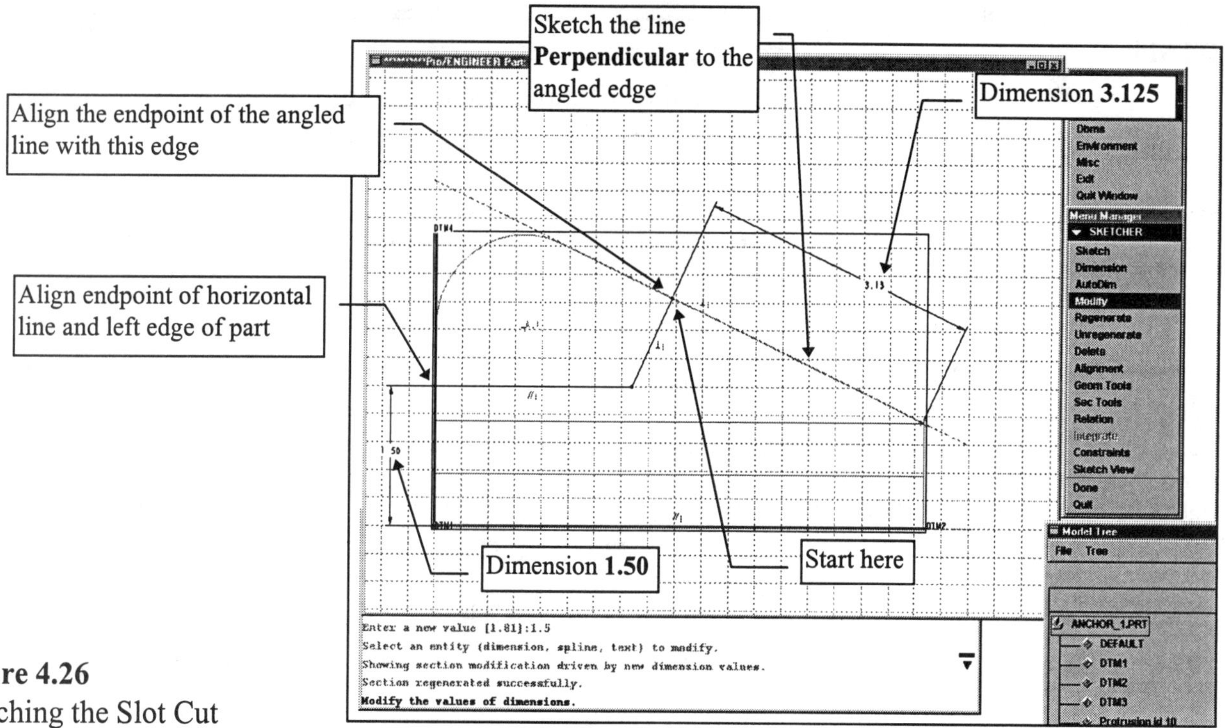

Figure 4.26
Sketching the Slot Cut

Complete the cut with the following commands:

Done ⇒ Okay (for the material removal side) **⇒ Blind ⇒ Done ⇒** (type the full width of **.750** for the slot at the prompt) **⇒ enter ⇒ OK ⇒ Done ⇒ View ⇒ Default** (shade the display as in Fig. 4.27) **⇒ Environment ⇒ Isometric ⇒ Done-Return**

The hole drilled in the angled surface appears to be aligned with **DTM4**. Upon closer inspection, it can be seen that the hole is a different distance from the edge (**.875** from **DTM3**) and is not in line with the slot and datum plane. Create the feature using a sketched hole (Fig. 4.28).

Dbms (**File**--PT/Modeler) ⇒
Save ⇒ **enter**
Purge ⇒ **enter** ⇒
Done-Return

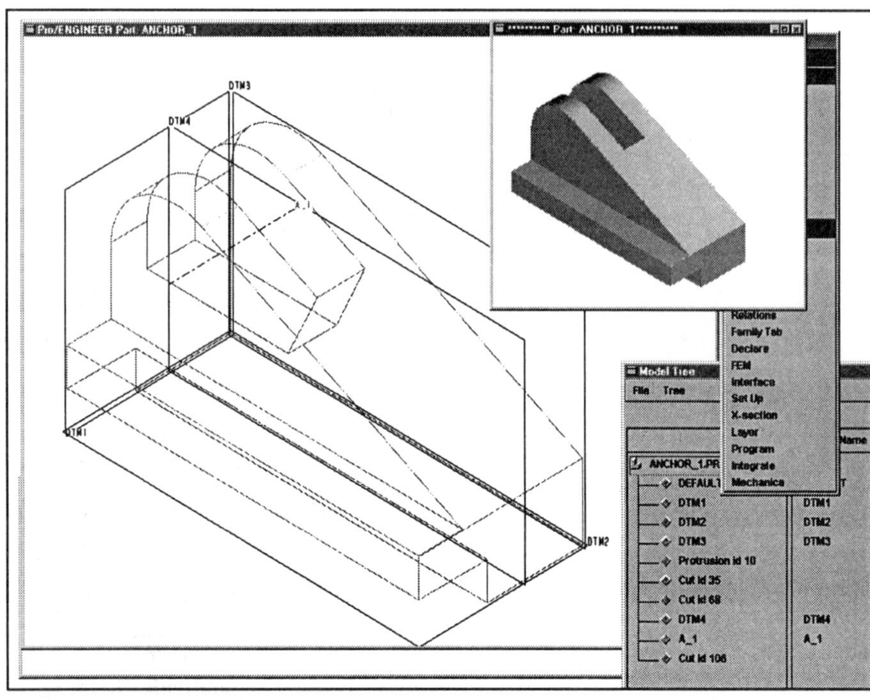

Figure 4.27
Cut Created on Both Sides of **DTM4**

Since the drill tip at the bottom of the hole needs to be modeled, the hole is created as a sketched hole (Fig. 4.28). Use the following commands:

Environment ⇒ ✓ **Grid Snap** ⇒ **Done-Return** ⇒ **Feature** ⇒ **Create** ⇒ **Hole** ⇒ **Sketch** ⇒ **Done** ⇒ **Linear** ⇒ **Done** ⇒ **Sketch** ⇒ **Line** ⇒ **Centerline** ⇒ **Vertical** (pick once on the sketch to create the centerline) ⇒ **Line** (pick five endpoints for the four lines to create the *closed section* on one side of the centerline) ⇒ **Regenerate** ⇒ **Dimension** (add the **1.125** depth, **62°** representing 1/2 of the drill tip angle, and a diameter of **1.00**) ⇒ **Regenerate** ⇒ **Modify** (Fig. 4.28)

HINT
Remember that the diameter dimension is created by picking the rightmost vertical line. Then pick the centerline, and finally pick the line again before placing the dimension on top.

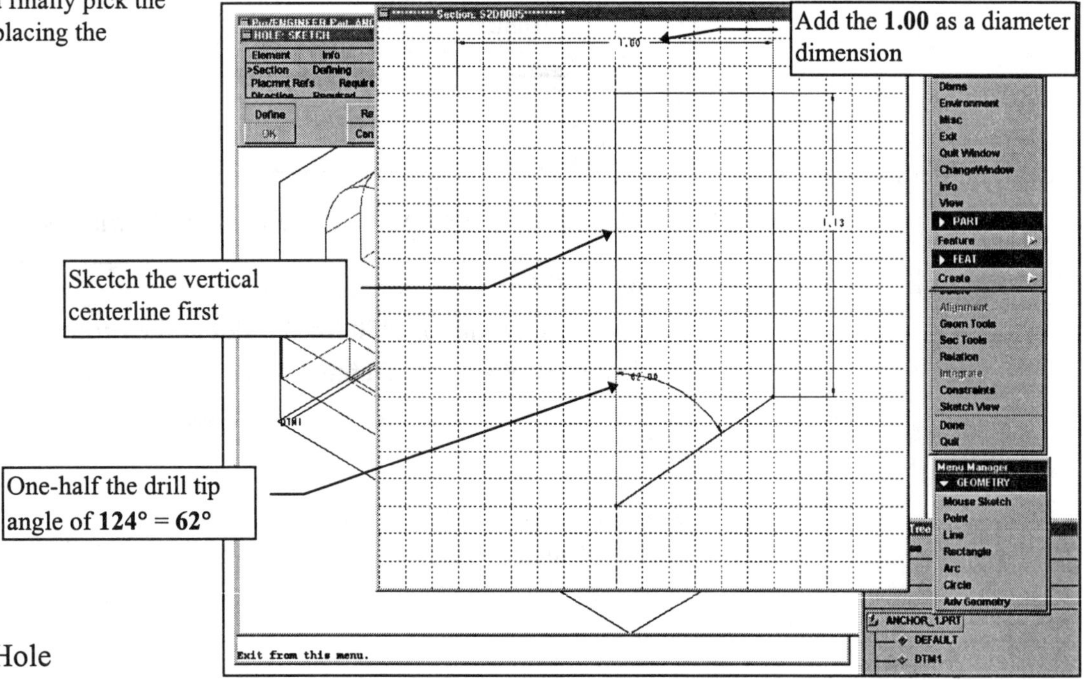

Figure 4.28
Sketching the Hole

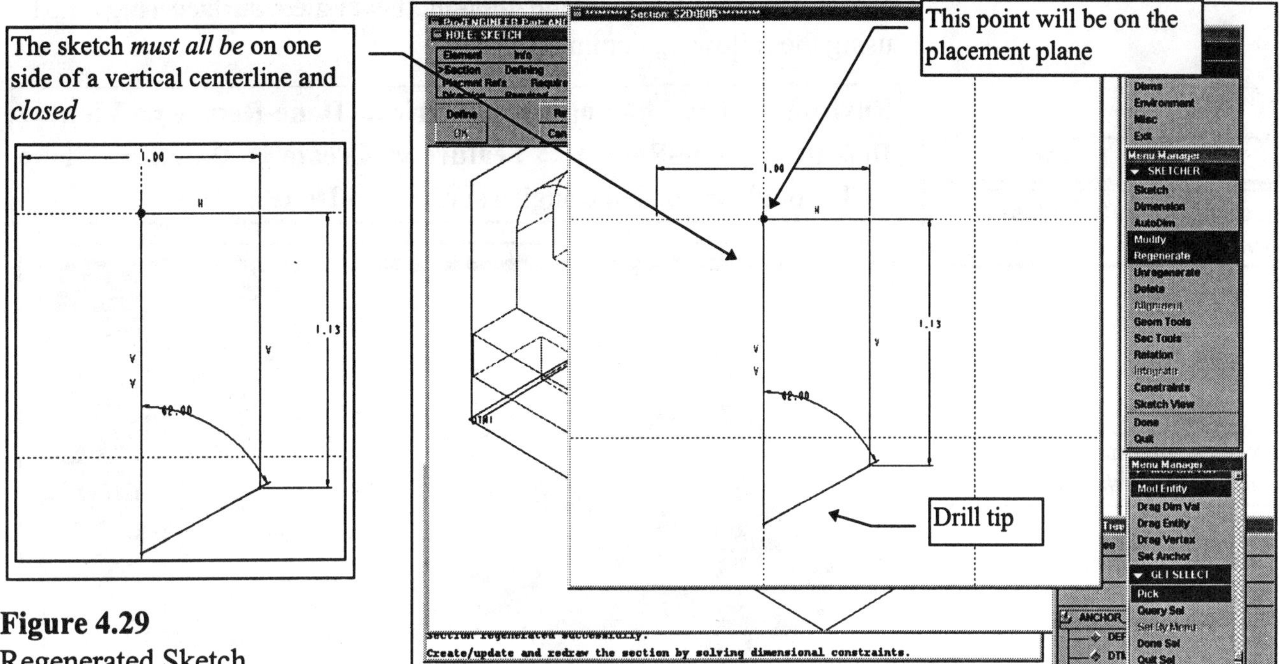

Figure 4.29
Regenerated Sketch

To complete the hole, choose the following commands:

Regenerate (Fig. 4.29) ⇒ **Done** ⇒ (pick the angled surface as the placement plane) ⇒ (select two edges to dimension from, pick the edge as shown in Fig. 4.30, and type **2.0625** at the prompt) ⇒ **enter** ⇒ (next, pick **DTM3** and type **.875** at the prompt) ⇒ **enter** ⇒ **OK** (from the dialog box) ⇒ **Done**

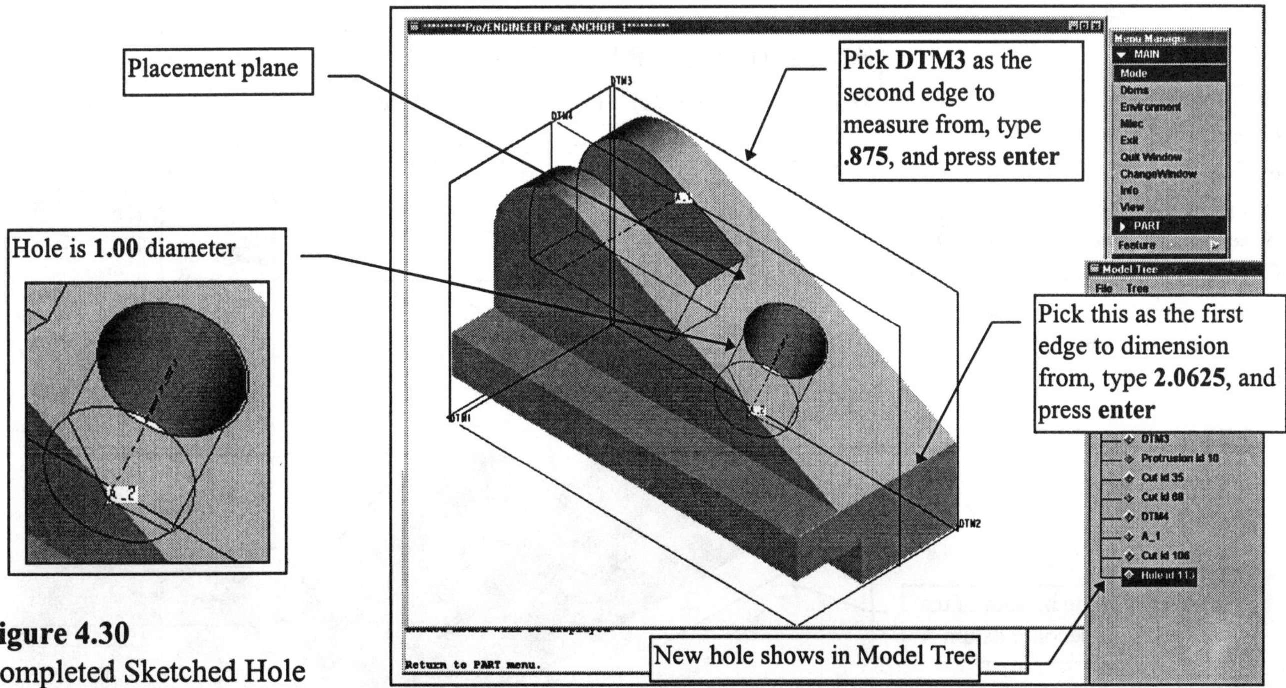

Figure 4.30
Completed Sketched Hole

A new datum plane will now be created that passes through the angled surface. All datum planes are to be *set* as geometric tolerancing features (basic) and added to a layer.

Create the datum plane though the angled surface (Fig. 4.31) using the following commands:

Environment ⇒ **Shading** ⇒ **Isometric** ⇒ **Done-Return** ⇒ **View** ⇒ **Default** ⇒ **Done-Return** ⇒ **Feature** ⇒ **Create** ⇒ **Datum** ⇒ **Plane** ⇒ **Through** ⇒ (pick the angled surface) ⇒ **Done** ⇒ **Done**

PT/Modeler™

⇒ **Feature** ⇒ **Datum Plane** ⇒ **Through etc.**

Pick the angled surface to create the datum plane *through*

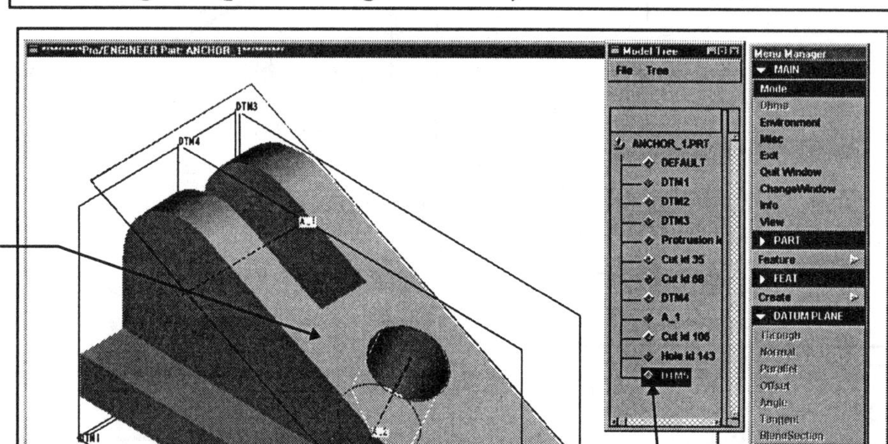

Figure 4.31
DTM5 Created Through the Angled Surface

Set (Fig. 4.32) and rename the datum planes as geometric tolerancing features using the following commands:

Set Up (from PART menu) ⇒ **Geom Tol** ⇒ **Set Datum** ⇒ (pick **DTM2** from Model Tree) ⇒ (change name from **DTM2** to **A**) ⇒ **OK**

HINT

Datums used in geometric tolerancing (basic):

A (primary--three-point contact),
B (secondary--two-point contact),
C (tertiary--one-point contact),

should be established on your ***DIPS*** before starting the part model!

DTM2 is on the bottom of the part and will become datum **A**

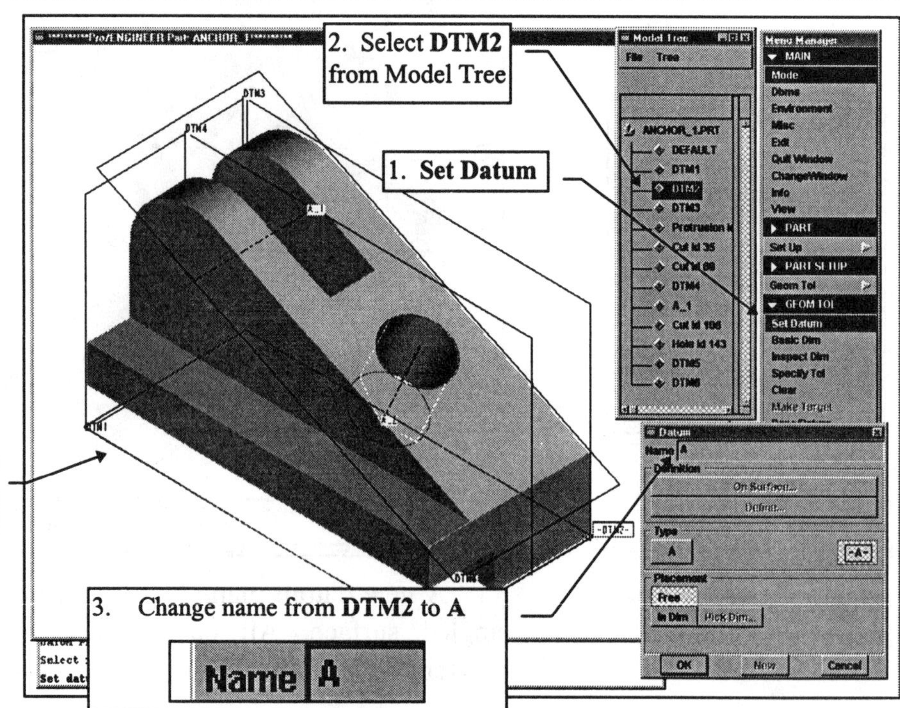

Figure 4.32
Changing Datum Names

Repeat to complete the renaming and setting of all five datum planes (Fig. 4.33). Move the positions of the names of the datums by choosing the following commands (Fig. 4.33):

Done/Return ⇒ **Done** ⇒ **Modify** (from the PART menu) ⇒ **Move Datum** ⇒ (pick the datum name, edge, or name from the Model Tree) ⇒ (pick a new position on the screen for the name) ⇒ **Done**

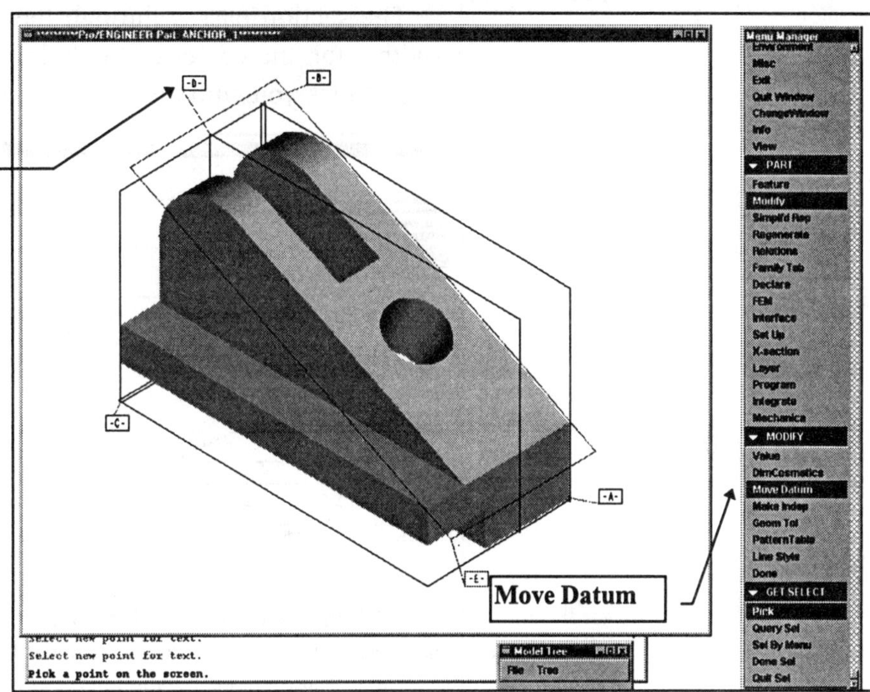

Figure 4.33
Moving Datum Names

Next you will create a section to be used when you are detailing the model in **Drawing Mode**. The section will pass through the part lengthwise using datum plane **D**. The section will be named **A** and will show as **Section A-A** when you are detailing the view in **Drawing Mode**. Figure 4.34 shows the section.

HINT
Create sections in **Part Mode** to be used later in **Drawing Mode** when you are detailing the part model.

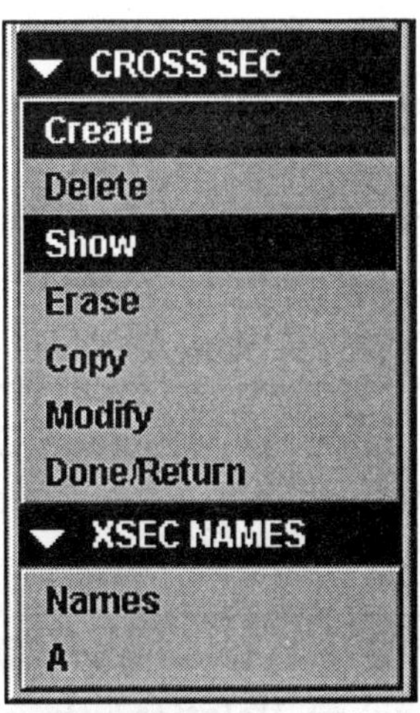

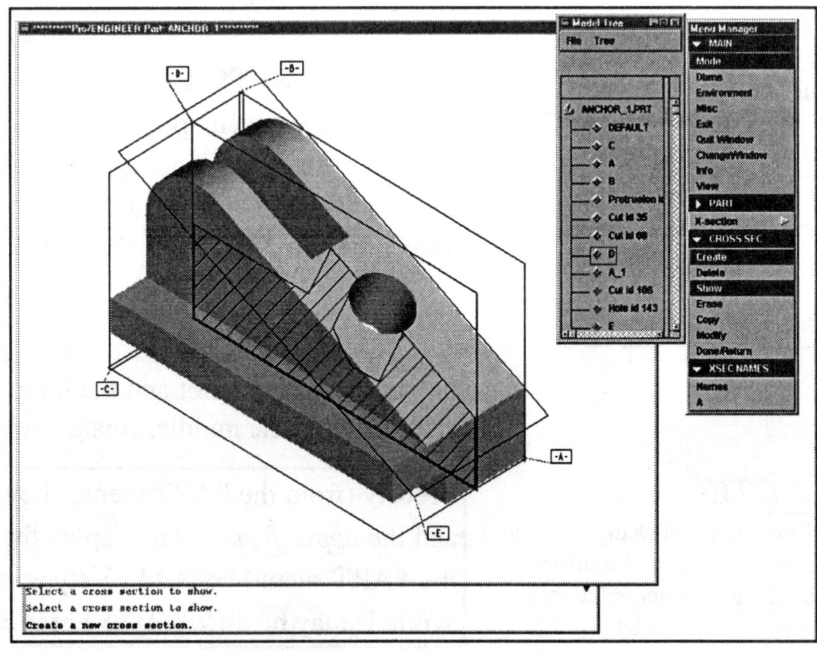

Figure 4.34
Creating a Section

Choose the following commands to create the section shown in Figure 4.34:

> **X-section** (from the PART menu) ⇒ **Create** ⇒ **Planar** ⇒ **Single** ⇒ **Done** ⇒ (enter the **NAME** for the cross section at the prompt: **A**) ⇒ **enter** ⇒ (select the planar surface of datum plane: **D**) ⇒ **Done/Return**

The section passes through the slot and the hole, but it doesn't pass through the center of the hole. Your "boss" has a "suggestion" and provides you with the following ECO (Fig. 4.35):

ECO

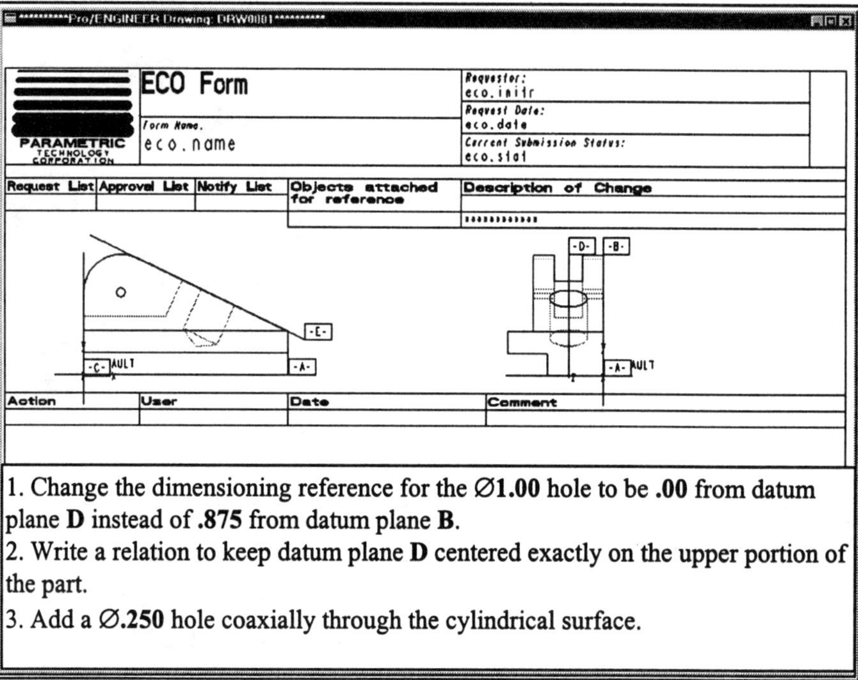

1. Change the dimensioning reference for the Ø**1.00** hole to be **.00** from datum plane **D** instead of **.875** from datum plane **B**.
2. Write a relation to keep datum plane **D** centered exactly on the upper portion of the part.
3. Add a Ø**.250** hole coaxially through the cylindrical surface.

Figure 4.35
ECO

Choose the following commands to change the dimensioning reference for the hole:

PT/Modeler™

Feature Oper ⇒ Redefine etc.

> **Feature** ⇒ **Redefine** ⇒ (pick the hole) ⇒ (pick **Placmnt Refs** from the **HOLE:SKETCH** dialog box) ⇒ **Define** ⇒ **Same Ref** (to keep the same placement plane) ⇒ **Same Ref** (to keep the same edge reference for dimensioning) ⇒ **Alternate** (pick datum **D** from the Model Tree or the part model to change the second reference--if asked, **"Align feature to reference?** [N]" press **enter)** ⇒ (type **0** at the prompt, as shown in Fig. 4.36) ⇒ **enter** ⇒ **OK** ⇒ **Done**

If you answer yes to this question (**Y**), the hole will be aligned to datum **D** and you will not be prompted for a dimension value.

The hole (and the slot) is now a ***child*** of datum **D**. If datum **D** moves, so will the slot and the hole. In order to ensure that the plane stays through the middle, create a relation to control its location:

> **Modify** (from the PART menu, shown in Fig. 4.37) ⇒ (pick datum **D** and the *upper front cut* to display the dimensions) ⇒ **Relations** (from the PART menu) ⇒ **Add** ⇒ (type **d18=d12/2** at the prompt--your **d** symbols may be different) ⇒ **enter** ⇒ **enter** ⇒ **Done**

NOTE

The dimension will change from its numerical value to its parameter value. The parameter value may be different on your model.

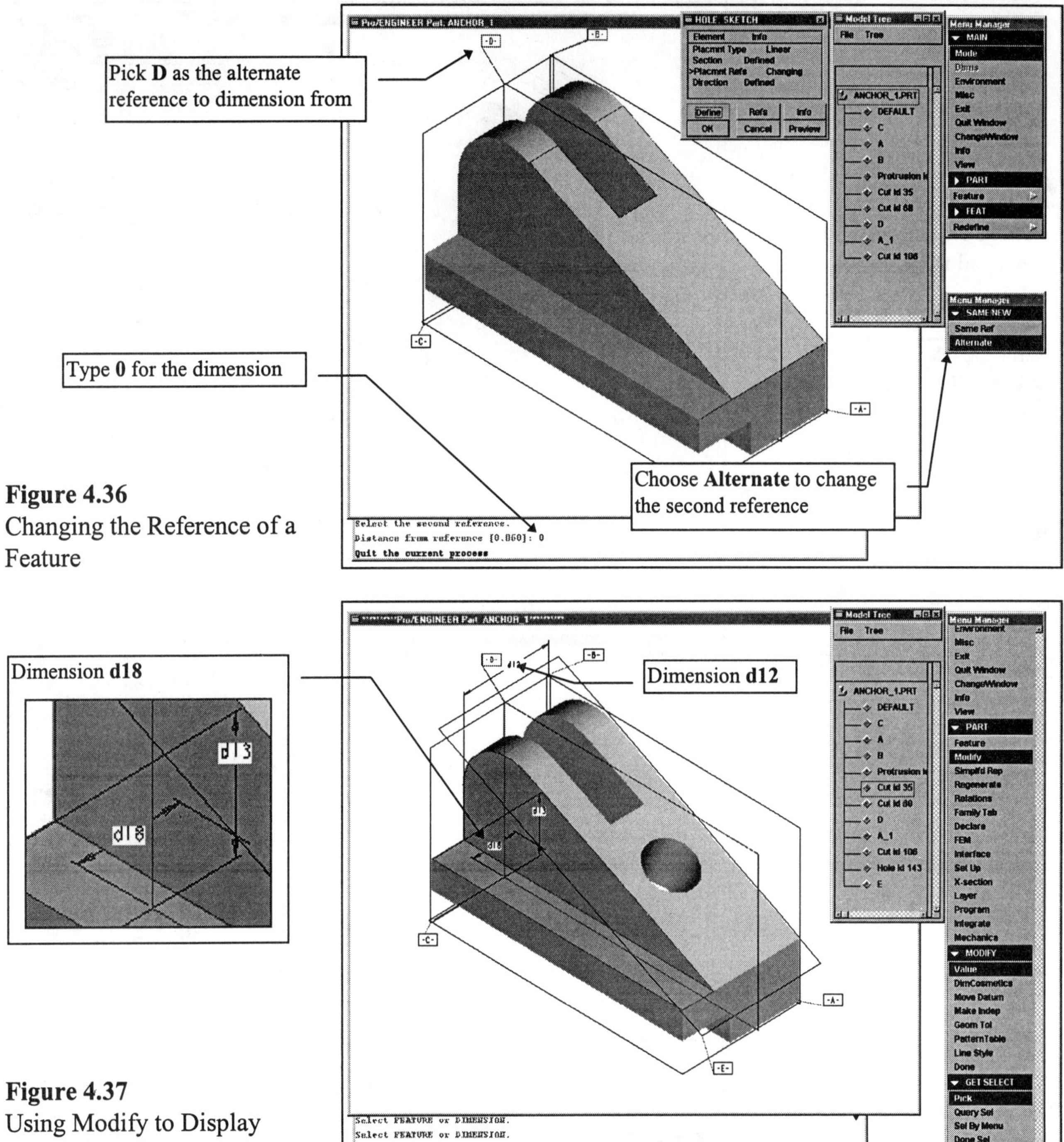

Figure 4.36
Changing the Reference of a Feature

Figure 4.37
Using Modify to Display Dimension of Features

The relation states that the distance (**d18**) from datum **B** to datum **D** will be one-half the value of the distance from datum **B** to the cut surface (**d12**). If the thickness of the upper portion of the part (**1.875**) changes, datum **D** will remain centered, as will the slot cut and the **1.00** inch diameter hole. If your *dimension symbols* do not show on the screen after you select them during the **Modify** command (Fig. 4.37), pick **Switch Dim** under the **Relations** command (Fig. 4.38). To see the new relation, select **Show Rel** before completing the command (Fig. 4.39) or choose **Relations** ⇒ **Show Rel** ⇒ (type **q** to quit the INFORMATION WINDOW) ⇒ **Done**.

NOTE
Relations will be covered in more detail in Lesson 8.

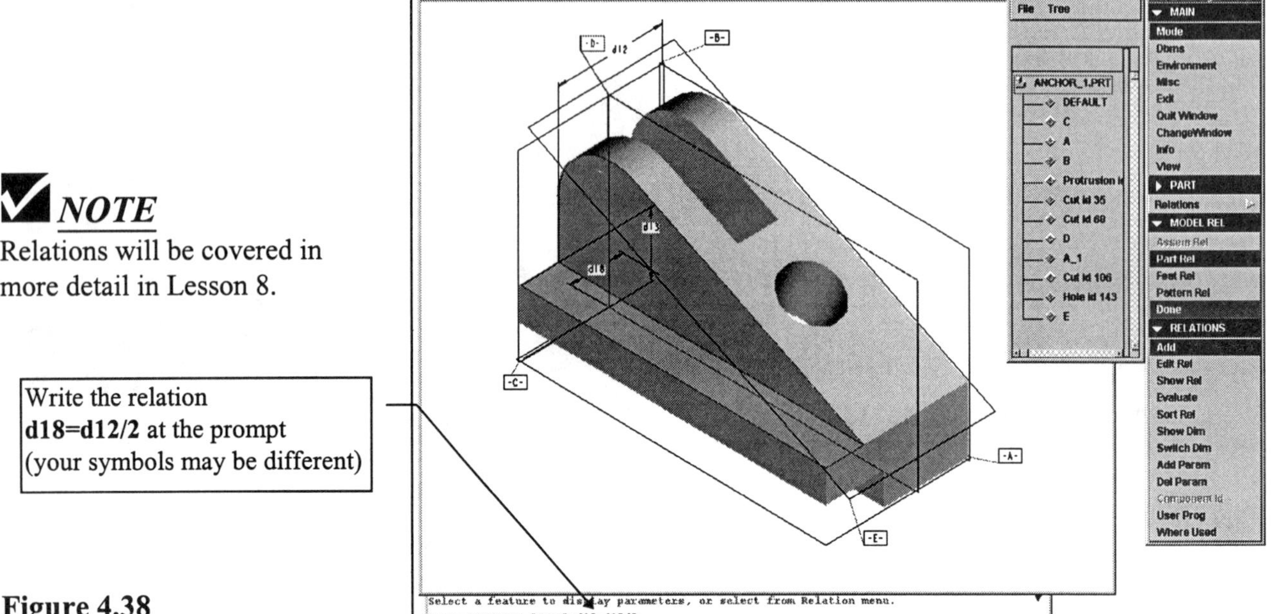

Figure 4.38
Writing a Relation

INFORMATION WINDOW

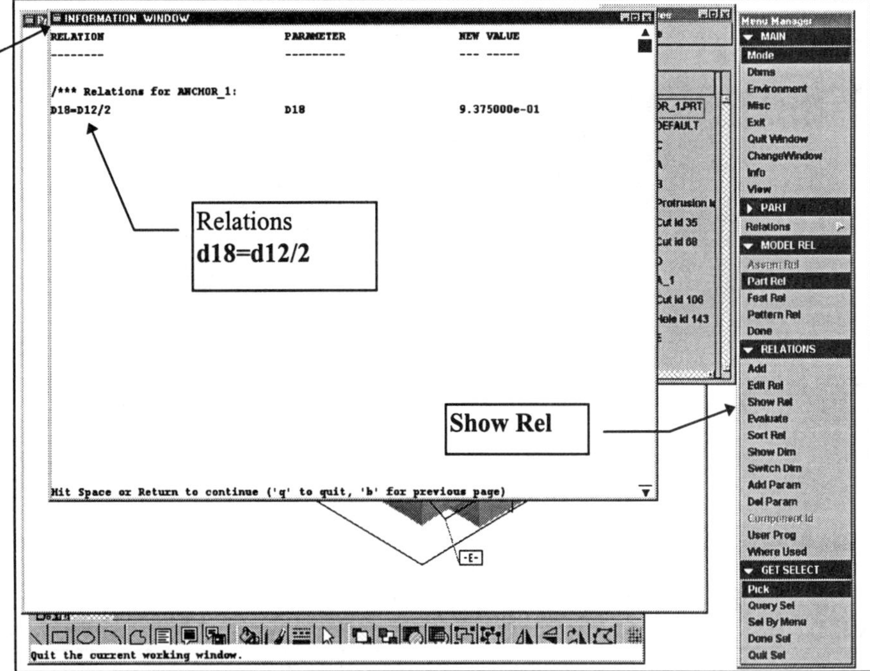

Figure 4.39
Showing a Relation in an INFORMATION WINDOW

The last feature to create is a **.250** inch diameter hole to be placed coaxially with **A_1**. Choose the following commands:

Dbms (**File**--PT/Modeler) ⇒
Save ⇒ **enter**
Purge ⇒ **enter** ⇒
Done-Return

Feature ⇒ **Create** ⇒ **Hole** ⇒ **Straight** ⇒ **Done** ⇒ **Coaxial** ⇒ **Done** ⇒ (pick on the axis line **A_1**) ⇒ (select the placement plane--datum **D**) ⇒ **Okay** ⇒ **Both Sides** ⇒ **Done** ⇒ **Thru All** ⇒ **Done** ⇒ **Thru All** ⇒ **Done** ⇒ (enter the diameter of **.250** at the prompt) ⇒ **enter** ⇒ **Preview** (Fig. 4.40) ⇒ **OK** ⇒ **Done**

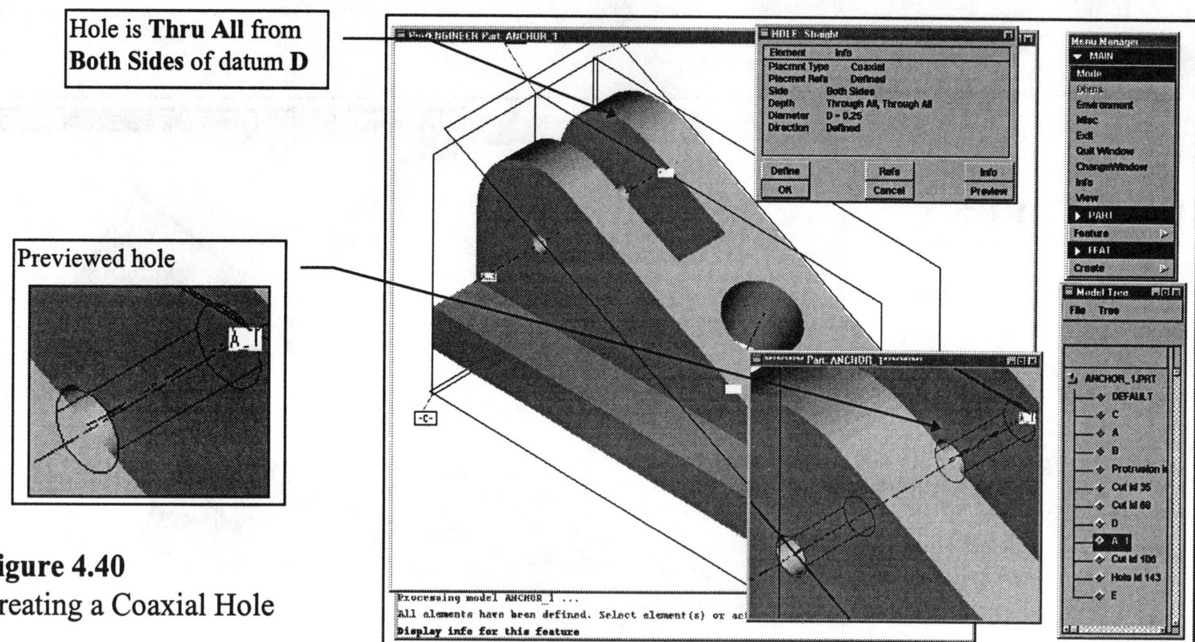

Figure 4.40
Creating a Coaxial Hole

Create a layer and add the datum planes to it (reread Section 8, Layers, at this time), using the following commands:

Dbms (File--PT/Modeler) ⇒ **Save** ⇒ **enter** **Purge** ⇒ **enter** ⇒ **Done-Return**

Layer (from the **PART** menu) ⇒ **Setup Layer** ⇒ **Create** ⇒ (type name of new layer at prompt: **DATUM_LAYER**) ⇒ **enter** ⇒ **enter** ⇒ **Set Items** ⇒ **Add Items** ⇒ (check **✓DATUM_LAYER**) ⇒ **Done Sel** ⇒ **Datum Plane** (from LAYER OBJ menu) ⇒ **Sel By Menu** ⇒ **Name** ⇒ (highlight all, or pick **Sel All--A, B, C, D, E**) ⇒ **Ok** ⇒ **Done Sel** ⇒ **Done/Return** ⇒ **Info** (from MAIN menu) ⇒ **Layer Info** ⇒ **Disp Status** ⇒ (check **✓DATUM_LAYER**) ⇒ **Done Sel** (Fig. 4.41) ⇒ (type **q**) ⇒ **Done/Return** (three times)

HINT

You can turn a layer off (**blank**) by:

Layer ⇒ **Set Display** ⇒ **Blank** ⇒ **✓ DATUM_LAYER** ⇒ **Done Sel** ⇒ **View** ⇒ **Repaint** ⇒ **Done/Return** (from LAYERS menu)

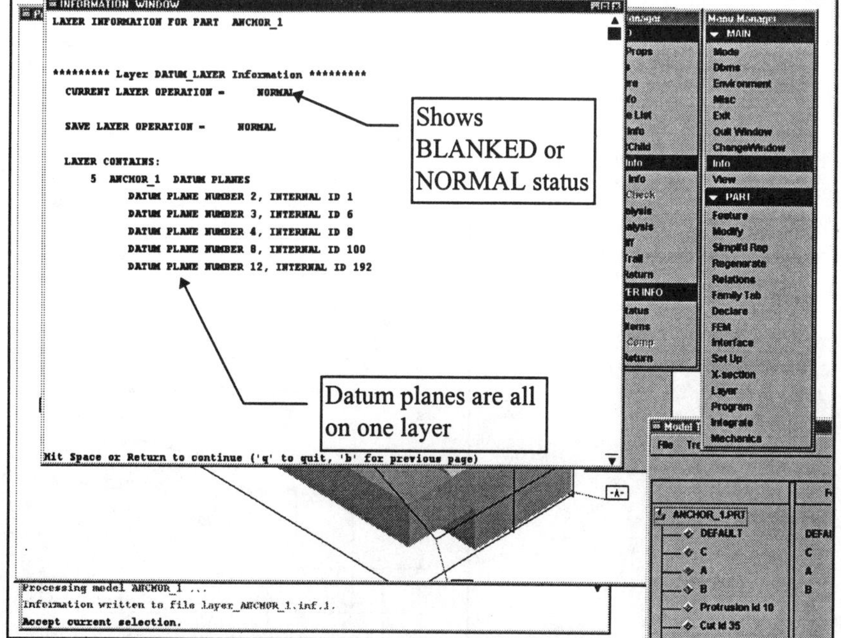

Figure 4.41
Layer Information

Lesson 4 Project

Angle Frame

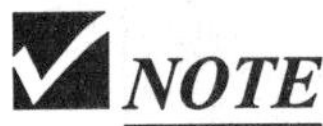

NOTE

Don't forget to set the units and the material (aluminum) for the part.

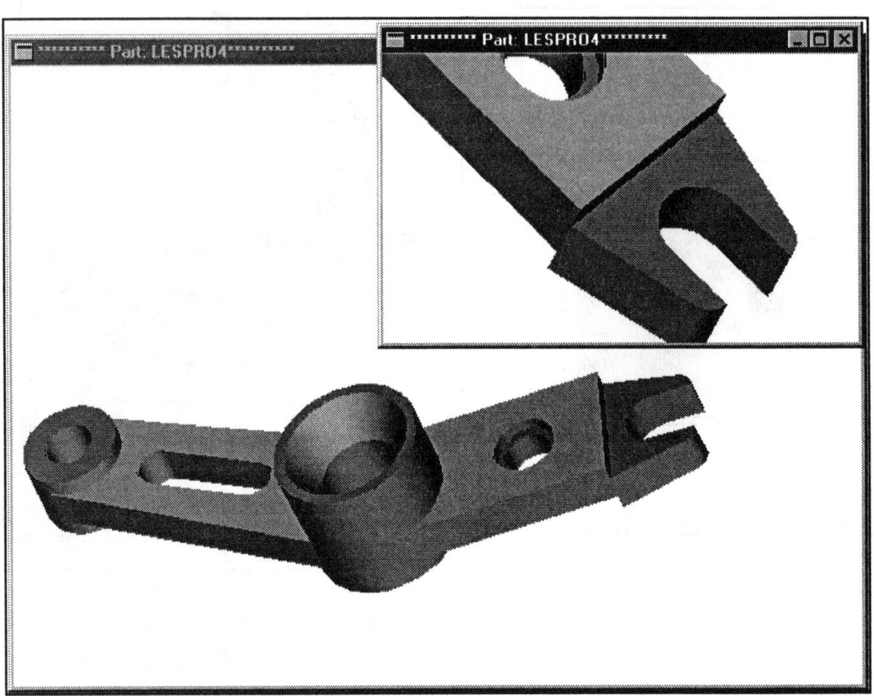

Figure 4.42
Angle Frame

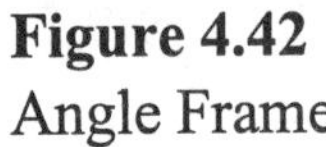

EGD REFERENCE

Engineering Graphics and Design with Graphical Analysis *or* **Fundamentals of Engineering Graphics and Design**
By L. Lamit and K. Kitto
See pages 311, 461, 548, and 929

Angle Frame

The fourth **lesson project** is a machined part that requires the use of a variety of datum planes and a layering scheme. You will also add a relation to control the depth of the large countersink hole at the part's center. Analyze the part and plan out the steps and features required to model it. Use the **DIPS** in Appendix D to establish a feature creation sequence before you start modeling. Create the part shown in Figures 4.42 through 4.51.

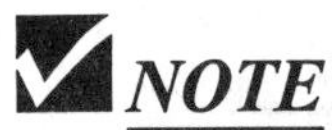

NOTE

Set the datums using **Geom Tol**, rename all three default datum planes to **A**, **B**, **C**, and so on. Set the datum planes on a separate layer.

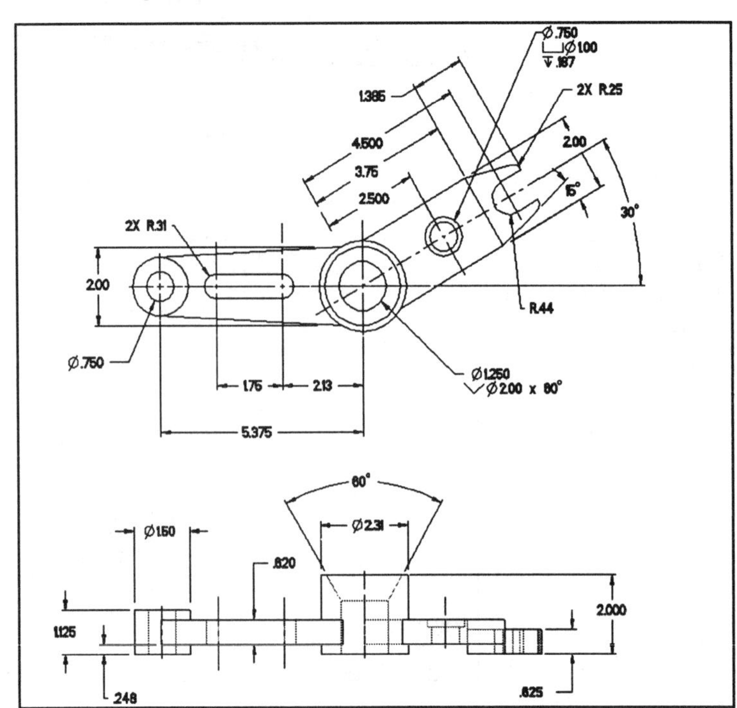

Figure 4.43
Angle Frame Drawing

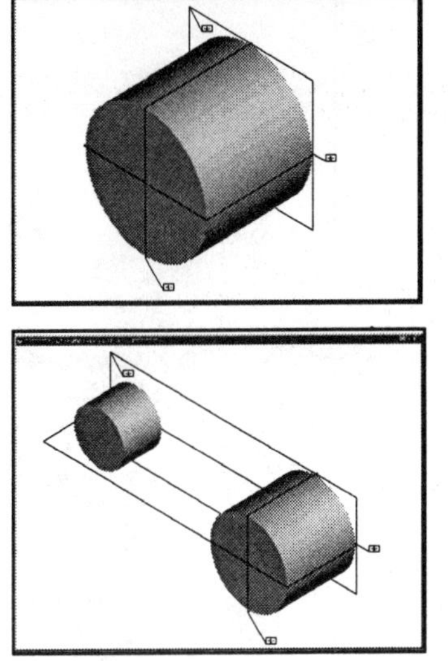

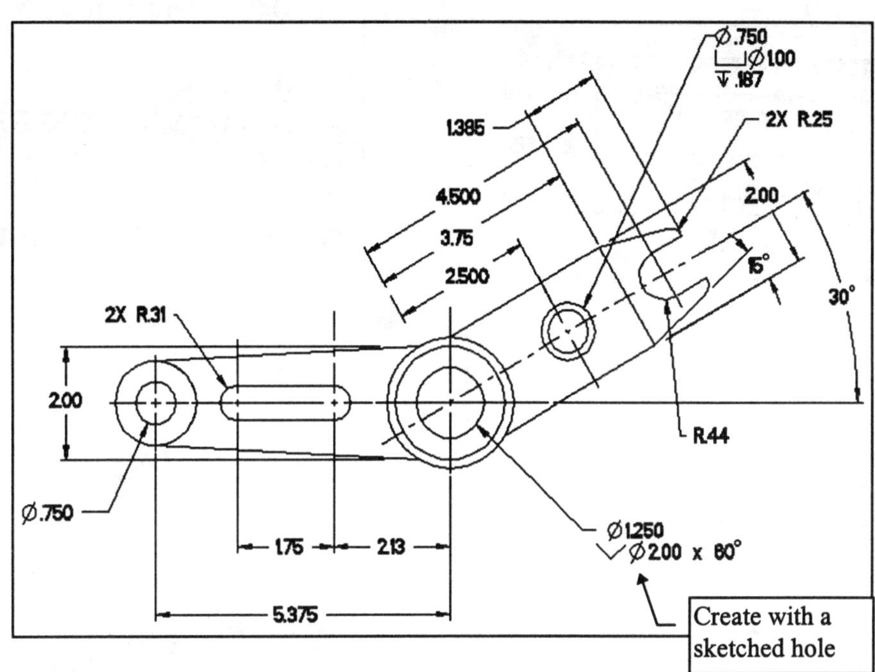

Figure 4.44
Angle Frame, Top View

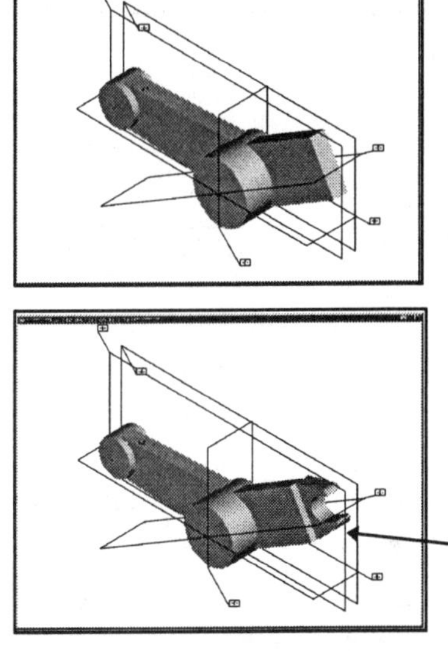

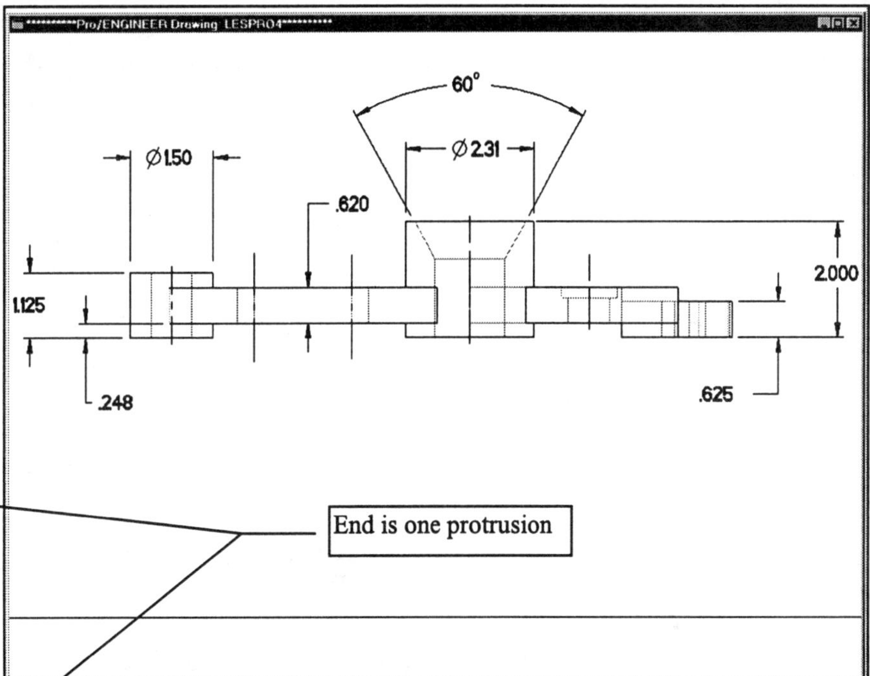

Figure 4.45
Angle Frame, Front View

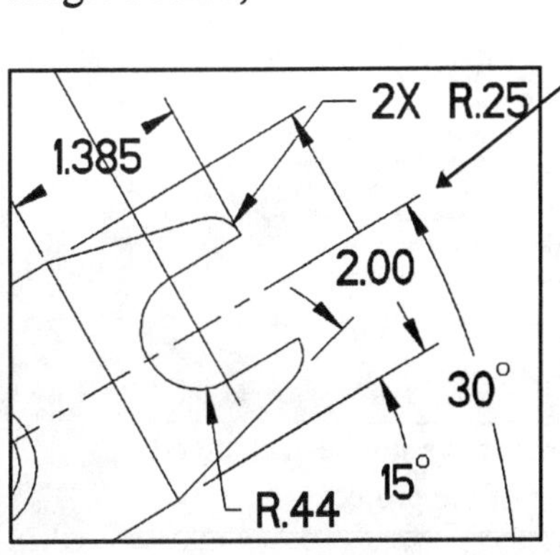

Create a layer for the datums and set them with the appropriate geometric tolerance names: **A**, **B**, **C**, and so on (see Fig. 4.49).

Create two *sections* through the Angle Frame to be used later in a Drawing Lesson. For the sections, use datum planes **B** and **E**, which pass vertically through the center of the part (Fig. 4.51). Name the cross sections **A (SECTION A-A)** and **B (SECTION B-B)**.

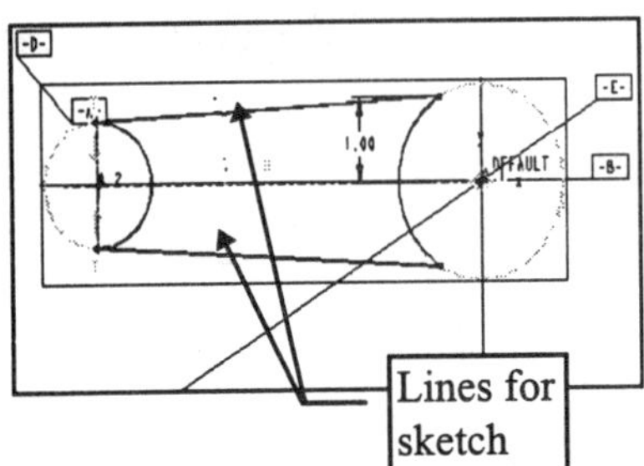

Datum and Section Naming
DTM1 - C
DTM2 - B - SECTION A-A
DTM3 - A
DTM4 - D
DTM5 - E - SECTION B-B

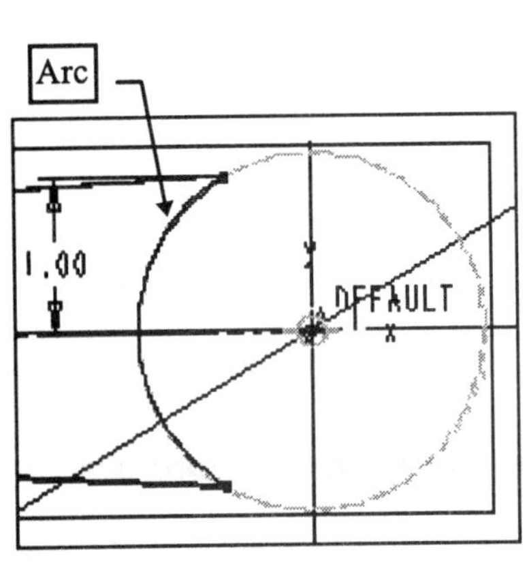

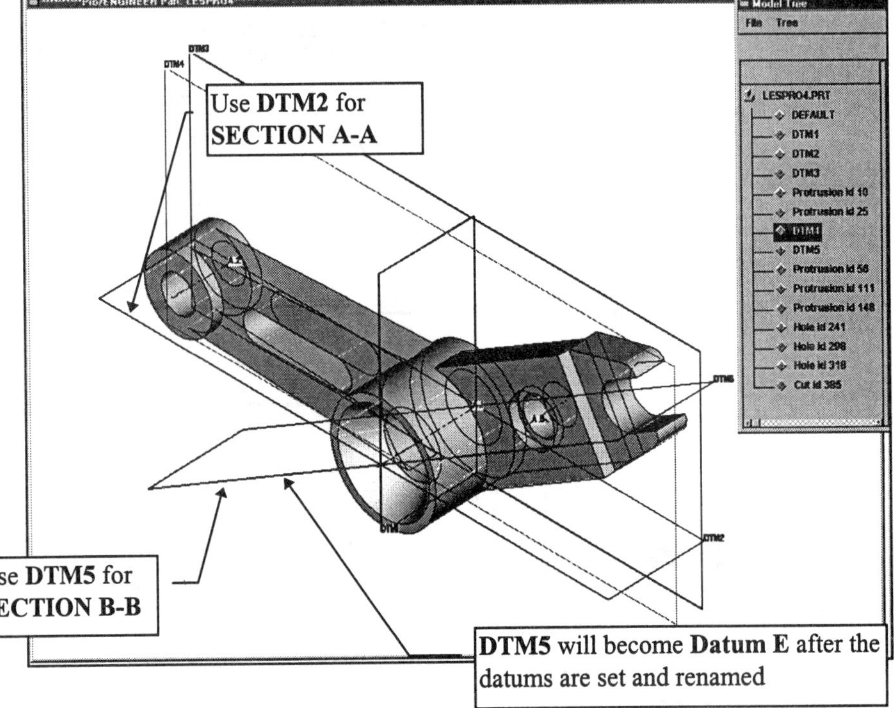

Figure 4.46
Datums

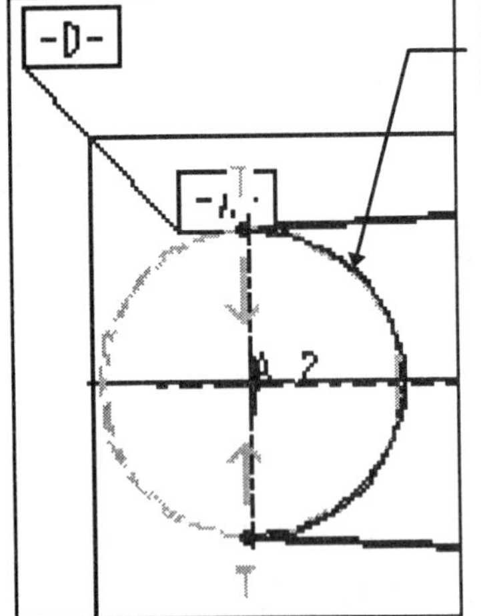

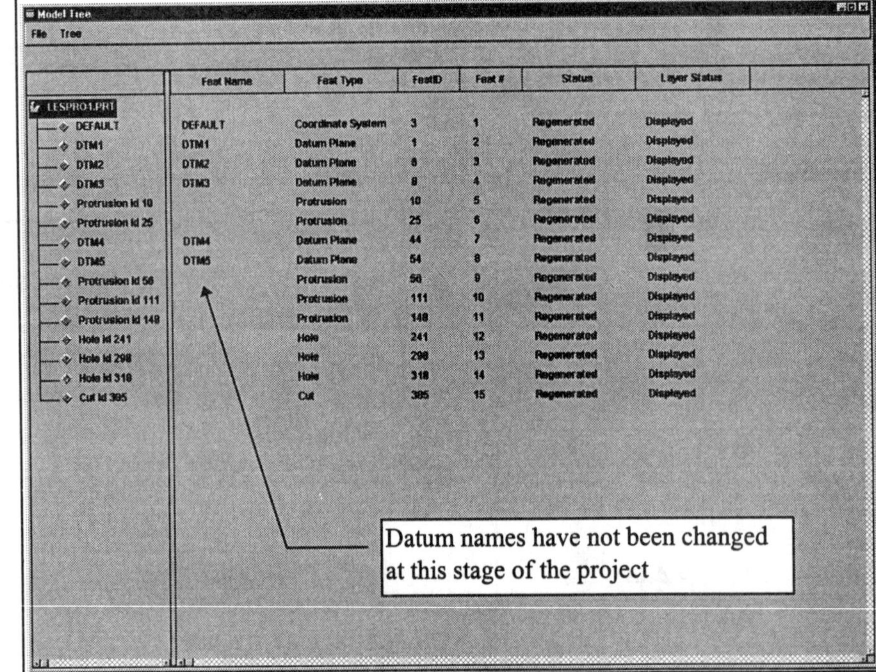

HINT
By clicking on **Tree** in the Model Tree, you can format, add, or remove columns of feature info about the model.

Figure 4.47
Model Tree

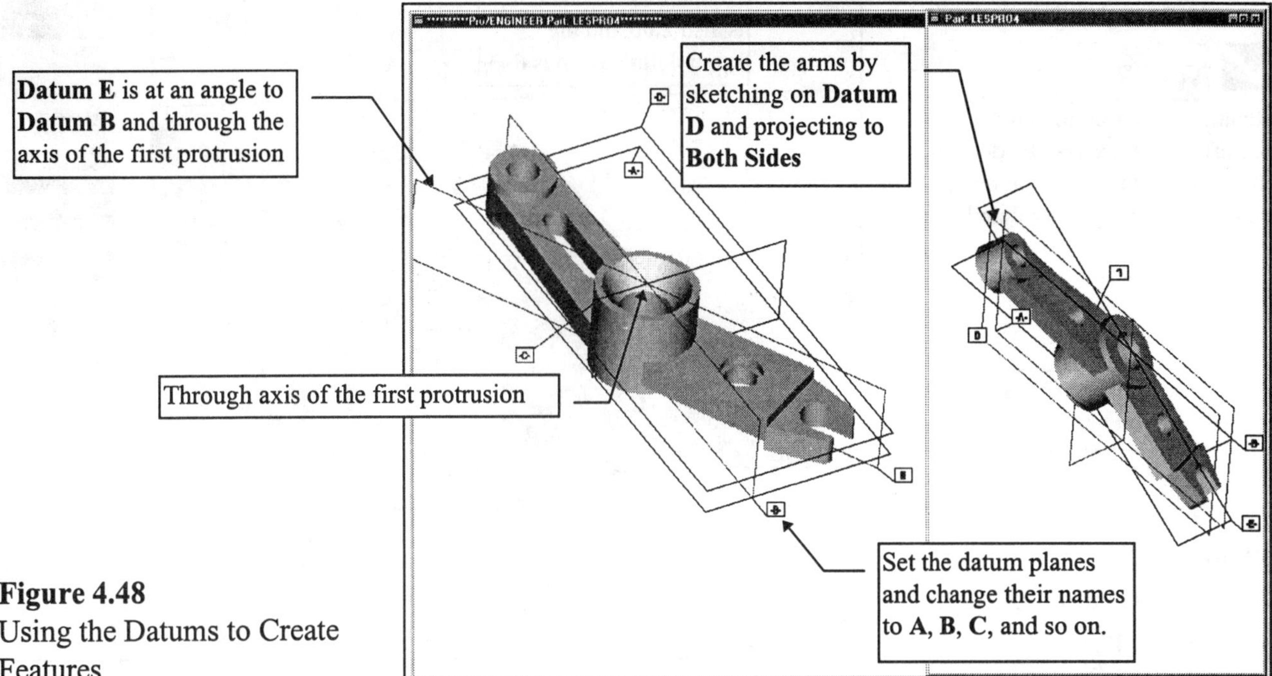

Figure 4.48
Using the Datums to Create Features

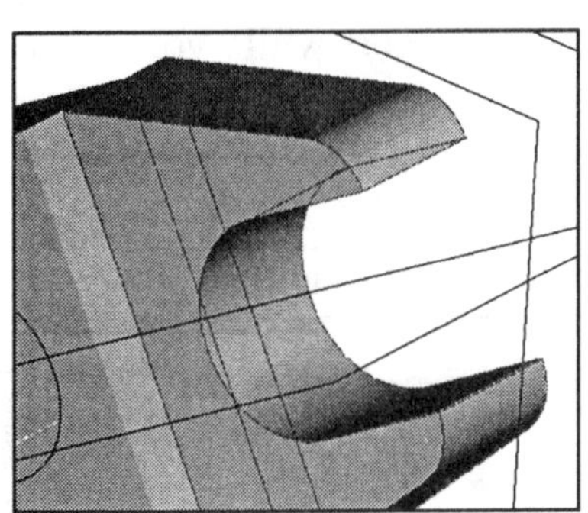

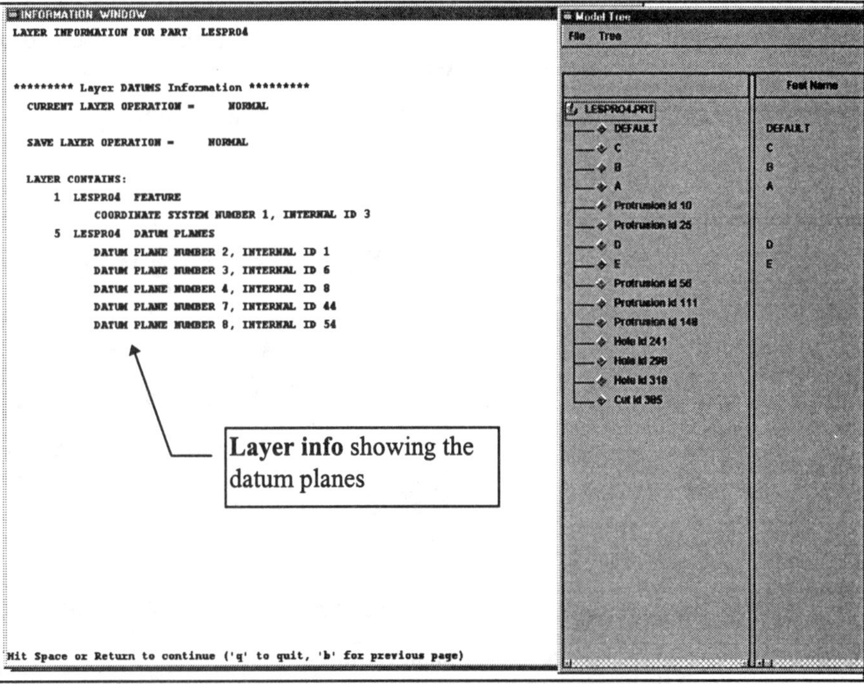

Figure 4.49
Layer Info

Modify the thickness of the boss from **2.00** to **2.50**. Note that the hole does not go through the part. Modify the boss back to the original design dimension of **2.00**. Add a relation to the hole that says the depth of the hole should be equal to the thickness of the boss **(d43=d6)**, as shown in Figure 4.50.

HINT
Your **d#** symbols will probably be different from the ones shown here.

Now change the thickness of the boss (original protrusion) to see that the hole still goes through the part. No matter what the boss thickness dimension changes to, the hole will always go completely through it. This relation controls the *design intent* of the hole.

NOTE

Relations are used to control features and preserve the design intent of the part. Lesson 8 will cover relations in more detail.

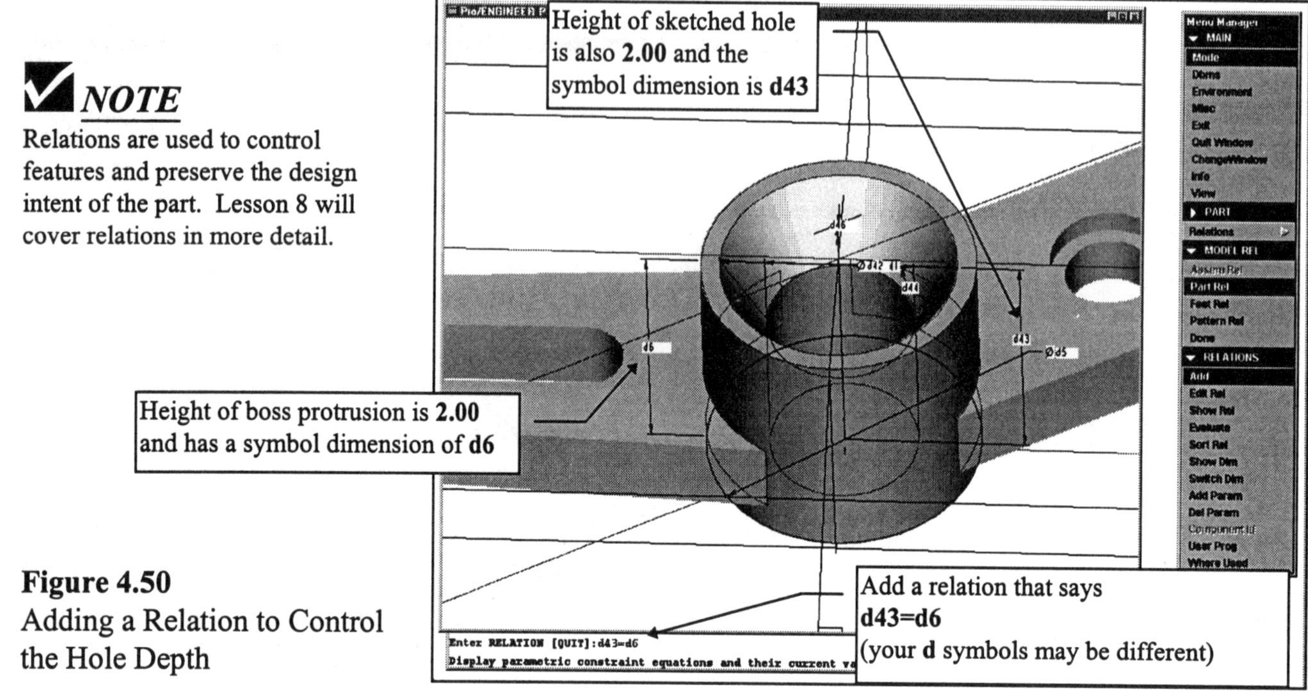

Figure 4.50
Adding a Relation to Control the Hole Depth

Create the sections required to describe the part while you are in **Part Mode** so they will be available for use when you are detailing the part in **Drawing Mode**.

Save as you model!
Dbms (File--PT/Modeler) ⇒
Save ⇒ **enter**
Purge ⇒ **enter** ⇒
Done-Return

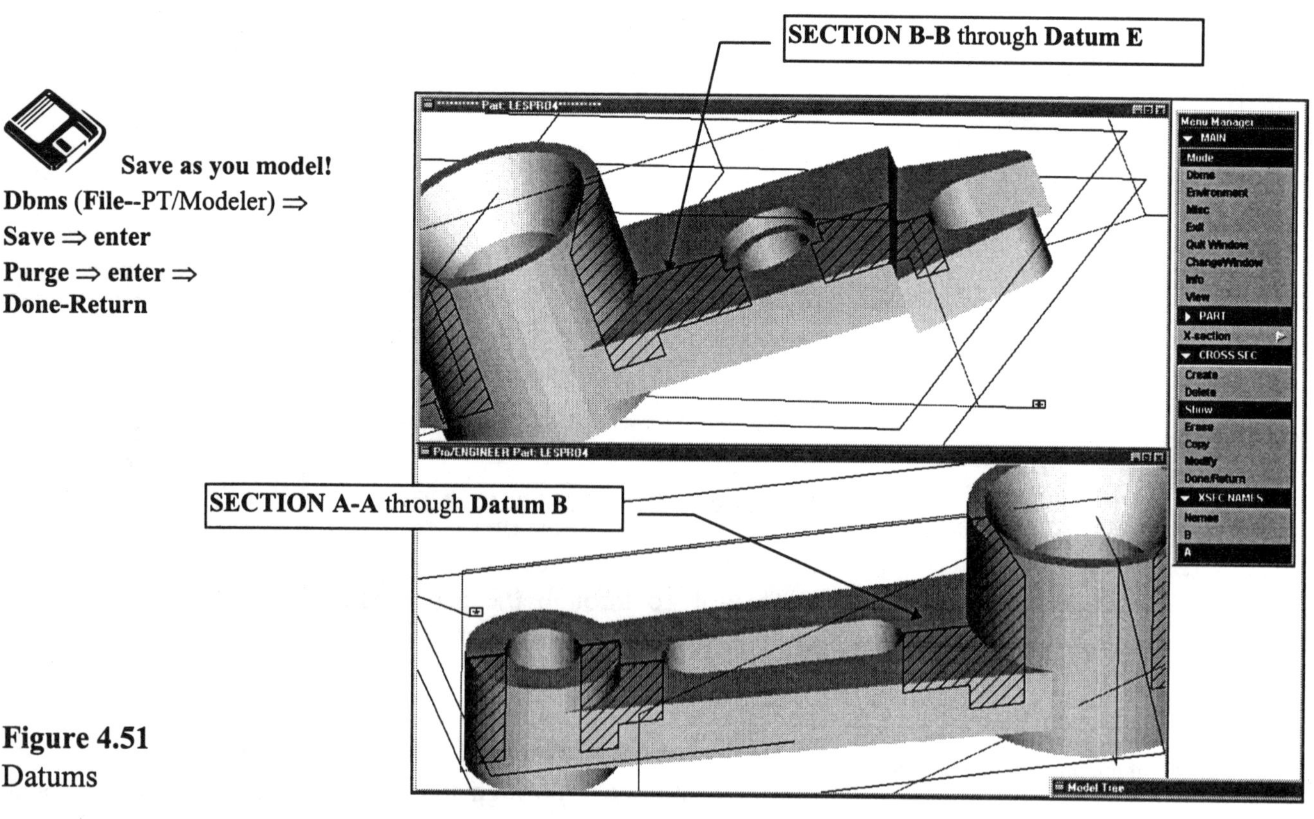

Figure 4.51
Datums

Lesson 5

Revolved Protrusions and Neck Cuts

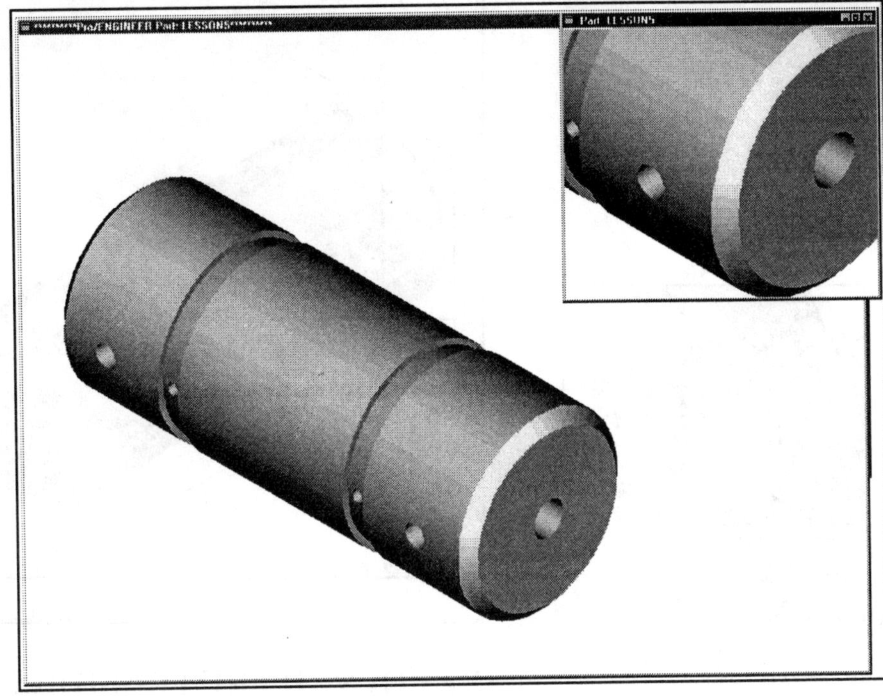

Figure 5.1
Pin

OBJECTIVES

EGD REFERENCE
Engineering Graphics and Design with Graphical Analysis *or* **Fundamentals of Engineering Graphics and Design**
by L. Lamit and K. Kitto
Read Chapters 8, 14, and 26
See pages 239, 486, and 956-957

1. **Create a simple revolved protrusion**

2. **Understand the angle options used to create revolved features**

3. **Use datums to locate holes**

4. **Cut necks in revolved protrusions**

5. **Create a conical revolved cut**

6. **Use the Info command to measure a revolved feature**

7. **Get a hard copy using the Interface command**

COAch™ for Pro/ENGINEER

If you have **COAch for Pro/ENGINEER** on your system, go to SEARCH and do the Segment shown in Figure 5.4.

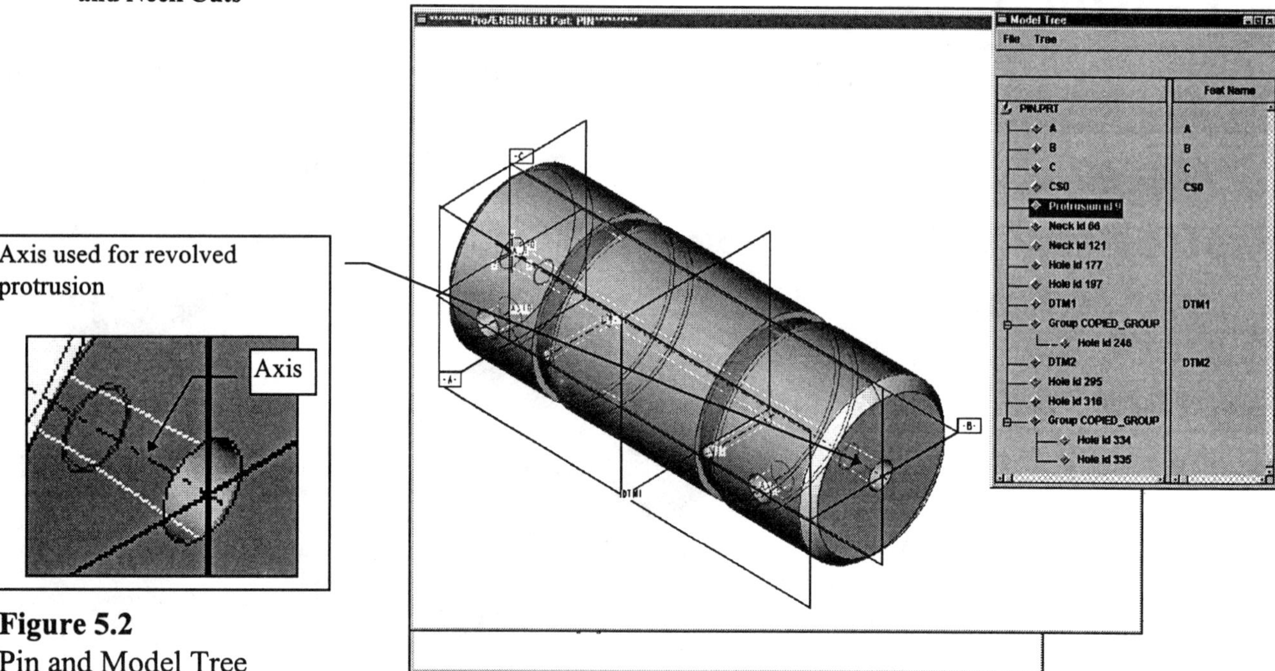

Figure 5.2
Pin and Model Tree

REVOLVED PROTRUSIONS AND NECK CUTS

The **revolve** option creates a feature by revolving the sketched section around a centerline from the sketching plane into the part (Fig. 5.1). You can have any number of centerlines in your sketch, but the first centerline will be the one used to rotate your section geometry.

When you are sketching the feature to be revolved, the first centerline sketched is the *axis of revolution* (Fig. 5.2). The section geometry must be closed and must lie on one side of this centerline (Fig. 5.3).

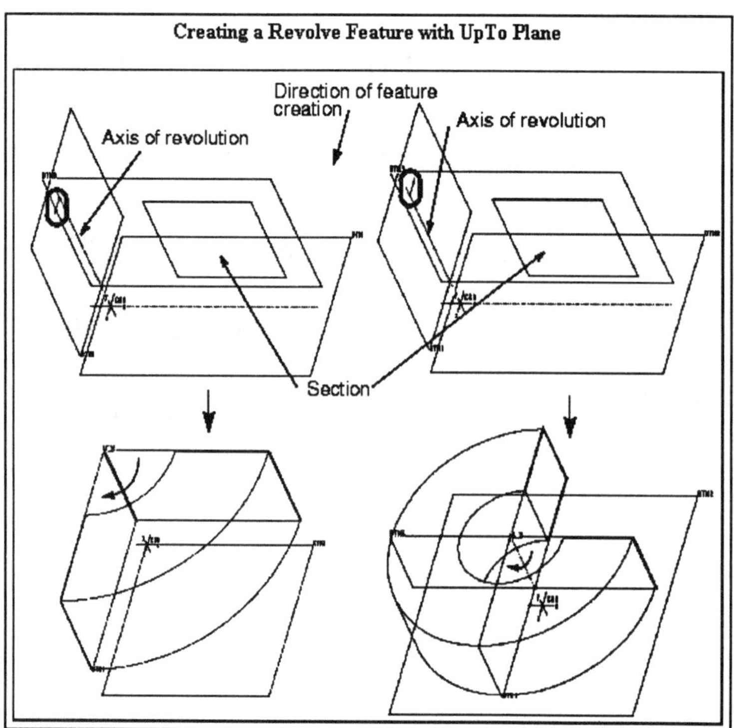

Figure 5.3
Online Documentation, Revolved Protrusions

A revolved feature can be created either entirely on one side of the sketching plane or symmetrically on both sides of the sketching plane. The **One Side** and **Both Sides** options are available for any but the first feature. If you choose **Both Sides,** the feature will be revolved symmetrically in each direction for one-half of the angle specified in the REV TO menu.

After successfully regenerating the revolved section, select **Done** and the REV TO menu appears. This menu allows you to specify the value of the feature's angle of revolution. You can choose the **Variable** option for a user-defined angle of revolution, or you can choose from one of four preset angles: **90, 180, 270,** and **360.**

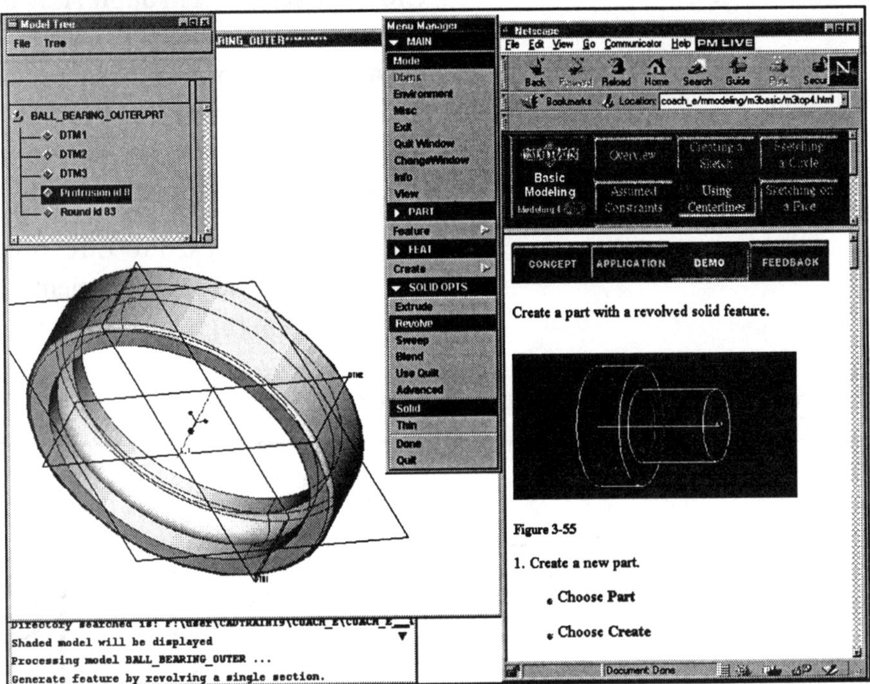

Figure 5.4
COAch for Pro/E, Basic Modeling (Using Centerlines)

If you choose **Variable**, the angle can be specified and modified after the section is created. This angle must be greater than **0°** and less than **360°**. The angle is controlled by a dimension that appears when you are modifying the part and in drawings. A corresponding dimension will not appear if a preset angle is chosen. The base feature of the pin was created with a **360°** revolved section.

If you have **COAch for Pro/ENGINEER** on your system, go to SEARCH and do the appropriate Segment, shown in Figure 5.4.

Neck Cuts

PT/Modeler™

PT/Modeler *does not* support the **Neck** command. Use a revolved **Cut** to accomplish the same task.

A **Neck** is a special type of *revolved slot* that creates a groove around a revolved part or feature. You always create a neck on a **Through/Axis** datum plane, and the sketch is inside the part. You might align both ends of the section to the revolved surface of the parent feature.

To create a neck:

1. Choose **Neck** from the SOLID menu.
2. Choose an option from the OPTIONS menu to specify the number of degrees in the revolution.
3. Create or select a **Through/Axis** datum plane as the sketching plane.
4. Create or select a reference/orientation plane
5. Sketch the centerline that becomes the axis of revolution.
6. Sketch the neck cross section open, with the ends aligned to the silhouette edge of the part or feature.

In creating a neck, Pro/E revolves the section around the part to the specified angle measurement, thereby removing the material inside the section (Fig. 5.5).

Figure 5.5 Online Documentation, Neck Features

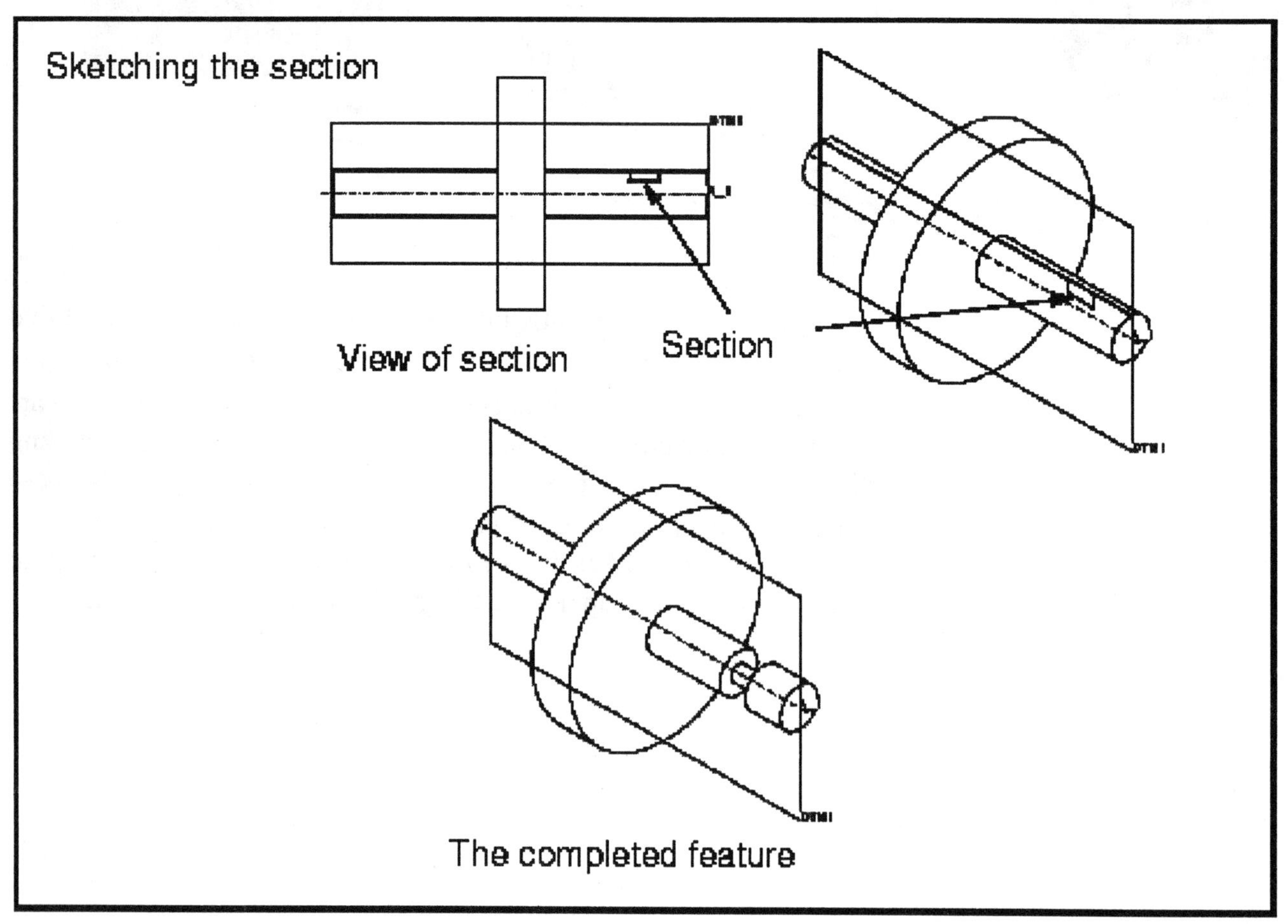

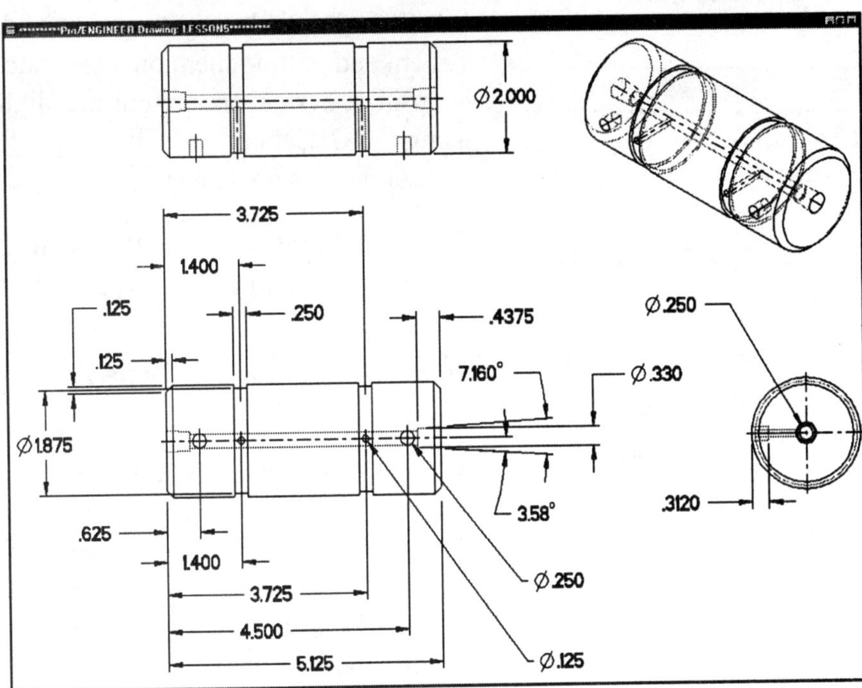

Figure 5.6
Pin Dimensions

Pin

The pin is an example of a part created by revolving one section about a centerline (Fig. 5.6). The pin was created as a **revolved protrusion.** The chamfers are created on the first revolved protrusion. The grooves were created with the **Neck** command. The holes were added using datum axes and a new datum plane.

The pin's complete geometry (with the exception of the holes) could have been created with one revolved protrusion. In general, this is poor design practice, since it limits the flexibility of modifying the geometry later in the design process. For most parts, the basic revolved shape should be the first protrusion, followed by the most important secondary features (cuts, protrusions, etc.). The holes required for the part are then created. Lastly, the rounds and chamfers are created where required. Choose **Part** from the MODE menu and choose the following commands:

Part ⇒ Create (type the part name, **Pin**) **⇒ enter**

SETUP AND ENVIRONMENT

Set Up ⇒ Units ⇒ Length ⇒ Inch ⇒ Done ⇒ Material ⇒ Define ⇒ (Type **Steel**, then press **enter**) ⇒ (table of material properties, change or add information) **⇒ File ⇒ Save ⇒ File ⇒ Exit ⇒ Assign ⇒** (pick **Steel**) **⇒ Accept ⇒ Done**

Environment ⇒ ✓Grid Snap
Hidden line Tan Dimmed

PT/Modeler™

Feature ⇒ Datum Plane ⇒ Coord Sys ⇒ Default ⇒ Done ⇒

Feature ⇒ Create ⇒ Datum ⇒ Plane ⇒ Default ⇒ Create ⇒ Datum ⇒ Coord Sys ⇒ Default ⇒ Done ⇒ Done

Start the Pin by creating a set of datum planes and a coordinate system and putting them on a separate layer. Note that the datum and coordinate system commands are slightly different from when you are using the **Offset** option. Choose the following commands to create and set datums on a layer:

PT/Modeler™

Layer ⇒ Setup ⇒ Create ⇒ (type **DATUM_LAYER**) **⇒ enter ⇒ enter ⇒ Set Items ⇒ Add Items ⇒** (**✓DATUM_LAYER**) **⇒ Done Sel**

Layer ⇒ Create ⇒ (type **DATUM_LAYER**) **⇒ enter ⇒ enter ⇒ Set Items ⇒ Add Items ⇒** (**✓DATUM_LAYER**) **⇒ Done Sel ⇒ Datum Plane ⇒ Sel By Menu ⇒ Name ⇒ Sel All ⇒ Ok ⇒ Done Sel ⇒ Done/Return ⇒ Done/Return**

The first protrusion is a revolved protrusion. Use the front view of the pin in Figure 5.6 to sketch the revolved protrusion's section in Figure 5.7. Create the chamfers and the main body of the pin with the first protrusion. In Lesson 6, you will see that chamfers, like rounds, are normally added near the end of the modeling sequence. The commands, sketch, references, and so on are very similar to those in the previous lessons:

PT/Modeler™

Feature ⇒ Protrusion ⇒ Revolve ⇒ Done ⇒ One Side ⇒ etc.

Feature ⇒ Create ⇒ Solid ⇒ Protrusion ⇒ Revolve ⇒ Done ⇒ One Side ⇒ Done ⇒ (pick **DTM3** as sketching plane) **⇒ Okay ⇒ Top** (pick **DTM2** as the horizontal reference) **⇒ View ⇒ Pan/Zoom ⇒ Pan** (pan the screen as shown) **⇒ Done-Return ⇒ Line ⇒ Centerline ⇒ Horizontal** (sketch the horizontal centerline *first*) **⇒ Mouse Sketch** (sketch the section geometry in Fig. 5.7) **⇒ Alignment** (align the horizontal centerline with **DTM2**, the horizontal line with **DTM2**, and the vertical left edge line with **DTM1**) **⇒ Regenerate** (Fig. 5.7) **⇒ Dimension** (Fig. 5.8) **⇒ Regenerate**

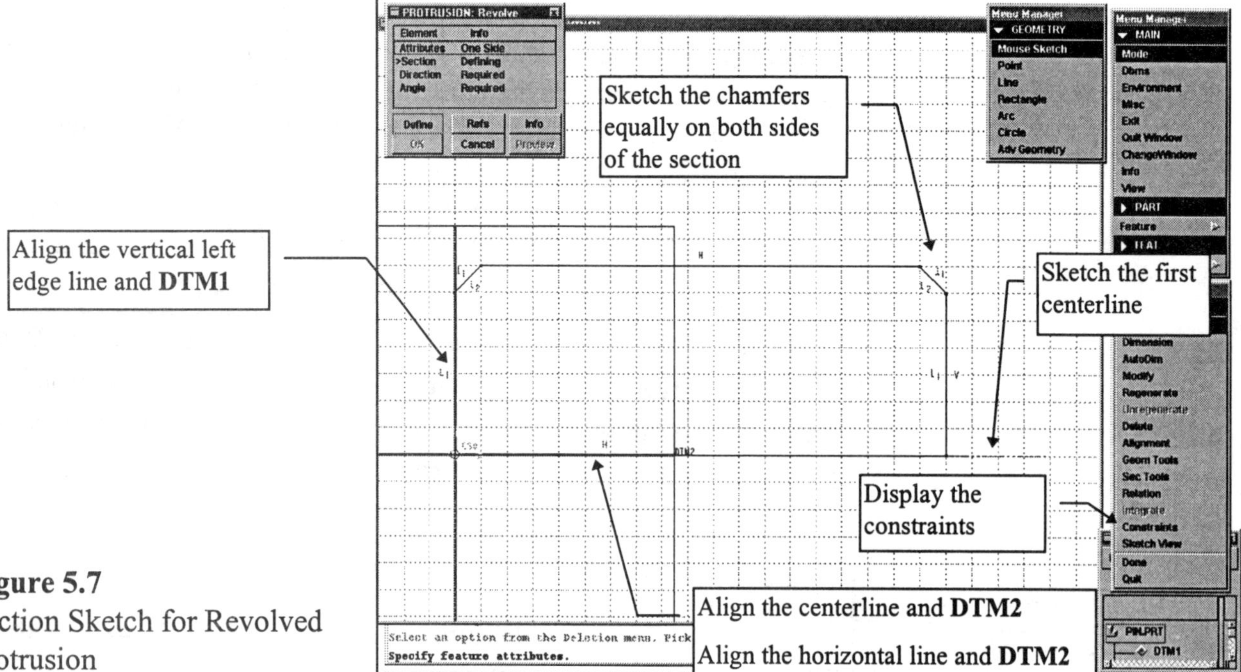

Figure 5.7
Section Sketch for Revolved Protrusion

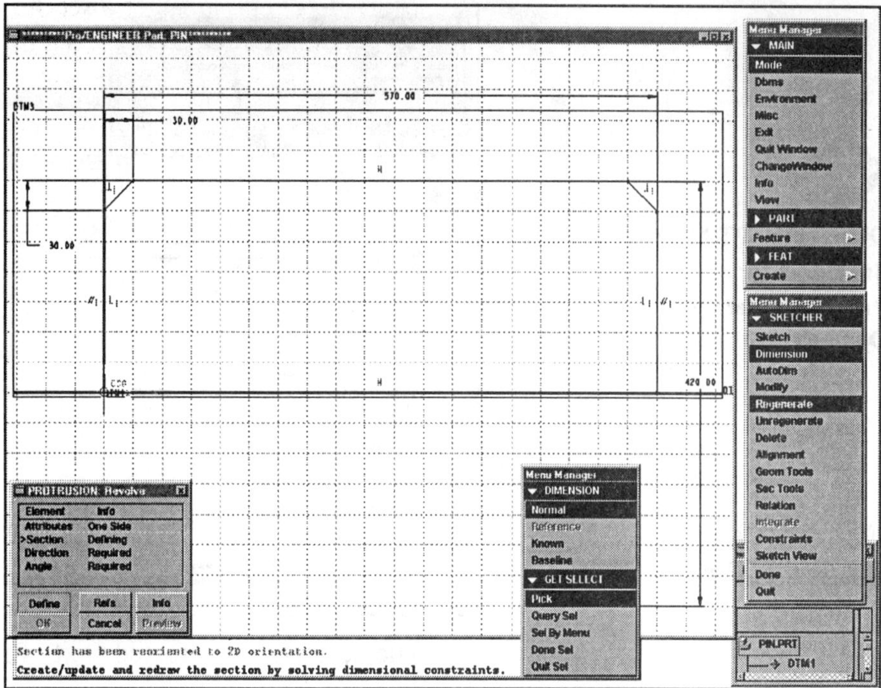

Figure 5.8
Dimensioned and Regenerated Section Sketch

Complete the sketch by choosing **Modify** (change the dimensions to the design values shown in Fig. 5.6) ⇒ **Regenerate**. If the section fails, **Unregenerate** *immediately*, then add and **Modify** the dimensions for the chamfer on the other side of the sketch and **Regenerate** again as in Figure 5.9.

HINT

If a section fails after you modify the dimensions, **Unregenerate** *immediately*.

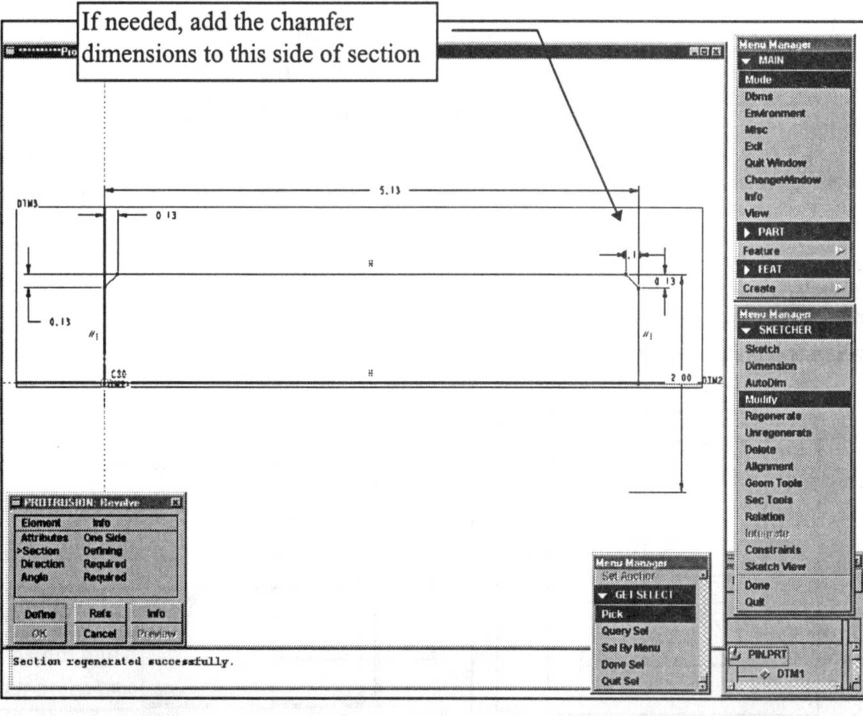

Figure 5.9
Regenerated Design Dimensions

Done ⇒ **360** (from the REV TO menu) ⇒ **Done** ⇒ **View** ⇒ **Default** ⇒ **Done-Return** ⇒ **Preview** (Fig. 5.10) ⇒ **OK** ⇒ **Done**

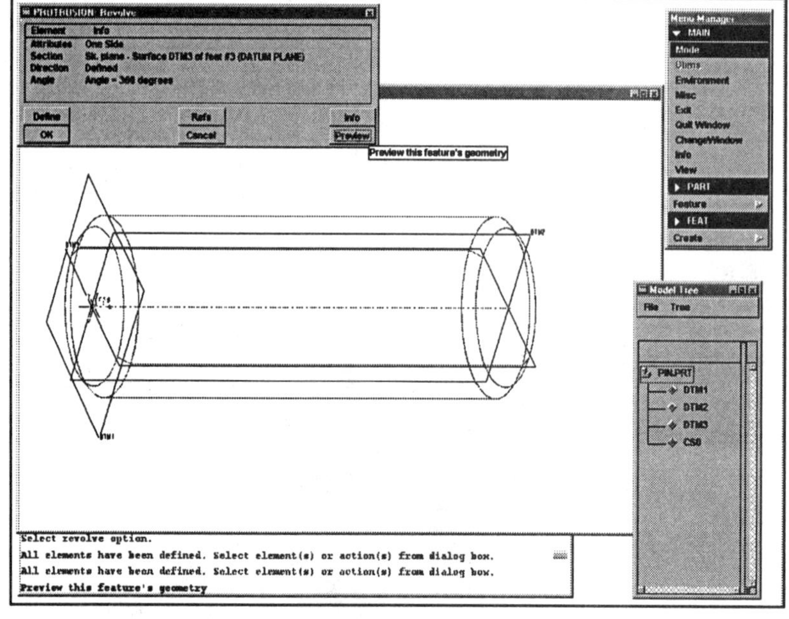

Dbms (**File**--PT/Modeler) ⇒
Save ⇒ **enter**
Purge ⇒ **enter** ⇒
Done-Return

Figure 5.10
Preview of Revolved Protrusion

The neck cuts are created next. Exaggerate the neck's size:

PT/Modeler™
Feature ⇒ **Cut** ⇒ **Revolve** ⇒ **Solid** ⇒ **Done** ⇒ (select **DTM3** as the sketching plane) ⇒ **Okay** ⇒ **Top** ⇒ (select **DTM2** as the reference plane) ⇒ **Sketch etc.**

Environment ⇒ □ **Grid Snap** ⇒ **Done-Return** ⇒ **Feature** ⇒ **Create** ⇒ **Neck** ⇒ **360** ⇒ **One Side** ⇒ **Done** ⇒ **Use Prev** ⇒ **Okay** ⇒ **Sketch** ⇒ **Line** ⇒ **Centerline** ⇒ **Horizontal** (sketch the centerline used to revolve the neck) ⇒ **Vertical** (sketch the vertical centerline to establish the middle of the neck cut) ⇒ **Sketch** ⇒ **Line** ⇒ **Vertical** (sketch the three lines of the neck) ⇒ **Regenerate** ⇒ **Alignment** (align the two endpoints of the vertical lines with the top of the part, and align the *horizontal* centerline with **DTM2**) ⇒ **Regenerate** ⇒ **Dimension** (Fig. 5.11) ⇒ **Regenerate** ⇒ **Modify** (change the sketch dimensions to the design dimensions as shown in Fig. 5.6) ⇒ **Regenerate** ⇒ **Done** ⇒ **Done**

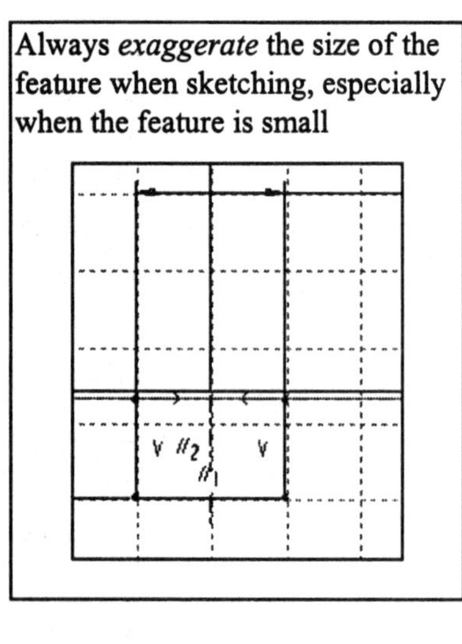

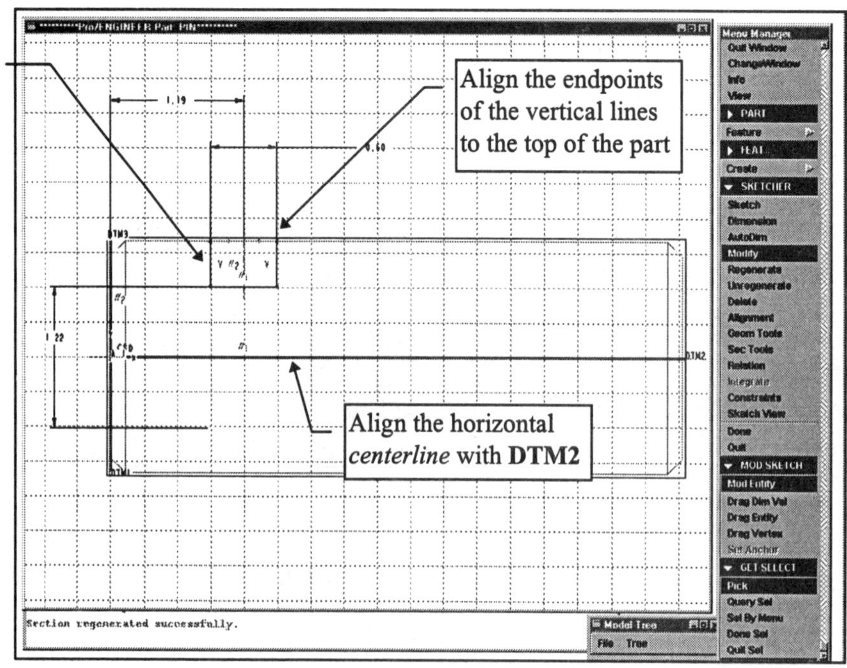

Figure 5.11
Neck Cut Sketch

The neck is now complete. Note that the neck command does not have a dialog box and that it is quicker to complete than a revolved protrusion (or a *revolved cut*, which is the same thing as a *neck*). Change the view, and shade and color the part (Fig. 5.12).

Figure 5.12
Completed Neck

Create the other neck cut using the same steps (Fig. 5.13). Try sketching with the shading on, as shown in Figure 5.14. Rotate the part to see the necks, as in Figure 5.14. Color the neck cuts differently from the base part. In industry, the neck would have been copied instead of created again, but it is good to practice repeating commands at this stage of your understanding of Pro/E.

Keep shading on when sketching

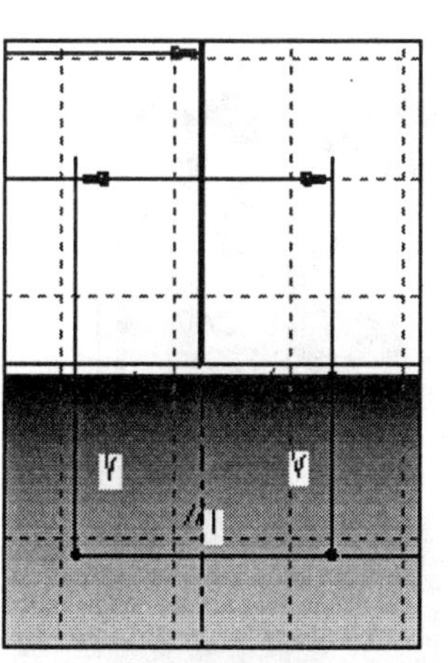

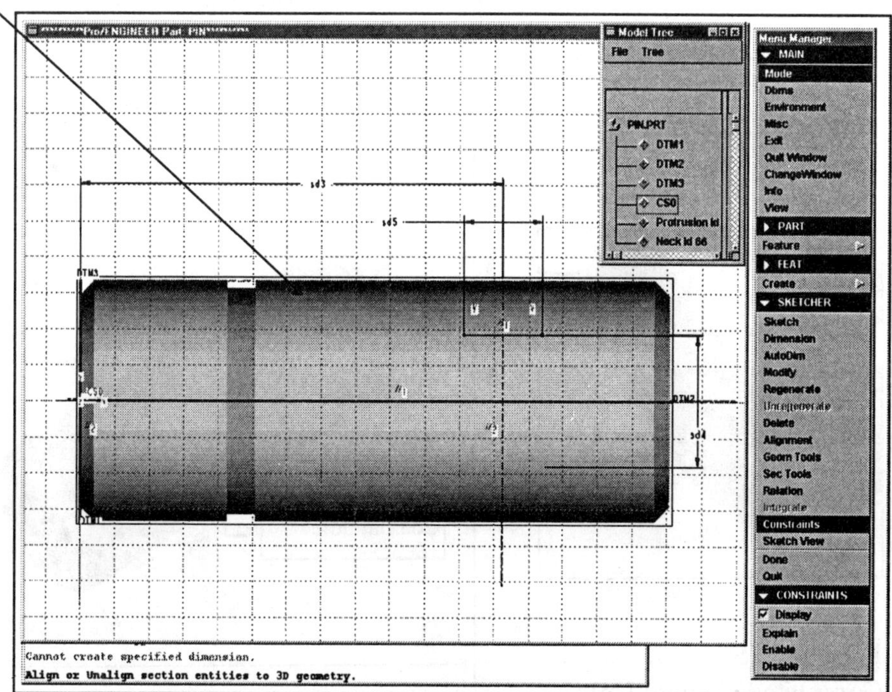

Figure 5.13
Second Neck with Shading

NOTE

In the example, we have renamed the datum planes:

DTM1 = datum **A**
DTM2 = datum **B**
DTM3 = datum **C**

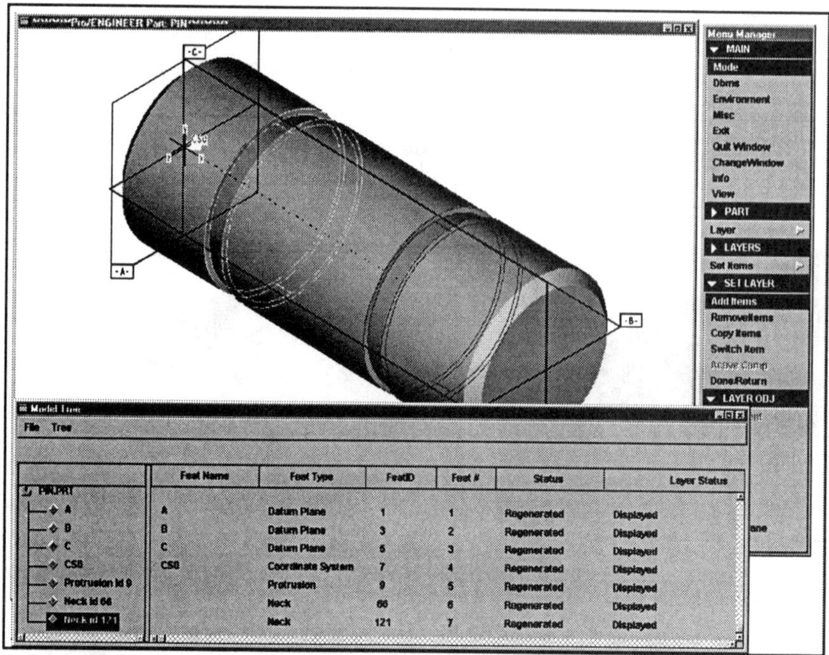

Figure 5.14
Completed Neck

Note that the datums have been *set* and the coordinate system and the datums have been put on a layer.

The remaining features are all holes. The first hole to create is the **∅.250** hole through the center of the pin (Fig. 5.15). Choose the following commands:

Feature ⇒ **Create** ⇒ **Hole** ⇒ **Straight** ⇒ **Done** ⇒ **Coaxial** ⇒ **Done** ⇒ **Query Sel** (pick on axis **A_1** with left mouse button and accept with middle mouse button) ⇒ (pick the placement plane datum **A**, which was **DTM1**) ⇒ **Flip** (if necessary, pick **Flip** a couple of times to see the direction of the arrow; it must point toward the part) ⇒ **Okay** ⇒ **One Side** ⇒ **Done** ⇒ **Thru All** ⇒ **Done** ⇒ (type the diameter at the prompt: **.250**) ⇒ **enter** ⇒ **OK** ⇒ **Done**

Datum plane **A** (**DTM1**) is the placement plane

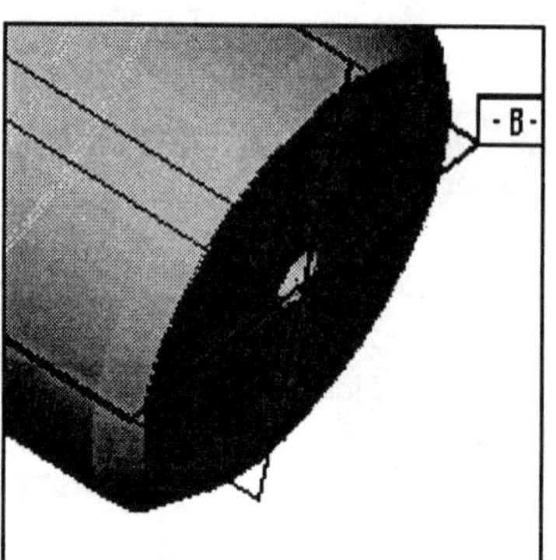

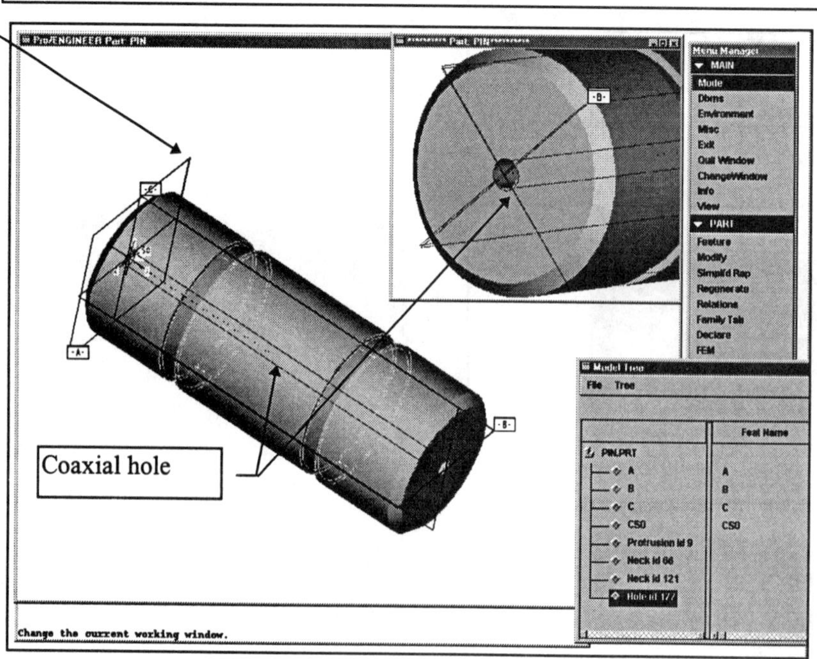

Figure 5.15
Completed Coaxial Hole

The pin has a conical cut (hole) at both ends that is coaxial with the **∅.250** hole and axis **A_1**. Make the conical feature with a *sketched hole* placed on the datum plane **A** side of the pin (Fig. 5.16):

Environment ⇒ **✓Grid Snap** ⇒ **Done-Return** ⇒ **Feature** ⇒ **Create** ⇒ **Hole** ⇒ **Sketch** ⇒ **Done** ⇒ **Coaxial** ⇒ **Done** ⇒ **Sketch** ⇒ **Line** ⇒ **Centerline** ⇒ **Vertical** (pick once) ⇒ **Sketch** ⇒ **Line** ⇒ **2Points** ⇒ (sketch the four lines of the *closed* section on one side of the centerline) ⇒ **Regenerate** ⇒ **Dimension** ⇒ **Regenerate** ⇒ **Modify** ⇒ **Regenerate** (Fig. 5.16) ⇒ **Done** ⇒ (pick axis **A_1)** ⇒ (pick datum **A** as the placement plane) ⇒ **Flip** ⇒ **Okay** ⇒ **Preview** (Fig. 5.17) ⇒ **OK** ⇒ **Done**

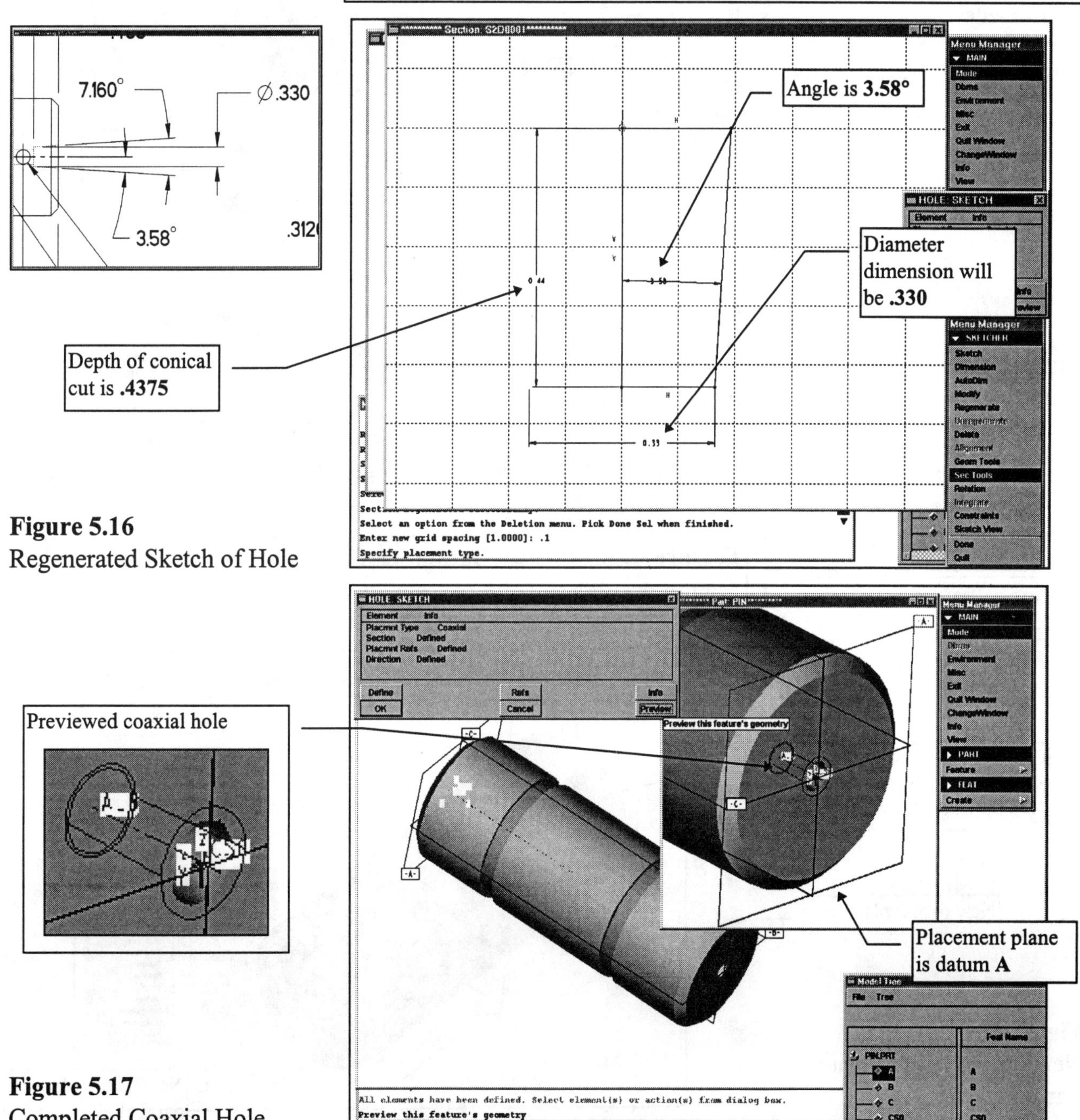

Figure 5.16
Regenerated Sketch of Hole

Figure 5.17
Completed Coaxial Hole

The conical hole is on both sides of the pin. Copy and mirror the hole using the following commands:

PT/Modeler™

Feature Oper ⇒ Copy etc.

Feature ⇒ Copy ⇒ Mirror ⇒ Dependent ⇒ Done ⇒ (select the conical hole to be mirrored; use **Query Sel** or rotate the model to see the hole) **⇒ Accept ⇒ Done Sel ⇒ Done ⇒ Make Datum ⇒ Offset** ⇒ (pick datum plane **A** to offset from) **⇒ Enter Value** (type **5.125/2** at the prompt) **⇒ enter ⇒ Done** (Fig. 5.18) **⇒ Done**

Dbms (File--PT/Modeler) ⇒
Save ⇒ enter
Purge ⇒ enter ⇒
Done-Return

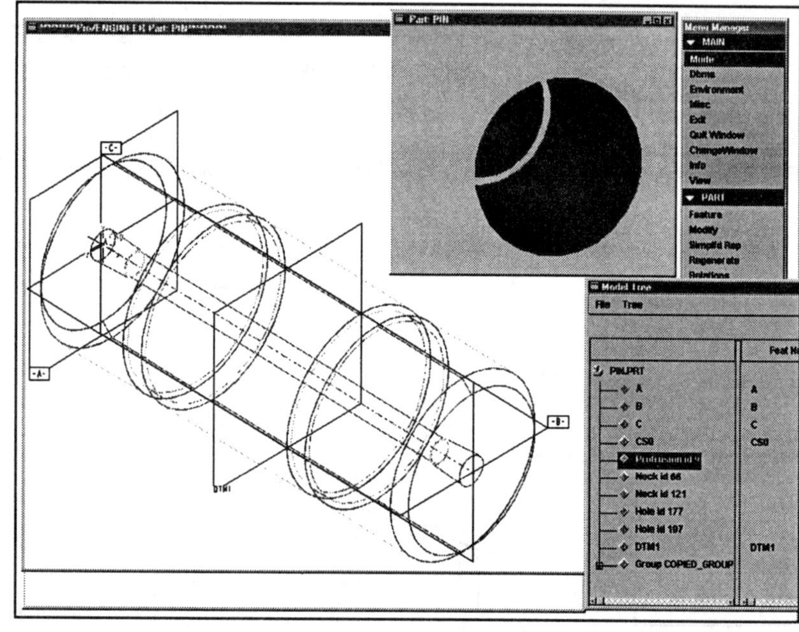

Figure 5.18
Mirrored Conical Hole

You just created your first ***datum-on-the-fly*** (**Make Datum**).

Create a new datum plane tangent to the pin's outside diameter and parallel to datum plane **C**:

PT/Modeler™

Feature ⇒ Datum Plane ⇒ Tangent etc.

Feature ⇒ Create ⇒ Datum ⇒ Plane ⇒ Tangent (pick the left front of the pin's cylinder) **⇒ Parallel** (pick datum plane **C**, as shown in Fig. 5.19) **⇒ Done ⇒ Done**

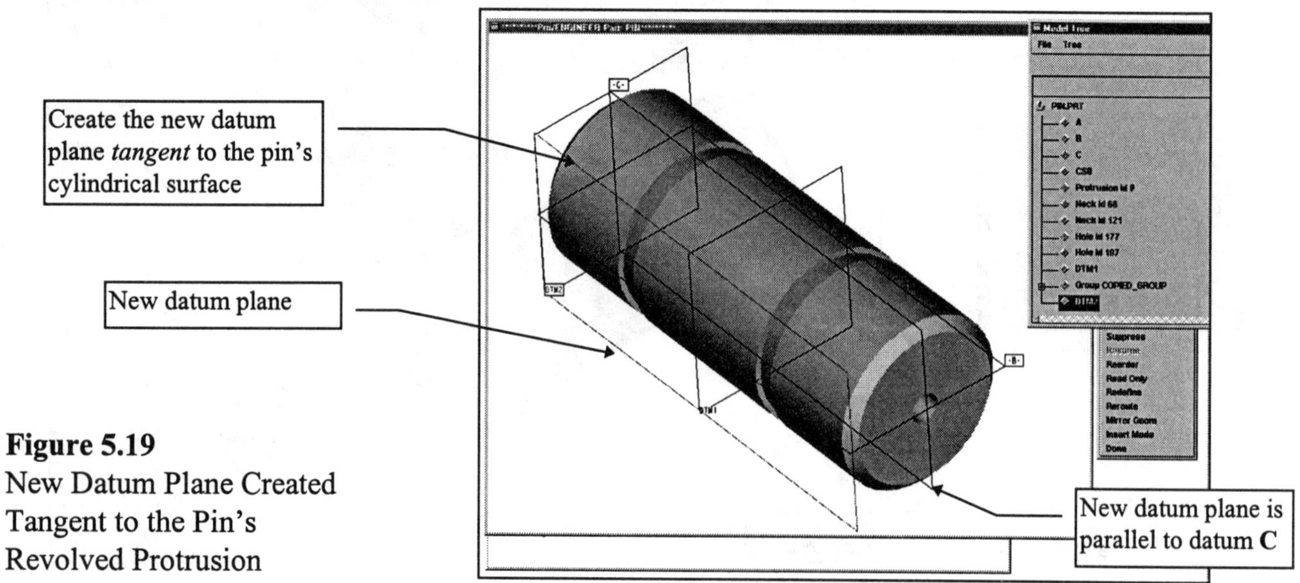

Figure 5.19
New Datum Plane Created Tangent to the Pin's Revolved Protrusion

Using this new datum plane, create the two **∅.125** holes using the new datum plane as the placement plane and the **∅.250** coaxial hole running through the pin as the ending surface.

Menu Manager
▼ PLACEMENT
Linear
Radial
Coaxial
On Point
Done
Quit

Feature ⇒ Create ⇒ Hole ⇒ Straight ⇒ Done ⇒ Linear ⇒ Done ⇒ (select the new datum plane as the placement plane) ⇒ (select the feature placement location on the datum plane; pick anywhere on the plane) ⇒ **Okay** (to accept the feature creation direction) ⇒ (select two edges to dimension from; pick datum **B** and type **0.00** at the prompt) ⇒ **enter** ⇒ (select datum **A** as the second reference and type **3.725** at the prompt) ⇒ **enter** ⇒ **One Side** ⇒ **Done** ⇒ **UpTo Surface** ⇒ **Done** ⇒ (pick the **∅.250** hole by using **Query Sel** to filter through the part to get to the hole) ⇒ **Accept** ⇒ (type **.125** as the hole's diameter) ⇒ **enter** ⇒ **OK** (from dialog box) ⇒ **Done**

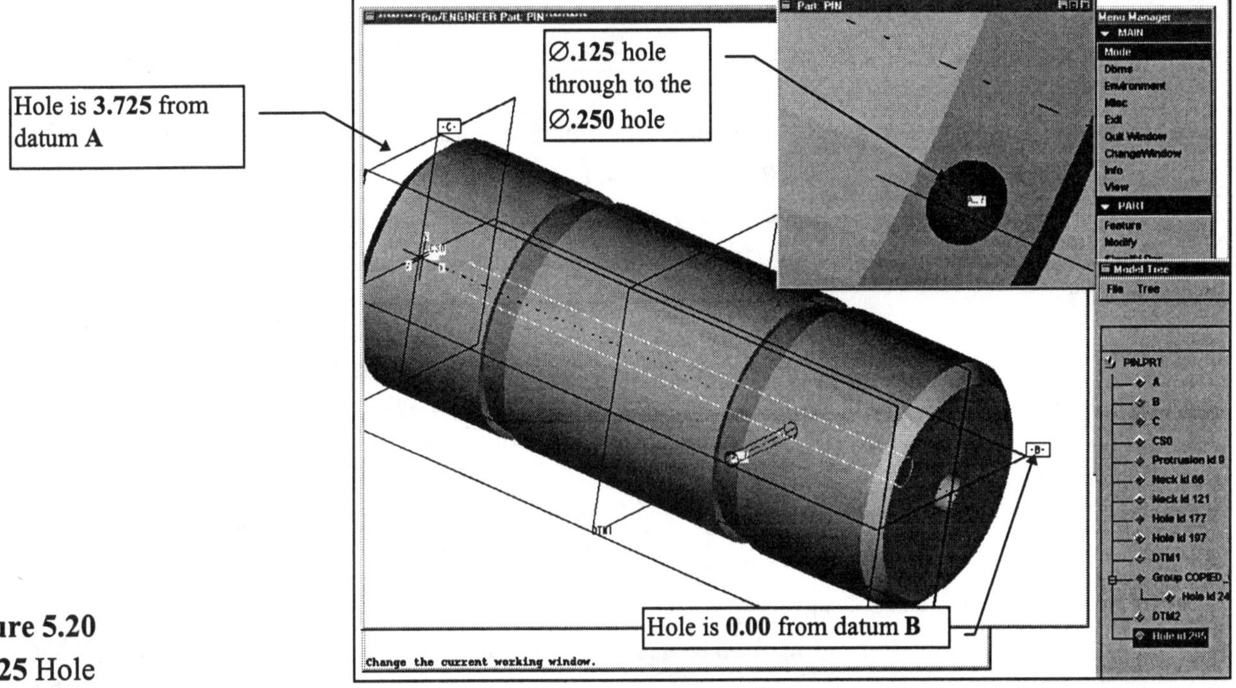

Figure 5.20
∅.125 Hole

Now create the **∅.250** hole on the part's side (you will mirror and copy both holes later). The command is exactly the same as the previous command, for the **∅.125** hole, except that the diameter is **.250**, its distance from datum plane **A** is **4.50**, and it is a *blind* hole with a depth of **.3120** (Fig. 5.21).

The holes can now be copied and mirrored about the same datum plane that was used for mirroring the conical hole (Fig. 5.6). Use the following commands:

PT/Modeler™

Feature Oper ⇒ Copy ⇒ Mirror ⇒ Dependent etc.

Feature ⇒ Copy ⇒ Mirror ⇒ Dependent ⇒ Done ⇒ (select both the **∅.125** and the **∅.250** holes just created) ⇒ **Done Sel ⇒ Done** ⇒ (pick **DTM1** as the plane to mirror about) ⇒ **Done** (Fig. 5.21)

Dbms (**File**--PT/Modeler) ⇒
Save ⇒ **enter**
Purge ⇒ **enter** ⇒
Done-Return

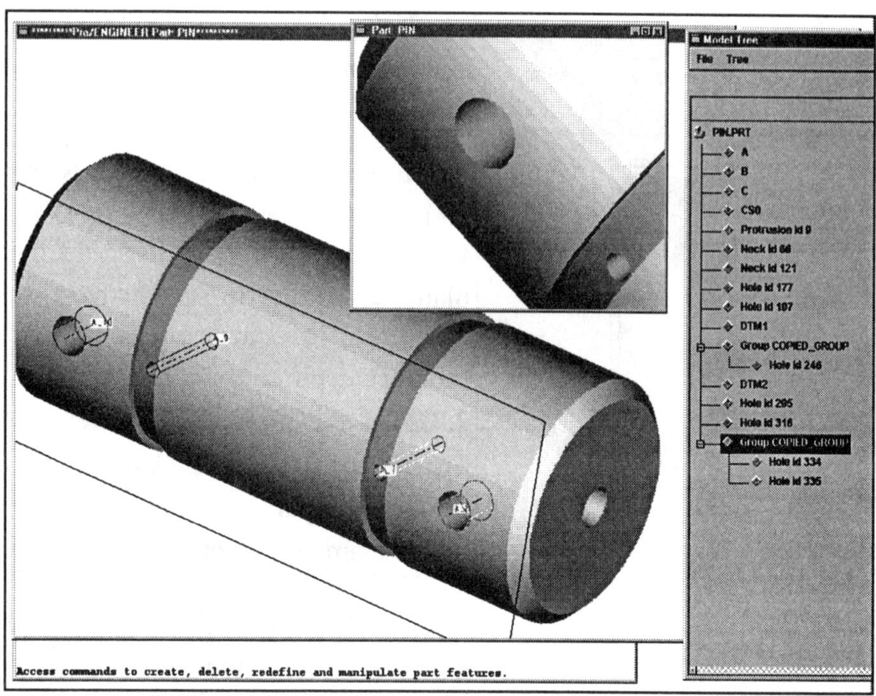

Figure 5.21
Mirrored Holes

PT/Modeler™
choose **Plotter** from the PART MENU

To plot your part or drawing, choose **Interface** ⇒ (from the PART menu) ⇒ **Export** ⇒ **Plotter** (select your plotter or printer). Since every system is different and there is a wide variety of printers and plotters available, you must ask your system manager or instructor for help in plotting. Read Section 7 for more information on plotting.

To get information about your part, choose **Info** (from the MAIN menu) ⇒ **Feat Info** (pick the revolved protrusion from the Model Tree as shown in Fig. 5.22) ⇒ **Done/Return**.

? Pro/HELP

Highlight the **Interface** command, pick with your right mouse button, and choose ?GetHelp

To close the INFORMATION WINDOW

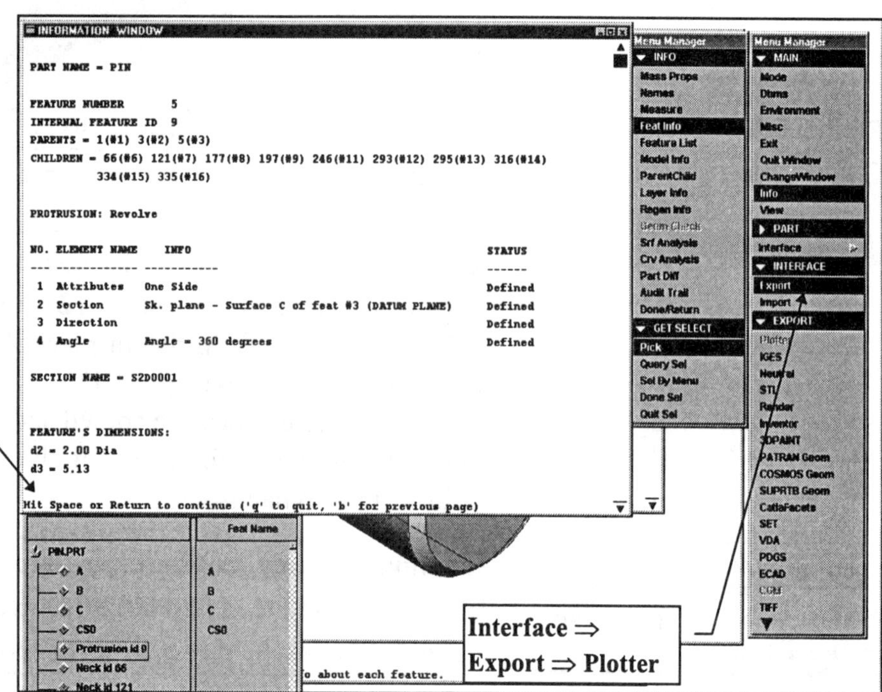

Figure 5.22
Feature Information

Lesson 5 Project

Clamp Foot and Clamp Swivel

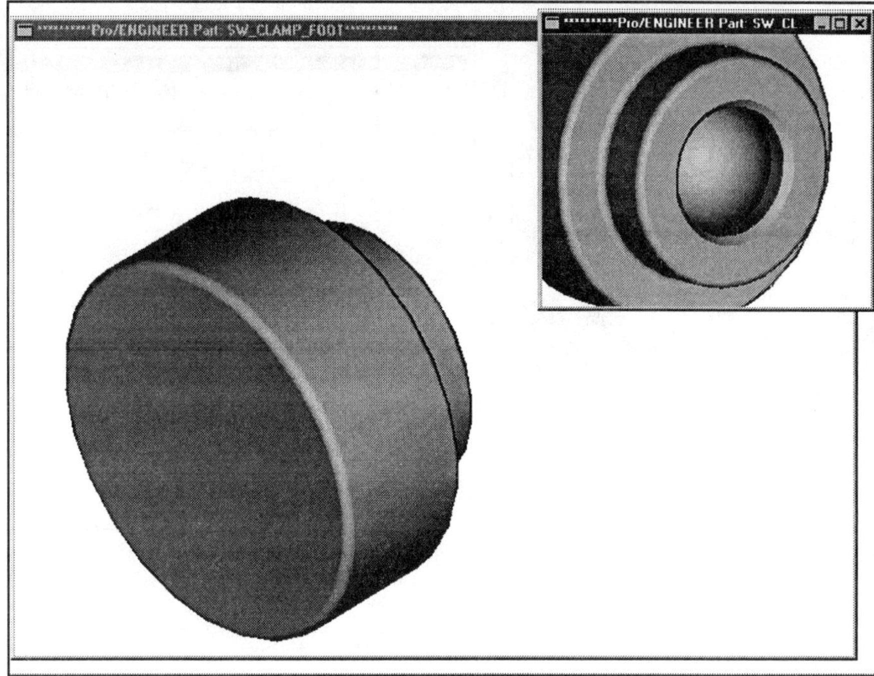

Figure 5.23
Clamp Foot

Clamp Foot and Clamp Swivel

Two lesson projects are provided in Lesson 5. You will use both of these parts in Lessons 14 and 15 when creating an assembly. Both the foot and the swivel are simple revolved protrusions. The Foot is nylon and the Swivel is steel. The Swivel fits inside the Foot.

Analyze each part and plan out the steps and features required to model it. Use the **DIPS** in Appendix D to establish a feature creation sequence before the start of modeling. Remember to set up the environment, establish datum planes, and set them on layers.

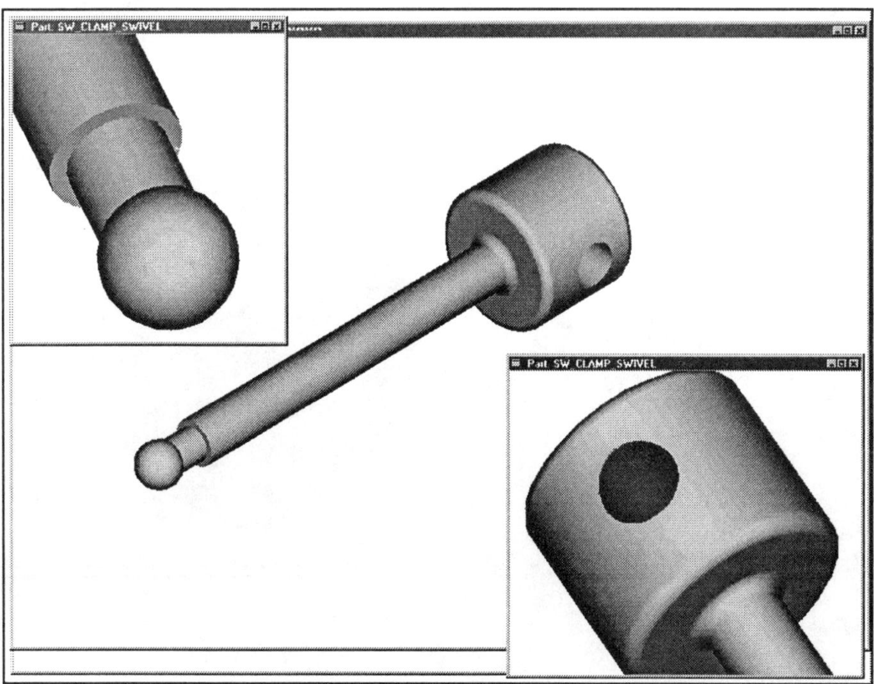

Figure 5.24
Clamp Swivel

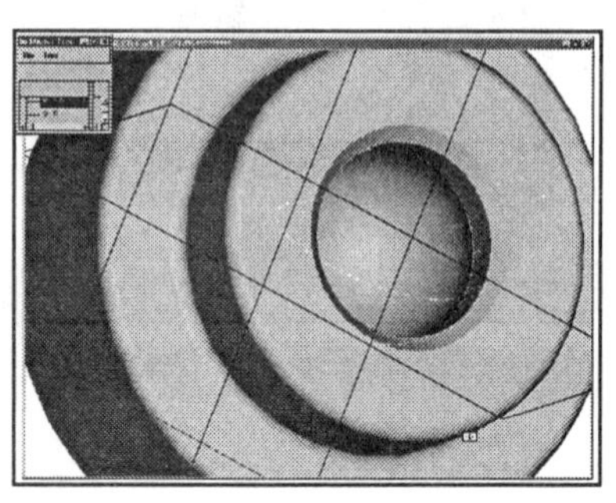

Figure 5.25
Clamp Foot Dimensions

Create the two parts (Figs. 5.23 through 5.30) with revolved protrusions using datum **C (DTM1)** as the sketching plane. Create the internal cut on the **Foot** with a *revolved cut*. Add the rounds on both parts at the end of the modeling process; do not include them on the first revolved protrusions.

Dbms (**File**--PT/Modeler) ⇒
Save ⇒ **enter**
Purge ⇒ **enter** ⇒
Done-Return

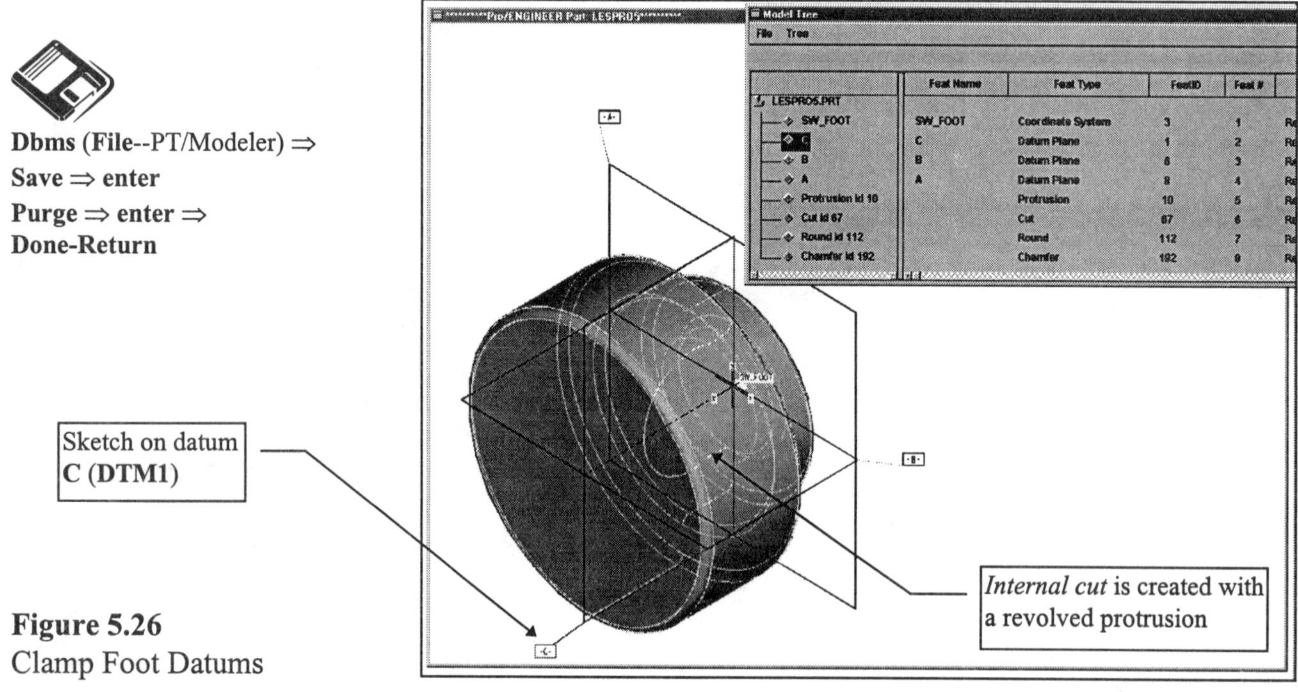

Figure 5.26
Clamp Foot Datums

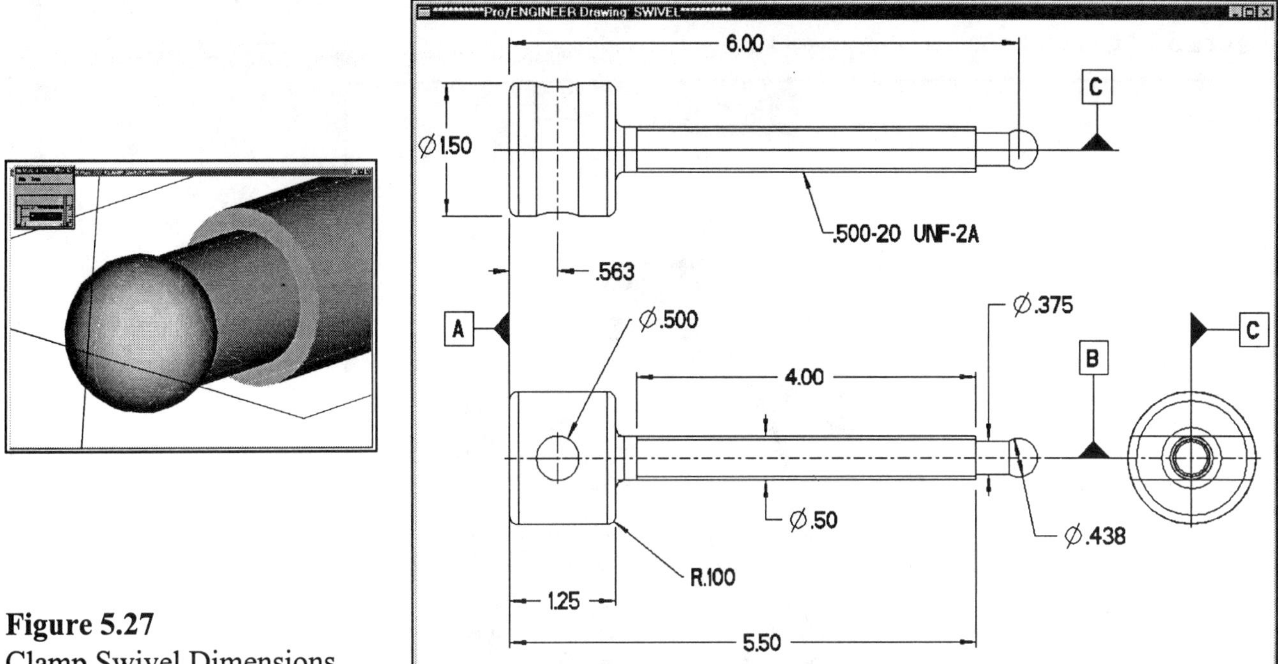

Figure 5.27
Clamp Swivel Dimensions

Figure 5.28 Clamp Swivel Datums

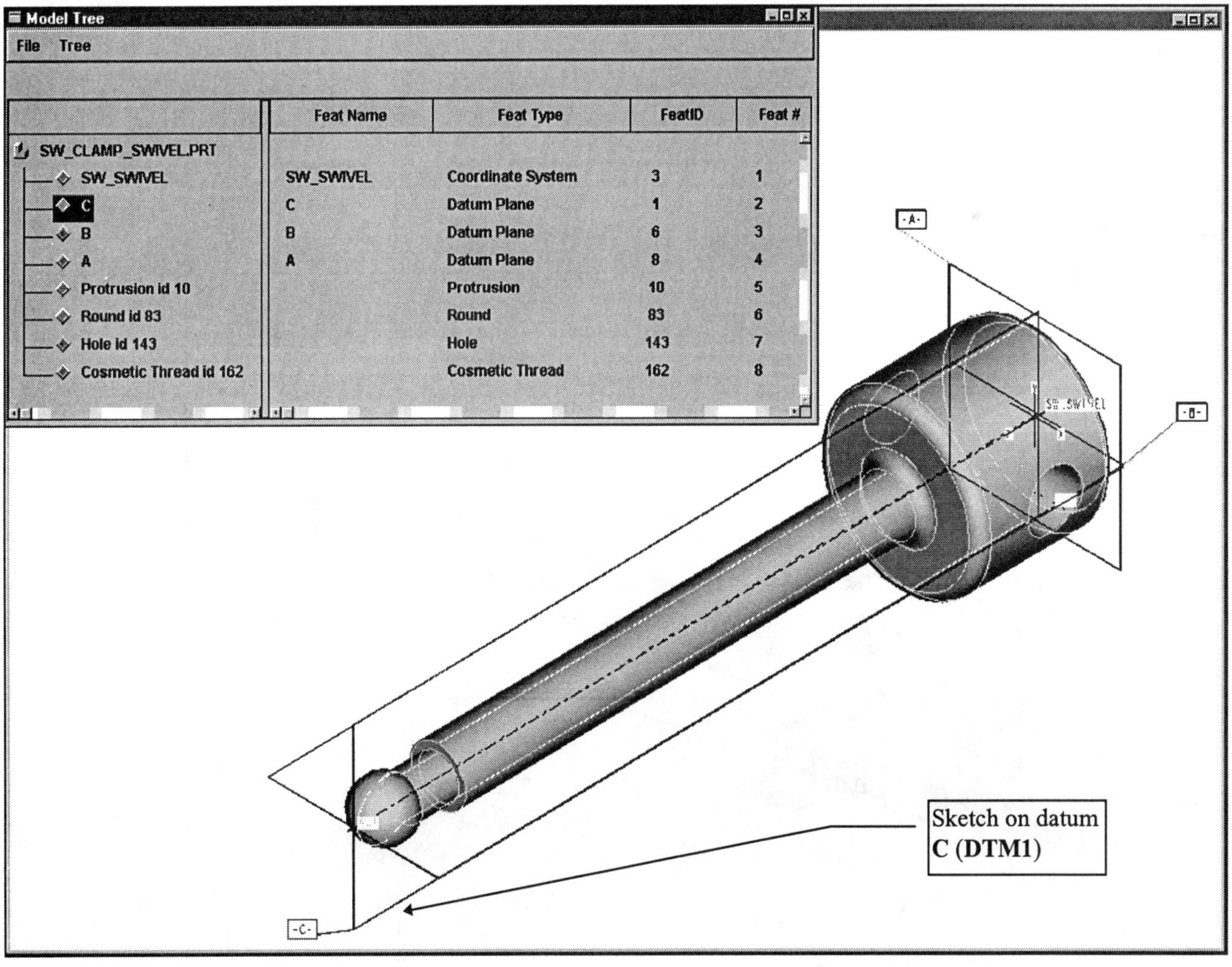

Model Tree

File Tree

SW_CLAMP_SWIVEL.PRT

	Feat Name	Feat Type	FeatID	Feat #
SW_SWIVEL	SW_SWIVEL	Coordinate System	3	1
C	C	Datum Plane	1	2
B	B	Datum Plane	6	3
A	A	Datum Plane	8	4
Protrusion id 10		Protrusion	10	5
Round id 83		Round	83	6
Hole id 143		Hole	143	7
Cosmetic Thread id 162		Cosmetic Thread	162	8

-A-

-B-

-C-

SW_SWIVEL

Sketch on datum **C (DTM1)**

Figure 5.29 Clamp Foot

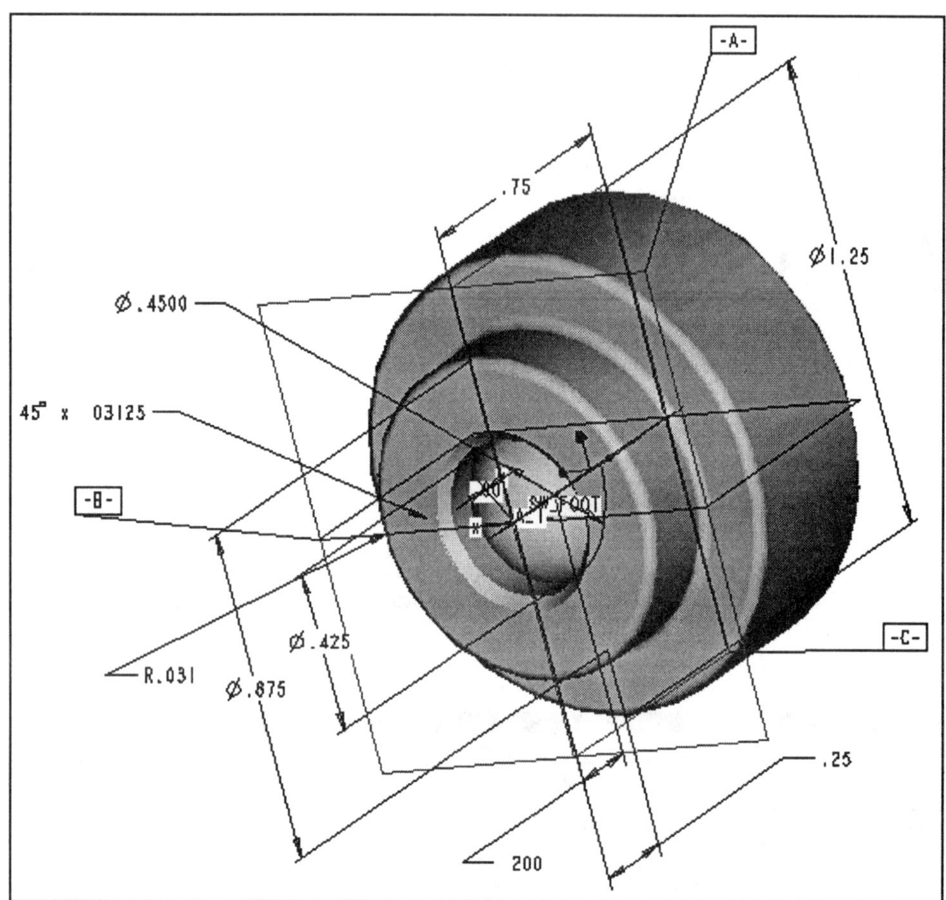

Figure 5.30 Clamp Swivel

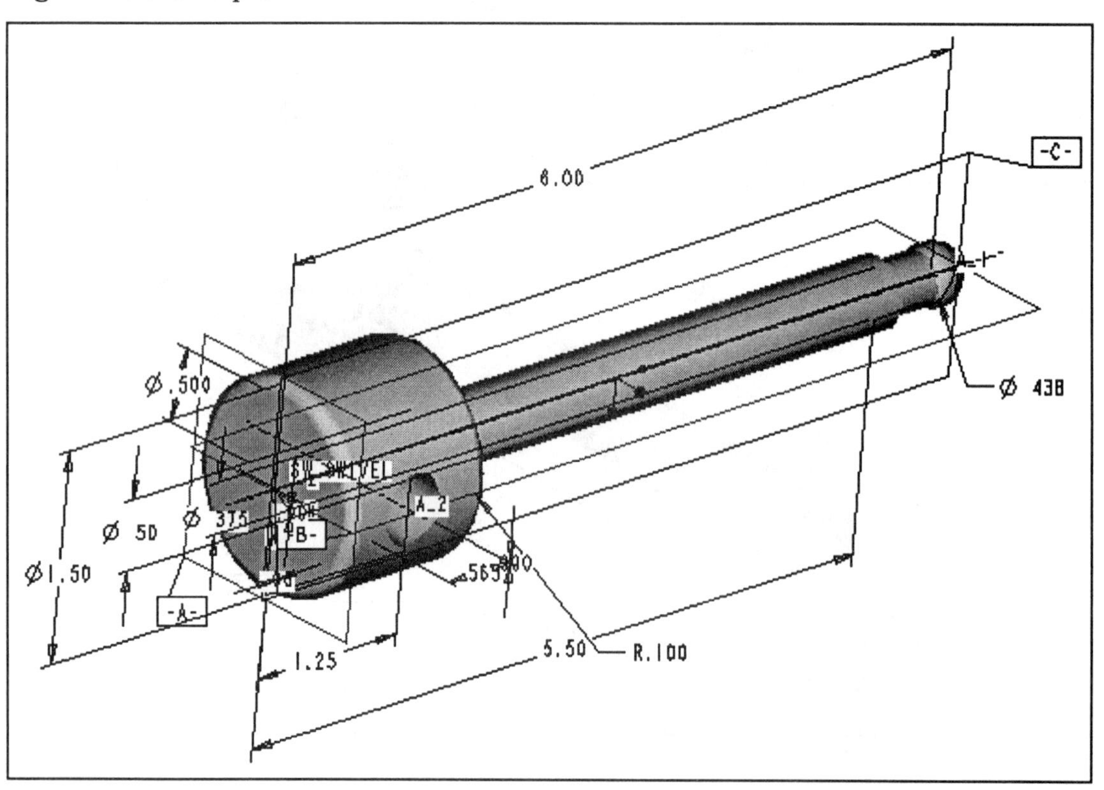

Lesson 6

Chamfers and Cosmetic Threads

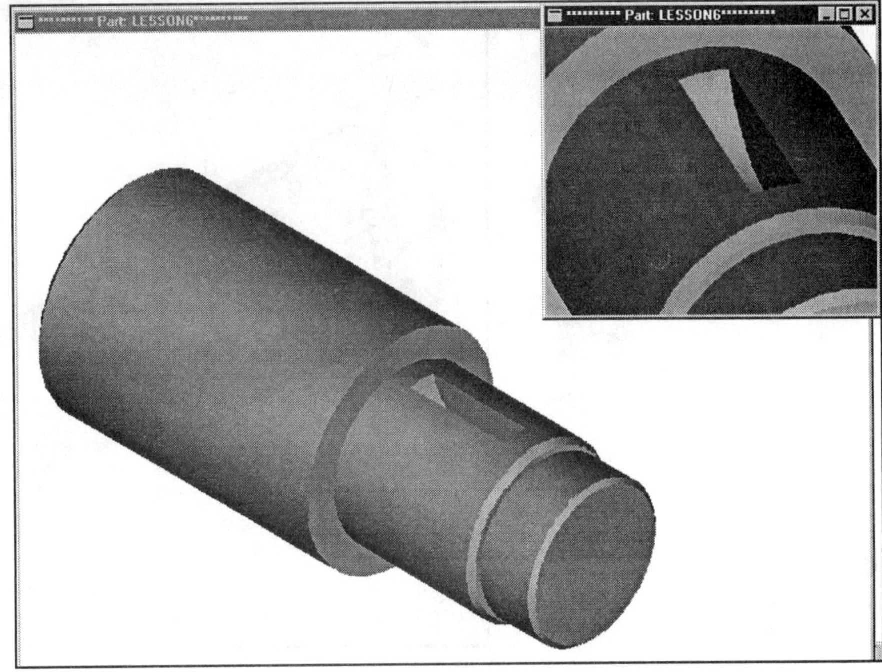

Figure 6.1
Cylinder Rod

OBJECTIVES

1. **Create simple chamfers along part edges**
2. **Learn how to sketch in 3D**
3. **Create cosmetic threads**
4. **Complete tabular information for threads**
5. **Get information on existing cosmetic threads**
6. **Dynamically modify sketch dimension values**

EGD REFERENCE

Engineering Graphics and Design with Graphical Analysis *or* **Fundamentals of Engineering Graphics and Design**
by L. Lamit and K. Kitto
Read Chapters 14 and17
See pages 497, 540-541, and 674-678

COAch™ for Pro/ENGINEER

If you have **COAch for Pro/ENGINEER** on your system, go to SEARCH and do the Segment shown in Figure 6.3.

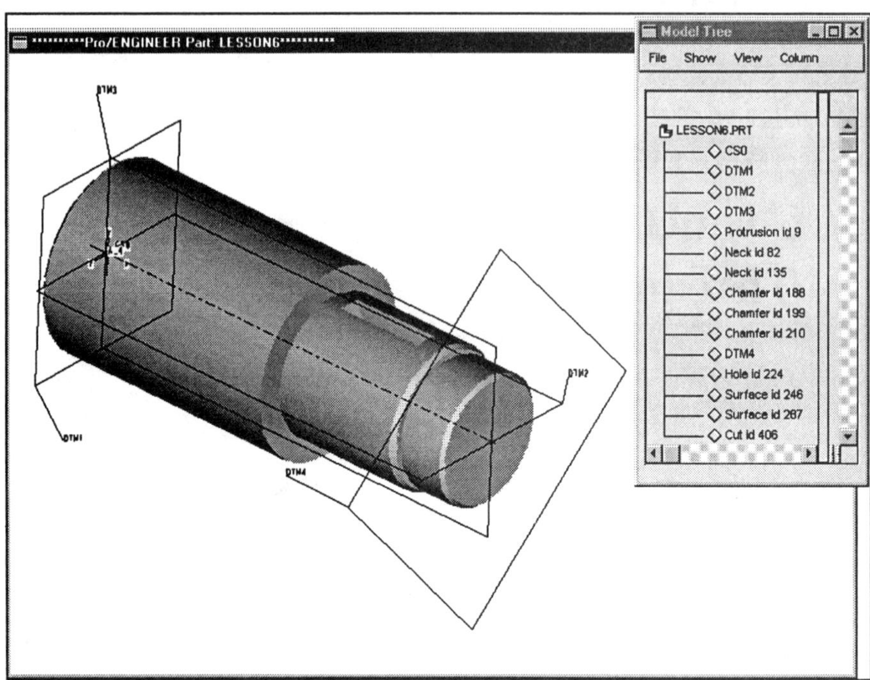

Figure 6.2
Cylinder Rod with Datums and Model Tree

CHAMFERS AND COSMETIC THREADS

A variety of geometric shapes and constructions are accomplished with a CAD system using parametric modeling. For instance, **chamfers** are created at selected edges of the part (Figs. 6.1 and 6.2). Chamfers are *pick-and-place* features (Fig. 6.3).

Threads are usually a *cosmetic feature* representing the *nominal diameter* or the *root diameter* of the thread. Information can be embedded in the feature. Threads show as a unique color (magenta). By putting cosmetic threads on a separate layer, you can display, blank, or suppress them as required.

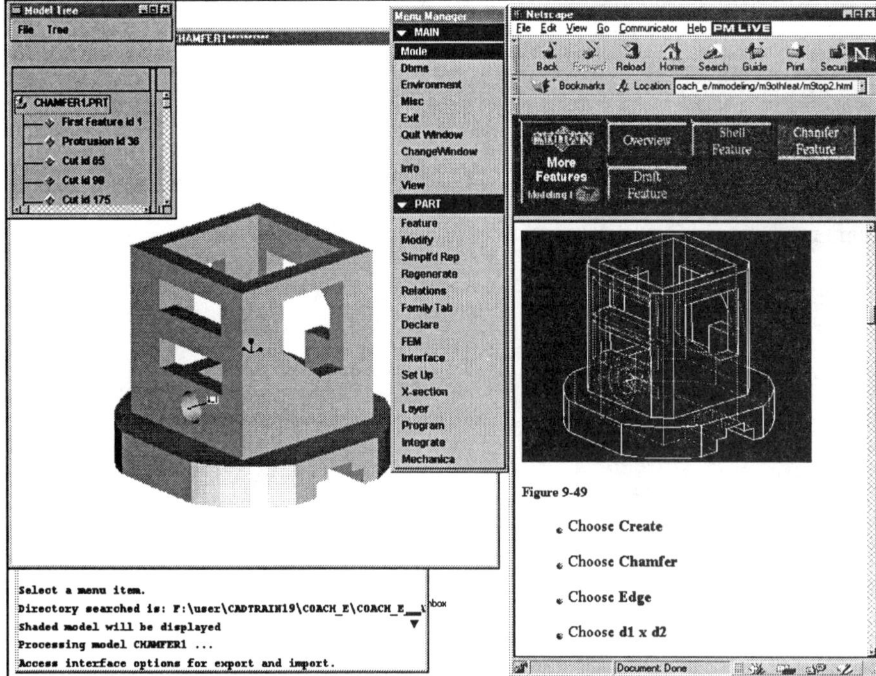

Figure 6.3
COAch for Pro/E, More Features (Chamfer Feature)

Chamfers

Chamfers are created between abutting edges of two surfaces on the solid model. An **edge chamfer** removes a flat section of material from a selected edge to create a beveled surface between the two original surfaces common to that edge (Fig. 6.4). Multiple edges can be selected.

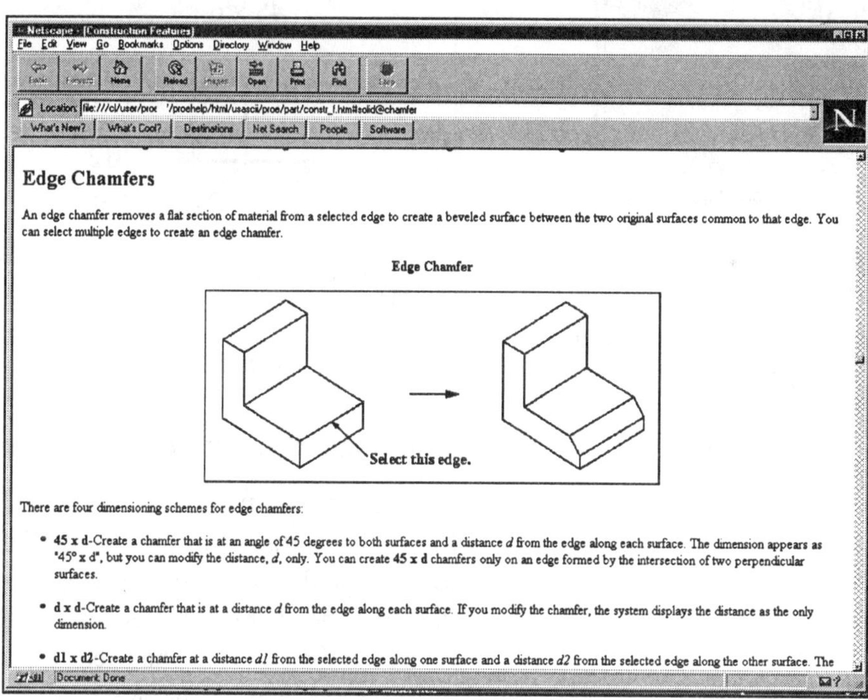

Figure 6.4
Online Documentation, Edge Chamfers

There are four dimensioning schemes for edge chamfers, as shown in Figure 6.5:

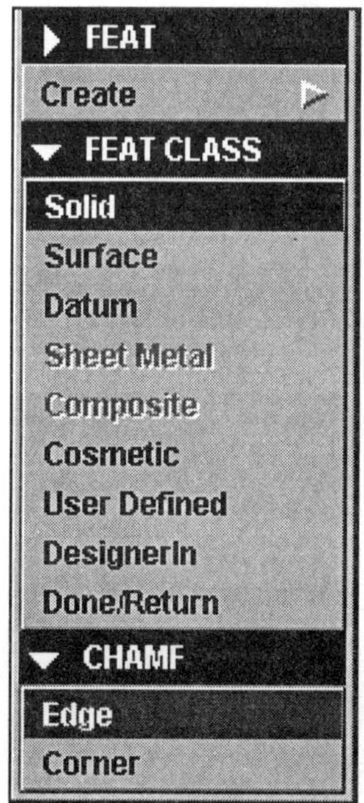

45 x d Creates a chamfer that is at an angle of **45°** to both surfaces and a distance **d** from the edge along each surface. The distance is the only dimension to appear when modified. **45 x d** chamfers can be created only on a edge formed by the intersection of two *perpendicular* surfaces.

d x d Creates a chamfer that is a distance **d** from the edge along each surface. The distance is the only dimension to appear when modified.

d1 x d2 Creates a chamfer at a distance **d1** from the selected edge along one surface and a distance **d2** from the selected edge along the other surface. Both distances appear along their respective surfaces when modified.

Ang x d Creates a chamfer at a distance **d** from the selected edge along one adjacent surface at a specified angle to that surface.

The dimensioning schemes appear as options in the SCHEME menu. The SCHEME menu appears after **Edge** is chosen from the CHAMF menu.

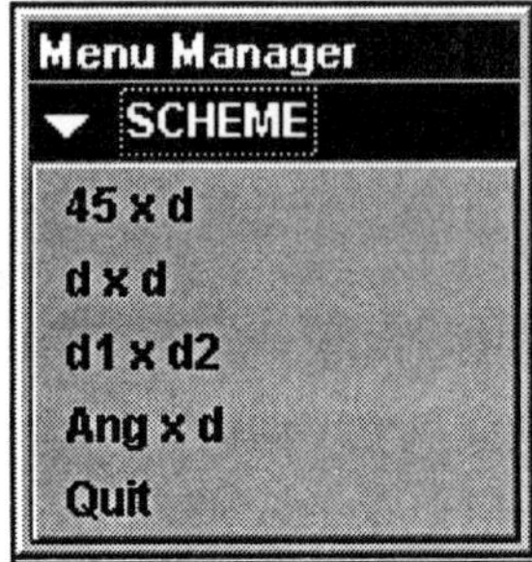

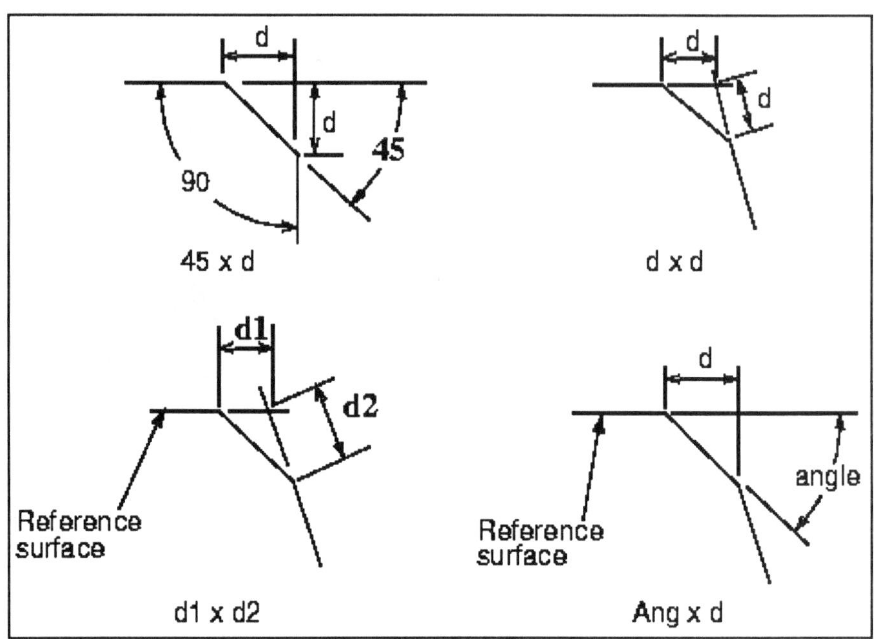

Figure 6.5
Online Documentation, Chamfer Dimensioning Schemes

To create a **45 x d** and **d x d** edge chamfer:

1. Choose **Chamfer** from the SOLID menu.
2. Choose **Edge** from the CHAMF menu.
3. Choose the **45 x d** or **d x d** option.
4. Enter the chamfer dimension.
5. Select the edges to chamfer. Remember that for a **45 x d** edge chamfer, the surfaces bounding an edge must be at **90°** to each other.

To create a **d1 x d2** chamfer:

1. Choose **Chamfer** from the SOLID menu.
2. Choose **Edge** from the CHAMF menu.
3. Choose the **d1 x d2** option.
4. Input a distance along a surface to be selected.
5. Input a second distance
6. Pick the surface along which the first distance will be measured; pick the edge to chamfer.

To create an **Ang x d** chamfer:

1. Choose **Chamfer** from the SOLID menu.
2. Choose **Edge** from the CHAMF menu.
3. Choose the **Ang x d** option.
4. Input distance.
5. Input an angle from a surface to be selected.
6. Select the reference surface from which the values will be measured.
7. Pick the edge(s) to chamfer.

CHAMFER: Edge Chamfer

Element	Info
>Scheme	Defining
Ref Surface	Required
Edge Refs	Required

Define Refs Info
OK Cancel Preview

To create a corner chamfer (Fig. 6.6):

1. Choose **Chamfer** from the SOLID menu, then choose **Corner** from the CHAMF menu.
2. Select the corner you want to chamfer.
3. Pro/E displays the PICK/ENTER menu, which allows you to specify the location of the chamfer vertex on the highlighted edge. The PICK/ENTER menu options are as follows:

   Pick Point Pick a point on the highlighted edge to define the chamfer distance along that edge.

   Enter input Type in a value for the chamfer distance along the highlighted edge.

4. Pick or enter values to describe the chamfer lengths along the edge. After you have selected the first vertex, Pro/E highlights the other edges, one at a time, so you can place the other two vertices.
5. Select the **OK** button in the dialog box.

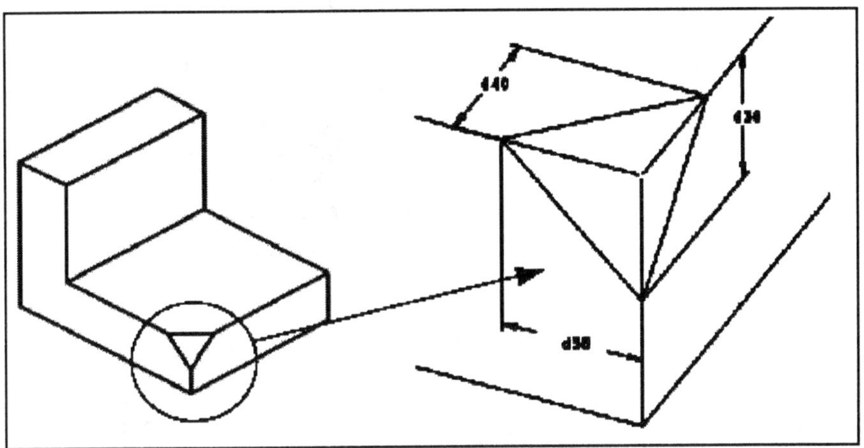

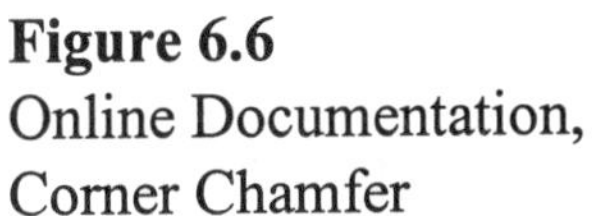

Figure 6.6
Online Documentation, Corner Chamfer

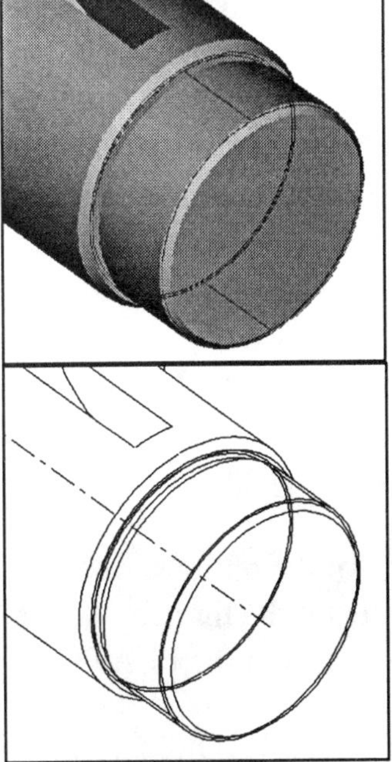

Threads

Cosmetic threads (Fig. 6.7) are displayed with magenta lines and circles. Cosmetic threads can be external or internal, blind or through. In the Rod part, one end has external blind threads and the opposite end has internal blind threads.

A thread has a set of supported parameters that can be defined at its creation or later, when the thread is added.

The following parameters can be defined for a thread:

PARAMETER DESCRIPTION	PARAMETER NAME	PARAMETER VALUE
Thread major diameter	MAJOR_DIAMETER	Number
Threads per inch (1/pitch)	THREADS_PER_INCH	Number
Thread form	THREAD_FORM	String
Thread class	CLASS	Number
Thread placement (A-external, B-internal)	PLACEMENT	A or B
Thread is Metric	METRIC	TRUE/FALSE

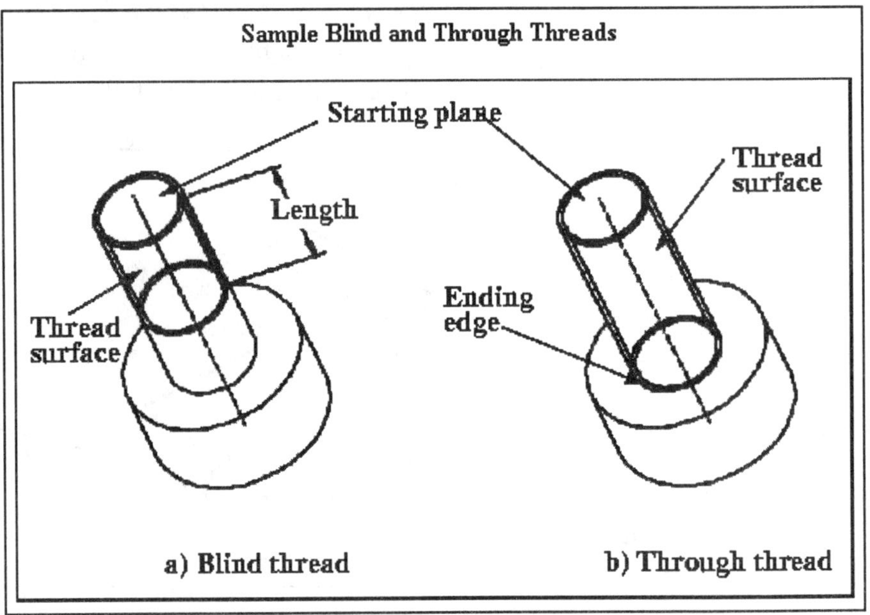

Figure 6.7
Online Documentation, Cosmetic Threads

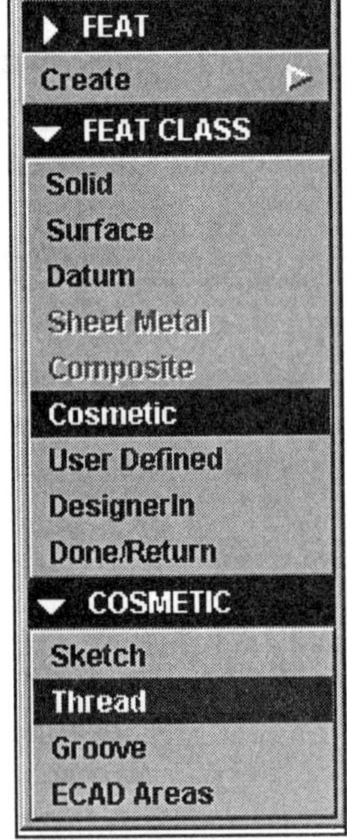

To create a cosmetic thread feature:

1. Choose **Create ⇒ Cosmetic ⇒ Thread.**
2. Pick the circular internal or external thread surface at the prompt.
3. Pro/E automatically knows whether the threads are internal or external, based on the feature selected.
Select the thread start surface, then **Flip** or **Okay** for the direction as needed.
4. From the SPEC TO menu pick **Blind, UpTo Pnt/Vtx, UpTo Curve**, or **UpTo Surface** and then **Done.**
5. Follow the prompts, which differ depending on Step 4.
6. Enter the diameter at the prompt.
7. From the FEAT PARAM menu, select one of the options. In general you will be picking **Mod Params** to enter the thread parameters into a file using Pro/TABLE. Select from the FEAT PARAM menu:

Retrieve Retrieves a previously created and saved thread file.

Save After completing the table, to save your thread file for use later.

Mod Params Modifies thread parameters in the Pro/TABLE environment (Fig. 6.8).

Show Displays a set of thread parameters in the INFORMATION WINDOW.

Done/Return To complete the process, exit from this menu.

8. When finished, choose **Done/Return** to continue, then **OK**.

Pro/TABLE TM Release 19.0 (c) 1988-95 by Parametric Technology Corp

File Edit View Format Help

! Thread parameters and callouts

! Thread parameters and callouts	
! Thread Major Diameter	
MAJOR_DIAMETER	2.75
! Threads Per Inch (Pitch)	
THREADS_PER_INCH	16
! Thread Form	
FORM	UNF
! Thread Class	
CLASS	2
! Thread Placement (A=external, B=internal)	
PLACEMENT	A
! Thread Is Metric (True or False)	
METRIC	FALSE

Figure 6.8
Using Pro/TABLE to Input Thread Parameters

Thread parameters can be manipulated like other user-defined parameters: they can be added, modified, deleted, or displayed using menu options.

The following information was extracted using the commands **Info** ⇒ **Feat Info** ⇒ (select the thread). Information similar to the following appears in a window:

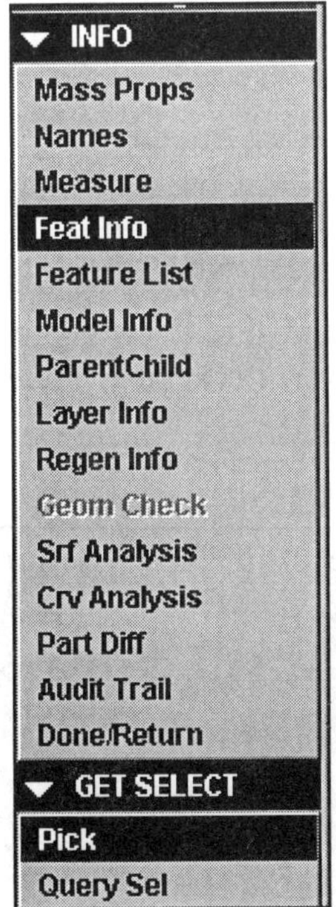

PART NAME =	ROD
FEATURE NUMBER	13
INTERNAL FEATURE ID	246
PARENTS =	9(#5) 224(#12)
TYPE THREAD	
FORM =	360 DEG. REVOLVED
SECTION NAME =	S2D0003
OPEN SECTION	
FEATURE DIMENSIONS:	
d28(d11) =	1.25
d29(d12) =	.63 Dia.
MAJOR_DIAMETER	.625
THREADS_PER_INCH	11
FORM	UNC
CLASS	2A

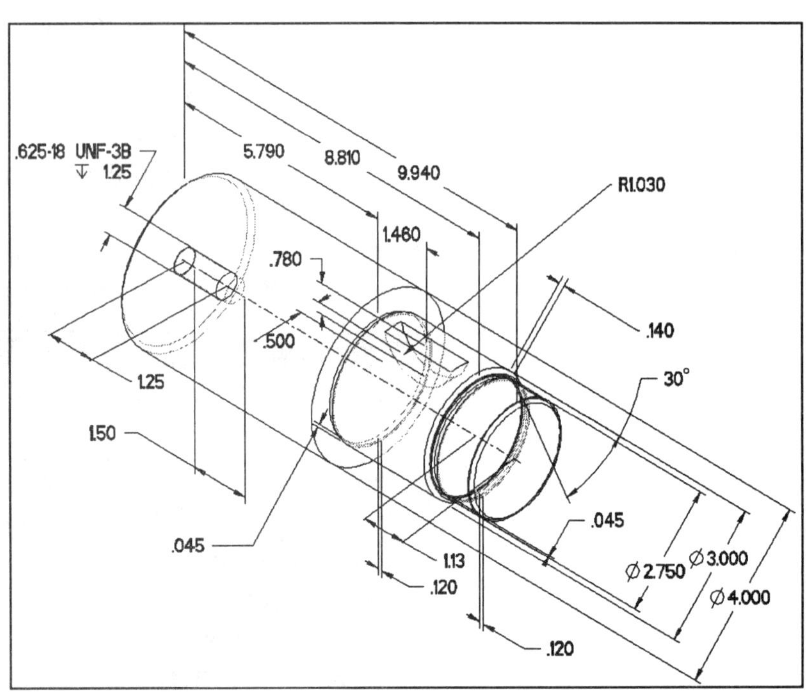

Figure 6.9
Cylinder Rod Dimensions
(Since this is a part model shown in default view, with dimensions displayed using **Modify**, not a dimensioned pictorial, many of the dimensions are rounded. Drawing dimensions for the cosmetic threads and construction datum planes are also displayed.)
Do not use this illustration for creating the part model. Use the step-by-step information provided.

Cylinder Rod

The **Cylinder Rod** is modeled by creating a revolved protrusion, similar to the Pin in Lesson 5. The geometry of the revolved feature is shown in Figure 6.9 (also see Fig. 6.38). After the revolved protrusion (base feature) is created, the necks (reliefs), the chamfers, the key seat, and the tap drill hole are modeled. In this lesson, we will create our first revolved protrusion by sketching in 3D.

Three chamfers are required for this part. The **45 x d** option was used to chamfer the left side (**∅4.00**) of the part (**45° x .125**). A **45° x .09** chamfer is added to the right side (**∅2.75**) of the Rod and a **30° x .14** chamfer is used on the **∅3.00** step of the Rod at the relief. Two necks are required; both are **.120 x .045 DEEP**.

The cosmetic threads for the external threaded shaft end and the internal hole threads are added last. They are created by specifying the minor or major diameter (for external or internal threads, respectively), starting plane (for the external threads; a **DTM4** was used for the external threads starting plane), and thread length or ending edge. The internal threaded hole is **.625-16 UNF-3B** by **1.25 DEEP.** The external threaded shaft is **2.75-16 UN-2A**.

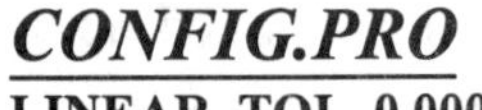
CONFIG.PRO

LINEAR_TOL_0.000

SKETCHER_DEC_PLACES 3

SKETCHER_DISPLAY_ CONSTRAINTS YES

DEFAULT_DEC_PLACES 3

NOTE

SETUP AND ENVIRONMENT blocks are for your information only; they are not meant to be input at this time.

SETUP AND ENVIRONMENT

Set Up ⇒ **Units** ⇒ **Length** ⇒ **Inch** ⇒ **Done** ⇒ **Material** ⇒ **Define** ⇒ (Type **Aluminum**, then press **enter**) ⇒ (table of material properties, change or add information) ⇒ **File** ⇒ **Save** ⇒ **File** ⇒ **Exit** ⇒ **Assign** ⇒ (pick **Aluminum**) ⇒ **Accept** ⇒ **Done** **Environment** ⇒ ✓**Grid Snap** □ **SketStart2D** (we will be sketching in 3D) **Hidden line Tan Dimmed**

Start the part with the usual commands (Fig. 6.10):

PT/Modeler™
Feature ⇒ Datum Plane ⇒ Coord Sys ⇒ Default ⇒ Done ⇒

Part ⇒ Create ⇒ CYLINDER_ROD ⇒ enter
Feature ⇒ Create ⇒ Datum ⇒ Plane ⇒ Default ⇒ Create ⇒ Datum ⇒ Coord Sys ⇒ Default ⇒ Done ⇒ Done

HINT

Set Up and **Environment** are set *after* Part or Assembly mode has been entered.

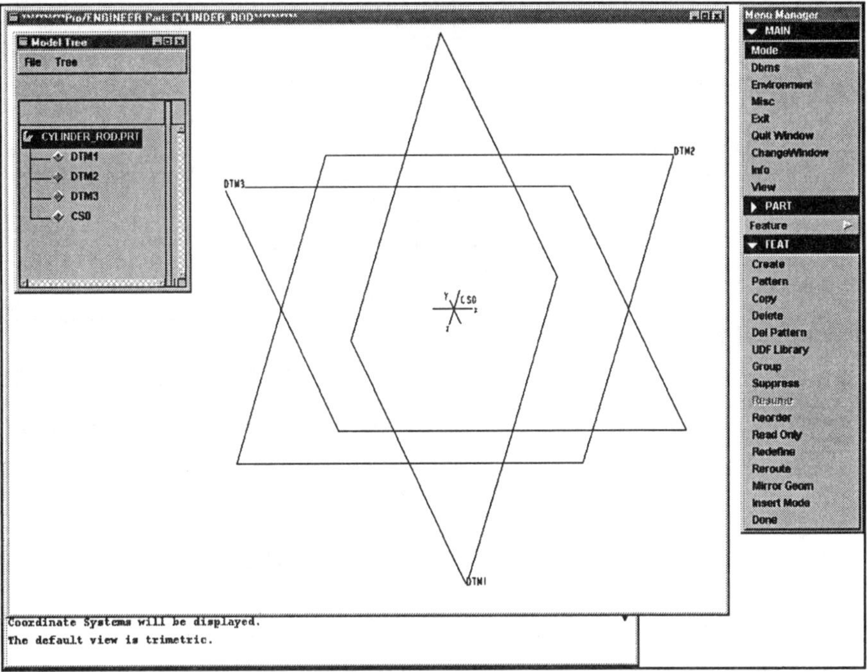

Figure 6.10
Cylinder Rod Coordinate System and Datum Planes

Create a layer for the datum planes and the coordinate system:

Layer ⇒ Create ⇒ (type **DATUM_LAYER) ⇒ enter ⇒ enter ⇒ Set Items ⇒ Add Items ⇒ (✓DATUM_LAYER) ⇒ Done Sel ⇒ Feature ⇒ Sel By Menu ⇒ Name ⇒ Sel All ⇒ Ok ⇒ Done Sel ⇒ Done/Return ⇒ Done/Return**

The first protrusion is a revolved feature, and you will be sketching the section in 3D for the first time:

PT/Modeler™
Feature ⇒ Protrusion etc.

Feature ⇒ Create ⇒ Solid ⇒ Protrusion ⇒ Revolve ⇒ Done ⇒ One Side ⇒ Done ⇒ (pick **DTM3** as sketching plane) **⇒ Okay ⇒ Top** (pick **DTM2** as horizontal reference) **⇒ View ⇒ Pan/Zoom ⇒ Pan** [pan the screen (Fig. 6.11)] **⇒ Done-Return ⇒ Line ⇒ Centerline ⇒ Horizontal** (sketch the horizontal centerline *first*) **⇒ Sketch ⇒ Mouse Sketch** [sketch the section geometry (Fig. 6.11)] **⇒ Regenerate ⇒ Alignment** (align the horizontal centerline with **DTM2**, the horizontal line with **DTM2**, and the left side vertical line with **DTM1**, as in Fig. 6.12) **⇒ Regenerate ⇒ Dimension** [use dimension scheme (Fig. 6.13)] **⇒ Regenerate ⇒ Modify** (modify to design dimensions shown in Figs. 6.9 and 6.38) **⇒ Regenerate**

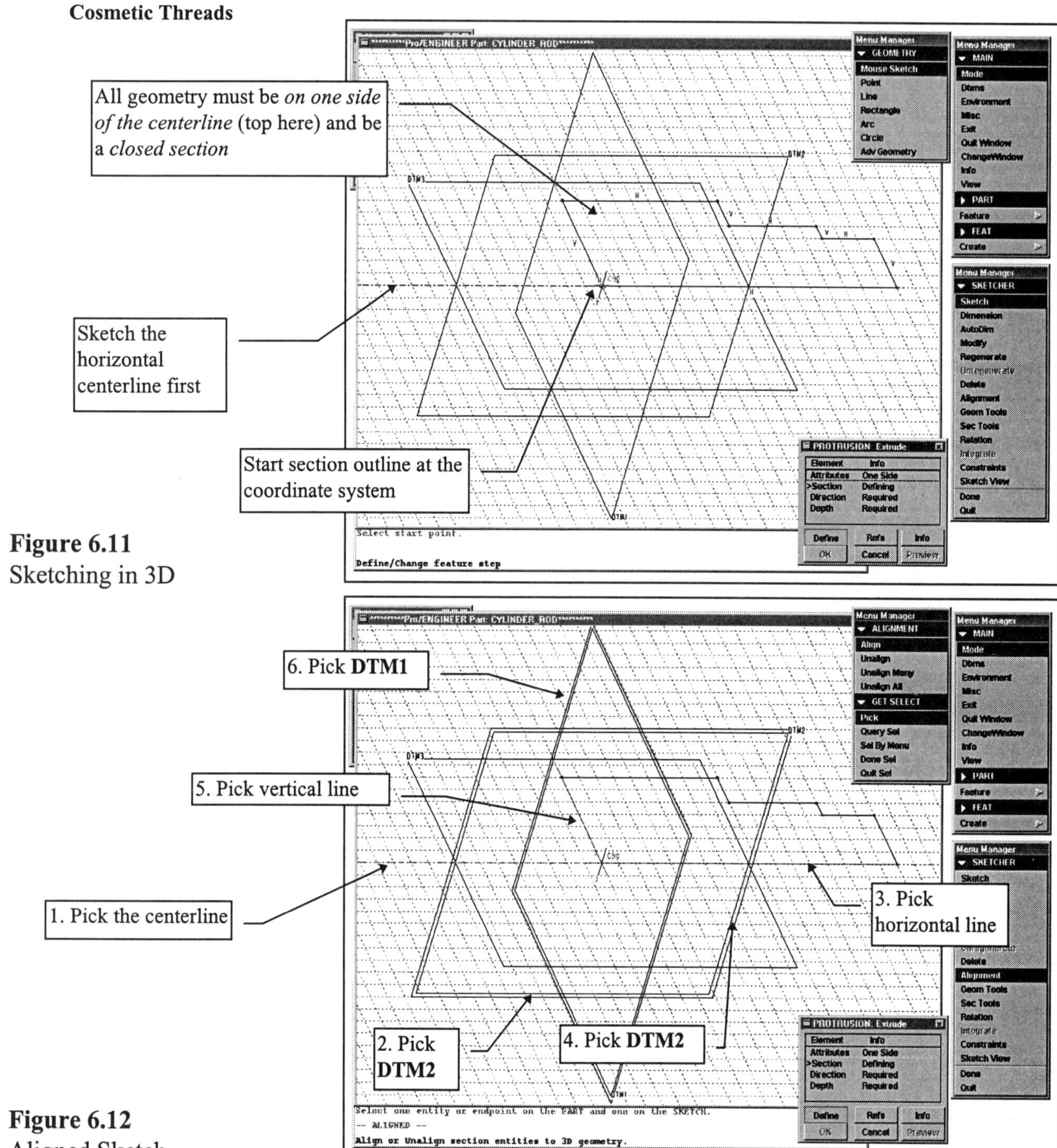

Figure 6.11
Sketching in 3D

Figure 6.12
Aligned Sketch

HINT

You can toggle between sketching in 3D and 2D. If you want to go to 2D, choose **Sketch View**.

You can dynamically change any dimension on the sketch. This capability is available for 3D and 2D sketching. After the sketch has been regenerated with the design dimensions (Fig. 6.9), choose the following commands:

Modify ⇒ **Drag Dim Val** ⇒ (pick the **9.94** dimension) ⇒ **Done Sel** ⇒ (the **Modify Dims** Scale Menu will appear on the screen) ⇒ (move the mouse pointer into the region under the title "**Linear**," pick once with the left mouse button, slide the linear indicator to the right until the dimension is about **12.25**, as shown in Fig. 6.14, then pick once more with the left mouse button to finish)

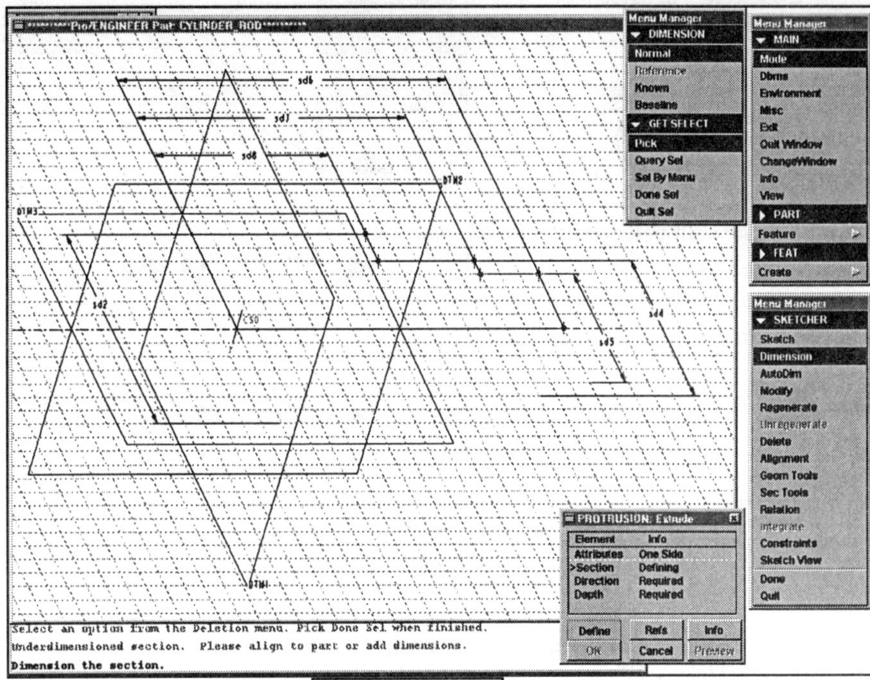

Figure 6.13
Section with Sketch Dimensions

Slide the bar to the right to increase the dimension size

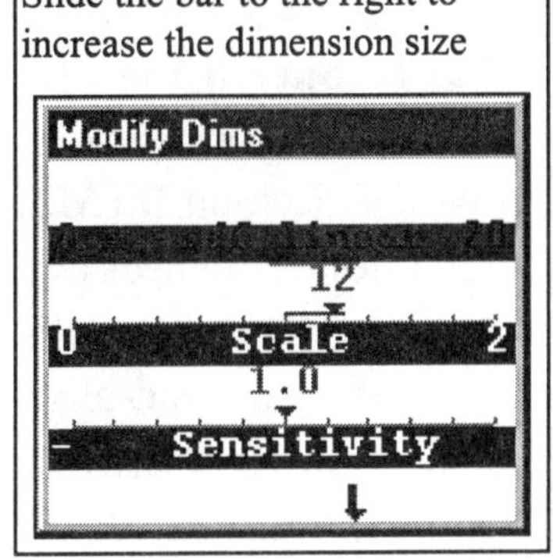

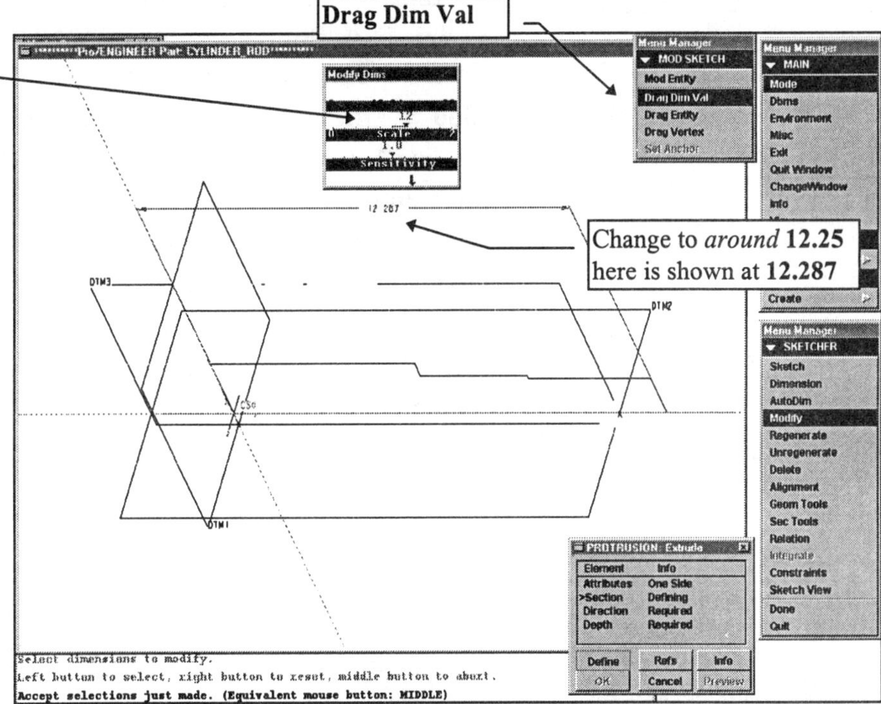

Figure 6.14
Dynamically Changing a Dimension Value

The sketch will dynamically change with the length value. After experimenting with this capability, **Regenerate** the sketch again, **Modify** the length back to the design dimension of **9.94**, and **Regenerate** once more. Complete the protrusion by choosing the following commands:

Done ⇒ **360** (from the REV TO menu) ⇒ **Done** ⇒ **OK** ⇒ **View** ⇒ **Cosmetic** ⇒ **Shade** ⇒ **Display** ⇒ **Done-Return** (Fig. 6.15) ⇒ **Done**

Dbms (File--PT/Modeler) ⇒ **Save** ⇒ **enter** **Purge** ⇒ **enter** ⇒ **Done-Return**

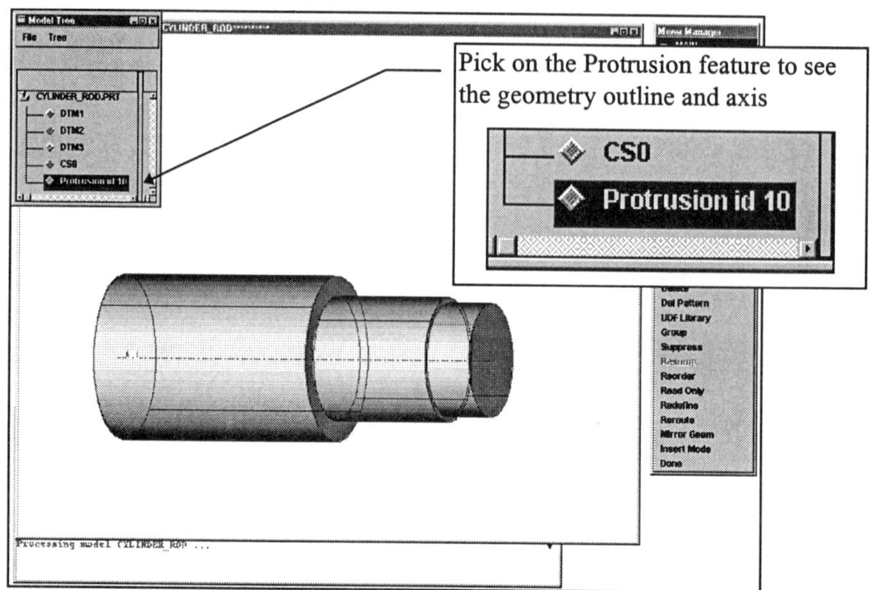

Figure 6.15
Revolved Protrusion

The next two features are neck cuts similar to those for the Pin in Lesson 5:

PT/Modeler™
As stated previously, PT/Modeler does not support the **Neck** command. Use a revolved **Cut** instead.

View ⇒ **Repaint** ⇒ **Done-Return** ⇒ **Environment** ⇒ **Shading** ⇒ □ **Grid Snap** ⇒ **Done-Return** ⇒ **Feature** ⇒ **Create** ⇒ **Neck** ⇒ **360** ⇒ **One Side** ⇒ **Done** ⇒ **Use Prev** ⇒ **Okay** ⇒ **Sketch View** (to go into a 2D view of the section) ⇒ **Sketch** ⇒ **Line** ⇒ **Centerline** ⇒ **Horizontal** (sketch the centerline used to revolve the neck on **DTM2**) ⇒ **Sketch** ⇒ **Line** ⇒ **Vertical** (sketch the three lines of the neck) ⇒ **Regenerate** ⇒ **Alignment** (align the centerline and **DTM2**, align the two endpoints of the vertical lines with the top of the part, and align the left *vertical* line with the part's vertical step, as in Fig. 6.16) ⇒ **Dimension** (Fig. 6.16) ⇒ **Regenerate** ⇒ **Modify** (change *sketch* dimensions to *design* dimensions) ⇒ **Regenerate** ⇒ **Done** ⇒ **Done**

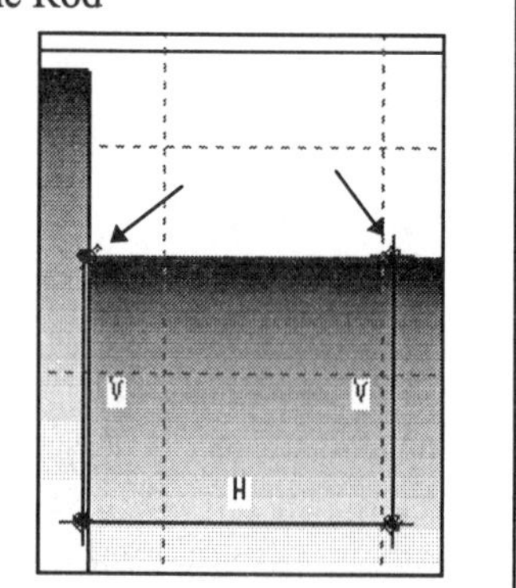

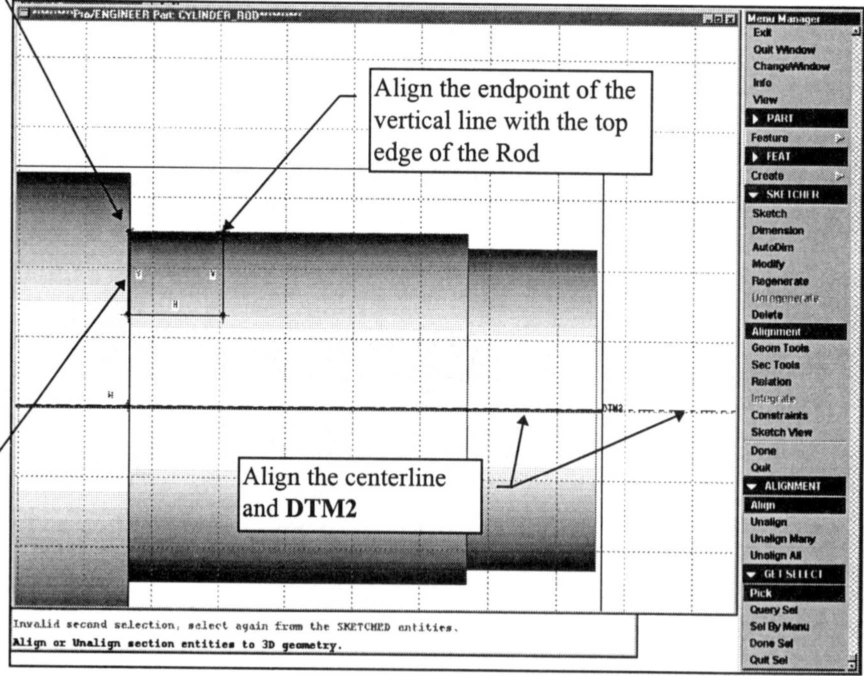

Figure 6.16
Aligning the Neck Cut

HINT

Always exaggerate the size of the sketch features. **Dimension** the sketch and **Regenerate**, then **Modify** to the design sizes. ***Never*** modify the sketch dimensions before you get a successful regeneration!

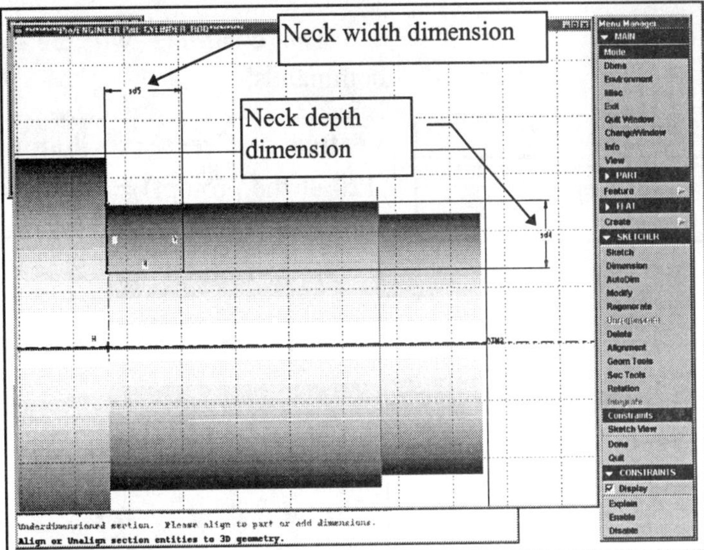

Figure 6.17
Dimensioning the **.120 X .045** Neck

After the first neck is successfully created (Fig. 6.18), model the second neck using the same procedure (Fig. 6.19).

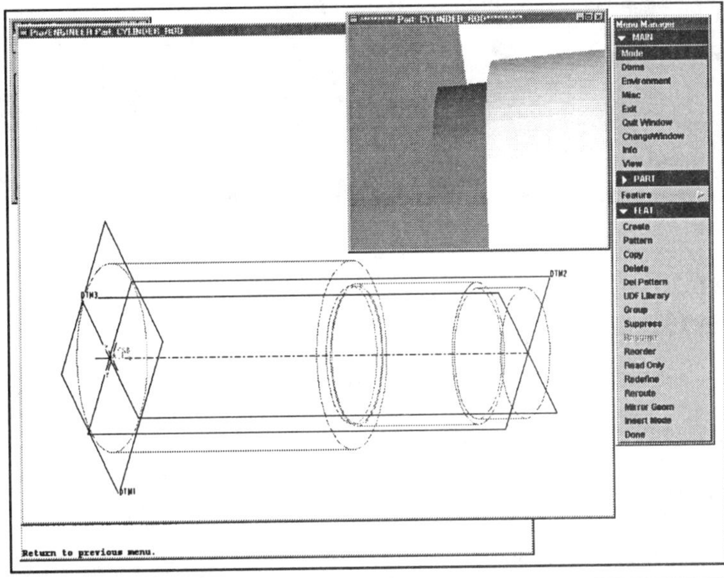

Figure 6.18
First Neck (revolved **Cut** with Pt/Modeler)

Dbms (File--PT/Modeler) ⇒
Save ⇒ **enter**
Purge ⇒ **enter** ⇒
Done-Return

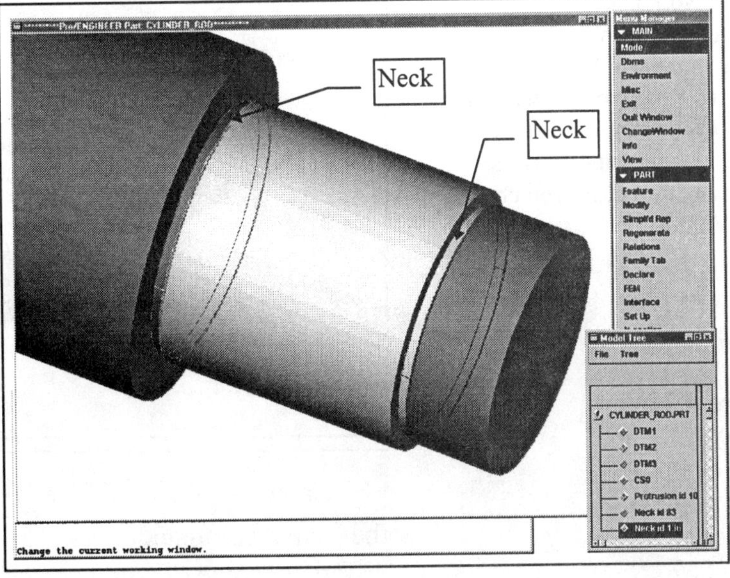

Figure 6.19
Both Necks Modeled Using the Same Dimensions

The chamfers will be created next. Choose the following commands:

PT/Modeler™
Feature ⇒ Chamfer ⇒ 45 x d **etc.**

Feature ⇒ Create ⇒ Chamfer ⇒ Solid ⇒ Edge ⇒ 45 x d ⇒ (type **.120** at the prompt) **⇒ enter ⇒** (select the edge to be chamfered as shown in Figs. 6.20 and 6.21) **⇒ Done Sel ⇒ Done Refs ⇒ Preview ⇒ OK ⇒ Done**

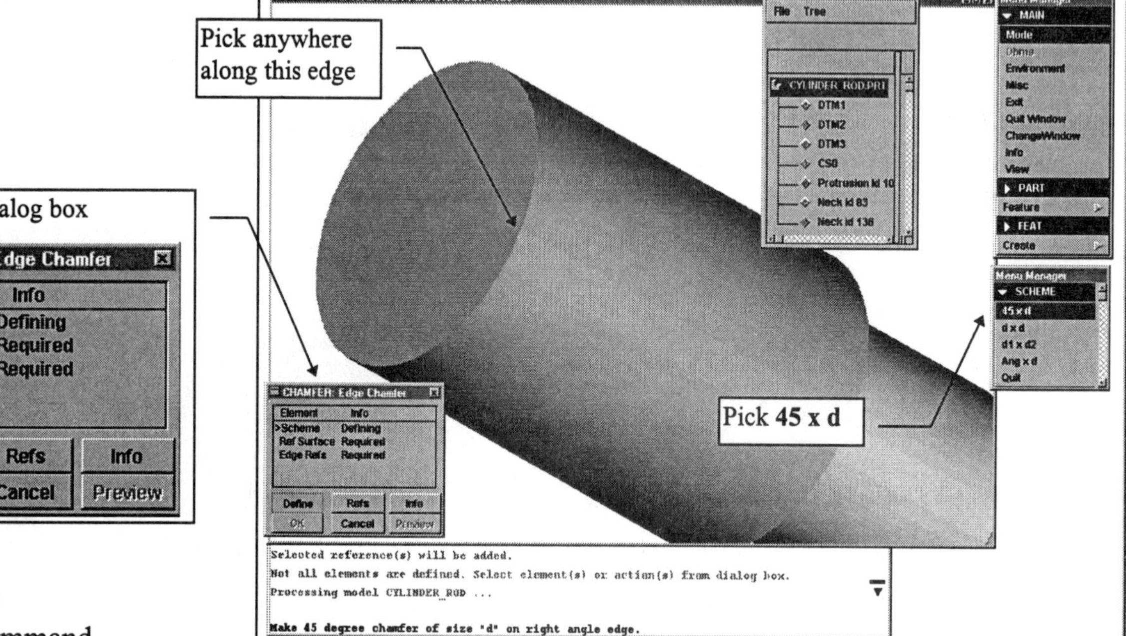

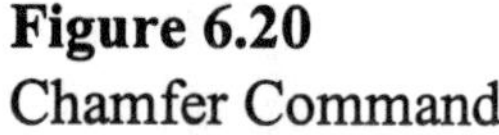

Figure 6.20
Chamfer Command

HINT

Open new window (**View ⇒ New Window**) to view your part from different angles and zoom states. You can work in either window by choosing **ChangeWindow** and picking in the window you wish to work. To close the new window, choose **Quit Window**. Make sure that the smaller window is the active window before you choose **Quit Window!**

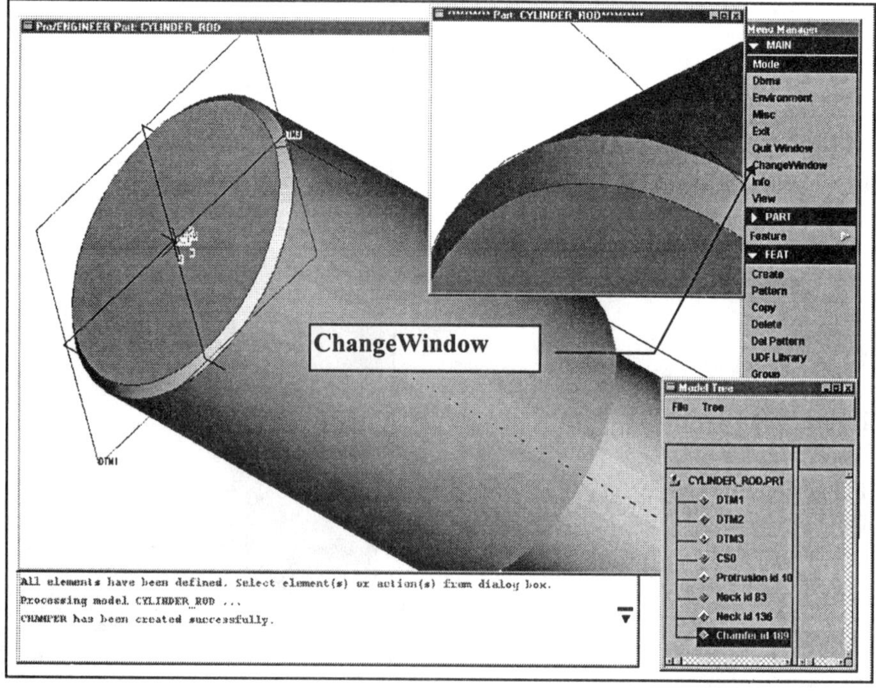

Figure 6.21
Chamfer

Create the **45° X .09** chamfer on the **∅2.75** side of the part using the same command. The **30° X .14** chamfer requires a slightly different choice of commands (Fig. 6.22):

PT/Modeler™
Feature ⇒ Chamfer ⇒ Ang x d etc.

Feature ⇒ Create ⇒ Chamfer ⇒ Solid ⇒ Edge ⇒ Ang x d ⇒ (type **.14** when prompted for distance) ⇒ **enter** ⇒ (type **30** when prompted for angle) ⇒ **enter** ⇒ (select the cylinder's revolved surface as the reference surface) ⇒ (select the edge to be chamfered, see Fig. 6.22) ⇒ **Done Sel ⇒ Done Refs ⇒ OK** (Fig. 6.23) ⇒ **Done**

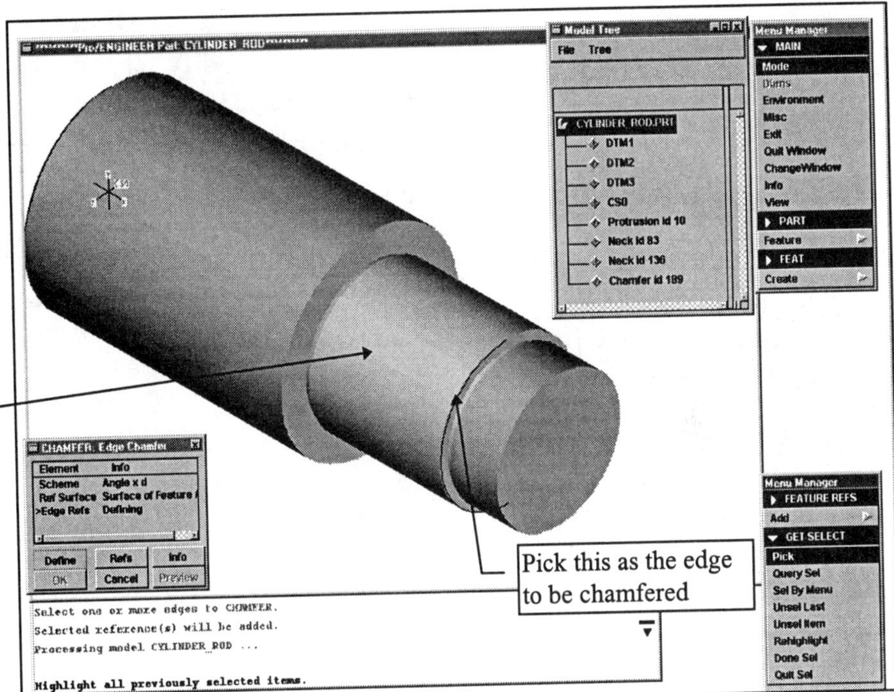

Figure 6.22
Creating the **30° X .14** Chamfer

Dbms (File--PT/Modeler) ⇒ Save ⇒ enter
Purge ⇒ enter ⇒ Done-Return

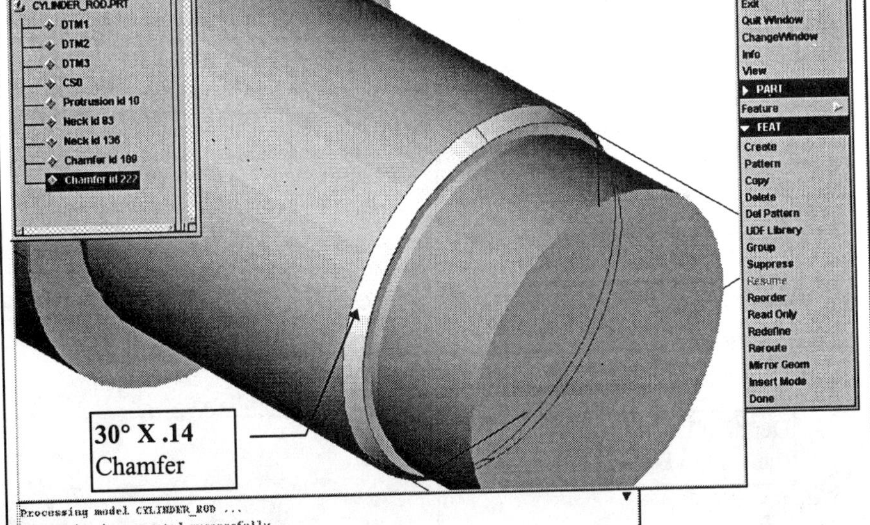

Figure 6.23
Chamfer **30° X .14**

Create the key seat using an extruded cut:

PT/Modeler™

Feature ⇒ Cut etc.

Feature ⇒ **Create** ⇒ **Cut** ⇒ **Extrude** ⇒ **Solid** ⇒ **Done** ⇒ **Both Sides** ⇒ **Done** ⇒ **Use Prev** ⇒ **Okay** ⇒ **Sketch View** ⇒ **Arc** ⇒ **Ctr/Ends** ⇒ (sketch the arc as in Fig. 6.24) ⇒ **Sketch** ⇒ **Line** ⇒ **Centerline** ⇒ **Vertical** (sketch the vertical centerline through the arc's center point, as in Fig. 6.25) ⇒ **Regenerate** ⇒ **Alignment** (align the endpoints of the arc to the upper edge of the part, as in Fig. 6.25) ⇒ **Regenerate** ⇒ **Dimension** (add the three dimensions) ⇒ **Regenerate** ⇒ **Modify** (change the dimensions to the design values of **.780** deep, **R1.030**, and **1.460** from the edge, as shown in Fig. 6.26) ⇒ **Regenerate** ⇒ **Done** ⇒ **Okay** ⇒ **Blind** ⇒ **Done** ⇒ (type **.500** at the prompt for the width of the keyseat) ⇒ **enter** ⇒ **Preview** ⇒ **OK** (Fig. 6.27) ⇒ **Done** ⇒ **View** ⇒ **Default** ⇒ **Done-Return**

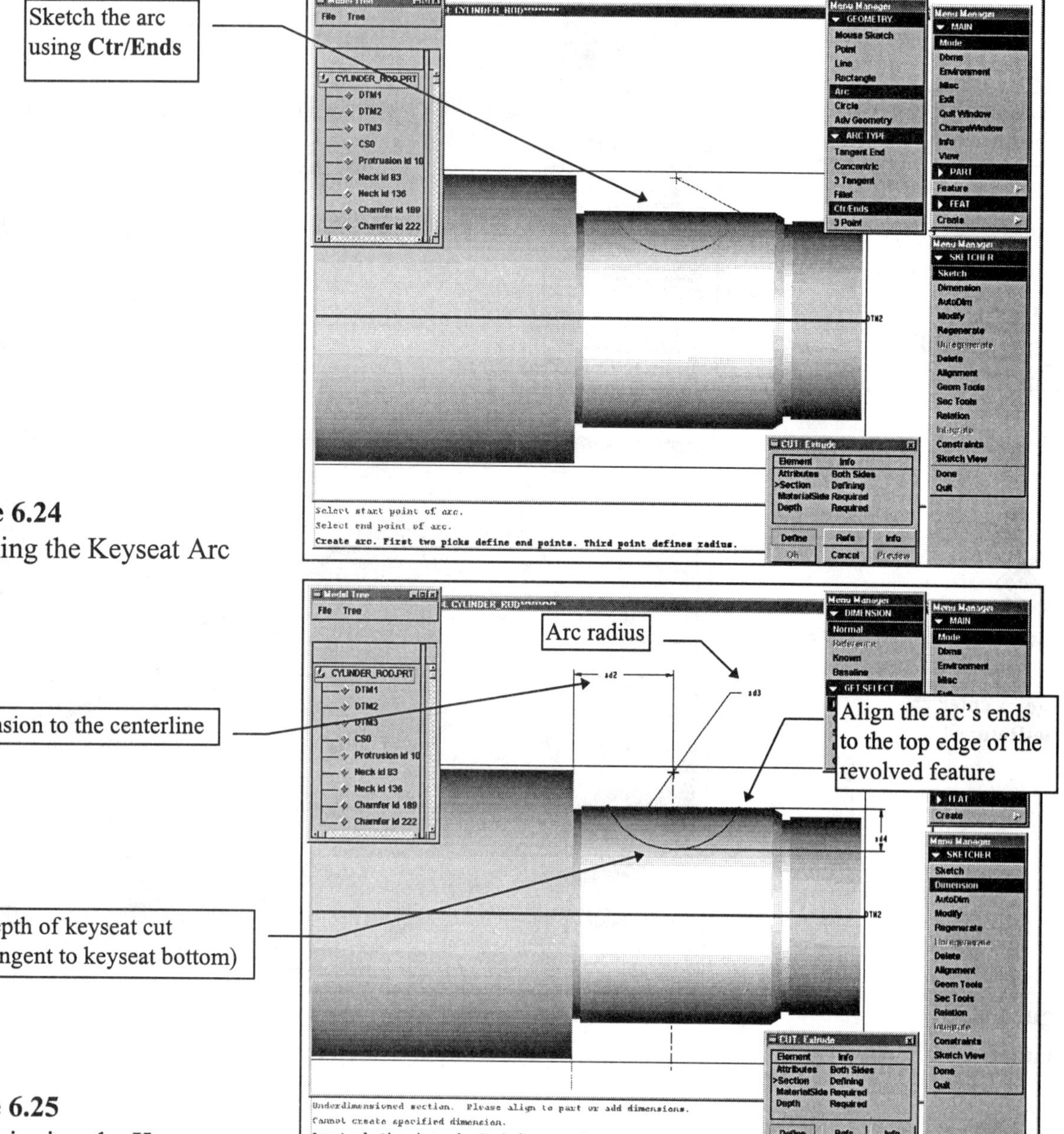

Figure 6.24
Sketching the Keyseat Arc

Figure 6.25
Dimensioning the Keyseat

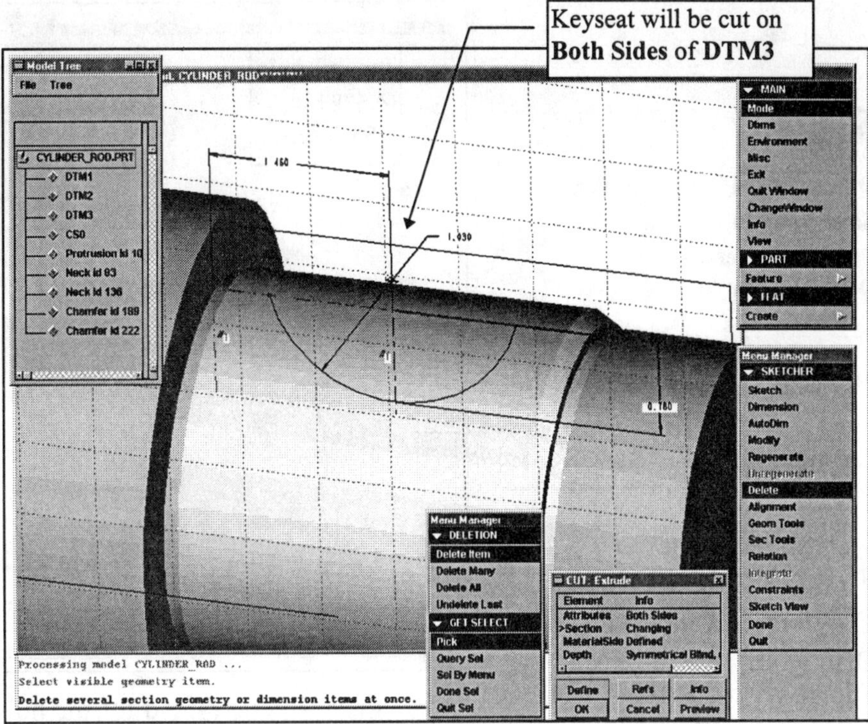

Figure 6.26
Keyseat Dimensions

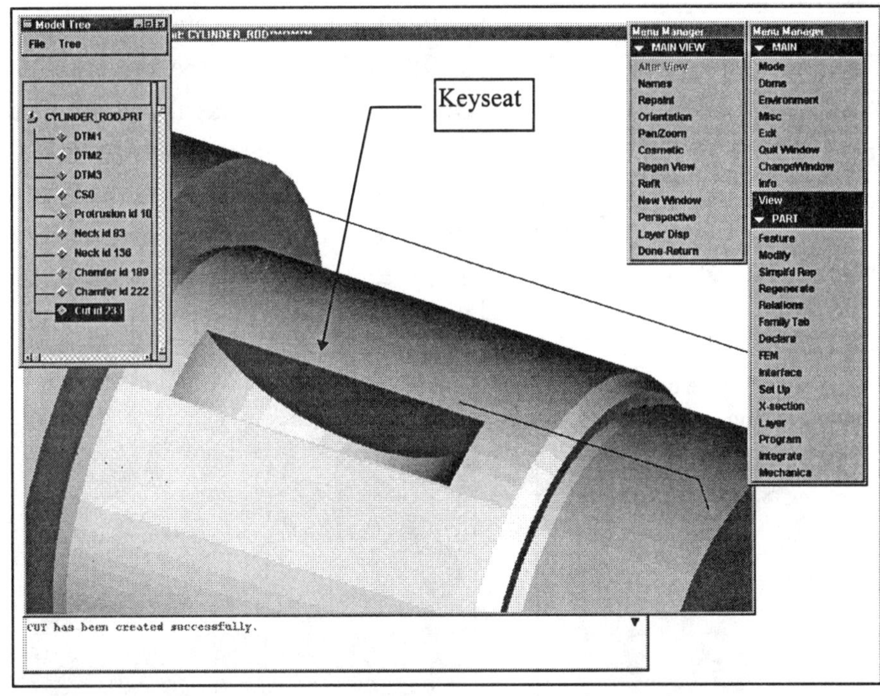

Figure 6.27
Keyseat

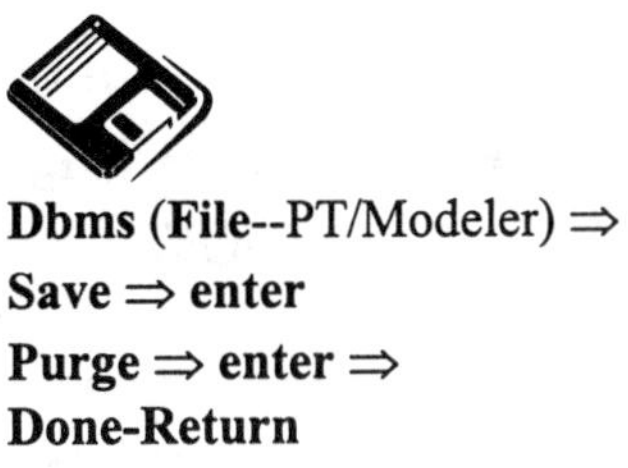

Dbms (File--PT/Modeler) ⇒ Save ⇒ enter
Purge ⇒ enter ⇒ Done-Return

At this stage of your understanding of Pro/E, you should be able to add the tap drill hole of **∅.578 X 1.50 DEEP** into the large diameter end of the part. Use coaxial for the sketched hole placement (Fig. 6.28).

The last features to be added to the Cylinder Rod will be cosmetic threads that will establish the thread size, type, class, and so on for the internal tap drill hole and the external thread that needs to be created on the **∅2.75** side of the rod.

Dbms (File--PT/Modeler) ⇒
Save ⇒ enter
Purge ⇒ enter ⇒
Done-Return

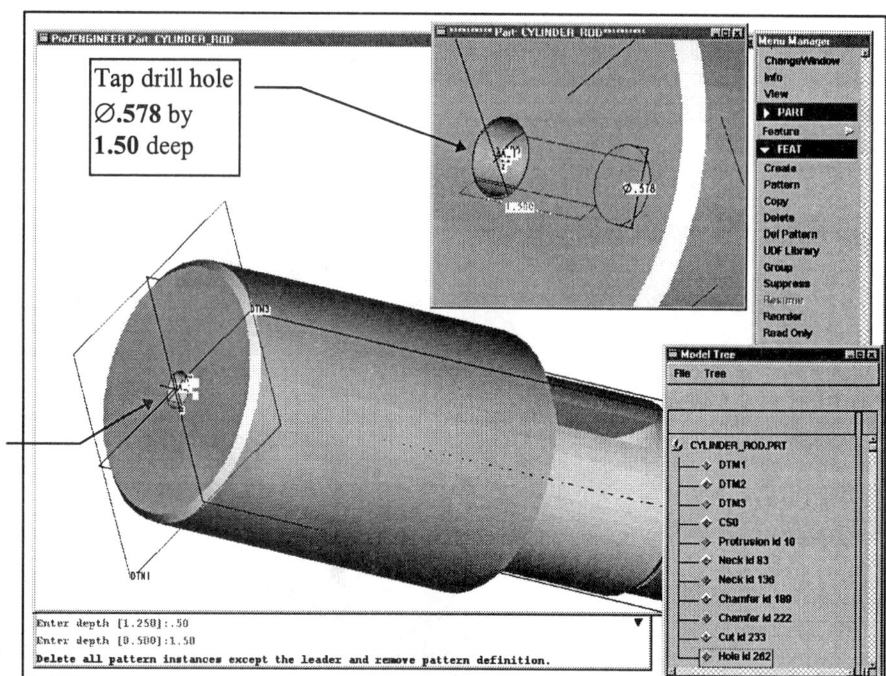

Figure 6.28
Ø.578 by **1.50** Deep Tap Drill Hole

Create the cosmetic thread for the hole with the following commands:

PT/Modeler™

Feature ⇒ Cosm Thread etc.

Feature ⇒ Create ⇒ Cosmetic ⇒ Thread ⇒ (pick the inside hole surface) ⇒ (pick the starting surface for the thread feature as in Fig. 6.29) ⇒ **Okay** (for the direction of feature creation) ⇒ **Blind ⇒ Done** ⇒ (type **1.25** as the depth of the cosmetic thread) ⇒ **enter** ⇒ (type **.625** as the thread diameter) ⇒ **enter** ⇒ **Mod Params** ⇒ (Pro/TABLE displays the thread options as shown in Fig. 6.30; fill in the table information) ⇒ **File ⇒ Save ⇒ File ⇒ Exit ⇒ Show** (Fig. 6.31) ⇒ type **q** (to quit the INFORMATION WINDOW) ⇒ **Done/Return ⇒ Preview** (Fig. 6.32) ⇒ **OK ⇒ Done**

HINT
You can select feature geometry directly from the screen or from the Model Tree.

Arrow points in the direction of feature creation

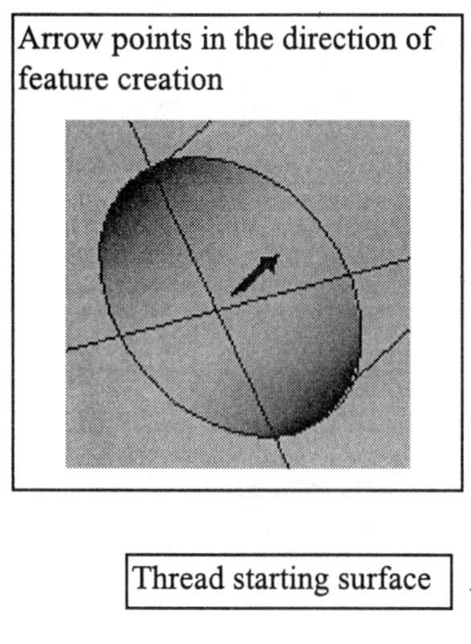

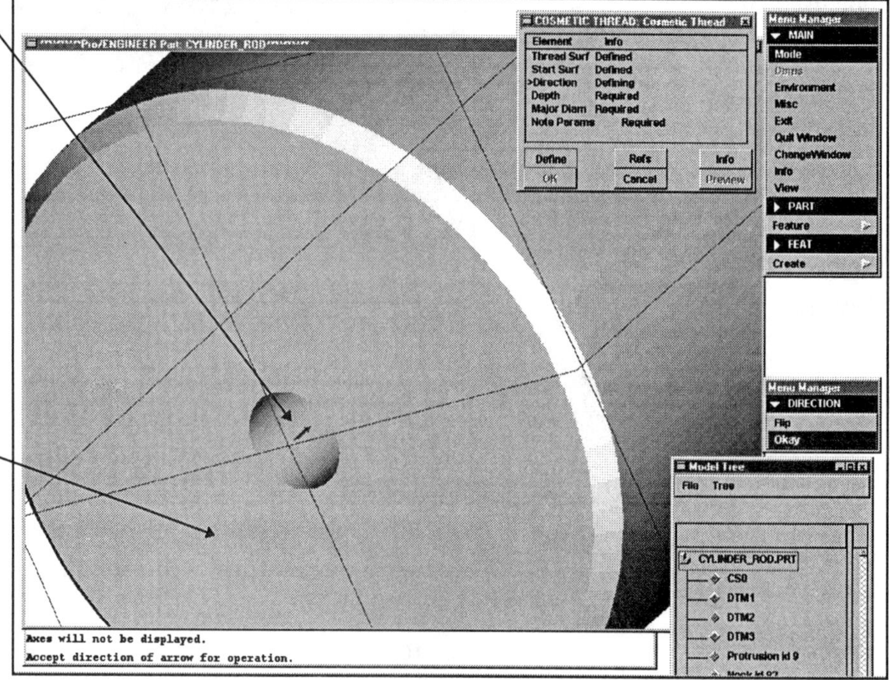

Figure 6.29
Creating a Cosmetic Thread

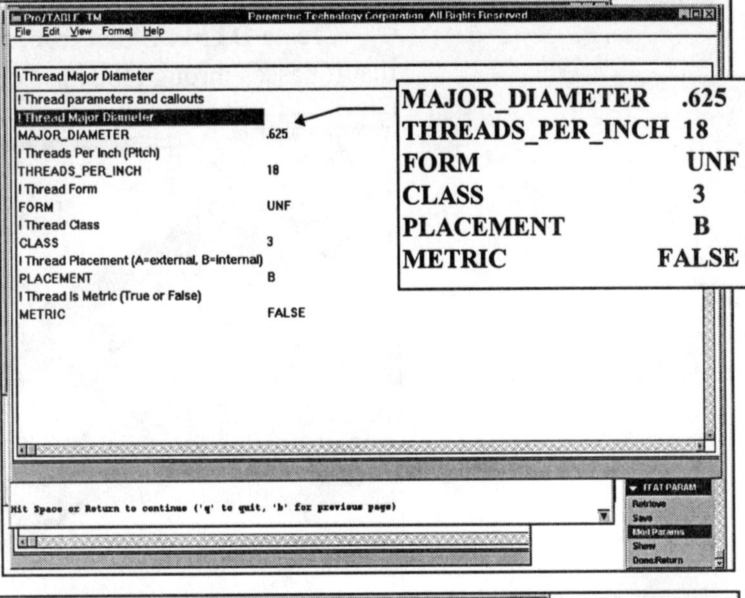

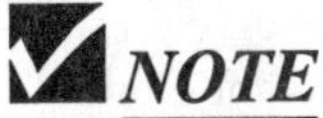

NOTE

Always **Save** the table information before exiting.

Figure 6.30
Thread Table

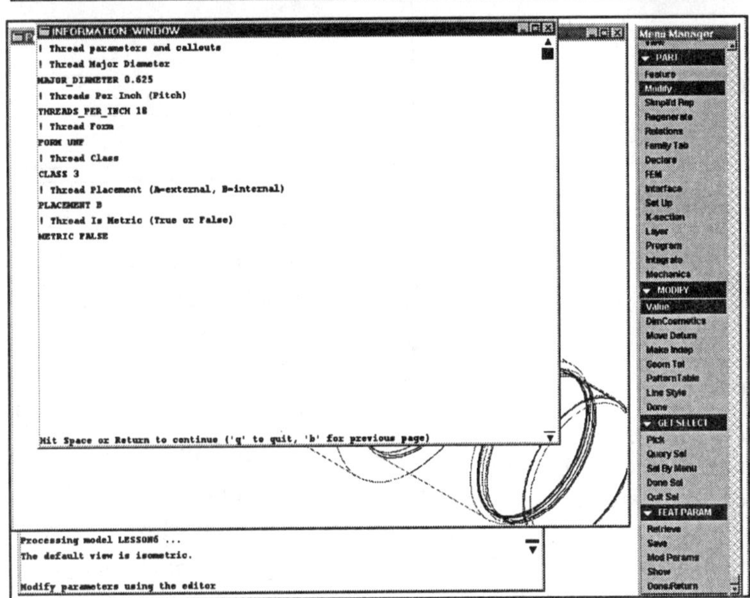

HINT

Turn off the shading and set **Hidden line** as the default; otherwise, you will not be able to see the cosmetic threads.

Figure 6.31
Showing the Feature Parameters

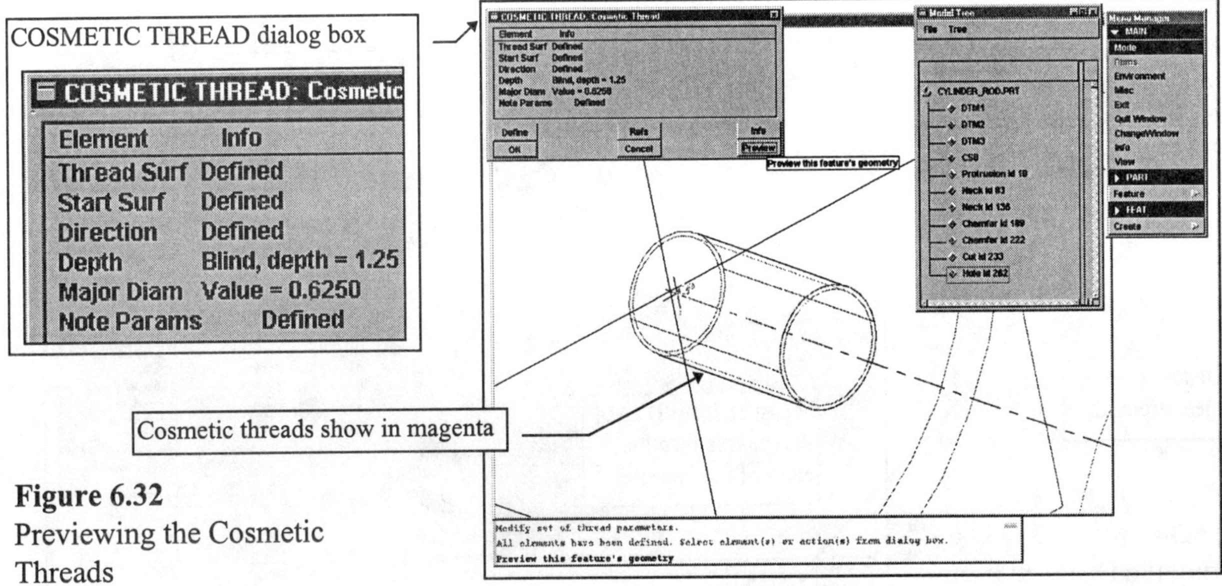

Figure 6.32
Previewing the Cosmetic Threads

Create **DTM4** offset from the end of the Rod a distance of **.09** so that it passes through the base of the chamfer (Fig. 6.33)

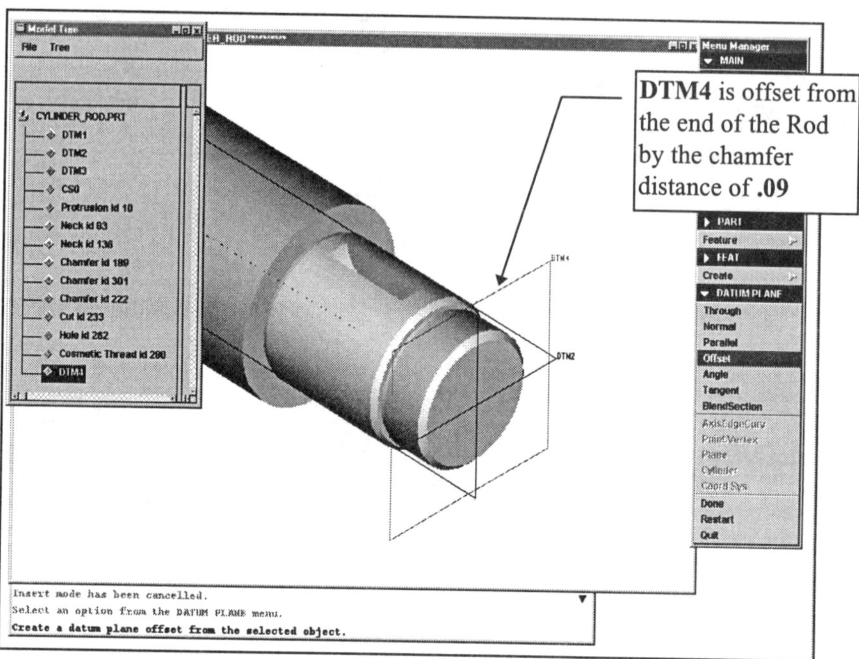

Figure 6.33
DTM4 Offset from the Rod End by **.09**

Measure the distance from **DTM4** to the lip of the neck cut before creating the cosmetic thread, as shown in Figure 6.34. Now create the other cosmetic thread, on the ∅**2.75** surface. The thread will start on **DTM4** and have a depth of **.920000** (choose **Flip** if the arrow points away from the part and then **Okay**). The Thread Table information (Fig. 6.35) will show the diameter as **2.75**, but the magenta-colored cylinder representing the thread on your screen is **2.6875** in diameter (Fig. 6.36). The thread diameter will be smaller than the nominal thread size of **2.75** (Fig. 6.37), since you are representing the *root diameter* of the thread (**2.6875**).

HINT

Internal cosmetic threads represent the **nominal diameter**. ***External cosmetic*** threads represent the **root diameter**. After creating the cosmetic thread, you must edit the thread table using Pro/TABLE. The diameter for a cosmetic internal thread will stay the same on the table, but the thread size of an external thread must be changed to the nominal size from the shown root diameter.

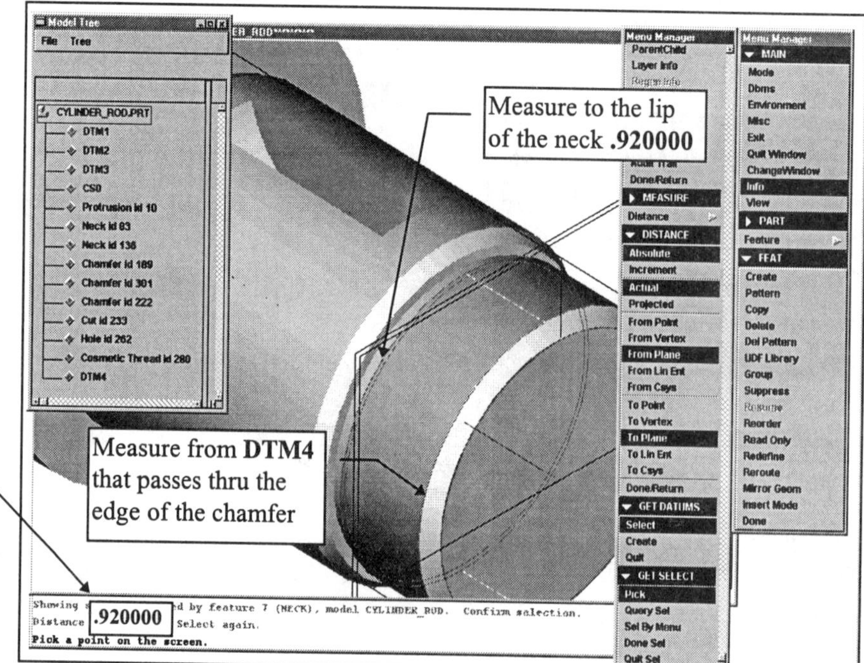

Figure 6.34
Measuring the Distance from **DTM4** to the Lip of the Neck

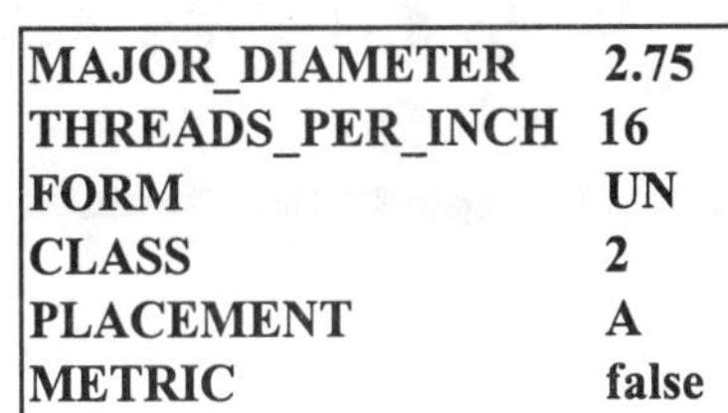

MAJOR_DIAMETER	2.75
THREADS_PER_INCH	16
FORM	UN
CLASS	2
PLACEMENT	A
METRIC	false

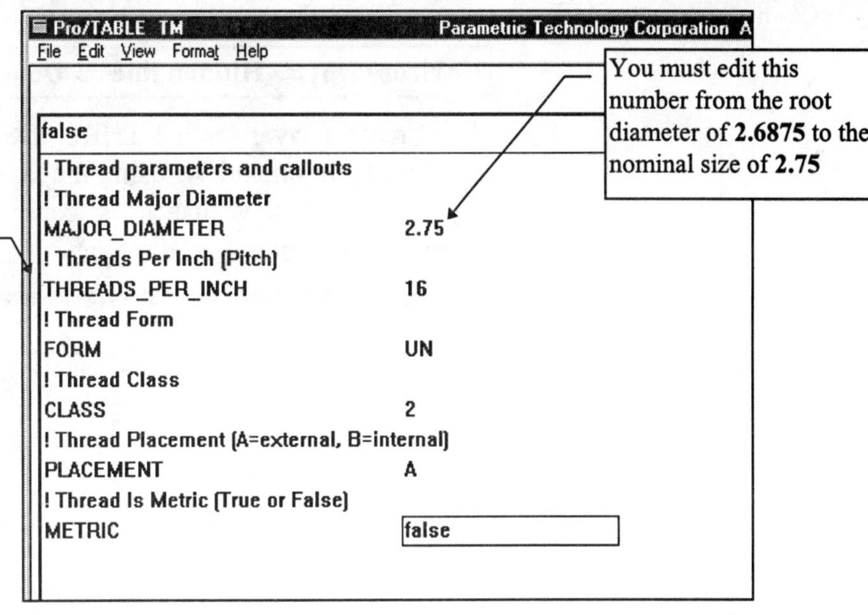

Figure 6.35
Thread Table for
2.75-16 UN-2A

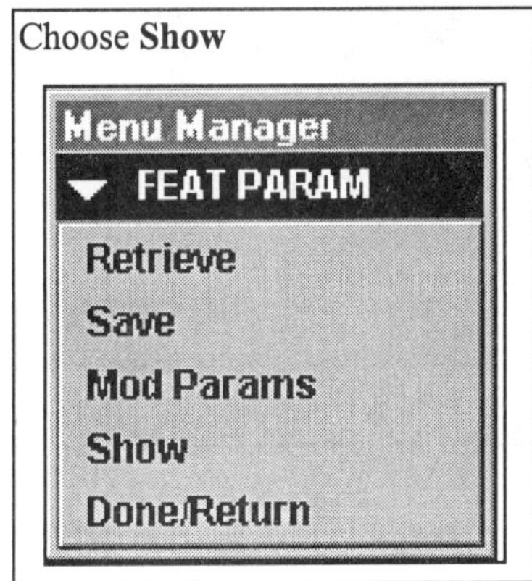

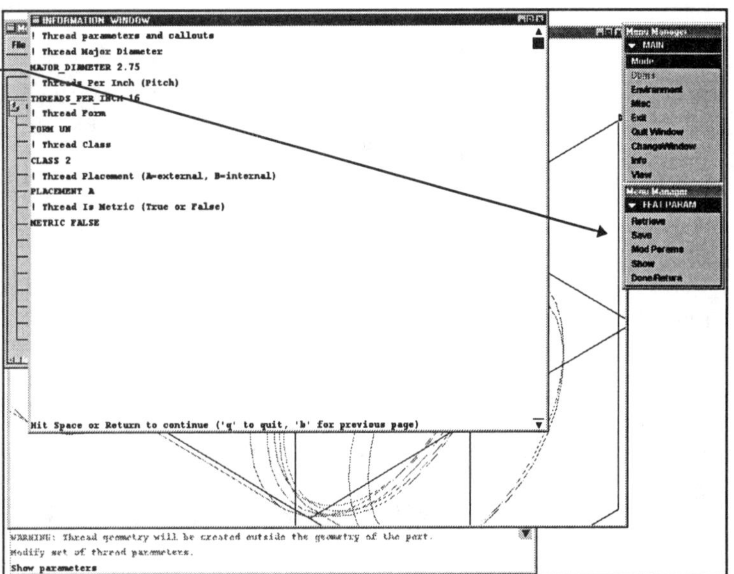

Figure 6.36
Showing Threads in the INFORMATION WINDOW

Dbms (File--PT/Modeler) ⇒
Save ⇒ **enter**
Purge ⇒ **enter** ⇒
Done-Return

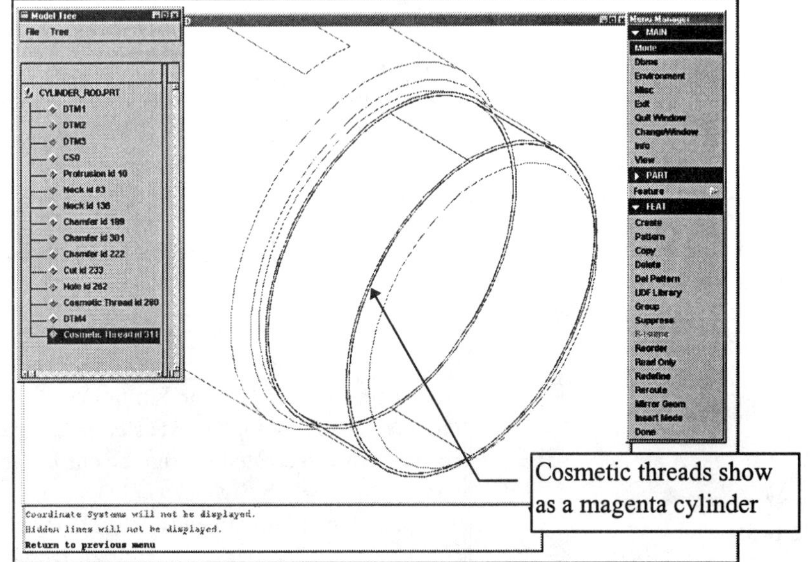

Figure 6.37
External Cosmetic Threads

Environment ⇒ Hidden line ⇒ Done-Return

Create a layer (called THREADS) for both threads so that you can turn them on and off using the layer display. Save and purge the Cylinder Rod now that it is complete. Figure 6.38 shows the completed Cylinder Rod drawing.

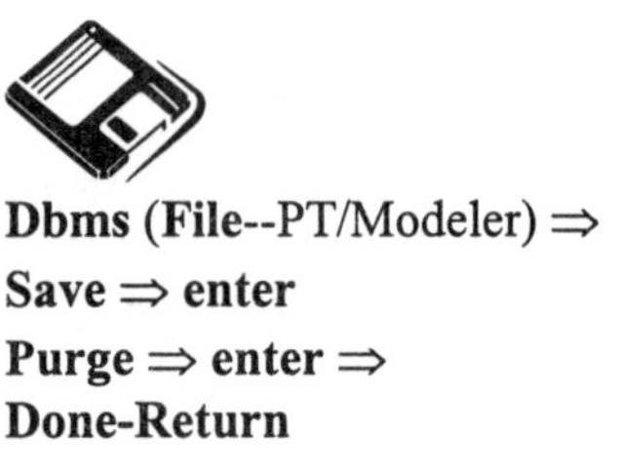

Dbms (File--PT/Modeler) ⇒
Save ⇒ enter
Purge ⇒ enter ⇒
Done-Return

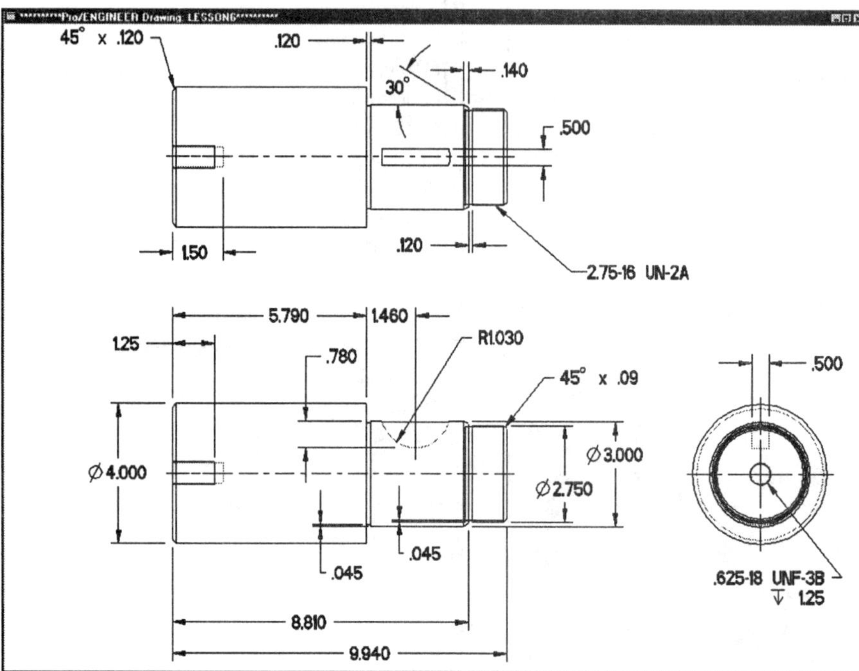

Figure 6.38
Cylinder Rod Drawing

Before going on to the Lesson 6 Project, complete the **ECO** shown in Figure 6.39.

ECO

NOTE

Look up the geometry sizes for the pipe thread in your **Machinery's Handbook**, a drafting text, or **Engineering Graphics and Design** by L. Lamit and K. Kitto, Chapter 17, Threads and Fasteners See pages 661-662

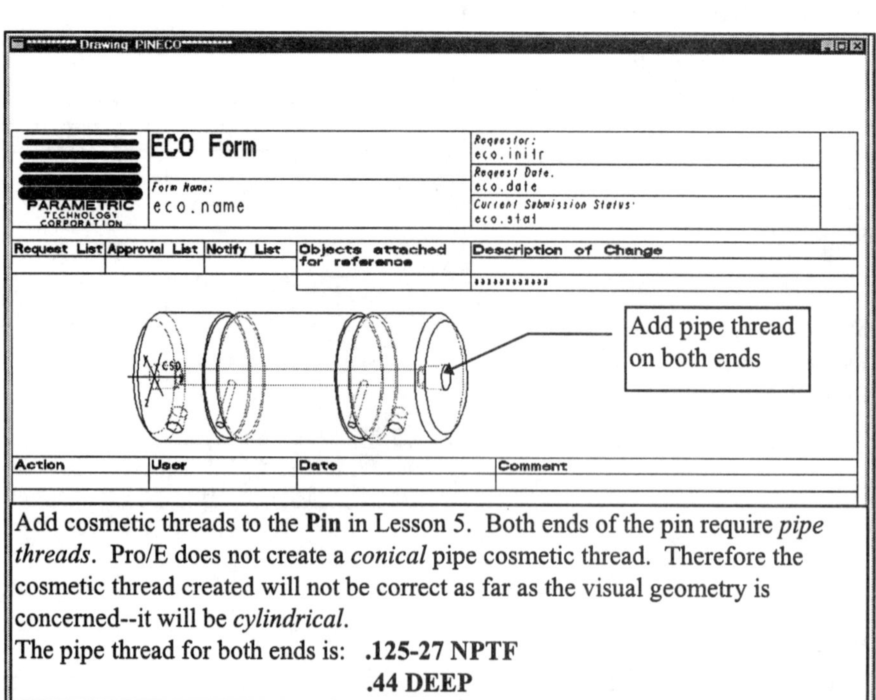

Add cosmetic threads to the **Pin** in Lesson 5. Both ends of the pin require *pipe threads*. Pro/E does not create a *conical* pipe cosmetic thread. Therefore the cosmetic thread created will not be correct as far as the visual geometry is concerned--it will be *cylindrical*.
The pipe thread for both ends is: **.125-27 NPTF**
.44 DEEP

Figure 6.39
ECO for the Pin in Lesson 5

Lesson 6 Project

Clamp Ball and Coupling Shaft

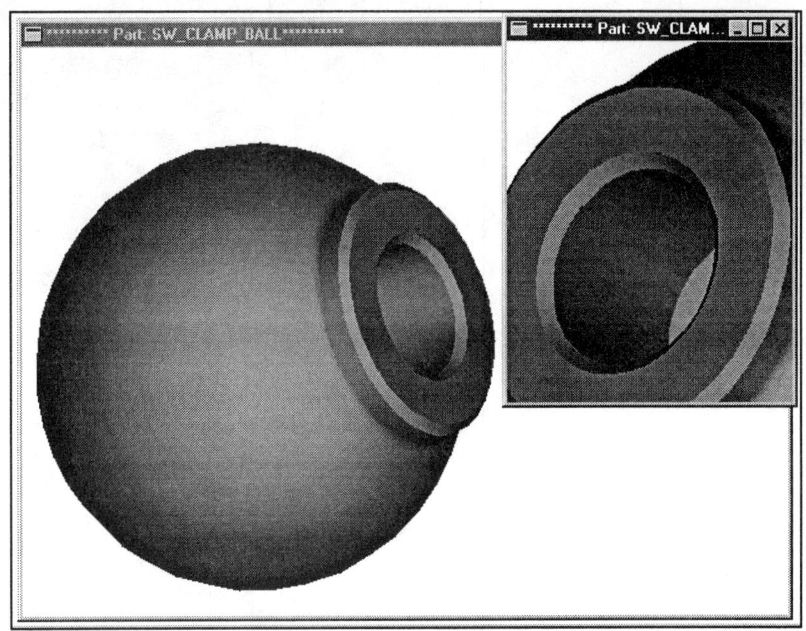

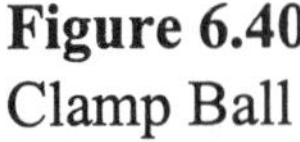
Figure 6.40
Clamp Ball

Clamp Ball and Coupling Shaft

As with Lesson 5, two lesson projects are provided here in Lesson 6 (Figs. 6.40 through 6.53). You will use both parts in Lessons 14 and 15 when creating different assemblies. Both the **Clamp Ball** (decimal inch) and the **Coupling Shaft** (SI units) are revolved protrusions. The Clamp Ball is simpler and easier to complete. The Ball is *black plastic* and the Shaft is *steel.* Create all cosmetic threads required on each part. The two parts are used on different assemblies.

Analyze the parts and plan out the steps and features required to model them. Use the **DIPS** in Appendix D to establish a feature creation sequence before the start of modeling. Remember to set up the environment, establish datum planes, and set them on layers.

EGD REFERENCE
Engineering Graphics and Design with Graphical Analysis *or* **Fundamentals of Engineering Graphics and Design**
by L. Lamit and K. Kitto
Read Chapter 23
See pages 865-866

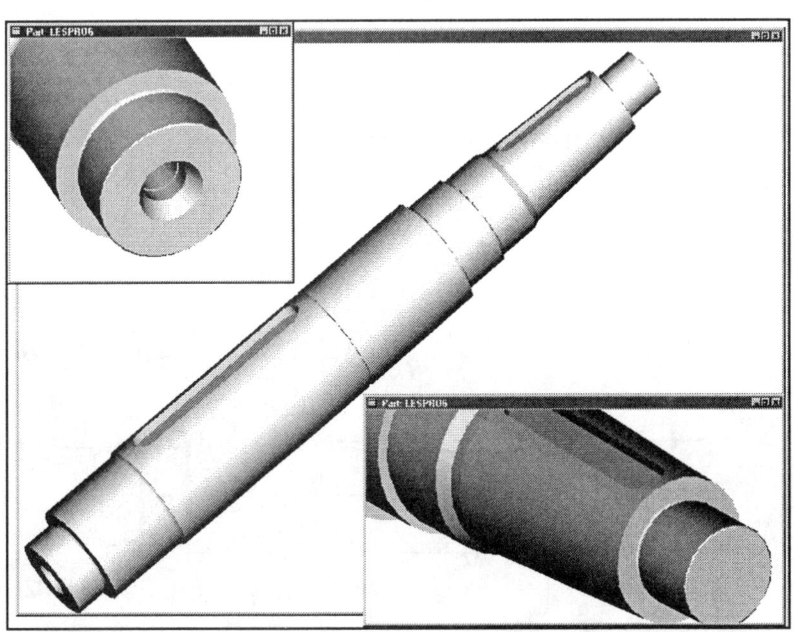

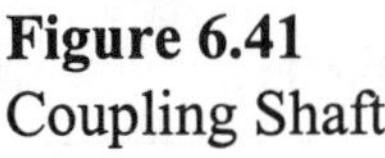
Figure 6.41
Coupling Shaft

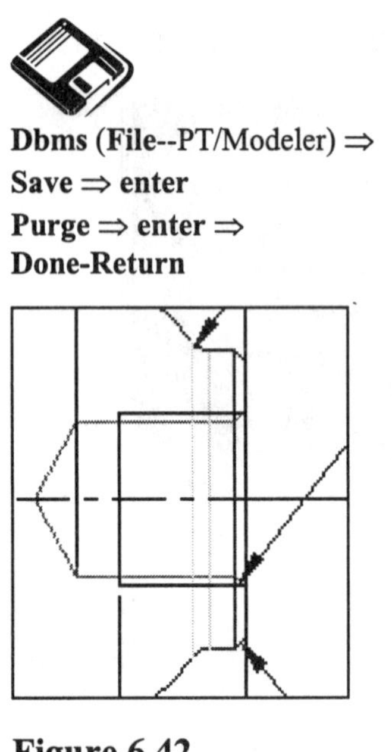

Dbms (**File**--PT/Modeler) ⇒
Save ⇒ **enter**
Purge ⇒ **enter** ⇒
Done-Return

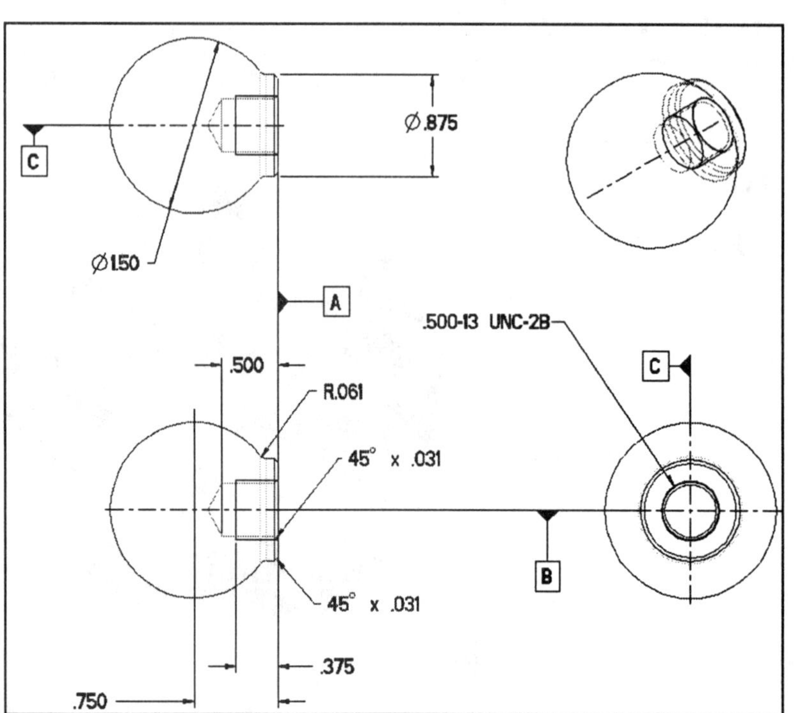

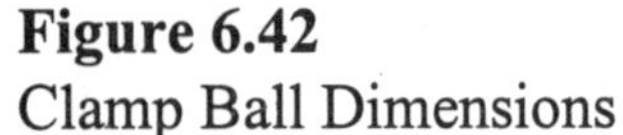

Figure 6.42
Clamp Ball Dimensions

Figure 6.43 Coupling Shaft Drawing, Sheet One

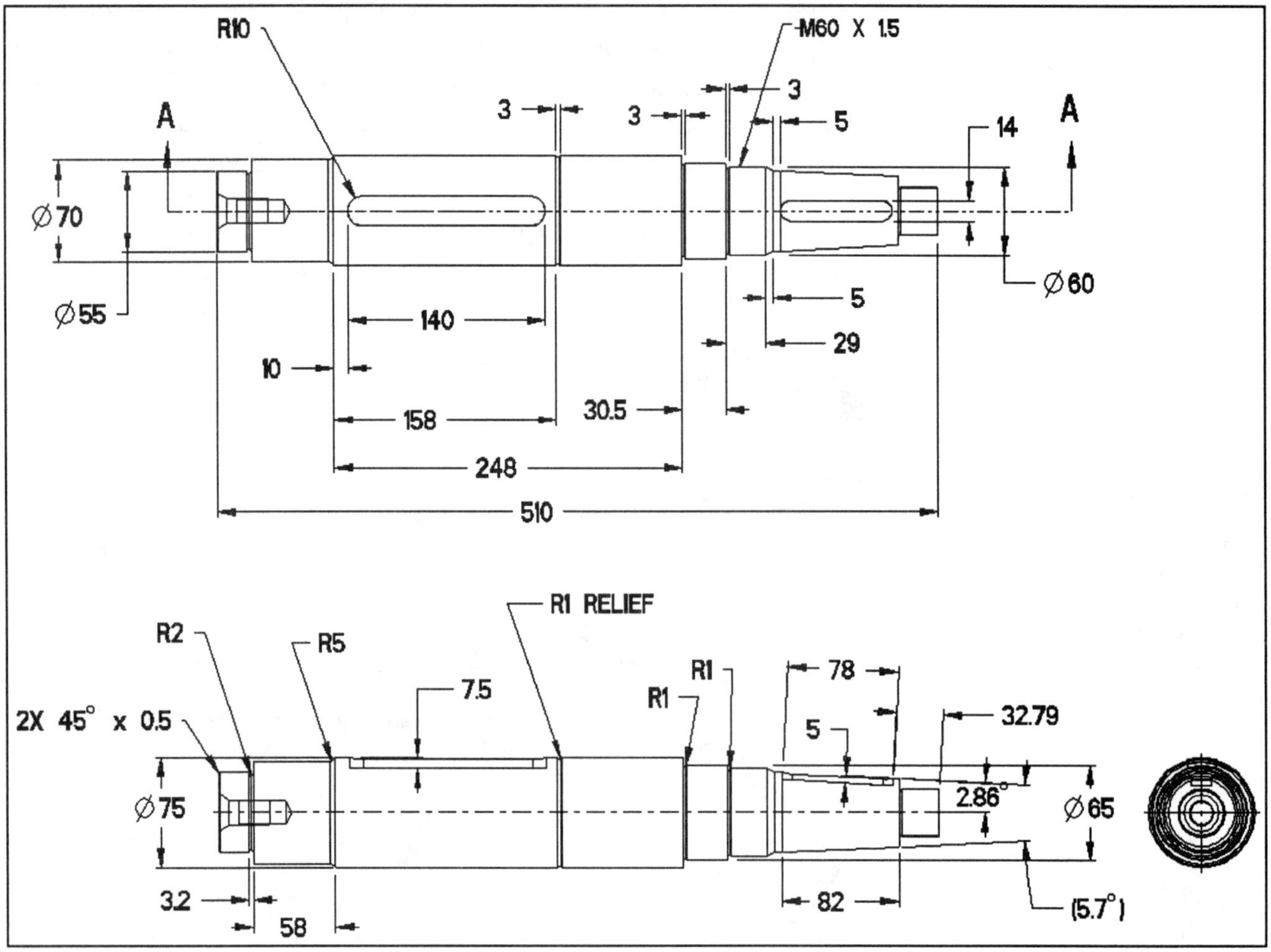

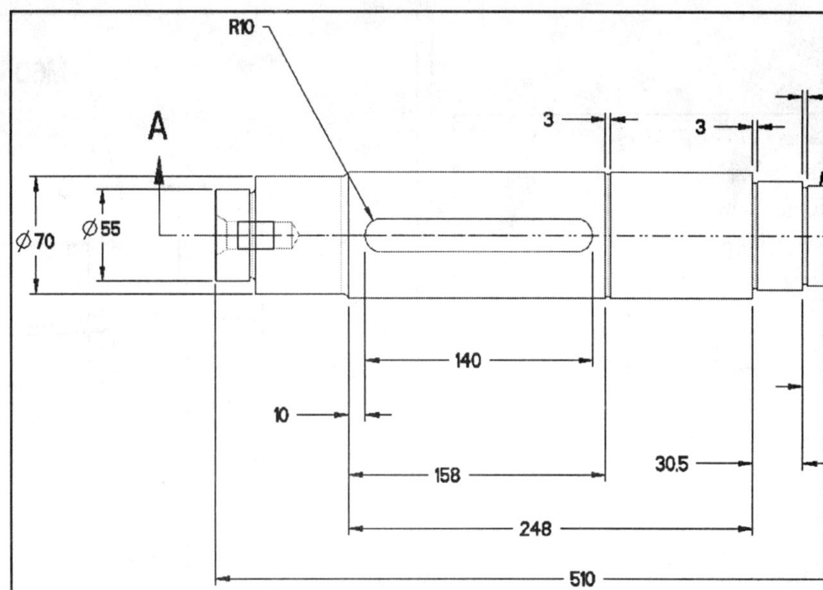

Figure 6.44
Coupling Shaft Drawing,
Top View Left Side

Figure 6.45
Coupling Shaft Drawing,
Sheet Two

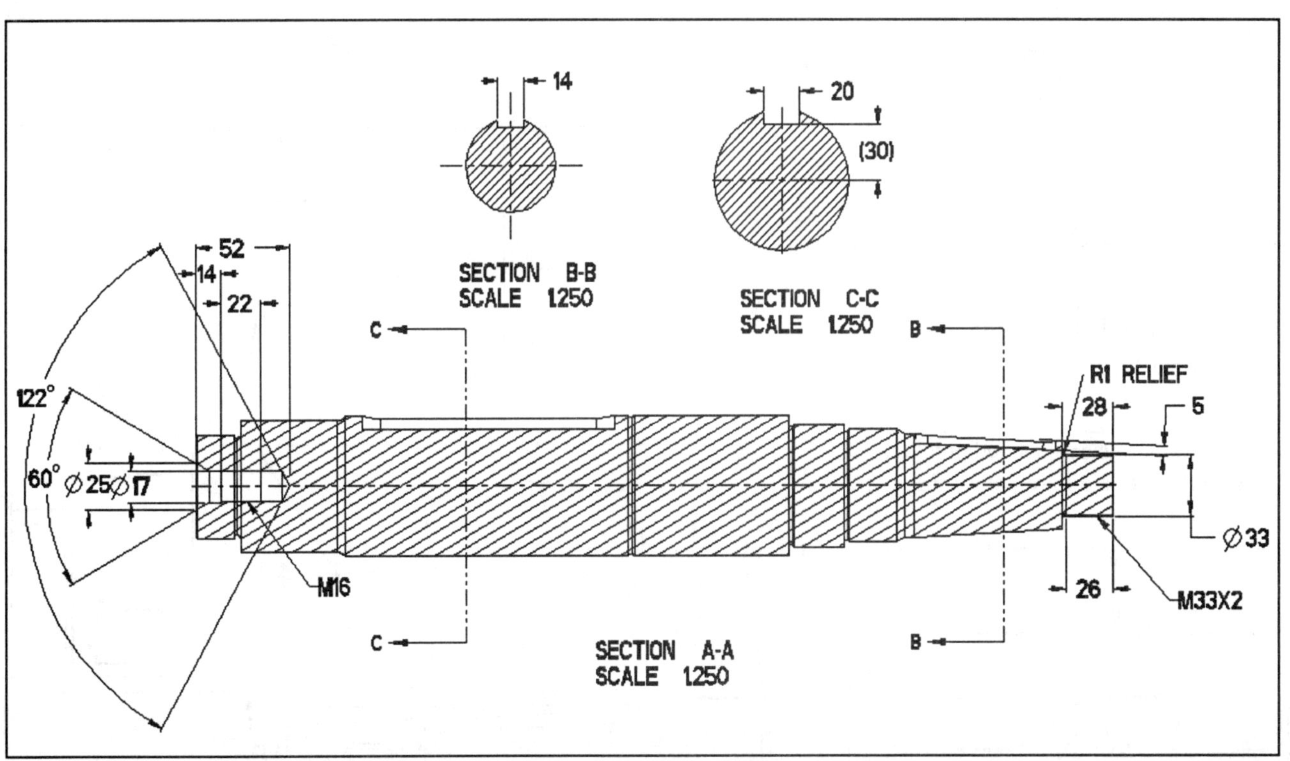

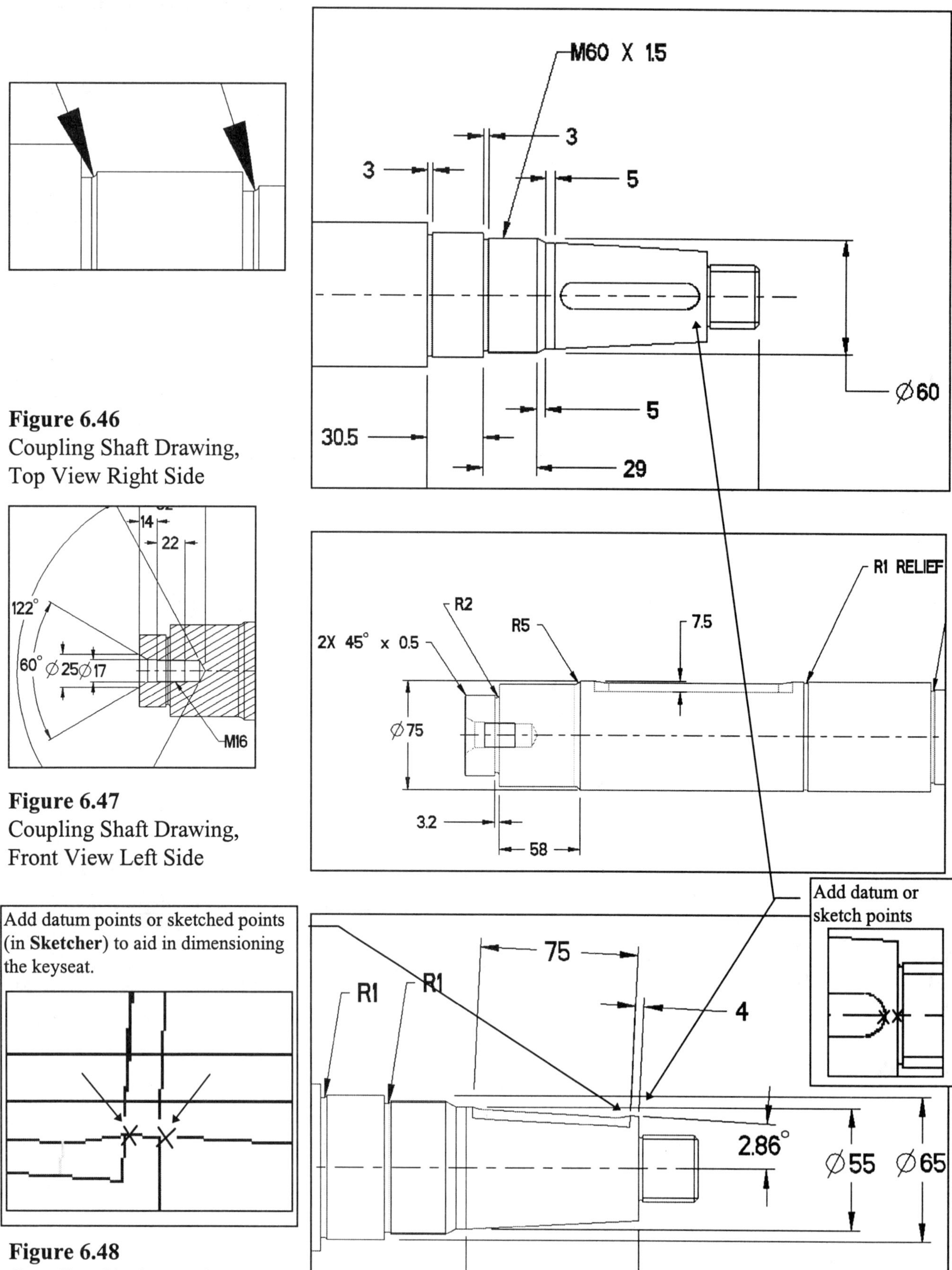

Figure 6.46
Coupling Shaft Drawing, Top View Right Side

Figure 6.47
Coupling Shaft Drawing, Front View Left Side

Figure 6.48
Coupling Shaft Drawing, Front View Right Side

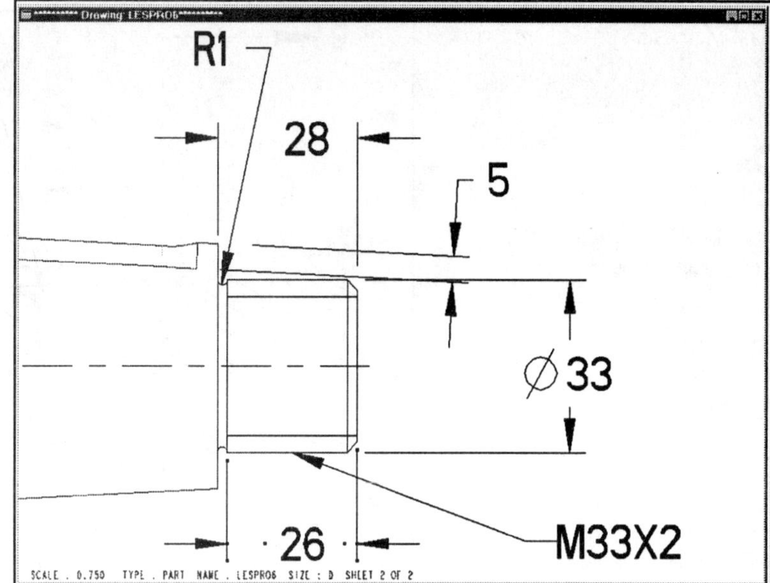

Figure 6.49
Coupling Shaft Drawing,
M33 X 2 Threads

Dbms (File--PT/Modeler) ⇒
Save ⇒ enter
Purge ⇒ enter ⇒
Done-Return

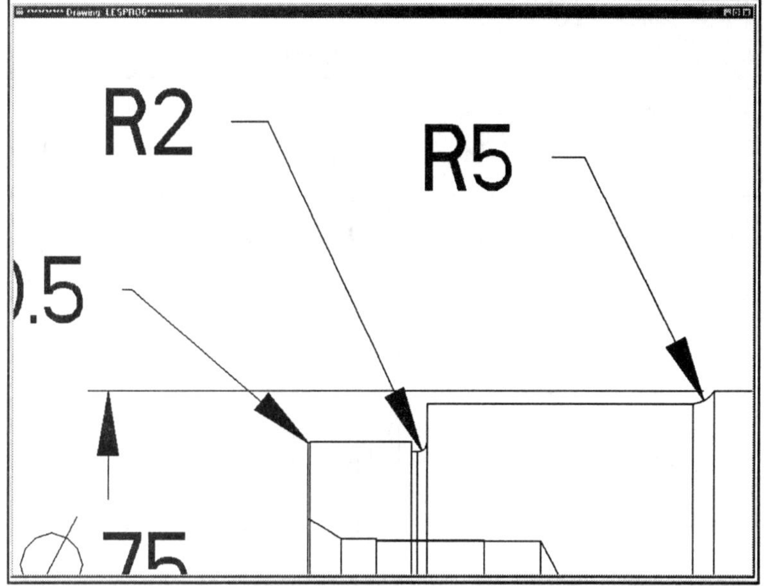

Figure 6.50
Coupling Shaft Drawing,
Reliefs

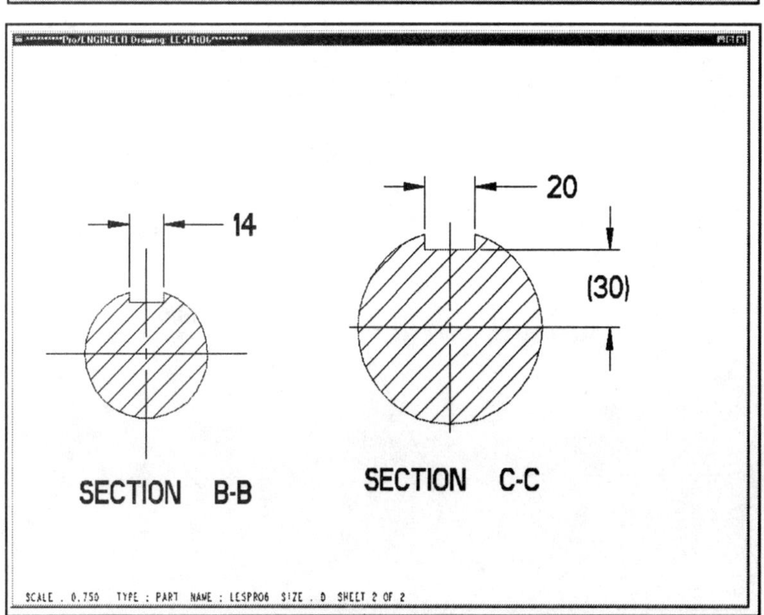

Figure 6.51
Coupling Shaft Drawing,
Sheet Two,
SECTION B-B and
SECTION C-C

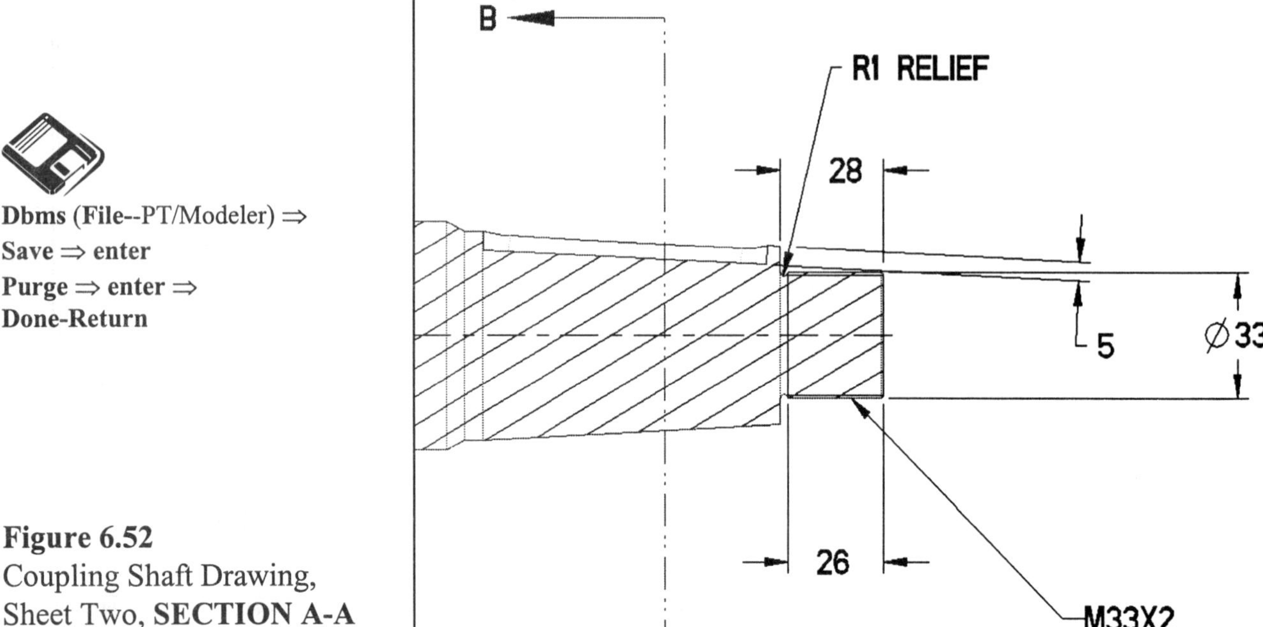

Dbms (**File**--PT/Modeler) ⇒
Save ⇒ **enter**
Purge ⇒ **enter** ⇒
Done-Return

Figure 6.52
Coupling Shaft Drawing, Sheet Two, **SECTION A-A** Right Side

Figure 6.53
Coupling Shaft Drawing, Sheet Two, **SECTION A-A** Left Side

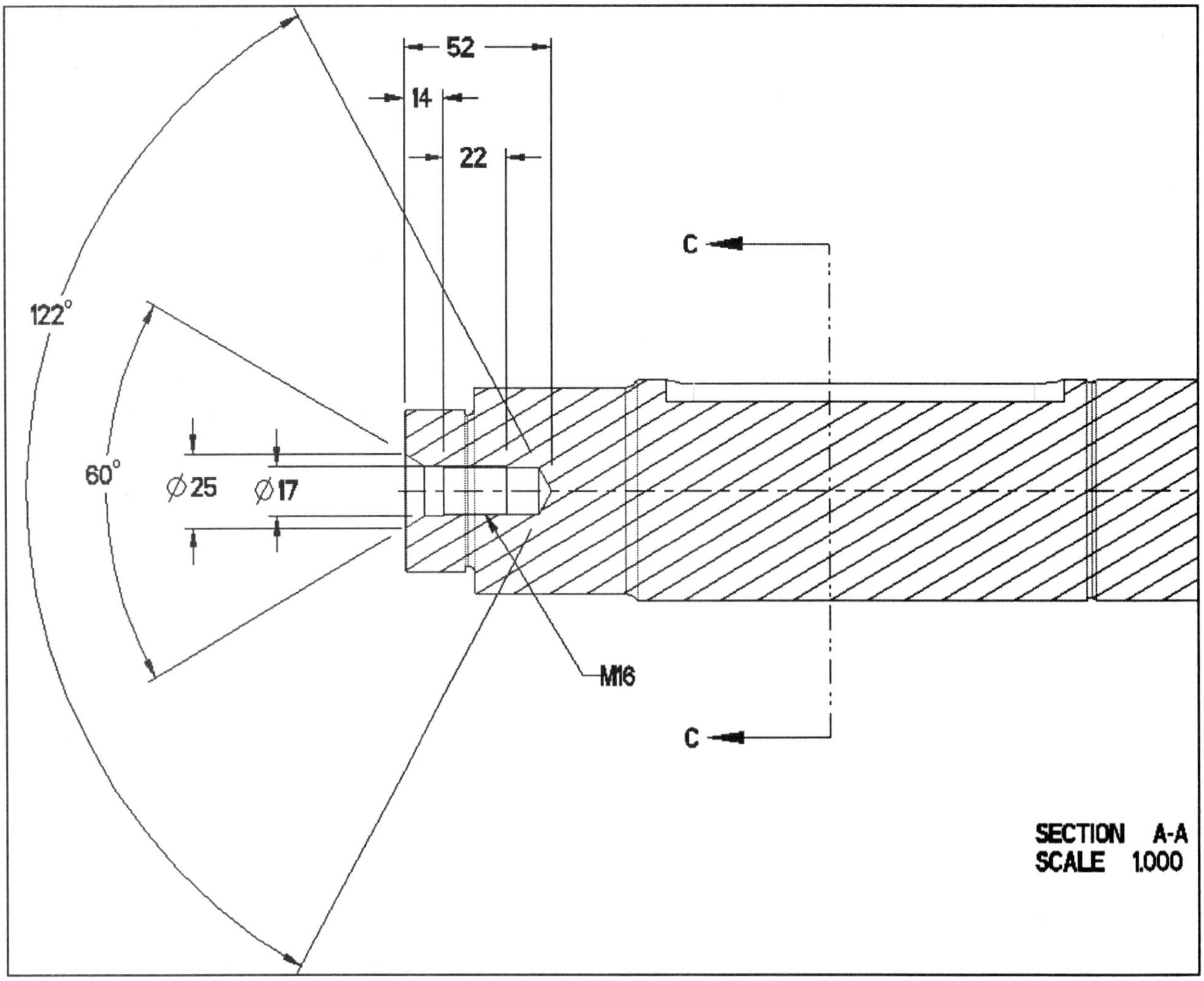

Lesson 7

Pattern and Group

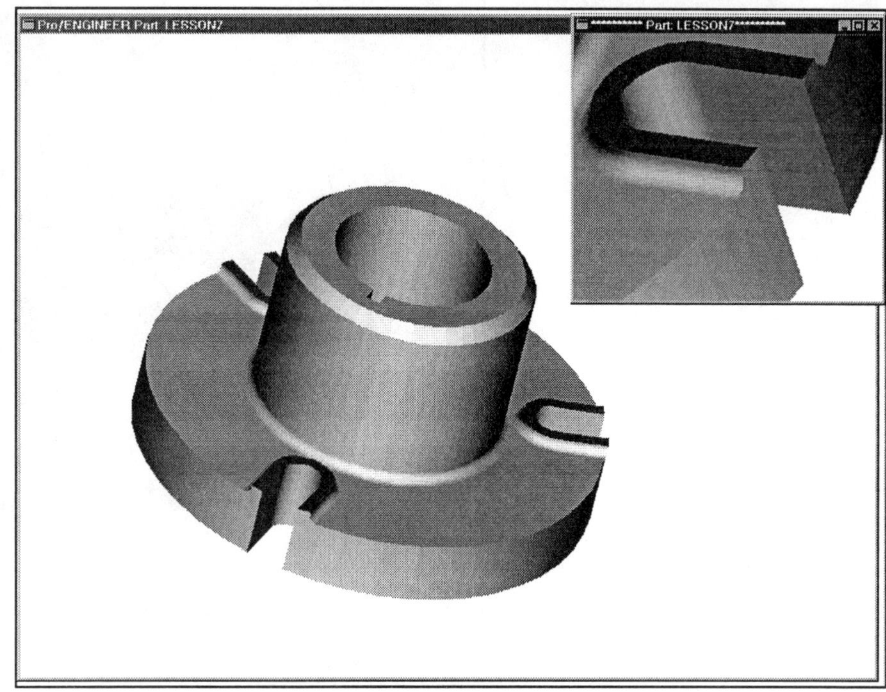

Figure 7.1
Post Reel

OBJECTIVES

1. Change dimension cosmetics and move part dimensions

2. Use offset edge to create sketch entities from existing geometry

3. Pattern features

4. Create and manipulate grouped features

5. Understand how to pattern groups

EGD REFERENCE
Engineering Graphics and Design with Graphical Analysis *or* **Fundamentals of Engineering Graphics and Design**
by L. Lamit and K. Kitto
Read Chapter 10
See page 296

COAch™ for Pro/ENGINEER

If you have **COAch for Pro/ENGINEER** on your system, go to SEARCH and do the Segment shown in Figure 7.3.

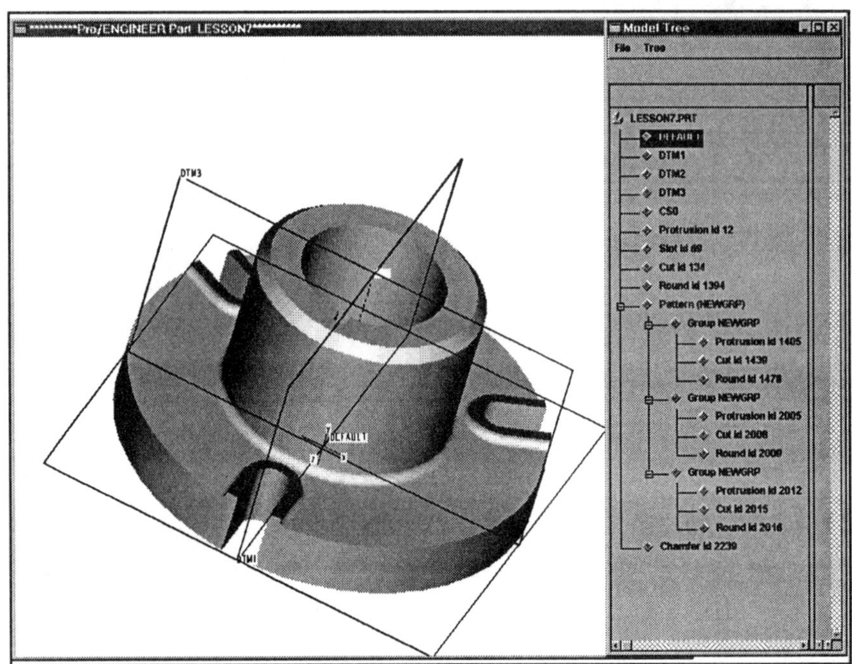

Figure 7.2
Post Reel Showing Datum Planes and Coordinate System

PATTERN AND GROUP

To create multiple features from a single feature (or group of features), the **Pattern** option can be used. After it is created, a pattern behaves as if it were a single feature. When you create a pattern, you create instances (copies) of the selected feature (or group of features). The **Group** option unites a series of features together.

Creating a pattern is a quick way to reproduce a feature or a set of features that are related and grouped for easy manipulation. Manipulating a pattern may be more advantageous for you than operating on individual features.

Figure 7.3
COAch for Pro/ENGINEER, Duplicating Features (Circular Pattern)

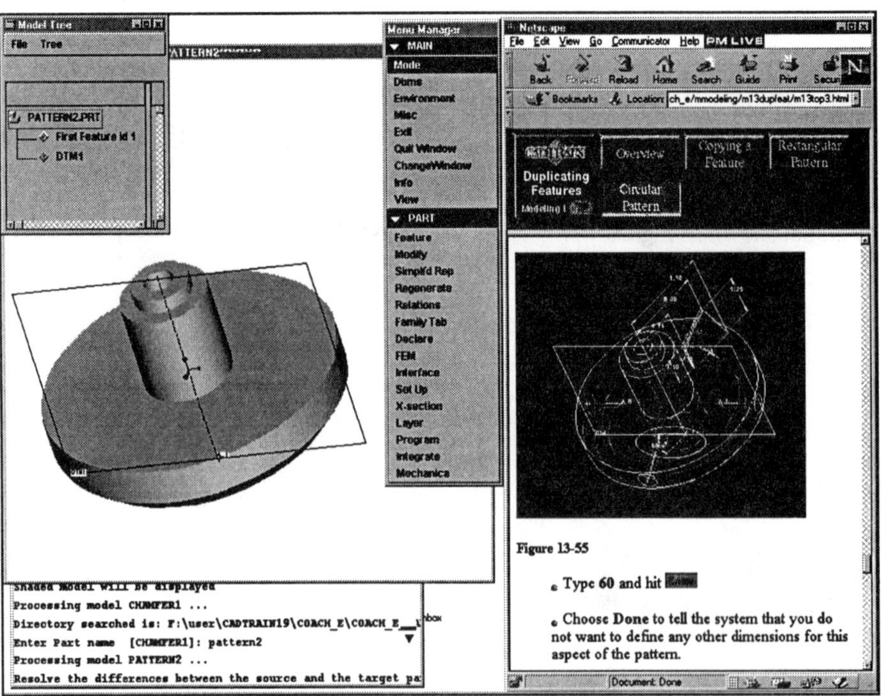

Group

When you create a *local group*, you must select the features in the consecutive sequential order of the regeneration list. A quick way to do this is to select the intended group by range. If there are features between the specified features in the regeneration list, Pro/E asks if you want to group all the features in between.

Features that are already in other groups cannot be grouped again. To create a local group, do the following:

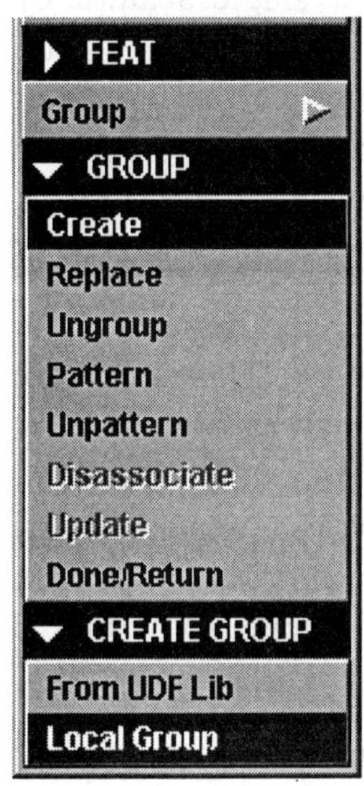

1. Choose **Create** from the GROUP menu and **Local Group** from the CREATE GROUP menu.
2. Select each feature to include in the local group

   or

3. Answer **yes** to the prompt asking if you want to group all the features in the between. If you answer **no** to this prompt, Pro/E does not create the local group.

Pattern

Modifying patterns is more efficient than modifying individual features. In a pattern, when you change the dimensions of the original feature, the whole pattern will be updated automatically. A pattern is parametrically controlled. Therefore, a pattern can be modified by changing pattern parameters, such as the number of instances, the spacing between instances, and the original feature dimensions.

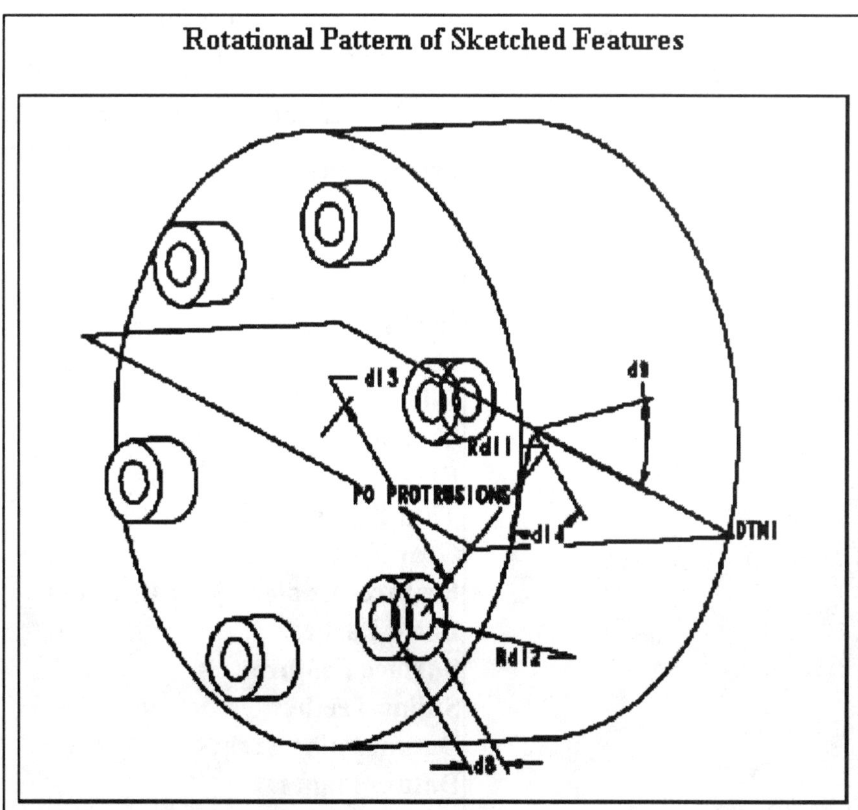

Figure 7.4
Online Documentation, Patterns

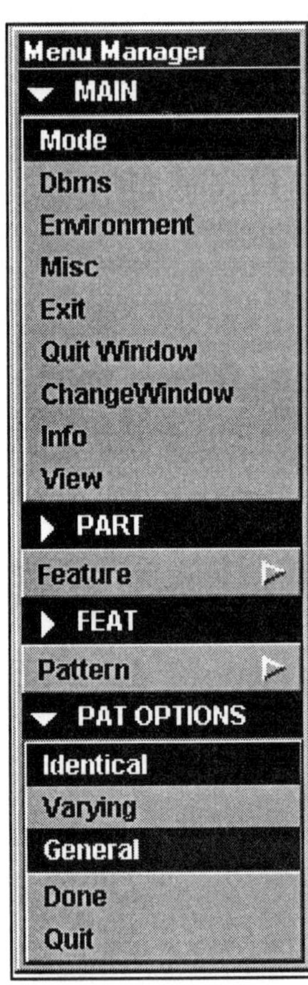

When you create a pattern, Pro/E assumes it is a "single" feature. Creating a pattern is a quick way to reproduce a feature. Patterning is an easier and more effective way to perform a single operation on the multiple features contained in a pattern, rather than on the individual features. For example, you can easily suppress a pattern or add it to a layer.

You can pattern most features using the FEAT menu's **Pattern** option. A thin feature "remembers" the surface to which it is attached and patterns to that surface.

Pro/E allows you to pattern a single feature only. However, you can pattern several features as if they were a single feature by arranging them in a **Local Group**, then patterning the group. After the pattern is created, you can unpattern and ungroup the instances, then make them independently modifiable using the option **Make Indep.** There are two ways to pattern a feature using the PRO PAT TYPE menu:

Dim Pattern Controls the pattern using driving dimensions to determine the incremental changes to the pattern. This is the case for all of the pattern.

Ref Pattern Controls the pattern by referencing another pattern. The dimension pattern must exist before you can create the next pattern type. For example, counterbore one hole in the pattern, then have Pro/E pattern the counterbore. It automatically makes a reference pattern.

When you are working with features or components for which it does not make sense to have both **Dim Pattern** and **Ref Pattern**, Pro/E does not display the PRO PAT TYPE menu.

The features that can be patterned are:

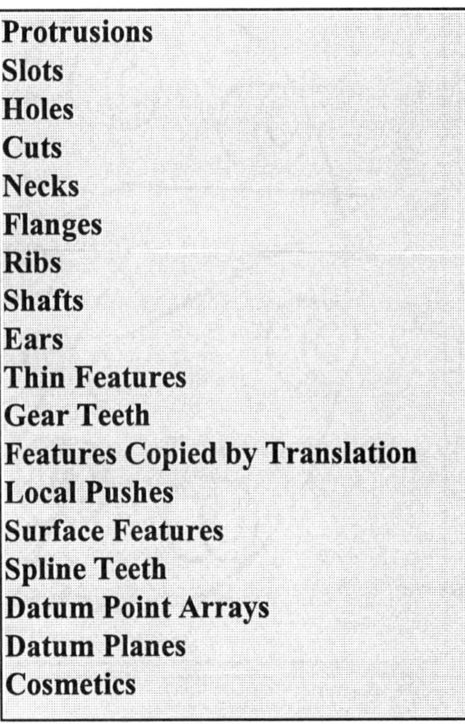
Protrusions
Slots
Holes
Cuts
Necks
Flanges
Ribs
Shafts
Ears
Thin Features
Gear Teeth
Features Copied by Translation
Local Pushes
Surface Features
Spline Teeth
Datum Point Arrays
Datum Planes
Cosmetics

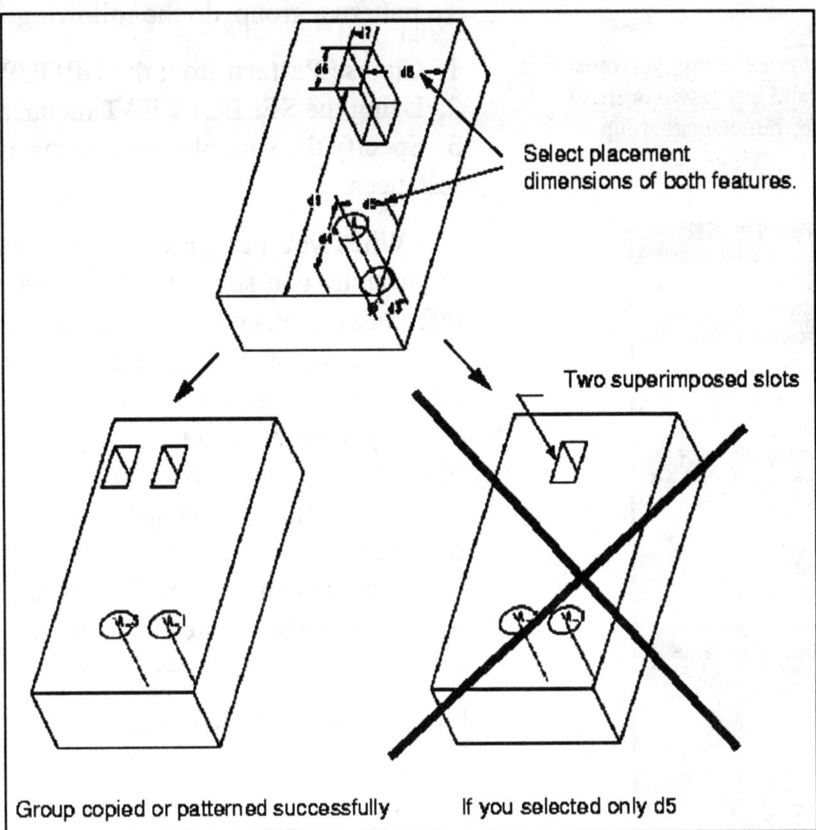

Figure 7.5
Online Documentation,
Patterns

Patterning a Group

You can pattern groups created from UDFs (user-defined features) and local groups using the GROUP menu option **Pattern**. This option differs from the FEAT menu option **Pattern** in that the GROUP menu option treats an entire group as a single entity. The FEAT menu PATTERN option is used to pattern one feature at a time.

You can select all the dimensions in the selected group, except those used to create a feature pattern within the group, as incremental dimensions. When you create a patterned group, one member represents the whole group. When regenerating, however, Pro/E regenerates all the features individually.

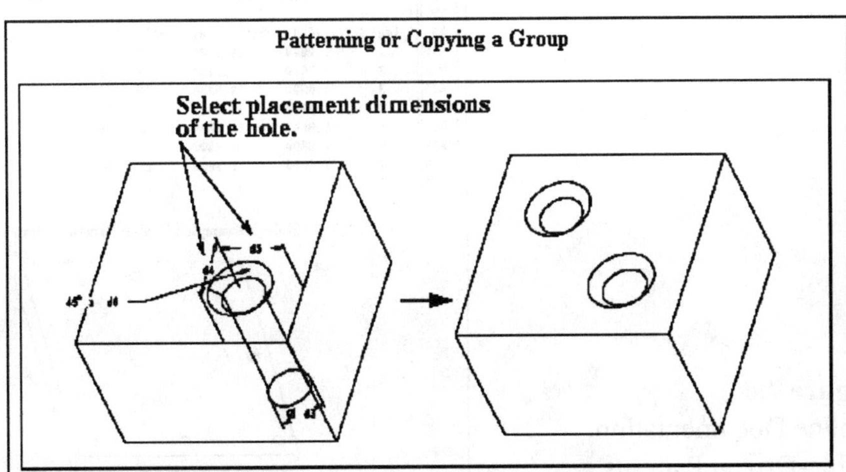

Figure 7.6
Online Documentation,
Patterning Groups

HINT

To pattern a group, you must first name and group two or more features into a local group.

Figure 7.7
Online Documentation, Modifying Patterns

To pattern a group, do the following:

1. Choose **Pattern** from the GROUP menu.
2. Using the SELECT FEAT menu, select the group to be patterned.
3. Specify the variable dimensions, increments, and number of instances.

When you pattern or copy a group, be careful which placement dimensions you select to increment or vary. If a feature in a group references another for placement (for example, a chamfer references the edge of a hole), you need to change only the placement dimensions of the referenced feature.

If you place features in a group separately, you must change the placement dimensions of ***each member,*** otherwise, features with unchanged dimensions will have several copies superimposed onto one another.

Patterns can be modified as shown in Figure 7.7. Table-driven patterns can also be created (Fig. 7.8).

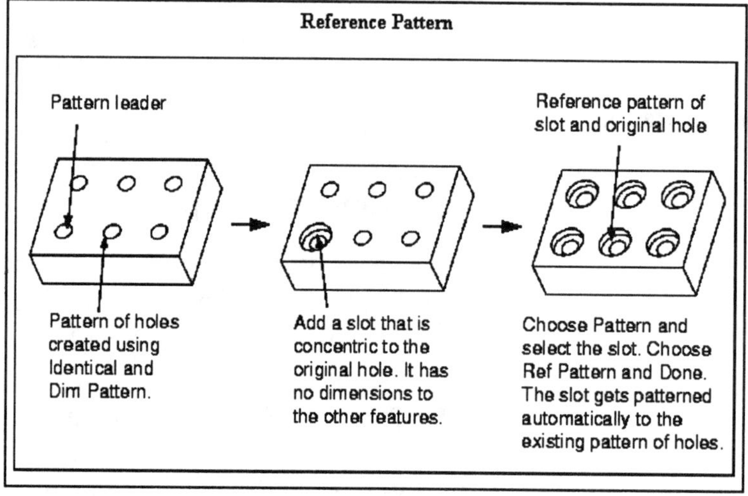

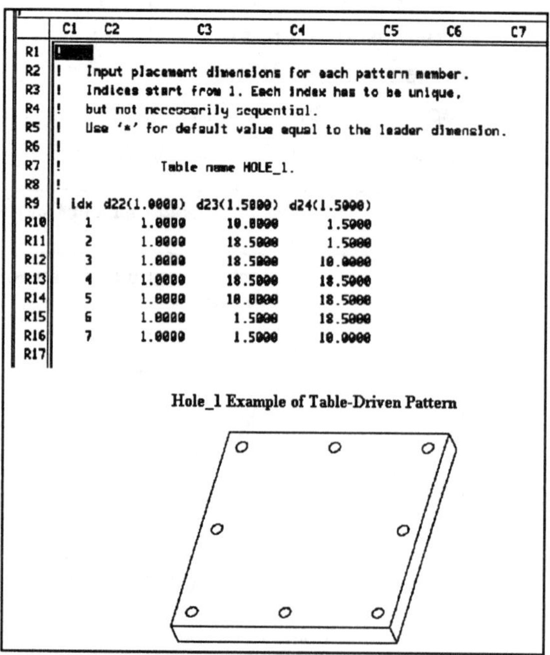

	C1	C2	C3	C4
R1				
R2	!	Input placement dimensions for each pattern member.		
R3	!	Indices start from 1. Each index has to be unique,		
R4	!	but not neccessarily sequential.		
R5	!	Use '*' for default value equal to the leader dimension.		
R6	!			
R7	!	Table name HOLE_1.		
R8	!			
R9	! idx	d22(1.0000)	d23(1.5000)	d24(1.5000)
R10	1	1.0000	10.0000	1.5000
R11	2	1.0000	18.5000	1.5000
R12	3	1.0000	18.5000	10.0000
R13	4	1.0000	18.5000	18.5000
R14	5	1.0000	10.0000	18.5000
R15	6	1.0000	1.5000	18.5000
R16	7	1.0000	1.5000	10.0000
R17				

Hole_1 Example of Table-Driven Pattern

Figure 7.8
Online Documentation, Table-Driven Patterns

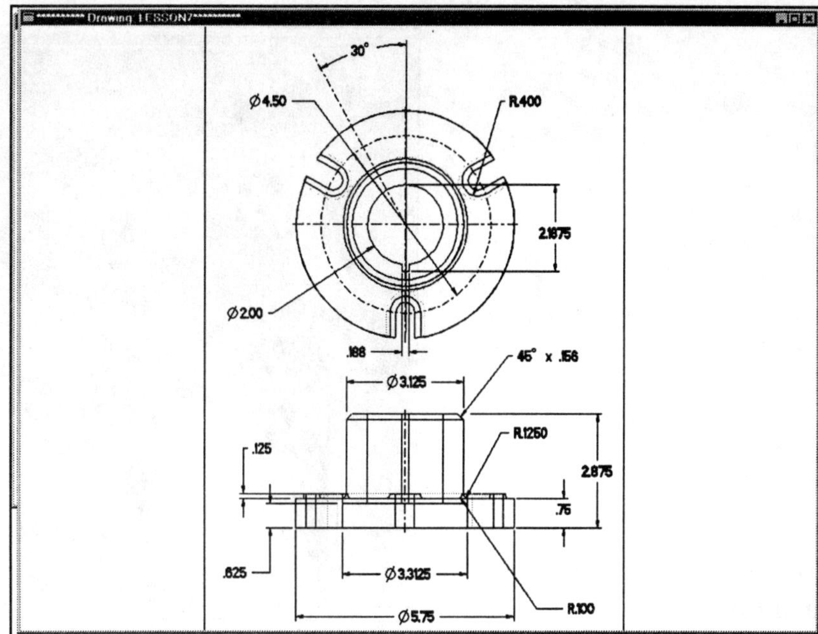

Figure 7.9
Post Reel Detail

Post Reel

The **Post Reel** (Fig. 7.9) is created with a *revolved protrusion*, as in Lesson 5 and Lesson 6. The internal geometry of the Post Reel can be created with a *revolved cut* instead of two holes of differing diameters or a sketched hole. The chamfers and the rounds are simple pick-and-place features. One boss and one slot (Fig. 7.10) are created using a datum on the fly (**Make Datum**), then grouped and patterned to complete the part. A detailed set of instructions will be supplied only for the boss and slot, since the other geometry is similar to that in previous lessons. The dimensions for the part are provided in Figures 7.9 through 7.12. Set up the part using the following commands:

Part ⇒ Create ⇒ POST_REEL ⇒ enter
Feature ⇒ Create ⇒ Datum ⇒ Plane ⇒ Default ⇒ Create ⇒ Datum ⇒ Coord Sys ⇒ Default ⇒ Done ⇒ Done

PT/Modeler™

Feature ⇒ Datum Plane ⇒ Coord Sys ⇒ Default ⇒ Done ⇒

Layer ⇒ Create ⇒ (type **DATUM_LAYER**) **⇒ enter ⇒ enter ⇒ Set Items ⇒ Add Items ⇒ (✓DATUM_LAYER) ⇒ Done Sel ⇒ Datum Plane ⇒ Sel By Menu ⇒ Name ⇒ Sel All ⇒ Ok ⇒ Done Sel ⇒ Done/Return ⇒ Done/Return**

SETUP AND ENVIRONMENT

Set Up ⇒ Units ⇒ Length ⇒ Inch ⇒ Done ⇒ Material ⇒ Define ⇒ (type **Steel_1020,** then press **enter**) **⇒** (table of material properties, change or add information) **⇒ File ⇒ Save ⇒ File ⇒ Exit ⇒ Assign ⇒** (pick **Steel_1020**) **⇒ Accept ⇒ Done**
Environment ⇒ ✓Grid Snap
☐ **SketStart2D** (sketch in 3D)
Hidden line Tan Dimmed

Figure 7.10
Slot and Boss

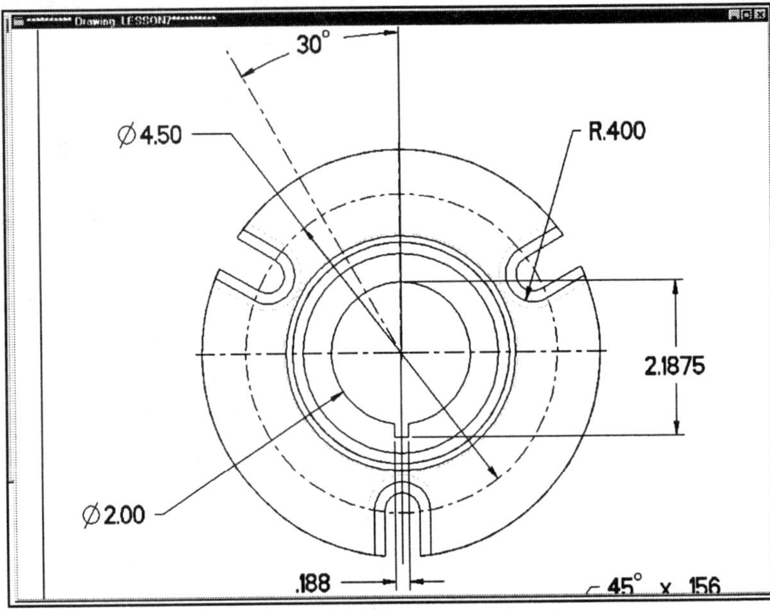

Figure 7.11
Post Reel Drawing,
Top View

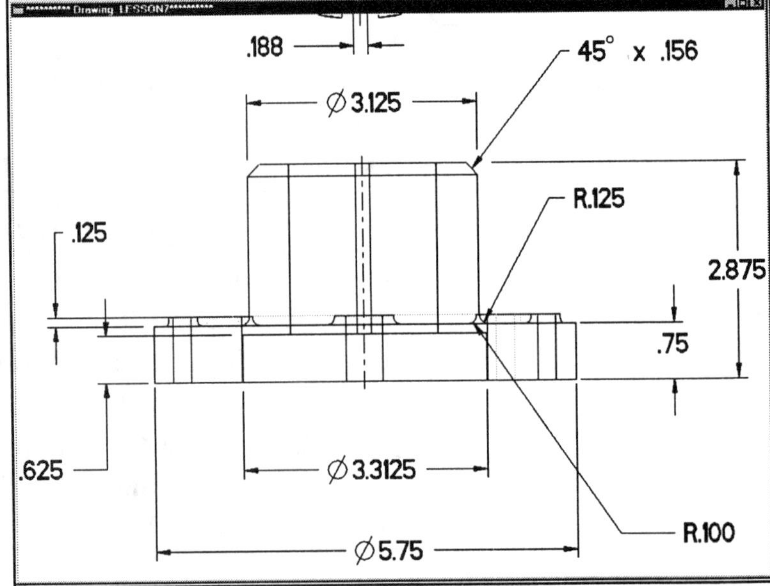

Figure 7.12
Post Reel Drawing,
Front View

The first protrusion is revolved and consists of the flangelike geometry shown in Figure 7.13. Sketch on **DTM3,** and revolve the section **360°** about a *vertical axis.*

HINT

Remember to create a palette of colors and use them on the model as required.

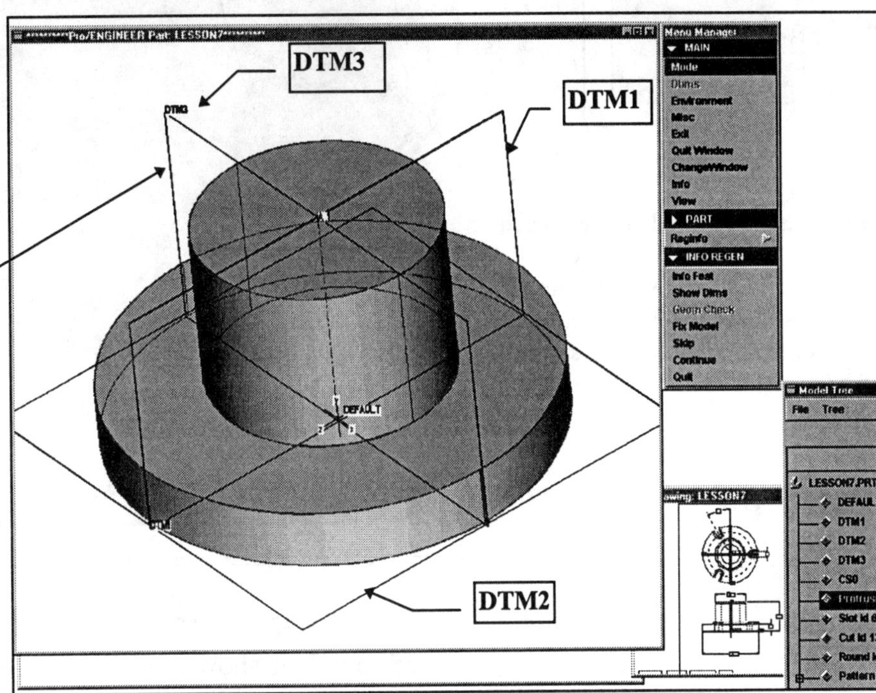

Figure 7.13
Revolved Protrusion

The second feature will be the internal cut shown in Figure 7.14. You can use the previous sketching plane and references.

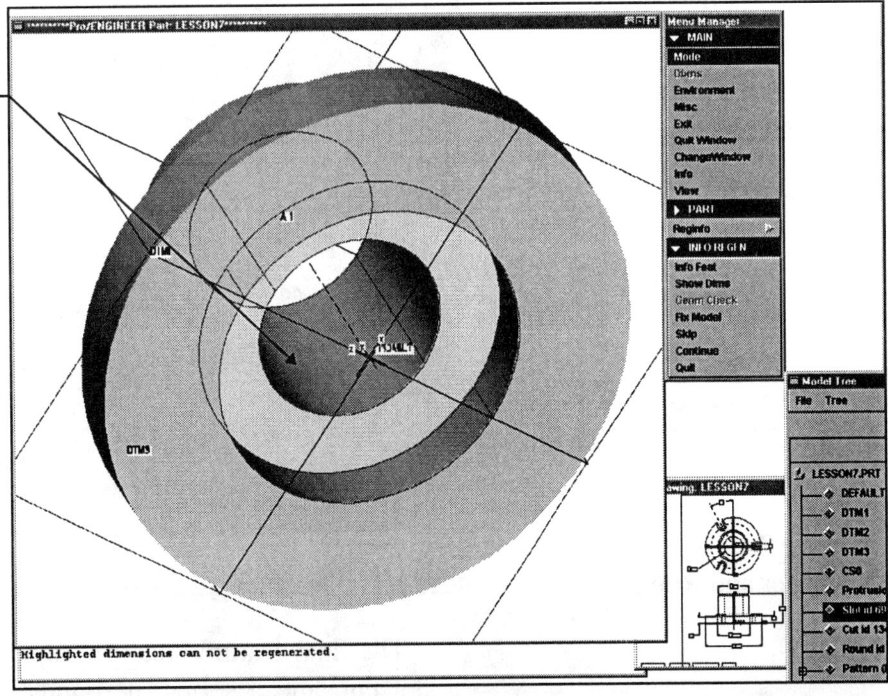

Figure 7.14
Revolved Cut

Dbms (File--PT/Modeler) **⇒**
Save ⇒ enter
Purge ⇒ enter ⇒
Done-Return

The keyseat can be created with a cut from the end of the Post Reel using three lines sketched on **DTM2** (or the top face), or can be created with one line from the side, sketching on **DTM1** and projecting to both sides (Fig. 7.15). Be careful to use the proper dimensioning scheme.

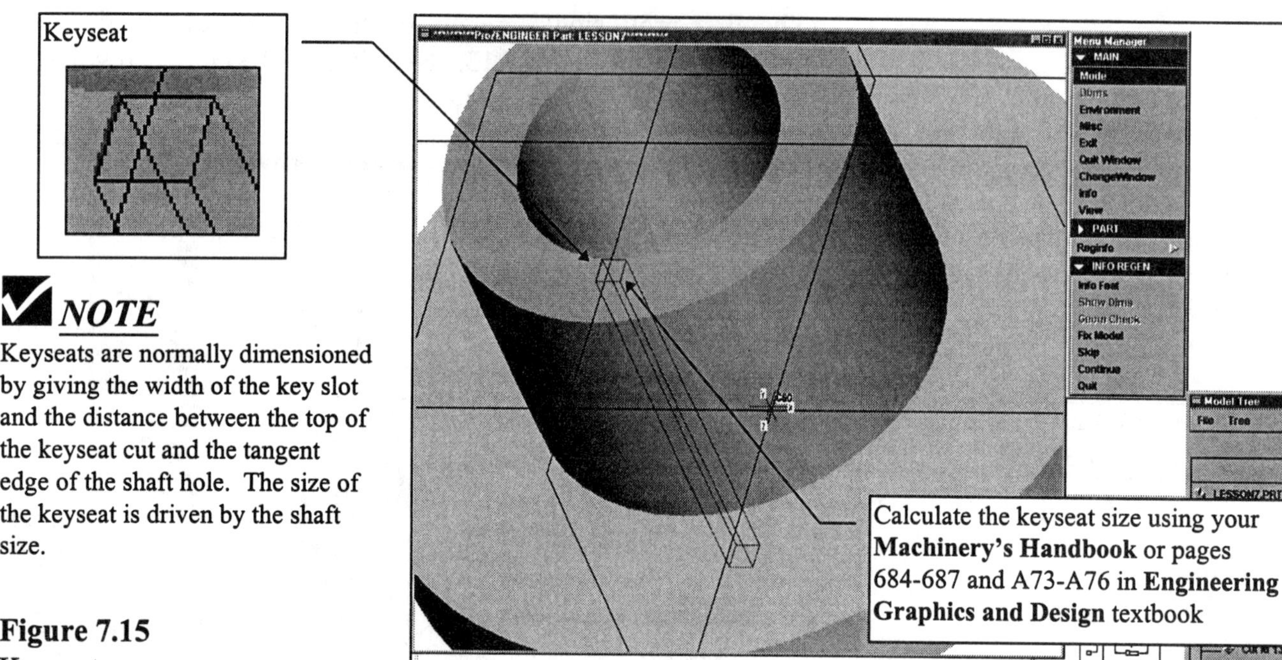

NOTE

Keyseats are normally dimensioned by giving the width of the key slot and the distance between the top of the keyseat cut and the tangent edge of the shaft hole. The size of the keyseat is driven by the shaft size.

Figure 7.15
Keyseat

Add the **45° X .156** chamfer to the top of the Post Reel and the **R.125** round, as shown in Figure 7.16.

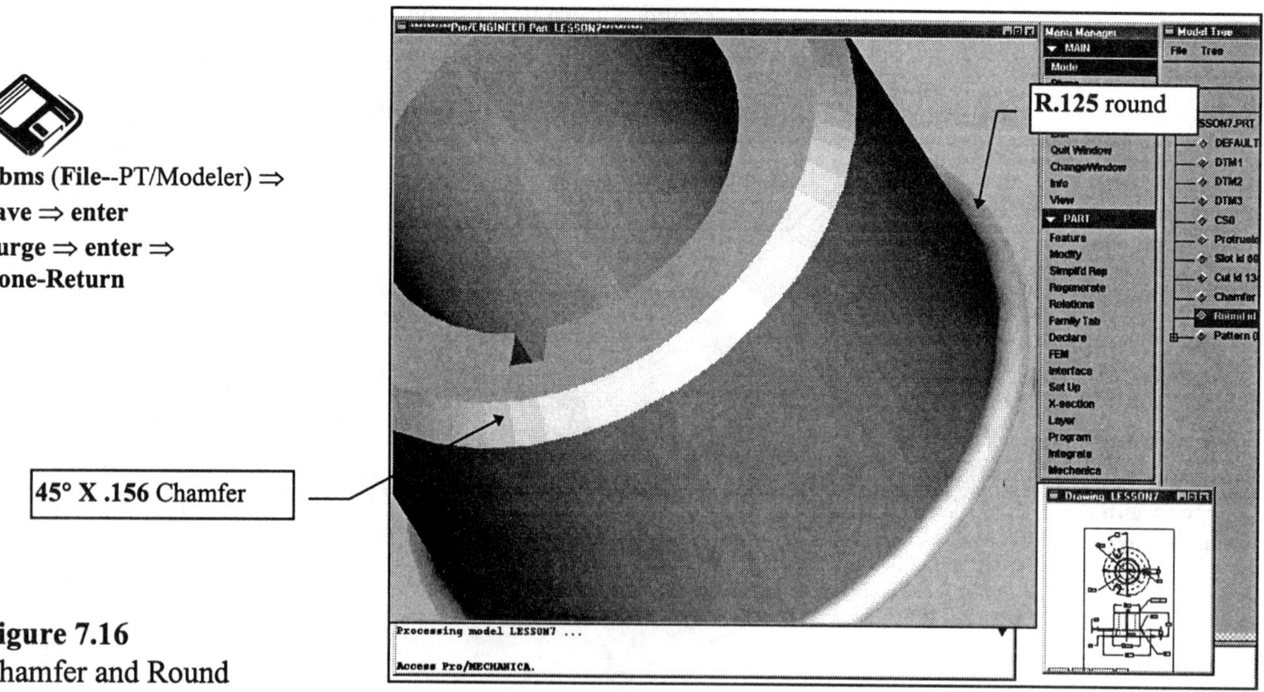

Dbms (File--PT/Modeler) ⇒
Save ⇒ enter
Purge ⇒ enter ⇒
Done-Return

Figure 7.16
Chamfer and Round

The final feature creation sequence consists of a boss protrusion, a slot cut, and a series of rounds. The three types of features are grouped and the group is then patterned. The protrusion and the slot must be created with a *datum-on-the-fly* (**Make Datum** within the SETUP PLANE menu).

Create the boss protrusion using the following commands:

PT/Modeler™
Feature ⇒ Protrusion etc.

Create ⇒ Protrusion ⇒ Extrude ⇒ Solid ⇒ Done ⇒ One Side ⇒ Done ⇒ (pick top of the flange as the placement surface, as in Fig. 7.17) **⇒ Okay** (for the direction of feature creation) **⇒ Bottom ⇒ Make Datum ⇒ Through** (pick axis **A_1**) **⇒ Angle** (pick **DTM3**) **⇒ Done ⇒ Enter Value** (type **30** at the prompt) **⇒ enter** (Fig. 7.18)

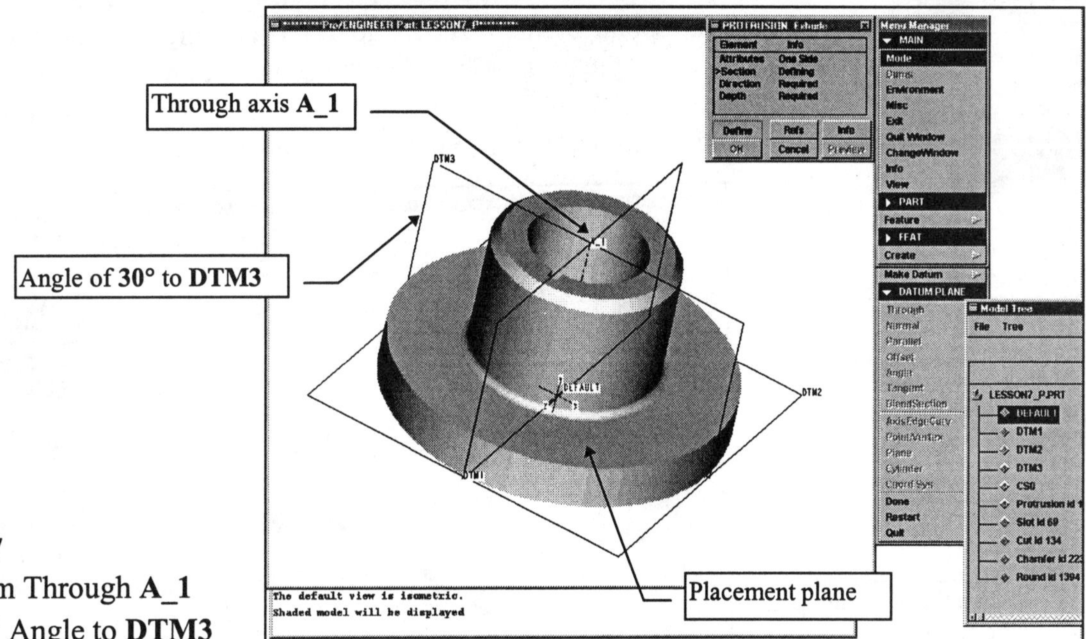

Figure 7.17
Make Datum Through **A_1** and at a **30°** Angle to **DTM3**

Turn on the grid snap and change the grid spacing:

PT/Modeler™
Sec Tools ⇒ Grid ⇒ Params ⇒ X Spacing ⇒ .20 ⇒ enter ⇒ Y Spacing ⇒ .20 ⇒ enter

Environment ⇒ ✓ Grid Snap ⇒ Done-Return ⇒ Sec Tools ⇒ Sec Environ ⇒ Grid ⇒ Params ⇒ X&Y Spacing ⇒ (type **.20** at the prompt) **⇒ enter ⇒** (zoom and pan the sketch as in Fig. 7.19)

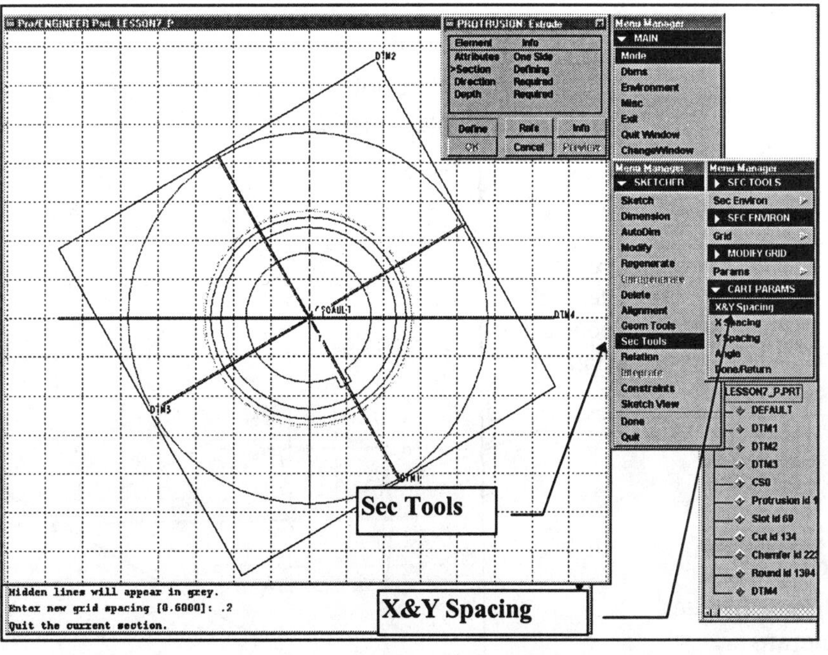

Figure 7.18
Sec Tools

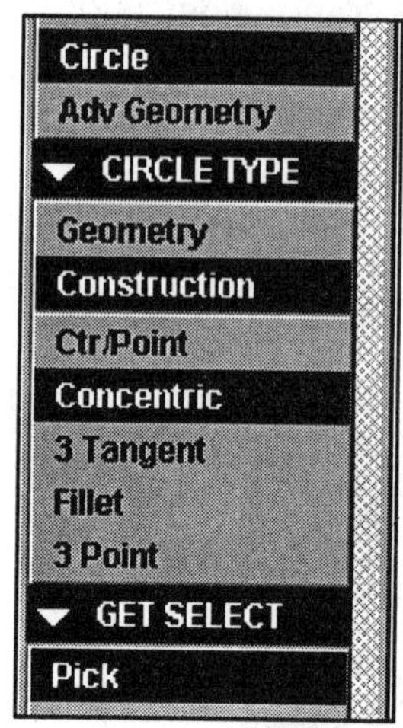

Sketch the two horizontal lines and the two arcs. The section must be *closed*. Create a construction circle to locate the center of the boss arc as shown in Figure 7.19:

> **Sketch** ⇒ **Circle** ⇒ **Construction** ⇒ **Concentric** ⇒ (pick any of the circles composing the part and then pick the center of the boss arc twice) ⇒ (sketch the remaining geometry) ⇒ **Regenerate**

Align the endpoints of the line to the flange curve and the centerpoint of the boss arc to **DTM4**. Align the arc to the outside circle of the part. Also align the construction circle with the center of the part (**DEFAULT** or **CS0**). Dimension the boss protrusion with an arc radius and the diameter of the construction circle.

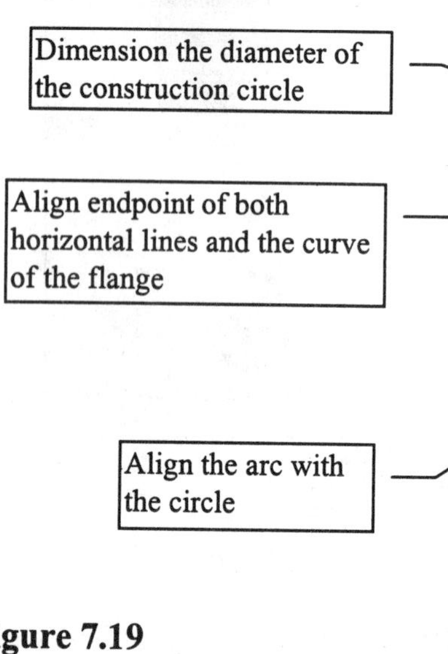

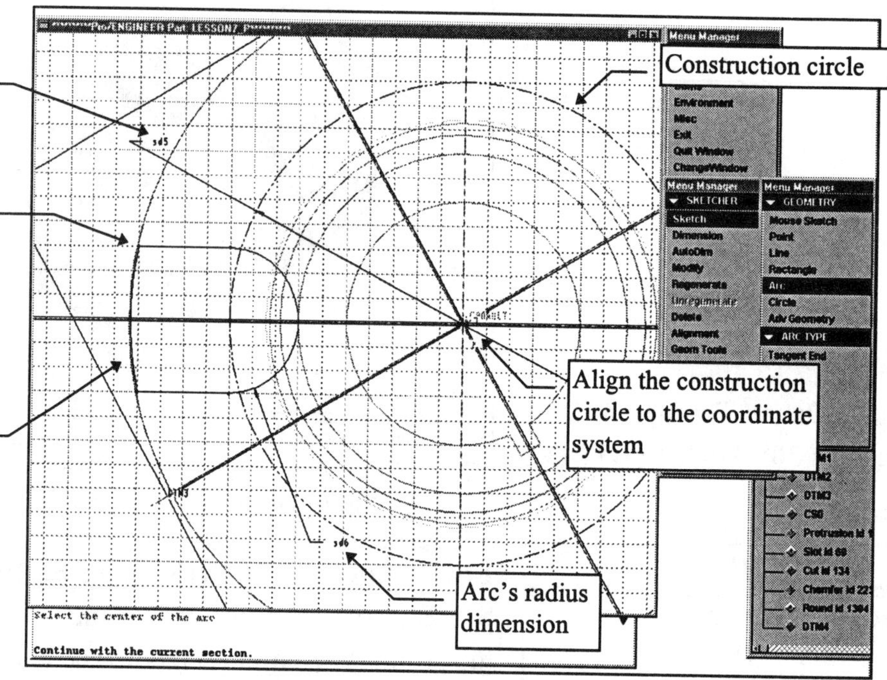

Figure 7.19
Sketch the Arc and Lines to Form the Boss Protrusion

If the sketch does not regenerate, add a horizontal dimension from the center of the boss arc to the center of the part, *or* add a sketch point at the intersection of the construction circle and the arc's center (Fig. 7.20).

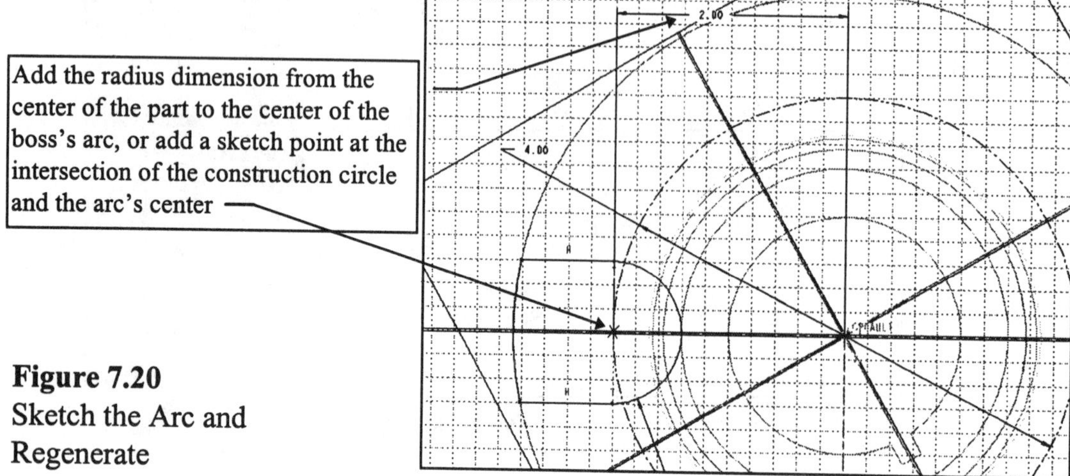

Figure 7.20
Sketch the Arc and Regenerate

After the sketch regenerates successfully, **Modify** the dimensions and **Regenerate** the sketch as shown in Figure 7.21.

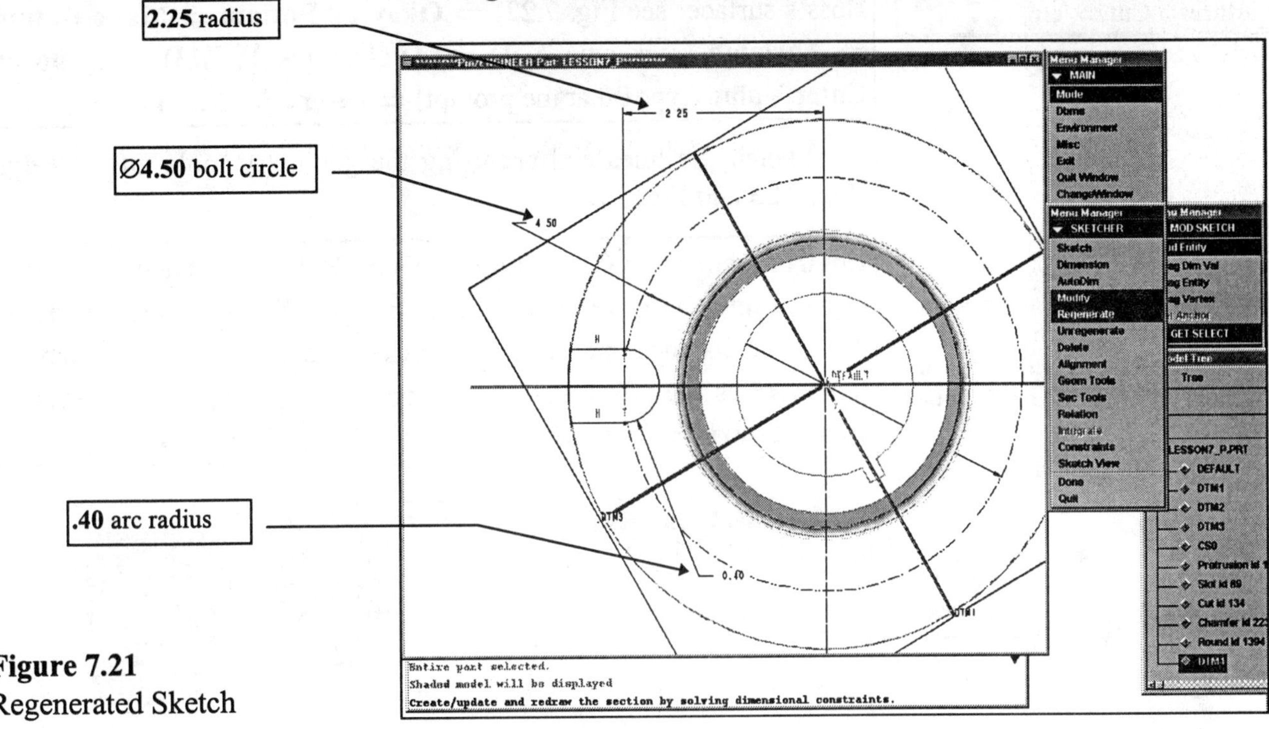

Figure 7.21
Regenerated Sketch

Complete the protrusion (Fig. 7.22) with the following commands:

Done ⇒ **Blind** ⇒ **Done** ⇒ (enter the depth of **.125**) ⇒ **enter** ⇒ **View** ⇒ **Default** ⇒ **Done-Return** ⇒ **Preview** ⇒ **OK** ⇒ **Done**

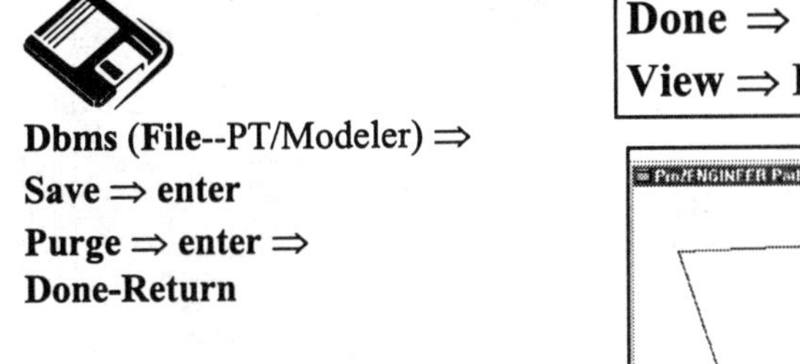

Dbms (File--PT/Modeler) ⇒
Save ⇒ **enter**
Purge ⇒ **enter** ⇒
Done-Return

Pick the top surface of the *boss* for the cut placement plane

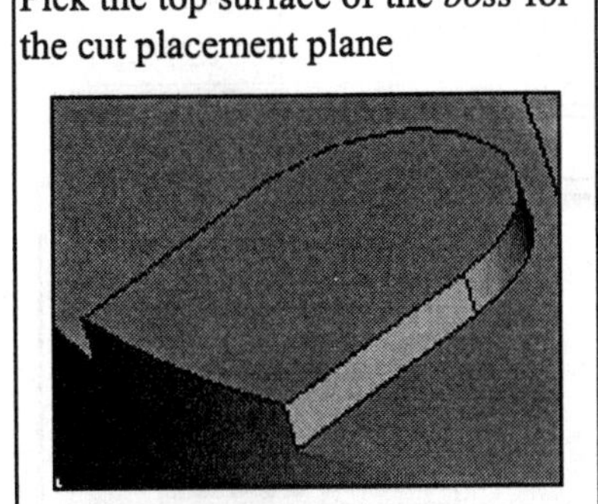

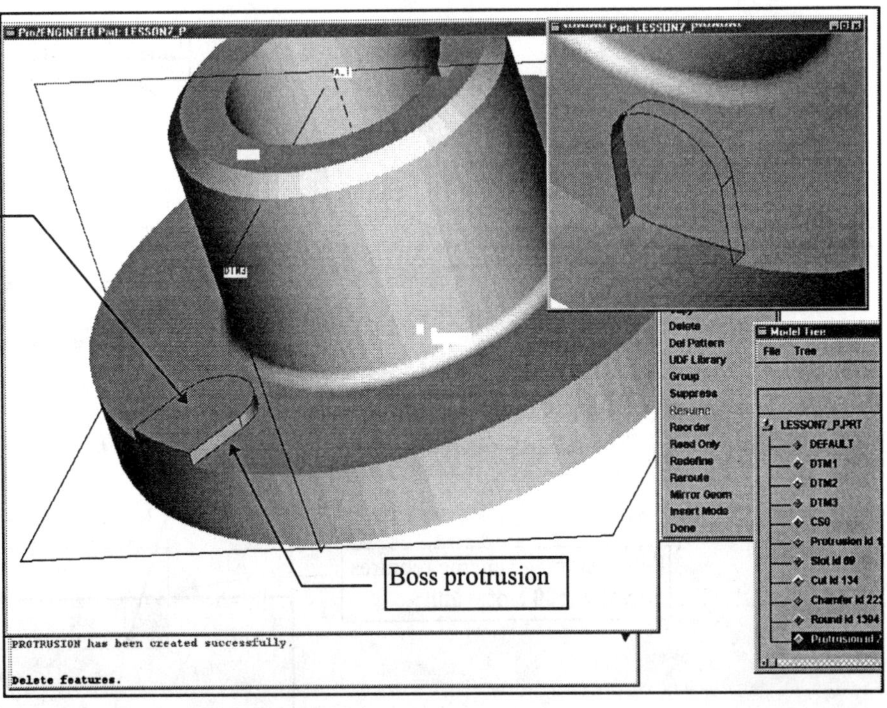

Figure 7.22
Regenerated Sketch

The next feature is a cut through the boss. The boss surface will be the placement plane (Fig. 7.22). The reference/orientation plane is created using **Make Datum**:

PT/Modeler™
Feature ⇒ Cut ⇒ etc.

Feature ⇒ Create ⇒ Cut ⇒ Done ⇒ Done ⇒ (pick the top of the Boss's surface; see Fig. 7.22) **⇒ Okay ⇒ Bottom ⇒ Make Datum ⇒ Through** (pick axis **A_1**) **⇒ Angle** (pick **DTM3**) **⇒ Done ⇒ Enter Value** (type **30** at the prompt) **⇒ enter** (Fig. 7.23)

Sketch the lines and arc using the geometry tools **Offset Edge** (Fig. 7.23 and Fig. 7.24):

HINT

Type a *plus* **.15** or a *minus* **.15** based on the direction of the arrow showing the side on which the geometry will be created. Draw the lines and the arc to the inside of the boss.

Environment ⇒ □ **Grid Snap ⇒ Done-Return ⇒ Geom Tools ⇒ Offset Edge ⇒** (pick the lower line) **⇒ Done/Return ⇒** (type **-.15** as the offset distance) **⇒ enter ⇒** (pick the arc) **⇒ Done/Return ⇒** (type **-.15** as the offset distance) **⇒ enter ⇒** (pick the upper line) **⇒ Done/Return ⇒** (type **-.15** as the offset distance) **⇒ enter**

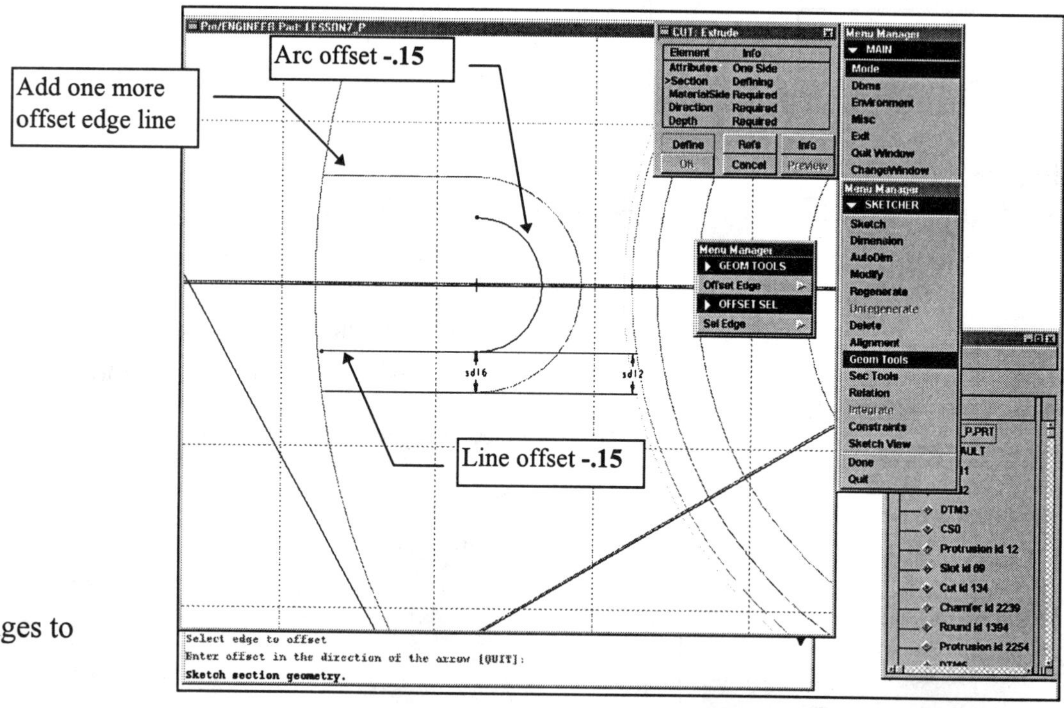

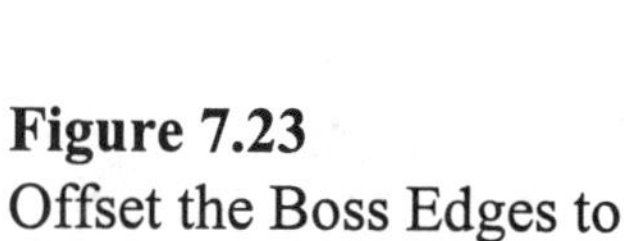

Figure 7.23
Offset the Boss Edges to Form the Slot

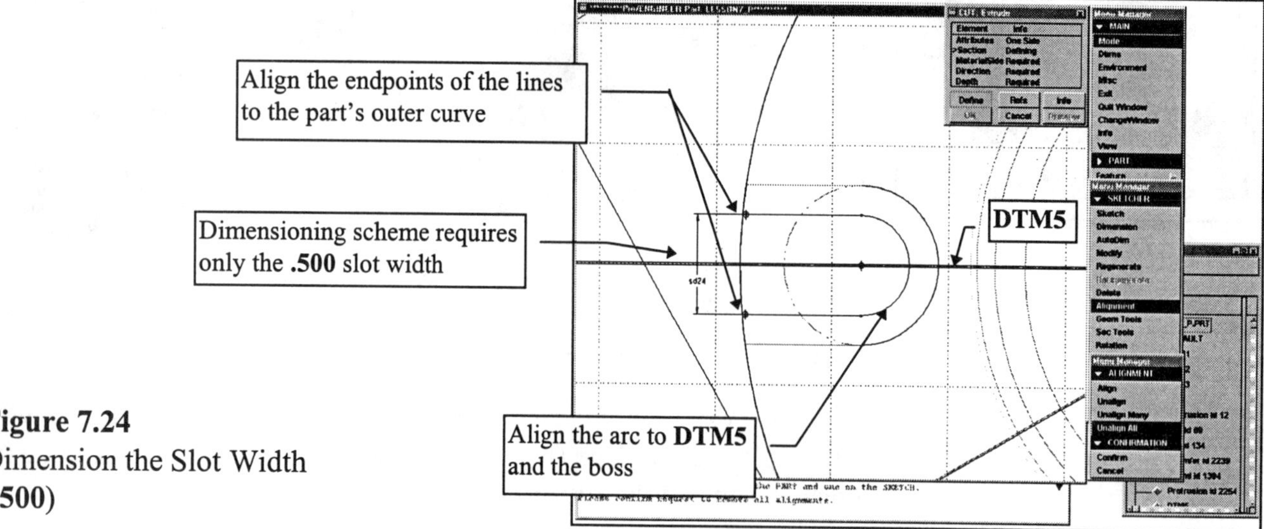

Figure 7.24
Dimension the Slot Width (.500)

Delete all three offset dimensions (**sd**) and repaint the screen to see the slot sketch geometry (Figs. 7.24 and 7.25). Align the endpoints of the horizontal lines to the outer curve of the part. Align the arc to the boss arc and to **DTM5**. **Regenerate** and add the dimension for the slot width. **Modify** the width (**.500**). *There is no need to modify the value if it was created at the correct design size.*

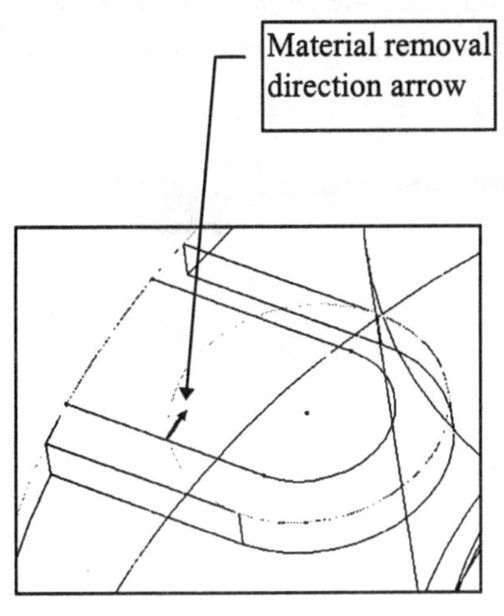

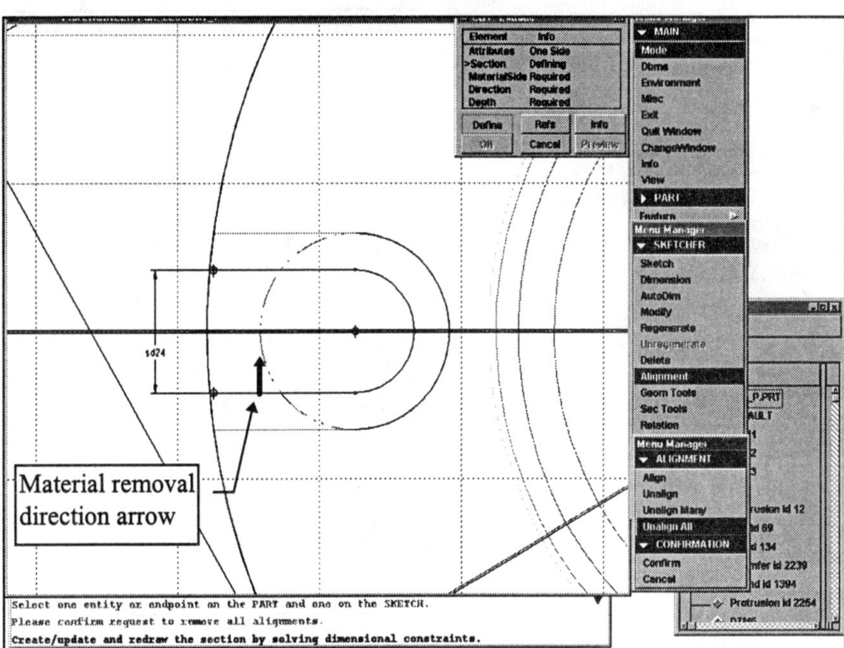

Figure 7.25
Regenerated Section Showing the Material Removal Side of Cut

Regenerate the sketch, choose **Done**, and choose **Okay** for the material removal direction. Make sure the direction of cut is toward the part, not out into space. The depth will be **Thru All**. Change the sketch orientation by rotating or using the default view. If the cut does not intersect the part, pick **Direction** and **Define** from the dialog box and flip the arrow, then choose **Okay, OK,** and **Done** (Fig. 7.26)

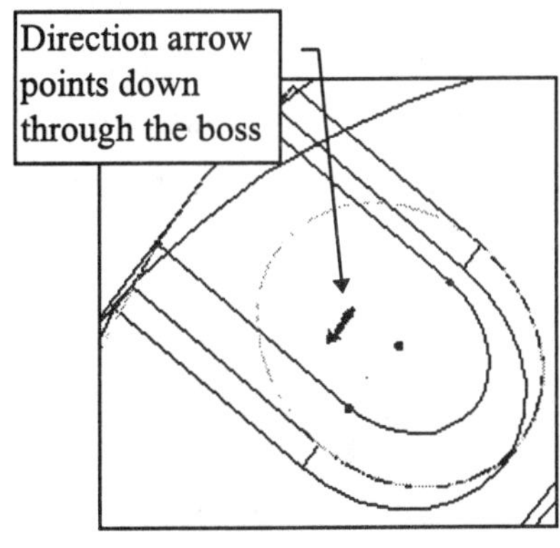

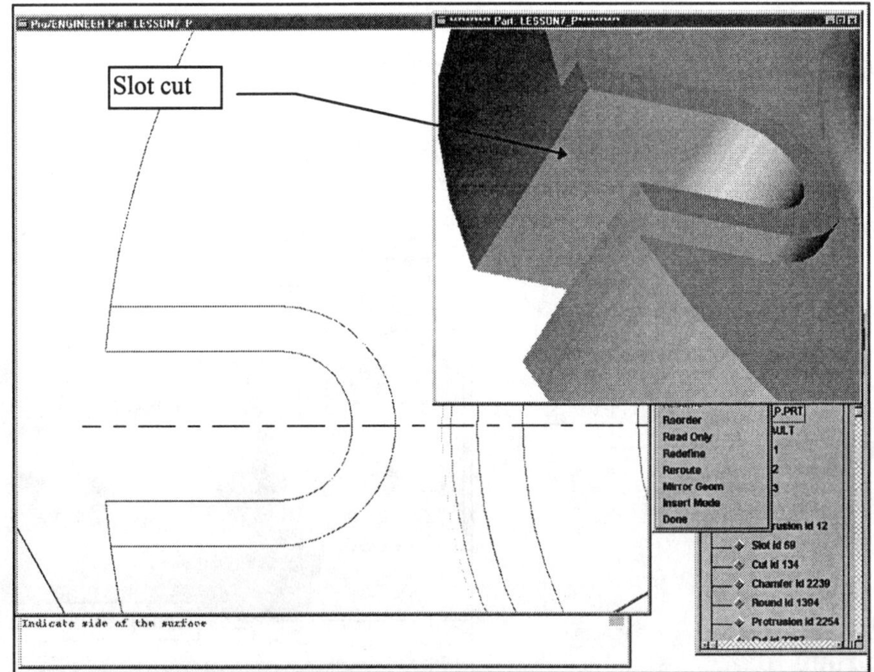

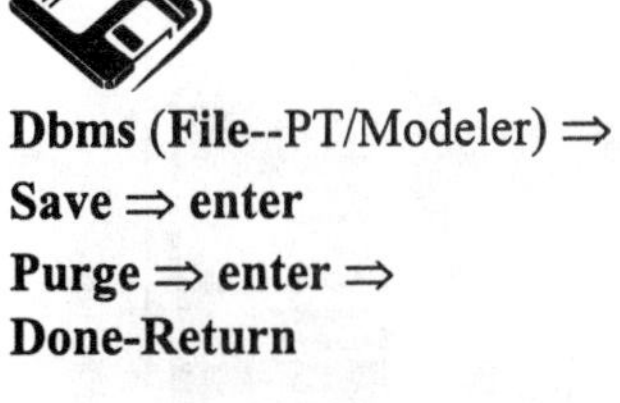

Dbms (**File**--PT/Modeler) ⇒
Save ⇒ enter
Purge ⇒ enter ⇒
Done-Return

Figure 7.26
Slot Cut

Create a set of rounds around the boss, as shown in Figure 7.27:

PT/Modeler™
Feature ⇒ Round ⇒ Done etc.

Feature ⇒ Create ⇒ Round ⇒ Simple ⇒ Done ⇒ Done ⇒ Tangnt Chain ⇒ (pick the line at the edge of the boss; both lines and the arc are highlighted) **⇒ Done Sel ⇒ Done ⇒** (type the radius value of **.10** at the prompt) **⇒ enter ⇒ Preview ⇒ OK ⇒ Done**

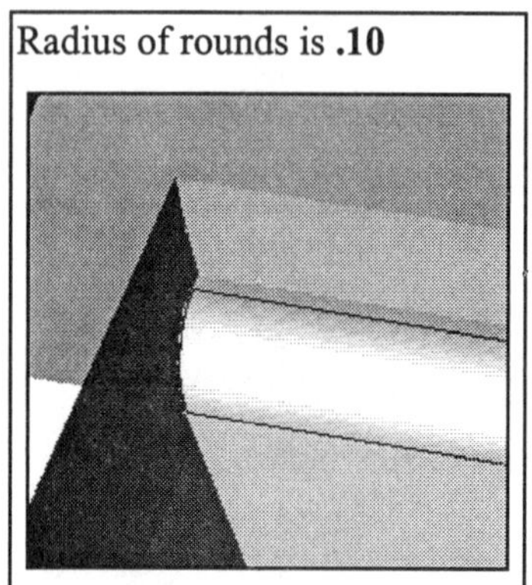

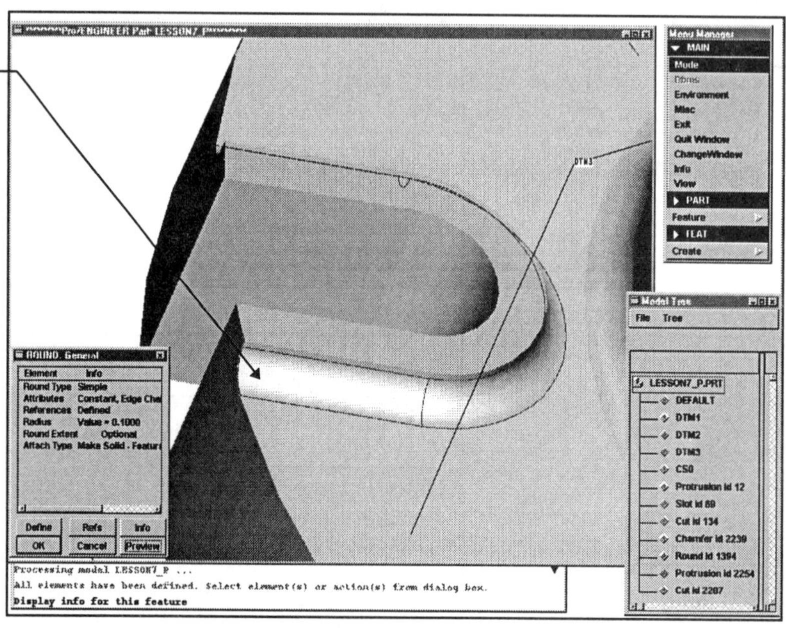

Figure 7.27
Rounds

Next we will group (Fig. 7.28) the three features together using the following commands:

PT/Modeler™
Feature Oper ⇒ Group ⇒ etc.

Feature ⇒ Group ⇒ Create ⇒ Local Group ⇒ (enter the group name at the prompt: **BOSS**) **⇒ enter ⇒** (select the features for the group: the *slot*, the *boss protrusion*, and the *rounds*) **⇒ Done Sel ⇒ Done ⇒** (Pro/E will respond with "**Group BOSS has been created**") **⇒ Done/Return ⇒ Done**

Dbms (**File**--PT/Modeler) **⇒**
Save ⇒ enter
Purge ⇒ enter ⇒
Done-Return

Group **BOSS**

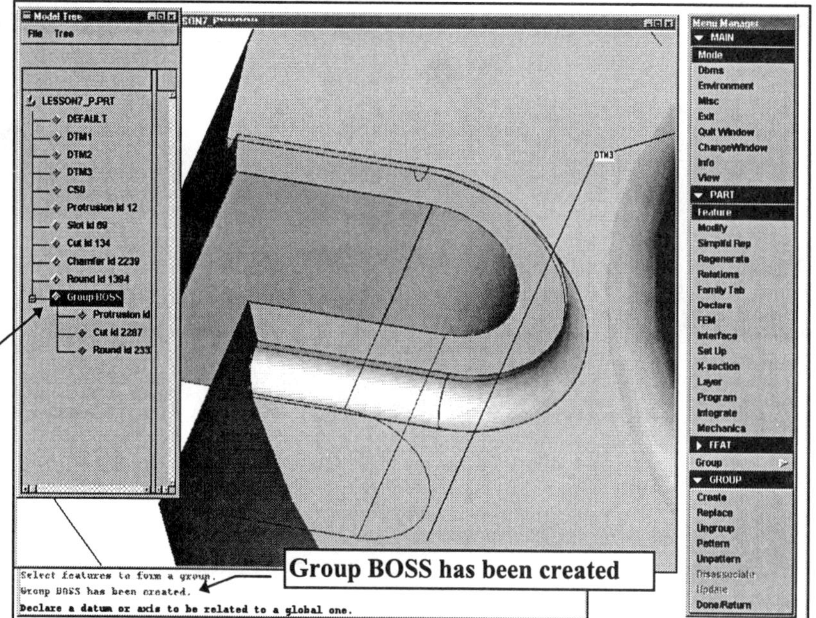

Figure 7.28
Group

Since the dimensions for the features that you will be using to pattern the group are the same for the angle of the boss protrusion and for the angle of the cut, you need to move the dimension. This must be done before the group is patterned (Fig. 7.29). Use the following commands:

PT/Modeler™
Modify ⇒ (pick the boss) ⇒ **Move Dim** ⇒ **etc.**

Modify ⇒ (pick the boss) ⇒ **DimCosmetics** ⇒ **Move Dim** ⇒ (pick the original position of the **30°** angle) ⇒ (pick the new position for the angle dimension)

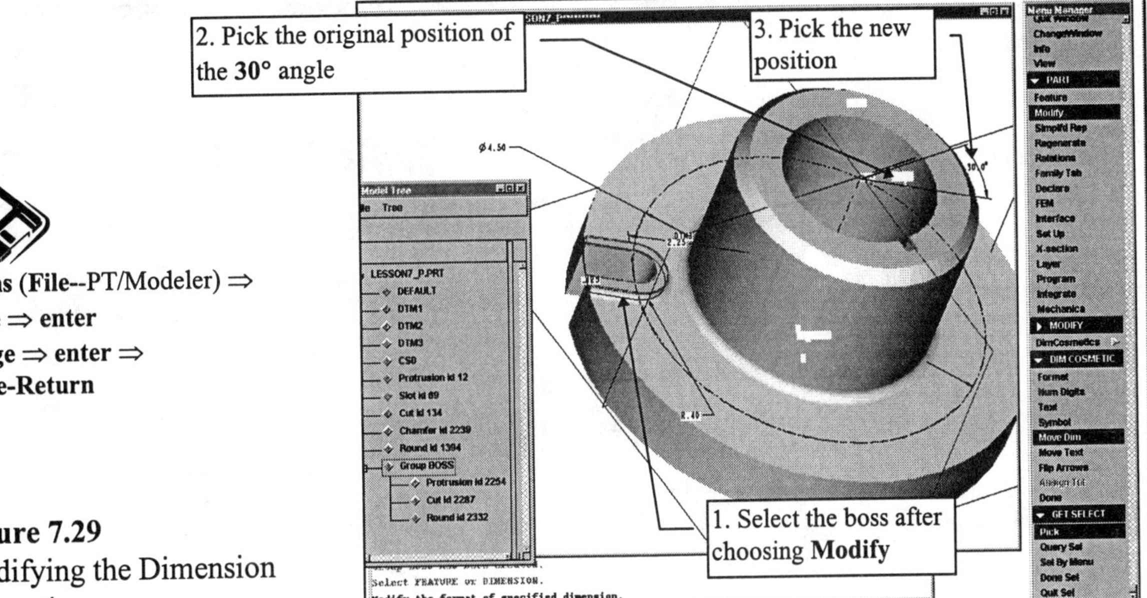

Dbms (File--PT/Modeler) ⇒
Save ⇒ **enter**
Purge ⇒ **enter** ⇒
Done-Return

Figure 7.29
Modifying the Dimension Cosmetics

Finally, the BOSS group needs to patterned. To pattern a group, you need to choose **Pattern** from the GROUP menu choices, not from the FEAT menu. Choose the following commands (Fig. 7.30):

PT/Modeler™
Feature Oper ⇒ **Group** ⇒ **etc.**

Feature ⇒ **Group** ⇒ **Pattern** ⇒ (select the group **BOSS** from the Model Tree; notice that two **30°** angle dimensions show on the screen, one for the slot and another for the protrusion)

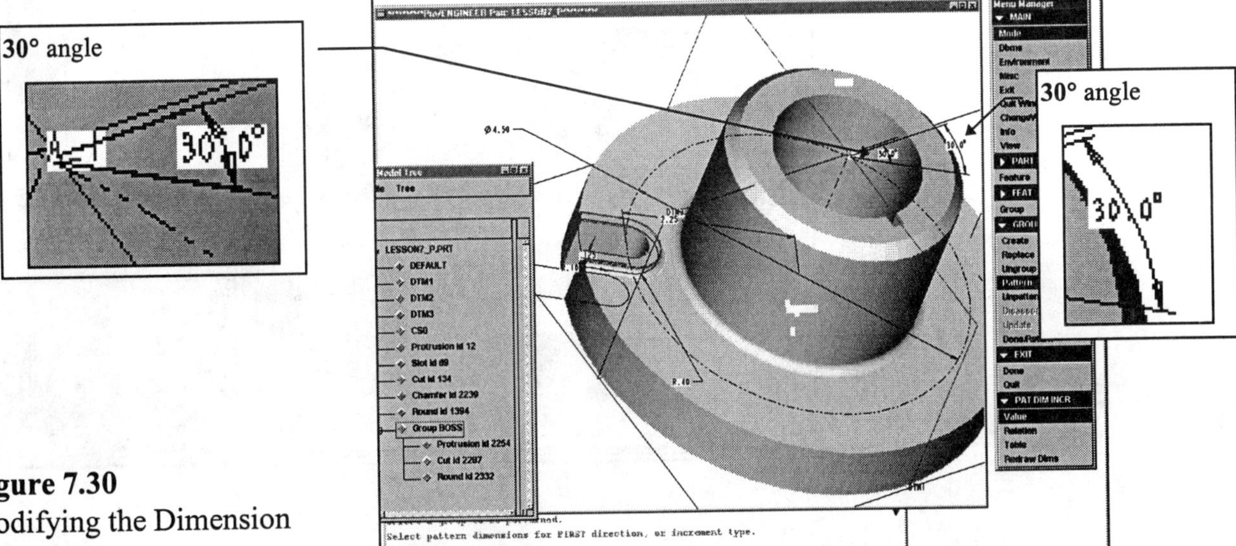

Figure 7.30
Modifying the Dimension Cosmetics

Pro/E will respond with "**Select pattern dimensions for the FIRST direction, or increment type.**" Continue with the following commands (Fig. 7.30):

(pick one of the **30°** angles) ⇒ (type **120**) ⇒ **enter** ⇒ (pick the other **30°** angle) ⇒ (type **120**) ⇒ **enter** ⇒ **Done** ⇒ (Pro/E responds with "**Enter TOTAL number of instances in this direction (including original):**") ⇒ (type **3**) ⇒ **enter** ⇒ **Done** (Figs. 7.31 and 7.32)

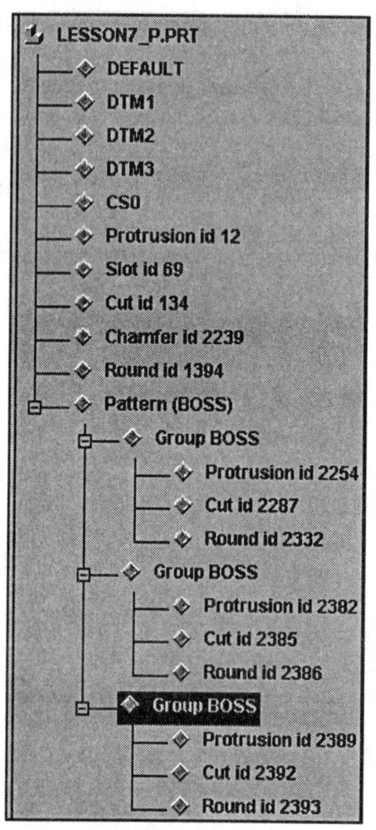

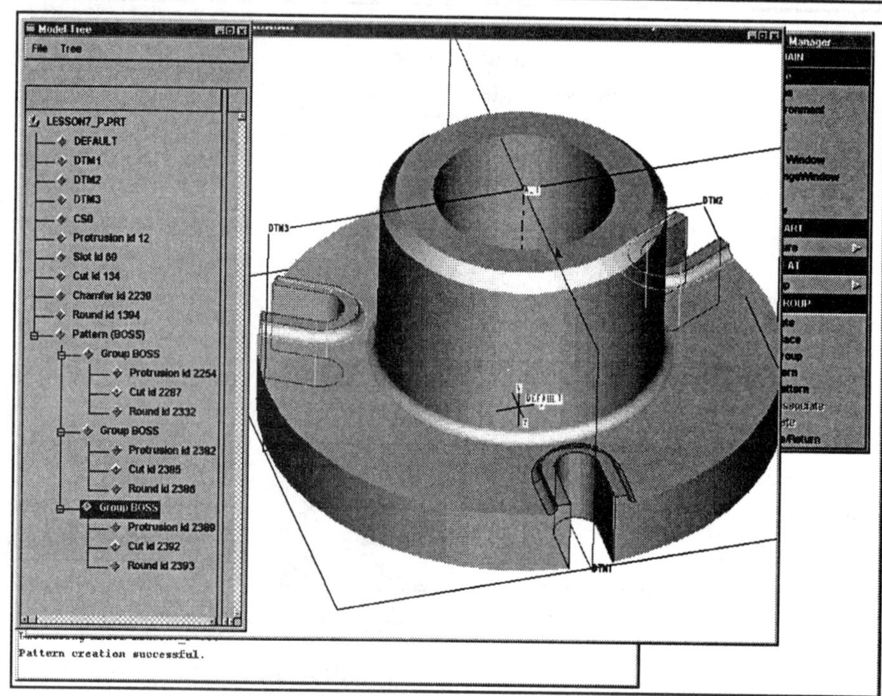

Figure 7.31
Patterning the Group

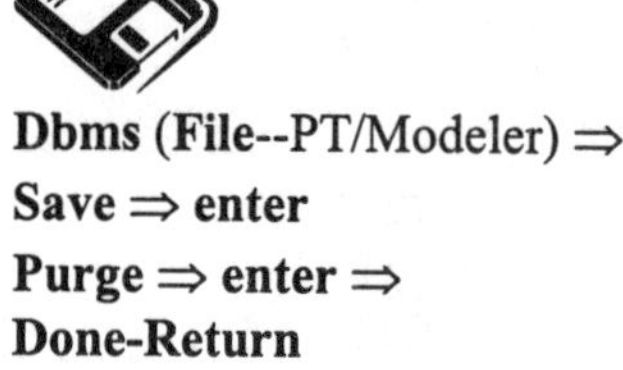

Dbms (File--PT/Modeler) ⇒
Save ⇒ **enter**
Purge ⇒ **enter** ⇒
Done-Return

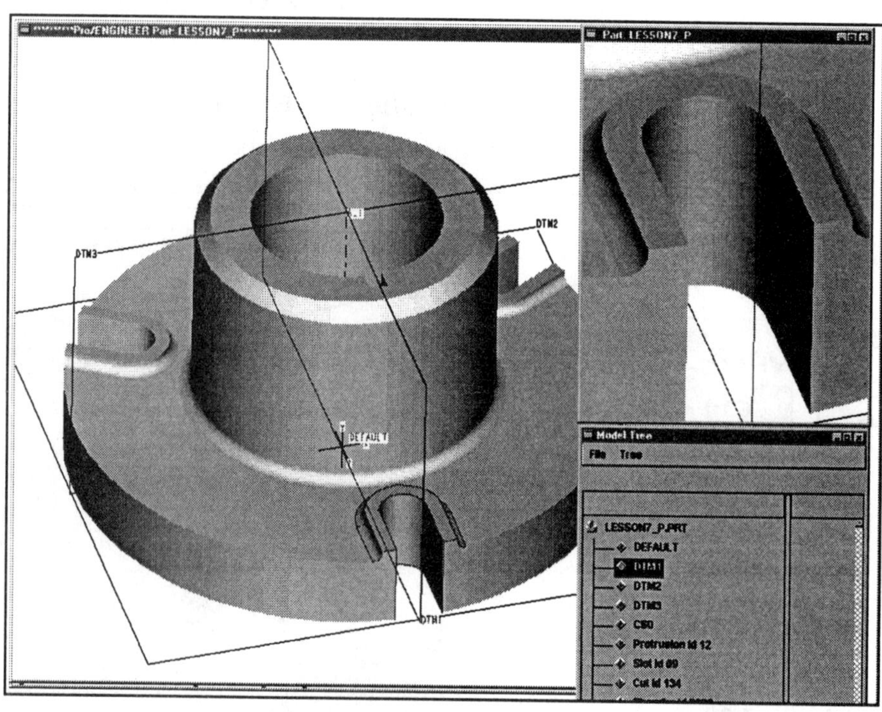

Figure 7.32
Completed Post Reel

Lesson 7 Project

Taper Coupling

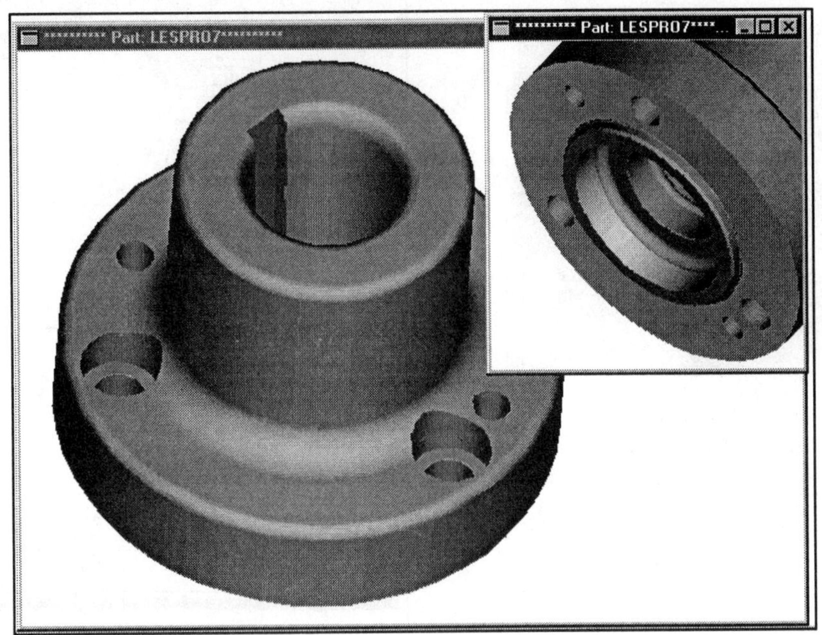

Figure 7.33
Taper Coupling

Taper Coupling

EGD REFERENCE
Engineering Graphics and Design with Graphical Analysis *or* **Fundamentals of Engineering Graphics and Design**
by L. Lamit and K. Kitto
Read Chapter 23
See pages 865-866

The seventh **lesson project** is a machined part that requires commands similar to the **Post Reel**. Create the part shown in Figures 7.33 through 7.47. This part will be used in Lessons 14 and 15.

At this stage in your understanding of Pro/E, you should be able to analyze the part and plan out the steps and features required to model it. Use the DIPS in Appendix D to plan out the feature creation and the parent-child relationships for the part. The coupling will be used for an assembly in Lessons 14 and 15. The machined face of the coupling mates with and is fastened to a similar surface when assembled. Plan your geometric tolerancing requirements accordingly, and set the datums to anticipate the mating surfaces.

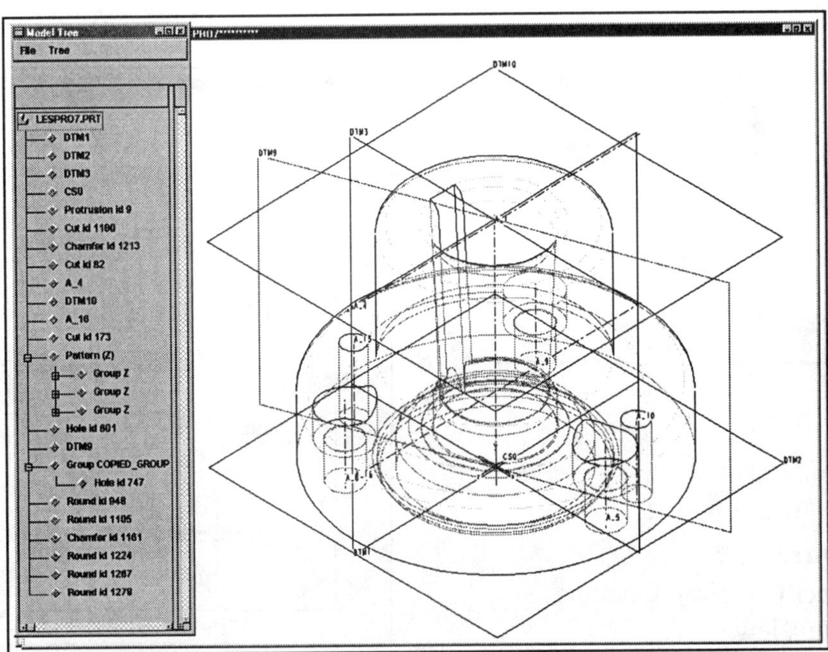

Figure 7.34
Taper Coupling Model with Datum Planes

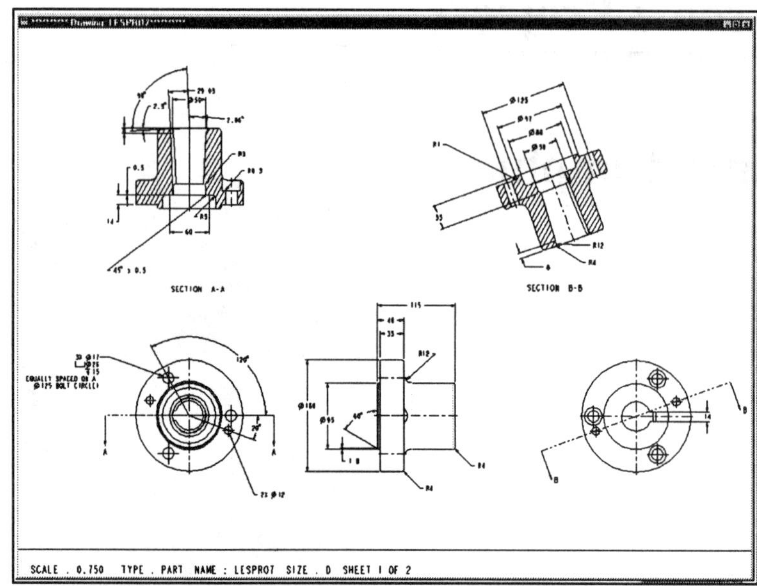

Figure 7.35
Taper Coupling Drawing

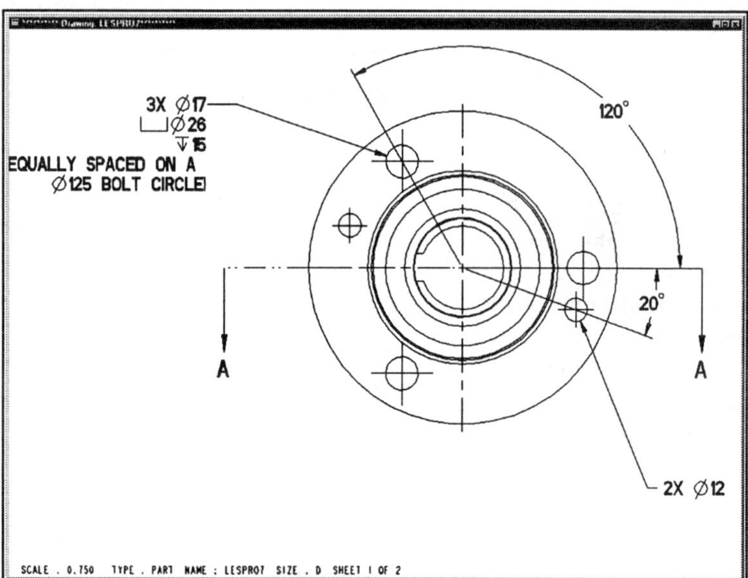

Figure 7.36
Taper Coupling Drawing, Bottom

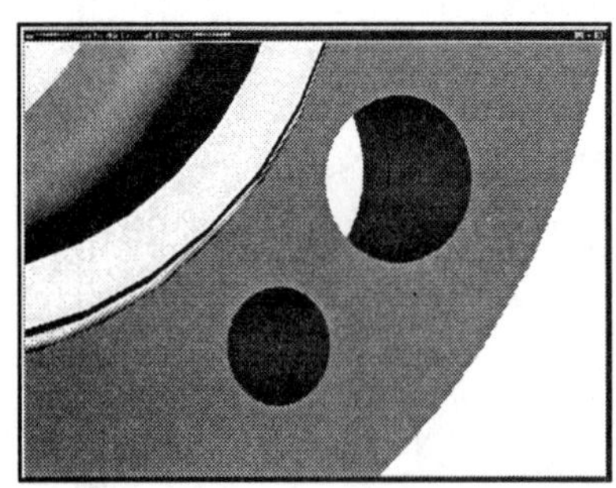

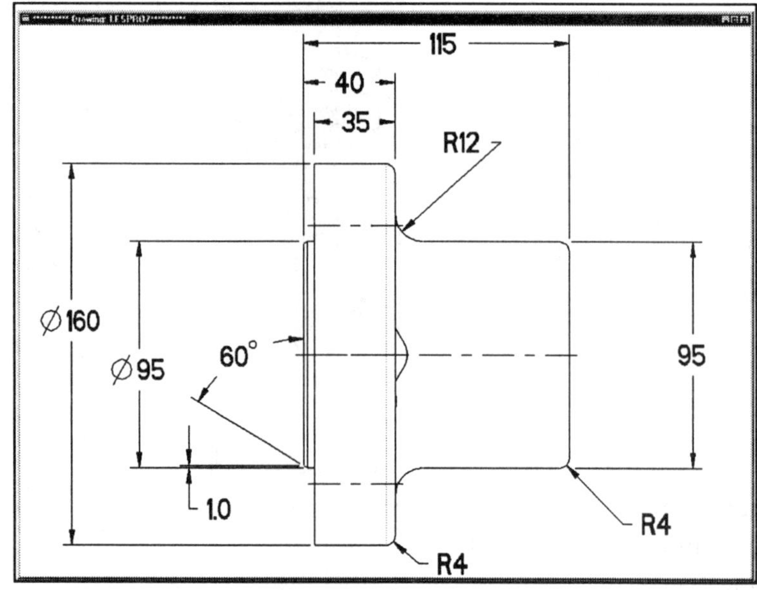

Figure 7.37
Taper Coupling Drawing, Side View

After completing Lesson 8, write a relation that will keep this dimension equal to the depth of the counterbore plus the radius of the large round (**R12**).

Dim=15+Radius
d18=d9+d6

Your dim symbols will differ.

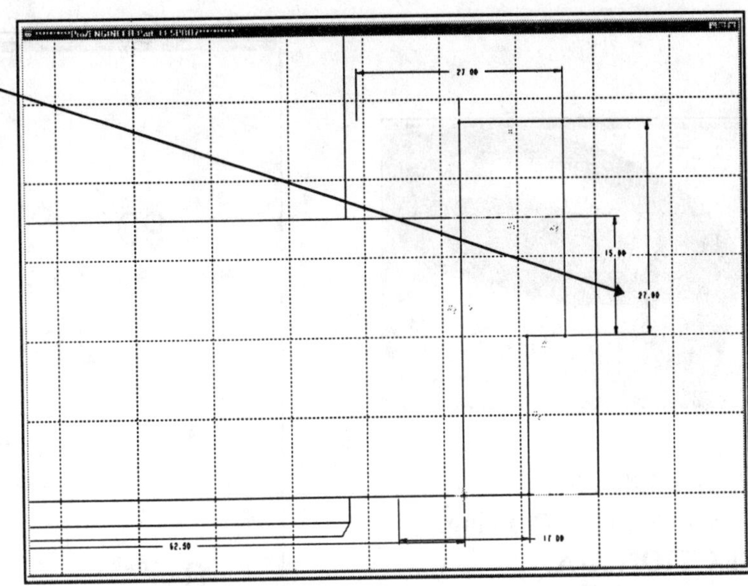

Figure 7.38
Taper Coupling Section, Counterbore

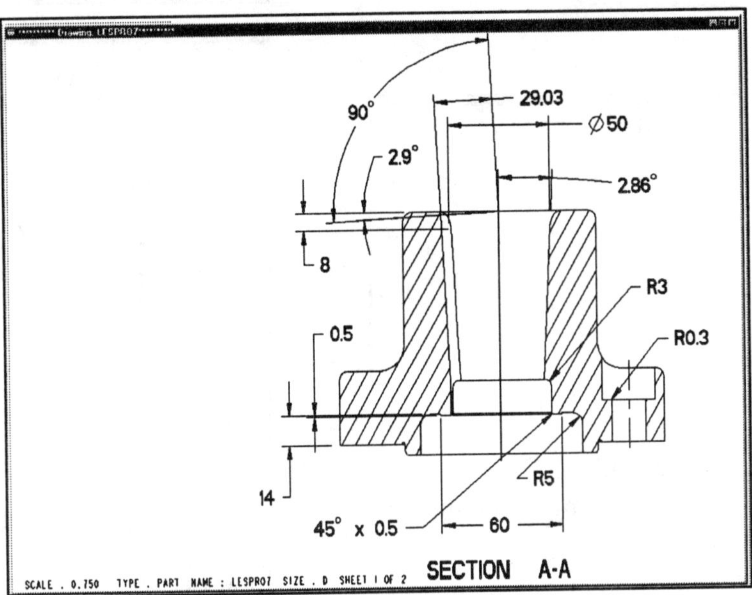

Figure 7.39
Taper Coupling Drawing, **SECTION A-A**

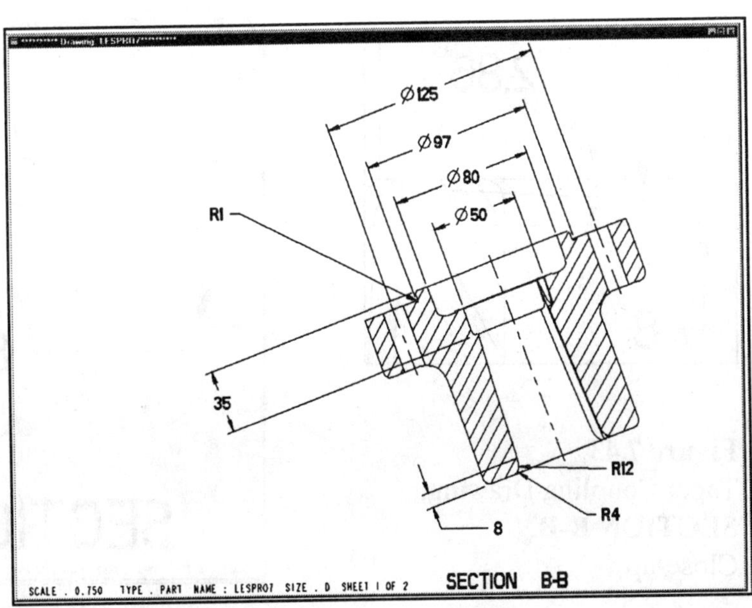

Figure 7.40
Taper Coupling Drawing, **SECTION B-B**

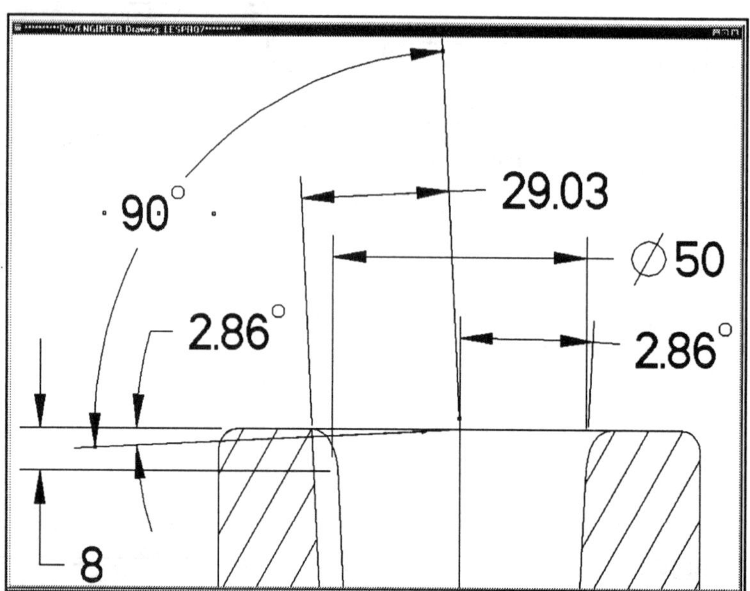

Figure 7.41
Taper Coupling Drawing,
SECTION A-A,
Taper Angle

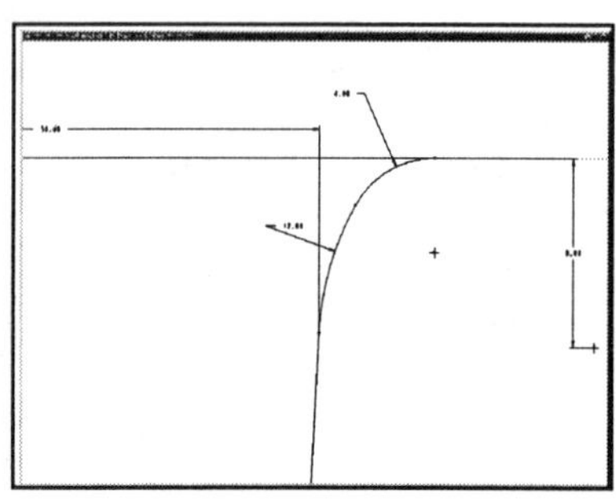

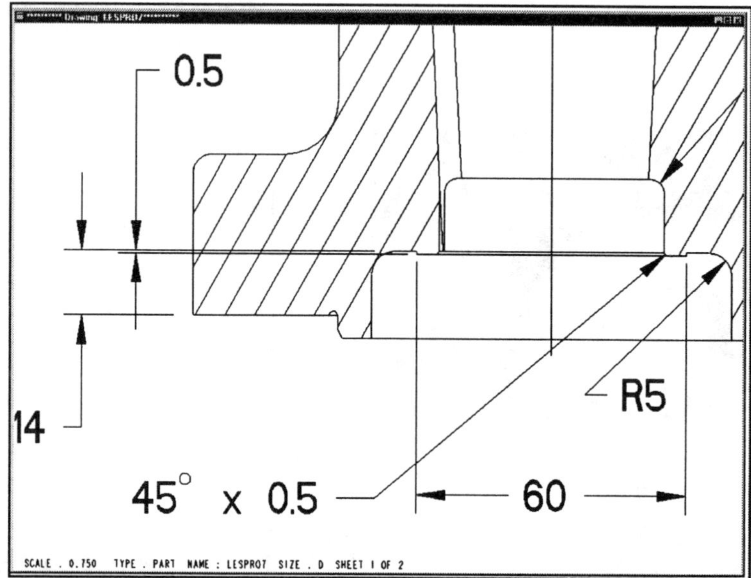

Figure 7.42
Taper Coupling Drawing,
SECTION A-A,
Close-up

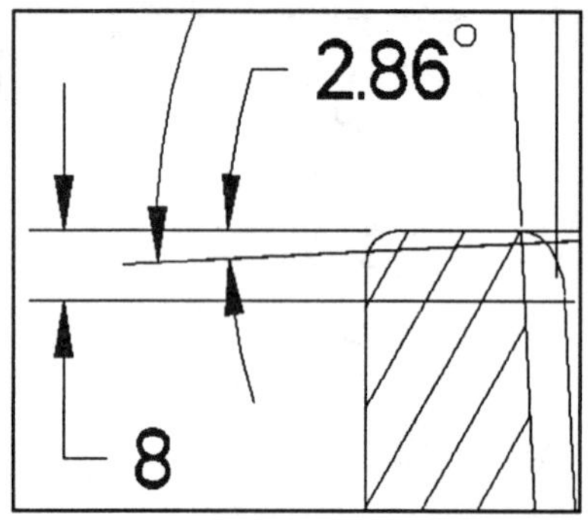

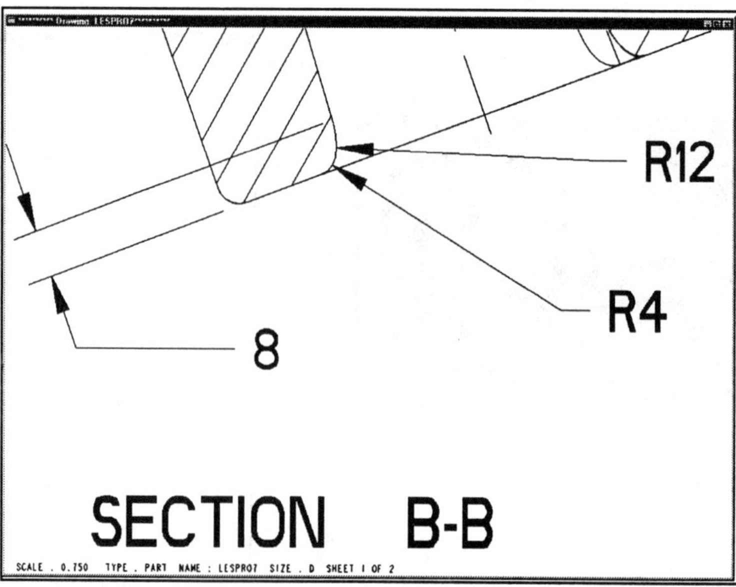

Figure 7.43
Taper Coupling Drawing,
SECTION B-B,
Close-up

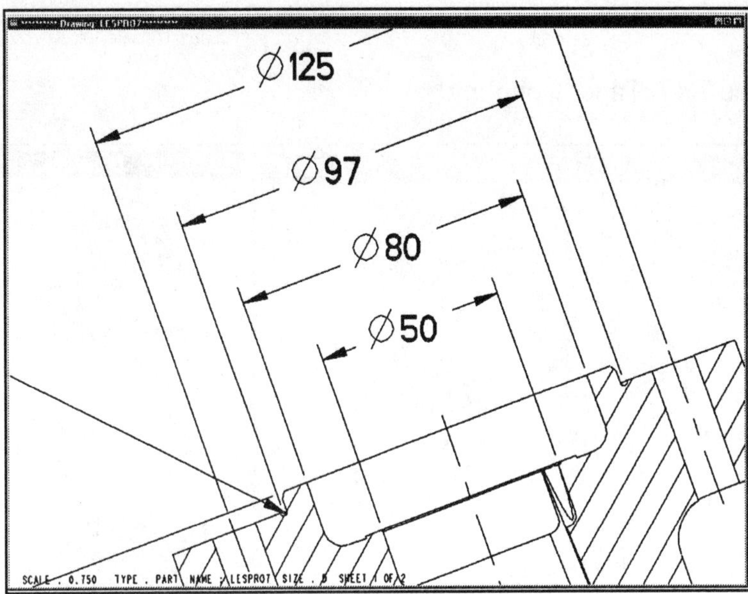

Figure 7.44
Taper Coupling Drawing,
SECTION B-B,
Mating Diameters

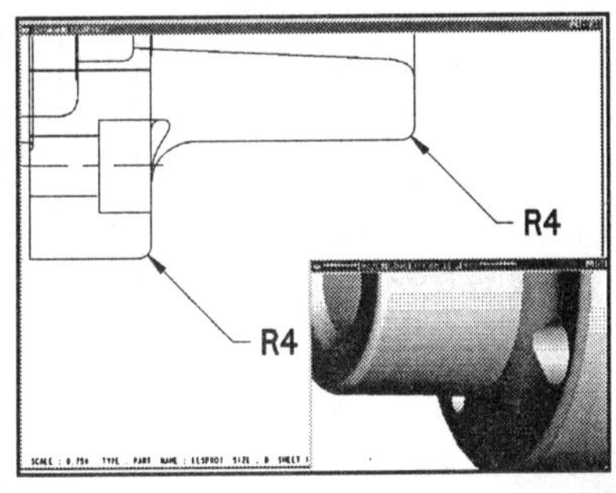

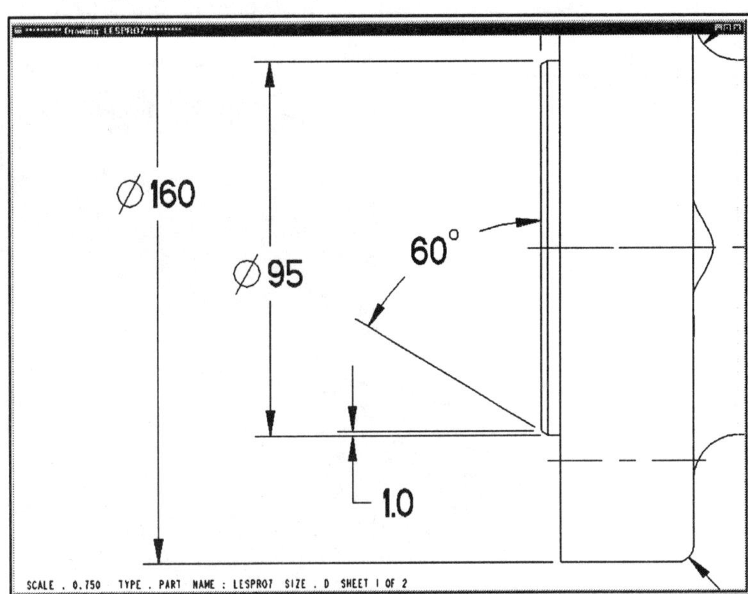

Figure 7.45
Taper Coupling Drawing,
Side View, Close-up

Dbms (File--PT/Modeler) ⇒
Save ⇒ enter
Purge ⇒ enter ⇒
Done-Return

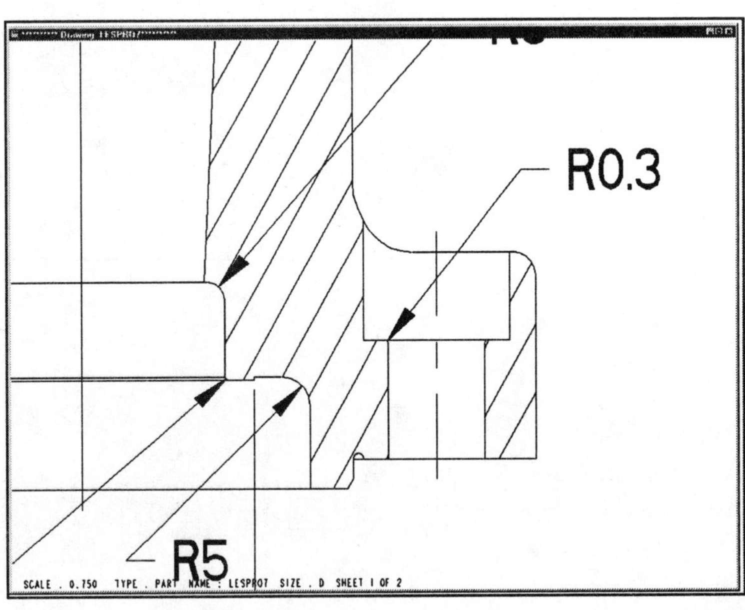

Figure 7.46
Taper Coupling Drawing,
SECTION A-A,
Close-up of Radii

Figure 7.47 Taper Coupling

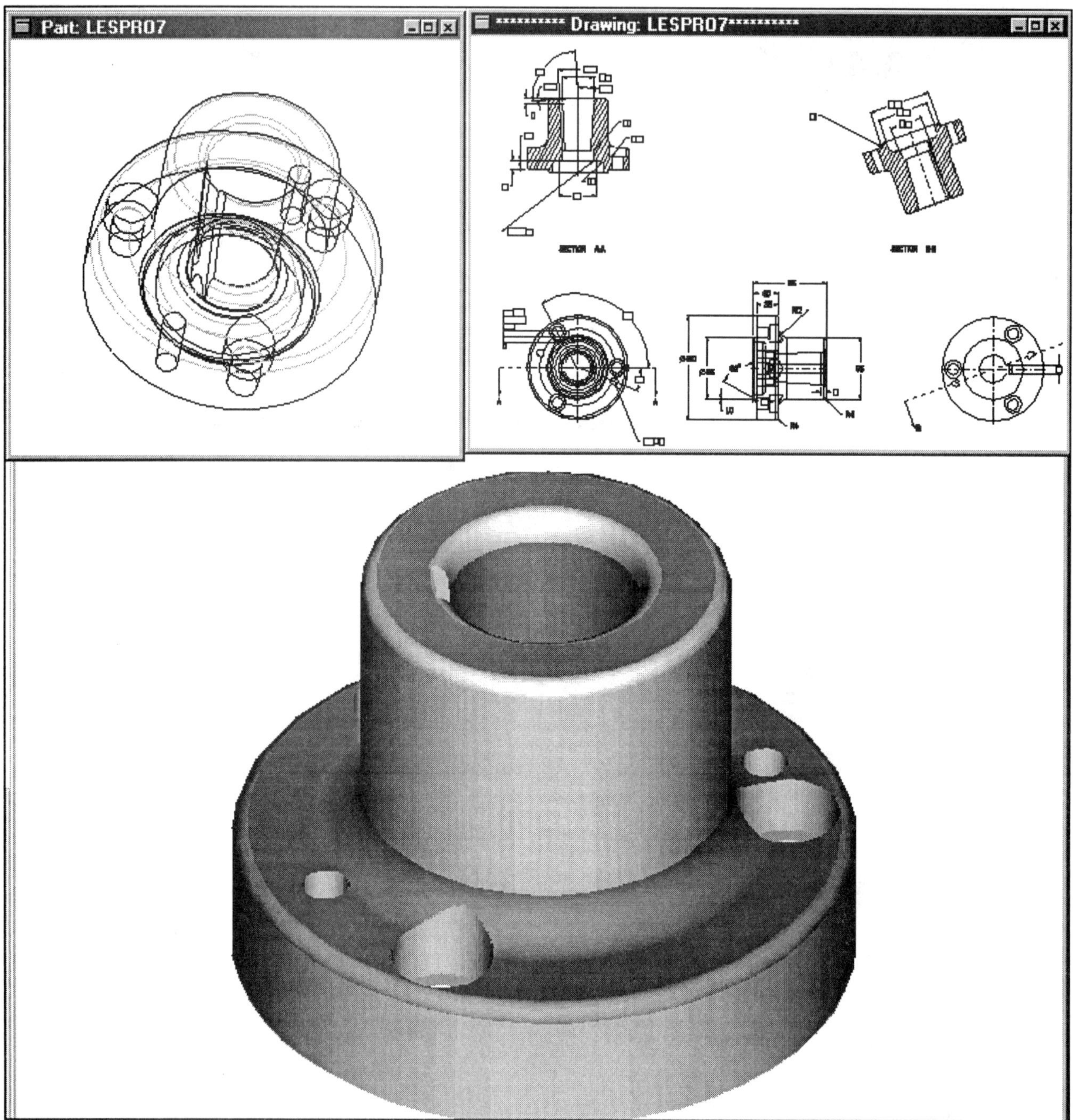

Lesson 8

Ribs, Relations, and Failures

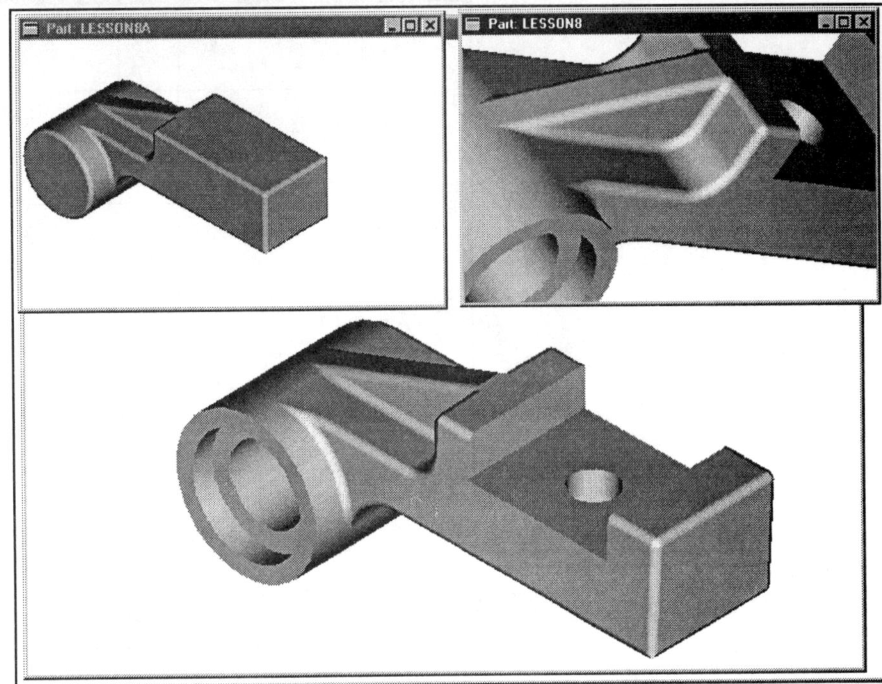

Figure 8.1
Adjustable Guide, Casting, and Machine Part

OBJECTIVES

1. **Understand parameters and relations**
2. **Create straight ribs**
3. **Troubleshoot and resolve failures**
4. **Write relations to control features**
5. **Create a manufacturing model**
6. **Model a workpiece and a design part to be used in Pro/MANUFACTURING**
7. **Establish parameters for a part**
8. **Create equality and comparison relations**
9. **Understand the four types of parameter symbols**

EGD REFERENCE
Engineering Graphics and Design with Graphical Analysis *or* **Fundamentals of Engineering Graphics and Design**
by L. Lamit and K. Kitto
Read Chapters 11 and 14
See pages 498 and 852

COAch™ for Pro/ENGINEER
If you have **COAch for Pro/ENGINEER** on your system, go to SEARCH and do the Segments shown in Figures 8.3, 8.6, and 8.9.

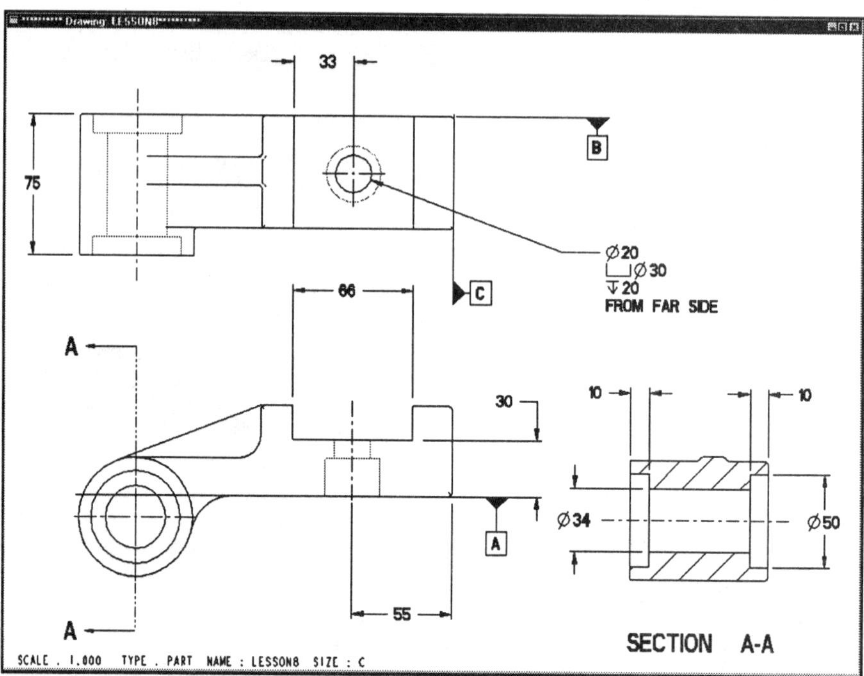

Figure 8.2
Adjustable Guide Machining Drawing

RIBS, RELATIONS, AND FAILURES

A **Rib** is a special type of protrusion designed to create a thin fin or web that is attached to a part (Figs. 8.1 through 8.4).

Relations are equations written between symbolic dimensions and parameters. By writing relations between dimensions in a part or an assembly, the effects of modifications can be controlled.

Failures happen when the model cannot be regenerated. You need to know how to avoid and how to resolve part and assembly failures.

PT/Modeler™

PT/Modeler does not support the **Rib** command. Model a protrusion on both sides of a new datum plane.

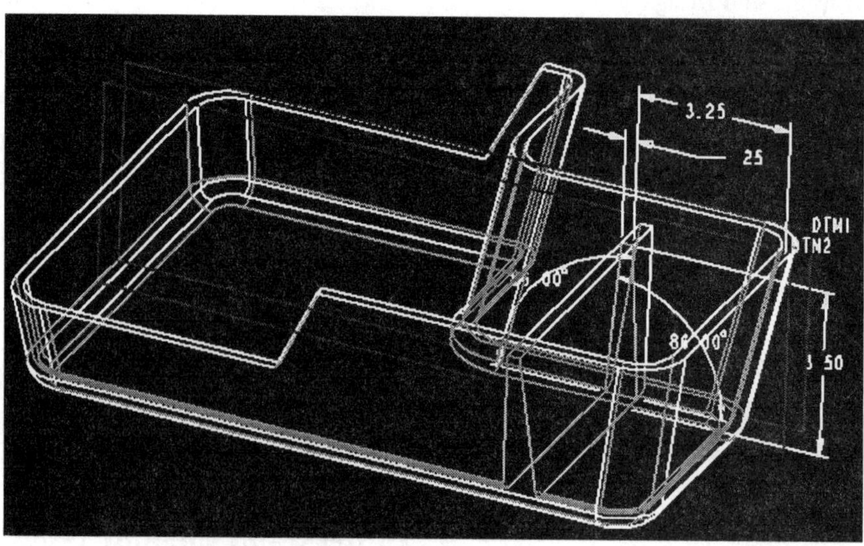

Figure 8.3
COAch for Pro/E, More on Modeling

A rib is always sketched from a side view, and it grows symmetrically about the sketching plane. Because of the way ribs are attached to the parent geometry, they are always sketched as open sections. A rib must "see" material everywhere it attaches to the part; otherwise, it becomes an unattached feature. There are two types of ribs, *straight* and *rotational*. We will confine our discussion to straight ribs. For more information on ribs and, in particular, rotational ribs, use online documentation (see Fig. 8.4).

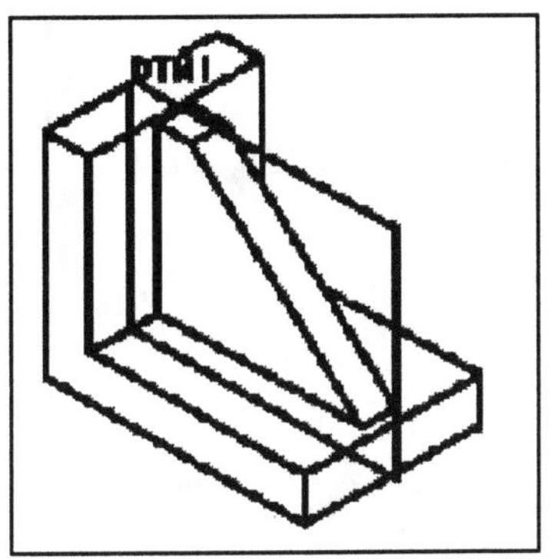

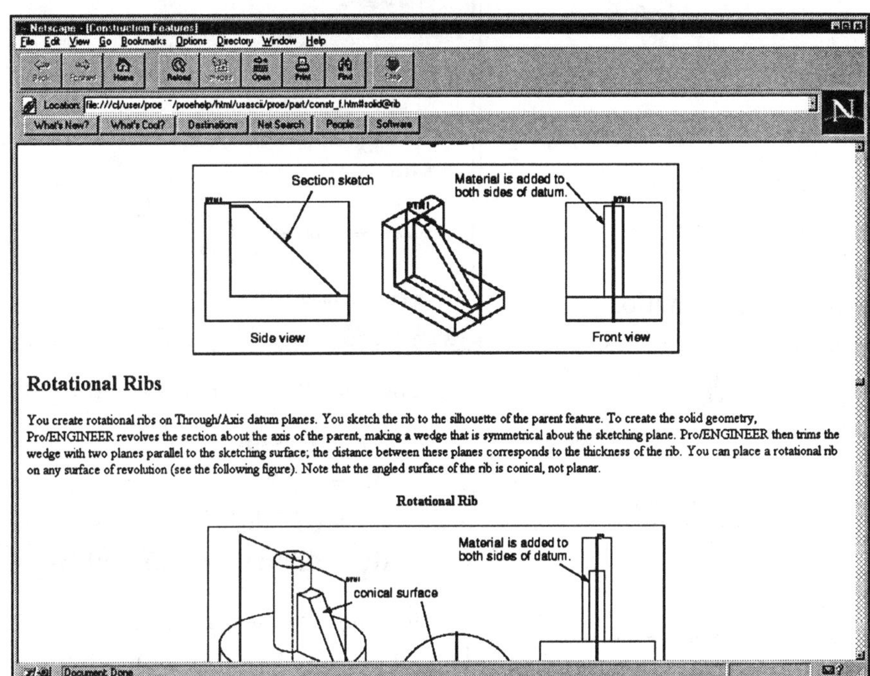

Figure 8.4
Online Documentation, Ribs

Straight Ribs

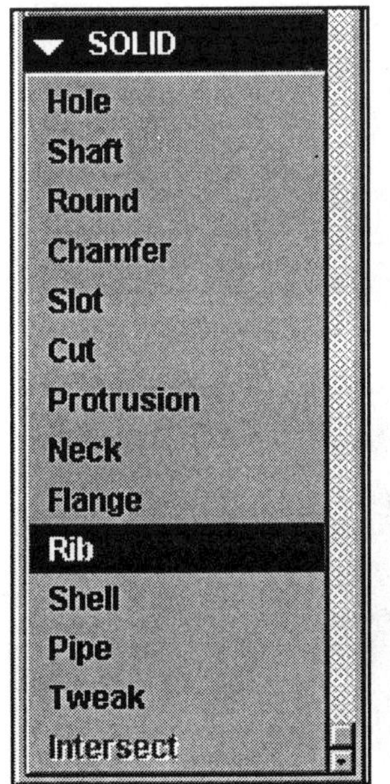

Ribs that are not created on **Through/Axis** datum planes are extruded symmetrically about the sketching plane. You must still sketch the ribs as open sections. Because you are sketching an open section, Pro/E may be uncertain about which side to add the rib on. Pro/E displays the DIRECTION menu after the rib section has been regenerated. Pro/E adds all material in the direction of the arrow. If the incorrect choice is made, correct the direction, pick **OK**, and enter the value for the thickness.

Relations

Relations (Figs. 8.5 and 8.6) can be used to provide a value for a dimension. But they can also be used to notify you when a condition has been violated, such as when a dimension exceeds a certain value.

There are two basic types of relations, *equality* and *comparison*. An *equality relation* equates a parameter on the left side of the equation to an expression on the right side. This type of relation is used for assigning values to dimensions and parameters.

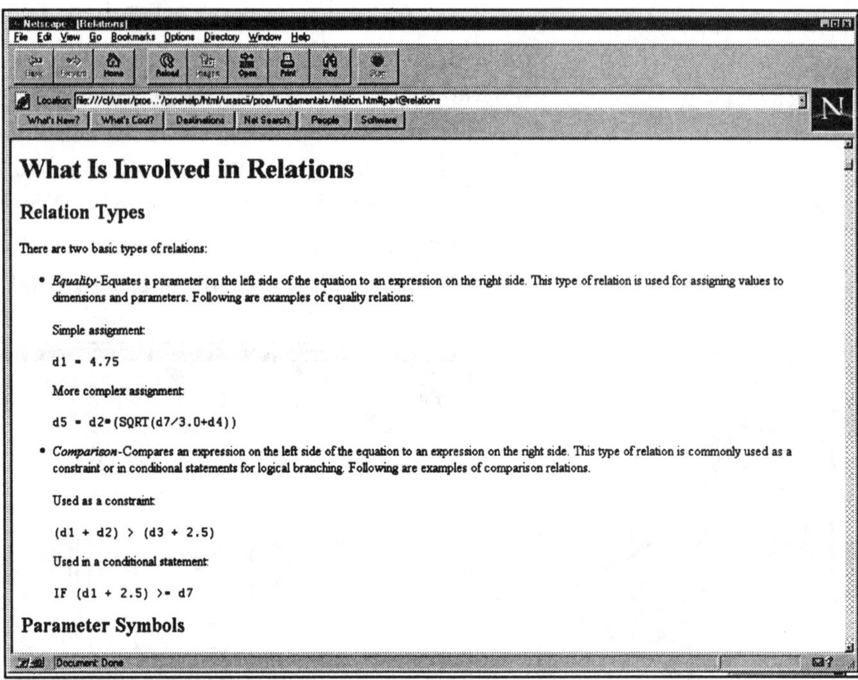

Figure 8.5
Online Documentation, Relations

The following are a few examples of equality relations:

d2 = 25.500
d8 = d4/2
d7 = d1+d6/2
d6 = d2*(sqrt(d7/3.0+d4))

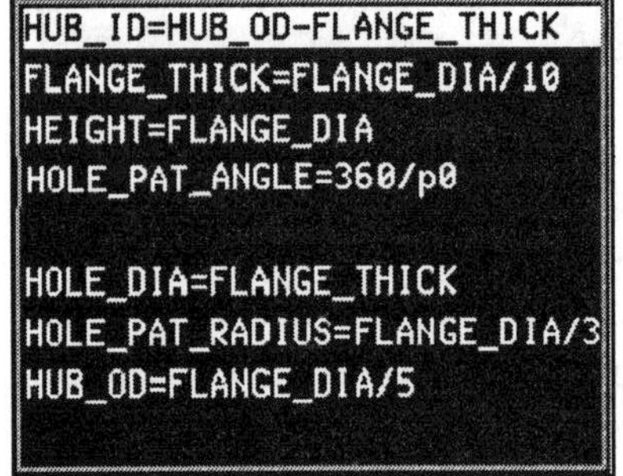

A *comparison relation* compares an expression on the left side of the equation to an expression on the right side. This type of relation is commonly used as a constraint or as a conditional statement for logical branching.

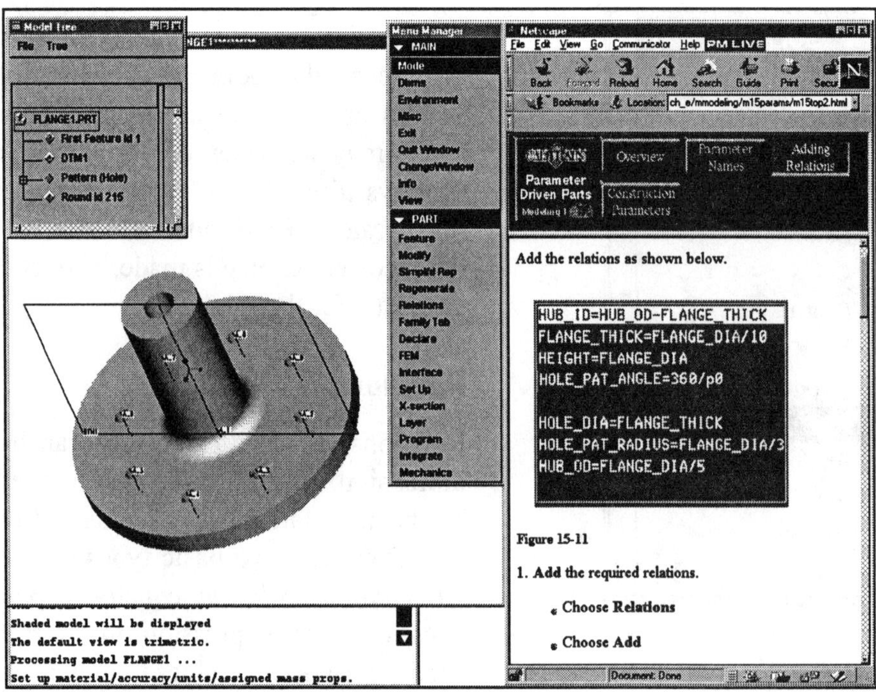

Figure 8.6
COAch for Pro/E, Parameter Driven Parts (Adding Relations)

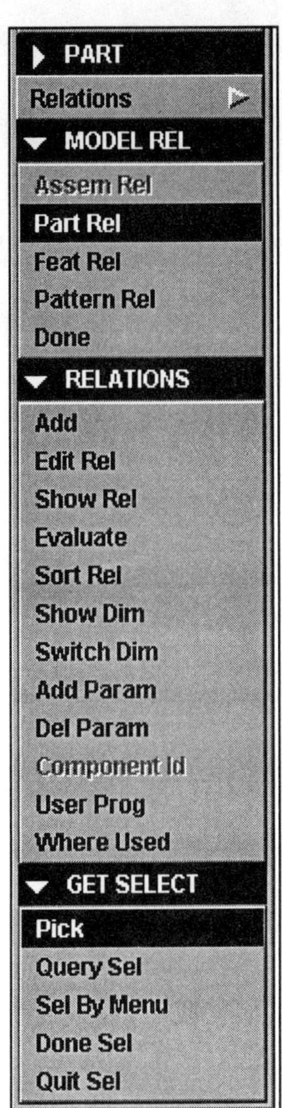

The following are examples of comparison relations:

d1 + d2 > (d3 + 5.5)	Used as a constraint
IF (d1 + 5.5) >= d7	Used in a conditional statement

Parameter Symbols

Four types of parameter symbols are used in relations:

Dimensions These are dimension symbols, such as **d8, d12.**
Tolerances These are parameters associated with ± symmetrical and plus-minus tolerance formats. These symbols appear when dimensions are switched from numeric to symbolic.
Number of Instances These are integer parameters for the number of instances in a direction of a pattern.
User Parameter These can be parameters defined by adding a parameter or a relation (e.g., **Volume = d3 * d4 * d5**).

Operators and Functions

The following operators and functions can be used in equations and conditional statements:

Arithmetic Operators

+ Plus, for addition
- Minus, for subtraction
/ Divided by, for division
* Times, for multiplication
^ Exponentiation
( ) Parentheses, for grouping; for example, **d10 = (d5 - d6)*d7**

Assignment Operator

The equals sign is an assignment operator that equates the two sides of an equation. Obviously, when it is used, the equation can have only a single parameter on the left side.

= Equal to

Comparison Operators

Comparison operators are used wherever a TRUE/FALSE value can be returned. For example, the equation **d1 >= .625** will return TRUE whenever **d1** is greater than or equal to **.625**. It will return FALSE whenever **d1** is less than **.625**.

The following comparison operators can be used:

=   Equal to
>   Greater than
>=   Greater than or equal to
!=, < >, ~ =   Not equal to
<   Less than
<=   Less than or equal to
| Or
&   And
~, !   Not

The "|", "&", "!", and "~" operators extend the use of comparison relations by allowing several conditions to be set in a single statement.

Functions

Mathematical Functions Relations may also include the following mathematical functions:

cos ( )   cosine
tan ( )   tangent
sin ( )   sine
sqrt ( )   square root
asin ( )   arcsine
acos ( )   arccosine
atan ( )   arctangent
sinh ( )   hyperbolic sine
cosh ( )   hyperbolic cosine
tanh ( )   hyperbolic tangent

Failures

Sometimes model geometry cannot be constructed because features that have been modified or created conflict with or invalidate other features. For example, this can happen when the following occurs:

* A protrusion is created that is unattached and has a one-sided edge.
* New features are created that are unattached and have one-sided edges.
* A feature is resumed that now conflicts with another feature (such as having two chamfers on the same edge).
* The intersection of features is no longer valid because dimensional changes have moved the intersecting surfaces.
* A relation constraint has been violated.

Resolving Feature Failures During Creation/Redefinition

Depending on the type of environment used to create a feature (e.g., whether the feature uses the dialog box interface--see Fig. 8.7), Pro/E handles feature failures that may occur during feature creation or redefinition in two different ways:

* *For features that use the dialog box interface* If the feature fails after you choose **OK** or **Preview,** the **Resolve** button appears in the feature creation dialog box. You can either stay in the dialog box environment and redefine feature elements with the **Define** button, or click on the **Resolve** button to access the Resolve environment so you can obtain diagnostics or make changes to other parts of the model.

* *For features that do not use the dialog box interface* If the feature fails, Pro/E brings up the FEAT FAILED menu.

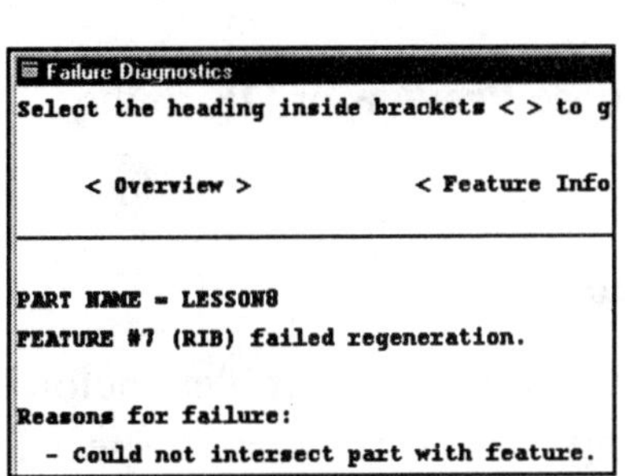

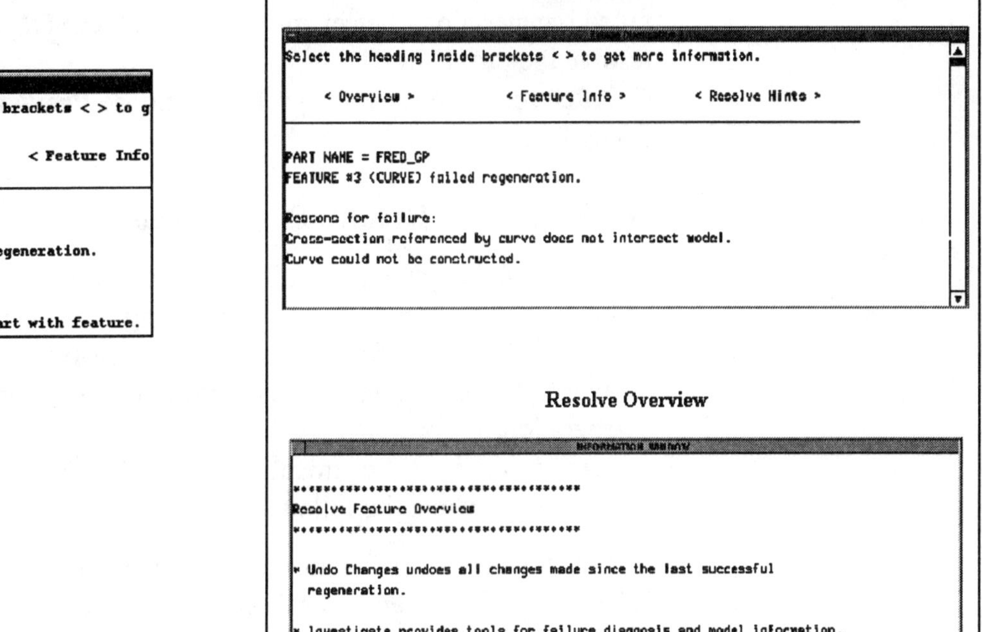

Figure 8.7
Online Documentation, Feature Failure Window

Using the FEAT FAILED Menu

If a feature fails during creation and it does not use the dialog box interface, Pro/E displays the FEAT FAILED menu, with the following options:

Redefine the feature.

Show Ref Display the SHOW REF menu so you can see the references of the failed feature. Pro/E displays the reference number in the MESSAGE WINDOW.

Geom Check for problems with overlapping geometry, misalignment, and so on. This command may be *dimmed.* If a shell, offset surface, or thickened surface fails, Pro/E stores information about the surfaces that could not be offset. The GEOM CHECK menu displays a list of features with failed geometry and a **Restore** command.

Feat Info Get information about the feature.

Using the Resolve Button in the Dialog Box

If a feature fails after you choose **Preview** or **OK** from a dialog box, the **Resolve** button appears in the dialog box, enabling you to enter the "fix model" environment. To resolve a feature failure by using the **Resolve** environment, follow these steps:

1. After a feature creation fails, choose **Resolve** from the dialog box to access the "fix model" environment.
2. The INFORMATION WINDOW appears, listing features that failed regeneration. Select an option in the RESOLVE FEAT menu to resolve the problem.
3. After you have fixed the problem, choose **Preview** or **OK** in the dialog box.

Working in the Resolve Environment

When a model regeneration fails, you must resolve the problem before continuing with normal model processing. Pro/E provides a special error resolution environment (the **Resolve** environment) for recovering from changes that have caused the model to fail regeneration.

As soon as a regeneration fails, Pro/E enters the **Resolve** environment, where the following occurs:

* The **Dbms** command is unavailable and the model cannot be saved.
* The failed feature and all subsequent features remain unregenerated. The current model displays only the regenerated features as they were at the last successful regeneration. Pro/E displays a message that indicates the problem in the MESSAGE WINDOW.
* Pro/E displays the RESOLVE FEAT menu and the failed-feature diagnostic window.

The **Resolve** environment (Fig. 8.8) allows you to do the following:

* Undo all the changes made since the last successful regeneration.
* Diagnose the cause of the model failure.
* Fix the problems within this special environment while using standard part or assembly functionality.
* Attempt a quick fix of the problems using shortcuts for performing standard operations on the failed feature, including **Redefine, Reroute, Suppress** (for parts), **Freeze** (for assemblies), **Clip Suppress,** and **Delete.**

For both diagnosing and fixing the problem, you can choose to work on the current (failed) model or on the backup model. The backup model shows all features in their preregenerated state, and can be used to modify or restore dimensions of the features that are not displayed in the current (failed) model.

Using the RESOLVE FEAT Menu

The RESOLVE FEAT menu options are as follows:

Undo Changes Undo the changes that caused the failed regeneration attempt, and return to the last successfully regenerated model. Pro/E displays the CONFIRMATION menu.
Investigate Investigate the cause of the regeneration failure using the INVESTIGATE submenu.
Fix Model Enter the environment to fix the cause of the regeneration failure.
Quick Fix Use the QUICK FIX menu to immediately perform the specified option on the current model. Use these options:

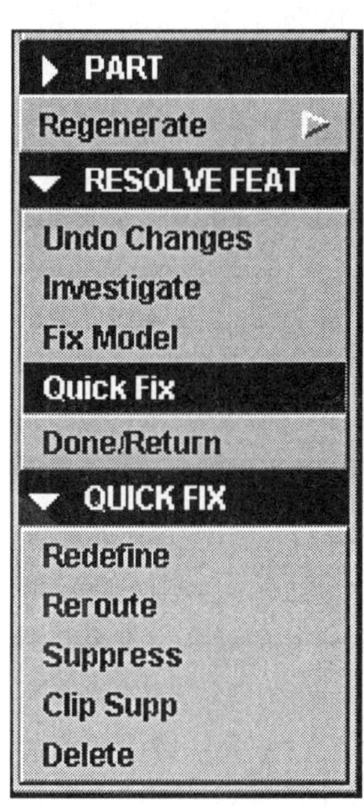

- **Redefine** Redefine the failed feature.
- **Reroute** Reroute the failed feature.
- **Suppress** Suppress the failed feature and its children.
- **Clip Supp** Suppress the failed feature and all the features after it.
- **Delete** Delete the failed feature and its children.

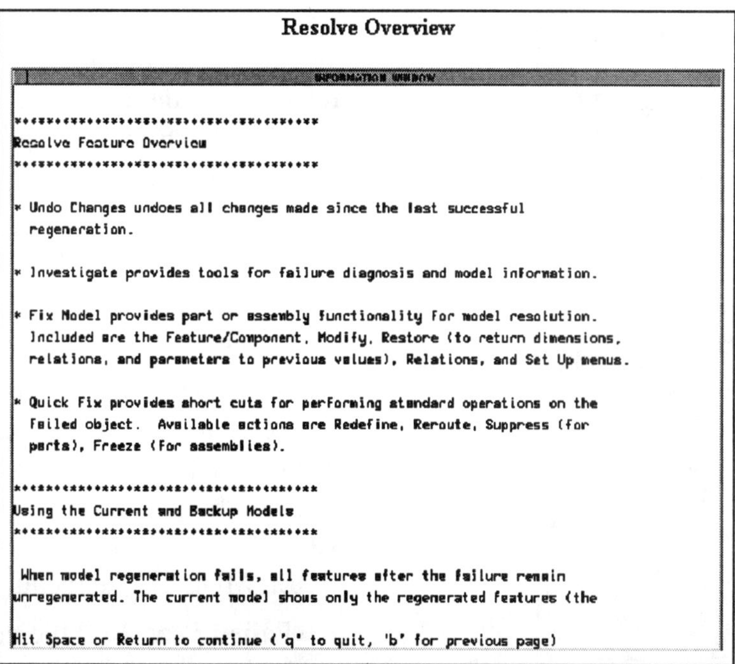
Resolve Overview

INFORMATION WINDOW

```
*************************************
Resolve Feature Overview
*************************************

* Undo Changes undoes all changes made since the last successful
  regeneration.

* Investigate provides tools for failure diagnosis and model information.

* Fix Model provides part or assembly functionality for model resolution.
  Included are the Feature/Component, Modify, Restore (to return dimensions,
  relations, and parameters to previous values), Relations, and Set Up menus.

* Quick Fix provides short cuts for performing standard operations on the
  failed object.  Available actions are Redefine, Reroute, Suppress (for
  parts), Freeze (for assemblies).

*************************************
Using the Current and Backup Models
*************************************

 When model regeneration fails, all features after the failure remain
unregenerated. The current model shows only the regenerated features (the

Hit Space or Return to continue ('q' to quit, 'b' for previous page)
```

Figure 8.8
Online Documentation, Resolve Overview

If you choose the **Investigate** option, Pro/E displays the INVESTIGATE menu, which has the following options:

Current Modl Perform operations on the current (failed) model.
Backup Modl Perform operations on the backup model, displayed in a separate window (current model displayed).

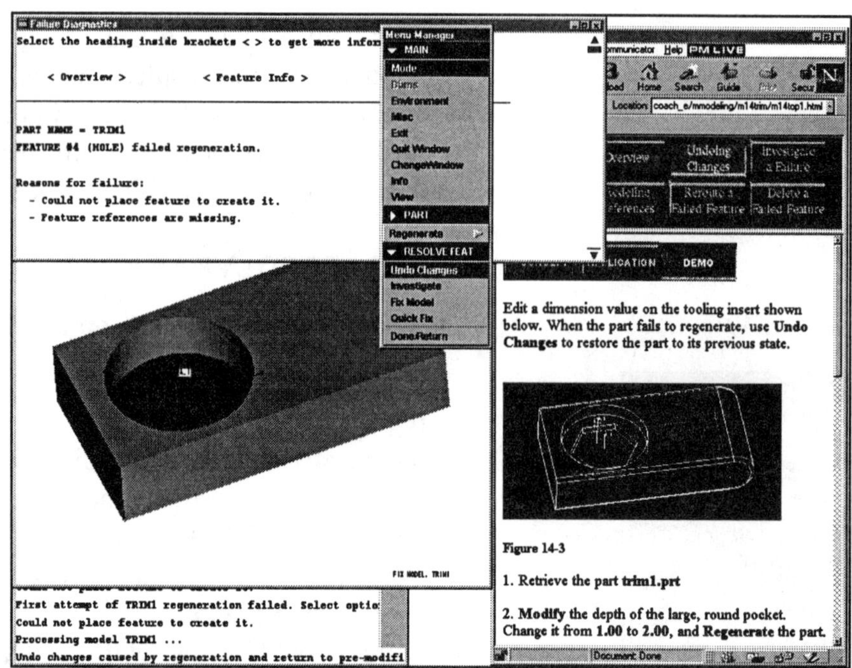

Figure 8.9
COAch for Pro/E, Resolve Feature Failures (Undoing Changes)

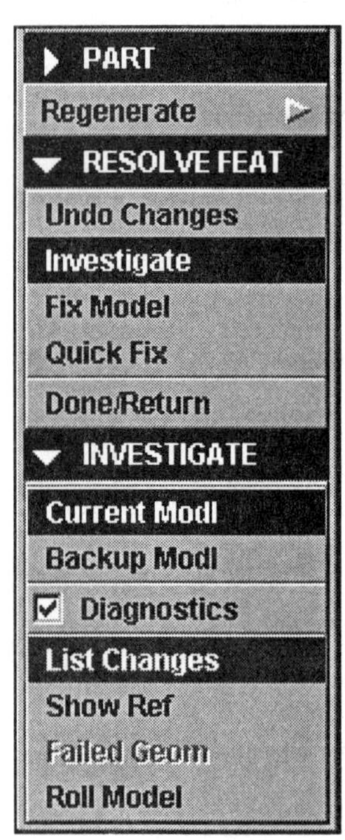

Diagnostics Toggle on or off the display of the failed-feature diagnostic window (Fig. 8.9).

List Changes Show the modified dimensions in the Main Window and in a preregenerated model window (Review Window), if available. Also, a table is displayed that lists all the modifications and changes.

Show Ref Display the SHOW REF menu to show all the references for the failed feature in the models, in both the Review Window and the Main Window. Pro/E highlights the first reference in the reference color (such as magenta), and displays the SHOW REF menu, which lists the following options:

> **Next** Highlight the next reference.
> **Previous** Highlight the previous reference.
> **Info** Display an INFORMATION WINDOW that provides information about the entity and the feature to which it belongs.

Failed Geom Display the invalid geometry of the failed feature. This command may be unavailable. The FAILED GEOM menu displays a list of features with failed geometry and a **Restore** command.

Roll Model Roll the model back to the option selected in the ROLL MDL TO submenu. The options are as follows:

> **Failed Feat** Roll the model back to the failed feature (for the backup model only).
> **Before Fail** Roll the model back to the feature just before the failed feature.
> **Last Success** Roll the model back to the state it was in at the end of the last successful feature regeneration.
> **Specify** Roll the model back to the specified feature.

If you choose the **Fix Model** option, Pro/E displays the FIX MODEL menu, which includes the following options:

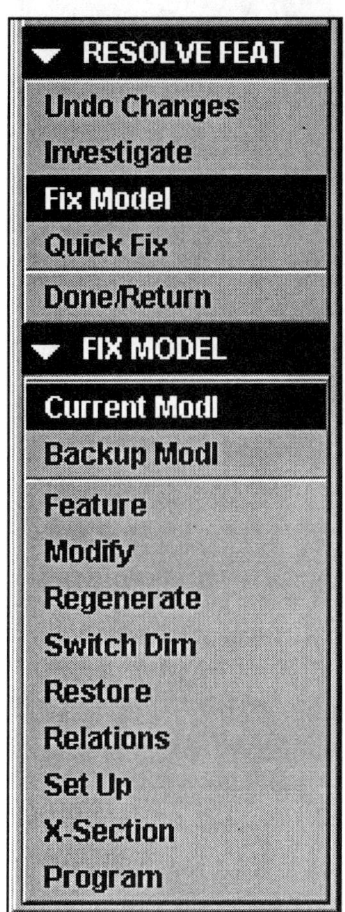

Current Modl Perform operations on the current active (failed) model.
Backup Modl Perform operations on the backup model, displayed in a separate window from the current model in the active window.
Feature Perform feature operations on the model using the standard FEAT menu. Pro/E displays the CONFIRMATION menu, so you can confirm or cancel the request only if the **Undo Changes** option is not possible. However, the **Undo Changes** option is not always possible if you used the **Regen Backup** option in the ENVIRONMENT menu.
Modify Modify dimensions using the standard MODIFY menu.
Regenerate Regenerate the model.
Switch Dim Switch the dimension display from symbols to values, or vice versa.
Restore Display the RESTORE menu so you can restore dimensions, parameters, relations, or all of these to their values prior to the failure. The RESTORE menu options include the following:

All Changes Restore all the changed items.
Dimensions Restore the dimensions.
Parameters Restore the parameters.
Relations Restore the relations.

Relations Add, delete, or modify relations, as necessary, to be able to regenerate the model, using the MODEL REL and RELATIONS menus.
Set Up Display the standard PART SETUP menu to perform additional part setup procedures.
X-Section Create, modify, or delete a cross-sectional view using the CROSS SEC menu.
Program Access Pro/PROGRAM capabilities using the PROGRAM menu.

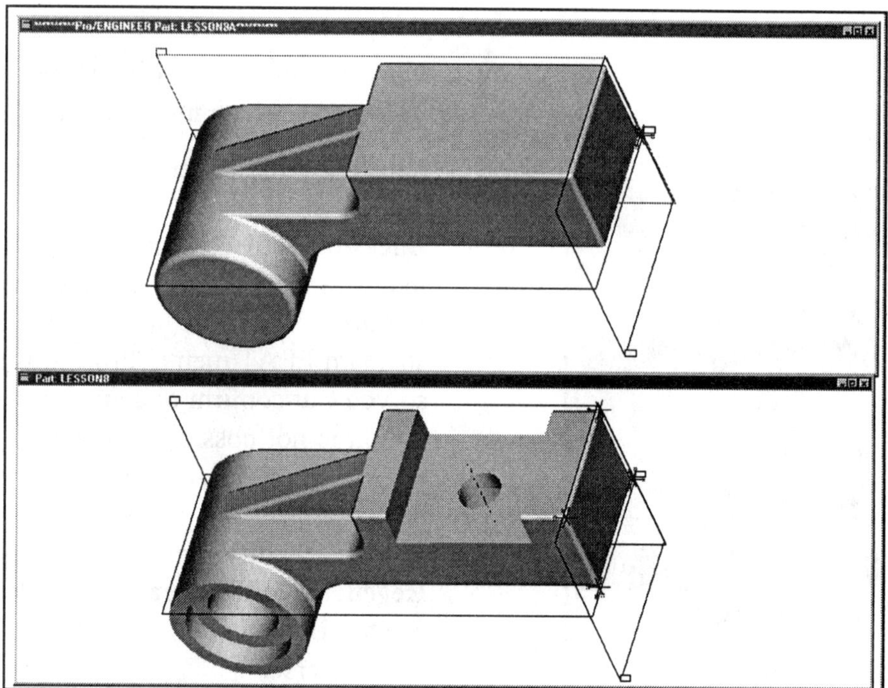

Figure 8.10
Adjustable Guide, Casting, and Machined Part

Adjustable Guide

Model the casting first (Fig. 8.10, top) and save it when it is completed. After saving the casting under a different name, use the part model to create the machined **Adjustable Guide** (Fig. 8.10, bottom). By having a *casting* (which is called a ***workpiece*** in **Pro/MANUFACTURING**) and a separate but almost identical *machined part* (which is called a ***design part*** in Pro/MANUFACTURING), you can create an operation for machining and an NC sequence. During the manufacturing process, you merge the workpiece into the design part and create a ***manufacturing model*** (Fig. 8.11). The difference between the two files is the difference between the volume of the casting and the volume of the machined part. The removed volume can be seen as *material removal* when you are doing an **NC Check** operation on the manufacturing model. The manufacturing model is *green;* the cuts completed in **NC Check** show as *yellow* until the machine tool reaches the design size, and then they show *magenta*. If the machining process gouges the part, the gouge will be displayed as *cyan*. The cutter location can also be displayed as an animated machining process, as shown in Figure 8.12.

The rib created in the casting model will have a **relation** added to it to control its location. The relation will keep the rib centered on the rectangular side of the part.

The Adjustable Guide is a simple part, so the process of describing step-by-step procedures will start with the creation of the rib. The rounds are added late in the modeling process, but will cause the model to fail in most cases. The process of fixing the part so that the rounds do not make the regeneration fail is also described. The machined version of the part can be finished with two cuts, a sketched counterbore hole, and a thru hole with two counterbores.

From this lesson on, you will be required to set up your parts without step-by-step commands.

Set up the Adjustable Guide:

- Material = Steel
- Units = Millimeters
- Default Datum Planes
- Default Coordinate System
- Layers = **DATUM_PLANES**
- Set datums with Geom Tol
- Rename datums **A**, **B**, and **C** as shown in the figures
- Hidden line
- Tan Dimmed
- ✓Grid Snap

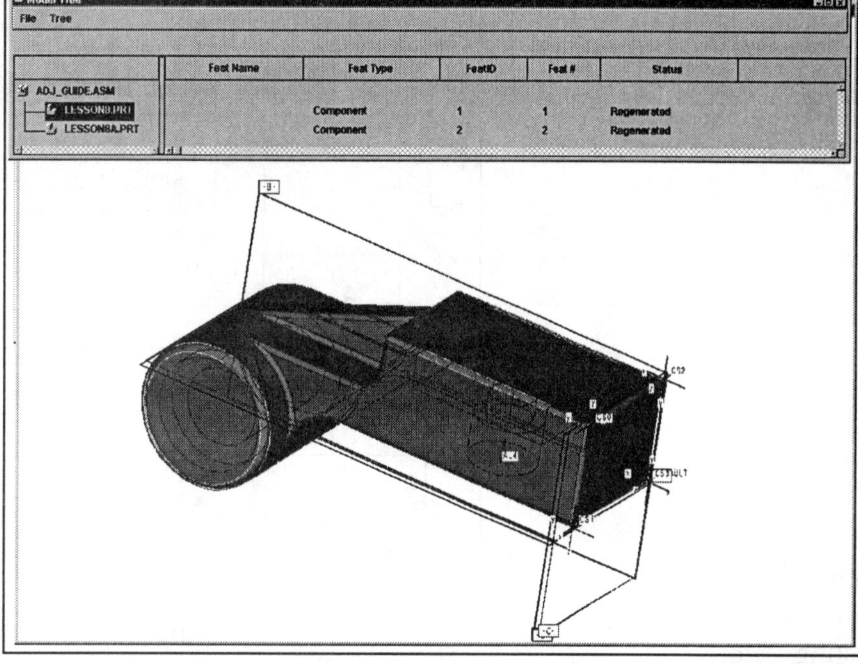

HINT

* Create the part with **ADJ_GUIDE_MACHINE** as the part name.
* Model only the casting dimensions.
* Save the part, using **Save As,** with the name of **ADJ_GUIDE_CAST**.
* Continue modeling the machined features.
* Save the part under its original name.

Figure 8.11
Manufacturing Model

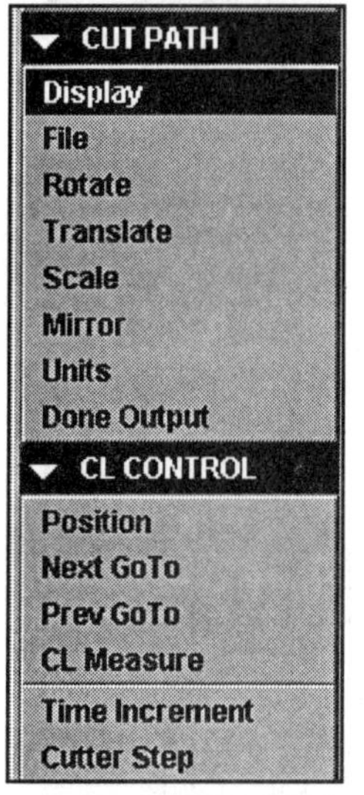

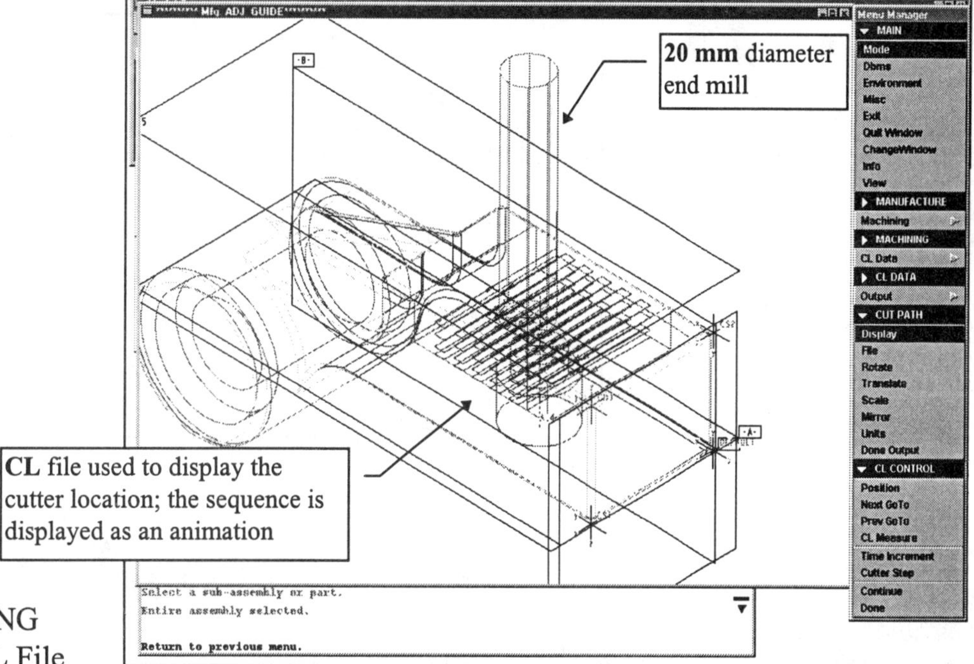

Figure 8.12
Pro/MANUFACTURING NC Sequence and a **CL** File

Start the process by creating a part called **ADJ_GUIDE_MACHINE**. Use Figure 8.13 for the modeling of the Adjustable Guide casting and Figure 8.14 for the Adjustable Guide machined part. Remember to create the casting first and save it using **Dbms** ⇒ **Save As** ⇒ **enter** ⇒ (type a new name, such as **ADJ_GUIDE_CAST)** ⇒ **enter** (to save). Continue working on the part by adding the machined features as required by the design. In this edition of this text, we will not be covering Pro/MANUFACTURING, but if your company or school has the manufacturing module, you may wish to create a manufacturing model and machine it with the assistance of your instructor.

NOTE:
ALL RADII TO BE R2 UNLESS OTHERWISE NOTED

Figure 8.13
Adjustable Guide, Casting Detail

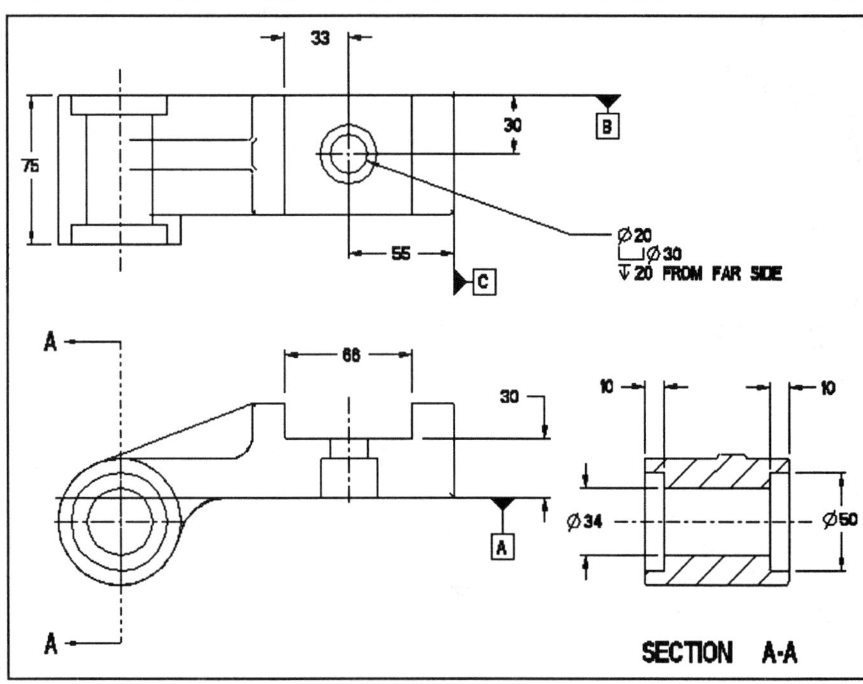

Figure 8.14
Adjustable Guide, Machining Detail

Use the same datum planes shown in the following figures. The sketching plane for the part is **DTM3** and is on the back side of the part (datum **B** in Fig. 8.15). Datum **A** was originally **DTM2** and runs along the bottom of the part. A *jig and fixture* assembly will be used to hold, locate, and clamp the part for machining. Datum **C** was originally **DTM1** and runs along the right side of the part.

The first protrusion is shown in Figure 8.15. The circular protrusion is extended in Figure 8.16. The first round set is created in Figure 8.17. This round is needed to create a tangent rib. Normally, all rounds would be added last. Create the round by choosing the following commands:

PT/Modeler™
Feature ⇒ Round ⇒ Constant ⇒ etc.

Feature ⇒ Create ⇒ Round ⇒ Simple ⇒ Done ⇒ Constant ⇒ Edge Chain ⇒ Done ⇒ Surf Chain ⇒ (pick the top surface of the part as shown in Fig. 8.17) **⇒ Select All ⇒ Done Sel ⇒ Done ⇒** (type **2**) **⇒ enter ⇒ Preview ⇒ OK ⇒ Done**

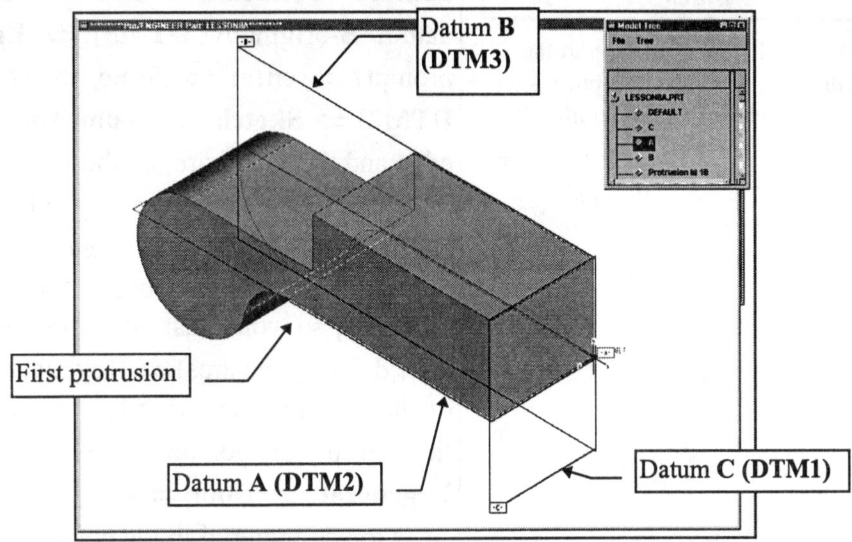

Figure 8.15
First Protrusion

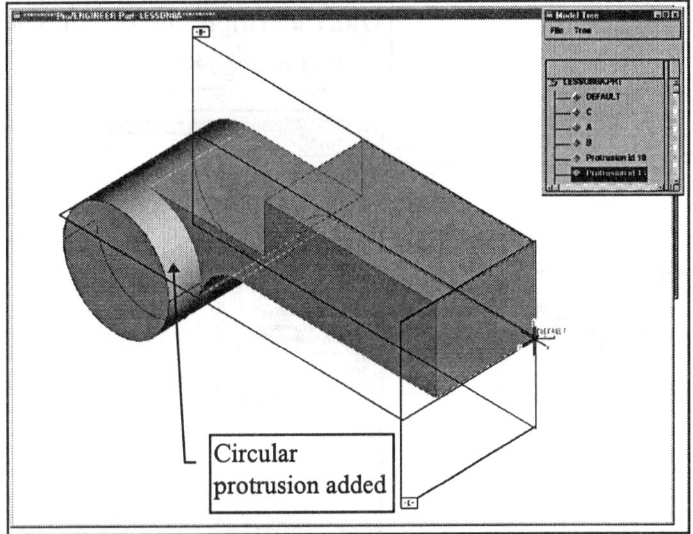

Figure 8.16
Circular Protrusion Added to the Model

Dbms (File--PT/Modeler) **⇒**
Save ⇒ enter
Purge ⇒ enter ⇒
Done-Return

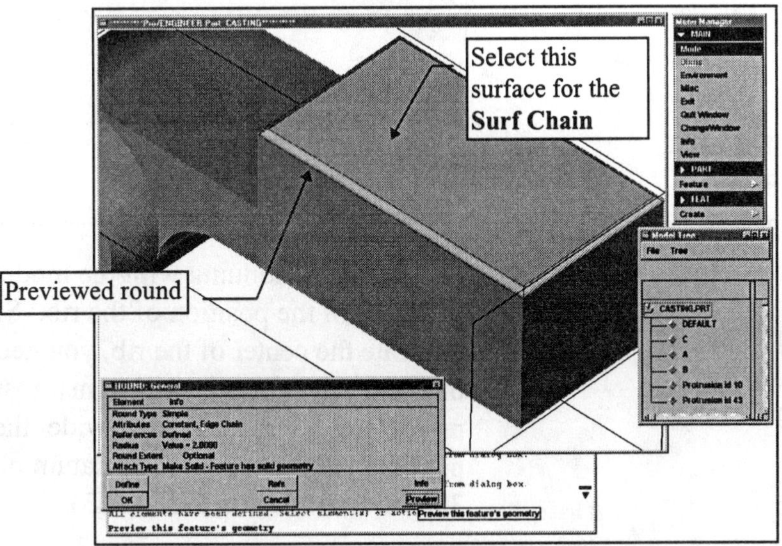

Figure 8.17
R2 Rounds Added to the Top Surface of the Part

The rib will be constructed using a datum on the fly (**Make Datum**). The sketch of the rib will require just one tangent line between the large circular protrusion and the round just created. Use the following commands:

PT/Modeler™

PT/Modeler does not support the **Rib** command. Model a protrusion on both sides of a new datum plane.

Feature ⇒ **Create** ⇒ **Rib** ⇒ **Make Datum** ⇒ **Offset** ⇒ (pick datum **B**-originally **DTM3**) ⇒ **Enter Value** ⇒ (type **30** at the prompt) ⇒ **enter** ⇒ **Done** ⇒ **Top** ⇒ (pick **datum A**--originally **DTM2**) ⇒ **Sketch** ⇒ **Geom Tools** ⇒ **Use Edge** (pick the circular edge and the round arc, as shown in Fig. 8.18) ⇒ **Done Sel** ⇒ **Sketch** ⇒ **Line** ⇒ **2 Tangent** ⇒ (pick the circle and the arc at the approximate points of tangency, as shown in Fig. 8.19) ⇒ **Delete** (delete the arcs--Pro/E broke the arc and the circle at the points of tangency, so you must zoom in and make sure only the line is left after deleting the arcs) ⇒ **Regenerate** ⇒ **Alignment** ⇒ (pick the end of the line and the circle for one alignment and the other end of the line and the arc for the second alignment, as shown in Fig. 8.19) ⇒ **Regenerate** ⇒ **Done** ⇒ **Flip** (flip the arrow to point toward the part so as to add material in the correct direction, as shown in Fig. 8.20) ⇒ **Okay** ⇒ (input the rib thickness by typing **15** at the prompt) ⇒ **enter** ⇒ **Done** (Fig. 8.21)

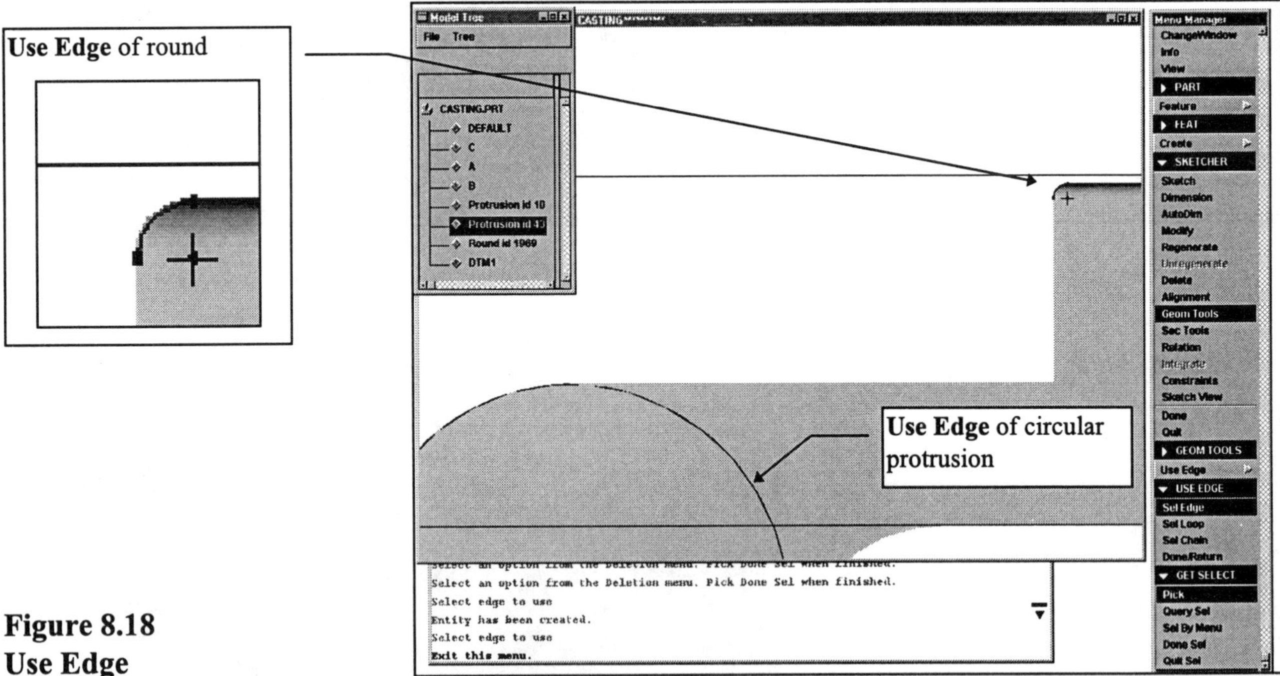

Figure 8.18
Use Edge

Before continuing with the modeling process, write a relation that will control the position of the rib. Since you used a datum on the fly to locate the center of the rib, you need to write a relation that will say that the rib's center will remain at the center of the rectangular protrusion no matter how wide the protrusion becomes during a possible **ECO** change. The datum plane used to sketch on was offset **30 mm** from **datum B (DTM3)**.

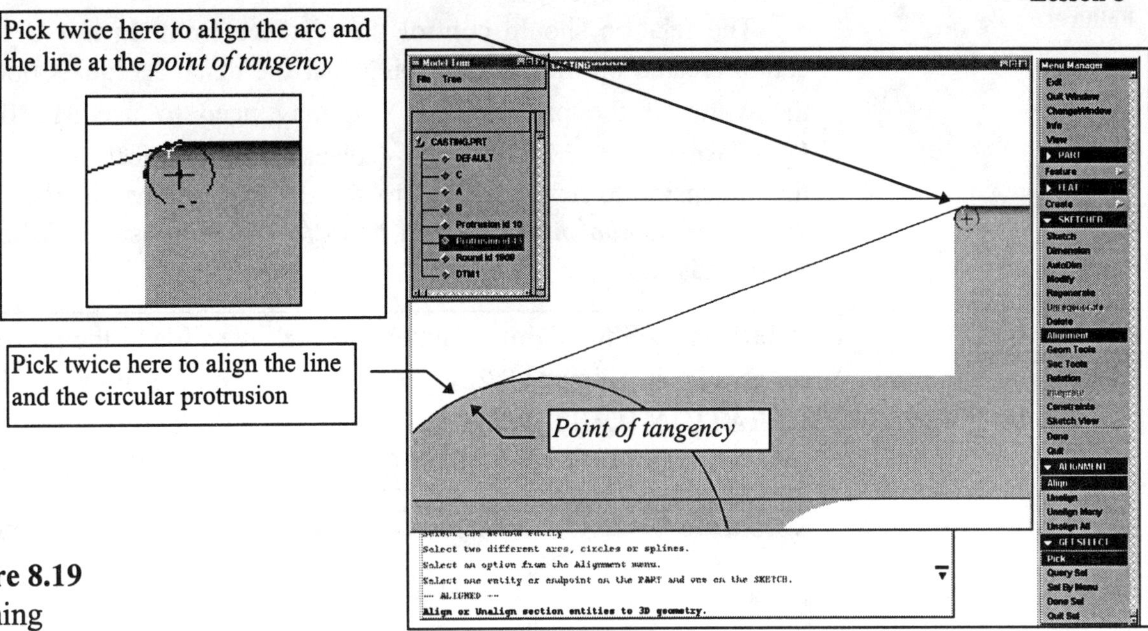

Figure 8.19
Aligning

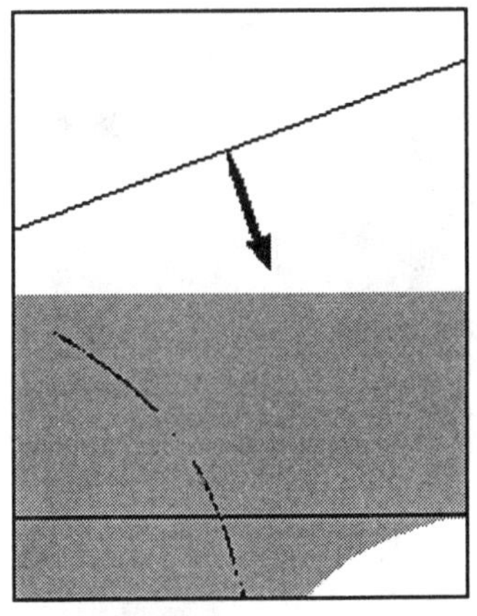

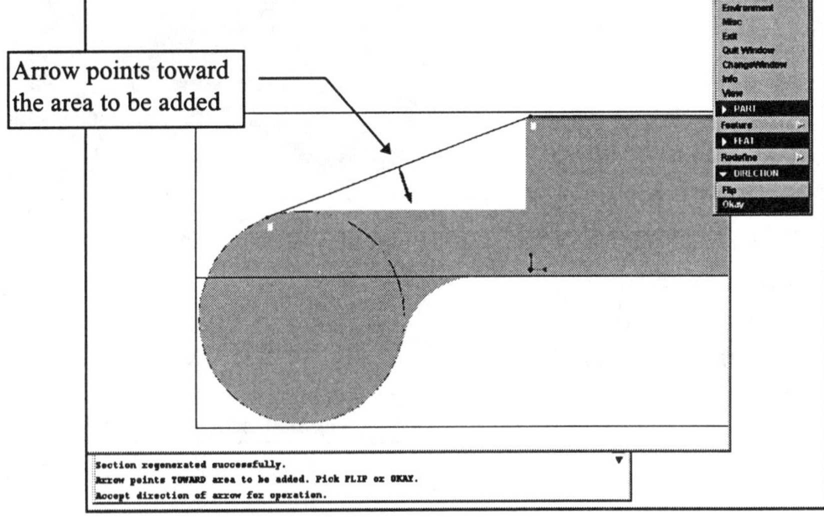

Figure 8.20
Arrow Points Toward the Area to Be Added

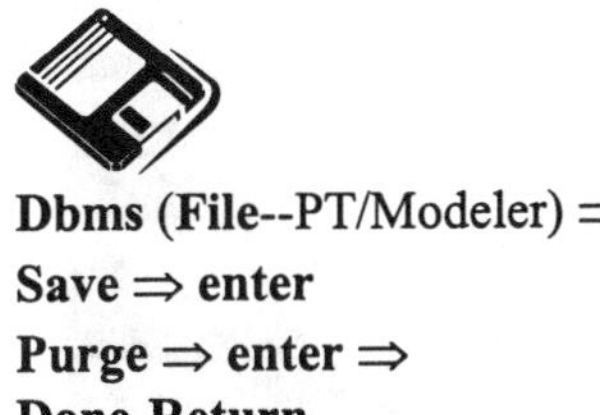

Dbms (File--PT/Modeler) ⇒
Save ⇒ enter
Purge ⇒ enter ⇒
Done-Return

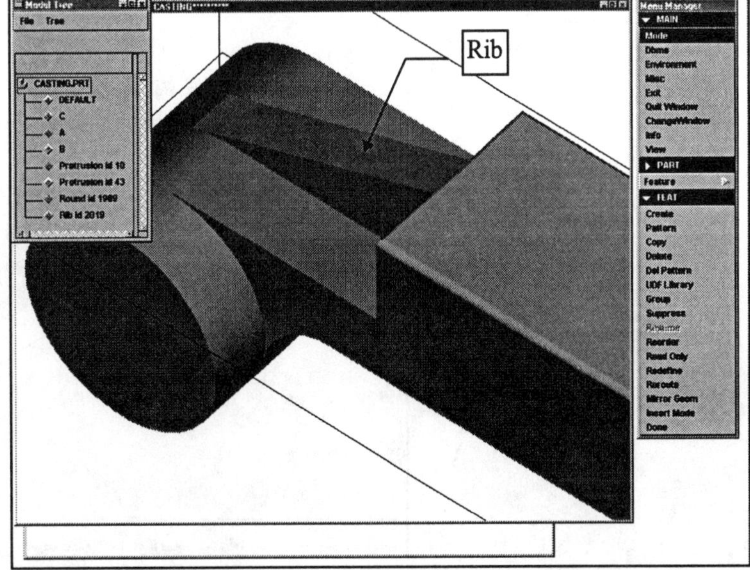

Figure 8.21
Completed Rib

The relation should control the distance from **datum B** to the datum created on the fly by saying that the distance equals one-half the width of the protrusion. You may need to use **Modify** ⇒ **DimCosmetics** ⇒ **Move Dim** ⇒ (select the desired feature and move the dimension to new position) to see the dimensions clearly. ***Your dimension symbol numbers will be different.*** Choose the following commands:

PT/Modeler™

Relations ⇒ (pick the rib and the protrusion) ⇒ **Switch Dim** ⇒ **Add etc.**

Relations ⇒ **Show Dim** ⇒ **enter** ⇒ (pick the rib and the protrusion as shown in Figure 8.22) ⇒ **Add** ⇒ (at the prompt, **Enter RELATION [QUIT]: d100=d11/2**) ⇒ **enter** ⇒ **enter** ⇒ **Show Rel** (Fig. 8.23) ⇒ (type **q** to quit INFORMATION WINDOW) ⇒ **Done**

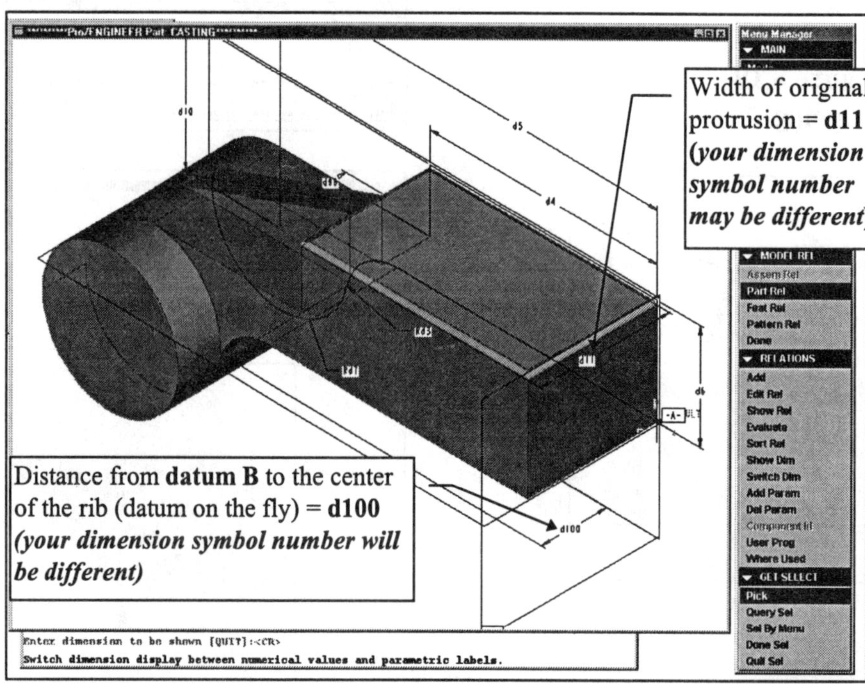

Figure 8.22
Using **Relations** ⇒ **Show Dim** to View the Dimension Symbols for the Center of the Rib and the Width of the Protrusion

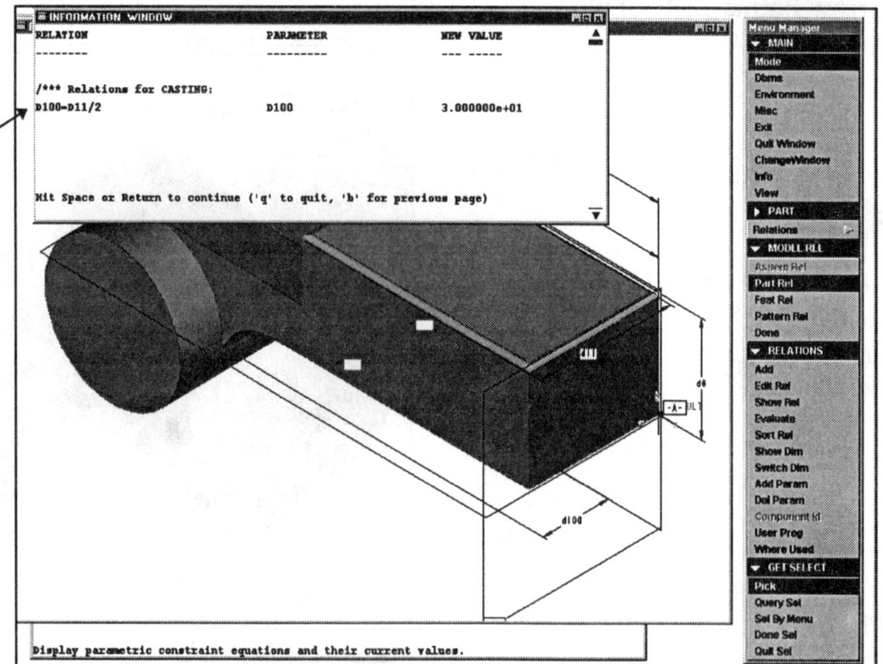

Figure 8.23
Show Relation

Create the rounds for the rest of the part. If you pick too many edges using **One By One**, you may get a failure, as shown in Figure 8.24, where the **Feature Info** choice was selected after the failure to explain the error. Redo the round, working one small set of connected edges at a time using **One By One** and **Tangnt Chain**, as shown in Figure 8.25.

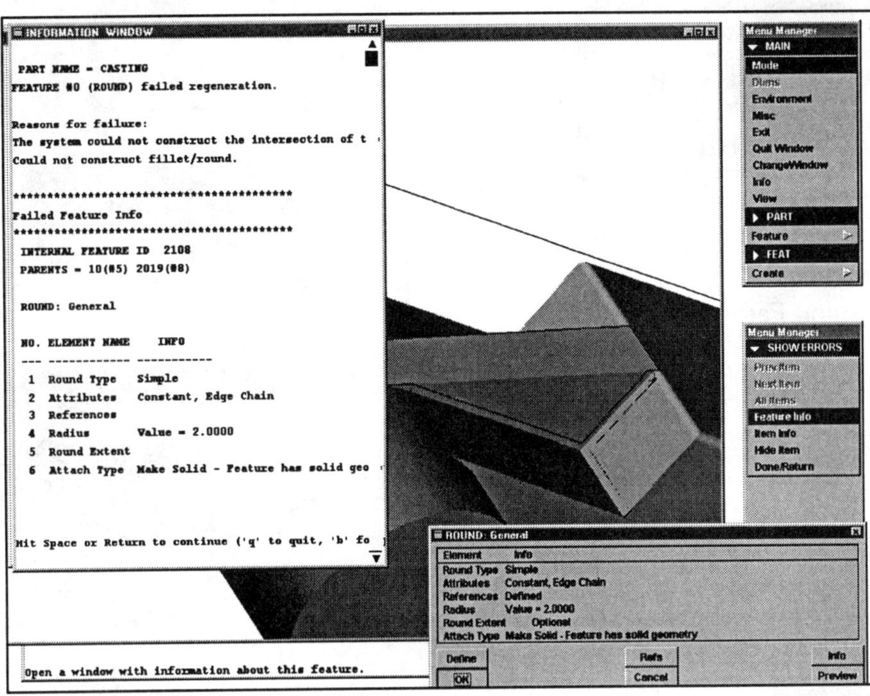

Figure 8.24
Feature Info During Failed Round

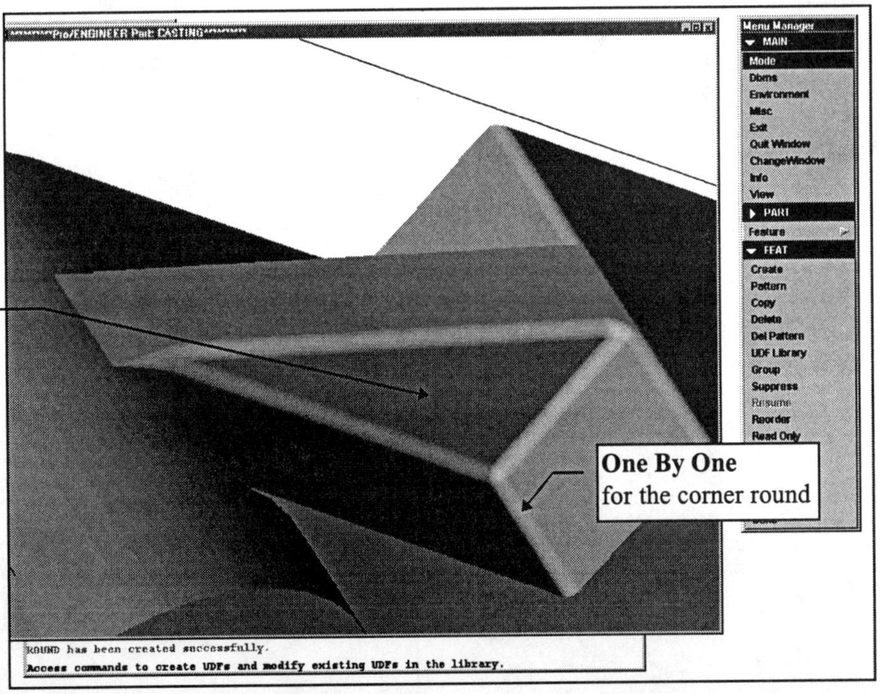

Use **Tangent Chain** for remaining rounds that use this surface

One By One for the corner round

ROUND has been created successfully.
Access commands to create UDFs and modify existing UDFs in the library.

Figure 8.25
Rounds Created Successfully Using **One By One** and **Tangnt Chain**

If you get a failure as shown in Figure 8.26, choose **Undo Changes ⇒ Confirm** and try another sequence of picks for the round selection. If you use **Quick Fix ⇒ Delete ⇒ Delete All** instead, you may remove other rounds that you want to keep.

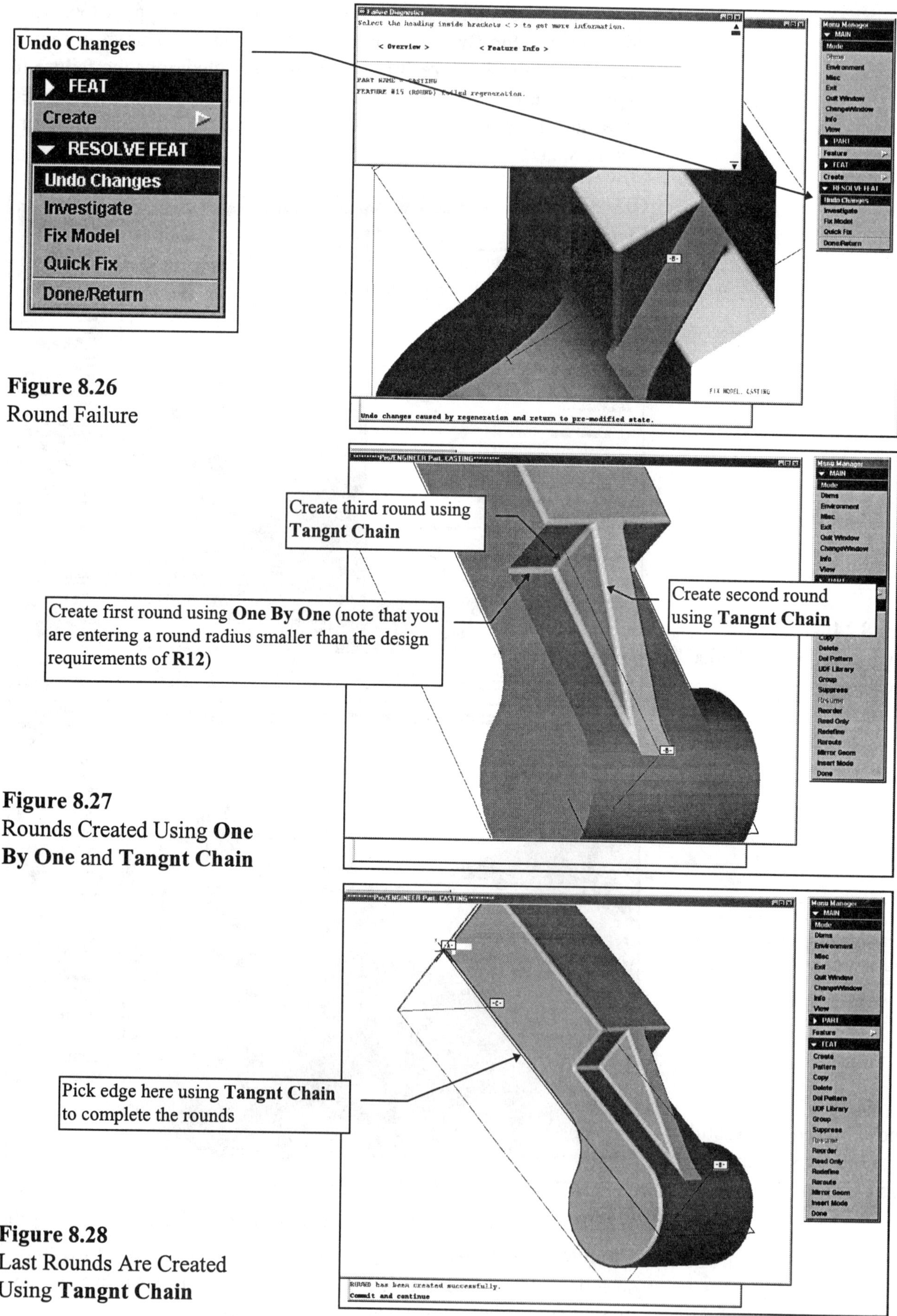

Figure 8.26
Round Failure

Figure 8.27
Rounds Created Using **One By One** and **Tangnt Chain**

Figure 8.28
Last Rounds Are Created Using **Tangnt Chain**

It will take some trial and error (Fig. 8.27 and 8.28), but it is possible to get all the rounds done correctly, as shown in Figure 8.29.

Dbms (**File**--PT/Modeler) ⇒
Save ⇒ **enter**
Purge ⇒ **enter** ⇒
Done-Return

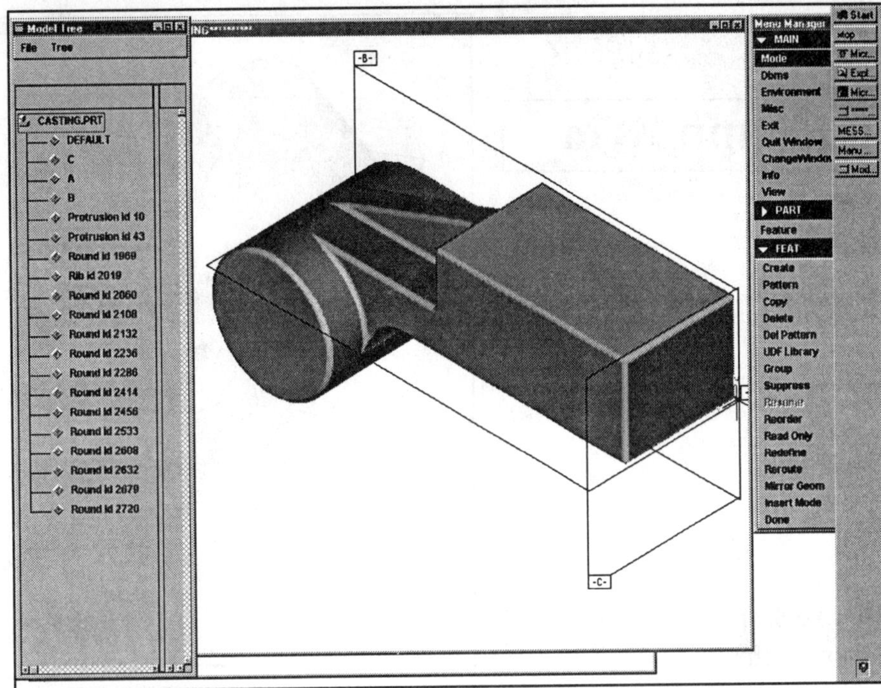

Figure 8.29
Completed Casting Part
(***Workpiece***)

Modify the **R2** round to the design size of **R12** for the feature shown in Figure 8.30. The radius for the round must be updated on the casting and machine part to reflect the design requirements.

Use **Dbms** ⇒ **Save As** ⇒ **enter** ⇒ (type new name for the casting **ADJ_GUIDE_CAST**) ⇒ **enter** ⇒ **Done-Return** to save the casting separate from the part file on which you will continue to model. Complete the machined part using the dimensions shown in Figure 8.14.

Dbms (**File**--PT/Modeler) ⇒
Save ⇒ **enter**
Purge ⇒ **enter** ⇒
Done-Return

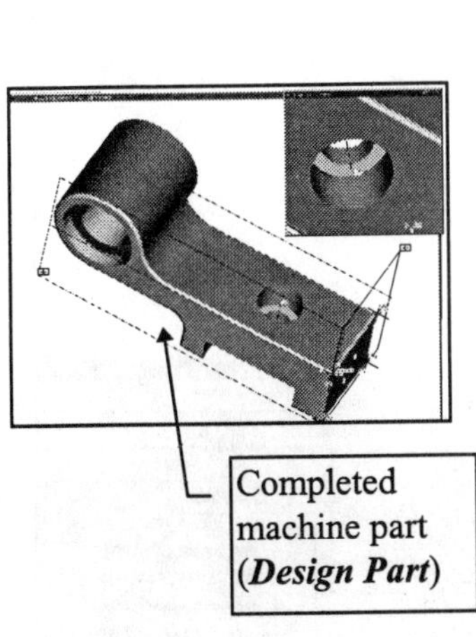

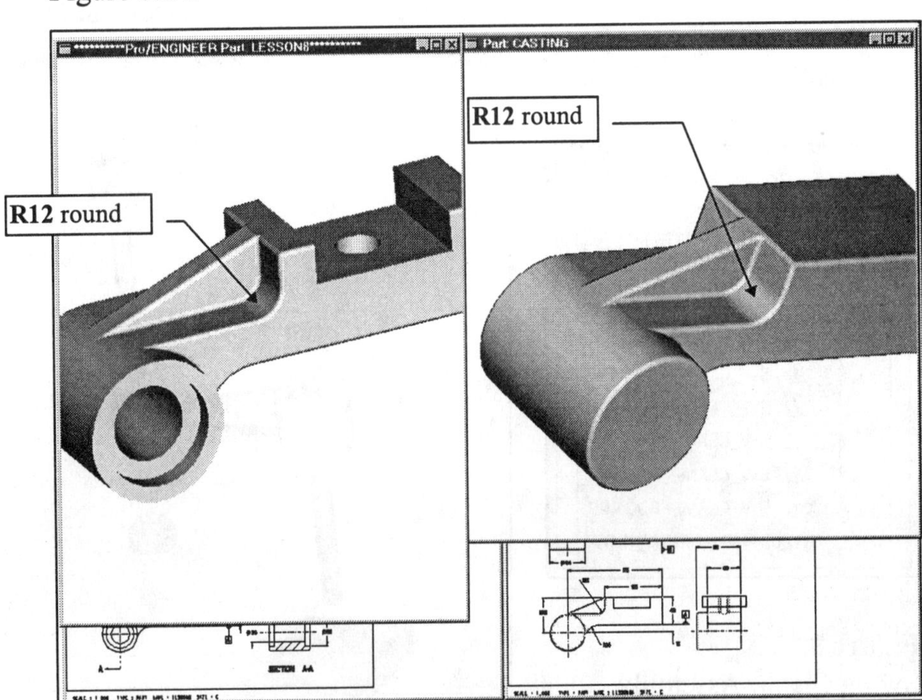

Figure 8.30
Modify the Round to **R12**

Lesson 8 Project

Clamp Arm

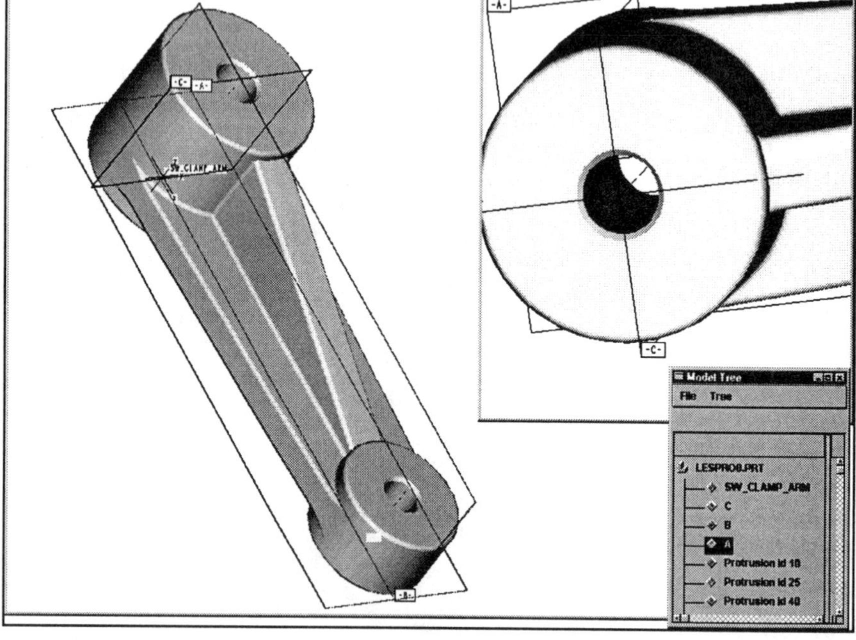

Figure 8.31
Clamp Arm

Clamp Arm

The eighth **lesson project** is a cast part that requires commands similar to those for the **Adjustable Guide**. Create the part shown in Figures 8.31 through 8.37. Analyze the part and plan out the steps and features required to model it. Use the DIPS in Appendix D to plan out the feature creation sequence and the parent-child relationships for the part. Note that the Clamp Arm is used for an assembly (Fig. 8.32) in Lessons 14 and 15. Create two versions of the Arm, one with all cast surfaces and one with machined ends.

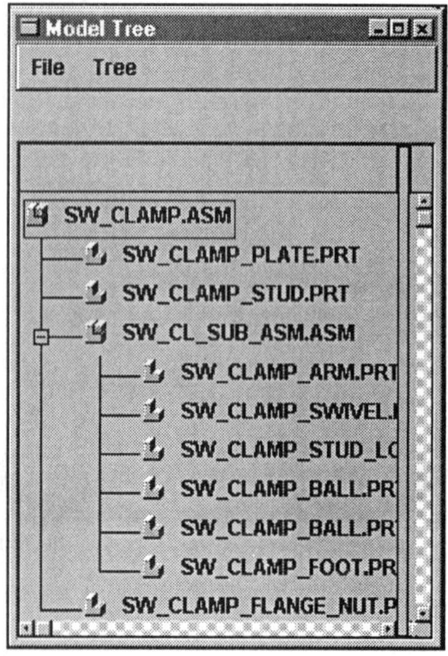

Figure 8.32
Swing Clamp Assembly

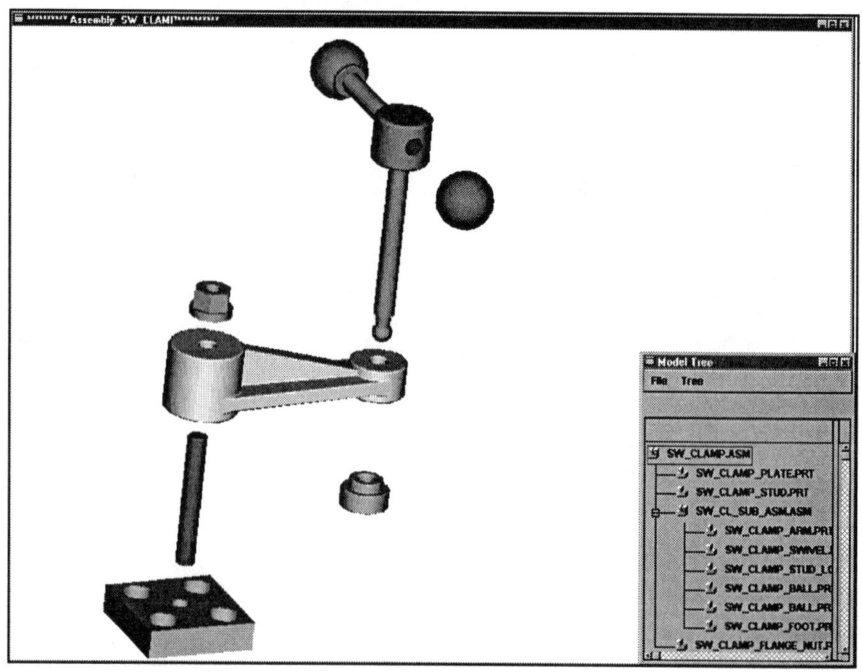

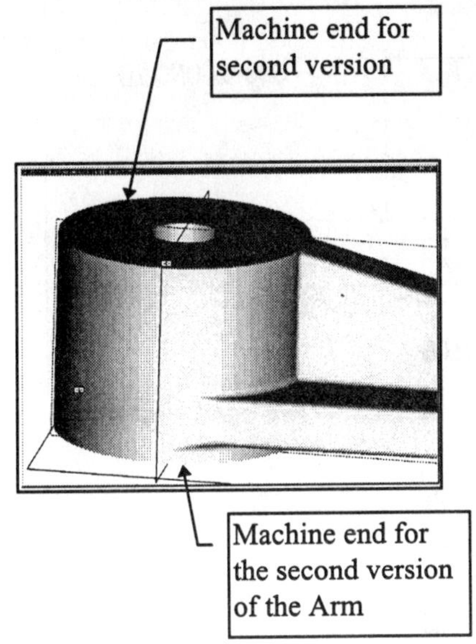

Machine end for second version

Machine end for the second version of the Arm

Save a casting version of the part (no holes or machined surfaces) and a machined version. Include cosmetic threads where required. Machine the top and bottom of both round protrusions (**.10** off either end).

Write a relation to control the position of the horizontal ribs (Fig. 8.33). They must remain at the center of the smaller circular end.

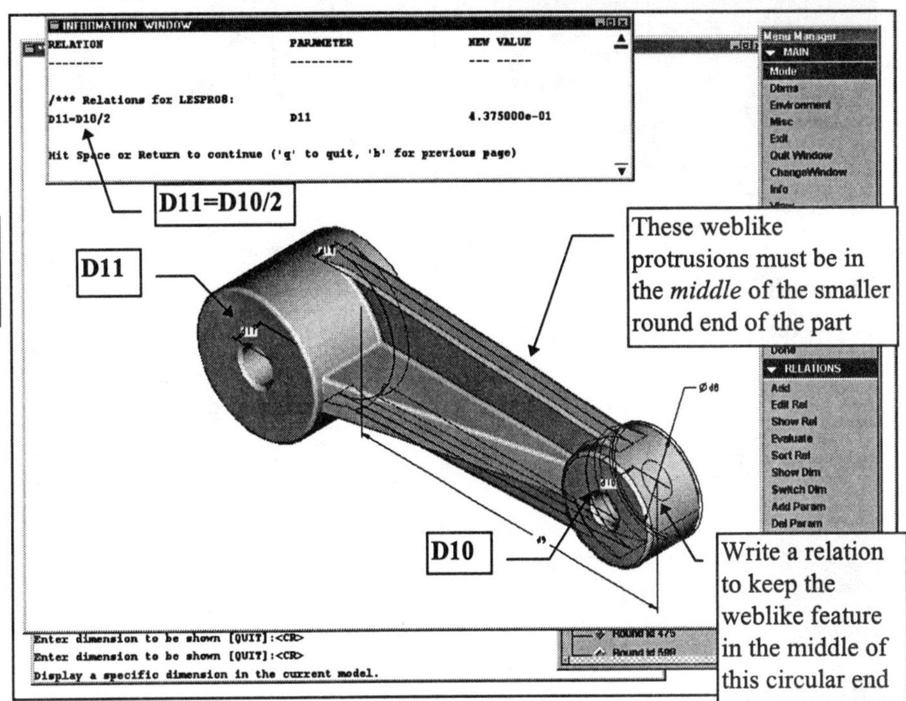

Figure 8.33
Relation

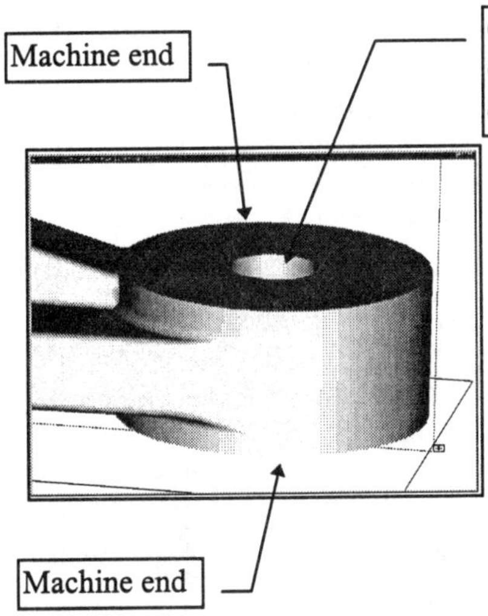

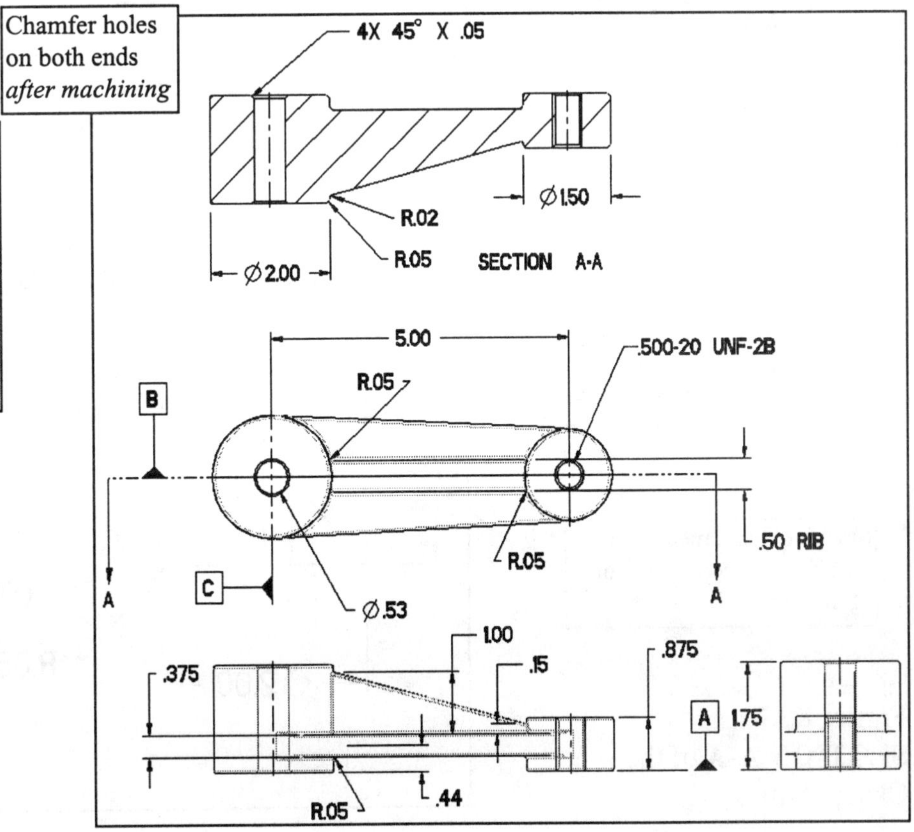

Machine end

Chamfer holes on both ends *after machining*

Machine end

4X 45° X .05

R.02

R.05

SECTION A-A

∅1.50

∅2.00

5.00

.500-20 UNF-2B

R.05

.50 RIB

R.05

∅.53

1.00

.15

.875

.375

1.75

.44

R.05

Figure 8.34
Detail of Clamp Arm
(ROUNDS ARE R.02 UNLESS OTHERWISE NOTED)

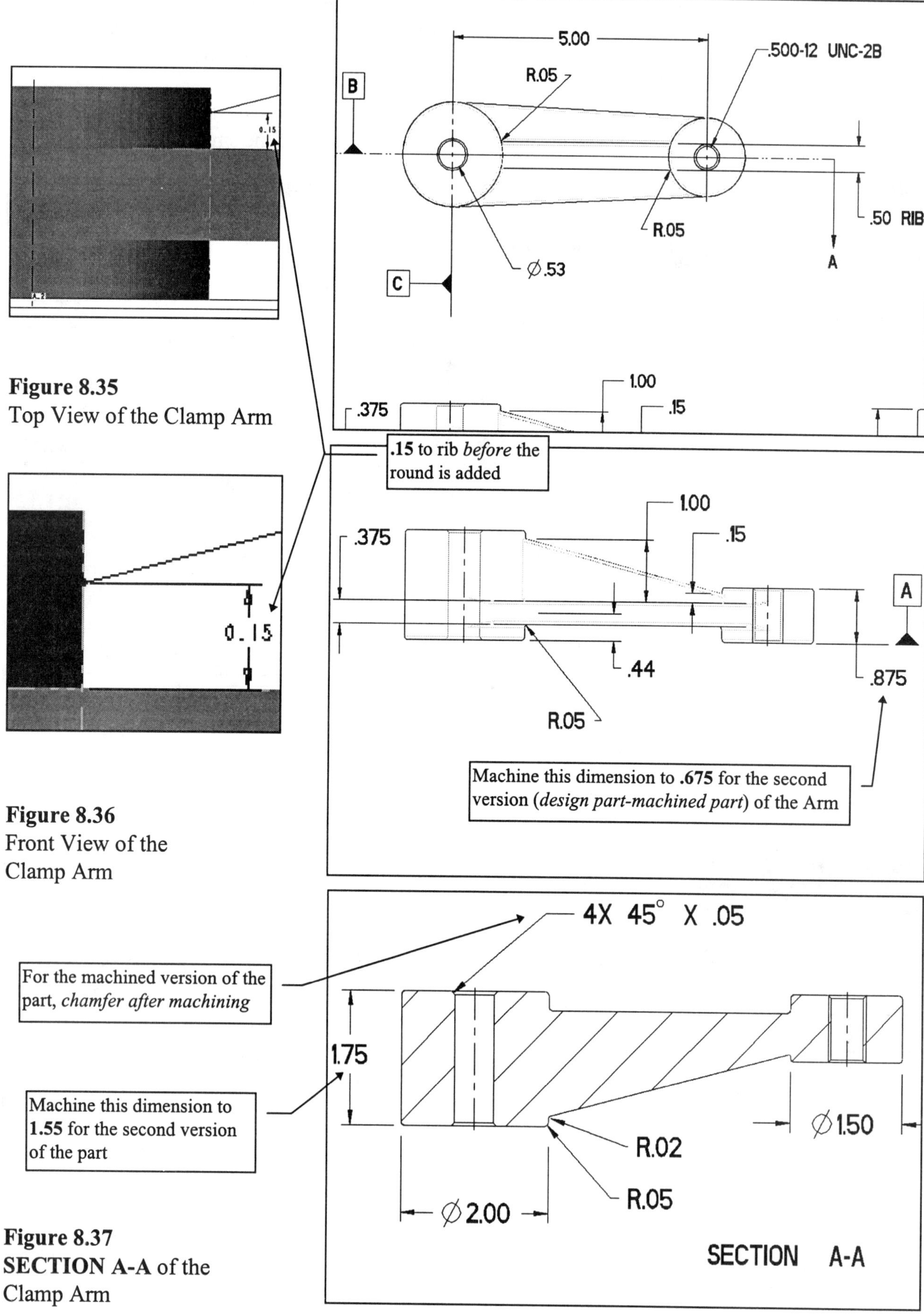

Figure 8.35
Top View of the Clamp Arm

Figure 8.36
Front View of the Clamp Arm

Figure 8.37
SECTION A-A of the Clamp Arm

Lesson 9

Drafts, Suppress, and Text Protrusions

Figure 9.1
Enclosure

OBJECTIVES

1. Create draft features
2. Create text protrusions on parts
3. Shell a part
4. Suppress features to decrease regeneration time
5. Resume a set of suppressed features

EGD REFERENCE

Engineering Graphics and Design with Graphical Analysis *or* **Fundamentals of Engineering Graphics and Design**
by L. Lamit and K. Kitto
Read Chapter 14
See pages 493-498

COAch™ for Pro/ENGINEER

If you have **COAch for Pro/ENGINEER** on your system, go to SEARCH and do the Segment shown in Figure 9.3.

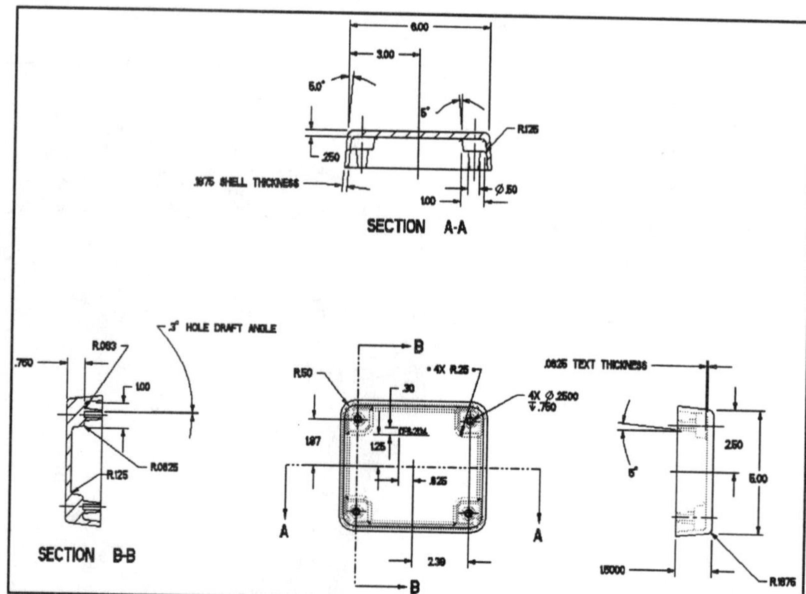

Figure 9.2
Enclosure Dimensions

DRAFTS, SUPPRESS, AND TEXT PROTRUSIONS

The **Draft** feature adds a draft angle between surfaces. A wide variety of parts incorporate drafts in their design. Casting, injection mold, and die parts normally have drafted surfaces. The ENCLOSURE in Figure 9.1 and Figure 9.2 is a plastic injection-molded part.

Suppressing features by using the **Suppress** command temporarily removes them from regeneration. Suppressed features can be "unsuppressed" **(Resume)** at any time. It is sometimes convenient to suppress text protrusions and rounds to speed up regeneration of the model.

Text can be included in a sketch for extruded protrusions and cuts, trimming surfaces, and cosmetic features. To decrease regeneration time of the model, text can be suppressed after it is created. Text can also be drafted.

Figure 9.3
COAch for Pro/E, More Features (Draft Features)

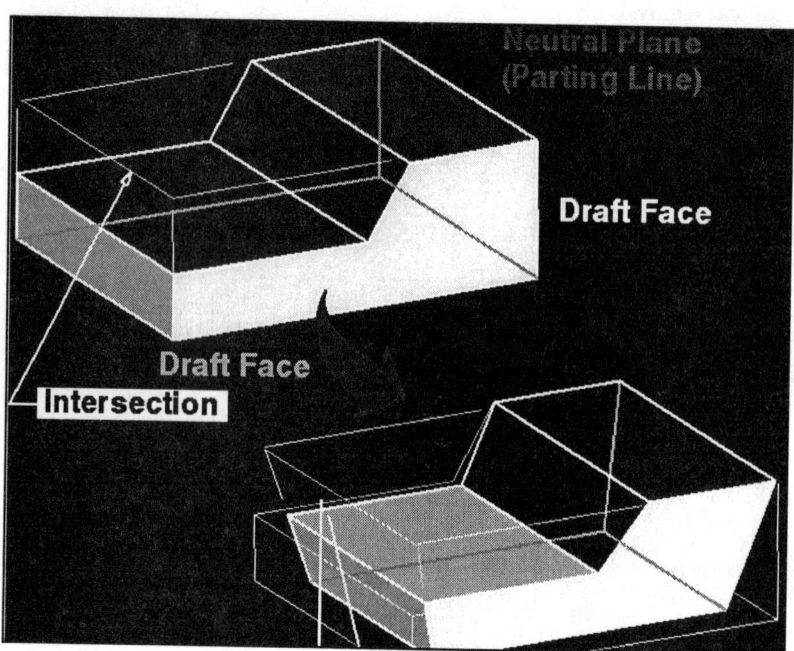

Drafts

The **Draft** feature adds a draft angle between **-15°** and **+15°** to individual surfaces or to a series of selected planar surfaces (Fig. 9.3). The following terminology is used in draft creation:

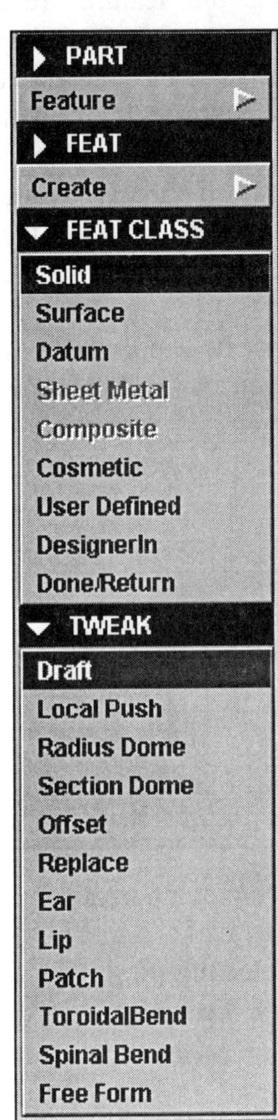

Draft surfaces are the selected surfaces of the model designated for drafting.

Neutral plane defines the pivot plane (Fig. 9.4) where the draft angle is **0°** (zero). The intersection of the neutral plane and drafted surfaces defines the *axis of rotation.*

Neutral curve defines the curve on the surfaces to be drafted where the draft angle is **0°** (zero). Drafted surfaces are rotated about the neutral curve.

Draft direction, or **reference direction**, indicates the direction in which material is added. The draft direction is defined as a normal to the reference plane that you specify; it is shown on the screen as a *red arrow.* The normal also indicates the direction from which the draft angle is measured.

Draft angle is measured between the draft direction and the resulting drafted surfaces. If the draft surfaces are split, you can define two independent angles for each portion of the draft. The draft angles must be within the range of **-15°** to **+15°**.

Direction of rotation defines how surfaces are rotated with respect to the neutral plane or neutral curve.

Split areas of the draft allow you to divide draft surfaces so different draft angles can be applied.

When creating drafts, you can draft only surfaces that are formed by tabulated cylinders or planes. The reference plane used to establish the draft direction must be parallel to the neutral plane.

Surfaces with fillets around the edge boundary cannot be drafted. Therefore, draft the surfaces first, then fillet the edges as required.

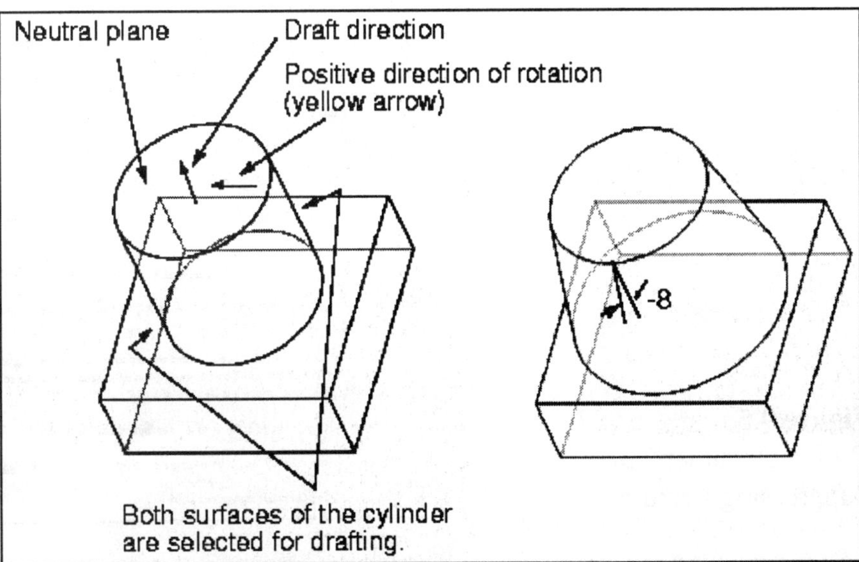

Figure 9.4
Online Documentation, Drafts

Suppressing and Resuming Features

Suppressing a feature is similar to removing the feature from regeneration temporarily (Fig. 9.5). You can "unsuppress" (**Resume**) suppressed features at any time. Features on a part can be suppressed to simplify the part model and decrease regeneration time. For example, while you work on one end of a shaft, it may be desirable to suppress features on the other end of the shaft. Similarly, while working on a complex assembly, you can suppress some of the features and components for which the detail is not essential to the current assembly process.

Unlike other features, the base feature cannot be suppressed. If you are not satisfied with your base feature, you can redefine the section of the feature, or delete it and start over again.

Suppressing Features

To suppress features, do the following:

1. Choose **Suppress** from the FEAT menu. Pro/E displays the SELECT FEAT and GET SELECT menus.
2. Choose one of the following options from the DELETE/SUPP menu:

   Normal Suppress the selected feature and all its children.
   Clip Suppress the selected feature and all the features that follow.
   Unrelated Suppress any feature that is not a child or parent of the selected feature.
3. Select a feature to suppress by picking on it, selecting from the Model Tree, specifying a *range*, entering its *feature number* or *identifier*, or using *layers*.

Suppressing Features

How to suppress features

1. Choose **Suppress** from the FEAT menu. The system displays the SELECT FEAT and GET SELECT menus.
2. Choose one of the following options from the DELETE/SUPP menu:
   - **Normal**-Suppress the selected feature and all its children.
   - **Clip**-Suppress the selected feature and all the features that follow.
   - **Unrelated**-Suppress any feature other than the selected ones and their parents.
3. Select a feature to suppress by picking on it, selecting from the tree tool, specifying a range, entering its feature number or identifier, or using layers (for more information on layers, see Layer Functions in the *Fundamentals* manual).
4. If any children are present and are not currently selected, the system highlights them in blue and displays the CHILD menu. Select one of the following options:
   - **Show Ref**-Show the reference identifier and highlight the reference geometry for each reference of the highlighted child. step through the references using **Next** and **Previous**. You can also obtain information about the reference, showing the reference identifier and the total number of references, and what type of reference it is (feature or entity).
   - **Reroute**-Reroute the references of the highlighted child feature to break the parent-child relationship (see Rerouting Features).
   - **Mod Scheme**-Modify the dimensioning scheme of the child (see Redefining Dimensioning Schemes).
   - **Suppress**-Suppress the highlighted child.

Figure 9.5
Online Documentation, Suppressing Features

You can use **Suppress** and **Resume** to simplify the part before inserting features such as text protrusions. Also, you may wish to suppress the text protrusion if there is other work to be done on the part. Text protrusions take time to regenerate, since they increase the file size considerably.

Text Protrusions

When you are modeling, **text** can be included in a sketch for extruded protrusions and cuts, trimming surfaces, and cosmetic features (Fig. 9.6). The characters that are in an extruded feature must use the font **font3d** for Pro/E. For cosmetic features, any font may be used, and this is done by modifying the text after creating the sketch.

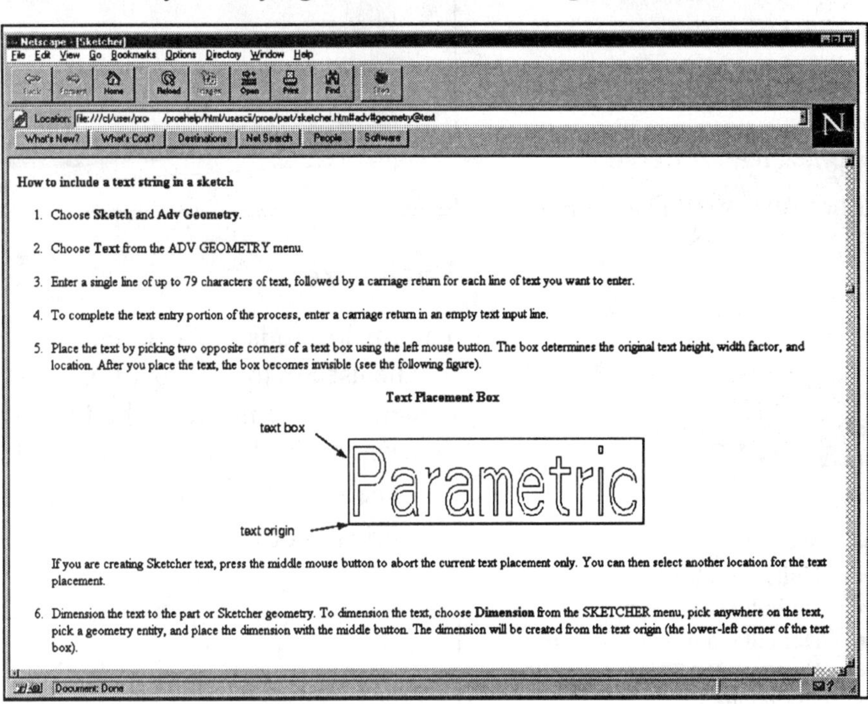

How to include a text string in a sketch

1. Choose **Sketch** and **Adv Geometry**.
2. Choose **Text** from the ADV GEOMETRY menu.
3. Enter a single line of up to 79 characters of text, followed by a carriage return for each line of text you want to enter.
4. To complete the text entry portion of the process, enter a carriage return in an empty text input line.
5. Place the text by picking two opposite corners of a text box using the left mouse button. The box determines the original text height, width factor, and location. After you place the text, the box becomes invisible (see the following figure).

Text Placement Box

If you are creating Sketcher text, press the middle mouse button to abort the current text placement only. You can then select another location for the text placement.

6. Dimension the text to the part or Sketcher geometry. To dimension the text, choose **Dimension** from the SKETCHER menu, pick anywhere on the text, pick a geometry entity, and place the dimension with the middle button. The dimension will be created from the text origin (the lower-left corner of the text box).

Figure 9.6
Online Documentation, Text Protrusions

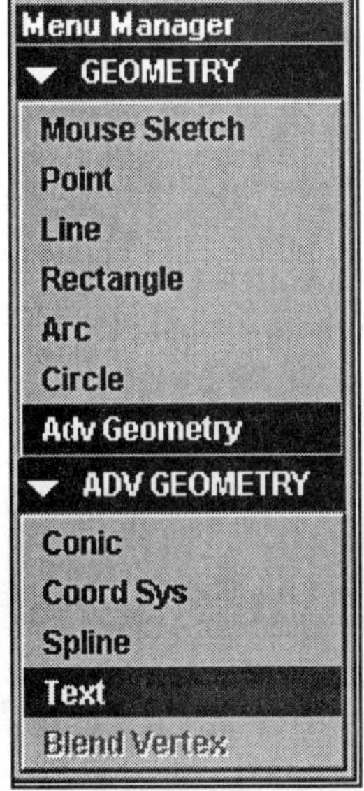

To include a text entry in a sketch:

1. Choose **Sketch ⇒ Adv Geometry.**
2. Choose **Text** from the ADV GEOMETRY menu.
3. Enter a single line of up to 79 characters of text. For the Enclosure, the part lot number, **CFS-2134**, was added to the model, as shown in Figure 9.1 and Figure 9.2.
4. Place the text by picking two opposite corners of a text box. The box determines the original text height, width factor, and location. After the text is placed, the box becomes invisible.
5. Dimension the text to the part or Sketcher geometry. To dimension the text, choose **Dimension** from the SKETCHER menu, pick anywhere on the text, pick a geometry entity, and place the dimension. The dimension will be created from the text origin (the lower left corner of the text box).

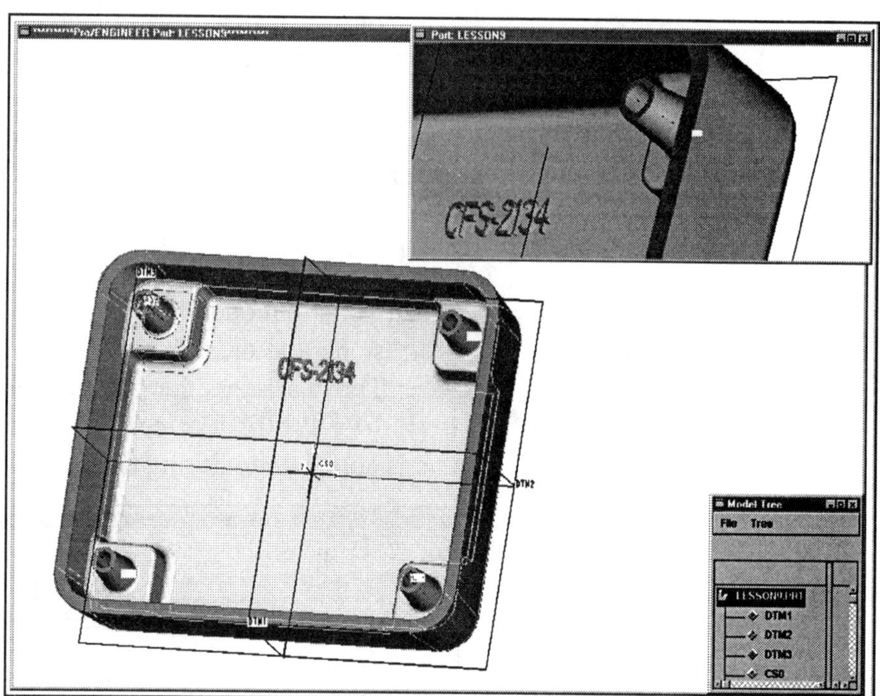

Figure 9.7
Enclosure with Datum Planes

Enclosure

The Enclosure is a plastic injection-molded part. A variety of drafts will be used in the design of this part. A *raised text protrusion* will be modeled on the inside of the Enclosure, as shown in Figure 9.7. The dimensions for the part are provided in Figures 9.8 through 9.13.

The first protrusion has a draft angle of **5°**, as well as the pad and cylindrical protrusions. The holes have a **.3°** draft angle.

NOTE

Set up the Enclosure:

- Material = Plastic
- Units = Inches
- Default Datum Planes
- Default Coordinate System
- Layers = **DATUM_LAYER**
- Hidden line
- Tan Dimmed
- ✓Grid Snap

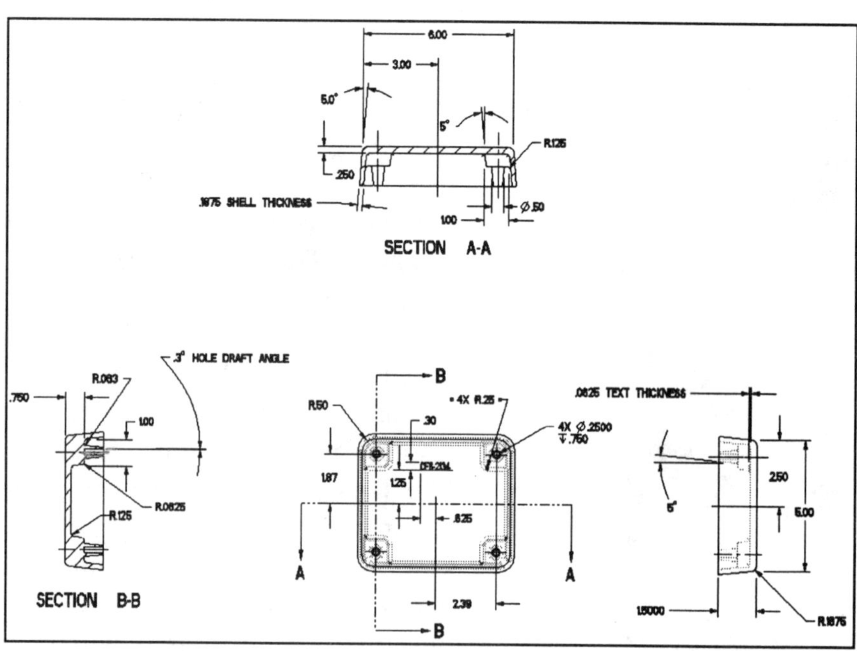

Figure 9.8
Enclosure Detail Drawing

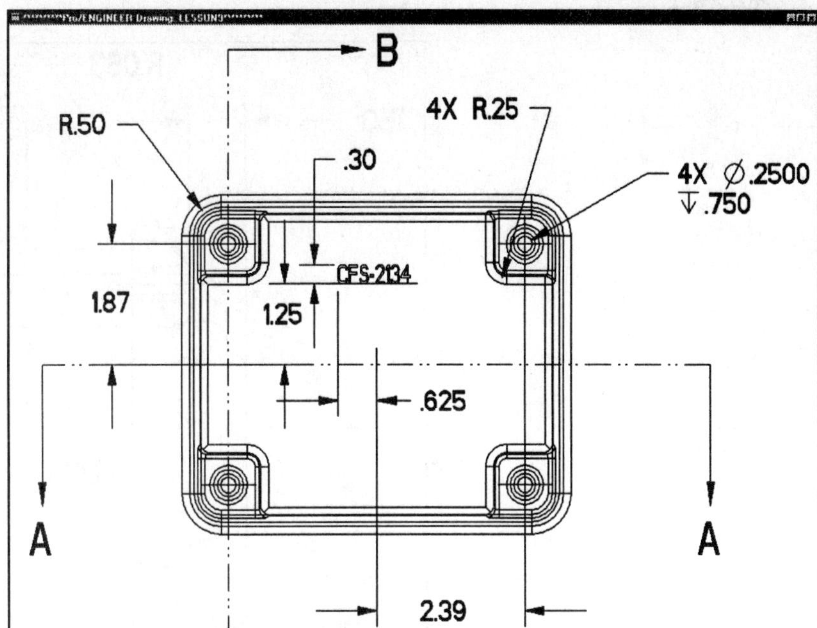

Figure 9.9
Enclosure Drawing,
Front View

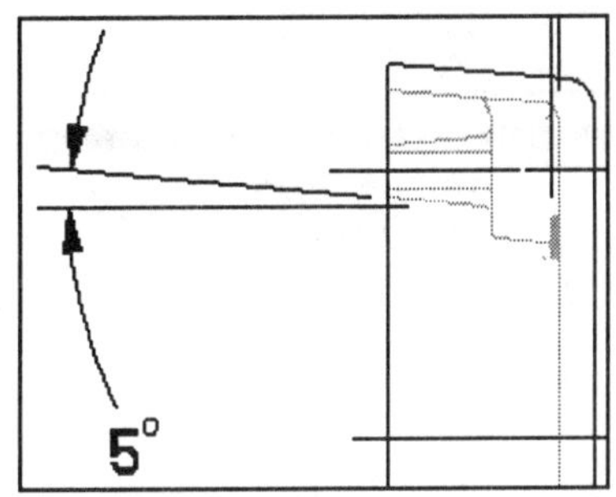

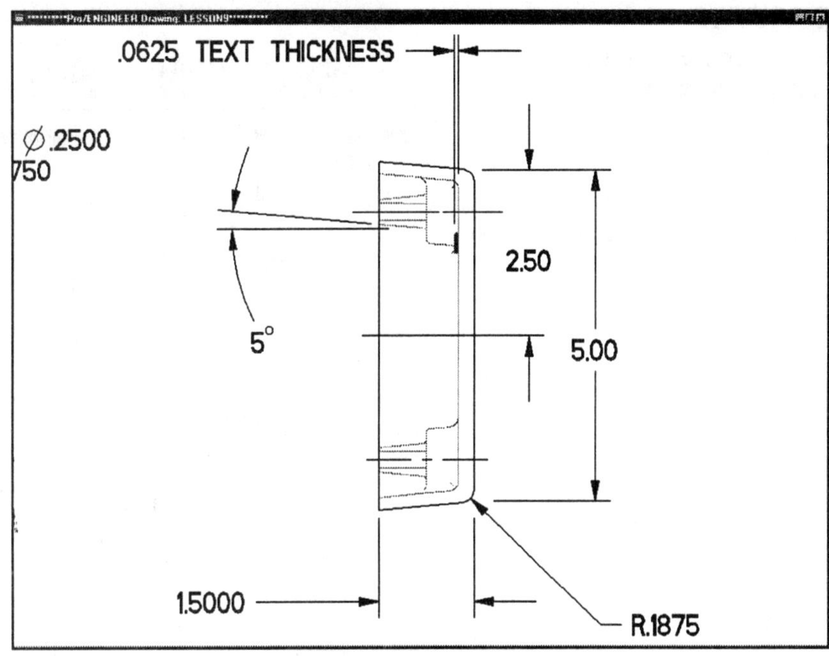

Figure 9.10
Enclosure Drawing,
Right Side View

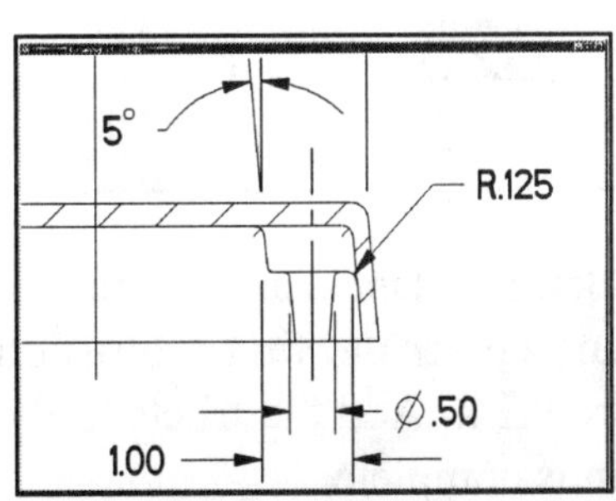

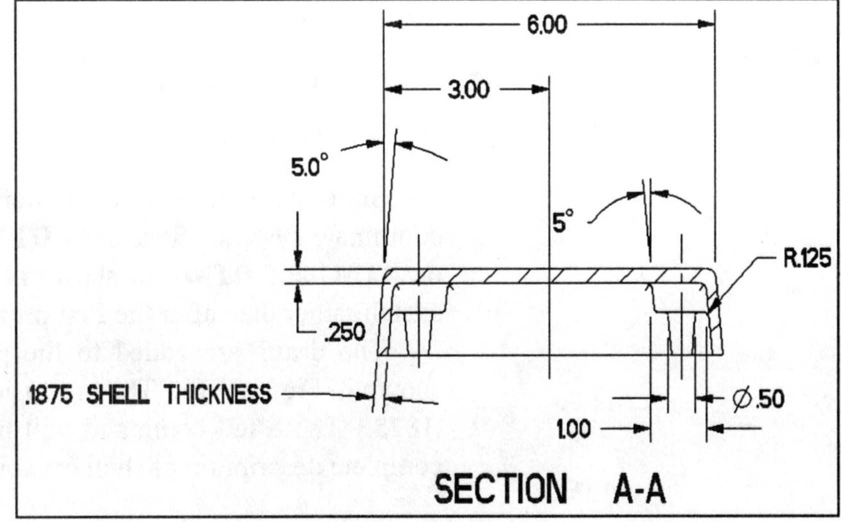

Figure 9.11
Enclosure Drawing,
SECTION A-A

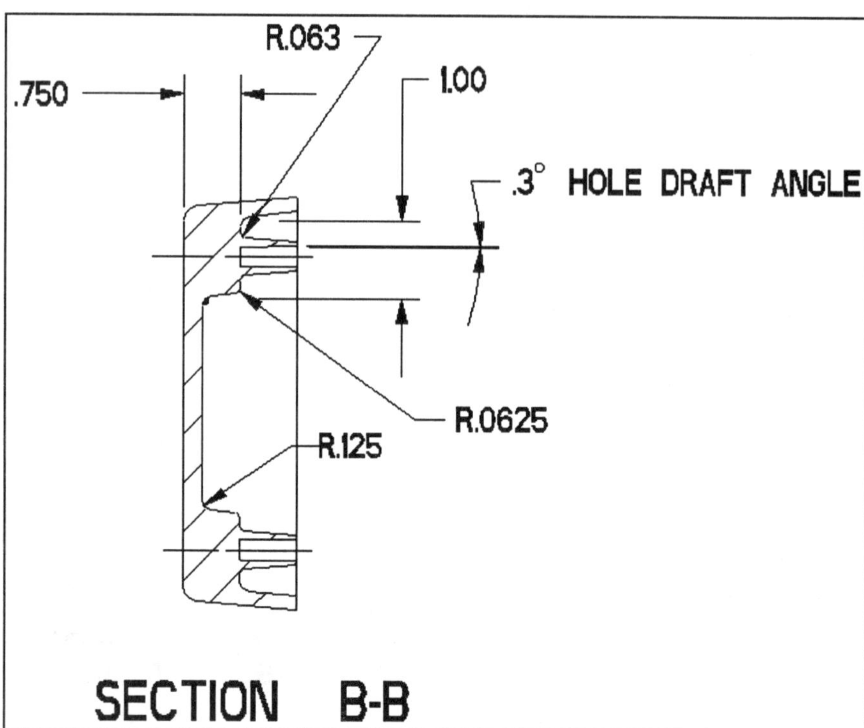

Figure 9.12
Enclosure Drawing,
SECTION B-B

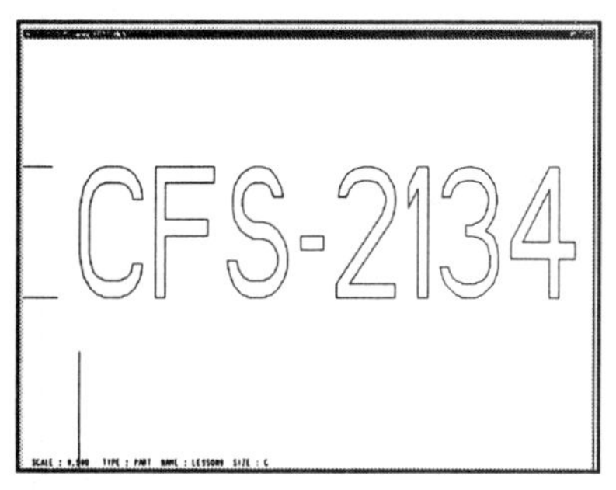

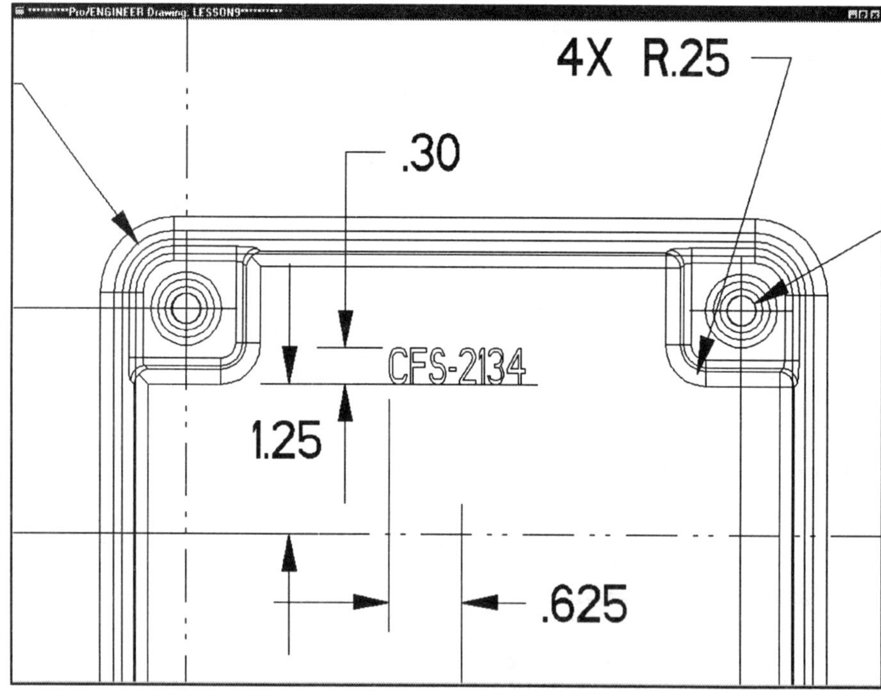

Figure 9.13
Enclosure Drawing,
Text Protrusion Location

Start the part with three default datum planes and a default coordinate system. Sketch on **DTM3**, and center the first protrusion on **DTM1** and **DTM2**, as shown in Figure 9.14. Add the fillets to the sketch rather than after the first protrusion is complete.

The drafts are added to the protrusion before it is shelled, as shown in Figure 9.15. The protrusion will be shelled to a thickness of **.1875**. The **Shell** command will be introduced in this lesson, but a complete description of shell capabilities is provided in Lesson 10.

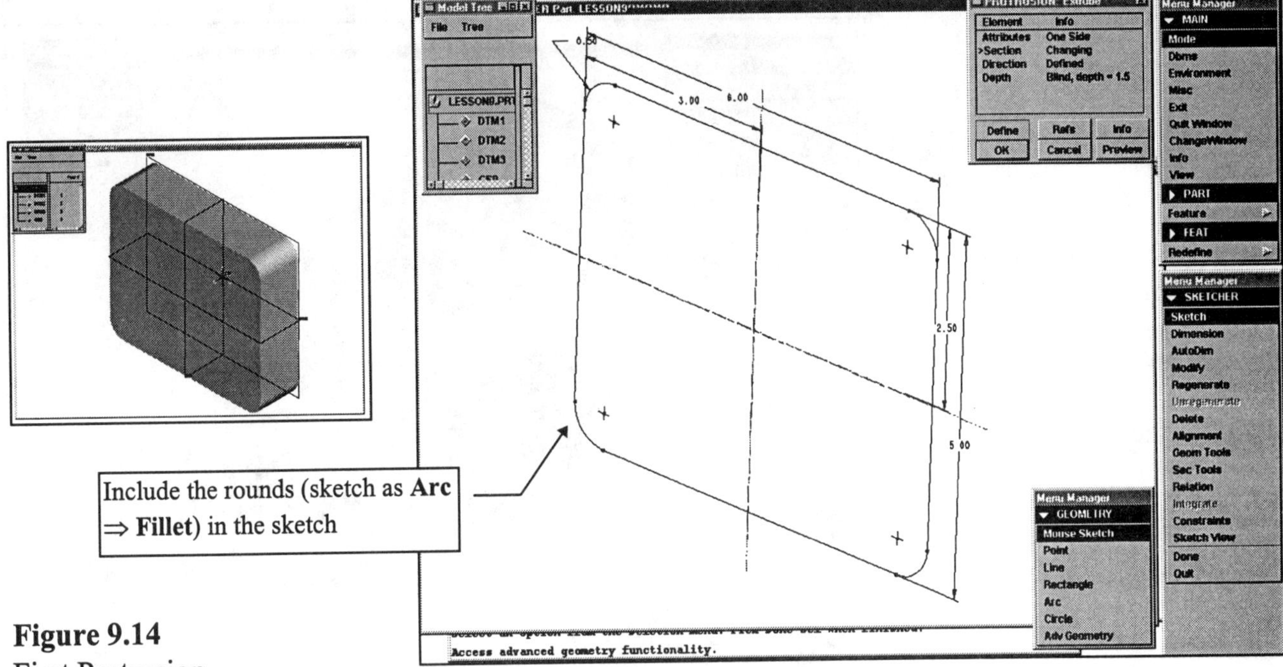

Figure 9.14
First Protrusion

Create the draft for the lateral surfaces of the protrusion using the following commands:

PT/Modeler™
Feature ⇒ **Draft** ⇒ **No Split** ⇒ **Done** ⇒ **etc.**

Feature ⇒ **Create** ⇒ **Tweak** ⇒ **Draft** ⇒ **Neutral Pln** ⇒ **Done** ⇒ **No Split** ⇒ **Constant** ⇒ **Done** ⇒ **Include** ⇒ **Indiv Surfs** ⇒ (pick all ***eight*** *lateral surfaces* of the protrusion, as shown in Fig. 9.15) ⇒ **Done Sel** ⇒ **Done** ⇒ (select **DTM3** as the neutral plane) ⇒ (select **DTM3** as the plane the direction will be perpendicular to, as shown in Fig. 9.15) ⇒ (specify the draft angle by typing **5** at the prompt) ⇒ **enter** ⇒ **Preview** (Fig. 9.16) ⇒ **OK** ⇒ **Done**

NOTE

The *neutral plane* and the *plane the direction will be perpendicular to* are often the same plane. For the draft created here, **DTM3** is used for both.

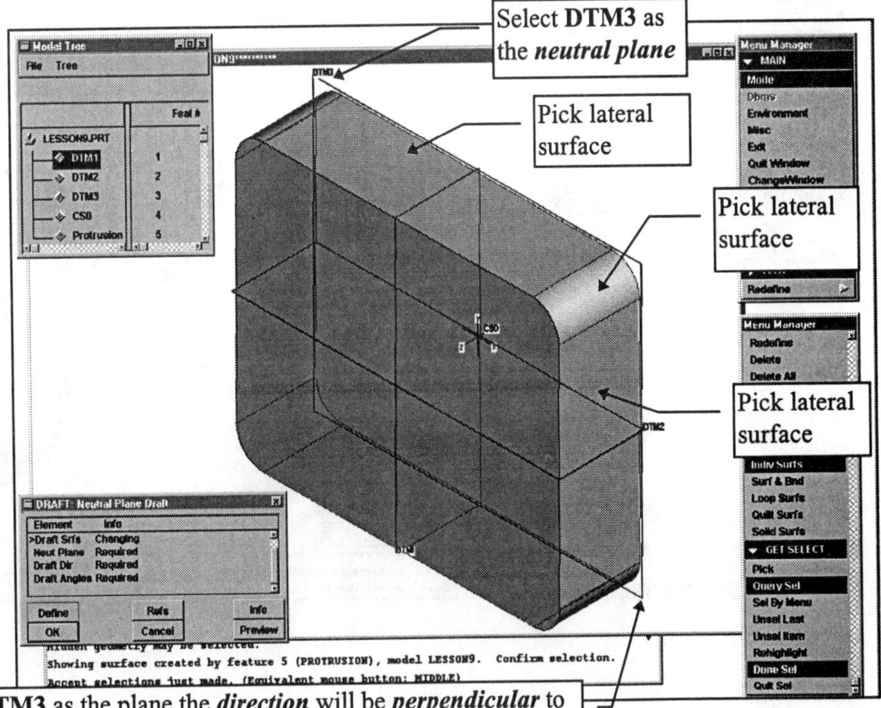

Figure 9.15
Draft

Dbms (**File--PT/Modeler**) ⇒
Save ⇒ **enter**
Purge ⇒ **enter** ⇒
Done-Return

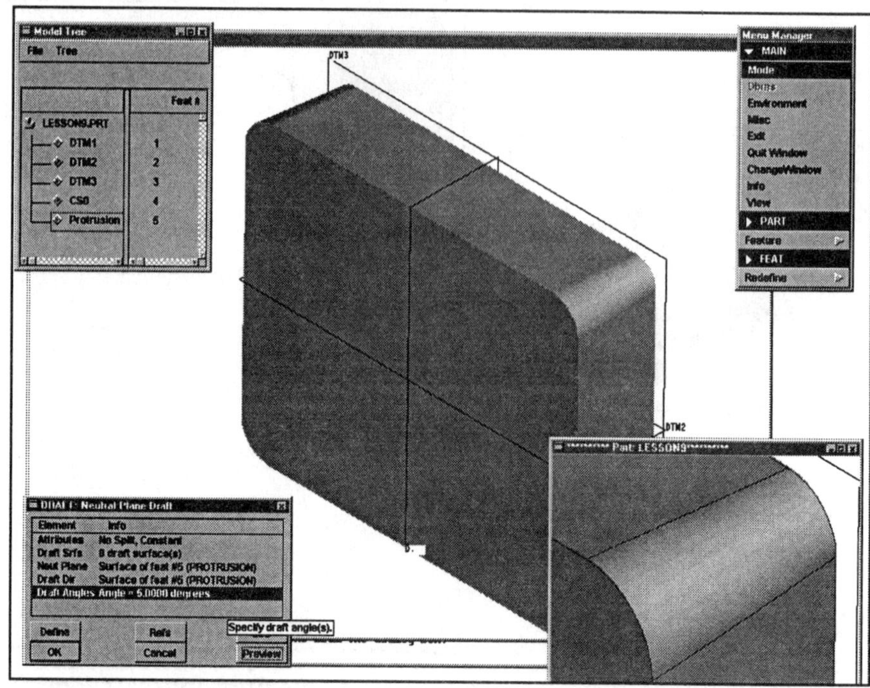

Figure 9.16
Preview of Draft

Next you will shell the part using the following commands:

PT/Modeler™
Feature ⇒ **Shell** ⇒ **etc.**

Feature ⇒ **Create** ⇒ **Shell** ⇒ (select the surface to remove by picking the vertical face, as shown in Fig. 9.17) ⇒ **Done Sel** ⇒ **Done Refs** ⇒ (enter the thickness by typing **.1875** at the prompt) ⇒ **enter** ⇒ **Preview** ⇒ **OK** (Fig. 9.18) ⇒ **Done**

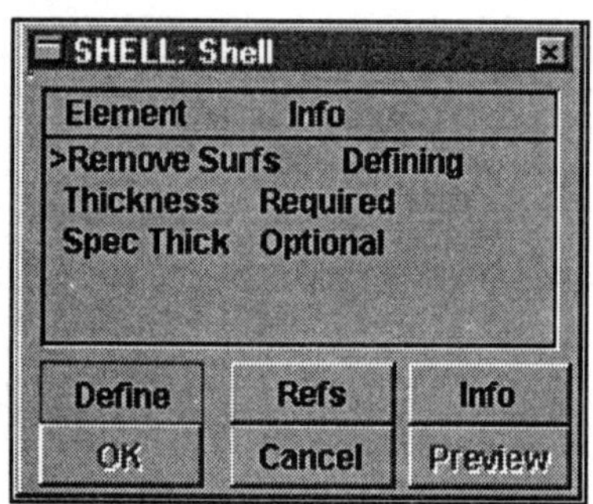

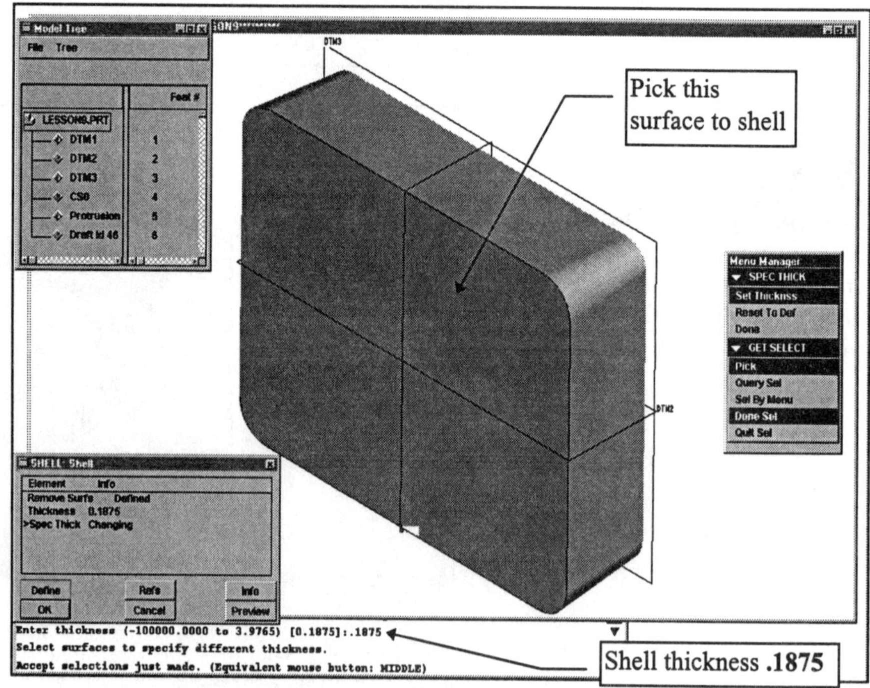

Figure 9.17
Selecting the Surface to Shell

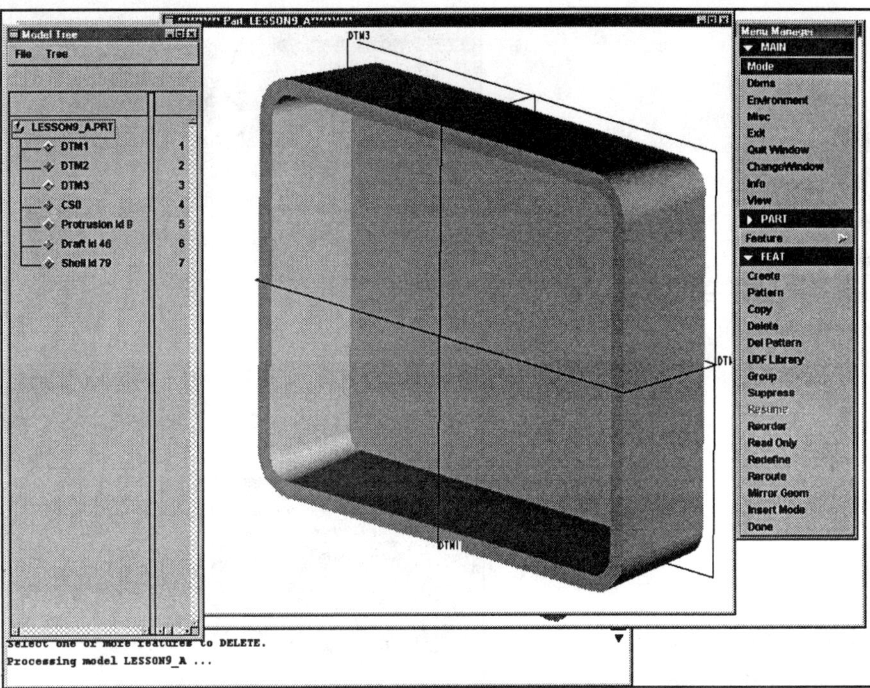

Figure 9.18
Shelled Part

Sketch the raised pedestal-like protrusion on **DTM3**, as shown in Figure 9.19. In most cases, the protrusion would be sketched on the inside of the shelled surface, but for this design we will project from the datum plane. Use the existing *internal* edge and the dimensioning scheme shown in the figure and in the detail drawings provided earlier. Figure 9.20 shows the completed protrusion.

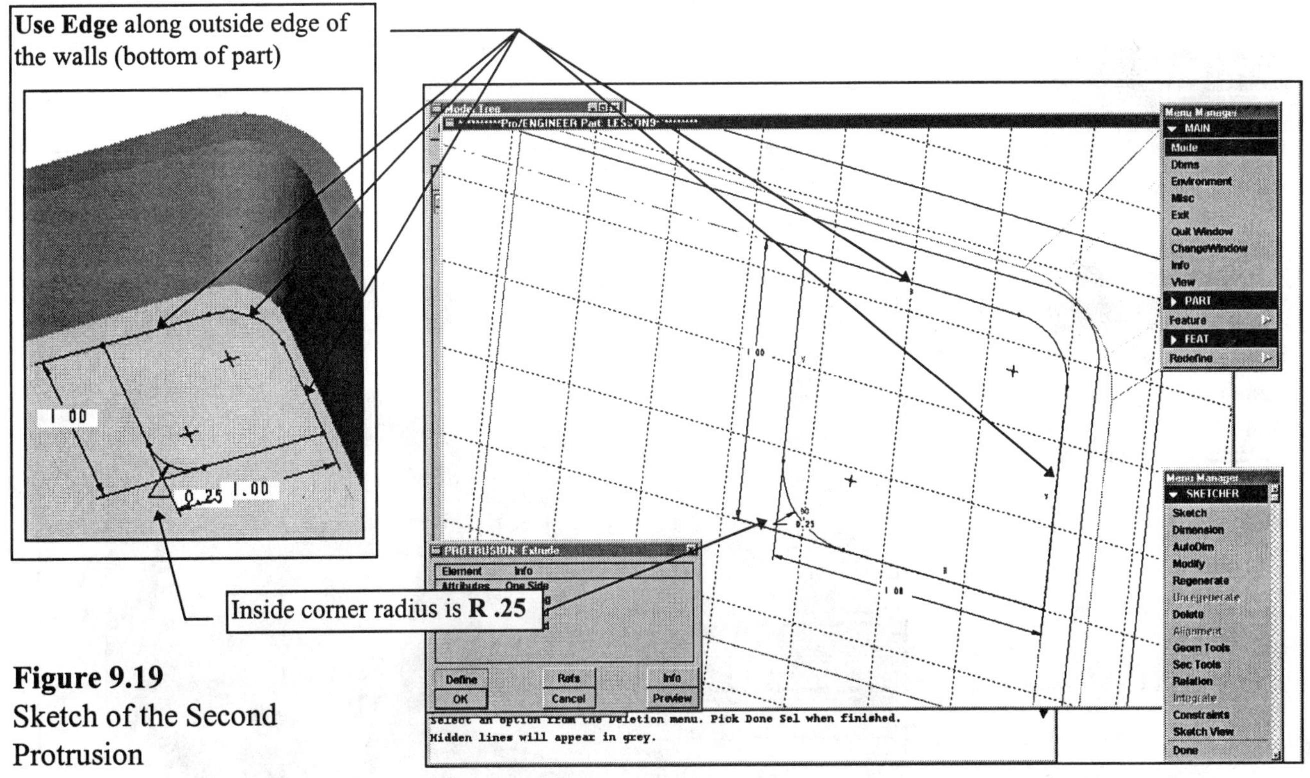

Figure 9.19
Sketch of the Second Protrusion

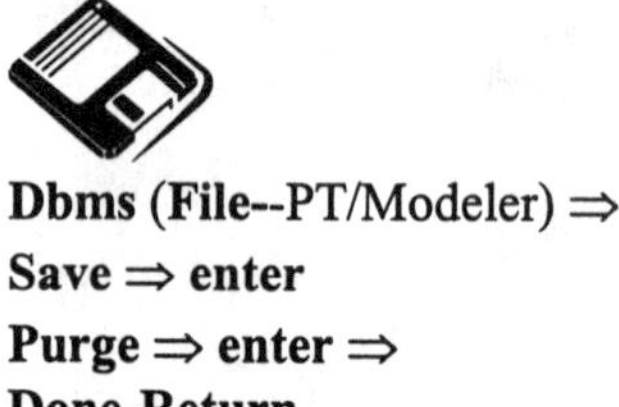

Dbms (**File**--PT/Modeler) ⇒ **Save** ⇒ **enter** **Purge** ⇒ **enter** ⇒ **Done-Return**

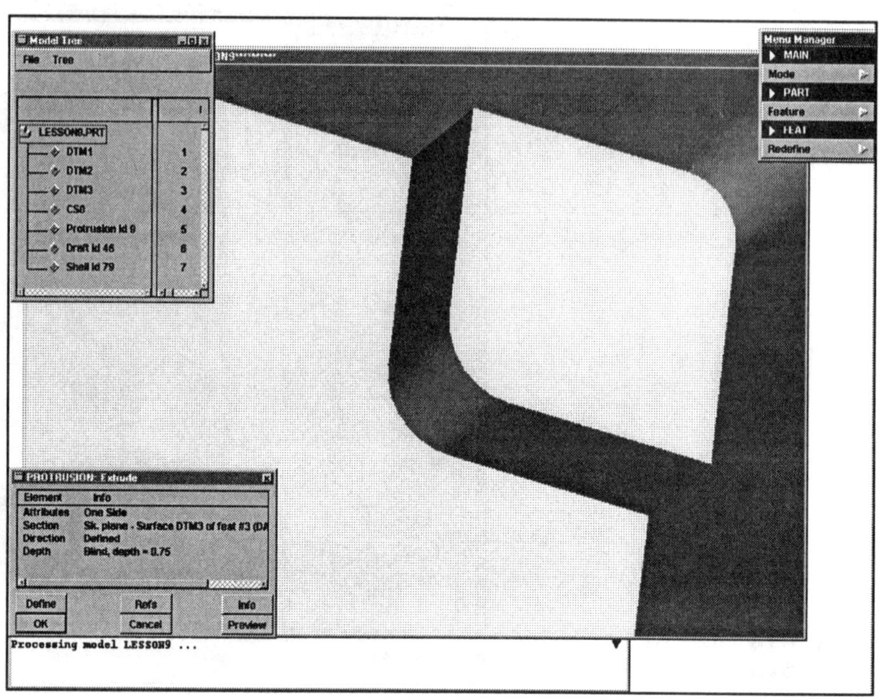

Figure 9.20
Second Protrusion

Use the following commands to draft the lateral surfaces of the second protrusion:

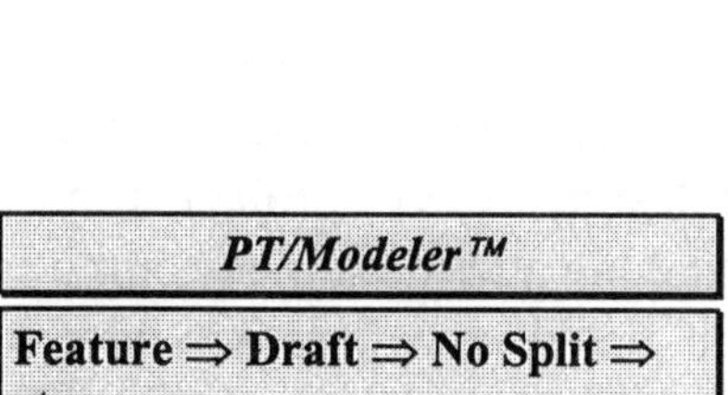

Feature ⇒ Create ⇒ Tweak ⇒ Draft ⇒ Neutral Pln ⇒ Done ⇒ No Split ⇒ Constant ⇒ Done ⇒ Include ⇒ Indiv Surfs ⇒ (pick all ***three lateral surfaces*** of the protrusion, as shown in Fig. 9.21) ⇒ **Done Sel ⇒ Done ⇒** (select **DTM3** as the neutral plane) ⇒ (select **DTM3** as the plane the direction will be perpendicular to) ⇒ (specify the draft angle by typing **-5** at the prompt) ⇒ **enter ⇒ Preview** (Fig. 9.21) ⇒ **OK ⇒ Done**

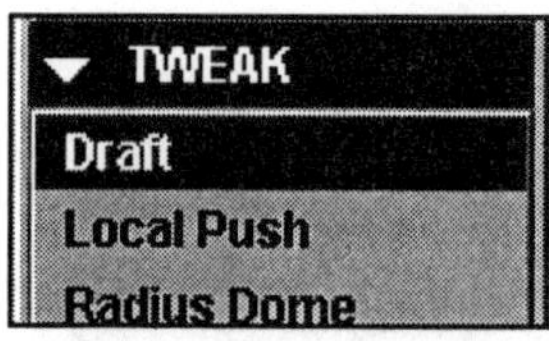

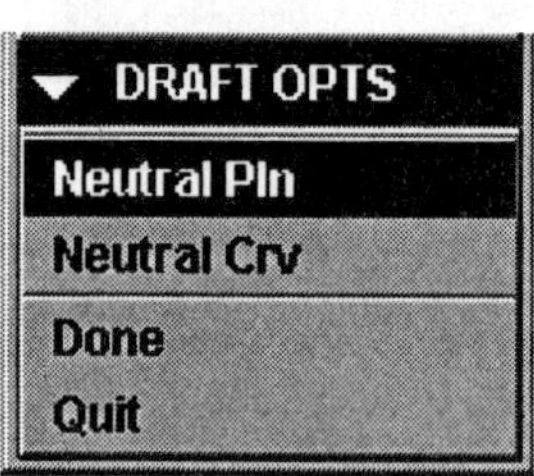

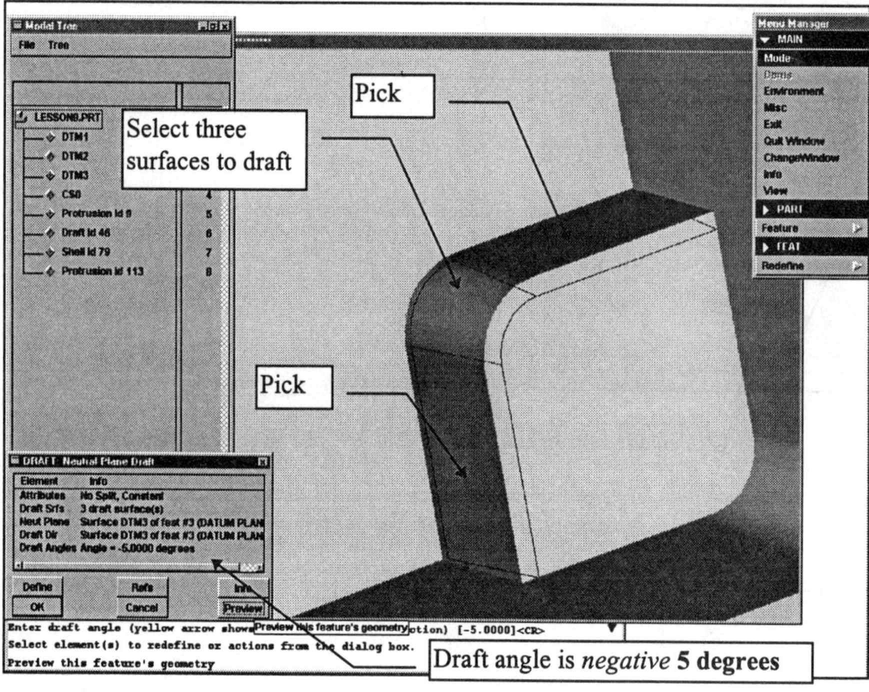

Figure 9.21
Preview of the Draft of the Second Protrusion

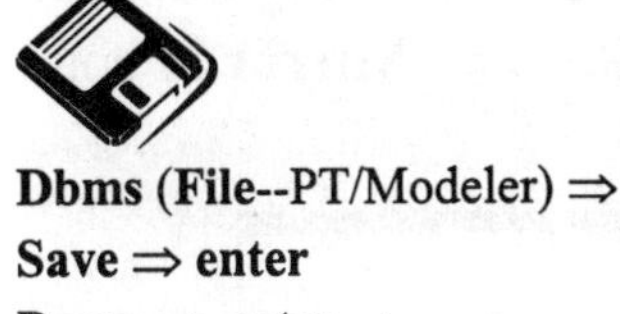

Dbms (File--PT/Modeler) ⇒
Save ⇒ **enter**
Purge ⇒ **enter** ⇒
Done-Return

Model the circular protrusion (Fig. 9.22). Draft the new protrusion at **-5°** (Fig. 9.23). Create a hole (Fig. 9.24), and add an internal draft of **-.3°**.

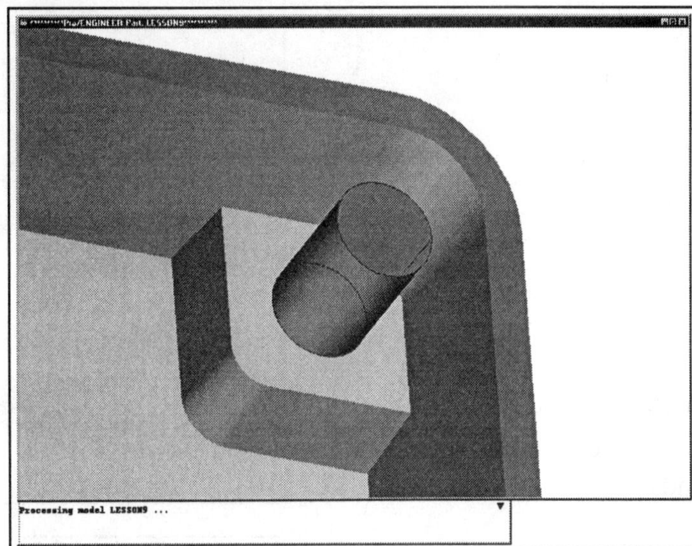

Figure 9.22
Circular Protrusion

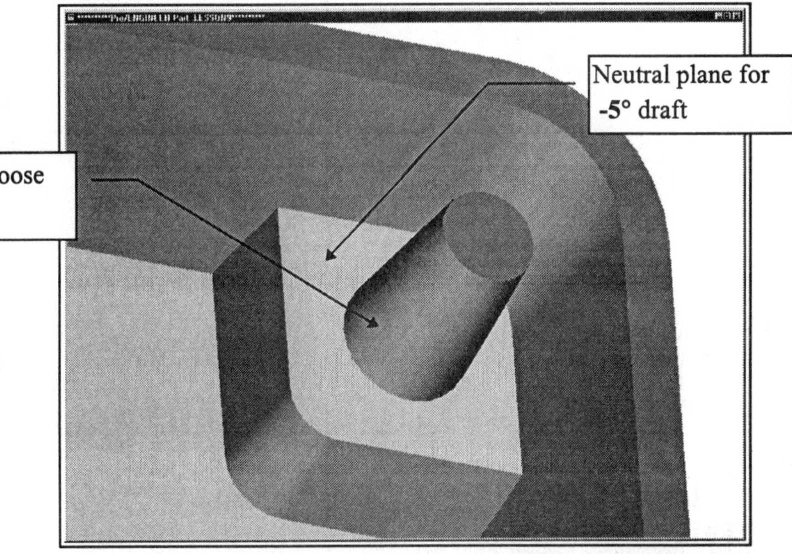

Figure 9.23
Draft

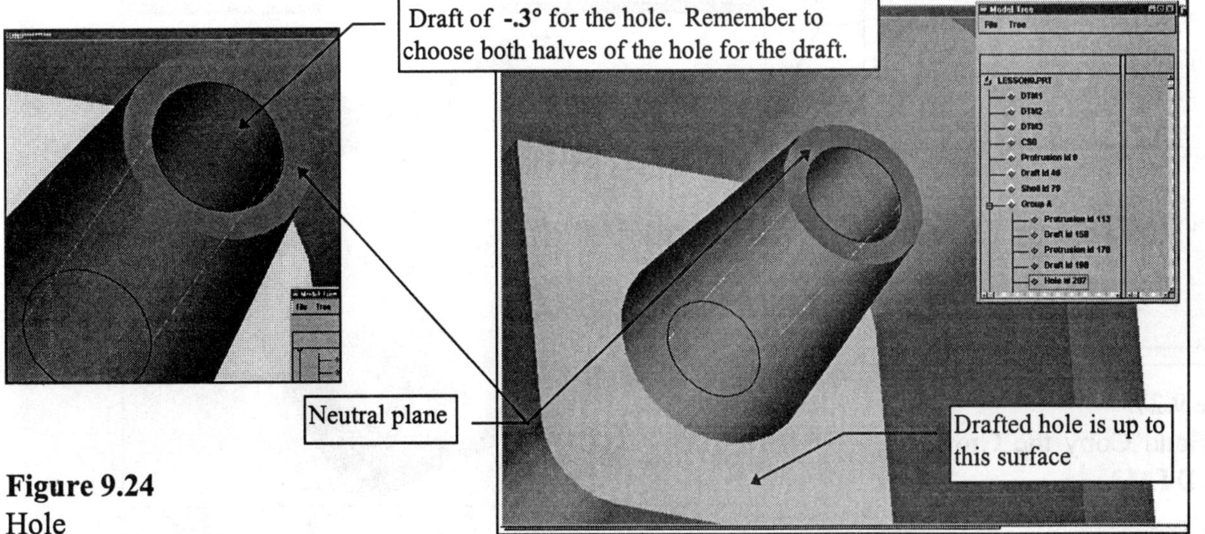

Figure 9.24
Hole

Create the rounds required for the part as shown in Figure 9.25. **Group** the protrusions, hole, and rounds and **Copy ⇒ Mirror** in both directions, as in Figures 9.26 and 9.27.

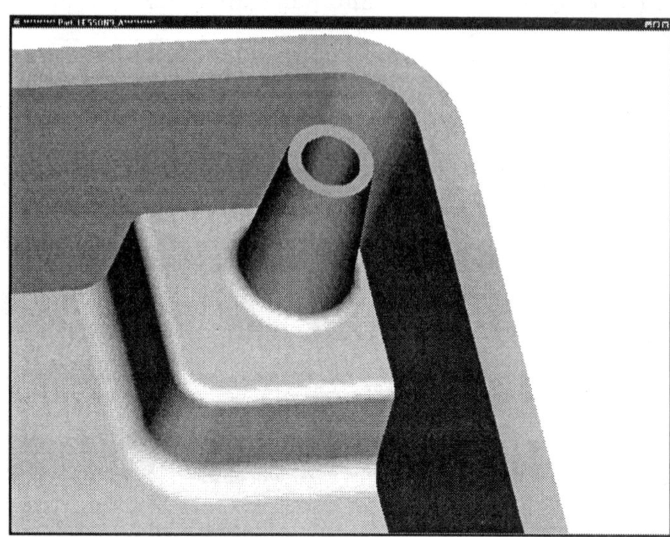

Figure 9.25
Group the Protrusions, Hole, and Rounds

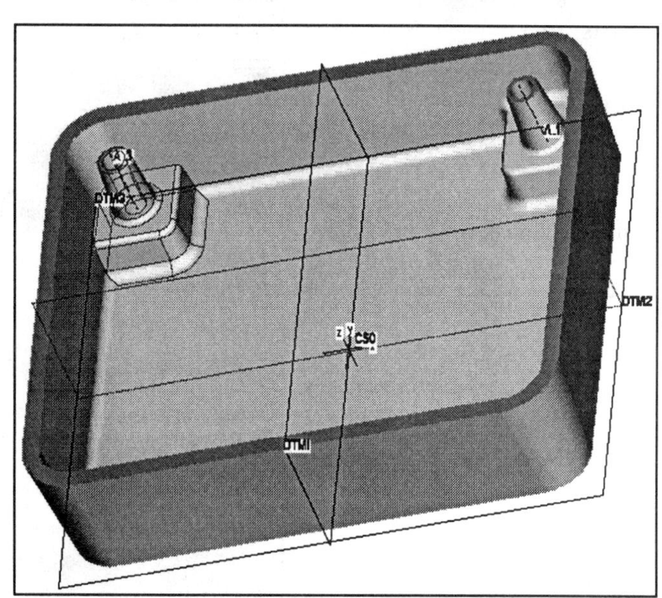

Figure 9.26
Copy ⇒ Mirror the **Group** About **DTM1**

Some rounds still need to be created

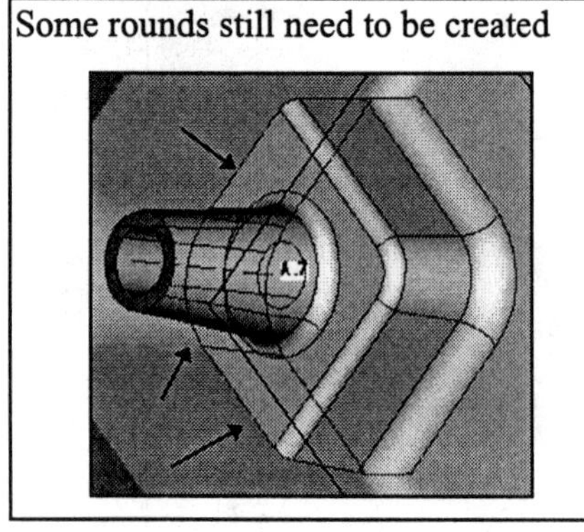

Figure 9.27
Mirror and Copy the Groups About **DTM2**

Create the missing rounds (Fig. 9.27) by using the following commands:

PT/Modeler™
Feature ⇒ Round ⇒ Surf-Surf⇒ etc.

Feature ⇒ Create ⇒ Round ⇒ Done ⇒ Surf-Surf ⇒ Done (pick the first surface) ⇒ (pick the second surface) ⇒ (enter the radius **.125** at the prompt) ⇒ **enter ⇒ OK** (Fig. 9.28) ⇒ **Done**

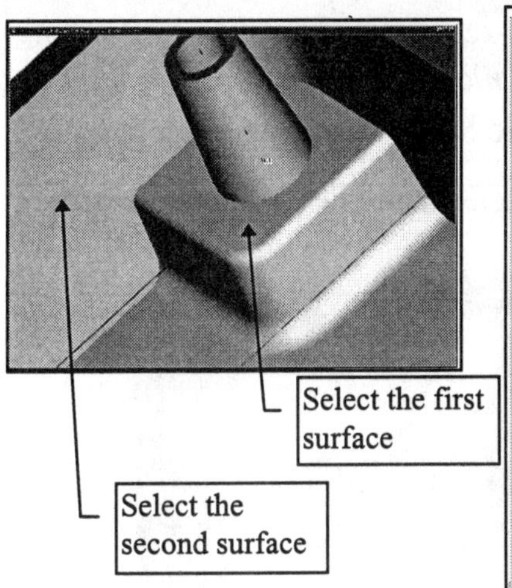

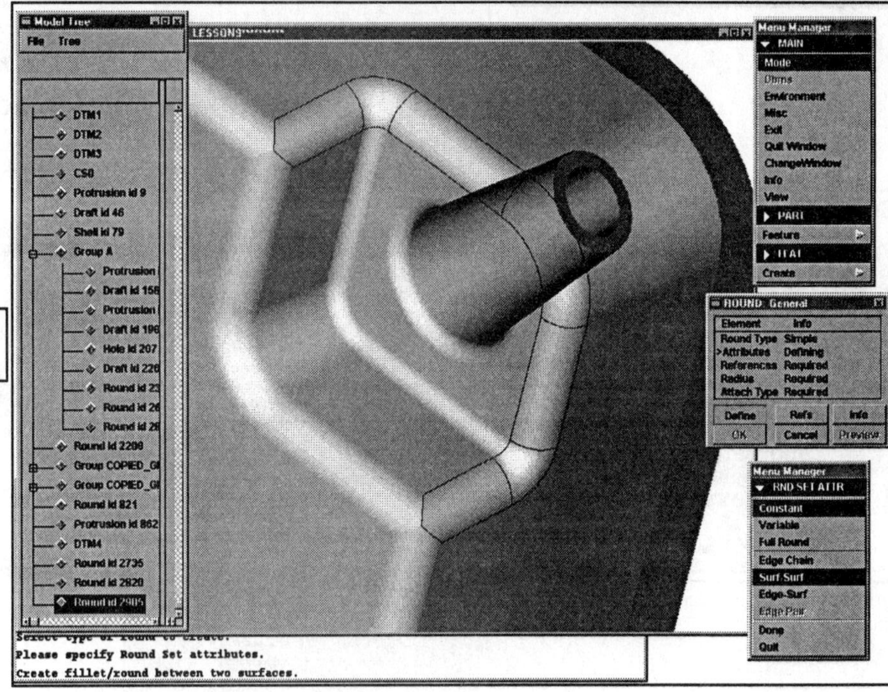

Figure 9.28
Create the Last Internal Rounds

Create the external round on the part, as shown in Figure 9.29. Try to **Reorder** the round to come before the shell. If the model fails, **Undo Changes ⇒ Confirm**.

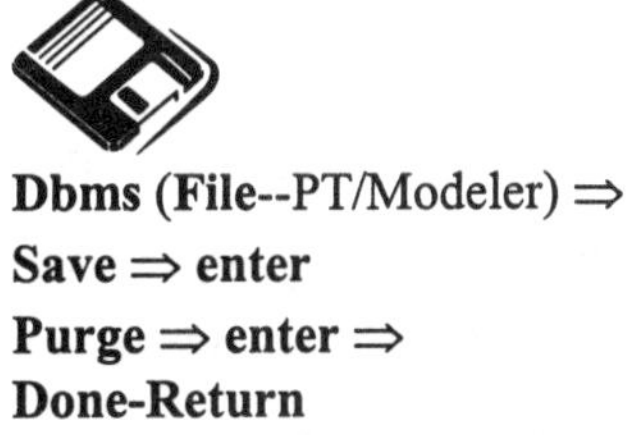

Dbms (File--PT/Modeler) ⇒
Save ⇒ enter
Purge ⇒ enter ⇒
Done-Return

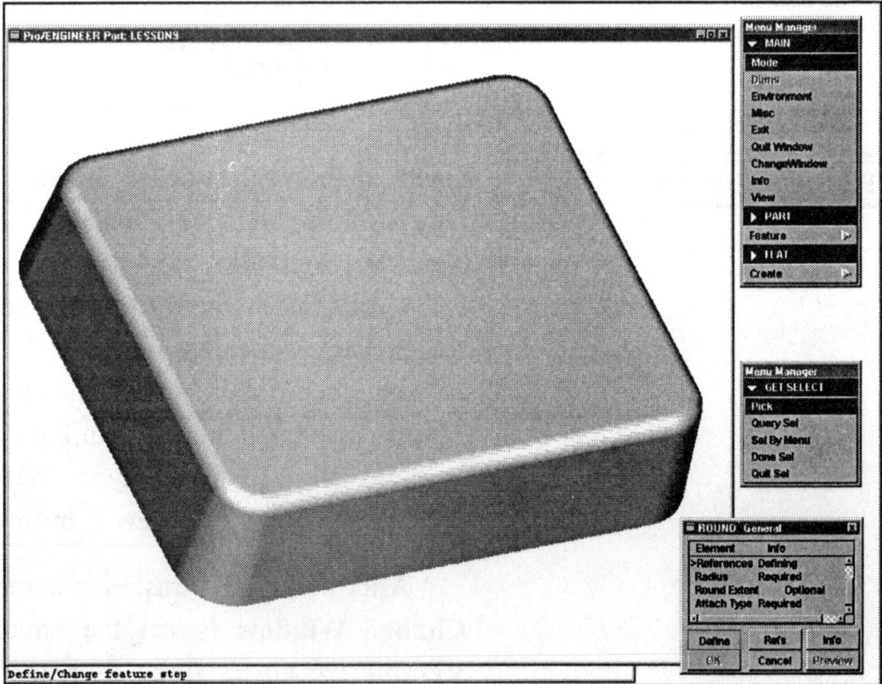

Figure 9.29
Create the Outside Round

Before creating the text protrusion, **Suppress** all the features after the shell command. Expand the Model Tree to include the feature number and status. Use the following commands to suppress the features:

PT/Modeler™

Feature Oper ⇒ Suppress etc.

Feature ⇒ Suppress ⇒ Range ⇒ (enter the number of the feature following the shell; type **8** at the prompt--*your number for this feature may be different*) ⇒ **enter ⇒** (enter regeneration number of ending feature; type **50** at the prompt--*your number for this feature may be different*) ⇒ **enter** (Fig. 9.30) ⇒ **Done Sel ⇒ Done ⇒ Done**

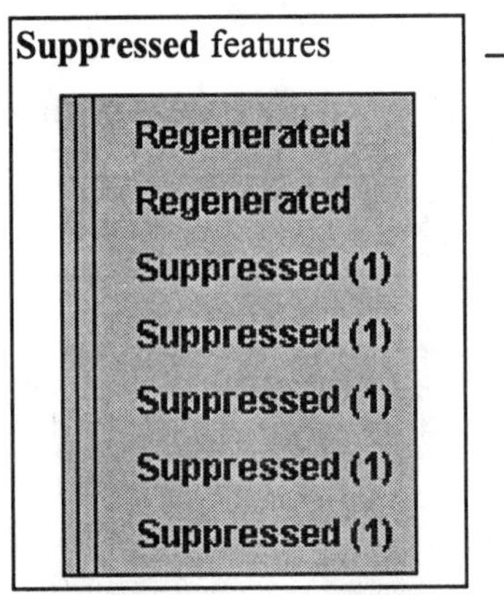

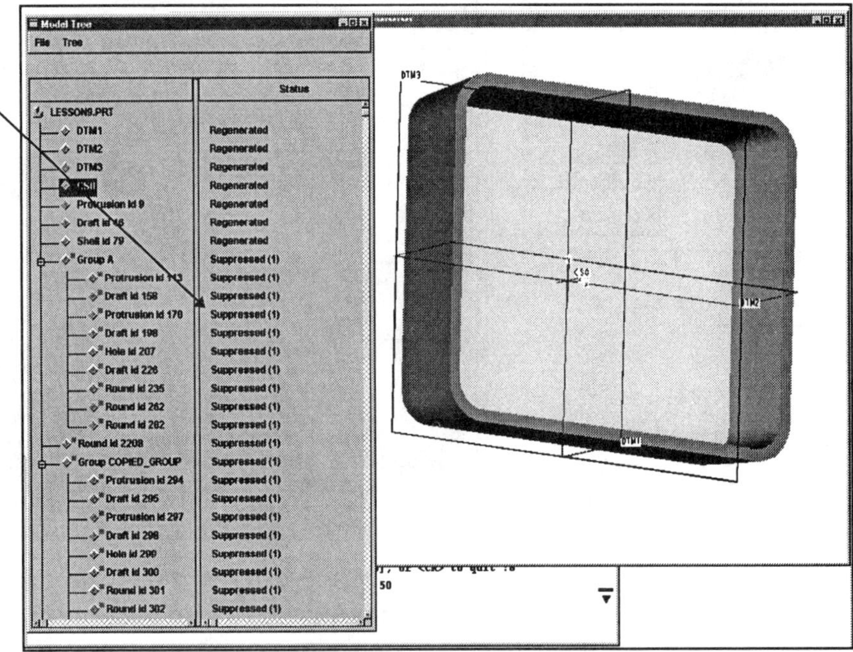

Figure 9.30
Suppressing Model Features

The regeneration time for your model will now be much shorter. Add the text protrusion shown in Figures 9.9 and 9.13 using the following commands:

PT/Modeler™

Feature ⇒ Protrusion ⇒ etc.

Feature ⇒ Create ⇒ Protrusion ⇒ Done ⇒ Done ⇒ (pick the inside of the enclosure for the sketching plane as in Fig. 9.31) ⇒ **Okay** (for direction) ⇒ **Top ⇒** (pick **DTM2**) ⇒ **Adv Geometry ⇒ Text ⇒** (enter **CFS-2134** at the prompt) ⇒ **enter** (Figs. 9.32 and 9.33) ⇒ (pick a box to place and size the text) ⇒ **enter ⇒ Regenerate ⇒ Dimension ⇒** (to locate the text, add the dimensions from the text to the datum planes as in Fig. 9.34) ⇒ **Regenerate ⇒ Modify** (use the dimensions in Figs. 9.9 and 9.34) ⇒ **Regenerate ⇒ Done ⇒ Blind ⇒ Done** (type **.0625** at the prompt) ⇒ **enter ⇒ OK ⇒ View ⇒ Default ⇒ New Window** (Fig. 9.35)

After the text protrusion is completed, choose **Quit Window ⇒ Change Window** (select the remaining window) ⇒ **Feature ⇒ Resume ⇒ All ⇒ Done** from the feature menu (Fig. 9.36). Try regenerating now. It will take longer. The enclosure is complete.

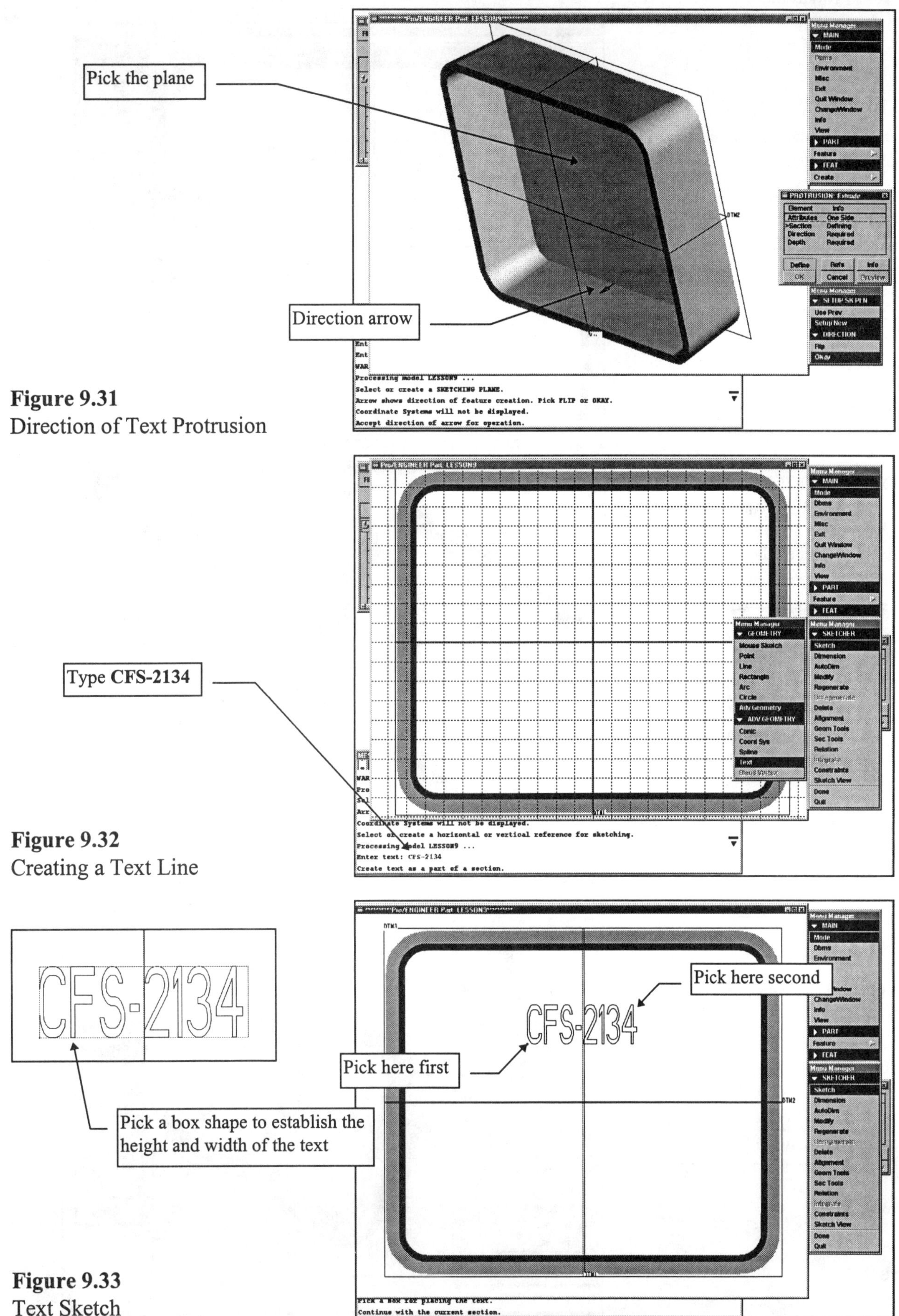

Figure 9.31
Direction of Text Protrusion

Figure 9.32
Creating a Text Line

Figure 9.33
Text Sketch

NOTE

To make the text clearly visible, it has been sketched larger here than in the detail drawings.

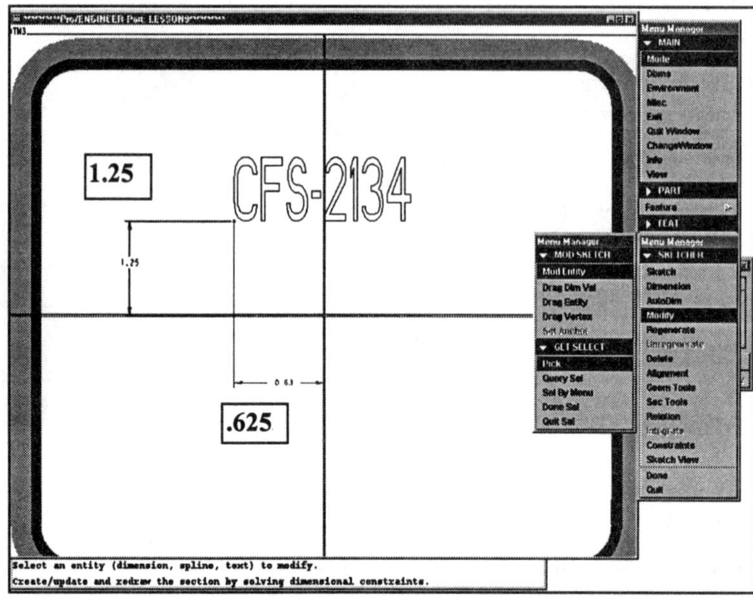

Figure 9.34
Dimension the Text Sketch

Try modifying the depth and location of the text protrusion: **Modify** ⇒ (pick the text) ⇒ (change the depth to **.25**) ⇒ **Regenerate** ⇒ (if you have not resumed the model do so now: **Resume** ⇒ **All** ⇒ **Done**)

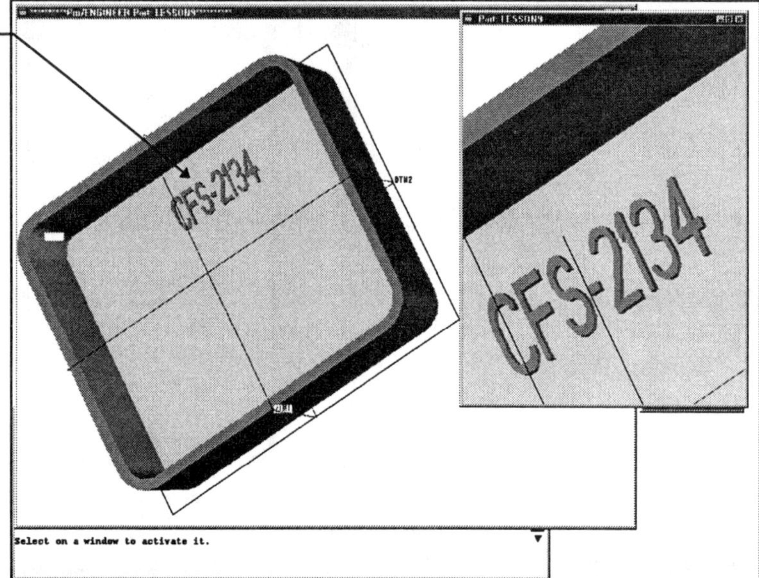

Figure 9.35
Text Protrusion

Dbms (File--PT/Modeler) ⇒
Save ⇒ **enter**
Purge ⇒ **enter** ⇒
Done-Return

Regenerated (**Resume**) features

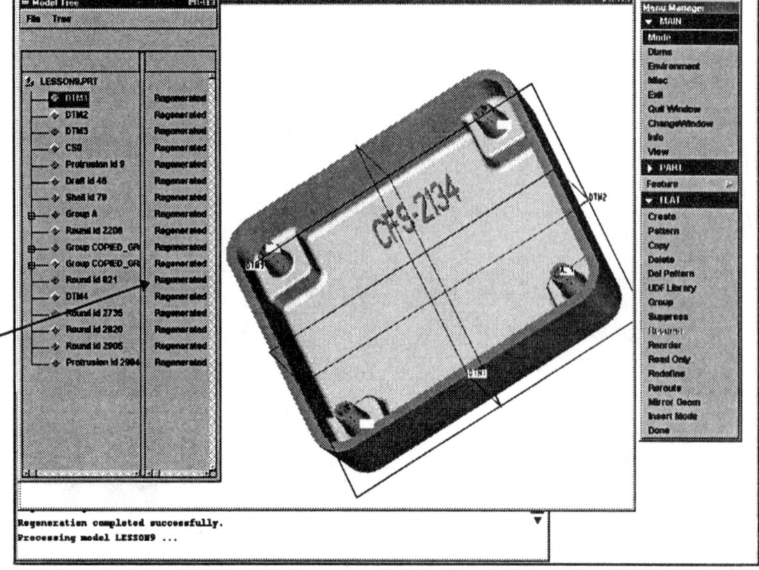

Figure 9.36
Resume

Lesson 9 Project

Cellular Phone Bottom

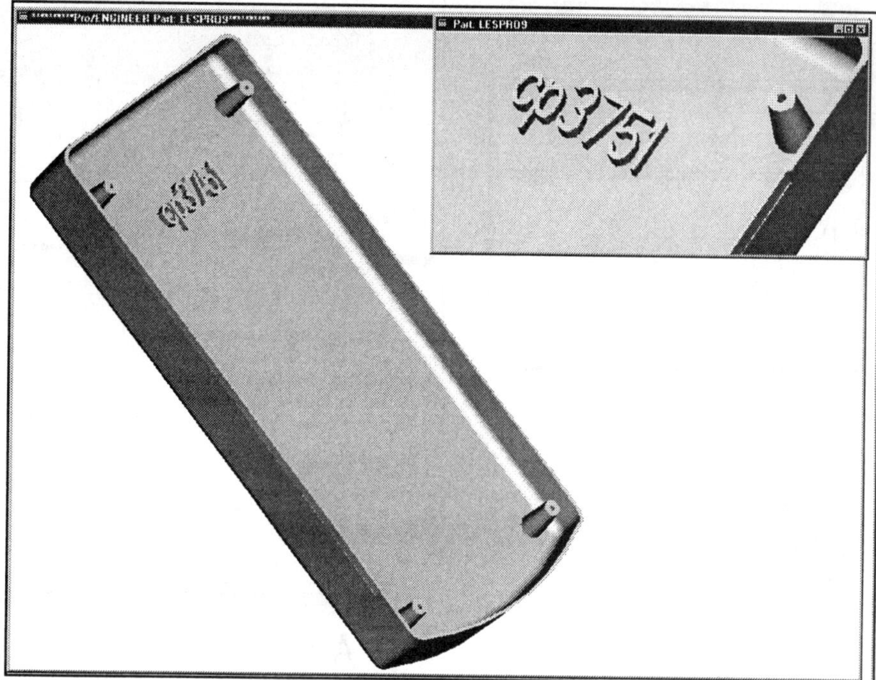

Figure 9.37
Cellular Phone Bottom

Cellular Phone Bottom

The ninth **lesson project** is a die-cast plastic part that requires commands similar to those for the **Enclosure**. Create the part shown in Figures 9.37 through 9.47. Analyze the part and plan out the steps and features required to model it. Use the DIPS in Appendix D to plan out the feature creation sequence and the parent-child relationships for the part. The top of the cellular phone is created in the Lesson 10 Project. Do not shell the Cellular Phone Bottom.

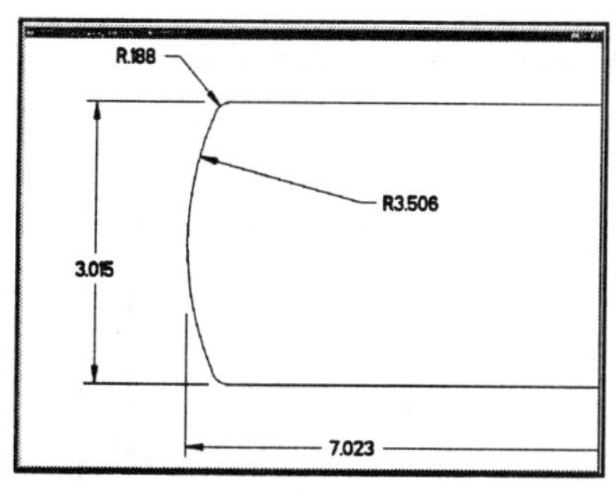

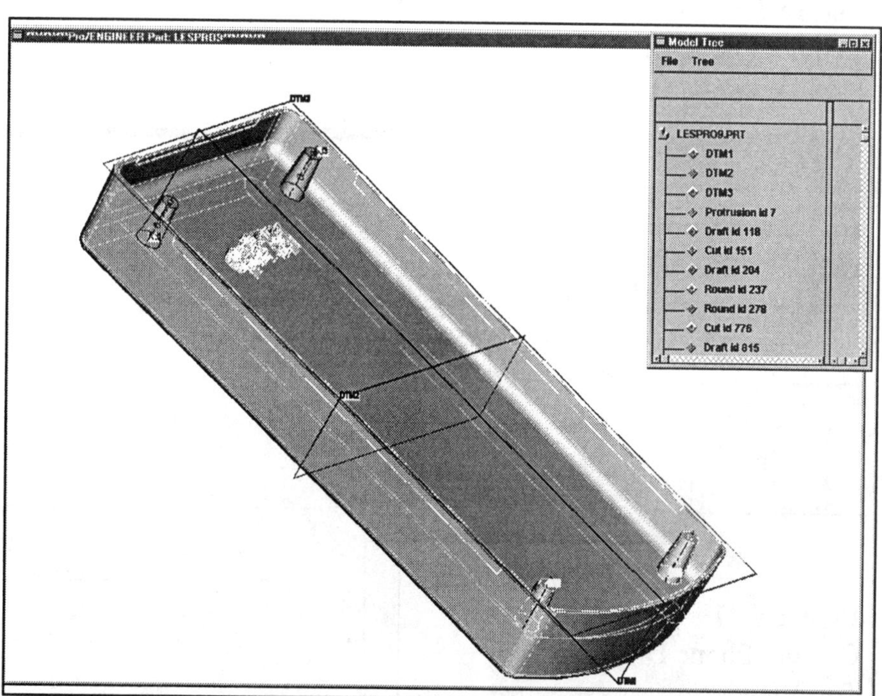

Figure 9.38
Cellular Phone Bottom
Showing Datum Planes

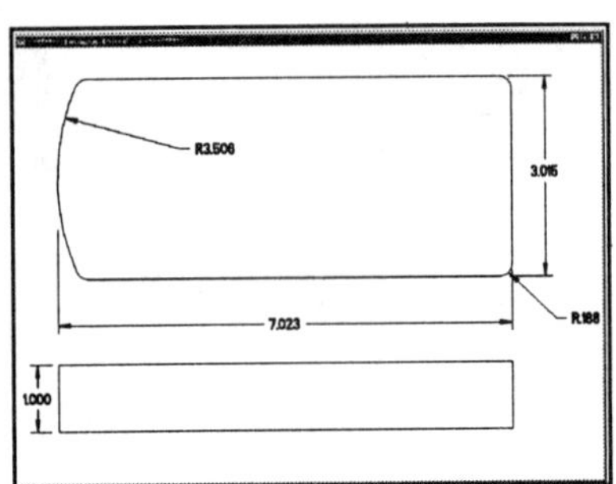

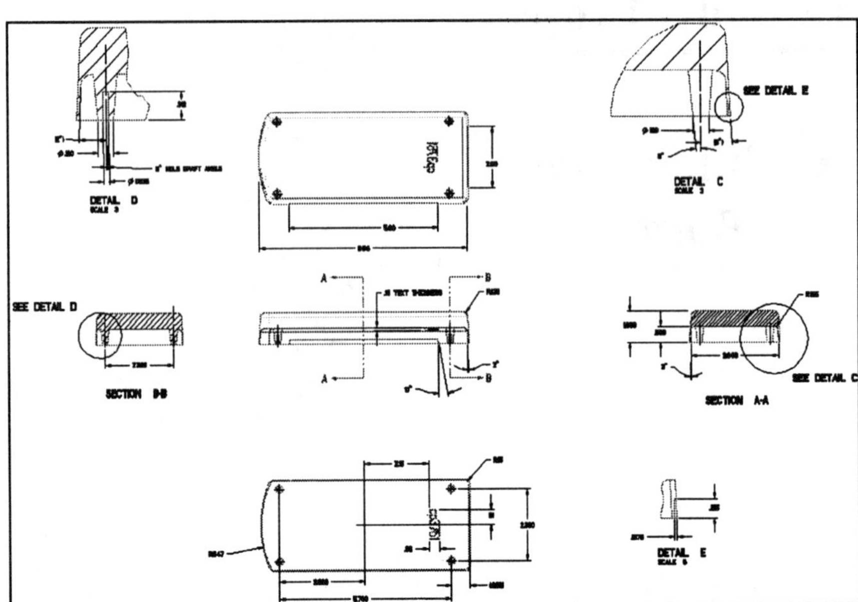

Figure 9.39
Cellular Phone Drawing

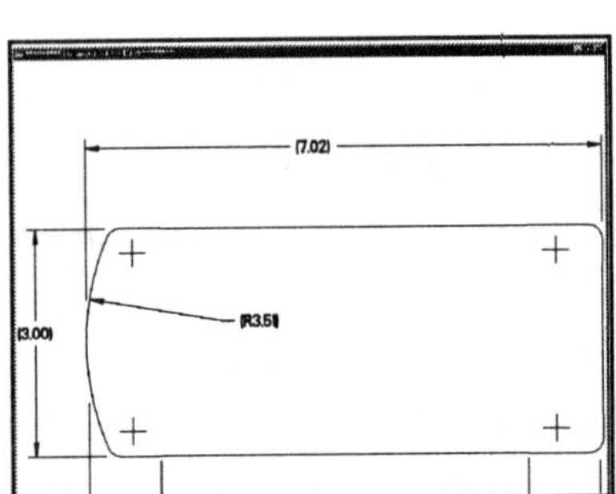

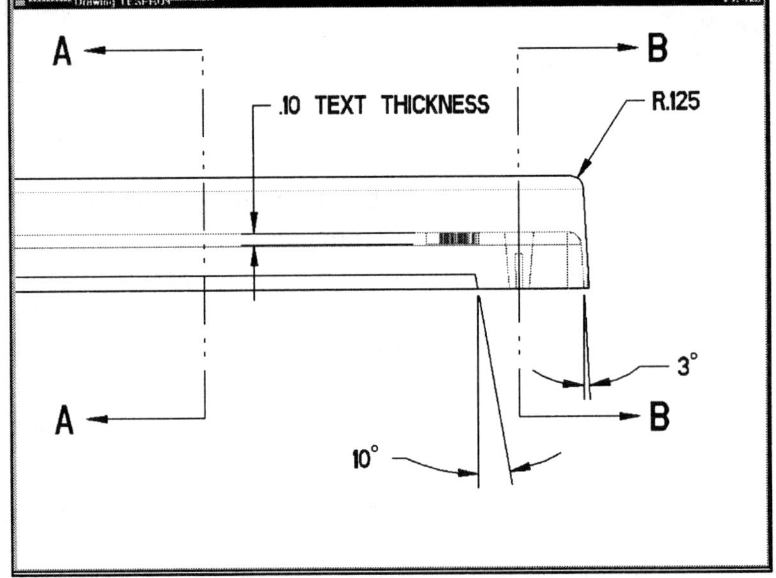

Figure 9.40
Cellular Phone Drawing, Front View

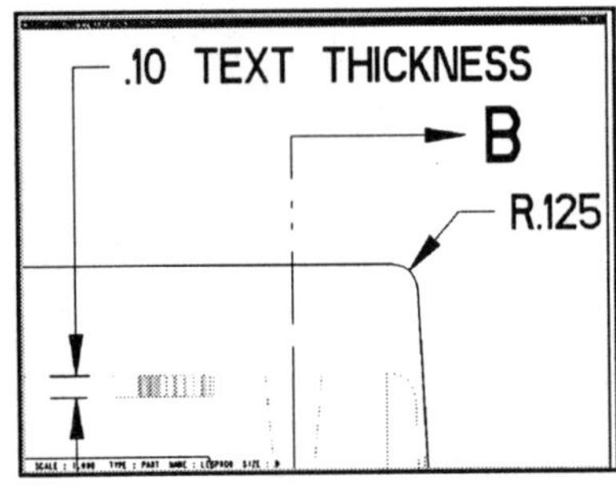

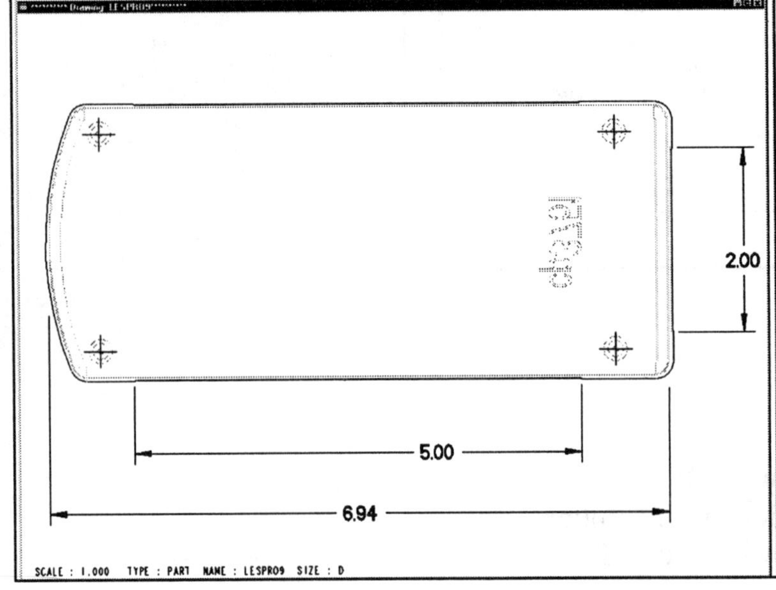

Figure 9.41
Cellular Phone Drawing, Top View

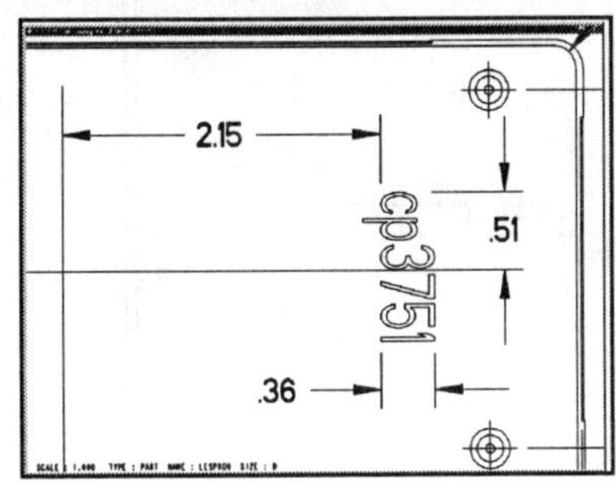

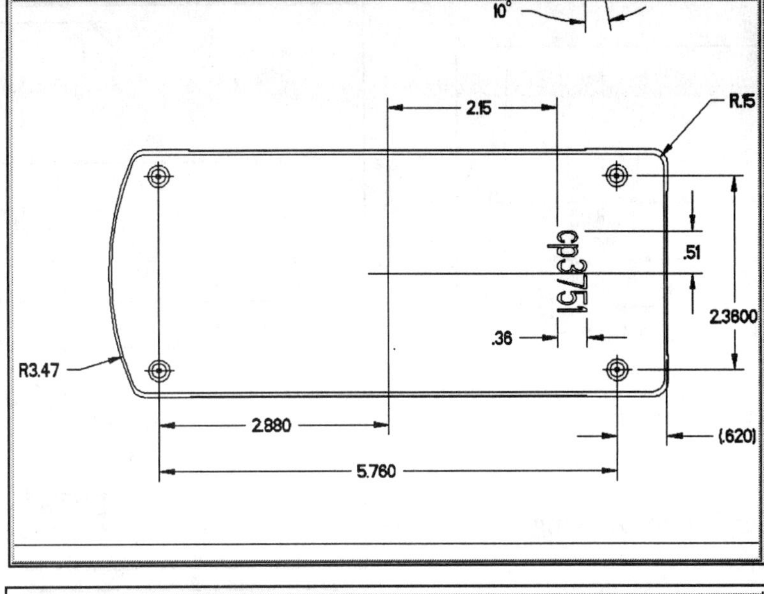

Figure 9.42
Cellular Phone Drawing,
Bottom View

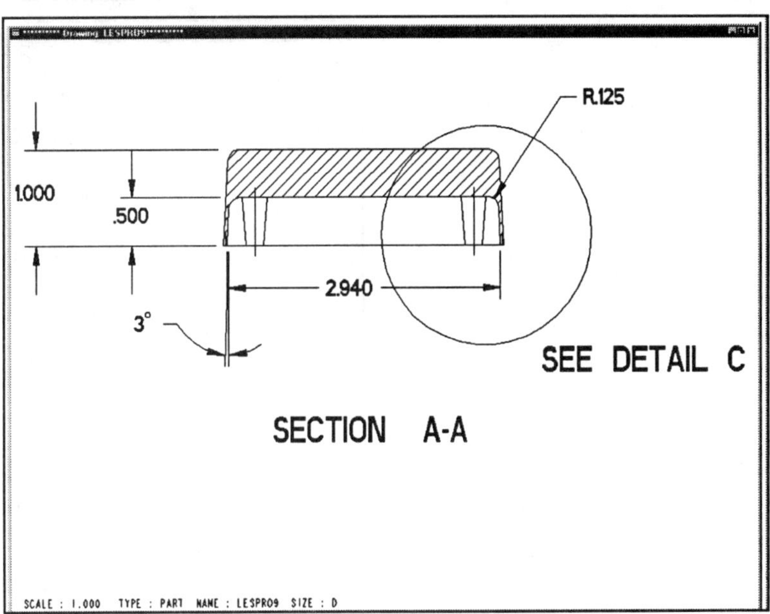

Figure 9.43
Cellular Phone Drawing,
SECTION A-A

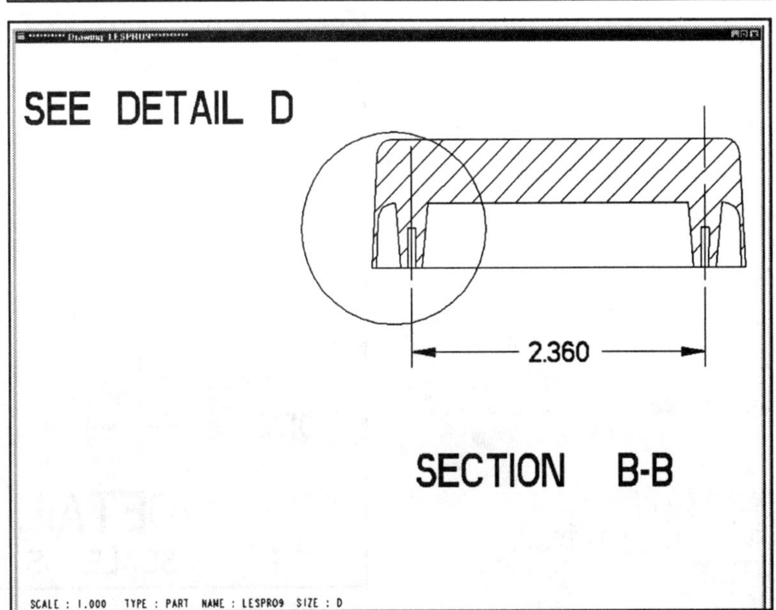

Figure 9.44
Cellular Phone Drawing,
SECTION B-B

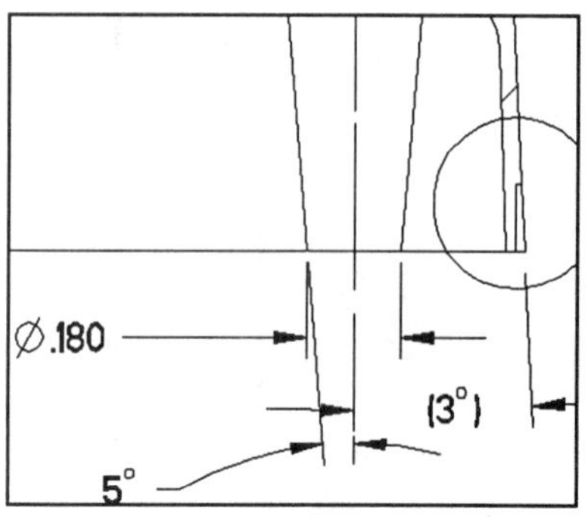

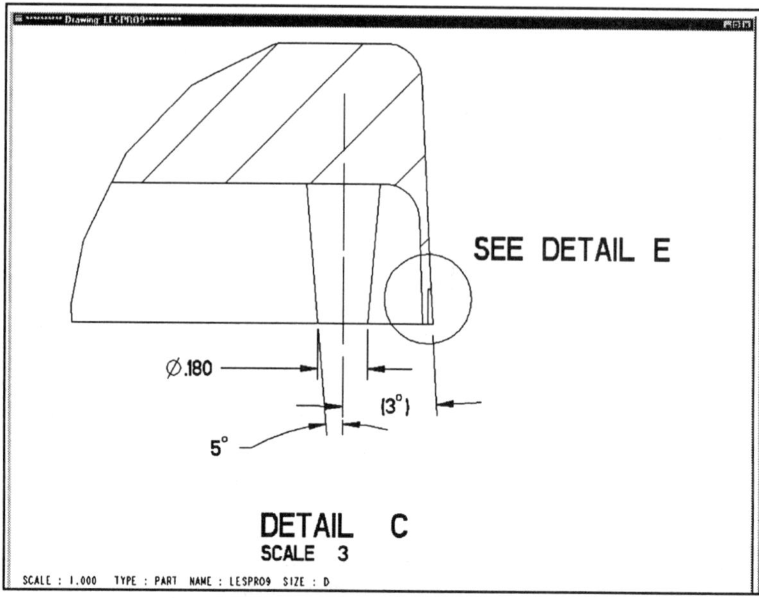

Figure 9.45
Cellular Phone Drawing,
DETAIL C

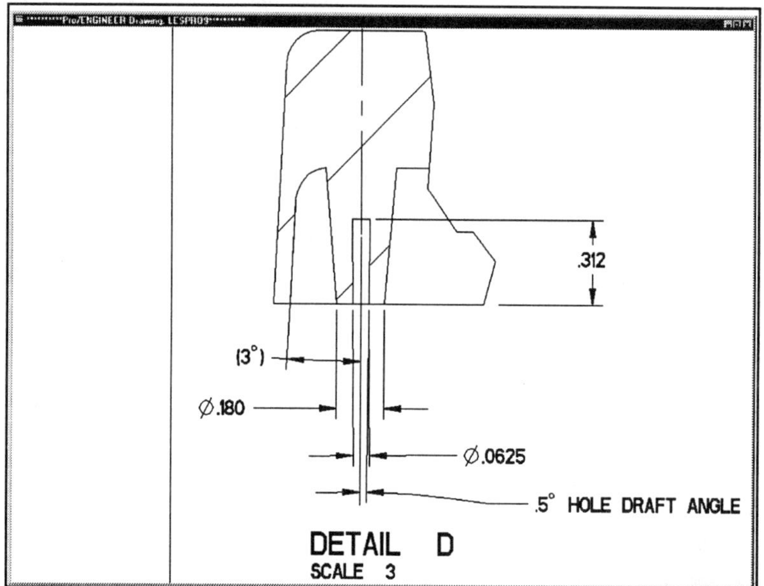

Figure 9.46
Cellular Phone Drawing,
DETAIL D

Dbms (File--PT/Modeler) ⇒
Save ⇒ **enter**
Purge ⇒ **enter** ⇒
Done-Return

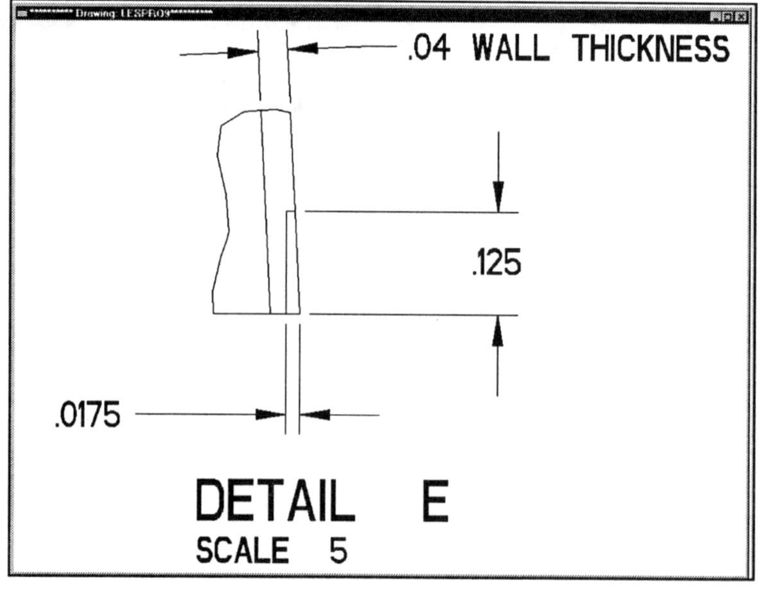

Figure 9.47
Cellular Phone Drawing,
DETAIL E

Lesson 10

Shell, Reorder, and Insert Mode

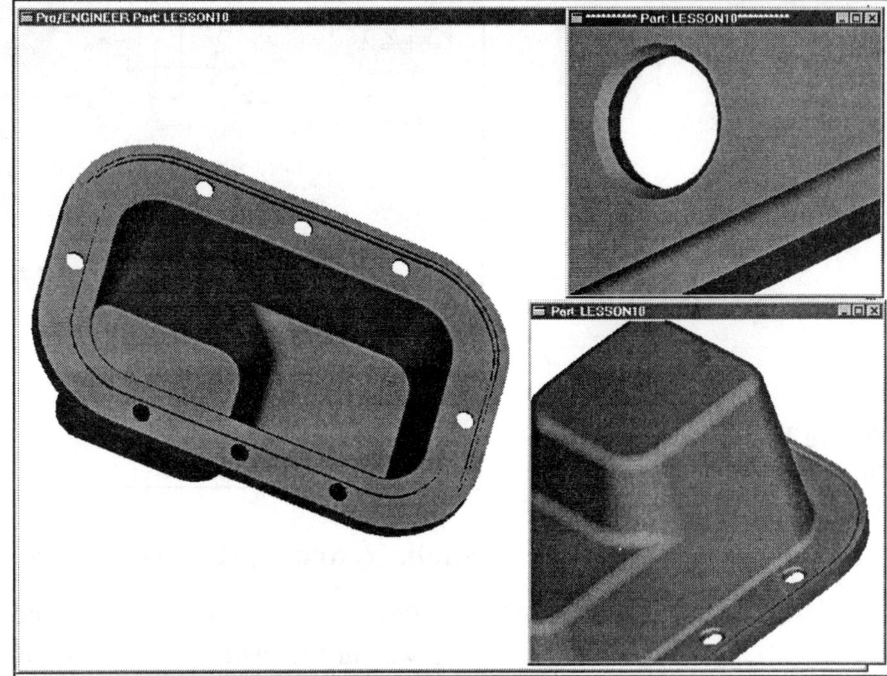

Figure 10.1
Oil Sink

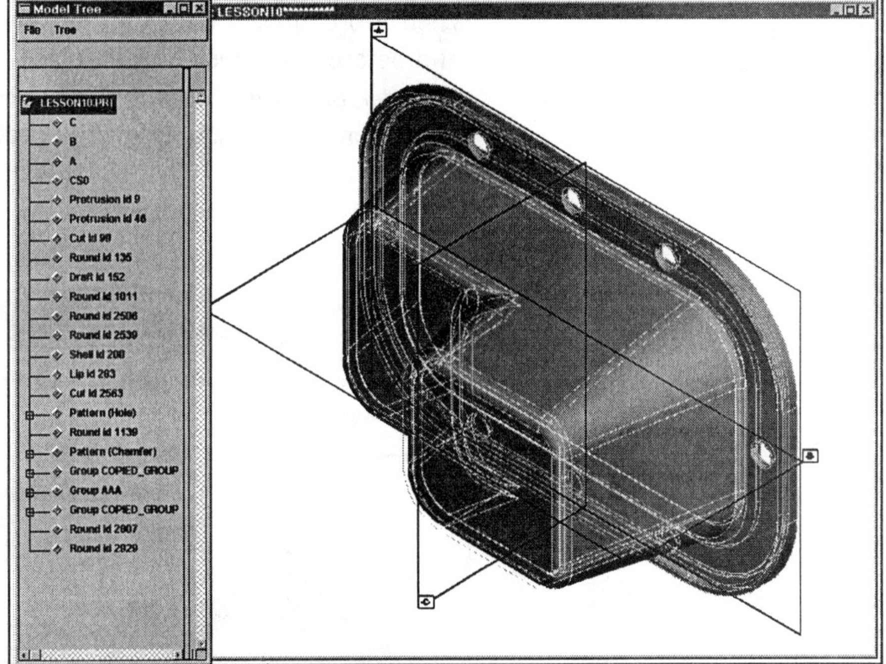

Figure 10.2
Oil Sink with Model Tree

EGD REFERENCE
Engineering Graphics and Design with Graphical Analysis *or* **Fundamentals of Engineering Graphics and Design**
by L. Lamit and K. Kitto
Read Chapter 14

OBJECTIVES

1. **Shell out a part**
2. **Alter the creation order of a feature**
3. **Insert a feature at a specific point in the design order**
4. **Create a lip on a part model**

COAch™ for Pro/ENGINEER

If you have **COAch for Pro/ENGINEER** on your system, go to SEARCH and do the Segments shown in Figures 10.4, 10.6, and 10.7.

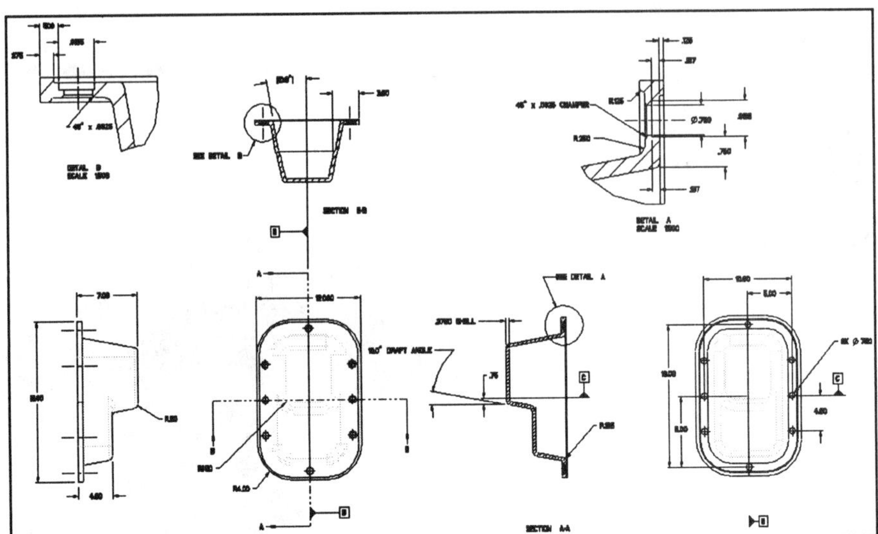

Figure 10.3
Oil Sink Dimensions

Shell, Reorder, And Insert Mode

The **Shell** option removes a surface or surfaces from a solid, then hollows out the inside of the solid, leaving a shell of a specified wall thickness, as in the **Oil Sink** shown in Figures 10.1 through 10.3. When Pro/E makes the shell, all the features that were added to the solid before you chose **Shell** are hollowed out (Fig. 10.4). Therefore, the *order of feature creation* is very important when you use **Shell**. You can alter the feature creation order by using the **Reorder** option. Another method of placing a feature at a specific place in the feature/design creation order is to use the **Insert Mode** option.

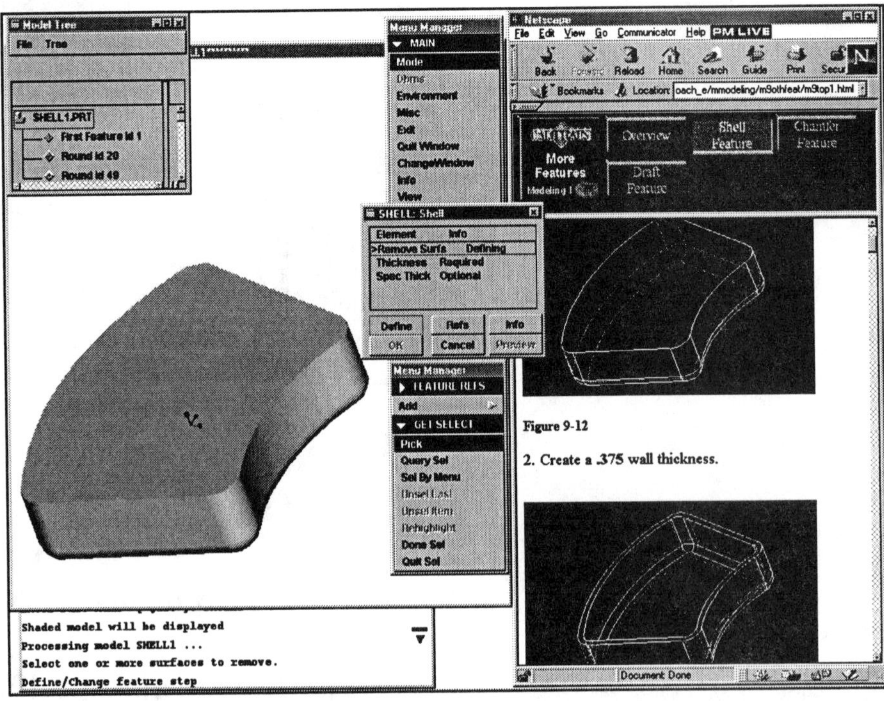

Figure 10.4
COAch for Pro/E, More Features (Shell Feature)

Creating Shells

To create a shell (Fig. 10.5), do the following:

1. Choose **Shell** from the SOLID menu.
2. Pro/E displays the feature creation dialog box. If you wish to, select the optional element **Spec Thick** to specify thickness individually. Select the **Define** button from the dialog box.
3. Select a surface or surfaces to be removed. When you have finished, choose **Done Refs** from the FEATURE REFS menu.
4. Enter the thickness of the wall. This thickness applies to all surfaces except those to which you assign a different thickness.
5. If you choose the **Spec Thick** element, Pro/E displays the SPEC THICK menu, which lists the following options:

Set Thickness Set thickness for the individual surface.
Reset to Def Reset the surfaces to the default thickness.

Choose **Set Thickness**. Select a surface and enter the thickness. Continue this process until you have specified all the surfaces you want. When you have finished, choose **Done** from the SPEC THICK menu.

6. To create the shell, select **OK** from the dialog box. If you entered a positive value for the thickness, material will be removed, leaving the shell thickness "inside" the part. However, if you entered a negative value, the shell thickness is added to the "outside" of the part.

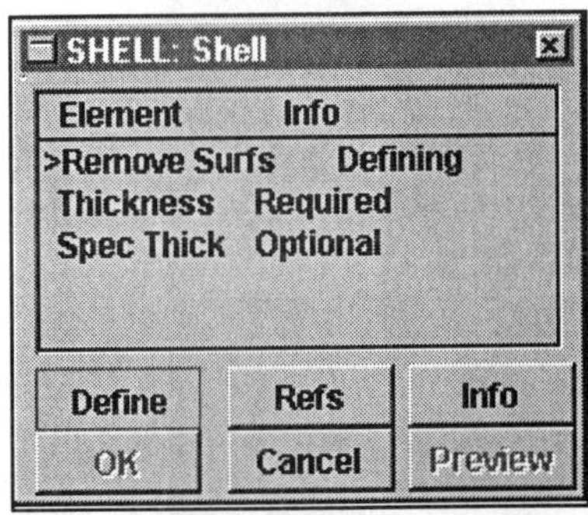

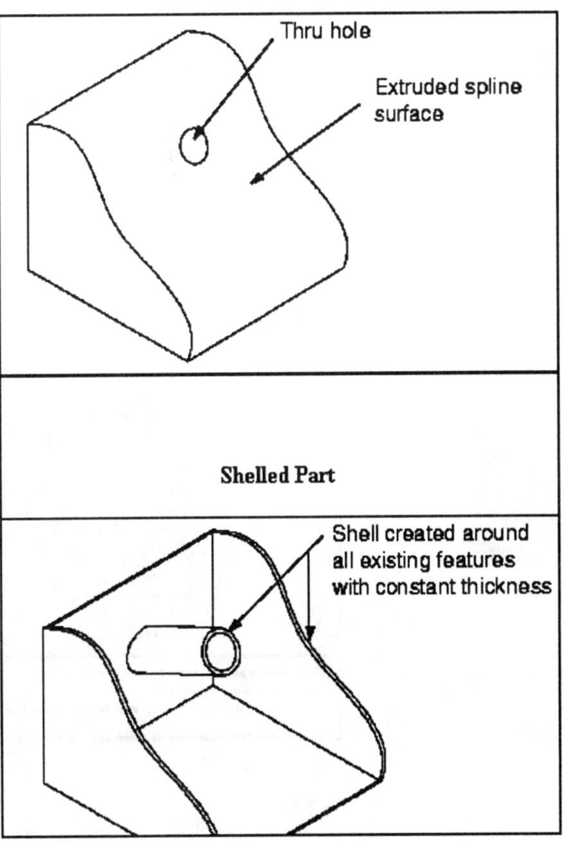

Figure 10.5
Online Documentation, Shell

Reordering Features

You can move features forward or backward in the feature creation (regeneration) order list, thus changing the order in which they are regenerated (Fig. 10.6). You can reorder multiple features in one operation, as long as these features appear in ***consecutive*** order.

Feature reorder *cannot* occur under the following conditions:

Parents cannot be moved so that their regeneration occurs after the regeneration of their children.
Children cannot be moved so that their regeneration occurs before the regeneration of their parents.

To reorder a feature, do the following:

1. Use the command sequence **Part, Feature, Reorder.**
2. Specify the selection method by choosing an option from the SELECT FEAT menu:

Select Select features to reorder by picking on the screen and/or from the tree tool. You can also choose **Sel By Menu** to enter the feature number. Choose **Done Sel** when finished selecting.
Layer Select all features from a layer by selecting the layer. Choose **Done Sel** from the LAYER SEL menu when finished.
Range Specify the range of features by entering the regeneration numbers of the starting and ending features.

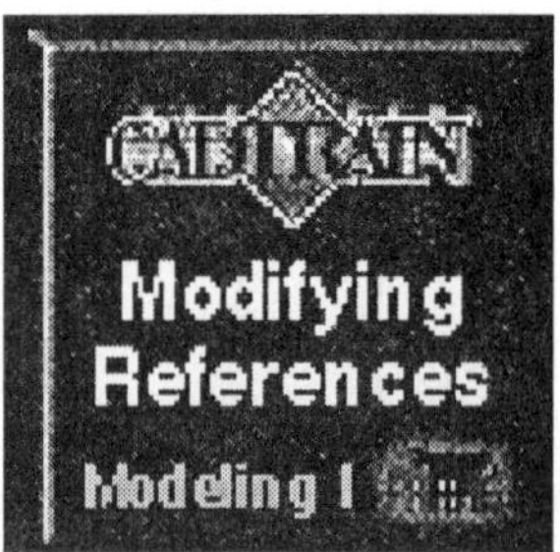

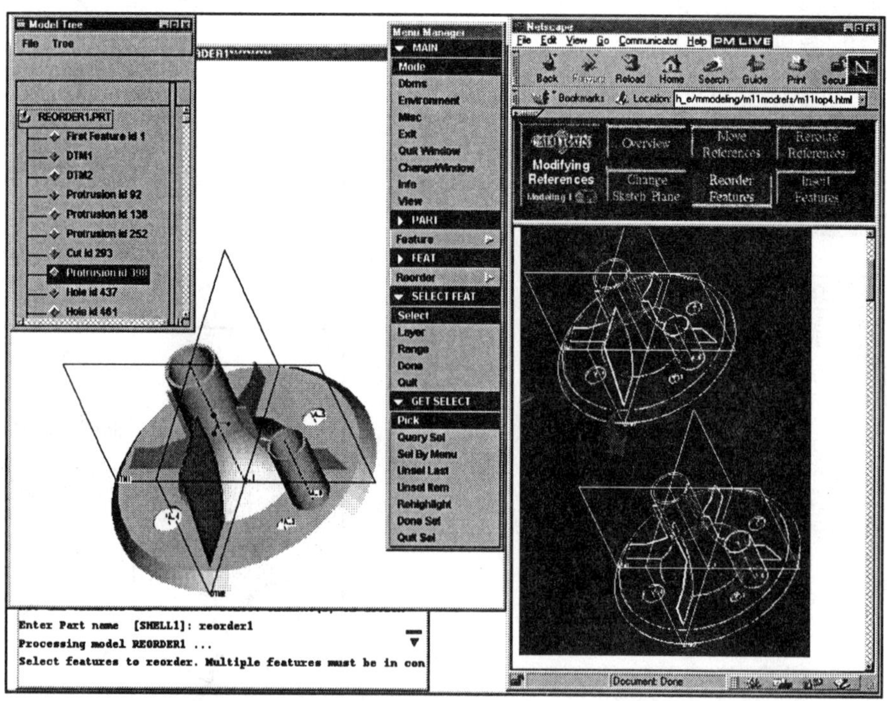

Figure 10.6
COAch for Pro/E, Modifying References (Reorder Features)

3. A Pro/E message lists the selected features for reorder and states the valid ranges for the new insertion point.
4. Choose **Done** from the SELECT FEAT menu.
5. Choose one of the options in the REORDER menu:

   Before Insert the feature before the insertion point feature.
   After Insert the feature after the insertion point feature.

Pick a feature indicating the insertion point, or choose **Sel By Menu** to enter the feature number.

Inserting Features

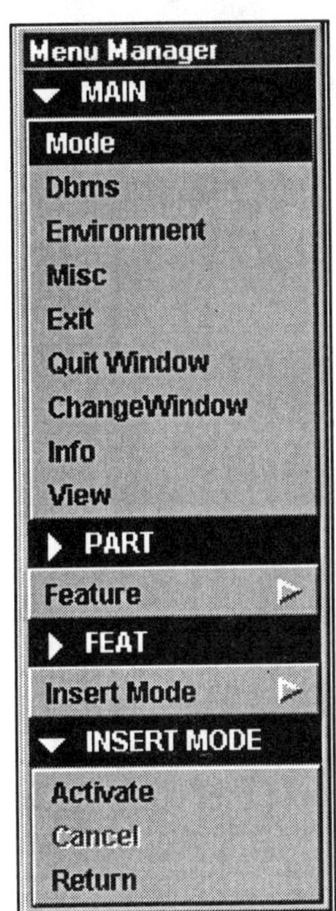

Normally, Pro/E adds a new feature after the last existing feature in the part, including suppressed features. **Insert Mode** (Fig. 10.7) allows you to add new features at any point in the feature sequence, except before the base feature or after the last feature.

To insert features, do the following:

1. Choose **Insert Mode** from the FEAT menu, then choose **Activate**.
2. Select a feature after which the new features will be inserted. All features after the selected one will be automatically suppressed.
3. Choose **Create** and create the new features as usual.
4. Cancel **Insert Mode** by choosing **Resume** from the FEAT menu and select to resume the features that were suppressed when you activated **Insert Mode**, or choose **Cancel** from the INSERT MODE menu. Pro/E asks you whether to resume the features that were suppressed when you activated **Insert Mode**, then automatically regenerates the part.

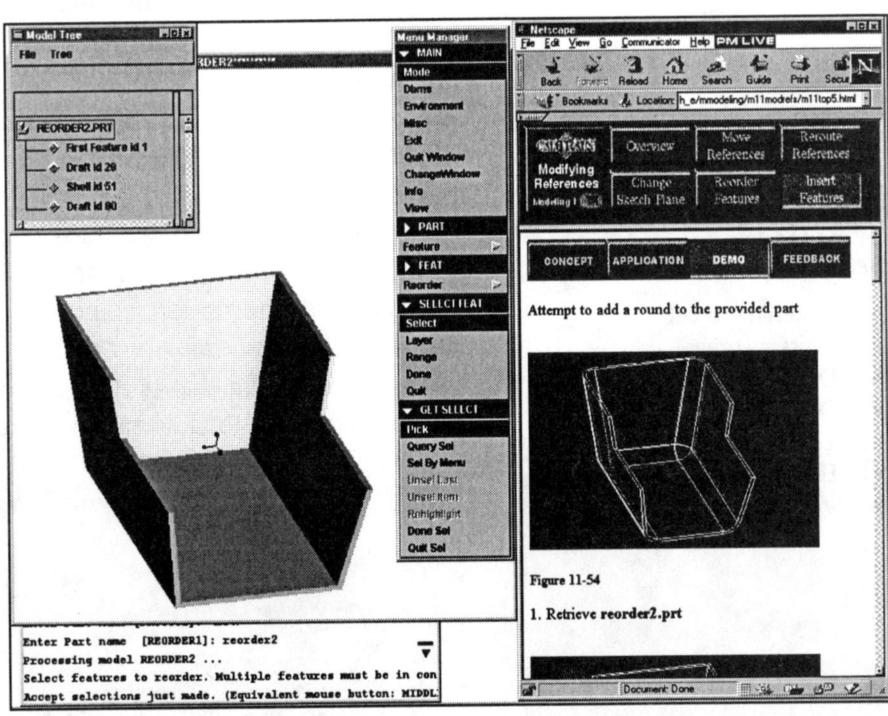

Figure 10.7
COAch for Pro/E, Modifying References (Insert Features)

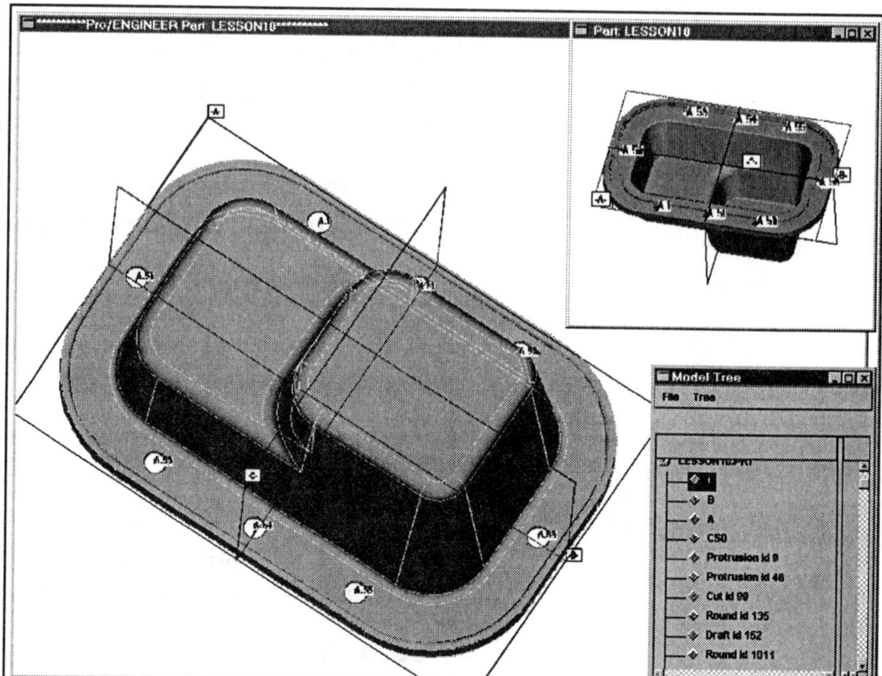

Figure 10.8
Oil Sink Showing Datum Planes and Coordinate System and Model Tree

Oil Sink

The **Oil Sink** (Fig. 10.8) requires the use of the **Shell** option. The shelling of a part should be done after the desired protrusions and most rounds have been modeled. This lesson part will have you create a protrusion, a cut, and a set of rounds. Some of the required rounds will be left off the part model on purpose. Pro/E's **Insert Mode** option enables you to insert a set of features at a previous stage in the design of the part. In other words, you can create a feature after or before a selected existing feature even if the whole model has been completed. You can also *move the order in which a feature was created* and therefore have subsequent features affect the reordered feature. A round created after a shell operation can be reordered to appear before the shell.

In this lesson, you will also insert a round or two before the existing shell feature using **Insert Mode**. The rounds will be shelled after the **Resume** option is picked, since the rounds now appears before the shell feature.

Another option that is new to you will also be introduced. The **Lip** option is used to add/remove material along an edge.

The first protrusion and cut for the Oil Sink are to be modeled by you using the dimensions provided in Figures 10.9 through 10.16.

Set up the Oil Sink:

- Material = Steel
- Units = Inches
- Default Datum Planes
- Default Coordinate System
- Layers = **DATUM_LAYER**
- Hidden line
- ✓Grid Snap

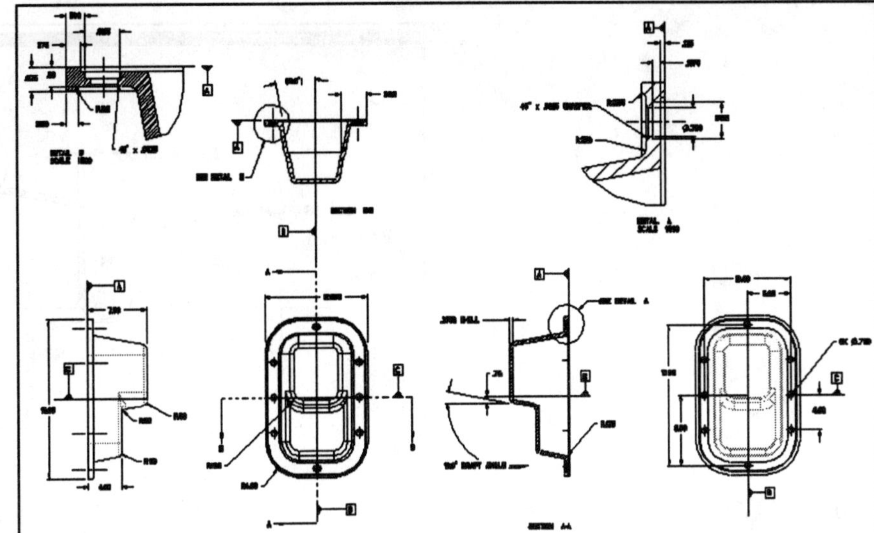

Figure 10.9
Oil Sink Detail Drawing

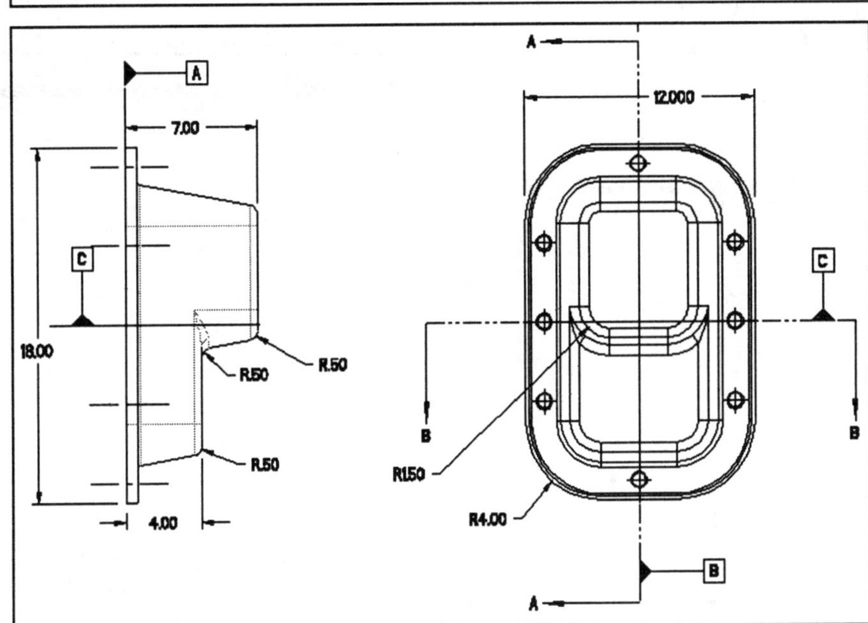

Figure 10.10
Oil Sink Drawing,
Front View

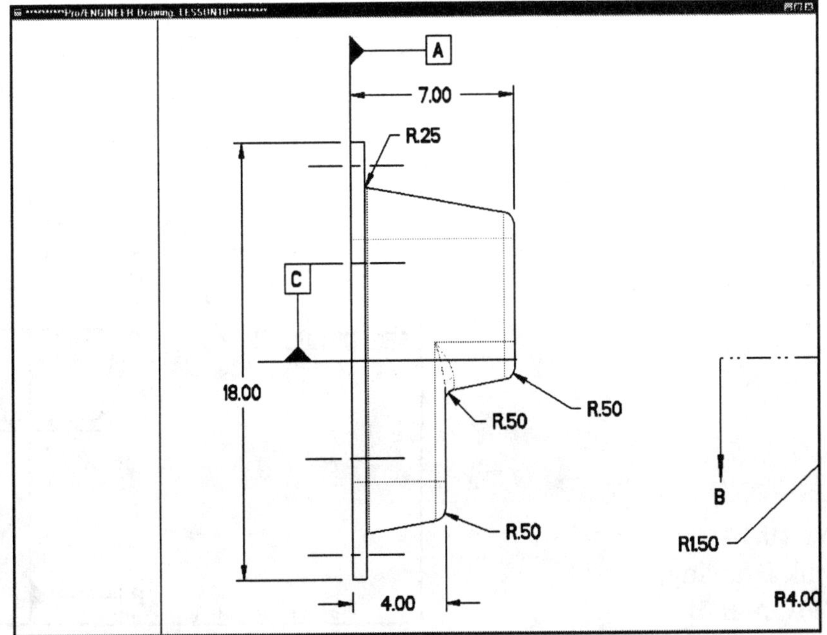

Figure 10.11
Oil Sink Drawing,
Left Side View

Figure 10.12
Oil Sink Drawing,
SECTION A-A

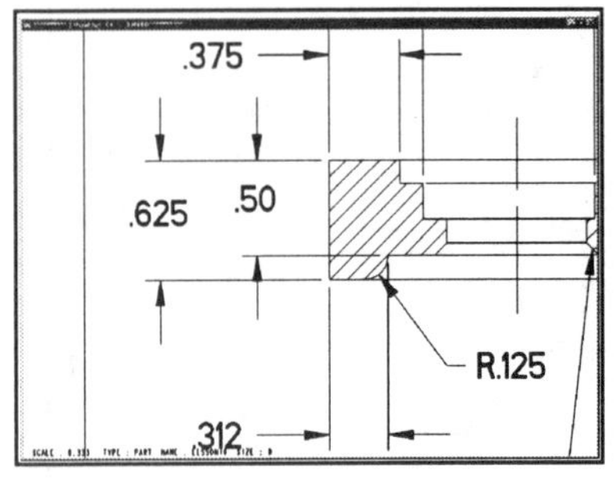

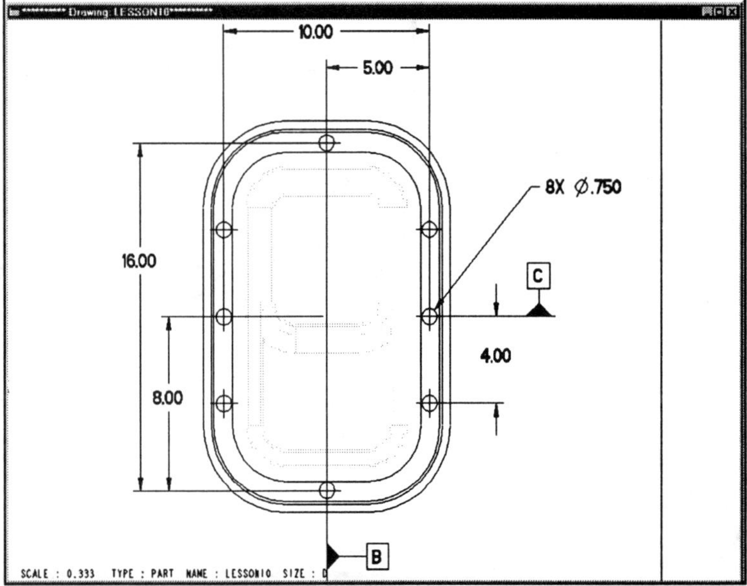

Figure 10.13
Oil Sink Drawing,
Bottom View

Figure 10.14
Oil Sink Drawing,
SECTION B-B

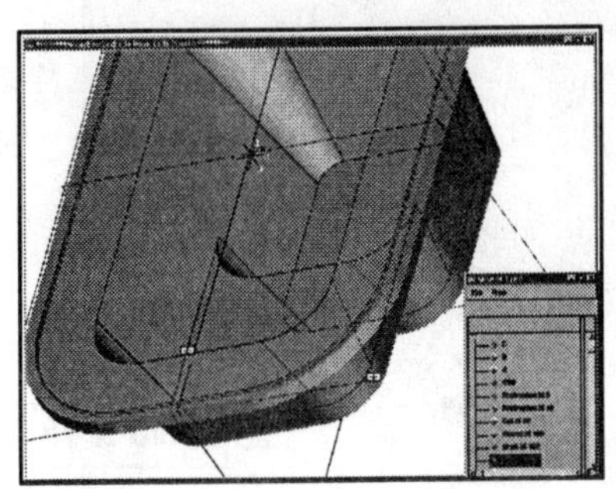

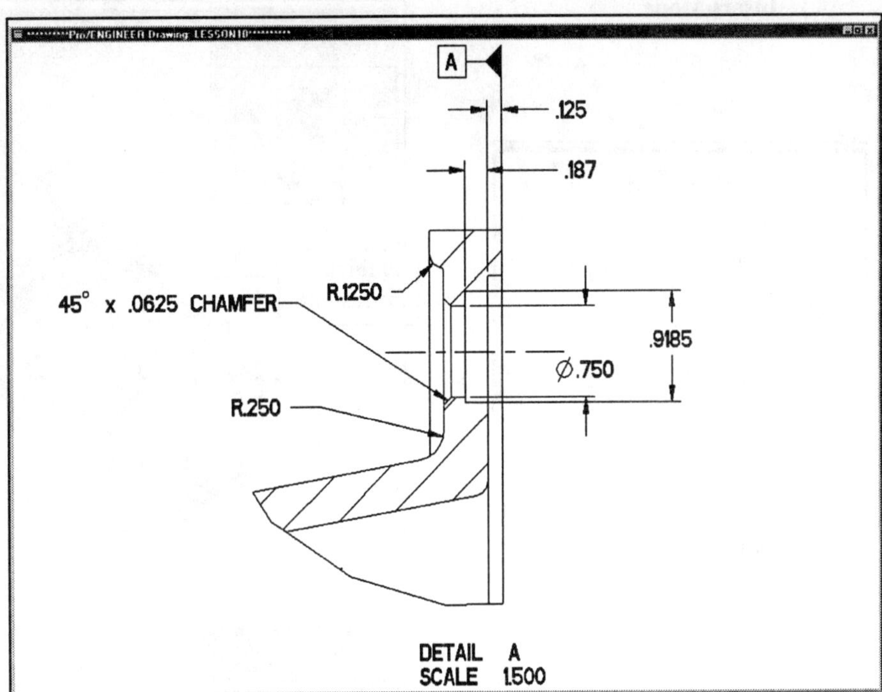

Figure 10.15
Oil Sink Drawing,
DETAIL A

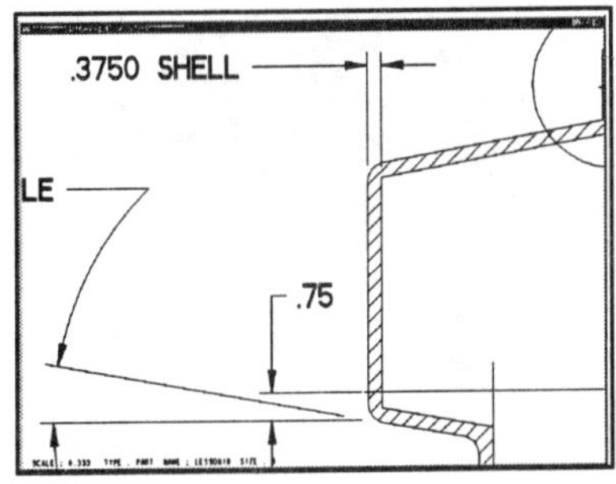

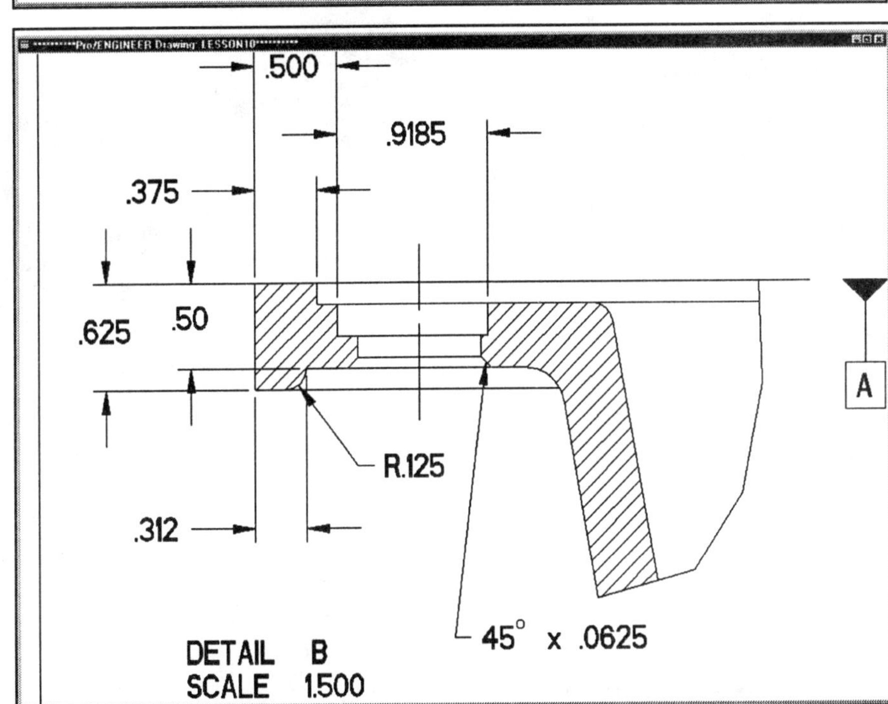

Figure 10.16
Oil Sink Drawing,
DETAIL B

After creating the default datum planes and coordinate system, model the first protrusion shown in Figure 10.17. The large protrusion forming the Oil Sink's tublike shape is modeled next (Fig. 10.18). In Figure 10.19, a cut is added to the model. The **R1.50** rounds are added next (Fig. 10.20). The lateral sides of the protrusion are drafted at a angle of **-10°**, as shown in Figure 10.21. Shell the part with the following commands (Fig. 10.22):

PT/Modeler™

Feature ⇒ Shell etc.

Feature ⇒ Create ⇒ Shell ⇒ (select one or more surfaces--pick the bottom surface) **⇒ Done Sel ⇒ Done Refs ⇒** (enter thickness of **.375** at the prompt) **⇒ enter ⇒ OK ⇒ Done**

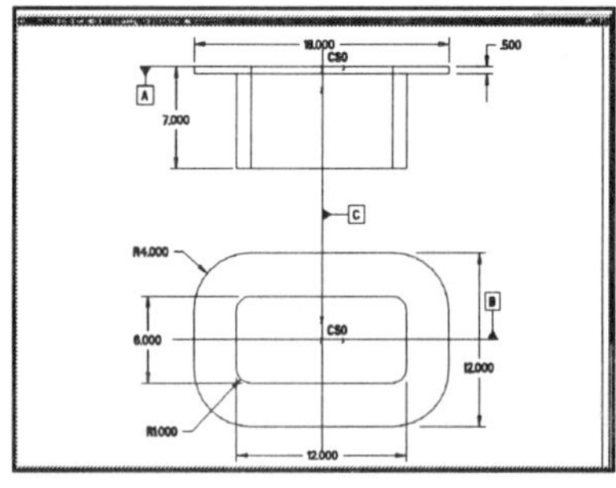

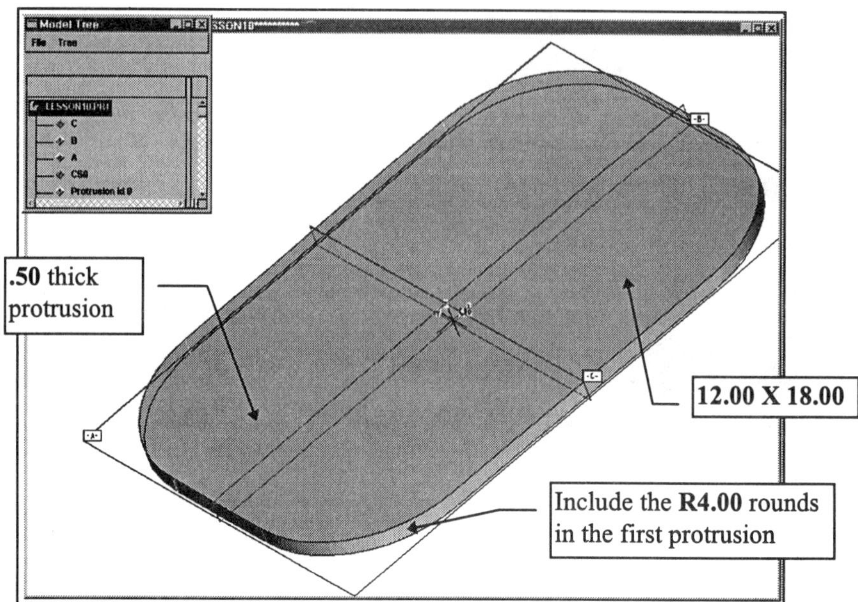

Figure 10.17
Oil Sink's First Protrusion

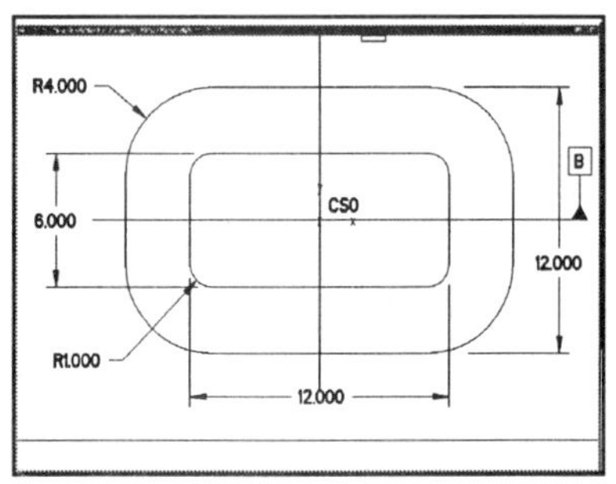

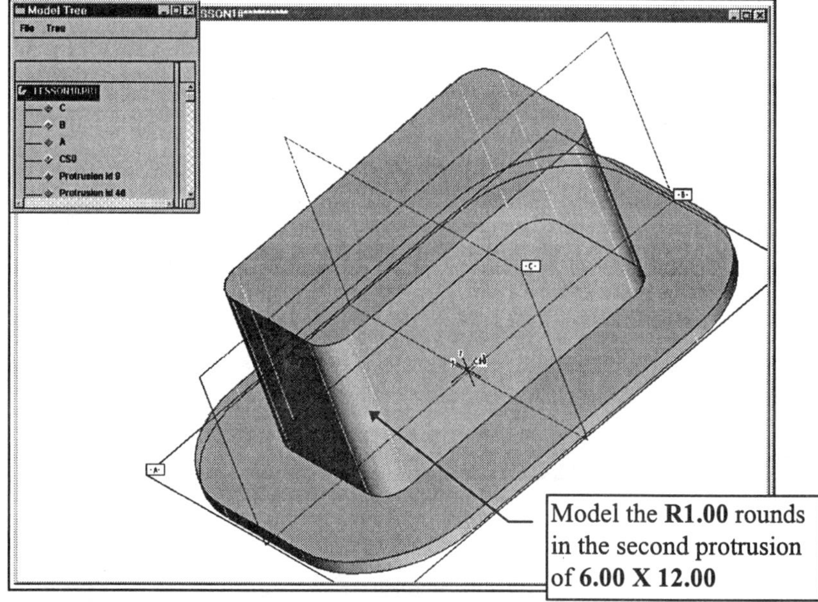

Figure 10.18
Oil Sink's Second Protrusion

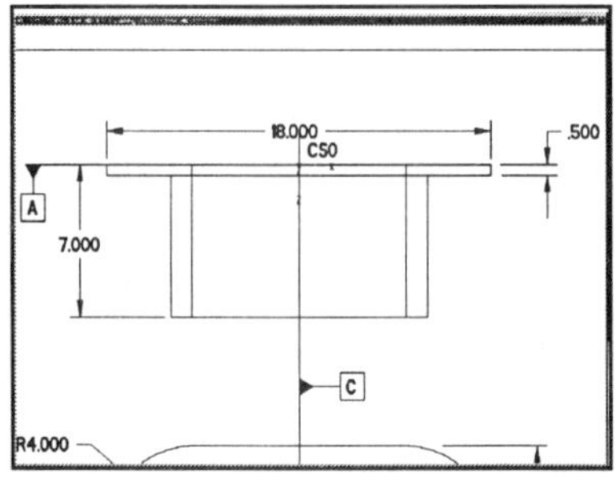

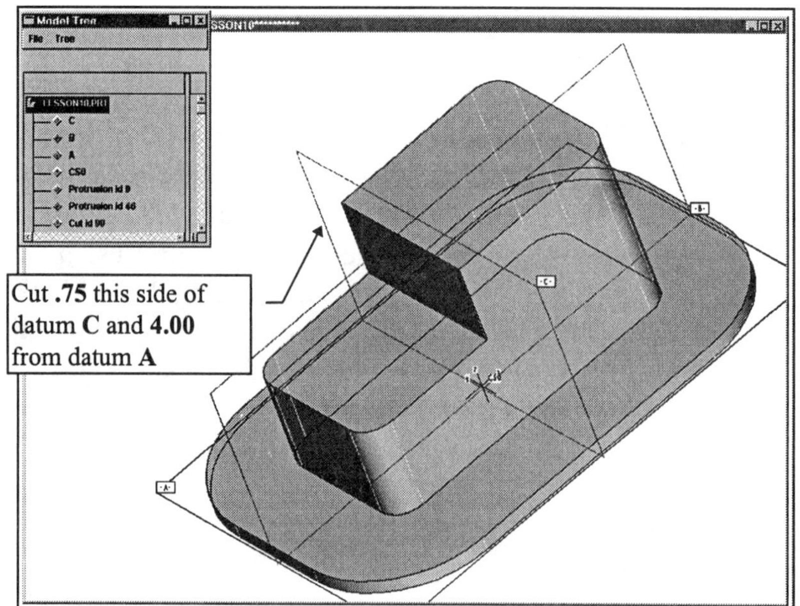

Figure 10.19
Oil Sink's First Cut

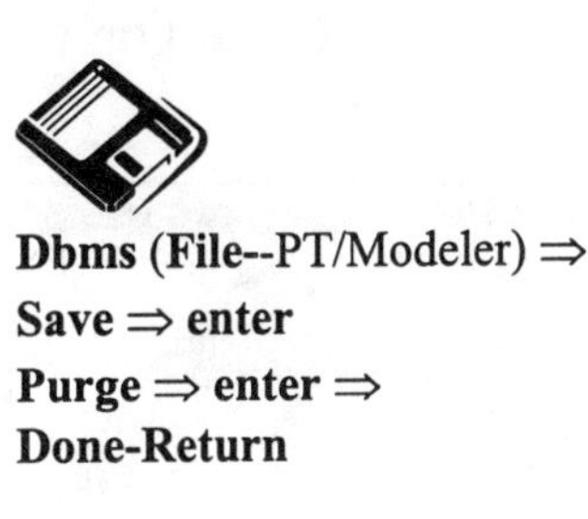

Dbms (File--PT/Modeler) ⇒
Save ⇒ **enter**
Purge ⇒ **enter** ⇒
Done-Return

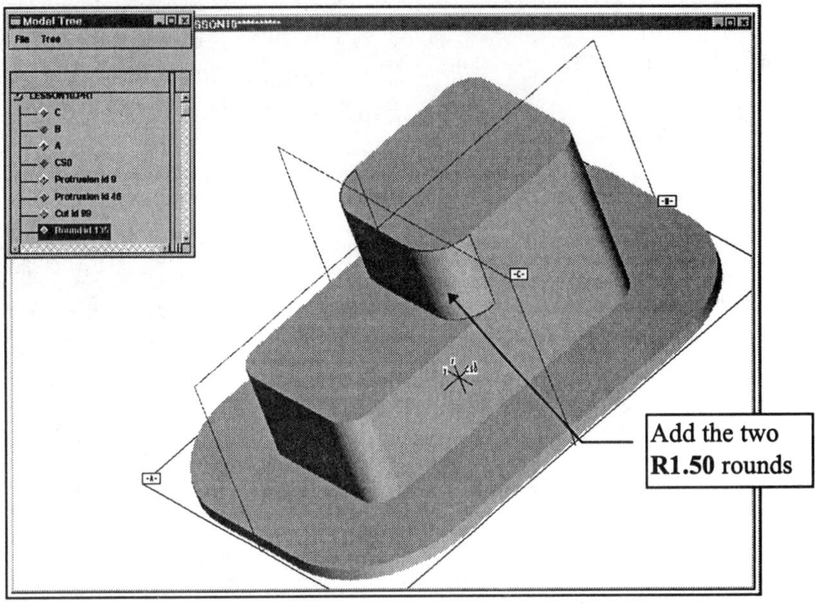

Add the two **R1.50** rounds

Figure 10.20
Add Rounds

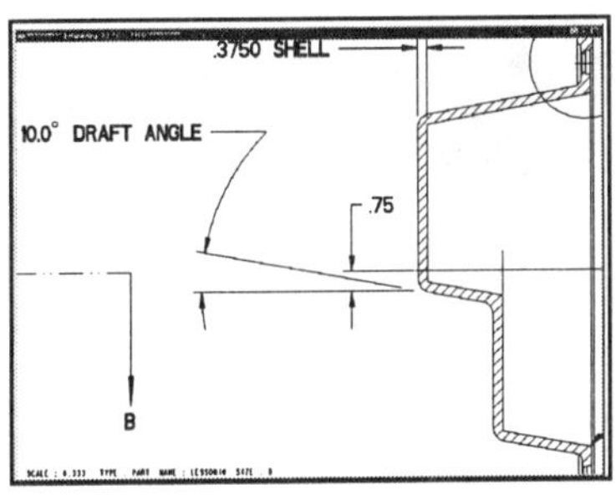

Neutral plane and plane used for *direction perpendicular to*

Draft

Draft

Draft

Draft

Draft

Draft

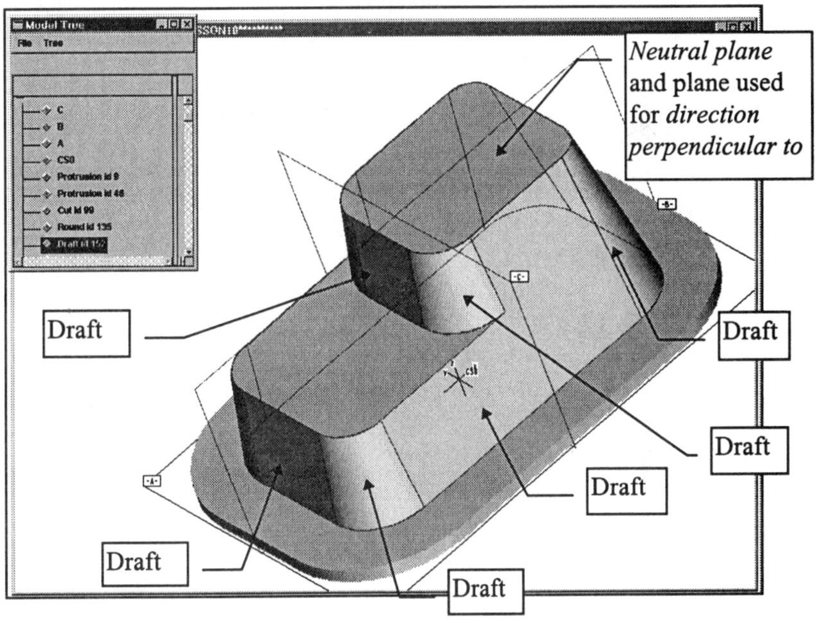

Figure 10.21
Draft *All* Upper Lateral Surfaces of the Oil Sink (**-10°**)

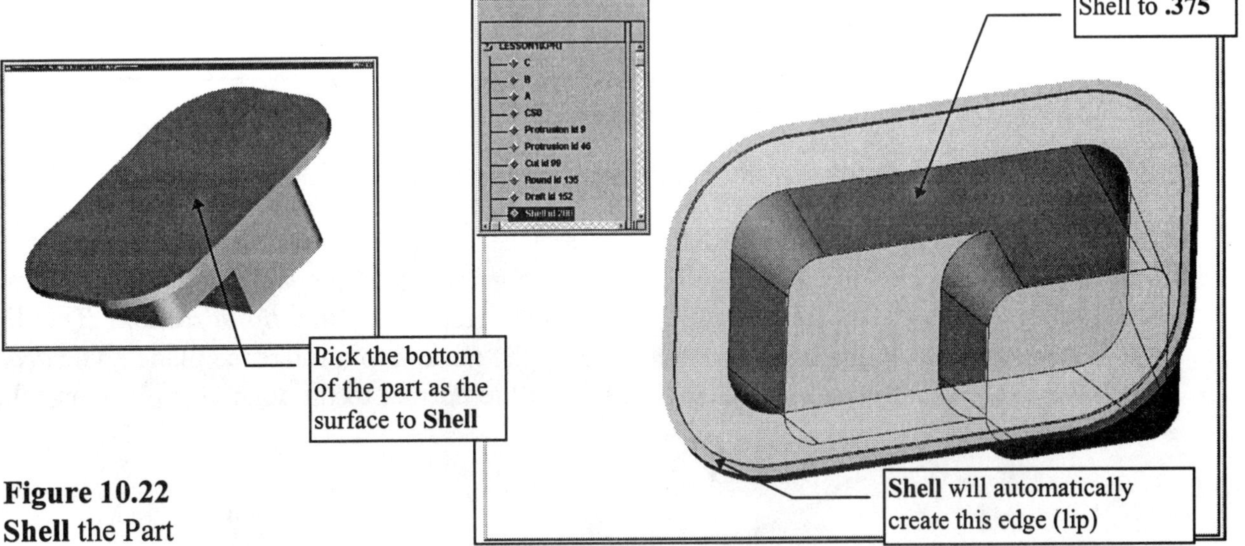

Figure 10.22
Shell the Part

The next feature you need to create is a **Lip** (Fig. 10.23). Go to the online documentation using Pro/HELP and read the section on the **Lip** feature (Fig. 10.24).

PT/Modeler ™

PT/Modeler does not support the **Lip** feature. Create a protrusion and then a draft, or wait until you complete Lesson 11 and then return to this lesson and create the lip with a **Sweep**.

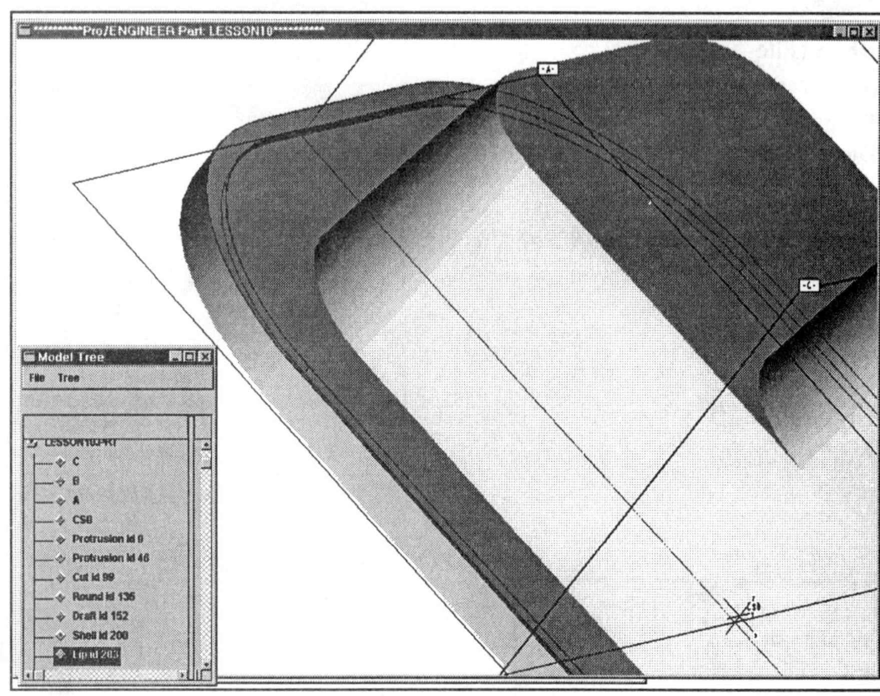

Figure 10.23
Lip

? Pro/HELP
Choose **Lip**

?GetHelp

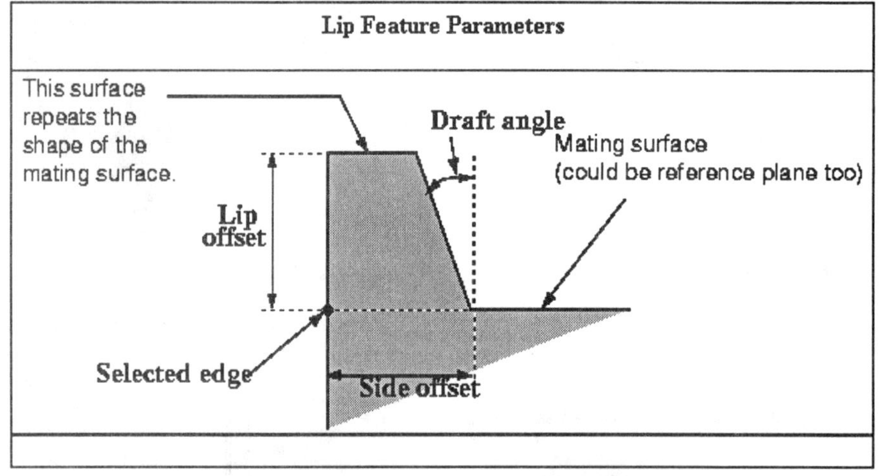

Figure 10.24
Pro/HELP for **Lip**

A **Lip** is constructed by offsetting the mating surface along the selected edges. The edges must form a continuous contour, either open or closed. Here, a closed contour is used. The top (or bottom) surface of the lip copies the geometry of the *mating surface:* you can draft the side surface with respect to the *lip direction*. The lip direction is determined by the normal to a reference plane. The *draft angle* is the angle between the normal to the reference plane and the side surface of the Lip.

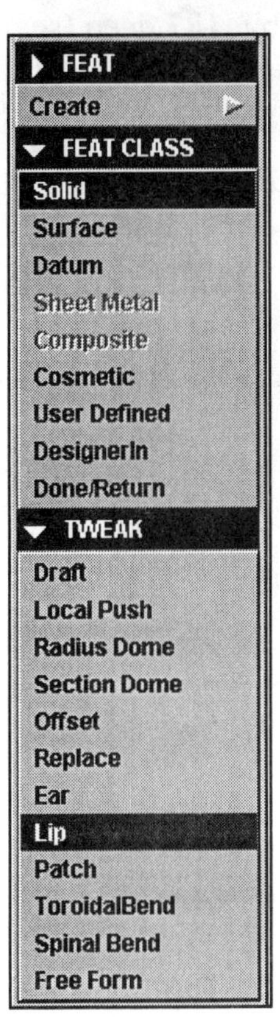

In most cases, the reference plane is coincident with the lip (mating) surface. You must select a separate reference plane in the following cases:

* The mating surface is not a plane.
* You want the lip creation direction not to be normal to the mating plane. The lip feature will then be distorted.

Create the lip with the following commands (Fig. 10.25):

Feature ⇒ **Create** ⇒ **Tweak** ⇒ **Lip** ⇒ **Chain** [pick the edge to select the mating surface (Fig. 10.25)] ⇒ **Done Sel** ⇒ **Done** ⇒ (pick the surface to be offset, here the mating surface) ⇒ (enter the lip offset from the mating surface: **.125**) ⇒ **enter** ⇒ (enter the *side offset distance* of **.312** from the selected edges to the *draft surfaces*) ⇒ **enter** ⇒ (select the *draft reference surface*--same as mating surface) ⇒ (enter the *draft angle*) ⇒ **enter** (for zero) ⇒ **Done** (Fig. 10.26)

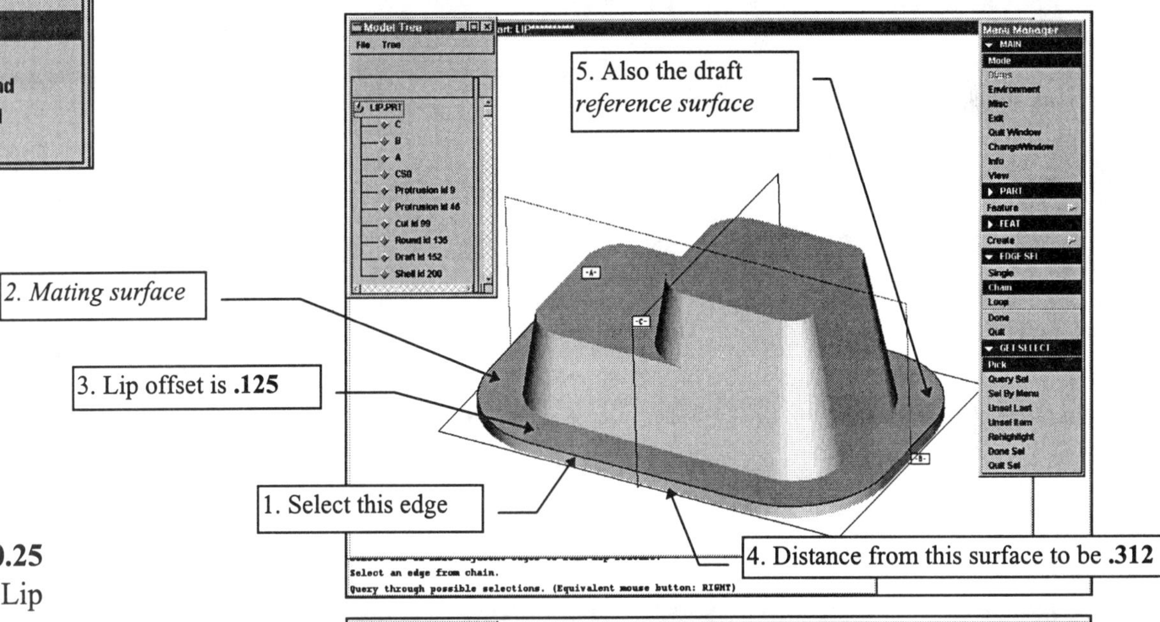

Figure 10.25
Adding a Lip

HINT

As of Pro/E 19, a **Lip** cannot be redefined.

.125 height by **.312** width lip with **0°** draft angle

Figure 10.26
Completed Lip

The next feature is a cut measuring **.9185** wide by **.187** deep (see Figs. 10.15 and 10.16). Figures 10.27 and 10.28 show the cut. Add the countersunk holes, chamfers, and the rounds (Fig. 10.29).

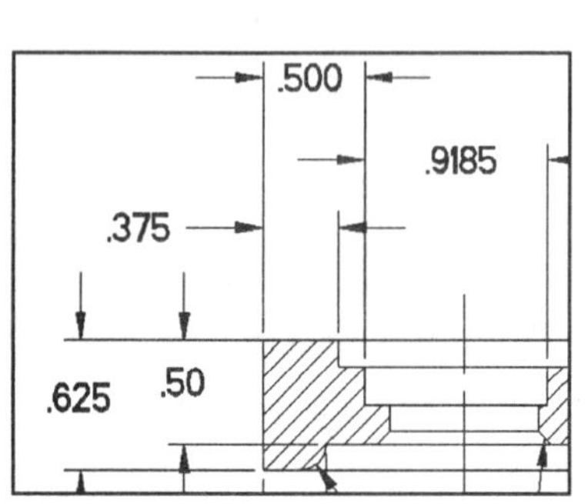

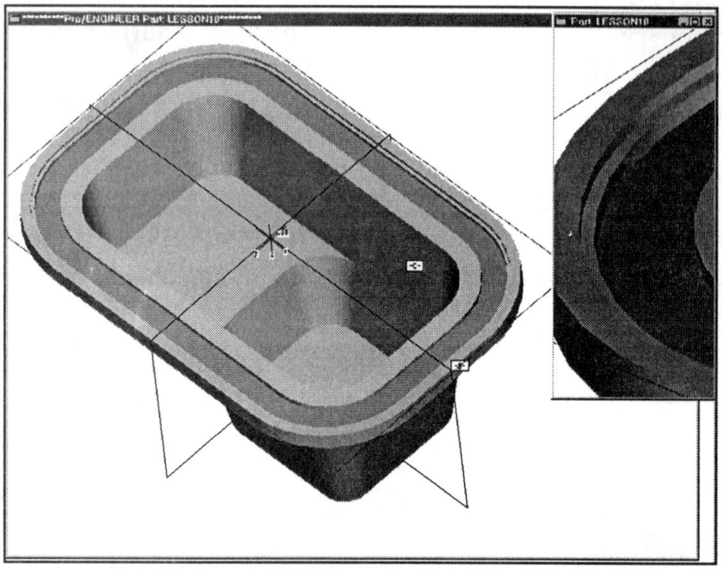

Figure 10.27
Cut

Dbms (**File**--PT/Modeler) ⇒
Save ⇒ **enter**
Purge ⇒ **enter** ⇒
Done-Return

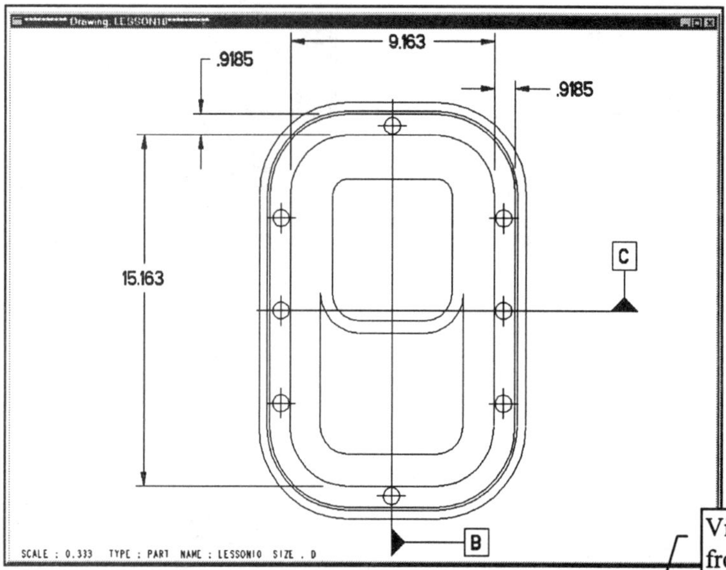

Figure 10.28
Cut Dimensions

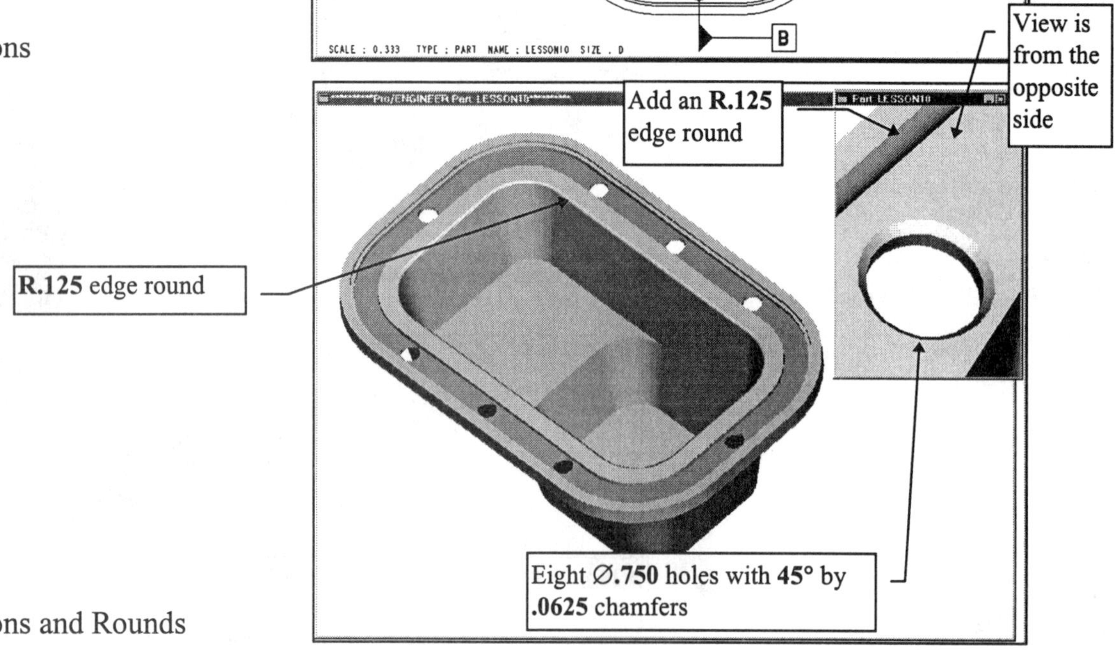

R.125 edge round

Add an **R.125** edge round

View is from the opposite side

Eight ∅**.750** holes with **45°** by **.0625** chamfers

Figure 10.29
Cut Dimensions and Rounds

The next series of features will purposely be created at the wrong stage of the project. Create an **R.50** round as shown in Figure 10.30. Since the design intent should have been to have a constant thickness for the part, shouldn't the round have been created before the shell? Use the **Reorder** option to change the position of this round in the design sequence. Your prompts will have different feature numbers. Choose the following commands:

PT/Modeler™
Feature Oper ⇒ **Reorder** etc.

Feature ⇒ **Reorder** ⇒ (pick the round to reorder) ⇒ **Done Sel** ⇒ **Done** ⇒ **Before** ⇒ (Pro/E will prompt with: ***Feature # can be inserted before features [10-26] or [30-34]. Please, select feature.***) ⇒ (from the Model Tree, select the **Shell id**) ⇒ **Done** (Fig. 10.31)

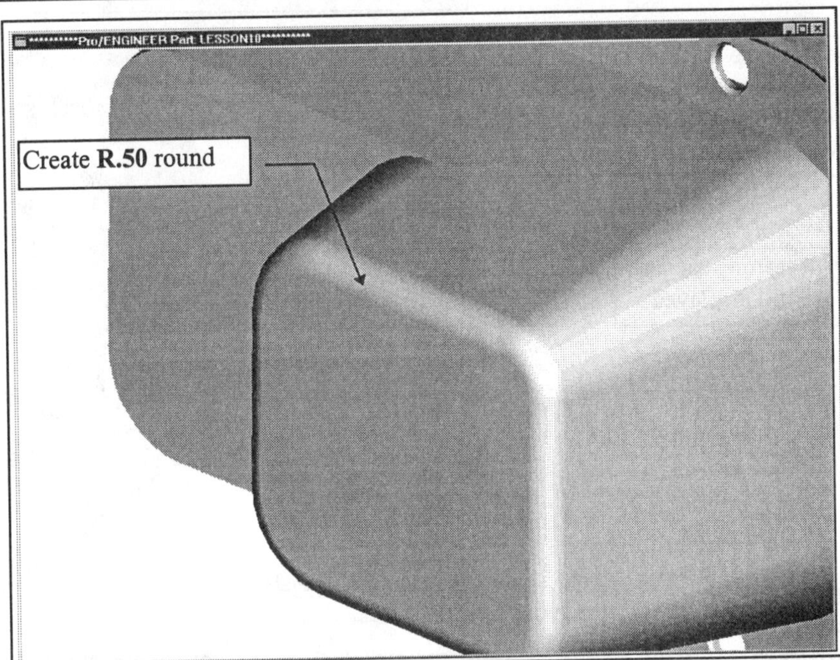

Figure 10.30
R.50 Round

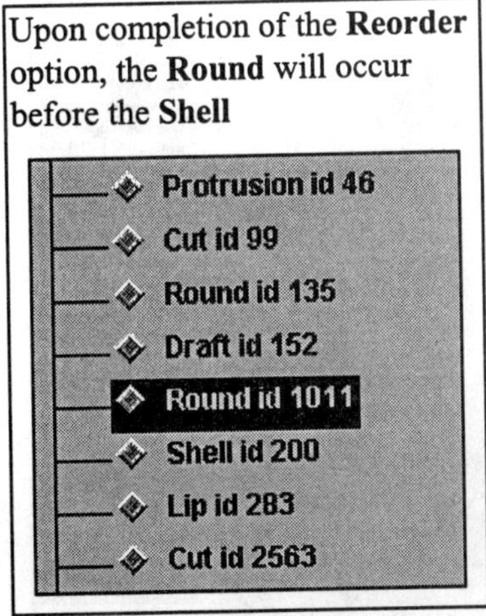

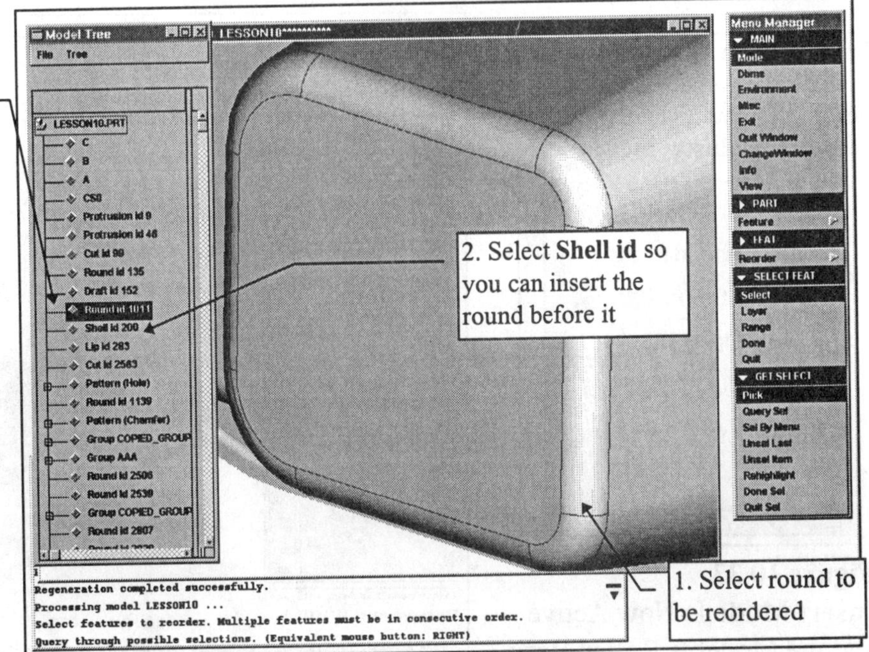

Figure 10.31
Reorder

The previous round was created at the wrong stage in the design sequence and then reordered. To eliminate the reordering of a feature, the **R.50** rounds will be created using **Insert Mode**. **Insert Mode** allows you to insert a feature at a previous stage of the design sequence. This is like going back in the past and doing something you wish you had done before--not possible with life, but with Pro/E less of a problem. Add the **R.25** round. Choose the following commands to enter **Insert Mode,** create the rounds, and then use **Resume** to return to the previous stage in the design (Fig. 10.32):

PT/Modeler™
Feature Oper ⇒ Insert Mode etc.

Feature ⇒ Insert Mode ⇒ Activate ⇒ (select feature to insert after; from the Model Tree, pick the round you just reordered, as shown in Fig. 10.33) **⇒ Done**

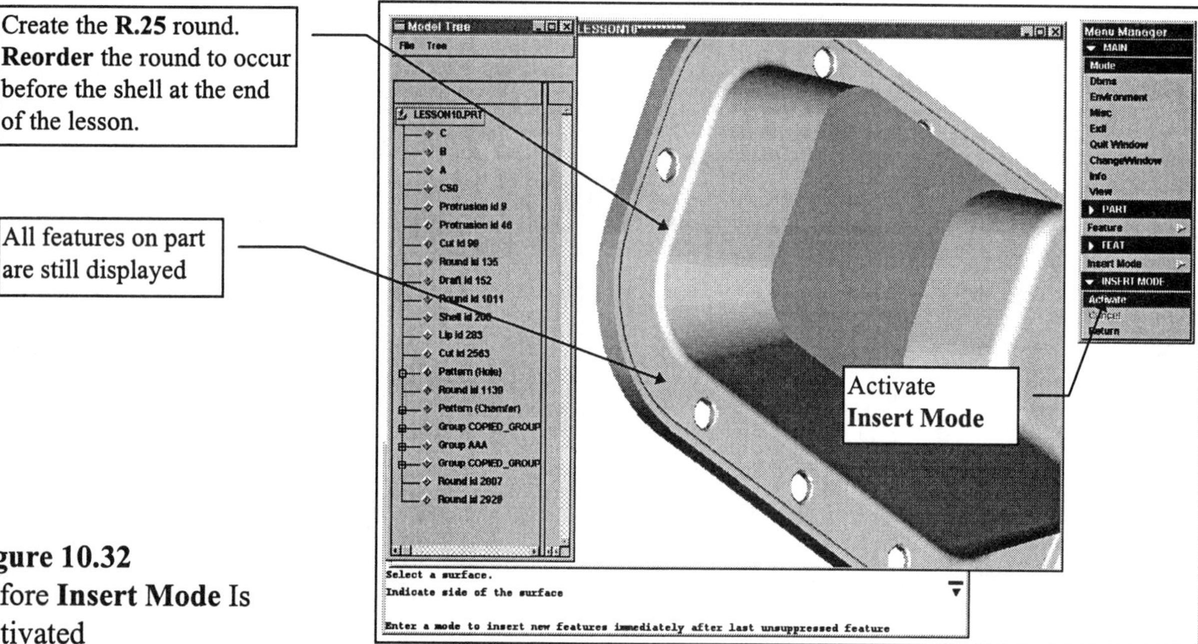

Figure 10.32
Before **Insert Mode** Is Activated

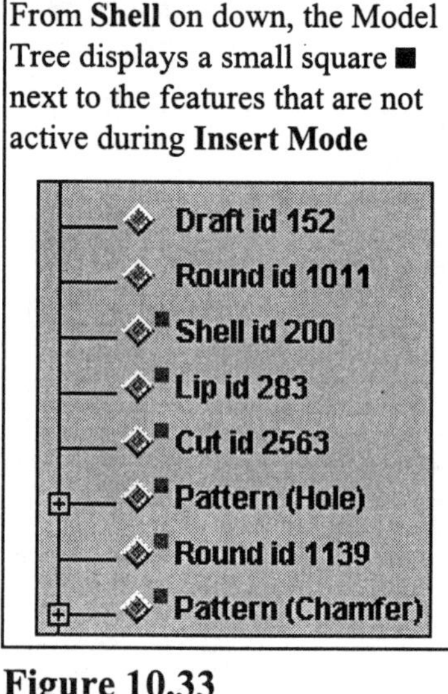

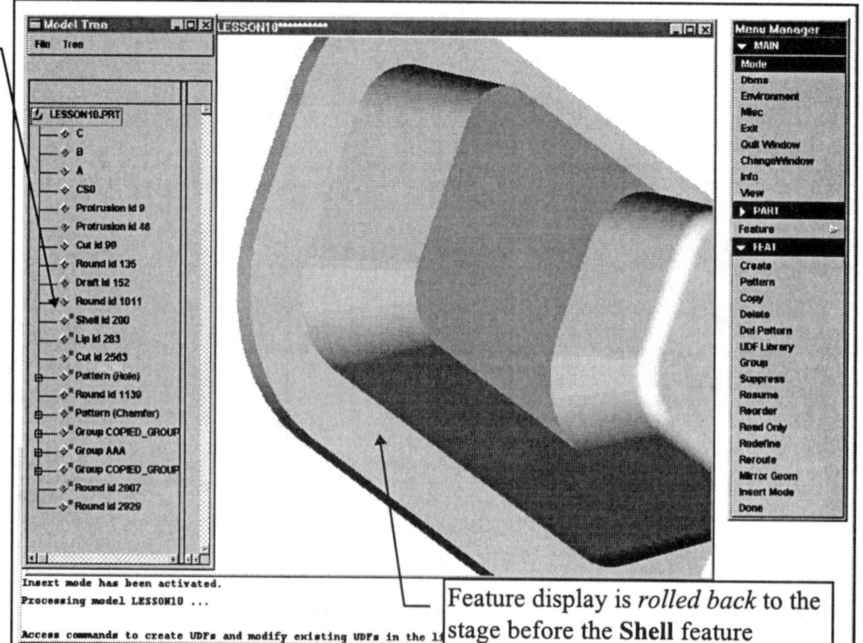

Figure 10.33
Insert Mode Is Now Active and the Model Is Rolled Back

The rounds can now be created using the following commands:

PT/Modeler™
Feature ⇒ Round ⇒ Done (pick the edge shown in Figure 10.34) ⇒ **Done etc.**

Feature ⇒ Create ⇒ Round ⇒ Simple ⇒ Done ⇒ Constant ⇒ Edge Chain ⇒ Done ⇒ Tangnt Chain ⇒ (pick the edge shown in Fig. 10.34) ⇒ **Done Sel ⇒ Done** ⇒ (enter the radius value of **.50** at the prompt) ⇒ **enter ⇒ OK ⇒ Done** ⇒ (*repeat the same commands* and pick the edge shown in Fig. 10.35) ⇒ **Feature ⇒ Resume ⇒ All** (Fig. 10.36) ⇒ **Done** (Fig. 10.37) ⇒ **Done**

NOTE

If the round fails, create the two rounds separately. Make sure you have picked all the correct edges.

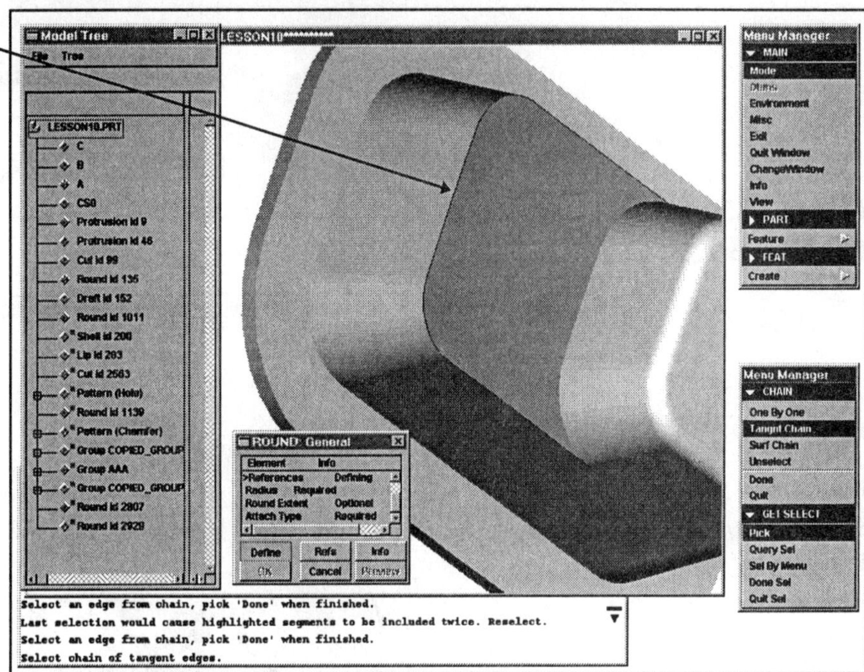

Figure 10.34
Create a Round

Dbms (File--PT/Modeler) ⇒
Save ⇒ enter
Purge ⇒ enter ⇒
Done-Return

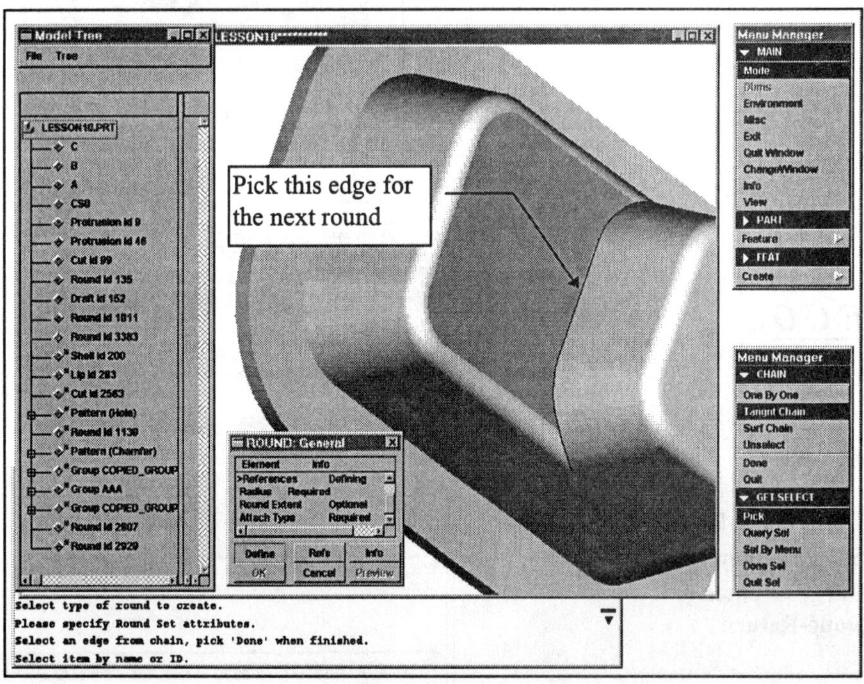

Figure 10.35
Create Another Round

The rounds are now in the proper design sequence for the shell feature to create a constant thickness. After the part is completed, save it under a new name and do the **ECO** (Fig. 10.38).

PT/Modeler™

(To cancel **Insert Mode**) **Feature Oper ⇒ Insert Mode ⇒ Cancel ⇒** (Type y at the prompt) **⇒ enter**

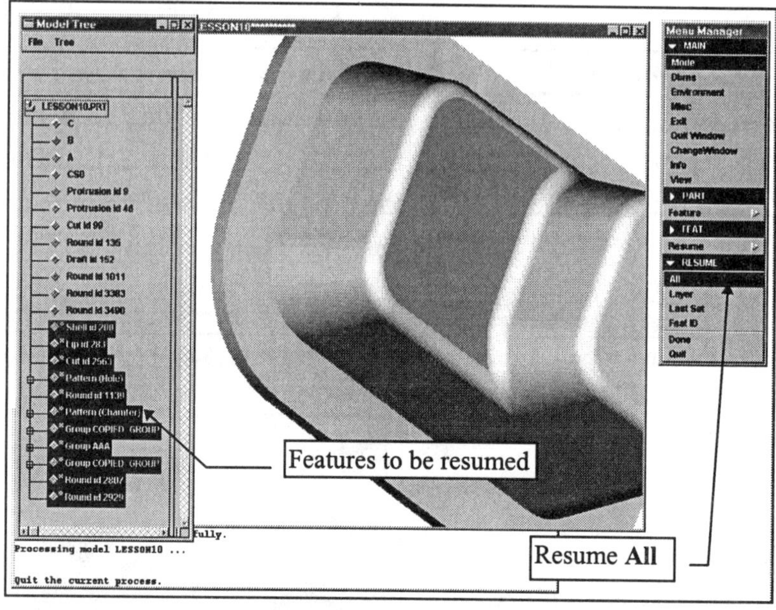

Figure 10.36
Resume ⇒ All

Dbms (File--PT/Modeler) ⇒
Save ⇒ enter
Purge ⇒ enter ⇒
Done-Return

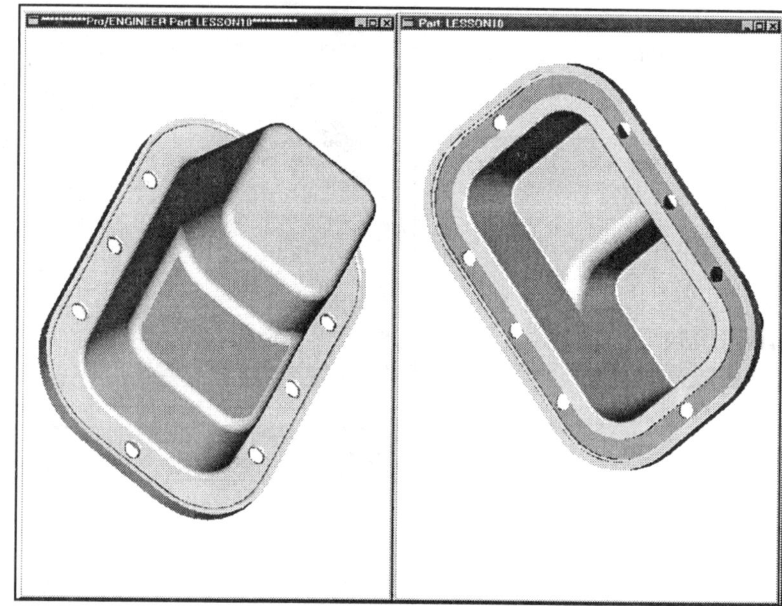

Figure 10.37
Resumed Part

ECO

Dbms (File--PT/Modeler) ⇒
Save ⇒ enter
Purge ⇒ enter ⇒
Done-Return

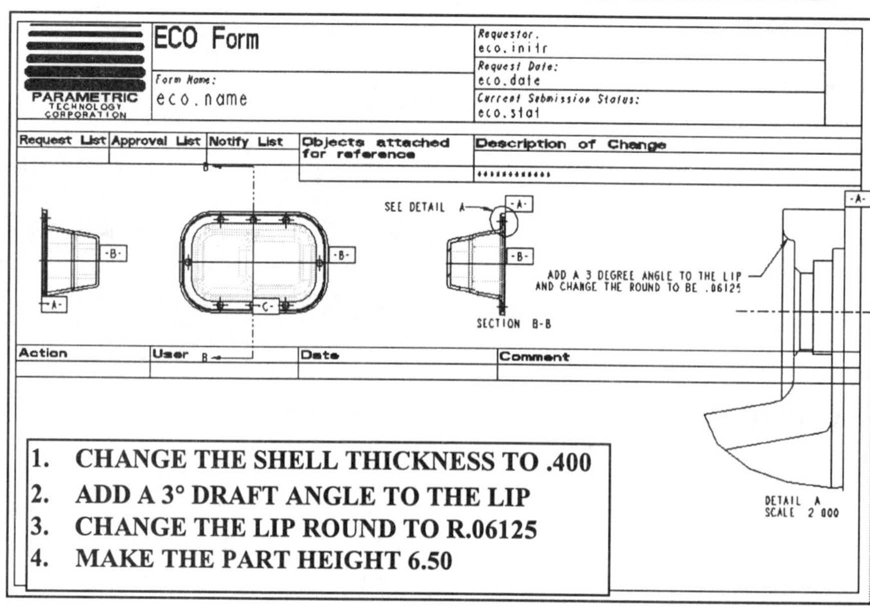

1. CHANGE THE SHELL THICKNESS TO .400
2. ADD A 3° DRAFT ANGLE TO THE LIP
3. CHANGE THE LIP ROUND TO R.06125
4. MAKE THE PART HEIGHT 6.50

Figure 10.38
ECO

Lesson 10 Project

Cellular Phone Top

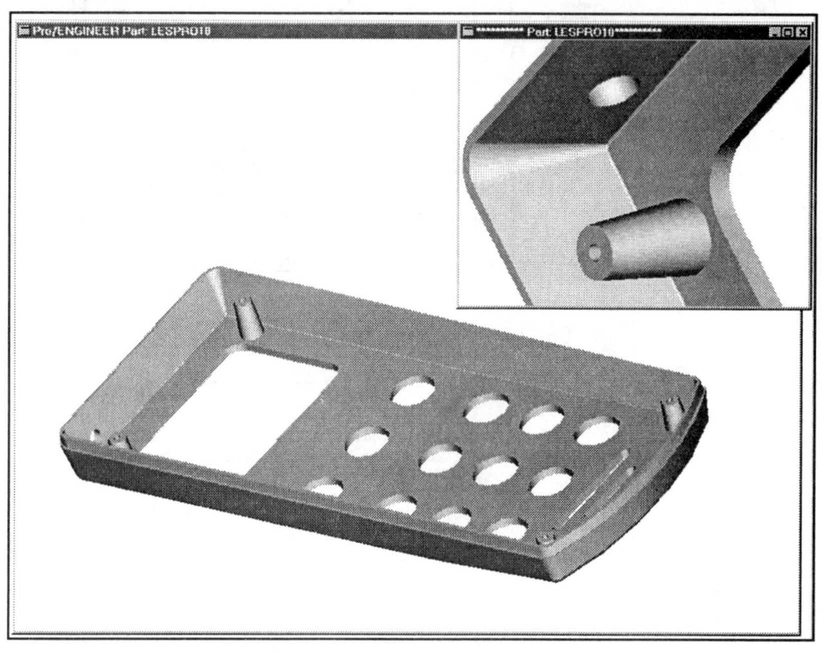

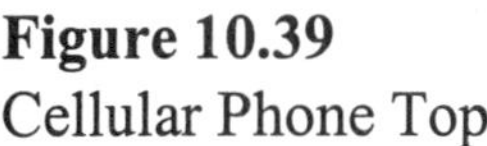

Figure 10.39
Cellular Phone Top

Cellular Phone Top

The Cellular Phone Top (Figs. 10.39 through 10.49) is one of two major components for a cellular phone. You created the other as a project in Lesson 9. The part is made of the same plastic as the Cellular Phone Bottom. If time permits, try to assemble the two pieces after completing Lesson 14, Assembly Constraints.

Analyze the part and plan out the steps and features required to model it. Use the **DIPS** in Appendix D to establish a feature creation sequence before the start of modeling. Remember to set up the environment, establish datum planes, and set them on a layer.

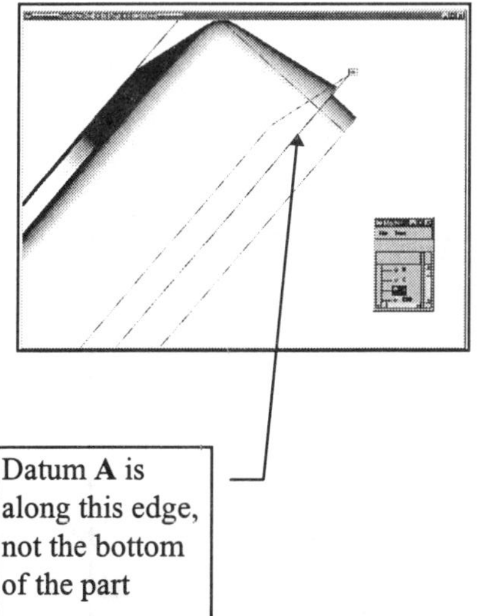

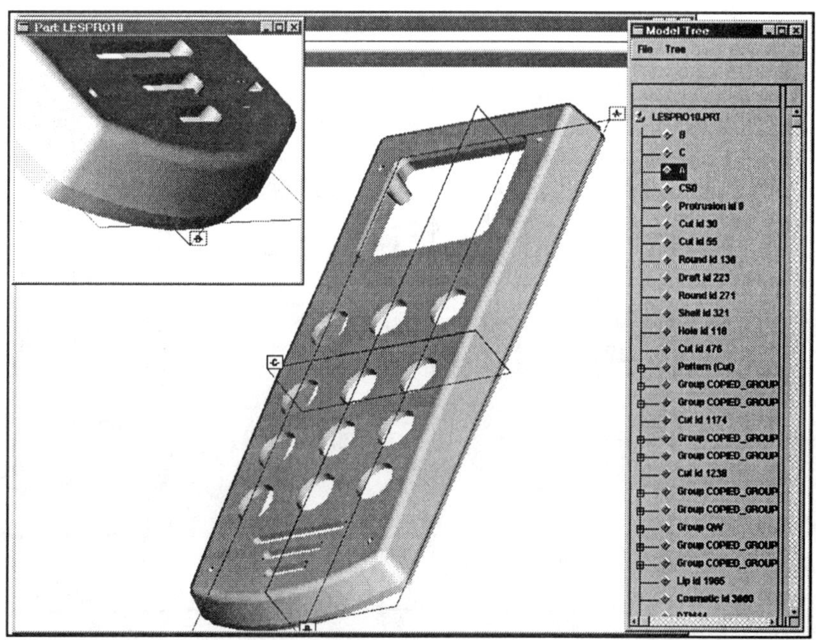

Figure 10.40
Cellular Phone Top Showing Datum Planes and Model Tree

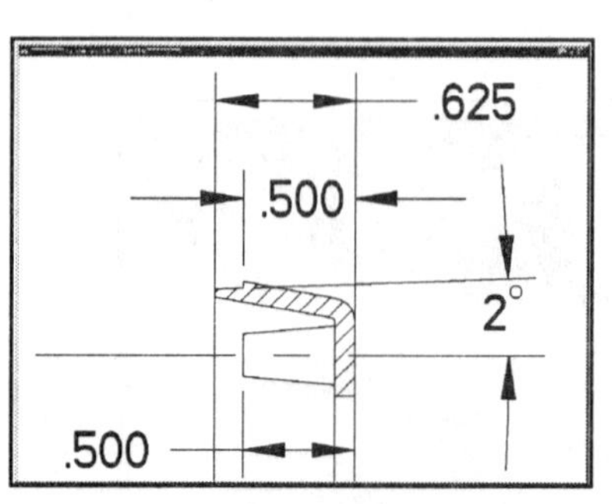

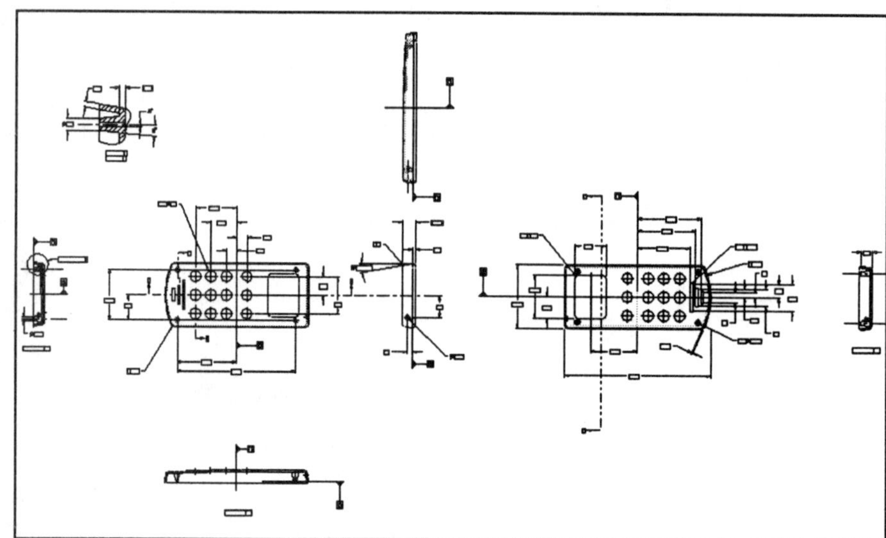

Figure 10.41
Cellular Phone Top, Detail Drawing

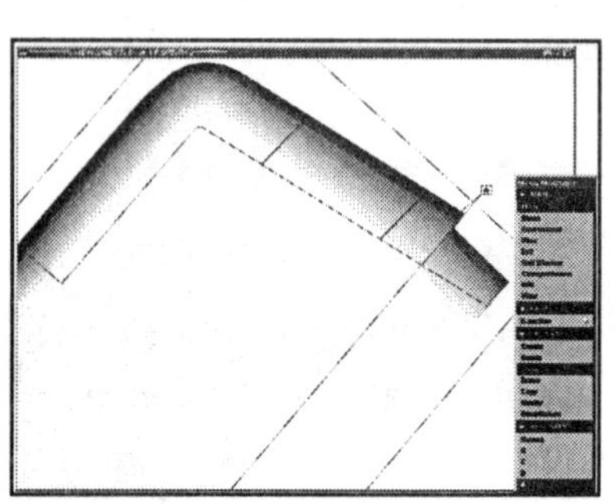

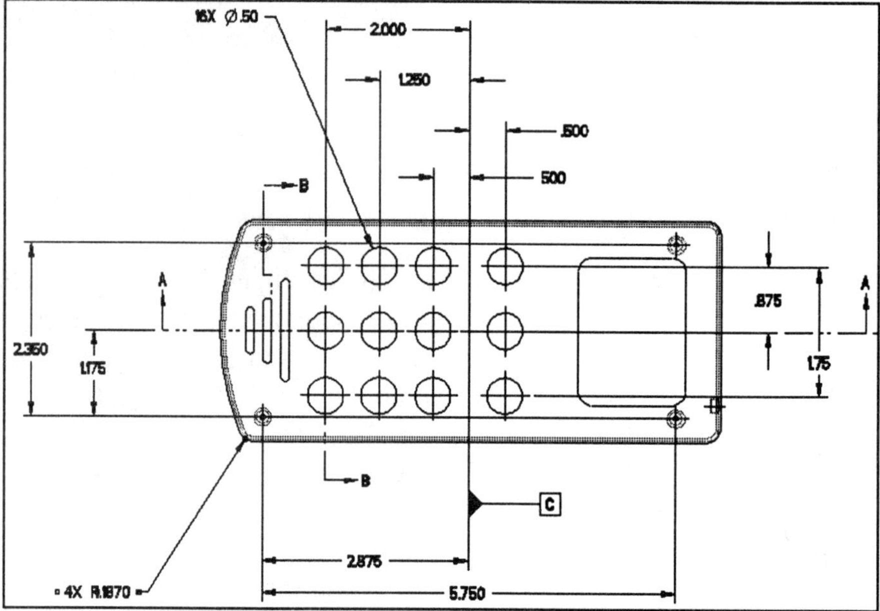

Figure 10.42
Cellular Phone Top, Front View

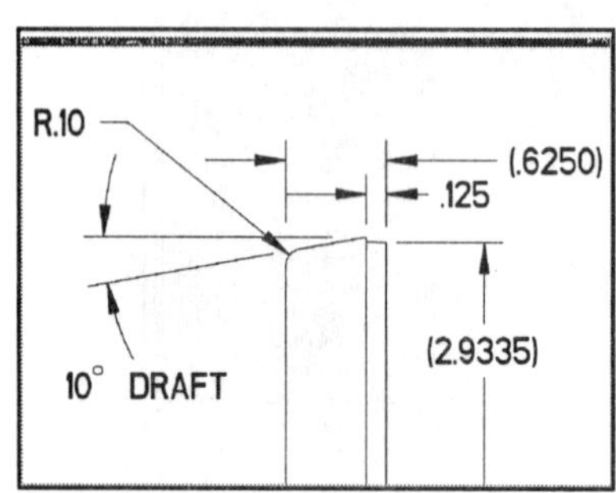

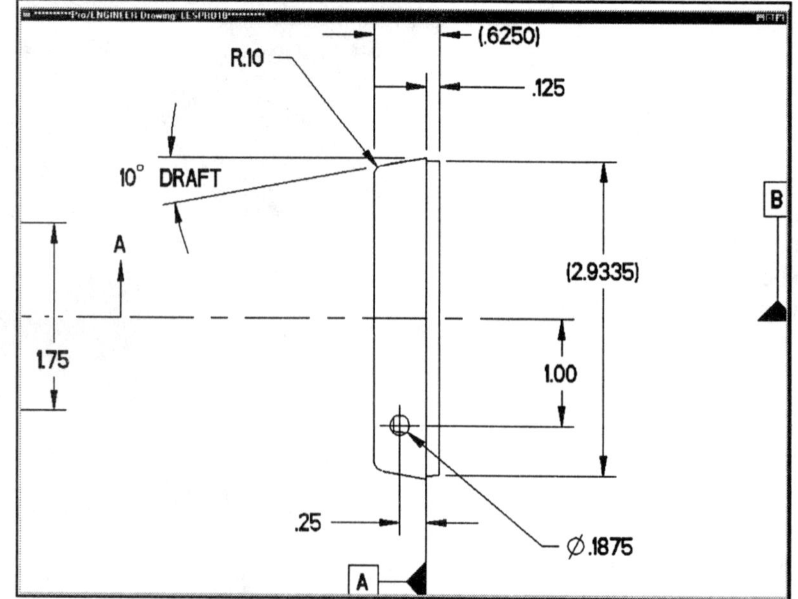

Figure 10.43
Cellular Phone Top, Right Side View

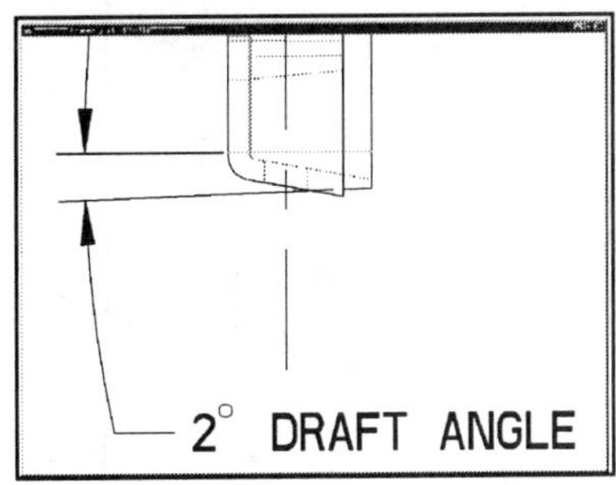

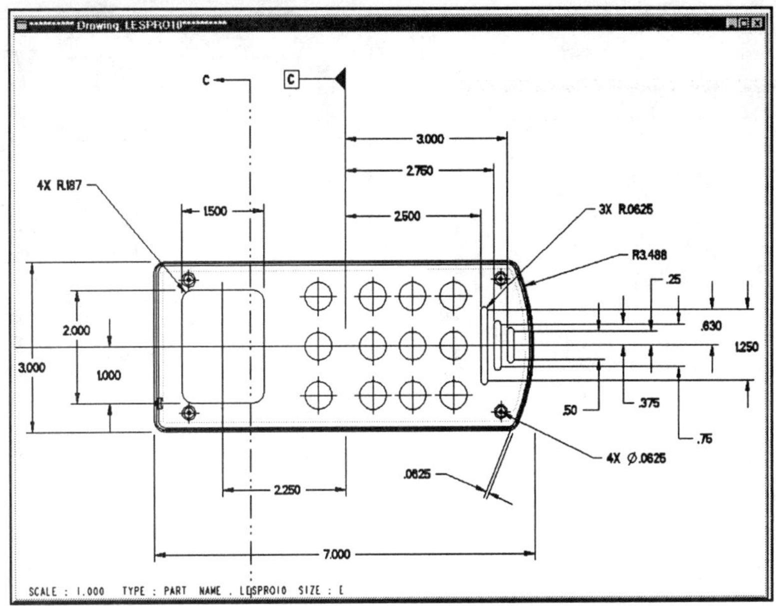

Figure 10.44
Cellular Phone Top,
Bottom View

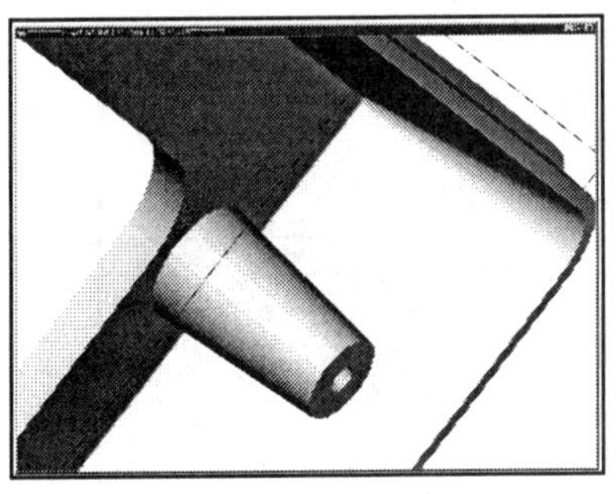

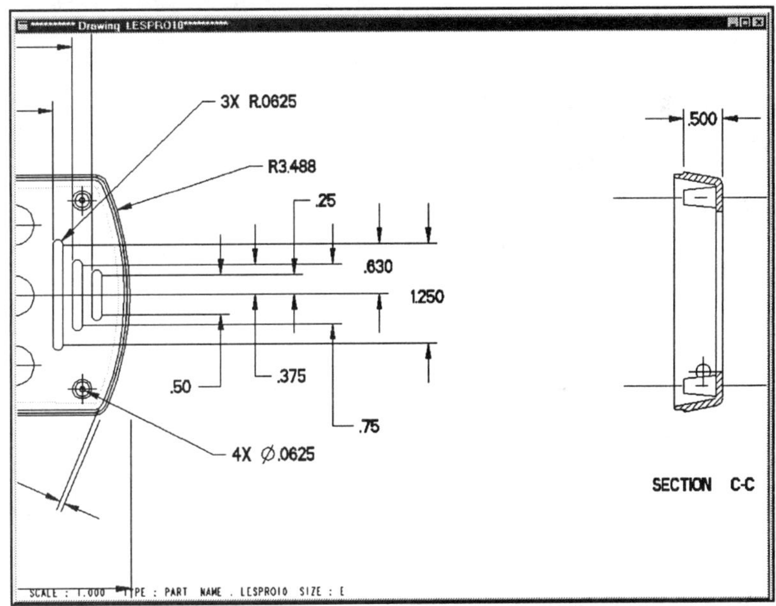

Figure 10.45
Cellular Phone Top,
SECTION C-C

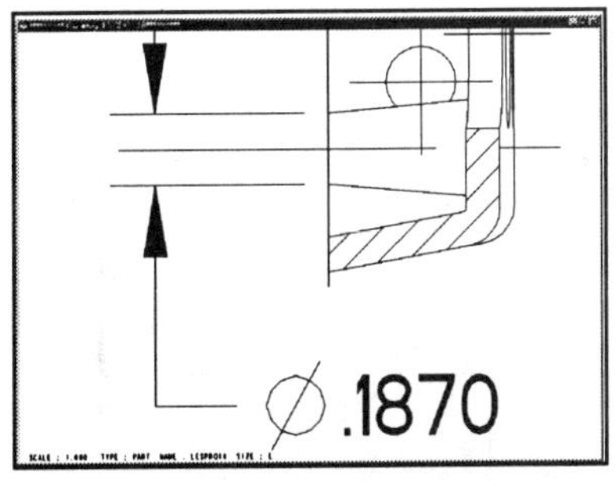

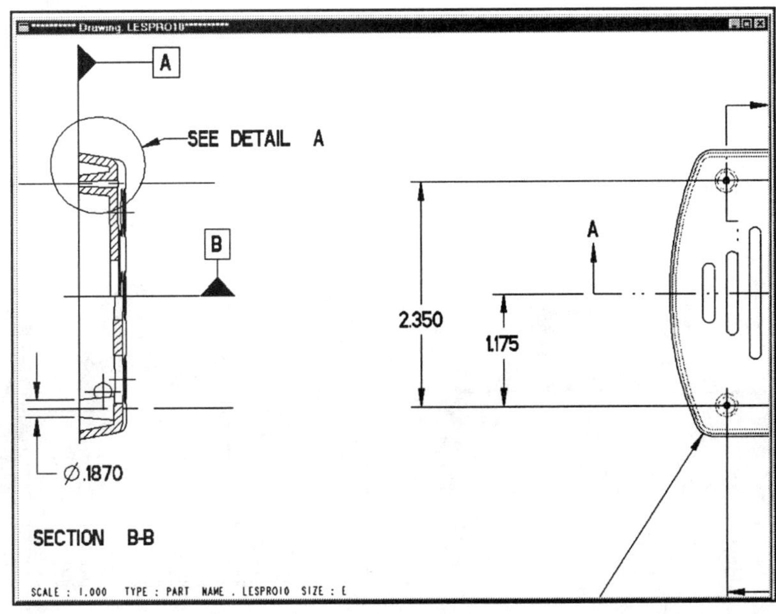

Figure 10.46
Cellular Phone Top,
SECTION B-B

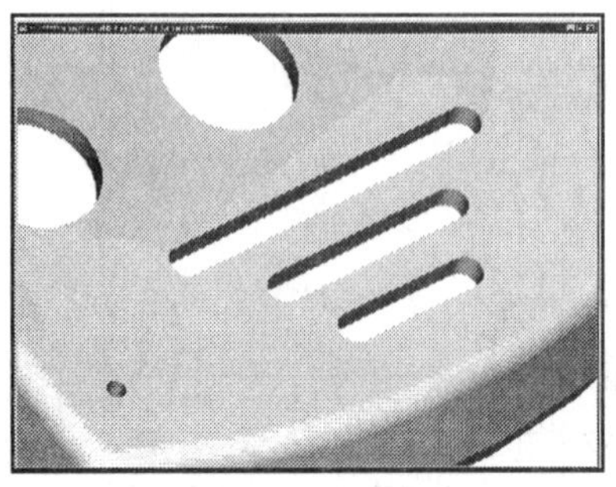

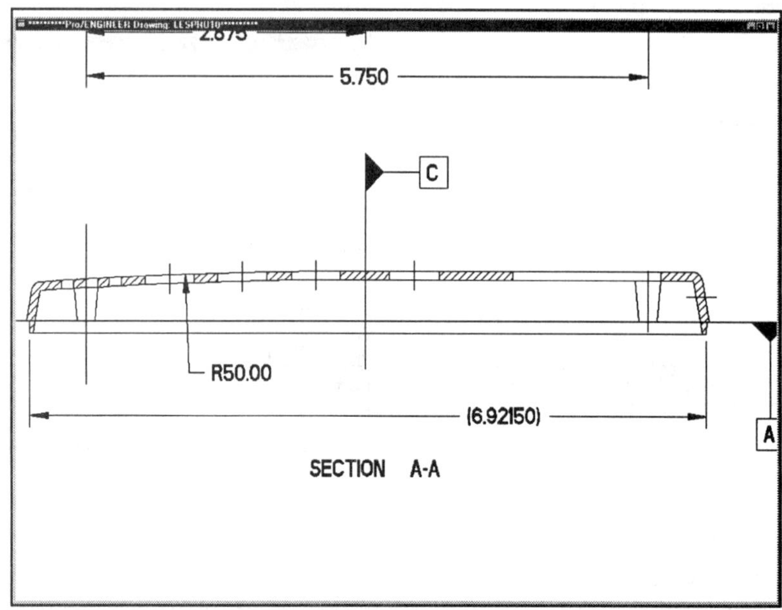

Figure 10.47
Cellular Phone Top,
SECTION A-A

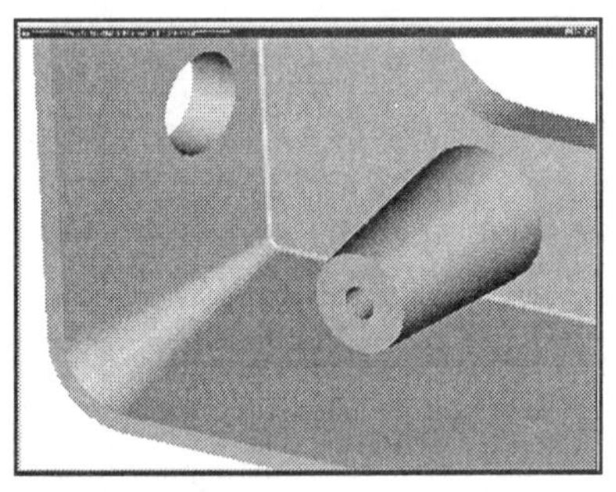

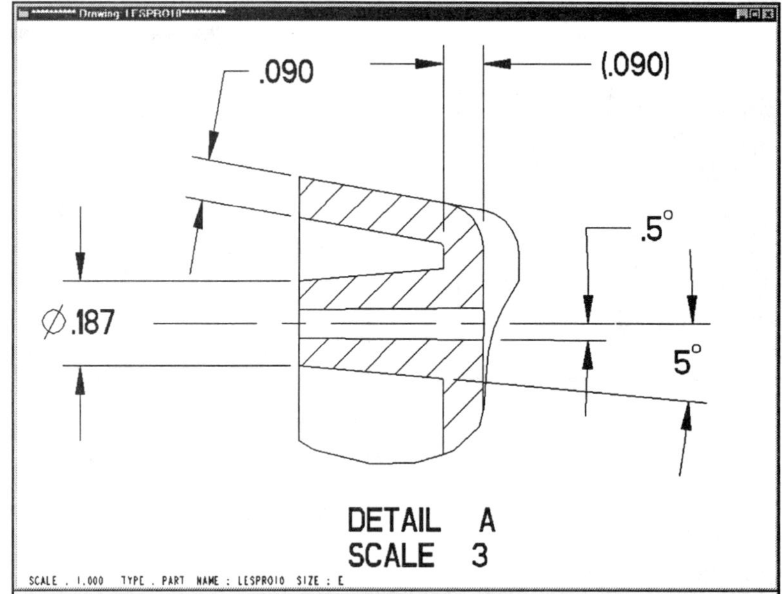

Figure 10.48
Cellular Phone Top,
DETAIL A

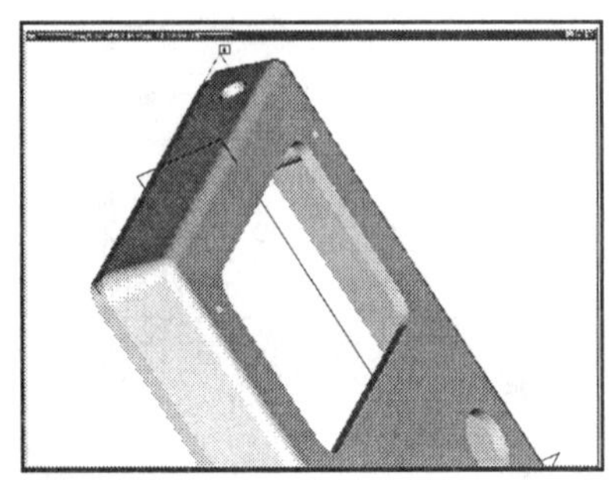

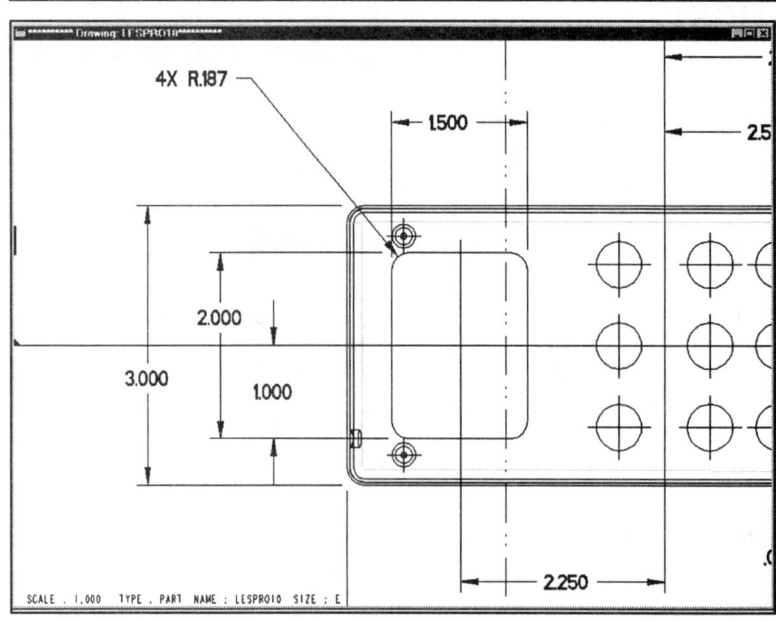

Figure 10.49
Cellular Phone Top,
Opening

Lesson 11

Sweeps

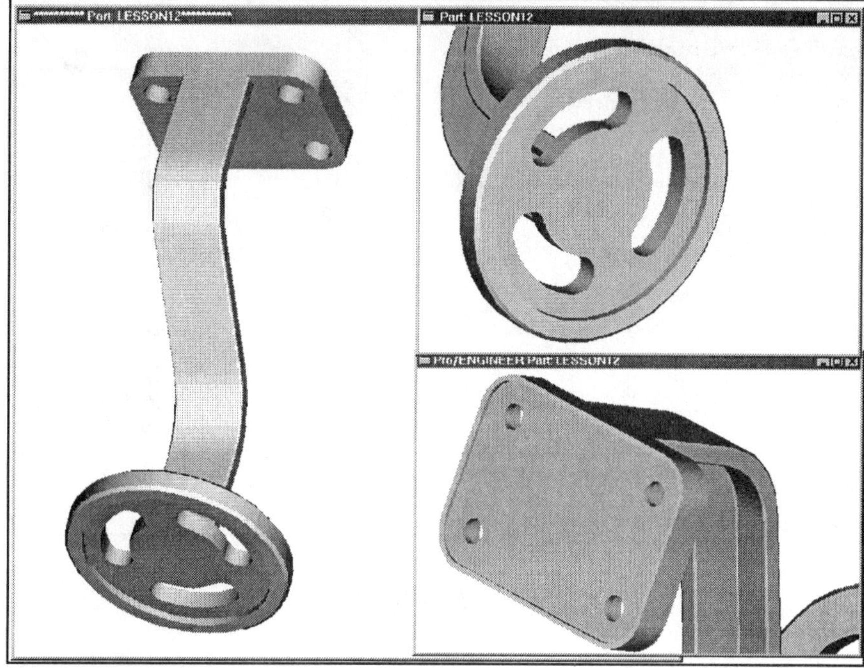

Figure 11.1
Bracket

OBJECTIVES

1. Create a constant-section swept feature

2. Sketch a trajectory for a sweep

3. Sketch and locate a sweep section

4. Understand the difference between adding and not adding inner faces

5. Be able to redefine a sweep

6. Understand the difference between a sketched and a selected trajectory

7. Create variable sweeps

EGD REFERENCE
Engineering Graphics and Design with Graphical Analysis *or* **Fundamentals of Engineering Graphics and Design**
by L. Lamit and K. Kitto
Read Chapter 20
See pages 378, 382

COAch™ for Pro/ENGINEER

If you have **COAch for Pro/ENGINEER** on your system, go to SEARCH and do the Segment shown in Figures 11.3 through 11.5.

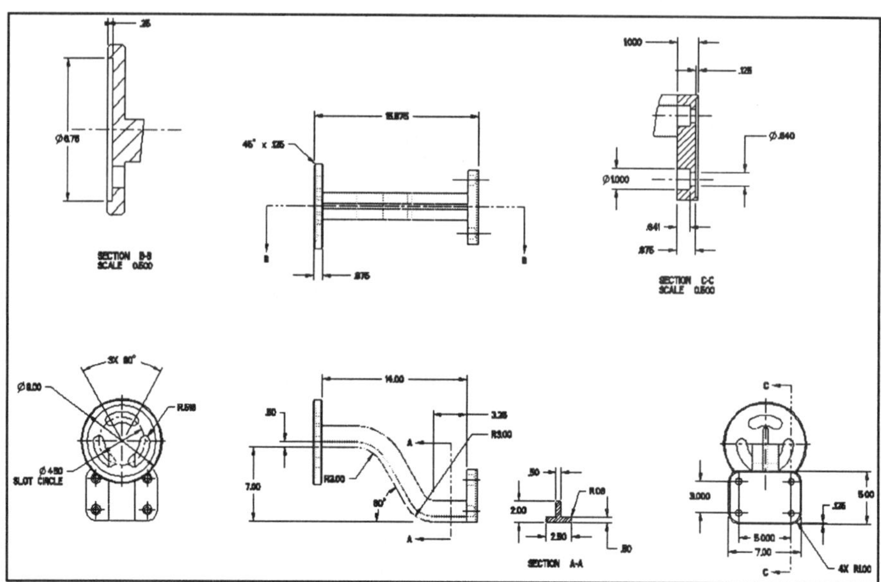

Figure 11.2
Bracket Detail

SWEEPS

A sweep is created by sketching or selecting a *trajectory,* then sketching a *section* to follow along it. The **Bracket** shown in Figures 11.1 and 11.2 uses a simple sweep in its design.

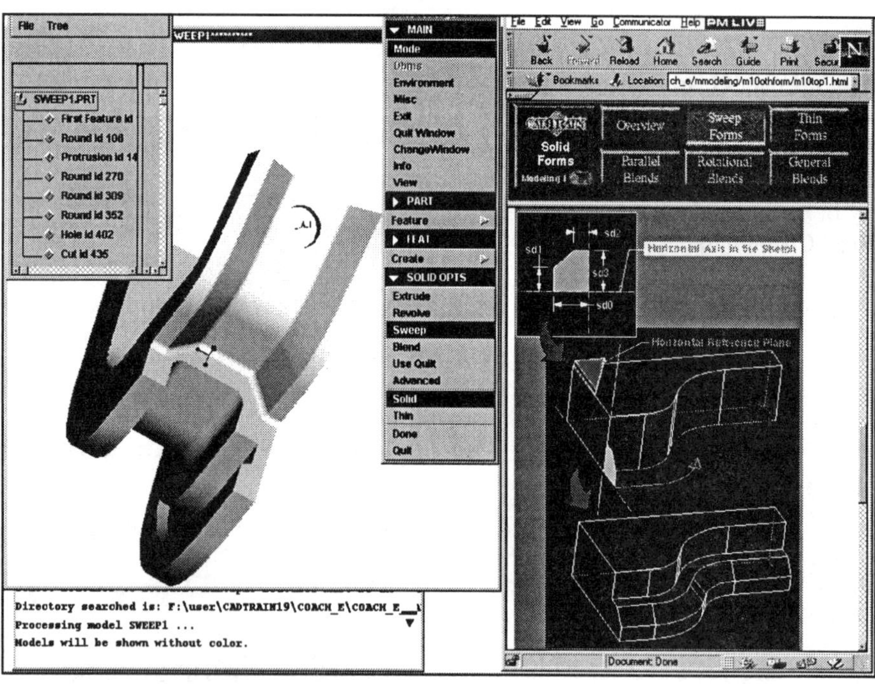

Figure 11.3
COAch for Pro/E, Solid Forms (Sweep Forms)

Defining a Trajectory

A *constant-section sweep* can use either trajectory geometry sketched at the time of feature creation or a trajectory (Fig. 11.3) made up of selected datum curves or edges. The trajectory must have adjacent reference surfaces or be planar (Fig. 11.4). When defining a sweep (Fig. 11.5), Pro/E checks the specified trajectory for validity and establishes normal surfaces. When ambiguity exists, Pro/E prompts you to select a normal surface.

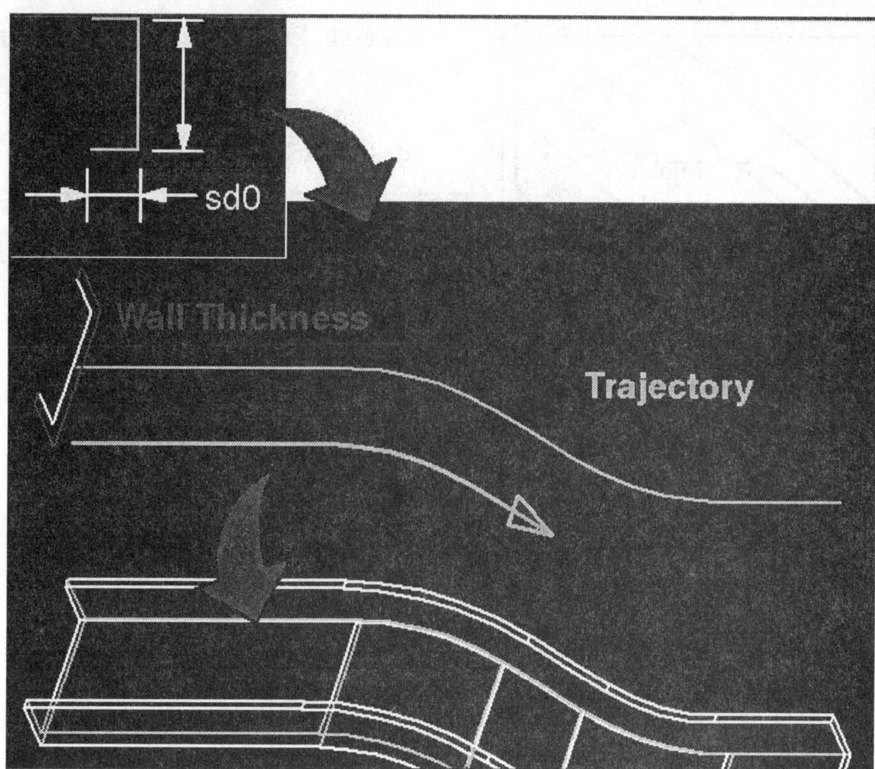

Figure 11.4
COAch for Pro/E, Solid Forms (Sweep Forms)

Creating a Swept Feature

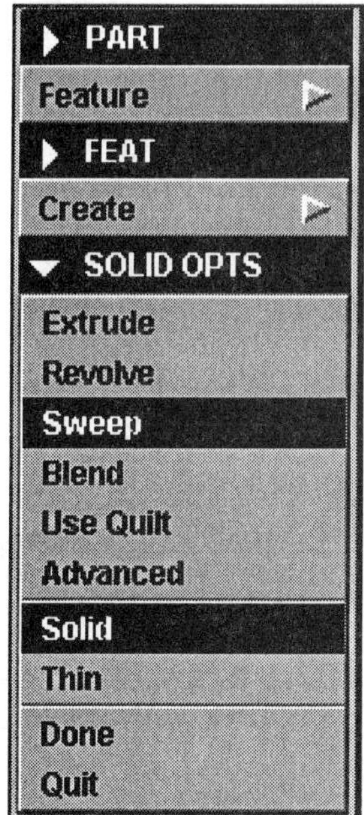

To create a sweep, do the following:

1. **Feature ⇒ Create ⇒ Solid ⇒ Protrusion.**
2. Choose **Sweep** and **Done** from the SOLID OPTS menu.
3. The feature creation dialog box for sweeps is displayed.

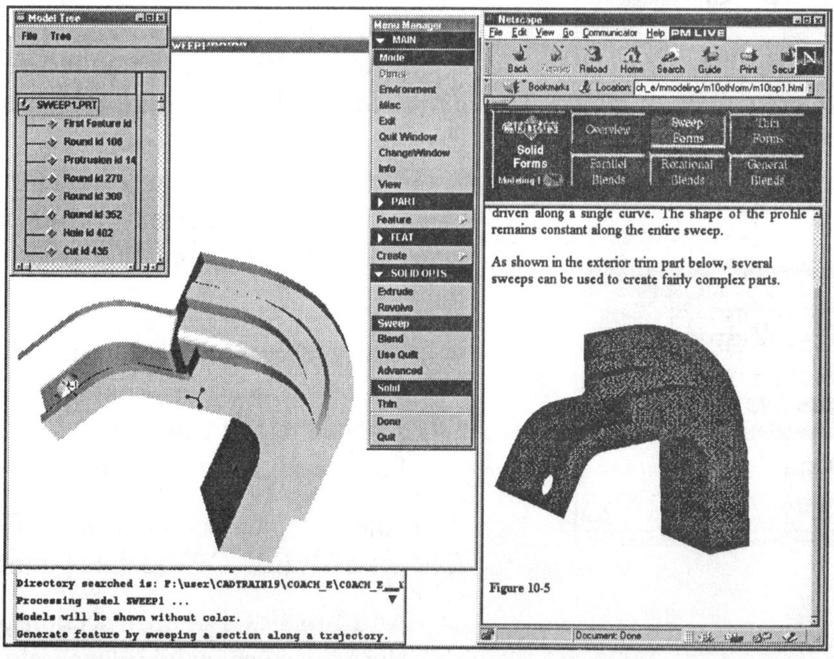

Figure 11.5
COAch for Pro/ENGINEER, Solid Forms (Sweep Forms)

4. Sketch or select an open or closed trajectory (Fig. 11.6), using a SWEEP TRAJ menu option. The following options are available:

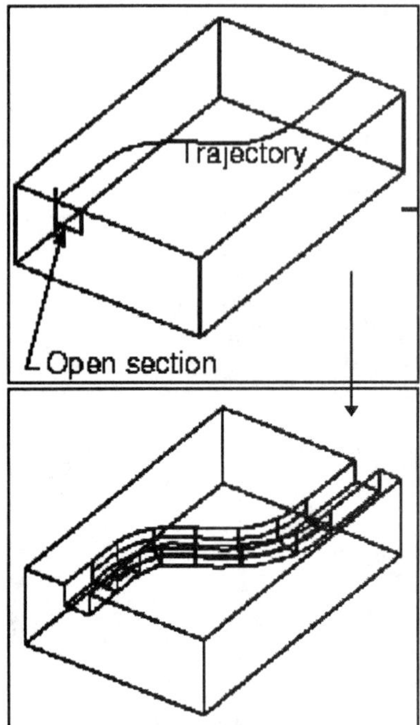

Sketch Traj Sketch the sweep trajectory using Sketcher mode.
Select Traj Select a chain of existing curves or edges as the sweep trajectory. The CHAIN menu allows you to select the desired trajectory.

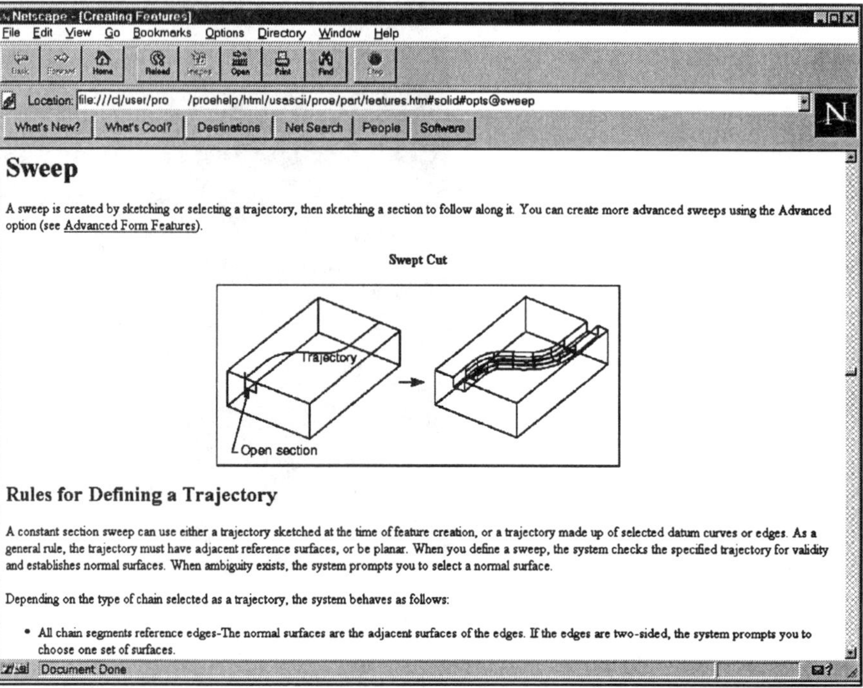

Sweep

A sweep is created by sketching or selecting a trajectory, then sketching a section to follow along it. You can create more advanced sweeps using the Advanced option (see Advanced Form Features).

Swept Cut

Rules for Defining a Trajectory

A constant section sweep can use either a trajectory sketched at the time of feature creation, or a trajectory made up of selected datum curves or edges. As a general rule, the trajectory must have adjacent reference surfaces, or be planar. When you define a sweep, the system checks the specified trajectory for validity and establishes normal surfaces. When ambiguity exists, the system prompts you to select a normal surface.

Depending on the type of chain selected as a trajectory, the system behaves as follows:

- All chain segments reference edges-The normal surfaces are the adjacent surfaces of the edges. If the edges are two-sided, the system prompts you to choose one set of surfaces.

Figure 11.6
Online Documentation, Sweeps

5. If the trajectory lies in more than one surface, such as a trajectory defined by a datum curve created using **Intr Surfs**, Pro/E prompts you to select a normal surface for the sweep cross section. Pro/E orients the **Y** axis of the cross section to be normal to this surface along the trajectory.

6. Create or retrieve the section to be swept along the trajectory, and dimension it relative to the *crosshairs* displayed on the trajectory. Choose **Done**.

7. If the trajectory is open (the start point and endpoint of the trajectory do not touch) and you are creating a solid sweep, choose an ATTRIBUTES menu option, then **Done**. The possible options are as follows:

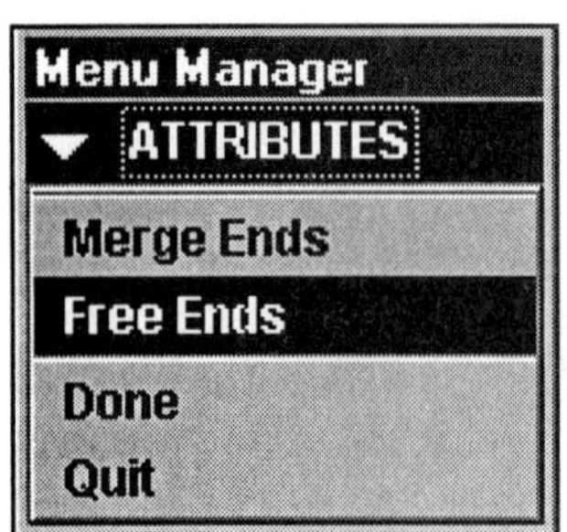

Merge Ends Merge the ends of the sweep, if possible, into the adjacent solid. For you to do this, the sweep endpoint must be attached to part geometry.
Free Ends Do not attach the sweep end to adjacent geometry.

8. If the sweep trajectory is closed (Fig. 11.7), choose one of the following SWEEP OPT menu options and then **Done**:

Add Inn Fcs For open sections, add top and bottom faces to close the swept solid (planar, closed trajectory, and open section). The resulting feature consists of surfaces created by sweeping the section and has two planar surfaces that cap the open ends.
No Inn Fcs Do not add top and bottom faces.

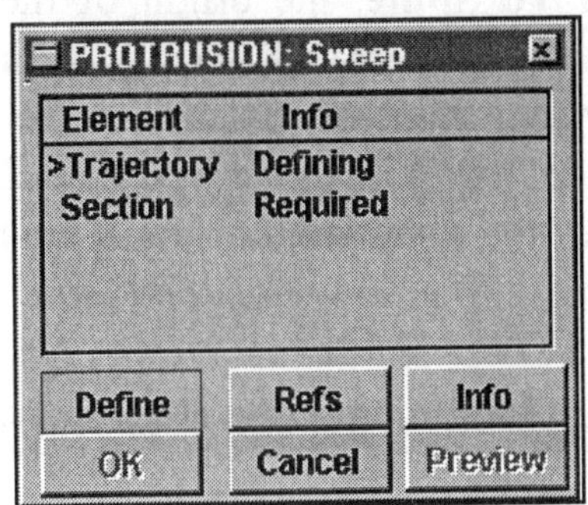

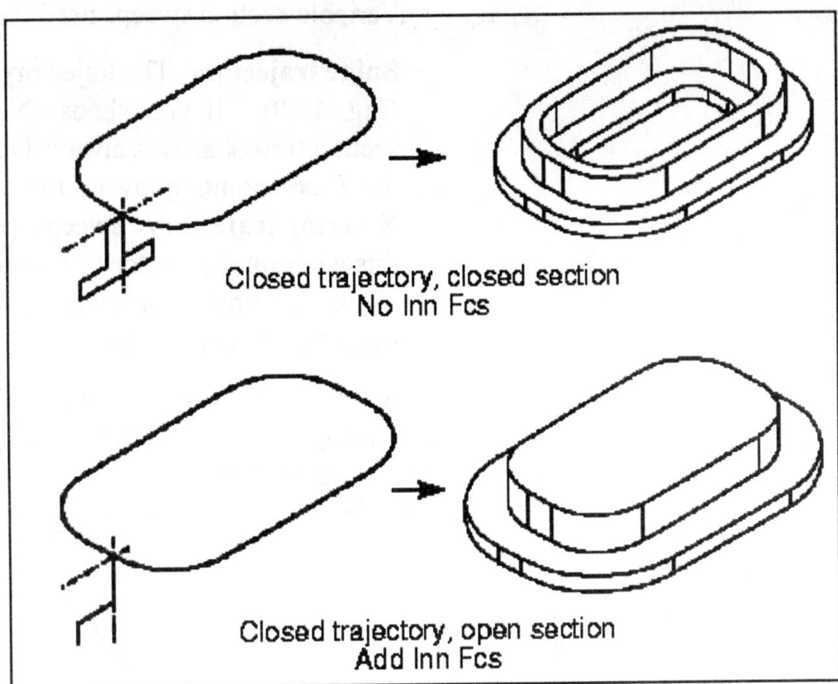

Figure 11.7
Online Documentation, Sweep Trajectories and Sections--**Add Inn Fcs** and **No Inn Fcs**

9. Choose **Flip**, if desired, then **Okay** from the DIRECTION menu to select the side on which to remove material for swept cuts.
10. Pro/E issues a message stating that all the elements have been defined. If you wish, select one of the buttons in the dialog box.
11. Select the **OK** button in the dialog box to create the sweep.

Redefine

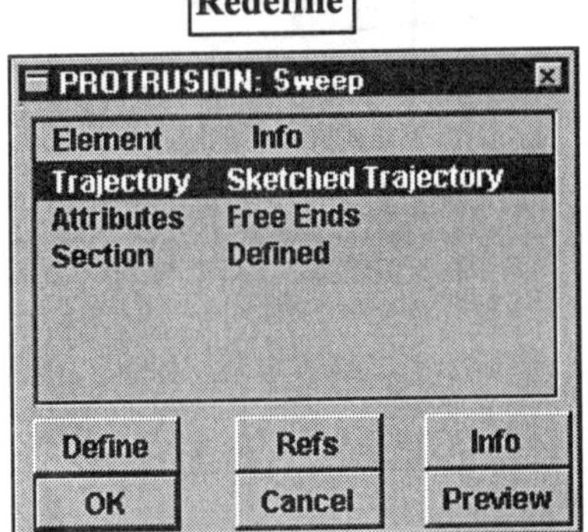

To redefine sweep sections or trajectories after the feature is created, choose **Redefine** and then select the sweep. Pro/E will display the feature creation dialog box. You can redefine the trajectory or section elements.

Variable-Section Sweeps

A solid sweep feature using one or more longitudinal trajectories and a single variable section can also be created (Fig. 11.8). The parameters of the section can vary as the section moves along the sweep trajectories. These sweeps are called **variable-section sweeps.**

Every variable-section sweep requires one *longitudinal "spine" trajectory*. You can define a sweep for which the **X** axis of the section follows the **X**-vector trajectory while remaining normal to the spine at all times as it sweeps along the spine (**Nrm to Spine**). To do so, you must also specify an **X**-vector trajectory to orient the section as it sweeps along the spine. The section plane is always normal to the spine trajectory at the point of their intersection. The **X** axis of each section's coordinate system is defined by the direction from the point of intersection of the plane and the spine to the point of intersection of the plane and the **X**-vector trajectory for that section. You can also define a variable-section sweep for which the **Y** axis of the section remains constant. The section will follow the spine such that it is normal to the selected pivot plane. The **X** axis and **Z** axis will still follow the spine and **X**-vector trajectories.

Variable-section sweeps need the following trajectories:

Spine trajectory The trajectory along which the section is swept (Fig. 11.9). If you choose **Nrm To Spine**, the origin of the section (crosshairs) is always located on the spine trajectory, with the **X** axis pointing toward the **X**-vector trajectory.

X-vector trajectory Sweeps created using **Nrm To Spine** need this additional trajectory. It defines the orientation of the **X** axis of the section coordinate system. *The X-vector and spine trajectories cannot intersect.*

After the direction is specified, Pro/E displays the VAR SEC SWP menu so you can define a trajectory. You can use a composite curve as a trajectory.

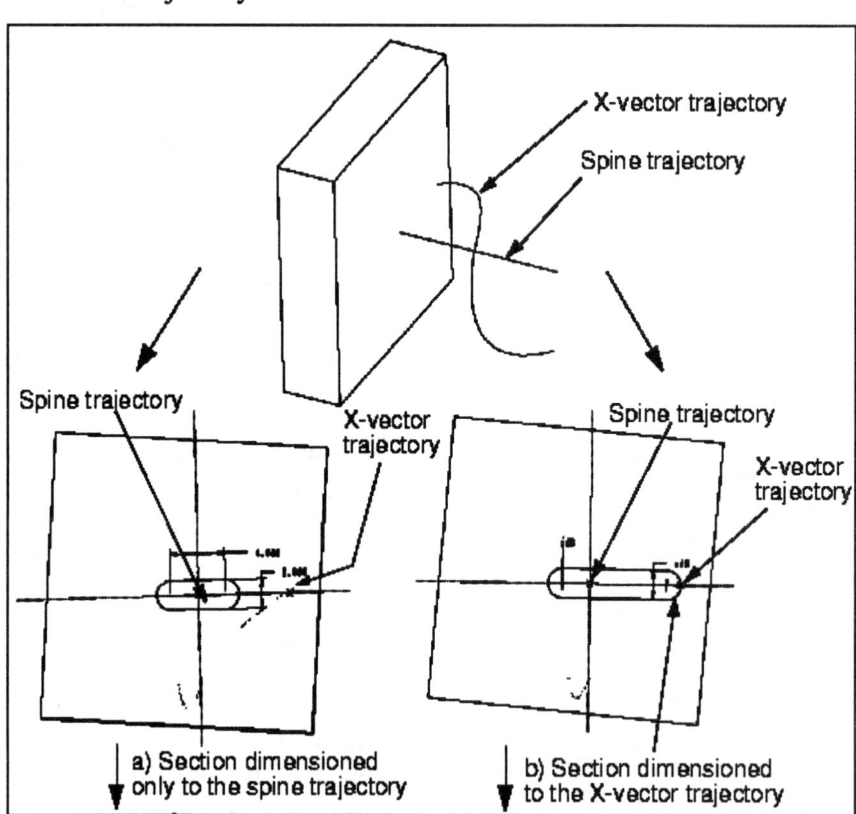

Figure 11.8
Online Documentation, Variable Section Sweeps

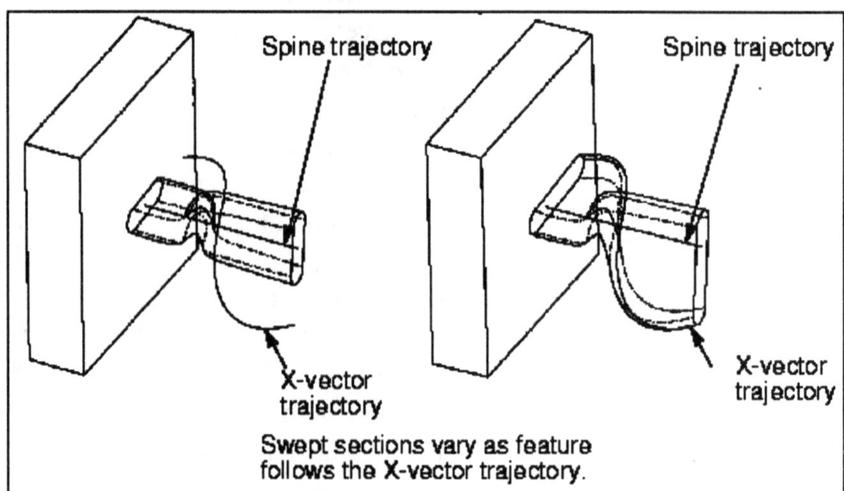

Figure 11.9
Online Documentation, Spine Trajectories

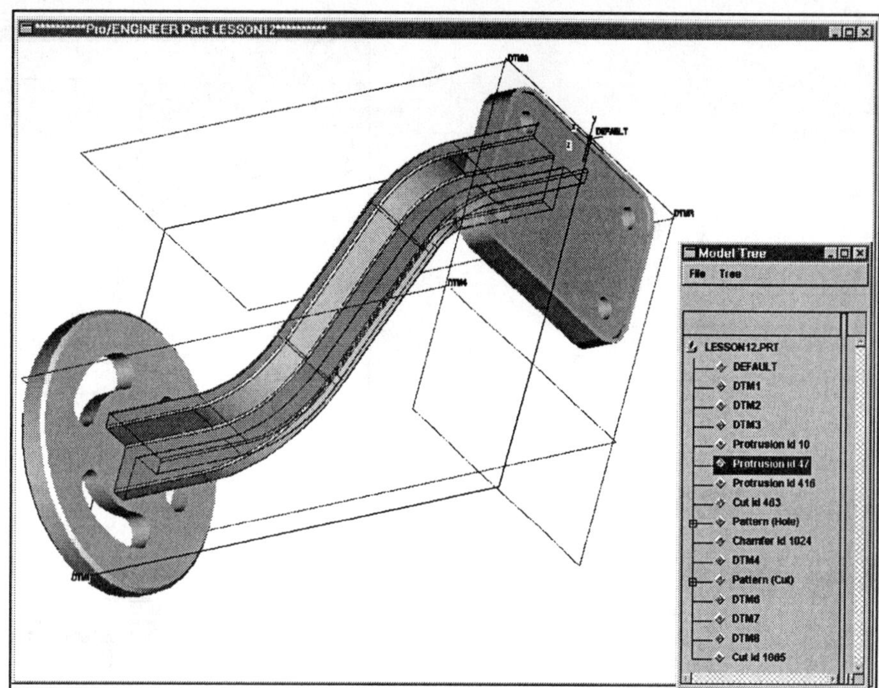

Figure 11.10
Bracket Showing Datum Planes and Coordinate System and Model Tree

Bracket

The **Bracket** (Figs. 11.10 through 11.39) requires the use of the **Sweep** command. The T-shaped section is swept along the selected *trajectory*. The protrusions on both sides of the swept feature are to be created with the dimensions given in Figures 11.11 through 11.17. Step-by-step commands are provided only for the sweep trajectory and its cross section.

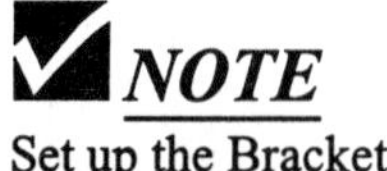

NOTE
Set up the Bracket:

- Material = Cast Iron
- Units = Inches
- Default Datum Planes
- Default Coordinate System
- Layers = **DATUM_LAYER**
- Hidden line
- ✓Grid Snap

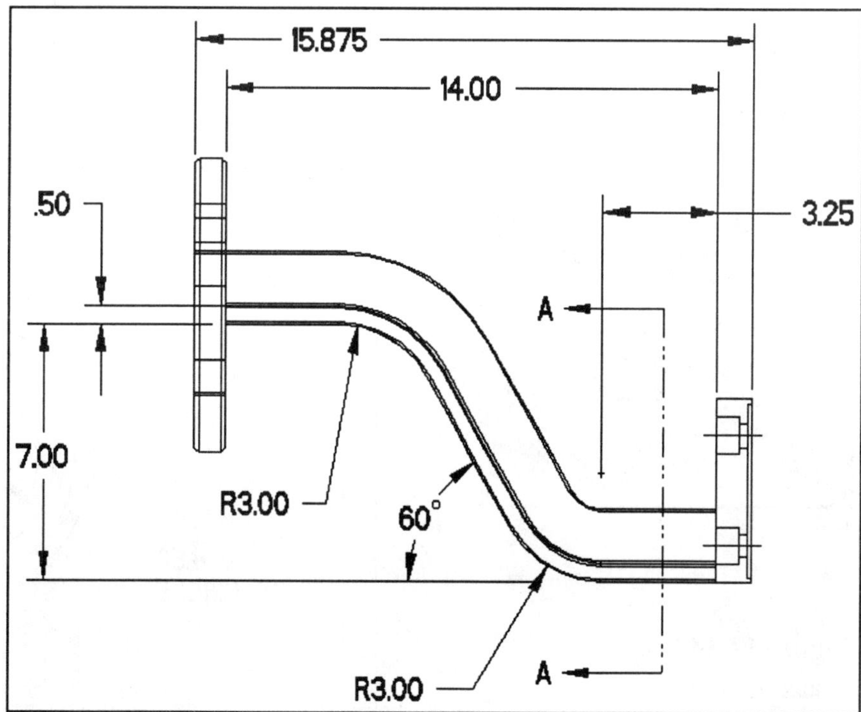

Figure 11.11
Bracket Drawing, Front View

15.875

45° X .125

.875

Figure 11.12
Bracket Drawing,
Top View

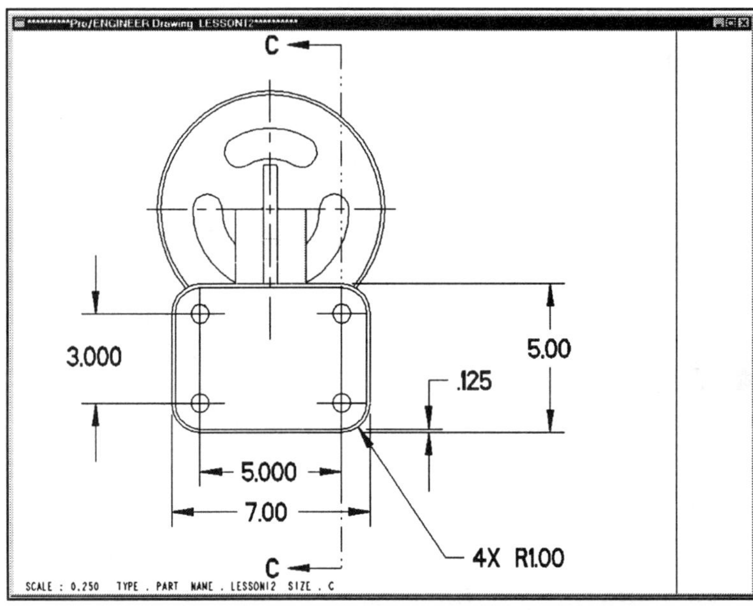

Figure 11.13
Bracket Drawing,
Right Side View

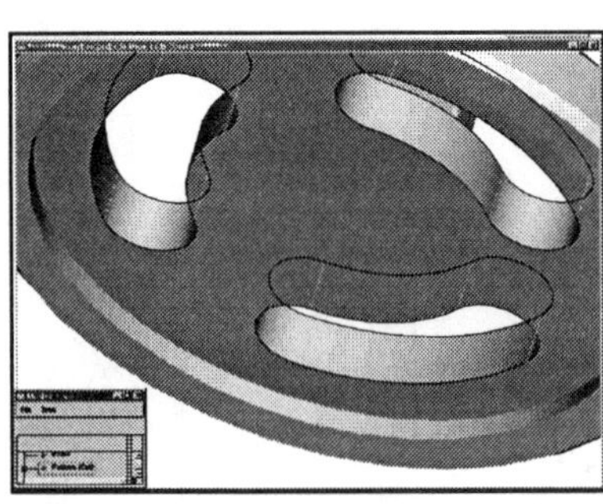

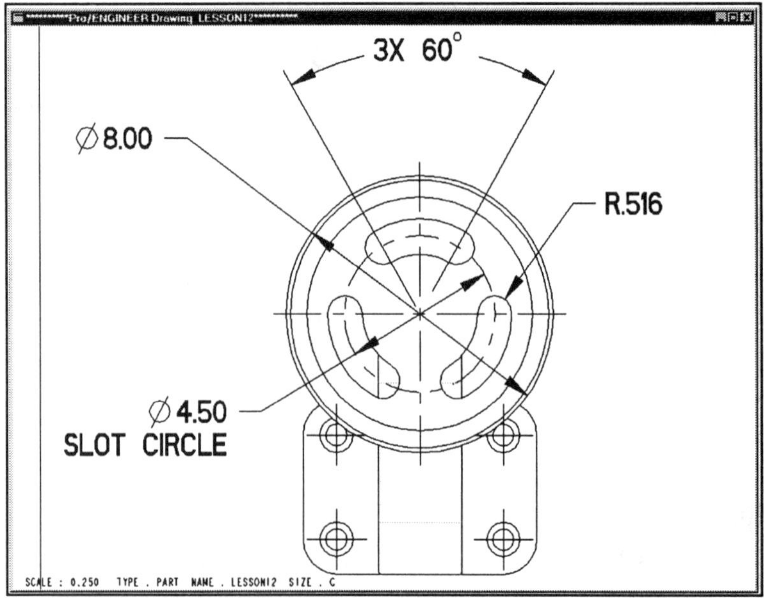

Figure 11.14
Bracket Drawing,
Left Side View

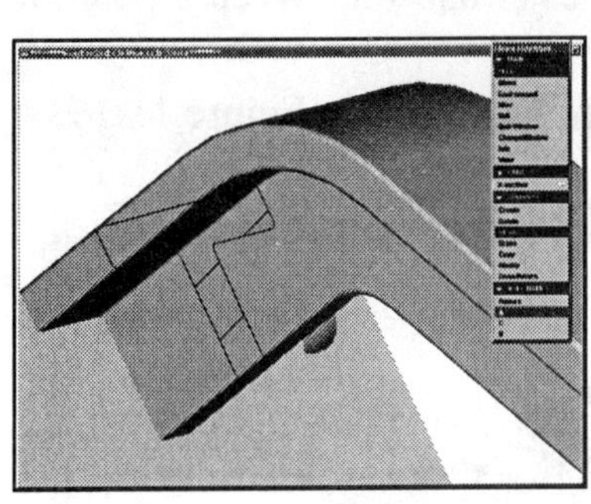

Figure 11.15
Bracket Drawing,
SECTION A-A

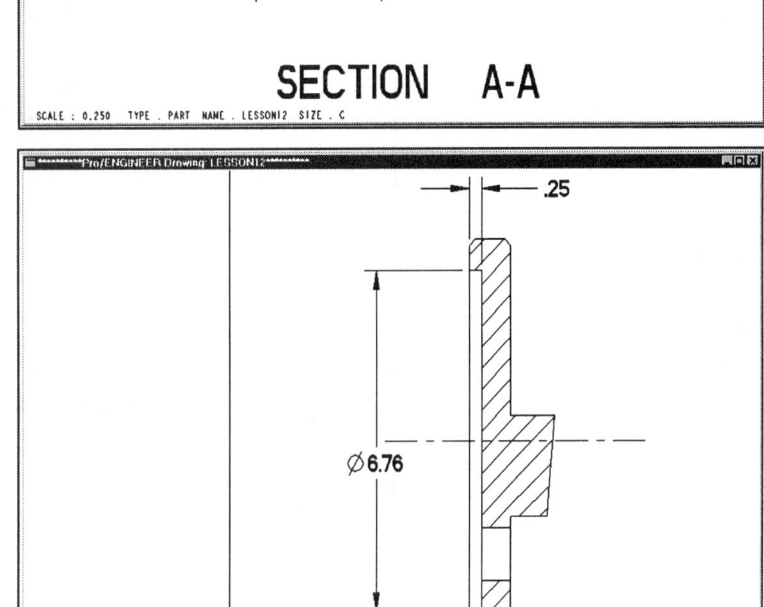

Figure 11.16
Bracket Drawing,
SECTION B-B

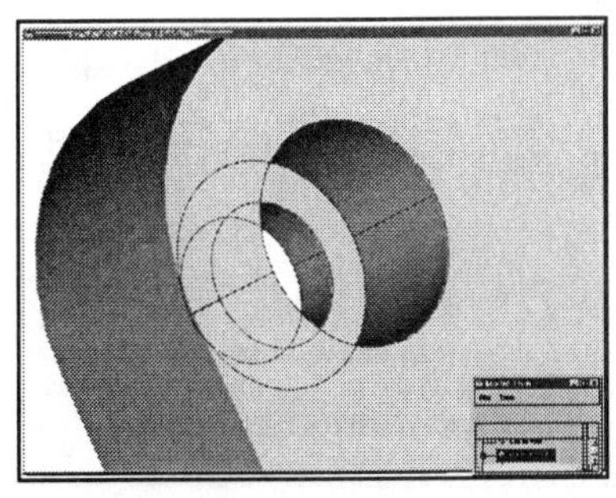

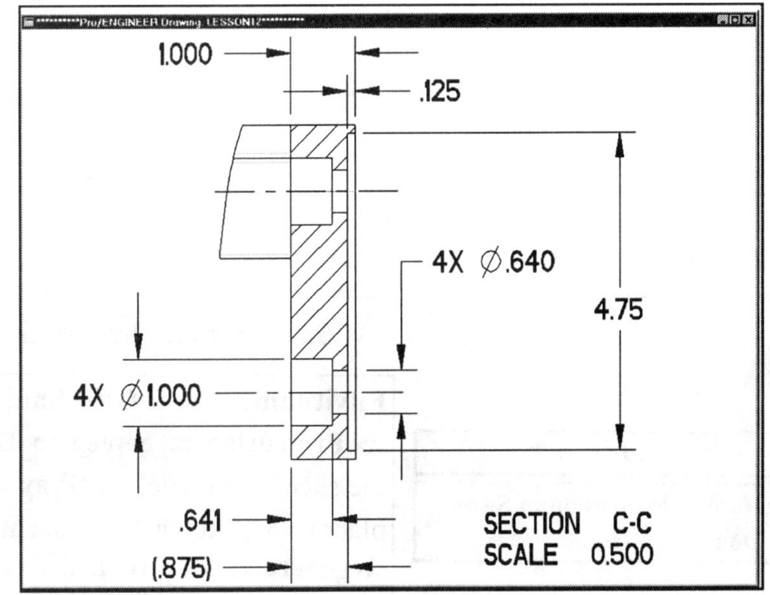

Figure 11.17
Bracket Drawing,
SECTION C-C

The Bracket is started by modeling the protrusion shown in Figure 11.18. This protrusion will be used to establish the sweep's position in space. Sketch the protrusion on **DTM3**.

The second protrusion is a sweep and is shown in Figure 11.19.

Dbms (**File**--PT/Modeler) ⇒ **Save** ⇒ **enter**
Purge ⇒ **enter** ⇒ **Done-Return**

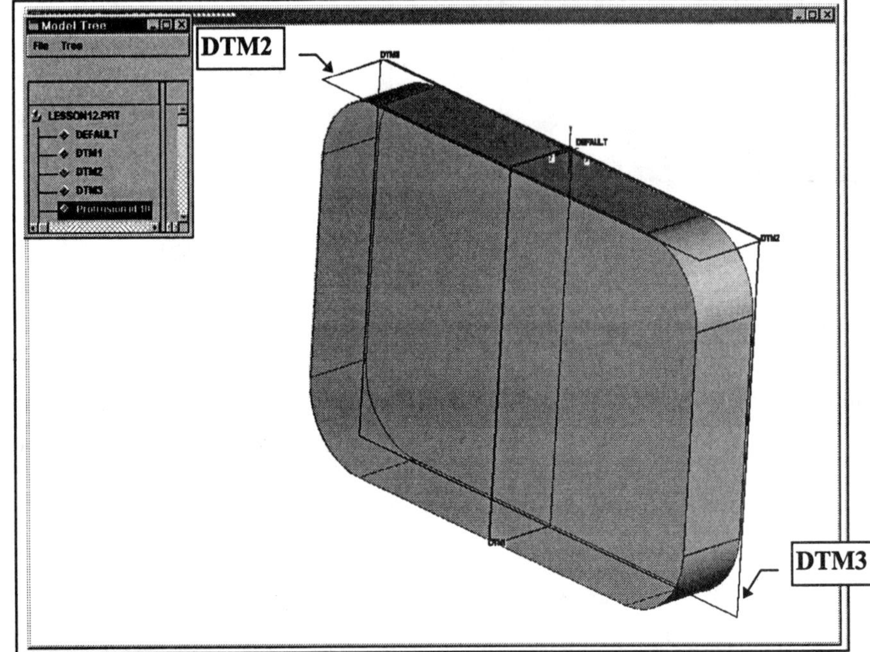

Figure 11.18
Bracket's First Protrusion

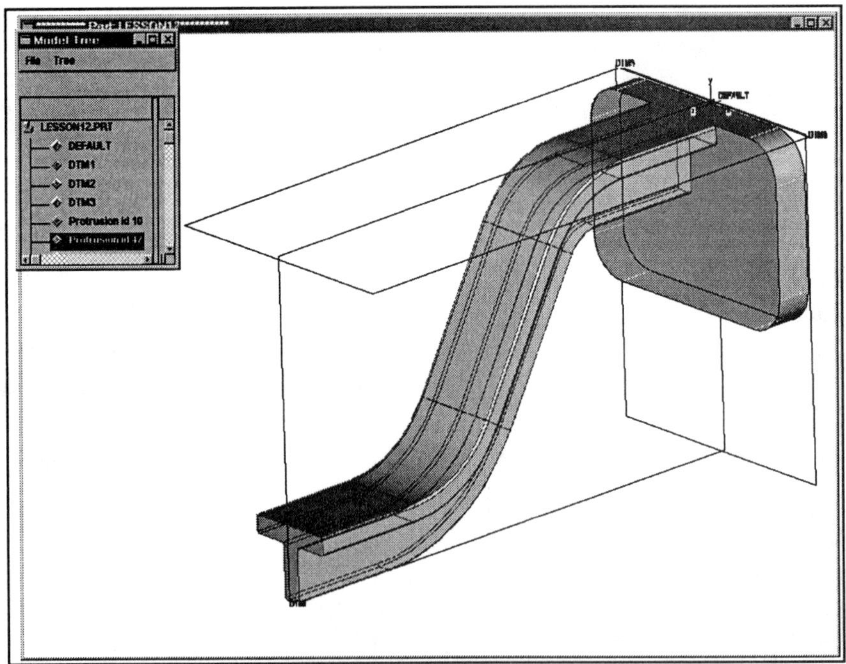

Figure 11.19
Swept Protrusion

Start the sweep by choosing the following commands:

PT/Modeler™
Feature ⇒ **Protrusion** ⇒ **Sweep** ⇒ **Done etc.**

Environment ⇒ □ **Grid Snap** ⇒ **Done-Return** ⇒ **Feature** ⇒ **Create** ⇒ **Protrusion** ⇒ **Sweep** ⇒ **Done** ⇒ **Sketch Traj** ⇒ (pick **DTM1** as the sketching plane) ⇒ **Okay** ⇒ **Top** ⇒ (pick **DTM2** as the orientation plane) ⇒ **(Sketch**, **Regenerate, Alignment, Regenerate, Dimension, Regenerate, Modify**, and **Regenerate** the trajectory as shown in Figs. 11.20 through 11.25)

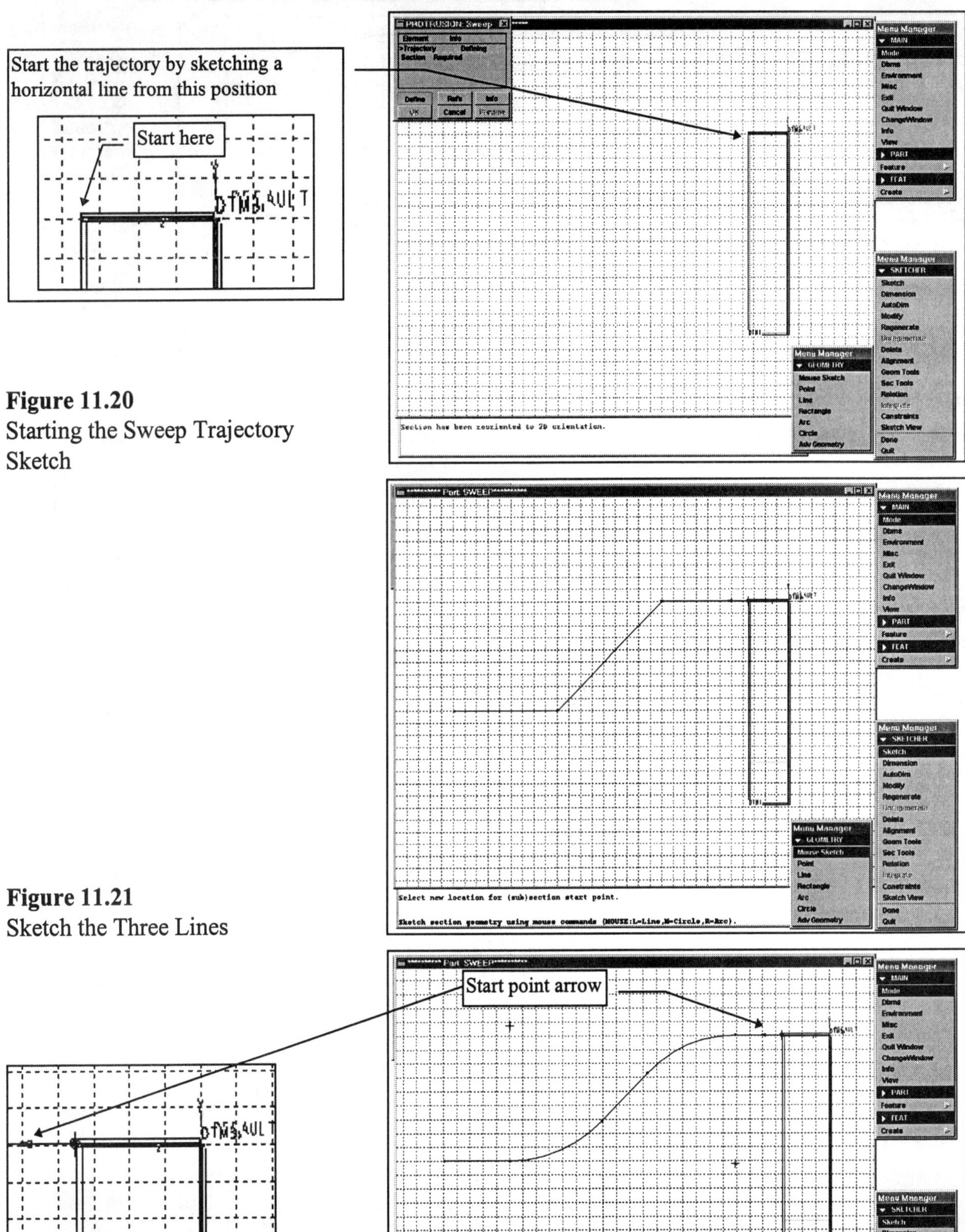

Figure 11.20
Starting the Sweep Trajectory Sketch

Figure 11.21
Sketch the Three Lines

Figure 11.22
Add the Arc Fillets

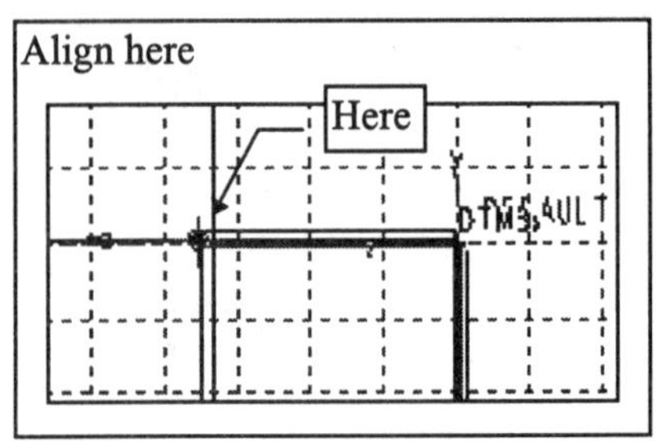

Figure 11.23
Alignment and **Dimension**

HINT

If the sketch fails to regenerate the first time try zooming out and regenerating and or zooming in and regenerating.

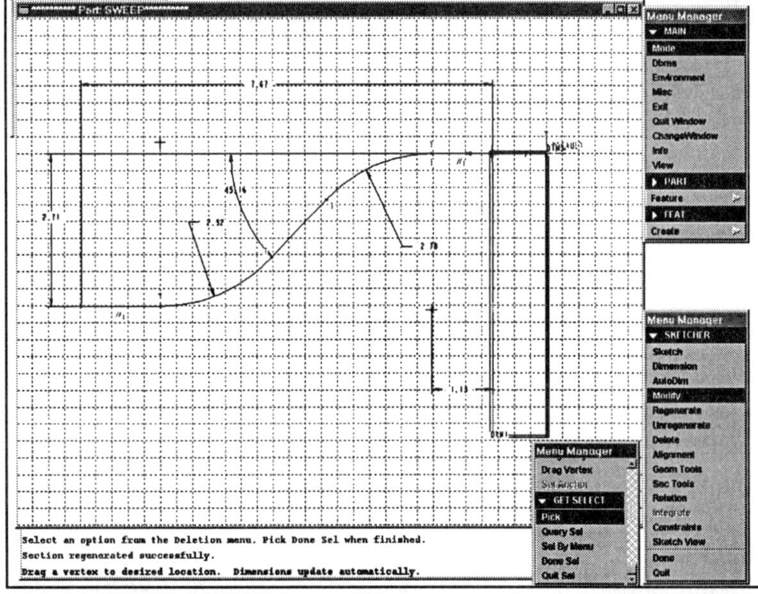

Figure 11.24
Regenerate the Sketch

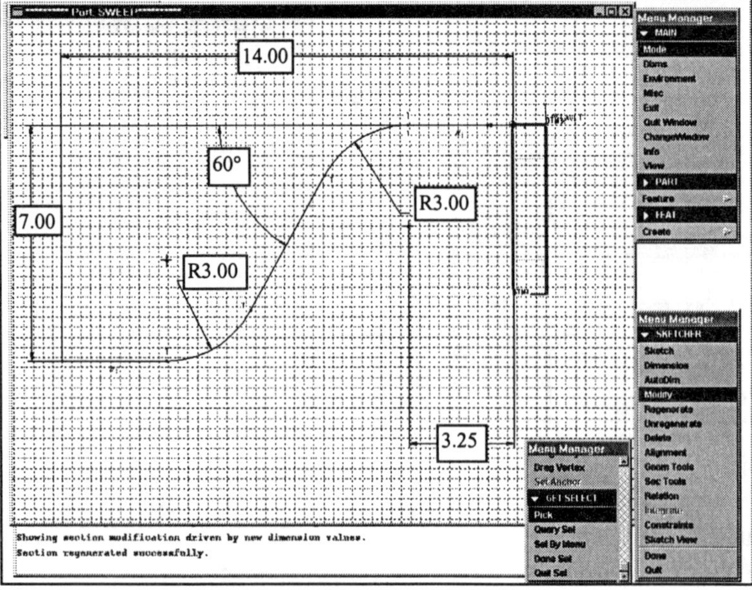

Figure 11.25
Modify and **Regenerate** the Sketch

After the trajectory has been regenerated, complete the sweep with the following commands, as shown in Figures 11.26 through 11.33:

Regenerate ⇒ Done ⇒ Free Ends ⇒ Done ⇒ (now sketch a section as shown in Figs. 11.26 and 11.27) ⇒ (add the *eight* fillets to the sketch) ⇒ **Regenerate ⇒ Alignment** (align the left vertical line of the sketch with **DTM2** and align the horizontal centerline to **DTM1**) ⇒ **Regenerate ⇒ Dimension ⇒ Regenerate ⇒ Modify ⇒ Regenerate ⇒ Done ⇒ Preview ⇒ OK ⇒ Done**

Sketch the section at the intersection of the *crosshairs*

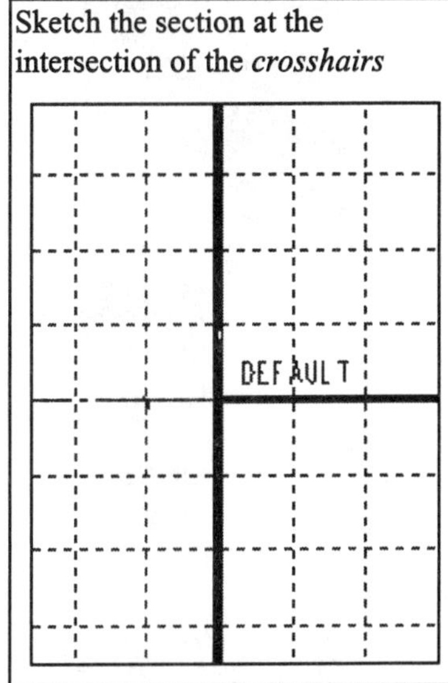

crosshairs

Section regenerated successfully.
Now sketch the cross-section.

Figure 11.26
Sketch the Section

HINT
Change the grid size:
Sketch ⇒ Sec Tools ⇒ Sec Environ ⇒ Grid ⇒ Params ⇒ X&Y Spacing ⇒ (type **.25** at prompt) ⇒ **enter**

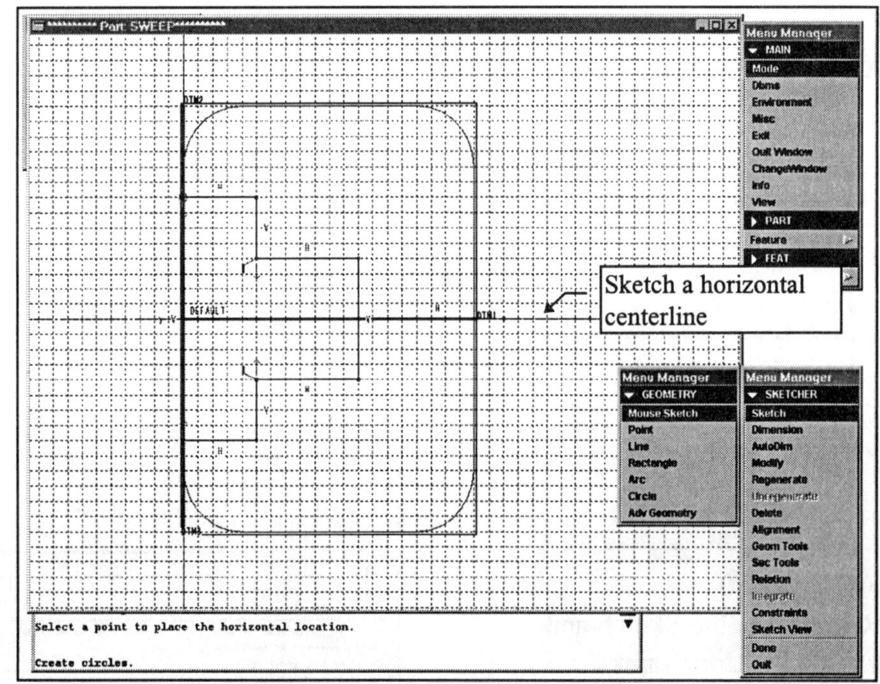

Figure 11.27
Sketch the Eight Lines of the Closed Section and a Centerline

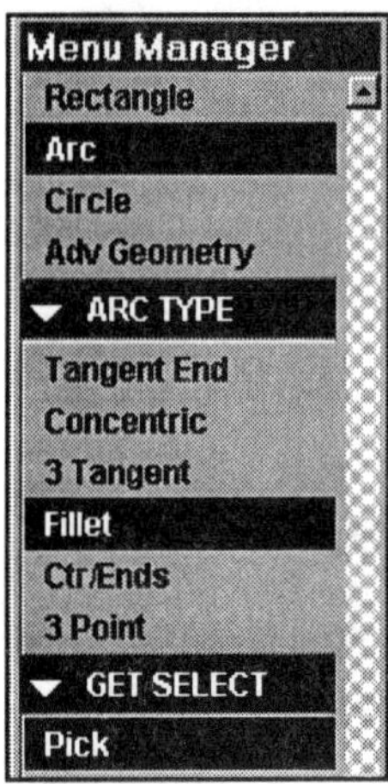

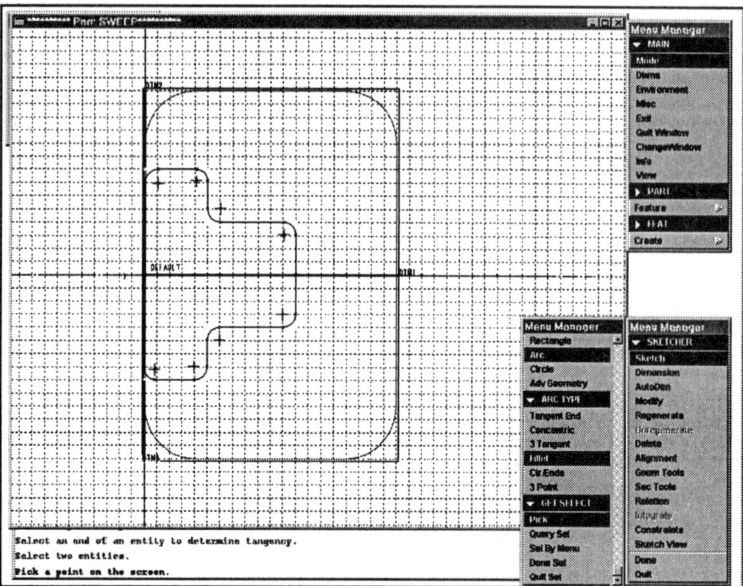

Figure 11.28
Sketch the Arc Fillets

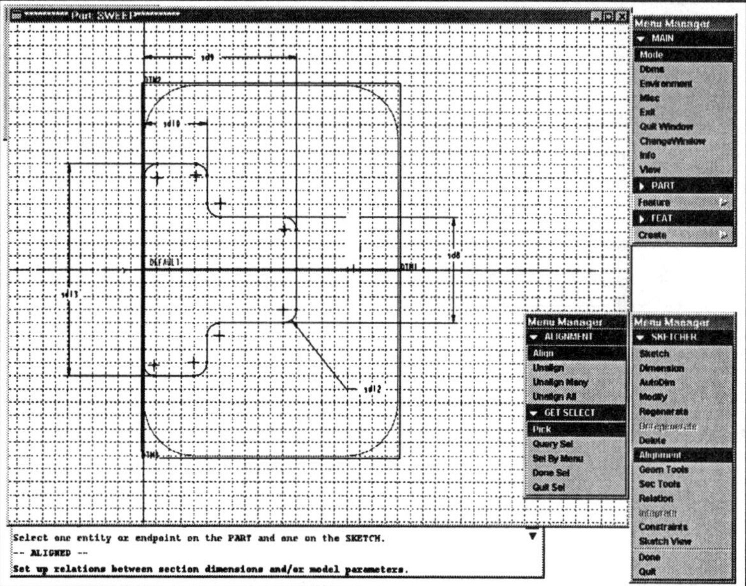

Figure 11.29
Alignment and **Dimension**

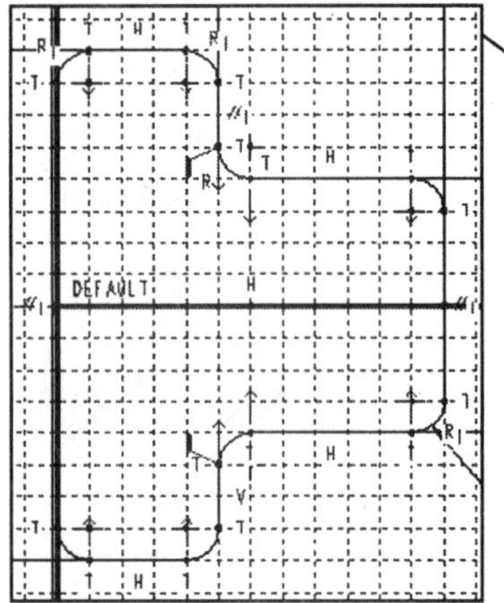

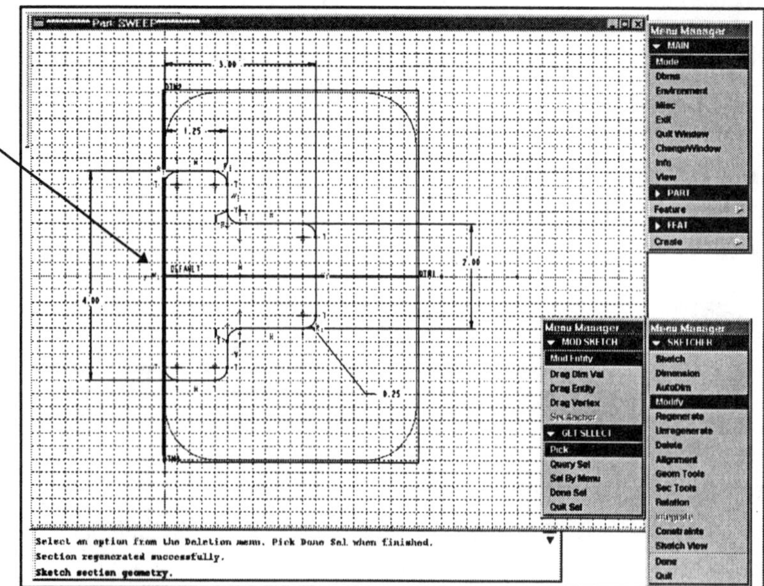

Figure 11.30
Regenerate the Sketch and Enable the Constraints

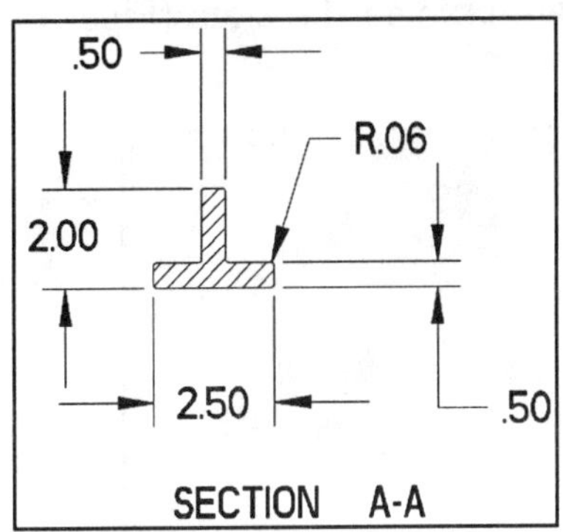

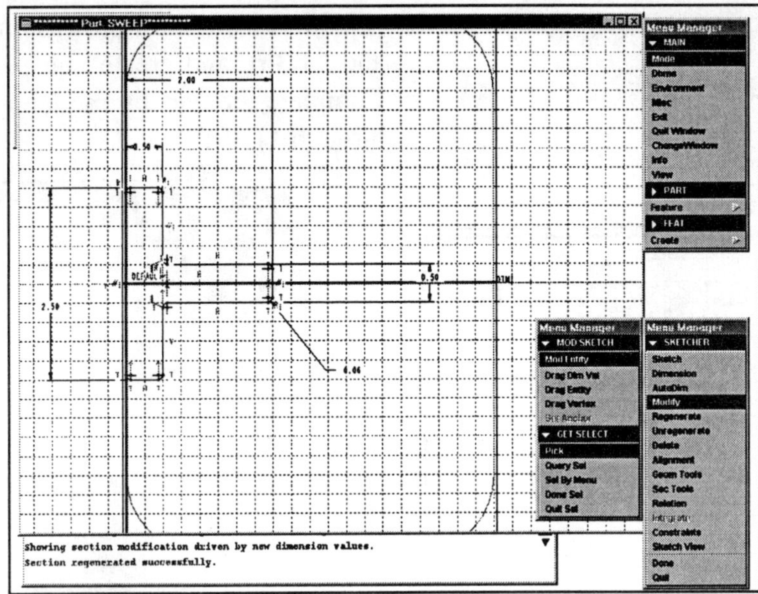

Figure 11.31
Modify and **Regenerate**

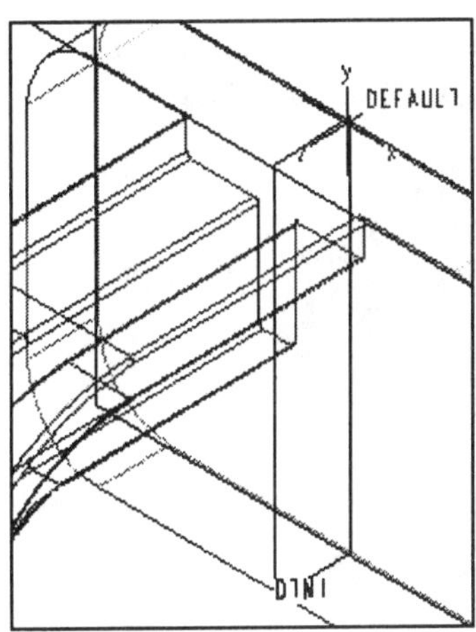

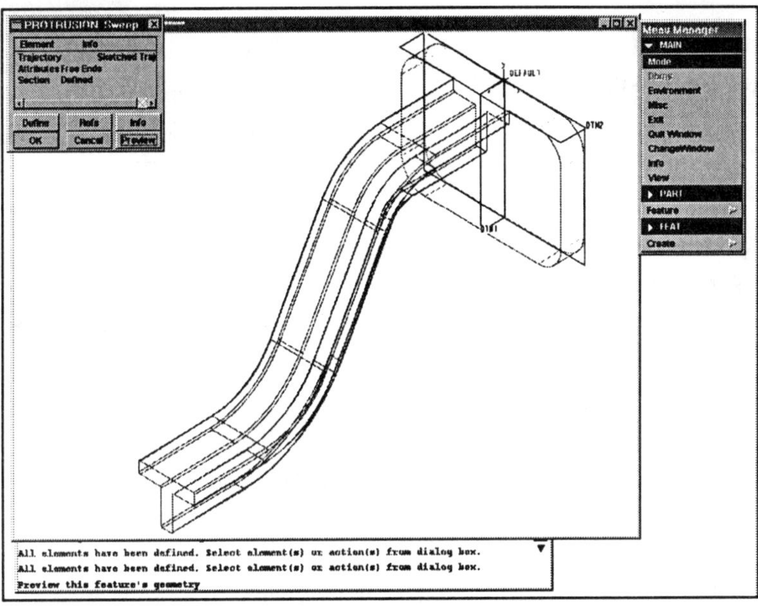

Figure 11.32
Preview

Dbms (File--PT/Modeler) ⇒
Save ⇒ **enter**
Purge ⇒ **enter** ⇒
Done-Return

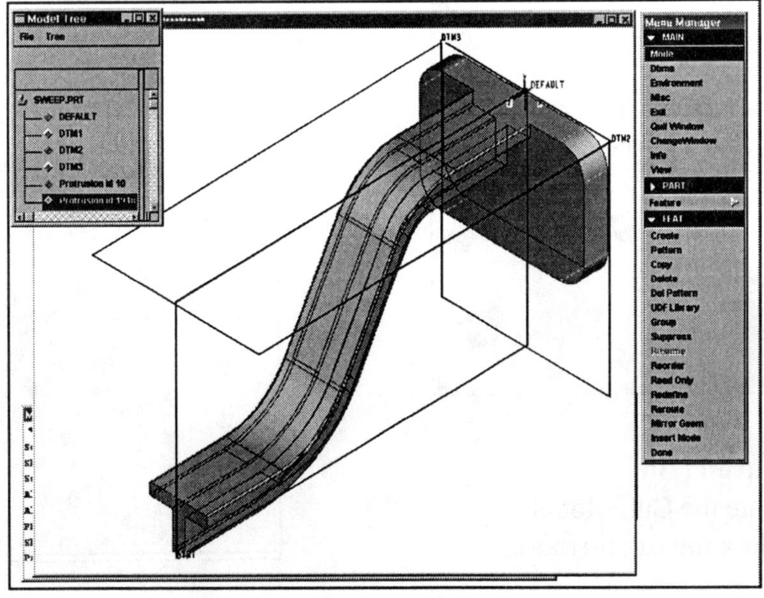

Figure 11.33
Completed Sweep

Model the third protrusion (Fig. 11.34), make the cuts (Figs. 11.35 and 11.36), and create and pattern the counterbore holes (Fig. 11.36). Complete the part by modeling the chamfers and the slots (Figs. 11.35 through 11.39).

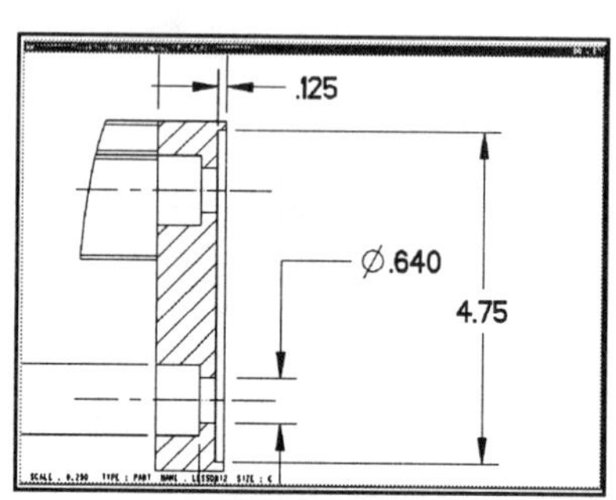

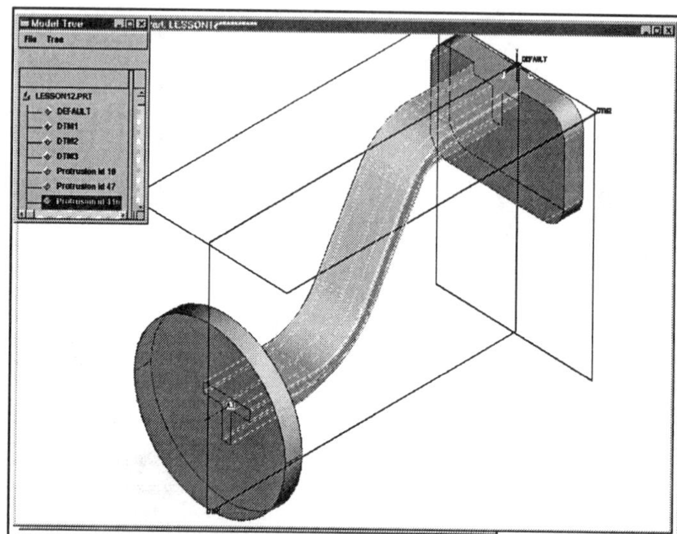

Figure 11.34
Third Protrusion

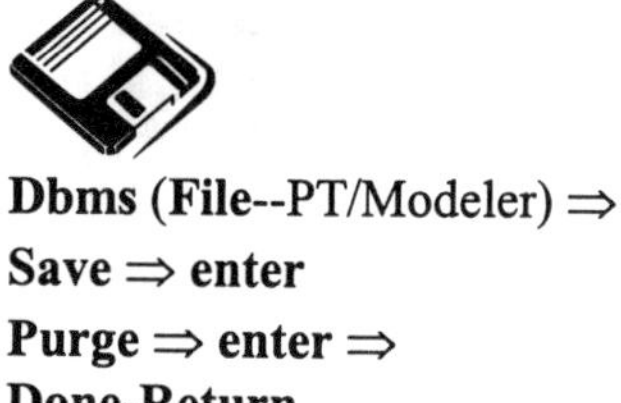

Dbms (File--PT/Modeler) ⇒
Save ⇒ **enter**
Purge ⇒ **enter** ⇒
Done-Return

Ø6.76 by **.250** deep cut

45° X .125 chamfer both sides

Figure 11.35
Add the Cuts and the Chamfers

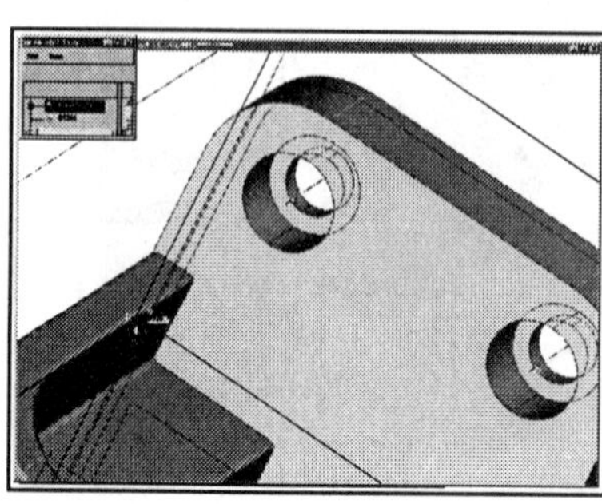

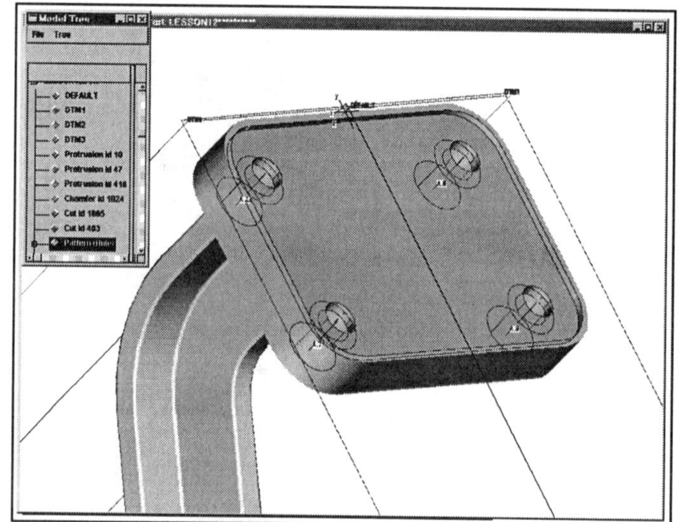

Figure 11.36
Create the Cut, Model and Pattern the Counterbores

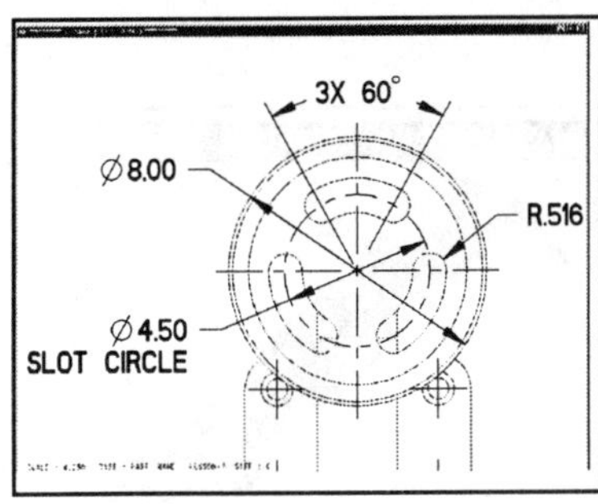

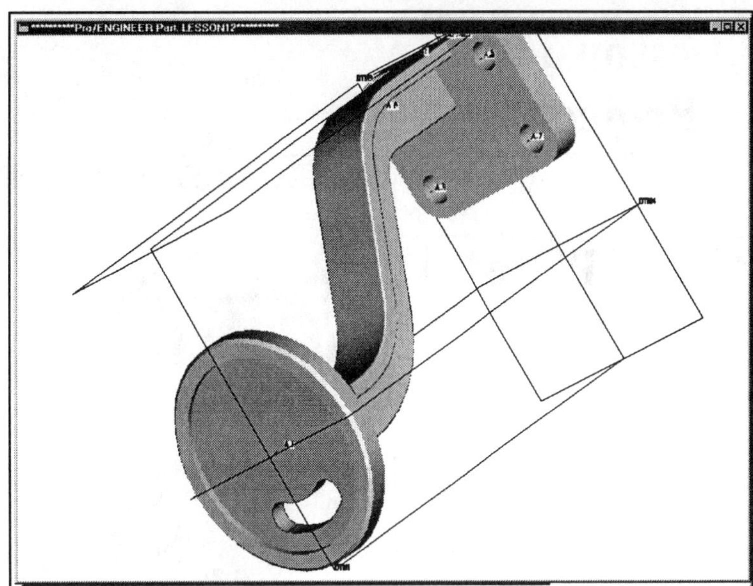

Figure 11.37
Create the Slot

NOTE

Use a datum-on-the-fly (**Make Datum**) when creating the first slot (see Lesson 7).

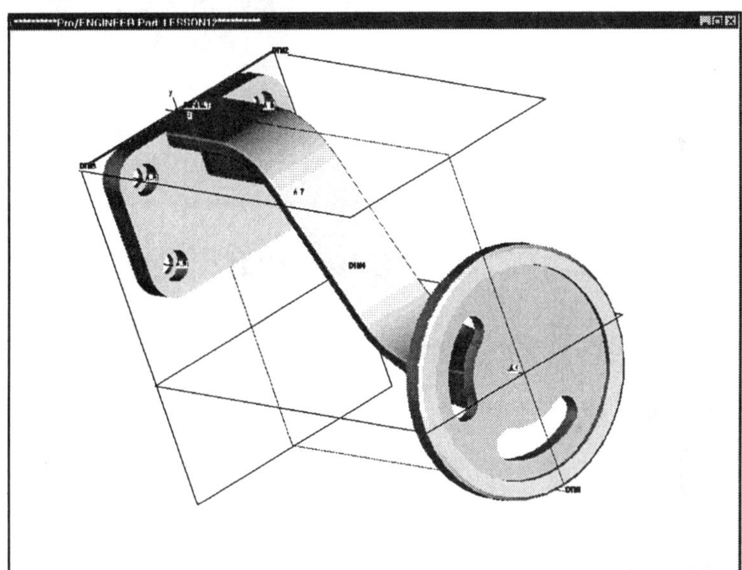

Figure 11.38
Pattern the Slots

Dbms (File--PT/Modeler) ⇒
Save ⇒ enter
Purge ⇒ enter ⇒
Done-Return

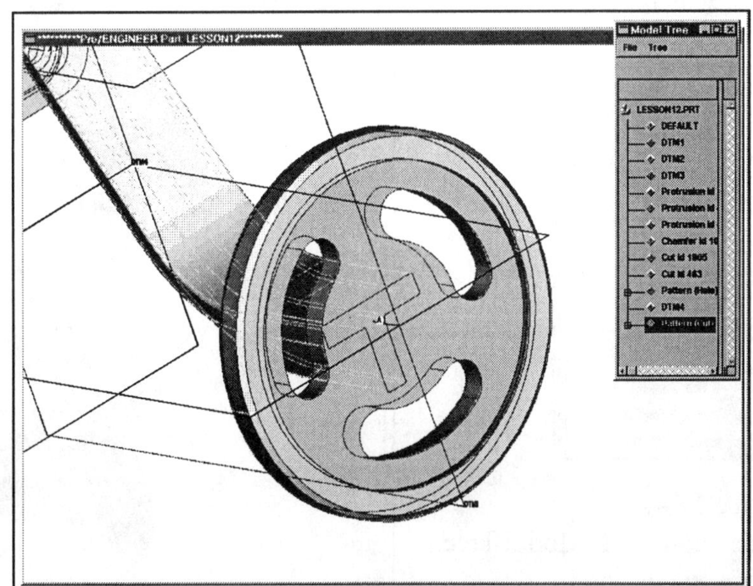

Figure 11.39
Completed Pattern and Part

Lesson 11 Project

Cover Plate

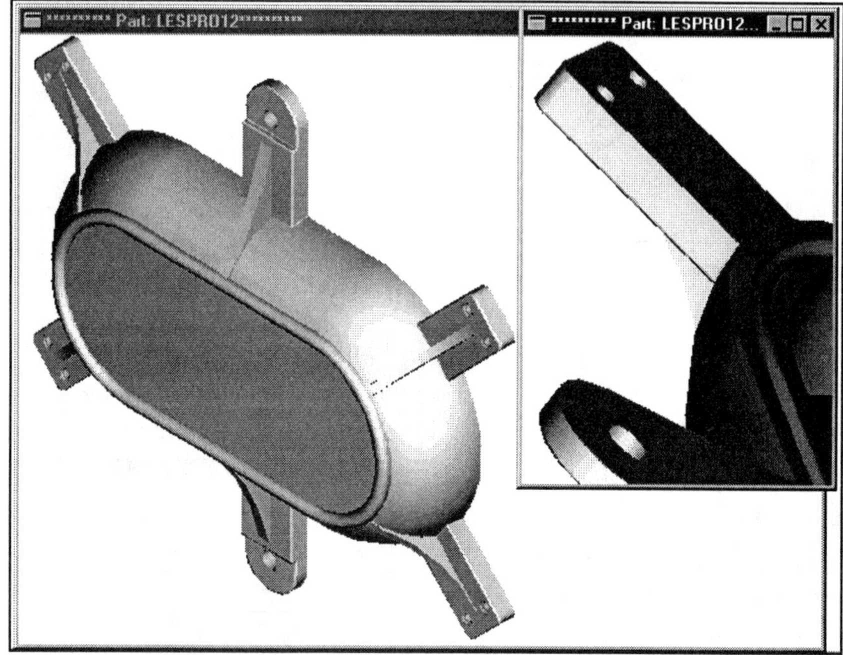

Figure 11.40
Cover Plate

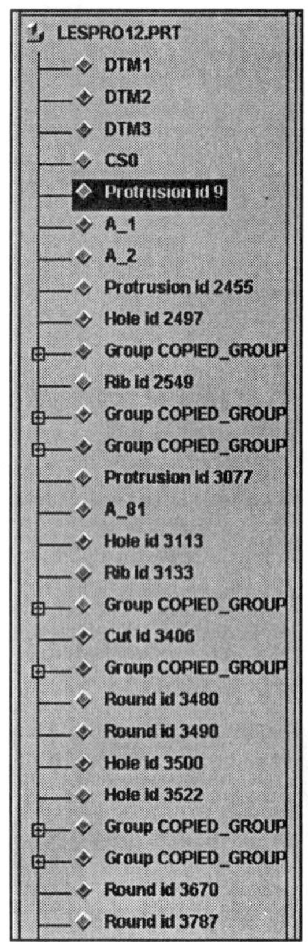

Figure 11.41
Cover Plate with Model Tree, Datum Planes, and Coordinate System

Cover Plate

The eleventh **lesson project** is a cast-iron part. Create the part shown in Figures 11.40 through 11.53. The sweep will have a *closed trajectory* with *inner faces included.* Analyze the part and plan out the steps and features required to model it. Use the DIPS in Appendix D to plan out the feature creation sequence and the parent-child relationships for the part. Add rounds on all nonmachined edges.

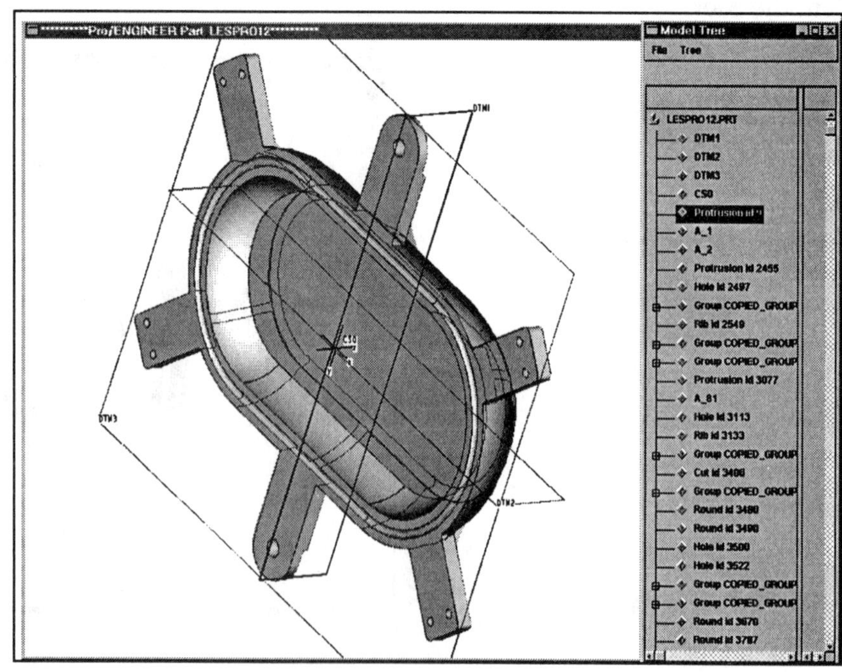

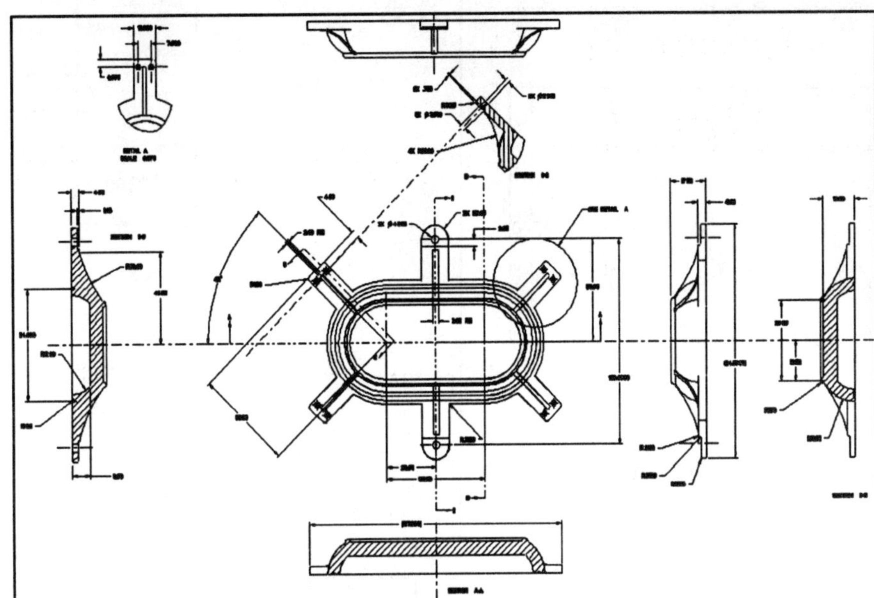

Figure 11.42
Cover Plate Detail Drawing, Sheet One

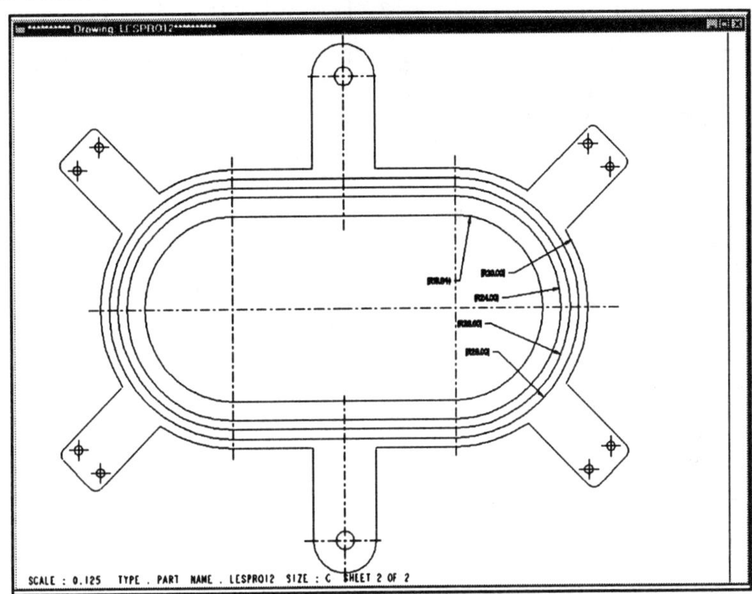

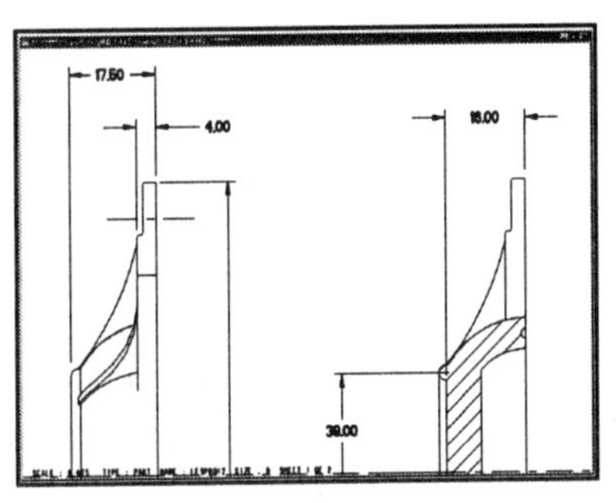

Figure 11.43
Cover Plate Drawing, Sheet Two, Bottom View

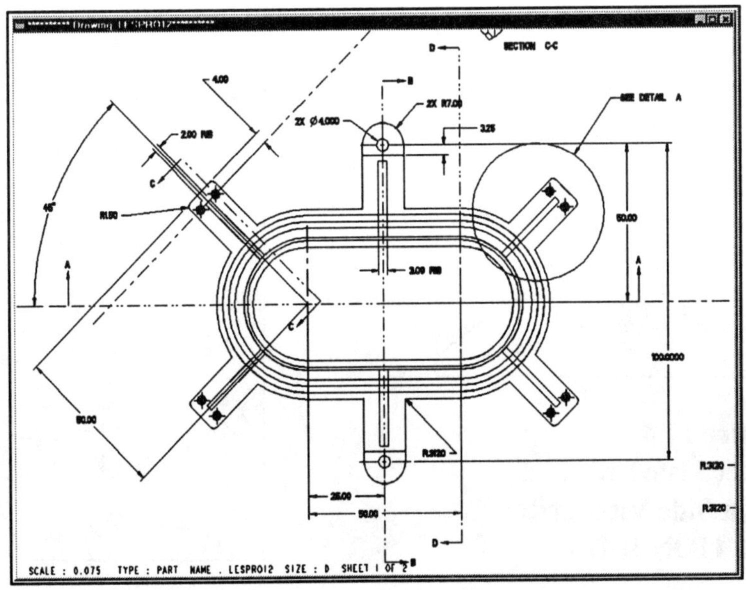

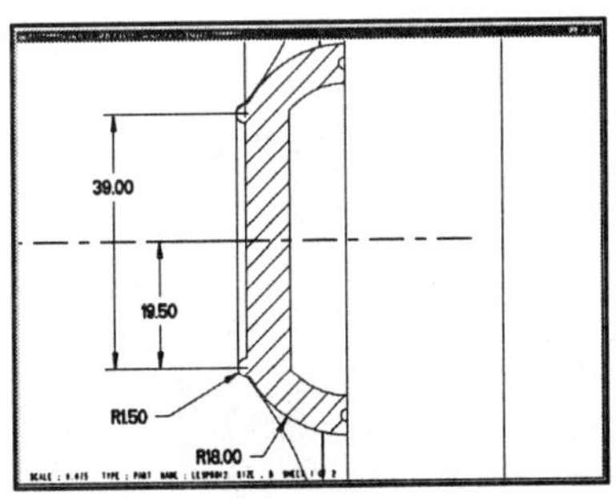

Figure 11.44
Cover Plate Drawing, Top View

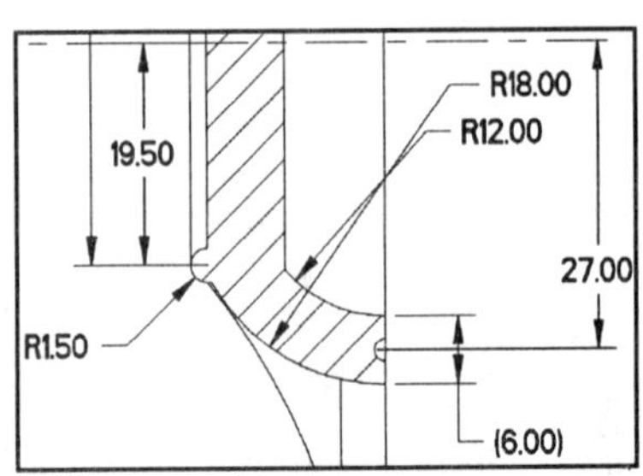

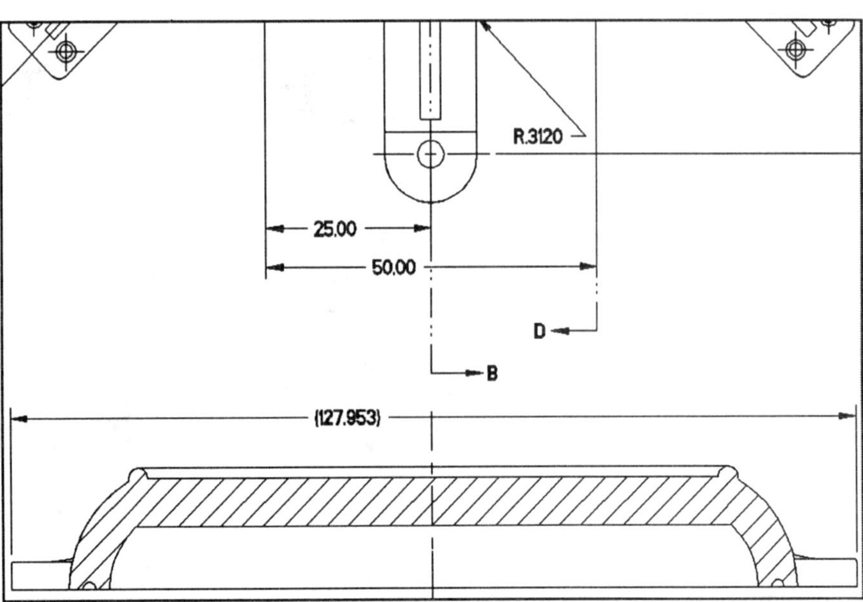

Figure 11.45
Cover Plate Drawing,
Front View,
SECTION A-A

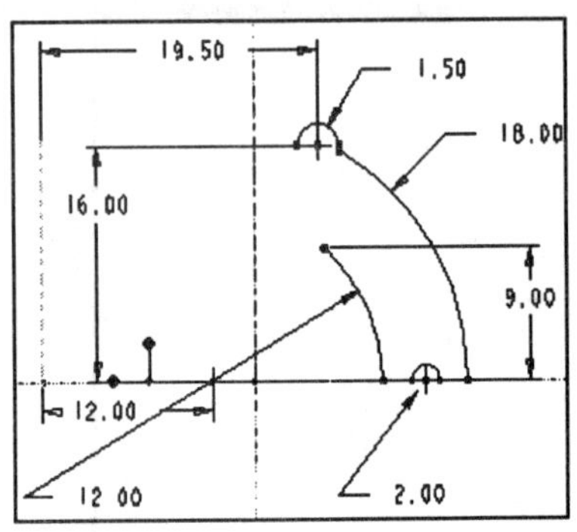

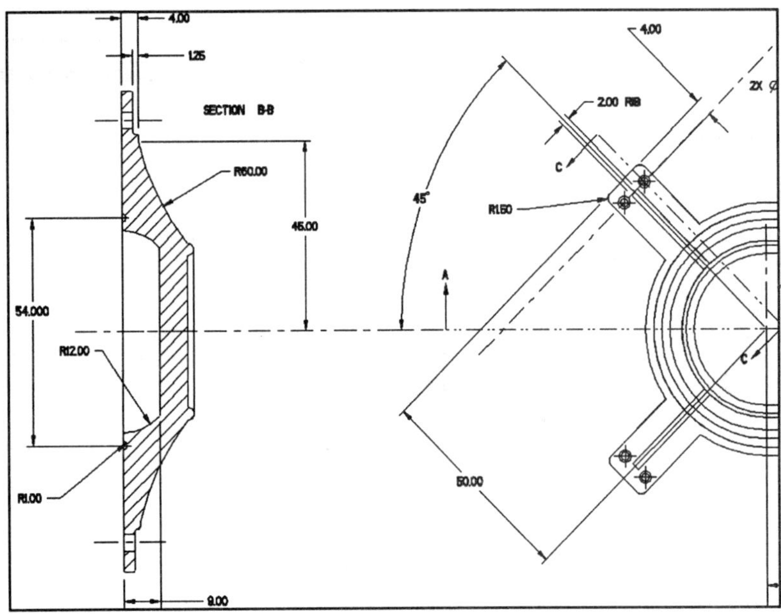

Figure 11.46
Cover Plate Drawing,
Left Side View,
SECTION B-B

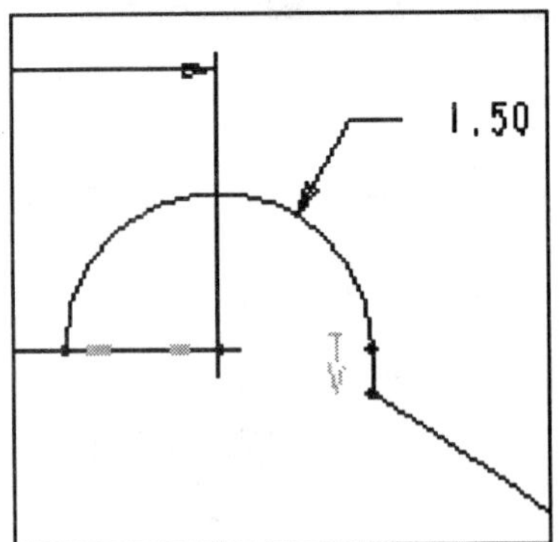

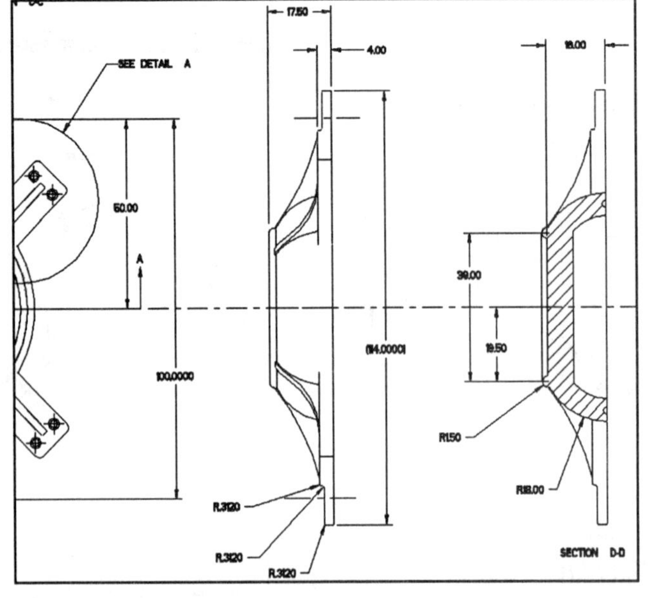

Figure 11.47
Cover Plate Drawing,
Right Side View and
SECTION D-D

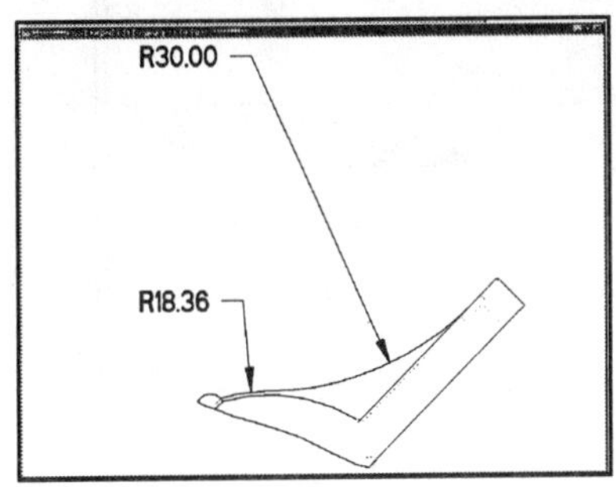

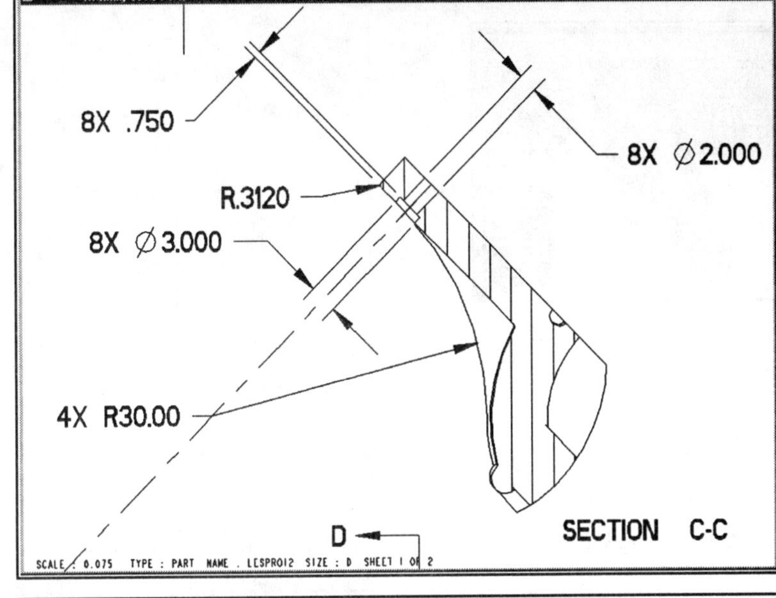

Figure 11.48
Cover Plate Drawing,
SECTION C-C

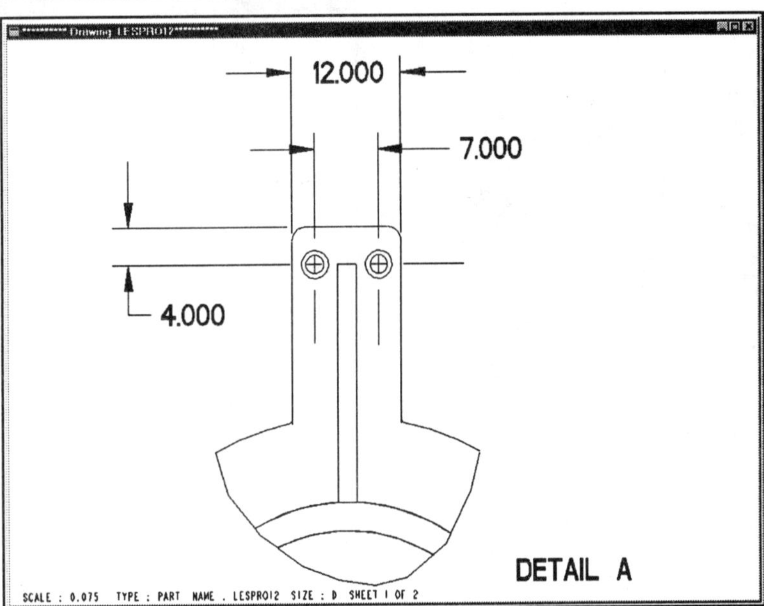

Figure 11.49
Cover Plate Drawing,
DETAIL A

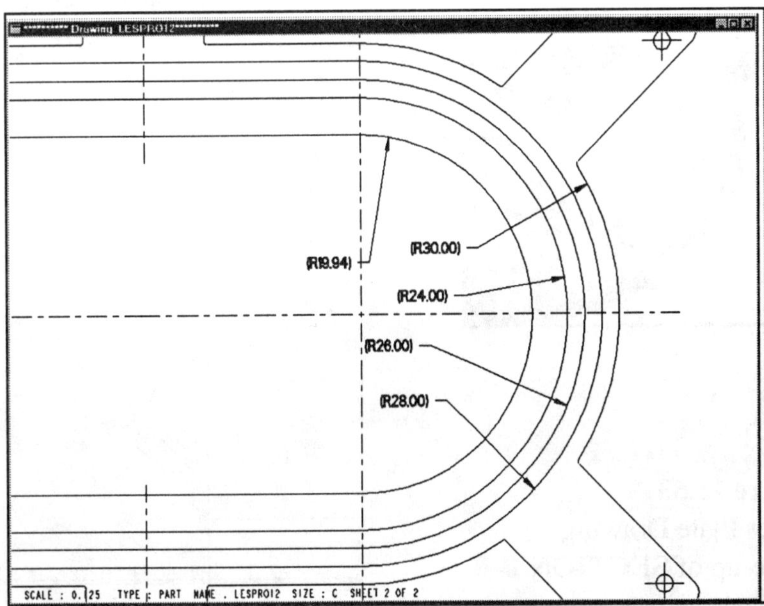

Figure 11.50
Cover Plate Drawing,
Sheet Two, Bottom View

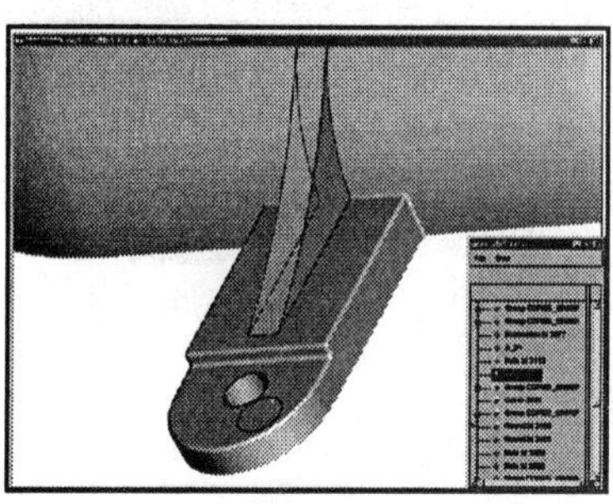

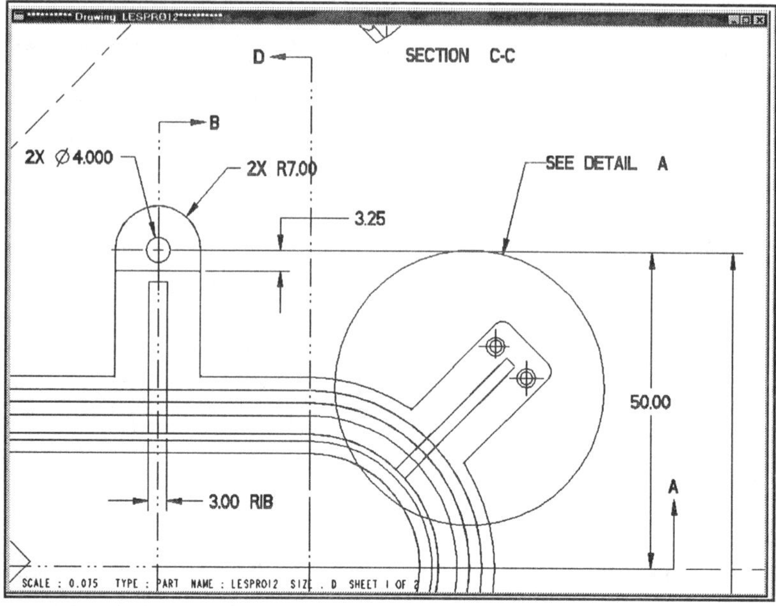

Figure 11.51
Cover Plate Drawing, Close-up of Top Right

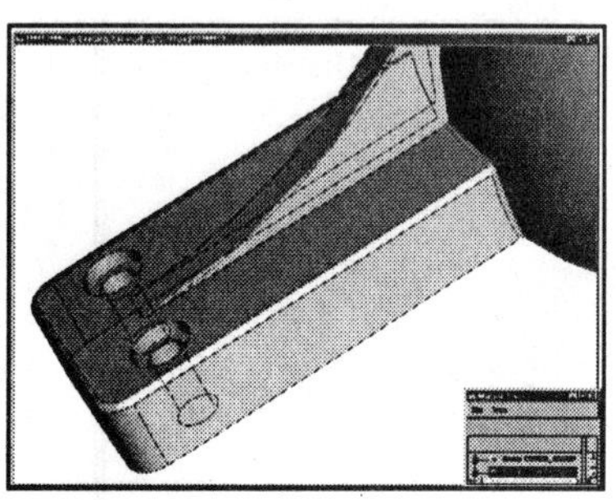

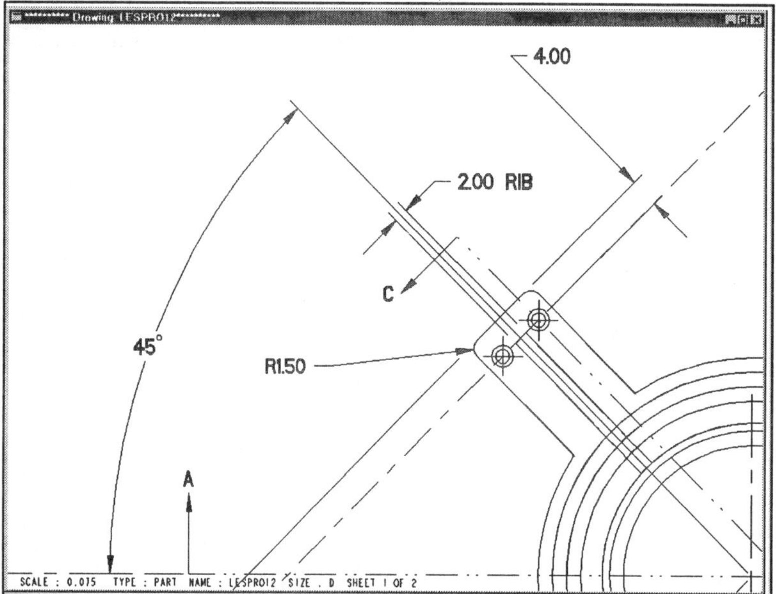

Figure 11.52
Cover Plate Drawing, Close-up of Top Left

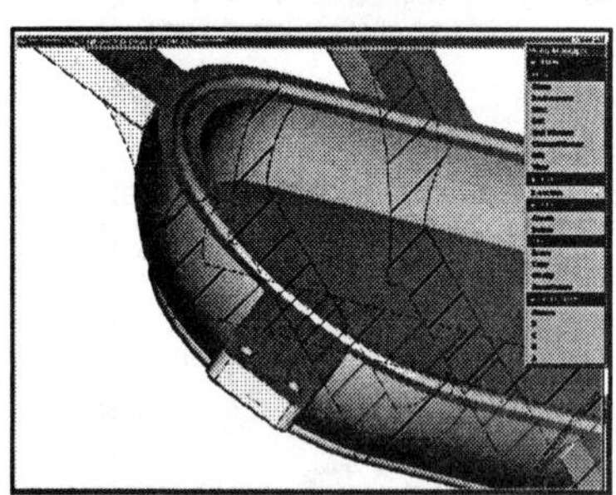

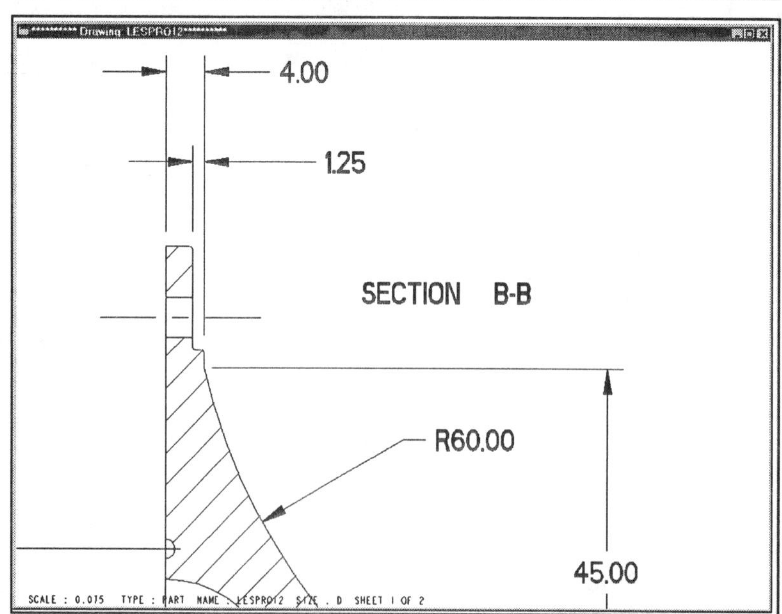

Figure 11.53
Cover Plate Drawing, Close-up of **SECTION B-B**

Lesson 12

Blends and Splines

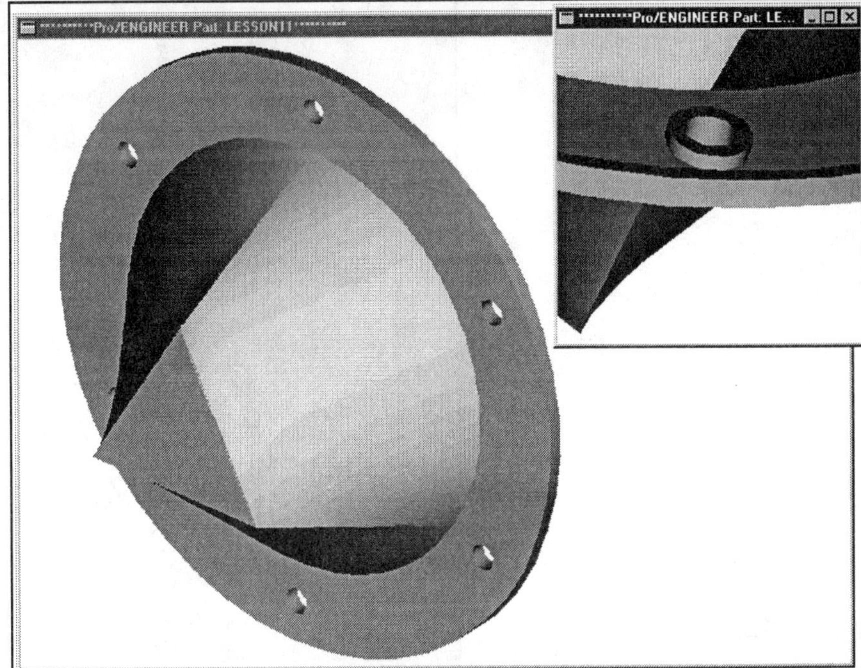

Figure 12.1
Cap

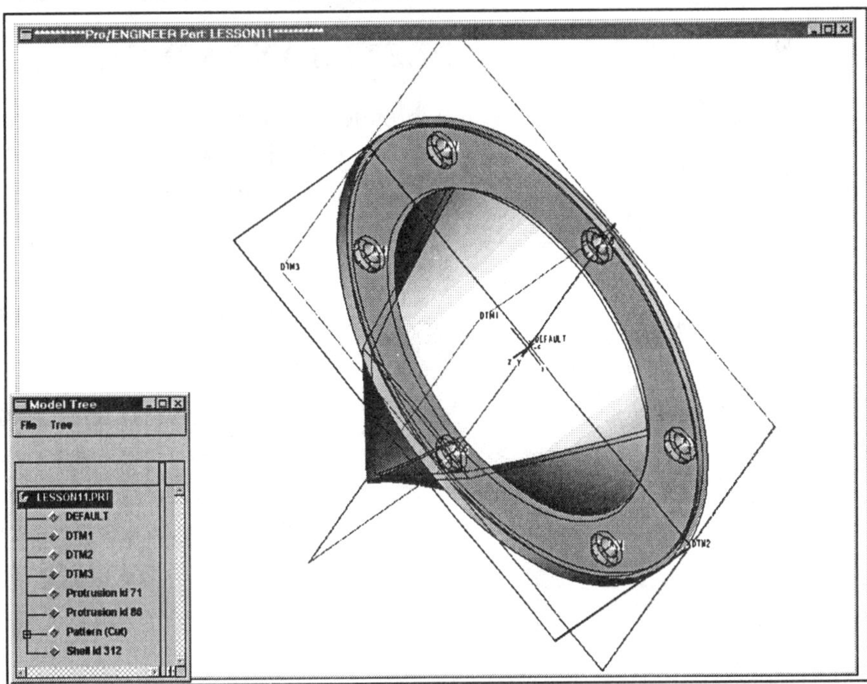

Figure 12.2
Cap with Datum Planes Displayed

✔ ***EGD REFERENCE***
Engineering Graphics and Design with Graphical Analysis *or* **Fundamentals of Engineering Graphics and Design**
by L. Lamit and K. Kitto
Read Chapter 12
See pages 364-366

COAch™ for Pro/ENGINEER

If you have **COAch for Pro/ENGINEER** on your system, go to SEARCH and do the Segment shown in Figure 12.5.

OBJECTIVES

1. **Create a parallel blend feature**
2. **Shell a blend feature**
3. **Create a spline and use it in a swept blend**
4. **Create a swept blend feature**
5. **Create sections in Part mode**

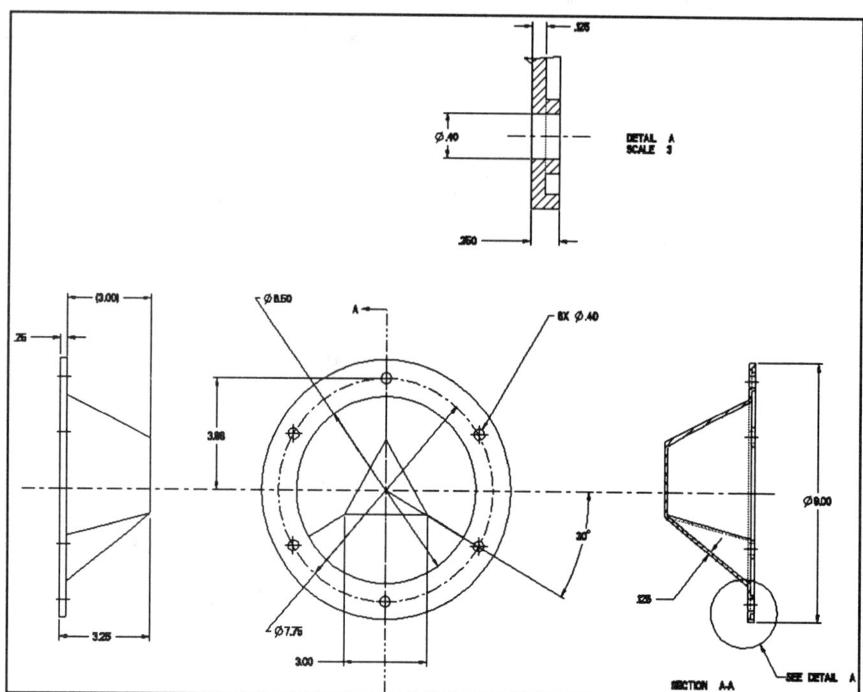

Figure 12.3
Cap Detail

BLENDS AND SPLINES

A blended feature consists of a series of at least two planar sections that are joined together at their edges with transitional surfaces to form a continuous feature. The Cap in Figures 12.1 through 12.3 uses a blend feature in its design. A **Blend** can be created as a **Parallel Blend** (Figs. 12.4 and 12.5), or you can construct a **Swept Blend**.

A **spline** is similar to an **irregular** curve and is used in a variety of industrial designs.

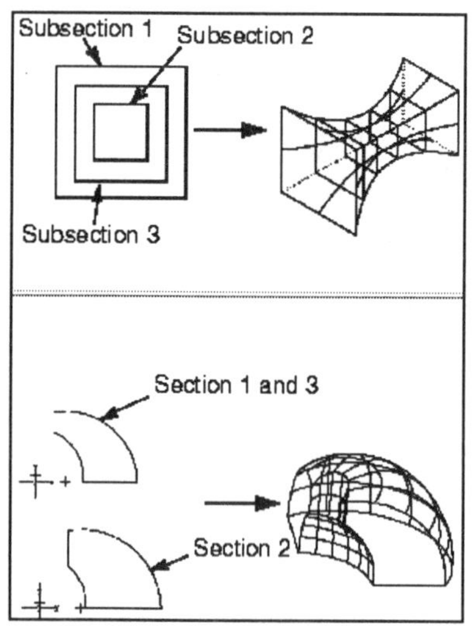

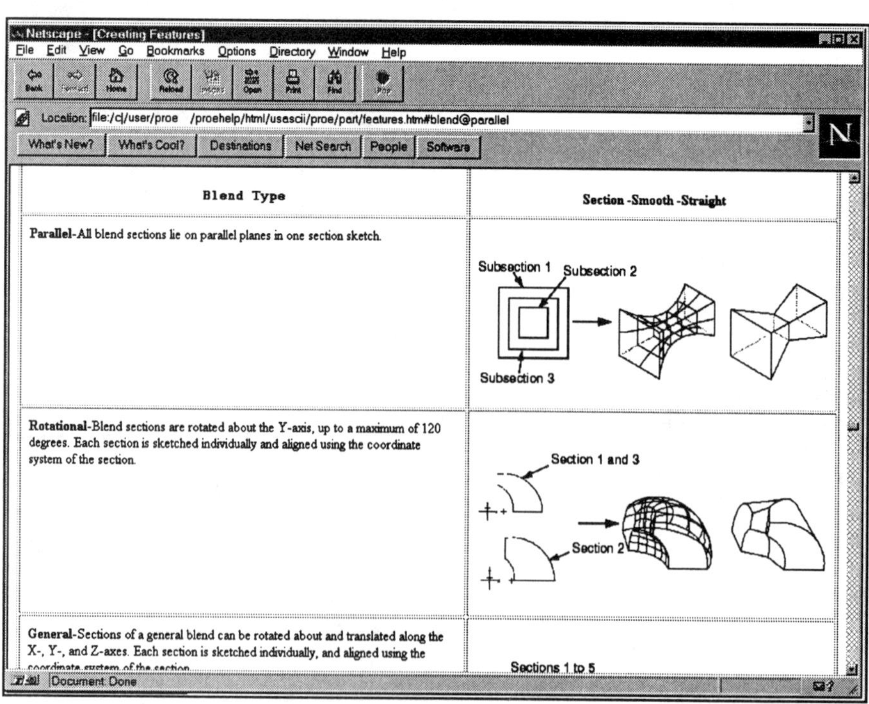

Figure 12.4
Online Documentation, Blends

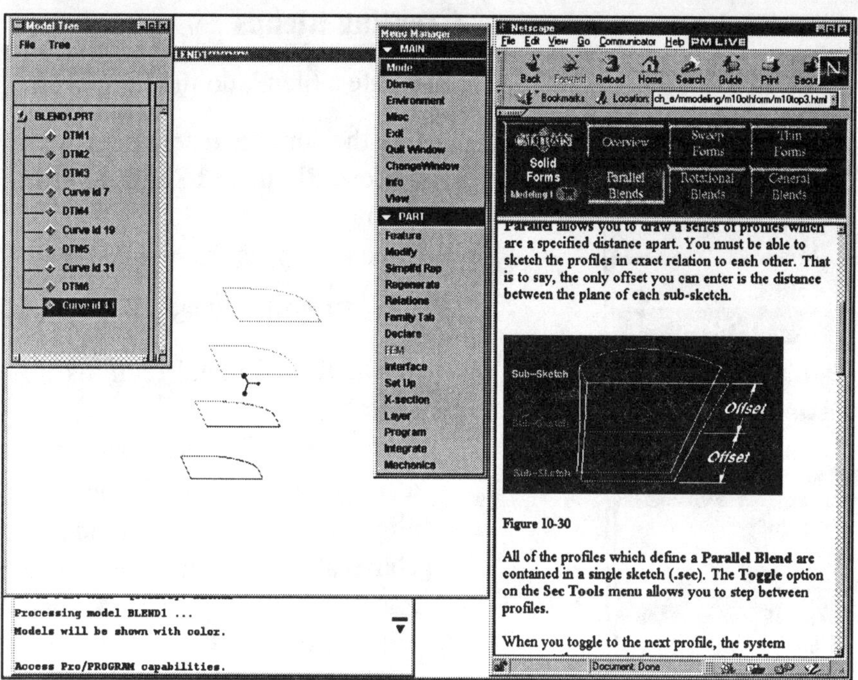

Figure 12.5
COAch for Pro/E, Solid Forms (Parallel Blends)

Blend Sections

Figure 12.5 shows a parallel blend for which the *section* consists of four *subsections*. Each segment in the subsection is matched with a segment in the following subsection; the blended surfaces are created between the corresponding segments.

HINT

For the most part, blends must have the same number of entities in each section. The only exception is a capped blend.

Starting Point of a Section

To create the transitional surfaces, Pro/E connects the *starting points* of the sections and continues to connect the vertices of the sections in a clockwise manner. By changing the starting point of a blend section, you can create blended surfaces that twist between the sections.

The default starting point is the first point sketched in the subsection. You can position the starting point to the endpoint of another segment by choosing the option **Start Point** from the SEC TOOLS menu and selecting the new position.

Smooth and Straight Attributes

Blends use one of the following transitional surface ATTRIBUTES menu options:

Straight Create a straight blend by connecting vertices of different subsections with straight lines. Edges of the sections are connected with ruled surfaces.

Smooth Create a smooth blend by connecting vertices of different subsections with smooth curves. Edges of the sections are connected with ruled (spline) surfaces.

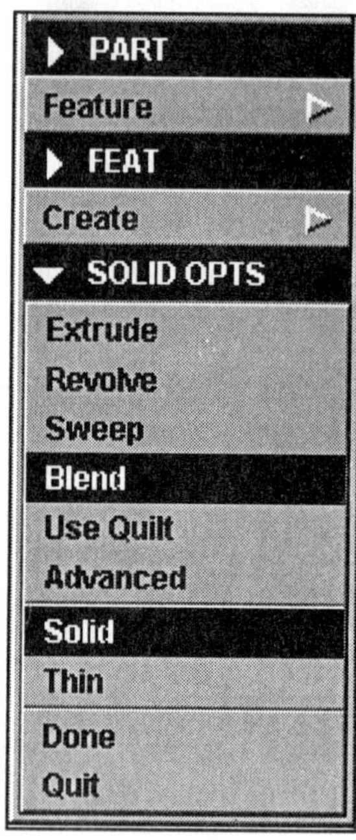

Creating Blends

To create a blend, do the following:

1. Use the command sequence **Feature**, **Create**, **Solid**, **Protrusion**.
2. Choose **Blend** and **Solid** or **Thin** from the SOLID OPTS menu, then **Done**.
3. Choose options from the BLEND OPTS menu, then **Done**.

The BLEND OPTS menu options are as follows:

Parallel All blend sections lie on parallel planes in one section sketch.
Rotational The blend sections are rotated about the **Y** axis, up to a maximum of **120°**. Each section is sketched individually and aligned using the coordinate system of the section.
General The sections of a general blend can be rotated about and translated along the **X**, **Y**, and **Z** axes. Sections are sketched individually and aligned using the coordinate system of the section.
Regular Sec The feature will use the regular sketching plane.
Project Sec The feature will use the projection of the section on the selected surface. This is used for parallel blends only.
Select Sec Select section entities (not available for parallel blends).
Sketch Sec Sketch section entities.

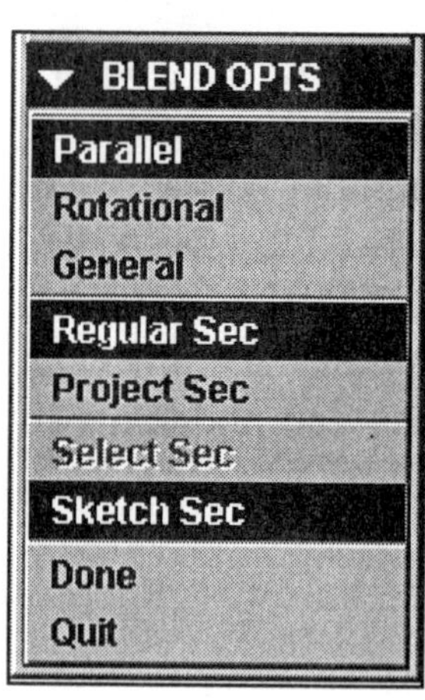

Parallel Blends

You create parallel blends (Fig. 12.6) using the **Parallel** option in the BLENDS OPTS menu. A parallel blend is created from a single section (Fig. 12.7) that contains multiple sketches, called *subsections*. A first or last subsection can be defined as a point or a blend vertex.

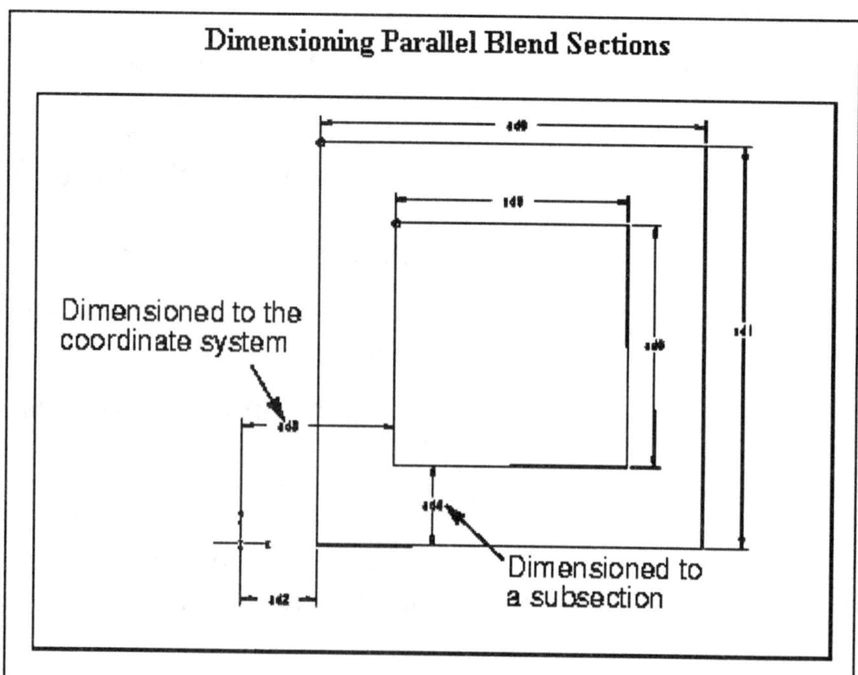

Figure 12.6
Online Documentation, Parallel Blends

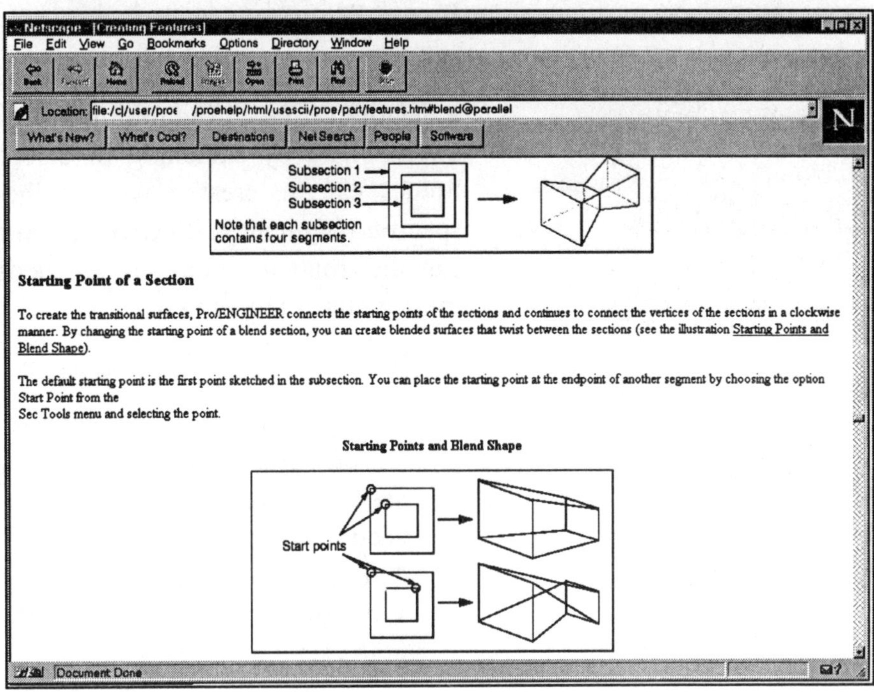

Starting Point of a Section

To create the transitional surfaces, Pro/ENGINEER connects the starting points of the sections and continues to connect the vertices of the sections in a clockwise manner. By changing the starting point of a blend section, you can create blended surfaces that twist between the sections (see the illustration Starting Points and Blend Shape).

The default starting point is the first point sketched in the subsection. You can place the starting point at the endpoint of another segment by choosing the option Start Point from the
Sec Tools menu and selecting the point.

Starting Points and Blend Shape

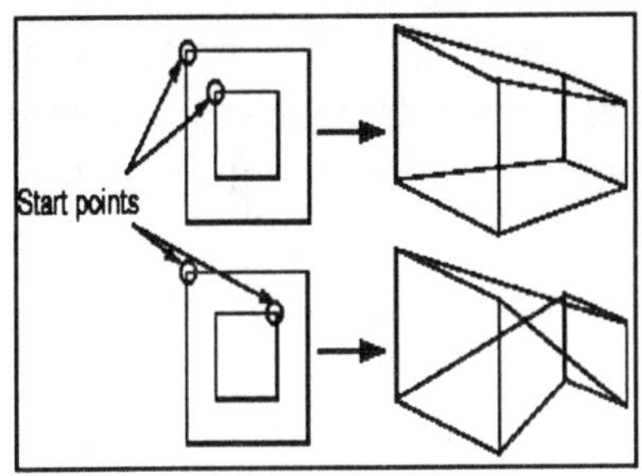

Figure 12.7
Online Documentation, Blend Sections

Creating a Parallel Blend

To create a parallel blend, do the following:

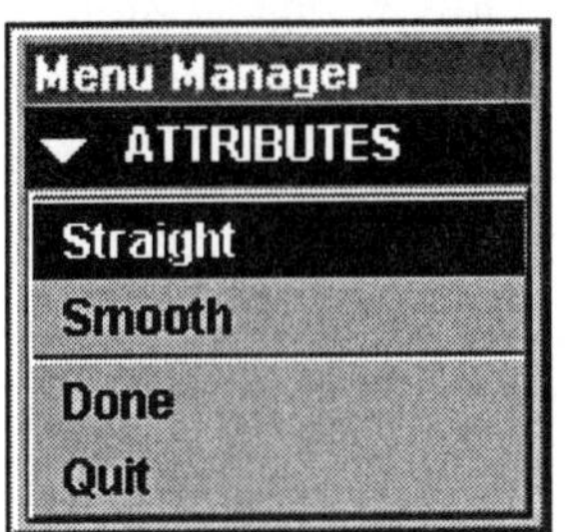

1. When you choose **Done** from the BLEND OPTS menu, Pro/E displays the feature creation dialog box and the ATTRIBUTES menu. Choose either **Straight** or **Smooth**.
2. Create the first subsection using the Sketcher. You determine the direction of feature creation as you set up the sketching plane. Dimension and regenerate each subsection sketch to ensure the validity of the dimensioning scheme. A parallel blend requires more than one subsection, so after successfully regenerating this section, choose **Sec Tools** from the SKETCHER menu.
3. Choose **Toggle** from the SEC TOOLS menu. The first subsection turns gray and becomes inactive.
4. Choose **Sketch** and sketch the second subsection. Make sure its starting point corresponds to the starting point of the first subsection in the manner that you intend. Dimension and regenerate it.
5. If you are sketching more than two subsections, choose **Toggle** repeatedly until all the current geometry is gray, then sketch the subsection. Repeat this step until all subsections have been sketched.
6. To modify an existing subsection, toggle through until the subsection you want is active. While you can place or move the starting point of a subsection only when it is active, you can modify the dimensions of any subsection at any time.
7. When you have sketched all the subsections and regenerated, choose **Done** from the SKETCHER menu. When prompted, enter the distances between each of the subsections.
8. Select the **OK** button to create the feature.

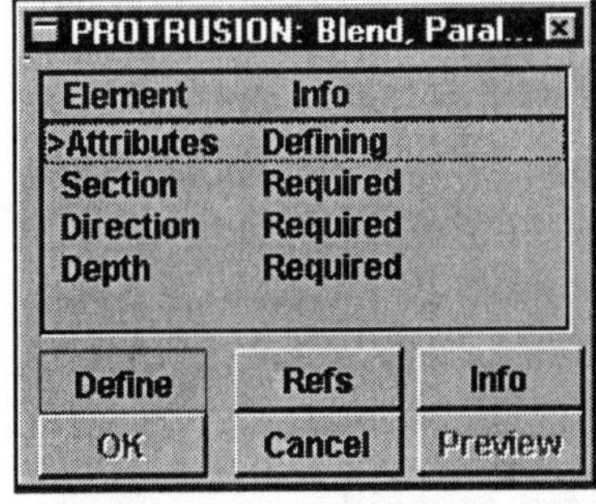

Swept Blends

A swept blend (Figs. 12.8 and 12.9) is created using a single trajectory (a spine) and multiple sections. You create the spine of the swept blend by sketching or selecting a datum curve or an edge. Spines can be created with splines. You sketch the sections at specified segment vertices or datum points on the spine. Each section can be rotated about the **Z** axis with respect to the section immediately preceding it.

Note the following restrictions:

* A section cannot be located at a sharp corner in the spine.
* For a closed trajectory profile, sections must be sketched at the start point and at least one other location. Pro/E uses the first section at the endpoint.
* For an open trajectory profile, you must create sections at the start point and endpoint. You cannot skip placement of a section at those points.
* Sections cannot be dimensioned to the model, because modifying the trajectory would invalidate those dimensions.
* A composite datum curve cannot be selected for defining sections of a swept blend (**Select Sec**). Instead, you must select one of the underlying datum curves or edges for which a composite curve is determined.
* If you choose **Pivot Dir** and **Select Sec**, all selected sections must lie in planes that are parallel to the pivot direction.
* You cannot use a nonplanar datum curve from an equation as a swept blend trajectory.

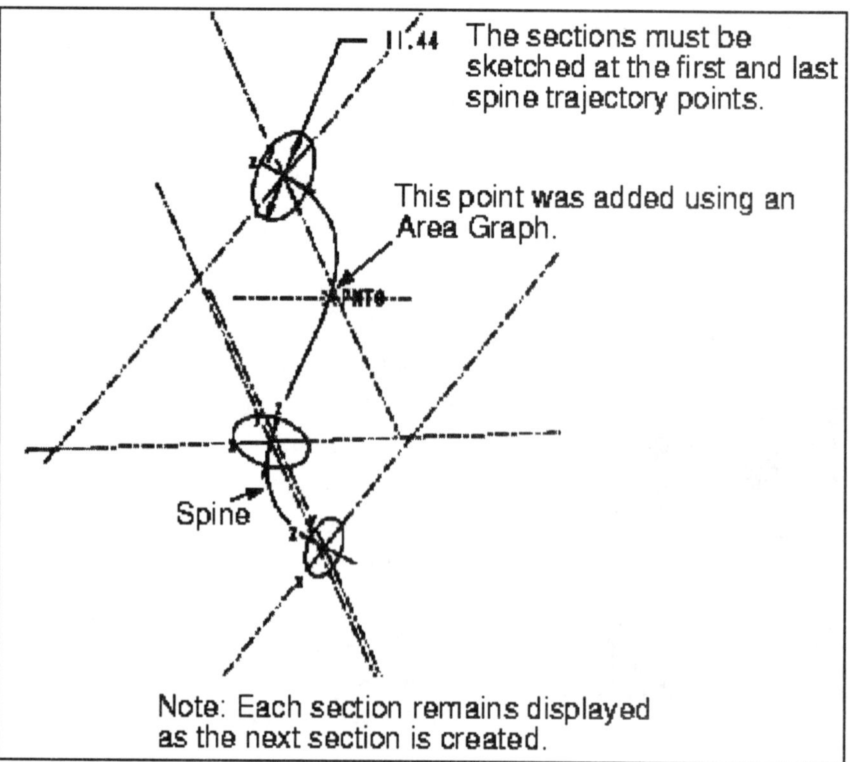

Figure 12.8
Online Documentation, Creating a Swept Blend

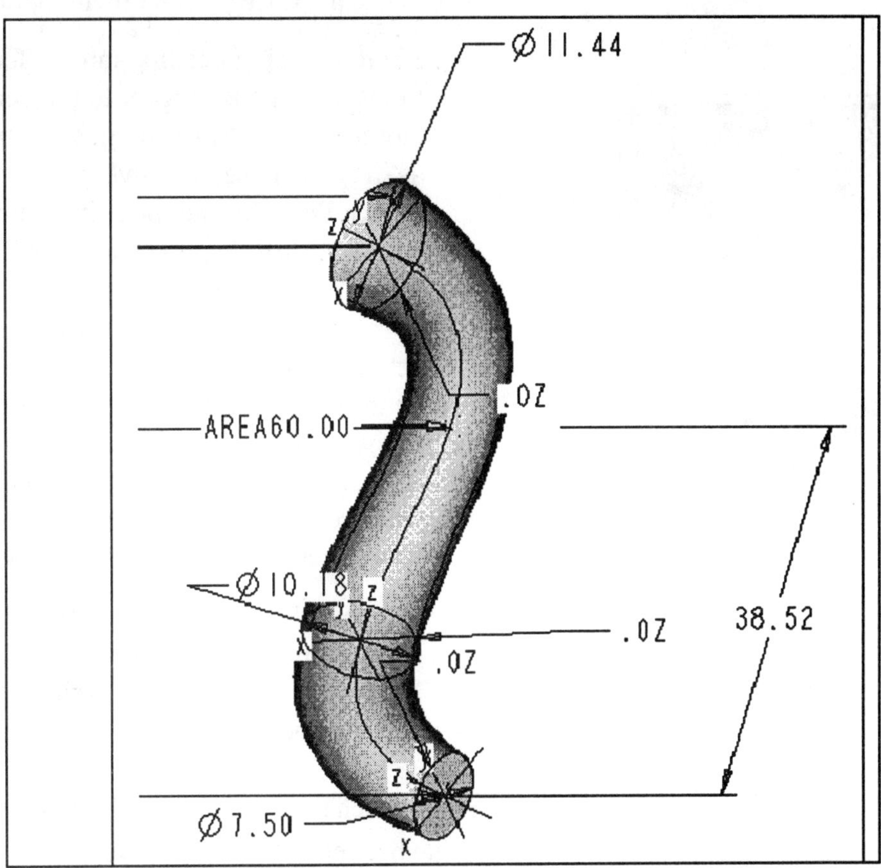

Figure 12.9
Online Documentation, Swept Blends

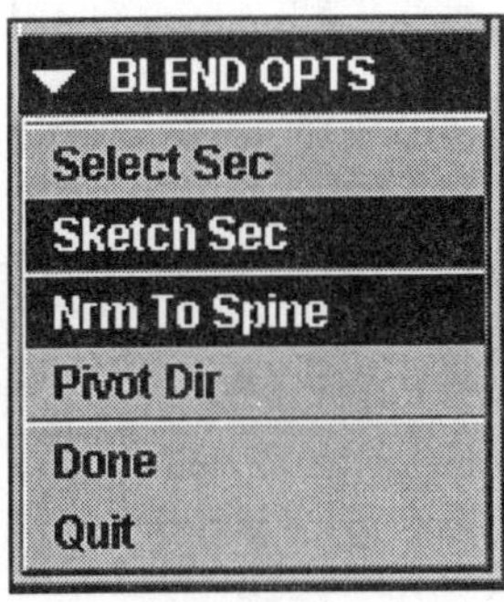

Creating a Swept Blend

To create a **Swept Blend** (Figs. 12.9 and 12.10), you can define the trajectory by sketching a trajectory or by selecting existing curves and edges and extending or trimming the first and last entity in the trajectory. Use the following procedure:

1. Choose **Advanced** from the SOLID OPTS menu and **Swept Blend** and **Done** from the ADV FEAT OPT menu.
2. Choose the desired options from the BLEND OPTS menu, presented in mutually exclusive pairs, then choose **Done** from the BLEND OPTS menu. The possible options are as follows:

> **Select Sec** Select existing curves or edges to define each section, using the CRV SKETCHER menu.
> **Sketch Sec** Sketch new section entities to define each section.
> **Nrm To Spine** The section plane remains normal to the spine trajectory.
> **Pivot Dir** The section plane will remain normal to the spine trajectory when viewed from a pivot direction. As the section plane sweeps along the trajectory, it always remains parallel to the pivot direction.

3. Choose a SWEEP TRAJ menu option. The options include:

> **Sketch Traj** Sketch a spine. The spine can have sharp corners (a discontinuous tangent to the curve), except at the endpoint of a closed curve. At nontangent vertices, Pro/E mitres the geometry as in constant section sweeps.
> **Select Traj** Define the spine trajectory using existing curves and edges. Pro/E displays the CHAIN menu. Choose **Select**, define the chain, then choose **Done**.

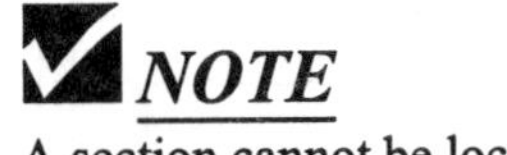

A section cannot be located at a sharp corner in the spine.

4. Sketch or select the trajectory of the spine.
5. Use the CONFIRM menu options to choose the points at which to define any additional sections. As appropriate to the defined trajectory, those points may include endpoints of spine entities, spline entity control points, and any existing datum points on spine entities (if you select a spine). The CONFIRM menu options are as follows:

> **Accept** Sketch or select a section at this highlighted location.
> **Next** Bypass this highlighted location and go to the next point.
> **Previous** Bypass this highlighted location and return to the previous point.

6. For each section, specify the rotation angle about the **Z** axis (with a value between **-120°** and **+120°**).
7. Select or sketch the entities for each section, depending on whether you choose **Select Sec** or **Sketch Sec**, respectively. Choose **Done** from the SKETCHER menu.
8. When all sections have been sketched or selected, unless you chose **Area Graph** or **Blend Control** element, select the **OK** button in the dialog box to generate the swept blend feature.

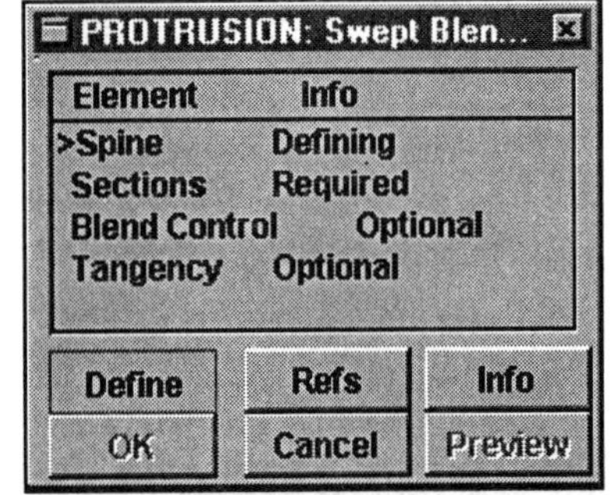

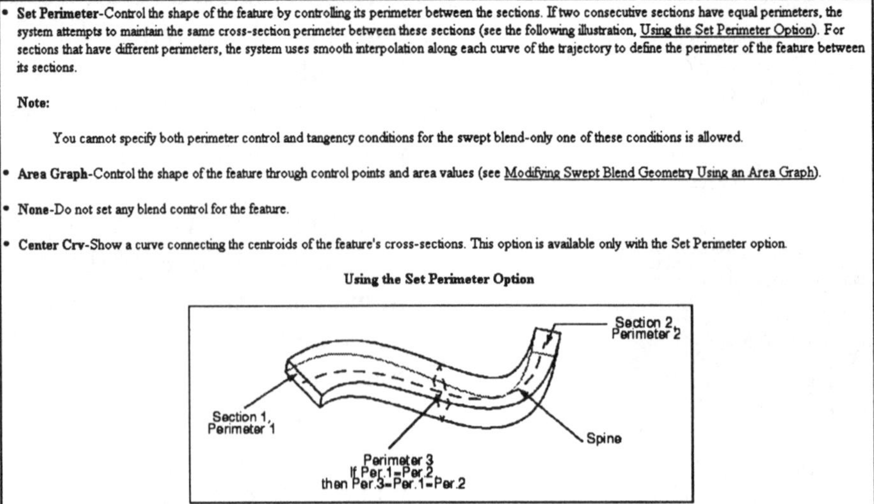

- **Set Perimeter**-Control the shape of the feature by controlling its perimeter between the sections. If two consecutive sections have equal perimeters, the system attempts to maintain the same cross-section perimeter between these sections (see the following illustration, Using the Set Perimeter Option). For sections that have different perimeters, the system uses smooth interpolation along each curve of the trajectory to define the perimeter of the feature between its sections.

  Note:

  You cannot specify both perimeter control and tangency conditions for the swept blend-only one of these conditions is allowed.

- **Area Graph**-Control the shape of the feature through control points and area values (see Modifying Swept Blend Geometry Using an Area Graph).
- **None**-Do not set any blend control for the feature.
- **Center Crv**-Show a curve connecting the centroids of the feature's cross-sections. This option is available only with the Set Perimeter option.

Using the Set Perimeter Option

Figure 12.10
Online Documentation, Spines, Set Perimeter Option

Splines

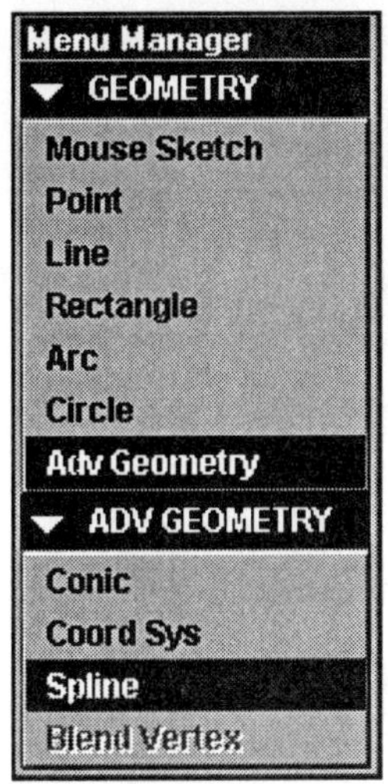

Sketching a **spline** is similar to drawing an **irregular** curve (Fig. 12.11). Splines (Fig. 12.12) are created by picking or sketching a series of specific points.

To create a spline:

1. Choose **Spline** from the ADV GEOMETRY menu.
2. The SPLINE MODE menu appears, with the following options:

   Sketch Points Create a spline by picking screen points for the spline to pass through.
   Select Points Create a spline by selecting existing Sketcher points. Once the point has been selected, there is no further link between the point and the spline.

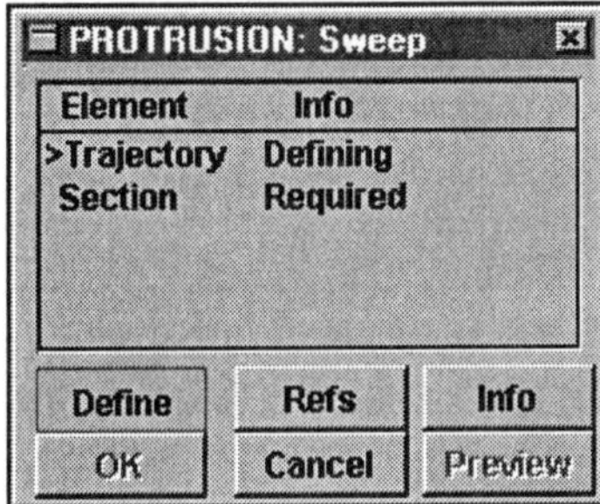

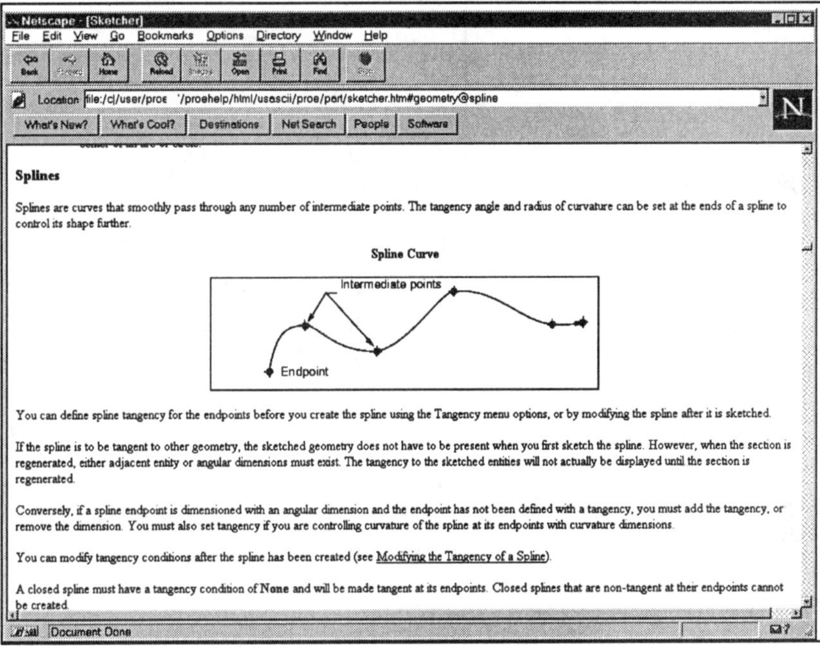

Splines

Splines are curves that smoothly pass through any number of intermediate points. The tangency angle and radius of curvature can be set at the ends of a spline to control its shape further.

Spline Curve

You can define spline tangency for the endpoints before you create the spline using the Tangency menu options, or by modifying the spline after it is sketched.

If the spline is to be tangent to other geometry, the sketched geometry does not have to be present when you first sketch the spline. However, when the section is regenerated, either adjacent entity or angular dimensions must exist. The tangency to the sketched entities will not actually be displayed until the section is regenerated.

Conversely, if a spline endpoint is dimensioned with an angular dimension and the endpoint has not been defined with a tangency, you must add the tangency, or remove the dimension. You must also set tangency if you are controlling curvature of the spline at its endpoints with curvature dimensions.

You can modify tangency conditions after the spline has been created (see Modifying the Tangency of a Spline).

A closed spline must have a tangency condition of **None** and will be made tangent at its endpoints. Closed splines that are non-tangent at their endpoints cannot be created.

Figure 12.11
Online Documentation, Splines

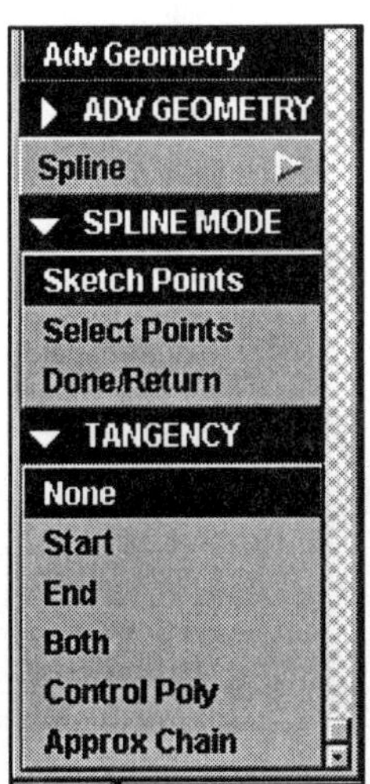

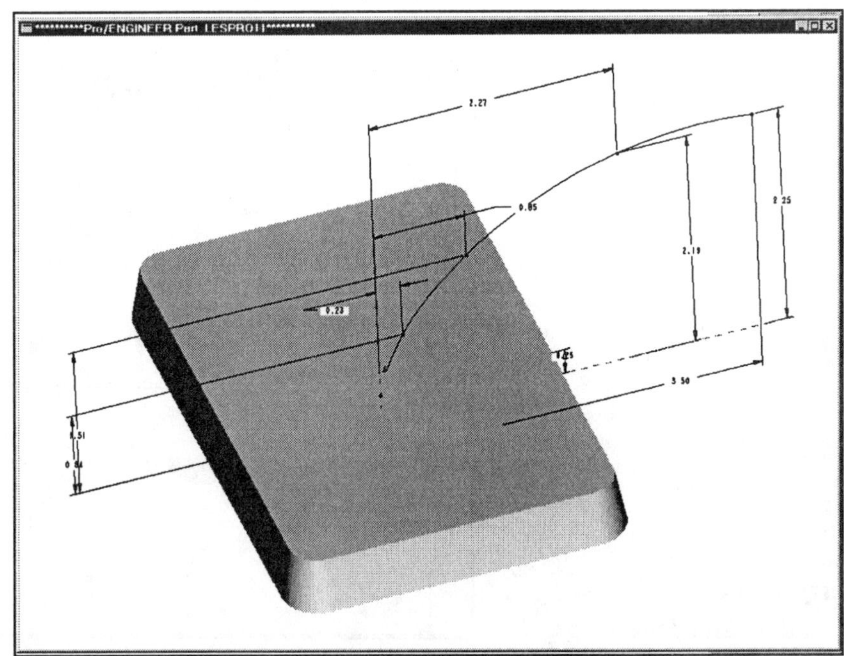

Figure 12.12
Spline

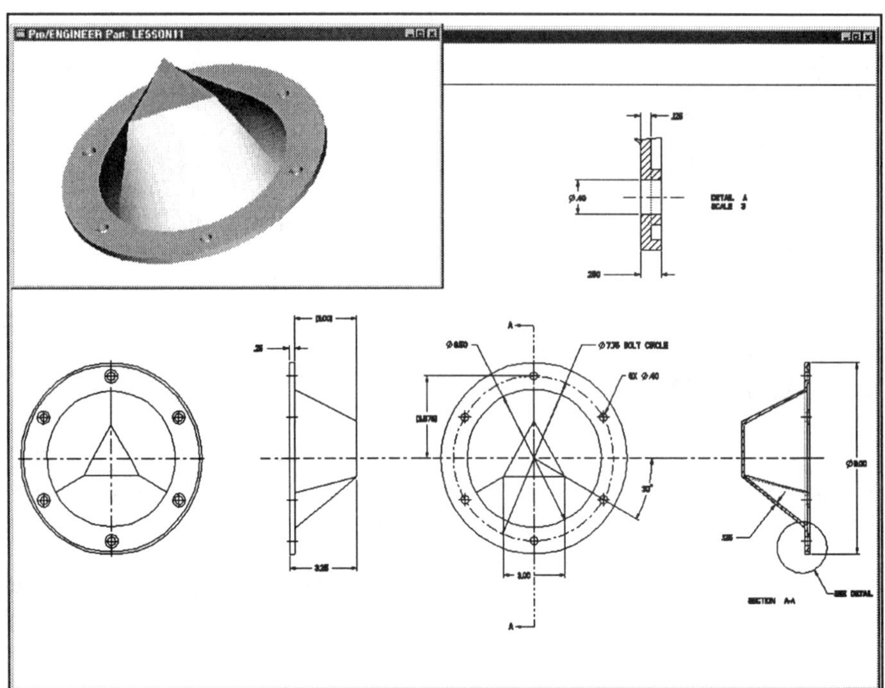

Figure 12.13
Cap Part and Detail Drawing

Cap

The **Cap** is a part created with a **Parallel Blend** (Figs. 12.13 through 12.19). The blend sections are a circle and a triangle. The circle is actually three equal arcs, since *the sections of a blend must have equal segments*.

The part is shelled as the last feature in its creation. The **Shell** command will create *bosses* around each hole as it hollows out the part.

For this part, a section will be created in Part mode to be used when you are detailing the Cap in the Drawing mode.

NOTE

Set up the Cap:

- Material = Plastic
- Units = Inches
- Default Datum Planes
- Default Coordinate System
- Layers = **DATUM_LAYER**
- Hidden line
- Tan Dimmed

✓ Grid Snap
✓ Display Tol

CONFIG.PRO

angular_tol	**0**
tol_mode	**plusminus**
sketcher_dec_places	**3**
default_dec_places	**3**

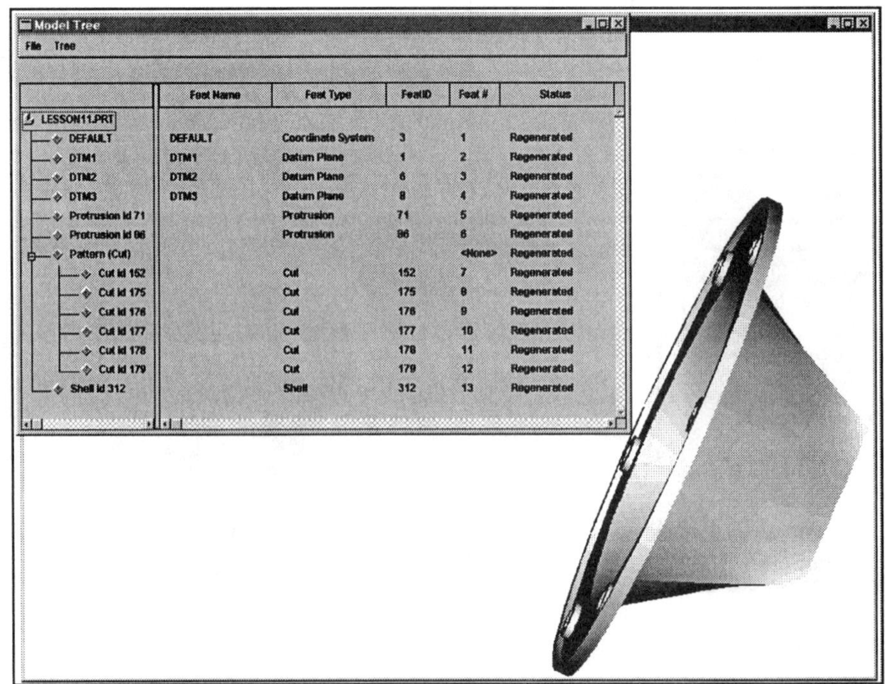

Figure 12.14
Cap and Model Tree

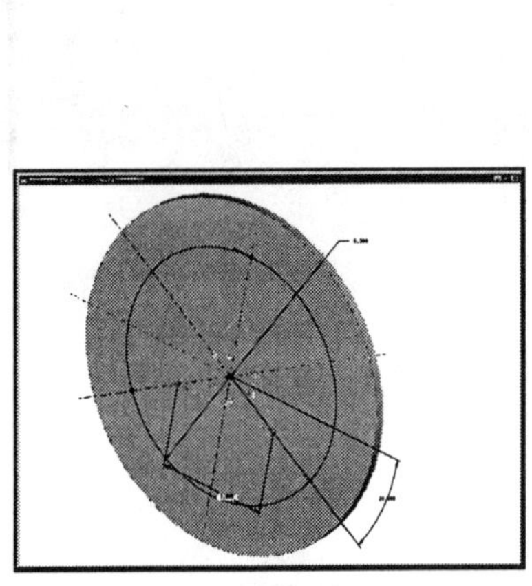

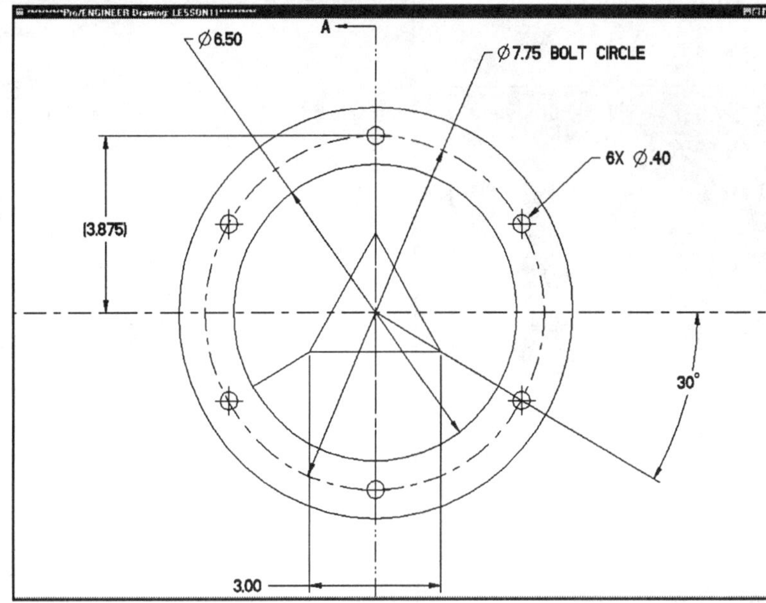

Figure 12.15
Cap, Top View

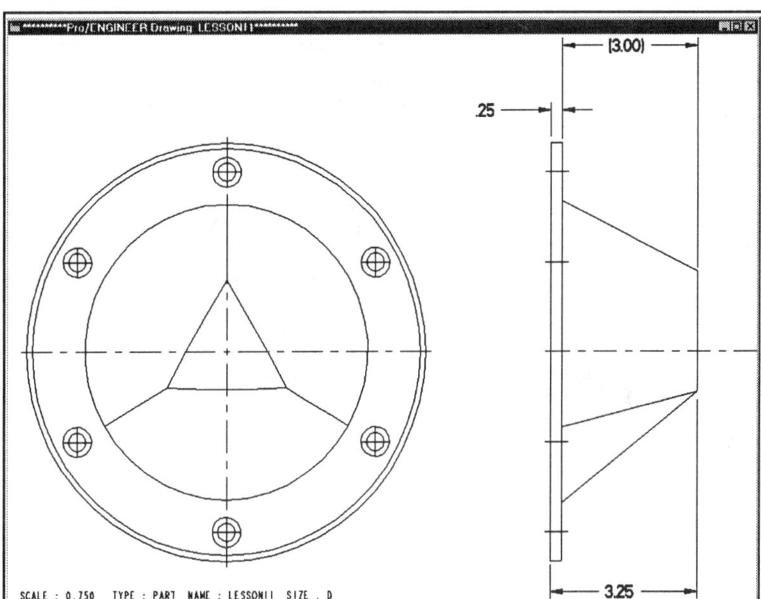

Figure 12.16
Cap, Left Side View, and Back View

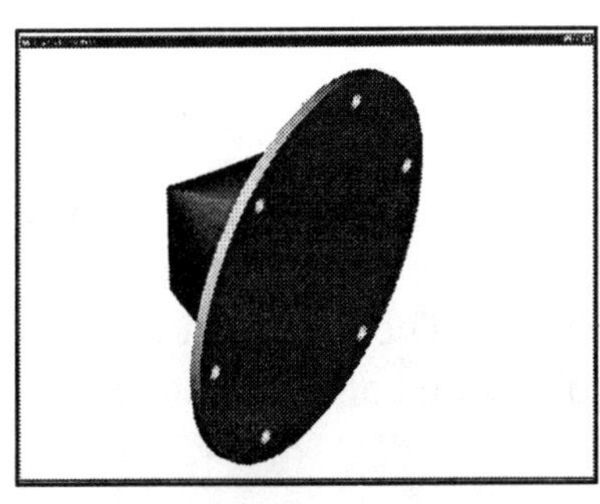

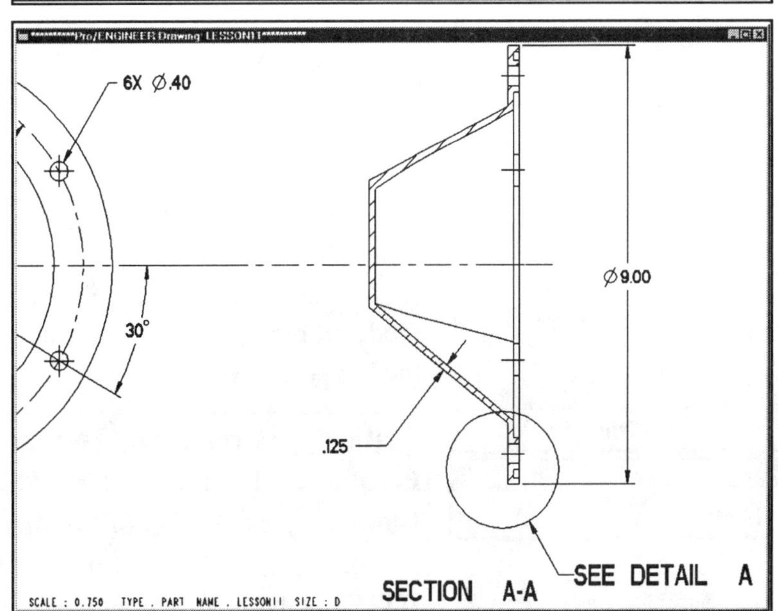

Figure 12.17
Cap, **SECTION A-A**

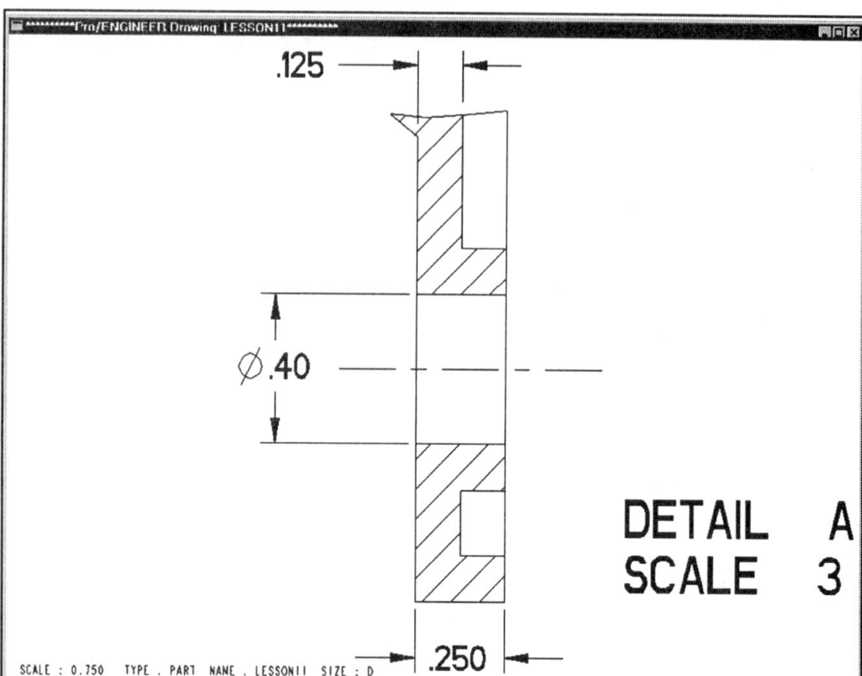

Figure 12.18
Cap, **DETAIL A**

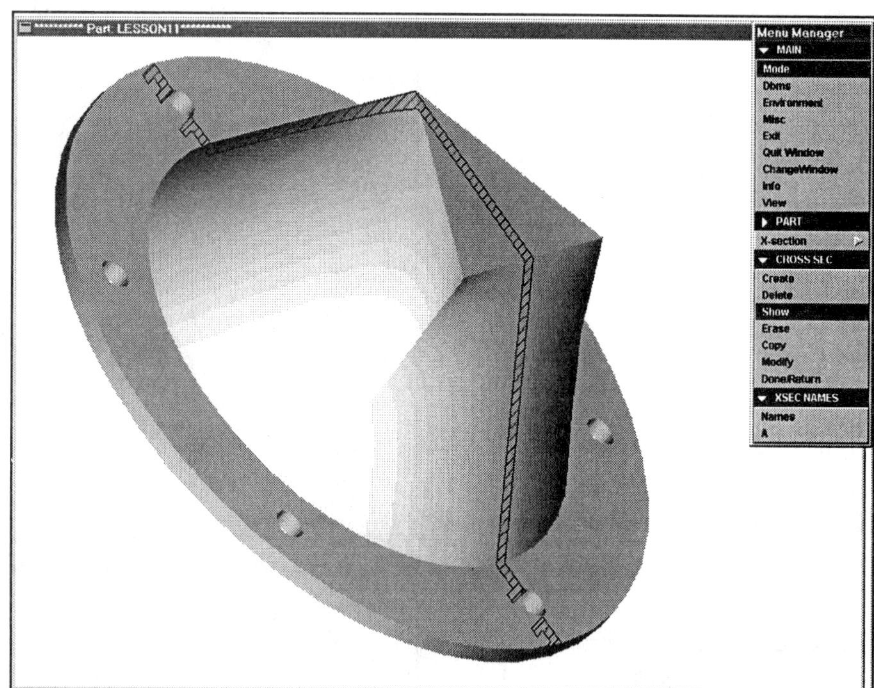

Figure 12.19
Cap, Part Section

Start the Cap by modeling the **∅9.00** by **.25** thick circular protrusion shown in Figure 12.20. Sketch the first protrusion on **DTM3** and centered on **DTM1** and **DTM2**. The **Blend** feature is modeled next (Fig. 12.21). Choose the following commands to create the blend:

PT/Modeler™

Feature ⇒ Protrusion ⇒ Blend ⇒ Solid etc.

Feature ⇒ Create ⇒ Protrusion ⇒ Blend ⇒ Solid ⇒ Done ⇒ Parallel ⇒ Regular Sec ⇒ Sketch Sec ⇒ Done ⇒ Straight ⇒ Done ⇒ (pick the top of the first protrusion as shown in Fig. 12.22) **⇒ Okay** (to confirm direction of feature creation) **⇒ Top ⇒** (choose **DTM2** as the orientation plane)

Dbms (File--PT/Modeler) ⇒
Save ⇒ enter
Purge ⇒ enter ⇒
Done-Return

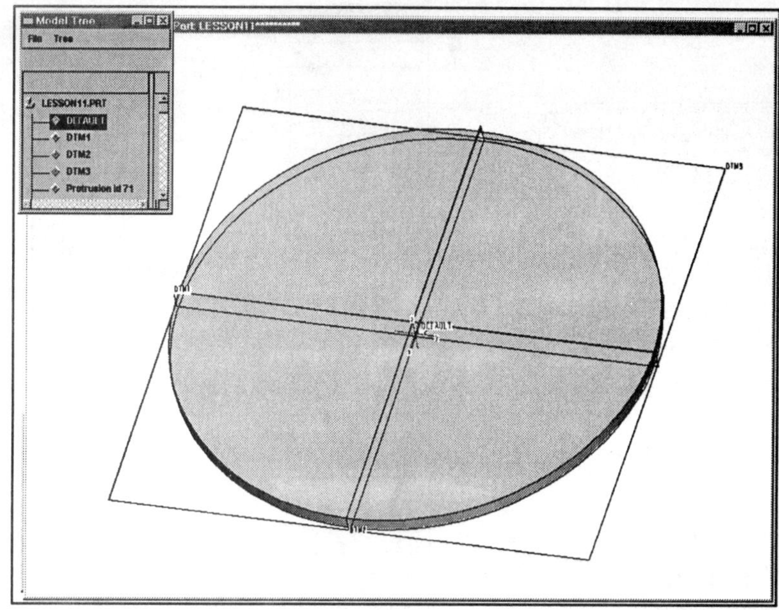

Figure 12.20
First Protrusion

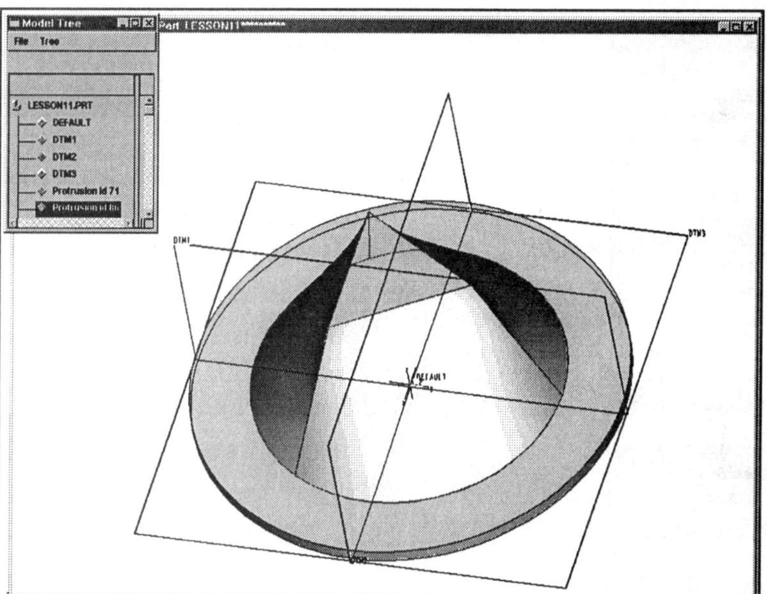

Figure 12.21
Blend Feature

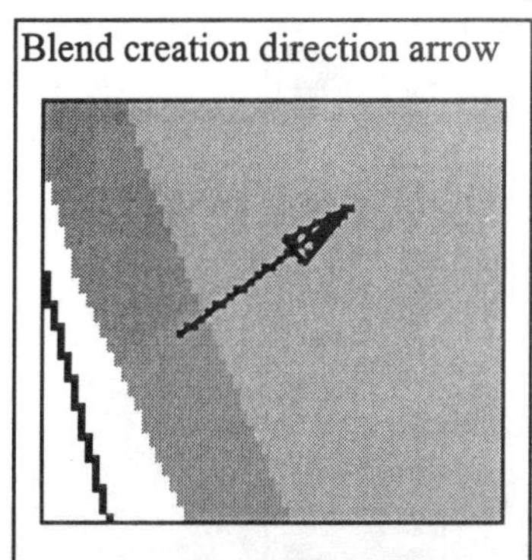

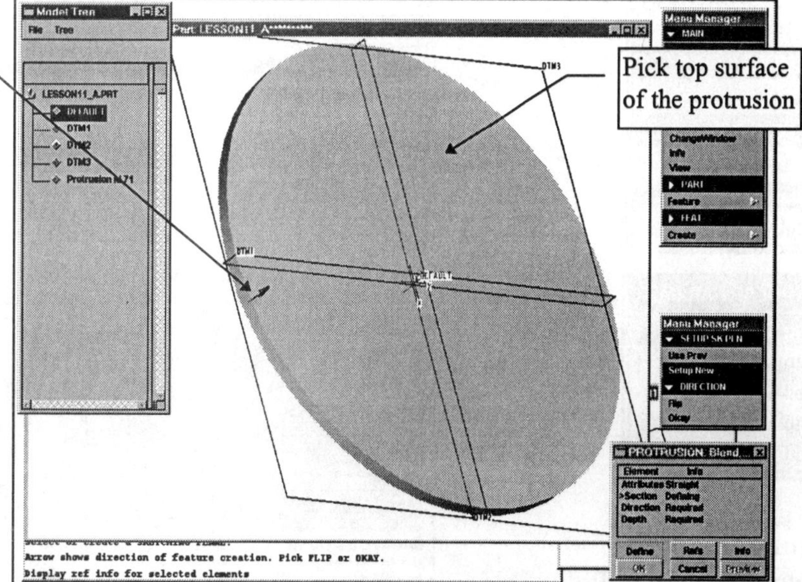

Figure 12.22
Blend Feature Starting Surface and Direction of Creation

Figure 12.23
Sketcher Showing Cartesian Grid

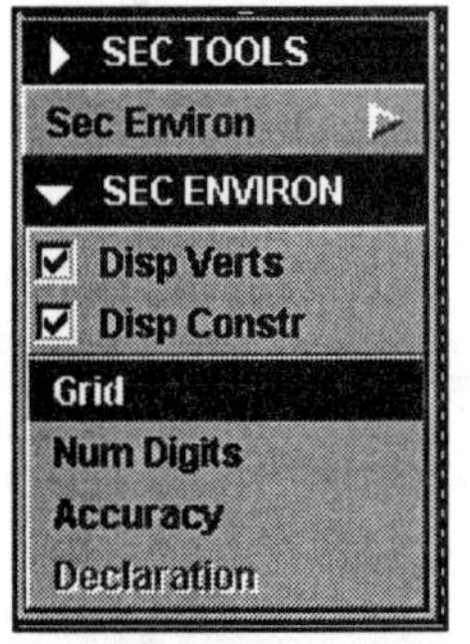

The section grid would be better utilized if it were a **Polar** grid rather than a **Cartesian** grid (Fig. 12.23). Change the grid type and size and continue with the blend commands:

Sec Tools ⇒ **Sec Environ** ⇒ **Grid** ⇒ **Type** ⇒ **Polar** ⇒ **Done/Return** ⇒ **Params** ⇒ **Ang Spacing** ⇒ **enter** (to accept the default of **30°**) ⇒ **Rad Spacing** ⇒ (type **.50** at the prompt) ⇒ **enter** ⇒ **Done/Return** (Fig. 12.24)

Sketch ⇒ **Arc** ⇒ **Ctr/Ends** ⇒ (sketch the first section of the blend by creating three equal **120°** arcs, sketch each arc in a counter clockwise direction) ⇒ **Regenerate** (Fig. 12.25) ⇒ **Alignment** (select each arc and the default coordinate system) ⇒ **Regenerate** ⇒ **Dimension** (add the diameter dimension) ⇒ **Regenerate** ⇒ **Sketch** (add two centerlines to locate the arcs' ends)

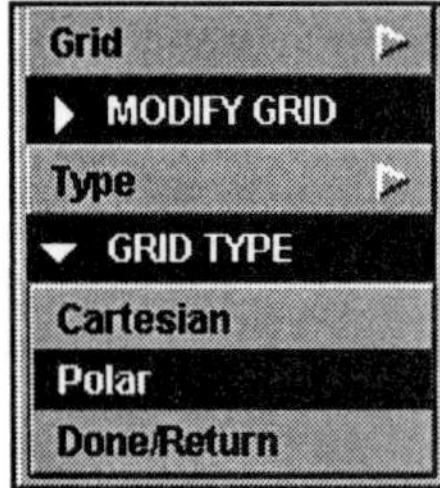

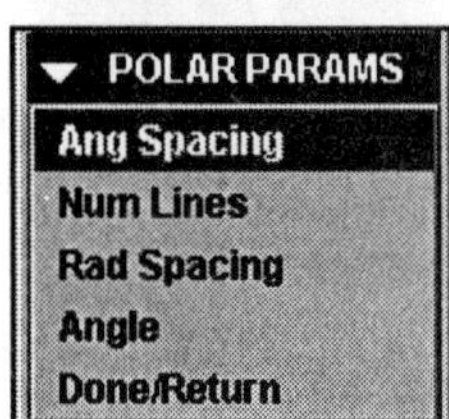

Figure 12.24
Sketcher Showing Polar Grid

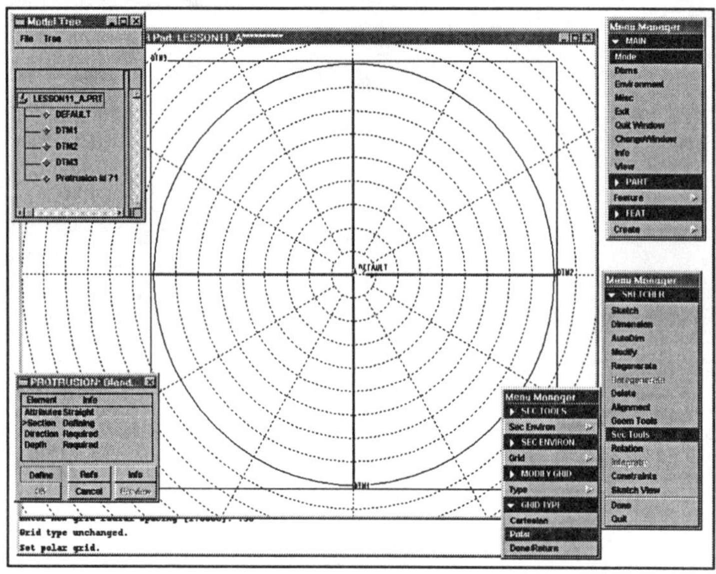

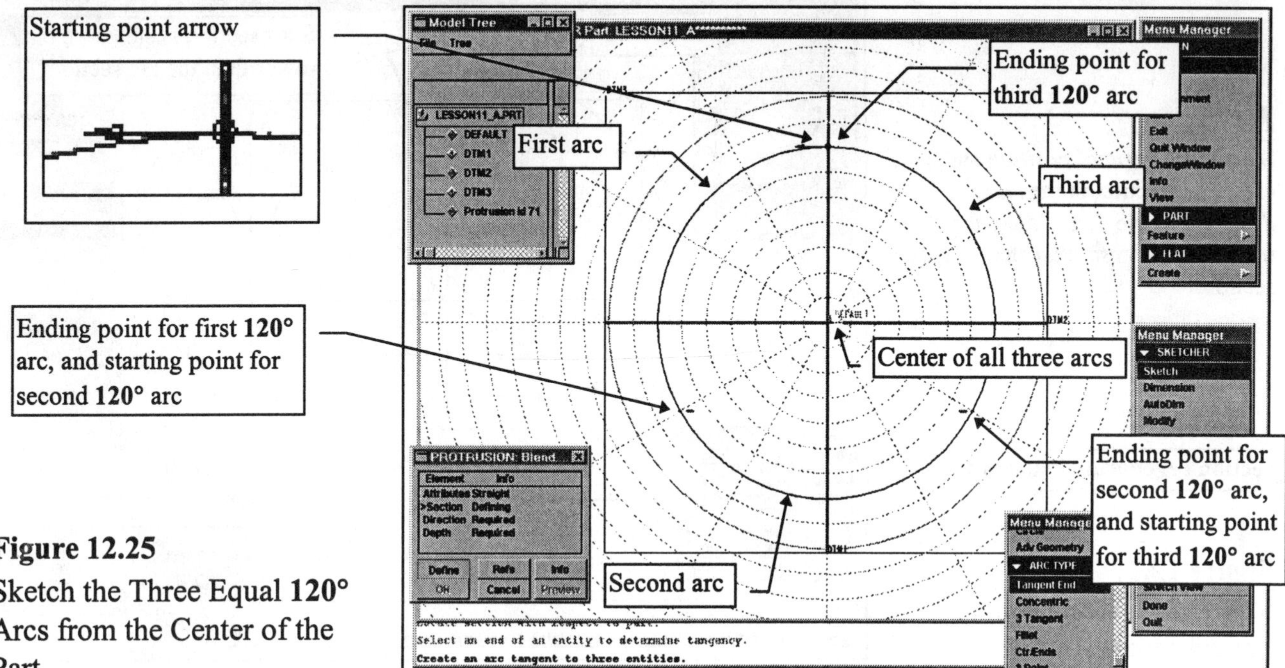

Figure 12.25
Sketch the Three Equal **120°** Arcs from the Center of the Part

NOTE

As you toggle between sections the active section shows in *cyan* (light blue), and the inactive section is *grayed*. You can toggle both directions. Remember that each section is separate and needs its own unique dimensioning scheme, as required by the design intent. Each section must have the same total number of entity segments. Any number of sections can be created depending on the design requirements. Here, only two sections are incorporated into the design, and each section has three segments: three arcs in one section and three lines in the other.

Two centerlines locating the arcs' ends are needed to get the sketch to regenerate. Continue with the commands:

Sketch ⇒ **Line** ⇒ **Centerline** ⇒ **Alignment** (align the end of the centerlines to the coordinate system) ⇒ **Dimension** [dimension the centerlines from the vertical datum plane with a **120°** angle (Fig. 12.26)] ⇒ **Regenerate** ⇒ **Sec Tools** ⇒ **Toggle** (to sketch the second parallel section) (the first section is *grayed out*) ⇒ **Sketch** ⇒ **Line** ⇒ (sketch the three lines of the triangle starting at **DTM1** and picking points *in the same direction* in which the arcs were created) ⇒ **Alignment** [align the first centerline to **DTM1** (Fig. 12.27)] ⇒

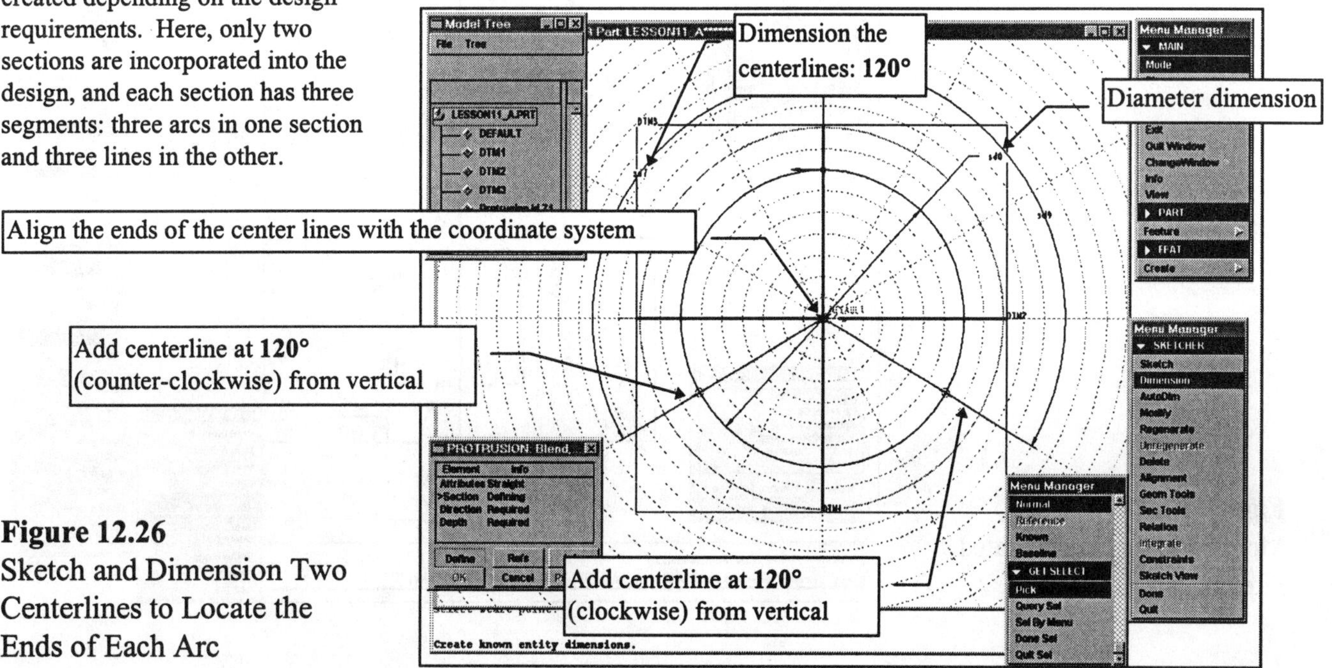

Figure 12.26
Sketch and Dimension Two Centerlines to Locate the Ends of Each Arc

HINT

Start at the *vertical position* and create the section in the *same direction* as the arcs. Make the triangle section *smaller* than the arc section.

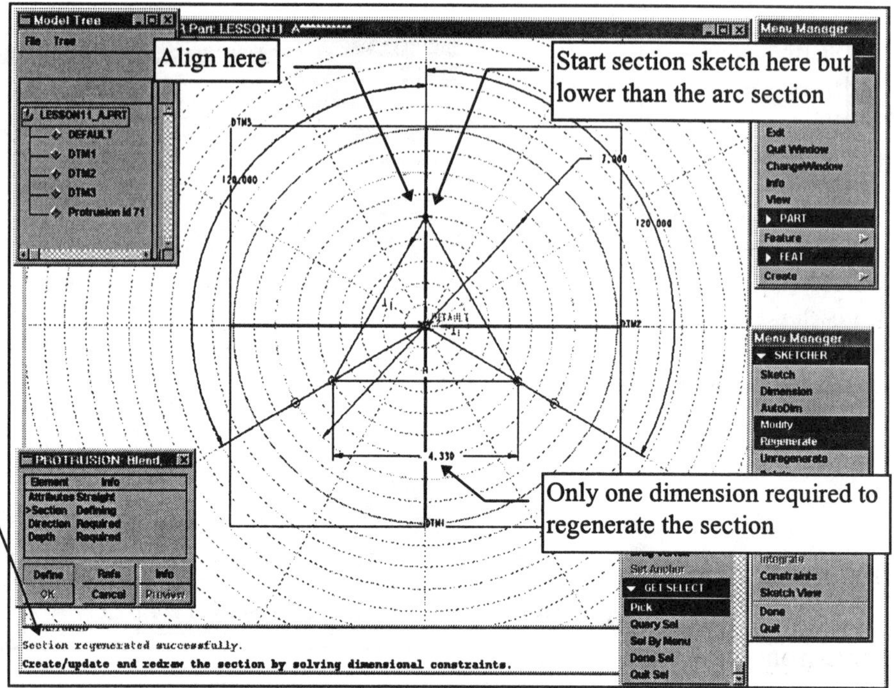

Section regenerated successfully

Figure 12.27
Sketch and Dimension the Three Lines

Regenerate ⇒ Dimension ⇒ Regenerate ⇒ Modify (change the diameter dimension to **7.75**, the leg of the triangle dimension to **3.00,** and the two angles to **120°**) **⇒ enter ⇒ Regenerate** (Fig. 12.28) **⇒ Done ⇒ View ⇒ Default ⇒ Done-Return ⇒ Blind ⇒ Done ⇒** (type **3.00** at the prompt) **⇒ enter ⇒ Preview ⇒ View ⇒ Cosmetic ⇒ Shade ⇒ Display ⇒ Done-Return** (Fig. 12.29) **⇒ OK ⇒ Done**

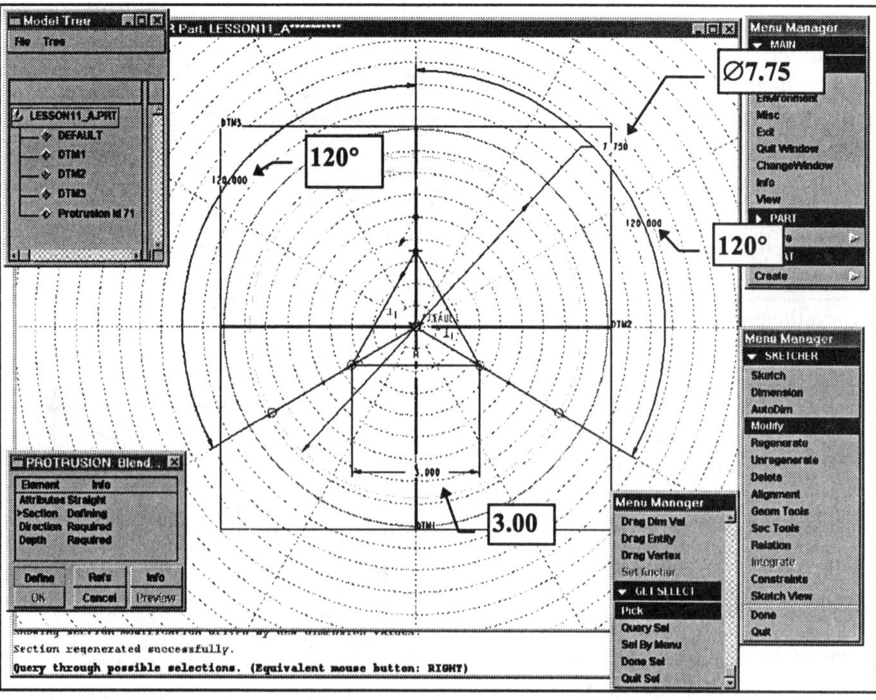

Figure 12.28
Modified and Regenerated Sketch

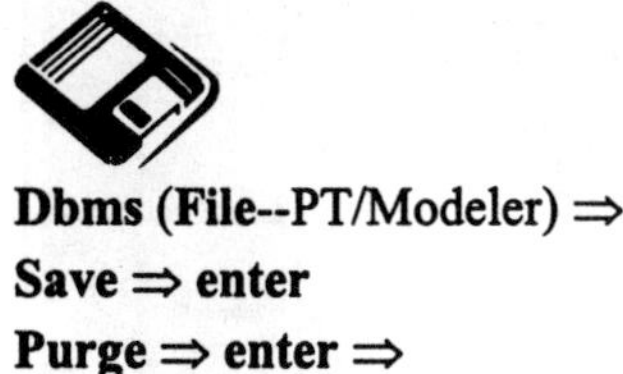

Dbms (File--PT/Modeler) ⇒
Save ⇒ enter
Purge ⇒ enter ⇒
Done-Return

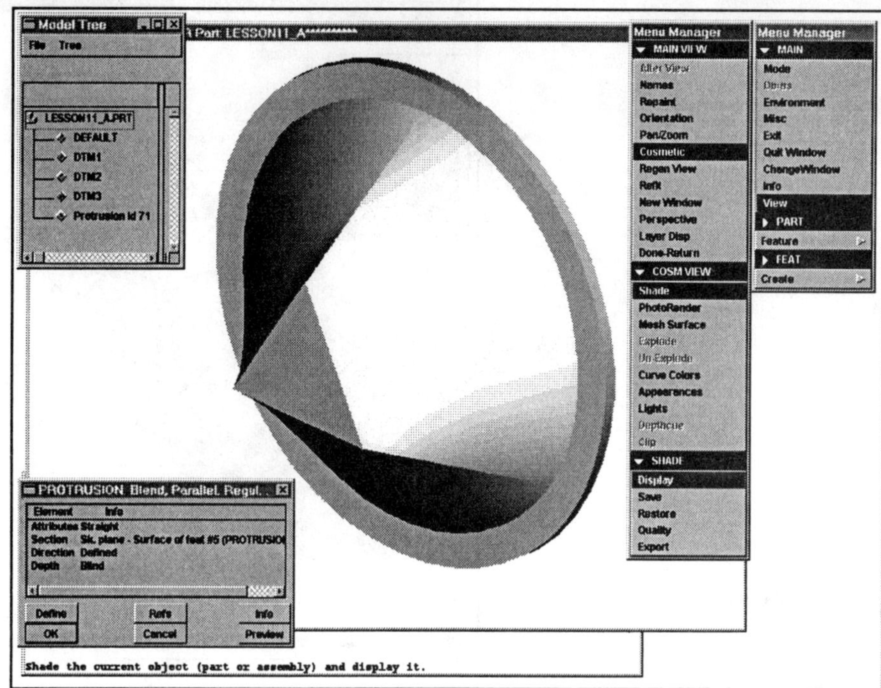

Figure 12.29
Shaded Preview of the Blend

You made a mistake (*we made a mistake*). The diameter of the blend was supposed to be **6.50,** not **7.75**, which is the diameter of the bolt circle.

Modify ⇒ (pick the blend protrusion from the Model Tree, as shown in Fig. 12.30) ⇒ (pick the **7.75** dimension) ⇒ (change the value to **6.50**; see Fig. 12.15) ⇒ **enter** ⇒ **Regenerate** (Fig. 12.31)

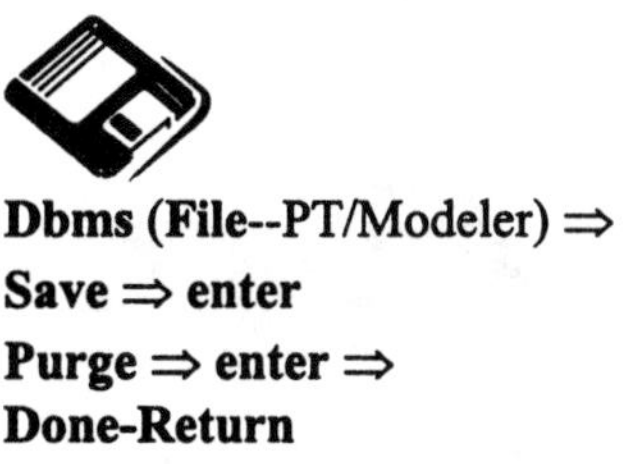

Dbms (File--PT/Modeler) ⇒
Save ⇒ enter
Purge ⇒ enter ⇒
Done-Return

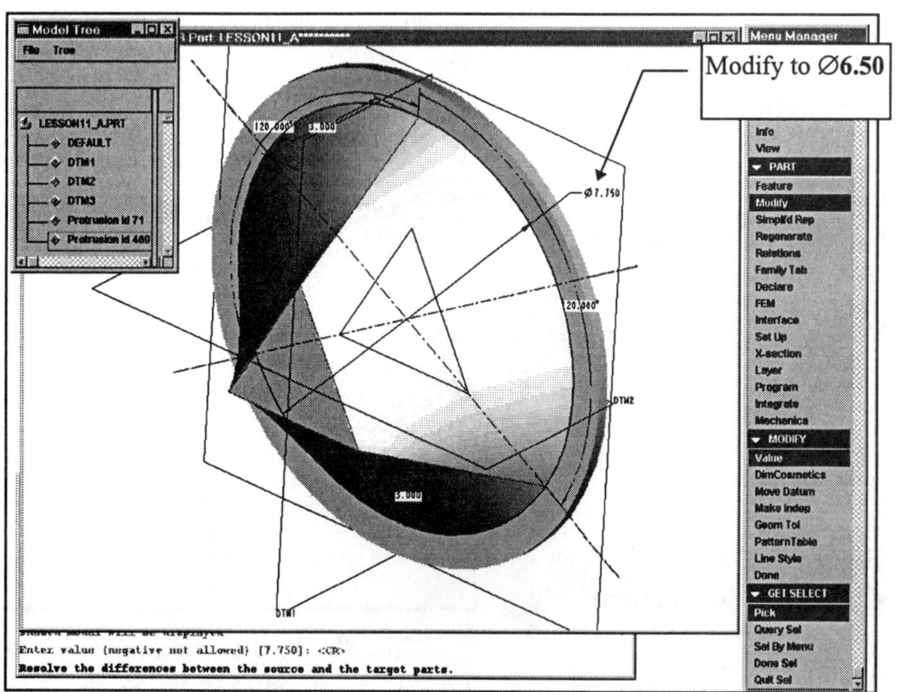

Figure 12.30
Modify ∅7.75 to ∅6.50

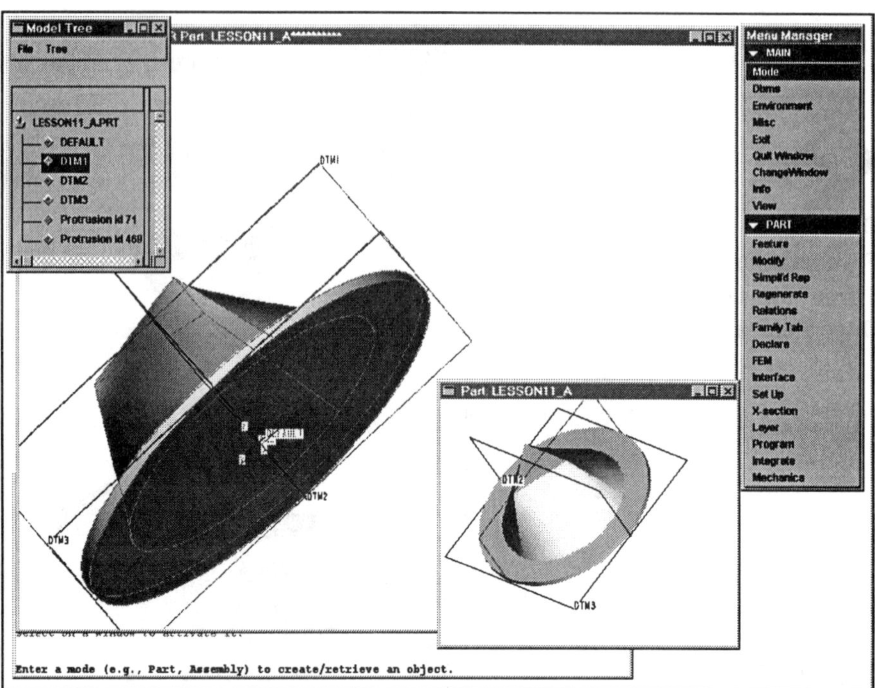

Figure 12.31
Completed Blend

Create and pattern the holes as shown in Figure 12.32. Next, **Shell** the part using the following commands:

PT/Modeler ™
Feature ⇒ Shell etc.

Feature ⇒ Create ⇒ Shell ⇒ (pick the bottom surface of the part as the surface to remove, as shown in Fig. 12.33) **⇒ Done Sel ⇒ Done Refs ⇒** (type a thickness of **.125** at the prompt) **⇒ enter ⇒ OK** (Fig. 12.34) **⇒ Done**

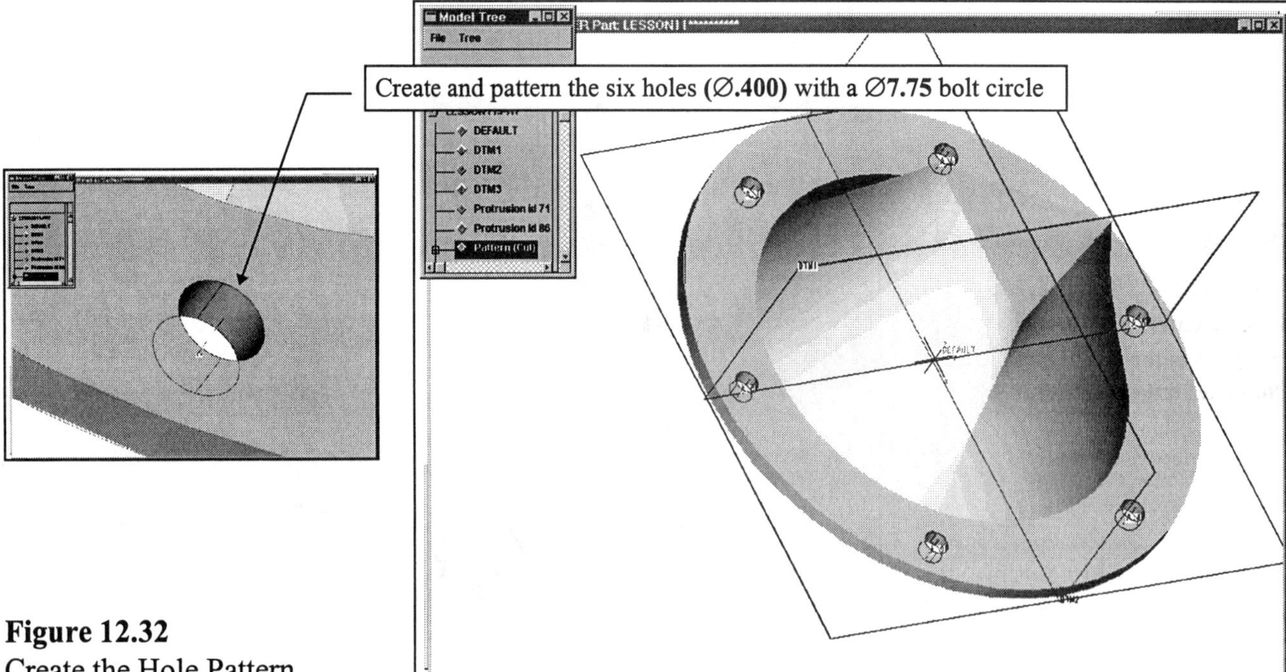

Figure 12.32
Create the Hole Pattern

The **Shell** will automatically create the bosses (Figs. 12.34 and 12.35), since the **.125** thickness is left around all previously created features. If the bosses were not desired, you would simply **Reorder** the holes to come after the shell feature.

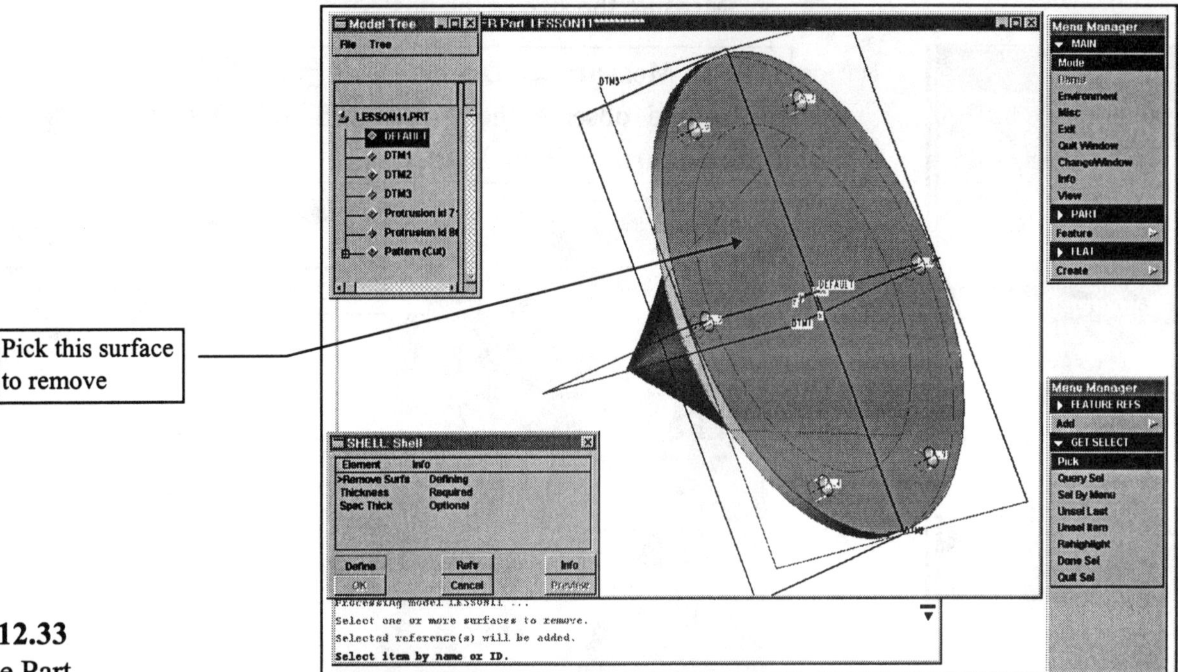

Figure 12.33
Shell the Part

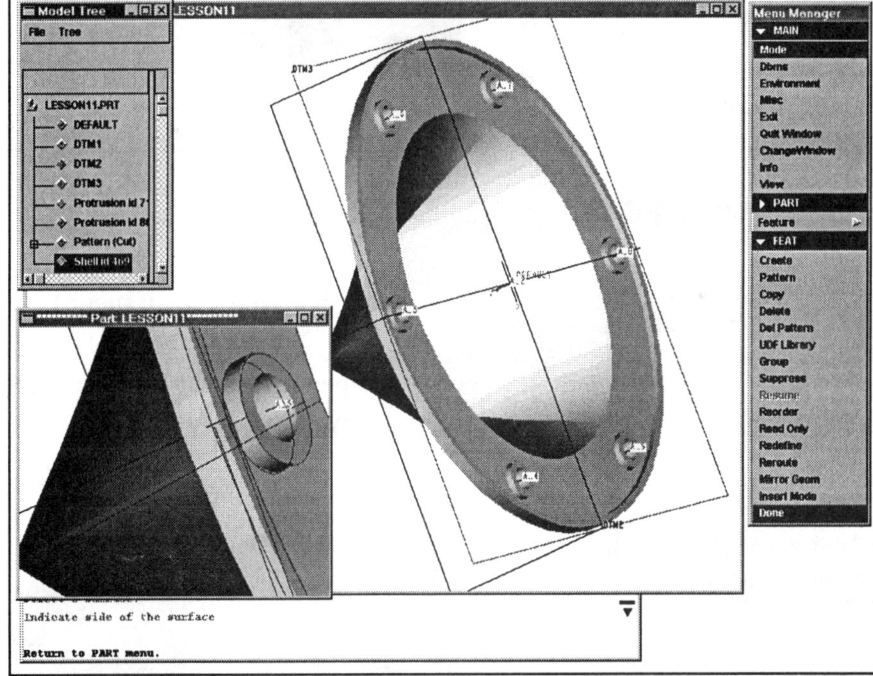

NOTE
The *bosses* around the holes are created automatically at **.250** larger than the holes:
(.250 + .400 = ∅.650)

Figure 12.34
Shelled Part

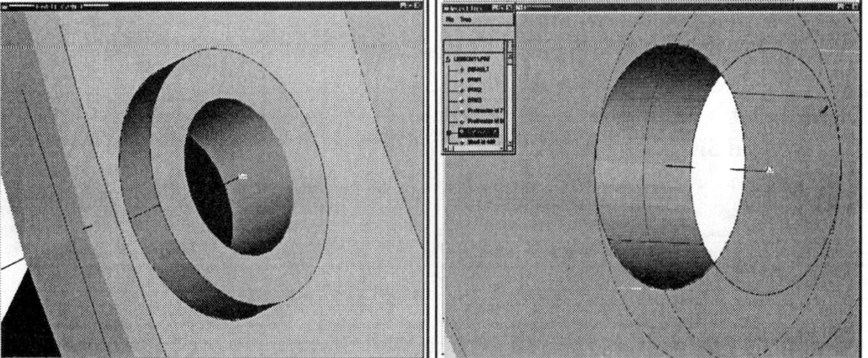

Figure 12.35
Hole and Boss Shown From Both Sides

Measure the size of the boss:

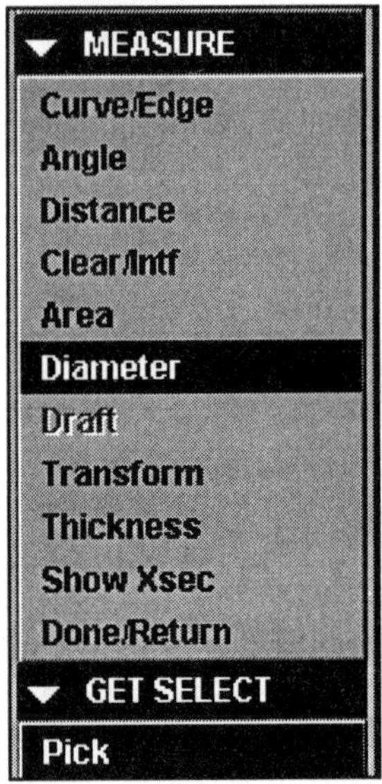

Info ⇒ **Measure** ⇒ **Diameter** ⇒ (pick the boss as shown in Fig. 12.36, and observe the MESSAGE WINDOW) ⇒ **Done Sel** ⇒ **Done/Return** ⇒ **Done/Return**

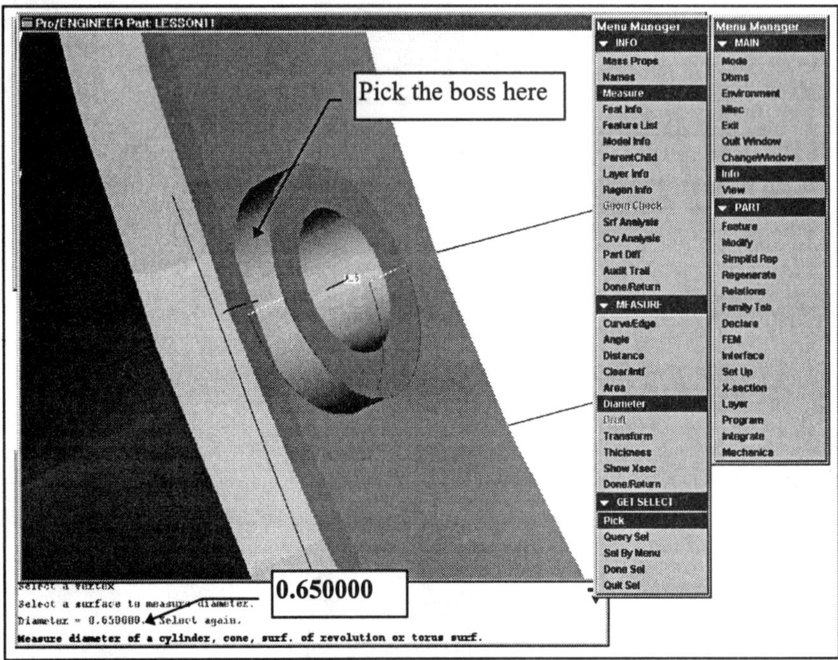

Figure 12.36
Measuring the Boss

Now create a section (Fig. 12.37) to be used in Drawing mode when you are detailing the Cap. Use the following commands:

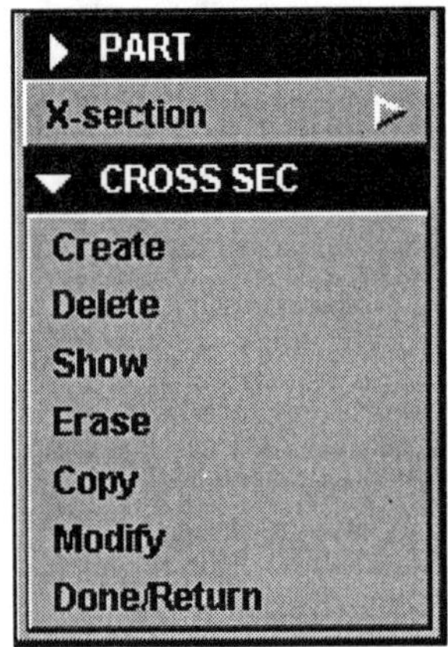

X-section (from the PART menu) ⇒ **Create** ⇒ **Planar** ⇒ **Single** ⇒ **Done** ⇒ (type **A** at the prompt) ⇒ **enter** ⇒ (select planar surface or datum plane; pick **DTM1** from the Model Tree, as shown in Fig. 12.37) ⇒ **Done/Return**

HINT

Type only *one letter* to establish a section name. Typing **A** will create a section called **SECTION A-A**. Typing *two* **A**'s will create a section called **SECTION AA-AA!**

Type *one* **A** at the prompt

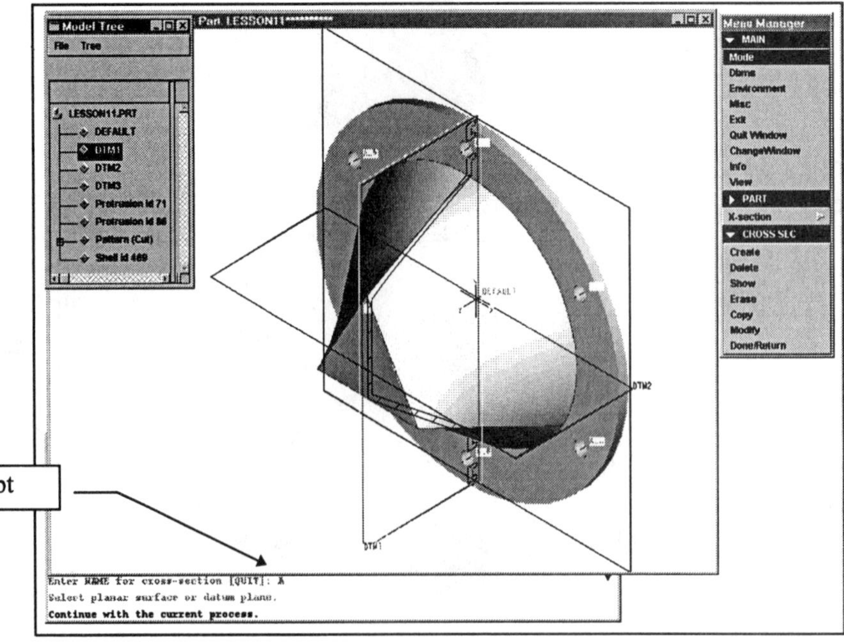

Figure 12.37
Creating a **X-section**

Save the Cap under a new name, and complete the **ECO** to establish a second part with different sizes and features.

E C O

Dbms (File**-PT/Modeler) ⇒ **Save As** ⇒ **enter** ⇒ (type **CAP_PART_ONE)** ⇒ **enter** ⇒ **Done-Return

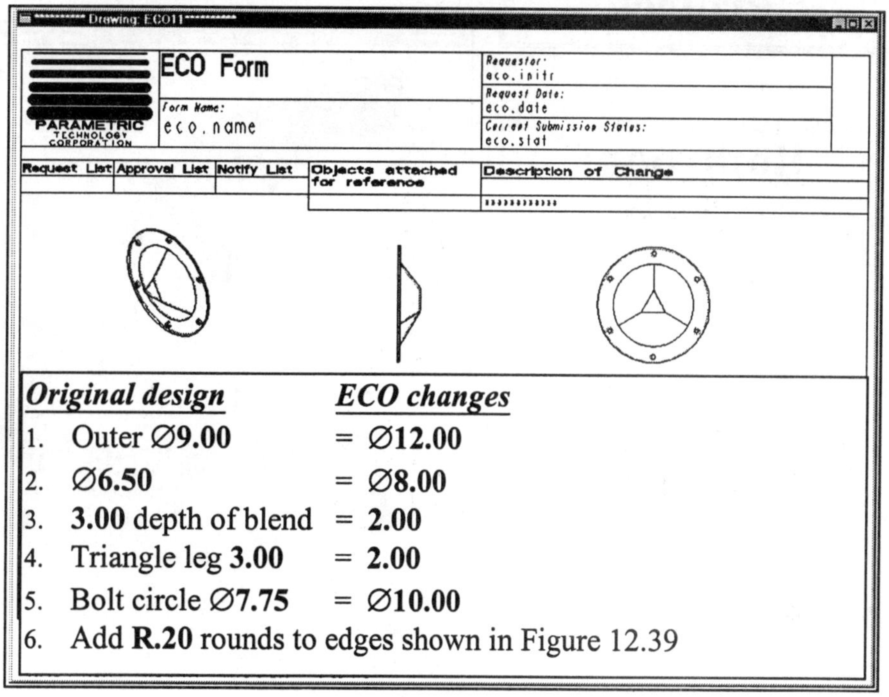

Figure 12.38
ECO

Drawing: ECO11

ECO Form

Form Name: eco.name

Requestor: eco.initr

Request Date: eco.date

Current Submission Status: eco.stat

PARAMETRIC TECHNOLOGY CORPORATION

Request List	Approval List	Notify List	Objects attached for reference	Description of Change

Original design	***ECO changes***
1. Outer **Ø9.00**	= **Ø12.00**
2. **Ø6.50**	= **Ø8.00**
3. **3.00** depth of blend	= **2.00**
4. Triangle leg **3.00**	= **2.00**
5. Bolt circle **Ø7.75**	= **Ø10.00**

6. Add **R.20** rounds to edges shown in Figure 12.39

Figure 12.39
ECO Changes to Model

Pro/ENGINEER Part: ECO11

*Add **R.20** rounds to edges*--Insert the rounds so as to be affected by the **Shell** option.

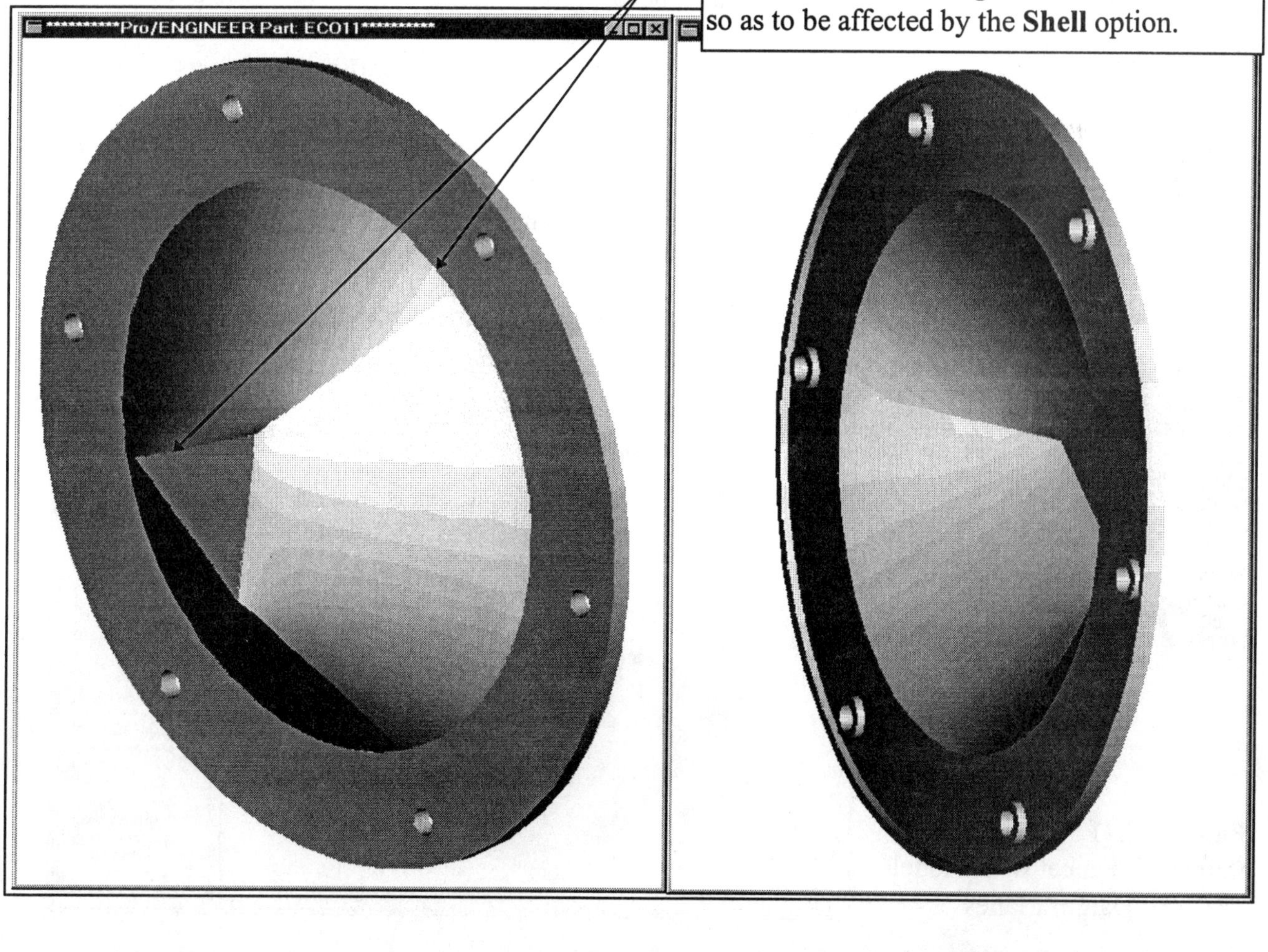

Lesson 12 Project

Bathroom Faucet

Figure 12.40
Bathroom Faucet

Bathroom Faucet

This is an advanced project. Since you have created over 20 parts using Pro/E, you should be able to use that knowledge to model the **Swept Blend** required to create the **Bathroom Faucet** (Figs. 12.40 through 12.82). Some instructions will accompany this lesson project, but you will be required to research online documentation (**Pro/HELP**) and learn more about **Splines** and **Swept Blends**.

After the model is complete, create a number of sections that can be used in Drawing mode when you are detailing the Faucet.

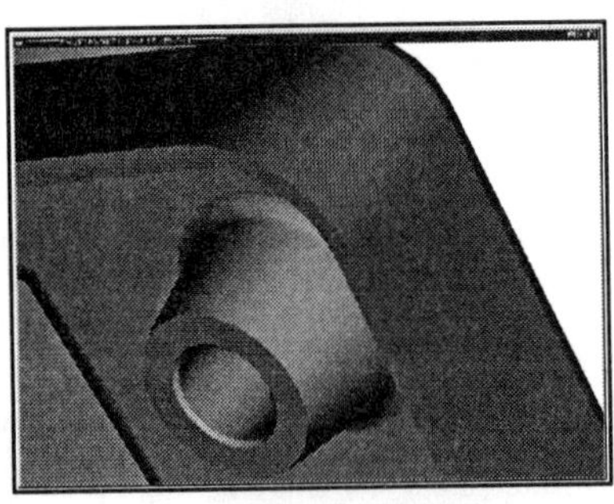

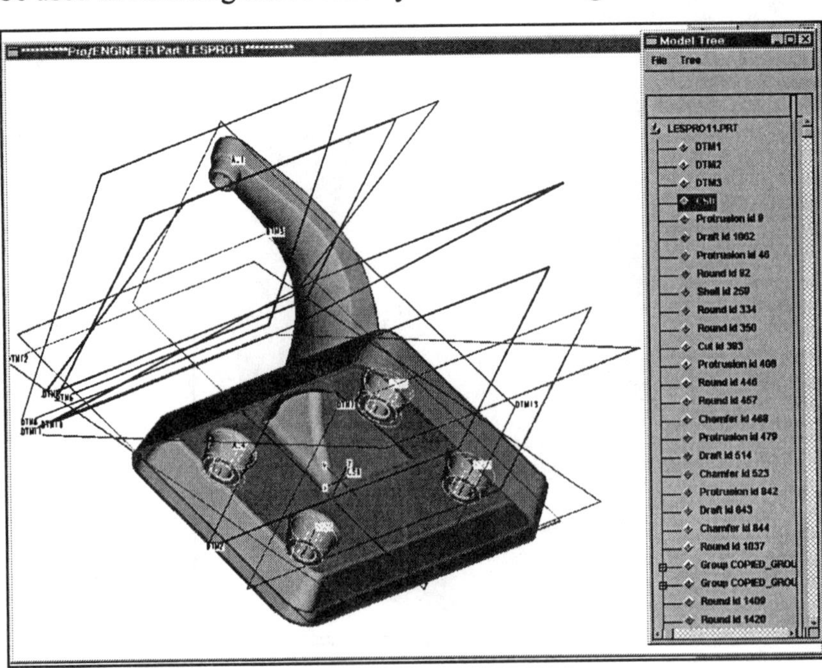

Figure 12.41
Bathroom Faucet with Model Tree and Datum Planes

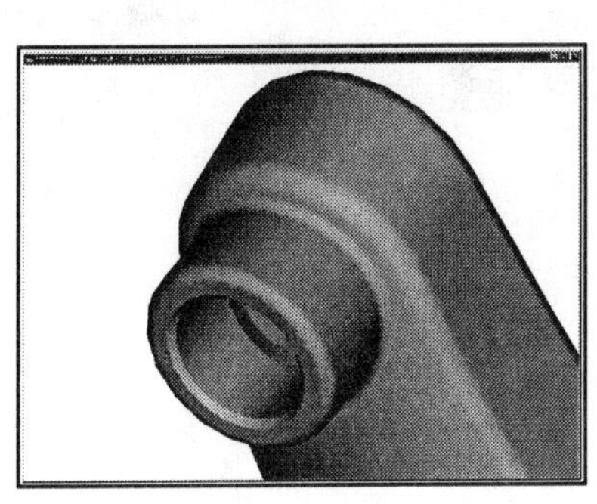

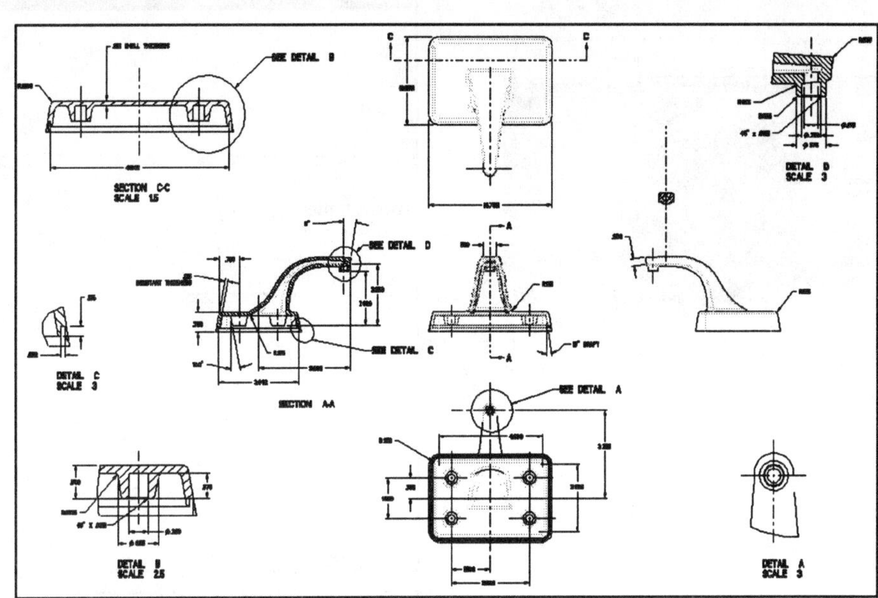

Figure 12.42
Bathroom Faucet,
Detail Drawing

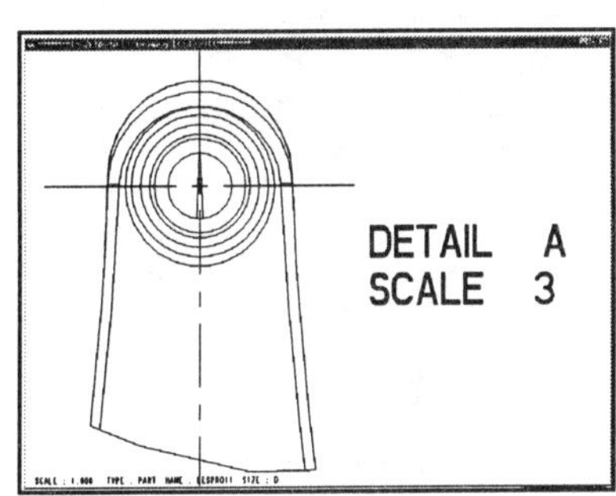

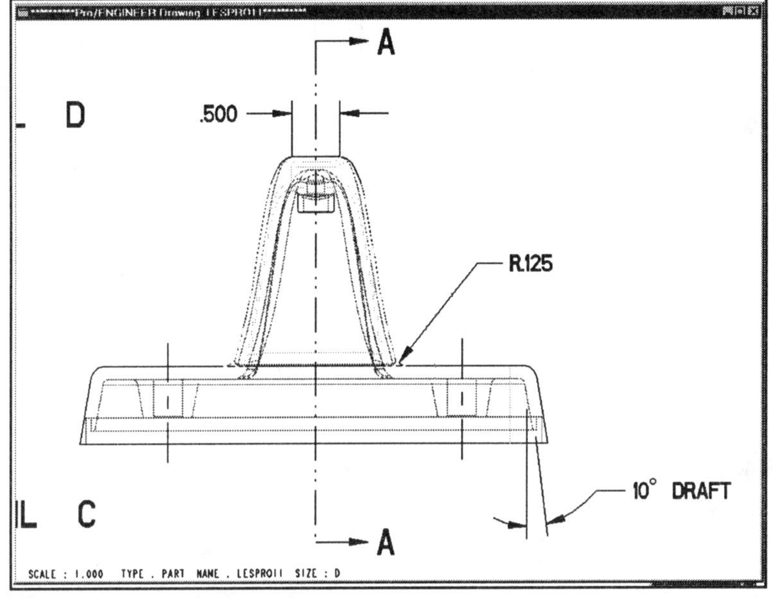

Figure 12.43
Bathroom Faucet Drawing,
Front View

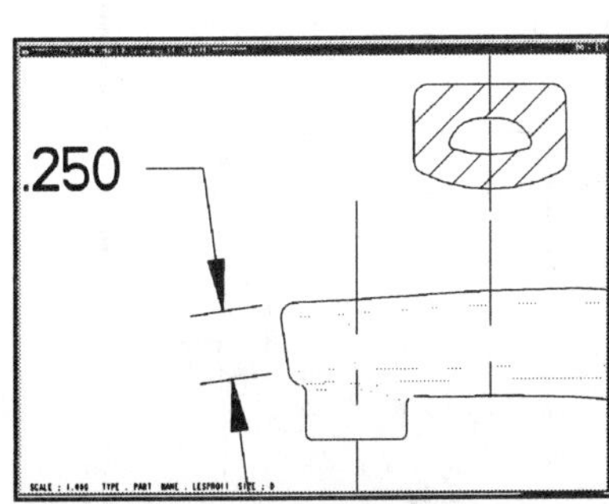

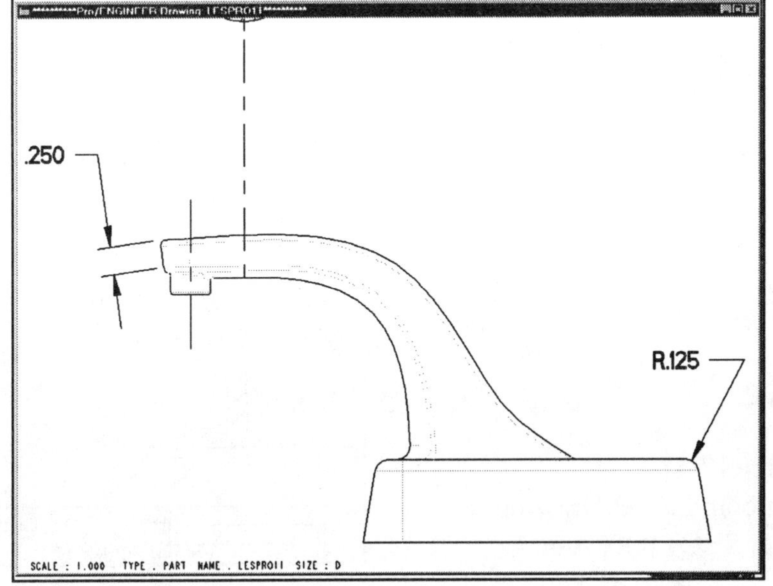

Figure 12.44
Bathroom Faucet Drawing,
Right Side View

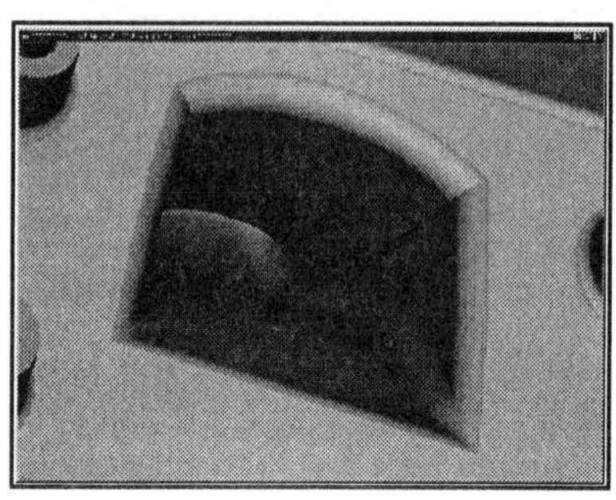

Figure 12.45
Bathroom Faucet Drawing,
SECTION A-A

Figure 12.46
Bathroom Faucet Drawing,
Lower **SECTION A-A**

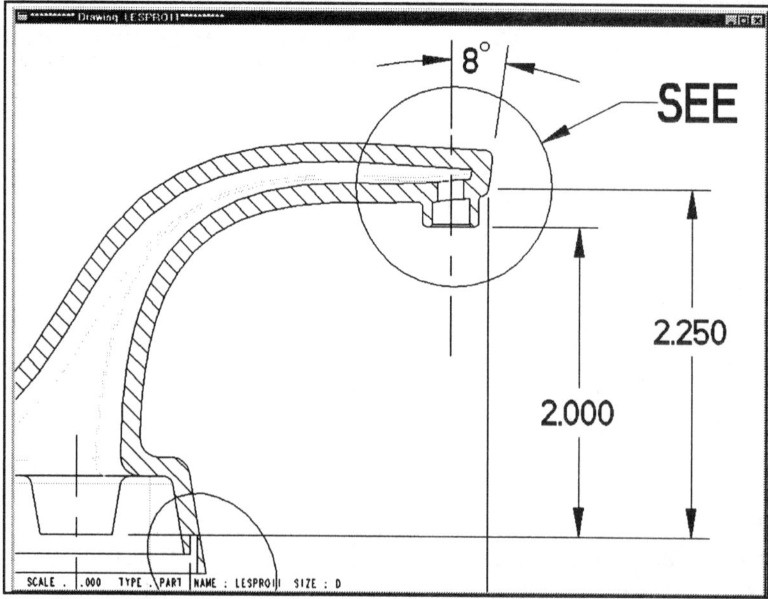

Figure 12.47
Bathroom Faucet Drawing,
Upper **SECTION A-A**

C
C
(3.265)
(4.765)
SCALE . 1.000 TYPE . PART NAME : LESPRO11 SIZE : D

Figure 12.48
Bathroom Faucet Drawing,
Top View

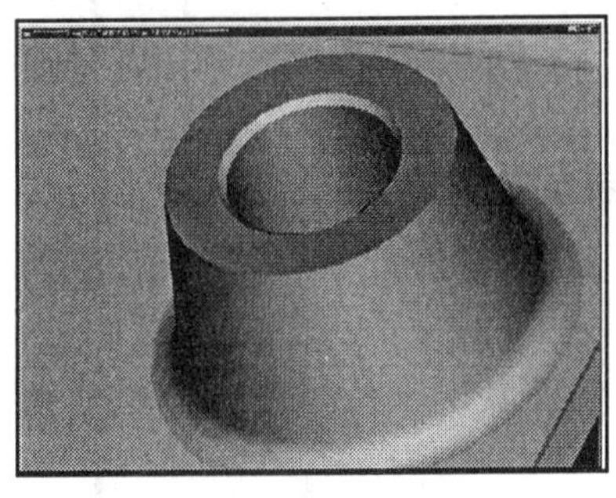

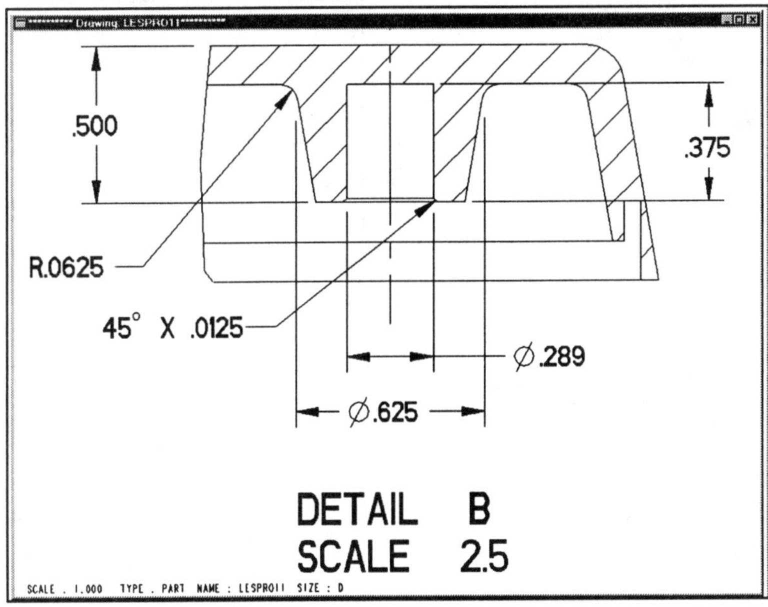

Figure 12.49
Bathroom Faucet Drawing,
DETAIL B

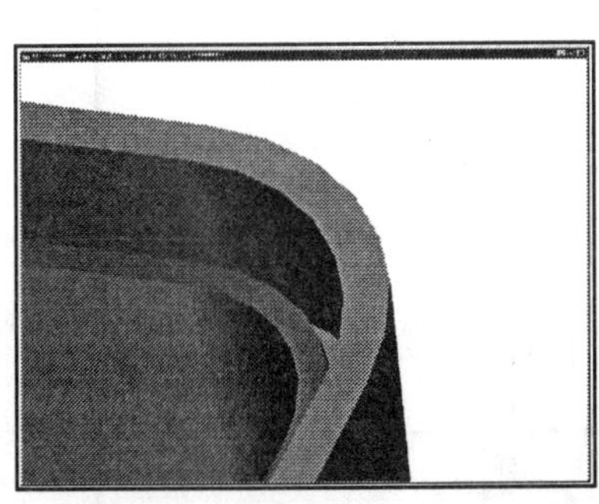

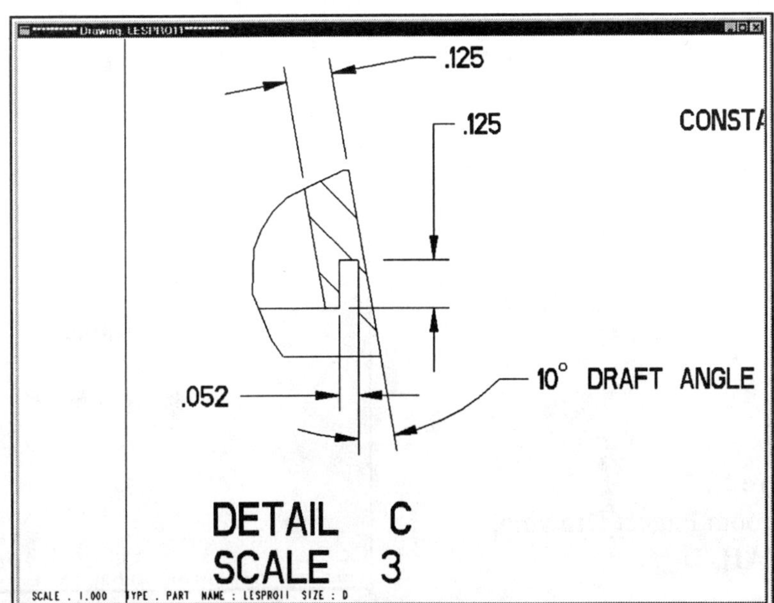

Figure 12.50
Bathroom Faucet Drawing,
DETAIL C

Figure 12.51
Bathroom Faucet Drawing,
SECTION C-C

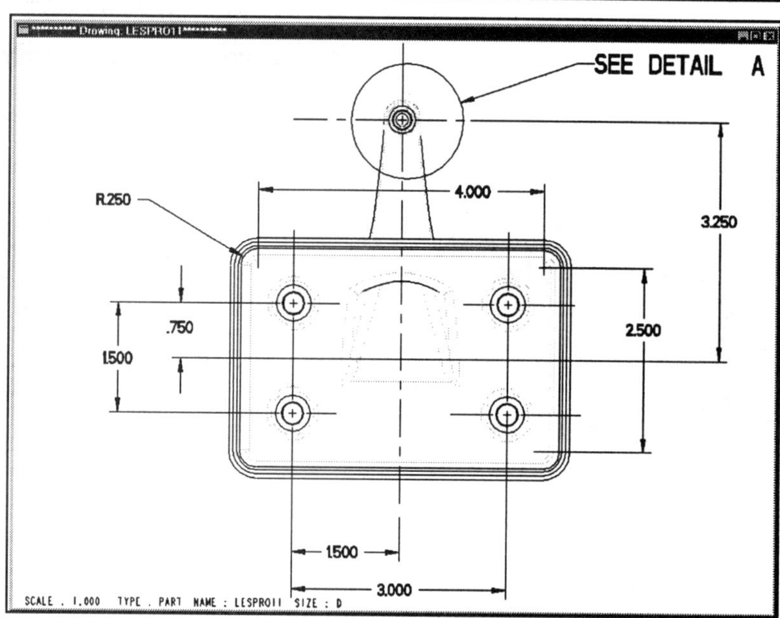

Figure 12.52
Bathroom Faucet Drawing,
Bottom View

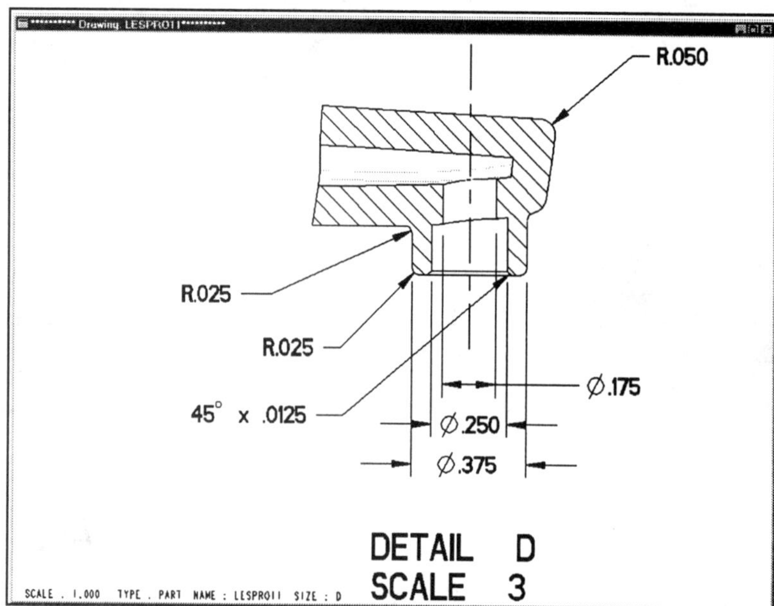

Figure 12.53
Bathroom Faucet Drawing,
DETAIL D

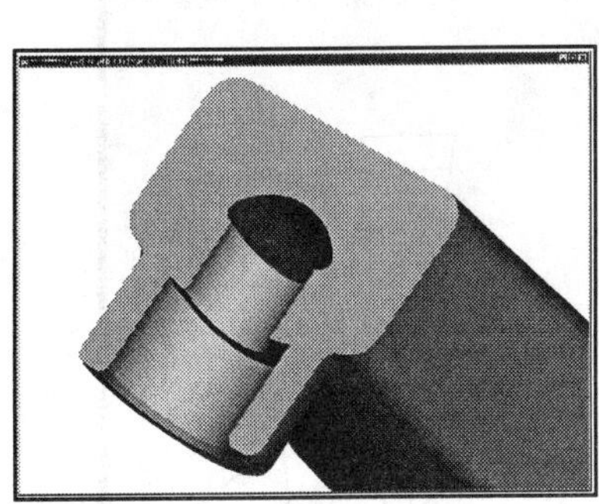

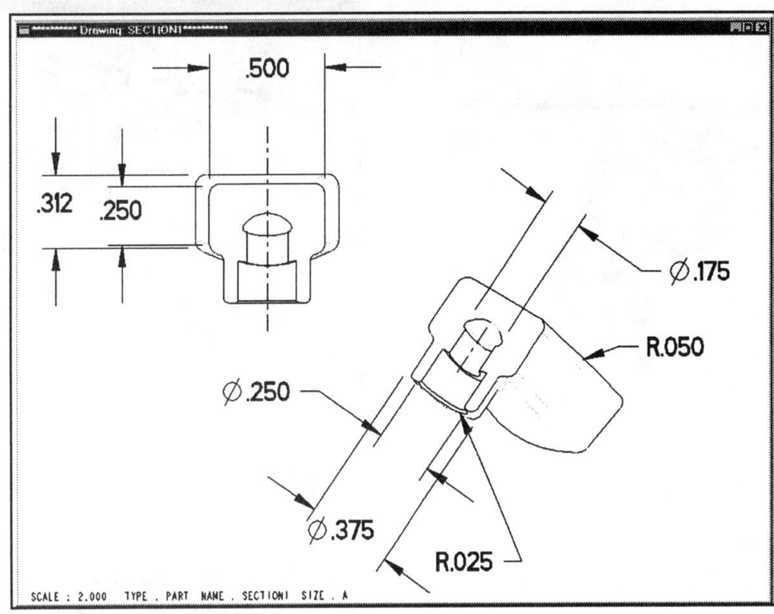

Figure 12.54
Bathroom Faucet Drawing, Spout Section

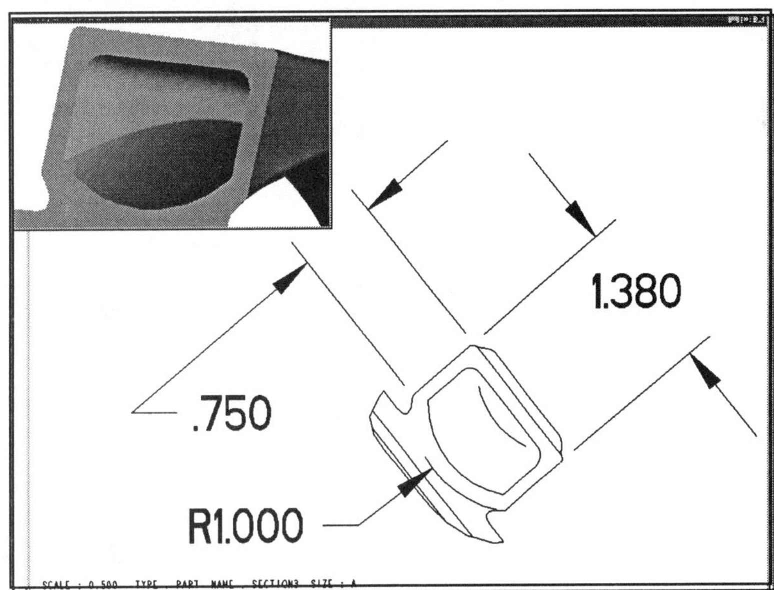

Figure 12.55
Bathroom Faucet Drawing, Second Blend Section

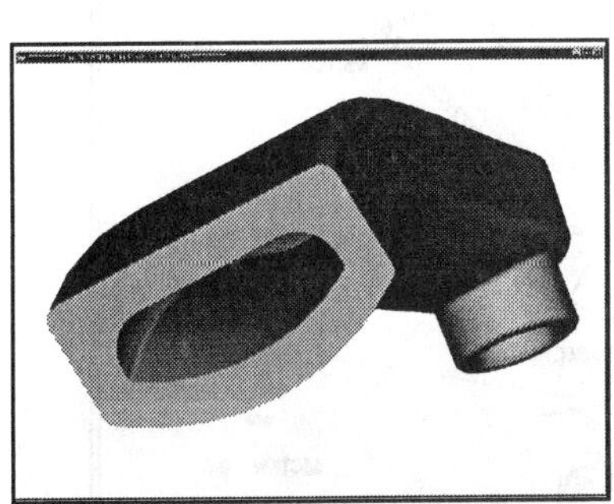

Figure 12.56
Bathroom Faucet Drawing, Third Blend Section

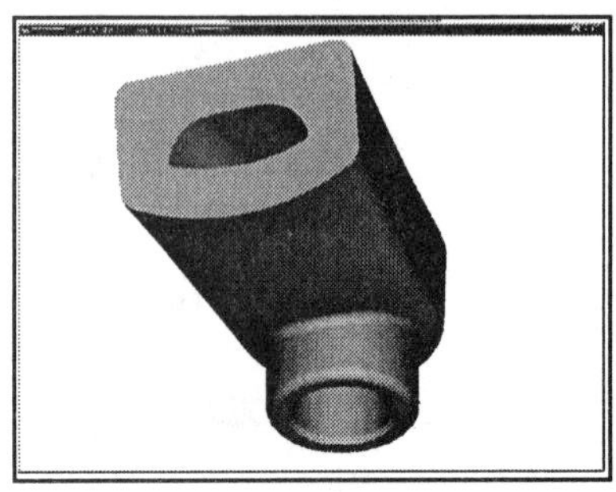

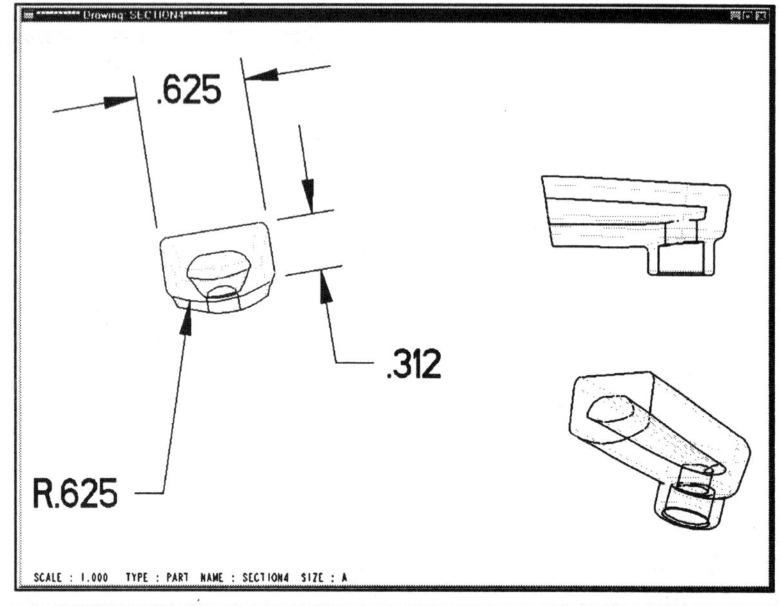

Figure 12.57
Bathroom Faucet Drawing,
Fourth Blend Section

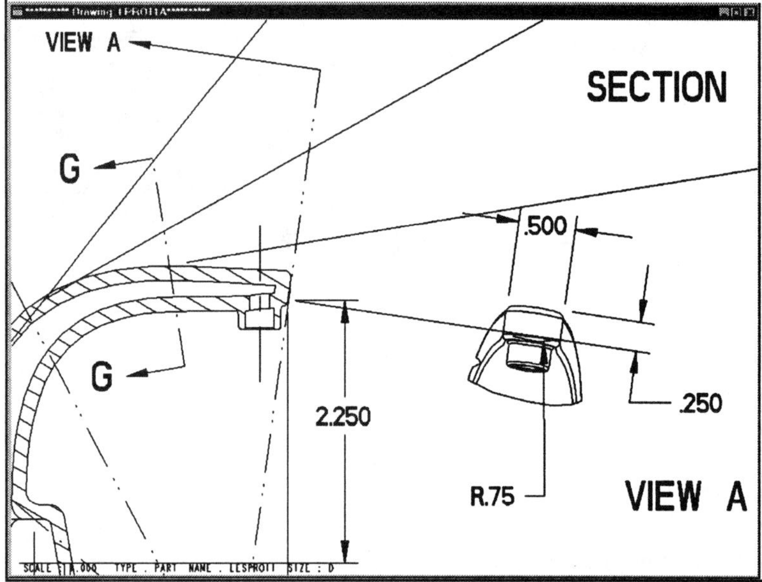

Figure 12.58
Bathroom Faucet Drawing,
Fifth Blend Section

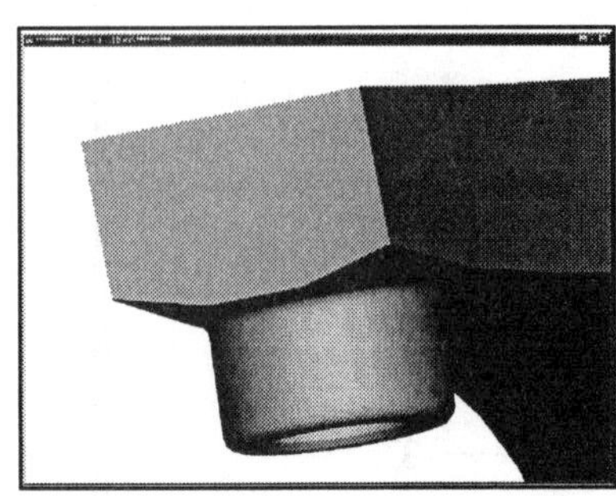

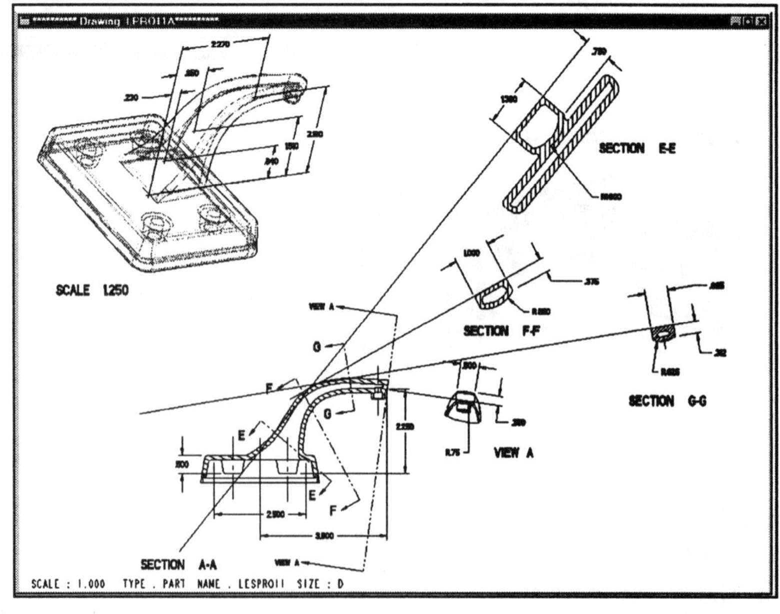

Figure 12.59
Bathroom Faucet Drawing,
Sections

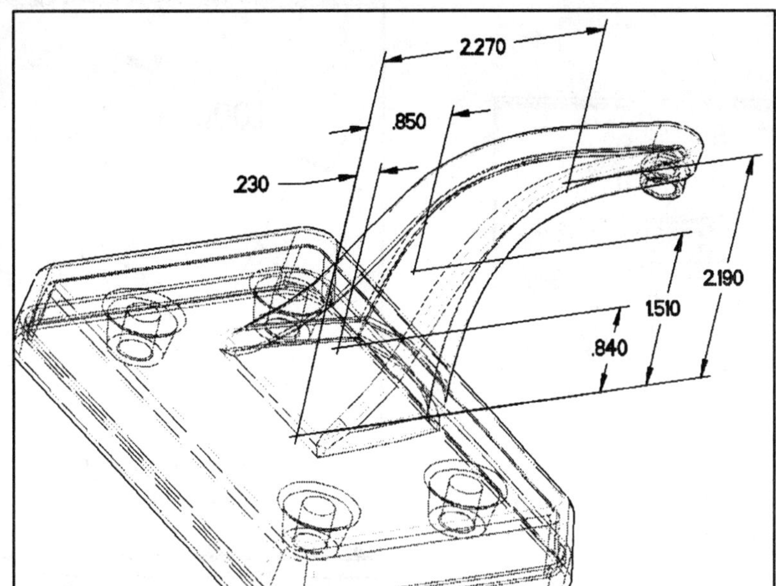

Figure 12.60
Bathroom Faucet Drawing, First Three Blend Section Locations

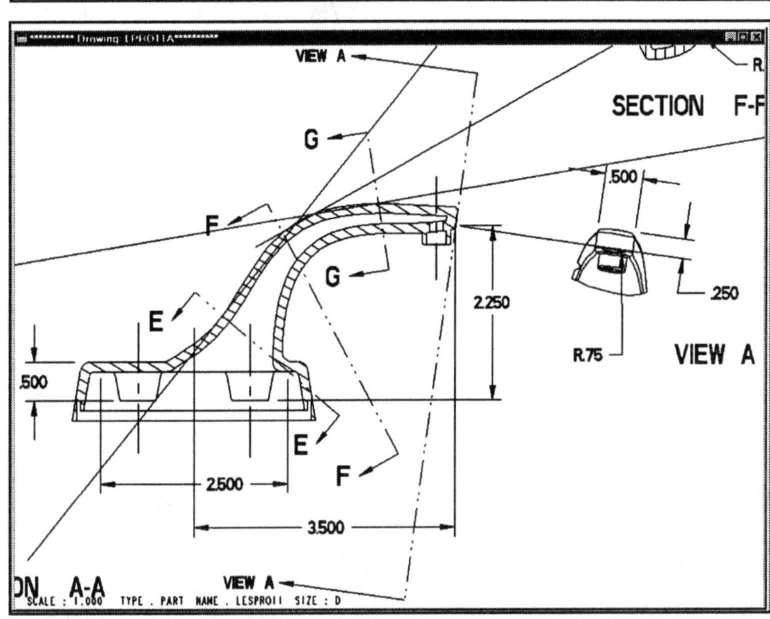

Figure 12.61
Bathroom Faucet Drawing, **SECTION A-A**

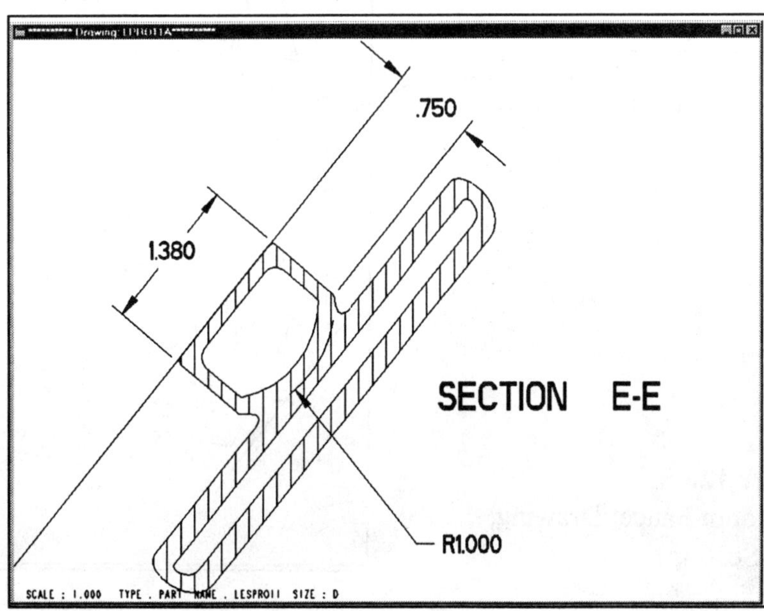

Figure 12.62
Bathroom Faucet Drawing, **SECTION E-E**

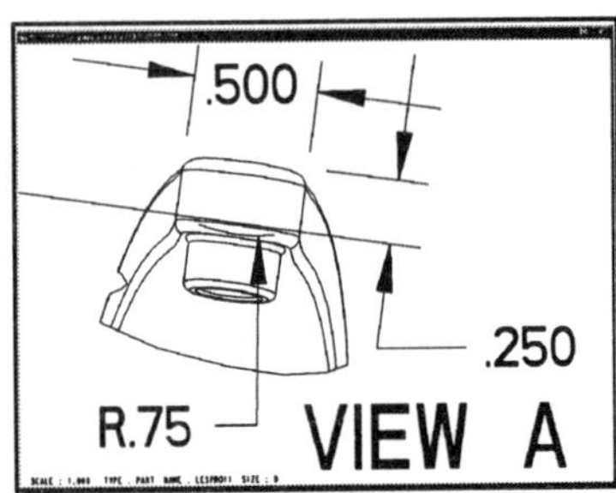

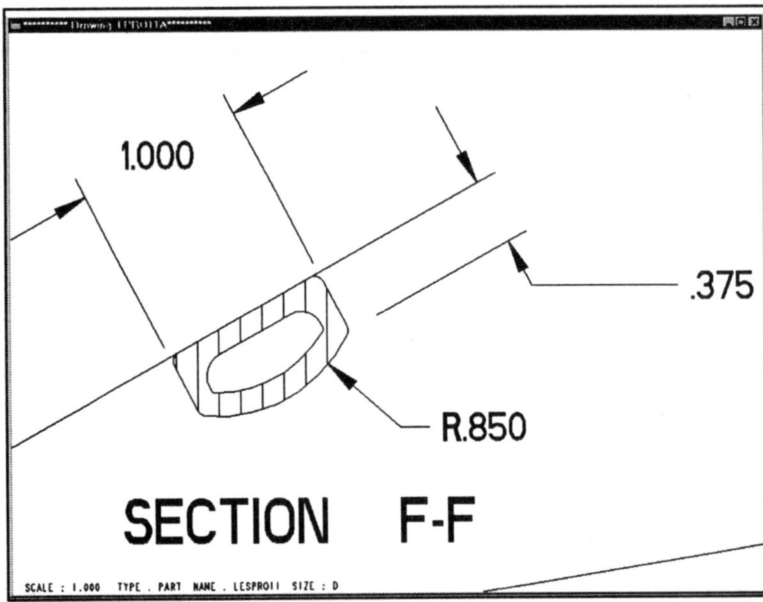

Figure 12.63
Bathroom Faucet Drawing,
SECTION F-F

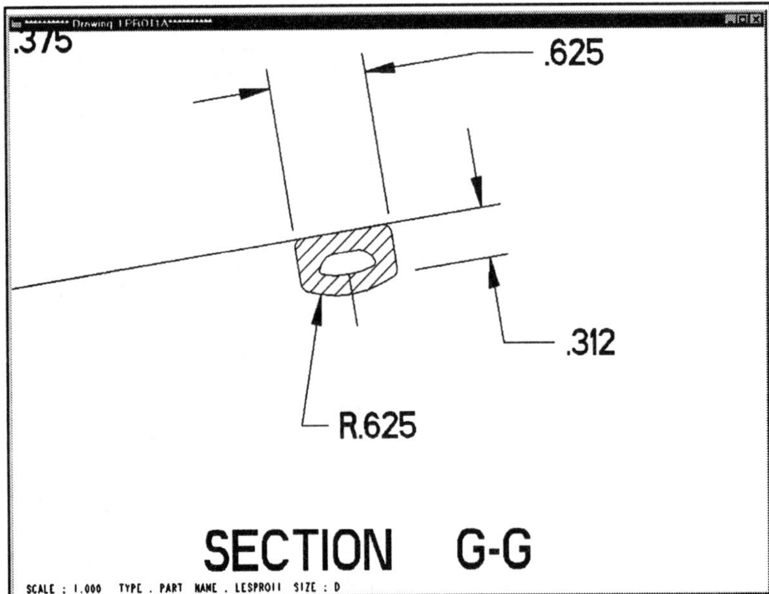

Figure 12.64
Bathroom Faucet Drawing,
SECTION G-G

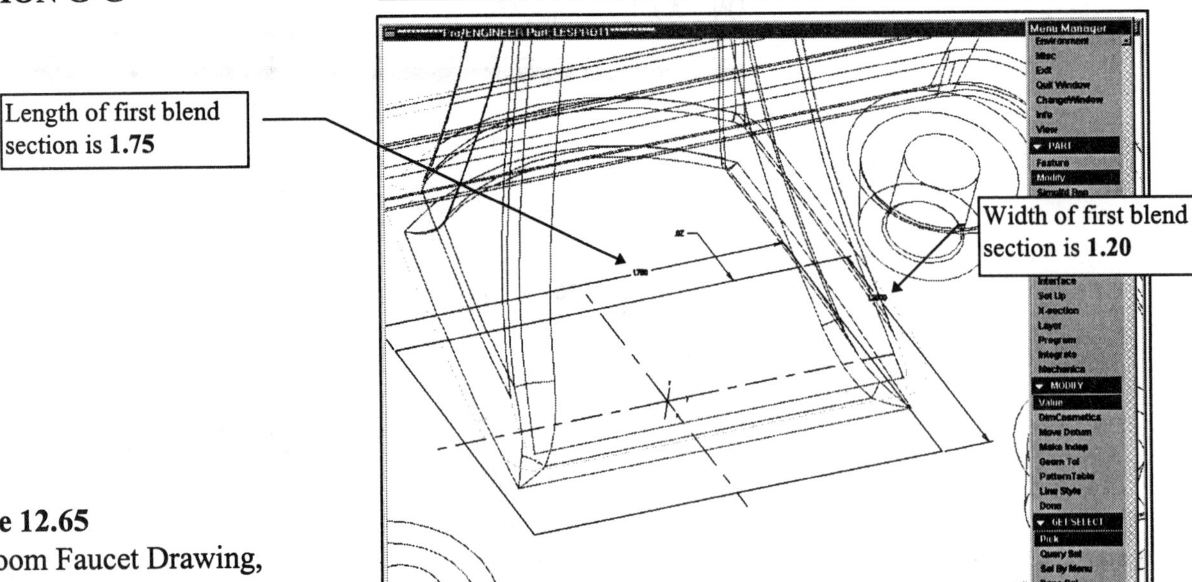

Figure 12.65
Bathroom Faucet Drawing,
First Section

Figure 12.66
Bathroom Faucet, First Protrusion and Draft

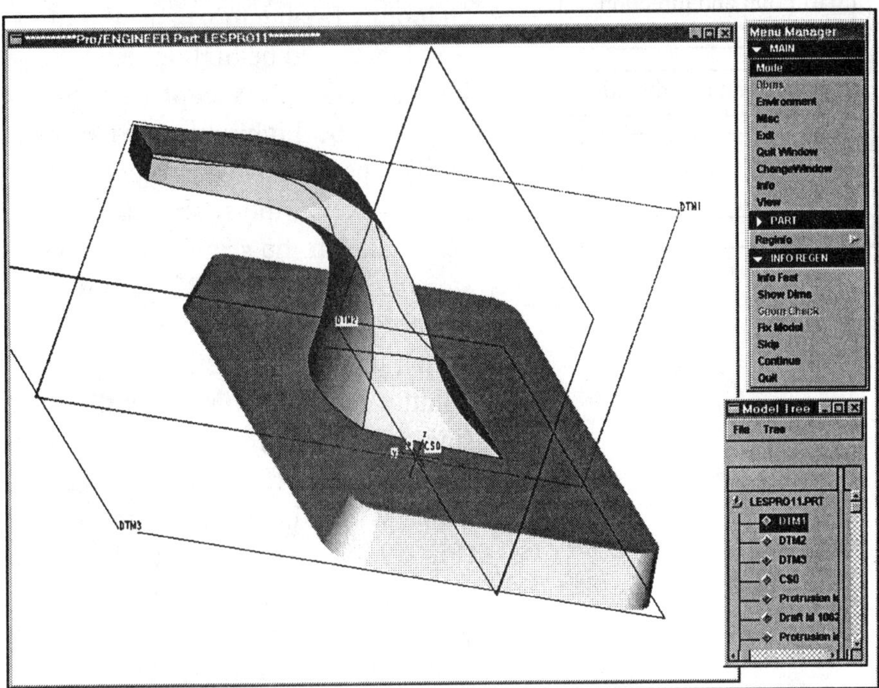

Figure 12.67
Bathroom Faucet, Swept Protrusion

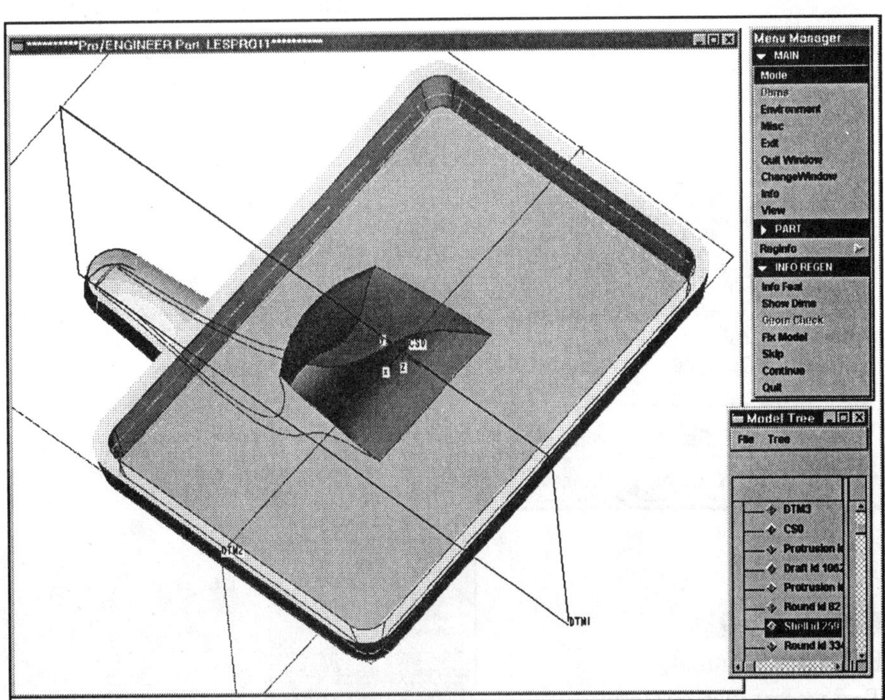

Figure 12.68
Bathroom Faucet, **Shell**

HINT

Each section in the Swept Blend has four entities. Keep the start point direction arrow of each section in the same region and facing the same direction.

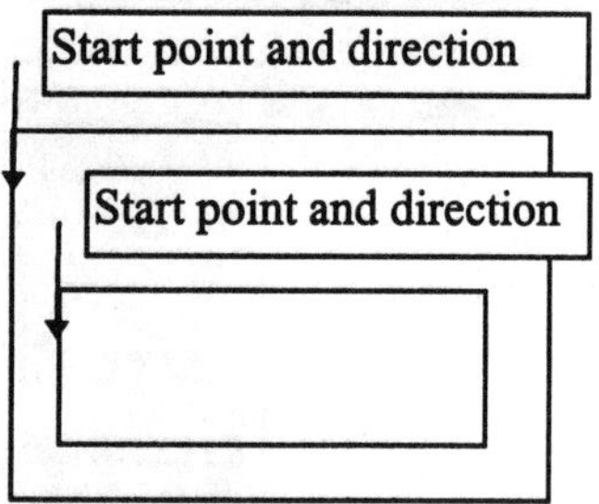

Create the first protrusion and draft as shown in Figure 12.66. Next, a **Swept Blend** will be used to create the geometry for the **Bathroom Faucet** as shown in Figures 12.69 through 12.82. The default options of **Sketch Sec** and **Nrm To Spine** will be used. When prompted for the trajectory, choose **Sketch Traj**. Sketch a trajectory (spine) with the **Spline** option in the **Adv Geometry** menu. Create a total of five points along the trajectory. These will locate the five sections of the blend. Before you begin to sketch any sections, you are prompted for where the sections are to be located. A total of five sections will be sketched for this protrusion. Two will be at the endpoints (mandatory) and the other three at the datum points. The first location to be highlighted will be the second point of the sketched trajectory. Select **Accept** from the CONFIRM menu. The next point will then be highlighted, so **Accept** this location. Choose **Accept** until all points have been accepted.

To sketch the first section, accept the default **Z**-axis rotation of zero. Sketch the section as shown in Figure 12.72. Be aware of the *start point* location. When finished with the first section, Pro/E will prompt you for the next **Z**-axis rotation. Accept the default value and sketch the second section (Fig. 12.73). Again, keep track of the start point; it must match up with the first section. Sketch the third section (Fig. 12.75), again using a **Z**-axis rotation value of zero. Sketch the fourth section (Fig. 12.77), again using a **Z**-axis rotation value of zero. When finished with the fourth section, sketch the fifth section (Fig. 12.79). Again, accept the default of zero for the **Z**-axis rotation.

The completed feature should look like Figure 12.81. **Preview** your feature elements before choosing **OK**, to check for twist in the swept blend because of start points that do not line up. If there is a problem, use the Sections element from the dialog box and choose **Define** to change the start points so that they are aligned.

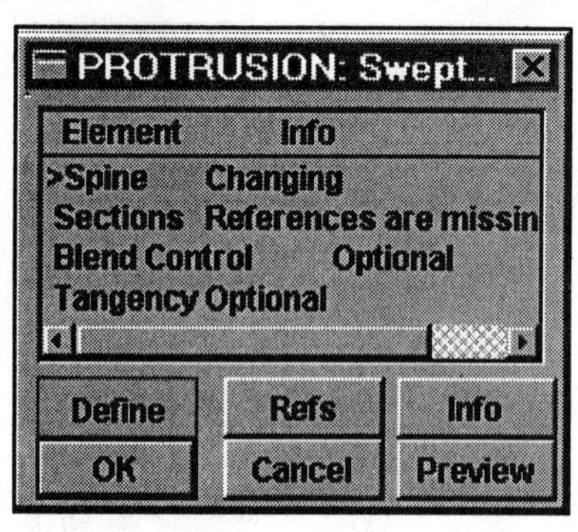

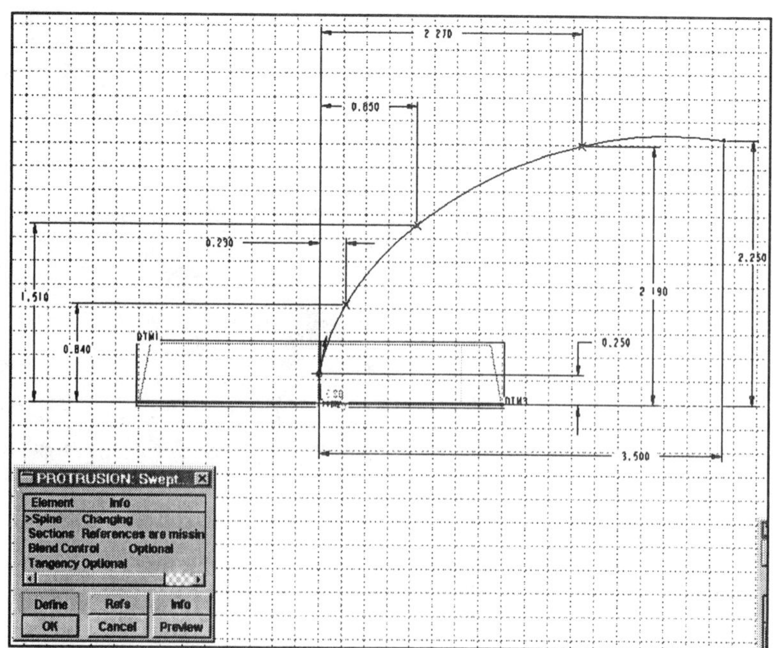

Figure 12.69
Bathroom Faucet, Trajectory Dimensions

Create the **Swept Blend** with the following commands:

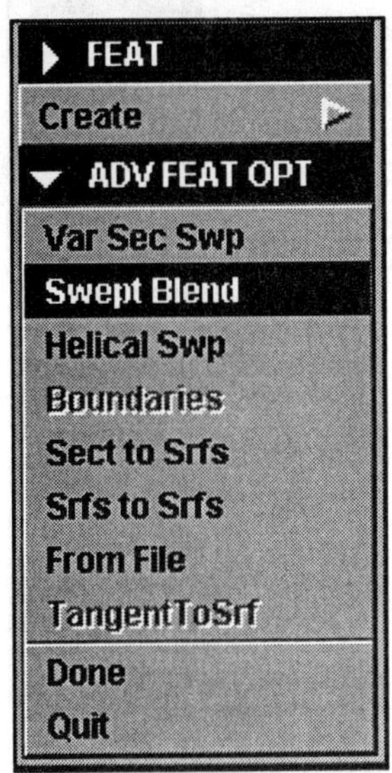

Feature ⇒ Create ⇒ Protrusion ⇒ Advanced ⇒ Solid ⇒ Done ⇒ Swept Blend ⇒ Done ⇒ Sketch Sec ⇒ Nrm To Spine ⇒ Done Sketch Traj ⇒ (pick **DTM1**) **⇒ Okay ⇒ Right ⇒** (pick **DTM2**) **⇒ Adv Geometry ⇒ Spline ⇒ Sketch Points ⇒** (sketch the trajectory as shown in Fig. 12.69) **⇒ Regenerate ⇒ Alignment** (align endpoint to **DTM2**) **⇒ Regenerate ⇒ Dimension** (pick the spline to display the points, dimension as shown in Fig. 12.69) **⇒ Regenerate ⇒ Modify ⇒ Regenerate ⇒ Done ⇒ Accept** (all three middle points will be highlighted one at a time as you accept them. The first and last points are automatically accepted) **⇒ enter** (accept the default of **0°**) **⇒** (sketch the first section centered about the light blue crosshairs--starting point--each section will have one) **⇒ Regenerate ⇒ Alignment ⇒ Regenerate ⇒ Dimension ⇒ Regenerate ⇒ Modify ⇒ Regenerate ⇒ Done ⇒** (continue creating the last four sections) **⇒ Done ⇒ Preview ⇒ OK**

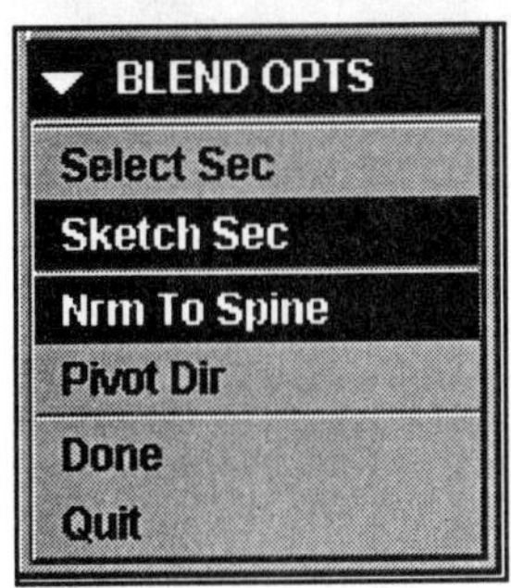

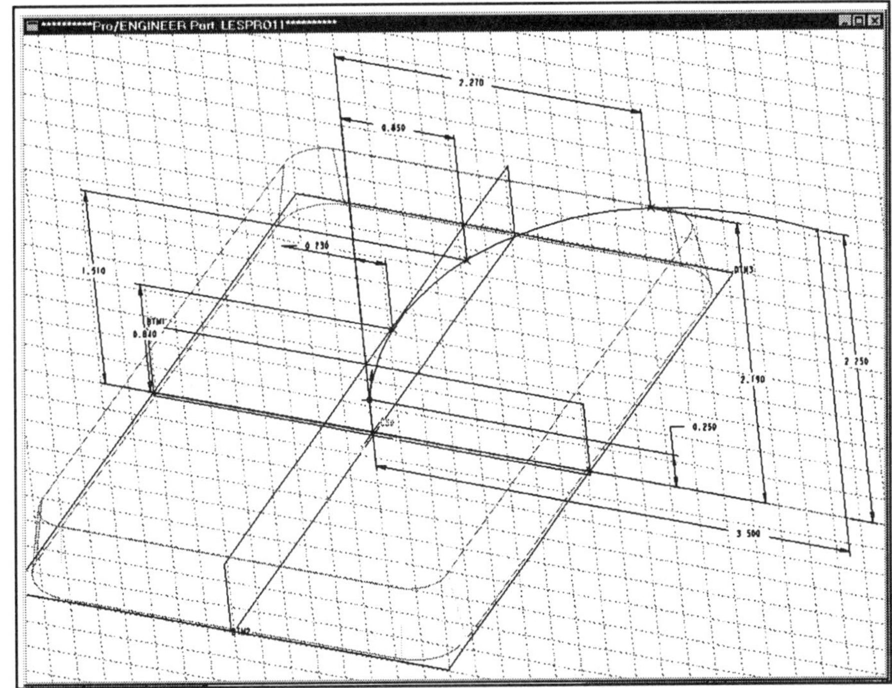

HINT

Create the first section approximately centered about its respective light blue crosshairs--starting point. Use centerlines when possible.

Figure 12.70
Bathroom Faucet,
First Section: Pictorial

The *first section* is **1.75 X 1.20,** shown in Figures 12.71 and 12.72. The *second section* is **1.380 X .75 X R1.00,** shown in Figures 12.55, 12.73, and 12.74. The *third section* is **1.00 X .375 X R.850,** shown in Figures 12.56, 12.75, and 12.76. The *fourth section* is **.625 X .312 X R.625,** shown in Figures 12.57, 12.77, and 12.78. The *fifth section* is **.500 X .25 X R.75,** shown in Figures 12.58, 12.79, and 12.80. Figure 12.81 shows the final part design.

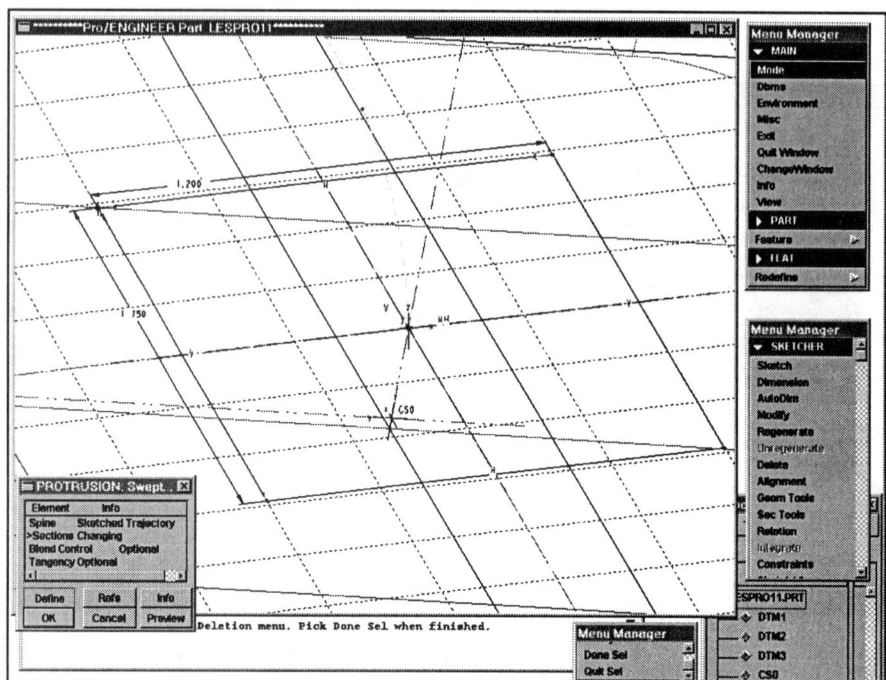

Figure 12.71
Bathroom Faucet,
Trajectory Dimensions

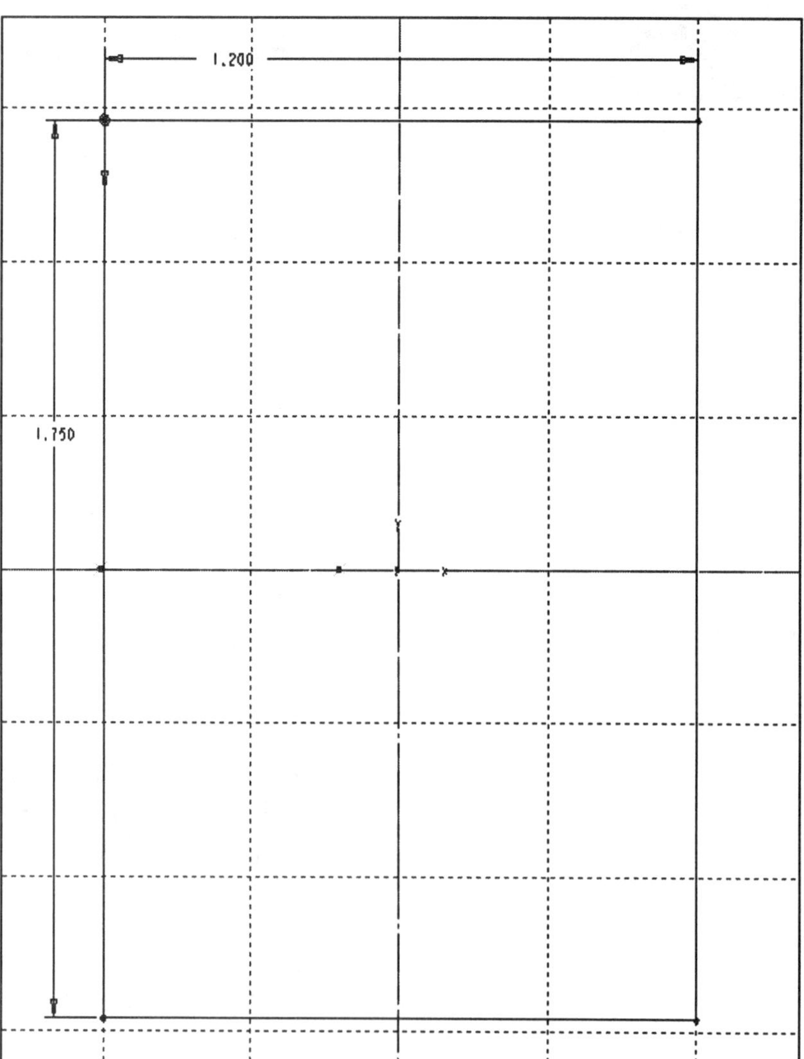

Figure 12.72
Bathroom Faucet,
First Section: **1.75 X 1.20**

HINT

For the second section, align the arc center to the light blue *crosshairs*--starting point.

Figure 12.73
Bathroom Faucet,
Second Section:
1.380 X .75 X R1.00

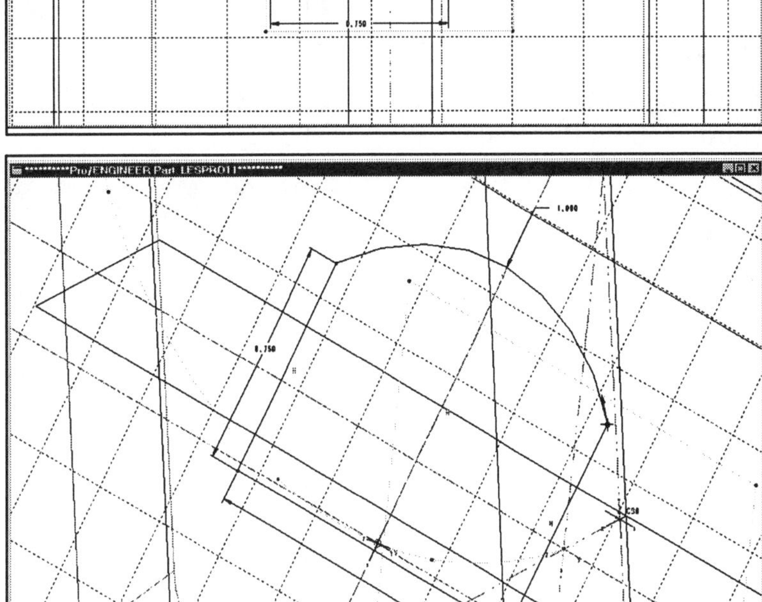

Figure 12.74
Bathroom Faucet,
Second Section: Pictorial

HINT

Note where all endpoints are in relation to the light blue *crosshairs* (starting point).

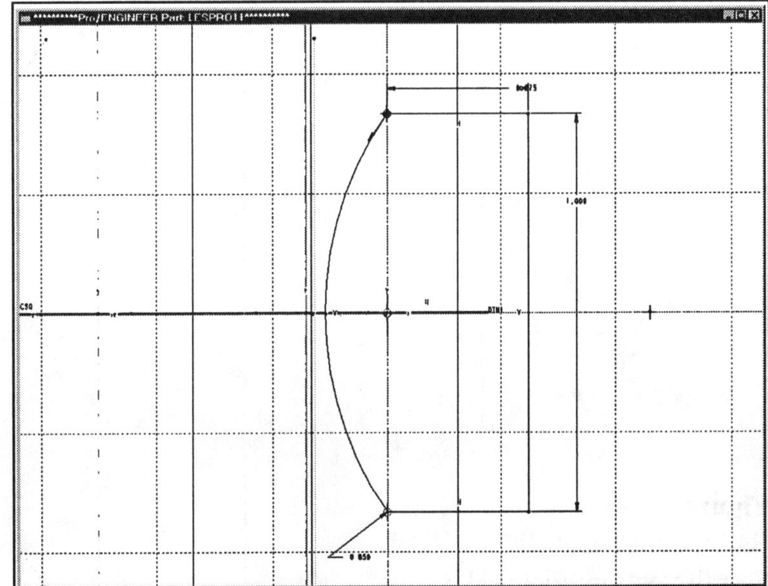

Figure 12.75
Bathroom Faucet,
Third Section:
1.00 X .375 X R.850

Figure 12.76
Bathroom Faucet,
Third Section: Pictorial

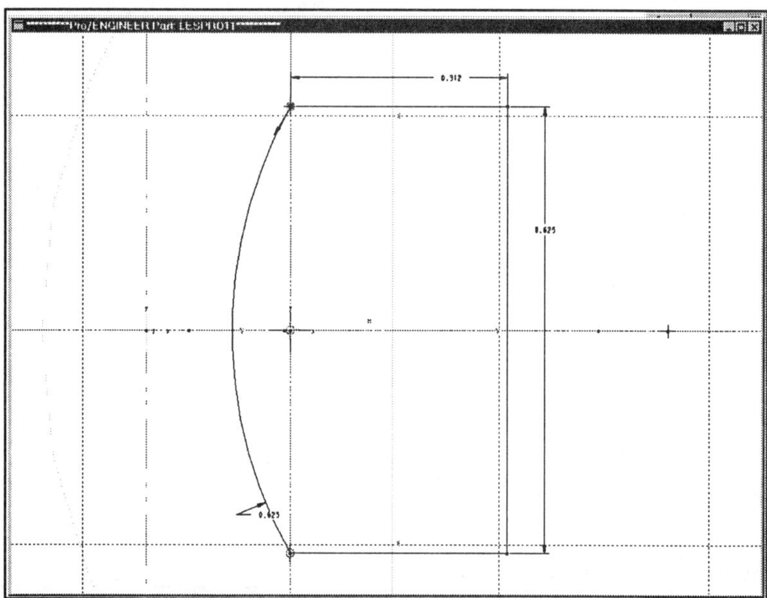

Figure 12.77
Bathroom Faucet
Fourth Section:
.625 X .312 X R.625

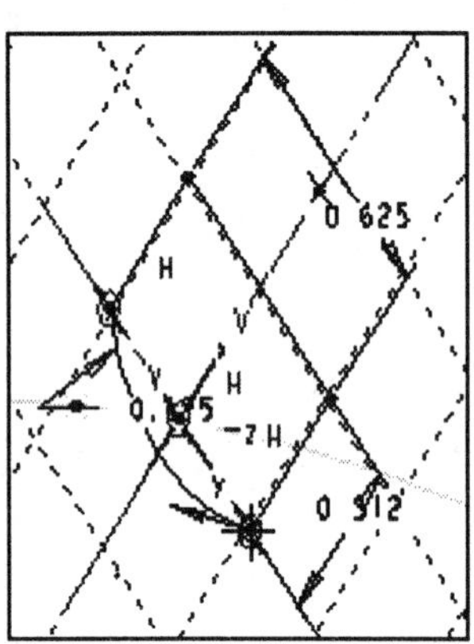

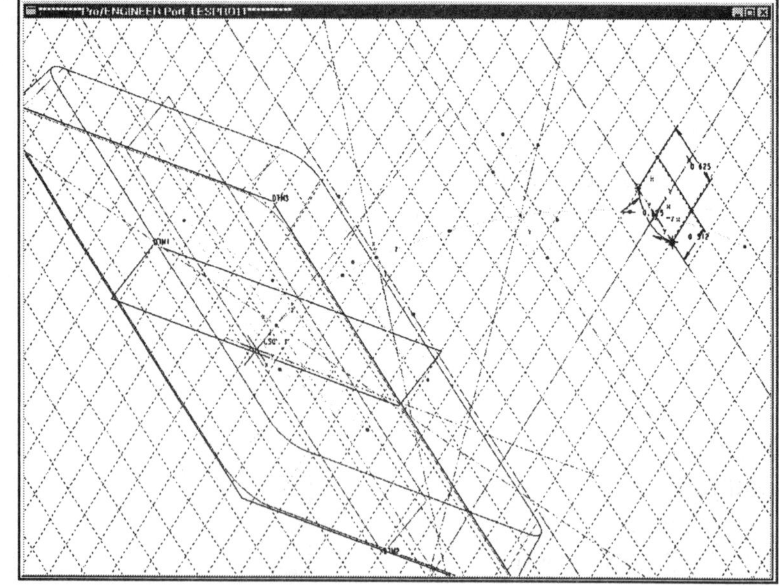

Figure 12.78
Bathroom Faucet,
Fourth Section: Pictorial

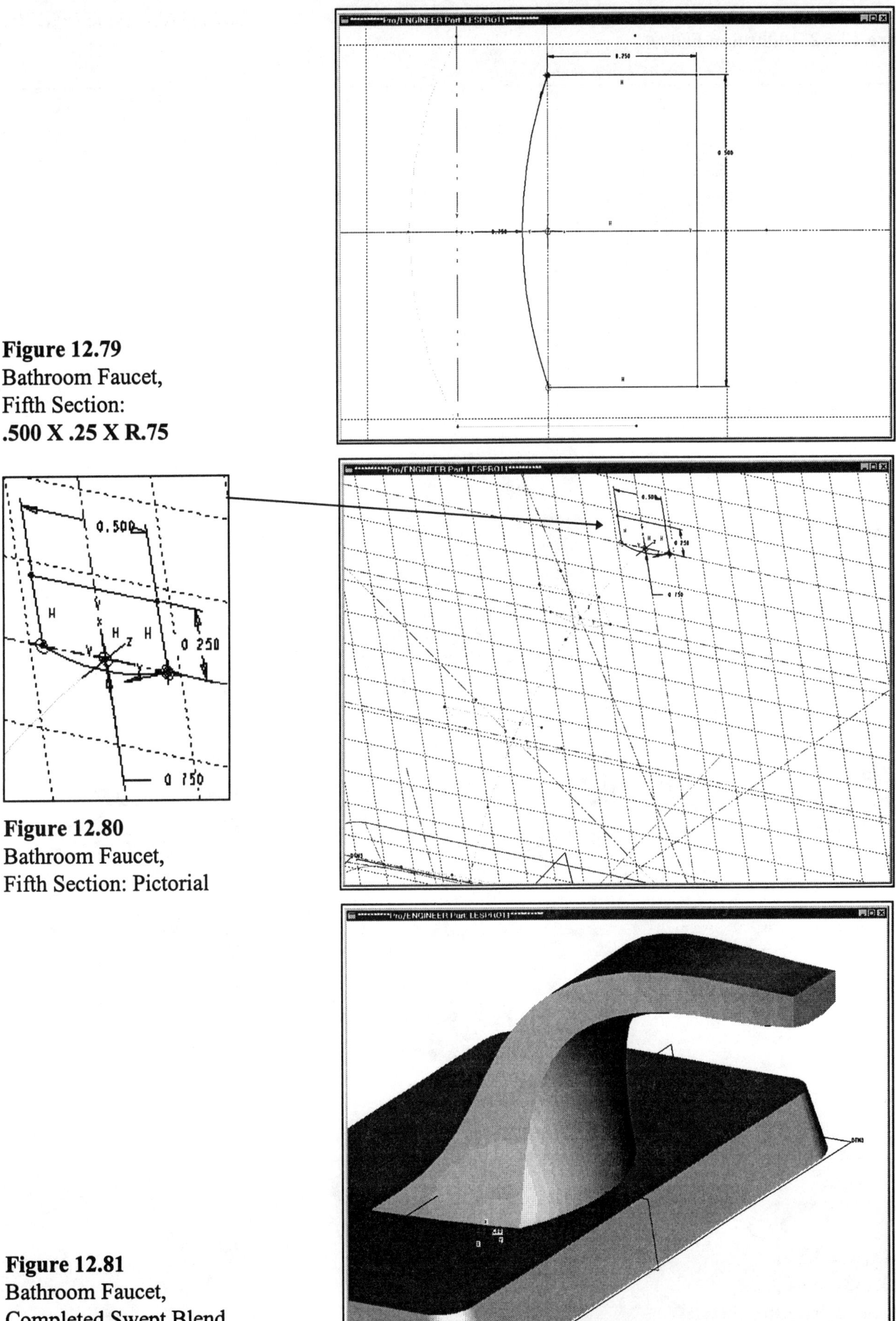

Figure 12.79
Bathroom Faucet,
Fifth Section:
.500 X .25 X R.75

Figure 12.80
Bathroom Faucet,
Fifth Section: Pictorial

Figure 12.81
Bathroom Faucet,
Completed Swept Blend

Figure 12.82 Bathroom Faucet

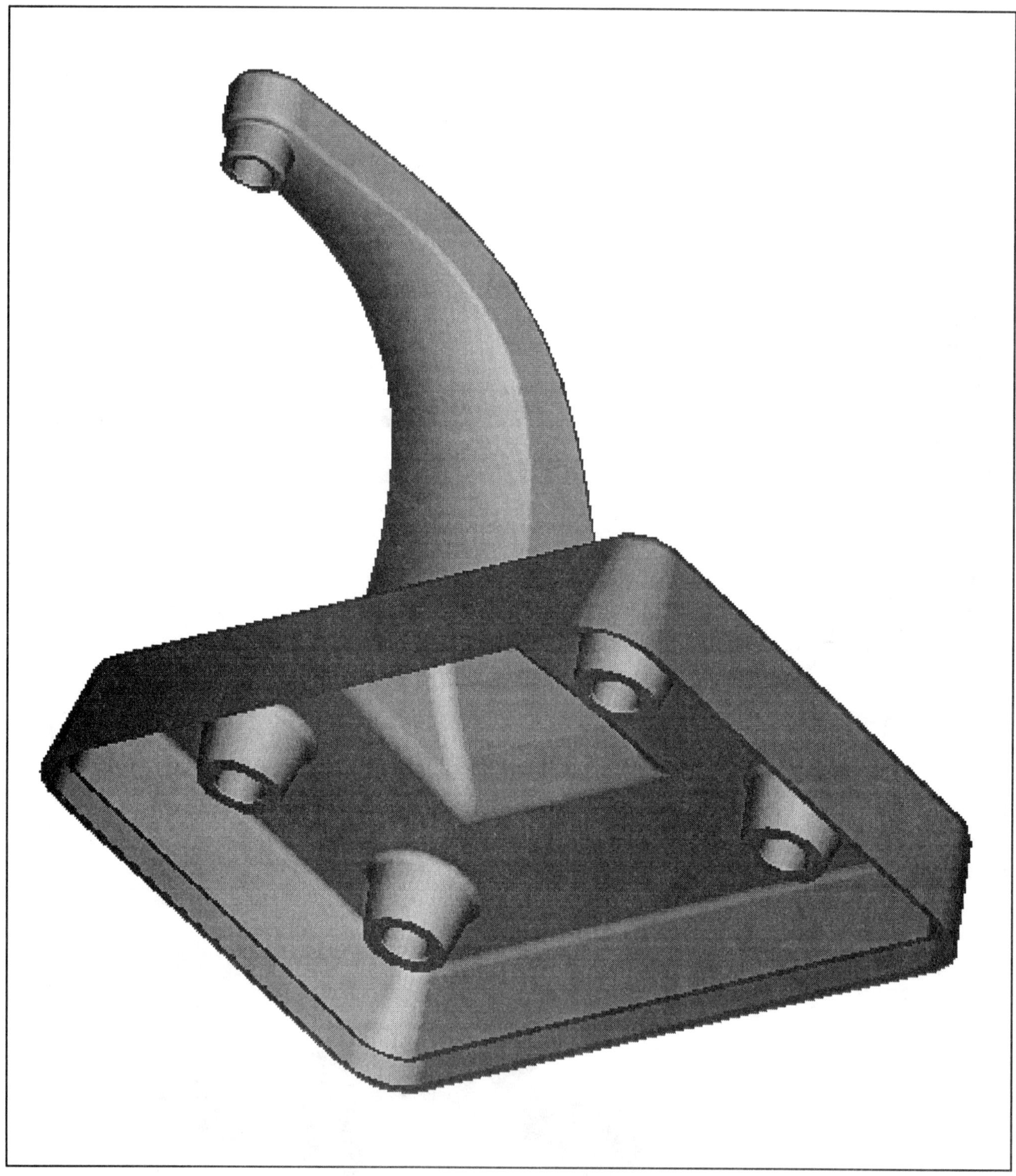

Lesson 13

Helical Sweeps

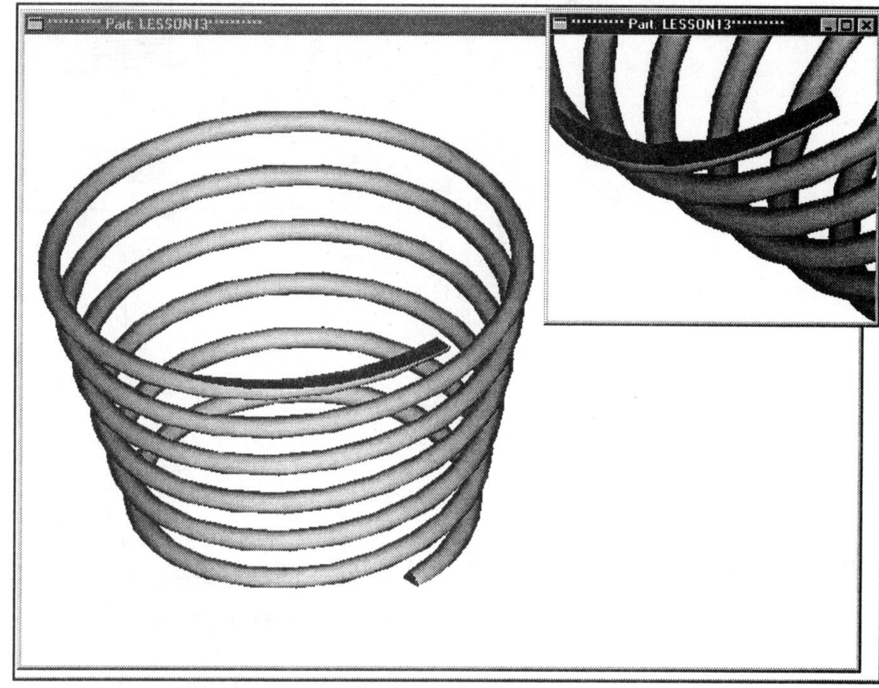

Figure 13.1
Helical Compression Spring

OBJECTIVES

1. **Create springs with a helical sweep**

2. **Model a helical compression spring**

3. **Use sweeps to create hooks on extension springs**

4. **Design an extension spring with a machine hook**

5. **Create plain ground ends on a spring**

6. **Model a convex spring**

EGD REFERENCE
Engineering Graphics and Design with Graphical Analysis *or* **Fundamentals of Engineering Graphics and Design**
by L. Lamit and K. Kitto
Read Chapter 18
See pages 702-704

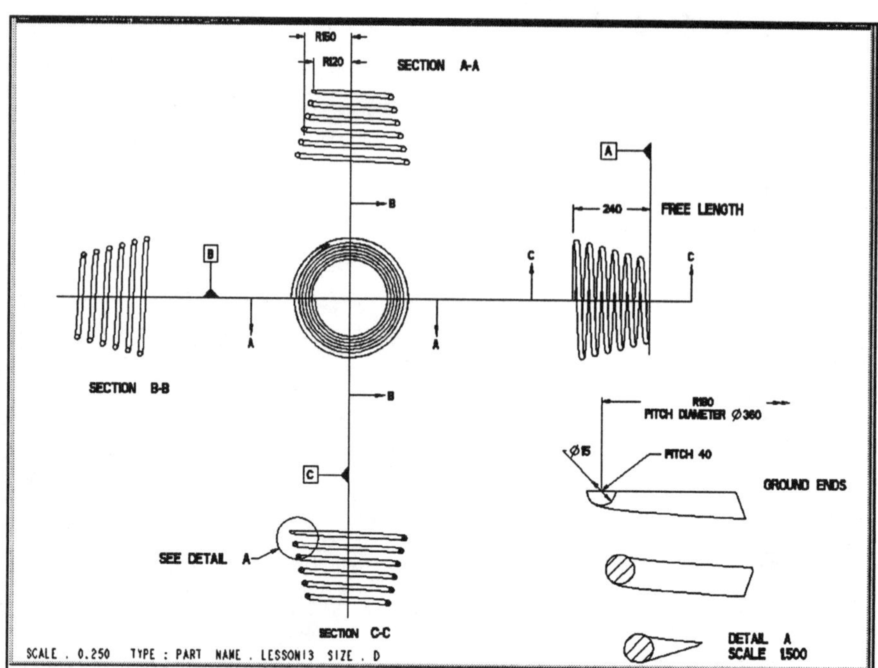

Figure 13.2
Helical Compression Spring Dimensions

HELICAL SWEEPS

A **helical sweep** (Figs. 13.1 through 13.4) is created by sweeping a section along a helical *trajectory*. The trajectory is defined by both the *profile* of the *surface of revolution* (which defines the distance from the section origin of the helical feature to its *axis of revolution*) and the *pitch* (the distance between coils). The trajectory and the surface of revolution are construction tools and do not appear in the resulting geometry.

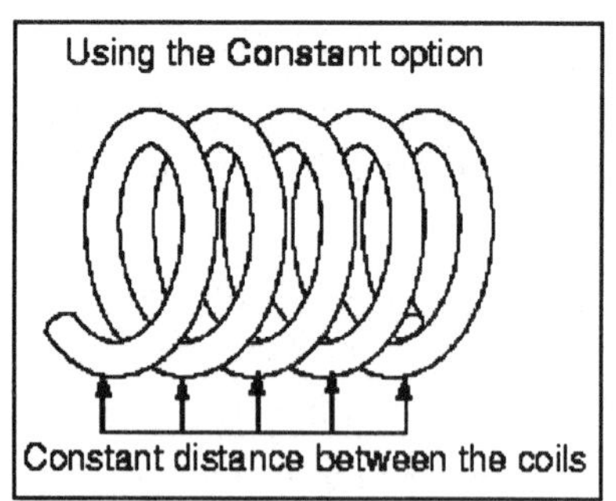

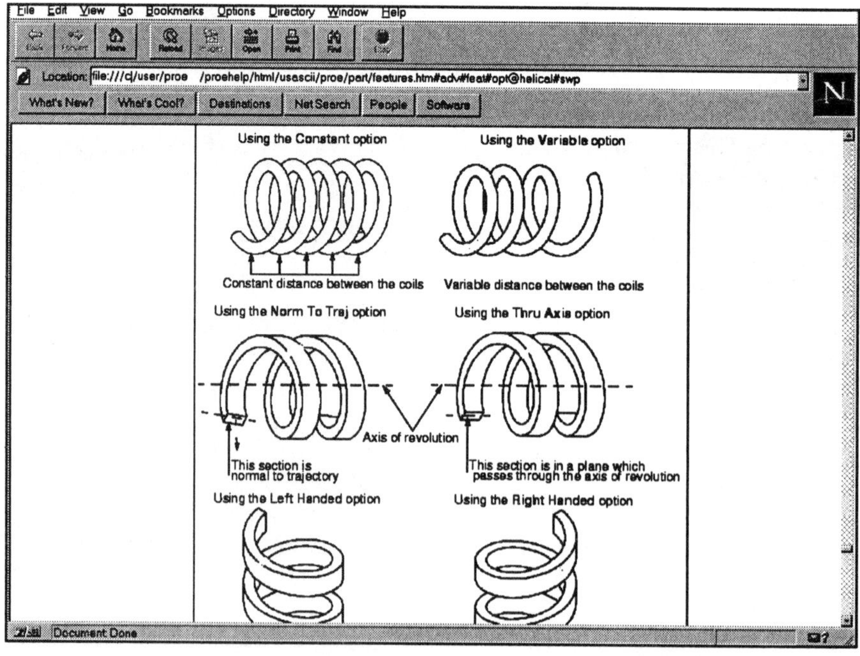

Figure 13.3
Online Documentation, Helical Sweeps

The **Helical Swp** option in the ADV FEAT OPT menu is available for both solid and surface features. Use the following ATTRIBUTES menu options, presented in mutually exclusive pairs, to define the helical sweep feature:

Constant The pitch is constant.
Variable The pitch is variable and defined by a graph.
Thru Axis The section lies in a plane that passes through the axis of revolution.
Norm To Traj The section is oriented normal to the trajectory (or surface of revolution).
Right Handed The trajectory is defined by the right-hand rule.
Left Handed The trajectory is defined by the left-hand rule.

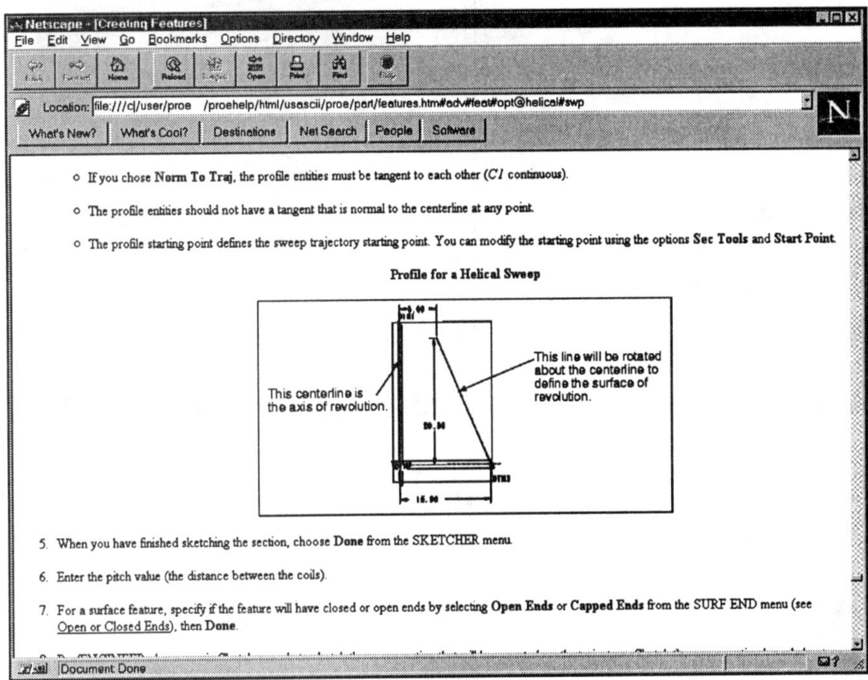

Netscape - [Creating Features]

File Edit View Go Bookmarks Options Directory Window Help

Location: file:///c|/user/proe /proehelp/html/usascii/proe/part/features.htm#advfeat#opt@helical#swp

What's New? What's Cool? Destinations Net Search People Software

- If you chose **Norm To Traj**, the profile entities must be tangent to each other (*C1* continuous).
- The profile entities should not have a tangent that is normal to the centerline at any point.
- The profile starting point defines the sweep trajectory starting point. You can modify the starting point using the options **Sec Tools** and **Start Point**.

Profile for a Helical Sweep

This centerline is the axis of revolution.

This line will be rotated about the centerline to define the surface of revolution.

5. When you have finished sketching the section, choose **Done** from the SKETCHER menu.
6. Enter the pitch value (the distance between the coils).
7. For a surface feature, specify if the feature will have closed or open ends by selecting **Open Ends** or **Capped Ends** from the SURF END menu (see Open or Closed Ends), then **Done**.

Document: Done

Figure 13.4
Online Documentation, Profile for a Helical Sweep

Constant Pitch Helical Sweeps

To create a helical sweep with a constant pitch value, choose the following command sequence:

1. Choose **Advanced ⇒ Done** from the SOLID OPTS menu, then **Helical Swp ⇒ Done.** Pro/E displays the feature creation dialog box.
2. Define the feature by selecting from the ATTRIBUTES menu, then choose **Done.**
3. Pro/E places you in Sketcher mode. Specify the sketching plane and its orientation, and the axis of revolution. Sketch, dimension, and regenerate the surface of revolution profile. The sketched entities must form an *open loop*. As with a revolved feature, you must *sketch a centerline* to define the **axis of revolution**.
4. Pro/E places you in another Sketcher orientation. Sketch, dimension, and regenerate the section profile to be swept along the trajectory.

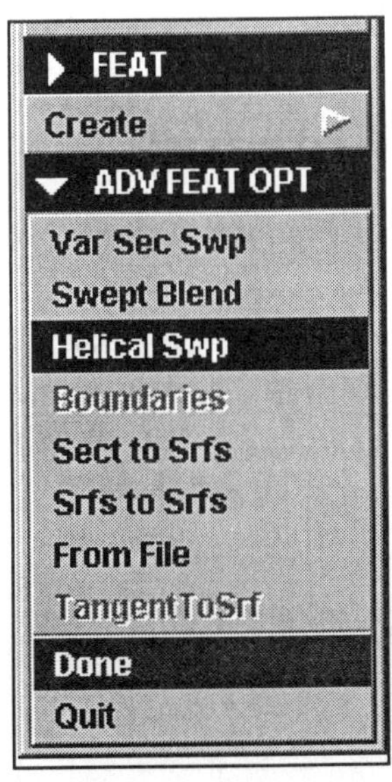

If you choose **Norm To Traj,** the profile entities must be tangent to each other (continuous). The profile entities should not have a tangent that is normal to the centerline at any point. The profile starting point defines the sweep trajectory starting point. Modify the starting point using the options **Sec Tools** and **Start Point**.

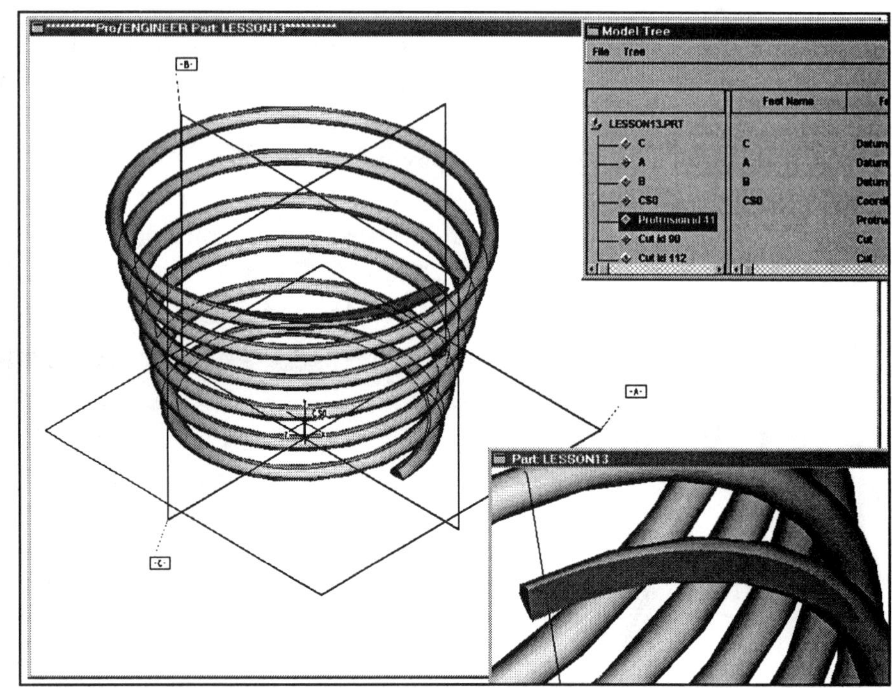

Figure 13.5
Helical Compression Spring with Datum Planes and Model Tree

Helical Compression Spring

Springs (Fig. 13.5) and other helical features are created with the **Helical Swp** command. A helical sweep is created by sweeping a *section* along a *trajectory* that lies in the *surface of revolution:* the trajectory is defined by both the *profile* of the surface of revolution and the distance between coils. The part for this lesson is a *constant-pitch right-handed helical compression spring with ground ends, a pitch of* ***40 mm****, and a wire diameter of* ***15 mm*** (Figs. 13.6 through 13.9).

 *NOTE*

Set up the Helical Compression Spring:

- Material = Spring Steel
- Units = Millimeters
- Default Datum Planes
- Default Coordinate System
- Layers = **DATUM_LAYER**
- Hidden line
- No Disp Tan
- ✓Grid Snap

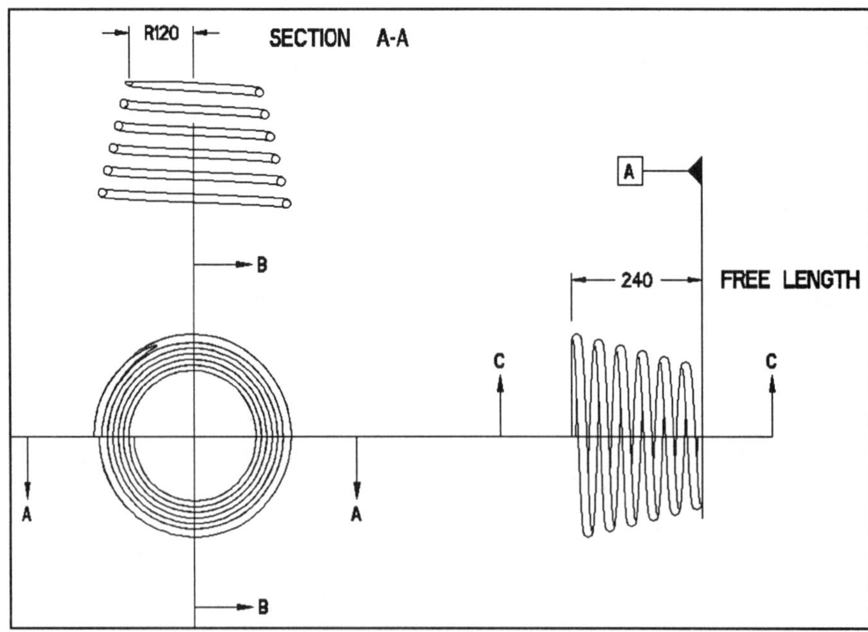

Figure 13.6
Helical Compression Spring

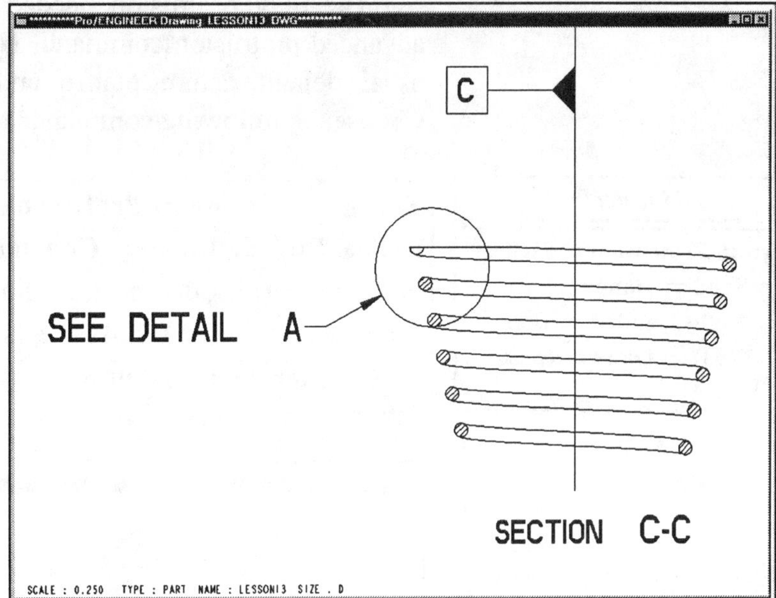

Figure 13.7
Helical Compression Spring Drawing, **SECTION C-C**

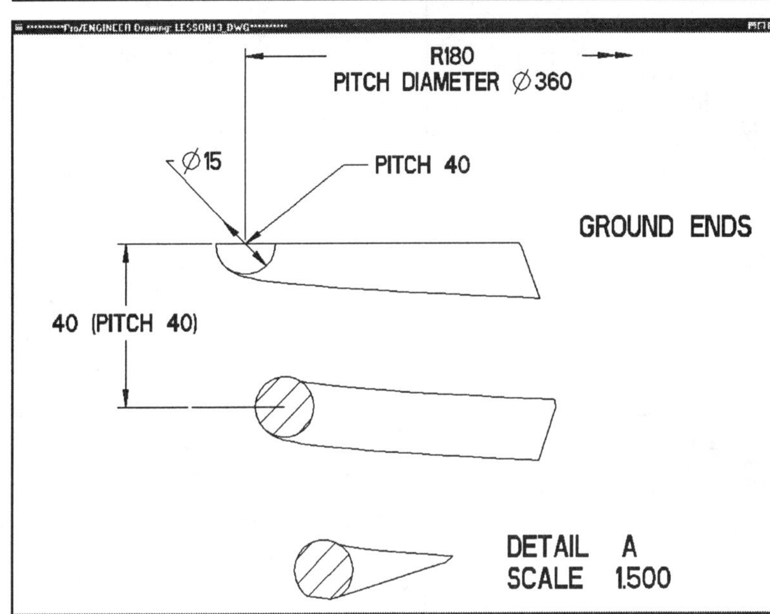

Figure 13.8
Helical Compression Spring Drawing, **DETAIL A**

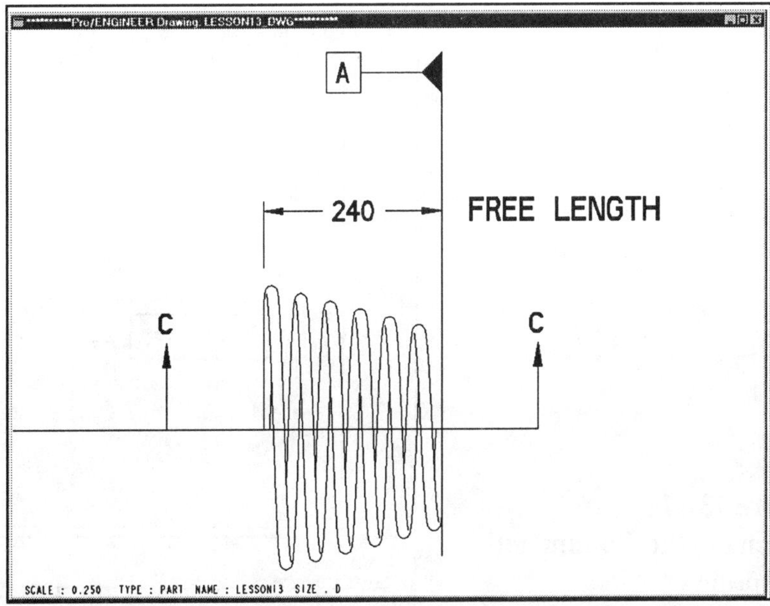

Figure 13.9
Helical Compression Spring Drawing, Right Side View

The only protrusion needed for the spring is created with an advanced protrusion command, **Helical Swp**. Start the part with the usual default datum planes and coordinate system (Fig. 13.10). Choose the following commands to create the first protrusion:

PT/Modeler™
Feature ⇒ Protrusion ⇒ Helical Swp ⇒ Solid ⇒ Done ⇒ Thru Axis ⇒ Right Handed ⇒ Done ⇒ (pick **DTM3**) **⇒ Okay ⇒** (pick **DTM2**) etc.

Feature ⇒ Create ⇒ Protrusion ⇒ Advanced ⇒ Solid ⇒ Done ⇒ Helical Swp ⇒ Done ⇒ Constant ⇒ Thru Axis ⇒ Right Handed ⇒ Done ⇒ (pick datum **B**, originally **DTM3**) **⇒ Okay ⇒ Top** (pick datum **A**, originally **DTM2**) ⇒ (sketch a line as shown in Fig. 13.11) **⇒ Regenerate ⇒ Alignment** (align the lower end of the line to datum **A**, originally **DTM2**)

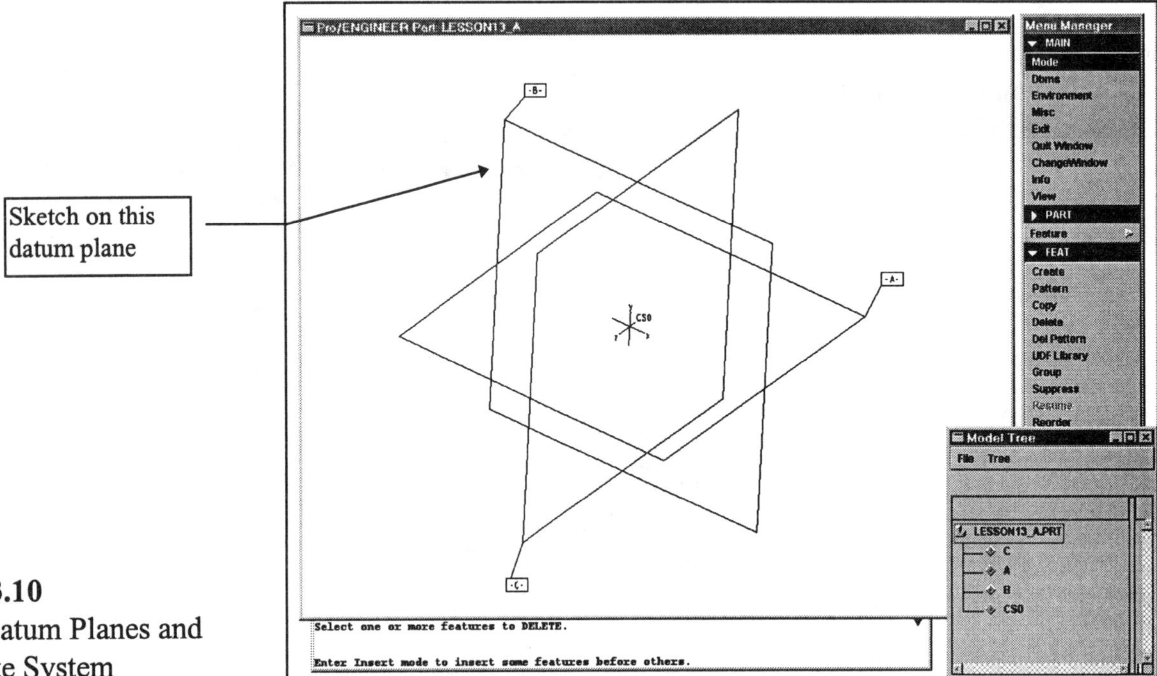

Figure 13.10
Default Datum Planes and Coordinate System

Sketch one line

Align here

Figure 13.11
Sketch the Profile Line for the Spring

> **Regenerate** ⇒ **Dimension** (add dimensions as shown in Fig. 13.12) ⇒ **Regenerate** ⇒ **Sketch** ⇒ **Line** ⇒ **Centerline** ⇒ **Vertical** ⇒ (add a vertical axis line along datum **C, DTM1**) ⇒ **Alignment** (align the axis line to datum **C, DTM1**) ⇒ **Regenerate** ⇒ **Modify** (change the values to the design sizes) ⇒ **Regenerate** ⇒ **Done** ⇒ (enter the pitch value **40** at the prompt, as shown in Fig. 13.13) ⇒ **enter**

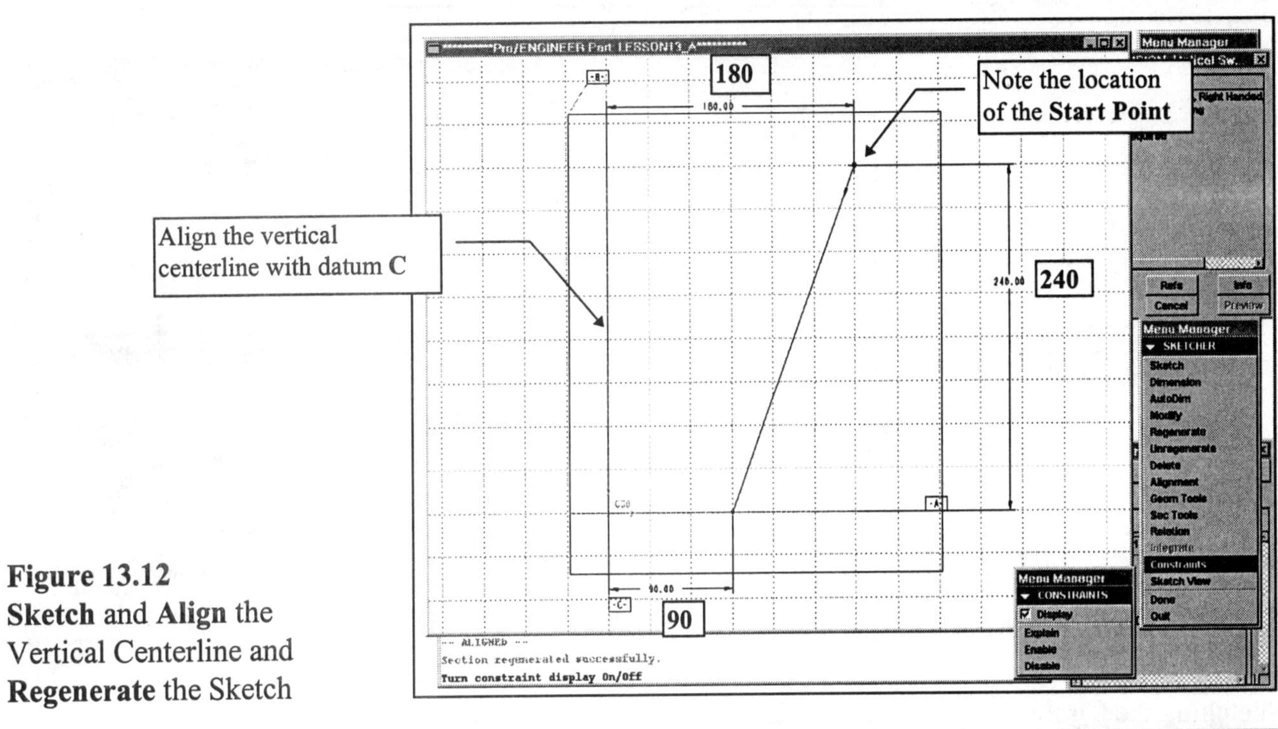

Figure 13.12
Sketch and **Align** the Vertical Centerline and **Regenerate** the Sketch

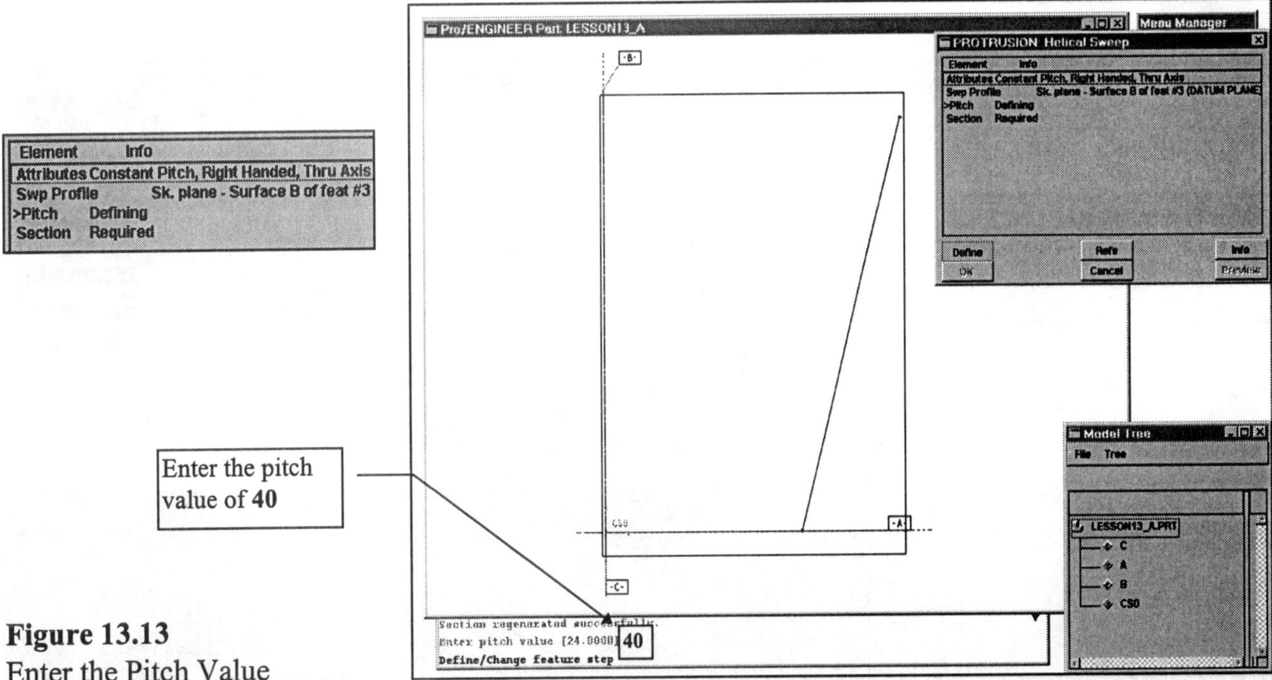

Figure 13.13
Enter the Pitch Value

Sketch the section geometry of the spring (here it is a circle):

Sketch ⇒ **Circle** ⇒ **Ctr/Point** (Fig. 13.14) ⇒ **Regenerate** ⇒ **Alignment** (align the center of circle with the *crosshairs*) ⇒ **Regenerate** ⇒ **Dimension** (add wire diameter) ⇒ **Regenerate** ⇒ **Modify** (type **15**) ⇒ **enter** ⇒ **Regenerate** ⇒ **Done** ⇒ **OK** ⇒ **View** ⇒ **Default** ⇒ **Cosmetic** ⇒ **Shade** ⇒ **Display** (Fig. 13.15)

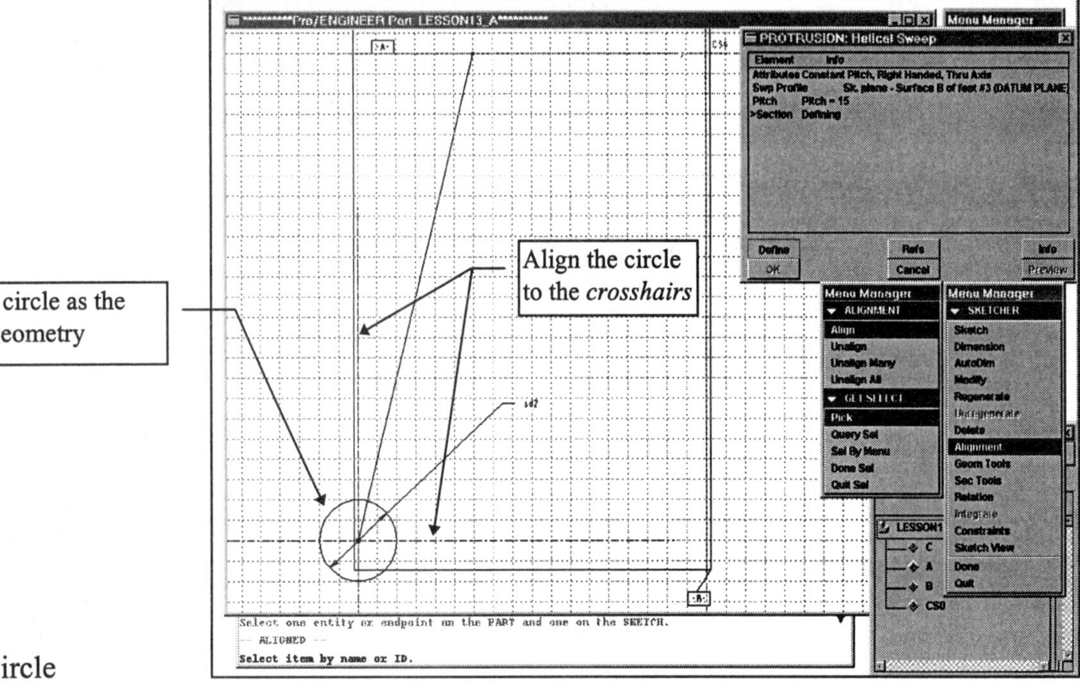

Figure 13.14
Sketching the Circle

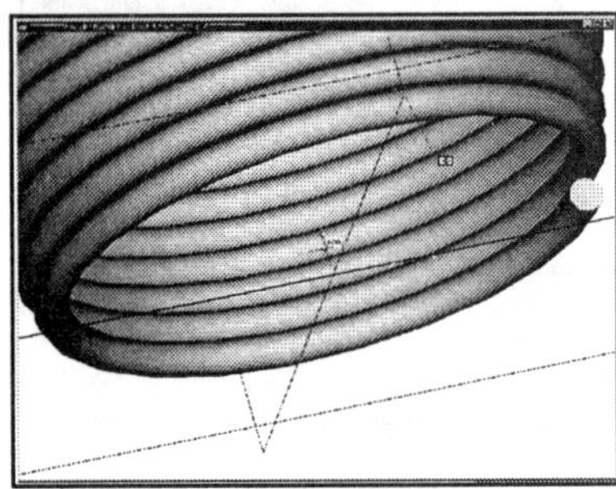

Dbms (File--PT/Modeler) ⇒
Save ⇒ **enter**
Purge ⇒ **enter** ⇒
Done-Return

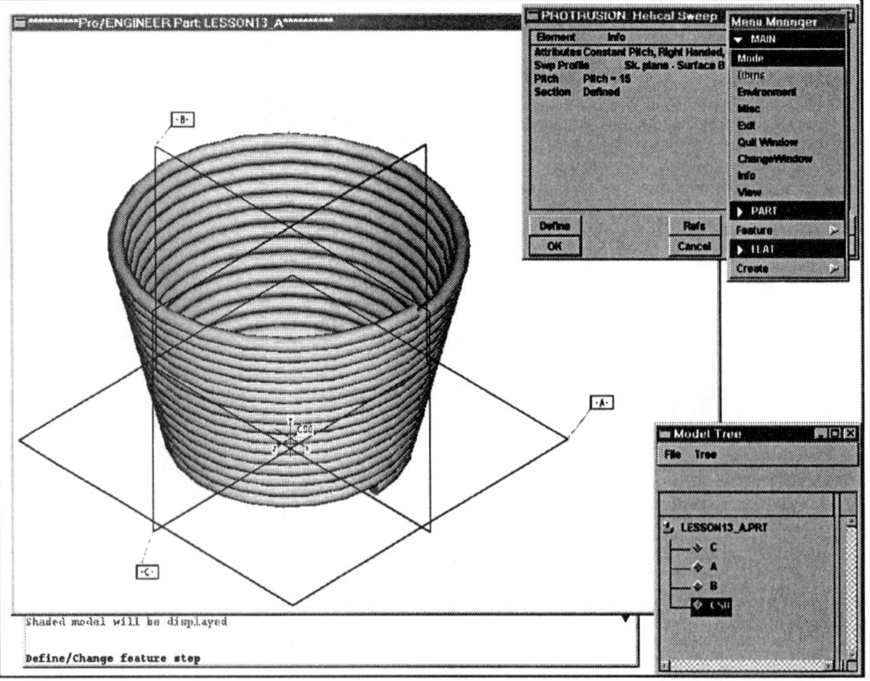

Figure 13.15
Completed Helical Sweep

The spring is almost complete (Fig. 13.16). Add the cut line to create the ground end as shown in Figure 13.17:

PT/Modeler™
Create ⇒ Cut etc.

Create ⇒ Cut ⇒ Extrude ⇒ Solid ⇒ Done ⇒ Both Sides ⇒ Done ⇒ Plane (pick datum **C, DTM1**) **⇒ Okay ⇒ Bottom** (pick datum **A, DTM2**) **⇒ Sketch ⇒ Line ⇒ Horizontal** (on **DTM2**) **⇒ Regenerate ⇒ Alignment** (align the line to datum **A, DTM2**, and the end of the line to datum **B, DTM3**) **⇒ Regenerate ⇒ Dimension ⇒ Regenerate** (there is no need to modify the dimension) **⇒ Done ⇒ Flip** (if necessary) **⇒ Okay ⇒ Thru All ⇒ Done ⇒ Thru All ⇒ Done ⇒ OK** (Fig. 13.18)

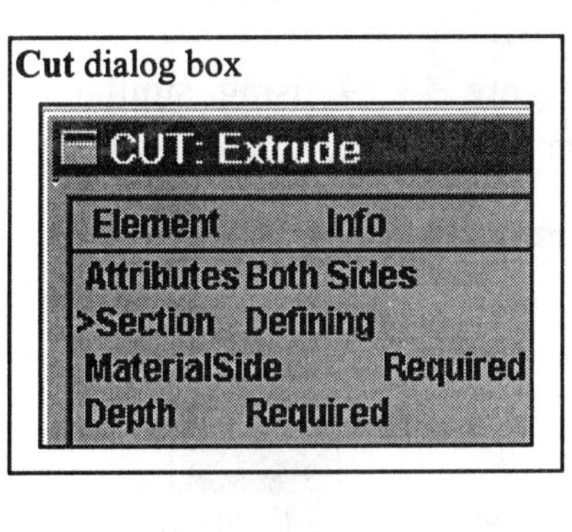

Figure 13.16
Creating a Cut

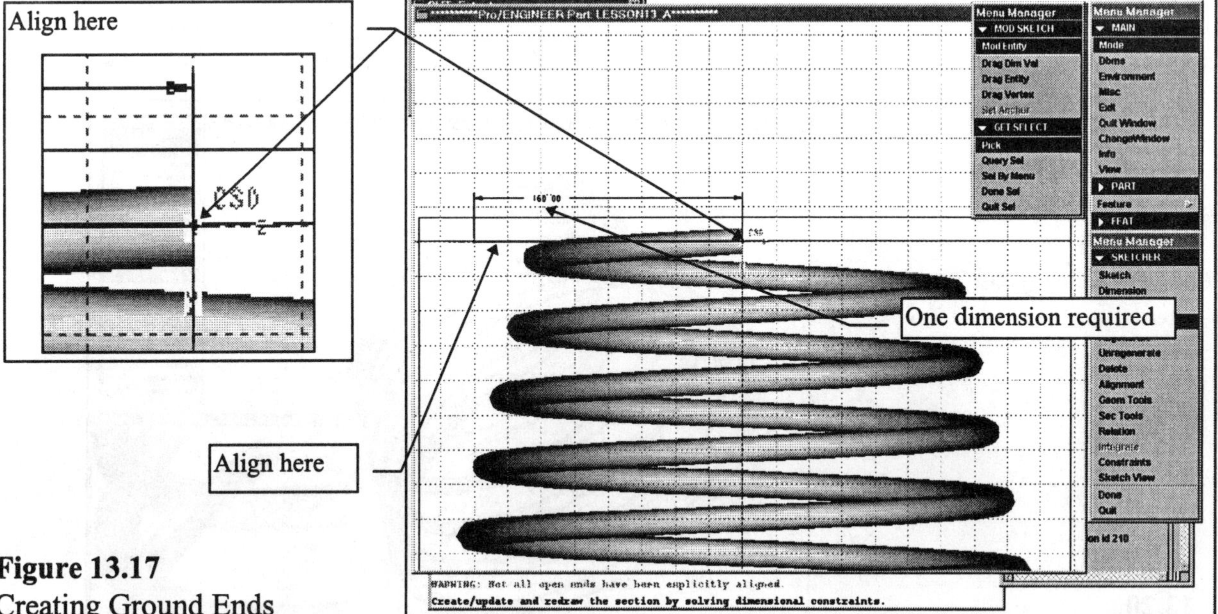

Figure 13.17
Creating Ground Ends

NOTE

For both Pro/ENGINEER and PT/Modeler you will get a message that says "**Not all open ends have been explicitly aligned.**" Ignore this warning and continue (Fig. 13.17).

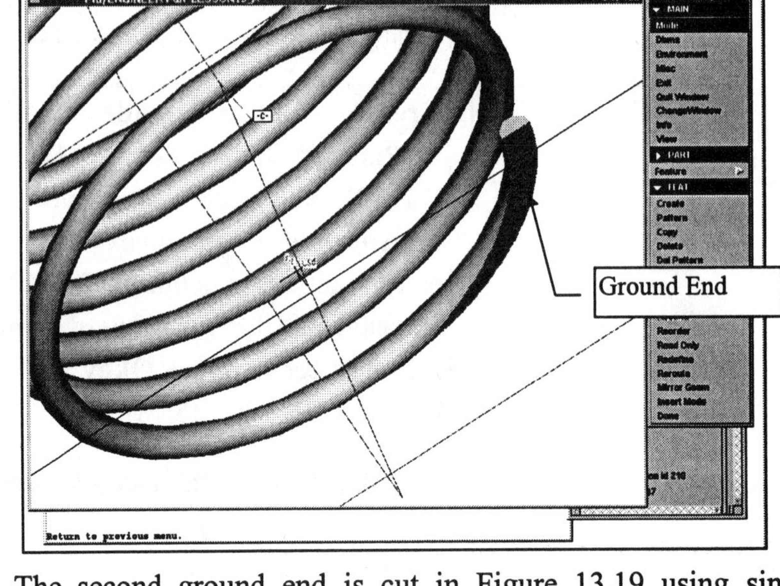

Figure 13.18
Completed Ground End

The second ground end is cut in Figure 13.19 using similar commands. The completed spring is shown in Figure 13.20.

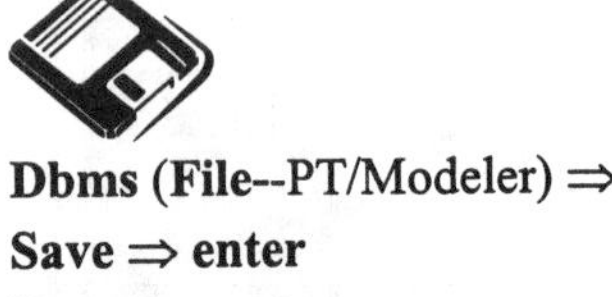

Dbms (File--PT/Modeler) ⇒
Save ⇒ **enter**
Purge ⇒ **enter** ⇒
Done-Return

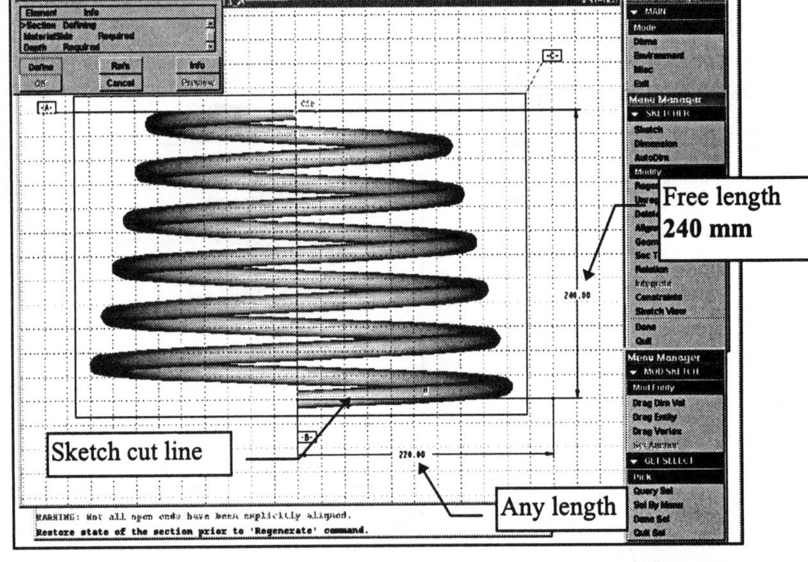

Figure 13.19
Creating the Second Ground End

Dbms (File--PT/Modeler) ⇒
Save ⇒ **enter**
Purge ⇒ **enter** ⇒
Done-Return

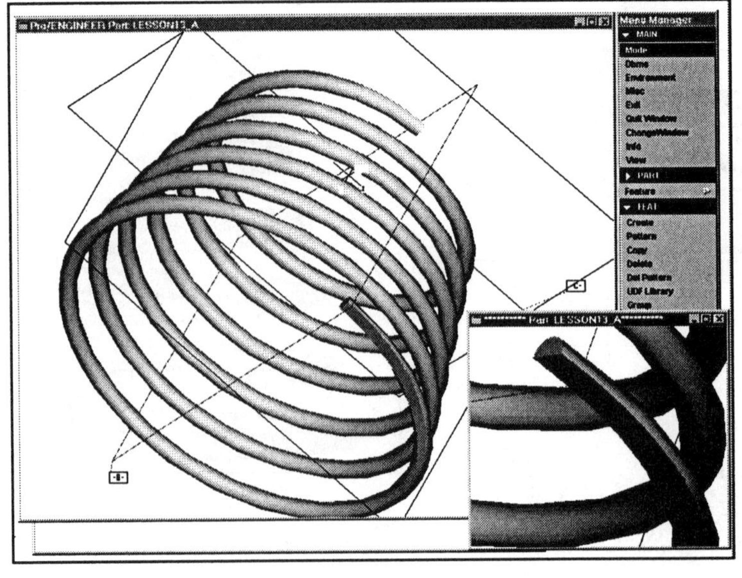

Figure 13.20
Completed Spring

Save the Helical Compression Spring by choosing the commands **Dbms ⇒ Save As ⇒ enter**, and giving it a different name, for example **HEL_COMP_SPR_GRND_ENDS**. Rename the file you are working on by using **Dbms ⇒ Rename ⇒ enter**, and give it a name such as **HEL_EXT_SPR_MACH_ENDS**. Figure 13.21 provides an **ECO** for the new spring. Delete the existing ground ends and modify the pitch to **10**. Figure 13.22 shows the wire diameter changed to **7.5** and the pitch changed to **10**. The ground ends have been deleted.

Complete the extension spring as shown in Figures 13.23 through 13.31. The free length is to be **120 mm**, the large radius is now **90 mm**, and the small radius is **60 mm**.

ECO

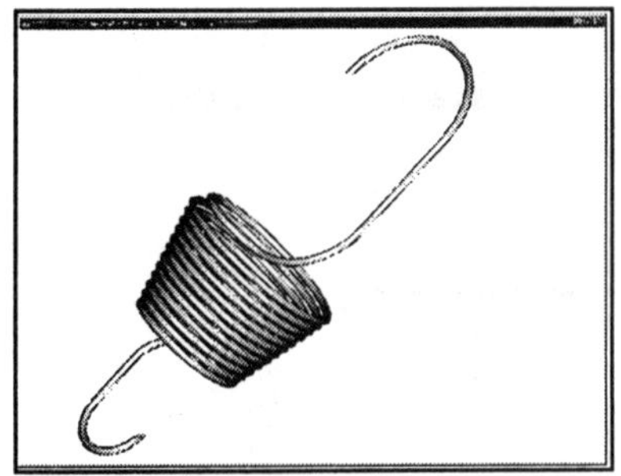

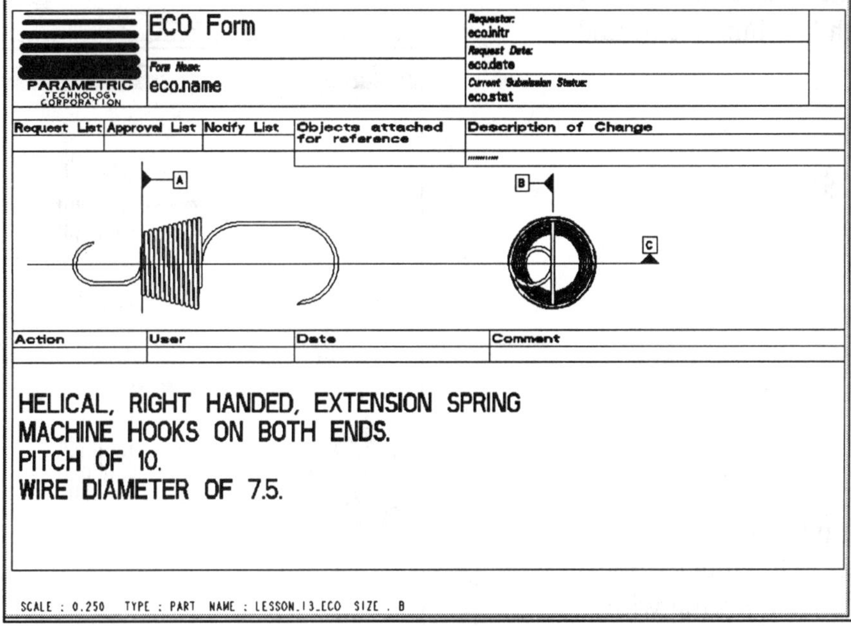

Figure 13.21
ECO to Create a Helical Extension Spring

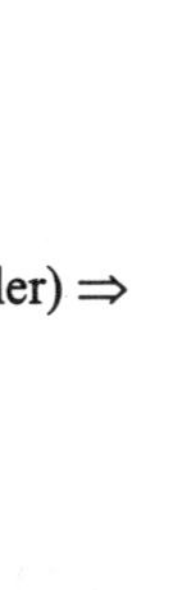

Dbms (File--PT/Modeler) ⇒
Save ⇒ enter
Purge ⇒ enter ⇒
Done-Return

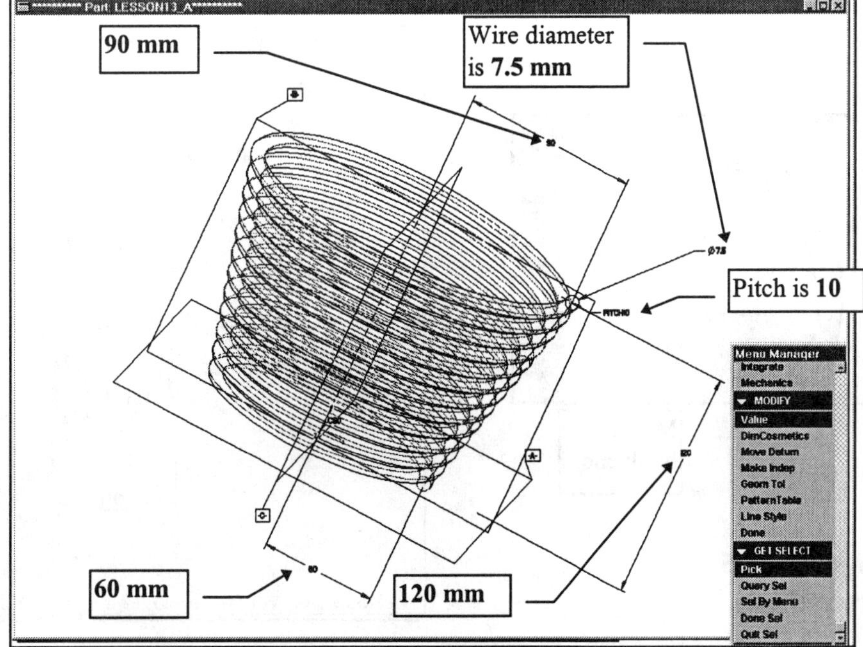

Figure 13.22
ECO Changes

Figure 13.23
Helical Extension Spring with Machine Hook Ends

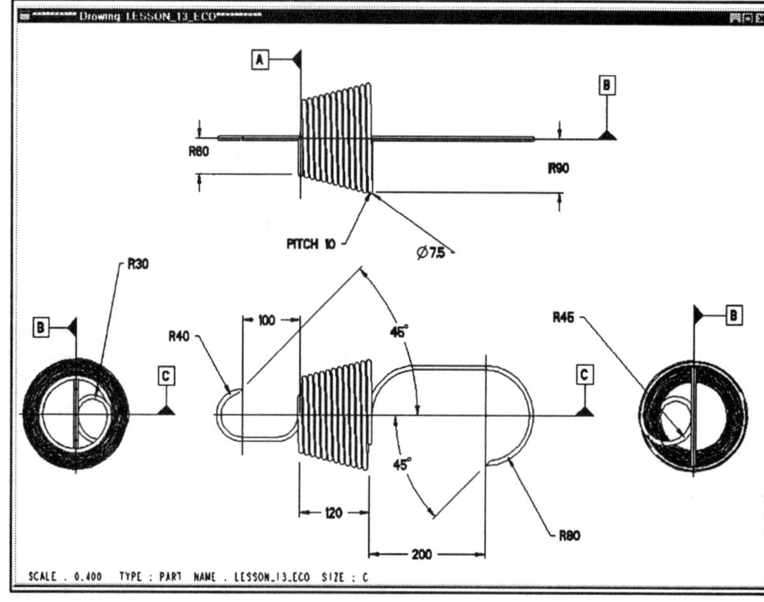

Figure 13.24
Detail Drawing of Helical Extension Spring with Machine Hook Ends

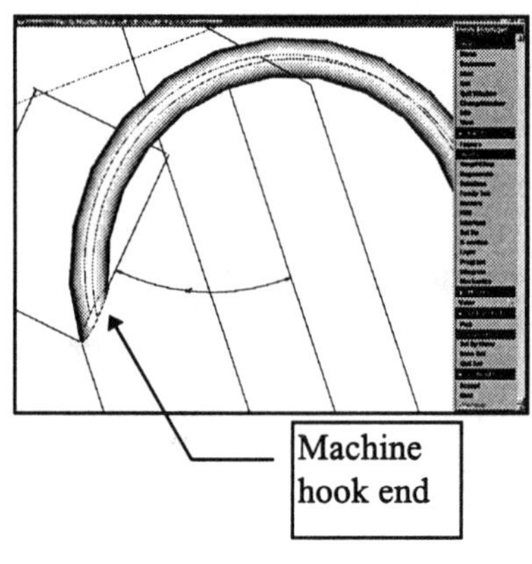

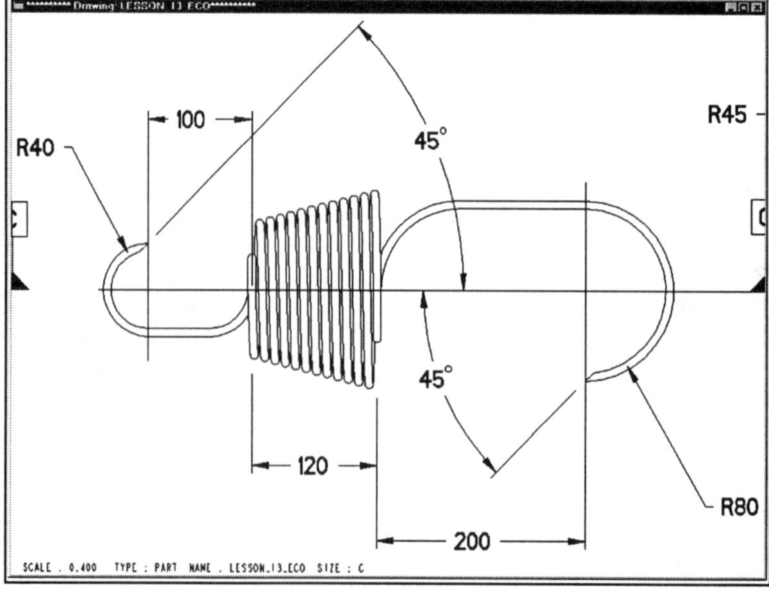

Figure 13.25
Front View

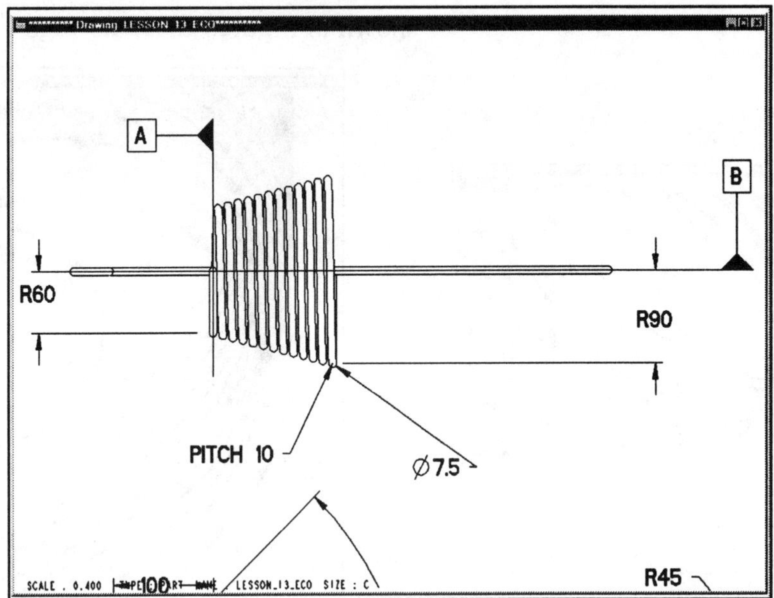

Figure 13.26
Top View

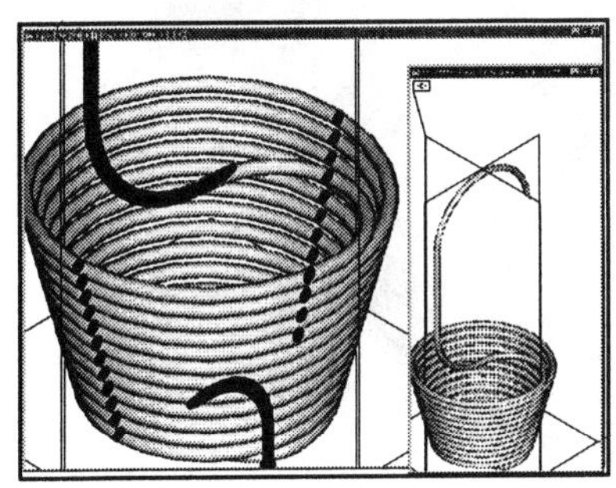

Figure 13.27
Right Side View

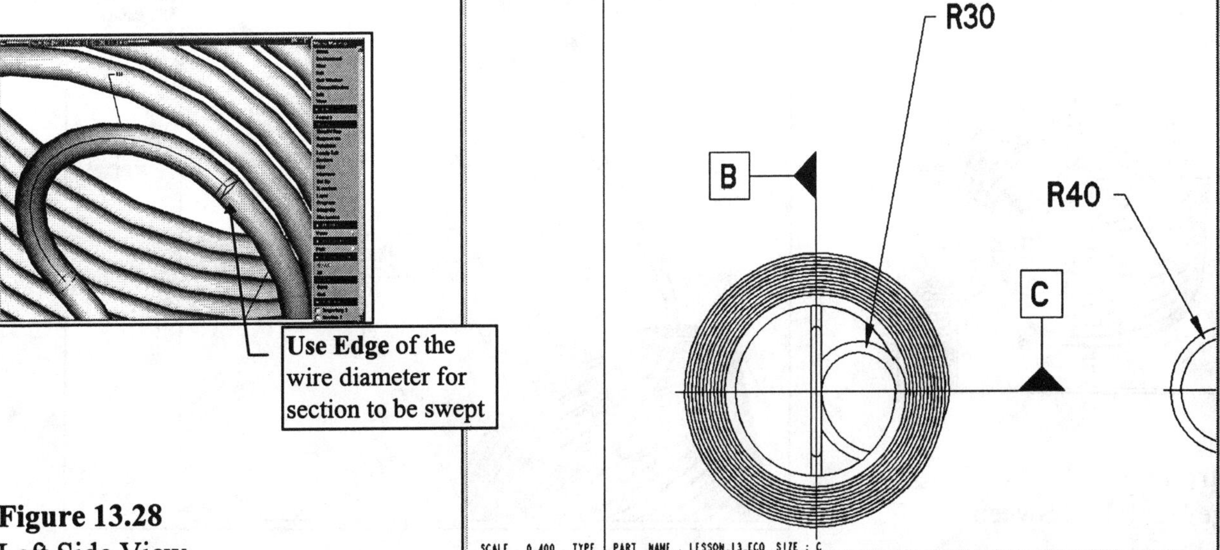

Figure 13.28
Left Side View

Create the machine hooks using simple sweeps and cuts, as shown in Figures 13.29 through 13.31.

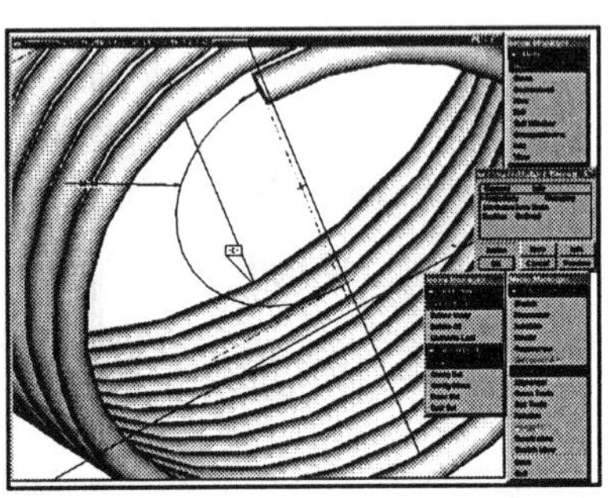

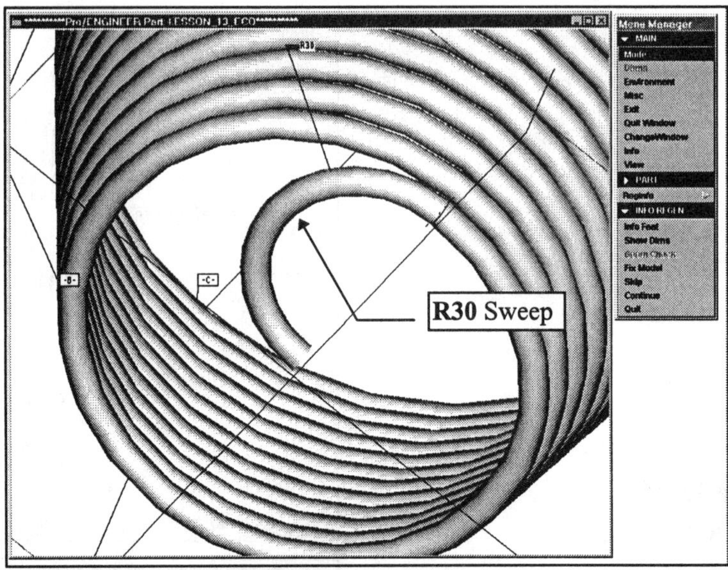

Figure 13.29
R30 Sweep

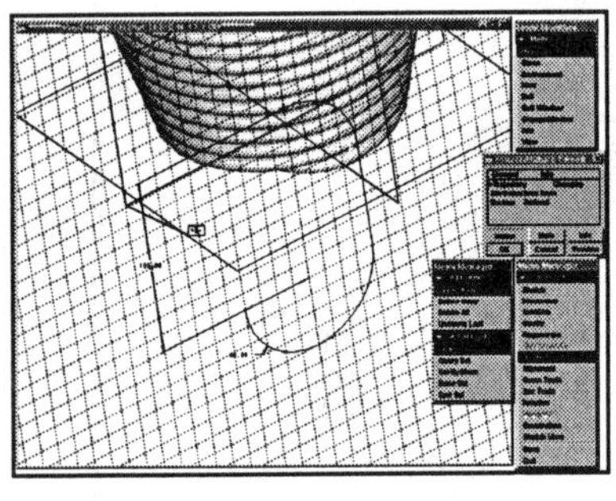

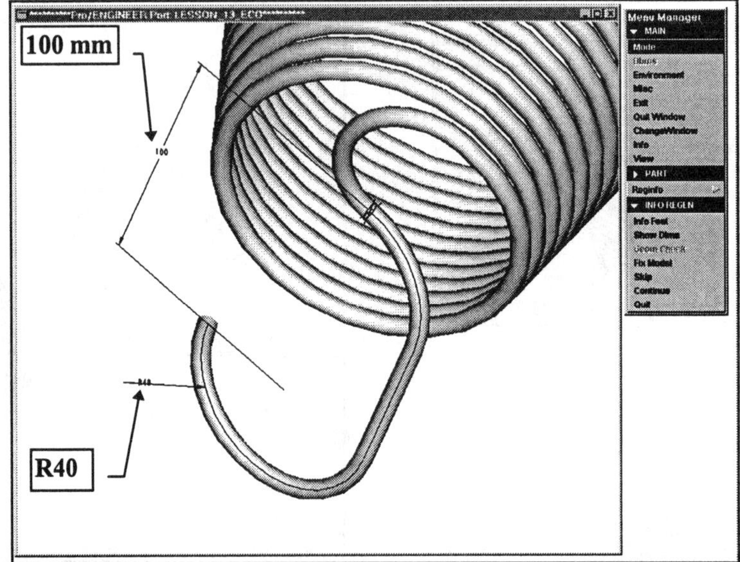

Figure 13.30
Small Hook End Sweep

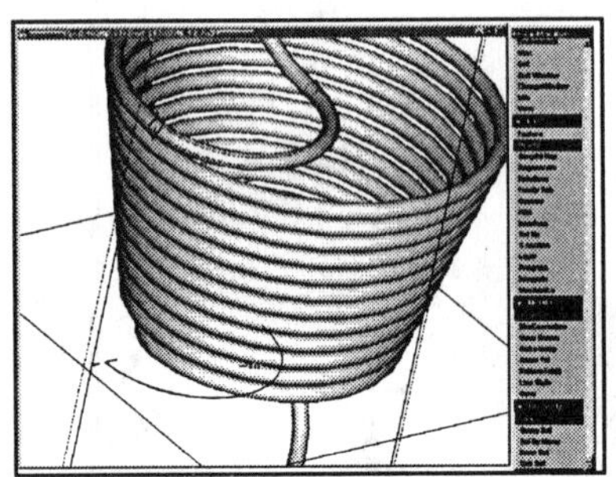

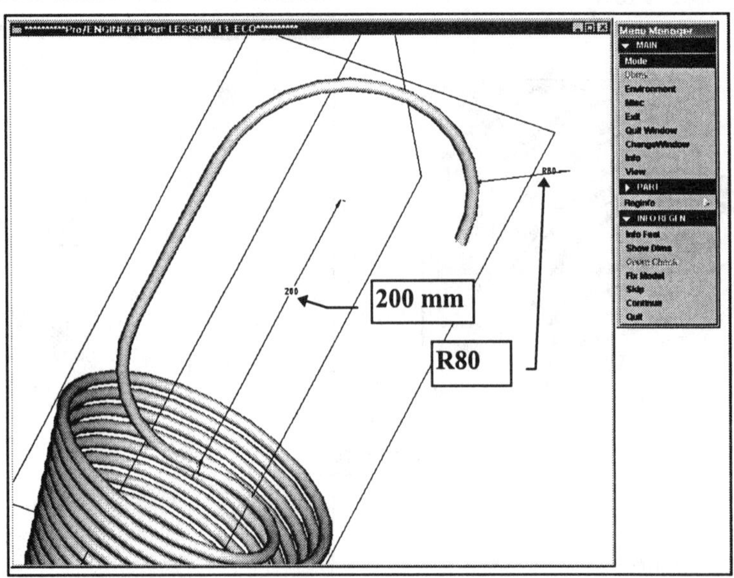

Figure 13.31
Large Hook End Sweep

Lesson 13 Project

Convex Compression Spring

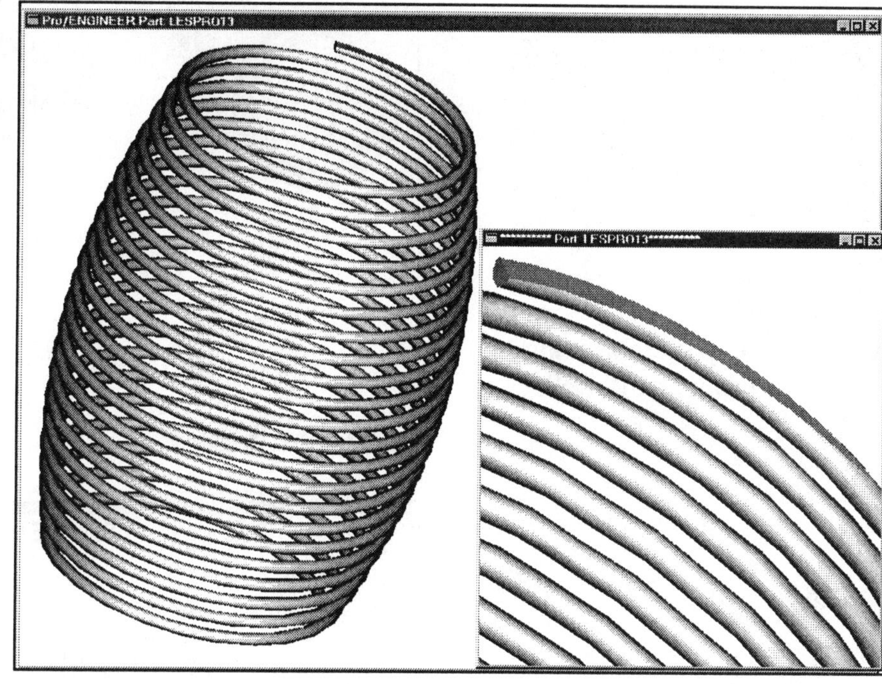

Figure 13.32
Convex Compression Spring

Convex Compression Spring

The last **lesson project** is a **Convex Compression Spring**. This project uses commands similar to those for the **Helical Compression Spring**. Create the part shown in Figures 13.32 through 13.39. Analyze the part and plan out the steps and features required to model it. Use the DIPS in Appendix D to plan out the feature creation sequence and the parent-child relationships for the part. The spring is made of *spring steel*.

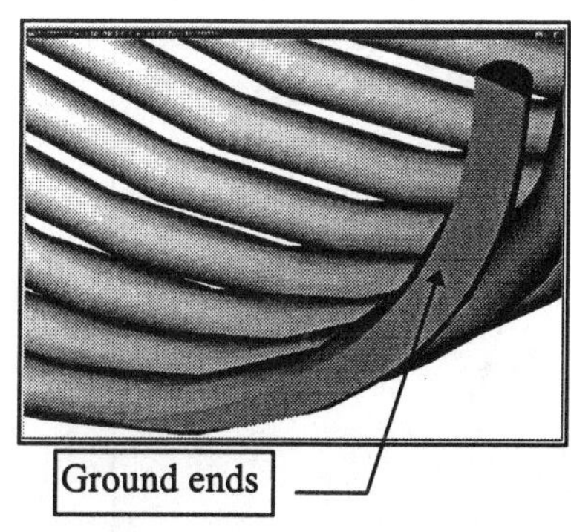

Ground ends

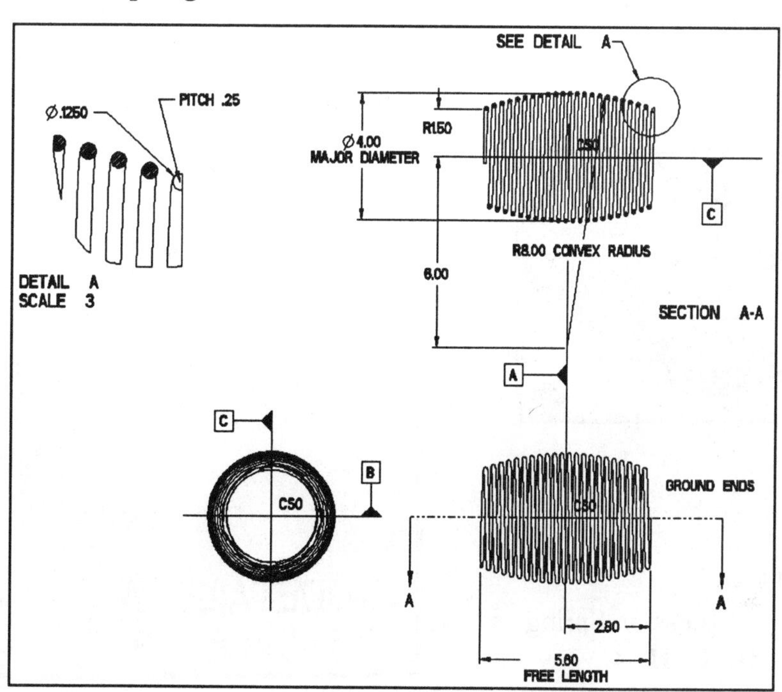

Figure 13.33
Convex Compression Spring, Detail Drawing

A
C
B
GROUND ENDS
A A
2.80
5.60
FREE LENGTH
SCALE . 0.750 TYPE : PART NAME . LESPRO13 SIZE . C

Figure 13.34
Convex Compression Spring Drawing, Front and Left Side Views

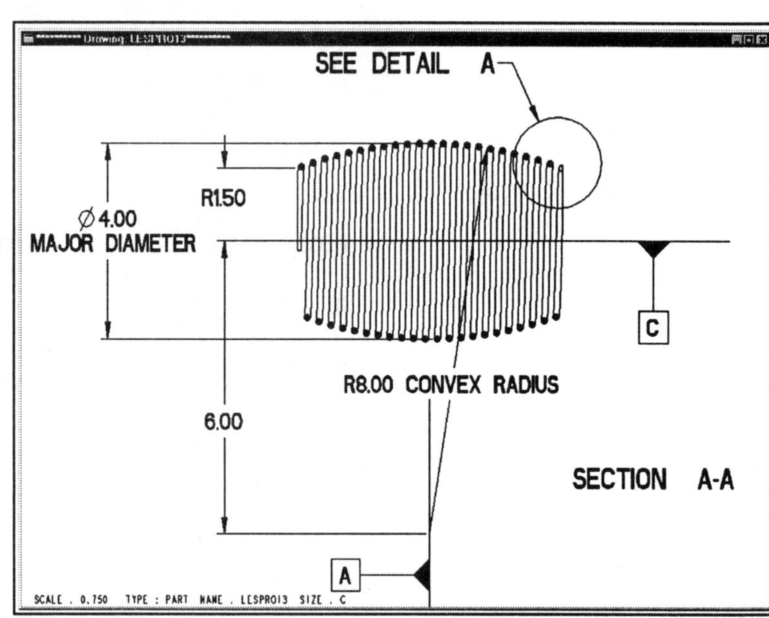

Figure 13.35
Convex Compression Spring Drawing, Top View

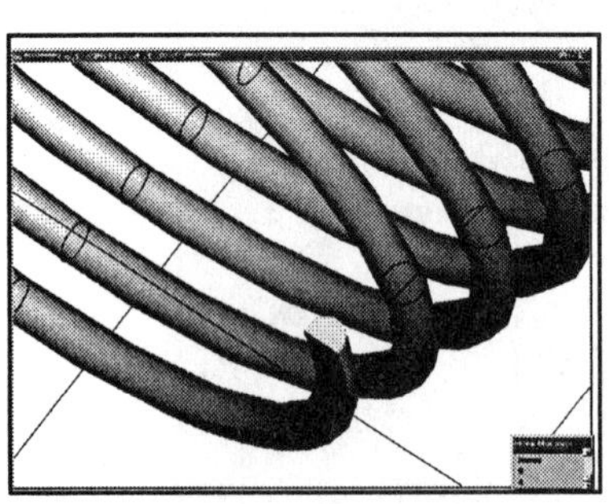

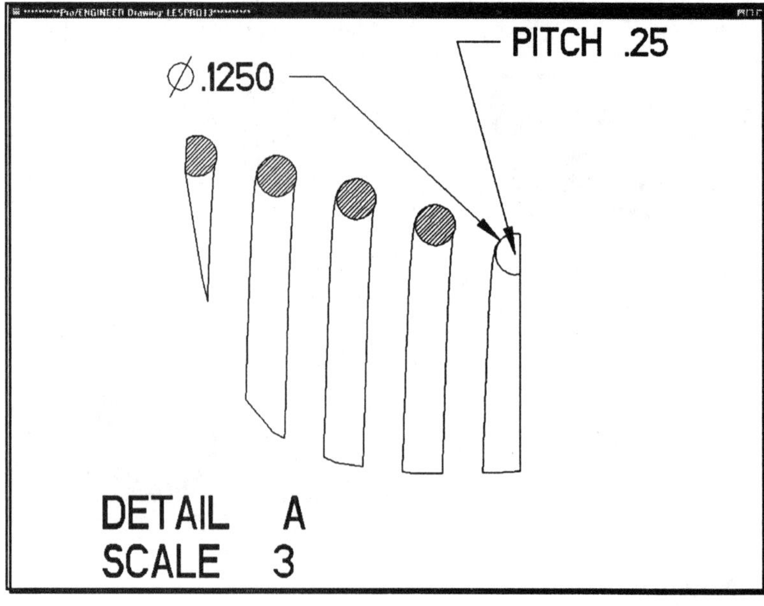

Figure 13.36
Convex Compression Spring Drawing, **DETAIL A**

Dbms (File--PT/Modeler) ⇒
Save ⇒ enter
Purge ⇒ enter ⇒
Done-Return

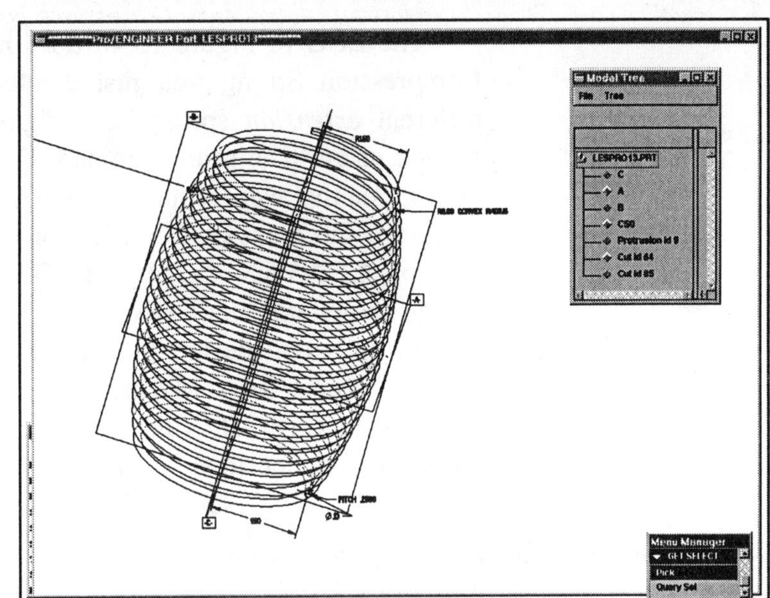

Figure 13.37
Convex Compression Spring Showing Datum Planes and Model Tree

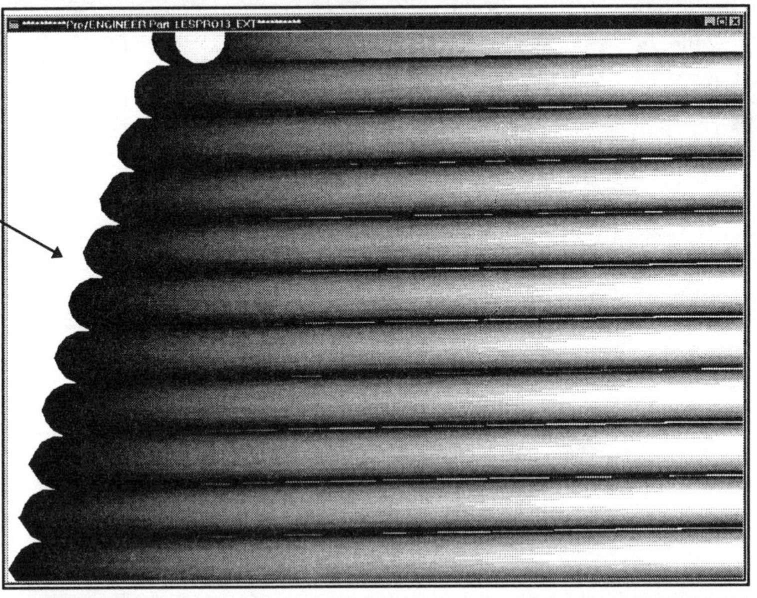

Figure 13.38
Convex Compression Spring, Pitch diameter **.125**

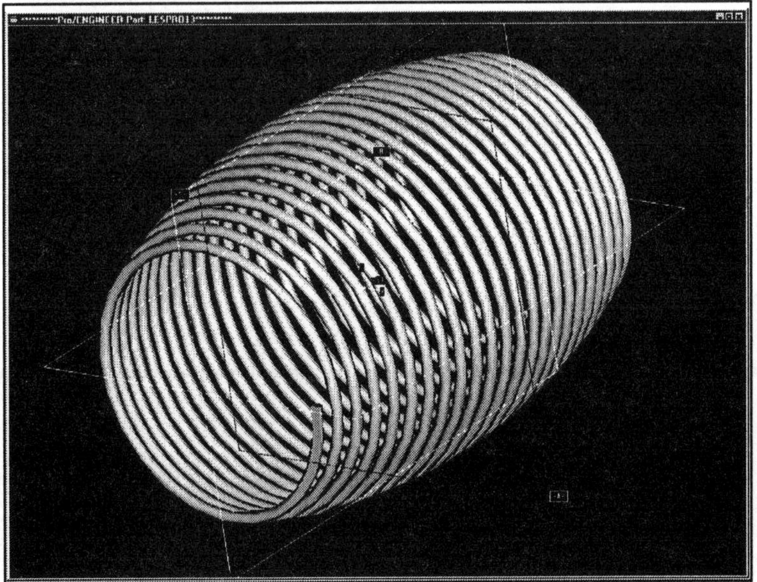

Figure 13.39
Completed Convex Compression Spring

The ECO in Figure 13.40 does *not* alter or change the Convex Compression Spring you just created. The ECO requests that a different *extension* spring be designed with hook ends instead of ground ends. The same size and dimensions are to be used for the new spring that were required in the Convex Compression Spring.

Save the Convex Compression Spring under a new name, for example **CONVEX_COM_SPR_GRND_ENDS**. Rename the active part to something like **CON_EXT_SPR_HOOK_ENDS**. Delete the ground ends and design machine hooks for both ends of the new spring. Refer to a Machinery's Handbook or your engineering graphics text (**EGD**, Chapter 18, Springs) for acceptable design options and dimensions for the hook ends.

Figure 13.40 ECO for New Convex *Extension* Spring

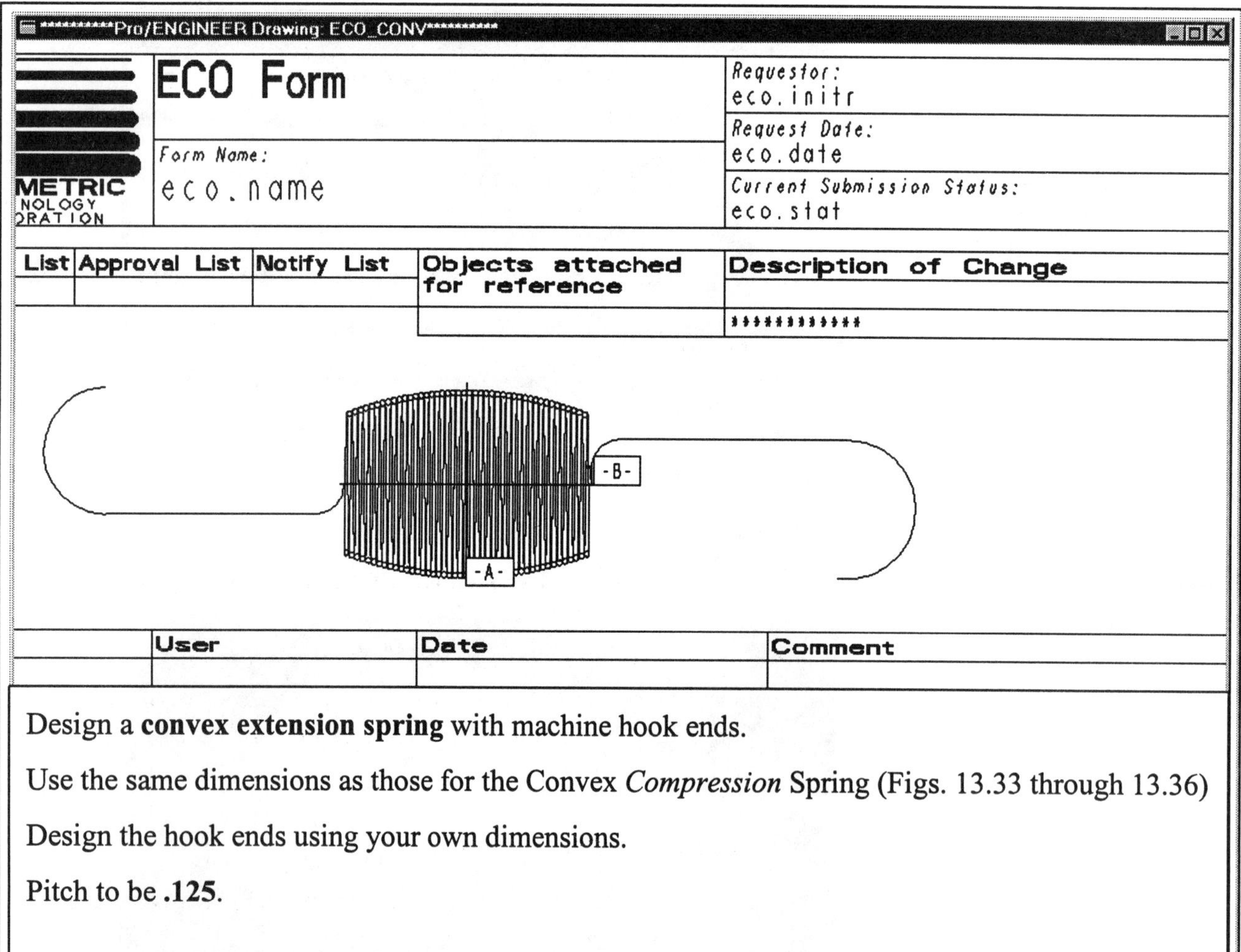

**********Pro/ENGINEER Drawing: ECO_CONV**********

METRIC
NOLOGY
ORATION

ECO Form

Form Name:
eco.name

Requestor:
eco.initr

Request Date:
eco.date

Current Submission Status:
eco.stat

List	Approval List	Notify List	Objects attached for reference	Description of Change
				************

	User	Date	Comment

Design a **convex extension spring** with machine hook ends.

Use the same dimensions as those for the Convex *Compression* Spring (Figs. 13.33 through 13.36)

Design the hook ends using your own dimensions.

Pitch to be **.125**.

Part Two

Assemblies

Lesson 14 Assembly Constraints
Lesson 15 Exploded Assemblies

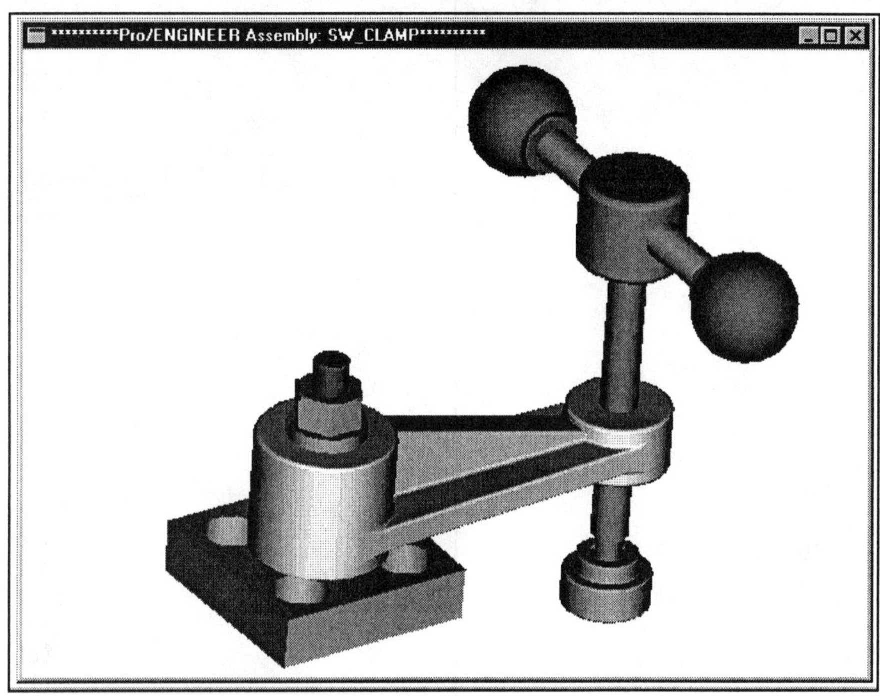

Swing Clamp Assembly

ASSEMBLIES

The **Assembly mode** allows you to place together component parts and subassemblies to form assemblies. Assemblies can then be modified, reoriented, analyzed, and documented. Assembly mode is used for the following functions:

* Placing components into assemblies (*Bottom-up* assembly design)
* Exploding views of assemblies
* Altering the display settings for individual components
* Designing in Assembly Mode (*Top-down* assembly design)
* Part modification, including feature construction
* Analysis of assemblies

With **Pro/ENGINEER** you can:

* Assemble component parts and subassemblies to form assemblies
* Delete or replace assembly components
* Modify assembly placement offsets, and create and modify assembly datum planes, coordinate systems, and sectional views
* Modify parts directly in assembly mode

* Get assembly engineering information, perform viewing and layer operations, create reference dimensions, and work with interfaces.

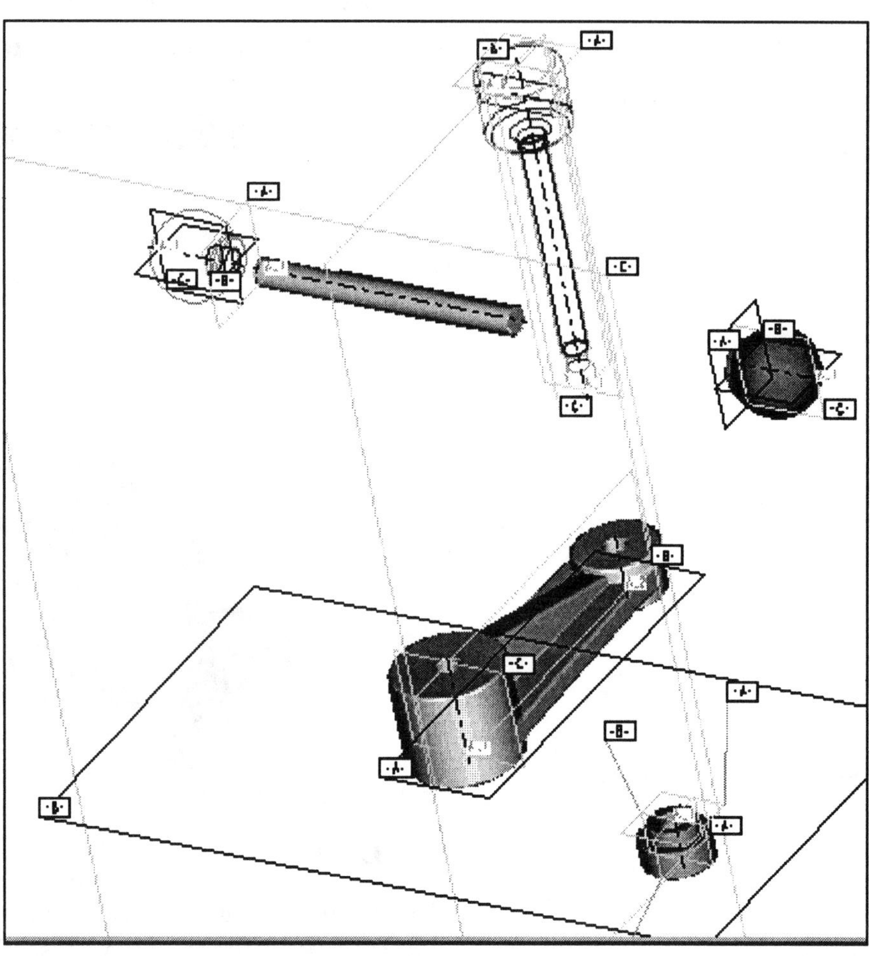

Exploded Swing Clamp Subassembly

With **Pro/ASSEMBLY** you can:

* Create new parts in Assembly mode
* Create sheet metal parts in Assembly mode
* Mirror parts in Assembly mode (create a new part)
* Replace components automatically by creating interchangeable groups. Create assembly features, existing only in Assembly mode and intersecting several components
* Create families of assemblies, using the family table
* Simplify the assembly representation
* Use the **Move** and **Copy** commands for assembly components
* Use Pro/PROGRAM to create design programs that allow users to respond to program prompts to alter the design model

The process of creating an assembly is accomplished by adding models (parts/subassemblies) to a base component (parent part/subassembly) using a variety of constraints. A **placement constraint** specifies the relative position of a pair of surfaces on two components. The **Mate**, **Align**, **Insert**, and **Orient** commands and their variations are used to accomplish this task.

Lesson 14

Assembly Constraints

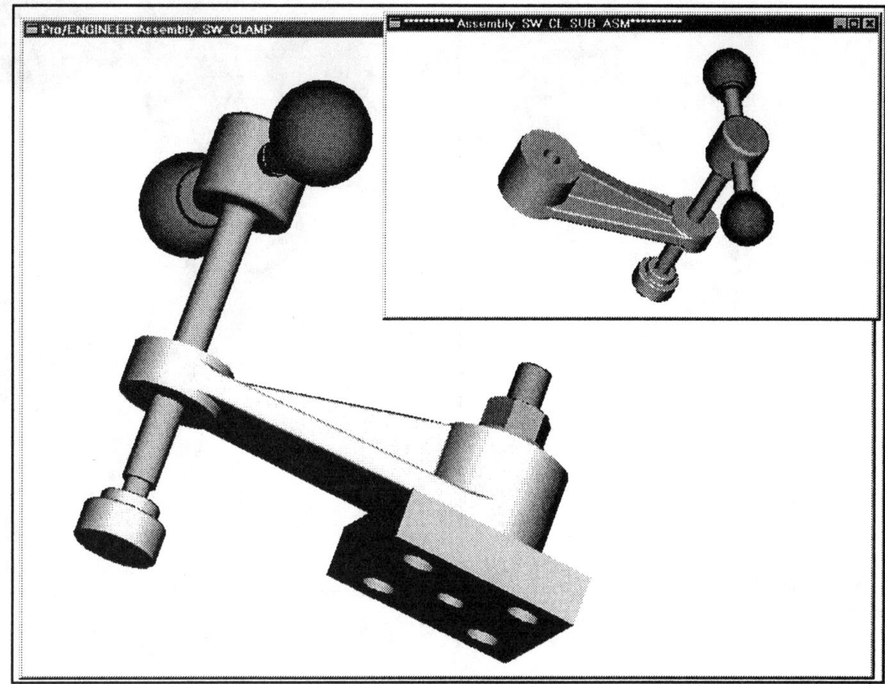

Figure 14.1
Swing Clamp

OBJECTIVES

1. **Assemble components to form an assembly**

2. **Create a subassembly**

3. **Understand and use a variety of assembly constraints**

4. **Redefine a component constraint**

5. **Modify a constraint**

6. **Check for clearance and interference**

☑ ***EGD REFERENCE***
Engineering Graphics and Design with Graphical Analysis *or* **Fundamentals of Engineering Graphics and Design**
by L. Lamit and K. Kitto
Read Chapter 23
See pages 832-838, 845-846

COAch ™ for Pro/ENGINEER

If you have **COAch for Pro/ENGINEER** on your system, go to SEARCH and do the Segments shown in Figures 14.3 through 14.5.

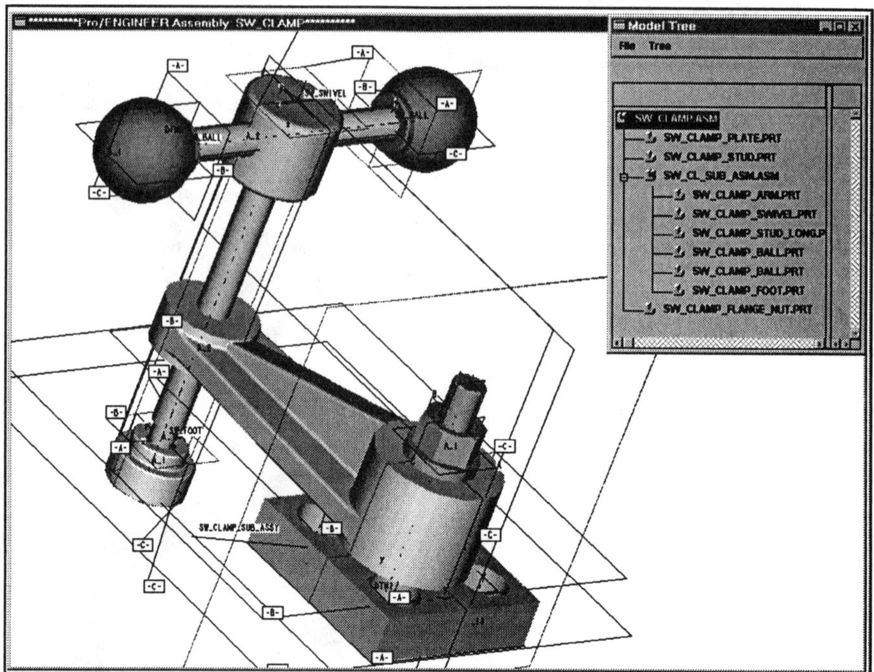

Figure 14.2
Swing Clamp with Model Tree

ASSEMBLY CONSTRAINTS

Assembly mode allows you to place together component parts and subassemblies to form an **assembly** (Figs. 14.1 and 14.2). Assemblies (Fig. 14.3) can be modified, reoriented, documented, or analyzed. An assembly can be assembled into another assembly, thereby becoming a **subassembly**.

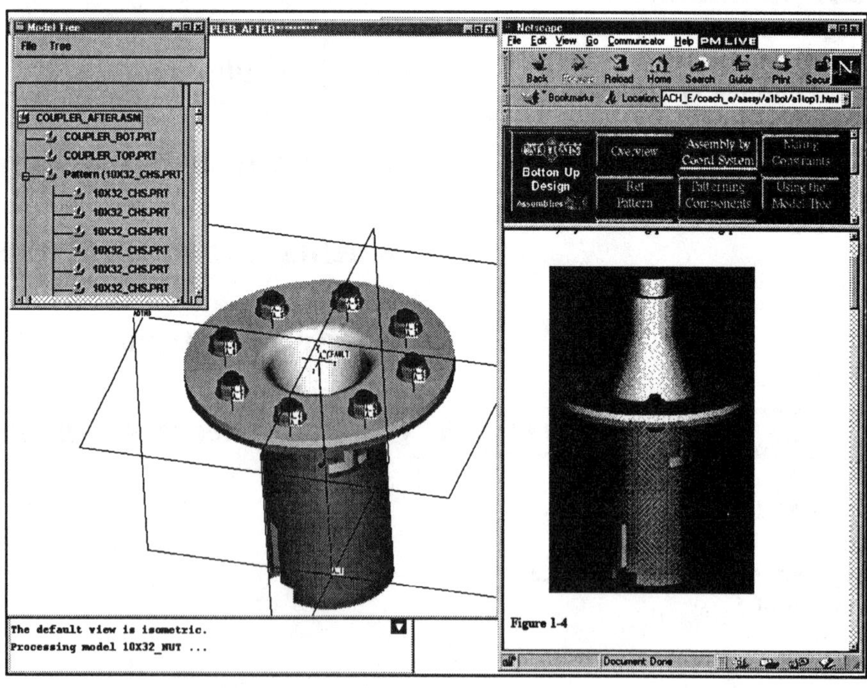

Figure 14.3
COAch for Pro/E, Assemblies, Bottom-Up Design (Assembly by Coord System)

Assembly mode is used for the following functions:

* Placing components in assemblies (Fig. 14.4)
* Exploding views of assemblies
* Part modification, including feature construction
* Analysis

With **Pro/E,** you can:

* Place together component parts and subassemblies to form assemblies.
* Remove or replace assembly components.
* Modify assembly placement offsets, and create and modify assembly datum planes, coordinate systems, and cross sections.
* Modify parts directly in Assembly mode.
* Get assembly engineering information, perform viewing and layer operations, create reference dimensions, and work with interfaces.

With **Pro/ASSEMBLY,** you can:

* Create new parts in Assembly mode.
* Create sheet metal parts in Assembly mode (using Pro/SHEETMETAL).
* Mirror parts in Assembly mode (create a new part).
* Replace components automatically by creating interchangeable groups. Create assembly features, existing only in Assembly mode, which intersect several components.
* Create families of assemblies, using the family table.
* Simplify the assembly representation.
* Use **Move** and **Copy** for assembly components.
* Use Pro/PROGRAM to create design programs that allow users to respond to program prompts to alter the design model.

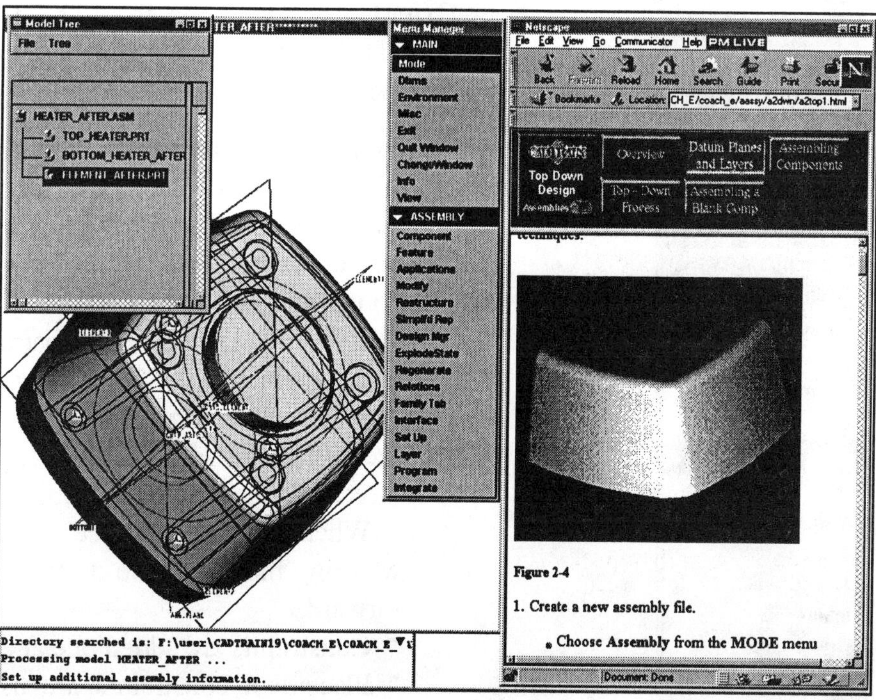

Figure 14.4
COAch for Pro/E, Assemblies, Top-Down Design (Datum Planes and Layers)

Assembling Components

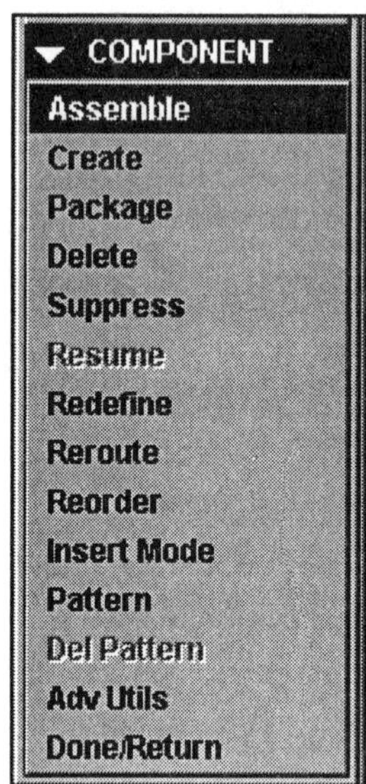

The process of creating an assembly involves adding components (parts/subassemblies) to a base component (parent part/subassembly) using a variety of constraints (Fig. 14.5). Components can also be created in Assembly mode using existing components as references.

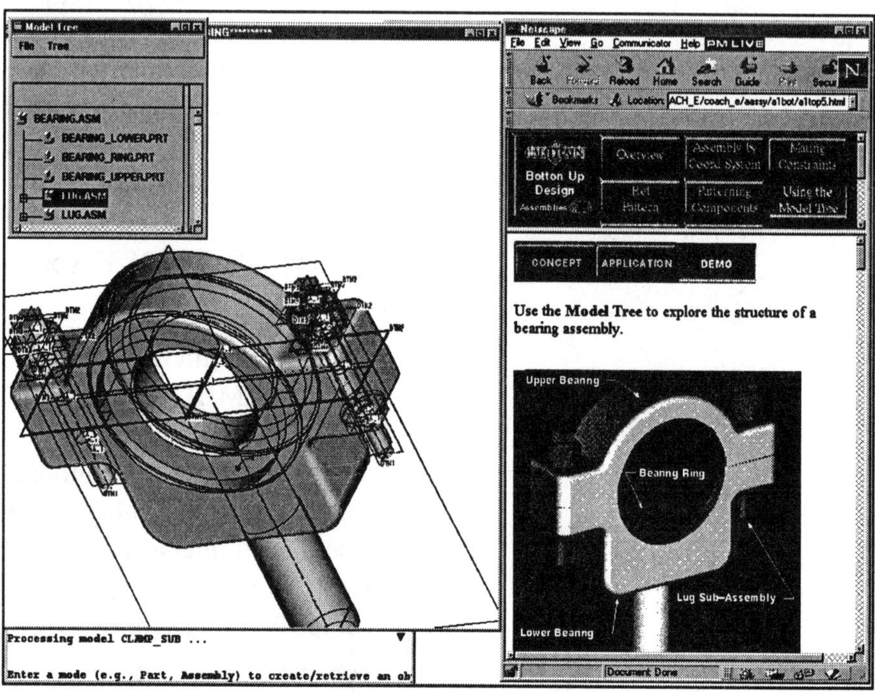

Figure 14.5
COAch for Pro/E, Assemblies, Bottom-Up Design (Using the Model Tree)

A **placement constraint** specifies the relative positions of a pair of references on two components. The **Mate**, **Align**, **Insert**, and **Orient** commands and their variations are used to accomplish this task. The general principles to apply during constraint placement are as follows:

* The two surfaces must be of the same type (for example, plane-plane, revolved-revolved). The term *revolved surface* means a surface created by revolving a section or by extruding an arc or a circle. Only the following surfaces are allowed: plane, cylinder, cone, torus, sphere.
* If you put a placement constraint on a datum plane, you should specify which side of it, yellow or red, you are going to use.
* When using **Mate Offset** or **Align Offset** you will be shown the positive offset direction. If you need an offset in the opposite direction, enter a negative value.
* When a model surface in selected in one window, another window may become hidden, and will need to be brought forward.
* Constraints are added one at a time.
* Placement constraints are used in combinations in order to specify placement and orientation completely. For example, one pair of surfaces may be constrained to mate, another pair to insert, and a third pair to orient.

Figure 14.6
Pro/HELP
Online Documentation,
Mate and **Mate Offset**

Mate (Fig. 14.6) is used to make two surfaces touch one another: coincident and facing each other. When you are using datums, this means that two yellow sides, or two red sides, will face each other. **Mate Offset** (Fig. 14.6) makes two planar surfaces parallel and facing each other. The offset value determines the distance between the two surfaces.

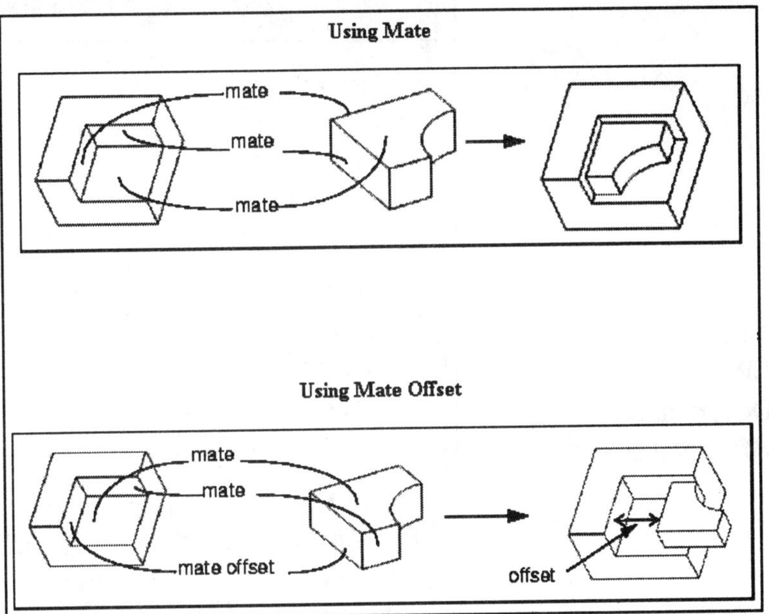

The **Align** command (Fig. 14.7) makes two planes coplanar: coincident and facing in the same direction. The **Align** constraint also aligns revolved surfaces or axes to make them coaxial. You can also align two datum points, vertices, or curve ends; selections on both parts must be of the same type (that is, if a datum point is selected on one part, only a datum point can be selected on another part). The **Align Offset** constraint (Fig. 14.8) aligns two planar surfaces at an offset: parallel and facing in the same direction.

Figure 14.7
Pro/HELP
Online Documentation,
Align

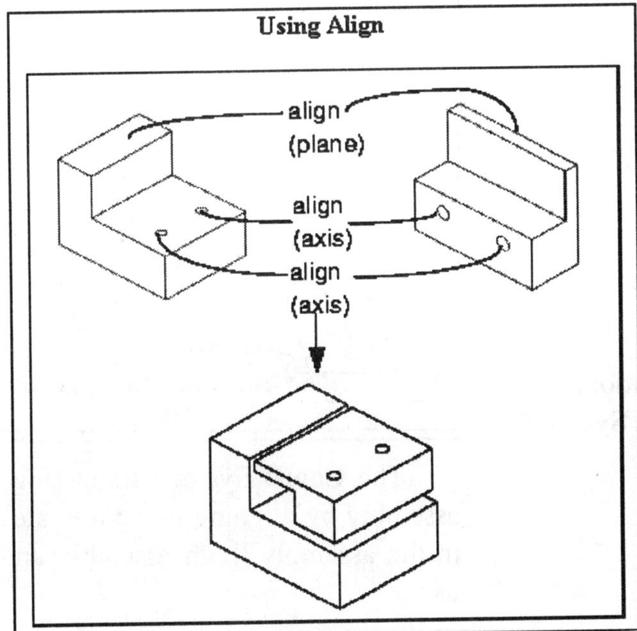

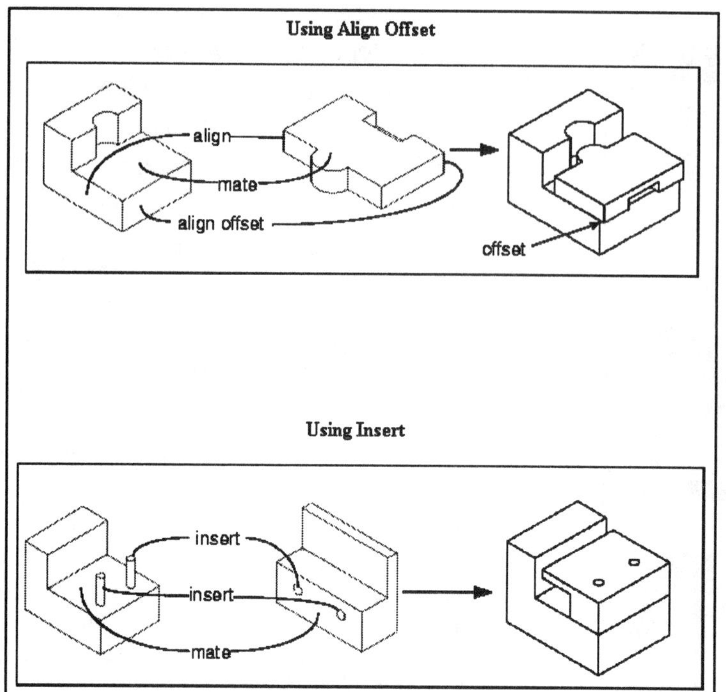

Figure 14.8
Pro/HELP
Online Documentation,
Align Offset and **Insert**

The **Insert** constraint (Fig. 14.8) inserts a "male" revolved surface into a "female" revolved surface, aligning axes.

The **Orient** constraint (Fig. 14.9) orients two planar surfaces so that they are parallel and facing in the same direction; offset is not specified.

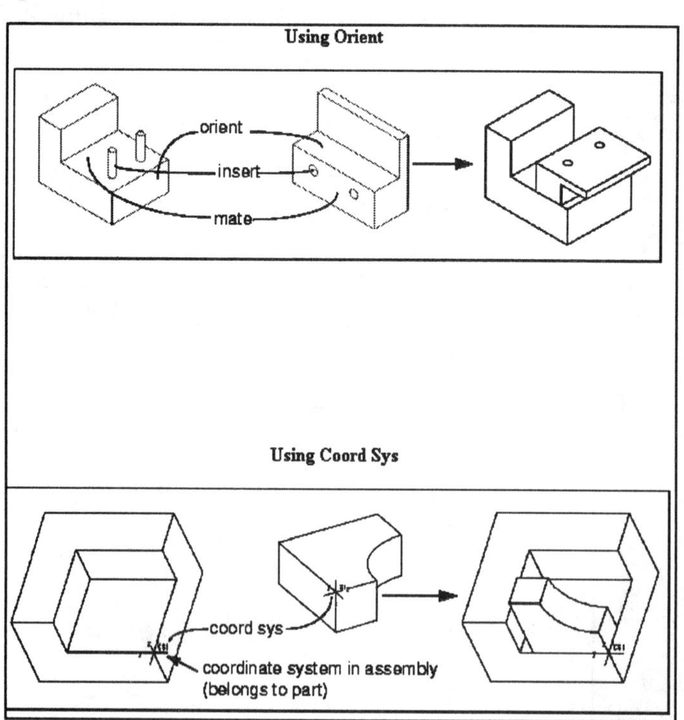

Figure 14.9
Pro/HELP
Online Documentation,
Orient and **Coord Sys**

The **Coord Sys** constraint (Fig. 14.9) places a component in an assembly by aligning its coordinate system with a coordinate system in the assembly (both assembly and part coordinate systems can be used).

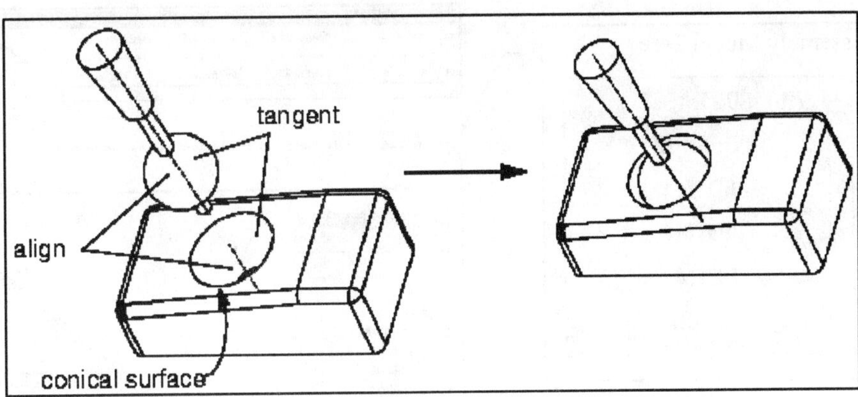

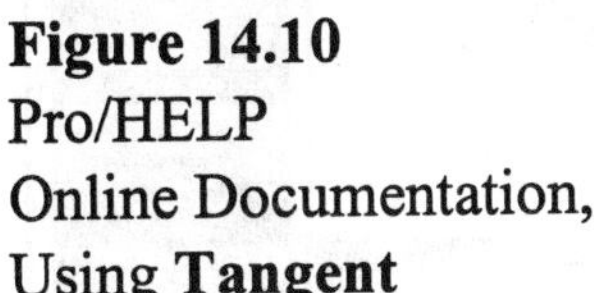

Figure 14.10
Pro/HELP
Online Documentation,
Using **Tangent**

Coordinate systems can be picked or selected by name from namelist menus. The components will be assembled by aligning the **X**, **Y**, and **Z** axes of the selected coordinate systems.

The **Tangent**, **Pnt On Srf**, and **Edge On Srf** constraints are used to control the contact of a surface at the tangency of another surface, at a point, or at an edge (Fig. 14.10). An example of the use of these placement options is the contact surface or point between a cam and its actuator.

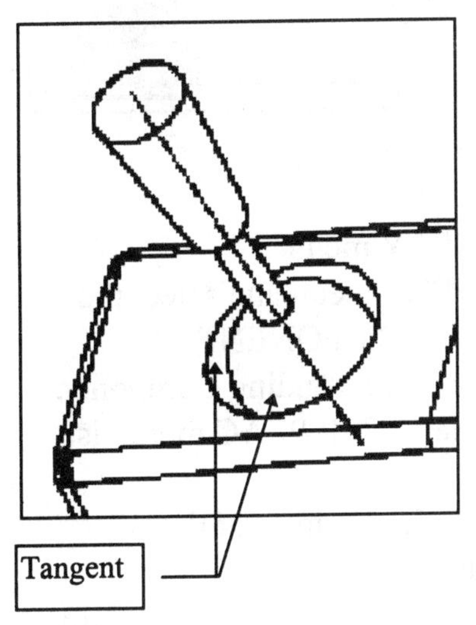

In most cases, a combination of constraints will be required. **Mate** and **Insert** are required to constrain the two parts shown in Figure 14.11. **Mate**, **Insert**, and **Orient** are another possibility, depending on the parts.

Assemblies can also be displayed *exploded*. The exploded components can be placed anywhere in 3-D space. An exploded cosmetic view of the assembly can be displayed with or without the reference planes, coordinate system, and hidden lines.

The completed assembly can be modified by redefining the constraints. The last stage of the project involves putting the assembly into Drawing mode and displaying the appropriate views and information.

Component Placement Window

When you choose **Assemble** or **Redefine** from the COMPONENT menu, a *Component Placement window* appears (Fig. 14.11 and below left). This window lists the *constraint number*, the *constraint type*, the corresponding *component* and *assembly references*, and the *offset value* (if any) of the component placement constraint.

Component Placement

No.	Type	Comp. Ref.	Assy. Ref.	Offset
1	Insert	Surface	Surface	0.000
2	Mate	Surface	Surface	0.000

Status: Component is fully constrained and can be placed.

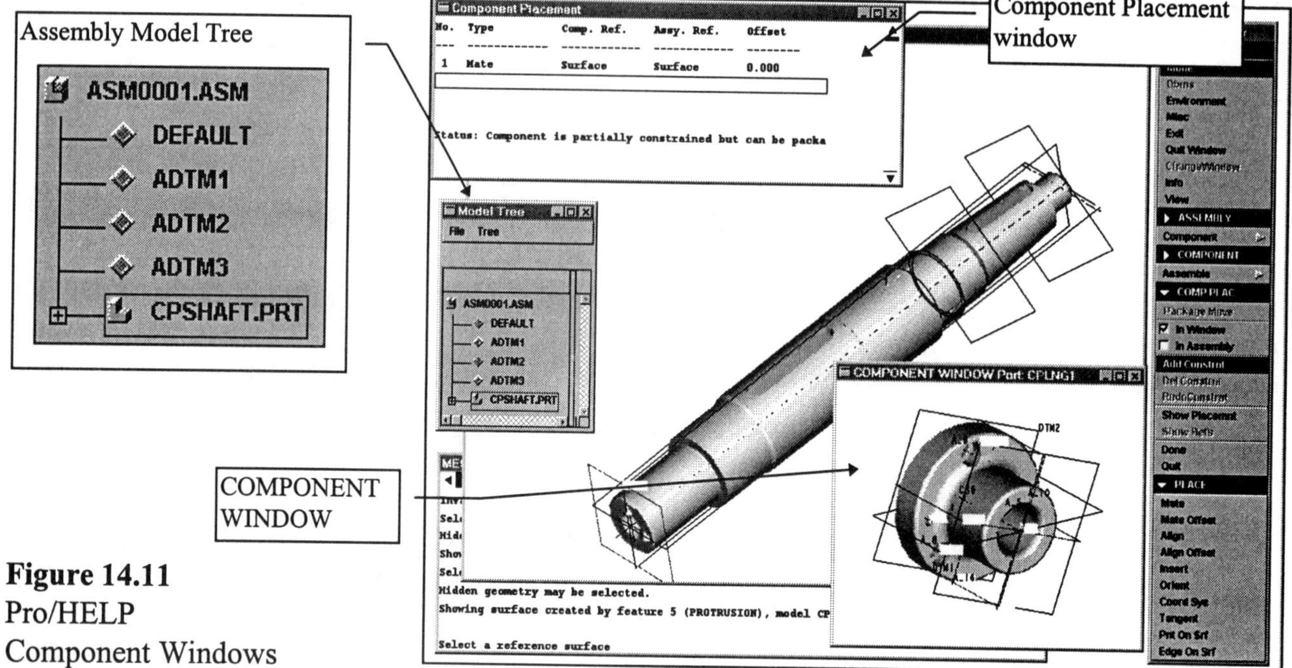

Figure 14.11
Pro/HELP
Component Windows

To assemble the component:

1. Choose **Component** from the ASSEMBLY menu.
2. Choose **Assemble** from the COMPONENT menu and select the component. The component appears in the COMPONENT WINDOW; the COMP PLAC menu and a corresponding Component Placement window are also displayed. The COMP PLAC menu lists the following options:

   Package Move Dynamically changes the position of the component without parametric constraints.

   In Window Shows the component in its own window while you modify its constraints.

   In Assembly Shows the component in the assembly window while you modify its constraints.

   Add Constrnt Adds a placement constraint for the component.

   Del Constrnt Deletes a placement constraint for the component.

   RedoConstrnt Changes a placement constraint for the component. The CONSTR REDEF menu contains the options:

   Type Redefines the constraint type (mate, align, orient).

   AssemblyRef Redefines the reference in the assembly.

   Comp Ref Redefines the reference on the placed component.

   Show Placemnt Shows the location of the component as it would be with the current placement constraints.

   Show Refs Highlights the current references.
3. Choose **Add Constrnt** from the COMP PLAC menu.
4. Define the placement constraints using the PLACE menu. As you do so, Pro/E automatically updates a line in the Component Placement window corresponding to the constraint.

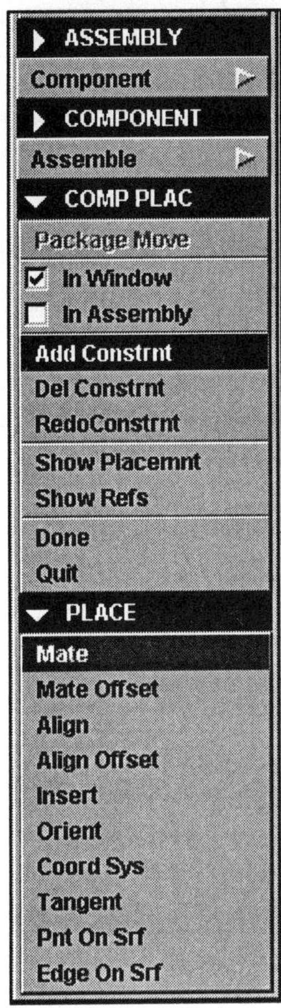

Figure 14.12
Status: **Component is fully constrained and can be placed.**

5. If the component is validly constrained, the Component Placement window displays the message **Status: Component is fully constrained and can be placed.** The MESSAGE WINDOW also displays the message **Component can now be placed.** (Fig. 14.12).
6. Choose **Done** from the COMP PLAC menu to complete the processing. If no component constraints are entered, the Component Placement window displays the message **Status: Component has no constraints but can be packaged** (Fig. 14.13). Pro/E allows you to continue placing the component, package the component, or **Quit**. If you choose to **Quit**, the insufficient constraints are erased and the component is not placed.

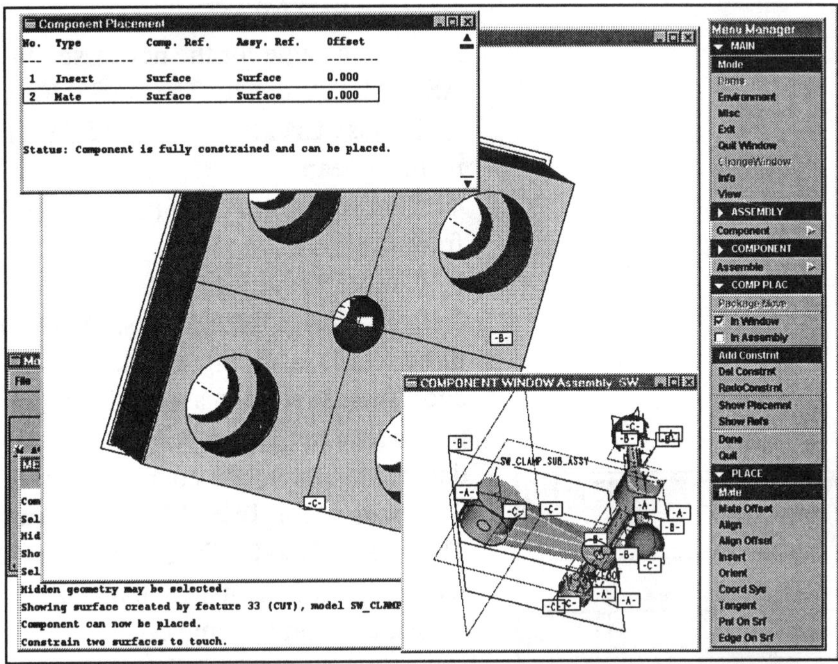

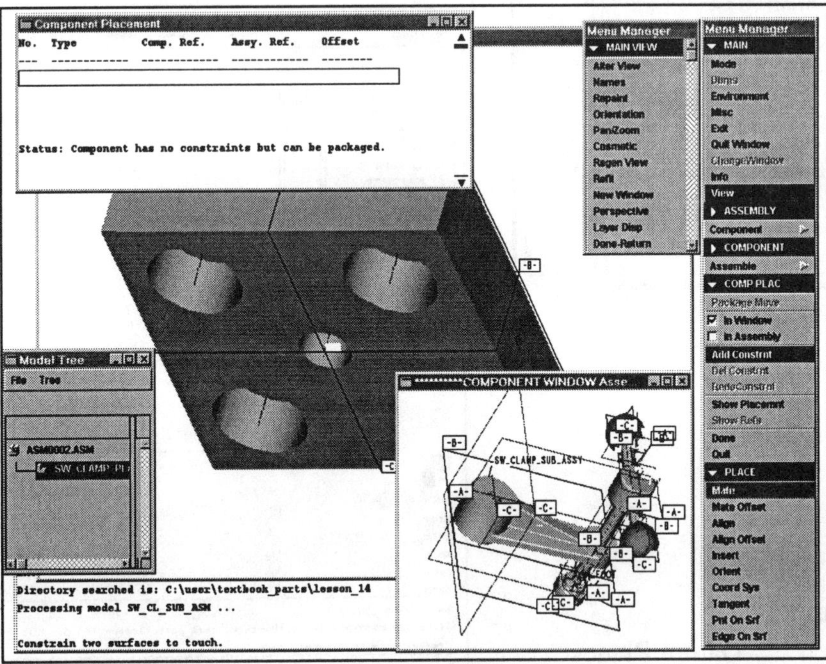

Figure 14.13
Status: **Component has no constraints but can be packaged.**

Datum Planes as the First Feature of an Assembly

When you create three default datum planes as the first features in an assembly, you can assemble a component with respect to these planes or create a part in Assembly mode as the first component. When you use datum planes as the first features, you create a more flexible design. You can *pattern* the first component you add, and you can *reorder* subsequent components to come before the first component (if the components are not children of the first component). To create a default datum plane assembly feature, follow these steps:

1. Choose **Assembly** from the MODE menu, then choose **Create** from the ENTERASSY menu.
2. Enter a name for the new assembly, then choose **Feature** from the ASSEMBLY menu. The ASSY FEAT menu appears.
3. Choose **Create** from the ASSY FEAT menu. The FEAT CLASS menu appears.
4. Choose **Datum** from the FEAT CLASS menu, then choose **Plane** from the DATUM menu.
5. Choose **Default** or **Offset** from the MENUDTM OPT menu. **Default** creates three planes; **Offset** creates a coordinate system and three planes, as in Figure 14.14. You can use the datums and/or a coordinate system to assemble other components.
6. Choose **Set Up** ⇒ **Name** ⇒ **Feature** to change the names of the datum planes and the coordinate system.
7. Choose **Modify** ⇒ **Move Datum** to reposition the new names.

The Model Tree for the assembly can be altered to show the assembly features, components, component features, assembly and component layers, and datum names.

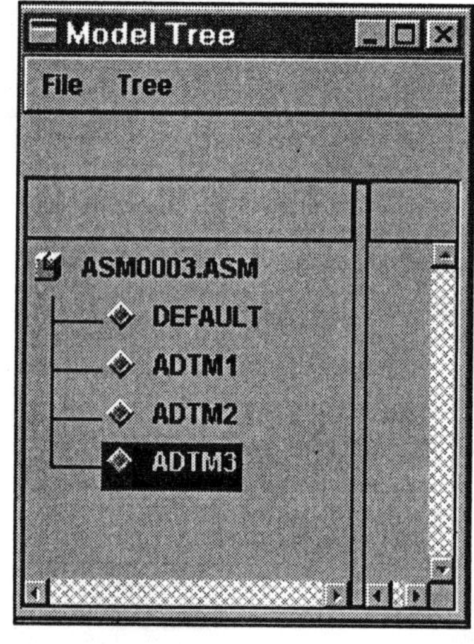

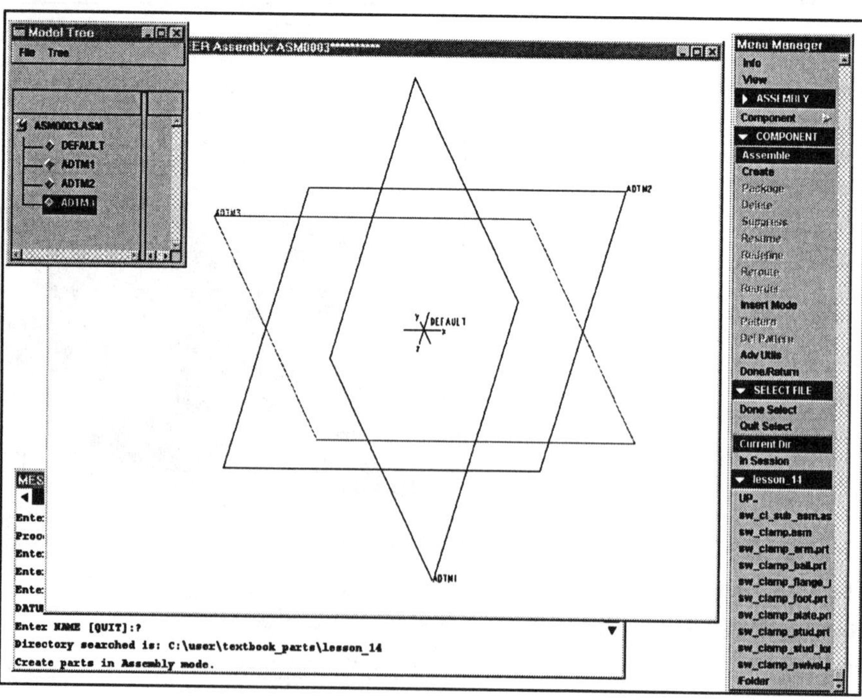

Figure 14.14
Assembly Datum Planes

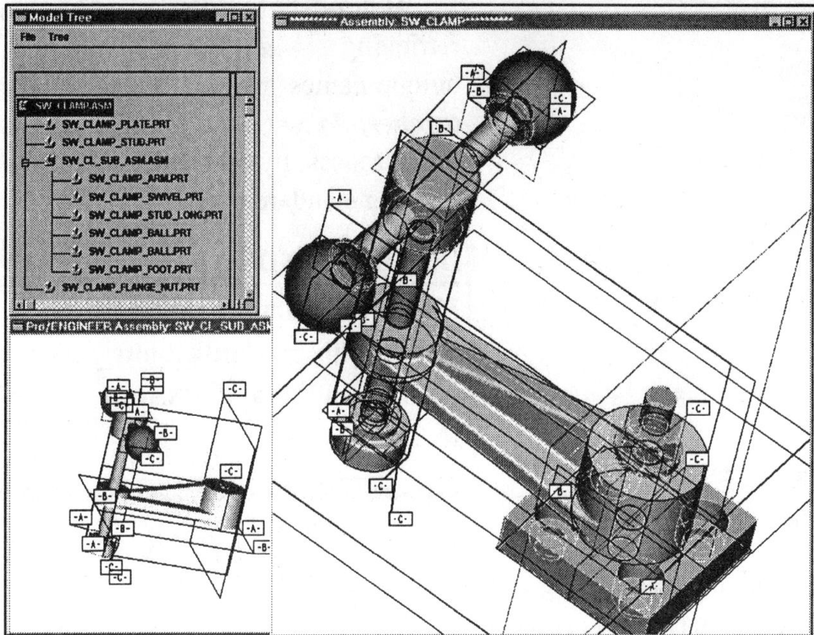

Figure 14.15
Swing Clamp Assembly and Subassembly with Model Tree and Datum Planes

Swing Clamp Assembly

Most of the parts required in this lesson are lesson projects from this text. The **Clamp Foot** and **Clamp Swivel** are from Lesson 5. The **Clamp Ball** is from Lesson 6, and the **Clamp Arm** was created in Lesson 8. If you have not modeled these parts previously, please do so before you start the following step-by-step instructions. The other parts required for the assembly (Fig. 14.15) are standard off-the-shelf hardware items that you can get from Pro/E by accessing the library. If your system does not have a Pro/LIBRARY license for the Basic and Manufacturing libraries, model the parts using the detail drawings provided here. The **Flange Nut**, the **3.50 Double-ended Stud**, and the **5.00 Double-ended Stud** are standard parts (Fig. 14.16).

NOTE

Set up the **Swing Clamp** Assembly:

- Units = Inches
- Datum Planes
- Default Coordinate System
- Layers = **ASM_DATUM_PLANE**
- Shading
- No Disp Tan
- □ Grid Snap

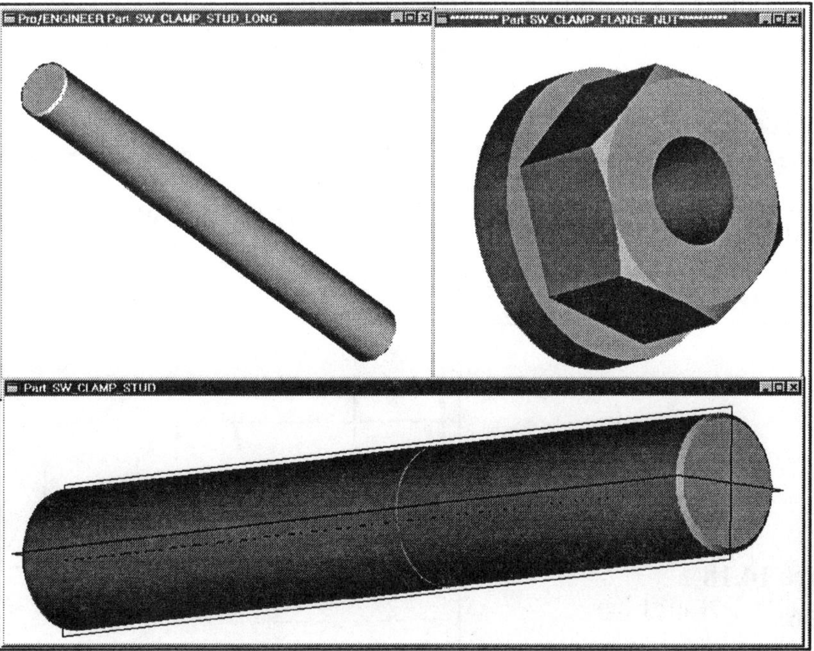

Figure 14.16
Double-Ended Studs and a Flange Nut

Before starting the assembly, you will be modeling each part or retrieving *standard parts* from the library and saving them under unique names in *your* directory. ***Unless instructed to do so by your teacher, do not use the library parts directly in the assembly.*** Start this process by choosing the following commands to access the existing standard double-ended studs in the library:

DOUBLE-ENDED STUD ∅.500 by 3.50 length

Mode ⇒ Part ⇒ Search/Retr ⇒ Pro/Library ⇒ /mfglib ⇒ /fixture_lib ⇒ /nuts_bolts_screws ⇒ st.prt ⇒ SelByParams ⇒ ✓d0,thread_dia ✓ d8,stud_length ⇒ Done Sel ⇒ .5000 ⇒ 3.500

INSTANCE = ST403 (Fig. 14.17)

NOTE

Save the library part with a new name:

Dbms (**File--**PT/Modeler) ⇒ **Save As ⇒ enter ⇒ STUD35 ⇒ enter ⇒ Done-Return**

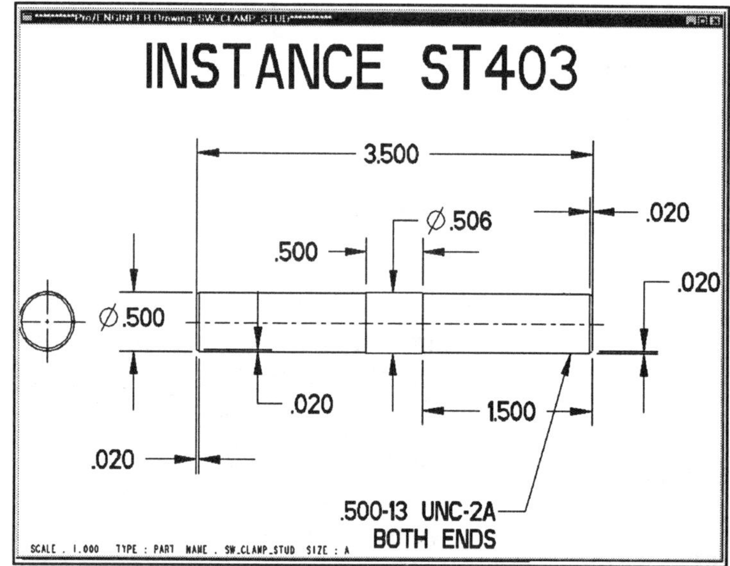

Figure 14.17
∅3.50
Double-Ended Stud

DOUBLE_ENDED STUD ∅.500 by 5.00 length

Mode ⇒ Part ⇒ Search/Retr ⇒ Pro/Library ⇒ /mfglib ⇒ /fixture_lib ⇒ /nuts_bolts_screws ⇒ st.prt ⇒ SelByParams ⇒ ✓d0,thread_dia ✓ d8,stud_length ⇒ Done Sel ⇒ .5000 ⇒ 5.00

INSTANCE = ST406 (Fig. 14.18)

NOTE

Save the library part with a new name:

Dbms (**File--**PT/Modeler) ⇒ **Save As ⇒ enter ⇒ STUD5 ⇒ enter ⇒ Done-Return**

INSTANCE ST406

5.000 · ∅.506 · .020 · .020 · ∅.500 · .020 · .020 · .500-13 UNC-2A BOTH ENDS · 2.000 · 1.500

SCALE : 1.000 TYPE : PART NAME : SW_CLAMP_STUD_LONG SIZE : A

Figure 14.18
∅5.00
Double-Ended Stud

The last standard item used for this assembly is a flange nut (Fig. 14.19). Use the following commands to access the part:

FLANGE NUT

Mode ⇒ Part ⇒ Search/Retr ⇒ Pro/Library ⇒ /mfglib ⇒ /fixture_lib ⇒ /nuts_bolts_screws ⇒ fn.prt ⇒ SelByParams ⇒ ✓ d4,thread_dia ⇒ Done Sel ⇒ .5000

INSTANCE = FN7

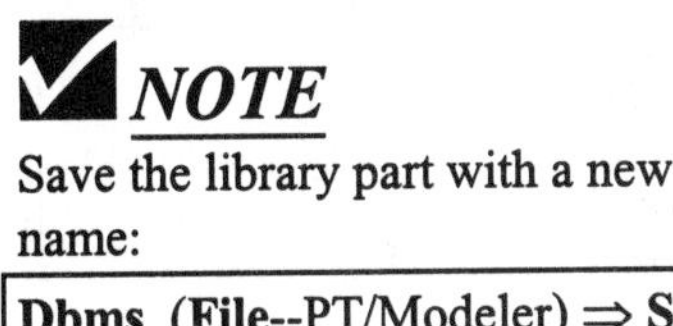

NOTE

Save the library part with a new name:

Dbms (File--PT/Modeler) ⇒ **Save As** ⇒ **enter** ⇒ **FLNGNUT** ⇒ **enter** ⇒ **Done-Return**

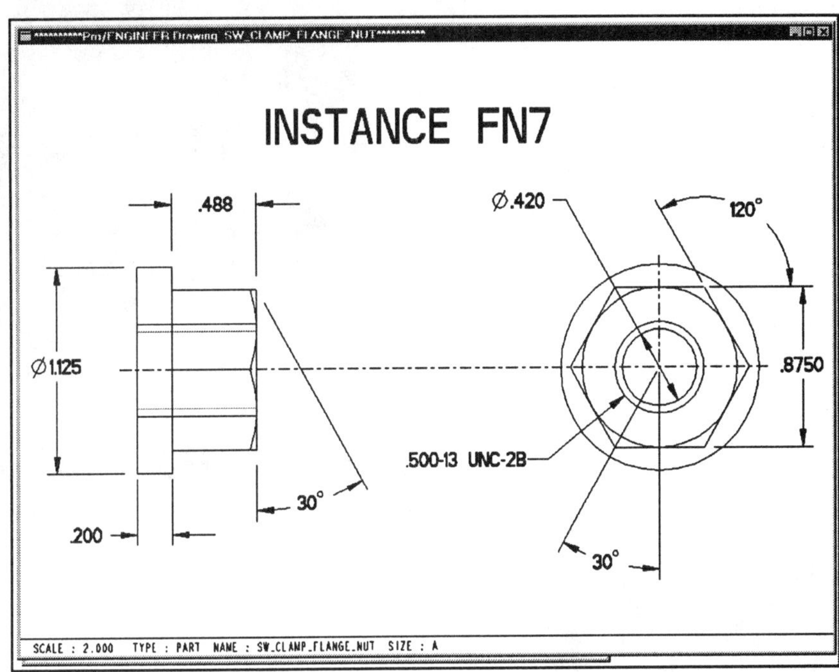

Figure 14.19
Flange Nut

Since you will be creating an assembly using the bottom-up design approach, all the components must be available before the assembling starts. *Bottom-up design* means that existing parts are assembled one by one until the assembly is complete. The assembly starts with a set of datum planes and a coordinate system. The parts are added to the datum features of the assembly. The sequence of assembly will determine the parent-child relationships between components.

Top-down design is the design of an assembly whose component parts are created in Assembly mode as the design unfolds. Some existing parts are available, such as standard components and a few modeled parts. The remaining design evolves during the assembly process.

Regardless of the design method, the assembly datums and coordinate system should be on their own separate *assembly layer*. Each part should also be placed on separate assembly layers; the part's datum features should already be on *part layers*.

The **Clamp Plate** component is the first component of the main assembly. The plate is a new part (Fig. 14.20). You must model the Plate before creating the assembly. Figure 14.21 provides the dimensions necessary to model the Plate.

Dbms (**File**--PT/Modeler) ⇒
Save ⇒ **enter**
Purge ⇒ **enter** ⇒
Done-Return

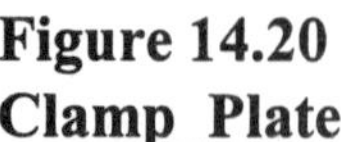

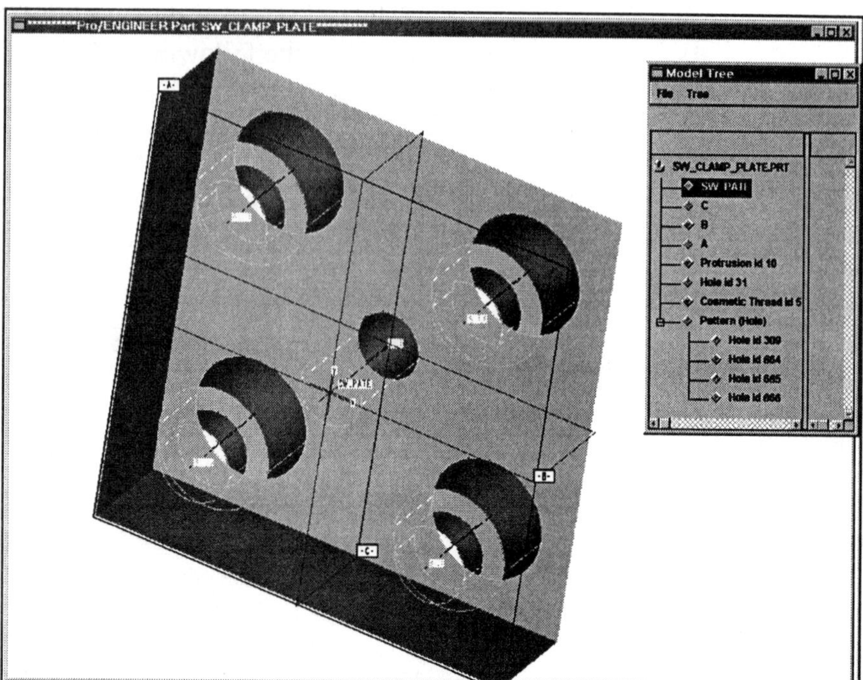

Figure 14.20
Clamp_Plate

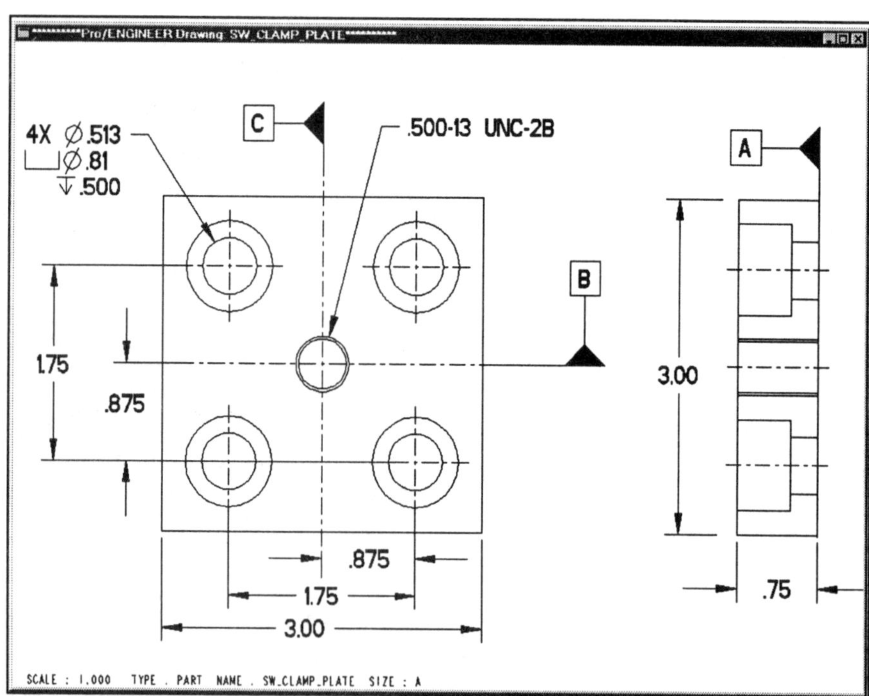

Figure 14.21
Clamp Plate Detail Drawing

You now have all nine parts (eight unique parts, two Ball components are used) required for the assembly. A subassembly will be created first. The main assembly is created second. The subassembly is assembled to the main assembly. Start the subassembly (Fig. 14.22) using the following commands:

PT/Modeler™
⇒ **Feature** ⇒ **Datum Plane** ⇒ **Coord Sys** ⇒ **Default** ⇒ **Done**

Assembly ⇒ **Create** ⇒ **CLAMP_SUBASSEMBLY** ⇒ **enter** ⇒ **Feature** ⇒ **Create** ⇒ **Datum** ⇒ **Plane** ⇒ **Offset** ⇒ **enter** ⇒ **enter** ⇒ **enter** ⇒ **Done/Return**

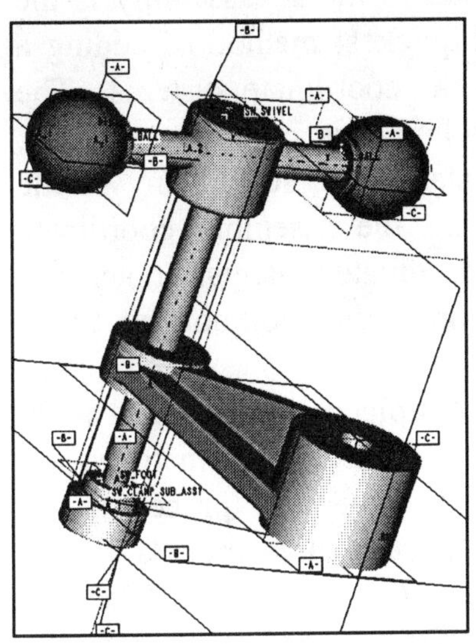

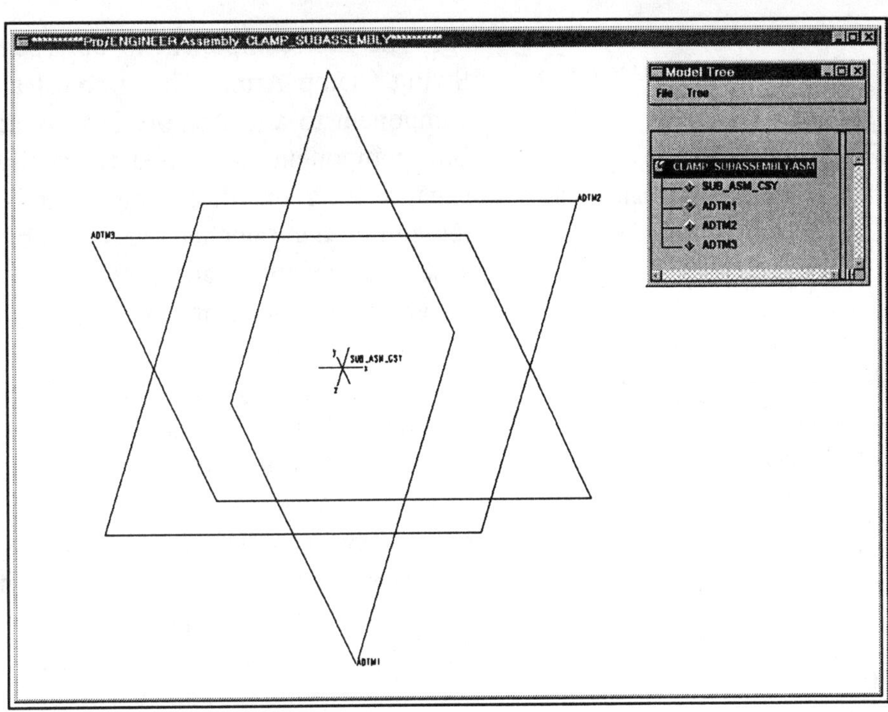

Figure 14.22
Subassembly Datums

Create a layer for the datum planes and the coordinate system (Fig. 14.23). The default datums will have default names, such as **ADTM1**, **ADTM2**, and **ADTM3,** provided by Pro/E. You can change the names using **Set Up ⇒ Name ⇒ Other**. The coordinate system's name has been changed in Figure 14.23 to **SUB_ASM_CSY**. The layer for the datum planes and coordinate system is **SUB_ASM_DATUMS**. You can create your own layering system, using unique names, or use the ones provided here.

Dbms (File--PT/Modeler) ⇒ Save ⇒ enter Purge ⇒ enter ⇒ Done-Return

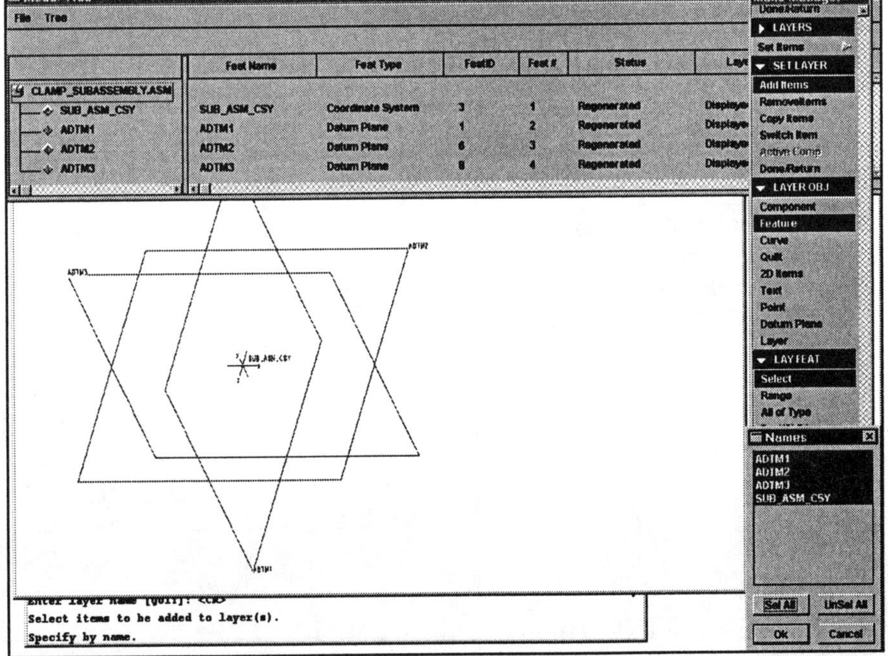

Figure 14.23
Subassembly Layers

The first component to be assembled to the subassembly is the Swing Clamp Arm. The simplest and quickest method of adding a component to an assembly is to match the coordinate systems. The first component assembled is usually where this *constraint* is used, because after the first component is established, few if any of the remaining components are assembled to the assembly coordinate system or, for that matter, other parts' coordinate systems. In general, different constraints are used when you are adding components to the assembly model.

HINT

On all parts and assemblies, **datum planes** will usually be the first feature.

The **Coord Sys** command (Fig. 14.24) places a component in an assembly by aligning its coordinate system with a coordinate system in the assembly. Before you start the assembly process, both coordinate systems should exist on their respective models (Pro/E refers to both assemblies and parts as *models* or *objects*). Make sure all your models are in the same working directory before you start the assembly process. If you named your models something that is not listed here, pick the appropriate part model as requested. After the component window appears, rotate the part model so you can see more clearly the features to be used for constraining. The window with the asterisks is the *active window:* ********MODEL_NAME********. When a feature in one window is selected, another window may become hidden.

PT/Modeler™

PT/Modeler does not open another window for the assembled component the way Pro/E does; after you select the component to assemble, PT/Modeler lets you drag the component to a convenient place in the window. Left-click to drop the component where you want it.

Choose the following commands (Figs. 14.25 and 14.26):

Component ⇒ **Assemble** ⇒ (enter a **?** from the keyboard) ⇒ **enter** ⇒ (pick the **CLAMP_ARM**--the machined part, not the casting--from the **Current Dir**) ⇒ **Coord Sys** ⇒ (pick the coordinate system on the part model and then the coordinate system on the assembly model) ⇒ **Done** ⇒ **Done/Return**

PT/Modeler™

Component ⇒ **Assemble** ⇒ **?** ⇒ **enter** ⇒ (select the component) ⇒ (drag the component to a convenient place in the main screen. Left-click to drop it)

Status: Component has no constraints but can be packaged.

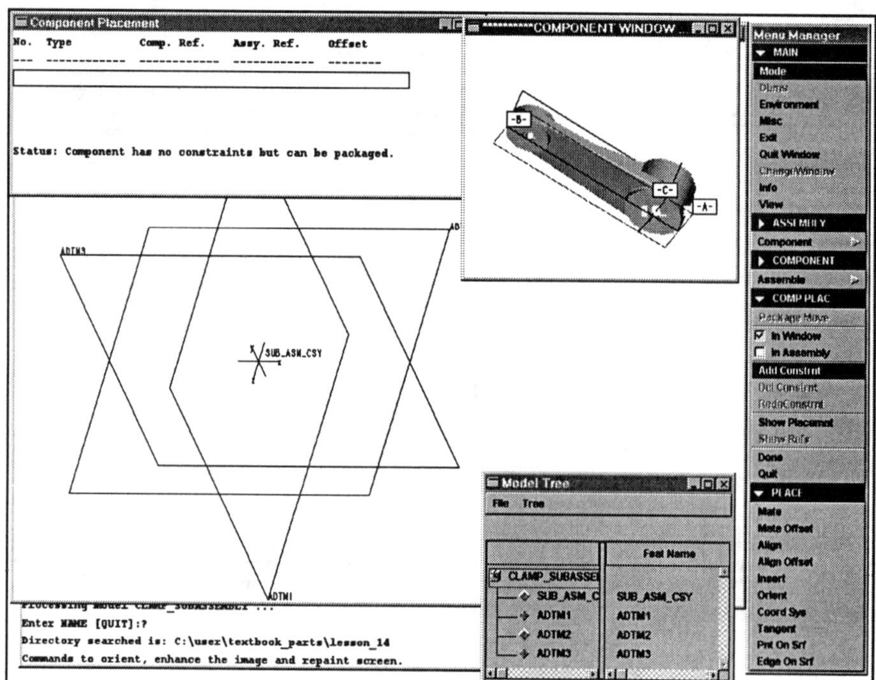

Figure 14.24
COMPONENT WINDOW and Component Placement Window

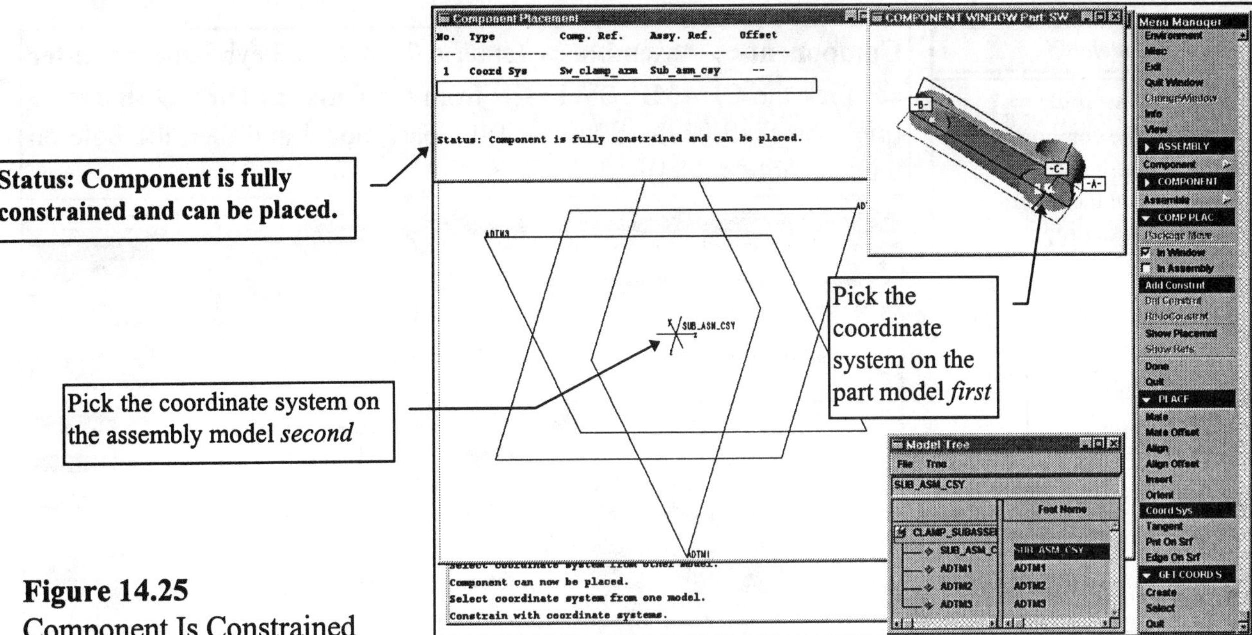

Figure 14.25
Component Is Constrained

Save the *assembly*
Dbms (File--PT/Modeler) ⇒
Save ⇒ **enter**
Purge ⇒ **enter** ⇒
Done-Return

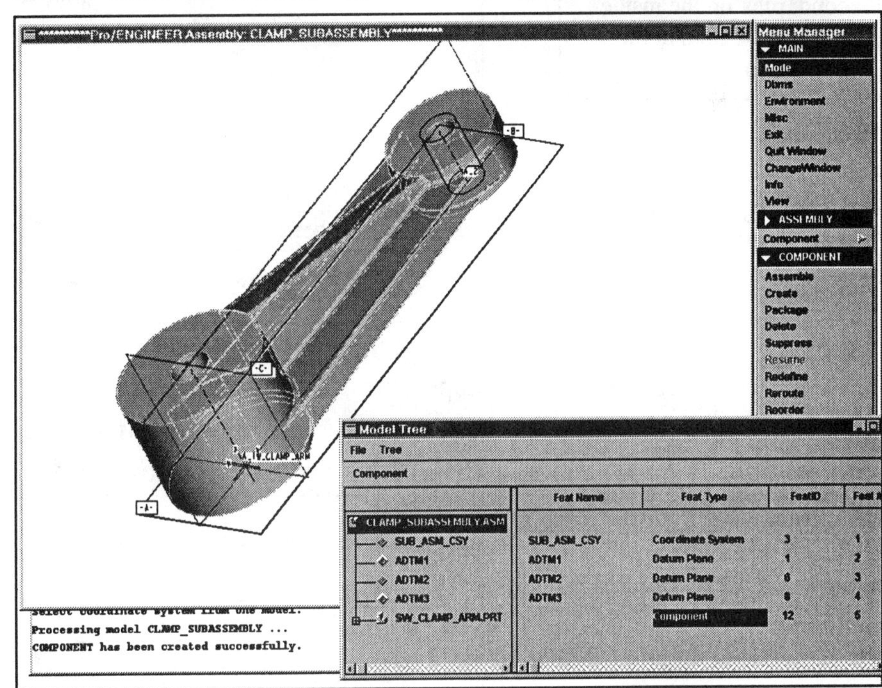

Figure 14.26
Clamp_Arm

The next component to be assembled is the **CLAMP_SWIVEL**. Two constraints will be used on this model: **Insert** and **Mate Offset**. *Placement constraints* are used to specify the relative position of a *pair of surfaces/references* between two components. The **Mate**, **Align**, **Insert**, and **Orient** commands are placement constraints. The two surfaces/references must be of the same type.

When using a datum plane as a placement constraint, specify which side to use, yellow or red. When using **Mate Offset** or **Align Offset**, enter the offset distance. The *offset direction* is displayed with a large arrow. If you need an offset in the opposite direction, enter a *negative value*. Choose the following commands to assemble the next part to the assembly model as shown in Figure 14.27:

PT/Modeler™

Component ⇒ Assemble ⇒ ? ⇒ enter ⇒ (select the component) ⇒ (drag the component to a convenient place in the main screen. Left-click to drop it)

Component ⇒ Assemble ⇒ (enter a **?** from the keyboard) **⇒ enter ⇒** (pick the **CLAMP_SWIVEL** from the **Current Dir**) **⇒ Insert ⇒** (pick the shaft of the Swivel on the part model and then the hole on the assembly model)

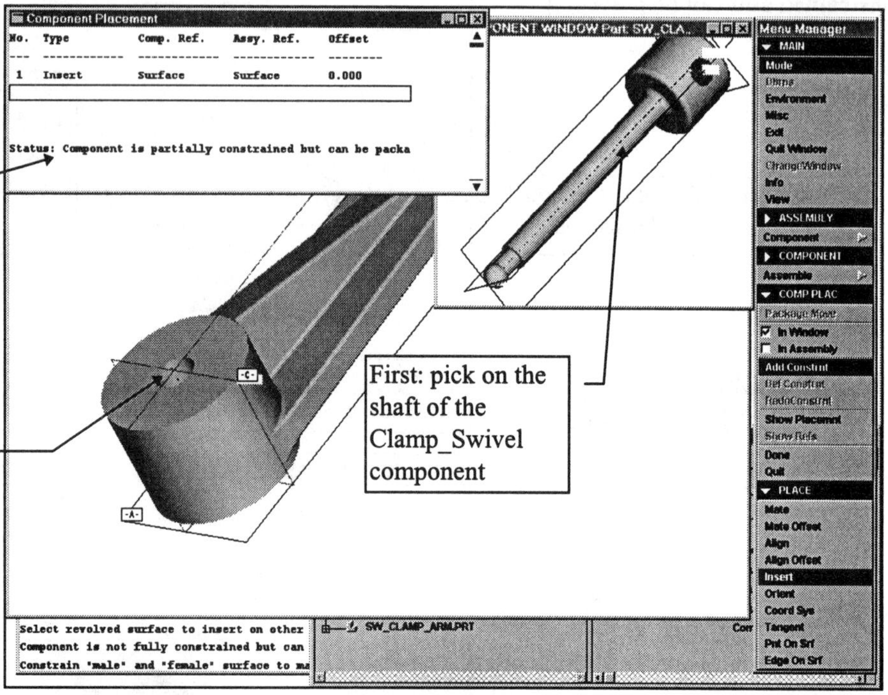

Component is only partially constrained

Second: pick on the inside surface of the hole in the Clamp_Arm

First: pick on the shaft of the Clamp_Swivel component

Figure 14.27
Assembling the **Clamp_Swivel**

The next constraint is **Mate Offset**:

Mate Offset ⇒ (pick the lower surface of the Swivel and then the upper surface on the assembly model, as shown in Fig. 14.28 and in Fig. 14.29) **⇒ [Offset (INCHs) in indicated direction [0.0000]:** (type) **2.00] ⇒ enter ⇒ Done ⇒ Done/Return**

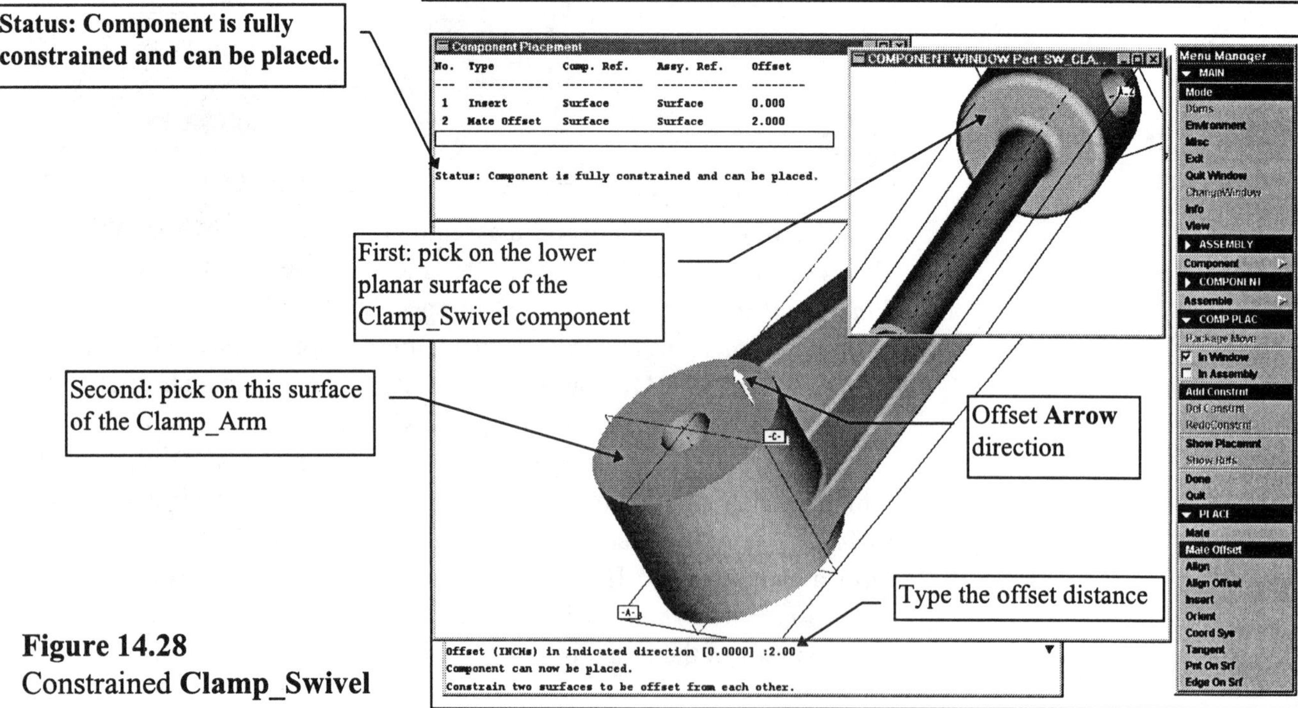

Figure 14.28
Constrained **Clamp_Swivel**

Dbms (File--PT/Modeler) ⇒
Save ⇒ **enter**
Purge ⇒ **enter** ⇒
Done-Return

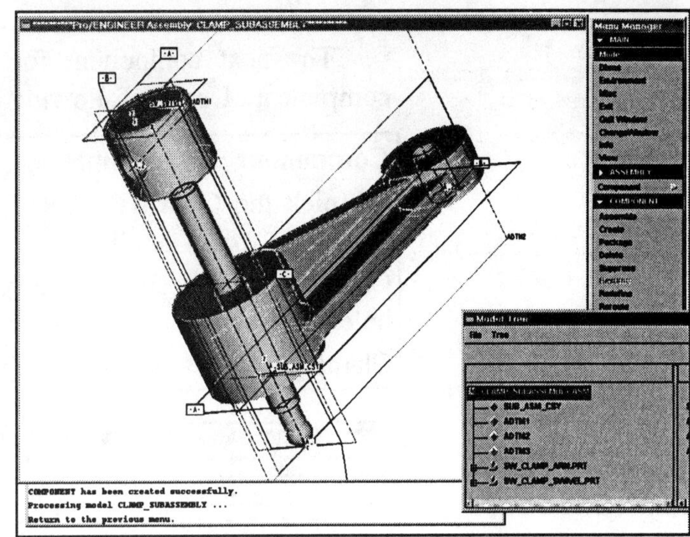

Figure 14.29
Subassembly with **Clamp_Arm** and **Clamp_Swivel**

Change the offset distance to **1.50** using the following commands:

PT/Modeler™
Modify Dim ⇒ **Value** ⇒ **etc.**

Modify ⇒ **Mod Assem** ⇒ **Modify Dim** ⇒ **Value** ⇒ (pick the Swivel) ⇒ (pick the **2.00** dimension from the model as in Fig. 14.30) ⇒ (type **1.50** at the prompt) ⇒ **enter** ⇒ **Done Sel** ⇒ **Done** ⇒ **Regenerate** ⇒ **Automatic** ⇒ **Done** ⇒ **Done/Return** (Fig. 14.31)

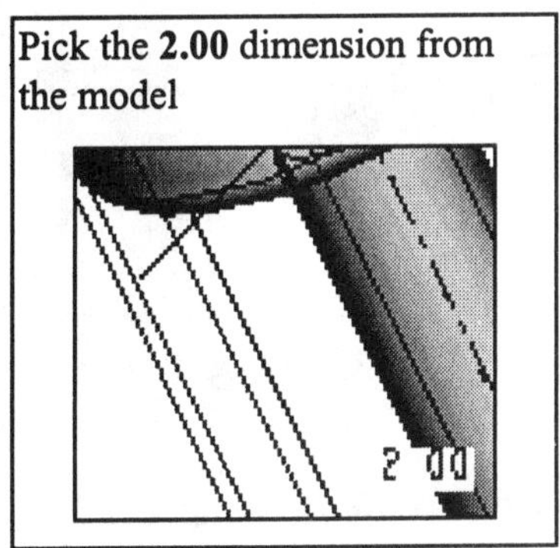

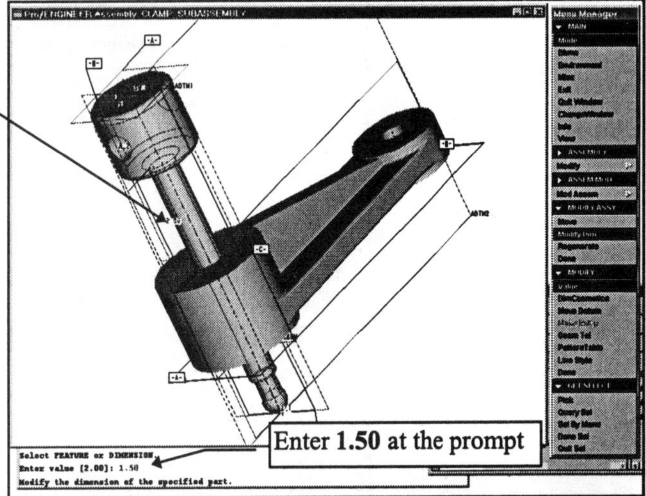

Figure 14.30
Modify the Offset Distance to **1.50**

Dbms (File--PT/Modeler) ⇒
Save ⇒ **enter**
Purge ⇒ **enter** ⇒
Done-Return

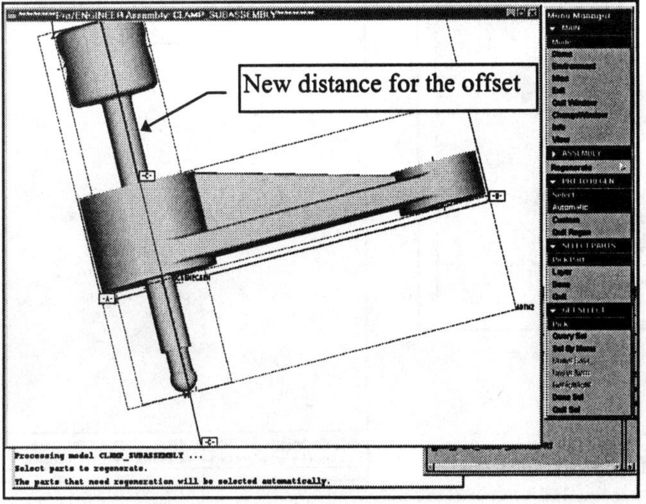

Figure 14.31
Regenerated Model with **1.50** as Offset

The next component for the subassembly is the Clamp_Foot component. Use the following commands:

Component ⇒ **Assemble** ⇒ (enter a **?** from the keyboard) ⇒ **enter** ⇒ (pick the **CLAMP_FOOT** from the **Current Dir**) ⇒ **Align** ⇒ (pick the *axis* of the Clamp_Foot and then the *axis* on the Clamp_Swivel as shown in Fig. 14.32) ⇒ **Mate** ⇒ (pick the spherical hole of the Clamp_Foot and then the spherical end on the Clamp_Swivel, as shown in Fig. 14.33) ⇒ **Done**

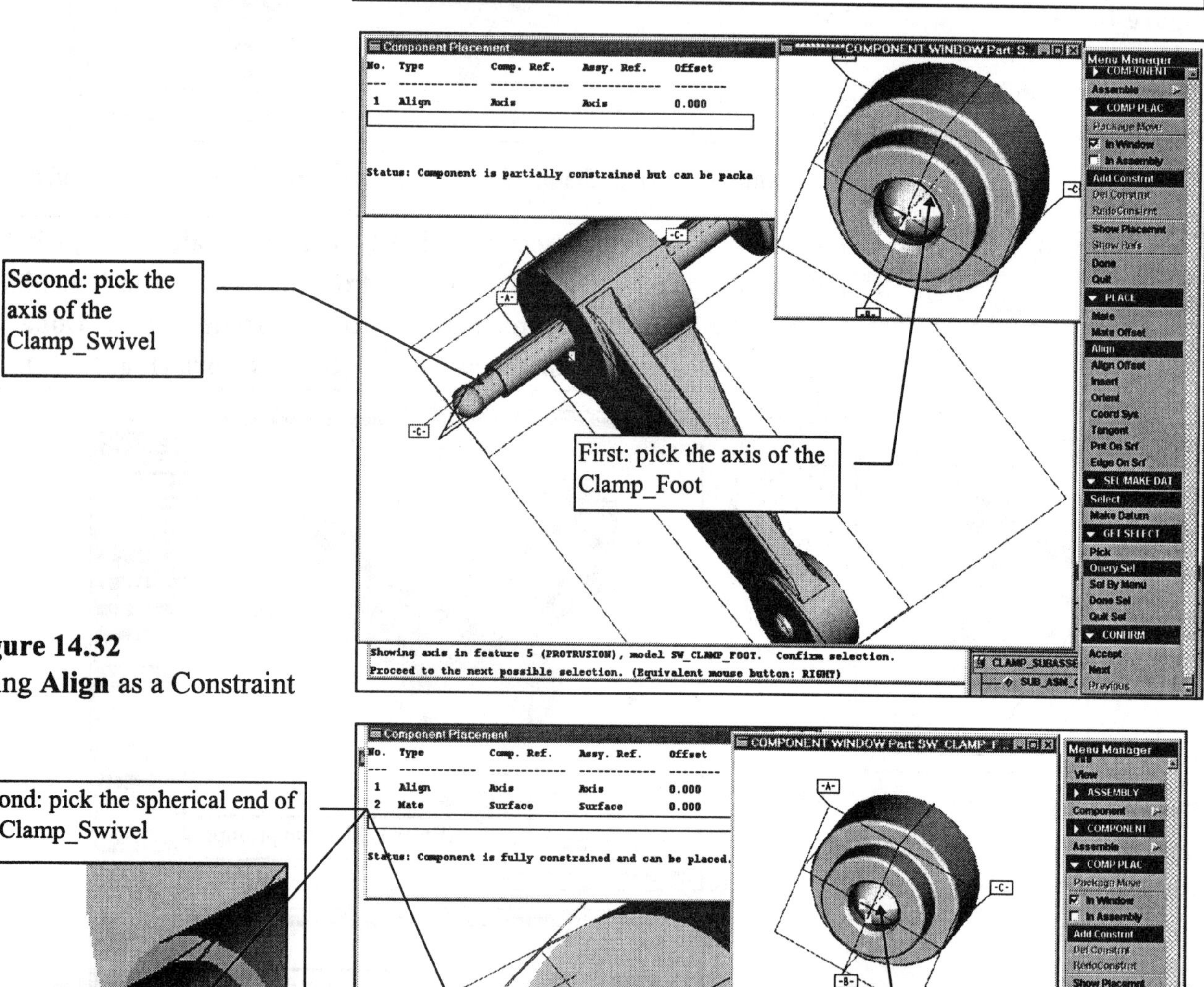

Figure 14.32
Using **Align** as a Constraint

Figure 14.33
Using **Mate** as a Constraint

Well, it doesn't look exactly right yet (Fig. 14.34). A third constraint needs to be added to orient the Foot correctly. Choose the following commands (Fig. 14.35):

> **Redefine** ⇒ (pick Clamp_Foot from the Model Tree) ⇒ **Add Constrnt** ⇒ **Orient** ⇒ (pick lower surface of the Clamp_Arm as in Fig. 14.36) ⇒ **Query Sel** ⇒ (pick top surface of the Clamp_Foot) ⇒ **Next** ⇒ **Accept** ⇒ **Done** ⇒ **Done/Return** (Fig. 14.37)

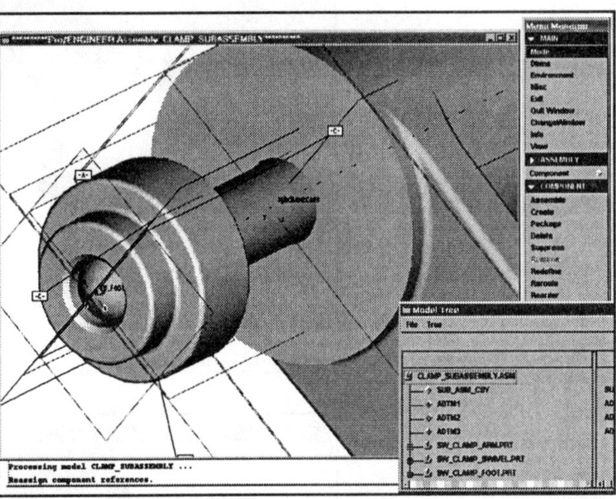

Figure 14.34
Foot Facing in the Wrong Direction

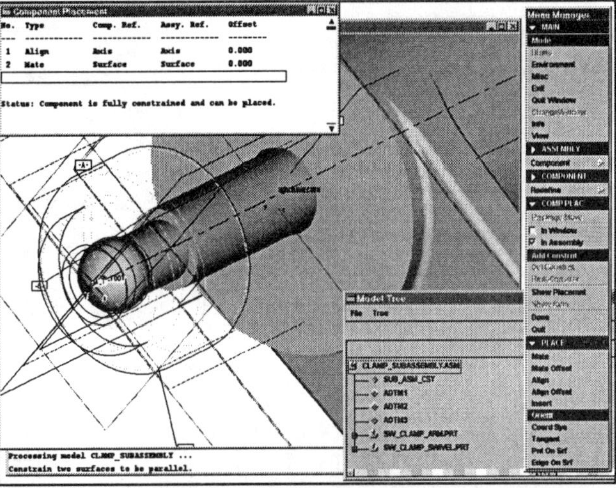

Figure 14.35
Using **Orient** as a Constraint

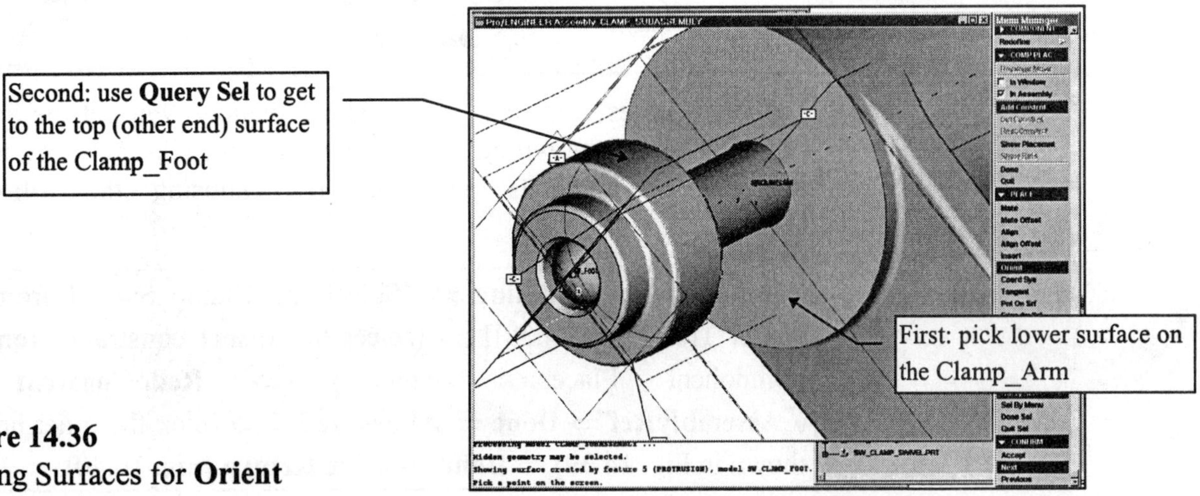

Second: use **Query Sel** to get to the top (other end) surface of the Clamp_Foot

First: pick lower surface on the Clamp_Arm

Figure 14.36
Picking Surfaces for **Orient**

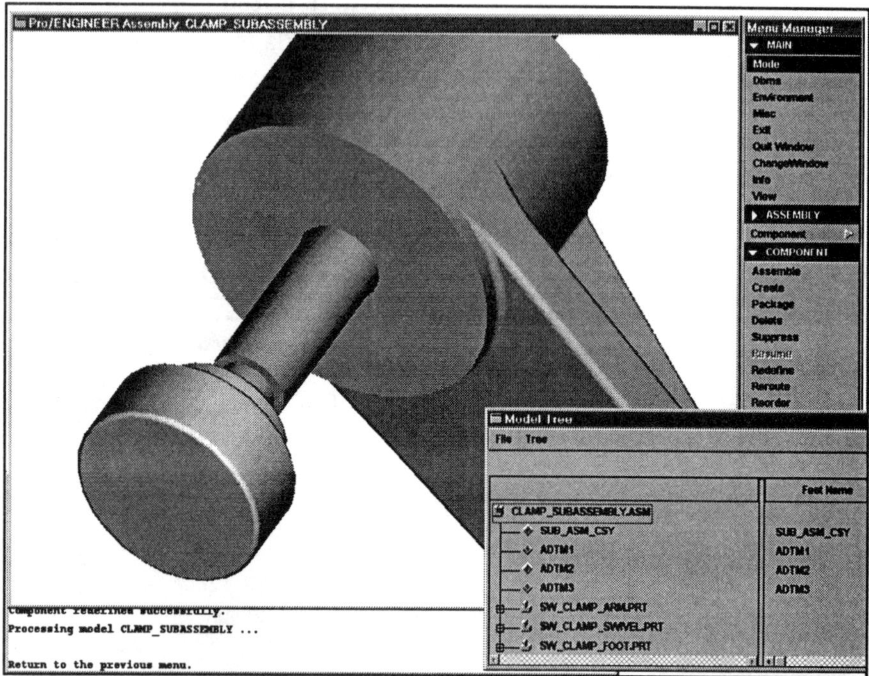

Figure 14.37
Redefined Foot

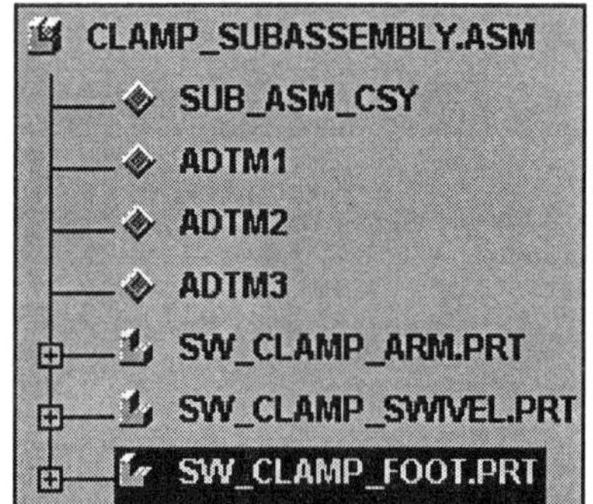

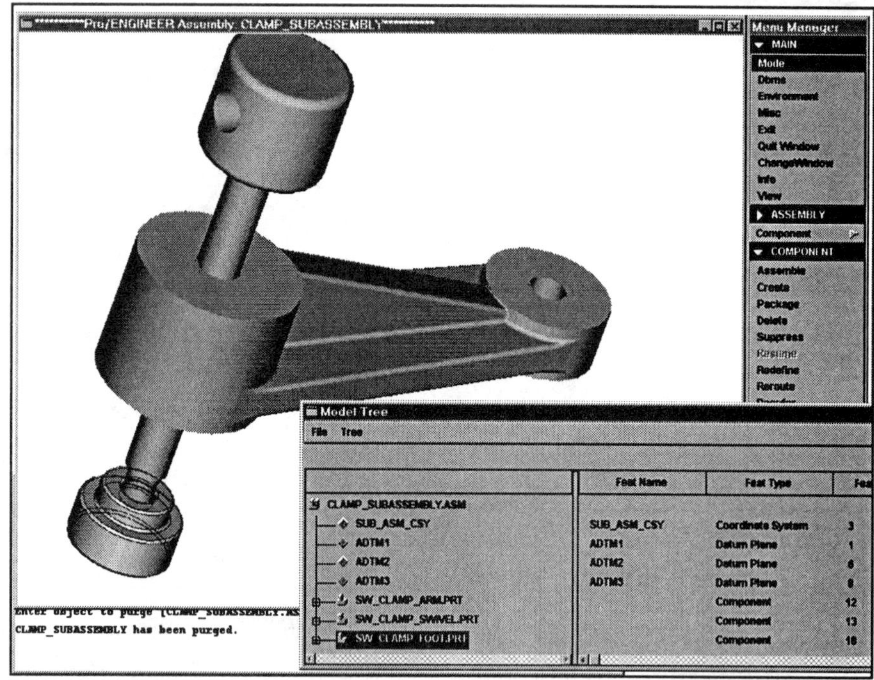

Figure 14.38
Picking Surfaces

Here comes your boss again. Unfortunately, *you* put the Swivel and the Foot in the wrong hole! Before the mistake gets any undue attention, redefine its placement by choosing the following commands:

Component ⇒ **Redefine** ⇒ [Select the Clamp_Swivel from the Model Tree (Fig. 14.38)] ⇒ (select the **Insert** constraint from the Component Placement window) ⇒ **RedoConstrnt** ⇒ **✓AssemblyRef** ⇒ **Done** ⇒ **AlternateRef** ⇒ (pick the other hole as shown in Fig. 14.39) ⇒ **Done** ⇒ **Done/Return** (Fig. 14.40)

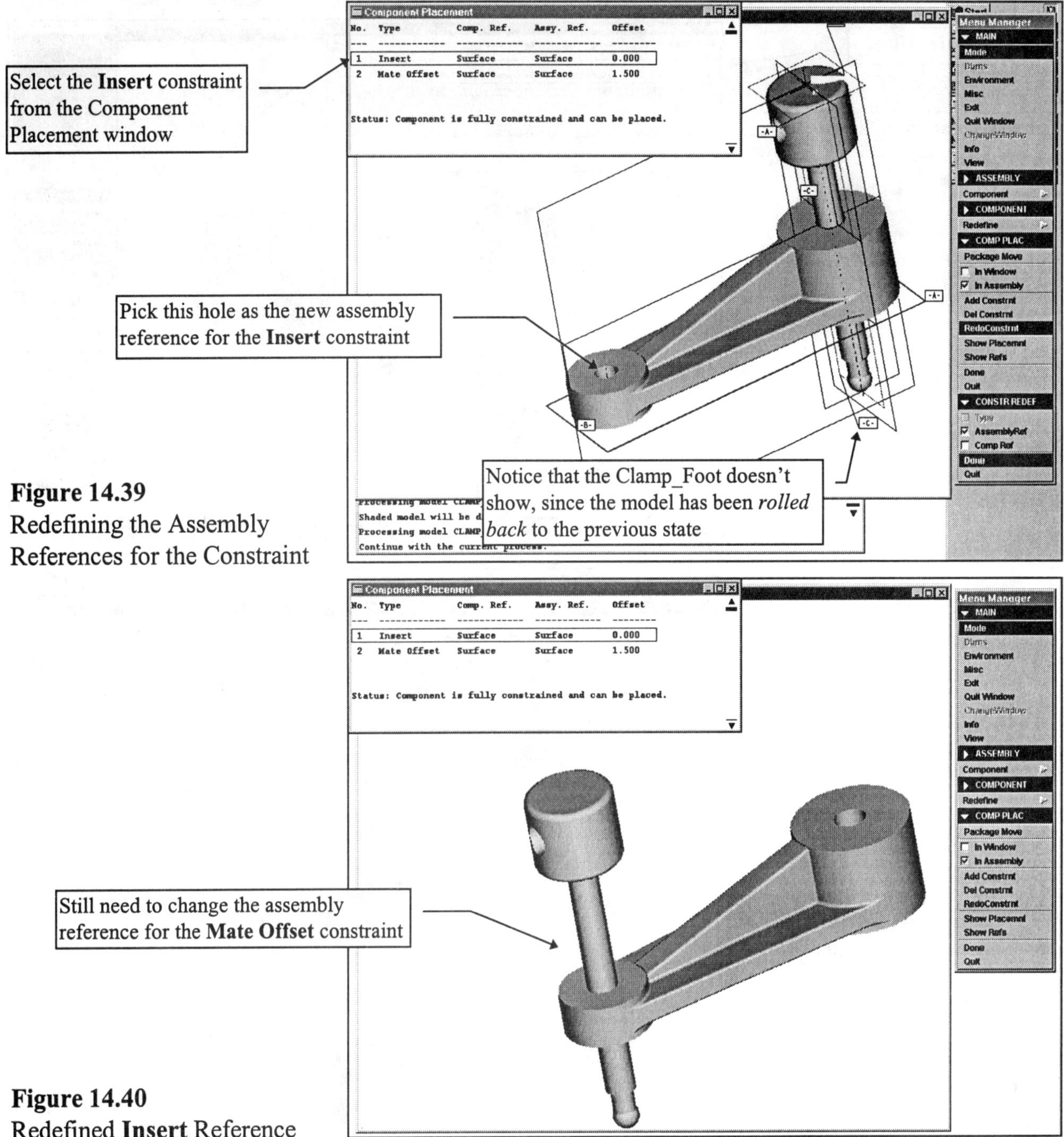

Figure 14.39
Redefining the Assembly References for the Constraint

Figure 14.40
Redefined **Insert** Reference

The Swivel is now in the correct hole, but the **Mate Offset** assembly reference is incorrect, since the bottom of the Swivel head is offset from the top surface of the large-diameter boss on the other side of the part. Complete the redefinition:

Component ⇒ **Redefine** ⇒ (pick the Clamp_Swivel from the Model Tree) ⇒ (select the **Mate Offset** constraint from the Component Placement window) ⇒ **RedoConstrnt** ⇒ **✓AssemblyRef** ⇒ **Done** ⇒ **AlternateRef** ⇒ (Select offset mating surface as shown in Fig. 14.41) ⇒ (Offset [INCHs] in indicated direction) ⇒ (type **2.25** at the prompt) ⇒ **enter** ⇒ **Done** ⇒ **Done/Return** (Fig. 14.42)

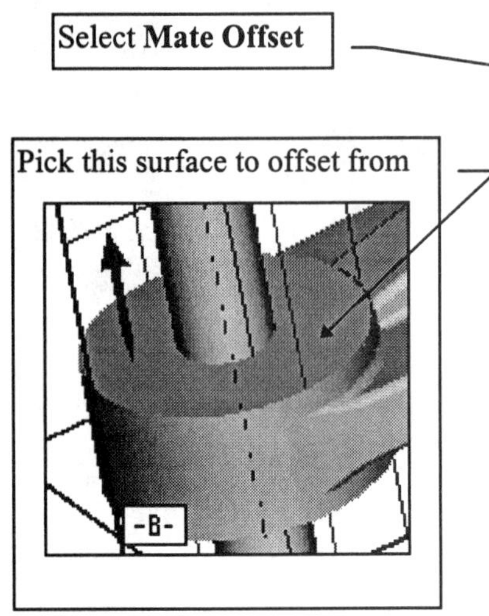

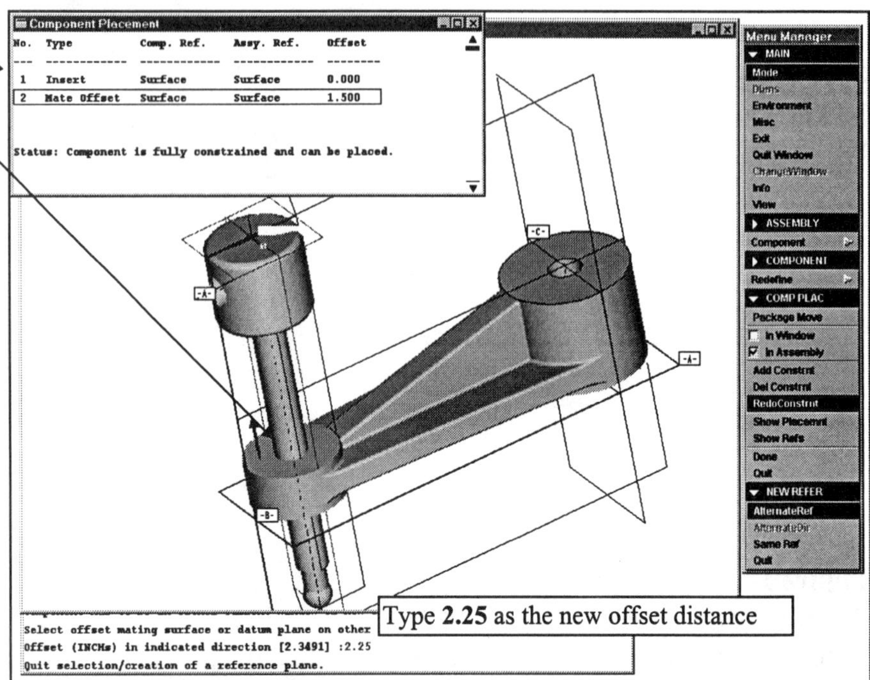

Figure 14.41
Redefined **Mate Offset** Reference

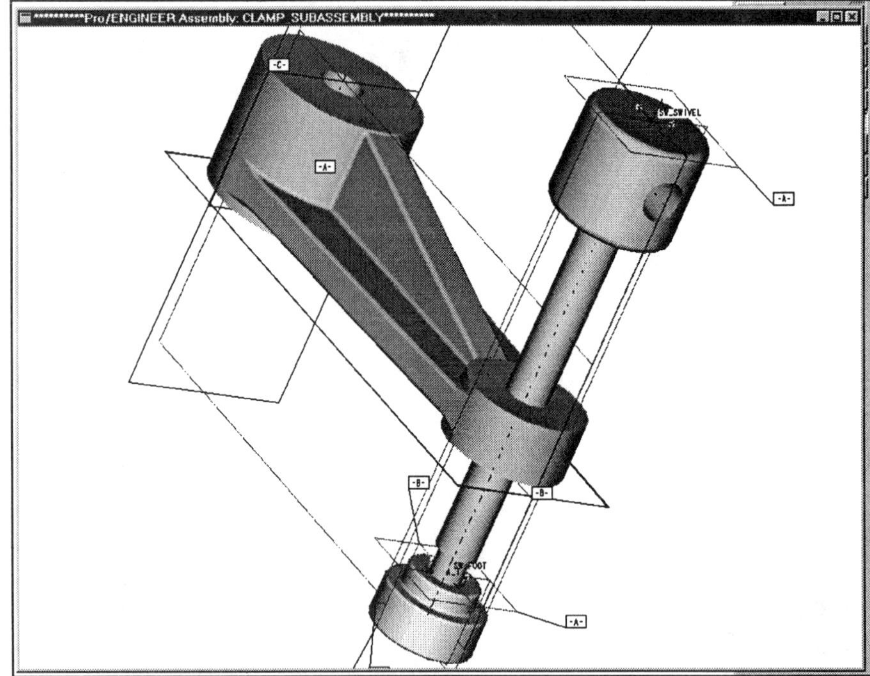

HINT
When the part was modeled, the datum planes were set as geometric tolerance features (Basic Datums) and renamed.

Figure 14.42
Completed Redefinition

As you can see, it is extremely easy to alter your model using **Modify** and **Redefine**. The next step in the assembly of the model is to assemble the **5.00** double-ended stud:

NOTE
Both **Feature ⇒ Redefine** and **Component ⇒ Redefine** are available. In this case, you want to redo the component placement, therefore use **Component ⇒ Redefine**.

Component ⇒ Assemble ⇒ (enter a **?** from the keyboard) **⇒ enter ⇒** (pick **STUD5** from the **Current Dir**) **⇒ Insert ⇒** (select the *revolved surface* of the Stud and then select the *revolved surface hole* on the Swivel as shown in Fig. 14.43) **⇒ Align Offset ⇒** (pick the end of the Stud and then pick the datum **C** of the Swivel from the Model Tree) **⇒ Yellow** (Fig. 14.44) **⇒** (type an offset of **5.00/2** at the prompt) **⇒ enter ⇒ Done ⇒ Done/Return** (Fig. 14.45)

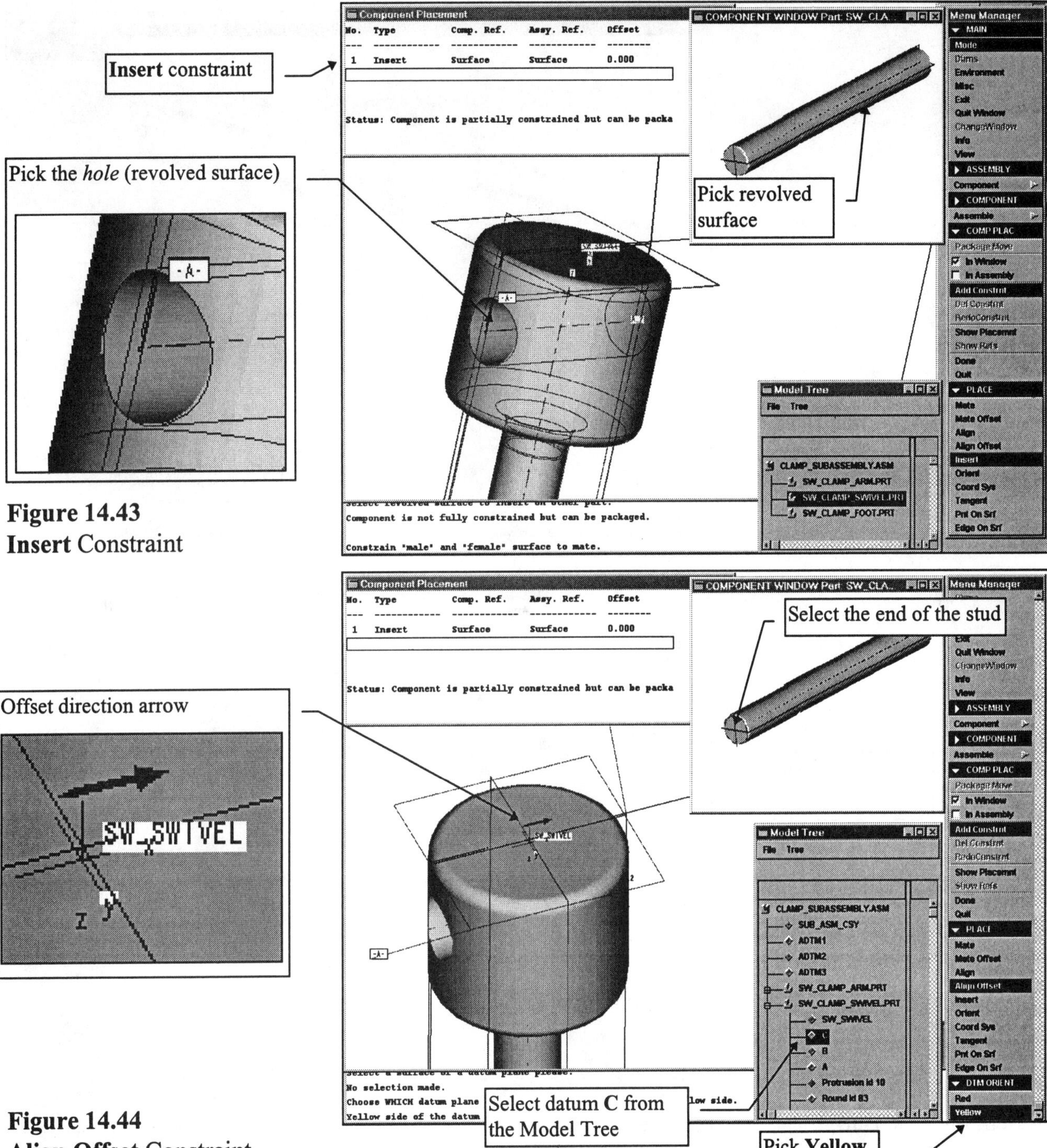

Figure 14.43
Insert Constraint

Figure 14.44
Align Offset Constraint

The two Ball handles are assembled next. Instructions are provided for assembling one Ball; you must assemble the other on your own. Use the following commands:

Component ⇒ **Assemble** ⇒ (enter a **?** from the keyboard) ⇒ **enter** ⇒ (pick the **CLAMP_BALL** from the **Current Dir**) ⇒ **Insert** ⇒ (pick the *revolved hole* of the Ball and then the *revolved surface* on the Stud, as shown in Fig. 14.46)

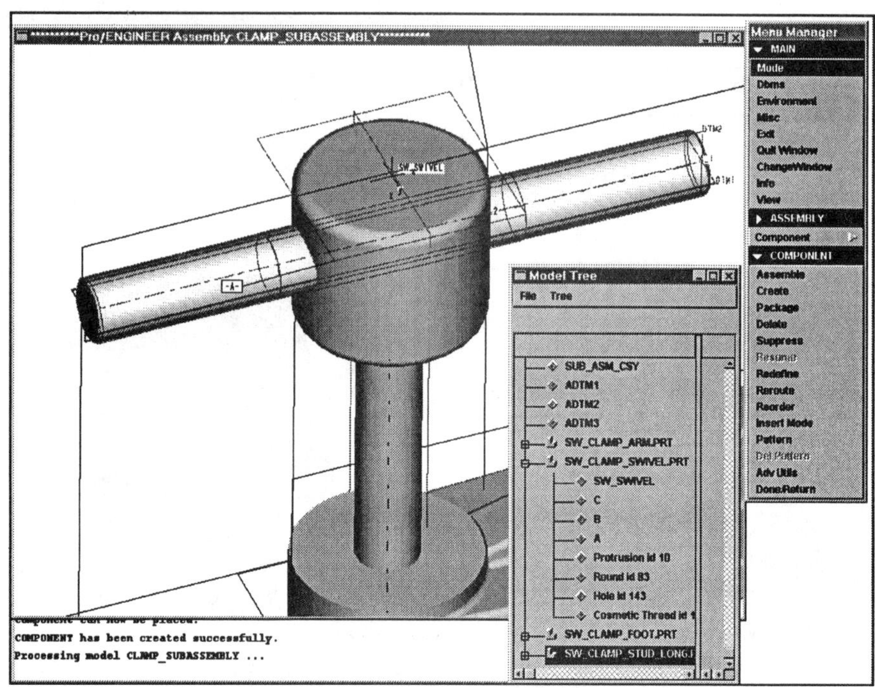

Figure 14.45
Assembled Double-Ended Stud

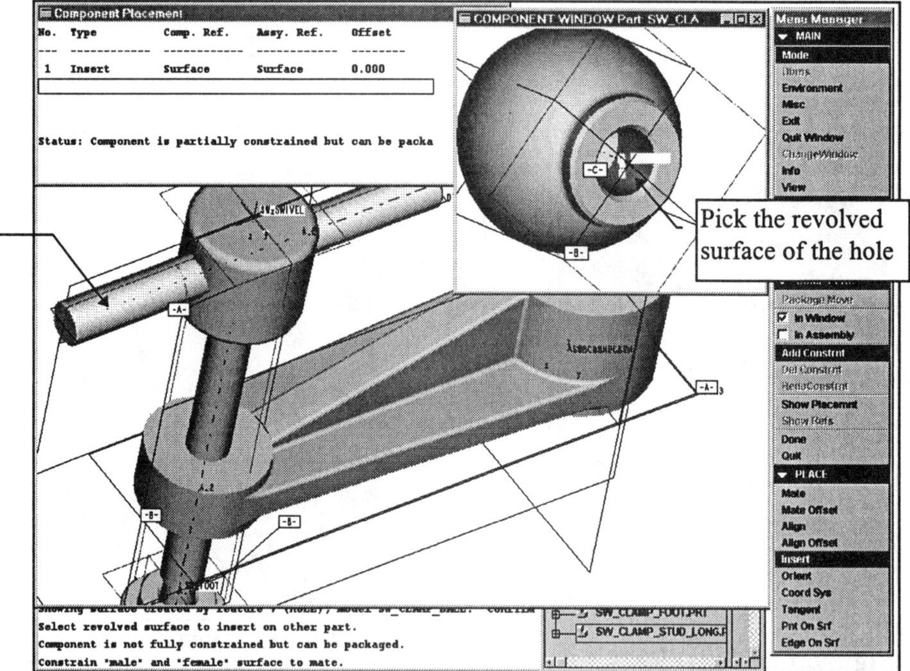

Figure 14.46
Assembling the Ball

Mate Offset ⇒ (pick the flat end of the Ball and then the end of the Stud, as shown in Fig. 14.47) ⇒ (Offset [INCHs] in indicated direction) ⇒ **(**type **-.500** at the prompt) ⇒ **enter** ⇒ **Done** ⇒ **Done/Return** (Fig. 14.48)

Assemble the second Ball as shown in Figure 14.49. The subassembly is now complete. **Save** and **Purge** the model. The subassembly will be added to the assembly in the next set of steps.

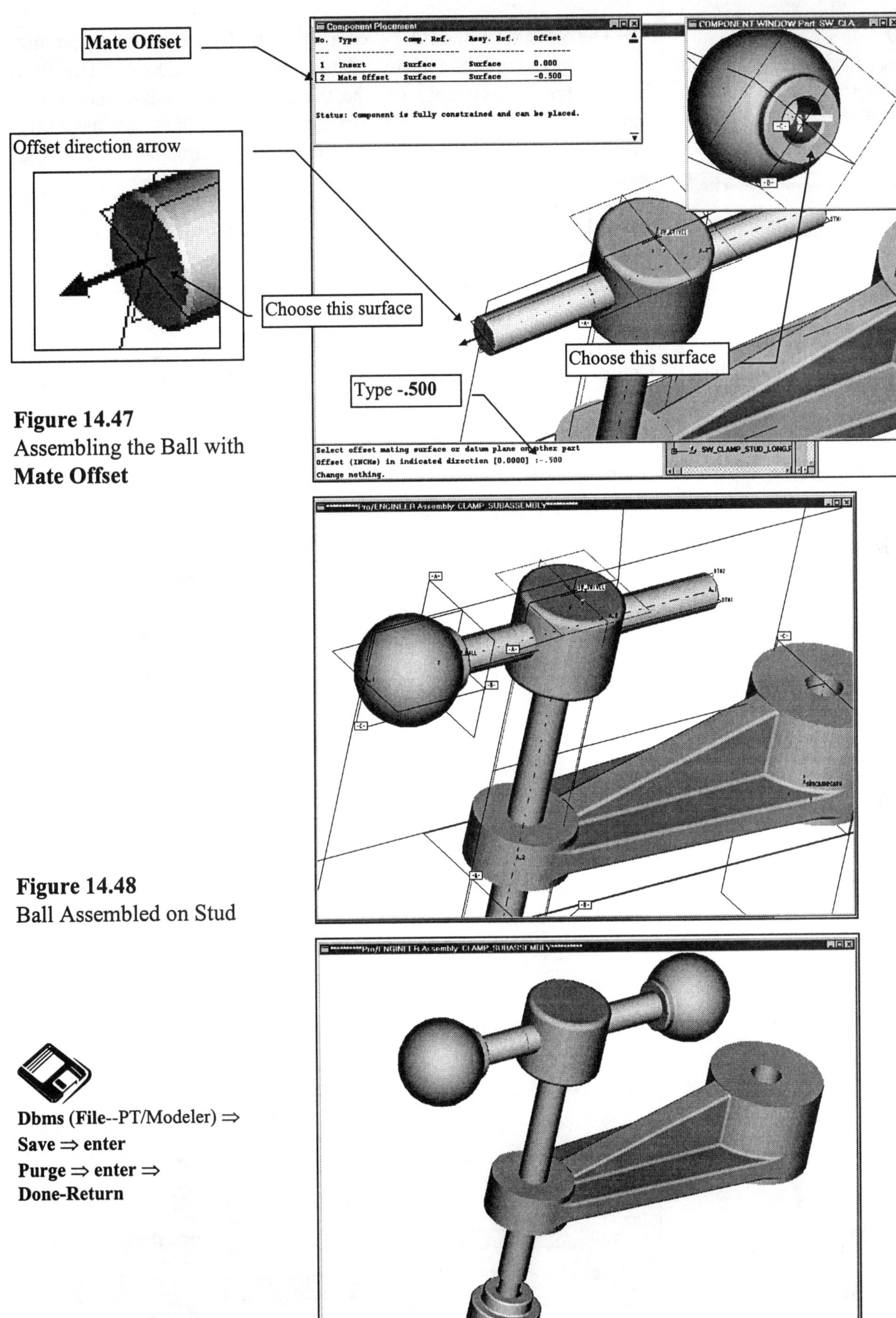

Figure 14.47
Assembling the Ball with **Mate Offset**

Figure 14.48
Ball Assembled on Stud

Dbms (File--PT/Modeler) ⇒
Save ⇒ enter
Purge ⇒ enter ⇒
Done-Return

Figure 14.49
Completed Subassembly

ECO

As a minor **ECO**, redefine the Ball component as offset **.44** from the end of the Shaft so that the Stud does not bottom out in the Ball's hole.

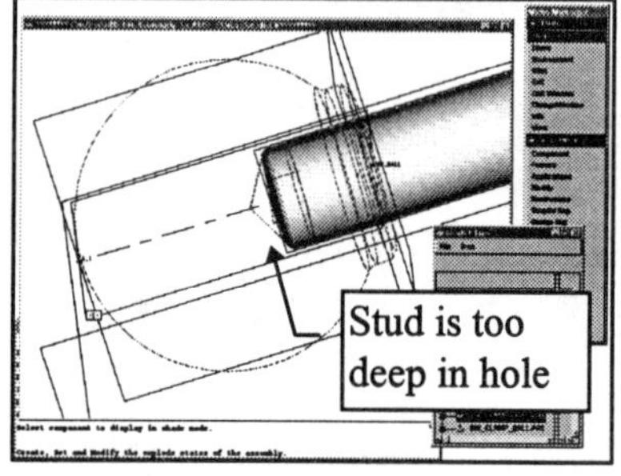

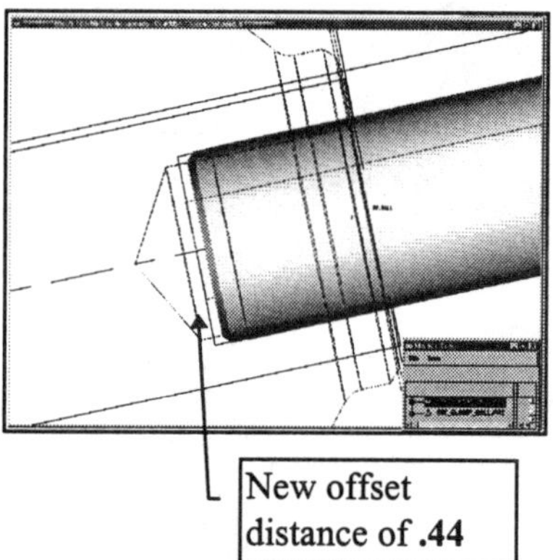

The assembly should be created with the same setups as the subassembly, including layers, datum names, and so on. The first features for the assembly will be the appropriate datum planes and coordinate system. The first part assembled on the main assembly will be the Clamp_Plate. Use the following commands to start the assembly:

Assembly ⇒ **Create** ⇒ **Clamp_Assembly** ⇒ **enter** ⇒ **Feature** ⇒ **Create** ⇒ **Datum** ⇒ **Plane** ⇒ **Offset** ⇒ **enter** ⇒ **enter** ⇒ **enter** ⇒ **Done/Return** ⇒ **Set Up** ⇒ **Geom Tol** ⇒ **Set Datum** (rename and set your datums) ⇒ **Done/Return** ⇒ **Done** (Fig. 14.50)

Component ⇒ **Assemble** ⇒ (enter a **?** from the keyboard) ⇒ **enter** ⇒ (pick the **CLAMP_PLATE** from the **Current Dir**) ⇒ **Coord Sys** (Fig. 14.51) ⇒ (pick the coordinate system on the Plate and then the coordinate system on the assembly) ⇒ **Done** ⇒ **Done/Return**

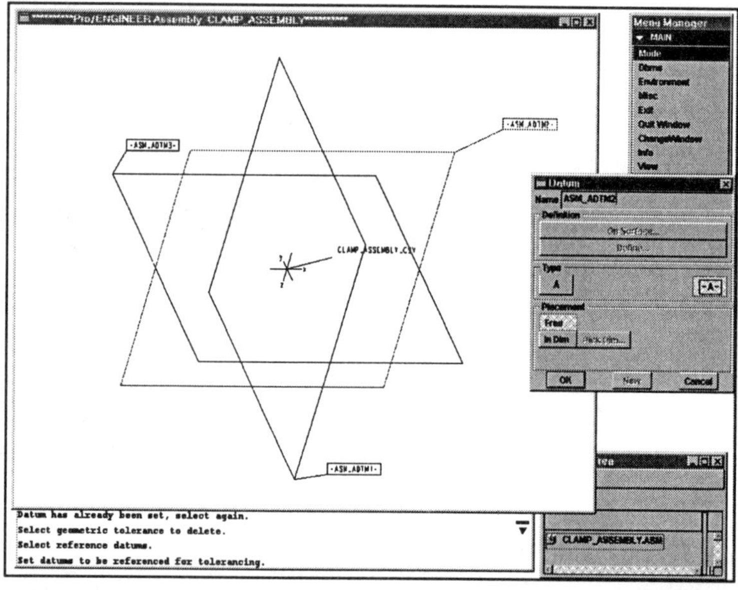

Figure 14.50
Clamp_Assembly
Datums and Coordinate System

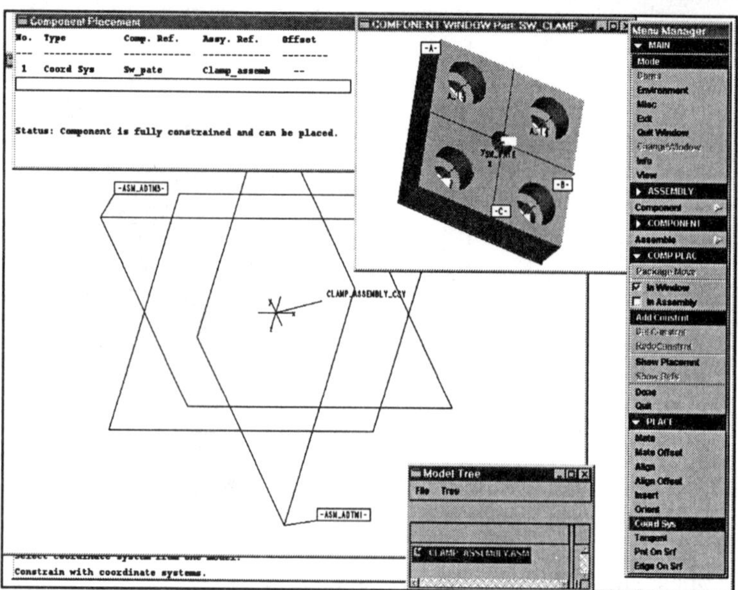

Figure 14.51
Use **Coord Sys** as the Only Constraint for the Plate

HINT

Your datum planes may have different names. Pick the *yellow side* of both datum planes. To **Orient**, pick a vertical datum on the assembly and subassembly.

Component ⇒ **Assemble** ⇒ (enter a ? from the keyboard) ⇒ **enter** ⇒ (pick the **CLAMP_SUBASSEMBLY** from the **Current Dir**) ⇒ **Mate** (Fig. 14.52) ⇒ (pick the top surface of the Plate and then the bottom circular planar surface on the Arm of the subassembly) ⇒ **Insert** ⇒ (pick the hole in the Plate and then the hole in the Arm) ⇒ **Orient** ⇒ [pick a vertical datum plane of the assembly (in this case **ASM_ADTM2**)] ⇒ **Yellow** ⇒ [pick the vertical datum plane **ADTM2** of the subassembly (Fig. 14.53)] ⇒ **Yellow** ⇒ **Done** ⇒ **Done/Return** (Fig. 14.54)

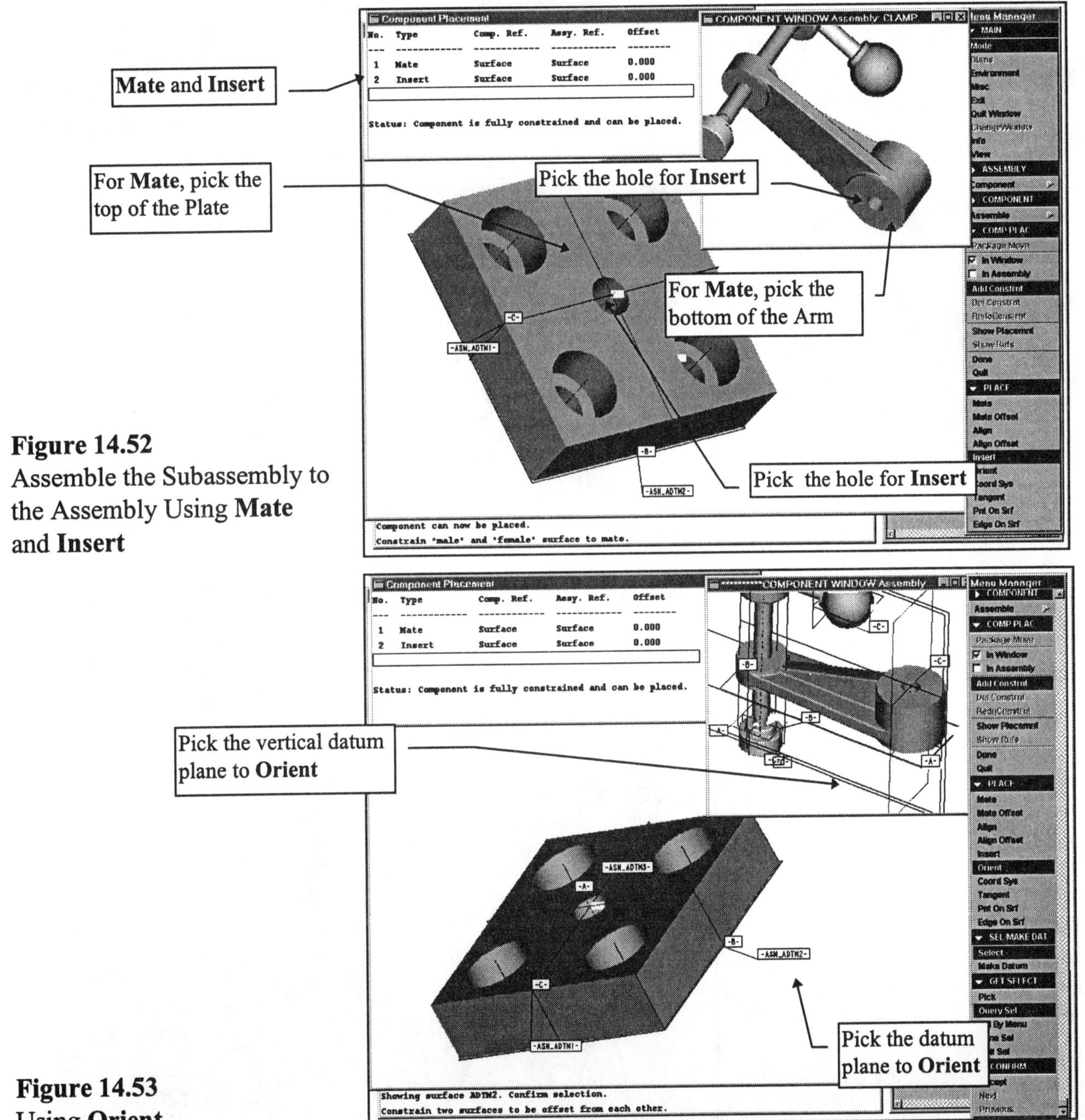

Figure 14.52
Assemble the Subassembly to the Assembly Using **Mate** and **Insert**

Figure 14.53
Using **Orient**

Assemble the **STUD35** and the Flange Nut (**FLNGNUT**) using constraints learned in this lesson (Figs. 14.55 and 14.56).

Dbms (**File**--PT/Modeler) ⇒
Save ⇒ **enter**
Purge ⇒ **enter** ⇒
Done-Return

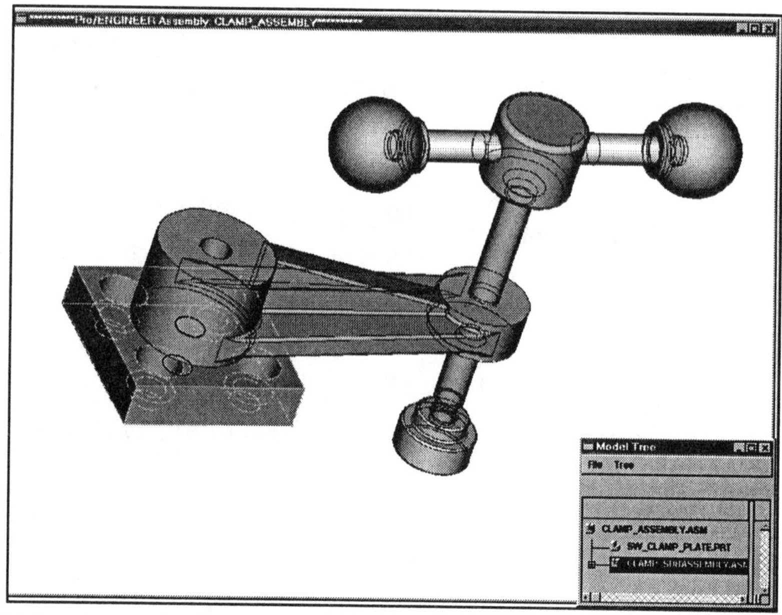

Figure 14.54
Assembled Plate and Subassembly

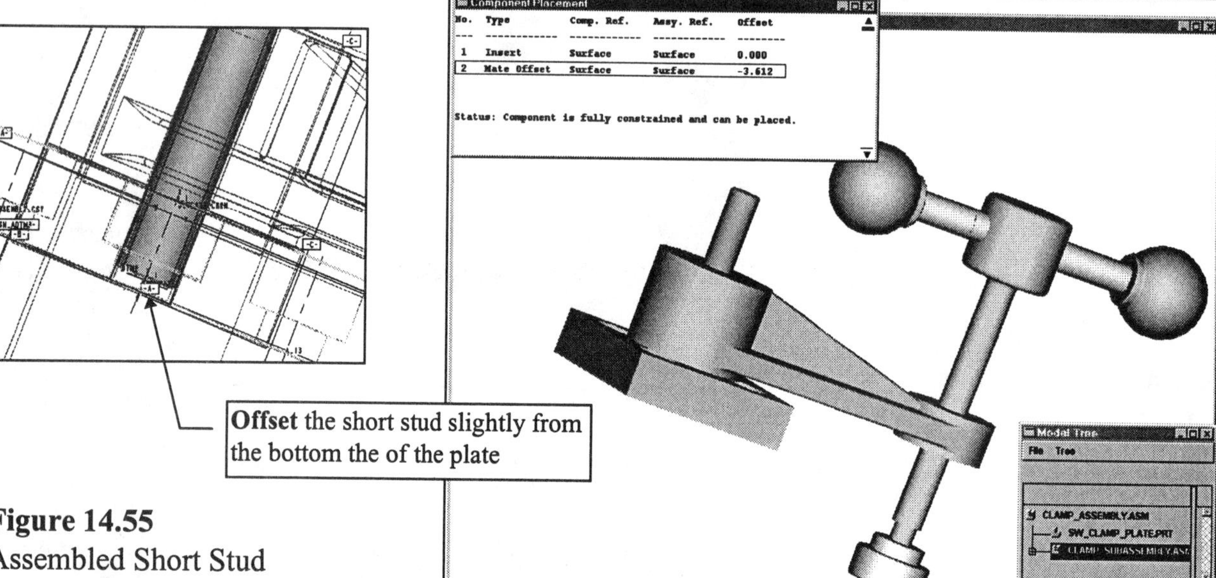

Figure 14.55
Assembled Short Stud

Flange Nut

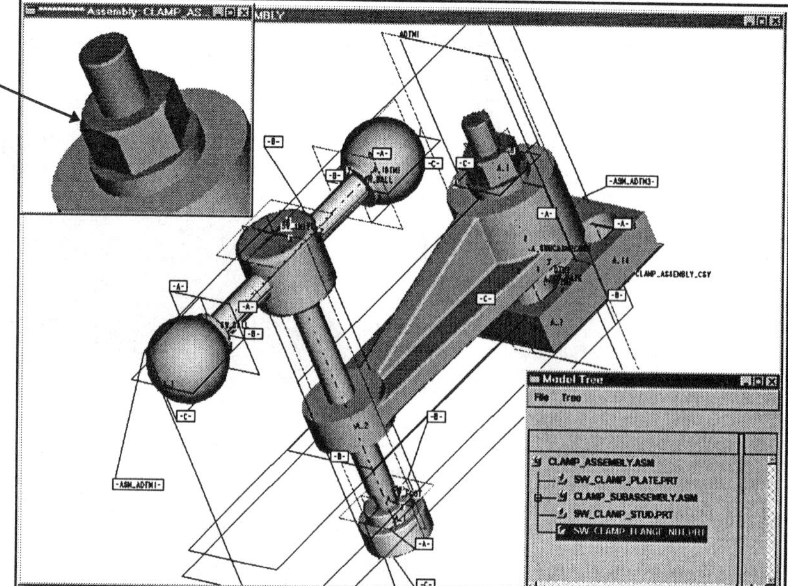

Dbms (**File**--PT/Modeler) ⇒
Save ⇒ **enter**
Purge ⇒ **enter** ⇒
Done-Return

Figure 14.56
Assembled Swing Clamp

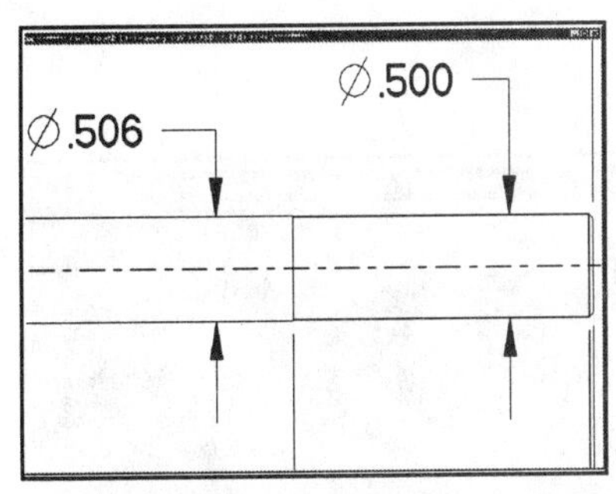

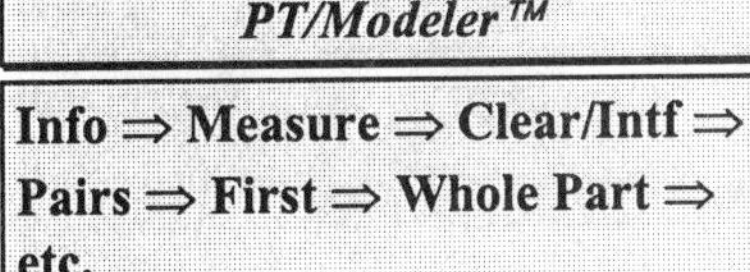

The last thing you will want to do before going on to the next lesson is check the assembly using the **Info** command (Fig. 14.57). If you look at the long **5.00** stud's detail drawing (see Fig. 14.18), you will see that the shaft diameter is greater than **.500** at the center of the stud and each end has a **.500-13 UNC** thread. The hole in the Swivel is **.500**. This means that there should be a slight interference between the two components. Check the clearance and interference between these components using the following commands (Fig. 14.57):

Info ⇒ Measure ⇒ Clear/Intf ⇒ Pairs ⇒ First ⇒ Whole Part ⇒ ExcludQuilts ⇒ (pick the Swivel) ⇒ (pick the **5.00** Stud)

The **Info** Command provides you with a statement as to the status of the two parts:

Interference detected. Volume of Interference is 0.0069.

The assembly is now complete. In the next lesson, you will learn how to move and rotate components in the assembly, establish views for use in Drawing mode, create exploded views of the assembly, generate a bill of materials, and change the component visibility.

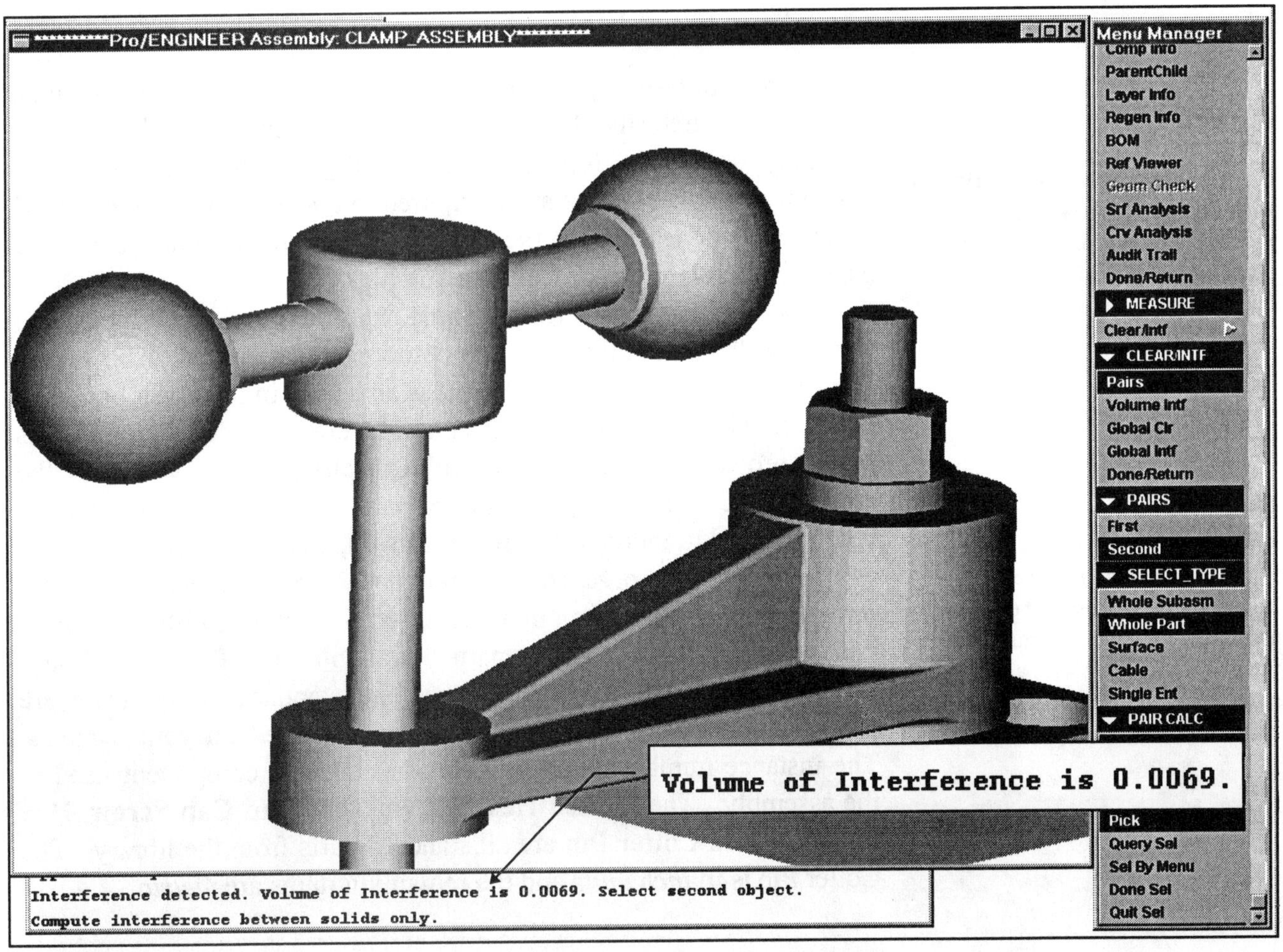

Figure 14.57
Info Command Used to Establish Clearance and Interference

Pro/ENGINEER Assembly: CLAMP_ASSEMBLY

Menu Manager
Comp Info
ParentChild
Layer Info
Regen Info
BOM
Ref Viewer
Geom Check
Srf Analysis
Crv Analysis
Audit Trail
Done/Return
MEASURE
Clear/Intf
CLEAR/INTF
Pairs
Volume Intf
Global Clr
Global Intf
Done/Return
PAIRS
First
Second
SELECT_TYPE
Whole Subasm
Whole Part
Surface
Cable
Single Ent
PAIR CALC
Pick
Query Sel
Sel By Menu
Done Sel
Quit Sel

Volume of Interference is 0.0069.

Interference detected. Volume of Interference is 0.0069. Select second object.
Compute interference between solids only.

Lesson 14 Project

Coupling Assembly

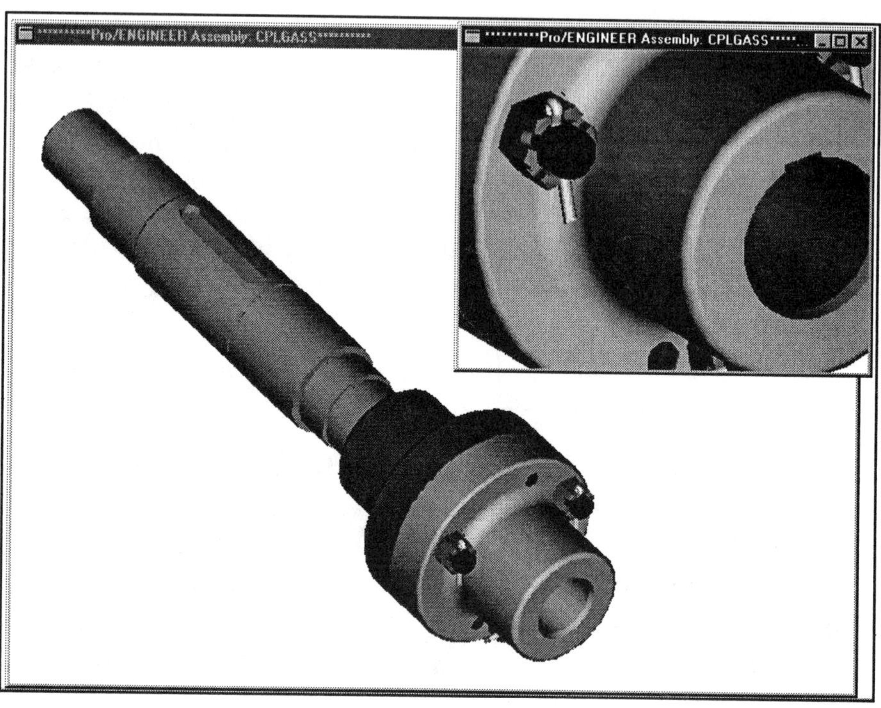

Figure 14.58
Coupling Assembly

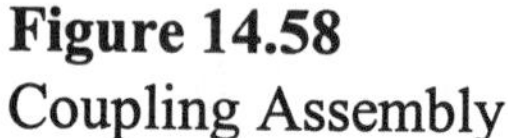

EGD REFERENCE
Engineering Graphics and Design with Graphical Analysis *or* **Fundamentals of Engineering Graphics and Design**
by L. Lamit and K. Kitto
Read Chapter 23
See pages 865-866

NOTE
For this project do not use the library parts directly in the assembly. Save each library part in your own directory with a new name and then use the new part names in the assembly.

Coupling Assembly

The fourteenth **lesson project** is an assembly that requires commands similar to those for the **Swing Clamp**. Model the parts and create the assembly shown in Figures 14.58 through 14.81. Analyze the assembly and plan out the steps required to assemble it. Use the DIPS in Appendix D to plan out the assembly component sequence and the parent-child relationships for the assembly.

You will use the **Coupling Shaft** from the Lesson 6 Project. The Coupling Shaft should be the first component assembled. The **Taper Coupling** from the Lesson 7 Project is also used in the assembly. The detail drawings for the second coupling are provided here, in this lesson project. Model this second **Coupling** *before* you start the assembly. Depending on the library parts available on your system, you may need to model the **Key**, the **Dowel**, and the **Washer**.

Since not all organizations purchase the libraries, details are provided for all the components required for the assembly, including the standard off-the-shelf parts available in Pro/E's library. Pro/LIBRARY commands to access the standard components are provided for those of you who have them loaded on your systems. The instance number is given for every standard component used in the assembly. The **Slotted Hex Nut**, **Socket Head Cap Screw**, **Hex Jam Nut**, and **Cotter Pin** are all standard parts from the library. The Cotter Pin is in *inch units* and the remaining items are *metric*.

Redesign the length of the threaded end to accommodate the washer and nut. **You** decide the new length based on the combined thickness of the two components.

Figure 14.59
Assembling the Coupling Shaft and Taper Coupling

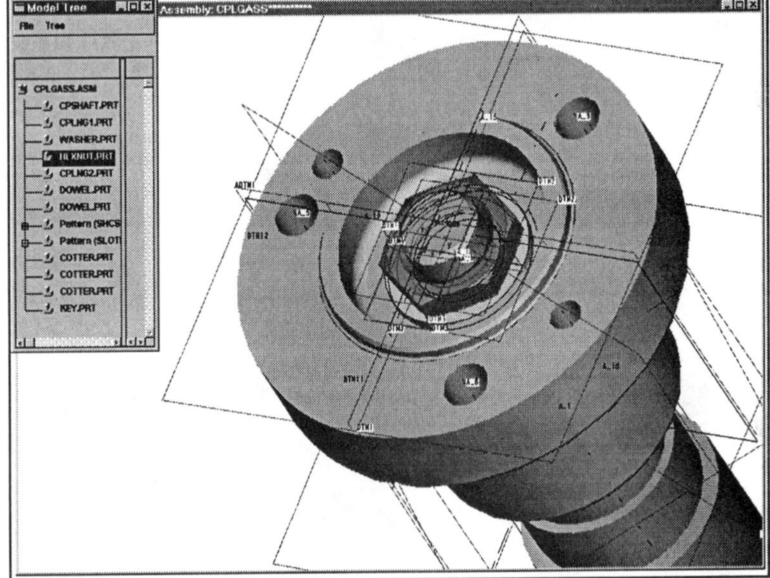

Figure 14.60
Hex Jam Nut and Washer

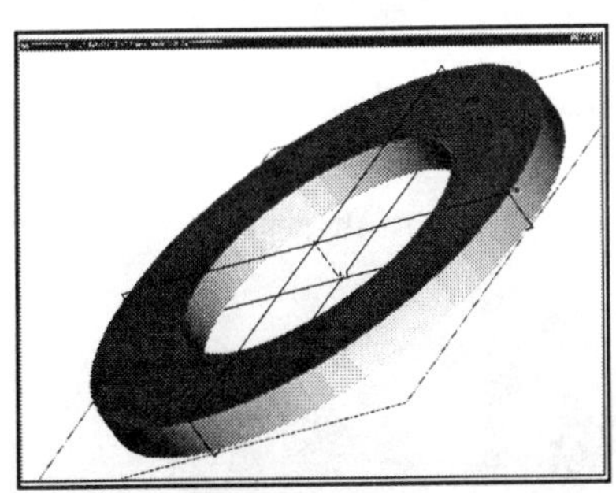

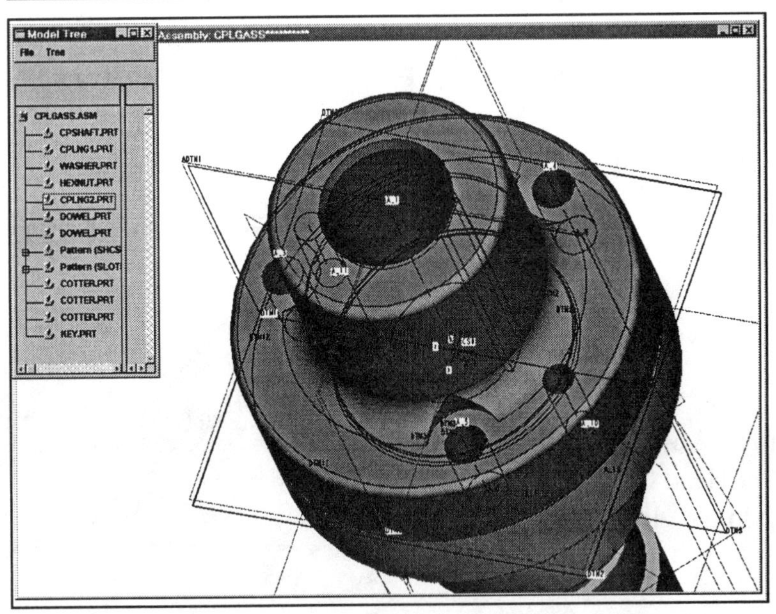

Figure 14.61
Second Coupling

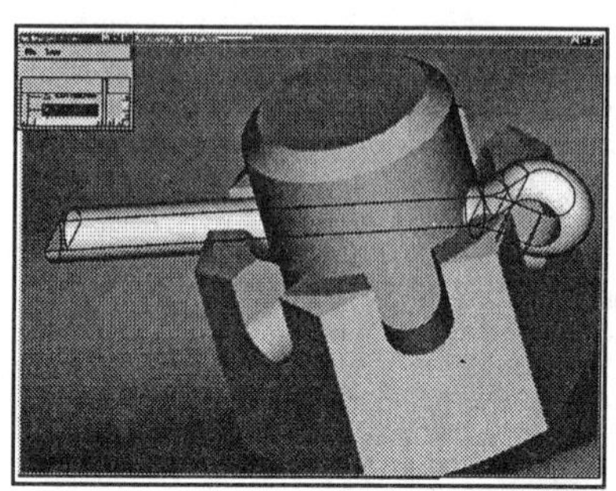

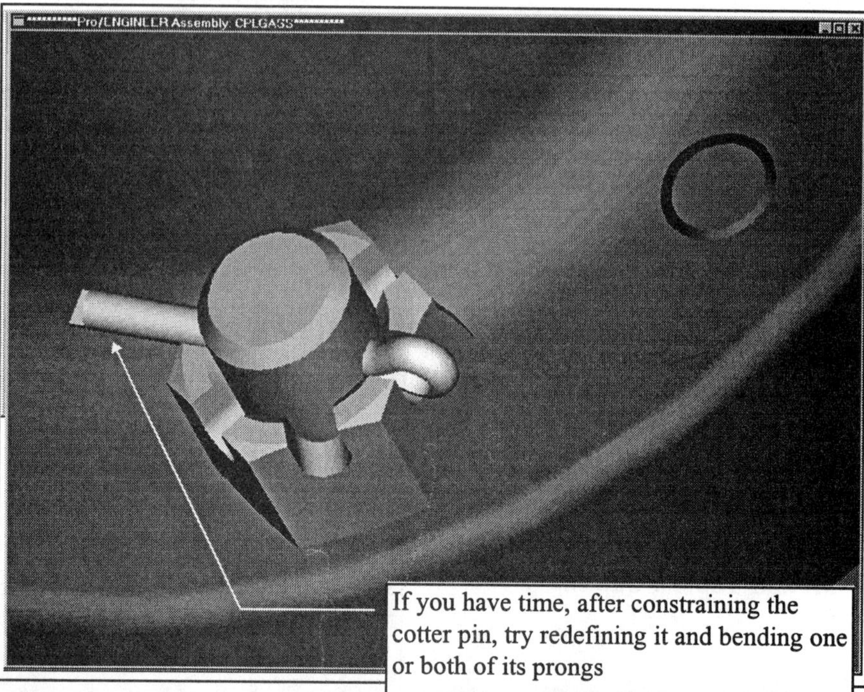

Figure 14.62
Dowel, Slotted Hex Nut, and Cotter Pin

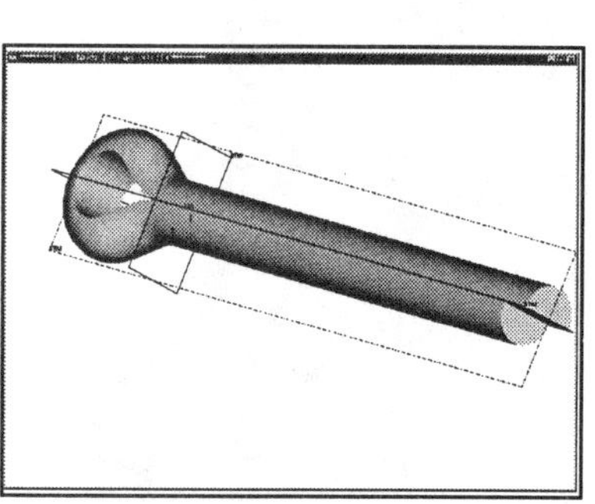

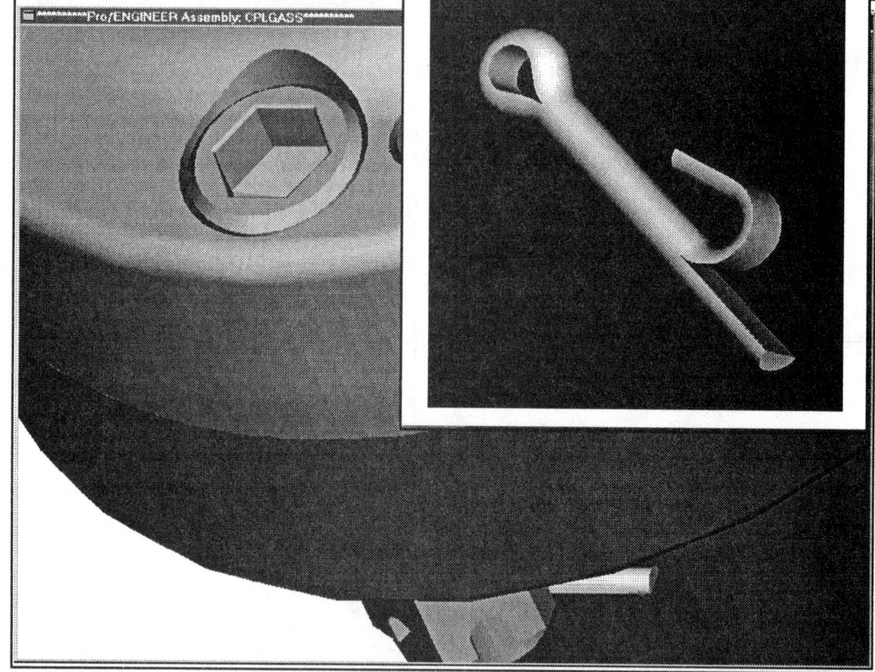

Figure 14.63
Socket Head Cap Screw

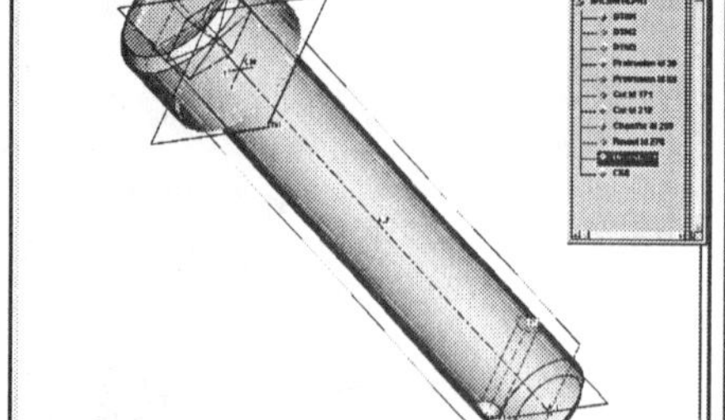

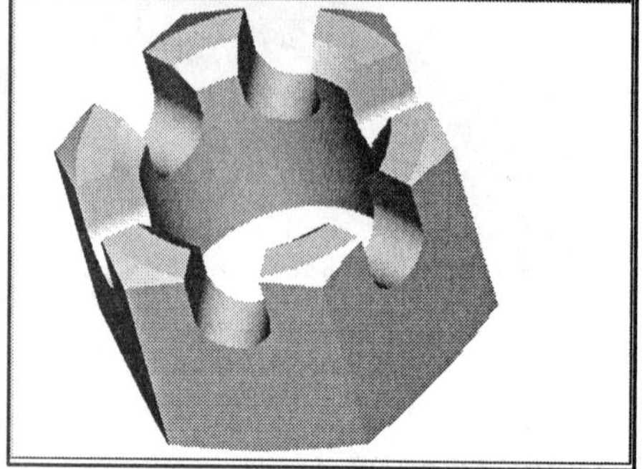

Figure 14.64
Socket Head Cap Screw and Slotted Hex Nut

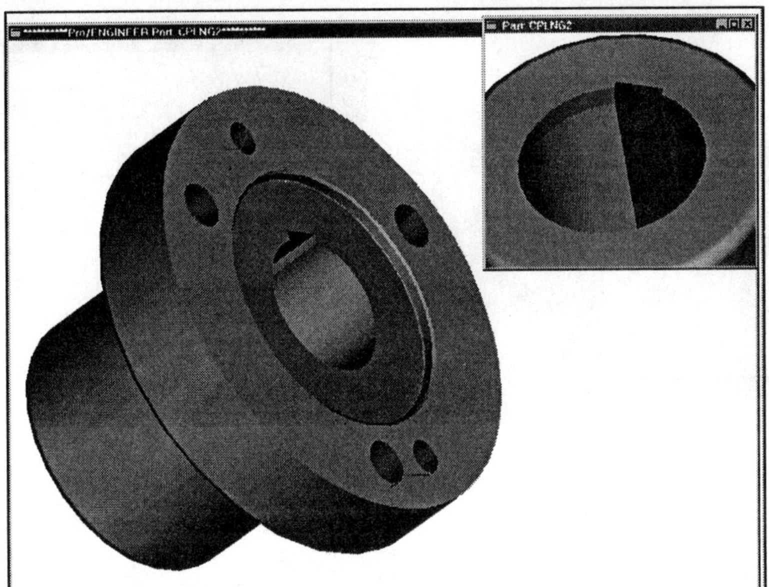

Figure 14.65
Second Coupling, Part Model

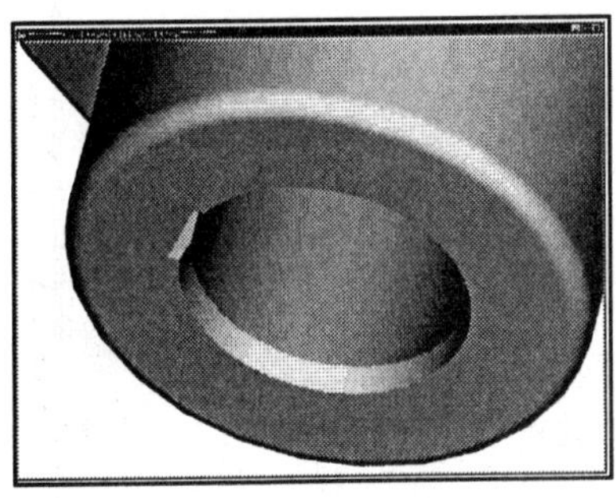

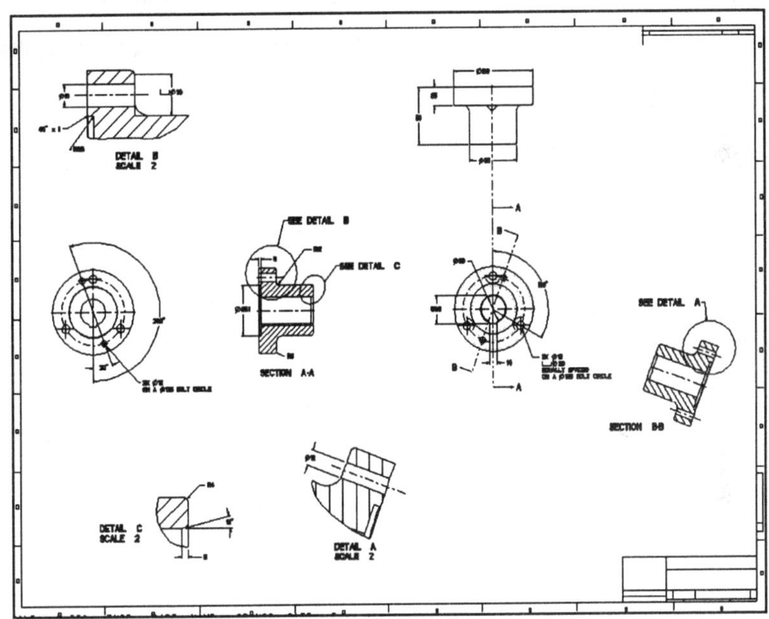

Figure 14.66
Second Coupling, Detail Drawing

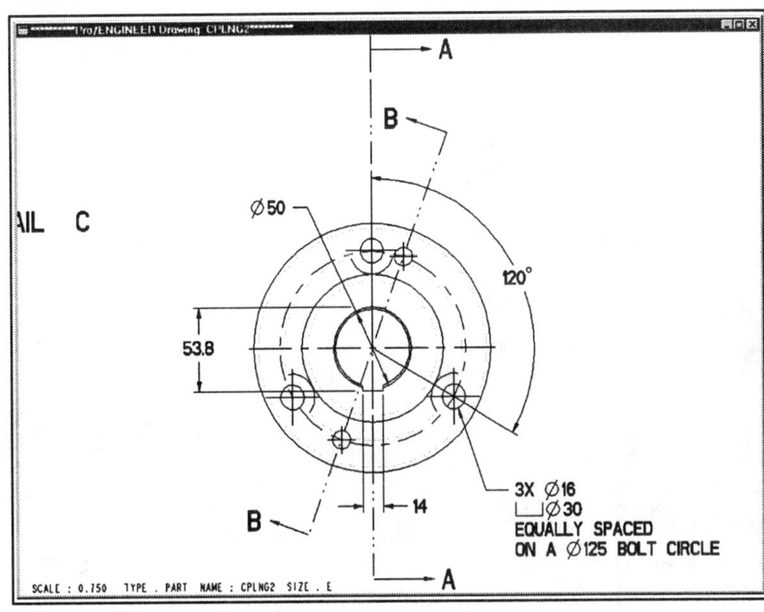

Figure 14.67
Second Coupling, Detail Drawing, Front View

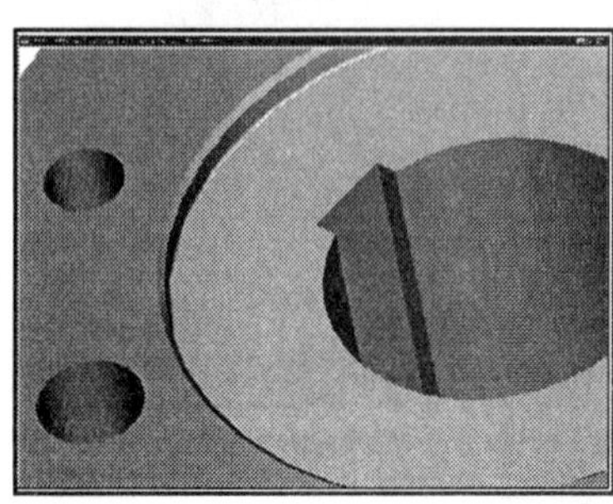

Figure 14.68
Second Coupling, Detail Drawing, Top View

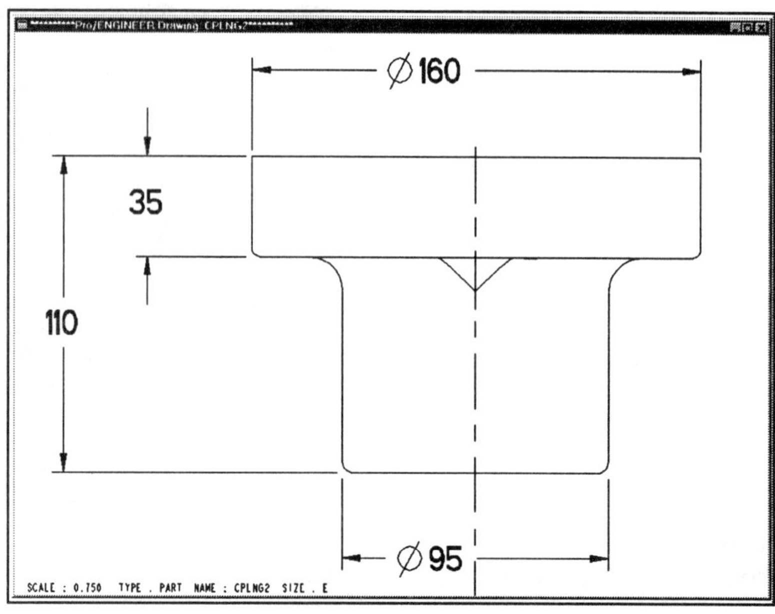

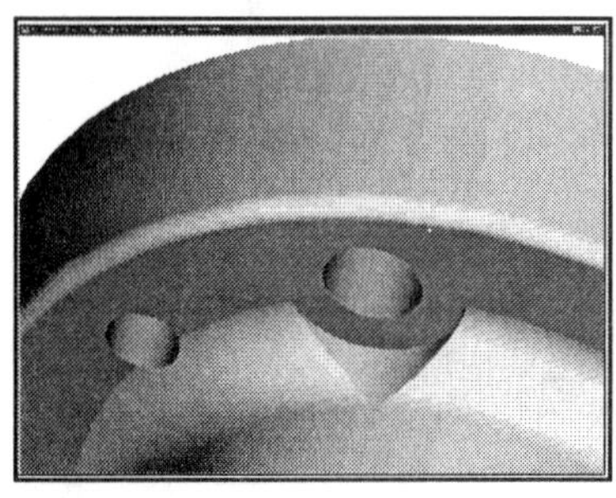

Figure 14.69
Second Coupling, Detail Drawing, **SECTION A-A**

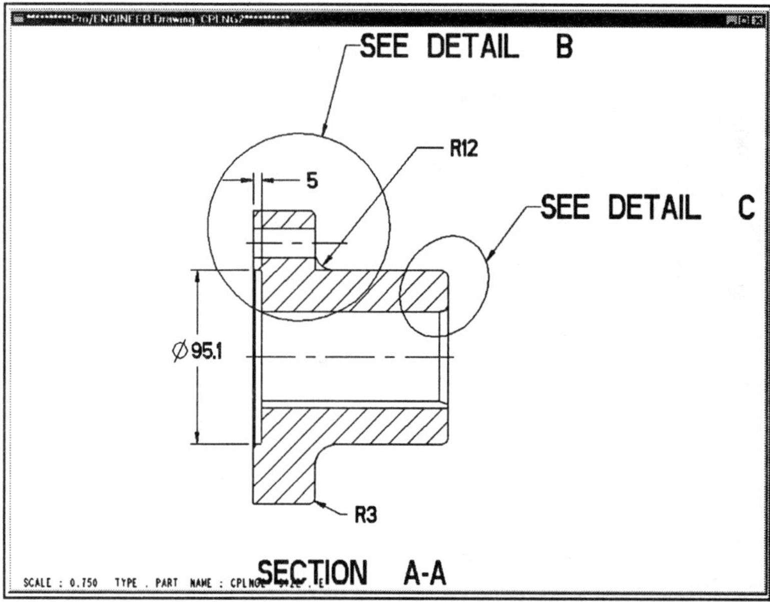

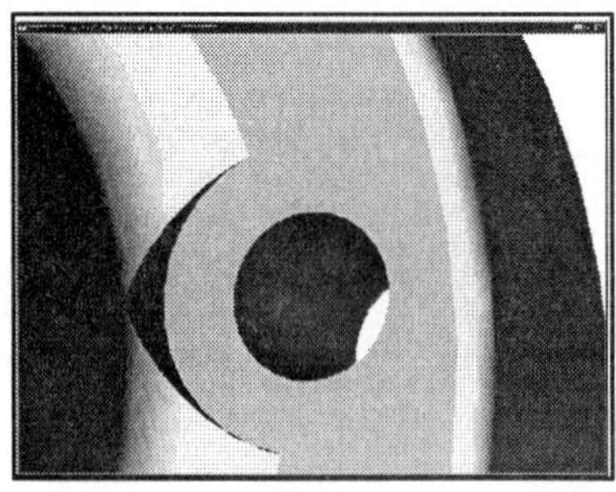

Figure 14.70
Second Coupling, Detail Drawing, Back View

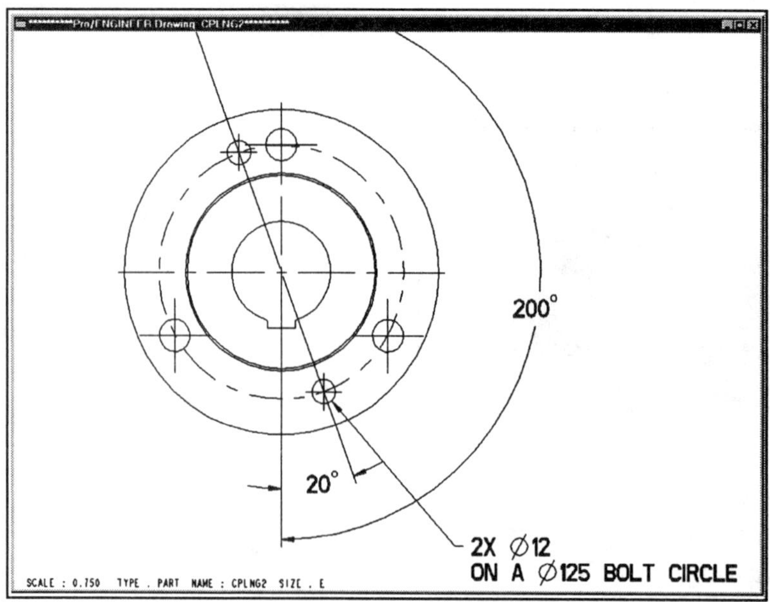

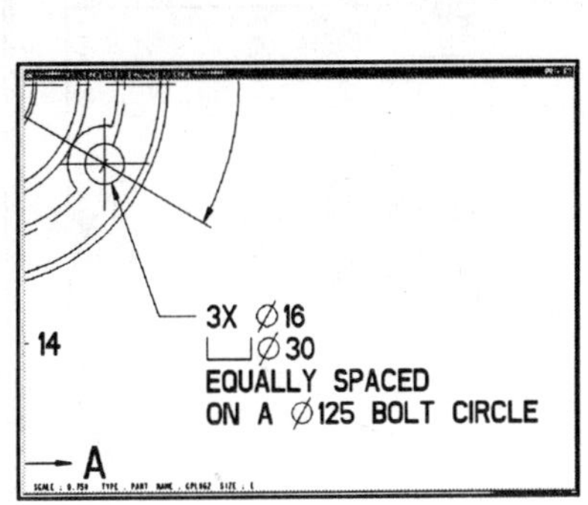

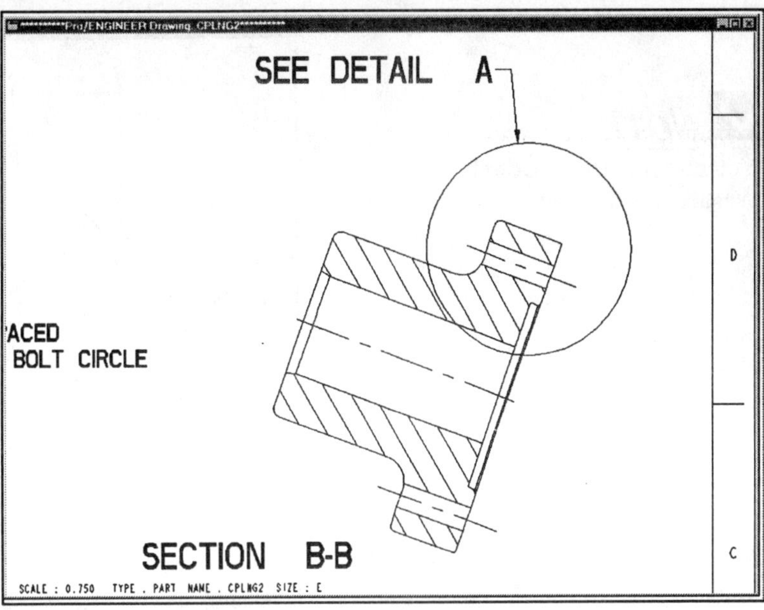

Figure 14.71
Second Coupling, Detail Drawing, **SECTION B-B**

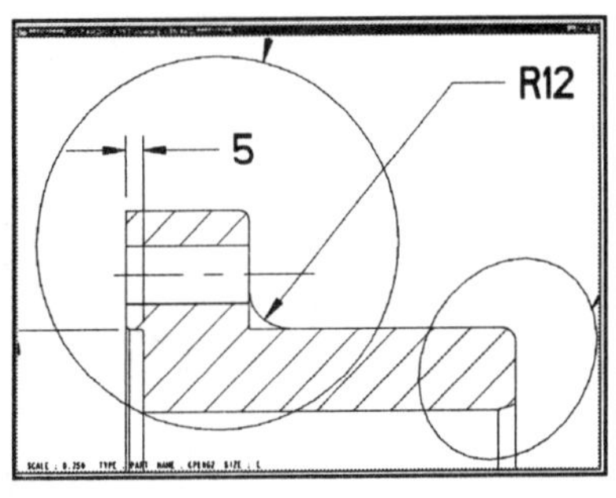

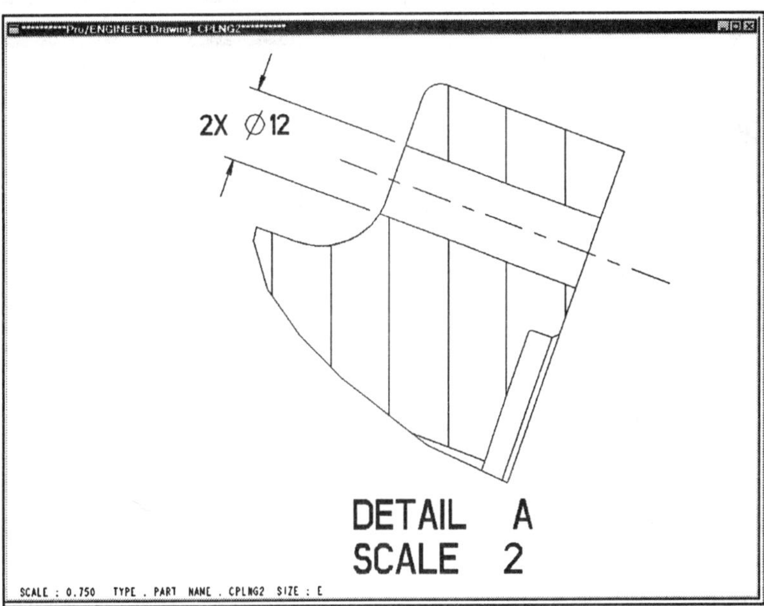

Figure 14.72
Second Coupling, Detail Drawing, **DETAIL A**

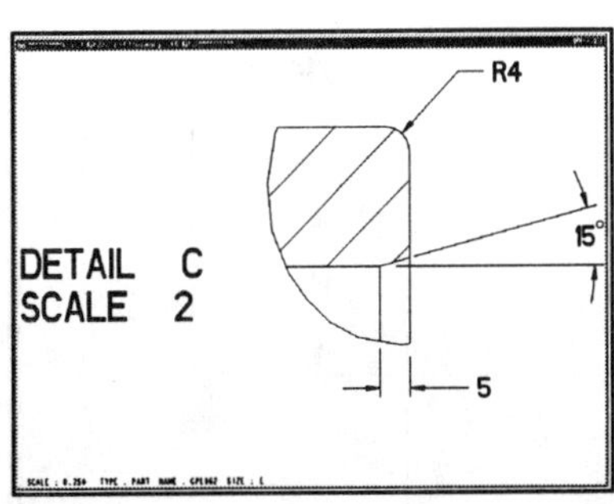

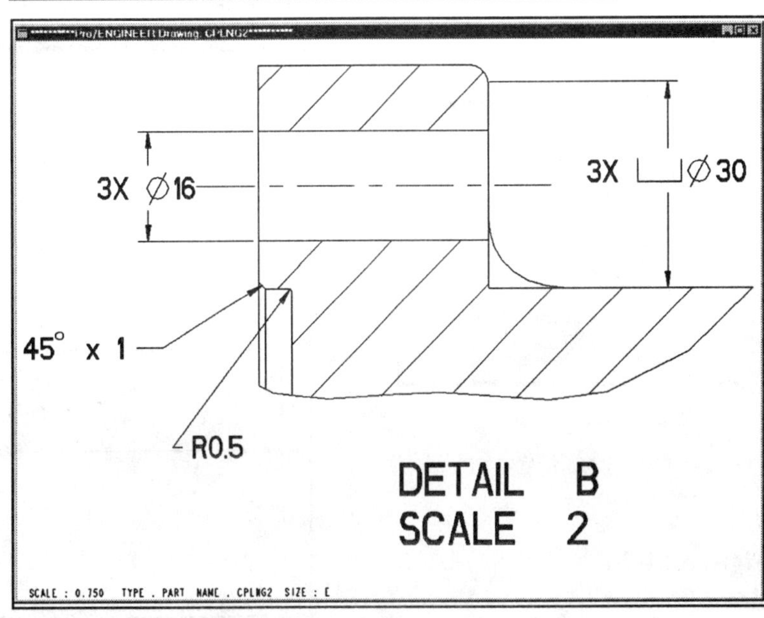

Figure 14.73
Second Coupling, Detail Drawing, **DETAIL B**

NOTE

You must model the Dowel, the Washer, and the Key.

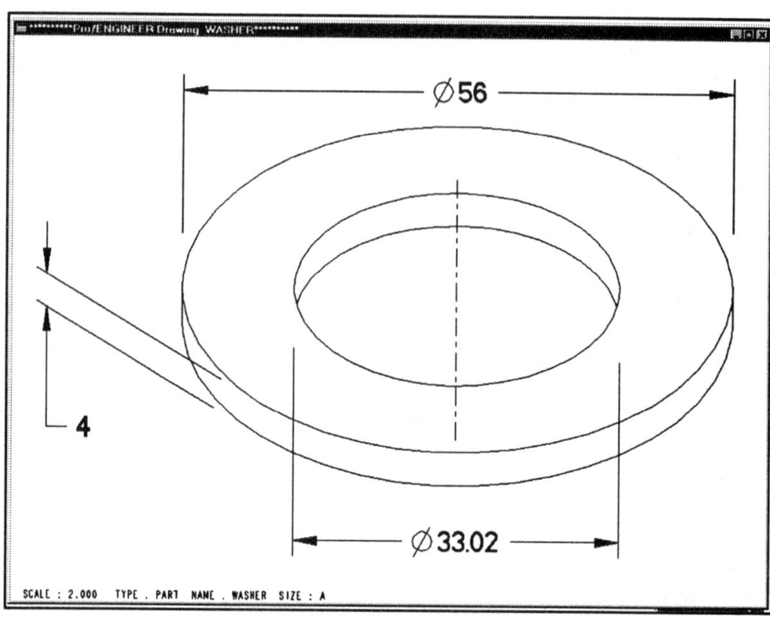

Figure 14.74
Washer

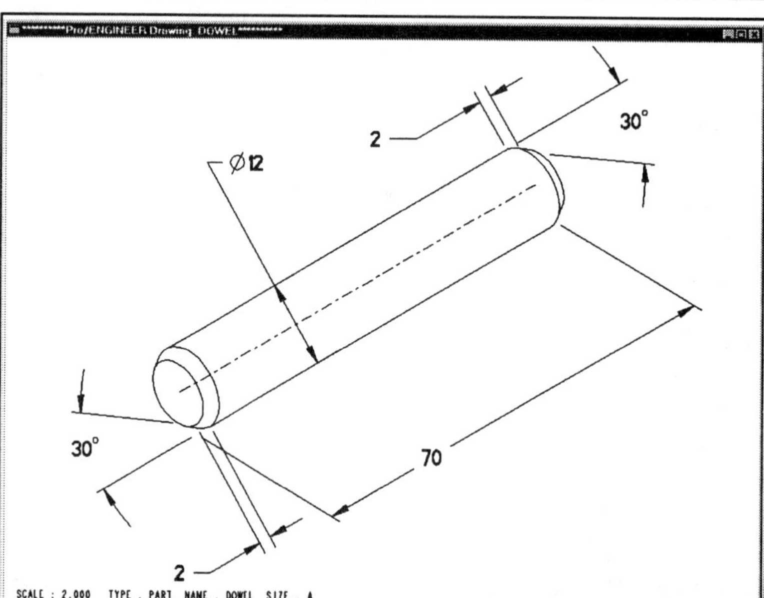

Figure 14.75
Dowel

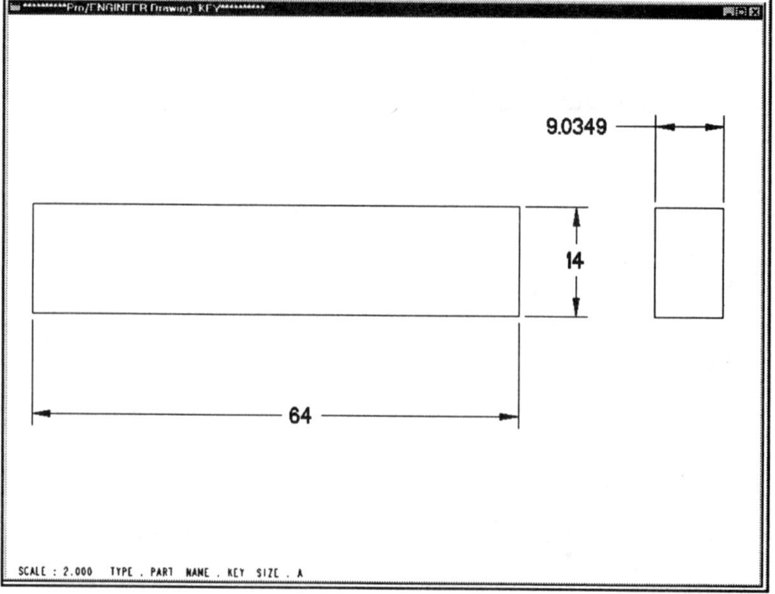

Figure 14.76
Key

SOCKET HEAD CAP SCREW

Part ⇒
Search/Retr ⇒
Pro/Library ⇒
/objlib ⇒ /metriclib ⇒
/sock_hd_scr ⇒ mscs.prt ⇒
SelByParams ⇒
✓NOM_SIZE_THR_PITCH ⇒
Done Sel ⇒ M16X2 ⇒
SelByParams ⇒
✓d5,length ⇒
Done Sel ⇒ 80.000

INSTANCE = MSCS1210

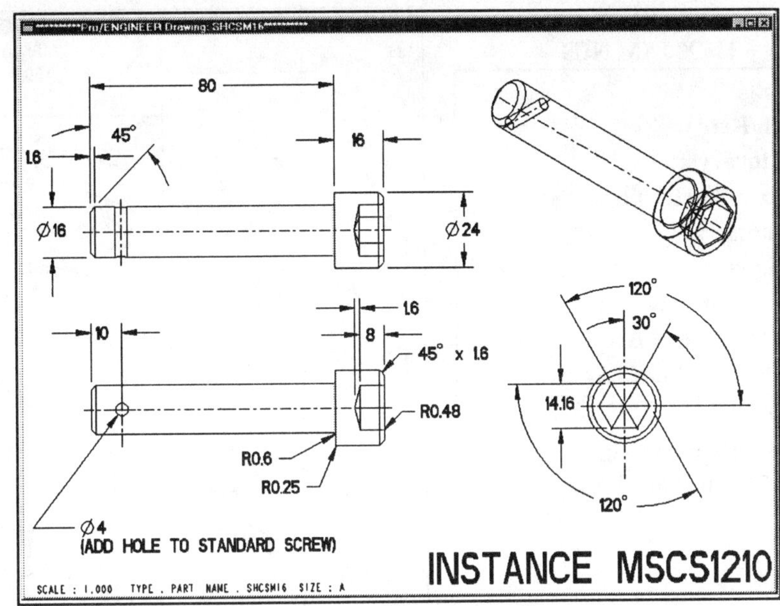

Figure 14.77
Socket Head Cap Screw

SLOTTED HEX NUT

Part ⇒
Search/Retr ⇒
Pro/Library ⇒
/objlib ⇒ /metriclib ⇒
/hex_nuts ⇒
mshn.prt ⇒
SelByParams ⇒
✓NOM_ DIA_THR_PITCH ⇒
Done Sel ⇒ M16X2

INSTANCE = MSHN07

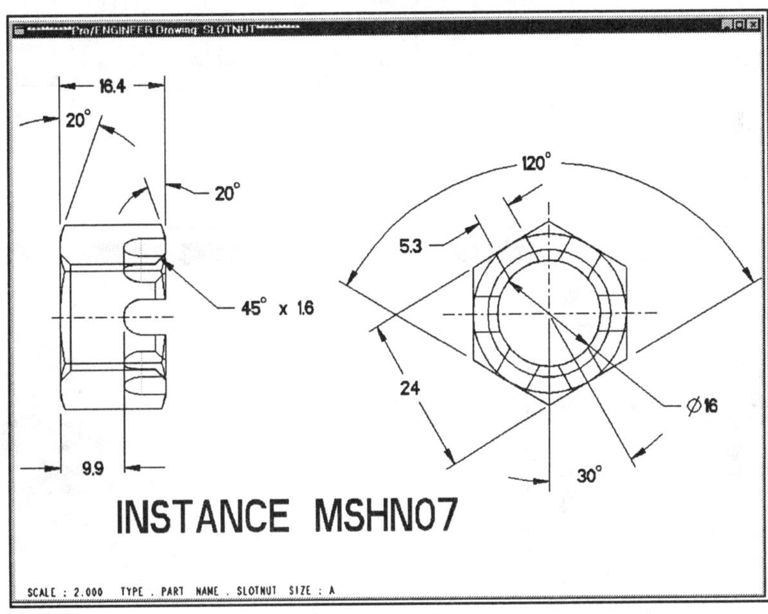

Figure 14.78
Slotted Hex Nut

COTTER PIN

Part ⇒
Search/Retr ⇒
Pro/Library ⇒
/objlib ⇒ /eng_part_lib ⇒
/cot_clvs_pin ⇒ Pina.prt ⇒
SelByParams ⇒
✓NOM_SIZE ⇒
Done Sel ⇒ .1562 ⇒
SelByParams ⇒
✓ d13,1 ⇒
Done Sel ⇒ 1.250

INSTANCE = PNA09L05

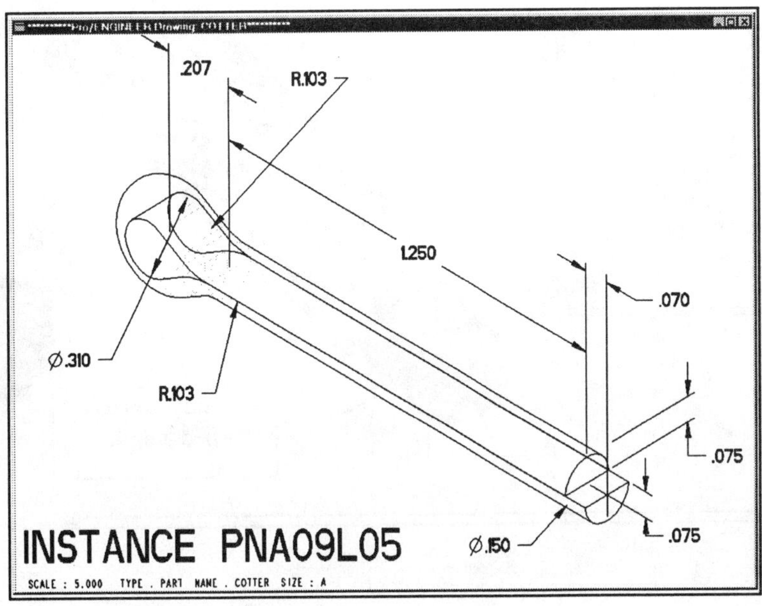

Figure 14.79
Cotter Pin

HEX JAM NUT

Part ⇒
Search/Retr ⇒
Pro/Library ⇒
/objlib ⇒ /metriclib ⇒
/hex_nuts ⇒
/mhjn.prt ⇒
SelByParams ⇒
✓NOM_SIZE_THR_PITCH ⇒
Done Sel ⇒ M30X3.5

INSTANCE = MHJN10

- **Modify the thickness of the nut to 10 mm**

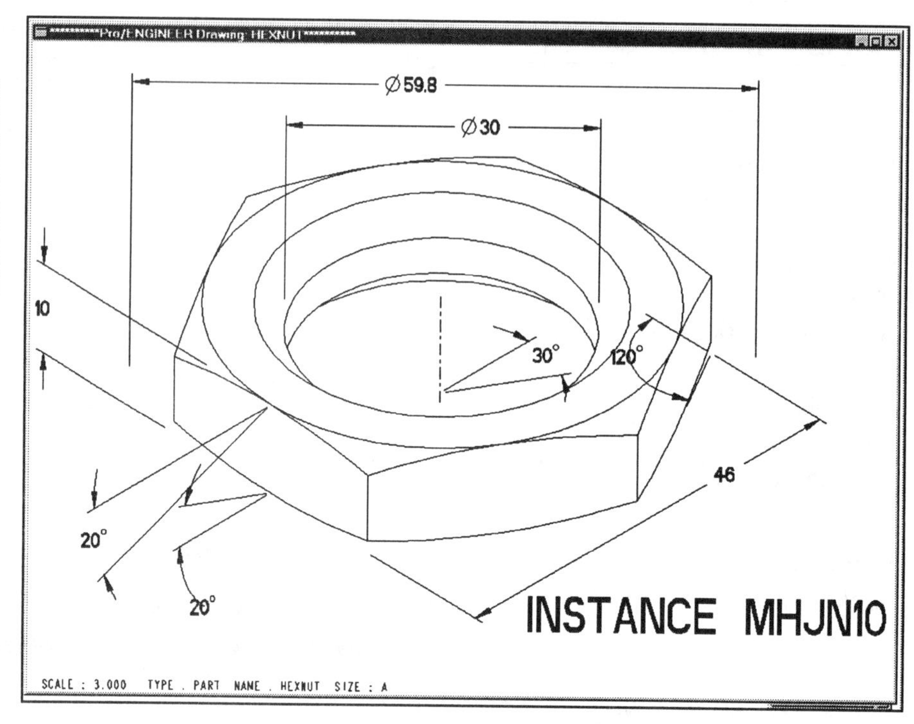

Figure 14.80
Hex Jam Nut

Figure 14.81
Coupling Assembly

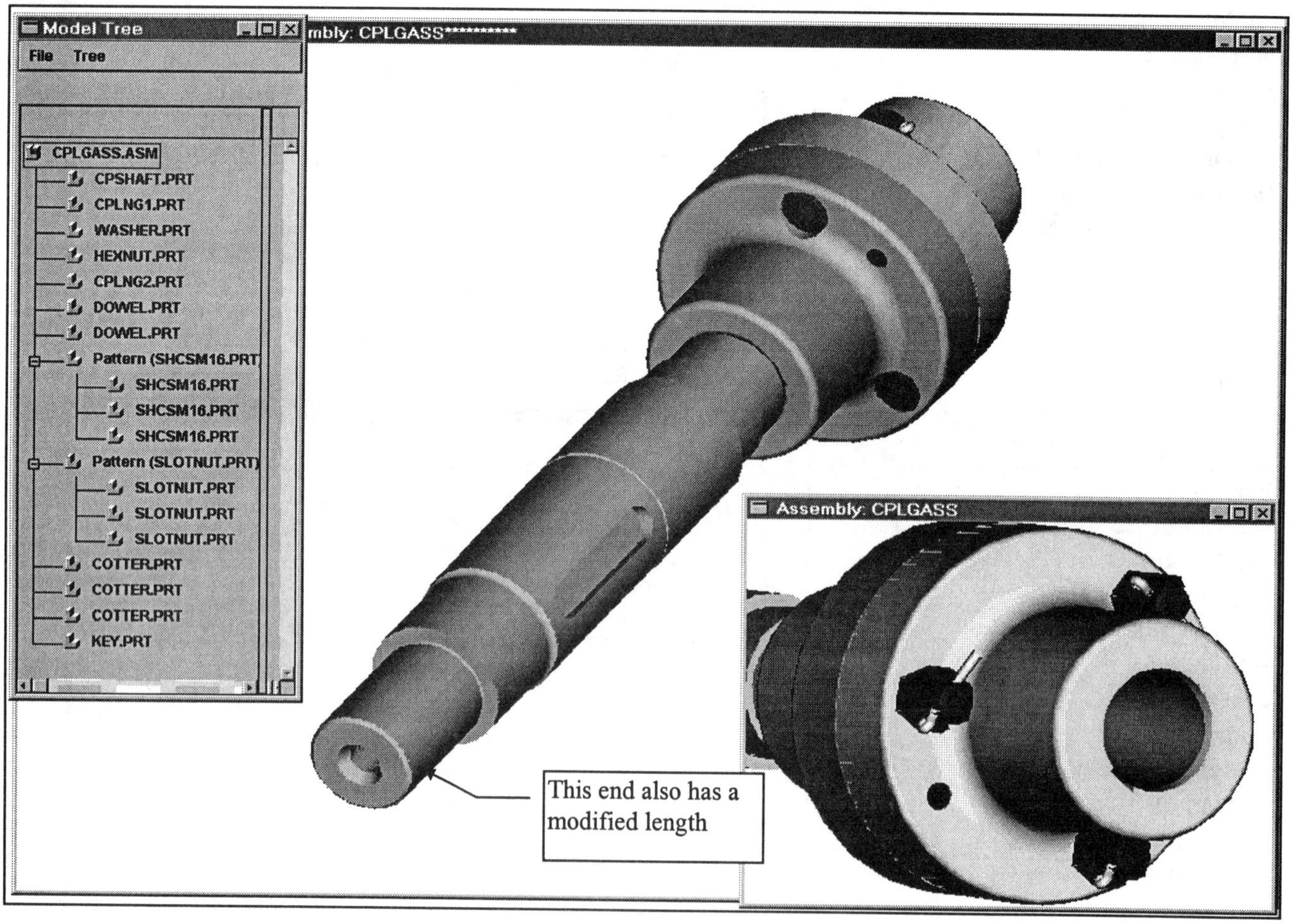

This end also has a modified length

Lesson 15

Exploded Assemblies

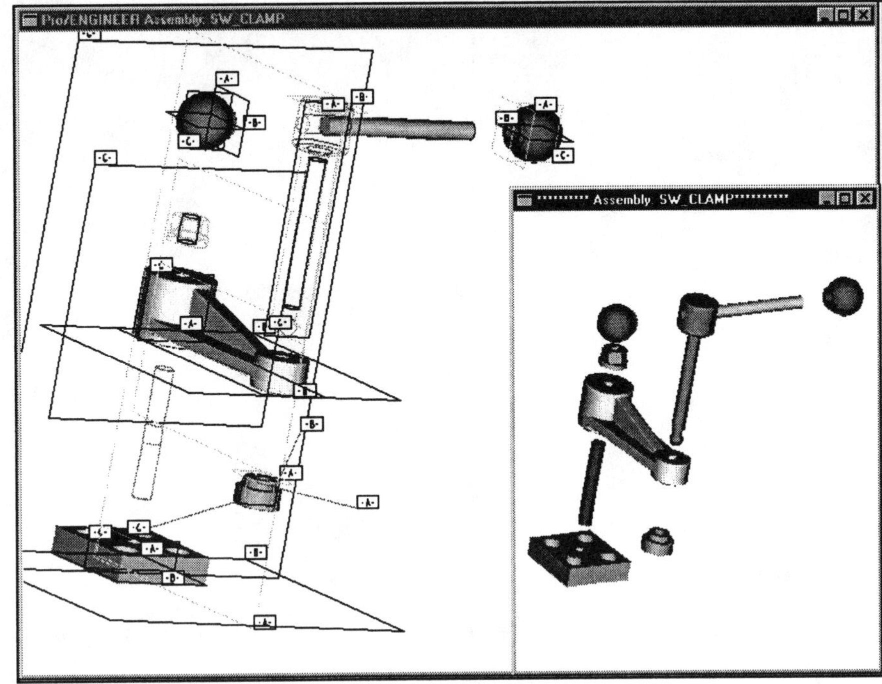

Figure 15.1
Exploded Swing Clamp

OBJECTIVES

1. **Create exploded views**
2. **Edit exploded views**
3. **Create unique component visibility settings**
4. **Move and rotate components in an assembly**
5. **Create shaded and pictorial views of an assembly**
6. **Save named views to use later in Drawing mode**

EGD REFERENCE
Engineering Graphics and Design with Graphical Analysis *or* **Fundamentals of Engineering Graphics and Design**
by L. Lamit and K. Kitto
Read Chapter 23
See pages 865-866

COAch™ for Pro/ENGINEER

If you have **COAch for Pro/ENGINEER** on your system, go to SEARCH and do the Segment shown in Figure 15.3.

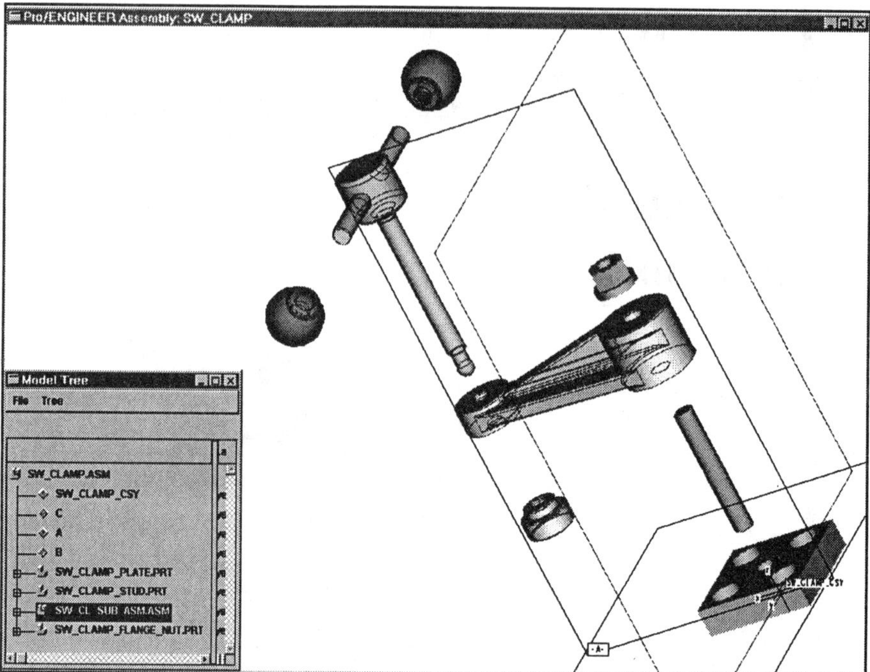

Figure 15.2
Exploded Swing Clamp, Model Tree

EXPLODED ASSEMBLIES

Pictorial illustrations such as exploded views are generated directly from the 3D model database (Figs. 15.1 through 15.7). The model can be displayed and oriented in any position. Each component in the assembly can have a different display type: wireframe, hidden line, no hidden, and shading. You can select and orient the part to provide the required view orientation to display the part from underneath or from any side or position. Perspective projections are made with selections from menus. The assembly can be spun around, reoriented, and even clipped to show the interior features. When assemblies are illustrated, you have the choice of displaying all components and subassemblies, or any combination of parts in the design.

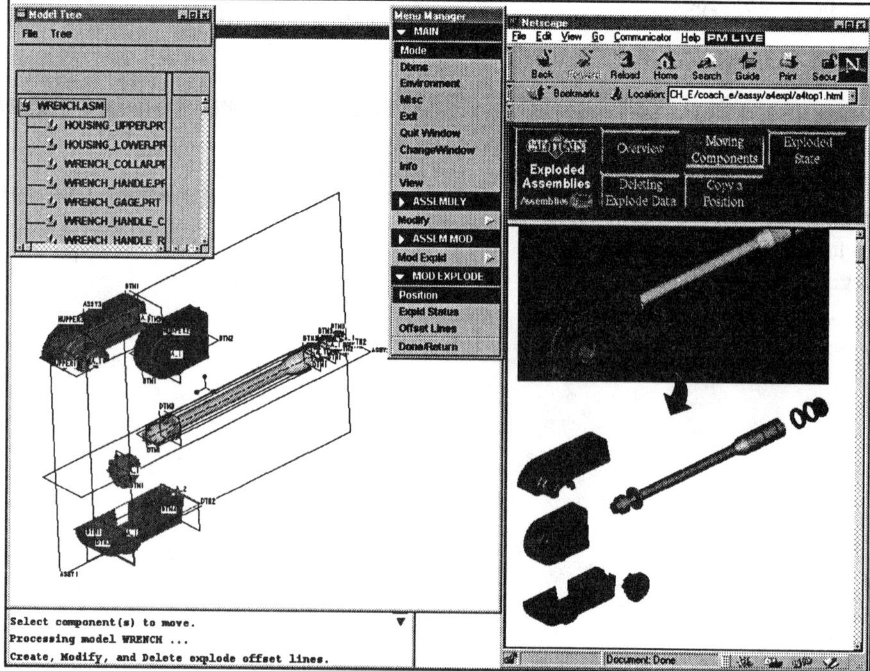

Figure 15.3
COAch for Pro/E, Exploded Assemblies (Moving Components)

Creating Exploded Views

> **NOTE**
> To use other than the default explode, you have the license Pro/PROCESS_ASM

Using the **ExplodeState** option in the ASSEMBLY menu, you can create an exploded view of an assembly. Exploding an assembly affects only the display of the assembly; it does not alter actual distances between components. Exploded states are created to define the exploded positions of all components. For each explode state, you can toggle the explode status of components, change the explode locations of components, and create explode offset lines. To access this functionality, choose the **ExplodeState** option in the ASSEMBLY menu.

You can define multiple explode states for each assembly, then explode the assembly using any of these explode states at any time. You can also set an explode state for each drawing view of an assembly. Pro/E gives each component a default explode position determined by the placement constraints. By default, the reference component of the explode is the parent assembly (top-level assembly or subassembly).

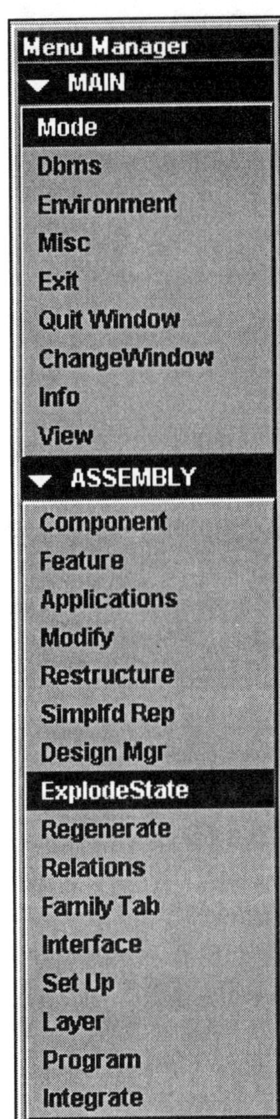

To explode components, you use a drag-and-drop user interface similar to the **package** functionality. You select one or more components and the motion reference, then drag the outlines to the desired positions. The component outlines drag along with the mouse cursor. You control the type of explode motion using a Preferences setting.

Two types of explode instructions can be added to a set of components. The children components follow the component being exploded or they do not follow it. Each explode instruction consists of a set of components, explode direction references, and dimensions that define the exploded position from the final (installed) position with respect to the explode direction references.

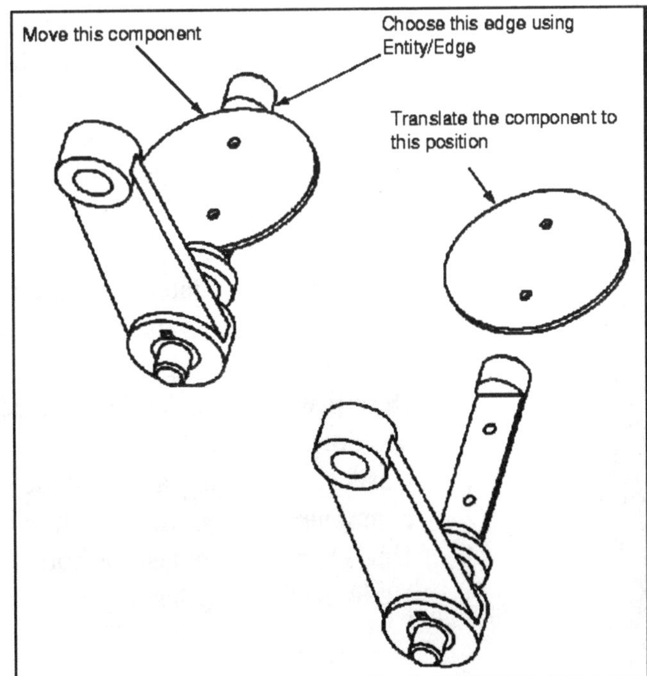

Figure 15.4
Online Documentation, Exploded Views

Creating New Exploded Views

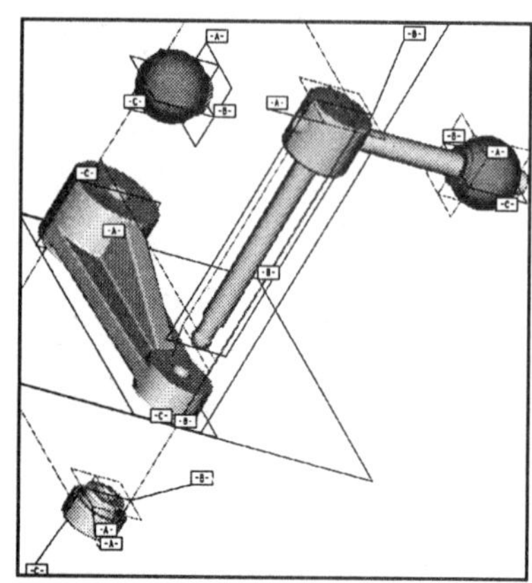

To create a new explode state:

1. Retrieve an assembly. Pro/E displays the Model Tree assembly tool.
2. Choose **ExplodeState** from the ASSEMBLY menu. The EXPLD STATE menu lists the following options:

 Create Creates a new explode state. Each time Pro/E creates a new state, it creates it with no components exploded.
 Set Current Displays the SEL STATE menu, listing available explode states.
 Copy Copies an existing explode state using the SEL STATE menu.
 Redefine Redefines an explode state using the MOD EXPLODE menu.
 Delete Deletes one or more explode states.
 List Lists all currently defined explode states in an INFORMATION WINDOW.

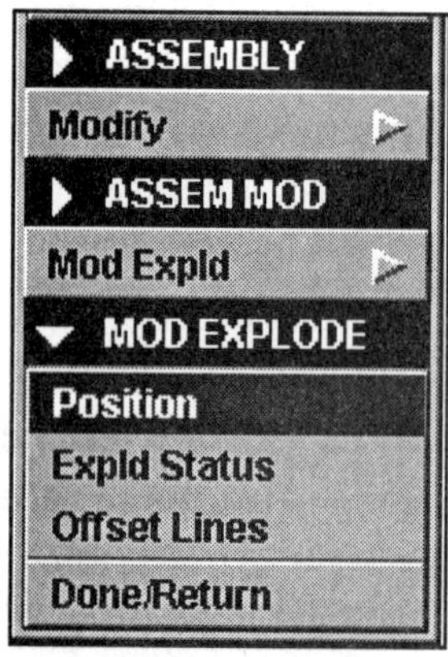

3. Choose **Create**, then enter a name for the state. The MTNPREF menu displays the following options:

 Preferences Sets up preferences for dragging packaged components.
 View Plane Uses the viewing plane as the reference plane (repositions the component in a plane that is parallel to it).
 Sel Plane Selects a plane other than the viewing plane as the reference plane (repositions the component in a plane that is parallel to it).
 Entity/Edge Selects an axis, straight edge, or datum curve (repositions the component in a line parallel to it).
 Plane Normal Selects a plane as the reference plane and repositions the component in a line that is normal to it.
 2 Points Picks two points or vertices (repositions the component in a line that is parallel to the line that connects them).
 C-sys Selects a coordinate system axis (repositions the component in the direction of it).
 Translate Translates the selected component.
 Copy Pos Copies the position instructions from one component to the selected components.
 DefaultExpld Places selected components in the default explode position.
 Reset Removes all explode instructions from the selected components, even the default explode, and resets their positions.
 Undo Undoes the last motion.
 Redo Redoes the last motion that was undone.

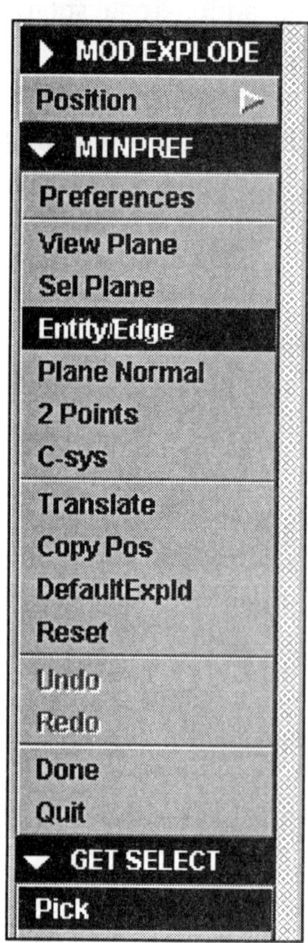

4. Choose **Preferences.** The MOTION PREFS menu displays the following options:

Trans Incr Specifies the motion increments for translational dragging. Displays the TRANS INCR menu. Choose **Enter** to enter an increment value for the translation, or choose one of the preset values displayed in the menu. To drag the component without apparent incrementing, choose **Smooth.**

Move Type Sets the type of movement (single or multiple components). Displays the MOVE TYPE menu with the following options:

Move One Moves one component at a time.

Move Many Moves multiple components simultaneously.

With Children Highlights the children of the selected component and moves the entire group as one unit.

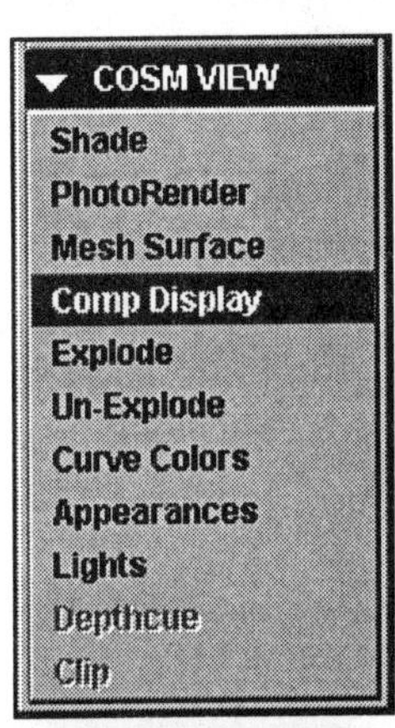

5. Choose an option from the MOVE TYPE menu, then specify the motion increments using the **Trans Incr** option. Choose **Done.**

6. Choose one of these options:

Copy Pos Copy the position of selected components.

DefaultExpld Place selected components in the default explode position.

Translate Specify the reference or direction for repositioning the exploded component by choosing **Translate** and one of the following options (the **Entity/Edge** option is highlighted):

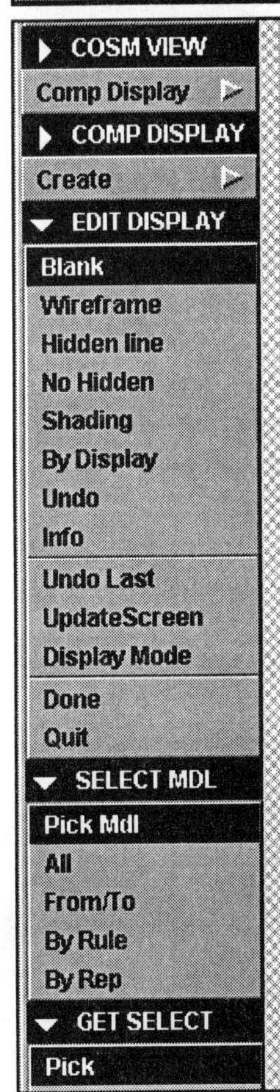

View Plane or **Sel Plane** Translation occurs about the component's drag origin in the same plane.

Entity/Edge Translation occurs about the reference line in the plane that is normal to it and contains the drag origin point.

Plane Normal Translation occurs about the component's drag origin in the plane that contains the drag origin point and is parallel to the reference plane.

2 Points Translation occurs about the reference line in the plane that is normal to it and contains the drag origin point.

C-sys Translation occurs about the axis in the plane that is normal to it and contains the drag origin point.

Setting Display Modes for Components

Using the **Comp Display** option in the COSMETIC menu, you can set different visualization (display) modes for components in an assembly. Wireframe, hidden line, no hidden, shaded, or blanked display modes can be assigned to components. The components will be displayed according to their display status in the current display state (that is, blanked, shaded, drawn in hidden line color, and so on). The setting in the ENVIRONMENT menu controls the display of unassigned components. As a result, they appear according to the current mode of the environment:

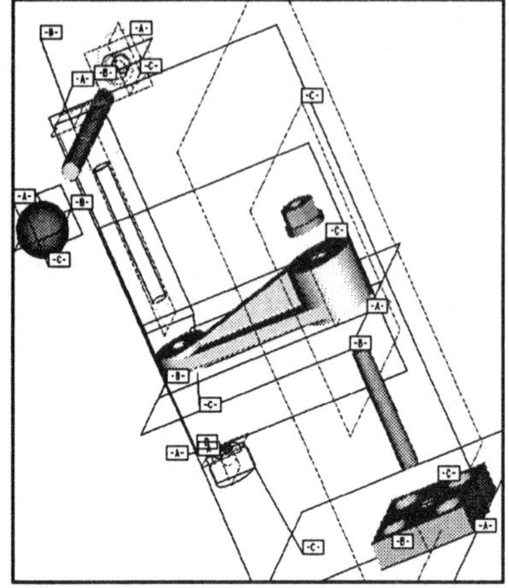

Wireframe Pro/E displays components in front of the assigned component in wireframe and shows all their edges in white (or the color they have) when you assign one component to display its hidden lines, leave the rest unassigned, or set the ENVIRONMENT menu to **Wireframe**. Pro/E displays the components behind the assigned part in wireframe, but does not display sections of edges masked by the hidden line component.

Hidden line If you set the ENVIRONMENT menu to **Hidden Line**, Pro/E displays all components and the hidden line component in the same way.

No Hidden If you set the ENVIRONMENT menu to **No Hidden**, Pro/E displays all components in front of the assigned components with no hidden lines, and displays the obscured edges of the hidden line component in gray and its visible edges in white. For components behind the assigned part, visible edges appear, but obscured ones do not. The edges of the other parts that are obscured by the hidden line component do not appear in gray.

Shading If you set the ENVIRONMENT menu to **Shading**, Pro/E displays all unassigned components as shaded, whether they are in front of or behind the hidden line component.

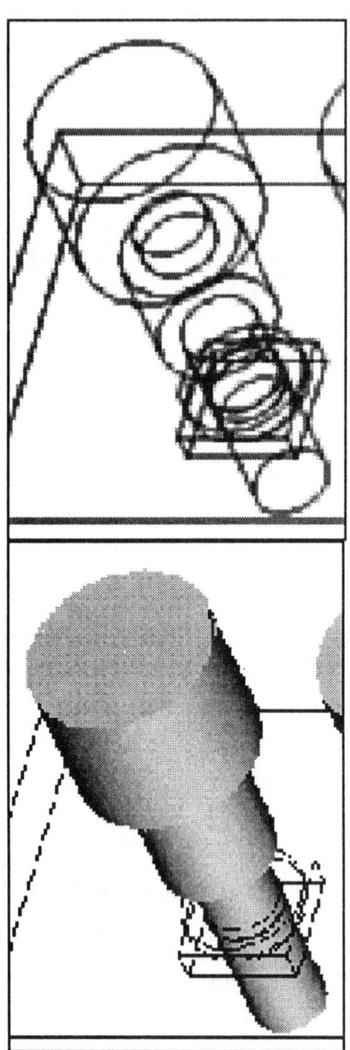

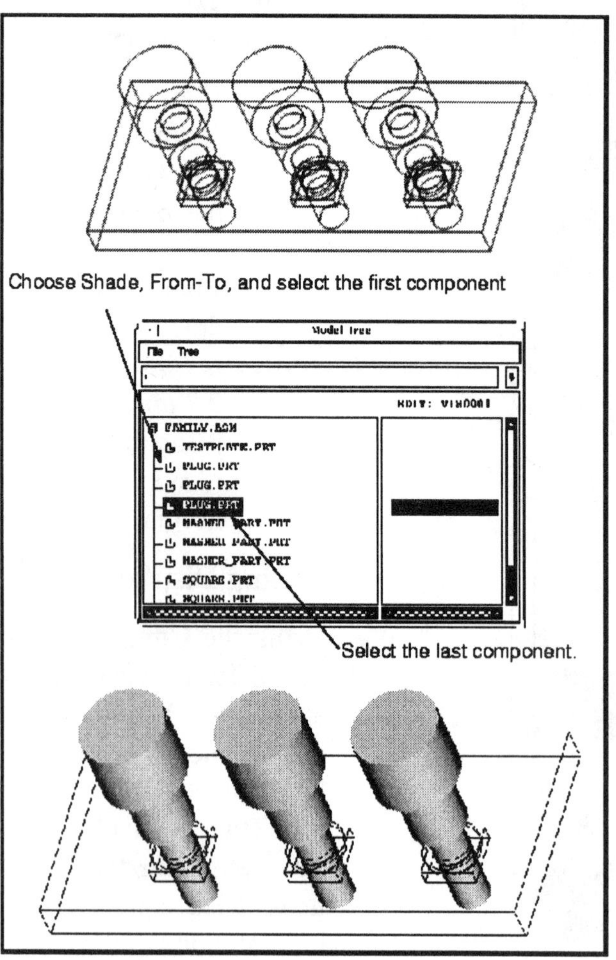

Figure 15.5
Online Documentation, Component Display Status

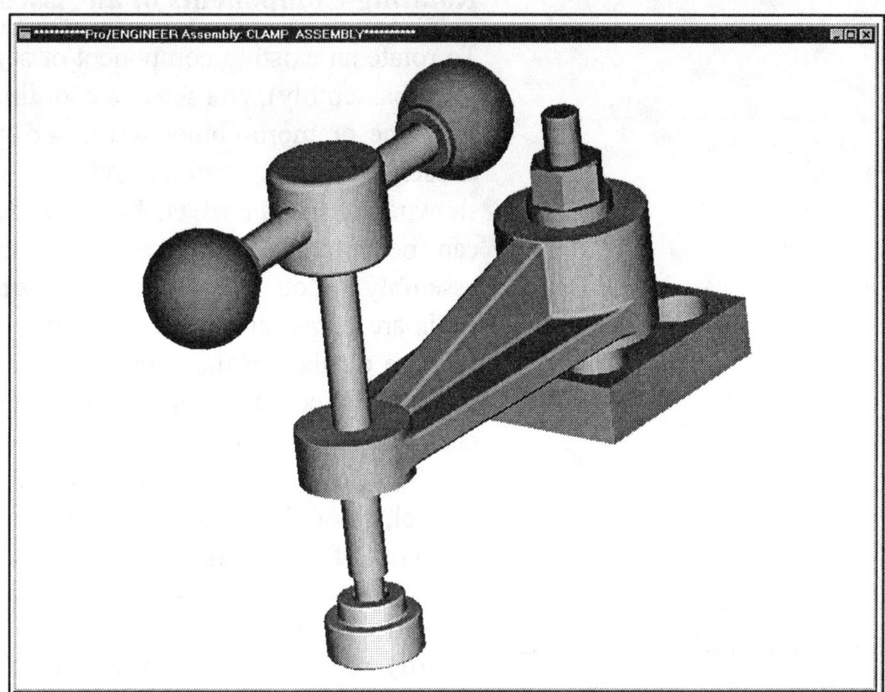

Figure 15.6
Swing Clamp Assembly, Original (default) State

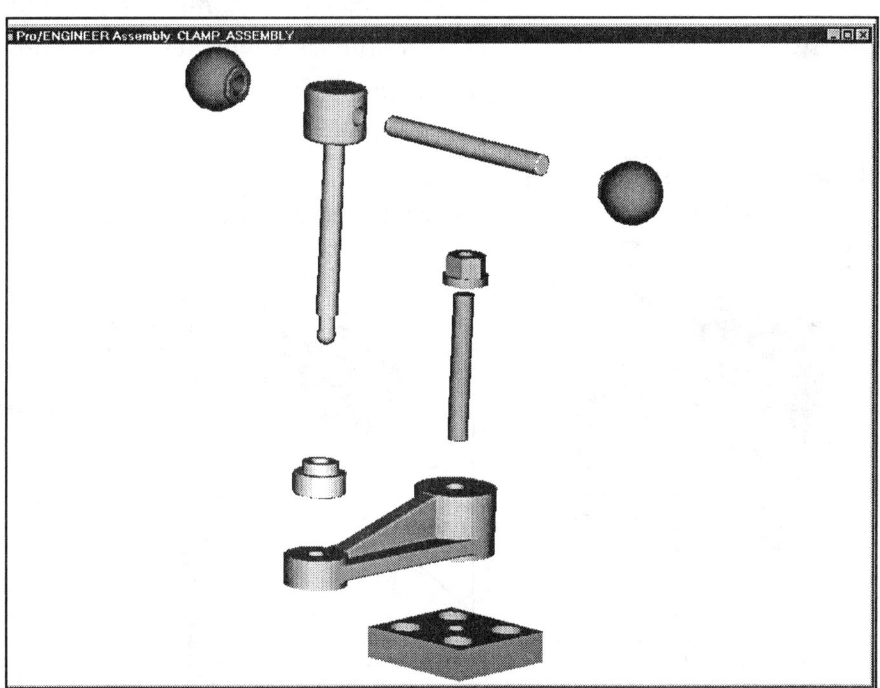

NOTE

Set up the **Swing Clamp** Assembly Exploded View:

- Units = Inches
- Shading
- No Disp Tan
- ☐ Grid Snap

Figure 15.7
Exploded Swing Clamp Assembly with Rotated Swivel

Exploded Swing Clamp

In this lesson, you will use the subassembly and assembly created in Lesson 14 to establish and save new views, exploded views, and views with component display states that differ from one another. You will also be required to move and rotate components of the assembly before cosmetically displaying the assembly in an exploded state. The creation and assembly of new components will not be required. A bill of materials will also be displayed using the **Info** command.

Rotating Components of an Assembly

To rotate an existing component or set of components of an assembly (or subassembly), you select a coordinate system to use as a reference, pick one or more components, and give the rotation angle about a chosen axis of a coordinate system. The **Swing Clamp** subassembly shown here has a **Swivel**, **Foot**, **Stud**, and two **Ball** components that can be rotated about the **Arm** during normal operation of the assembly. You will rotate these components so that the Stud and Balls are perpendicular to the Arm. This position looks better when you are displaying the assembly as exploded (and in its unexploded state). The components will be rotated in the subassembly, and the change will be propagated to the assembly.

The following sequence of commands are used to rotate the Swivel, Foot, Stud, and two Ball components about the Arm (Fig. 15.8) (all of these make up the *subassembly*):

PT/Modeler™

Command not supported

Assembly ⇒ **Search/Retr** ⇒ **CLAMP_SUBASSEMBLY** ⇒ **Modify** ⇒ **Mod Assem** ⇒ **Move** ⇒ (select the Swivel's coordinate system) ⇒ (select components to move, then pick the Swivel and the Foot *only*) ⇒ **Done Sel** ⇒ **Rotate** ⇒ **Z Axis** ⇒ (input the angle about the **Z** direction, **90°**) ⇒ **enter** ⇒ **Done Move** ⇒ **Regenerate** ⇒ **Automatic** ⇒ **Done** ⇒ **Done/Return**

Second: Select the Swivel's coordinate system

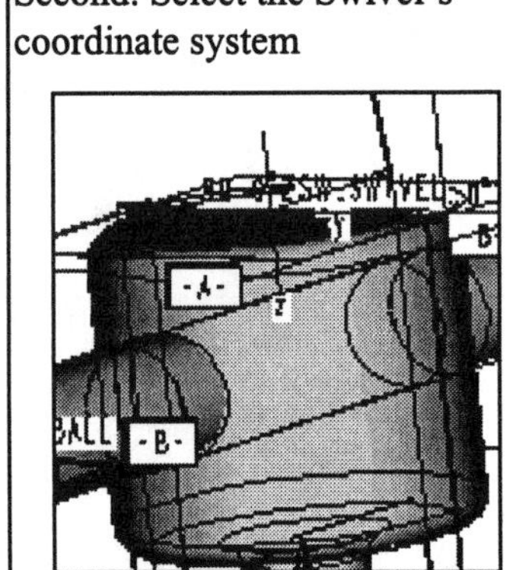

Third: Select the Swivel and the Foot

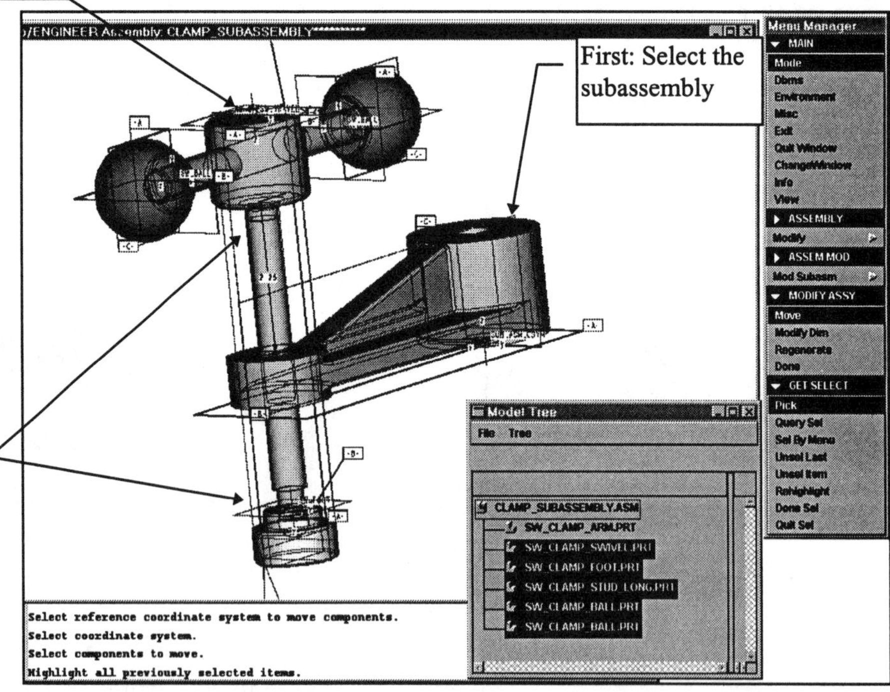

Figure 15.8
Modifying the Subassembly

The Swivel, Foot, Balls, and Stud are now rotated **90°** (Fig. 15.9). Capture this view with the following commands:

View ⇒ **Names** ⇒ **Save** ⇒ (enter NAME: **subpict1**) ⇒ **enter** ⇒ **Done-Return**

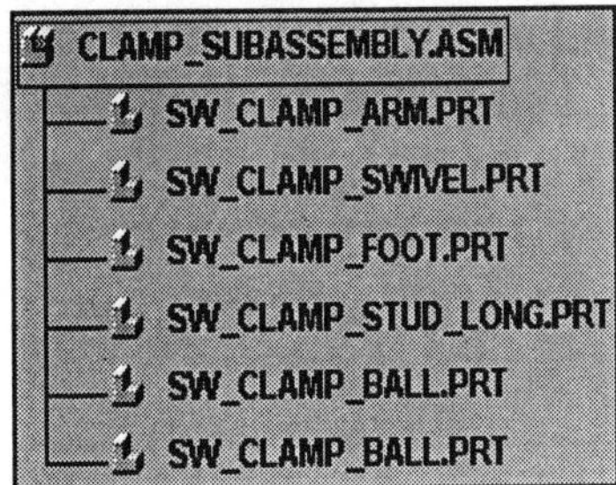

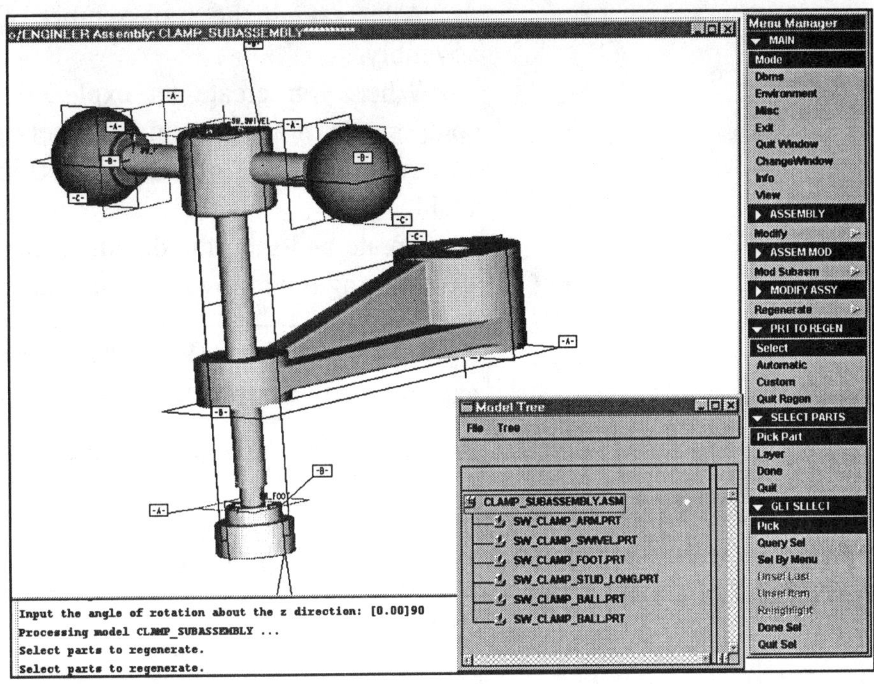

Figure 15.9
Subassembly with Rotated Components

Dbms (File--PT/Modeler) ⇒ Save ⇒ enter
Purge ⇒ enter ⇒ Done-Return

Views: Perspective, Exploded, and Altered Component Displayed State Views

Another type of view that can be created is the perspective view. You will now create a variety of cosmetically altered views. Cosmetic changes to the assembly do not affect the model itself, only the way it is displayed on the screen. Use the following commands to create a perspective (Fig. 15.10) and save the view with a unique name:

View ⇒ Perspective ⇒ Default ⇒ Names ⇒ Save ⇒ (enter NAME: **perspec) ⇒ enter ⇒ Done-Return**

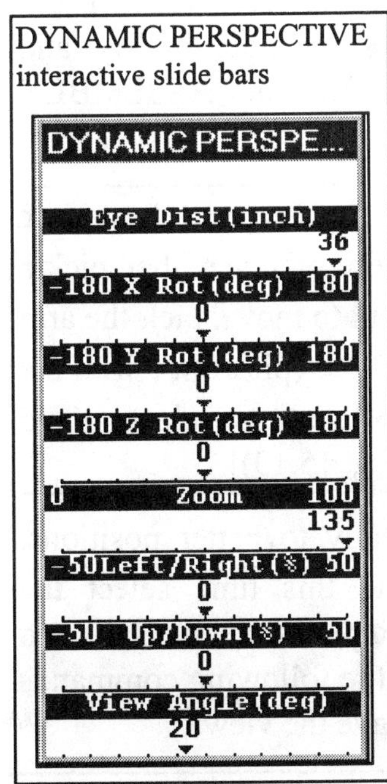

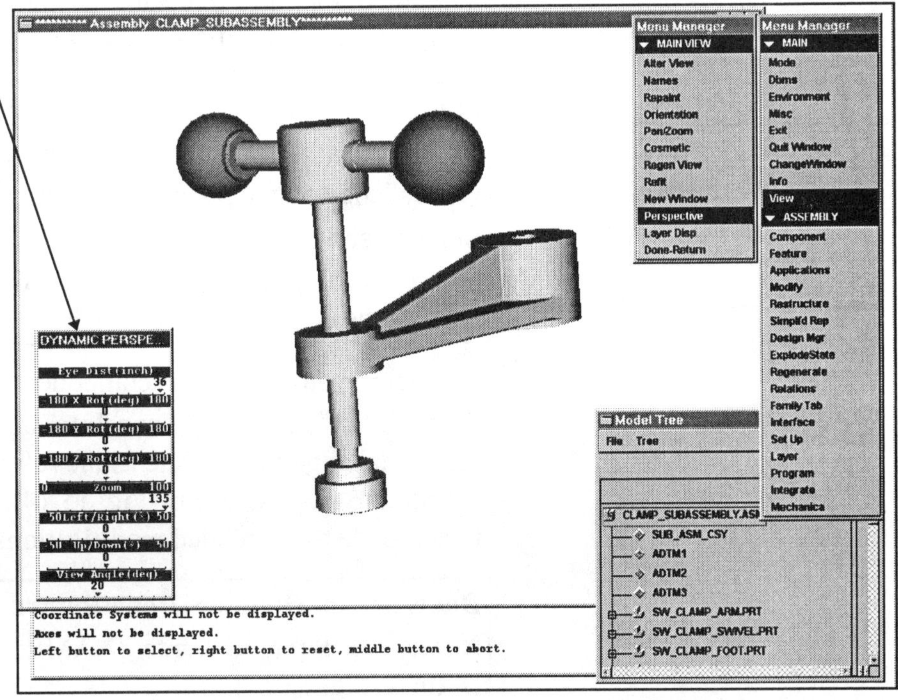

Figure 15.10
Perspective Assembly View

After saving the subassembly, **Quit Window** and retrieve the assembly.

When you create an exploded view, Pro/E moves apart the components of an assembly to a set default distance. Your exploded view will probably look different from the example shown in Figure 15.11.

Create and save the default exploded view of the assembly (Fig. 15.11) using the following commands:

View ⇒ **Cosmetic** ⇒ **Explode** ⇒ **Names** ⇒ **Save** ⇒ (enter NAME: **expldefault**) ⇒ **enter** ⇒ **Done-Return**

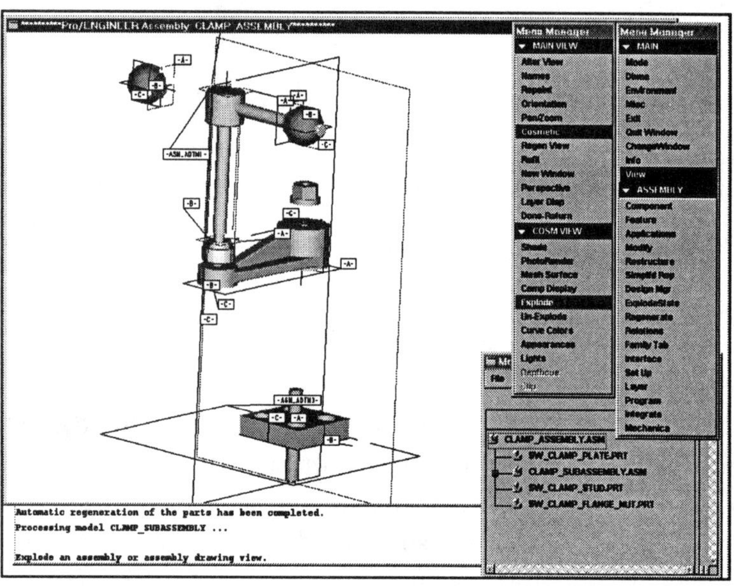

Figure 15.11
Default Exploded Assembly

Since the default exploded view is seldom perfect, Pro/E has a wide variety of commands available to adjust, change, modify, and reorient the exploded view. Using **Modify** from the ASSEMBLY menu, choose the following commands:

PT/Modeler™

Mod Explode ⇒ **Move** ⇒ (left mouse button to select component) ⇒ (move to new location, left mouse button to drop at new location)

Modify ⇒ **Modify Expld** ⇒ **Position** ⇒ **Entity/Edge** ⇒ **Translate** ⇒ (select an axis or straight edge as the motion reference, then pick a vertical edge on the plate) ⇒ [select component to move, pick the arm first and slide it to a new position (Fig. 15.12)] ⇒ (pick once with the left mouse button to place the component) ⇒ [continue selecting and moving the components you wish to adjust (Fig. 15.13)]

After you move the components vertically to better positions, pick **Entity/Edge** ⇒ **Translate** again, and this time select the horizontal edge of the plate as the reference edge. Move the Balls and the Stud to new positions (Fig. 15.14). Use the following commands to complete the exploded modifications and save the view:

Done Sel ⇒ **Done** ⇒ **Done/Return** ⇒ **Done/Return** ⇒ **View** ⇒ **Names** ⇒ **Save** ⇒ (enter NAME: **explod1**) ⇒ **enter** ⇒ **Done-Return** (Fig. 15.15)

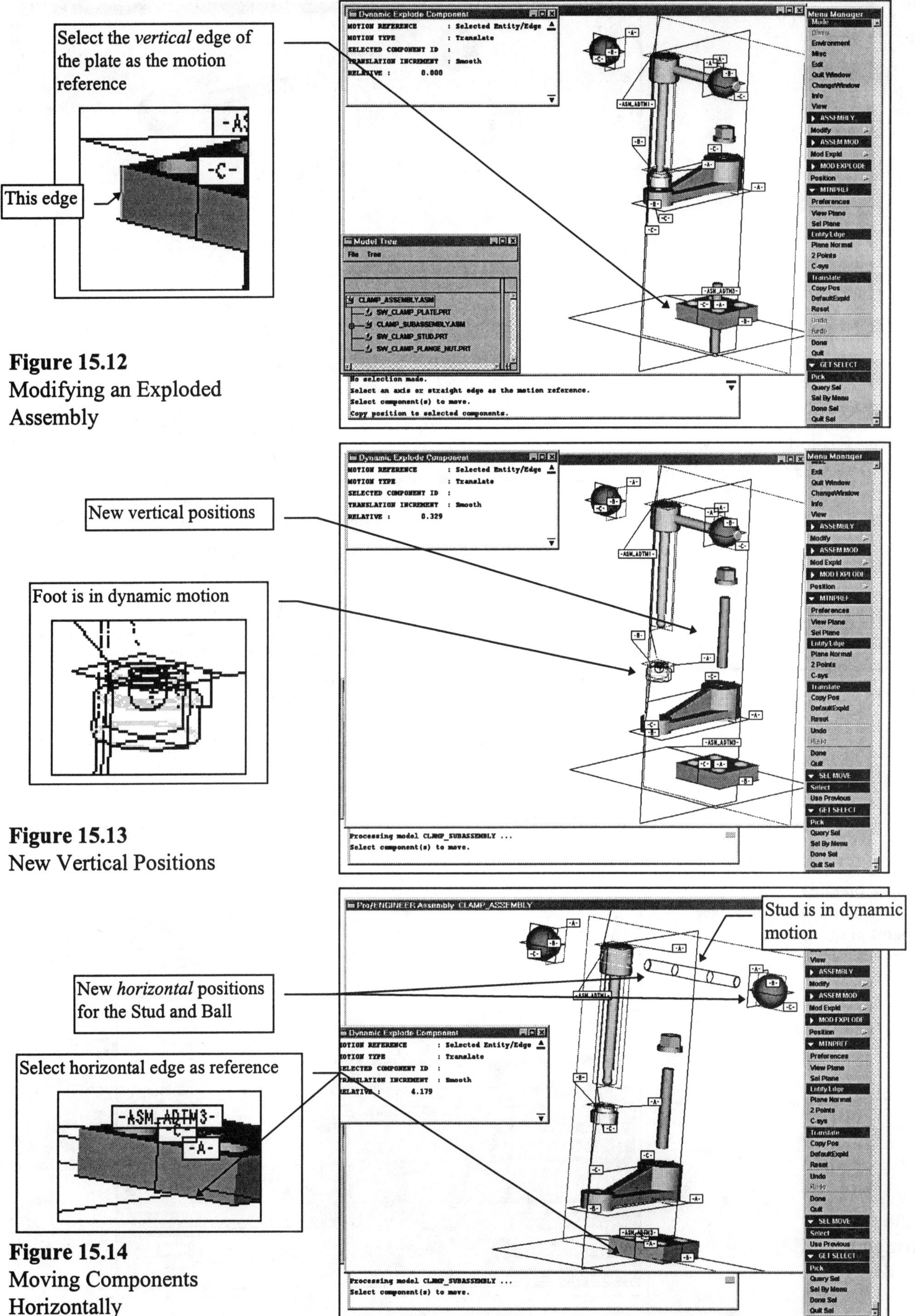

Figure 15.12
Modifying an Exploded Assembly

Figure 15.13
New Vertical Positions

Figure 15.14
Moving Components Horizontally

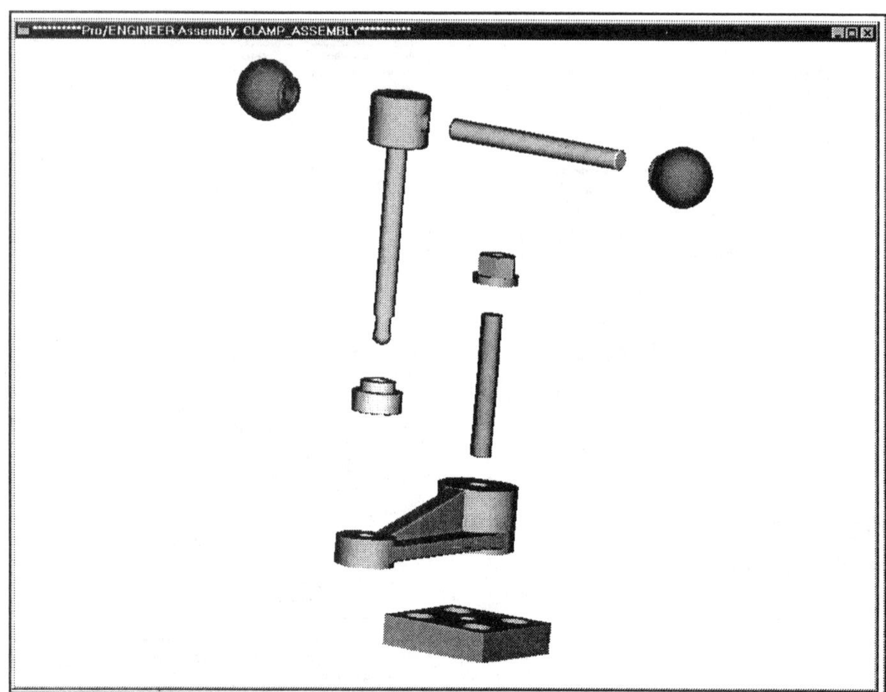

Figure 15.15
Save the Exploded View's New Component Positions

The components of an assembly (whether exploded or not) can be displayed individually with **Wireframe, Hidden line, No Hidden,** or **Shading** using **Comp Display**. Change the component display of the exploded view (Fig. 15.16) with the following commands:

PT/Modeler™

Does not support **Comp Display**

View ⇒ **Cosmetic** ⇒ **Comp Display** ⇒ **Create** ⇒ **(VIS0001:)** ⇒ **enter** (to accept the default) ⇒ **Hidden line** ⇒ (pick the Swivel) ⇒ **No Hidden** ⇒ (pick the right Ball) ⇒ **Wireframe** ⇒ (pick the Foot) ⇒ **Done Sel** ⇒ **Done** ⇒ **Set Current** ⇒ **✓VIS0001** ⇒ **Done** (Fig. 15.17)

HINT
Check **✓Shading** in the ENVIRONMENT menu and then **Repaint** to see the likeness of Figure 15.17.

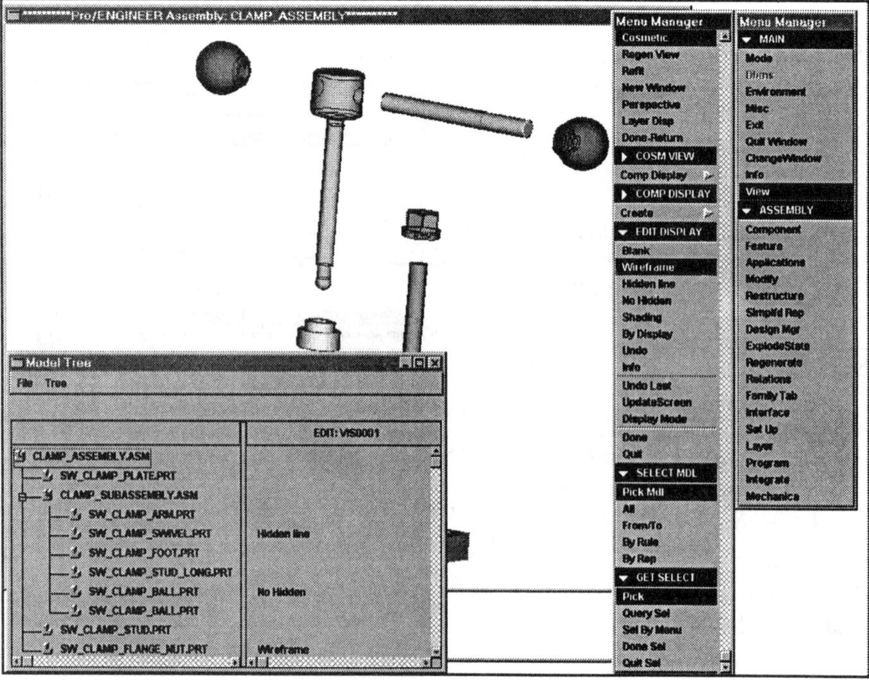

Figure 15.16
Changing the Component Display State

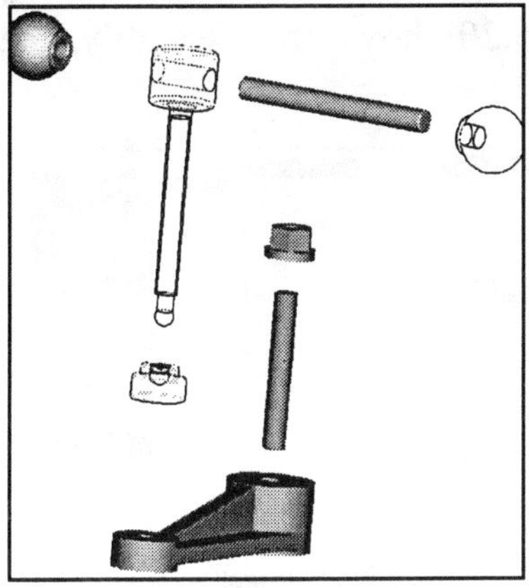

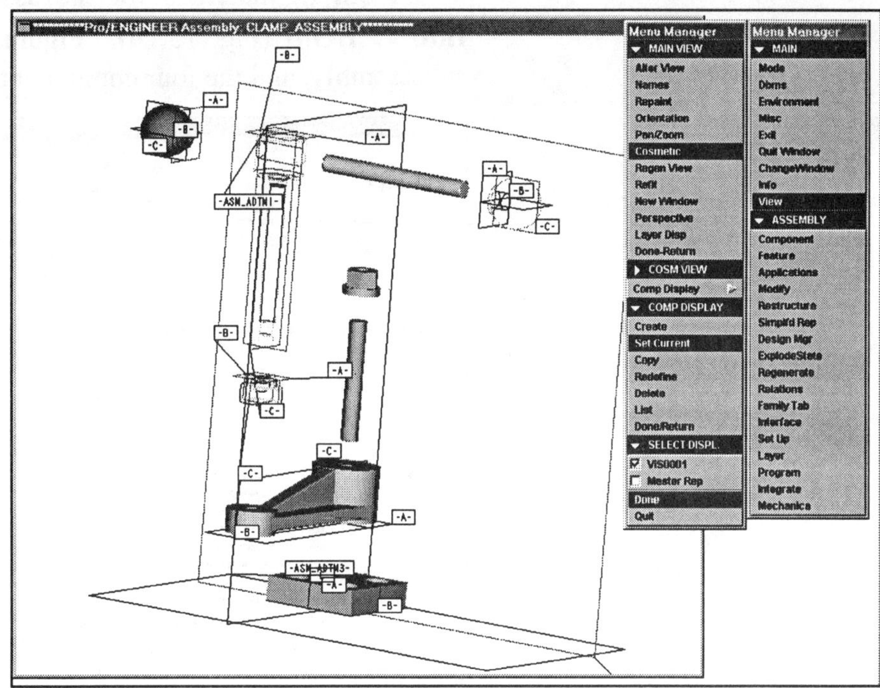

Figure 15.17
Component Display State Is Different for Individual Components

You can see the new component display states (Fig. 15.18) in the Model Tree by choosing the following:

PT/Modeler™

Does not support **Visual Modes**

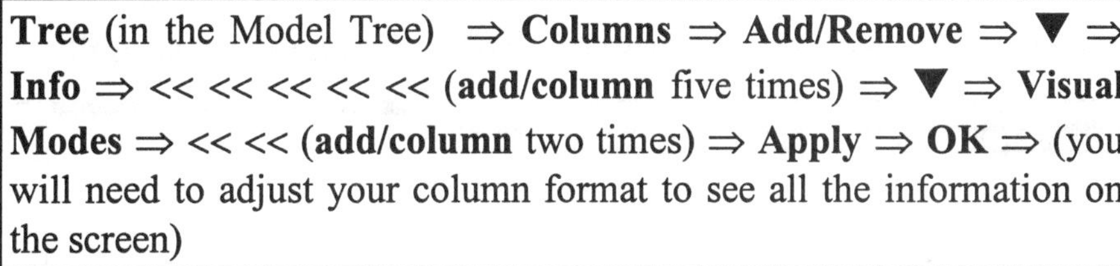

Tree (in the Model Tree) ⇒ **Columns** ⇒ **Add/Remove** ⇒ **▼** ⇒ **Info** ⇒ << << << << << (**add/column** five times) ⇒ **▼** ⇒ **Visual Modes** ⇒ << << (**add/column** two times) ⇒ **Apply** ⇒ **OK** ⇒ (you will need to adjust your column format to see all the information on the screen)

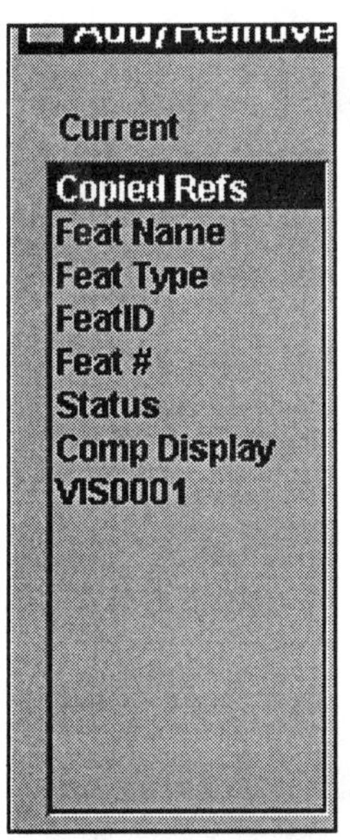

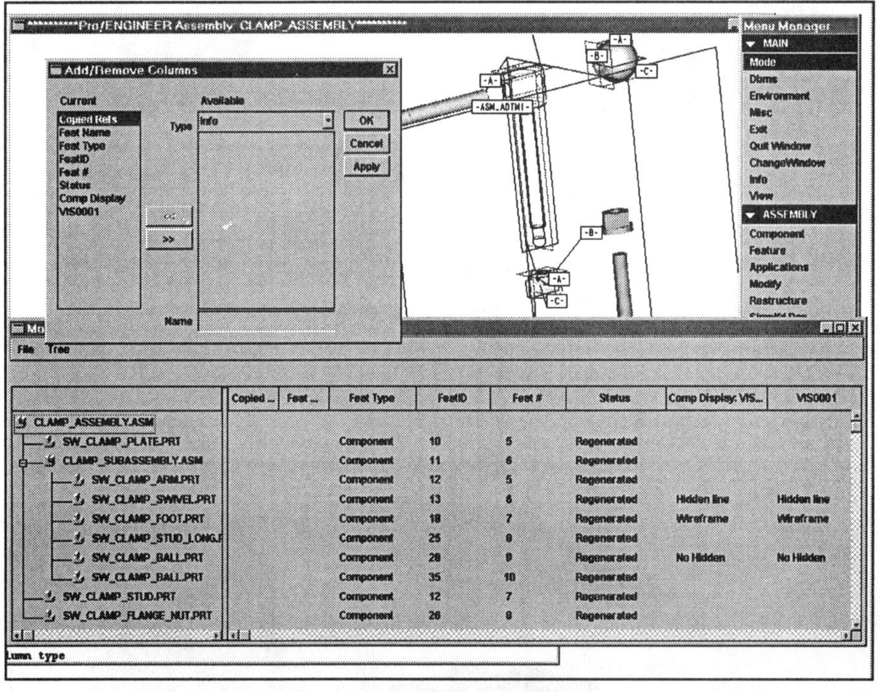

Figure 15.18
Components Display

There are thousands of views, exploded view capabilities, and display states. Experiment with Pro/E commands and enjoy yourself.

A bill of materials for the assembly can be seen by picking **Info ⇒ BOM** (Fig. 15.19). Figure 15.20 shows the assembly, the subassembly, and the four components.

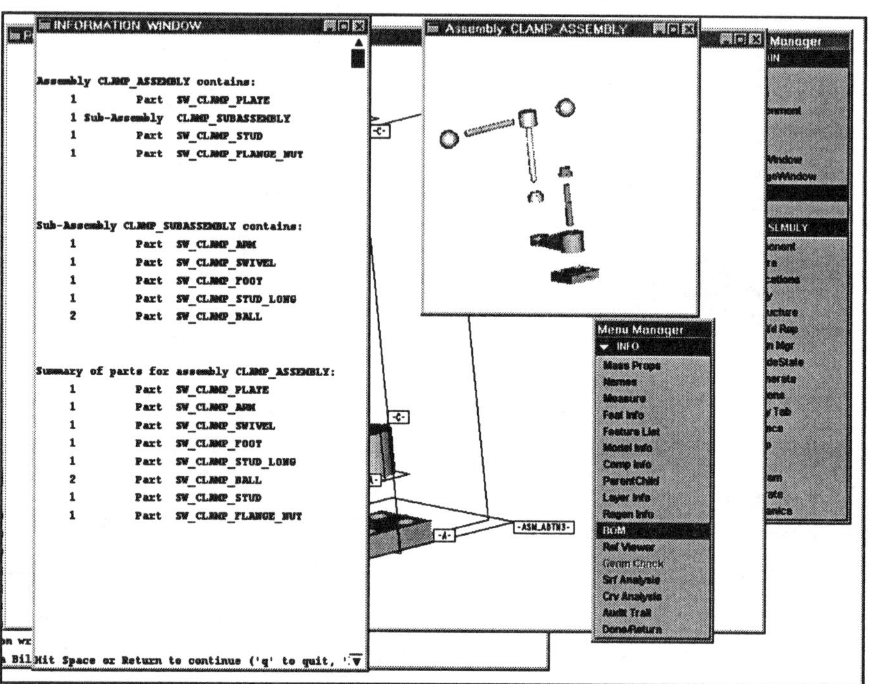

Figure 15.19
BOM

Figure 15.20 Assembly, Subassembly, Foot, Arm, Plate, and Ball

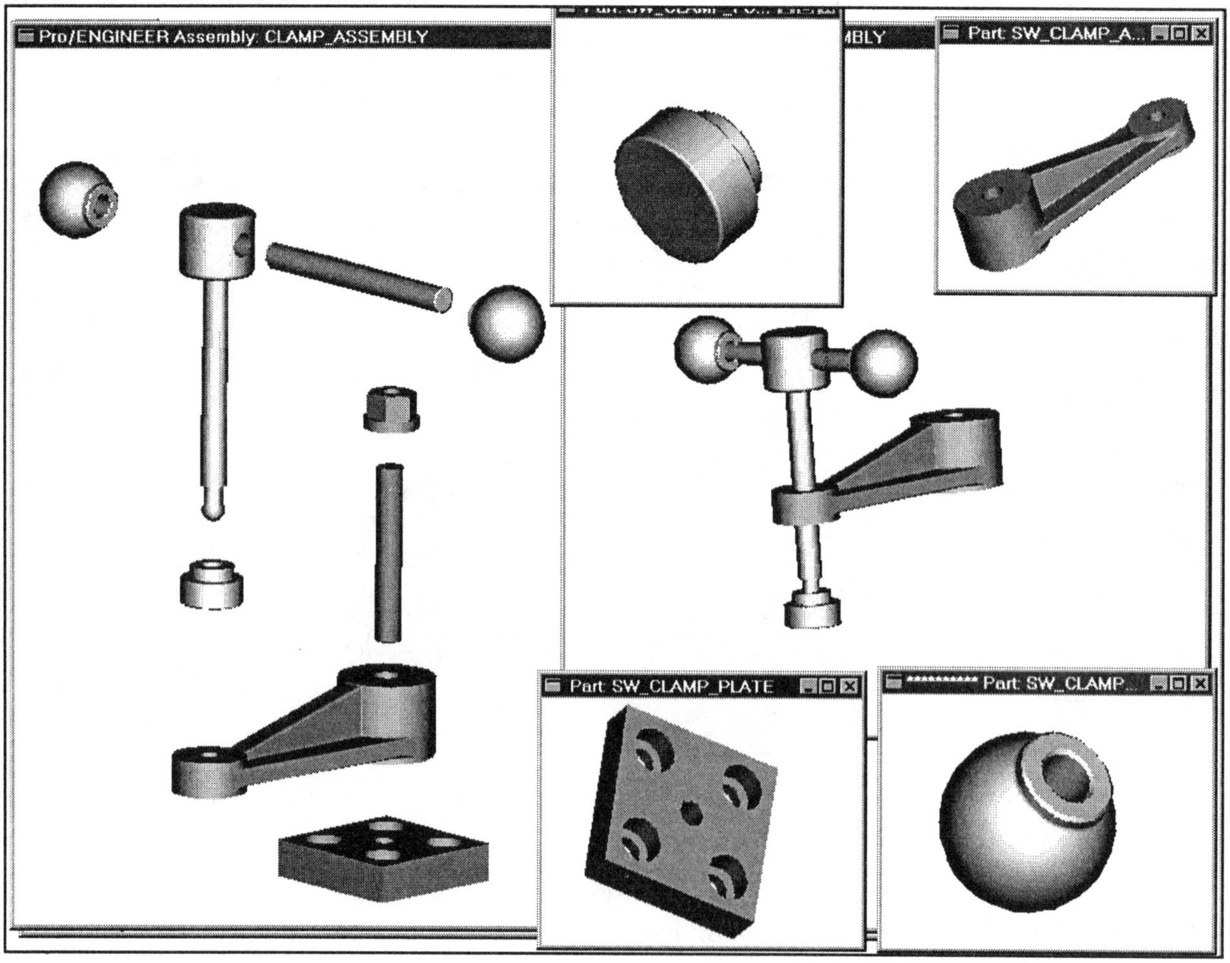

Lesson 15 Project

Exploded Coupling Assembly

☑ ***EGD REFERENCE***
Engineering Graphics and Design with Graphical Analysis *or* **Fundamentals of Engineering Graphics and Design**
by L. Lamit and K. Kitto
Read Chapter 23
See pages 865-866

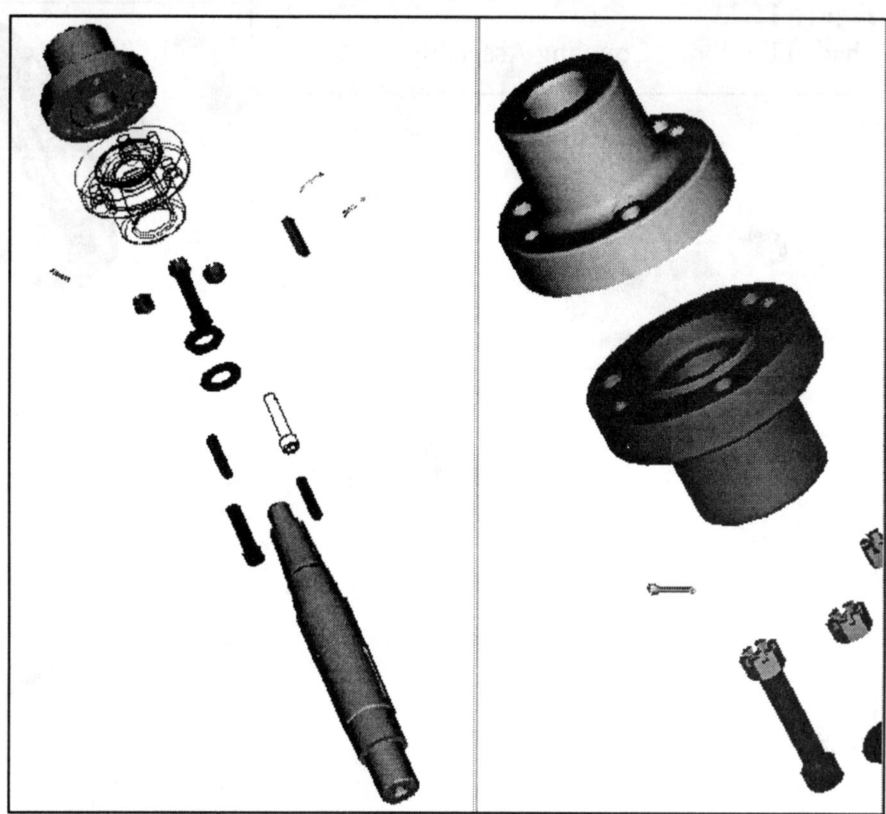

Figure 15.21
Exploded Coupling Assembly

Exploded Coupling Assembly

The fifteenth **lesson project** uses the assembly created in the Lesson 14 Project. An exploded view needs to be created and saved for use later in Drawing mode for the Lesson 20 Project. A variety of other views are suggested, including a section of the assembly, a perspective view, and an exploded view with a different component display variation. Each component should have its own color. If you did not color the components during the part creation, bring up each part in Part mode, define, and set the component with a color. Save three or four unique exploded states (view positions) and component display (visibility) states. You do not need to match the examples (Figs. 15.21 through 15.24) provided in this lesson project.

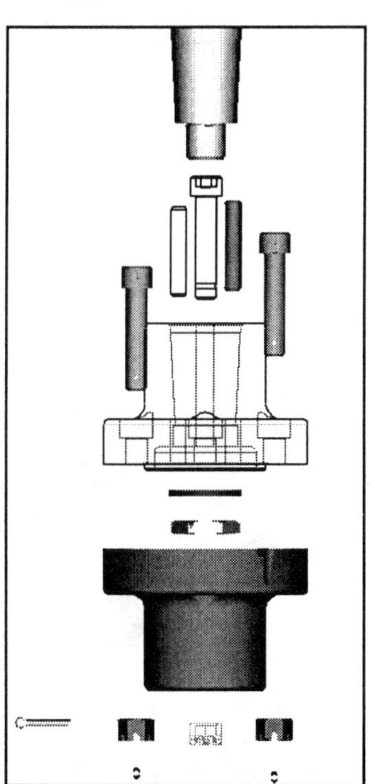

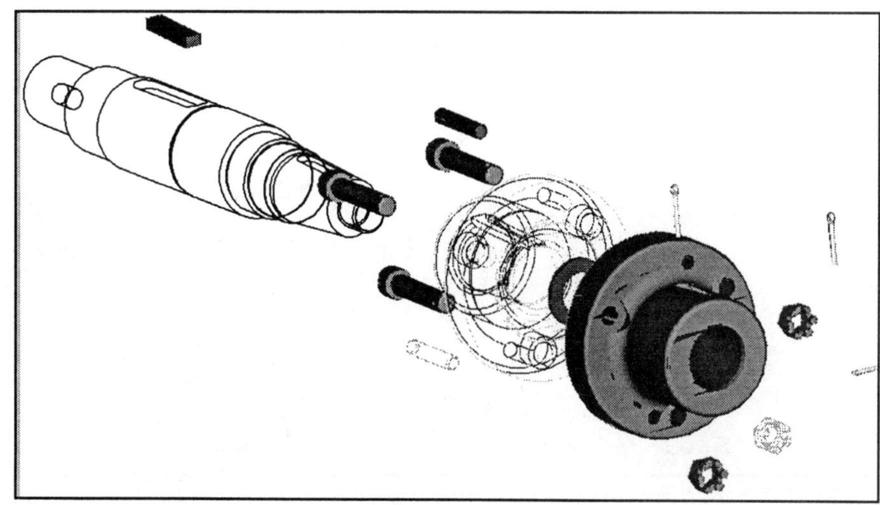

Figure 15.22
Perspective View of Exploded Coupling Assembly with a Variety of Component Display States

Figure 15.23
Shaded Exploded Coupling Assembly

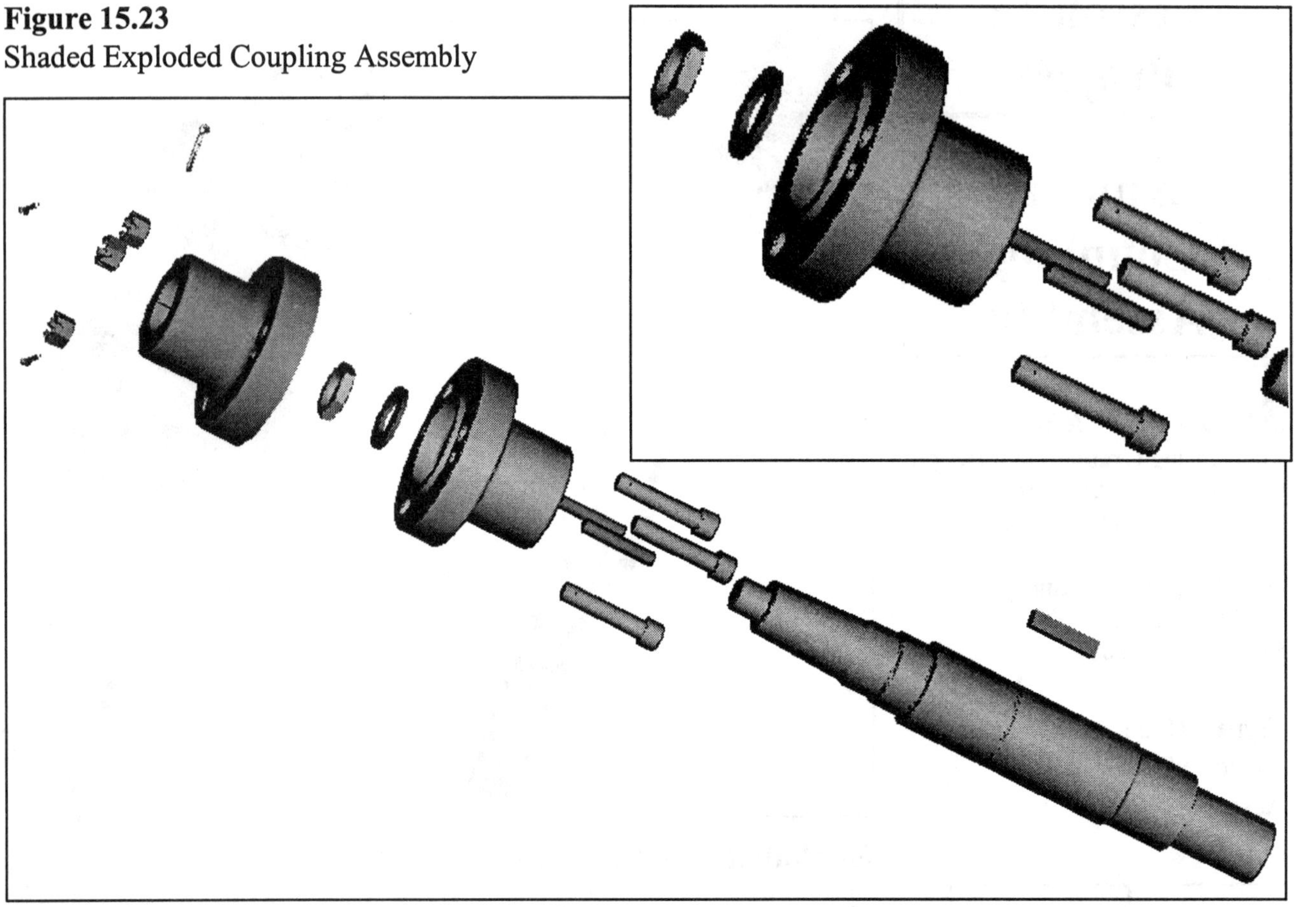

Figure 15.24
Exploded Coupling Assembly with Different Component Display States

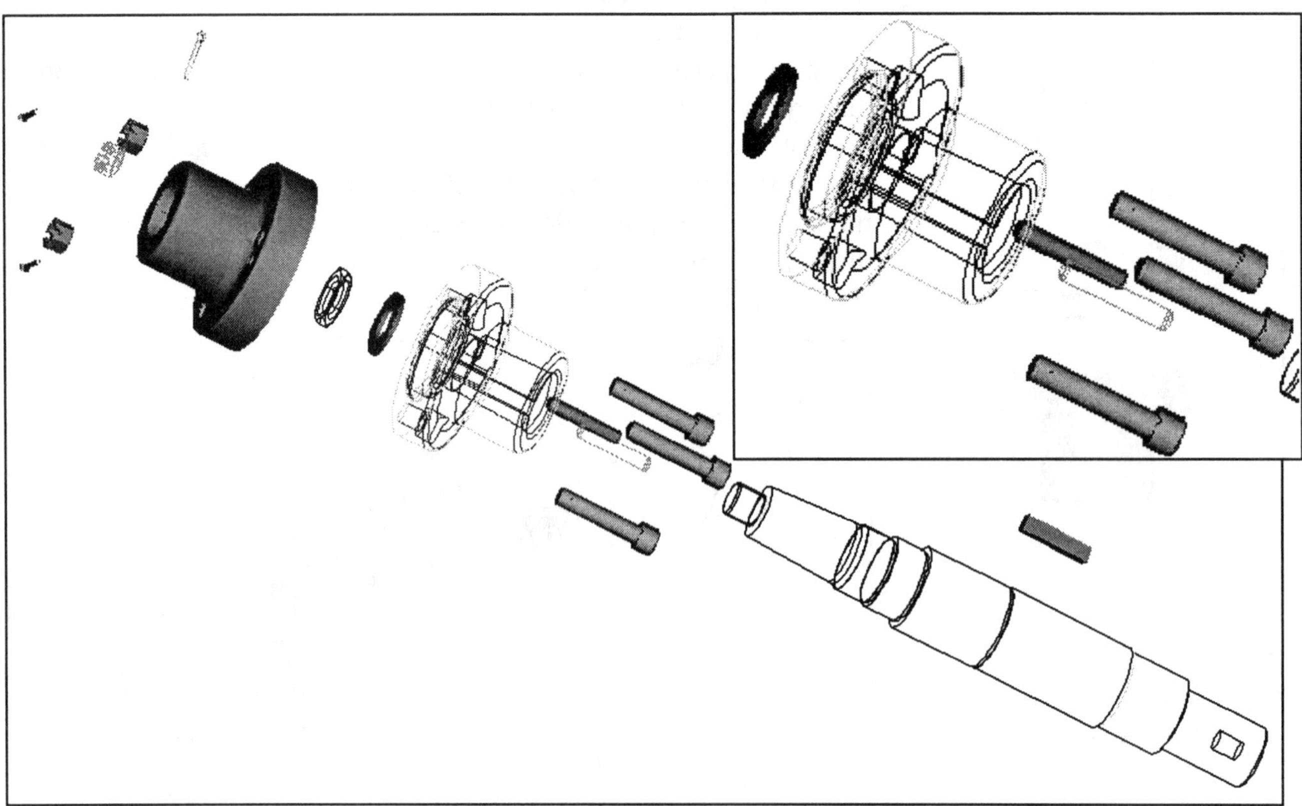

Part Three

Generating Drawings

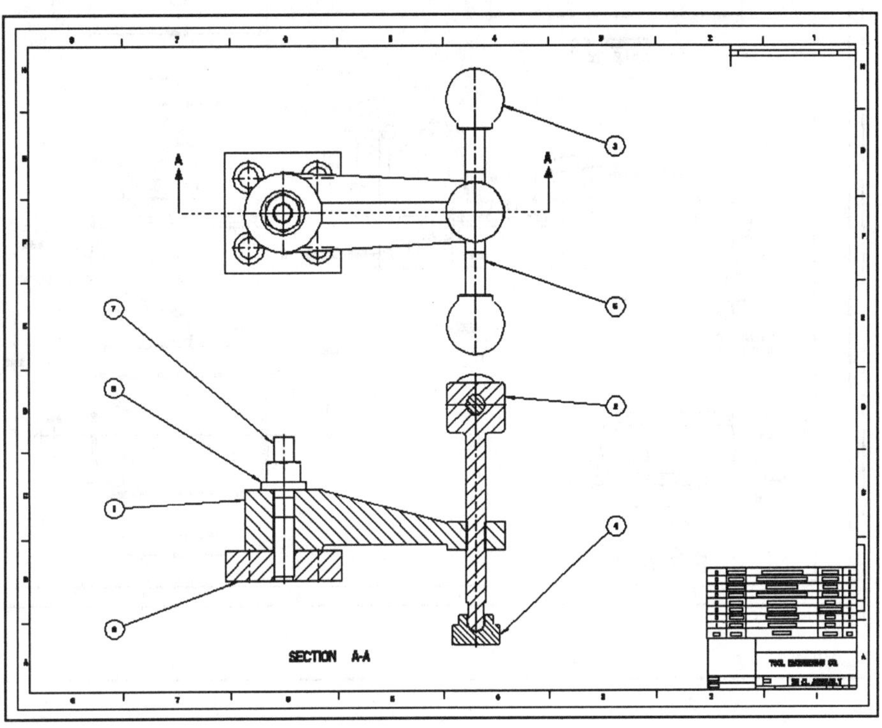

Assembly Drawing

GENERATING DRAWINGS

The drawing functionality in Pro/E is used to create annotated drawings of parts and assemblies. During drawing creation a variety of options are available, including:

* Add views of the part or assembly to the drawing
* Display existing design dimensions
* Create additional driven or reference dimensions
* Create and insert notes on the drawing
* Add views of additional parts or assemblies
* Add multiple sheets to the drawing
* Create a BOM and balloon the assembly
* Add draft entities to the drawing

Construction of Drawings

Drawings can be created of parts and assemblies. Drawings can be multiview or pictorial, and include section, auxiliary, detailed, exploded, and broken views. With Pro/DETAIL, ANSI, ISO, DIN, or JIS standard drawings can be created.

Detail Drawing

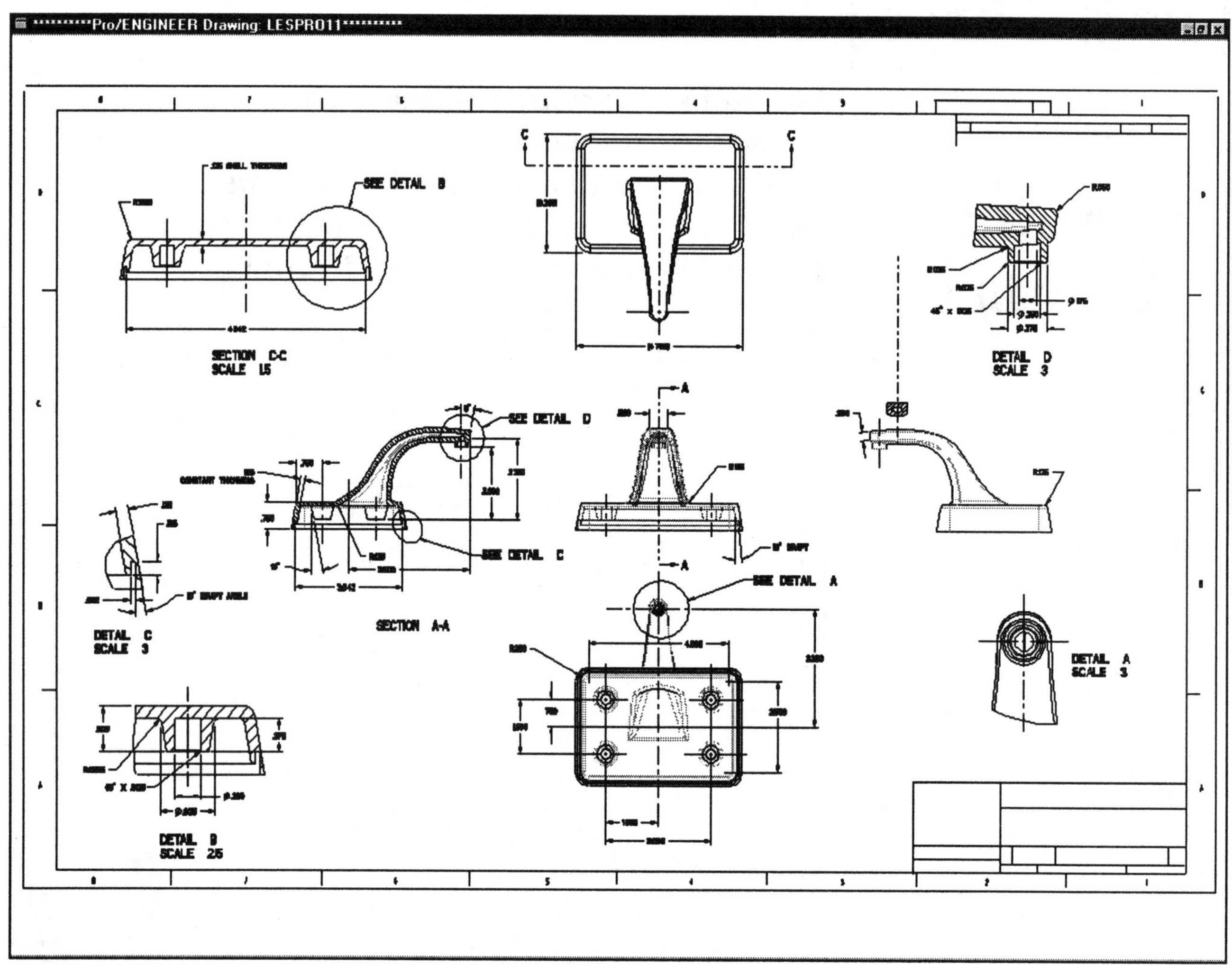

Lesson 16

Formats, Title Blocks, and Views

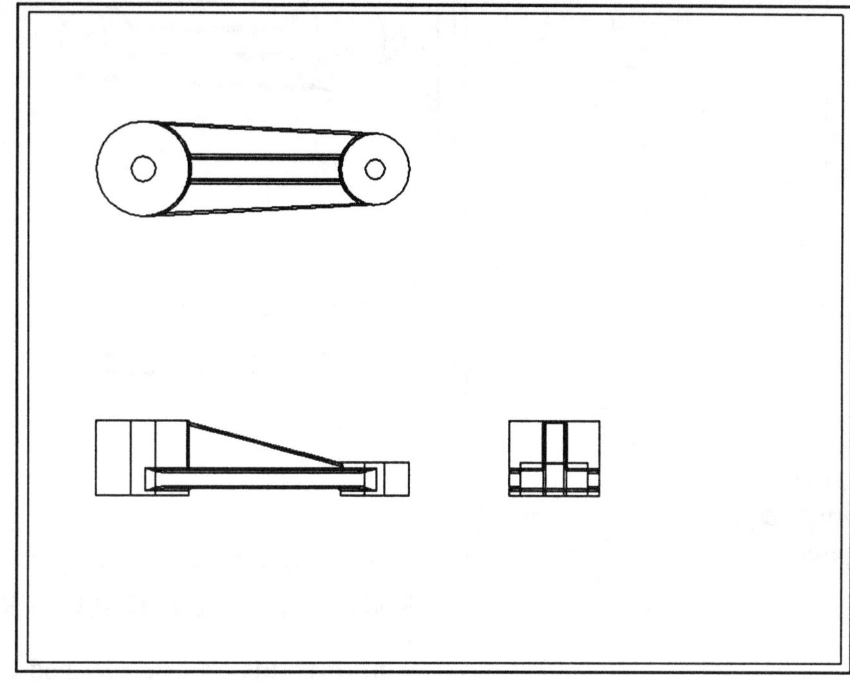

Figure 16.1
Clamp Arm Drawing
Without Drawing Format

EGD REFERENCE
Engineering Graphics and Design with Graphical Analysis *or* **Fundamentals of Engineering Graphics and Design**
by L. Lamit and K. Kitto
Read Chapters 10 and 23
See pages 314-316

COAch™ for Pro/ENGINEER

If you have **COAch for Pro/ENGINEER** on your system, go to SEARCH and do the Segments shown in the many figures taken from COAch in this lesson. Note that *COAch is not required to use this text.* Having access to CADTRAIN's COAch tutorials *will expedite* your mastering of Pro/ENGINEER and PT/Modeler.

OBJECTIVES

1. **Create drawings with views**
2. **Create and save title blocks and sheet formats**
3. **Change the scale of a view**
4. **Display views for detailing a project**
5. **Move, erase, and delete views**
6. **Specify a standard-format paper size and units**
7. **Retrieve formats from the format library**

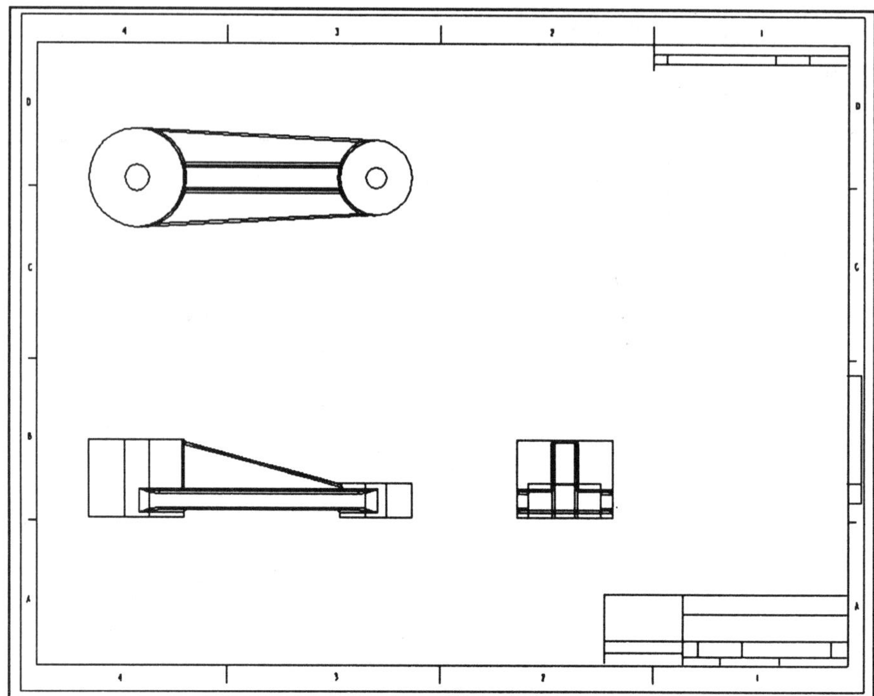

Figure 16.2
Formatted Clamp Arm Drawing

FORMATS, TITLE BLOCKS, AND VIEWS

Drawing formats are user-defined drawing sheet layouts (Figs. 16.1 and 16.2). A drawing format can be used in any number of drawings. It can also be modified or replaced in a Pro/E drawing at any time. **Title Blocks** are standard or sketched line entities that can contain parameters so that the part name, tolerances, scale, and so on will show when the drawing format is retrieved (Fig. 16.3). **Views** created by Pro/E are exactly like views constructed manually by a designer on paper. The same rules of projection are applied; the only difference is that you choose commands in Pro/E to create the views as needed.

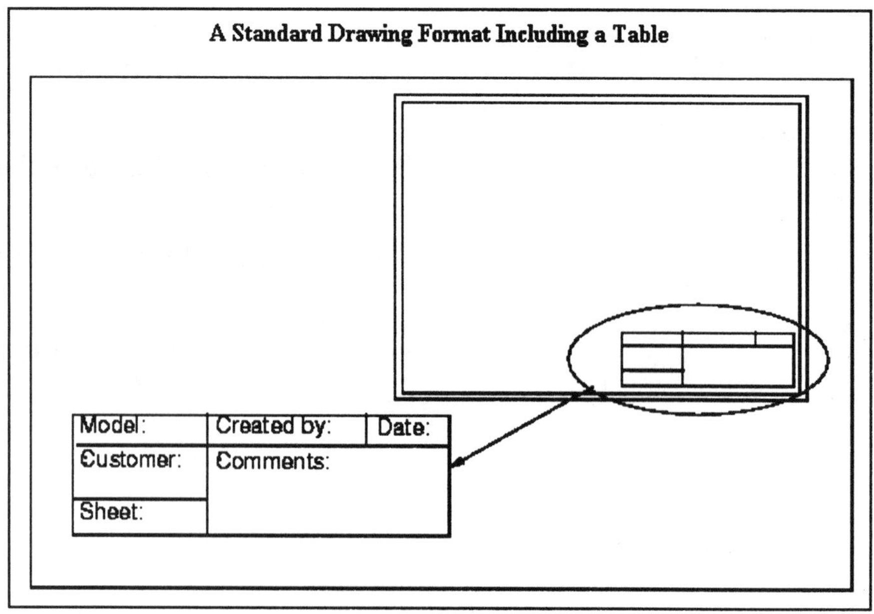

Figure 16.3
Online Documentation, Formats

Drawing Formats

There are two types of drawing formats: standard (Fig. 16.2) and sketched. **Drawing formats** consist of draft entities, not model entities. You can select the desired format size from a list of standard sizes **(A-F** and **A0-A4)** or create a new size by entering values for length and width.

Sketched formats created in Sketcher mode (Fig. 16.4) may be parametrically modified, enabling you to create nonstandard-size formats or families of formats.

Formats can be altered to include note text, symbols, tables, and drafting geometry, including drafting cross sections and filled areas.

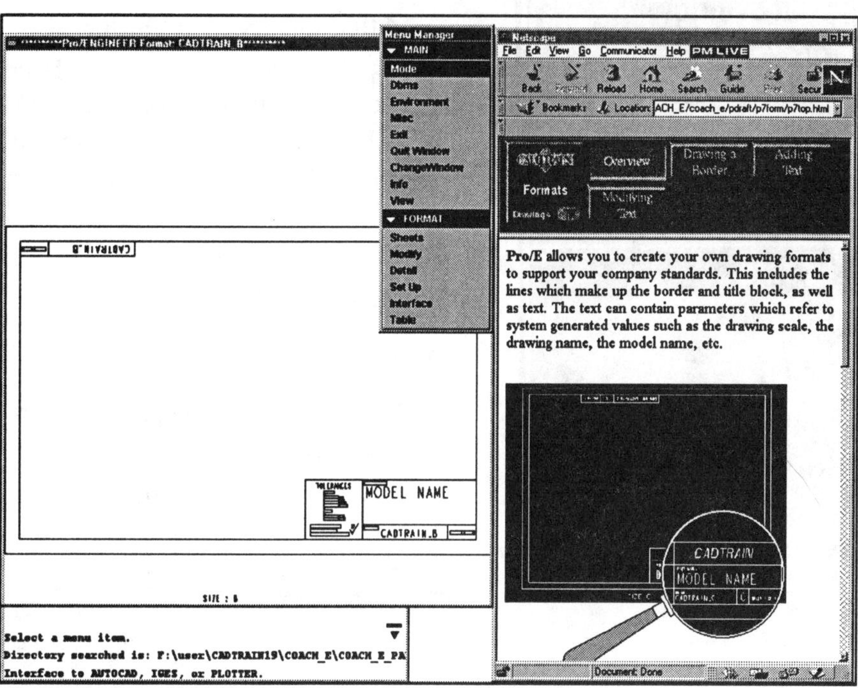

Figure 16.4
COAch for Pro/E, Drawings, Formats (Overview)

With Pro/E (Pro/DETAIL) you can do the following in **Format** mode:

* Create draft geometry, notes.
* Move, mirror, copy, group, translate, and intersect geometry.
* Use and modify the draft grid.
* Enter user attributes (Fig. 16.5).
* Create drawing tables.
* Use interface tools to create plot, DXF, SET, and IGES files.
* Import IGES, DXF, and SET files into the format.
* Create user-defined line styles.
* Create, use, and modify symbols.
* Include drafting cross sections in a format.

Whether you use a standard format or a sketched format, the format is added to a drawing which is created for a set of specified views of a parametric 3D model.

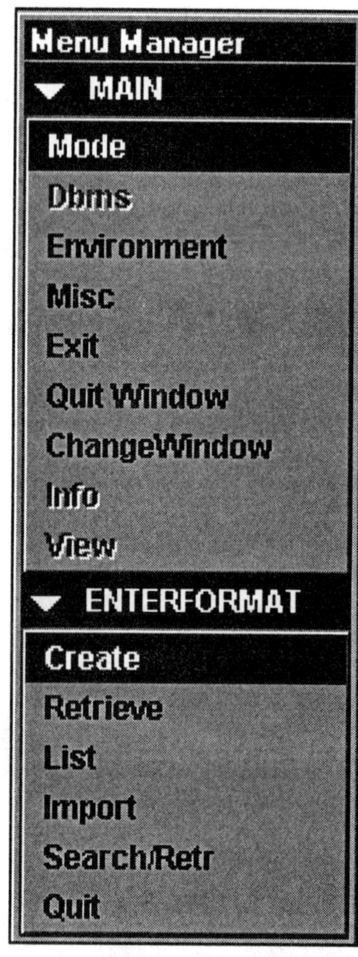

Retrieving Pro/E Formats from the Format Library

You can retrieve formats from a format library directory within Pro/E. To enter the format directory, use the **Format Dir** option in the SELECT FILE menu. This option is not available unless you have set the **pro_format_dir** configuration file option. To retrieve these formats:

1. Choose **Format** from the MODE menu.
2. Choose **Search/Retr** from the ENTERFORMAT menu.
3. Choose **Format Dir** from the SELECT FILE menu.
4. Choose a format from the displayed list.

After specifying a format size, you enter **Format** mode; a sheet outline appears on the screen. If necessary, turn on the drawing grid by choosing **Detail** from the FORMAT menu, **Modify** from the DETAIL menu, **Grid** from the MODIFY DRAW menu, and **Grid On** from the GRID MODIFY menu. The grid size depends on the units for the format.

Creating a Format

To Create a format choose:

1. Choose **Format** from the MODE menu. The ENTERFORMAT menu is displayed.
2. Choose **Create** from the ENTER menu, then enter a format name.

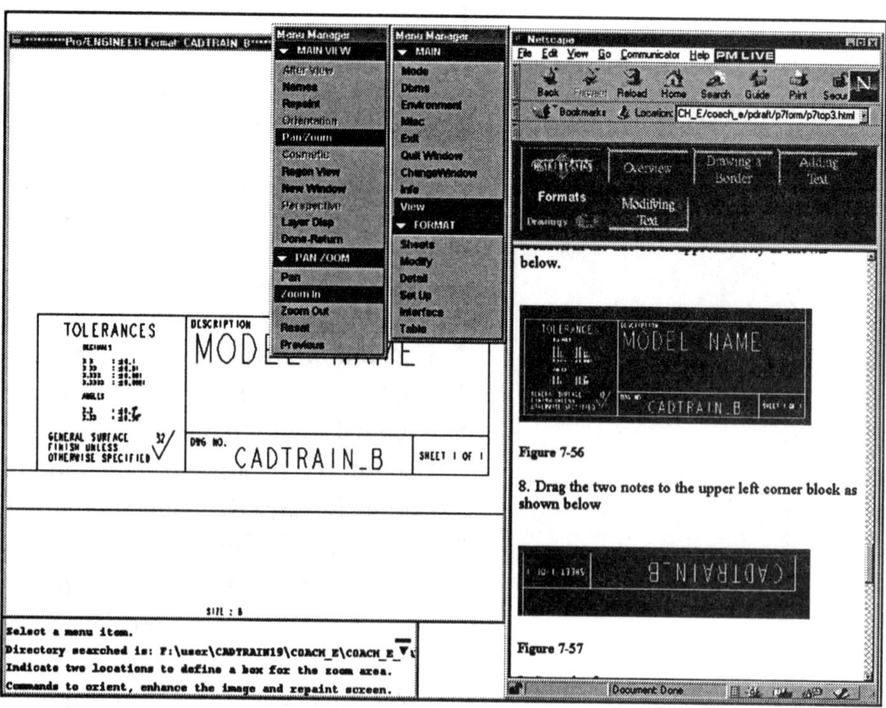

Figure 16.5
COAch for Pro/E, Drawings, Formats (Modifying Text)

Specifying the Format Size and Units

You establish the format size by selecting one from a list of standard format sizes or by creating your own size by entering values for the length and width of the format. Once you set the size, you must choose the units of the drawing format. The main grid spacing and format text units depend on the units you select for the format. To select the format units, choose **Inch** or **Millimeter** from the FORMAT UNITS menu. The text height units are in the units of the format: inches or millimeters. The major grid spacing is an inch if the units are inches, and a centimeter if the units are millimeters.

To Specify a Standard-Format Paper Size and Units:

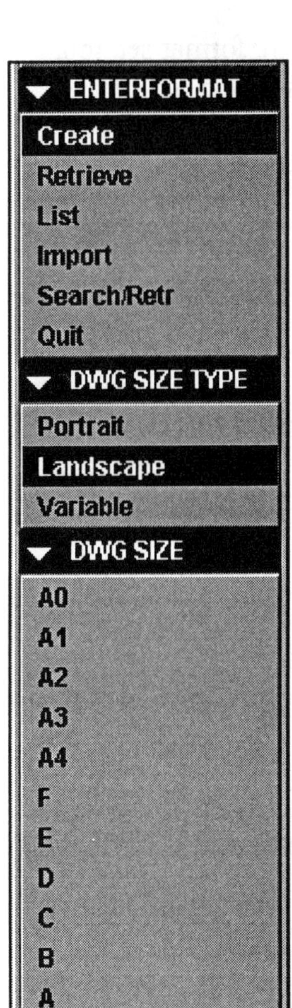

1. Choose an option from the DWG SIZE TYPE menu to determine the orientation of the format sheet:

   Portrait Uses the larger of the dimensions of the sheet size for the format's height; uses the smaller for the format's width.
   Landscape Uses the larger of the dimensions of the sheet size for the format's width; uses the smaller for the format's height.
   Variable Enter specific values for the height and width of the format.

2. After you select **Landscape** or **Portrait**, choose the desired standard size from the DWG SIZE menu.

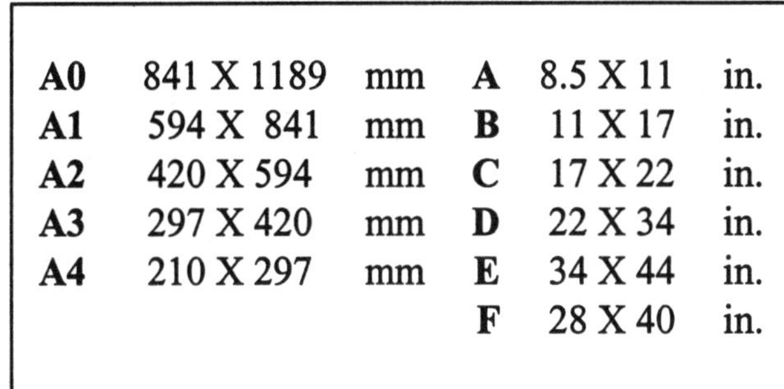

A0	841 X 1189	mm	**A**	8.5 X 11	in.
A1	594 X 841	mm	**B**	11 X 17	in.
A2	420 X 594	mm	**C**	17 X 22	in.
A3	297 X 420	mm	**D**	22 X 34	in.
A4	210 X 297	mm	**E**	34 X 44	in.
			F	28 X 40	in.

3. Choose the desired units (**Inch** or **Millimeter**) from the FORMAT UNITS menu. The units that you choose supersede those of the standard format size that you chose.

Sketching the Format

To create format geometry, you use draft geometry. To sketch draft geometry, choose **Detail** from the FORMAT menu, **Sketch** from the DETAIL menu, and any option from the DRAFT GEOM menu.

The sheet outline is the border of the standard drawing format you selected. Because it is the actual border, it may not show up on "pen plots" unless you use a paper size larger than the drawing size. Everything within the sheet outline border also plots, but you should make an allowance for the plotter's hold-down rollers.

Sketched Format

When you add a sketched format to a drawing, Pro/E aligns the lower left corner (the origin) of the format to the lower left corner (the origin) of the drawing, then centers all items in the drawing on the new sheet in the locations that correspond to their positions on the original sheet. If necessary, it adjusts the drawing scale, maintaining relative distances between items.

To create a sketched format (Fig. 16.6):

1. Choose **Sketcher** from the MODE menu.
2. Choose **Create**, then enter a new format name.
3. Sketch the boundary and dimension it.
4. Sketch the title block and dimension it appropriately.
5. After Pro/E successfully regenerates the drawing format section, save it using **Dbms** menu options. You cannot save text with a sketched format.

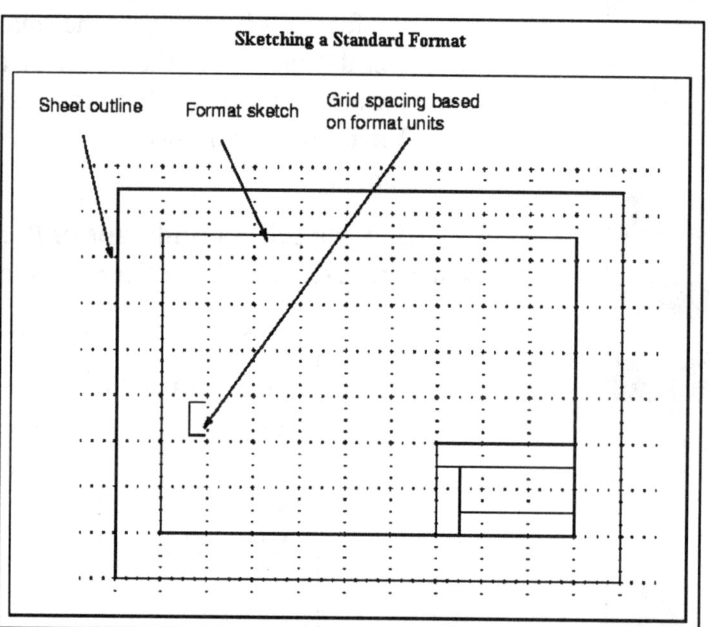

Figure 16.6
Online Documentation, Sketching Formats

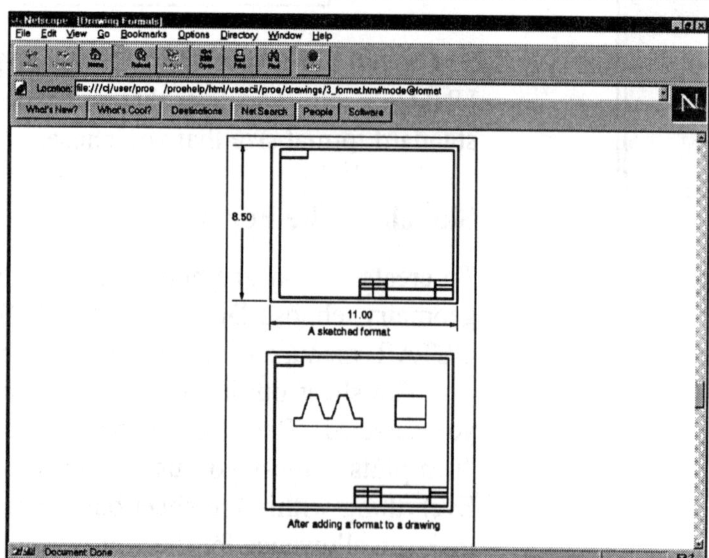

Figure 16.7
Online Documentation, Adding Views to a Sketched Format

Views

▼ DRAWING
Views
Sheets
Modify
Regenerate
Detail
Interface
Table
Layer
Set Up
Set Model
Overlay
Program
Integrate

A wide variety of views (see Fig. 16.7) can be derived from the parametric model. Among the most common are projection views. Pro/E creates projection views by looking to the left of, to the right of, above, and below the picked view location to determine the orientation of a projection view (Fig. 16.8). When conflicting view orientations are found, you are prompted to select the view that will be the parent view. A view will then be constructed from the selected view.

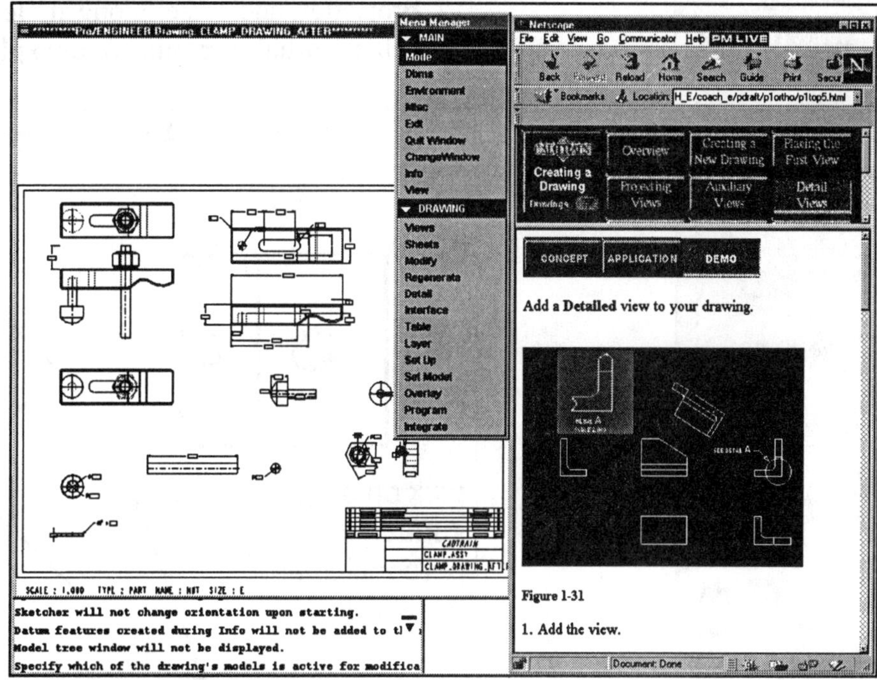

Figure 16.8
COAch for Pro/E, Creating a Drawing (Detail Views)

At the time when they are created, projection, auxiliary, detailed, and revolved views have the same representation and explosion offsets, if any, as their parent views. From that time onward, each view can be simplified, restored, and have its explosion distance modified without affecting the parent view. The only exception to this is detailed views, which will always be displayed with the same explosion distances and geometry as their parent views. Some view types that are available include:

Projection Creates a view that is developed from another view by projecting the geometry along a horizontal or vertical direction of viewing (orthographic projection). The projection type is specified by you in the drawing setup file, and can be based on third-angle (default) or first-angle rules.

Auxiliary Creates a view that is developed from another view by projecting the geometry at right angles to a selected surface or along an axis. The surface selected from the parent view must be perpendicular to the plane of the screen (Fig. 16.9).

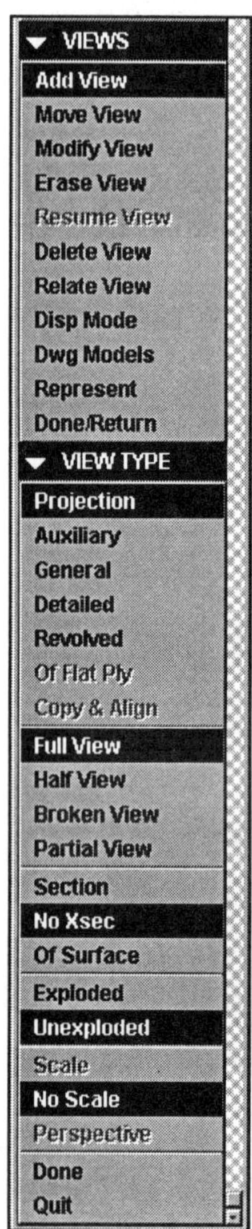

Figure 16.9
Online Documentation, Auxiliary Views

General Creates a view with no particular orientation or relationship to other views in the drawing. The model must first be oriented to the desired view orientation established by you.
Detailed Details a portion of the model appearing in another view. Its orientation is the same as that of the view it is created from, but its scale may be different so that the portion of the model being detailed can be better visualized.
Revolved Creates a planar area cross section from an existing view; the section is revolved **90°** around the cutting plane projection and offset along its length. A revolved view may be full or partial, exploded or unexploded.

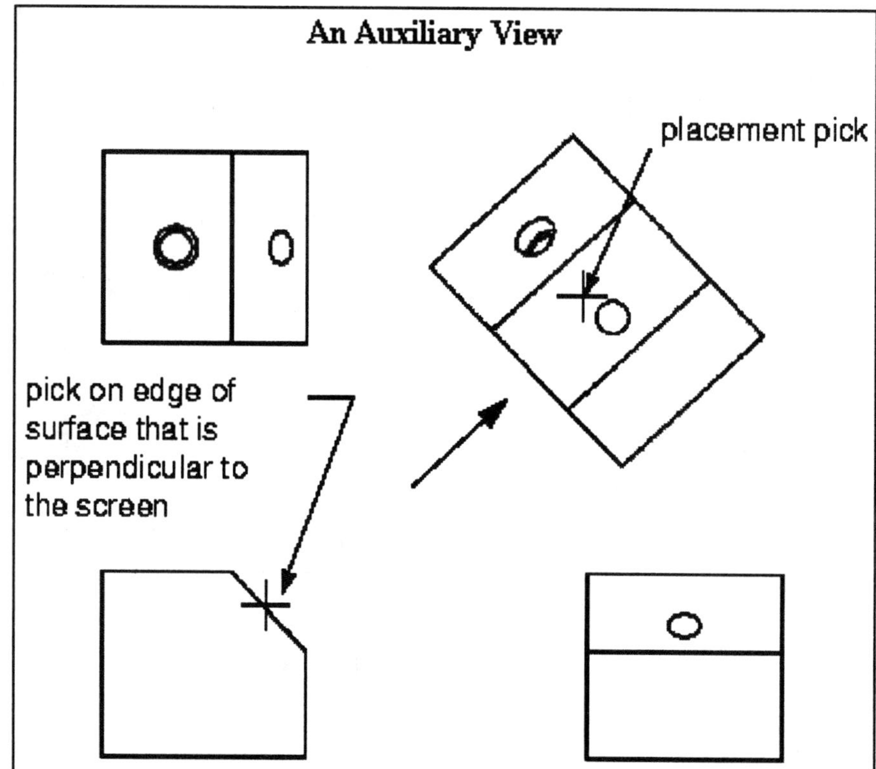

The view options that determine how much of the model is visible in the view are:

Full View Shows the model in its entirety.
Half View Removes a portion of the model from the view on one side of a cutting plane.
Broken View Removes a portion of the model from between two selected points, and closes the remaining two portions together within a specified distance.
Partial View Displays a portion of the model in a view within a closed boundary. The geometry appearing within the boundary is displayed; the geometry outside of it is removed.

XSEC TYPE
Full
Half
Local
Full & Local
Total Xsec
Area Xsec
Align Xsec
Total Align
Unfold Xsec
Total Unfold
Done
Quit

The options that determine whether the view is of a single surface or has a cross section are:

Section Displays an existing cross section of the view if the view orientation is such that the cross-sectional plane is parallel to the screen (Figs. 16.10 and 16.11).
No Xsec Indicates that no cross section is to be displayed.
Of Surface Displays a selected surface of a model in the view. The single-surface view can be of any view type except detailed.

The options that determine whether the view is scaled are:

Scale Allows you to create a view with an individual scale shown under the view. When a view is being created, Pro/E will prompt you for the scale value. This value can be modified later. General and detailed views can be scaled.
No Scale A view will be scaled automatically using a pre-defined scale value that will be shown in the lower left corner of the screen as "SCALE."
Perspective Creates a perspective general view.

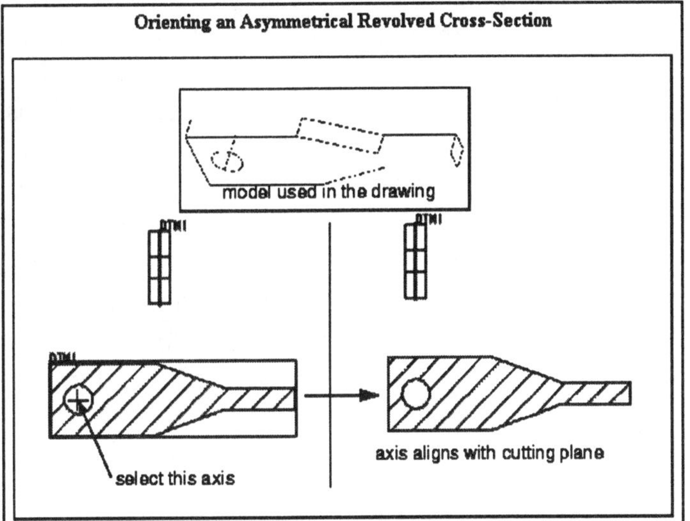

Figure 16.10
Online Documentation, Section Views

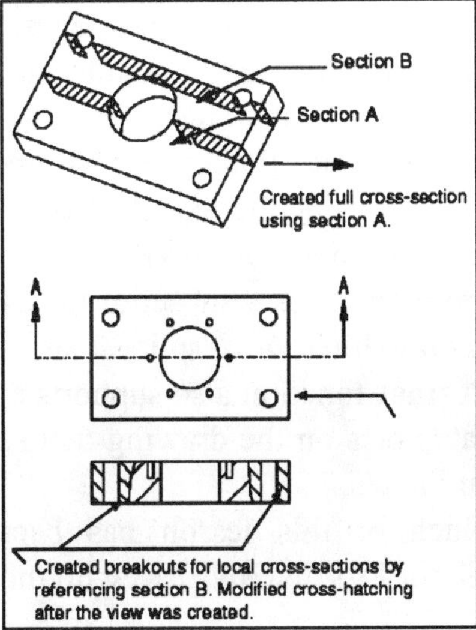

Figure 16.11
Online Documentation, Sections

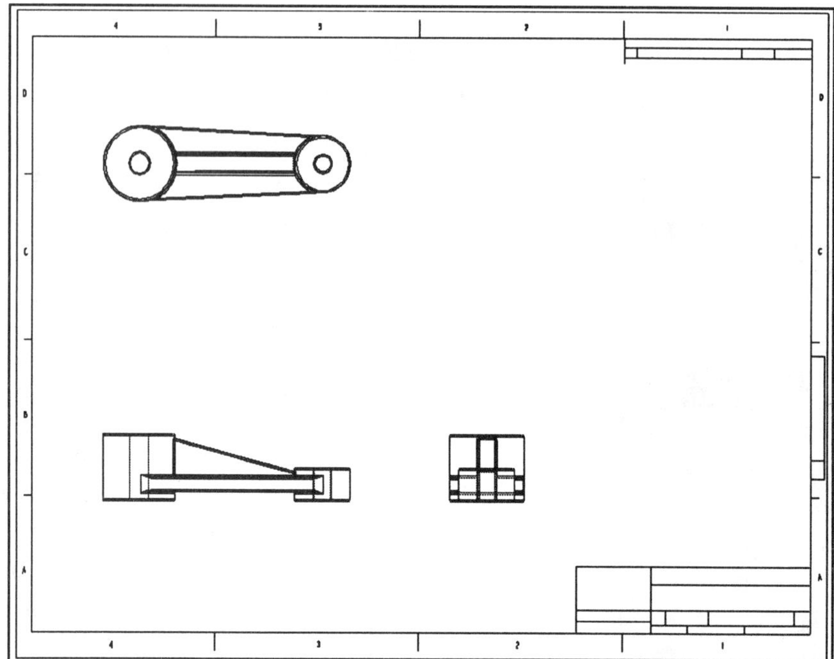

Figure 16.12
Clamp Arm Drawing on an ANSI Standard Drawing Format

Clamp Arm Drawing

Pro/E allows you to create your own drawing formats to support your company or school standards. This includes the lines that make up the border and title block, as well as text. The text can also contain parameter information that refers to Pro/E-generated values, such as the drawing scale, the drawing name, and the model name.

In most cases, you will use the standard formats that come with Pro/E. A standard "**C**" size format has been added to the drawing of the Clamp Arm part in Figure 16.12.

When you add a format to your drawing, Pro/E can write the appropriate notes based on information stored or defined in the model you use.

You can create two basic types of formats: a standard format, which is *locked* to a specific drawing size, or a sketched format, which is parametrically linked to the size of the drawing and thus changes as the drawing size changes.

This lesson deals with altering an existing standard format to match your format requirements. After you master the techniques required to make this type of format, you can easily extend these principles to using the Sketcher to draw the border outlines (to make a sketched parametric format).

Pro/E provides a subset of the 2D Drafting functionality to allow you to draw lines, arcs, splines, and so on to define a drawing border. The **Format** function also supports a complete range of text functions to create notes on the drawing that serve as title block information or standard notes.

Much of this lesson has been adapted from CADTRAIN's COAch for Pro/ENGINEER, with the company's permission.

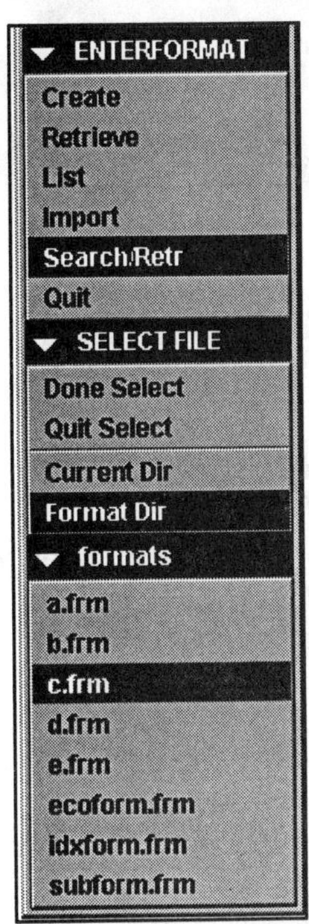

Altering Pro/E Formats from the Format Library

One of the easiest ways to create a drawing format is to use an existing format from the Pro/E format library, save it in your directory under a different name, and alter it as required. This method will keep the sheet an ANSI standard size and format with a standard title block. You can add parameters to the sheet so as to display the appropriate information in the title block and at other locations on the sheet where appropriate. Formats have an **.frm** extension and are read-only files.

Retrieve and save for later modification a "C" size format using the following commands (Fig. 16.13):

Mode ⇒ Format ⇒ Search/Retr ⇒ Format Dir ⇒ c.frm ⇒ Dbms ⇒ Save As ⇒ enter ⇒ (type a unique name for your format: **FORMAT_C) ⇒ enter ⇒ Quit Window ⇒ Format ⇒ Search/Retr ⇒** (pick the new format: **format_c.frm**)

You can make any number of personal formats using this method. Adding parameters to the sheet is the next step in creating your own library of formats that will automatically display the part's name, scale, tolerances, and so on.

Figure 16.13 FORMAT_C

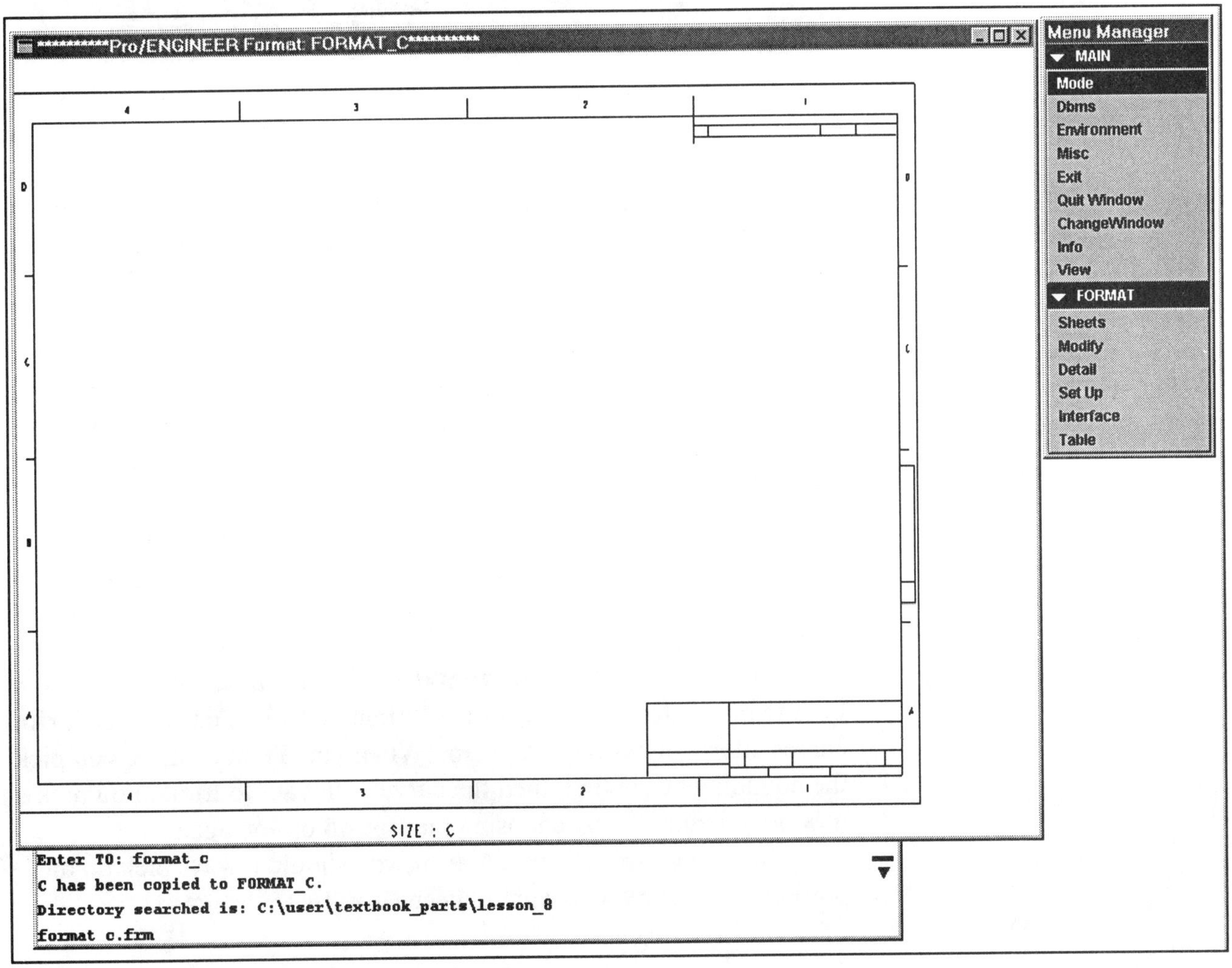

Sketching the Format (Using Draft Geometry)

This part of the lesson leads you through creating a "**C**" size, inches format (Fig. 16.14). You can substitute metric values to create the equivalent-size metric format if desired.

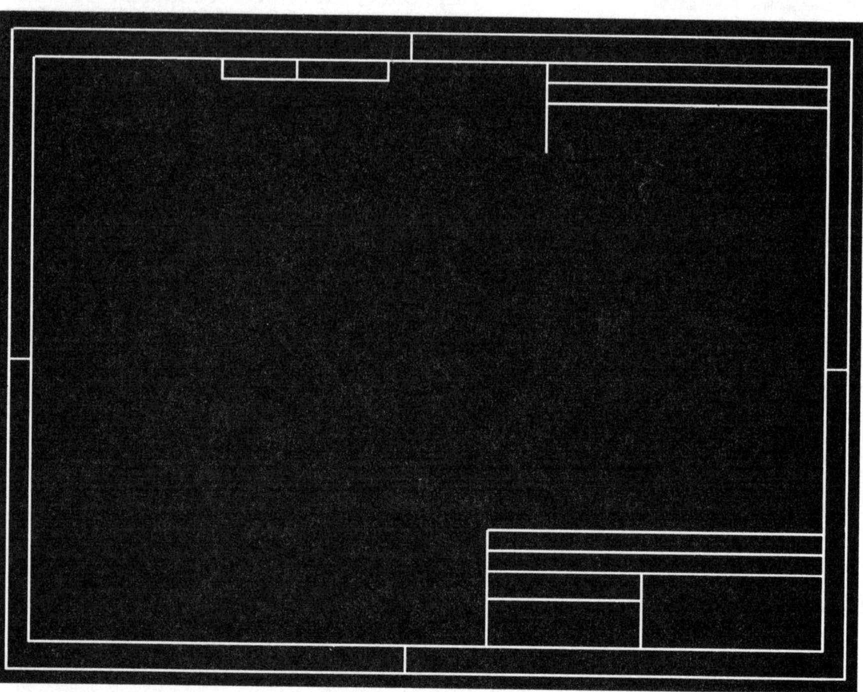

Figure 16.14
CADTRAIN, Production Drawing (Formats)

You can draw lines, arcs, splines, and so on using a drawing format to represent the border and title block. Even though the option to draw geometry is called "sketch," you need to be aware that the **Format** mode does not support the true Sketcher. No allowance is made for constraining the curves (entities) and regenerating them to a dimension-driven size. You **must** draw the curves exactly as they will appear on the format.

You use the basic **2D Drafting** functionality of Pro/E to create the geometry you need. Therefore, you do not have implied constraints such as horizontal or vertical lines. You also do not have implied endpoint connection (even if the curves are close to each other they are not connected).

You can, however, draw curves using the **On Entity** option, which will force picks near an endpoint to snap to that endpoint. You can also use the **Start Chain** option to make sure that endpoints connect.

You use either the **Horizontal** or the **Vertical** option to *force* lines to be created correctly for the format border. You can also **Trim** curves after you have drawn them. When you **Trim** a curve, you pick the bounding curve first, then the curve you want to trim. You pick a new bounding curve by choosing the **Bound** option again.

When you pick the curve to trim, you should always pick on the portion of the curve you want to keep (Fig. 16.15).

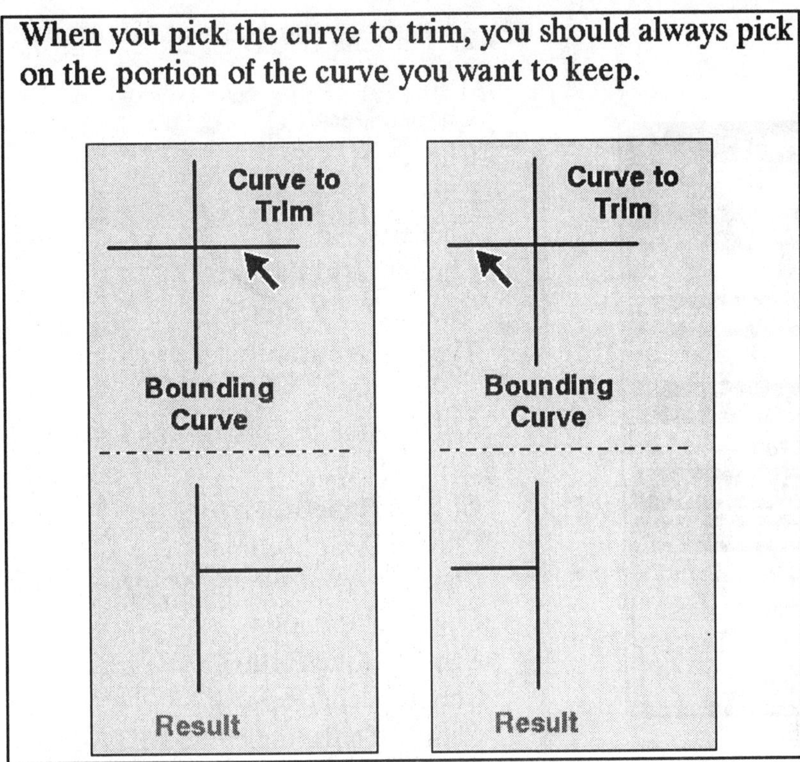

Figure 16.15
CADTRAIN, Production Drawing (Trimming)

Draw the lines of a standard "C" size format using the following commands (Fig. 16.16):

1. Create a new format file.

Choose **Format** from the **Mode** menu
Choose **Create**
Type **my_format_** and ***your initials*** and press **enter**.
Choose **C** from the list of sizes
Choose **Inch** as the format unit

HINT

The box that is displayed represents the edge of the paper. Many companies allow for about .500 inches inside this box for the border of the format.

Figure 16.16
CADTRAIN, Production Drawing (Edge of the Paper)

2. Turn the **GRID** on and change the spacing to **.50**.

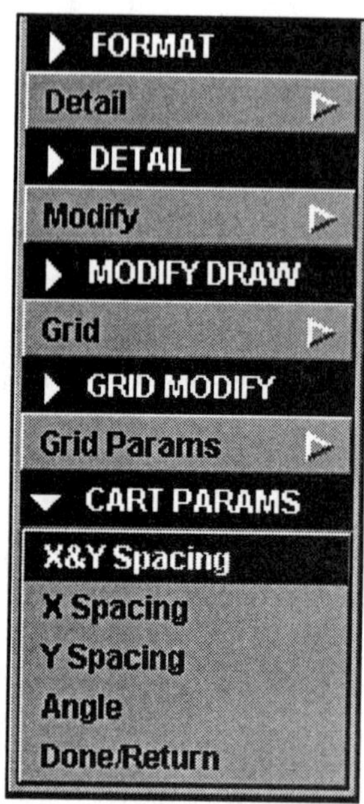

Choose **Detail**
Choose **Modify**
Choose **Grid**
Choose **Grid On**
Choose **Grid Params**
Choose **X&Y Spacing**
Type **.5** ⇒ **enter**

3. Draw the lines of the border (Fig. 16.17).

Choose **Done/Return**
Choose **Done/Return**
Choose **View** from the **MAIN** menu
Choose **Zoom Out**
Choose **Done/Return**
Choose **Done/Return**
Choose **Tools** (under the DETAIL menu)
Choose **Offset**
Choose **Ent Chain**
Select the four lines on the screen
Choose **Done Sel**
Type **-.5** ⇒ **enter** (the boundary is zero with negative defined as inward)
Choose **View** from the **MAIN** menu
Choose **Reset**
Choose **Done/Return**

Figure 16.17
CADTRAIN, Production Drawing (Four Lines of the Border)

4. Create the lines for the title block.

Select the second to bottom horizontal line using the **left** mouse button (Fig. 16.18)
Click the **middle** mouse button to finish selection
Type **-.5 ⇒ enter**
Select the line you just created with the **left** mouse button
Click the **middle** mouse button
Type **-1.5 ⇒ enter**
Select the second right vertical line as shown in Figure 16.18
Click the **middle** mouse button
Type **-1 ⇒ enter**
Select the line you just created
Click the **middle** mouse button
Type **-3 ⇒ enter**
Select the line you just created
Click the **middle** mouse button
Type **-2 ⇒ enter**

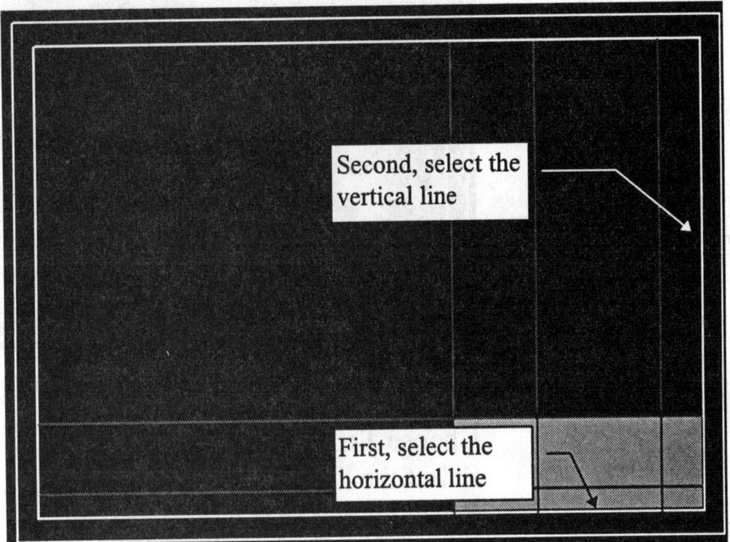

Figure 16.18
CADTRAIN, Production Drawing (Create Lines for the Title Block)

5. **Trim** the lines to form the finished title block (Fig. 16.19).

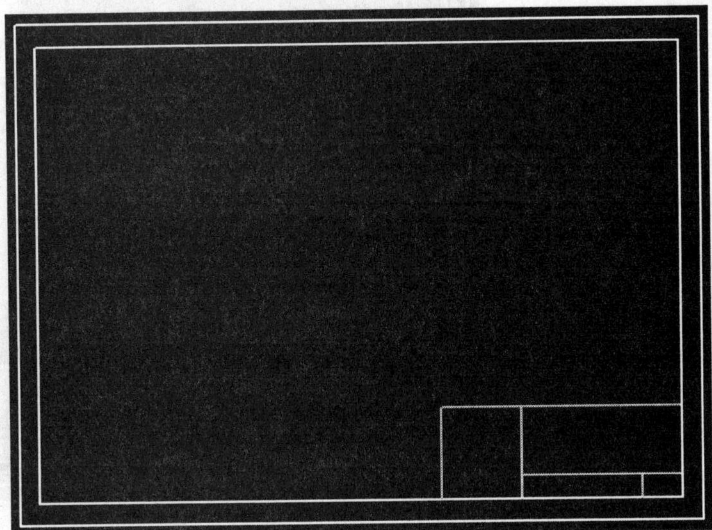

Figure 16.19
CADTRAIN, Production Drawing (Finished Title Block)

6. Draw the lines for the drawing number that is inverted at the top left corner of the format.

Choose **Offset** ⇒ **Ent Chain**
Select second the left vertical line (Fig. 16.20)
Click the **middle** mouse button
Type **-1** ⇒ **enter**
Select the line you just created
Click the **middle** mouse button
Type **-3** ⇒ **enter**
Select the horizontal line (Fig. 16.20)
Click the **middle** mouse button
Type **-.5** ⇒ **enter**

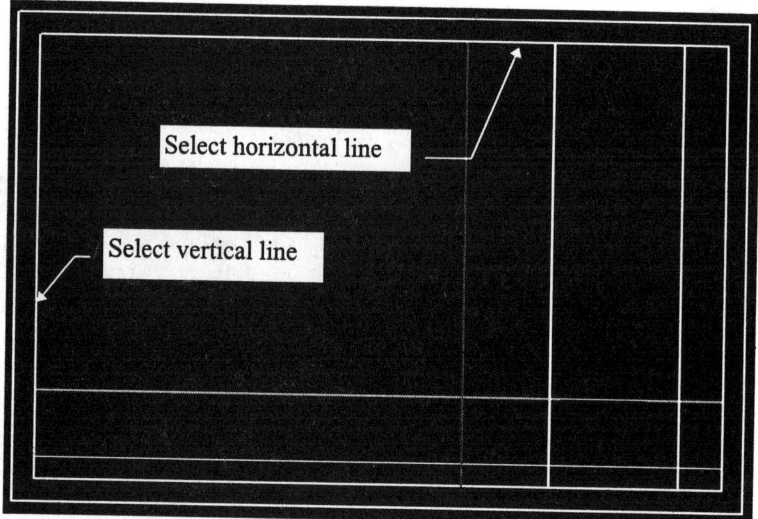

Figure 16.20
CADTRAIN, Production Drawing (Inverted Drawing Number)

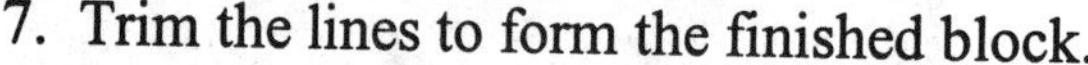

7. Trim the lines to form the finished block.

Choose **Trim**
Select the line as shown in Figure 16.21 as the bounding curve
Select the *two* trim curves as shown in Figure 16.22

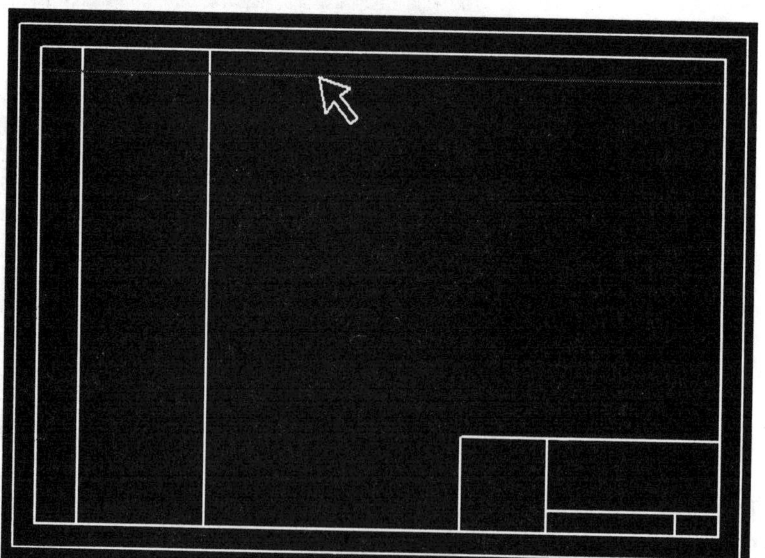

Figure 16.21
CADTRAIN, Production Drawing (Bounding Curve)

Choose **Bound**
Select the rightmost vertical line of the inverted drawing number block (Figs. 16.22 and 16.23) as the bounding curve
Select the trim curve as shown in Figure 16.23
Choose **Done/Return**
Choose **Done/Return**
Choose **View ⇒ Repaint ⇒ Done-Return**

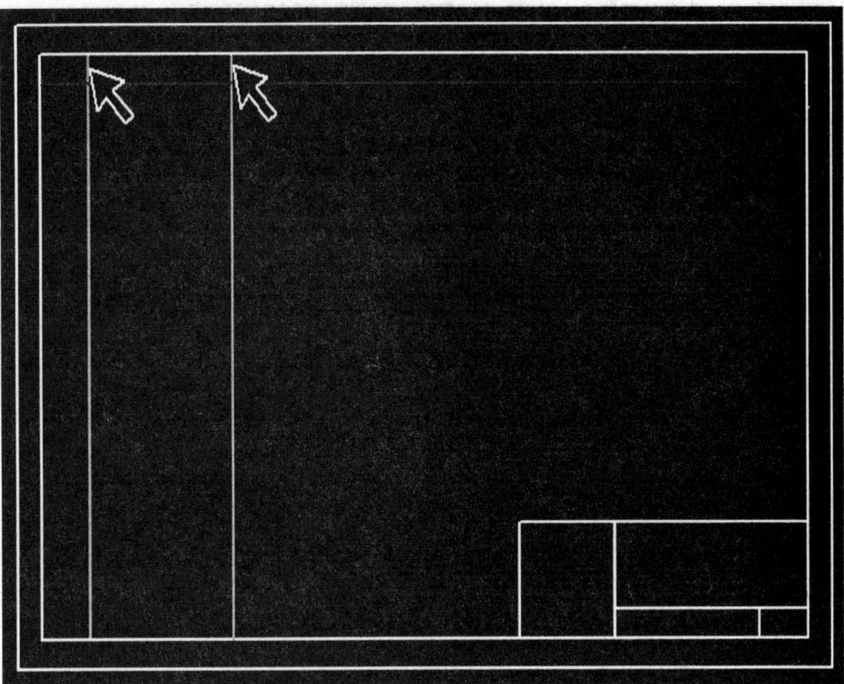

Figure 16.22
CADTRAIN, Production Drawing (Trimming the Block)

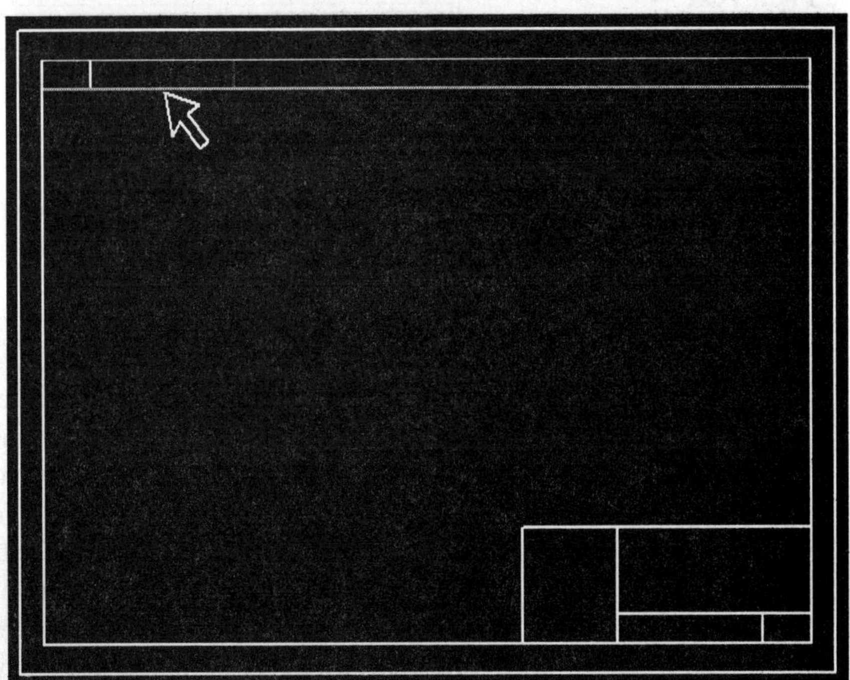

Figure 16.23
CADTRAIN, Production Drawing (Trimming the Block, Final Trim Curve)

Next, you will add *parameters* to the sheet.

Adding Text to the Title Block (Border and Format)

NOTE

If user defined parameters are not defined in the model (part or assembly), then the format will display the *parameter titles* created in the format mode, instead of converting the parameters to model values.

For every drawing, the title block is filled in with standard data or data unique to that design. **Parameters** for the title block information are created and saved in a format in **Format** mode. *Parameter text* will automatically reflect the design when the format is added to a drawing (in **Drawing** mode). Examples of *parameter text* would be tolerance standards, and sheet numbering (Fig. 16.24). Besides parameter text, *plain text* which does not change when the format is added can also be added in the format. An example of *plain text* on a format might be the company name (Fig. 16.24).

User defined parameters can be included in the format, but these user defined parameters must also be imbedded in the part or assembly model. After the model is inserted on the drawing, views made and placed, and the format is added to the drawing, then the user defined parameters will be read by the format parameters and the information displayed on the drawing. An example of a user defined parameter is the DESCRIPTION of a component used in a bill of materials (BOM) on an assembly drawing.

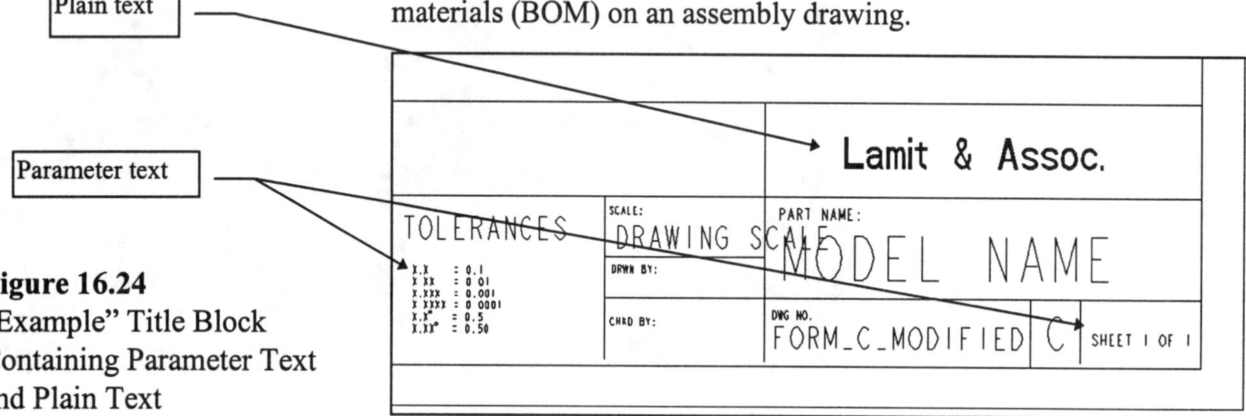

Figure 16.24
"Example" Title Block Containing Parameter Text and Plain Text

You can then create and place the notes almost anywhere on the format, since you will almost always want to alter and move the parameter and plain text after they have been created.

A format has its own parameter **Set Up** file. The **.dtl** entries listed apply to a *format*. A *drawing* **.dtl** file will have these entries and more (see Lesson 19). You may want to establish a standard **Set Up** file for use only with formats. This is especially true if your standards call for text parameters (font size, arrow size, etc.) that are different from those used when you detail a drawing.

You may also want to create items such as your company logo and any other special symbols you plan to use as symbols (e.g., projection angle and inspection symbols) before you create your formats. You can then add these symbols without having to redraw them for each format size (**A**, **B**, **C**, etc.).

You should also be aware that you can make copies of text as you create the format. However, when you make a copy, any parameters in the copy become simple text. This means that the copy will *not* utilize the parametric value.

Create the notes you need on the drawing (Fig. 16.25). The notes initially will all have the same text parameters; you can edit them later. You also do not need to place the notes anywhere special. You will find it easier to move them into place later in the process.

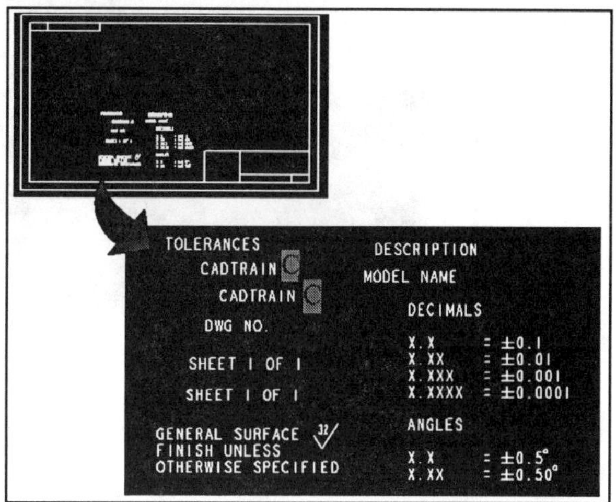

Figure 16.25
Creating Drawing Format Notes

By saving a personalized **.dtl** file, you can use the settings on other formats by retrieving the file.

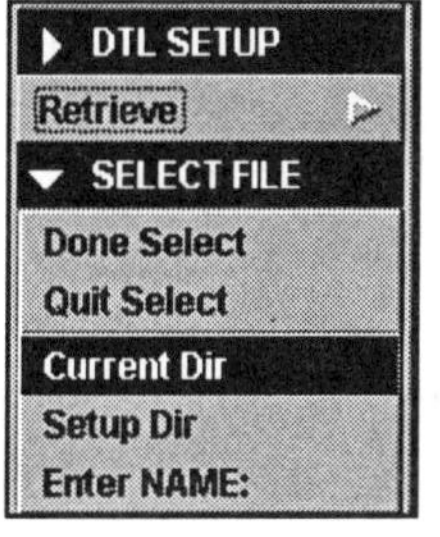

You will be working on one of the formats you previously created or one that you saved under a different name from the Pro/E format directory. When you save this file, it will become a format (**.frm**) and can be used to replace a blank format for a project or as the original format retrieved from your directory.

Retrieve the **cadtrain.dtl Set Up** file if you have CADTRAIN's COAch for Pro/ENGINEER installed on your system, or use a **.dtl** file provided by your instructor. You may also alter or create your own file (Fig. 16.26) by choosing:

Set Up ⇒ **Modify Val** ⇒ (edit the file as required) ⇒ **File** ⇒ **Save** ⇒ **File** ⇒ **Exit** ⇒ **Save** (enter filename) ⇒ **enter**

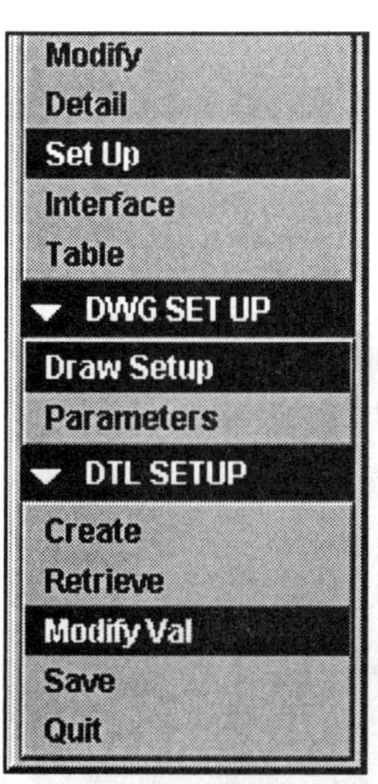

Figure 16.26
.dtl File

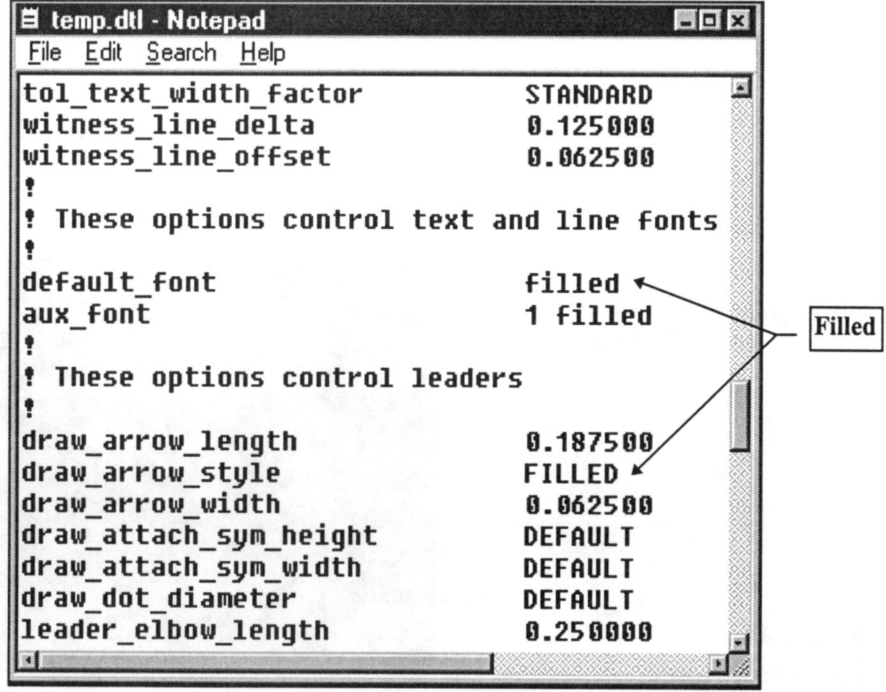

1. Create the plain text notes (Fig. 16.27).

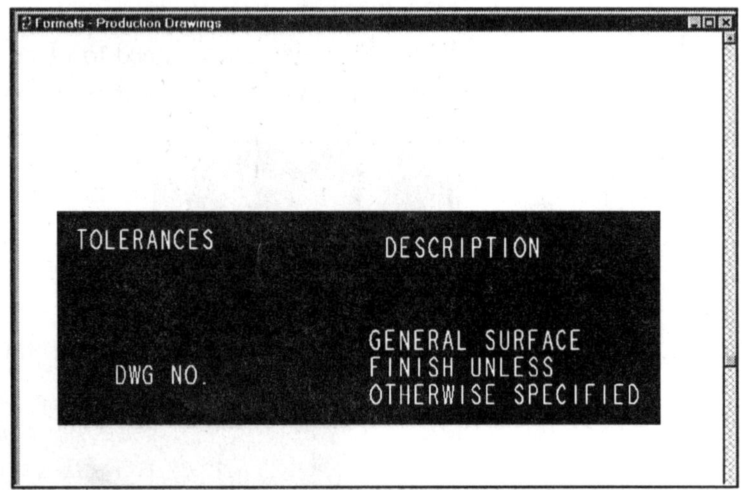

Figure 16.27
Plain Text Notes

Press your ***Caps Lock*** key to turn it on.

Choose **Detail**
Choose **Create**
Choose **Note**
Choose **Make Note**
Indicate the note origin anywhere on the screen (pick once with the *left* mouse button)
Type **TOLERANCES** and press **enter** twice
Choose **Make Note**
Indicate the note origin anywhere on the screen
Type **DESCRIPTION** and press **enter** twice
Choose **Make Note**
Indicate the note origin anywhere on the screen
Type **DWG NO.** and press **enter** twice
Choose **Make Note**
Indicate the note origin anywhere on the screen
Type **GENERAL SURFACE** and press **enter**
FINISH UNLESS and press **enter**
OTHERWISE SPECIFIED and press **enter** twice

NOTE
If you have CADTRAIN, retrieve the **cadtrain.dtl** Set Up file:
Set Up ⇒ **Retrieve** ⇒ (type **?**) ⇒ **enter** ⇒ (locate and select **cadtrain.dtl** file) ⇒ **Quit**

NOTE
A variety of Pro/E standard drawing and format parametric notes are available including:
&dwg_name
&todays_date
&scale
&type (drawing model type)
&format (format size)
&linear_tol_0_0 through
&linear_tol_0_000000
&angular_tol_0_0 through
&angular_tol_0_000000
¤t_sheet
&total_sheets
&det_scale (detail scale)

2. Create the tolerances note (Fig. 16.28).

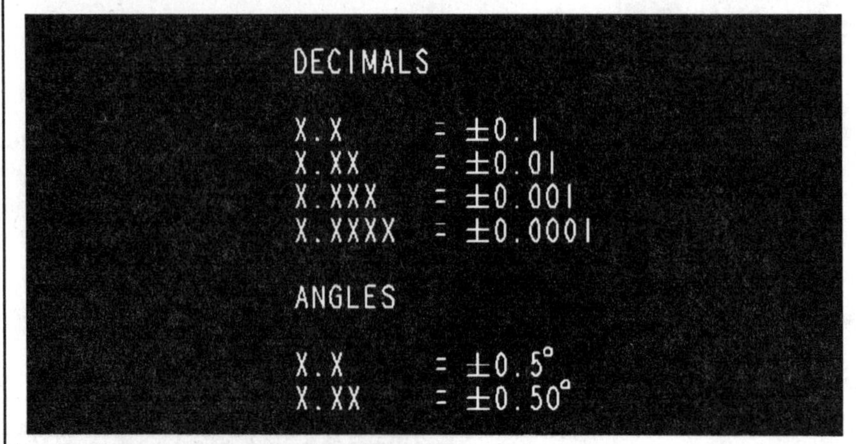

Figure 16.28
Tolerances Note

HINT
Set your font and arrow style to filled:

default_font **filled**
draw_arrow_style **filled**

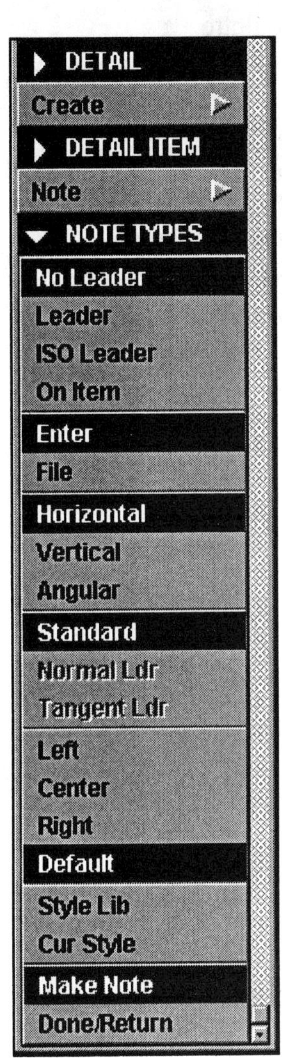

Choose **Make Note**
Indicate the note origin anywhere on the screen
Type **DECIMALS** ⇒ press **enter**

You now need a ***blank*** line. *If* you just press **enter** again, Pro/E will assume that you are finished making the note, which you are not.

Press the ***Spacebar*** and then press **enter**

The ***Spacebar*** counts as a character on the current line and makes Pro/E think that you have entered actual text. Therefore, **enter** is not taken to mean that you are finished making the note.

Now you must input the parameterized variables.

Press ***Caps Lock*** key to turn it off.

This is very important: *You cannot enter the name of a system parameter (such as **&linear_tol_0_0**) in capital letters.* Pro/E will not recognize it as a parameter.

Type **X.X**
Press the ***Spacebar*** five times
Type = ***Spacebar***
Choose the ***plus and minus*** symbol ± from the palette
Type **&linear_tol_0_0** ⇒ **enter**
Type **X.XX**
Press the ***Spacebar*** four times
Type = ***Spacebar***
Choose the ***plus and minus*** symbol ± from the palette
Type **&linear_tol_0_00** ⇒ **enter**
Type **X.XXX**
Press the ***Spacebar*** three times
Type = ***Spacebar***
Choose the ***plus and minus*** symbol ± from the palette
Type **&linear_tol_0_000** ⇒ **enter**
Type **X.XXXX**
Press the ***Spacebar*** twice
Type = ***Spacebar***
Choose the **plus and minus** symbol ± from the palette
Type **&linear_tol_0_0000** ⇒ **enter**

You now need another *blank* line. *If* you press **enter** again, Pro/E will assume that you are finished making the note, and you are not.

Press the ***Spacebar*** and press **enter**
Type **ANGLES** ⇒ **enter**
Press the ***Spacebar*** and press **enter**
Type **X.X**
Press the ***Spacebar*** five times
Type = ***Spacebar***
Choose the ***plus and minus*** symbol ± from the palette
Type **&angular_tol_0_0** and press **enter**

Type **X.XX**
Press the ***Spacebar*** four times
Type = ***Spacebar***
Choose the **plus and minus** symbol ± from the palette
Type **&angular_tol_0_00** and press **enter** (and press **enter** again to finish creating the note)

3. Create the note that will display the model name (Fig. 16.29) in the description area in the title block. Since there is no model at the moment, the text of the name of the parameter is displayed. ***When this format is placed on a drawing and a view is added (thus adding a model), this text will reflect the name of the model.***

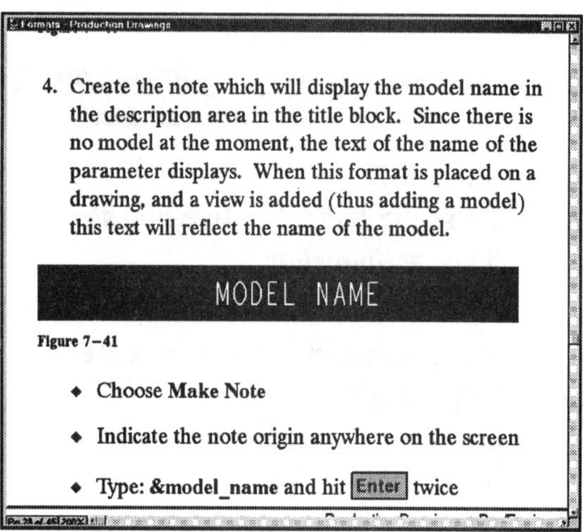
4. Create the note which will display the model name in the description area in the title block. Since there is no model at the moment, the text of the name of the parameter displays. When this format is placed on a drawing, and a view is added (thus adding a model) this text will reflect the name of the model.

MODEL NAME

Figure 7–41

- Choose **Make Note**
- Indicate the note origin anywhere on the screen
- Type: **&model_name** and hit Enter twice

Figure 16.29
Model Name

Choose **Make Note**
Indicate the note origin anywhere on the screen
Type **&model_name** and press **enter** twice

4. Create the note that will display the drawing name in the drawing number area in the title block (Fig. 16.30). Your version of this note will actually show the name of the format that is the current drawing name. When this format is placed on a drawing, the drawing name will replace this text.

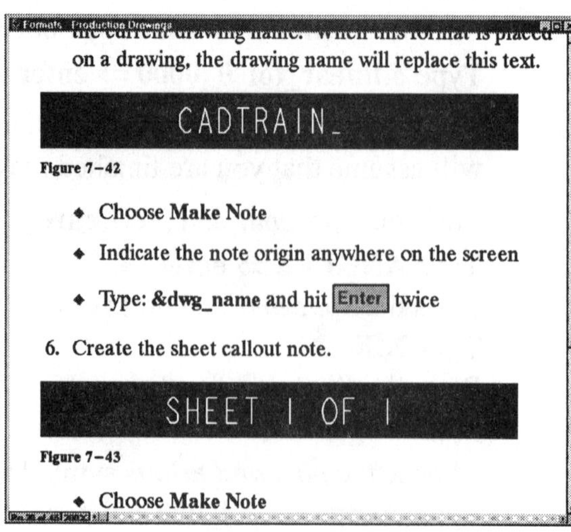
on a drawing, the drawing name will replace this text.

CADTRAIN_

Figure 7–42

- Choose **Make Note**
- Indicate the note origin anywhere on the screen
- Type: **&dwg_name** and hit Enter twice

6. Create the sheet callout note.

SHEET 1 OF 1

Figure 7–43

- Choose **Make Note**

Figure 16.30
Model Name and Sheet Callout

Choose **Make Note**
Indicate the note origin anywhere on the screen
Type **&dwg_name** and press **enter** twice

5. Create the sheet callout note (Fig. 16.30).

Choose **Make Note**
Indicate the note origin anywhere on the screen
Type **SHEET (*Spacebar*) ¤t_sheet (*Spacebar*) OF (*Spacebar*) &total_sheets** and press **enter** twice

6. Create another copy of the note that will display the drawing name (Fig. 16.30). This one will be rotated and will be placed in the box in the upper left corner of the drawing (see Fig. 16.25).

Choose **Make Note**
Indicate the note origin anywhere on the screen
Type **&dwg_name** and press **enter** twice

7. Create another copy of the sheet callout note (see Fig. 16.30).

Choose **Make Note**
Indicate the note origin anywhere on the screen
Type **SHEET (*Spacebar*) ¤t_sheet (*Spacebar*) OF (*Spacebar*) &total_sheets** and press **enter** twice

8. Add a finish symbol to go with the note you created earlier.

Choose **Done/Return**
Choose **Symbol**
Choose **Instance**

Pro/E displays the **Symbol** dialog box (Fig. 16.31).

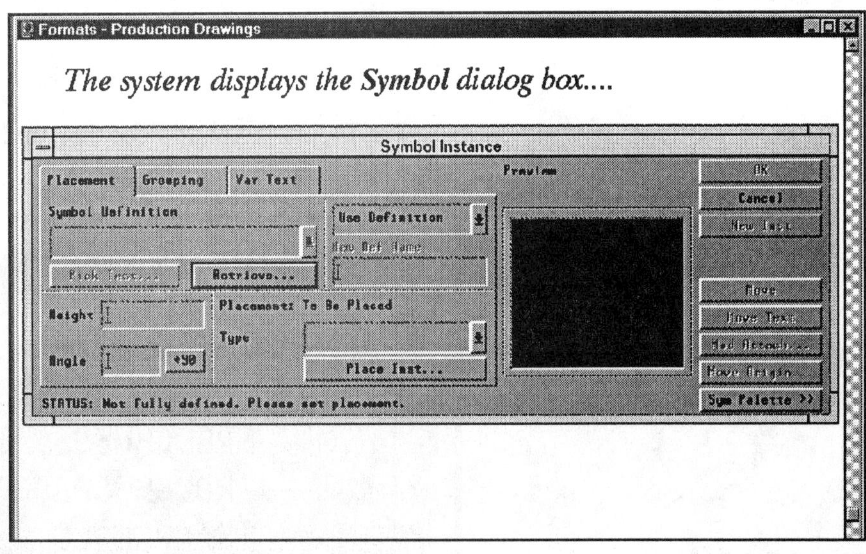

Figure 16.31
Complete Symbol Instance Dialog Box

Choose the **Retrieve** button in the dialog box (Fig. 16.32)
Choose **System Syms** from the menu, not the dialog box
Choose **/surftextsymlib** from the menu
Choose **surftexture.sym** from the menu
Highlight and type **.5** for the symbol **Height** in the dialog box
Choose the **Grouping** tab in the dialog box (Fig. 16.32)

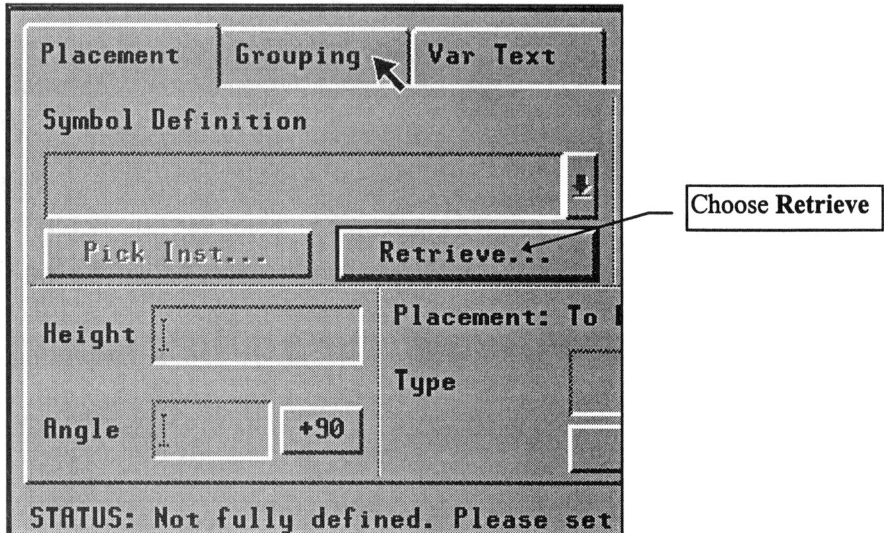

Figure 16.32
Left Side of Symbol Instance Dialog Box

Pro/E now changes the left side of the dialog box so it contains the settings for the finish symbol (Fig. 16.33).

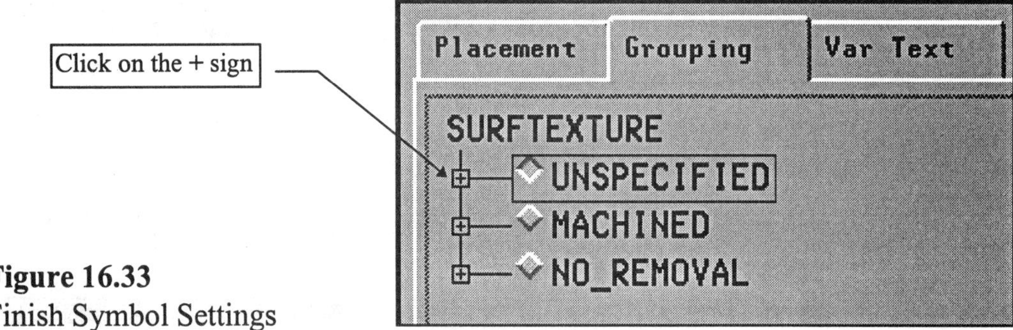

Figure 16.33
Finish Symbol Settings

Click on the + just to the left of the text **UNSPECIFIED**

Pro/E now expands the list under **UNSPECIFIED** to show the options available (Fig. 16.34).

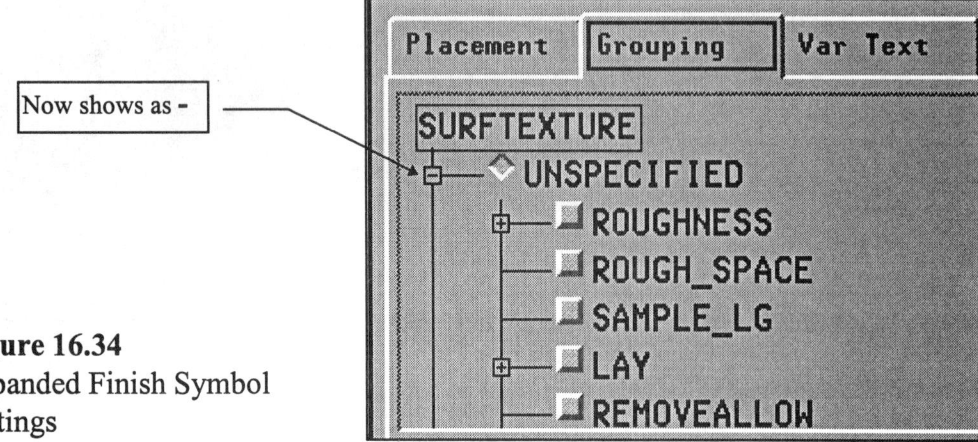

Figure 16.34
Expanded Finish Symbol Settings

Click on the toggle box □ just to the left of the **ROUGHNESS** text (Fig. 16.35).

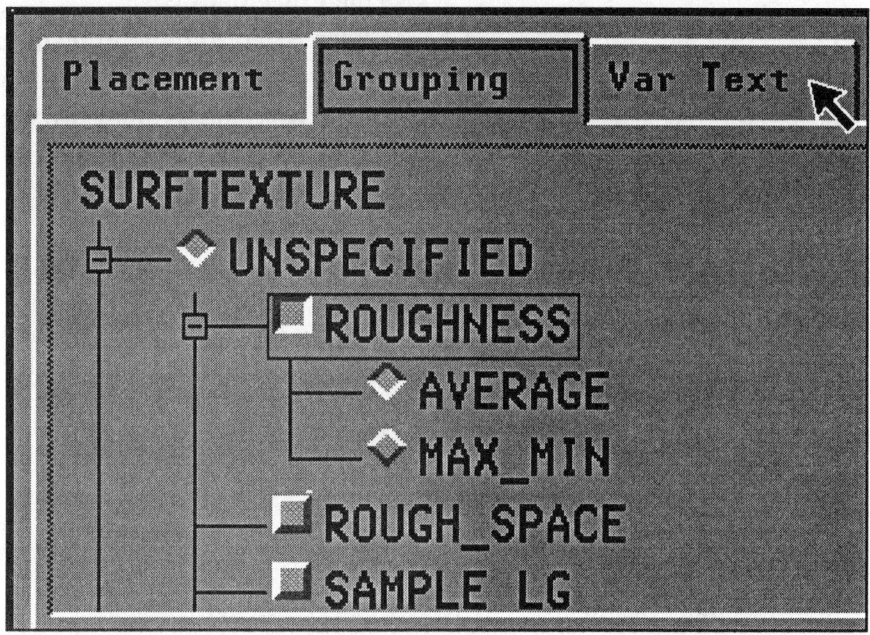

Figure 16.35
Expanded Finish Symbol Roughness

Choose the **Var Text** tab (Fig. 16.36)

Pro/E now changes the left side of the dialog box so it contains the variable text entries for the finish symbol (Fig. 16.36).

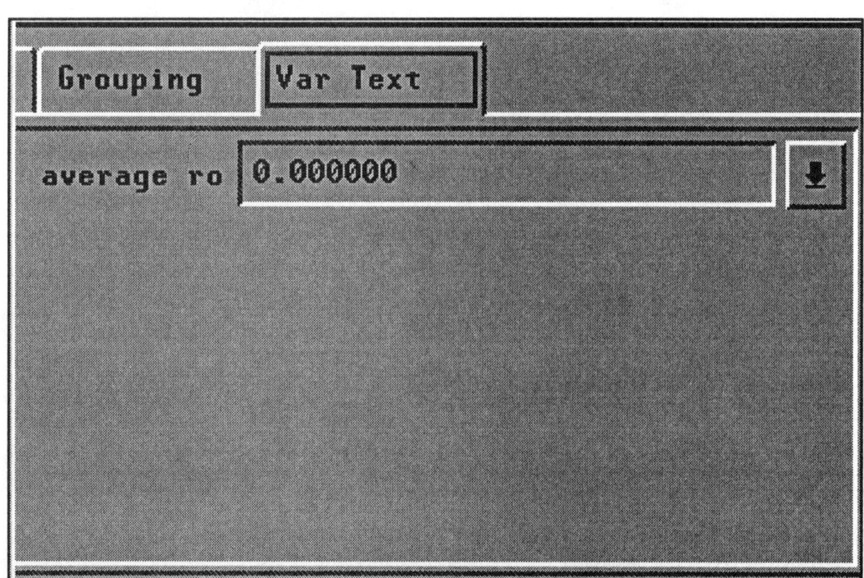

Figure 16.36
Finish Symbol Variable Text

If this symbol had been defined with a series of default parameters, you would be able to select them from an option menu. Since this symbol has the roughness value only as a single number, you need to type it. Figure 16.37 shows the Surface Texture dialog box.

Click in the text area containing **0.000000**, type **32.000000** for the roughness value, and press **enter**
Choose the **Placement** tab
Choose **Free Note** from the **Placement**: option menu
Indicate the origin of the symbol, anywhere on the screen
Choose **OK** (from the Symbol Instance dialog box as shown in Figure 16.37)
Choose **Done/Return**

HINT

If you did not get the correct roughness number:
Choose **Modify**
(from the DETAIL menu)
Choose **Value**
Pick the value "**0**"
Type **32** and press **enter**
Choose **Done Sel**
Choose **Done/Return**

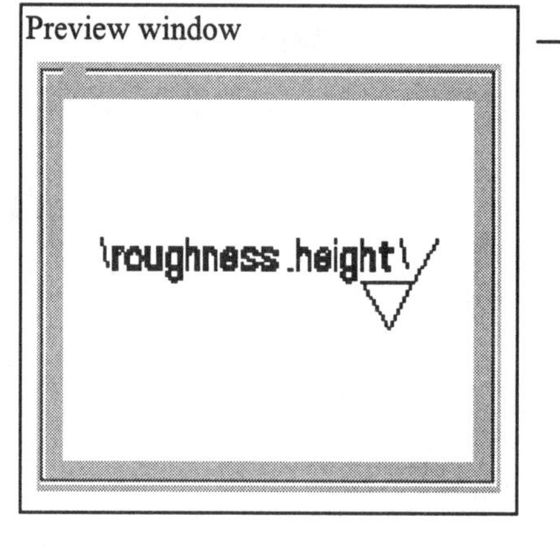

Figure 16.37
Symbol Instance: Surface Texture

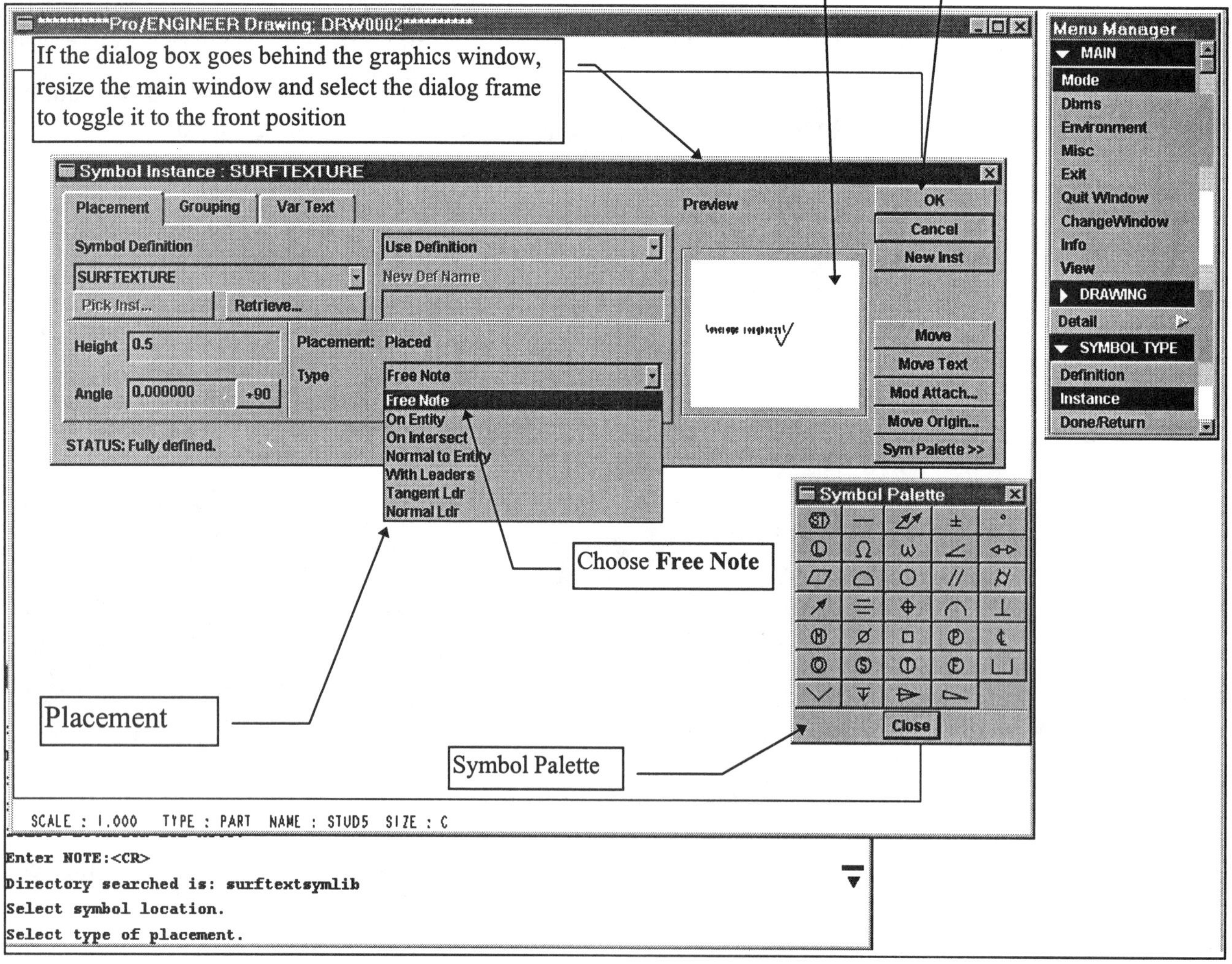

Pick **OK** when done

**********Pro/ENGINEER Drawing: DRW0002**********

If the dialog box goes behind the graphics window, resize the main window and select the dialog frame to toggle it to the front position

Symbol Instance : SURFTEXTURE
Placement | Grouping | Var Text
Symbol Definition: SURFTEXTURE
Use Definition
New Def Name
Pick Inst... | Retrieve...
Height 0.5
Angle 0.000000 +90
Placement: Placed
Type: Free Note
Free Note / On Entity / On Intersect / Normal to Entity / With Leaders / Tangent Ldr / Normal Ldr
STATUS: Fully defined.
Preview
OK | Cancel | New Inst | Move | Move Text | Mod Attach... | Move Origin... | Sym Palette >>

Menu Manager
MAIN: Mode, Dbms, Environment, Misc, Exit, Quit Window, ChangeWindow, Info, View
DRAWING: Detail
SYMBOL TYPE: Definition, Instance, Done/Return

Choose **Free Note**

Symbol Palette
Close

Placement

Symbol Palette

SCALE : 1.000 TYPE : PART NAME : STUD5 SIZE : C

```
Enter NOTE:<CR>
Directory searched is: surftextsymlib
Select symbol location.
Select type of placement.
```

Modifying Text

NOTE

You can **Modify Text** in either the **Format** Mode or the **Drawing** Mode. The **Format** Mode is demonstrated here.

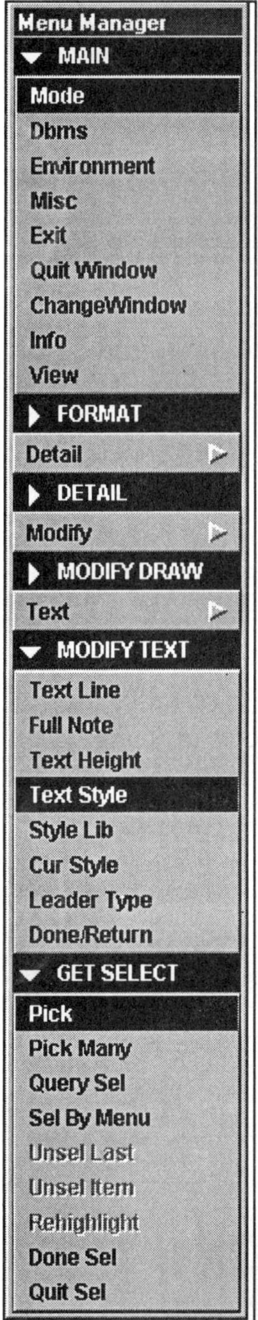

Figure 16.38
Text Style Dialog Box

After you create the text, you often need to modify the text parameters. The most common modification is to the text height. When you modify the parameters of text, Pro/E treats each text segment (**{1:text1}**) as a separate entity. This allows you to set the text height, for example, of different entity sets within a text string.

Since this would make selection quite time-consuming for complex notes, the **Pick Many** option allows you to pick the entire note using a rectangle.

You can also use the **Angle** option to rotate text. This allows you to write text upside down along the top of a format, or at **90°** or **270°** to make it parallel to the right or left border. The **Mirror** option allows you to change text so that it can be read from the back side of the vellum (when you plot it).

Modify the text height of the notes and place them at the proper location in the format.

1. Change the height of all of the small title text to **.09**.

Choose **Modify**
Choose **Text**
Choose **Text Style**
Select the notes: **DESCRIPTION**, the entire **GENERAL SURFACE** note, and **DWG NO.**
Click the *middle* mouse button to stop selecting notes to change

(Pro/E now displays the **Text Style** dialog box, which contains all of the parameters that control how text displays, as shown in Fig. 16.38.)

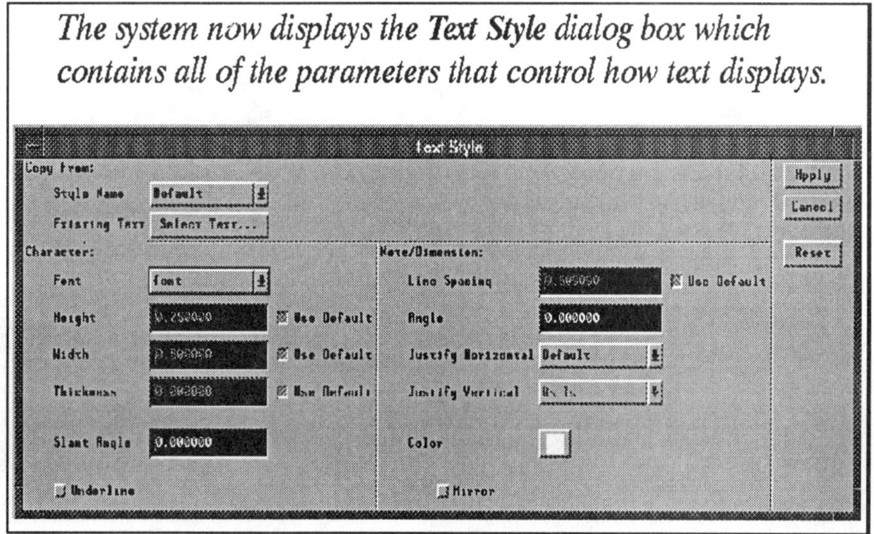

Click on the □ **Use Default** toggle box that is just to the right of the **Height** text area to select it (Fig. 16.39)

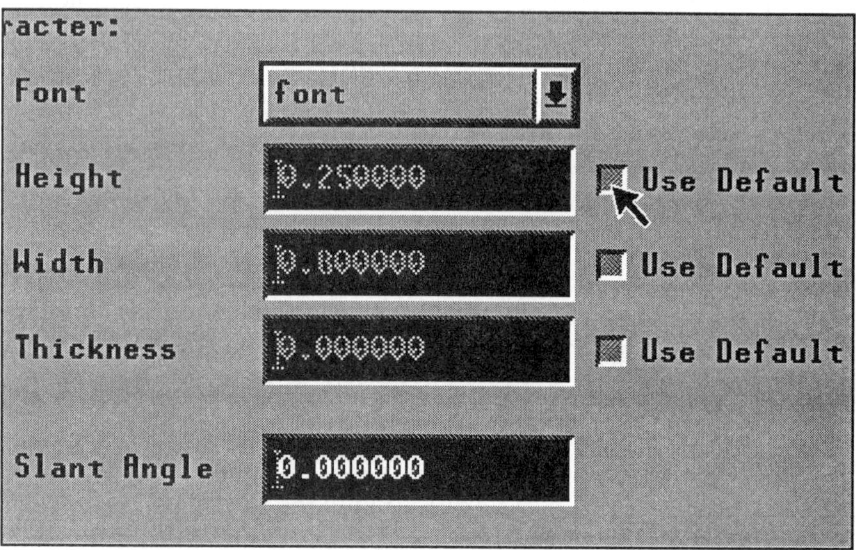

Figure 16.39
Text Height Use Default

Double-click in the **Height** text area and type **.09**
Choose **Apply**
Choose **Close**

2. Change the size of the two **SHEET** callouts to **.09**.

Choose **Pick Many**
Drag a box around both the texts **SHEET 1 OF 1** to select them
Click the *middle* mouse button to stop selecting notes to change
Click on the □ **Use Default** toggle box that is just to the right of the **Height** text area to select it (Fig. 16.40)

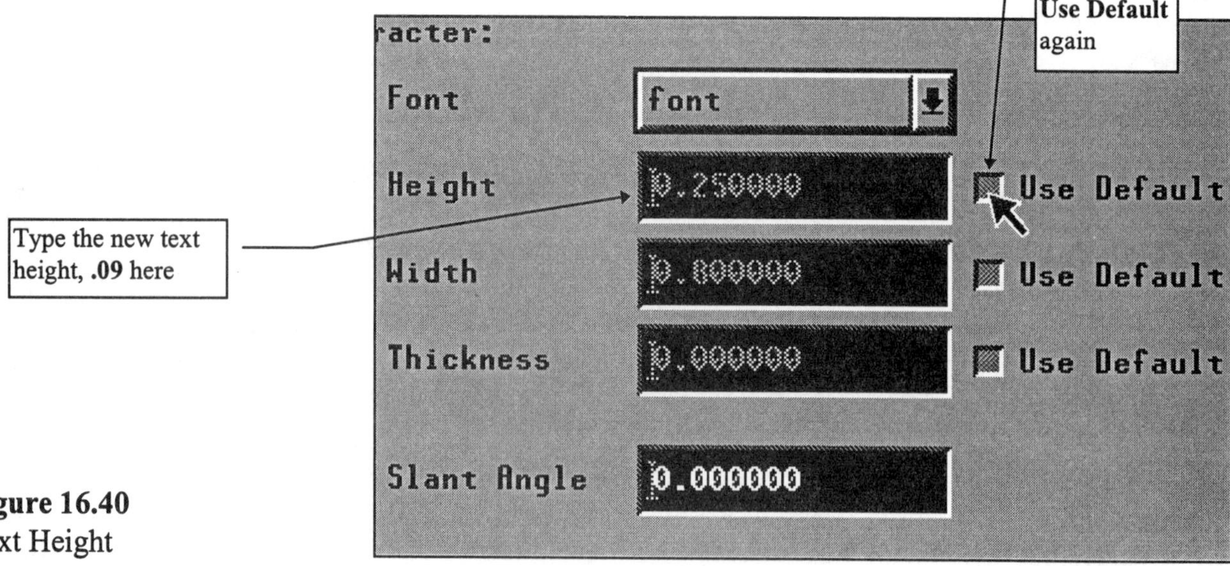

Figure 16.40
Text Height

Double-click in the **Height** text area and type **.09**
Choose **Apply**
Choose **Close**

3. Change the size of the two drawing number notes to **.25**.

Select the two notes that reflect the current name of the format **(MY_FORMAT...)**
Click the *middle* mouse button to stop selecting notes to change
Click on the □ **Use Default** toggle box that is just to the right of the **Height** text area to select it
Double-click in the **Height** text area and type **.25**
Choose **Apply**
Choose **Close**

4. Change the size of the tolerances note.

Choose **Pick Many**
Drag a box around the complete tolerances note (not the text **TOLERANCES)**
Click the *middle* mouse button to stop selecting notes to change
Click on the □ **Use Default** toggle box that is just to the right of the **Height** text area to select it
Double-click in the **Height** text area and type **.06**
Choose **Apply**
Choose **Close**

5. Change the size of the **MODEL NAME** text to **.375**.

6. Rotate one of the drawing name notes and one of the sheet callouts.

Select one of the notes that reflects the current name of the format **(MY_FORMAT)** and the sheet callouts note **(SHEET 1 OF 1)**
Click the *middle* mouse button to stop selecting notes to change
Double-click in the **Angle** text area (Fig. 16.41)

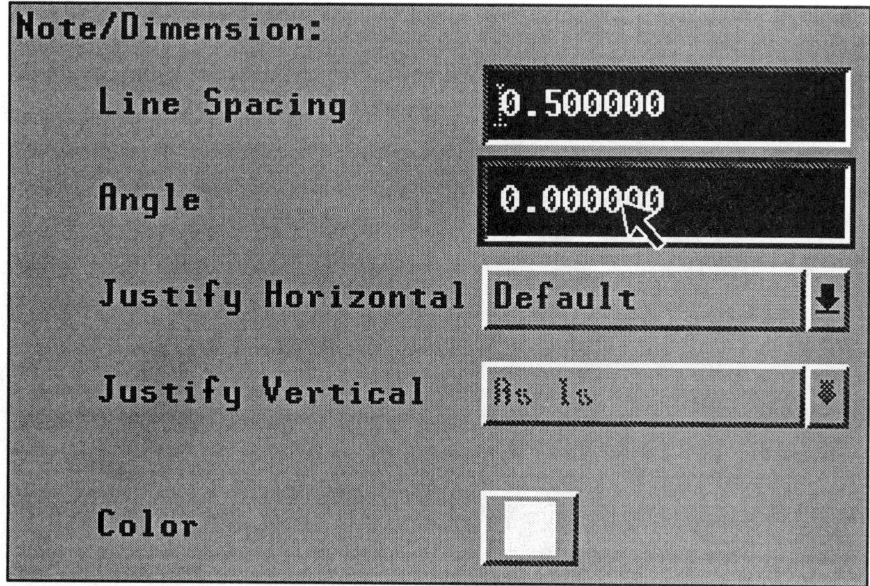

Figure 16.41
Text Angle

HINT

You can use the grid to line up and position text. Change the grid spacing to a convenient distance.

In the **Format** mode and the **Drawing** mode the **Grid** is 2D and aligned with the lower left-hand corner of the sheet. You can also change the **Origin** of the grid. First, check to see whether your **Grid Snap** is on or off in the **Environment** menu. Turn it on.

Second, to *see* the grid you must turn it on using **Modify**. Choose: **Modify ⇒ Grid ⇒ Grid On ⇒ Grid Params ⇒ X&Y Spacing ⇒** (type the new spacing value, the default is **.500**).

Type **180**
Choose **Apply**
Choose **Close**
Choose **Done/Return ⇒ Done/Return**

7. Use the **Detail ⇒ Move** option to drag the text to the proper location in the title block, approximately as shown in Figure 16.42.

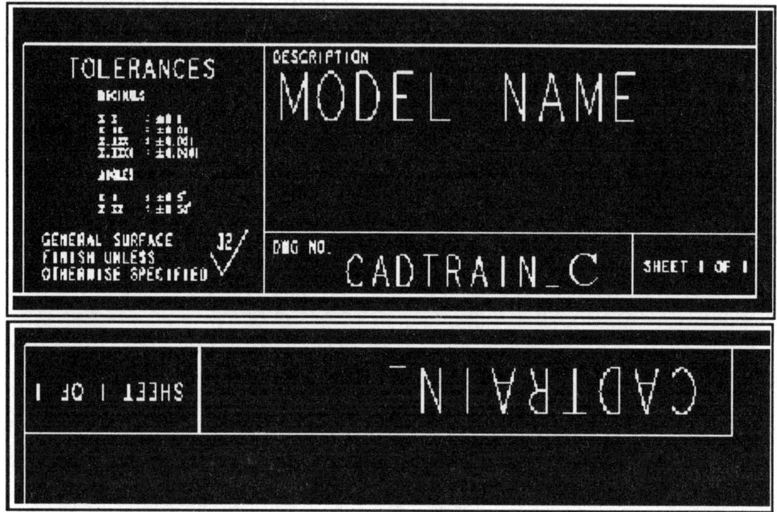

Figure 16.42
Moving the Text to the Title Block

8. Drag the two notes to the upper left corner block, as shown in Figure 16.42. Figure 16.43 shows another format that can be placed on a drawing.

9. Save the format.

Choose **Dbms**
Choose **Save**
Press **enter**
Choose **Done-Return**

Dbms (**File**--PT/Modeler) **⇒ Save ⇒ enter**
Purge ⇒ enter ⇒ Done-Return

Now you have a **"C"** size format with parameters. As the model is added to the first drawing view (Drawing mode), the format will "read" imbedded *parameter values* from any model with the same parameters.

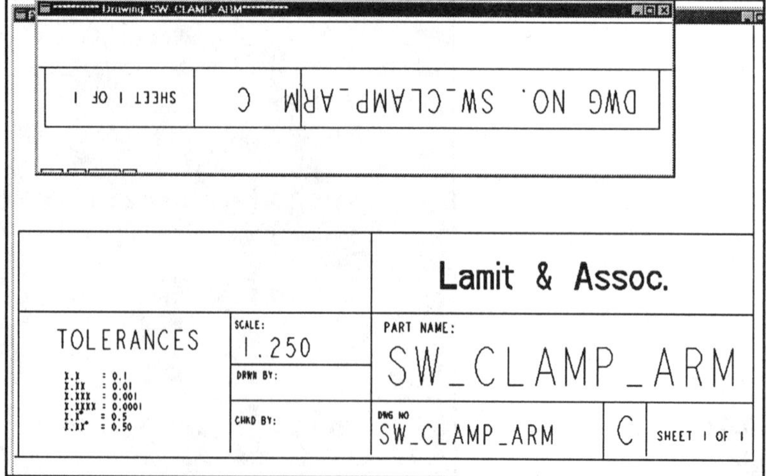

Figure 16.43
Company Formatted Sheet

Drawing Mode and Views

When you start a drawing, you must follow steps similar to creating a part or an assembly (Fig. 16.44). After the drawing is created, you must add the views required to detail the part.

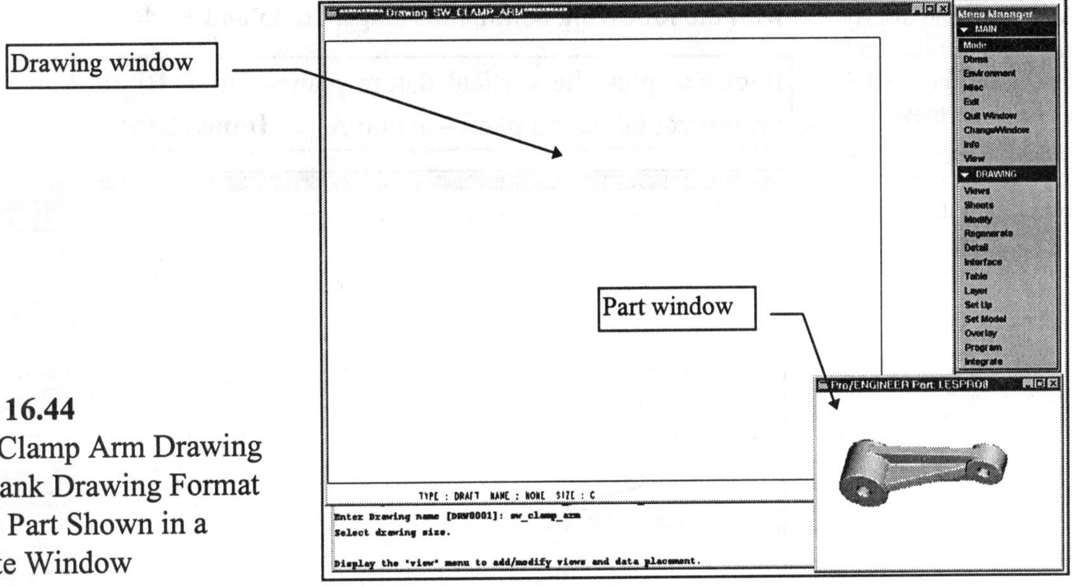

Figure 16.44
Swing Clamp Arm Drawing on a Blank Drawing Format and the Part Shown in a Separate Window

You must be in the same directory as your part. It is actually a good idea to bring up the part in a separate window. Choose the following commands (Fig. 16.45):

Mode ⇒ **Drawing** ⇒ **Create** ⇒ (type a name for your drawing: **SW_CLAMP_ARM**) ⇒ **enter** ⇒ (from the DWG SIZE menu, pick **C**) ⇒ **Mode** ⇒ **Part** ⇒ **Search/Retr** ⇒ (pick the name of the Lesson 8 Project) ⇒ (move and resize the Part window) ⇒ **Change Window** ⇒ (pick in Drawing Window) ⇒ **Views** ⇒ **enter** to accept the default (or **?** ⇒ **enter** ⇒ and choose the model name from the directory list) ⇒ **Done** (to accept the general default view) ⇒ [pick a position on the drawing for the first view (Fig. 16.45)]

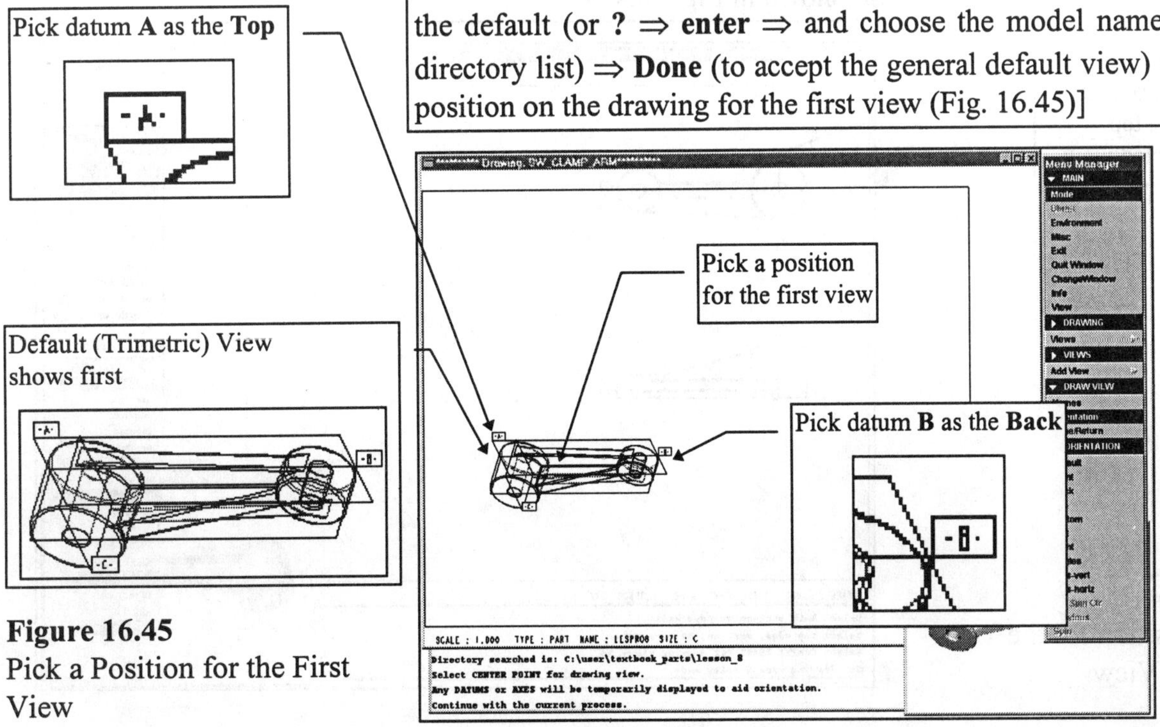

Pick datum **A** as the **Top**

Default (Trimetric) View shows first

Pick a position for the first view

Pick datum **B** as the **Back**

Figure 16.45
Pick a Position for the First View

The next step is to reorient the first (original) view, which is now displayed as a *general view* in the default orientation. Using the part's planar edges or datum planes, orient the view to display a *front view* of the part. This first view is important, since all other projected views are generated from it using orthographic projection. Continue with the following commands (Figs. 16.45 and 16.46):

NOTE

The original view can be any standard view you wish to display first. *This view does not have to be a standard front view*, though the front view is one of the most convenient since the top, bottom, right-side, and left-side views are readily projected from it.

> **Back** ⇒ (pick the vertical datum plane--datum **B**) ⇒ **Top** ⇒ (pick the horizontal datum plane--datum **A**) ⇒ **Done/Return**

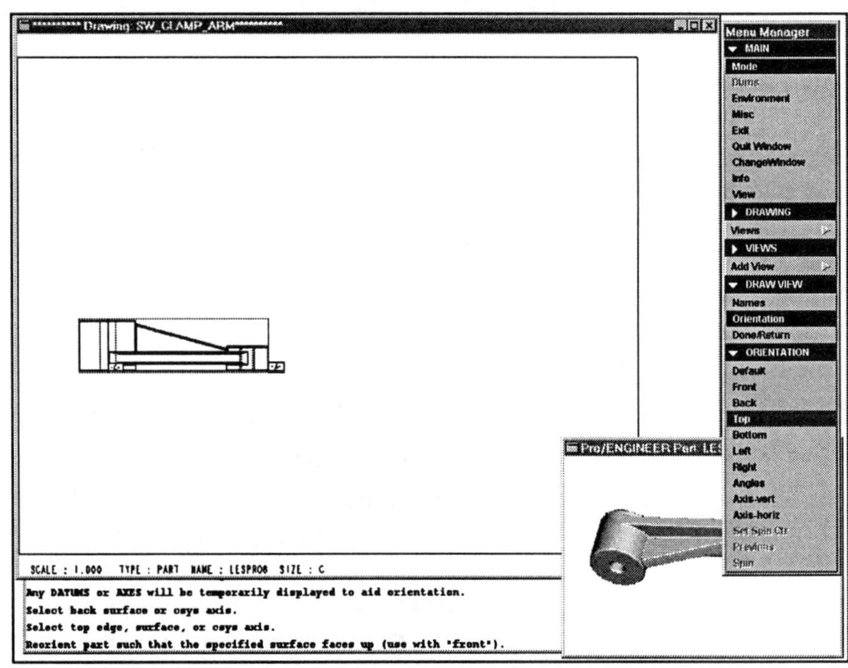

Figure 16.46
First View

> **Add View** ⇒ **Projection** ⇒ **Full View** ⇒ **No Xsec** ⇒ **No Scale** ⇒ **Done** ⇒ (pick a position near the place where the top view should go, as shown in Fig. 16.47)

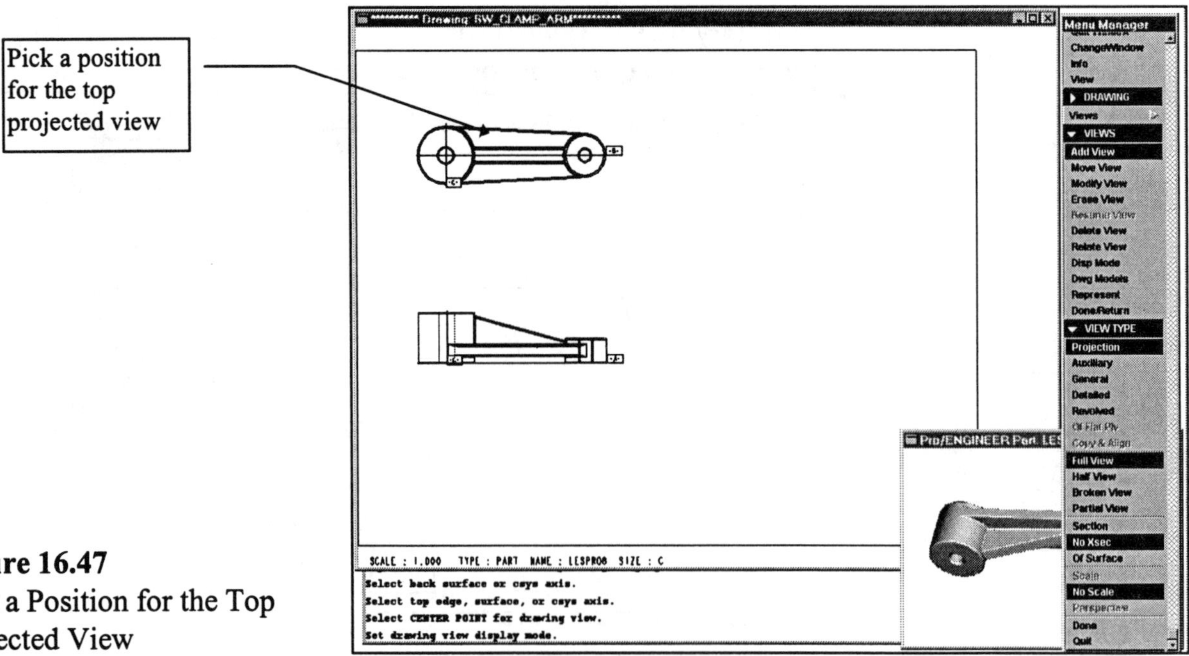

Figure 16.47
Pick a Position for the Top Projected View

Add View ⇒ **Done** (to accept defaults) ⇒ (pick the position for the right side view, as shown in Fig. 16.48)

Dbms (File--PT/Modeler) ⇒
Save ⇒ **enter**
Purge ⇒ **enter** ⇒
Done-Return
(this will save your drawing)

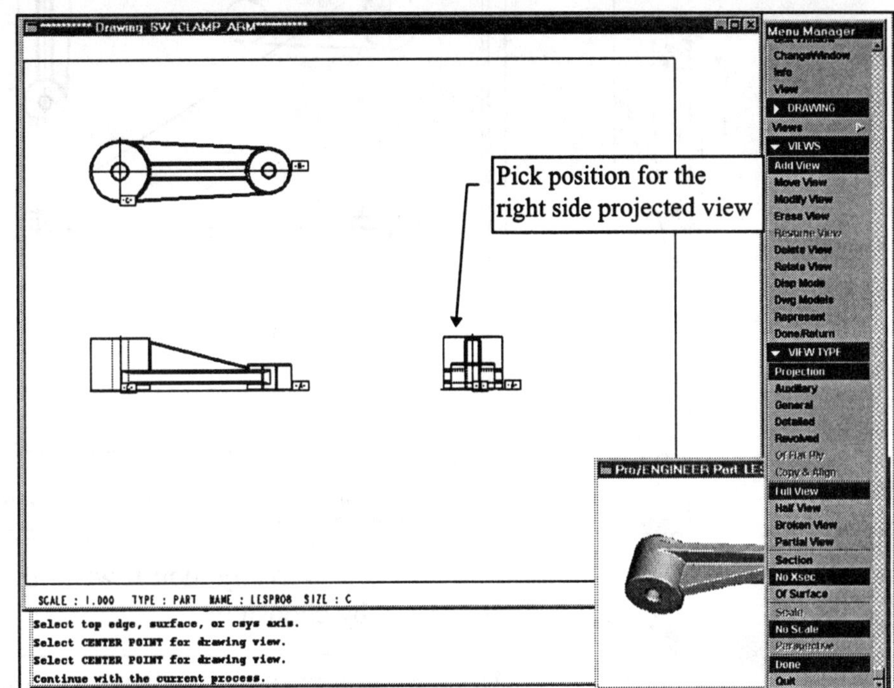

Figure 16.48
Pick a Position for the Right Side Projected View

Add another projected view (Figs. 16.49 and 16.50):

Add View ⇒ **Done** ⇒ (pick the position for the new view) (Pro/E will prompt you with the following: **Conflict in parent view exists. Select parent view for making the projection.**) ⇒ [Select the right side view (Fig. 16.50)]

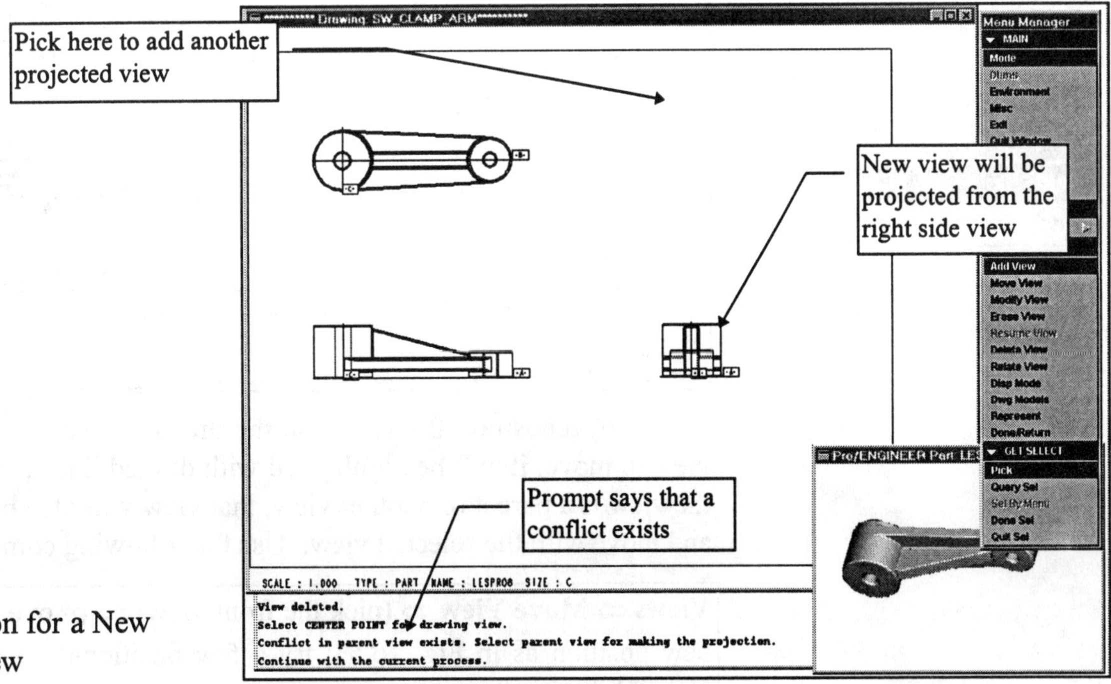

Figure 16.49
Pick a Position for a New Projected View

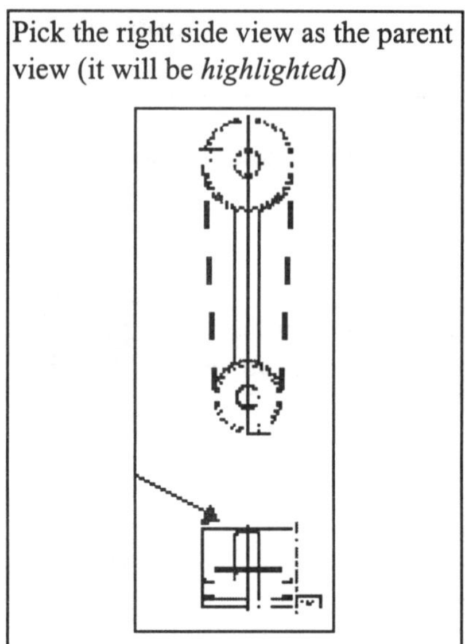

Figure 16.50
New Projected View

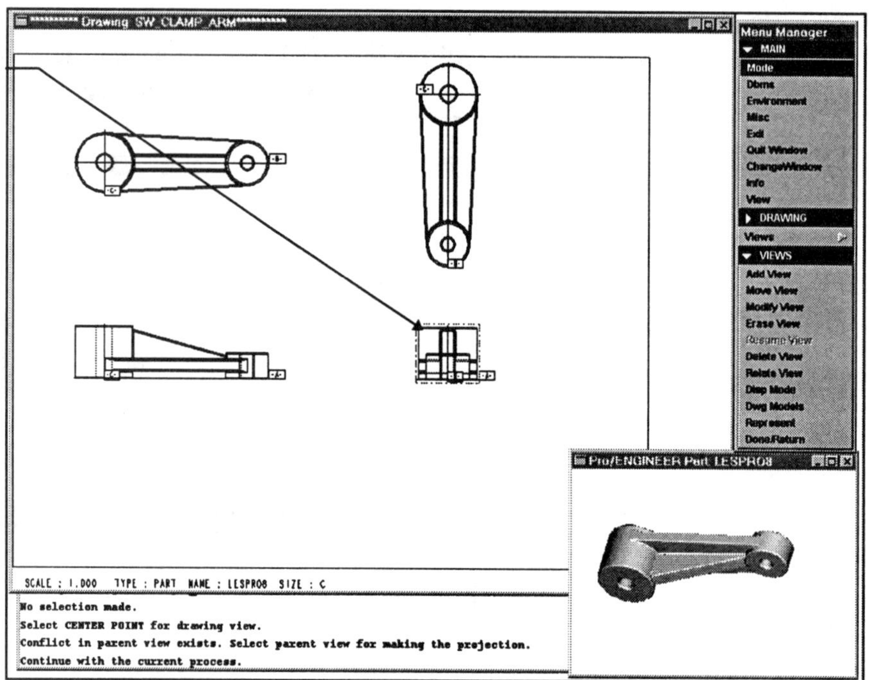

Since you really don't want the view that was just created, remove it by choosing:

Delete View ⇒ (pick the view) ⇒ **Confirm** (Fig. 16.51) ⇒ **Done Sel** ⇒ **Done/Return**

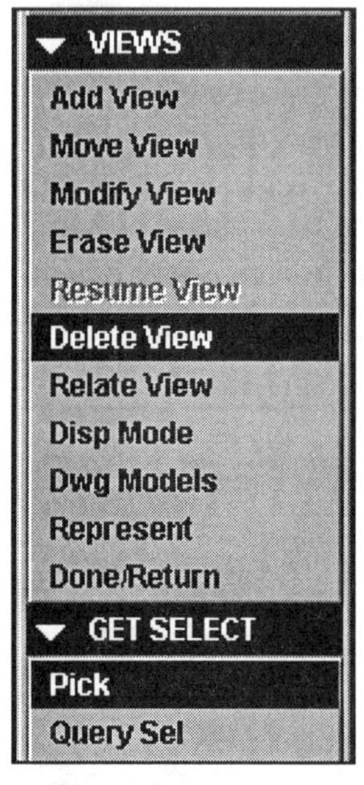

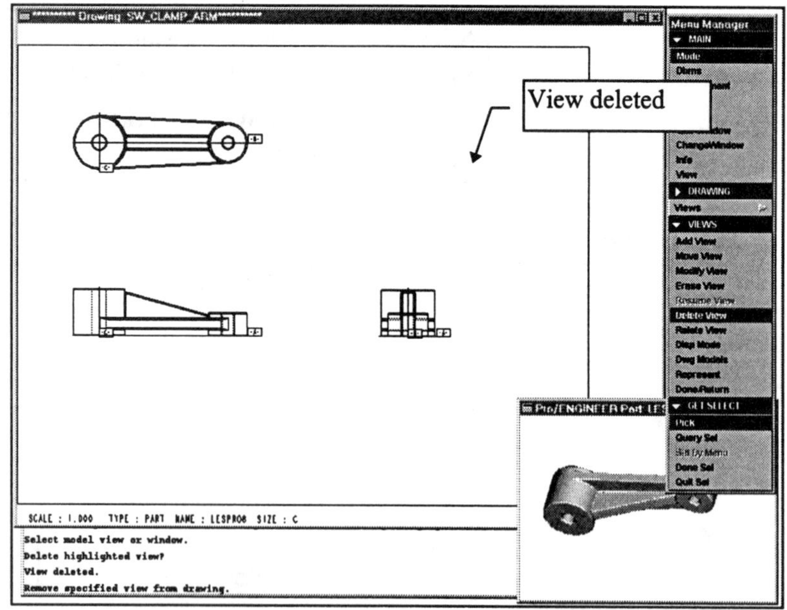

Figure 16.51
Delete the New View

Next, reposition the views on the drawing. After you select the view to move, it will be highlighted with dashed lines (Fig. 16.52). If the view is a parent of another view, that view will also be highlighted and move with the selected view. Use the following commands:

Views ⇒ **Move View** ⇒ (pick the front view) ⇒ (move the view to a new position as in Fig. 16.53; try a few positions) ⇒ **Done Sel** ⇒ **Done/Return**

Pick the front view to move (it will be *highlighted*)

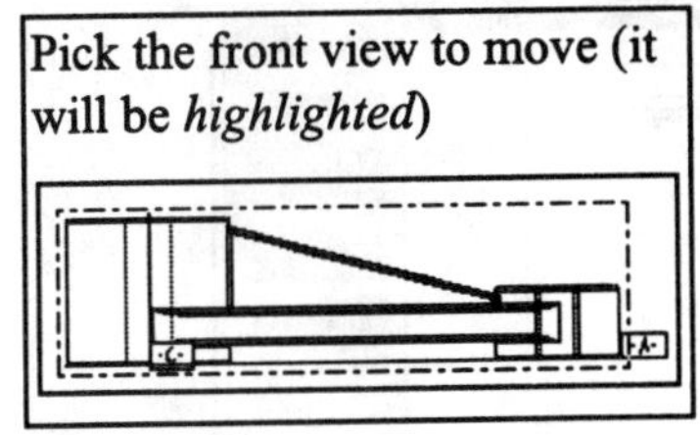

NOTE

The basic datum planes are displayed in the old ANSI standard style [icon].

You will learn how to change to the new ASME 1994 standard and display the datums correctly in the next lesson.

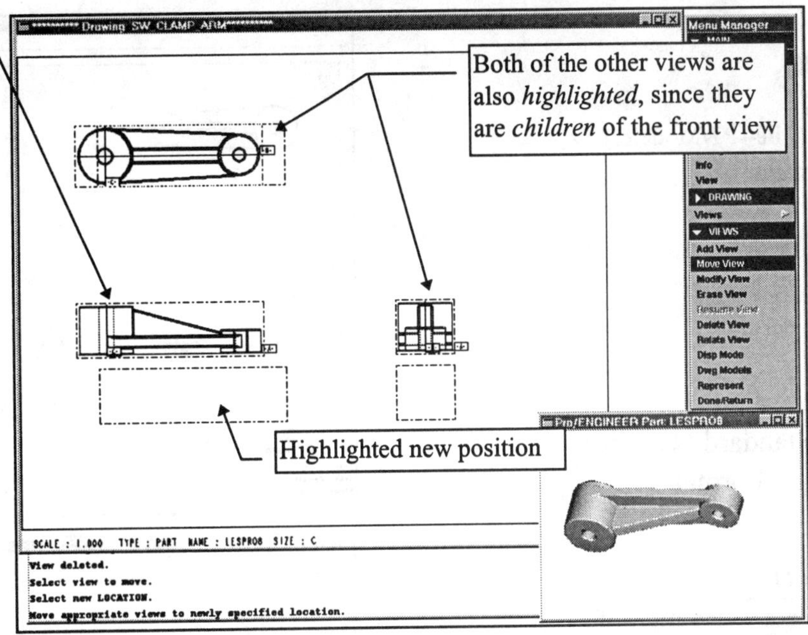

Figure 16.52
Moving Views

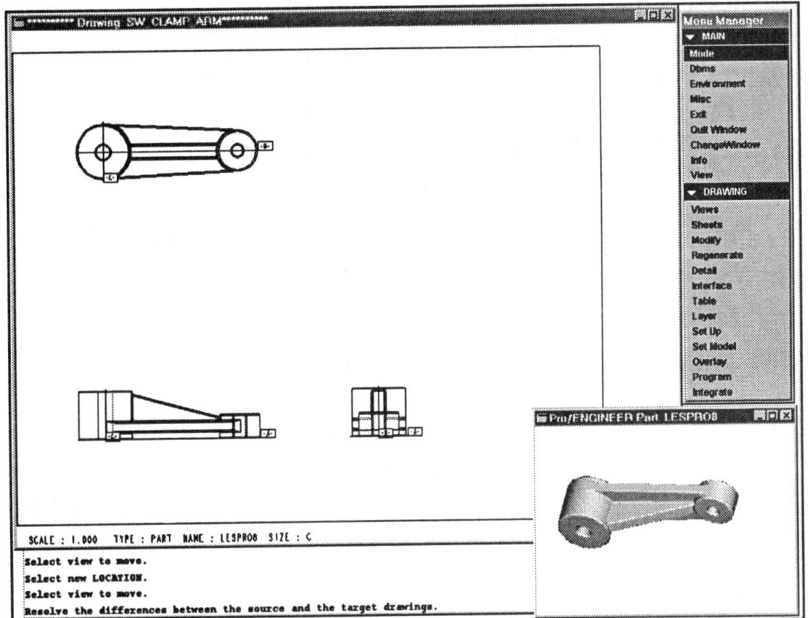

Dbms (**File**--PT/Modeler) ⇒ **Save** ⇒ **enter**
Purge ⇒ **enter** ⇒
Done-Return

Figure 16.53
New View Position

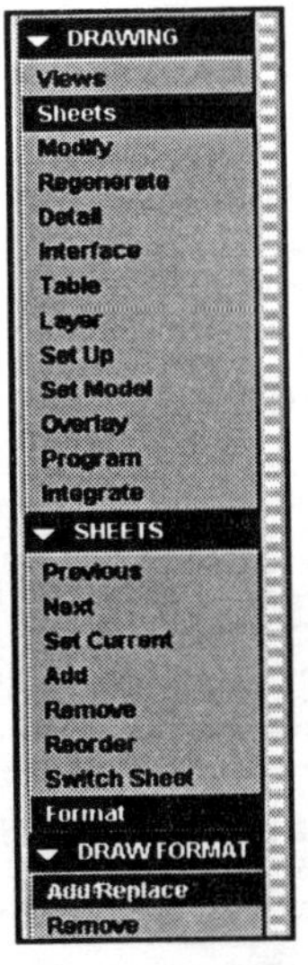

Now add to the drawing sheet a standard ANSI "**C**" size format using the following commands (Fig. 16.54):

Sheets ⇒ **Format** ⇒ **Add/Replace** ⇒ **?** ⇒ **enter** ⇒ **Format Dir** ⇒ (pick: **c.frm**) (notice there are no parameters)

Now change the drawing format to the one you previously created and saved as a **.frm** for use on your projects. This format should contain parameters you set up and saved with the format. The example in Figure 16.55 uses the CADTRAIN format and text parameters. Use the following commands:

Sheets ⇒ **Format** ⇒ **Add/Replace** ⇒ **?** ⇒ **enter** ⇒ **Current Dir** ⇒ (pick ***your_format_name*.frm**)

Dbms (File--PT/Modeler) ⇒
Save ⇒ enter
Purge ⇒ enter ⇒
Done-Return

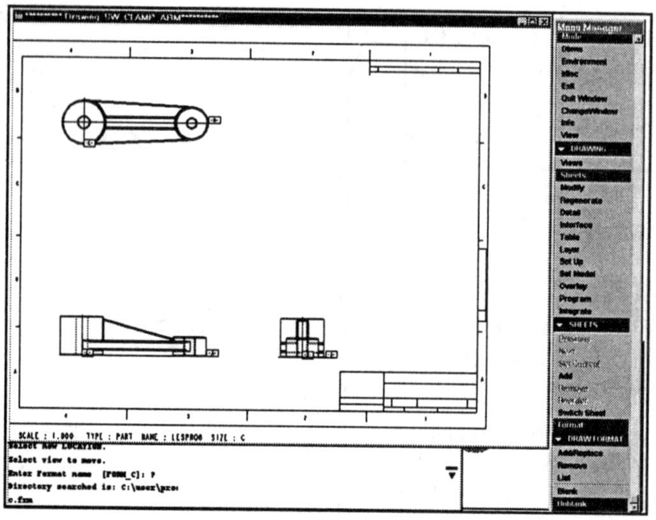

Figure 16.54
New Standard **"C"** Size Drawing Format

NOTE
Pro/E has a wide variety of capabilities that will allow you to tailor your drawing to the needs of your company or class.

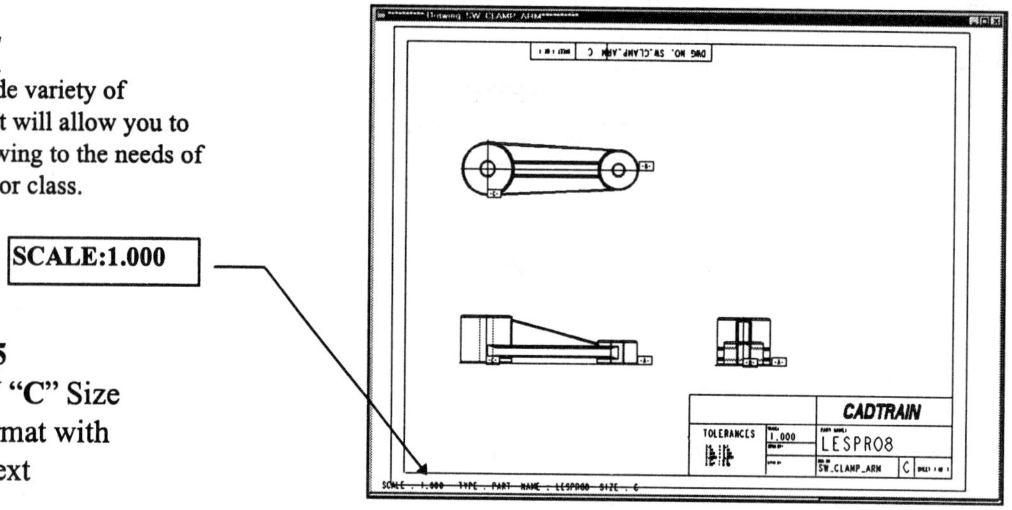

SCALE:1.000

Figure 16.55
CADTRAIN **"C"** Size Drawing Format with Parameter Text

To change the scale of the drawing (*not the part*), pick **Modify ⇒** (pick the **SCALE: 1.000** on the bottom left edge of the Drawing Window in Fig. 16.55) ⇒ (type **.75**) ⇒ **enter** (Fig. 16.56).

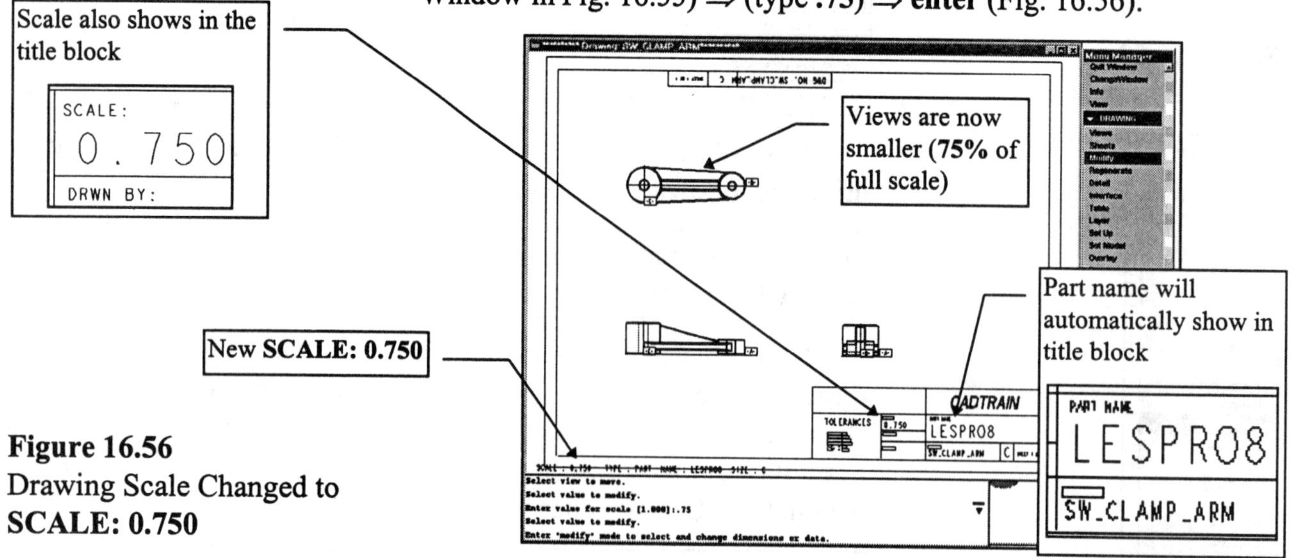

Figure 16.56
Drawing Scale Changed to **SCALE: 0.750**

You can now experiment with creating other drawings using this part or other parts, adding, moving, and erasing views, and creating, adding, and removing drawing formats.

Lesson 16 Project

Base Angle Drawing

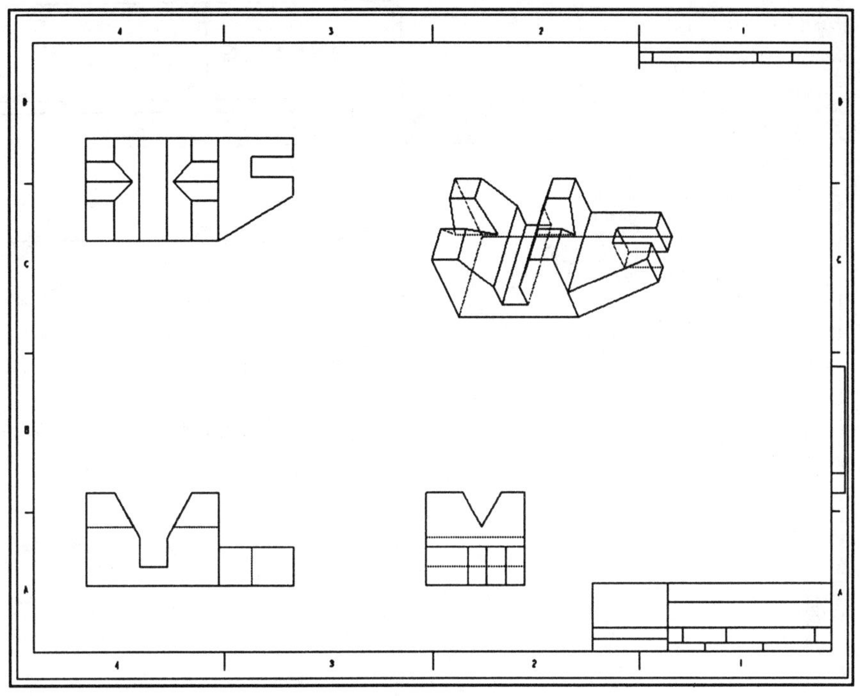

Figure 16.57
Base Angle Drawing Using ANSI (ASME) Standard Format

Base Angle Drawing

The first **lesson project** for Drawing mode will use the part modeled in Lesson 2. The drawing for the Base Angle (Fig. 16.57) is just one of many lesson parts and lesson projects that can be brought into Drawing mode and detailed.

Analyze the part and plan out the sheet size and the drawing views required to display its features for detailing (Fig. 16.58). Use the format created in this lesson. Figure 16.59 shows a **B** format.

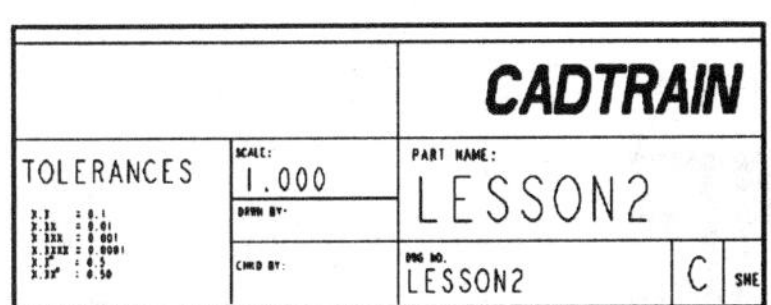

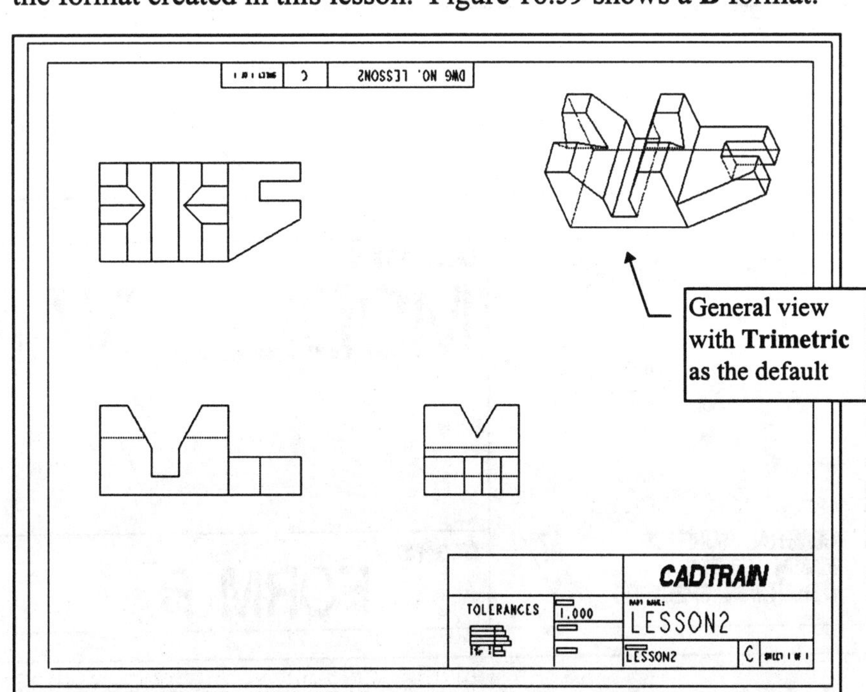

Figure 16.58
Base Angle Drawing Using CADTRAIN Format with Parameters

Figure 16.59 B Format

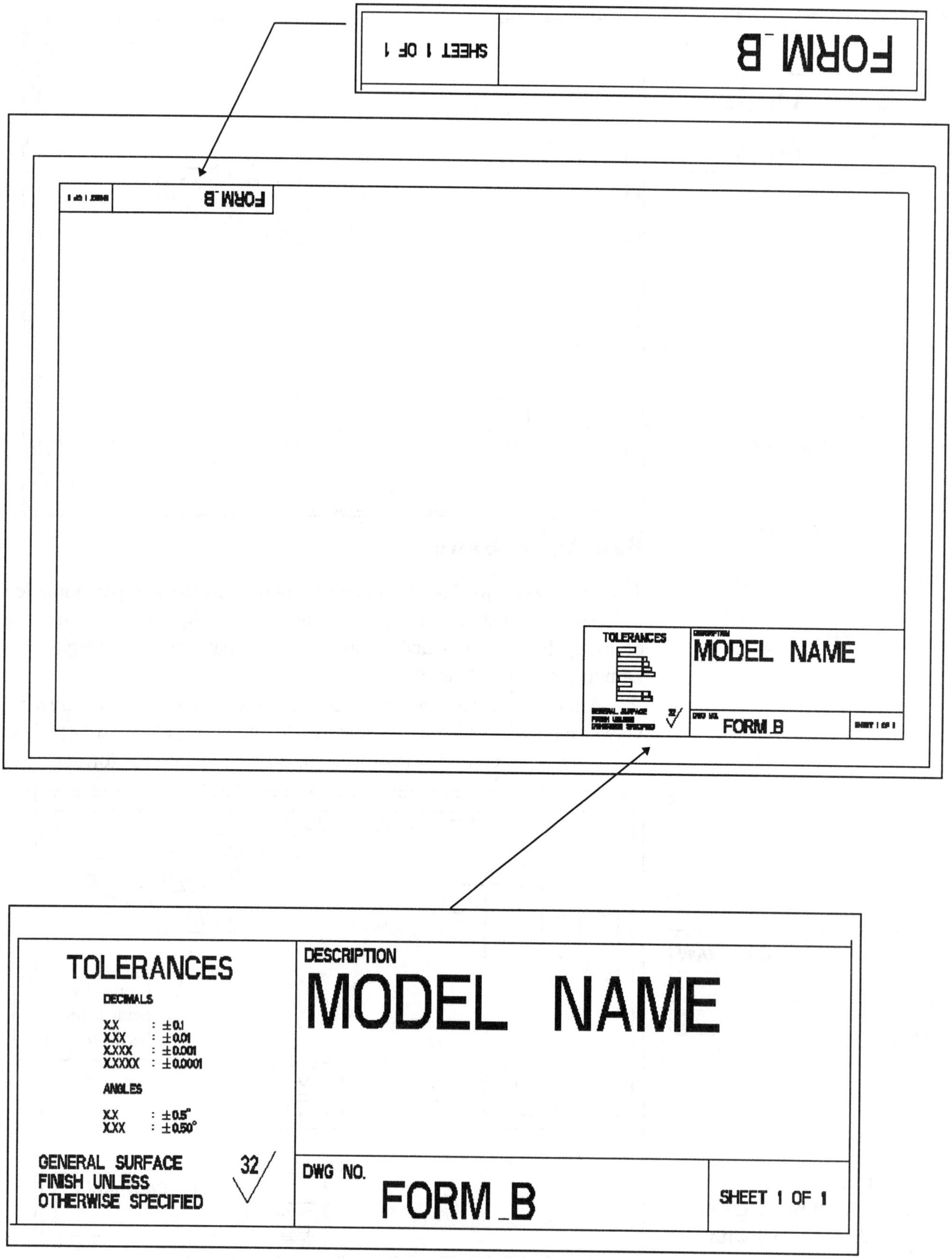

SHEET 1 OF 1 FORM_B

TOLERANCES MODEL NAME FORM_B

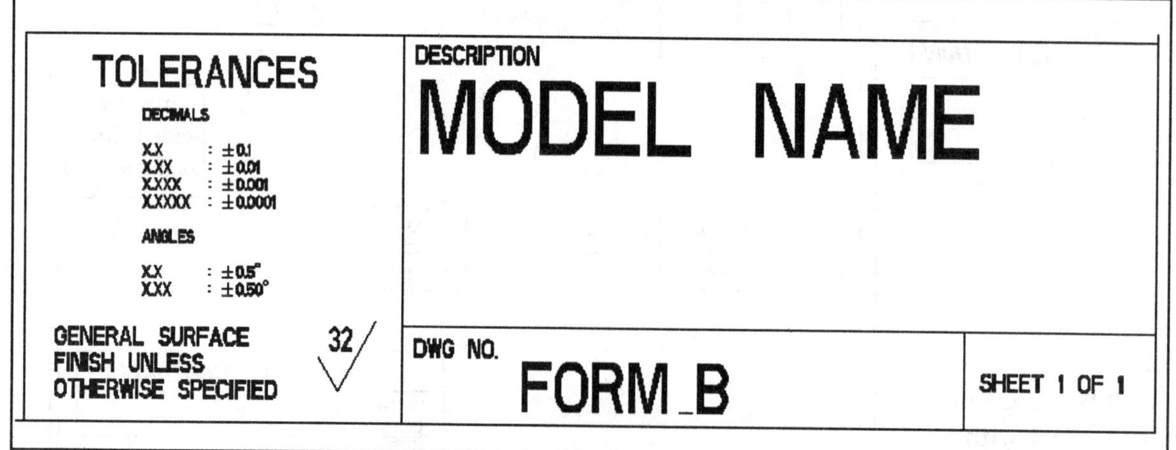

Lesson 17

Detailing

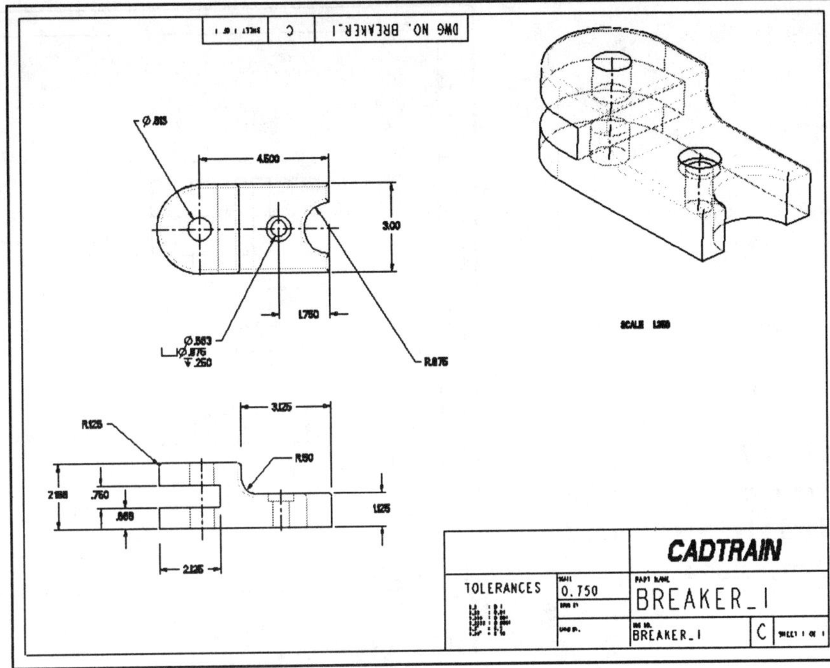

Figure 17.1
Breaker Drawing

OBJECTIVES

1. **Use ASME Y14.5 1994 standards to detail drawings**

2. **Dimension a part**

3. **Create and save .dtl files**

4. **Add geometric tolerancing information to a drawing**

5. **Use Pro/MARKUP to see checker changes**

6. **Move and modify dimensions**

EGD REFERENCE
Engineering Graphics and Design with Graphical Analysis *or* **Fundamentals of Engineering Graphics and Design**
by L. Lamit and K. Kitto
Read Chapters 15 and 16
See pages 321, 364, 556-559, and 622-626

COAch™ for Pro/ENGINEER

If you have **COAch for Pro/ENGINEER** on your system, go to SEARCH and do the Segment shown in Figures 17.3, 17.7, and 17.10.

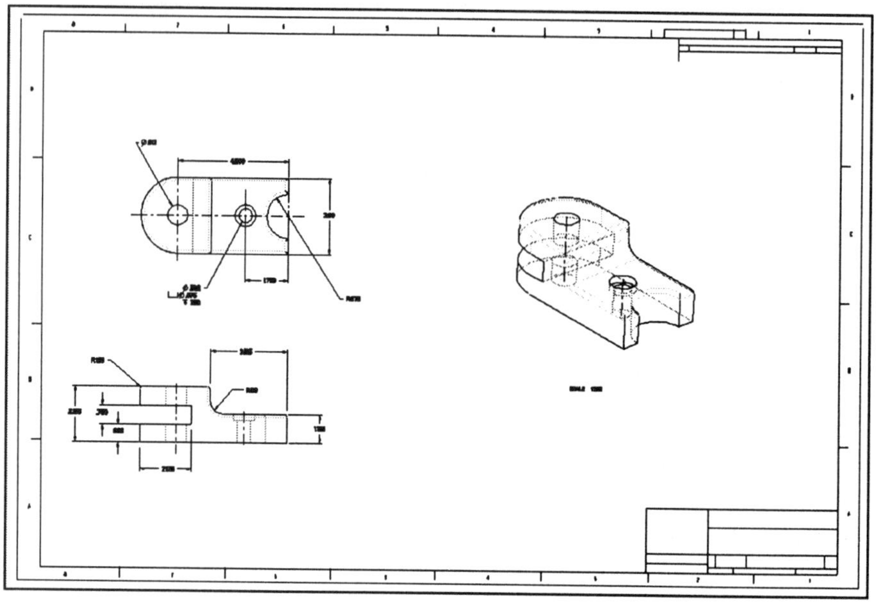

Figure 17.2
Breaker Drawing with ANSI Standard "**D**" Size Format and Dimensions

NOTE
Many of the concepts presented here (for parts) also apply to assemblies.

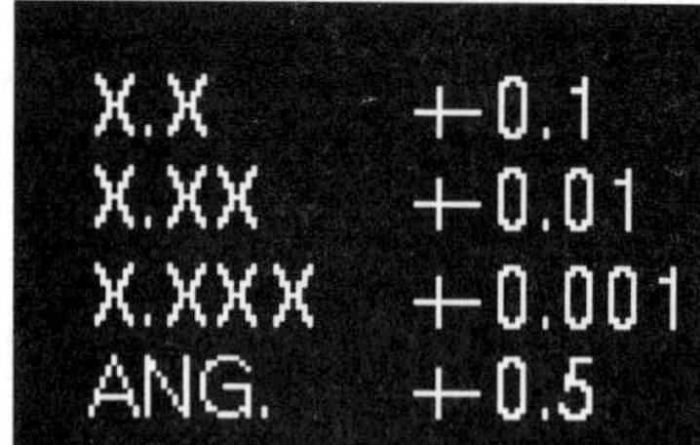

DETAILING

The purpose of an engineering drawing is to convey information so that the part can be manufactured correctly. Engineering drawings use dimensions and notes to convey this information (Fig. 17.1). Knowledge of the methods and practices of dimensioning and tolerancing is essential to the engineer or designer. The multiview projections of a model (part or assembly) provide a graphic representation of its shape (*shape description*). However, the drawing must also contain information that specifies size and other requirements.

Drawings are *annotated* with dimensions and notes. Dimensions must be provided between points, lines, or surfaces that are functionally related or to control relationships of other parts. With Pro/E, the **design intent** used in the original sequence of feature creation and the selection of dimensions used on the feature's sketch will determine the dimensions shown on the drawing. You need not create dimensions (unless desired), since the dimensions are already established during the modeling of the part. Dimensions created in Drawing mode to describe drafting features will not be *driving dimensions* (they will be *driven dimensions* that cannot be modified). *Only dimensions used in modeling the part drive the part feature database.*

Each dimension on a drawing has a **tolerance**, implied or specified. The general tolerance given in the title block is called a **general** or **sheet tolerance**. Specific tolerances are provided with each appropriate dimension. Together, the views, dimensions, and notes give the complete shape and size description of the part. Uniform practices for stating and interpreting dimensioning and tolerancing requirements were established in **ASME Y14.5 1994.**

Dimensioning

Views of a model (part or assembly) may be dimensioned in Drawing mode (Fig. 17.2). Pro/E displays dimensions in a view based on the way the part (or the assembly) was modeled. The dimension type is selected from options before showing the dimensions on the drawing (Figs. 17.3 and 17.4). Linear dimensions and ordinate dimensions are two of these options.

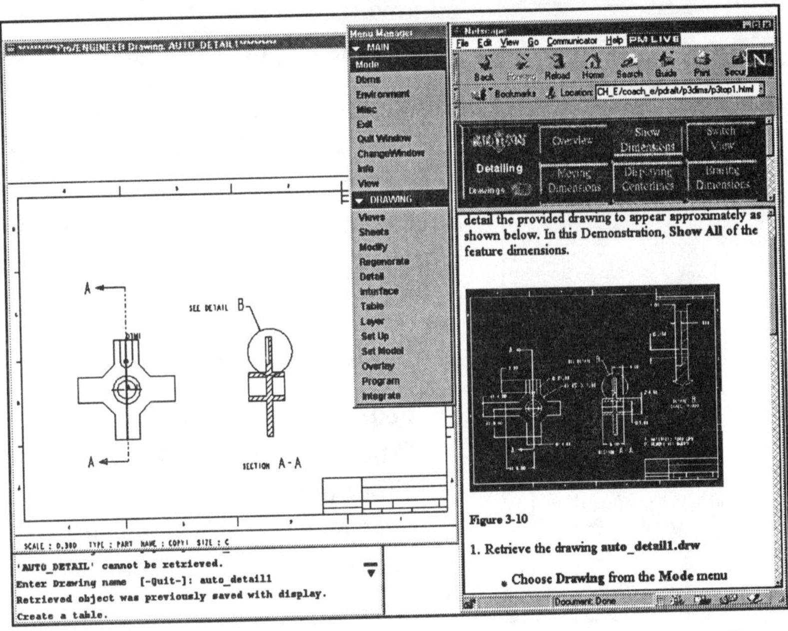

Figure 17.3
COAch for Pro/E, Detailing (Show Dimensions)

After a part's features are sketched, aligned, and dimensioned, you modify the dimension values to the exact sizes required for the design. The dimensioning scheme, the controlling features, the parent-child relationships, and the datums used to define and control the part features are determined as you design and model with Pro/E.

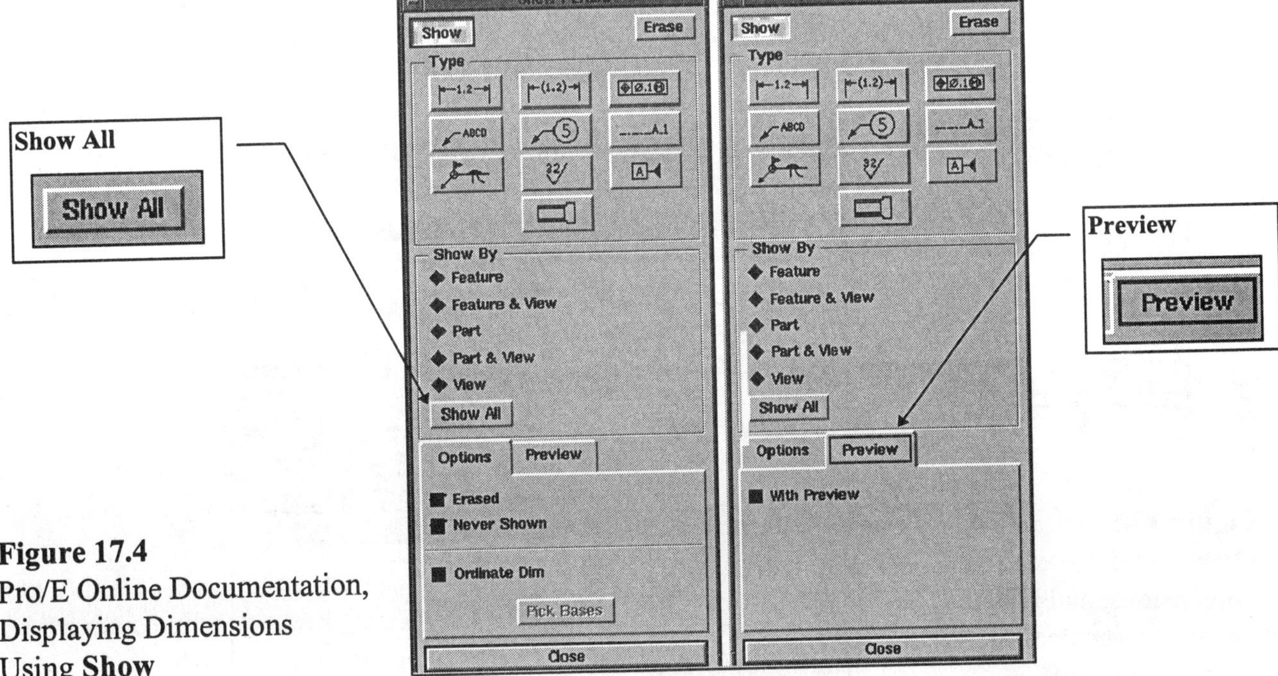

Figure 17.4
Pro/E Online Documentation, Displaying Dimensions Using **Show**

When detailing, you simply choose Pro/E commands to display views needed to describe the part and then display the dimensions used to model the part. These are the same dimensions used in the part's design. You cannot under dimension or over dimension the part, since Pro/E displays exactly what is required to model the part. Pro/E will not duplicate dimensions on a drawing. If a dimension is shown in one view, it will not be shown in another view. The dimension, however, can be switched to the other view via detailing options.

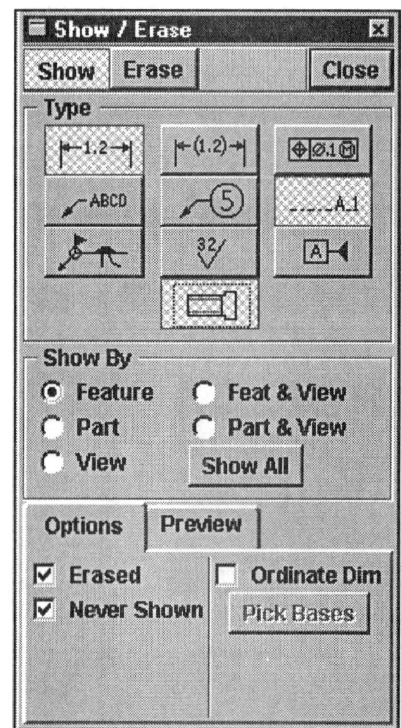

To display dimensions on a drawing:

1. Choose **Detail** from the DRAWING menu.
2. Choose **Show/Erase.** The **Show/Erase** dialog box is displayed. Select the dimension icon in the **Type** area. To specify the dimensions as ordinate, choose the **Ordinate Dim** check box at the bottom of the dialog box, the **Pick Bases** button highlights, then pick a *baseline*. In this case, the baseline must already exist.
3. Choose one of the following radio buttons:

Feature Show all the dimensions associated with a particular feature in the appropriate views. Select a feature to be dimensioned.
Feat & View Show all dimensions for a single feature in a single view. Select a feature in the view where the dimensions are to be displayed.
Part Show dimensions associated with a part.
Part & View Show dimensions associated with a part in a view.
View Show all the dimensions associated with a selected view. Select the view(s) you would like dimensioned.

To show all the dimensions for the current model, choose the **Show All** command button (Fig. 17.4). To *preview* what the drawing will look like when you make the changes, choose the **Preview** button. To specify the items that you wish to show on the drawing, choose the **Preview** tab and then select **With Preview** and one of the following command buttons: **Accept All**, **Erase All**, or **Select To Keep**. Figure 17.5 shows all dimensions.

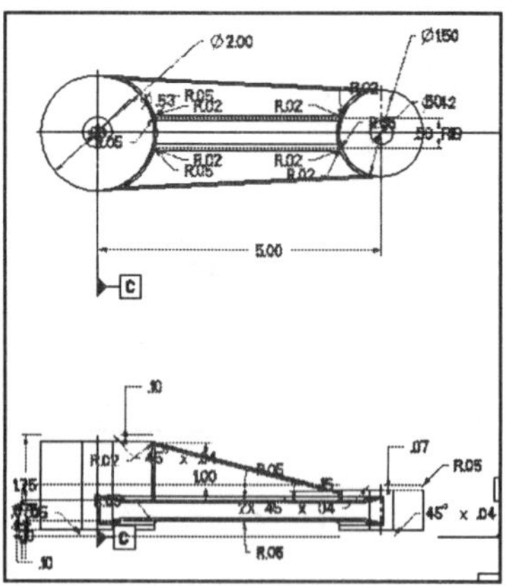

Figure 17.5
Drawing with Axes, Dimensions, and Datums Displayed (uncleaned, original position)

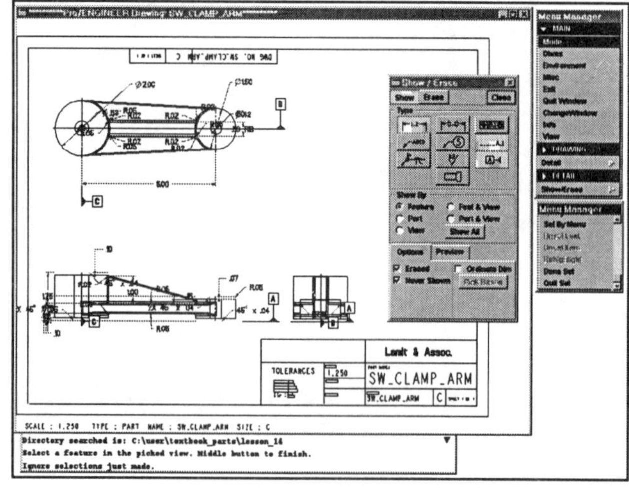

The **Clean Dims** option allows you to distribute standard and ordinate dimensions with equidistant spacing along witness lines, displaying them in a more orderly and readable fashion.

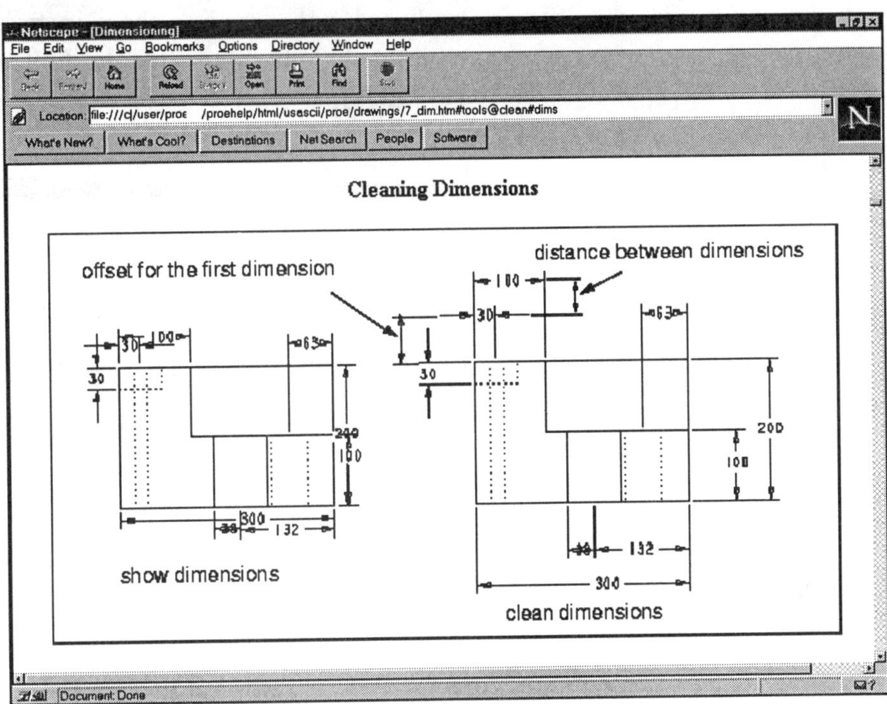

Figure 17.6
Pro/E Online Documentation, Cleaning Dimensions

To clean up the dimension display:

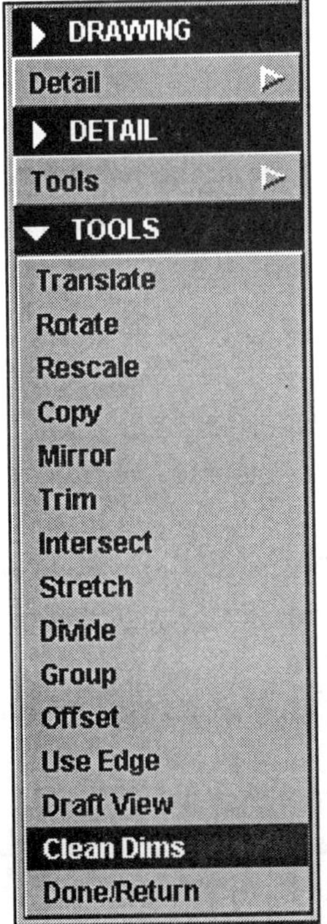

1. Select **Detail ⇒ Tools ⇒ Clean Dims**
2. Enter the offset value for the first dimension line (the one that is closest to the model).
3. Enter the distances between all other dimensions.
4. Select the view to be cleaned up by picking on the model.
5. Dimensions pertaining to the selected view are displayed with the specified spacing. In the event that the new display of dimensions is unsatisfactory, Pro/E will ask if you would like to move the dimensions back to their previous display. Type **(YES)** to restore.
6. Select another view to clean dimensions with the same offsets, or press the middle mouse button to quit the process.

The cleaned dimensions are usually not in the best positions for each dimension and note. After cleanup, the next step is to move and reposition the dimensions to create an ASME standard drawing.

Dimensions can be removed from the display by erasing them. Erasing dimensions does not delete them from the model (driving dimensions, "true part" dimensions, *cannot be deleted*, but reference dimensions and driven dimensions can). Dimensions that have been erased can be redisplayed via the **Show/Erase** dialog box.

Text and Notes on Drawings

Notes (Fig. 17.7) can be part of a dimension, attached to one or many edges on the model, or "free." You can add notes by typing from the keyboard or by reading them in from a text file. Notes are created with the default values (height, font, etc.) specified in the drawing **Set Up (.dtl)** file.

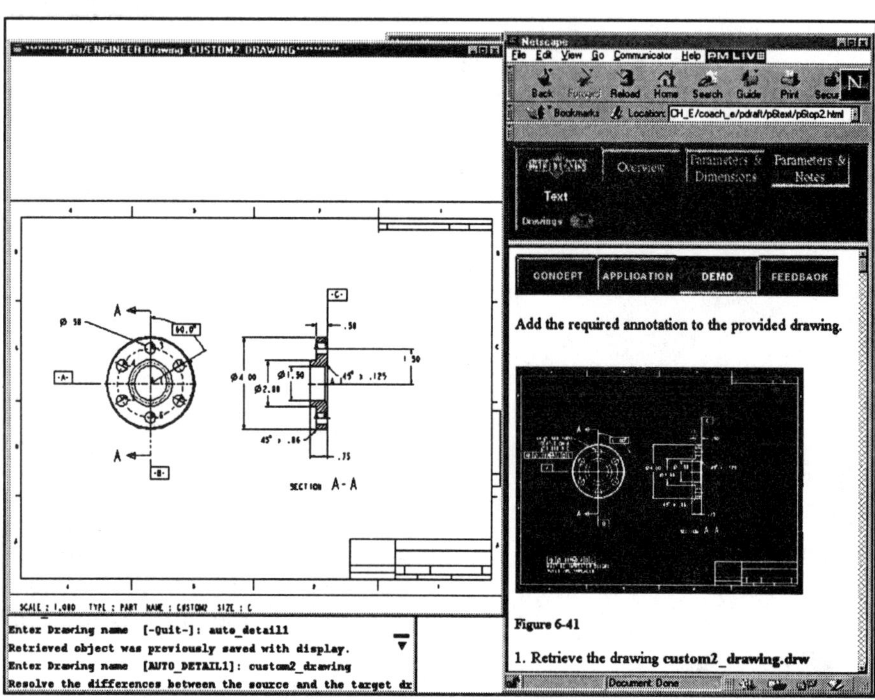

Figure 17.7
COAch for Pro/E, Drawings, Text (Parameters and Notes)

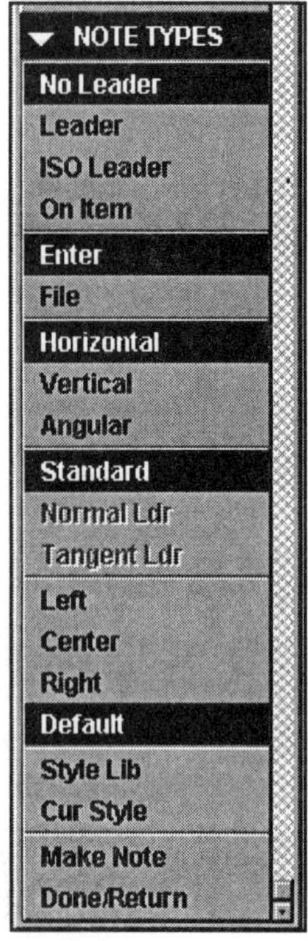

To add notes to the drawing, choose **Detail ⇒ Create ⇒ Note ⇒ Make Note**. Notes (Fig. 17.8) can be added with or without leaders, and the style of text can be modified via a dialog box (Fig. 17.9). The following options are available:

No Leader/Leader/On Item Create a note, with or without a leader, attached to an entity or not.
ISO Leader Create a note with ISO standard leader line.
Enter/File Enter the note from the keyboard, or read the note in from a text file.
Horizontal/Vertical/Angular Create a horizontal or vertical note, or enter an angular value between **0°** and **359°**.
Standard Create a note with multiple leaders.
Normal Ldr Create a note with a leader that is normal to an entity.
Tangent Ldr Create a note that is tangent to an entity.
Left/Center/Right/Default Create the note text as left-justified, center-justified, or right-justified. **Default** is left-justified.

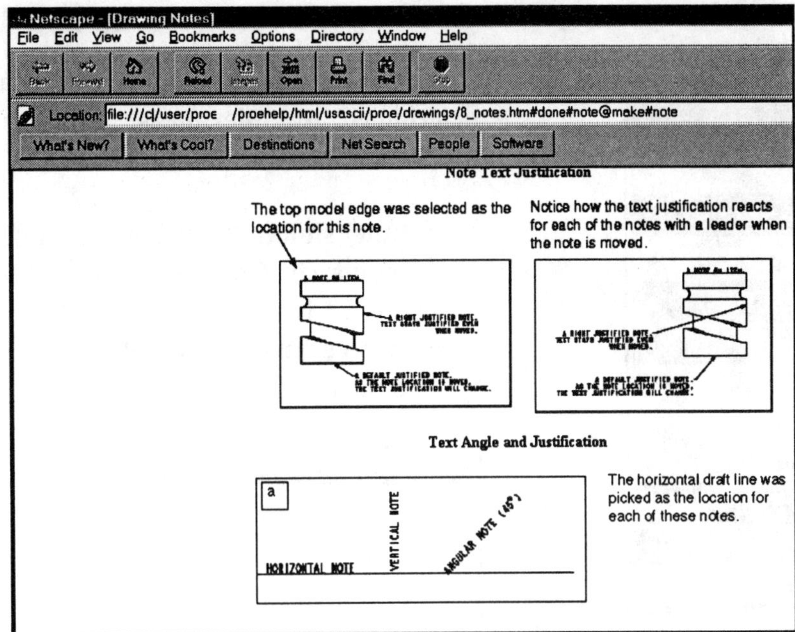

Figure 17.8
Pro/E Online Documentation, Notes

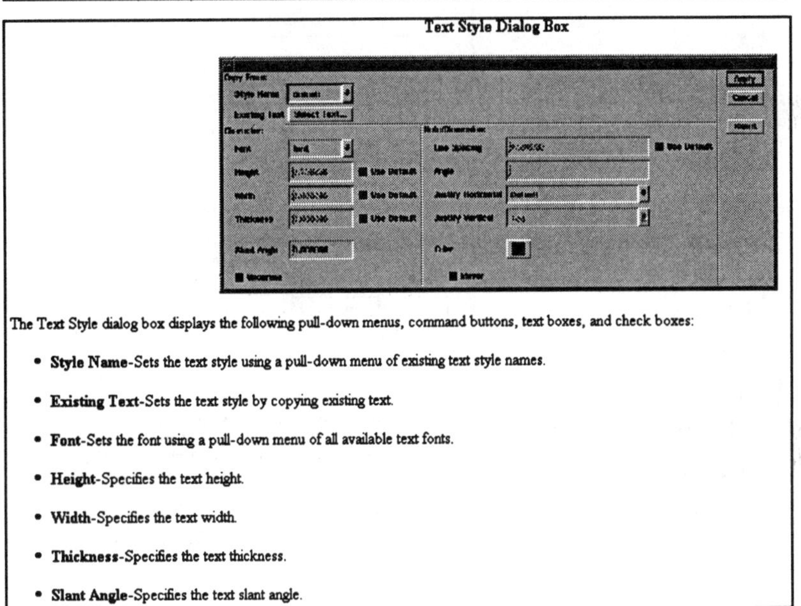

Text Style Dialog Box

The Text Style dialog box displays the following pull-down menus, command buttons, text boxes, and check boxes:

- **Style Name**-Sets the text style using a pull-down menu of existing text style names.
- **Existing Text**-Sets the text style by copying existing text.
- **Font**-Sets the font using a pull-down menu of all available text fonts.
- **Height**-Specifies the text height.
- **Width**-Specifies the text width.
- **Thickness**-Specifies the text thickness.
- **Slant Angle**-Specifies the text slant angle.

Figure 17.9
Pro/E Online Documentation, Text Style Dialog Box

Adding Text to a Dimension

The **Text** option from the MODIFY DRAW menu allows you to add text to a dimension value (for example, **DIAMETER**, **REF**, and **TYP**), as well as special symbols. You can also define your own special fonts and symbols. To add text:

1. Choose **Text** from the MODIFY DRAW menu.
2. Choose **Text Line** to modify one line of the note or dimension, or **Full Note** to modify the whole note or dimension.
3. Pick the dimension to which you will add the text.
4. Enter the line or lines of text. Each line must end with a carriage return **(enter)**.
5. To complete the text string, follow the line with an **enter**, or exit the editor when you are finished with the **Full Note**.

Geometric Dimensioning and Tolerancing

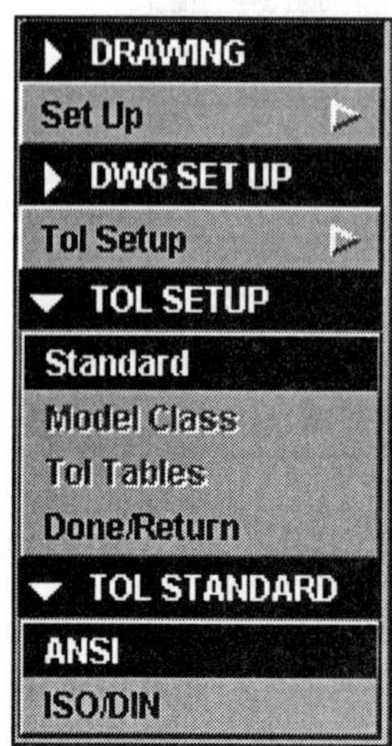

The manufacturing of parts and assemblies uses a degree of precision determined by **tolerances**. A typical parametric design system supports three types of tolerances:

Dimensional Specifies allowable variation of size.
Geometric Controls form, profile, orientation, and runout (Fig. 17.10).
Surface finish Controls the deviation of a part surface from its nominal value.

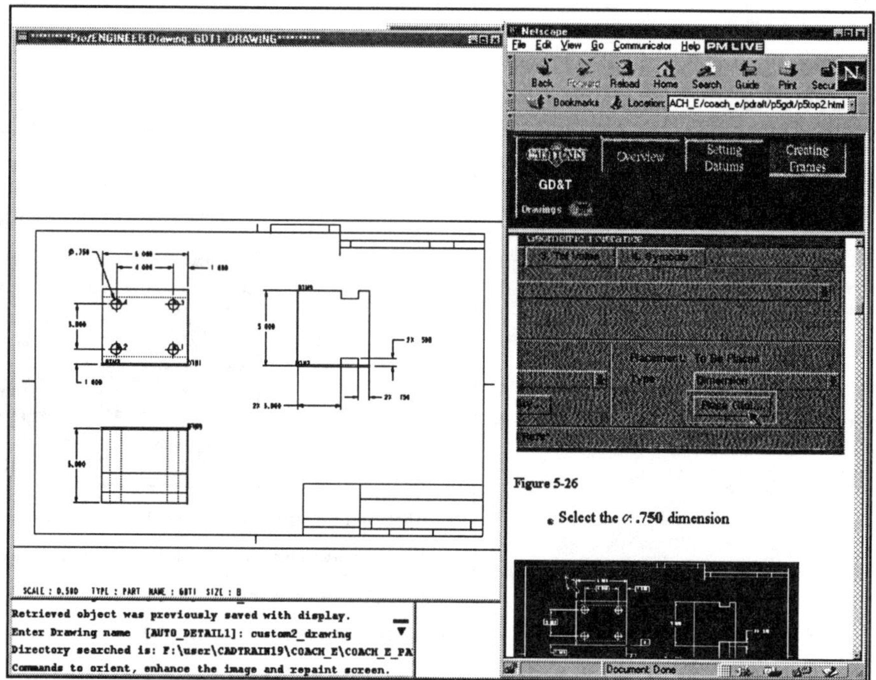

Figure 17.10
COAch for Pro/E, Drawings, GD&T (Creating Frames)

When you design a part, you specify dimensional tolerance, which means, the ***allowable variations in size***. All dimensions are controlled by tolerances, except "**basic**" dimensions, which for the purpose of reference are considered to be *exact*.

Dimensional tolerances on a drawing can be expressed in two forms:

* As **general tolerances** presented in a tolerance table. These apply to those dimensions that are displayed in nominal format, that is, without tolerances.
* As **individual tolerances** specified for individual dimensions.

You can use general tolerances given as defaults in a table or set individual tolerances by modifying default values of selected dimensions. Default tolerance values are used at the moment you start to create a model; therefore, *default tolerances must be set prior to creating geometry*.

Pro/E recognizes six decimal places for which you can specify the tolerance values. When you start to create a part, the table at the bottom of the window will display the current defaults for tolerances. If you have not specified tolerances, Pro/E defaults are assumed, and the table will look as follows:

*** x.x**	**±0.1**
*** x.xx**	**±0.01**
*** x.xxx**	**±0.001**
*** ANG.**	**±0.5**

You have a choice of displaying or blanking tolerances. If tolerances are not displayed, Pro/E still stores dimensions with their default tolerances. You can specify geometric tolerances, create "basic" dimensions, and set selected datums as reference datums for geometric tolerancing. ISO tolerances are generated from a table, as shown in Figure 17.11.

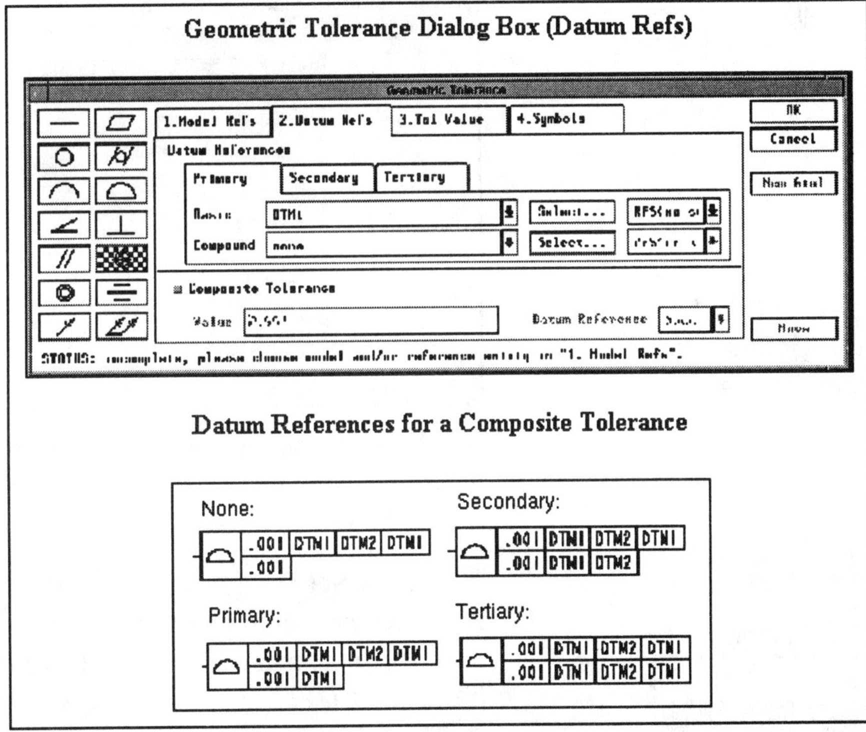

Figure 17.11
Pro/E Online Documentation, Geometric Tolerance Dialog Box

The available tolerance formats are:

Nominal Dimensions displayed without tolerances.
Limits Tolerances displayed as upper and lower limits.
Plus-Minus Tolerances displayed as nominal with plus-minus tolerance. The positive and negative values are independent.
±Symmetric Tolerances displayed as nominal with a single value for both the positive and the negative tolerance.
As Is Tolerances as is.

NOTE

Pro/DETAIL enables you to add geometric tolerances to the model from Drawing mode. Note that geometric tolerances can be added in Part or Drawing mode, but are reflected in all other modes. Geometric tolerances are treated by Pro/E as *annotations*, and they are always associated with the model. Unlike dimensional tolerances, *geometric tolerances do not have any effect on part geometry.*

Geometric tolerances provide a method for controlling the location, form, profile, orientation, and runout of features. You add geometric tolerances to the model from Part mode or Drawing mode. The geometric tolerances are treated by Pro/E as annotations, and they are always associated with the model. *Unlike dimensional tolerances, geometric tolerances do not have any effect on part geometry.*

When adding a geometric tolerance to the model, you can attach it to existing dimensions, edges, and existing geometric tolerances, or display them as notes without a leader.

Before you can reference a datum in a geometric tolerance, you must first indicate your intention by **setting** the datum (Fig. 17.12). Once a datum is set, hyphens are added before and after the datum name, and it is enclosed in a rectangle (using the old standards). The ASME Y14.5M 1994 standards display the datum name, as in Figures 17.13 and 17.14. You can change the name of a datum either before or after it has been set, by using the **Name** option in the **Set Up** menu from Part mode.

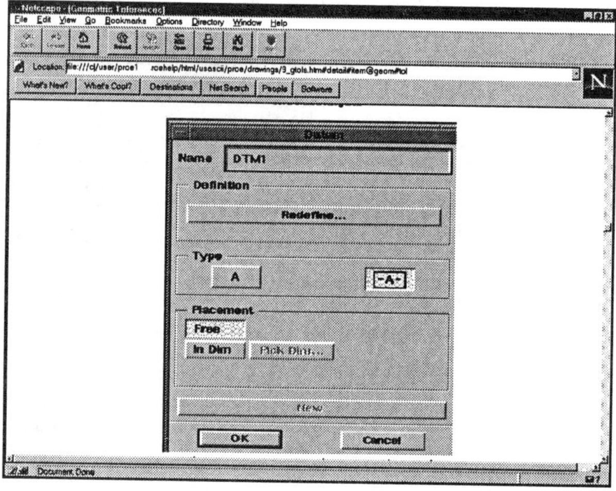

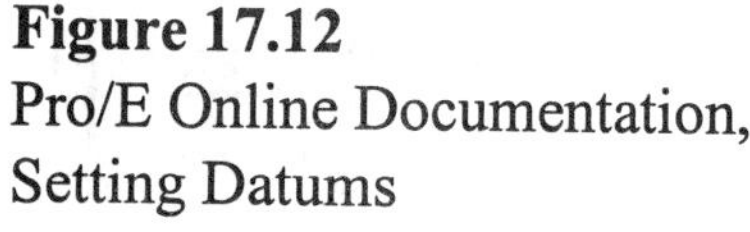

Figure 17.12
Pro/E Online Documentation, Setting Datums

You can choose any datum feature as a reference datum for a geometric tolerance. To set a reference datum in Drawing mode, choose **Detail ⇒ Create ⇒ Geom Tol**:

1. Choose **Set Datum** from the GEOM TOL menu.
2. Select the datum plane or axis to be set.
3. Change the name if desired. Change the type of feature control frame, then **OK**.
4. The datum is enclosed in a feature control frame.

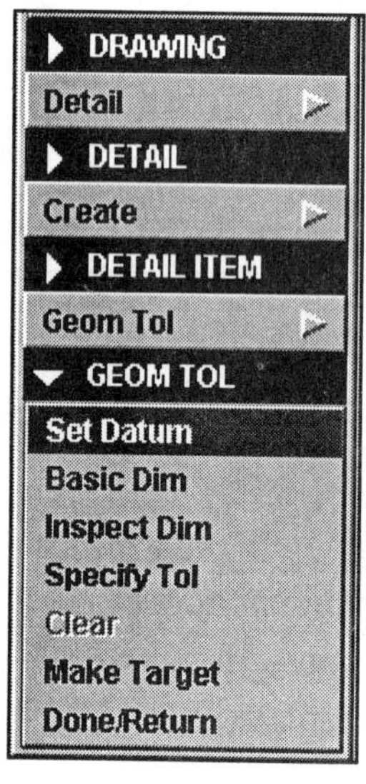

A geometric tolerance for individual features is specified by means of a **feature control frame** (a rectangle) divided into compartments containing the geometric tolerance symbol followed by the tolerance value. Where applicable, the tolerance is followed by a **material condition symbol**. Where a geometric tolerance is related to a datum, the reference datum name is placed in a compartment following the tolerance value. Where applicable, the reference datum name is followed by a material condition symbol.

For each class of tolerance, the types of tolerances available and the appropriate types of entities can be referenced. The available material condition symbols are shown in the dialog box (Fig. 17.11).

You are guided in the building of a geometric tolerance by Pro/E requests for each piece of required information. You respond by making menu choices, entering a tolerance value, and selecting entities and datums. As the tolerance is built, the choices are limited to those items that make sense in the context of the information you have already provided. For example, if the geometric characteristic is one that does not require a datum reference, you will not be prompted for one. Other checks are made to help prevent mistakes in the selection of entities and datums.

HINT

Change your **.dtl** file using **Set Up ⇒ Modify Val ⇒ gtol_datums STD_ANSI** change to **STD_ASME ⇒ File ⇒ Save ⇒ File ⇒ Exit** to see your set datums displayed with the correct ASME Y14.5 1994 symbology (Fig. 17.14).

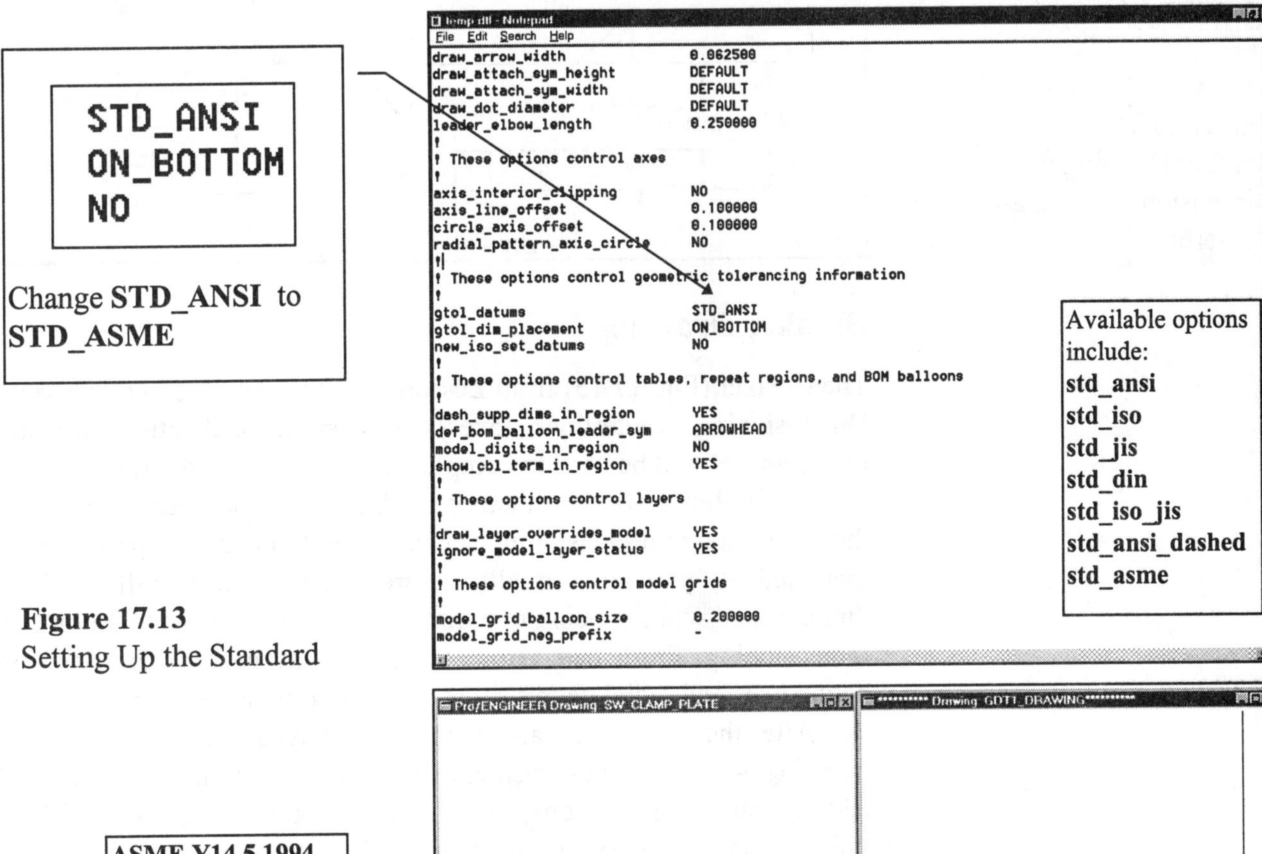

Figure 17.13
Setting Up the Standard

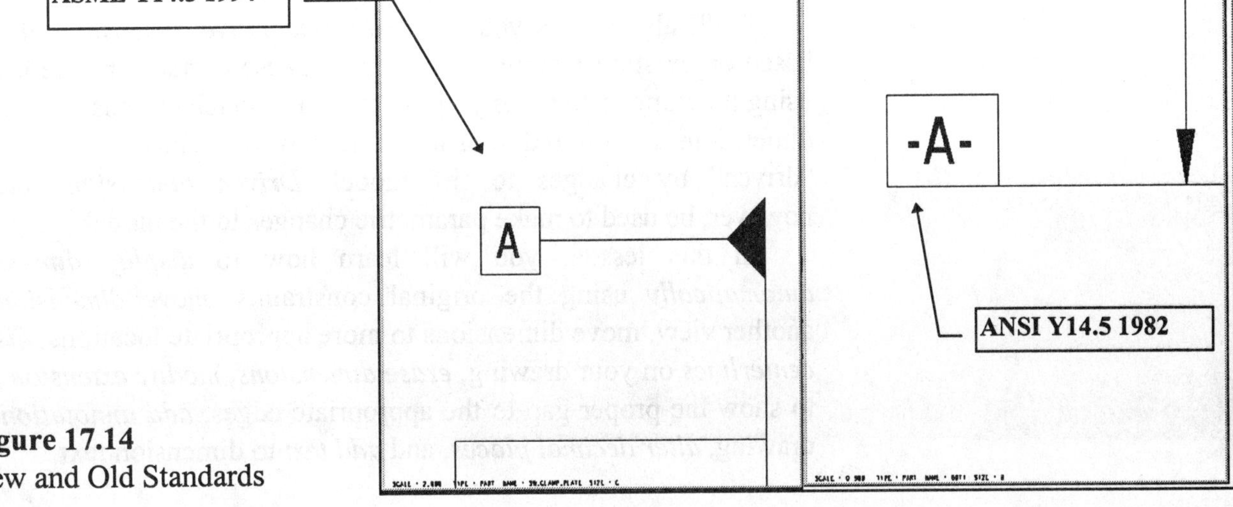

Figure 17.14
New and Old Standards

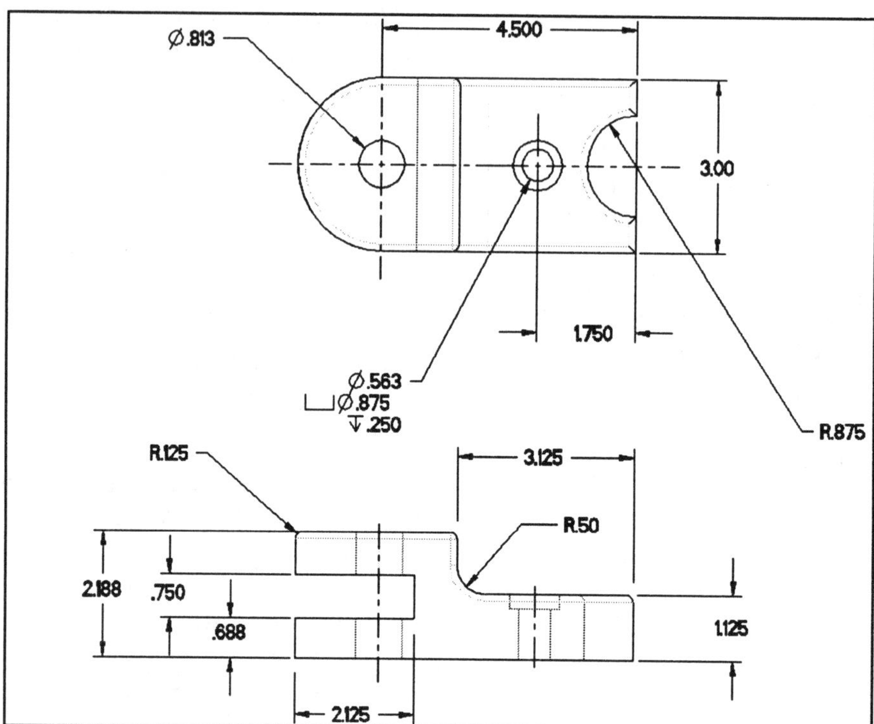

Figure 17.15
Breaker Drawing with Dimensions, Notes, and Centerlines

Breaker Drawing

The Breaker (Fig. 17.15) from Lesson 3 will be detailed in this lesson. Dimensioning, centerlines (axes), annotation, and other drawing requirements will be used to complete the detailing of the model.

Pro/E allows you to dimension a drawing automatically based on the design intent dimensioning scheme used to create the part. Using this method, you are virtually assured that the part will be fully dimensioned without being overdimensioned. In addition to automatic dimensioning, Pro/E can automatically display centerlines based on datum axes and radial patterns that were used to model the part.

After the dimensions and axes are displayed, the only remaining detailing work involves cosmetically cleaning up their display and adding other annotations, such as general notes, title block information, and required tolerancing information.

Pro/E also allows you to *create associative dimensions* that are based on existing feature constraints. These dimensions are created using the same techniques you use to create sketch dimensions. These dimensions are referred to as *driven* dimensions, since their display is "driven" by changes to the model. *Driven dimensions* cannot, however, be used to make parametric changes to the model.

In this lesson, you will learn how to *display dimensions automatically* using the original constraints, *move dimensions* to another view, move dimensions to more appropriate locations, *display centerlines* on your drawing, *erase dimensions*, *modify extension lines* to show the proper gap to the appropriate edges, *add annotation* to a drawing, *alter decimal places*, and *add text* to dimension text.

The model needs to be brought into Drawing mode and placed on a format. The format can be a blank format, a Pro/E-provided ANSI standard blank format, or a user-defined format created with parameters that will automatically display title block information and sheet callouts, as was shown in Lesson 16. Use the following commands:

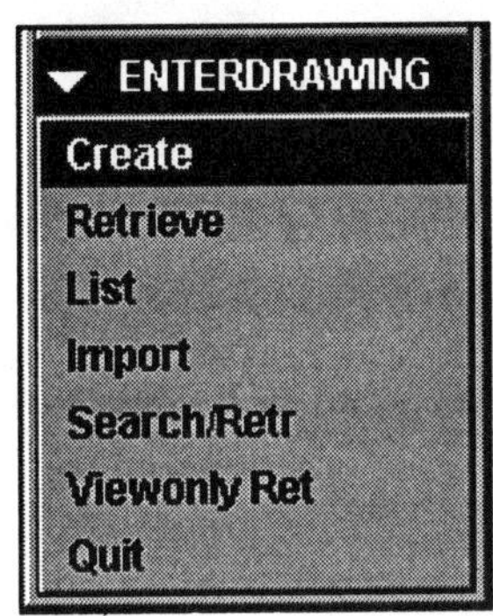

Mode ⇒ **Drawing** ⇒ **Create** ⇒ (**Breaker_dwg** or some other logical name) ⇒ **enter** ⇒ (from the DWG SIZE menu pick **C**) ⇒ **Mode** ⇒ **Part** ⇒ **Search/Retr** ⇒ (choose **Breaker** from the directory list) ⇒ (move and resize of the Part window) ⇒ **Change Window** ⇒ (pick in Drawing window) ⇒ **Views** ⇒ **enter** to accept the default (or **?** ⇒ **enter** ⇒ and choose the model name from the directory list) ⇒ **Done** (to accept the defaults) ⇒ [pick a place on the drawing for the first view (Fig. 17.16)] ⇒ **Front** ⇒ (pick **DTM3**) ⇒ **Top** ⇒ (pick **DTM2**) ⇒ **Done/Return** ⇒ **Add View** ⇒ **Done** (to accept defaults) ⇒ [pick in the area you wish to have a front view (Fig. 17.17)] ⇒ **Done/Return**

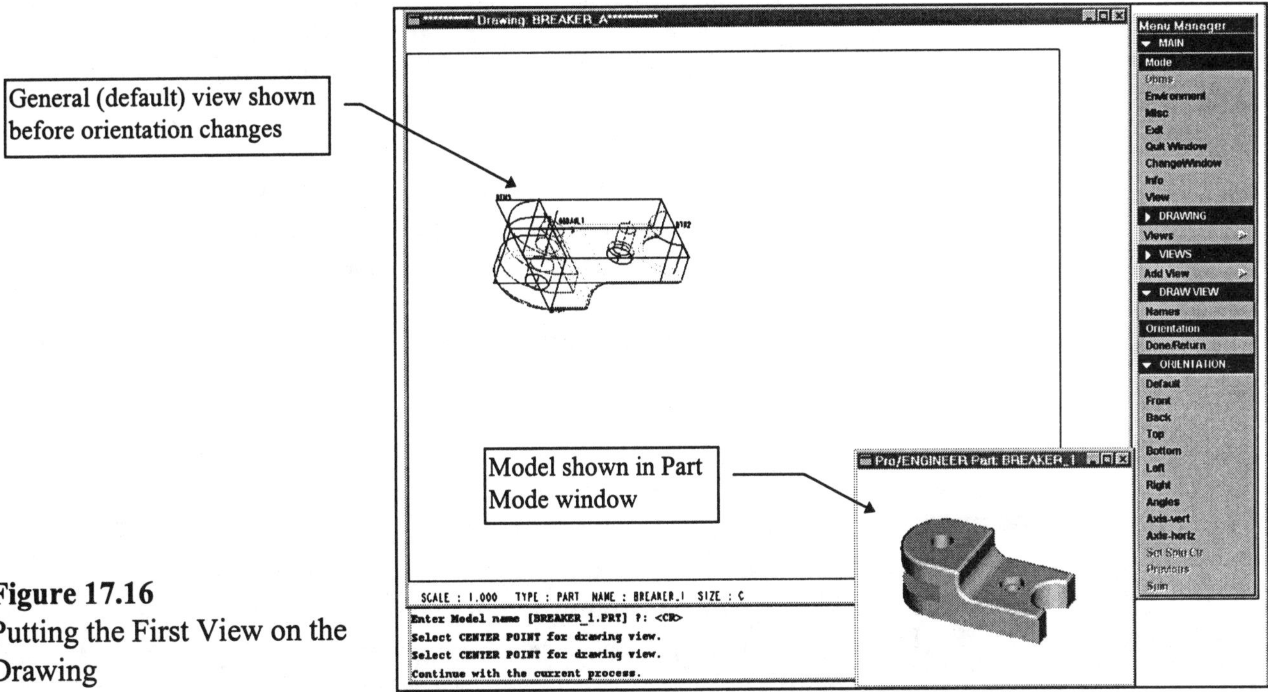

Figure 17.16
Putting the First View on the Drawing

After you have the two views on your drawing, use **Move View** to relocate them to better positions for detailing. Remember, you can reposition the views at any time in the detailing process, even after the dimensions are placed.

Dbms (**File**--PT/Modeler) ⇒
Save ⇒ enter
Purge ⇒ enter ⇒
Done-Return

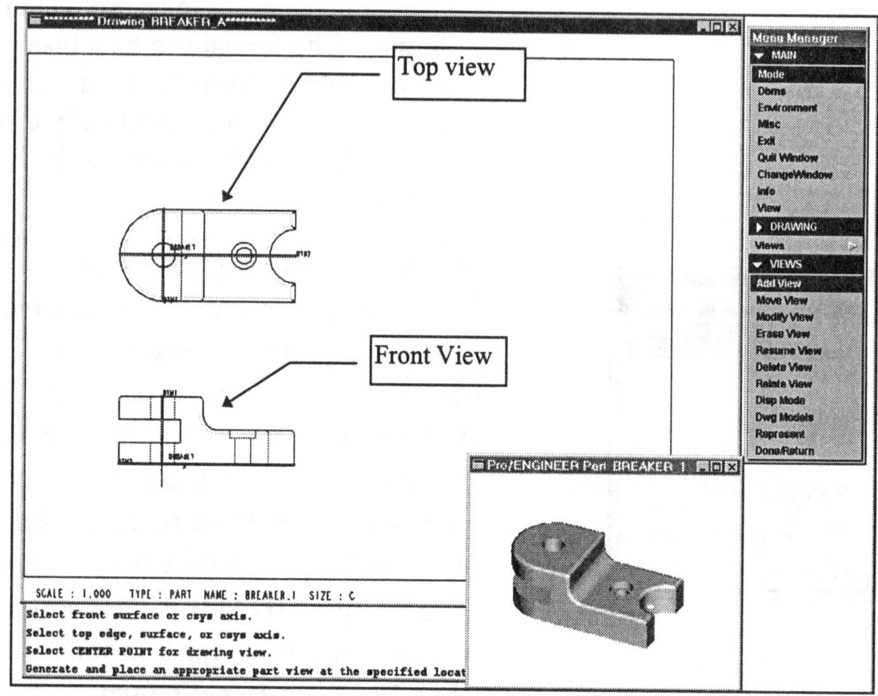

Figure 17.17
Putting the Second View on the Drawing

Showing Axes

Before adding dimensions Pro/E can automatically display centerlines at each location where there is a **Datum Axis**. The axis must be either parallel or perpendicular to the screen to be displayed.

If the axis is perpendicular to the screen, Pro/E displays a "crosshairs" centerline. If the axis is parallel to the screen, Pro/E displays a linear centerline. If the **Environment** option **Disp Axes** is off, the name of the axis is not displayed on the drawing. A crosshairs axis actually consists of four centerline segments. Each segment makes up one "leg" of the crosshairs. A linear axis actually consists of two centerline segments. Each segment makes up one "half" of the centerline. This distinction means very little, unless you need to trim the centerline segments because of interference with other drafting objects or edges.

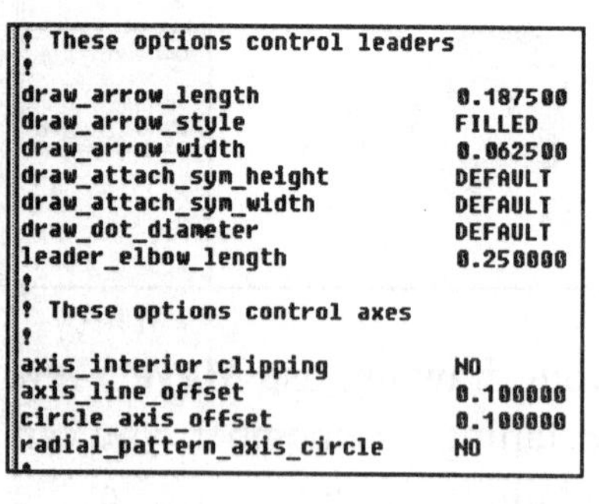

```
! These options control leaders
!
draw_arrow_length            0.187500
draw_arrow_style             FILLED
draw_arrow_width             0.062500
draw_attach_sym_height       DEFAULT
draw_attach_sym_width        DEFAULT
draw_dot_diameter            DEFAULT
leader_elbow_length          0.250000
!
! These options control axes
!
axis_interior_clipping       NO
axis_line_offset             0.100000
circle_axis_offset           0.100000
radial_pattern_axis_circle   NO
```

There are several **Set Up** file parameters that control the display of centerlines. The parameter **circle_axis_offset** controls the distance that centerlines extend past the circle they lie on. The parameter **axis_line_offset** controls the distance that centerlines extend past the end of a cylindrical face. The parameter **axis_interior_clipping** controls whether or not you can trim the interior segments of a centerline. If this parameter is set to **no**, you can shorten or lengthen the ends of only linear and crosshairs centerlines. If the parameter is set to **yes**, you can shorten or lengthen any end of any segment. The parameter **radial_pattern_axis_circle** will display a *bolt circle centerline.*

HINT

Always turn off most of the **Environment** options so that they do not clutter the drawing views:

□ **Disp DtmPln**
□ **Disp Points**
□ **Disp Axes**
□ **Disp Csys**
□ **Spin Center**
□ **Grid Snap**

To show the centerlines/axes for the Breaker, choose the following commands:

Detail ⇒ Show/Erase ⇒ A_1 (axis radio button on) **⇒ Show All ⇒ Yes** (Fig. 17.18) **⇒ Close ⇒ Done/Return**

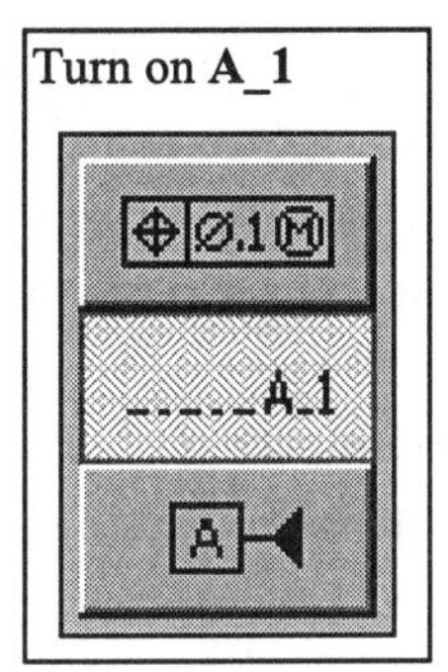

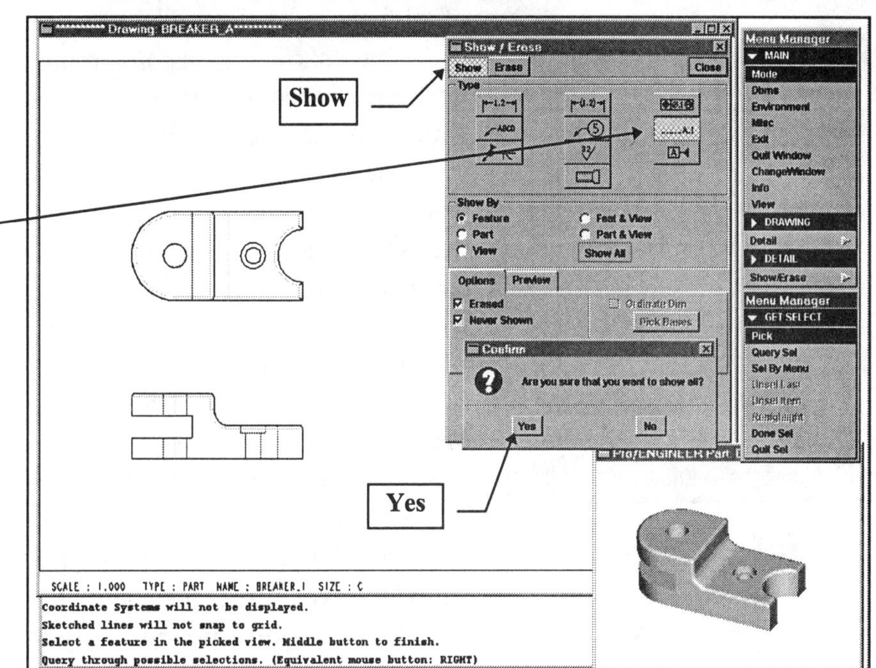

Figure 17.18
Displaying Axes

You can edit with a pop-up menu by *clicking the right mouse button while the cursor is in the main window* (when in Drawing mode), as shown in Figure 17.19. From the pop-up window, select **Modify Item**, then select (in this case) the axis to modify. The **Move/Activate** option is used to move the end of the axis. Modify the axes on both views to their required positions.

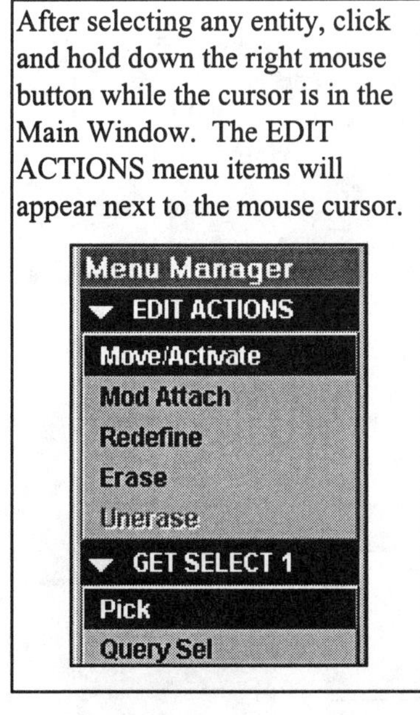

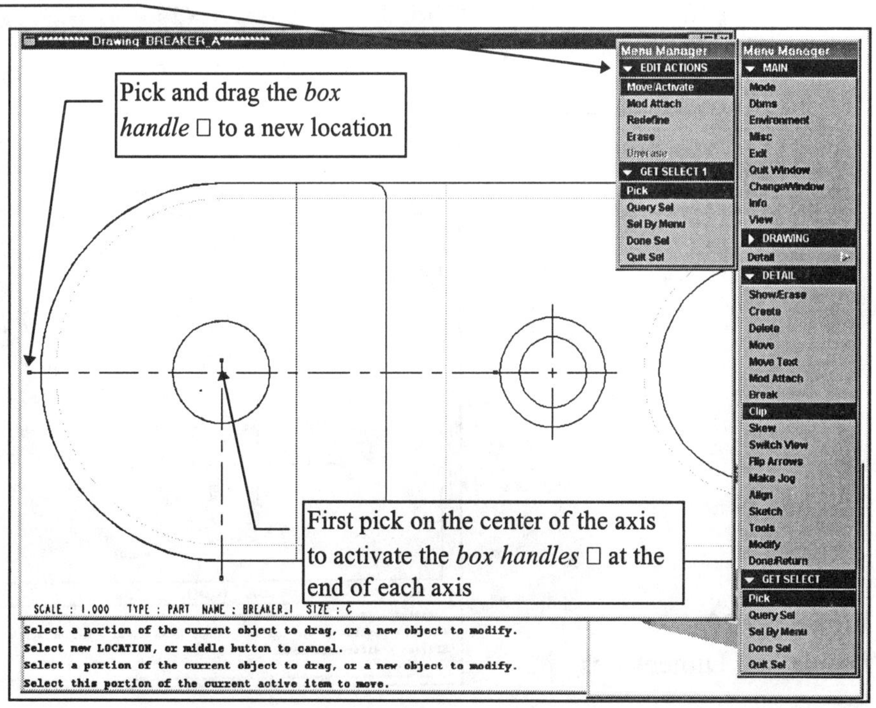

Figure 17.19
Modifying Axes

Show Dimensions

The **Show** option allows you to display dimensions based on the dimensional references and sketch dimensions that were used to create the model. You can show the dimensions of a feature, all the dimensions of all features in a particular view, or all the dimensions of all features on the model.

If you show dimensions in more than one view, and a feature can be dimensioned in more than one view, Pro/E attempts to decide which view is most appropriate to show the dimensions in.

Diameter dimensions are displayed differently based on the view where they are displayed. If a dimension is shown in a view where the cylinder axis is normal to the screen, the arrows are drawn to the circular edge of the cylinder. If the dimension is shown in a view where the cylinder axis is parallel to the screen, extension lines are added along the silhouette of the cylinder.

Changing the length of an axis can also be accomplished with the **Move** option. The **Move** option works on either end of the axis, based on which end is selected.

You can show dimensions, axes, datums, and so on at the same time, but for a complex part, the views quickly become cluttered. Showing axes first, then the dimensions, and then the cosmetic features gives you an opportunity to modify each drawing entity type separately to see the view requirements more clearly. Figure 17.20 illustrates the dimensions displayed on the drawing using the following commands (your drawing may differ slightly):

Detail ⇒ **Show/Erase** ⇒ **(Dimension** radio button on) ⇒ **Show All** ⇒ **Yes** (Fig. 17.20) ⇒ **Close** ⇒ **Done/Return**

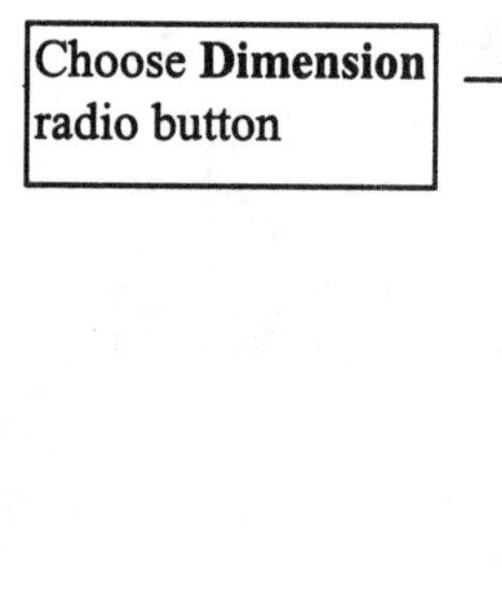

Figure 17.20
Displaying Dimensions

Before modifying the dimension locations and views, change the settings in the **.dtl** file using **default_font** to **filled** and **draw_arrow_style** to **filled**:

Set Up ⇒ **Modify Val** ⇒ (make the necessary changes) **File** ⇒ **Save** ⇒ **File** ⇒ **Exit** (Fig. 17.21) ⇒ **Quit** ⇒ **View** ⇒ **Repaint** ⇒ **Done-Return** (Fig. 17.22)

HINT

Always set your font and arrow style to filled:

default_font **filled**
draw_arrow_style **filled**

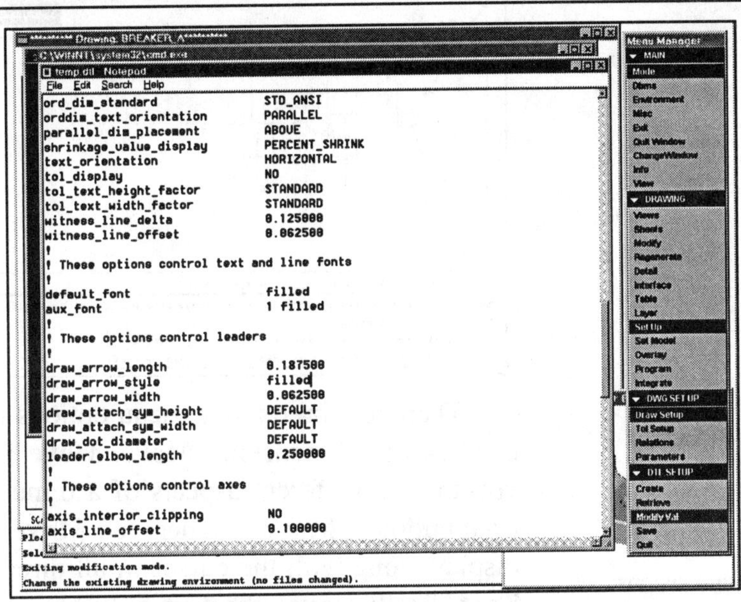

Figure 17.21
Set Up ⇒ Modify Val

Modifying Dimensions and Drawing Entities

In Drawing mode, you can access a pop-up menu item by clicking and holding down the *right* mouse button while the cursor is in the main window. From the pop-up window, select **Modify Item**, then select the item you wish to modify (Fig. 17.22). Another menu displays the actions that can be performed on this type of item. The **Move/Activate** option is the default. To move an item, simply select it and you will be in dynamic move.

HINT

Again, after selecting any entity, click and hold down the *right* mouse button while the cursor is in the Main Window. The EDIT ACTIONS menu will appear next to the mouse cursor.

The list of additional actions you can perform on the item is also available in the pop-up menu. To modify another item, simply select it, and the menu will display the actions appropriate to it.

This functionality reduces the number of menus you must navigate. Also, it eliminates the need to traverse the screen to reach the menus, because the pop-up window is available at all times. For example, while the **Show/Erase** dialog box is displayed, you can use the pop-up menu option to move dimensions into the correct locations without having to close the dialog box.

Move, Move Text, Clip, and **Skew** functionality are all available using **Move/Activate** (Fig. 17.23). The selection of which box handles determines how items are moved. For example, if you select the dimension text, you can move the text of the dimension and the leader line. Selection of the leader line allows you to clip that leader line. Picking the extension (witness) line allows you to clip that extension line, and picking the other end of the extension line allows you to skew the dimension.

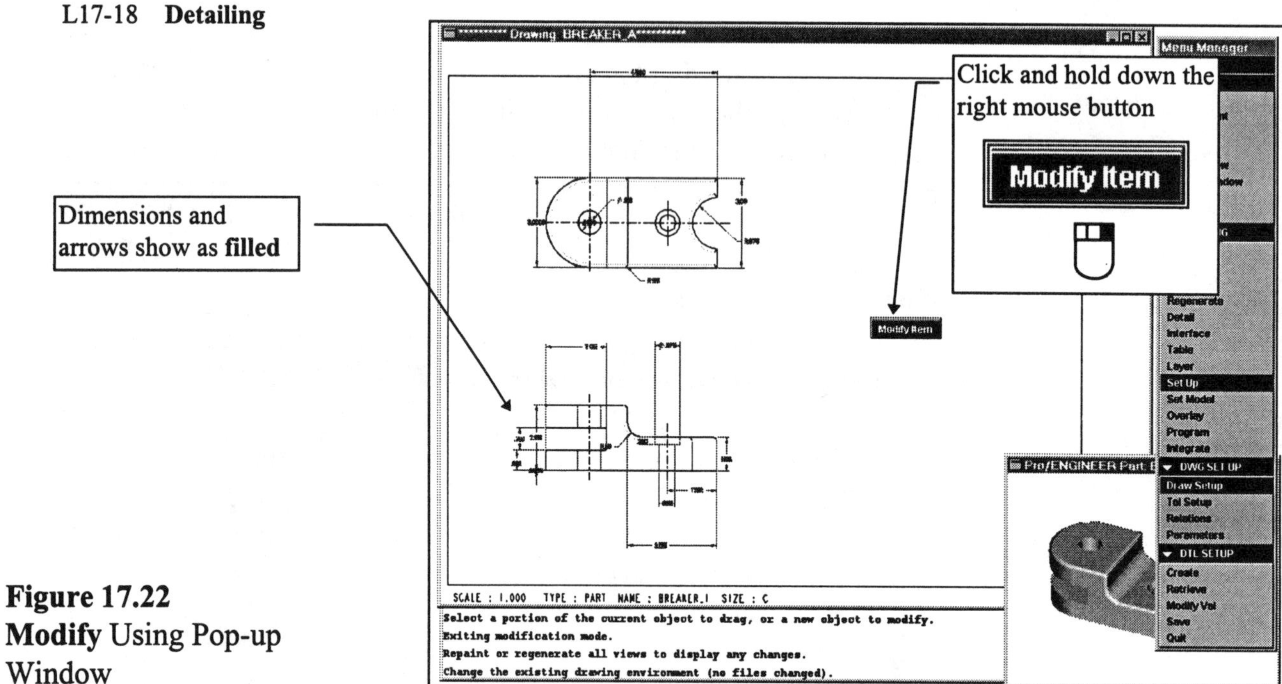

Figure 17.22
Modify Using Pop-up Window

There are three move menu options that allow you to change the position of a dimension. **Move, Move Text,** and **Mod Attach** allow you to alter different aspects of a dimension. The most commonly used option is **Move**. When you **Move** a dimension, Pro/E drags its display along with the cursor in all directions. When you use **Move Text**, Pro/E moves the text only and does not move the extension lines. When you use **Mod Attach,** select a note, geometric tolerance, symbol, or surface finish to modify. Select the desired reference options, then select the leader line you want to move, and then the new attach point. Reposition the dimensions and clean up the drawing. Do not concern yourself about dimensions in the wrong view at this time.

HINT
Before modifying dimensions, make your grid size very small or turn the grid off:
□ **Grid Snap**

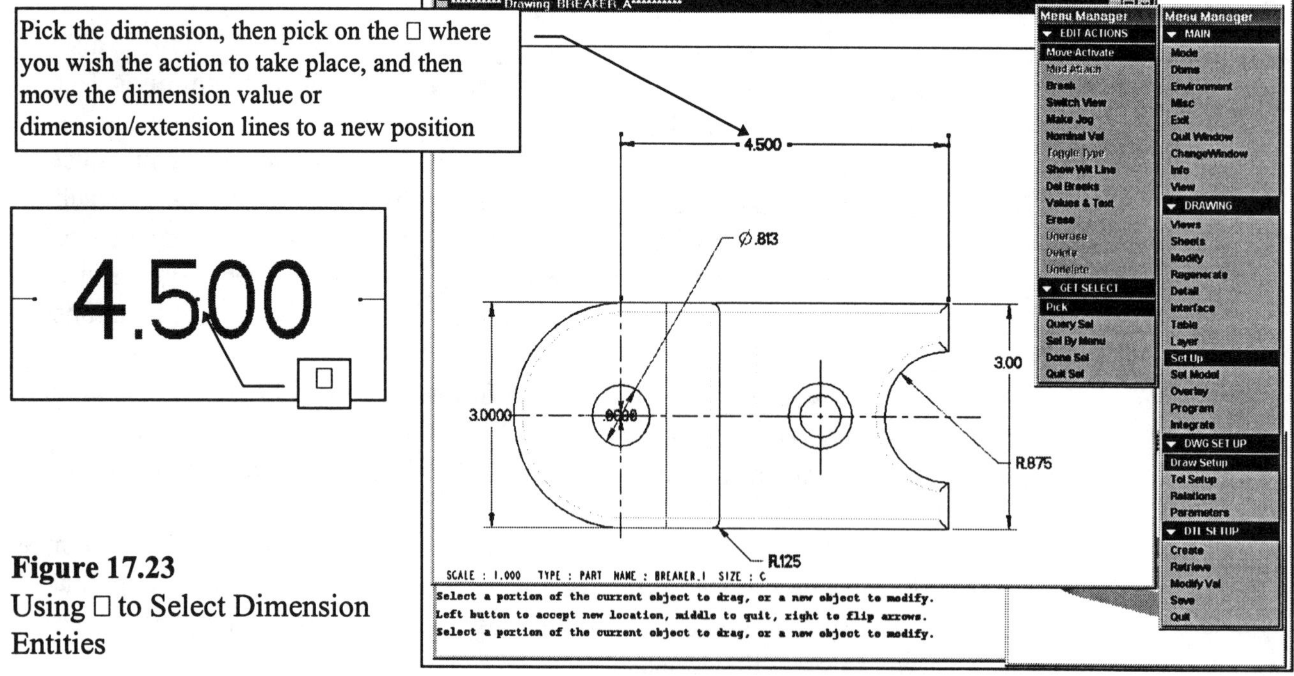

Figure 17.23
Using □ to Select Dimension Entities

Clipping Drafting Objects

The **Clip** option allows you to move the endpoints of many drafting object components, such as extension lines and centerlines. This option allows you to "pick up" the end of an extension line and drag it to a new length (Fig. 17.24). Of course, clipping can also be done with the pop-up option **Modify Item** menu.

Dbms (File--PT/Modeler) ⇒
Save ⇒ enter
Purge ⇒ enter ⇒
Done-Return
This will save the drawing.

One of the most common applications for **Clip** is to "regap" extension lines. When you **Show** feature dimensions, the extension lines are often gapped to a point on the part that is not appropriate for the view in which the dimension is displayed.

Clip allows you to drag the extension line end to a location that is visually more pleasing (and consistent with drafting standards). When you **Clip** the extension line, Pro/E remembers the clip distance and continues to apply it even if the model changes or if you **Erase** and redisplay the dimension.

NOTE

If you are clipping *multiple dimensions*, Pro/E "grabs" the extension line in each dimension that is closest to the cursor when you click the *middle* mouse button.

When you **Clip** extension lines, Pro/E prompts you to pick dimensions to clip. This allows you to clip several dimensions' extension lines together.

To stop picking dimensions, click the *middle* mouse button. Next, indicate which extension line(s) you want to clip, by moving the cursor near the desired extension line end. Pro/E then begins to drag the extension line(s). To **Clip** the line(s) after you move the cursor to the desired location, click the *left* mouse button. If you want to abort the **Clip**, click the *middle* mouse button.

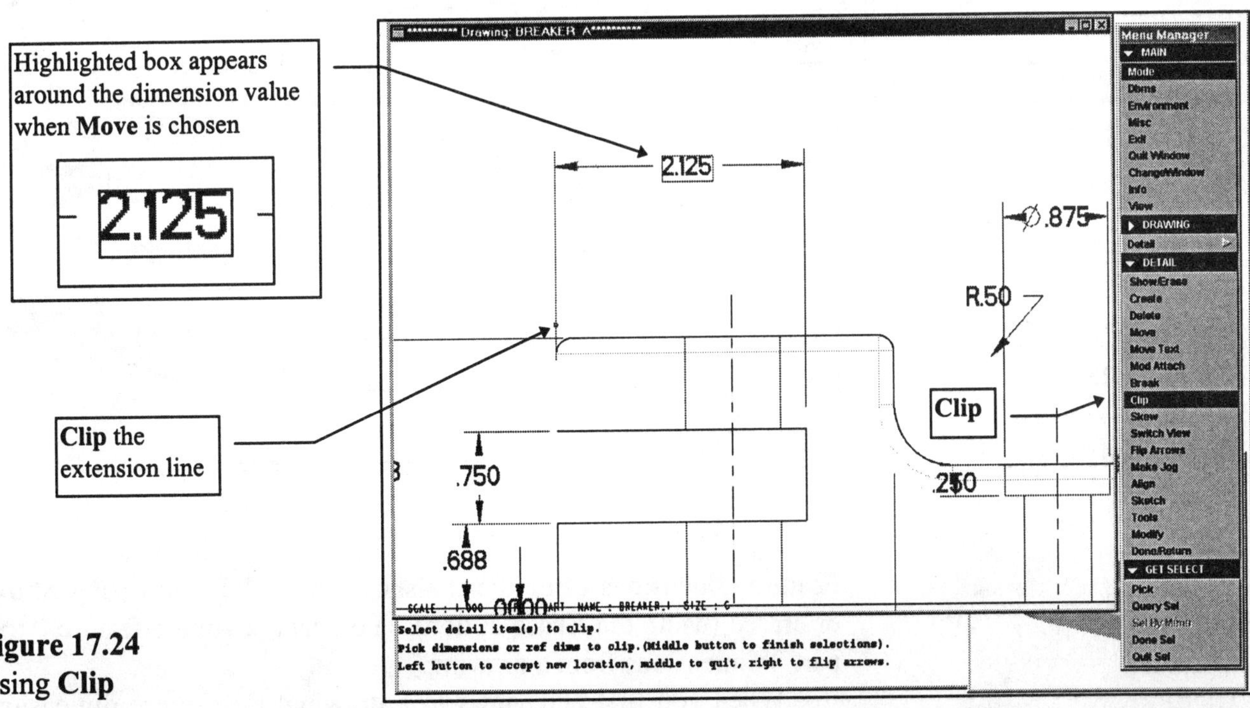

Figure 17.24
Using **Clip**

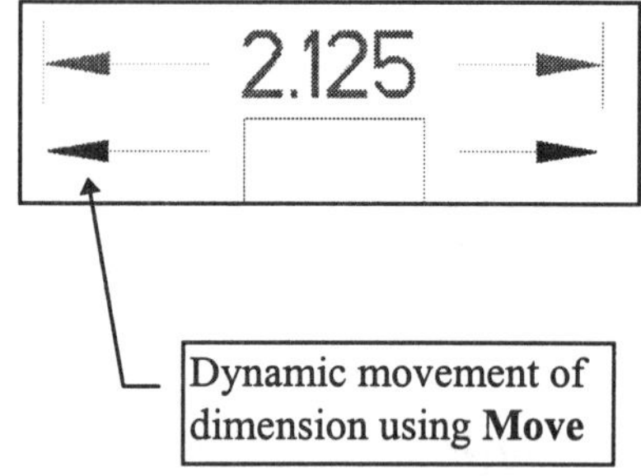

When you use **Move** to move a linear dimension, Pro/E adjusts the leader, arrows, and extension lines in proper relation to the text (which is what you are actually dragging).

If you place the text of a dimension between the extension lines and it really does not fit (because the extension lines are too close together), you may find the display of the arrows to be unacceptable. Move the origin of the text from one side of the extension lines to the other, Pro/E automatically flips the leader to the other side of the text.

When you move a radial dimension using **Move**, you can only move the text along the leader and the location where the arrow touches the arc, and thus alter the angle of the leader; however, the short line (stub) between the leader and the text stays at its original length. When you use **Move Text**, you can change only the length of the stub.

Besides moving dimensions, you may also need to switch the view that a dimension appears in after choosing **Show All**. In Figure 17.25, **Switch View** was chosen from the DETAIL menu (also available after choosing a dimension when the **Modify Item** option is selected from the pop-up window menu).

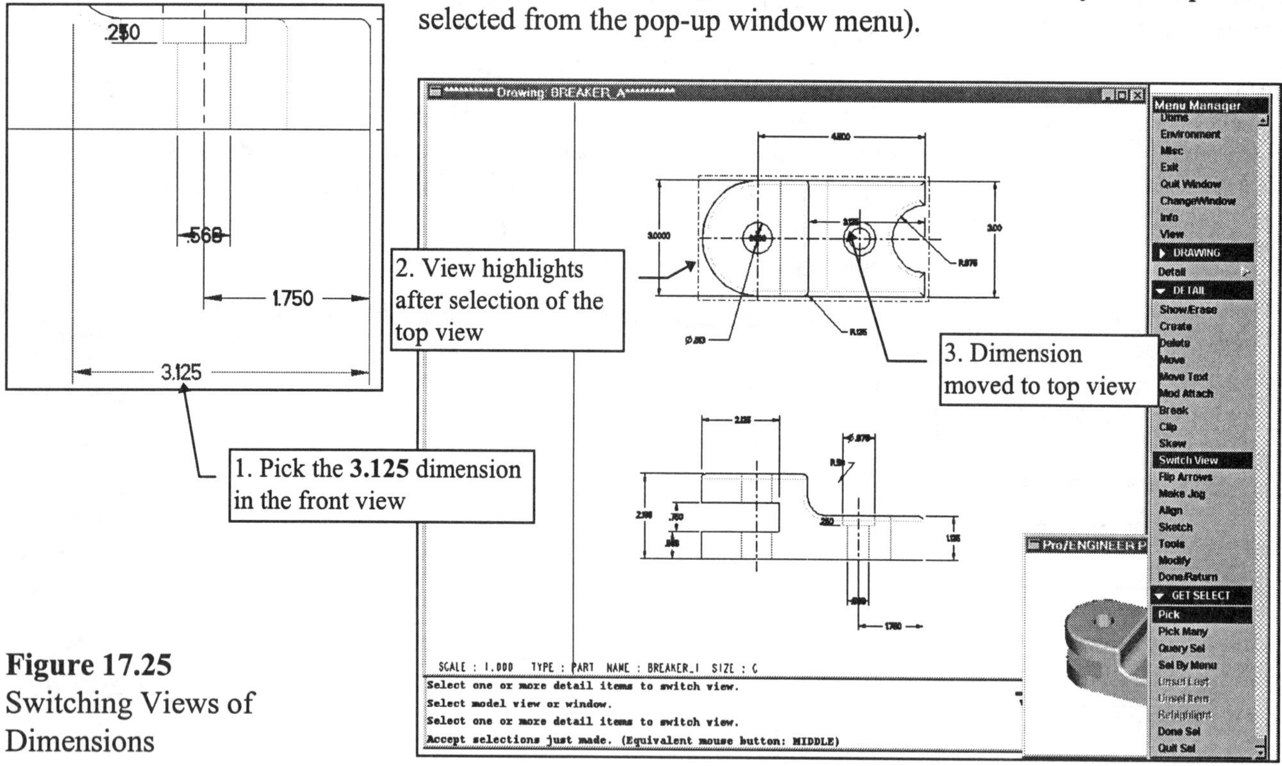

Figure 17.25
Switching Views of Dimensions

Erasing Feature Dimensions

Feature dimensions can either be shown on a drawing (using **Show**) or erased (using the **Erase** option). You cannot **Delete** feature ("true part") dimensions from a drawing.

When you first add views to a drawing, the feature dimensions are also present; they are simply in the **Erased** state. When you choose **Show/Erase** and the desired options, you are unerasing them.

Many times, when you **Show** feature dimensions, there are dimensions you do not need. This is especially true of dimensions of thickness and pattern number callouts. You may find that there are other dimensions you do not want shown.

If you want a dimension, but it appears in the wrong view, you do **not** want to erase it. The **Switch View** option allows you to move that dimension to the view where you want it displayed.

If, however, there are actually dimensions that you do not need, you can **Erase** them. When you choose **Erase** and the desired options, Pro/E prompts you to select the dimensions to erase. You use the *middle* mouse button to finish the selecting of the desired dimensions to erase.

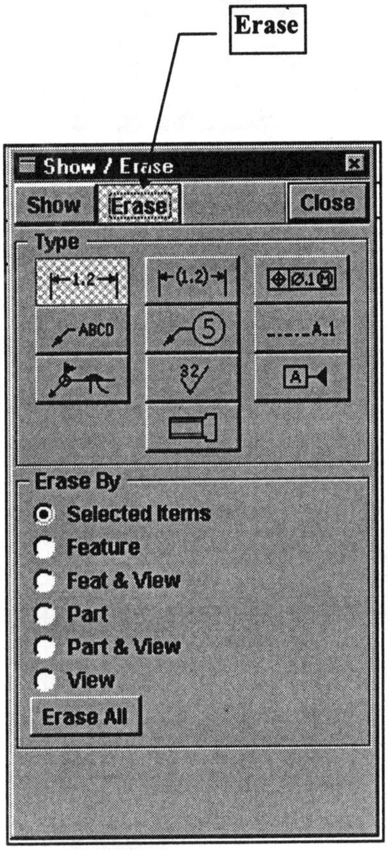

If you accidentally erase desired dimensions, you can redisplay them using the **Show** option. Unfortunately, the **Show** option does not know exactly which dimensions you want to redisplay (from the set of dimensions that you have erased). Therefore, you may have to redisplay several dimensions and re-erase those you really do not need. Using the **Feat & View** option with **Show** (instead of **Show All**) will minimize the number of dimensions you need to redisplay.

The **Erase** menu gives you a great deal of control over how you select the items to erase. You can choose to select a type of drafting object individually, or erase all objects of a type, in the drawing, in a view, and/or by feature.

If you choose to **Erase All** of an item, Pro/E prompts you to confirm the erasure. If you pick **No**, the erasure operation is aborted. If you pick **Yes**, the items are all erased.

The simplest way to erase drawing items is to use the **Show/Erase** dialog box, choose the **Type**, and pick the **Selected Items**. You may now erase items by using **Query Sel**.

Using the right mouse button, activating **Modify Item**, and picking the item to modify (as in Fig. 17.26, where the **.0000** dimension was chosen) will provide a choice of several actions, including **Erase**. Picking **Erase** will make the dimension *gray out*. You can continue picking items to erase. When the selections are complete, choose **Done Sel** to remove the items from the screen. This method, although faster, provides fewer choices.

When you use the **Create ⇒ Datum ⇒ Plane ⇒ Offset ⇒ enter ⇒ enter ⇒ enter ⇒** method to establish datum planes and a coordinate system in **Part** mode, you will also be creating three **.0000** dimensions that need erasing in **Drawing** mode. At this time, erase the **.0000** dimensions (Fig. 17.26) from the two views and move the dimensions off the faces of the views, per ASME standards.

Another aspect to change on the drawing is the number of digits displayed for individual dimensions (Fig. 17.27). In most cases, the number of digits is dependent on the tolerance for the dimension.

Erase

Show / Erase

Show | Erase | Close

Type

Erase By

Selected Items

Feature

Feat & View

Part

Part & View

View

Erase All

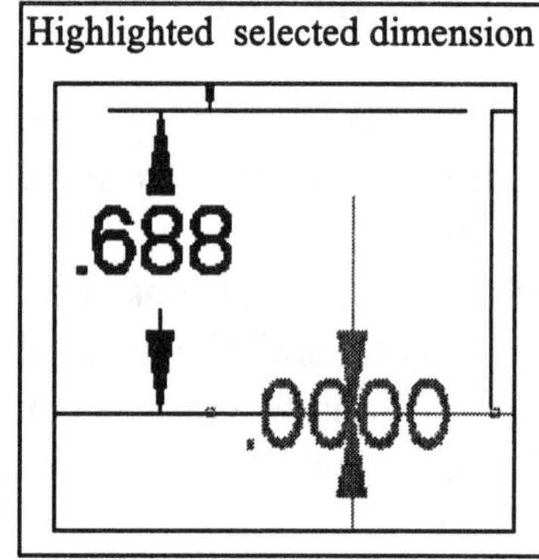

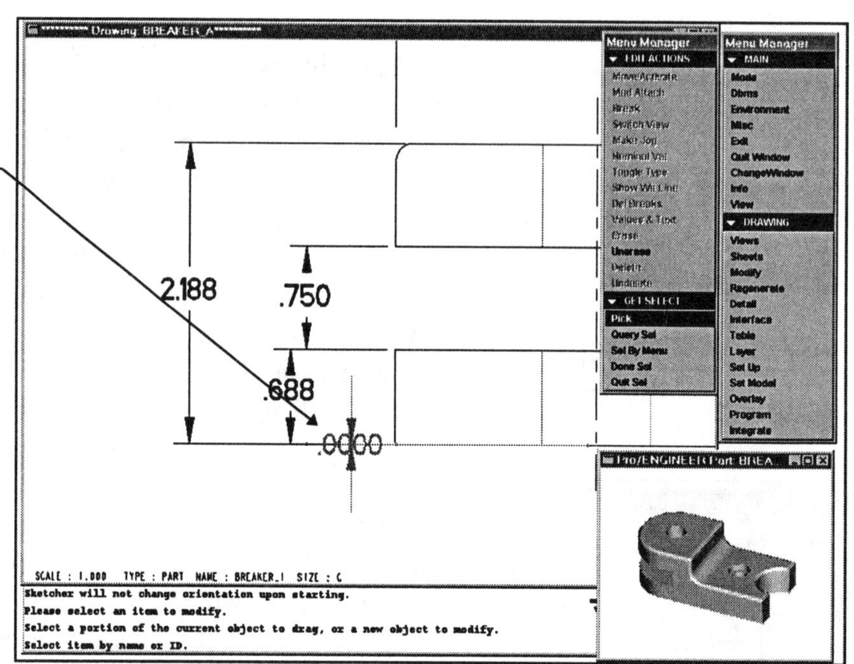

Figure 17.26
Using **Modify Item** to **Erase** a Dimension

HINT

Be careful: by changing the number of digits of a driving dimension the dimension color will change from yellow to white and will print bold (object line thickness). You will have to change the color back to yellow for the line thickness to print correctly (dimension line thickness). Also, as an example, a **.8125** driving dimension will become **.813** with three digits.

Change the number of digits for dimensions displayed with four decimal places to three decimals (Fig. 17.27) by choosing the commands:

Detail ⇒ **Modify** ⇒ **Num Digits** ⇒ (type **3** ⇒ **enter**) ⇒ (pick the dimension to alter) ⇒ **Done Sel** ⇒ **Done/Return**

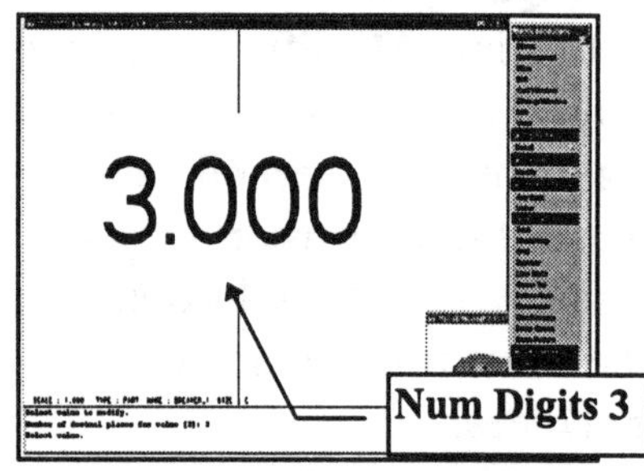

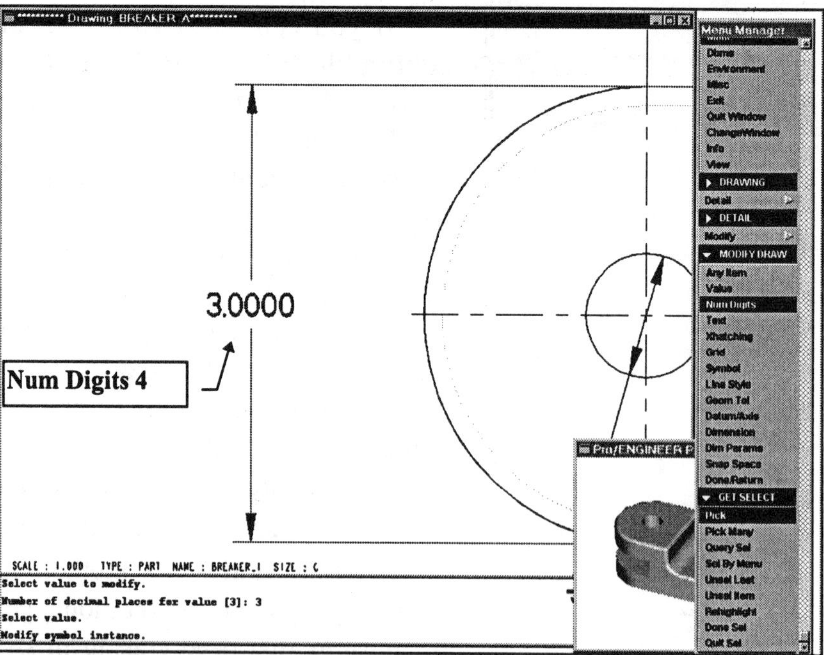

Figure 17.27
Changing the Number of Digits

The **Flip Arrows** option is used to change the dimension arrows from inside to outside arrows or from outside to inside arrows. Small dimension values will create arrows on top of the dimension value, and radial dimensions sometimes point to the wrong side of the arc, as in Figure 17.28, where the arc radius is dimensioned to the side of the arc that does not exist. All three dimensions shown in Figure 17.28 should be flipped.

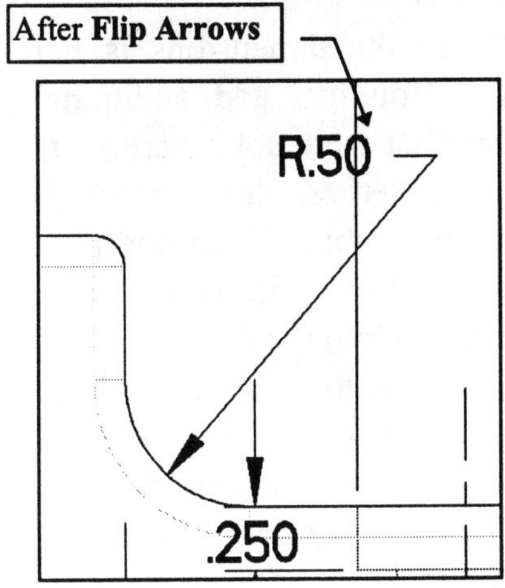

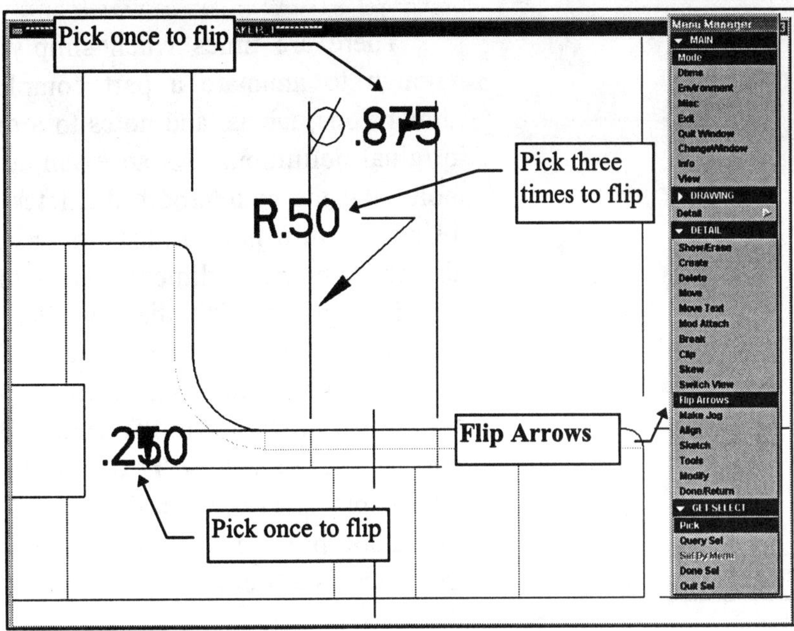

Figure 17.28
Flipping Arrows

To have diameter dimensions point to the outside of the circle with one arrowhead instead of across the diameter using two arrowheads, flip the arrows.

The dimensions are complete (Fig. 17.29), but you still need to make a number of changes to create a correctly dimensioned detail of the part. In some cases, dimensions need to be combined into notes and reference dimensions added to the drawing.

Figure 17.29
Dimensioned Drawing

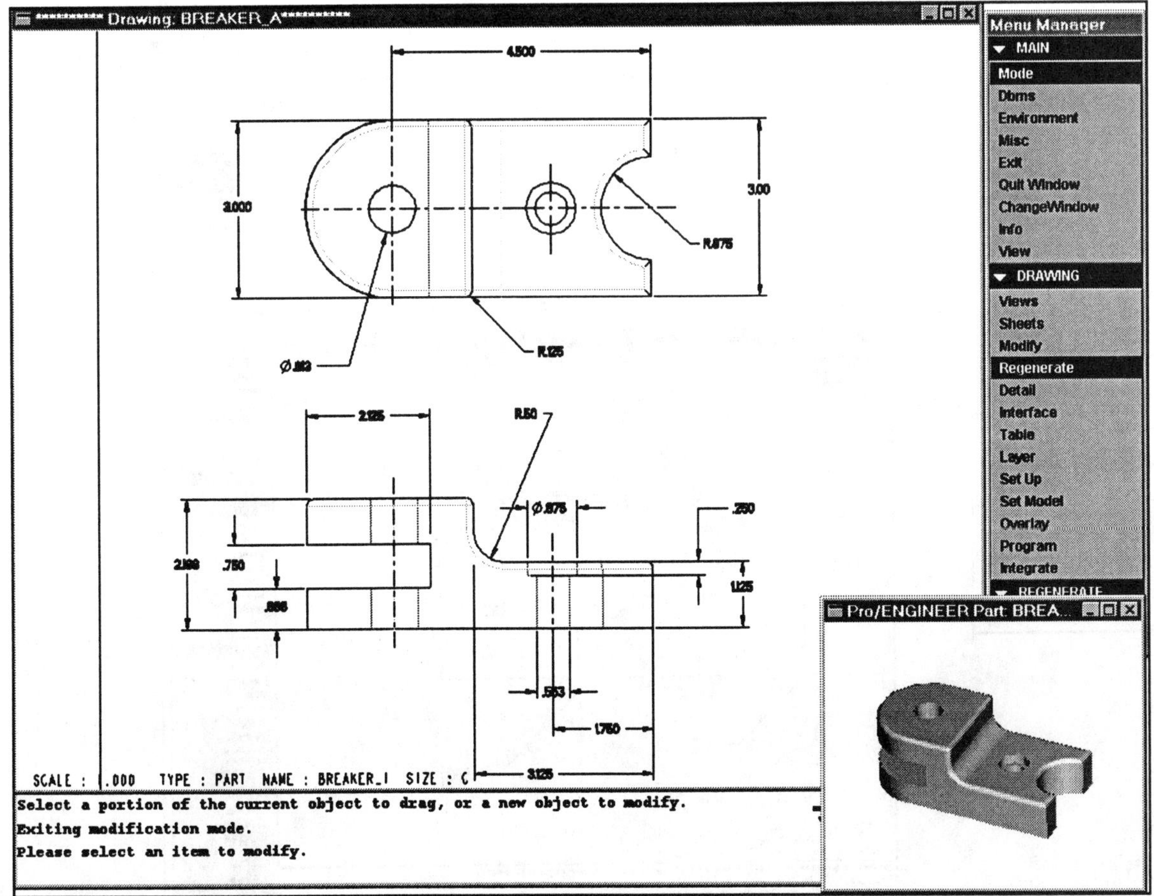

There are times when simply showing the dimensions is not enough to annotate a part completely. You can add additional dimensions, labels, and notes to a drawing that were not a part of its original definition. As an example, the counterbore depth, the thru hole, and the counterbore diameter need be combined into one note using the thru hole as the dimension to modify. The counterbore diameter and depth dimensions are then erased from the drawing.

In Figure 17.29, the diameter dimensions for the counterbore need to be switched to the top view. After the dimensions are switched, click and hold down the *right* mouse button anywhere on the screen to show **Modify Item**. Pick the **.563** diameter dimension and choose **Values & Text**, and the Modify Dimension dialog box will display, as shown in Figure 17.30. Select the **Dim Text** tab from the dialog box (Fig. 17.31). Remember, add the diameter symbol to the value as a prefix, and add the **.875** counterbore diameter and **.250** depth to the note.

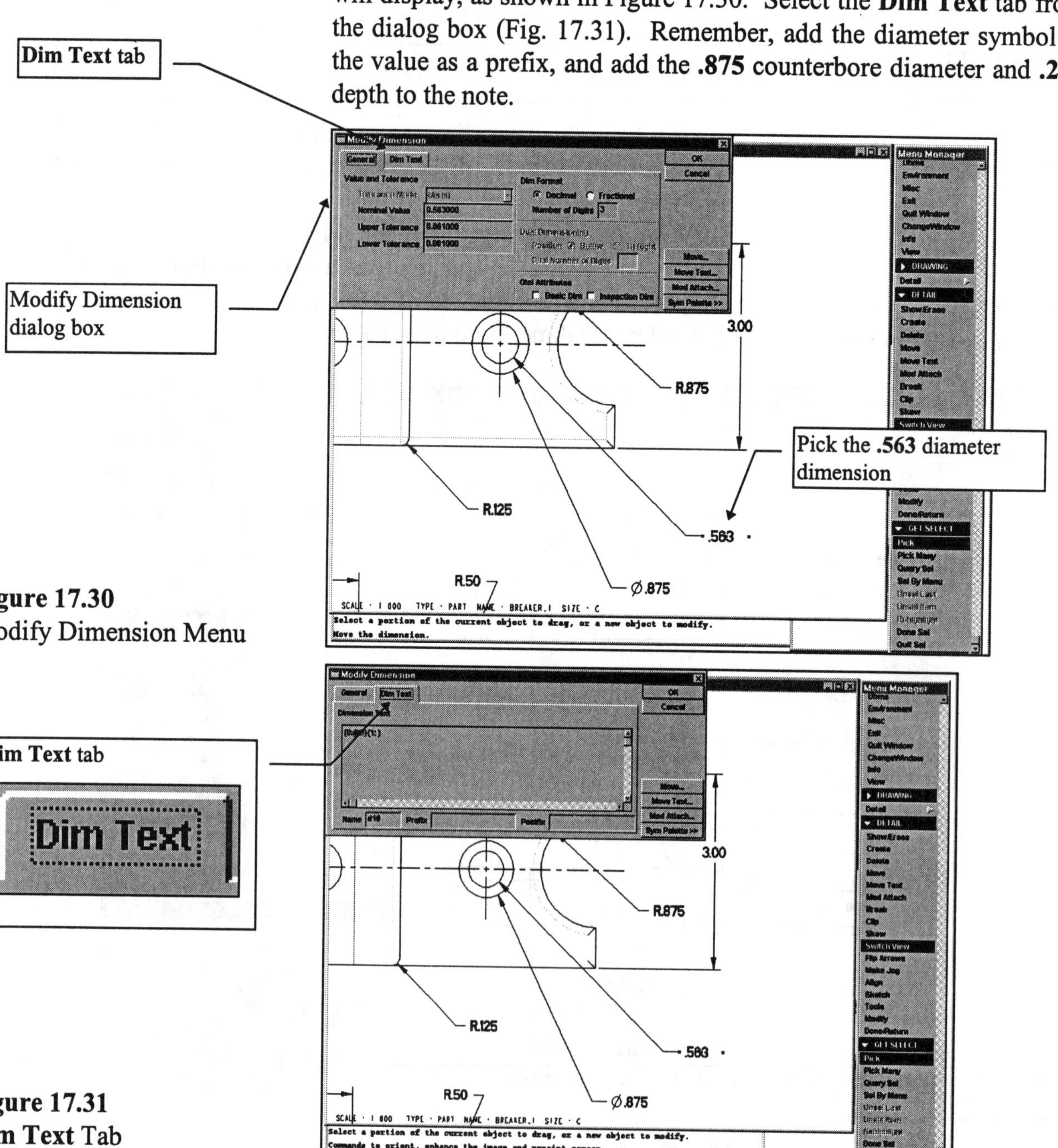

Figure 17.30
Modify Dimension Menu

Figure 17.31
Dim Text Tab

If instead of entering **.875** and **.250** as text, you enter **&** followed by the proper dim symbols (e.g., **d23**), the dimensions are then parameters that will reflect any modifications and changes. To see the dim symbols in Drawing mode, choose **Info ⇒ Switch Dim.** As an example:

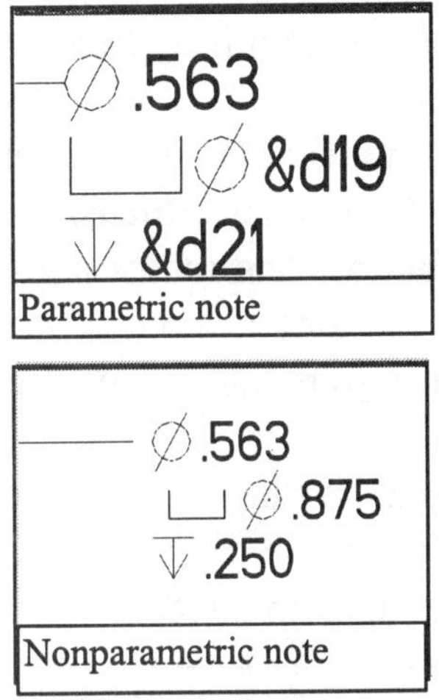

Choose the **Sym Palette** button, then place your cursor at the beginning of the text line in the dialog box window or in the **Prefix** sub-window, as shown in Figure 17.32. Pick the diameter symbol Ø from the palette. This will add the symbol at the beginning of the text line if it is not already there. Add a counterbore symbol on a new text line, along with the **Ø.875** dimension. To complete the note, add a third line with a depth symbol and the **.250** dimension. Erase the **Ø.875** and **.250** dimensions from the drawing (Fig. 17.33).

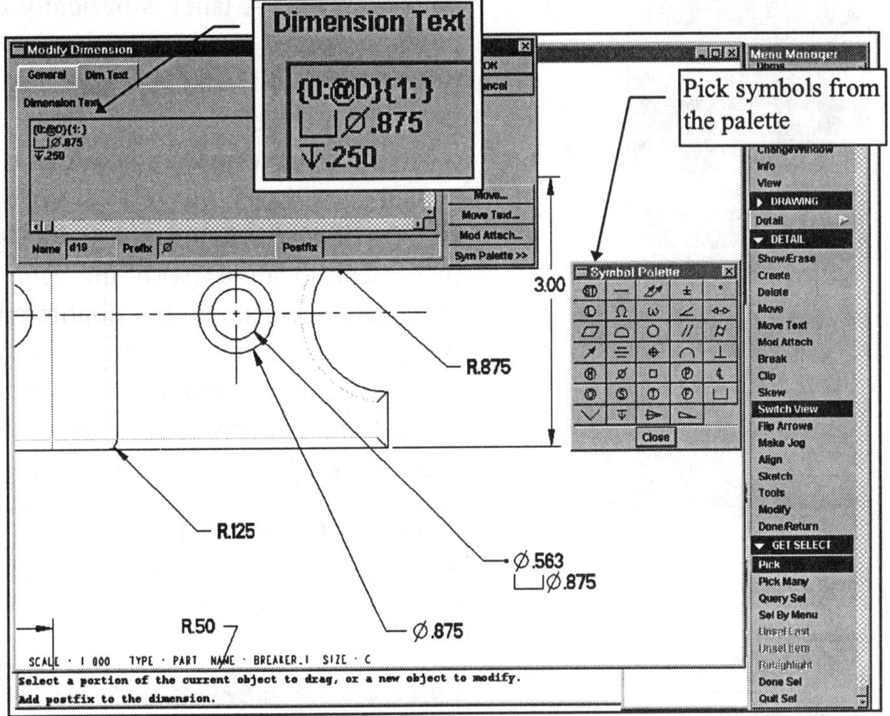

Figure 17.32
Modifying and Adding Text to a Note

HINT

Use **Info** and **Switch Dim** to show the dim symbols.

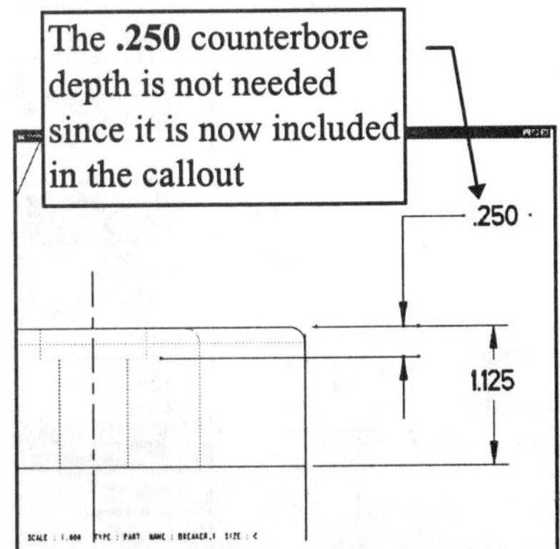

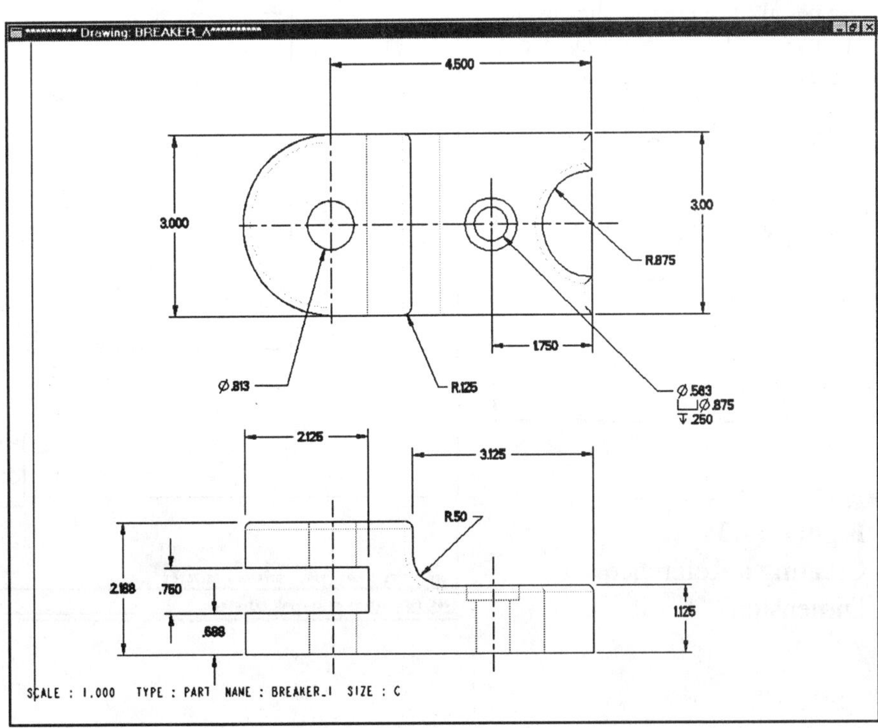

Figure 17.33
Erasing Unneeded Dimensions

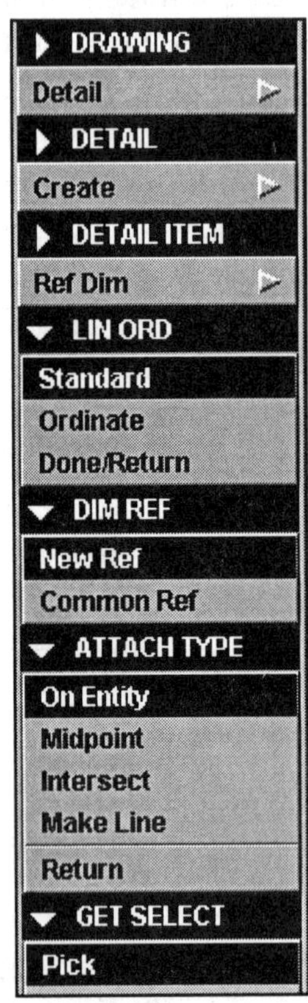

Dimensions that you add "manually" are called ***driven dimensions***. Driven dimensions change when the model changes, but they cannot be used to make changes to the model. Only dimensions that you place on the drawing using the **Show** option can ***drive*** changes to the model (and they are referred to as ***driving dimensions***).

Driven dimensions can be erased in the same manner as feature dimensions. In addition, however, driven dimensions can be deleted; this permanently removes them from the drawing database.

You can also create notes and labels to add to the annotation on your drawing. A label is basically a note that has a leader. Leaders can be attached to edges or drawn to positions in space. It is better practice to modify an existing diameter dimension than it is to create a note with a leader.

You create driven dimensions using the same techniques that are available in sketching. For example, you create parallel dimensions by selecting a linear edge (with the *left* mouse button) and indicating a placement location (with the *middle* mouse button). You create diameter dimensions by double-clicking on a circular edge and indicating a placement location.

Reference dimensions can also be added to a drawing. Reference dimensions are driven dimensions. Create a reference dimension to show the total width of the part using the following commands:

Detail ⇒ **Create** ⇒ **Ref Dim** ⇒ (pick both ends of the part in the front view and place the dimension below the view, as shown in Fig. 17.34) ⇒ **Done/Return**

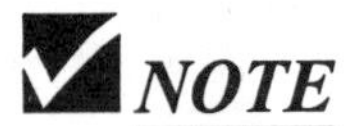

NOTE

ASME standards use parentheses around the reference dimension. The use of **REF** is not standard, but it is used by some companies.

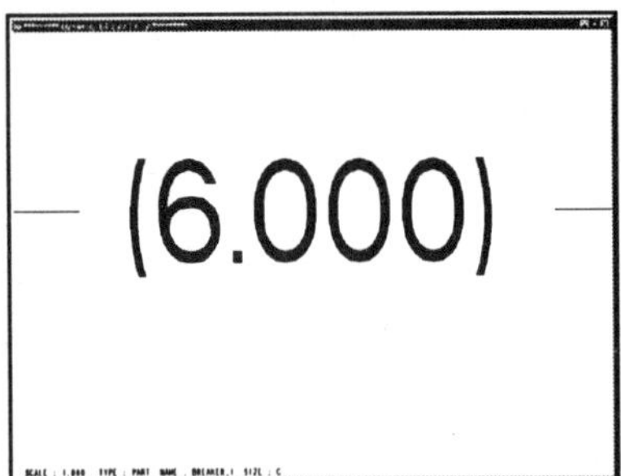

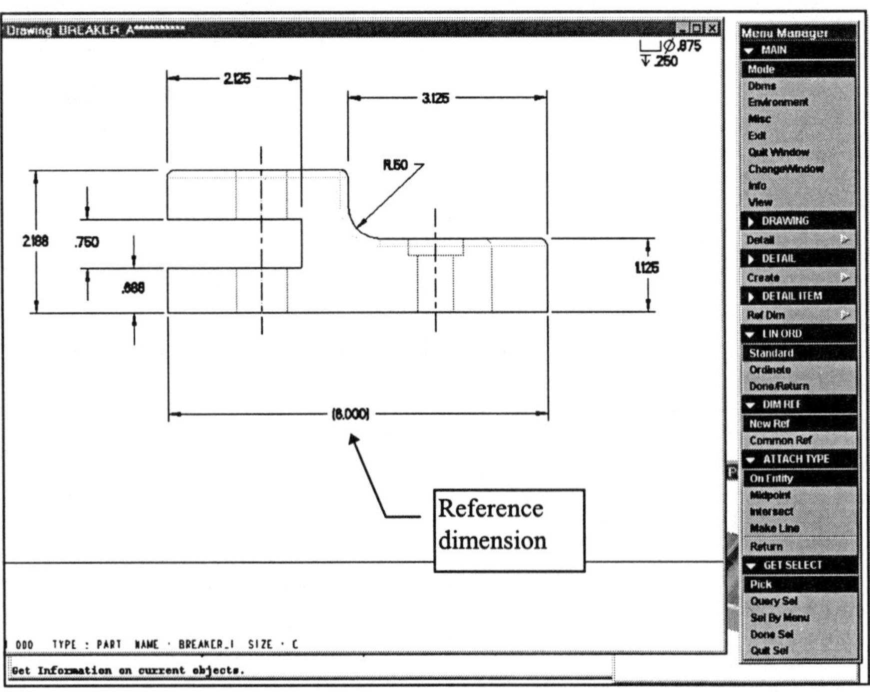

Figure 17.34
Creating a Reference Dimension

HINT

If you set your **Environment** to **Isometric**, your pictorial will be isometric instead of **Trimetric**.

Alphabetic characters are created in a font and sized such that they comply with standards for **Geometric Dimensioning and Tolerancing (GD&T).** These characters can be used for any purpose, but they are designed to be properly displayed in **GD&T** frames.

You may have noticed that up to this point, most of the text you enter is automatically capitalized by Pro/E. For example, you name a section by entering "**a**" but Pro/E displays it as "**A.**" When you enter text in labels, notes, or as text appended to a dimension, Pro/E displays exactly the case you enter. So if you enter "**material,**" Pro/E displays "**material**" on the drawing. If you want "**MATERIAL,**" you need to type "**MATERIAL.**"

Since the drawing is almost complete, let us add another general view (pictorial view) in the upper corner of the drawing using the following commands:

Views ⇒ **General** ⇒ **Done** ⇒ (pick in the upper right of the drawing, as shown in Fig. 17.35) ⇒ **Done/Return** ⇒ **Done/Return**

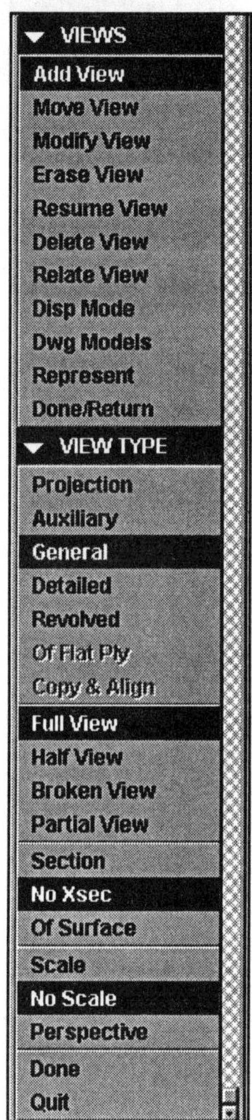

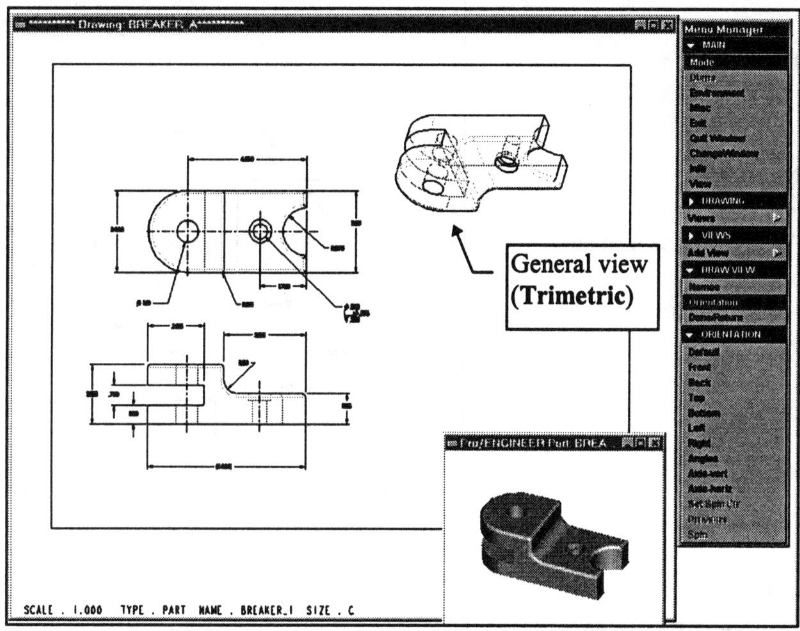

Figure 17.35
Adding a General View

Since you used a blank format to start the drawing, replace the format with a standard ANSI "**C**" size drawing format using the following commands (Fig. 17.36) :

Sheets ⇒ **Format** ⇒ **Add/Replace** ⇒ **?** ⇒ **enter** ⇒ **Format Dir** ⇒ (pick **c.frm**)

Also, try to use your personal format (Fig. 17.37):

Sheets ⇒ **Format** ⇒ **Add/Replace** ⇒ **?** ⇒ **enter** ⇒ **Current Dir** (or the directory where your formats are stored) ⇒ (pick ***your_format_name*.frm**)

Figure 17.36
ANSI Standard **"C"** Size Format

Dbms (File--PT/Modeler) ⇒ Save ⇒ enter
Purge ⇒ enter ⇒ Done-Return

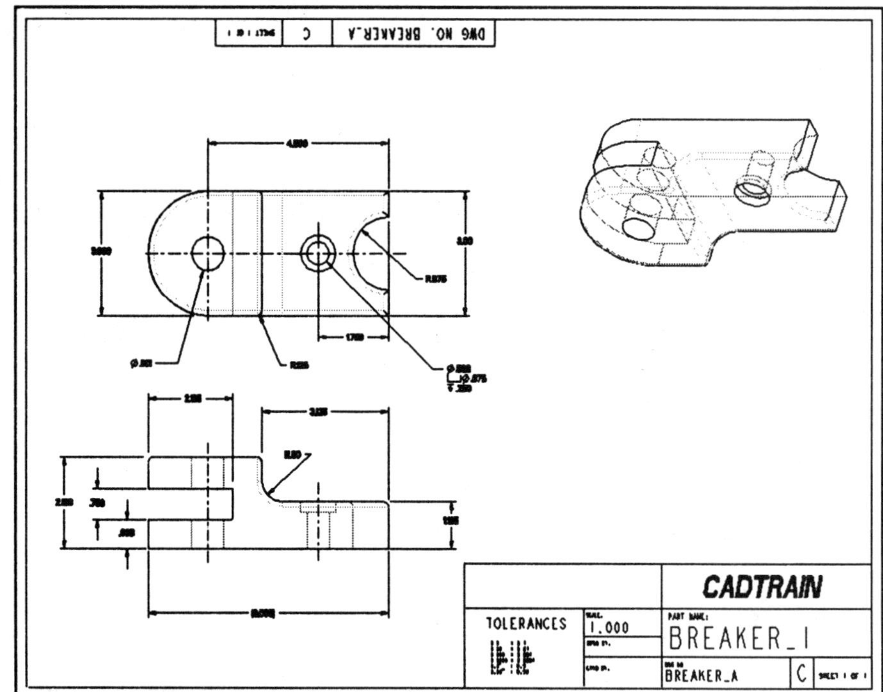

Figure 17.37
Your Personal Format (here, from CADTRAIN)

Each view of your model can have a different line display style: **Wireframe, Hidden Line, No Hidden,** or **Default**. The tangent display status can also be established independent of the environment default. **Tan Solid, No Disp Tan, Tan Ctrln, Tan Phantom, Tan Dimmed**, and **Tan Default** are available. Use the following commands to set the display of each view, starting with the general pictorial view:

Views ⇒ Disp Mode ⇒ View Disp ⇒ (pick the general view, as shown in Fig. 17.38) **⇒ Done Sel ⇒ No Hidden ⇒ Tan Solid ⇒ Done ⇒ Done Sel ⇒ Done/Return**

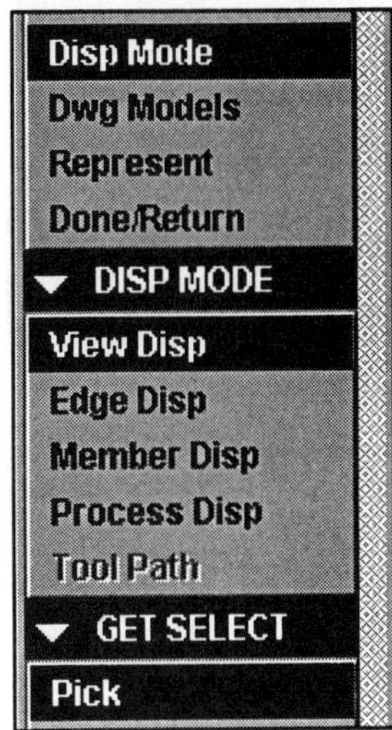

Figure 17.38
Setting the Display Status for a General View: **No Hidden** and **Tan Solid**

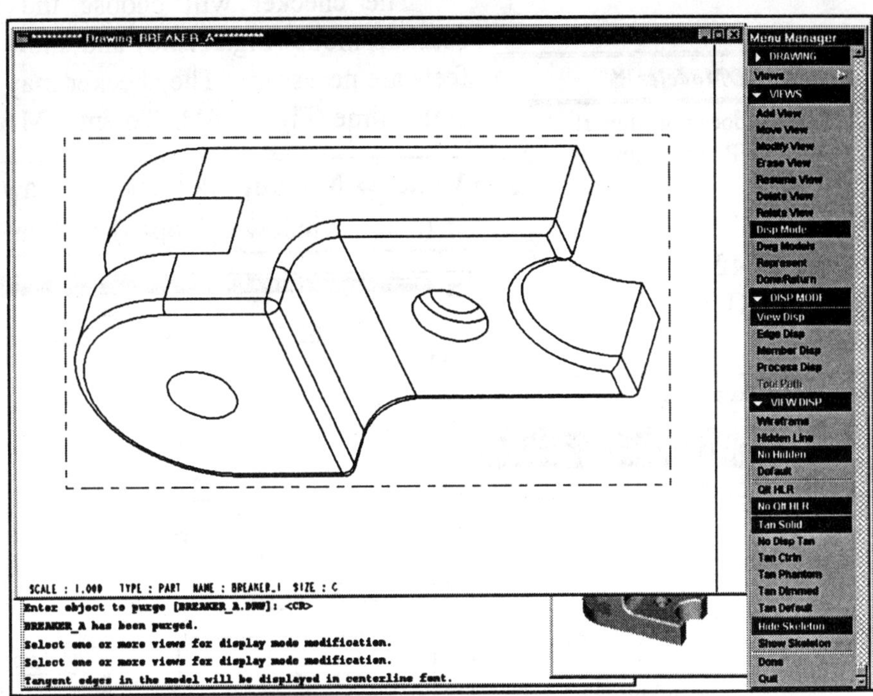

Figure 17.39
Setting the Display Status for the Top and Front Views: **Hidden Line** and **Tan Dimmed**

Set the front and top views to have **Hidden Line** and **Tan Dimmed** using the following commands (Fig. 17.39):

Views ⇒ **Disp Mode** ⇒ **View Disp** ⇒ (pick the top and front views) ⇒ **Done Sel** ⇒ **Hidden Line** ⇒ **Tan Dimmed** ⇒ **Done** ⇒ **Done Sel**

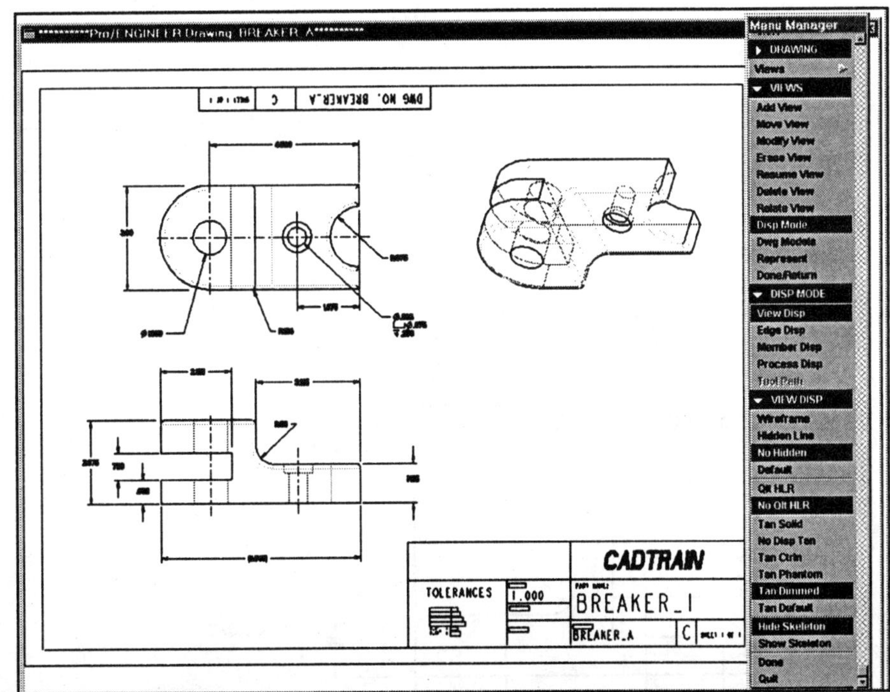

Since you believe the project is complete, print or plot out a copy and submit it to the checker (teacher, boss, design checker). The checker will use **Markup** mode to check and mark up your drawing.

PT/Modeler™

PT/Modeler does not support Pro/MARKUP at this time.

The checker will choose the following commands to enter Markup mode (Fig. 17.40) and show the **checker changes** he or she feels are necessary. The checker may also make some design changes at this time (Fig. 17.41). To enter Markup mode, choose:

Mode ⇒ **Markup** ⇒ (pick the drawing name used for the Breaker) ⇒ (a new window will open) ⇒ **New** ⇒ (type **Brk_Check**) ⇒ **enter**

Figure 17.40
Pro/MARKUP

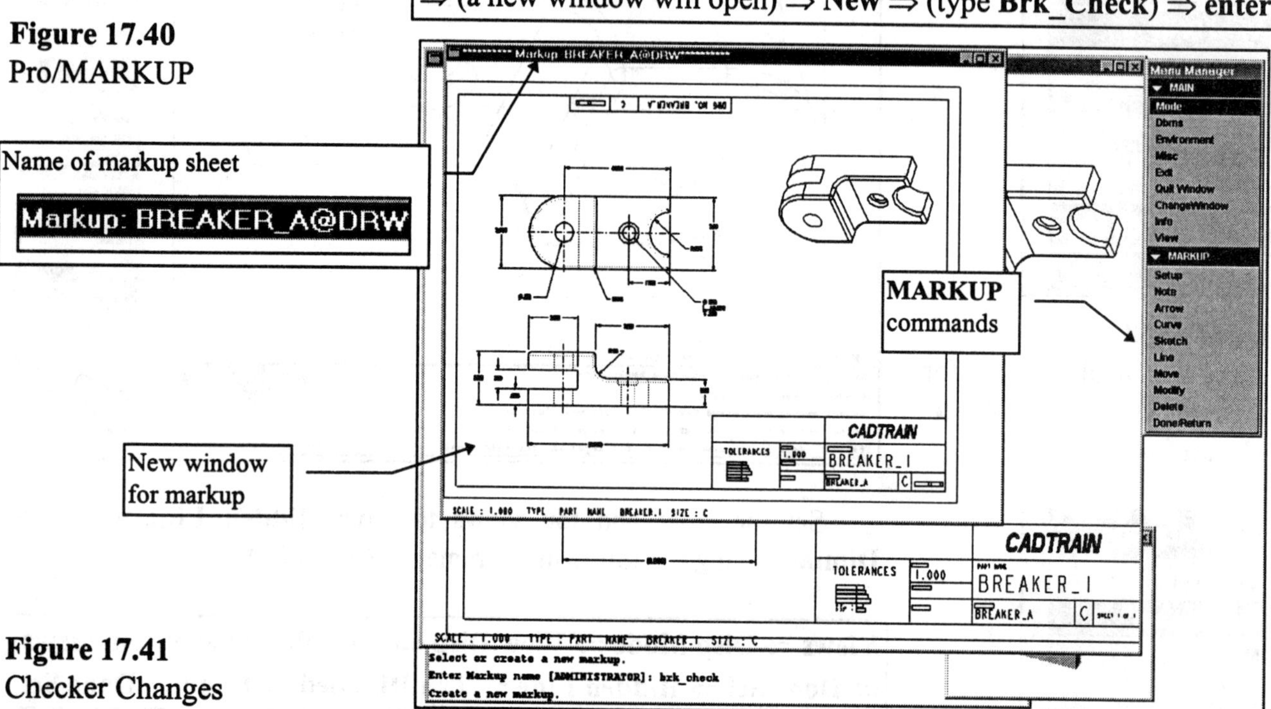

Figure 17.41
Checker Changes

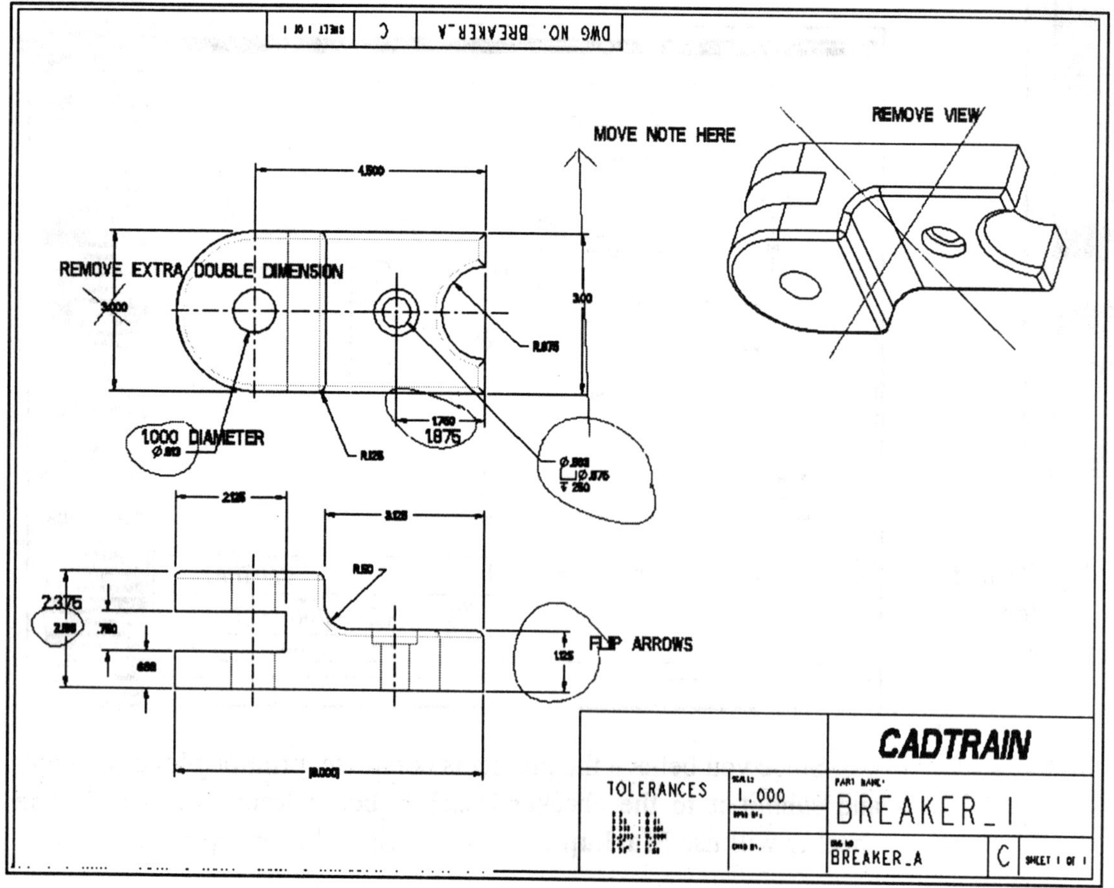

After the checker saves the changes, you can bring up the markup drawing and your Breaker detail drawing. By keeping the two files in session, you can work in your detail window and still see the markup drawing. Since you will also be making some design changes, it is a good idea to keep the part window active to see the changes propagated throughout the model, including the part, the drawing, and the assembly (if used in an assembly). All files will be updated with the new design changes after regeneration. If you are starting a new session, use the following commands (Fig. 17.42):

1. Mode ⇒ Drawing ⇒ Search/Retr ⇒ BREAKER
2. Mode ⇒ Part ⇒ Search/Retr ⇒ BREAKER
3. Mode ⇒ Markup ⇒ BREAKER ⇒ BRK_CHECK

Move the windows of each mode for easier viewing and work, as shown in Figure 17.42. Change the active window so it is the drawing, not the part or the markup.

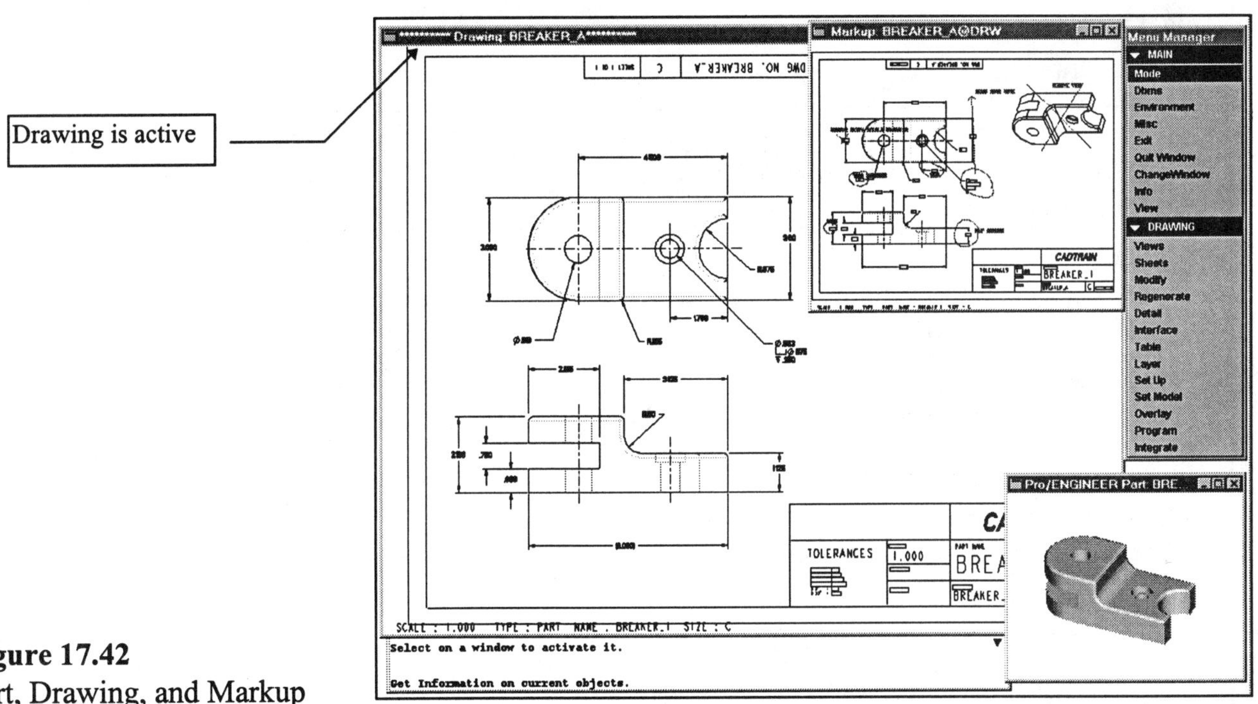

Figure 17.42
Part, Drawing, and Markup

Make the checker changes to the drawing (of course, you could also make the Part mode the active window and modify and regenerate the model there). Regardless of the mode or window you are working in, remember to **Regenerate** both the drawing *and* the part (Fig. 17.43).

The completed design can now be saved and plotted, as in Figure 17.44.

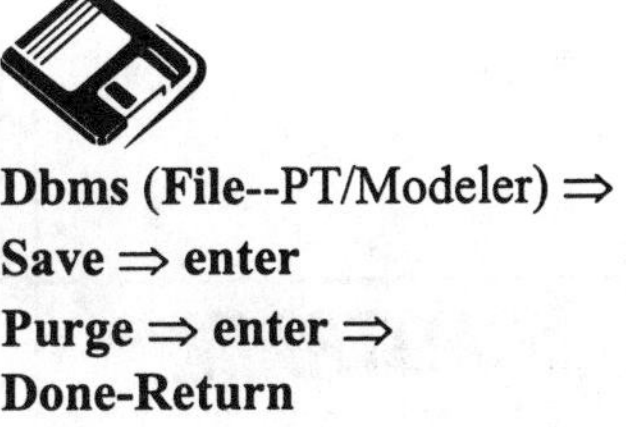

Dbms (**File--PT/Modeler**) **⇒**
Save ⇒ enter
Purge ⇒ enter ⇒
Done-Return

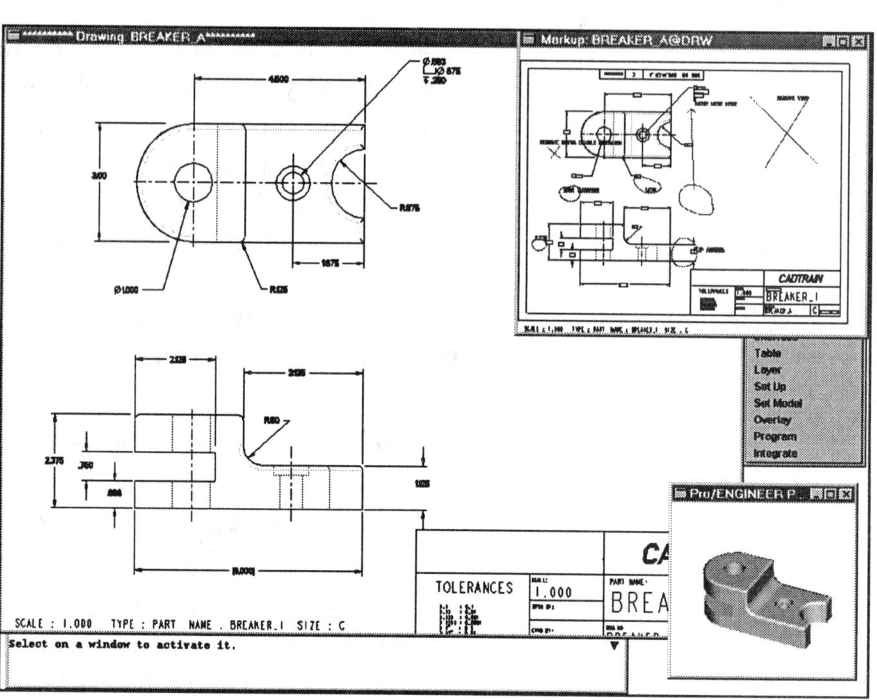

Figure 17.43
Regenerated Drawing and Part With Checker Changes

Figure 17.44
Completed Drawing

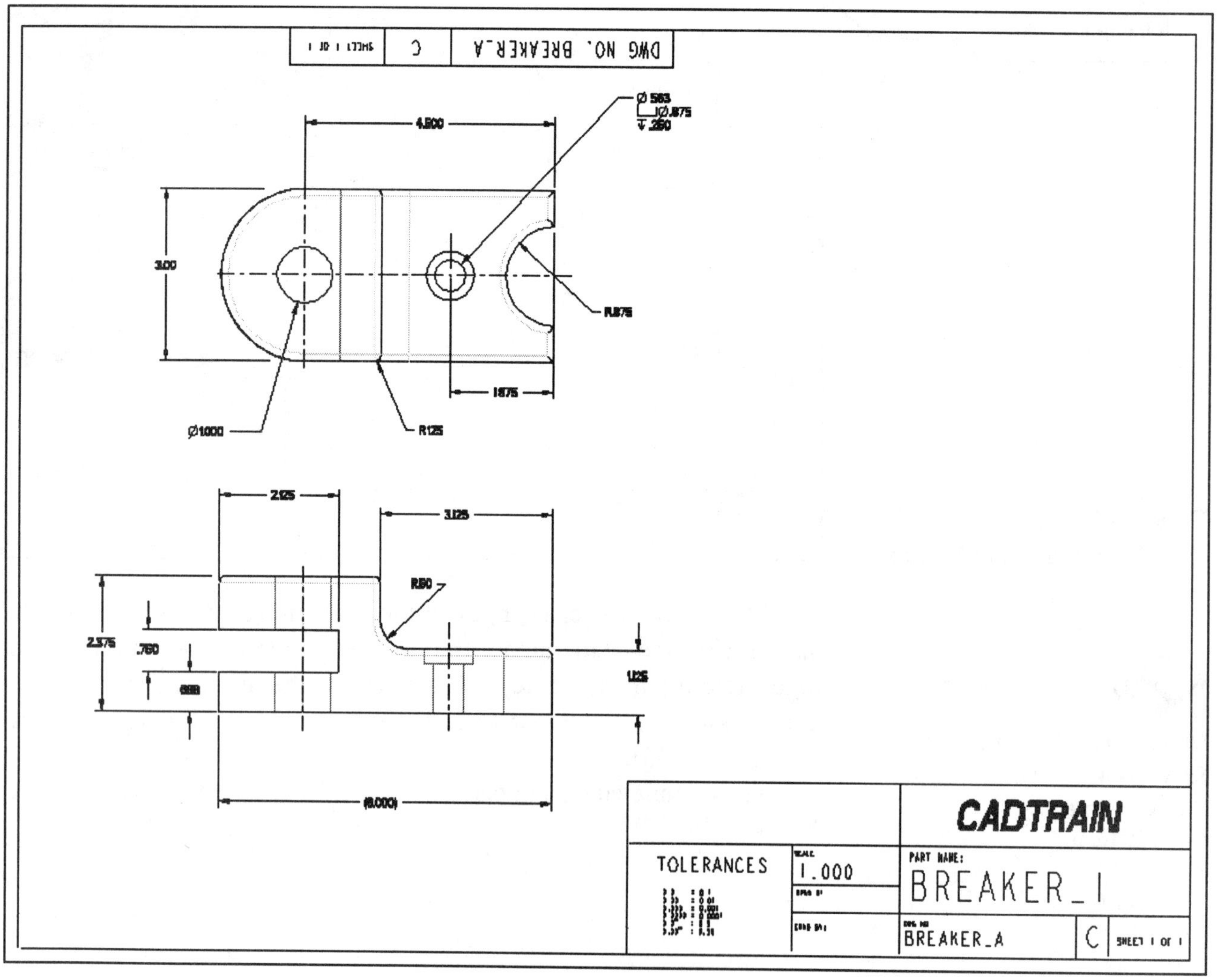

DWG NO. BREAKER_A
C
SHEET 1 OF 1

Ø.563
Ø.875
▽.250
4.500
3.00
R.875
1.875
Ø1.000
R.125
2.125
3.125
R.50
2.375
.750
.688
1.125
(8.000)

CADTRAIN
TOLERANCES
SCALE
1.000
PART NAME:
BREAKER_1
DWG NO
BREAKER_A
C
SHEET 1 OF 1

Lesson 17 Project

Cylinder Rod Drawing

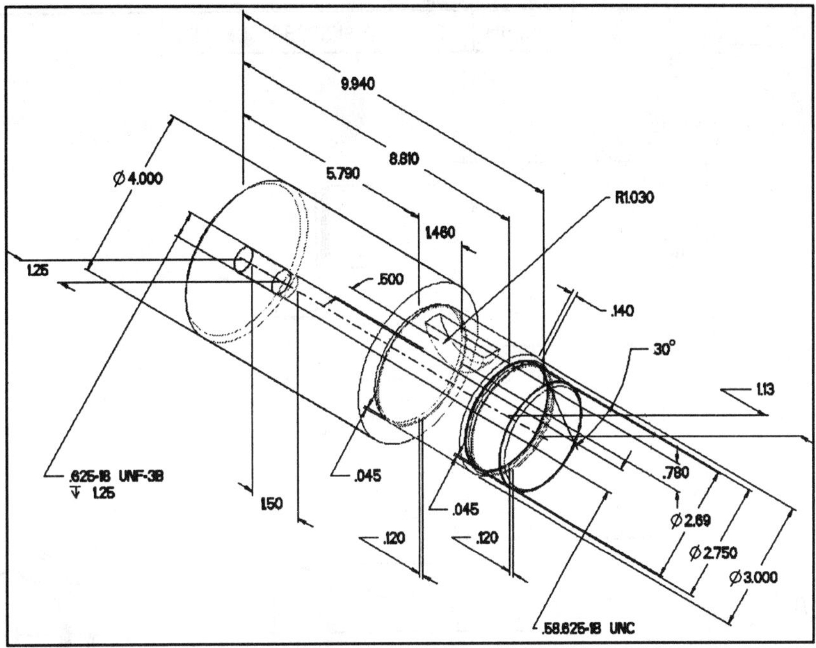

Figure 17.45
Cylinder Rod Pictorial Drawing

Cylinder Rod Drawing

The second **lesson project** for Drawing mode will use the part modeled in Lesson 6. The drawing for the Cylinder Rod (Fig. 17.45) is just one of many lesson parts and lesson projects that can be brought into Drawing mode and detailed.

Analyze the part and plan out the sheet size and the drawing views required to display the model's features for detailing (Figs. 17.46 and 17.47). Use the formats created in Lesson 16. Detail the part according to **ASME Y14.5 1994**.

EGD REFERENCE
Engineering Graphics and Design with Graphical Analysis *or* **Fundamentals of Engineering Graphics and Design**
by L. Lamit and K. Kitto
See pages 487 and 674-677

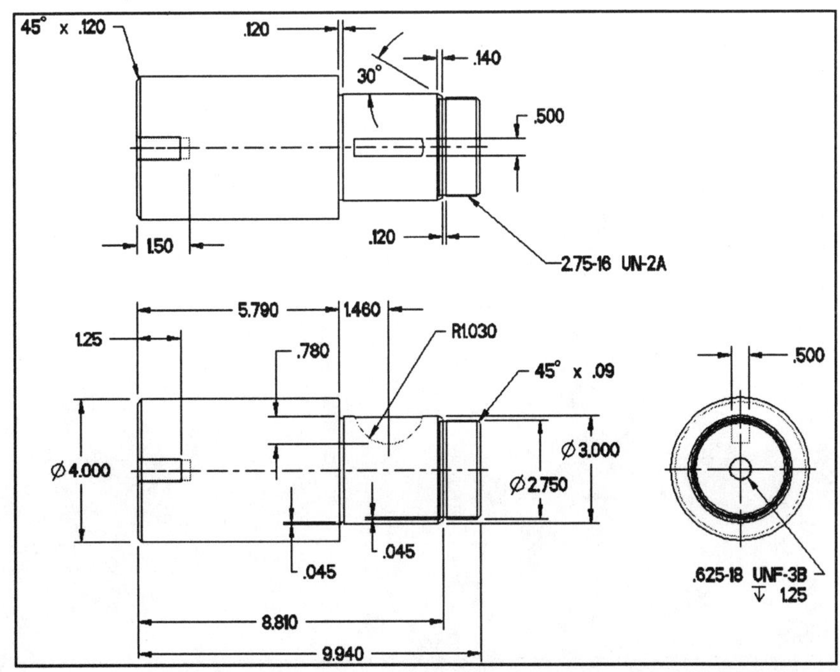

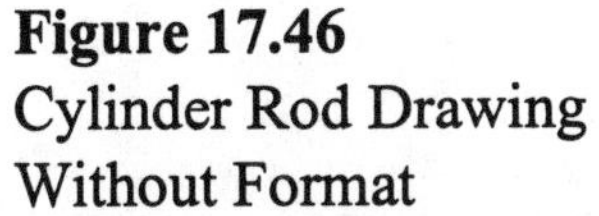

Figure 17.46
Cylinder Rod Drawing Without Format

Figure 17.47
Cylinder Rod Drawing With Format (before dimensioning)

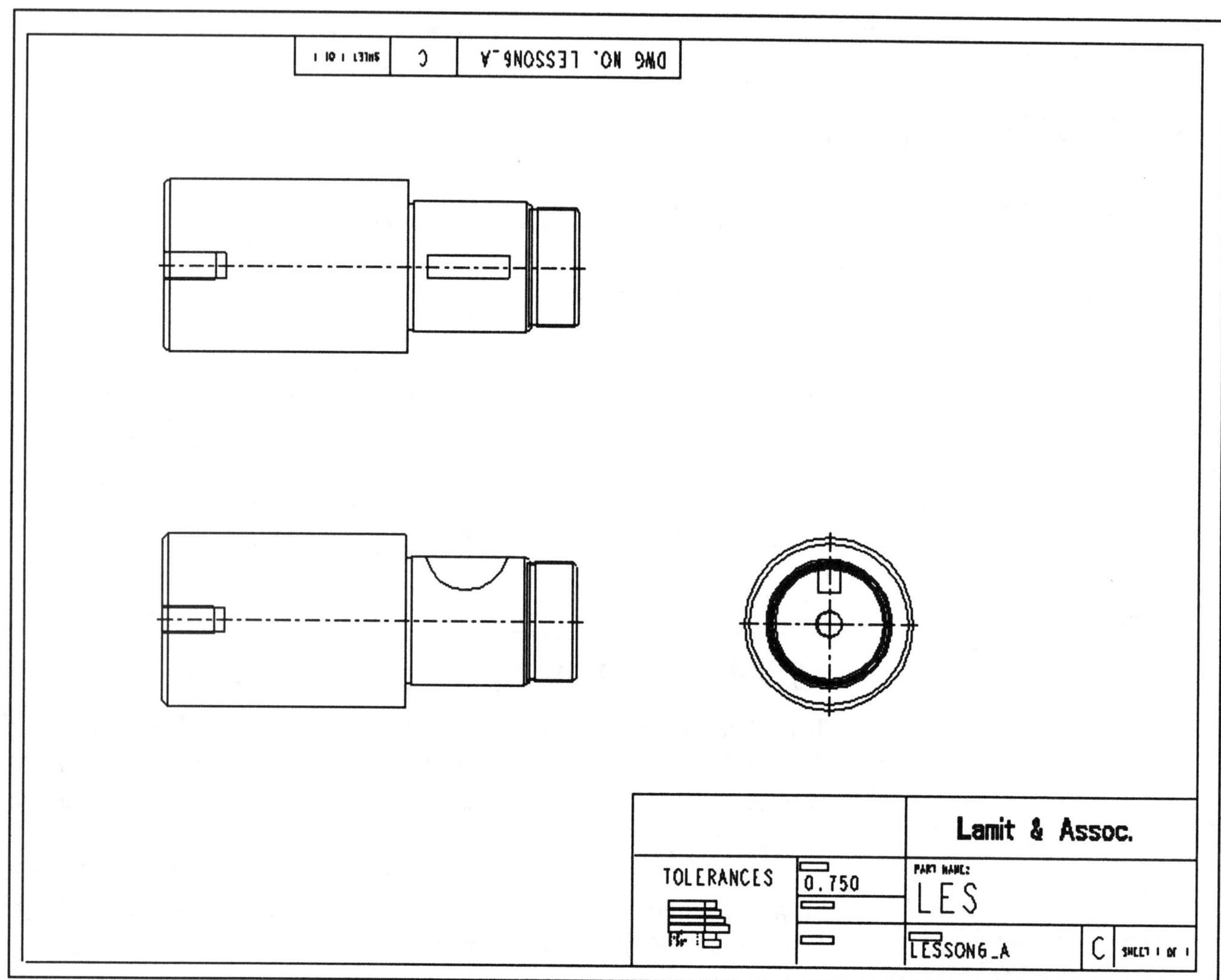

Lesson 18

Sections and Auxiliary Views

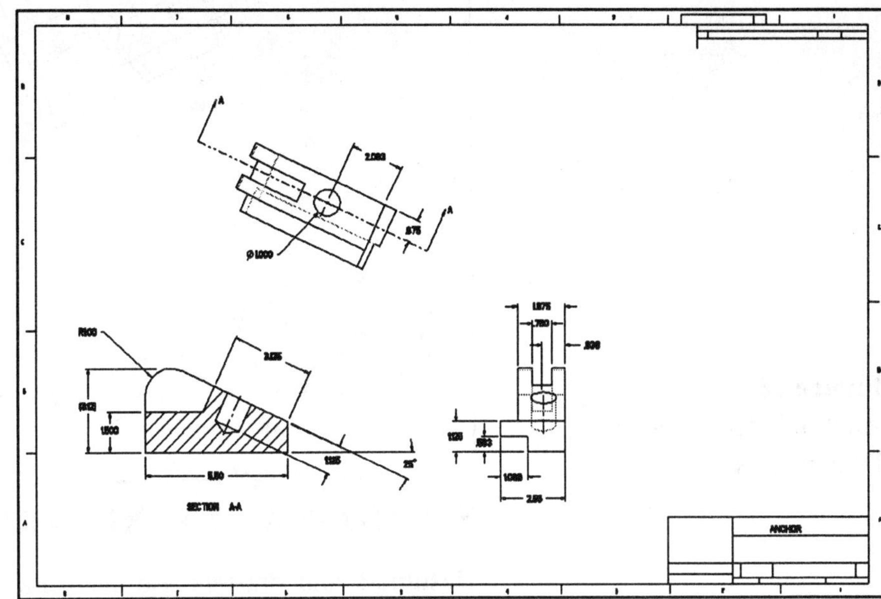

Figure 18.1
Anchor Drawing

OBJECTIVES

1. **Identify the need for sectional views to clarify interior features of a part**

2. **Establish a .dtl file to use when detailing and creating section drawings**

3. **Identify cutting planes and the resulting views**

4. **Create sections along datum planes**

5. **Detail section views**

6. **Develop the ability to produce auxiliary views**

7. **Create detail views**

8. **Create scaled detail views of complicated feature geometry**

9. **Apply standard drafting conventions and linetypes to illustrate interior features**

EGD REFERENCE
Engineering Graphics and Design with Graphical Analysis *or* **Fundamentals of Engineering Graphics and Design**
by L. Lamit and K. Kitto
Read Chapter 12
See pages 387-416, 549

COAch™ for Pro/ENGINEER

If you have **COAch for Pro/ENGINEER** on your system, go to SEARCH and do the appropriate Segment shown in Figures 18.3 through 18.6. A variety of CADTRAIN illustrations are also incorporated with the lesson.

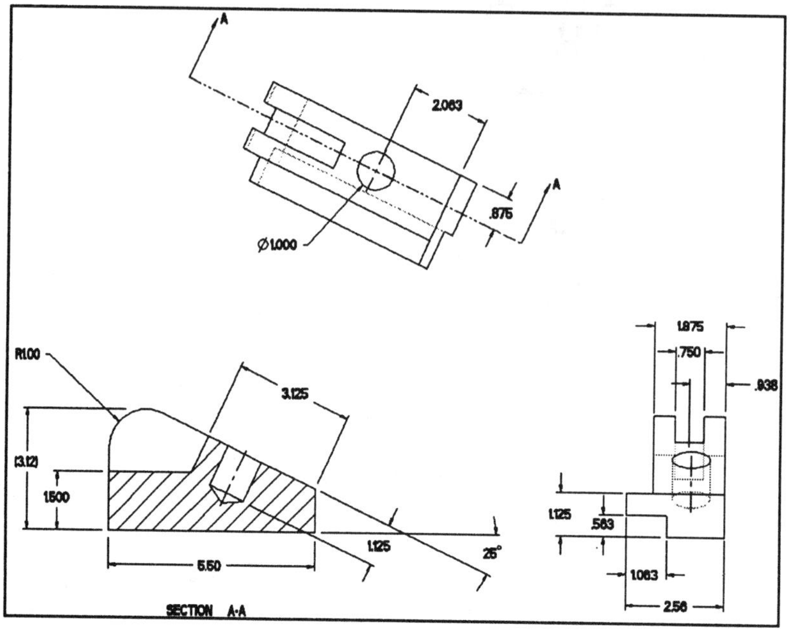

Figure 18.2
Anchor Drawing with Dimensions

SECTIONS AND AUXILIARY VIEWS

Designers and drafters use **sectional views**, also called **sections,** to clarify and dimension the internal construction of a part. Sections are needed for interior features that cannot be clearly described by hidden lines in conventional views (Figs. 18.1 and 18.2).

Auxiliary views are used to show the true shape/size of a feature or the relationship of part features that are not parallel to any of the principal planes of projection. Many parts have inclined surfaces and features that cannot be adequately displayed and described by using principal views alone. To provide a clearer description of these features, it is necessary to draw a view that will show the *true shape/size.* Besides showing features true size, auxiliary views are used to dimension features that are distorted in principal views and to solve a variety of engineering problems graphically (Fig. 18.3).

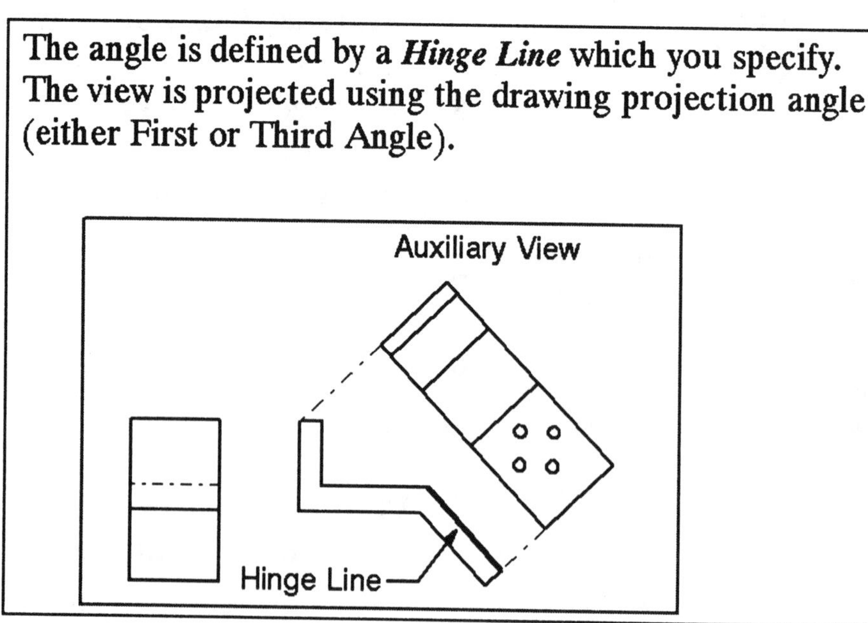

Figure 18.3
COAch for Pro/E, Drawings (Auxiliary Views)

Sections

A sectional view (Fig. 18.4) is obtained by passing an imaginary **cutting plane** through the part, perpendicular to the **line of sight**. The line of sight is the direction in which the part is viewed. The portion of the part between the cutting plane and the observer is "removed." The part's exposed solid surfaces are indicated by section lines. **Section lines** are uniformly spaced, angular lines drawn in proportion to the size of the drawing. There are many different types of section views.

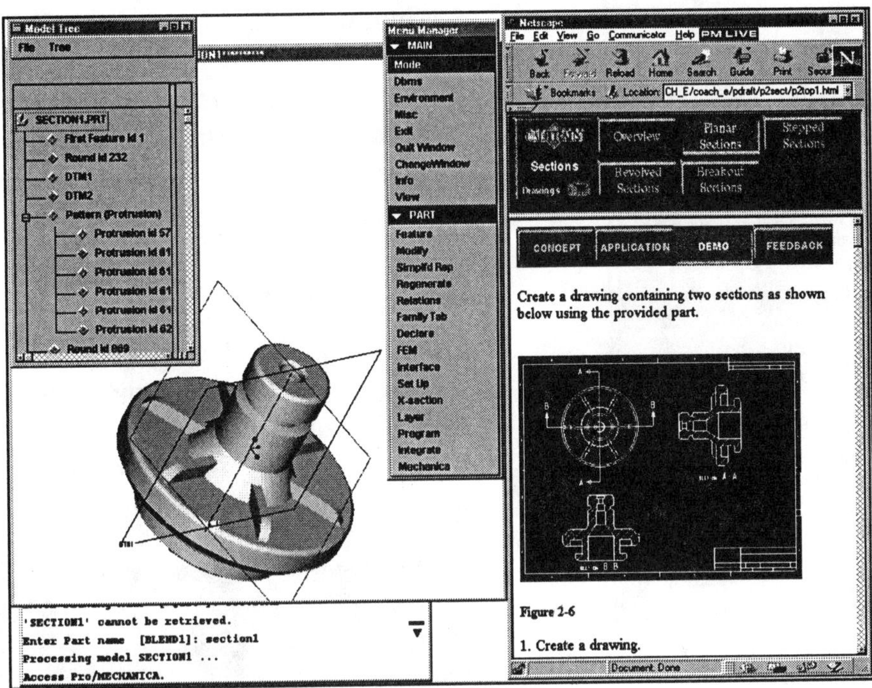

Figure 18.4
COAch for Pro/E, Sections (Planar Sections)

Sectional views are slices through a part or assembly and are valuable for opening up the part or assembly for displaying features and detailing in Drawing mode. Part sectional views can also be used to calculate sectional view mass properties. Each sectional view has its own unique name within the part or assembly, allowing any number of sectional views to be created and then retrieved for use in a drawing. A variety of standard (ANSI) section lining symbols (cross-hatch patterns) representing the type of material can be generated and displayed. You can create a variety of sectional view types:

* Standard planar sections of models (parts or assemblies)
* Offset sections of models (parts or assemblies)

Planar Sections

Planar sectional views are created along a datum plane. The datum may be established during the creation of the sectional view using the **Make Datum** option, or an existing plane may be selected.

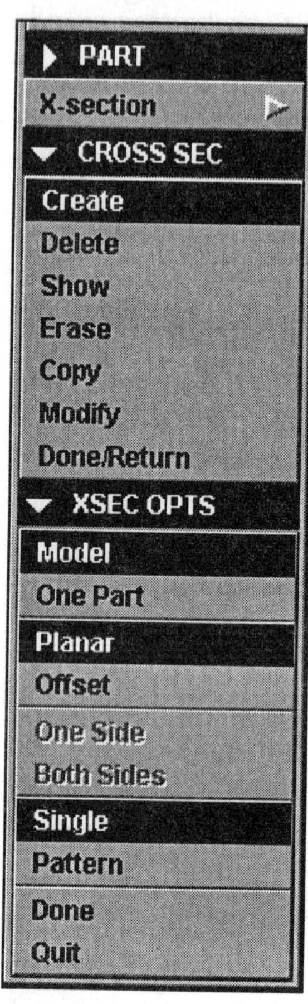

To create a planar sectional view of a part:

1. Choose **X-section** from the PART menu, then **Create** from the CROSS SEC menu.
2. Choose **Planar** from the XSEC OPTS menu, then **Done**.
3. Enter a name for the sectional view, then select, or make, the datum along which the section is to be generated.

To create a planar sectional view of an assembly:

1. Choose **Set Up** from the ASSEMBLY menu and **X-Section** from the ASSEM SETUP menu, then **Create** from the CROSS SEC menu.
2. Choose **Planar** from the XSEC OPTS menu, then **Done**.
3. Enter a name for the sectional view.
4. Select or create the **assembly** datum along which the section is to be generated.

Offset Sections

An **offset** sectional view (Fig. 18.5) is created by extruding a 2D section perpendicular to the sketching plane, just like creating an extruded cut but without removing any material. This type of sectional view is valuable for opening up the part to display several features with a single section.

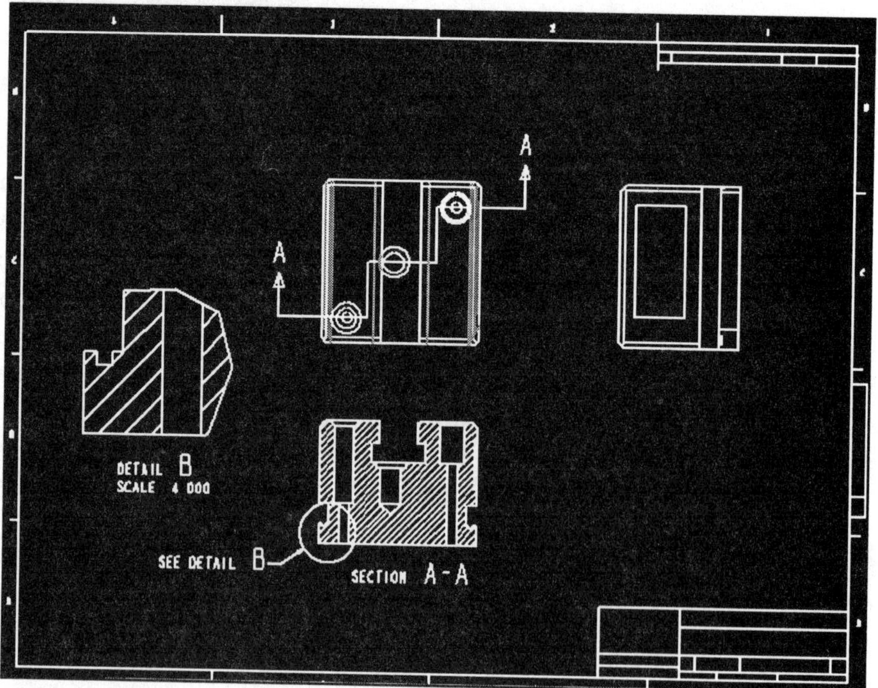

Figure 18.5
COAch for Pro/E, Sections (Stepped Sections)

The sketched section must be an *open section*. The first and last segments of the open section must be straight lines.

▼ XSEC ENTER
Create
Retrieve
▼ XSEC CREATE
Planar
Offset
One Side
Both Sides
Single
Pattern
Done
Quit

To create an offset section:

1. Choose **X-section** from the PART or ASSEM SETUP menu, then **Create** from the CROSS SEC menu.
2. Choose **Offset** and **One Side** or **Both Sides** from the XSEC OPTS menu, then **Done**.
3. Enter a name for the sectional view.
4. Answer the prompts for entering Sketcher. The sketching plane can be created using the **Make Datum** option.
5. **Sketch** the sectional view and **Dimension** it to the model. Choose **Done** when the section has been regenerated successfully.

Auxiliary Views

The proper selection of views, view orientation, and view alignment is determined by a part's features and its natural or assembled position. Normally, the front view is the primary view and the top view is obvious, based on the position of the part in space or when assembled. The choice of additional views is determined by the part's features (Fig. 18.6) and the minimum number of views necessary to describe the part and show its dimensions.

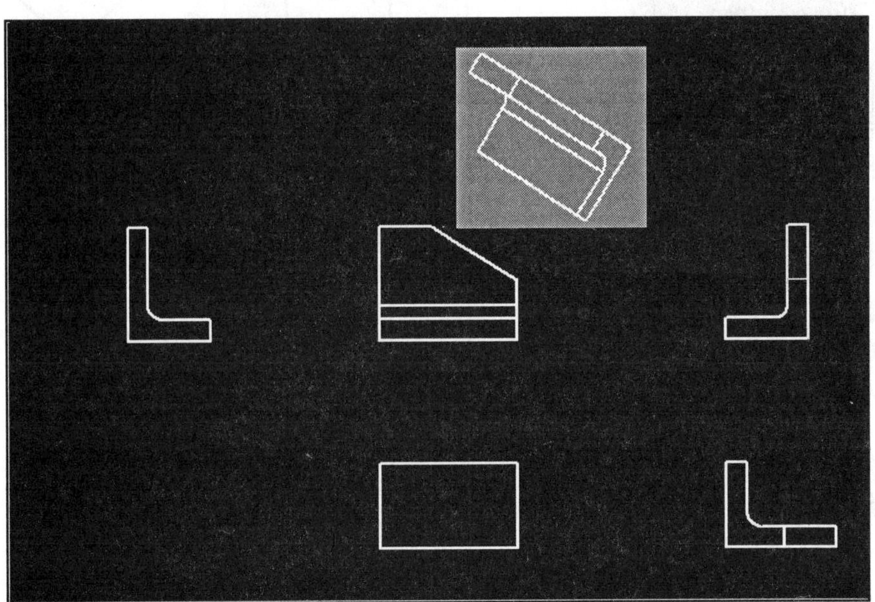

Figure 18.6
COAch for Pro/E, Auxiliary Views (Creating an Auxiliary View)

Auxiliary views are created by making a projection of the model perpendicular to a selected edge. They are normally used to discern the true size and shape of a planar surface on a part. An auxiliary view can be created from any other type of view. Auxiliary views can have arrows created for them that point back at the view(s) from which they were created.

To add an auxiliary view to the drawing, use the following command options:

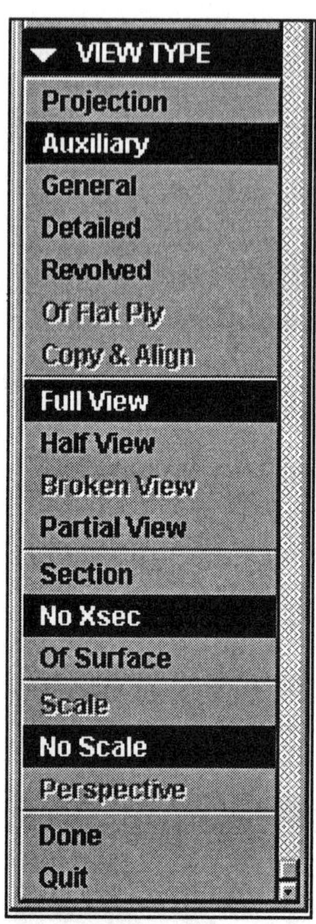

1. Choose **Auxiliary** and other available options from the VIEW TYPE menu.
2. Choose **Done** to accept the options, or **Quit** to quit the creation of a new view.
3. Pick the location of the new view on the drawing (Fig. 18.7).
4. Pick an edge of, or axis through, or datum plane of, the surface of the model in the view from which the auxiliary view will be developed. If the edge selected is from a view that has a pictorial (isometric-trimetric) orientation, the new view will be oriented as the base feature section was. Otherwise, the view will be oriented with the selected surface parallel to the plane of the drawing.

Detail views will also be discussed in this lesson (Fig. 18.8).

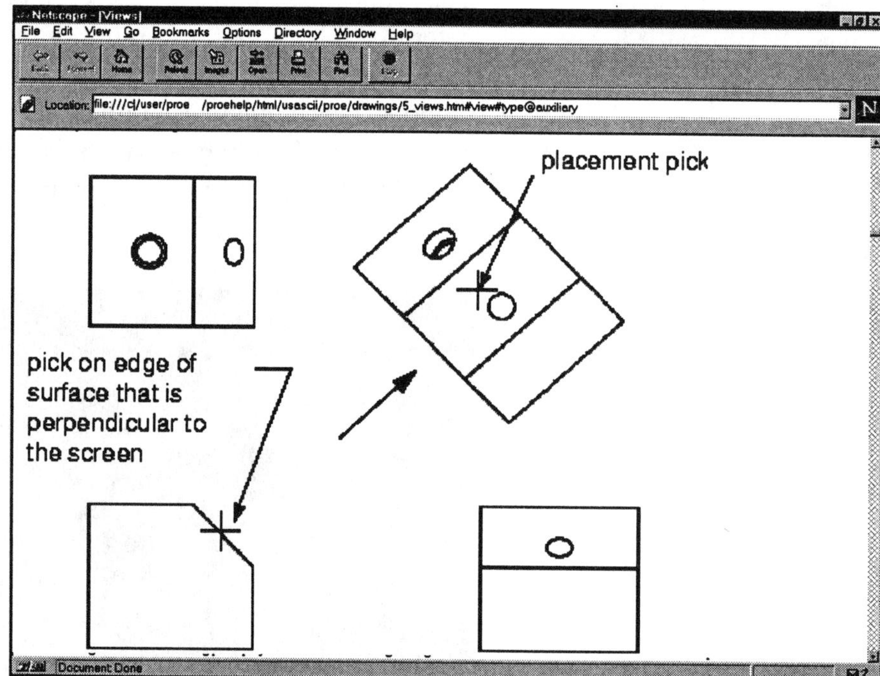

Figure 18.7
Online Documentation, Auxiliary Views

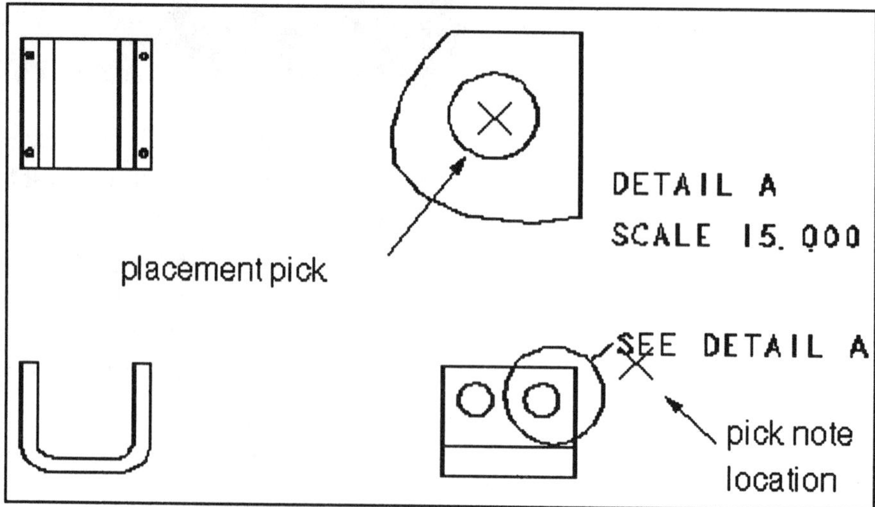

Figure 18.8
Online Documentation, Detailed View

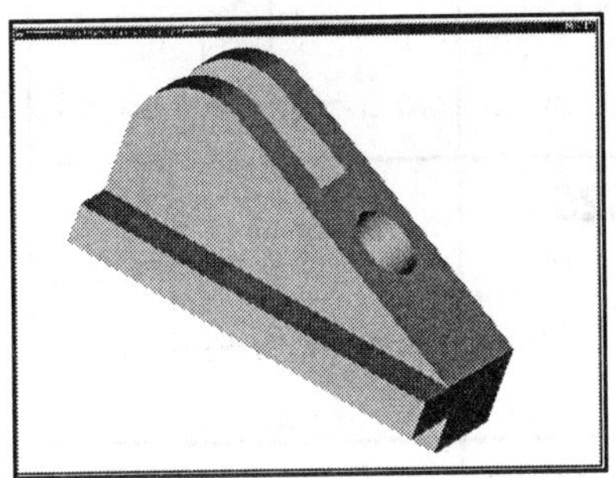

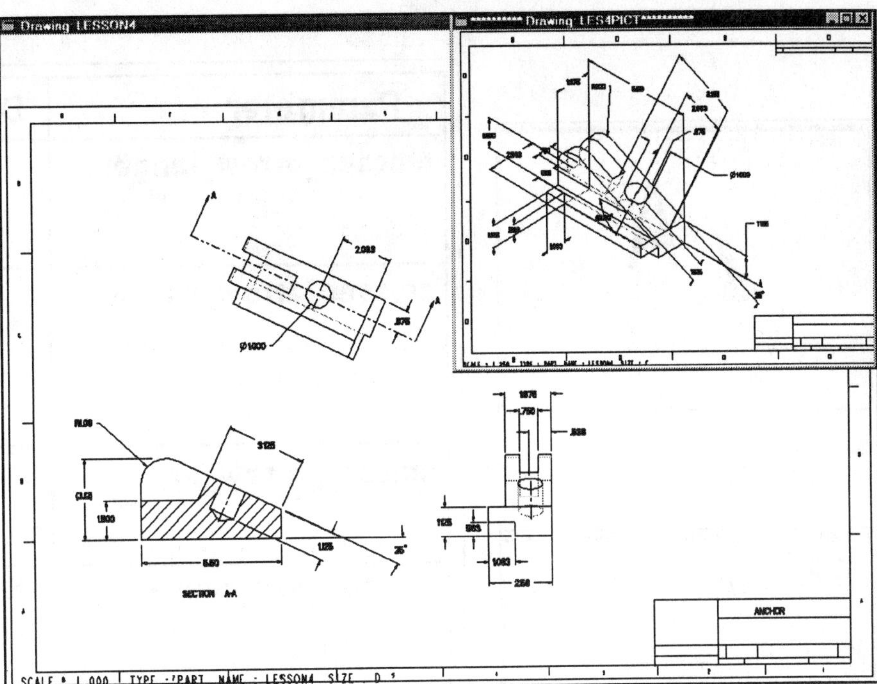

Figure 18.9
Anchor, Orthographic View Detail and Pictorial Detail

Anchor Drawing

Auxiliary views are created, using Pro/E, by making a projection of the model perpendicular to a selected edge or along an axis. They are normally used to discern the true size and shape of a planar surface on a part, such as the Anchor, which has an angled surface with a hole machined perpendicular to that surface.

Pro/E is able to create and display fully associative *section views* of solid models (Fig. 18.9). You can create a section view using an existing view on a drawing in Drawing mode or by creating a section in Part mode for retrieval on a drawing. As you were completing some of the lessons in this text, you were instructed to create sections in a number of lesson parts and lesson projects.

You create a section view while looking at the drawing by sketching (with the help of Pro/E) a section line (*Offset*) or by selecting an existing **Datum Plane** for the section line (*Planar*) to pass through.

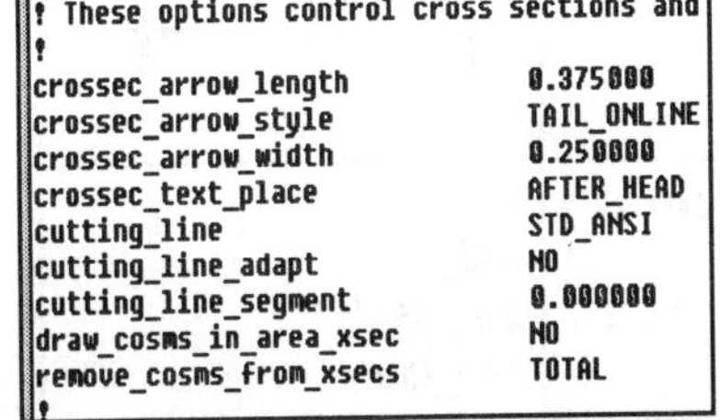

```
! These options control cross sections and
!
crossec_arrow_length          0.375000
crossec_arrow_style           TAIL_ONLINE
crossec_arrow_width           0.250000
crossec_text_place            AFTER_HEAD
cutting_line                  STD_ANSI
cutting_line_adapt            NO
cutting_line_segment          0.000000
draw_cosms_in_area_xsec       NO
remove_cosms_from_xsecs       TOTAL
!
```

The display of the *section line symbol* and the *section view text* is controlled by the **Section Line Display Parameters** in the **Drawing Set Up** file (**.dtl**).

The sectioning parameters apply to the two basic types of standards: the ANSI (ASME) Standards and the ISO/JIS/DIN Standards. Section line display and the manner in which the view titles (parameters) are created vary based on the standard you choose to use. See the **Pro/ENGINEER Drawing User's Guide**, Appendix A, for a listing of section line and view parameters. The pertinent parameters are shown in Figure 18.10. Figure 18.11 shows the **.dtl** file used by CADTRAIN for its sectioned drawings.

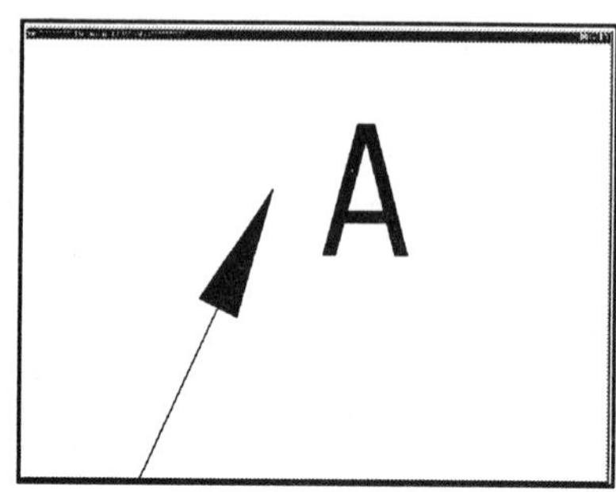

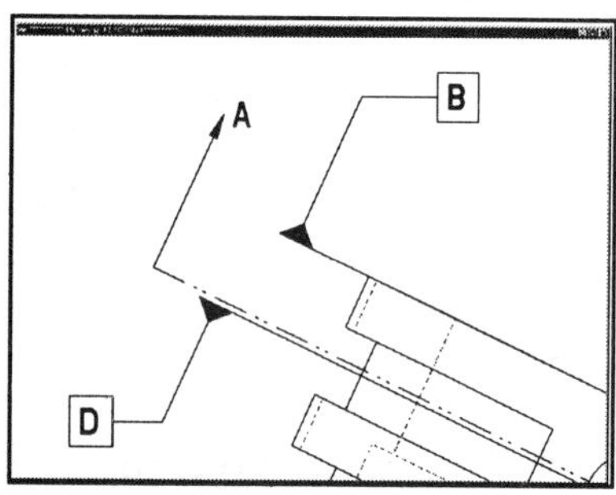

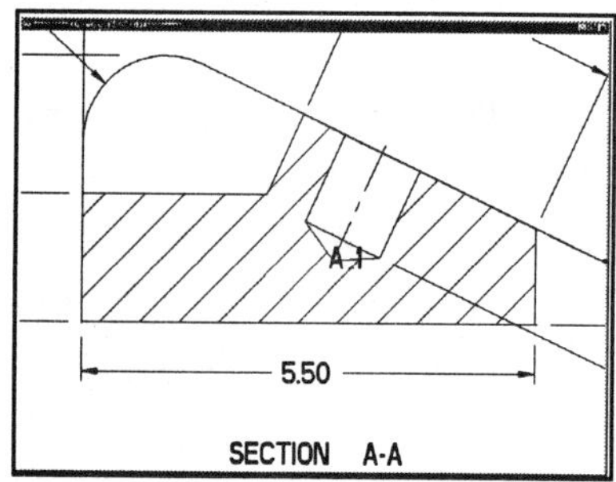

Figure 18.10
COAch for Pro/E, Section Line Symbol and the Section View Text Parameters

Parameter	Default	Hint
crossec_arrow_length	.187	
crossec_arrow_width	.0625	
crossec_arrow_style	tail_online	Tail Head
crossec_text_place	after_head	A after_head A before_tail A above_tail A above_line
cutting_line	std_ansi	std_ansi std_din std_jis std_iso std_ansi_dashed std_jis_alternate
cutting_line_segment	0	0 value
def_view_text_height	0	A
view_note	std_ansi	std_ansi std_din std_jis std_iso

Pro/TABLE

File Edit View Format Help

filled

	C1	C2	C3	C4	C5
R21	!				
R22	! These options control cross sections and their arrows				
R23	!				
R24	crossec_arrow_length	0.375000			
R25	crossec_arrow_style	TAIL_ONLINE			
R26	crossec_arrow_width	0.250000			
R27	crossec_text_place	AFTER_HEAD			
R28	cutting_line	STD_ANSI			
R29	cutting_line_adapt	NO			
R30	cutting_line_segment	0.000000			
R31	draw_cosms_in_area_xsec	NO			
R32	remove_cosms_from_xsecs	TOTAL			
R33	!				
R34	! These options control solids shown in views				
R35	!				
R36	datum_point_size	DEFAULT			
R37	datum_point_shape	CROSS			
R38	hlr_for_pipe_solid_cl	NO			
R39	hlr_for_threads	NO			
R40	location_radius	DEFAULT(2.)			
R41	mesh_surface_lines	ON			
R42	thread_standard	STD_ANSI			
R43	hidden_tangent_edges	DEFAULT			
R44	!				

Figure 18.11
.dtl File for Drawing Parameters

If you are going to create sections while you are in Drawing mode, you may require **Datum Planes**. Since most views on a drawing are orthographic, and since they cannot be changed with **Spin**, you may want to prepare the model with the necessary **Datum Planes** *before* you enter Drawing mode. You were asked to create a section of the Anchor in Lesson 4. If you did not, you now need to create a section that passes vertically, lengthwise, through the model in Part mode (Fig. 18.12), or to create a section in Drawing mode, as described next.

HINT

If you did not create a section when modeling the Anchor, use the following commands in Part mode:

X-Section ⇒ **Create** ⇒ **Done** ⇒ (type **A** for the section name) ⇒ **enter** ⇒ **Make Datum** ⇒ **Through** ⇒ (pick axis **A_1** of the hole) ⇒ **Parallel** ⇒ [pick **DTM3** (or **-B-**)] ⇒ **Done** ⇒ **Done/Return**

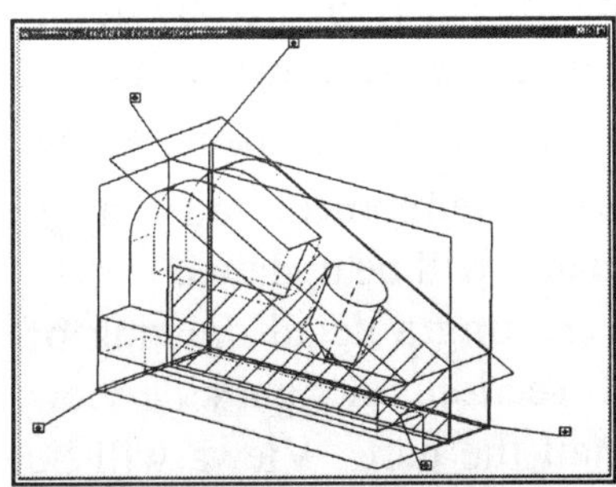

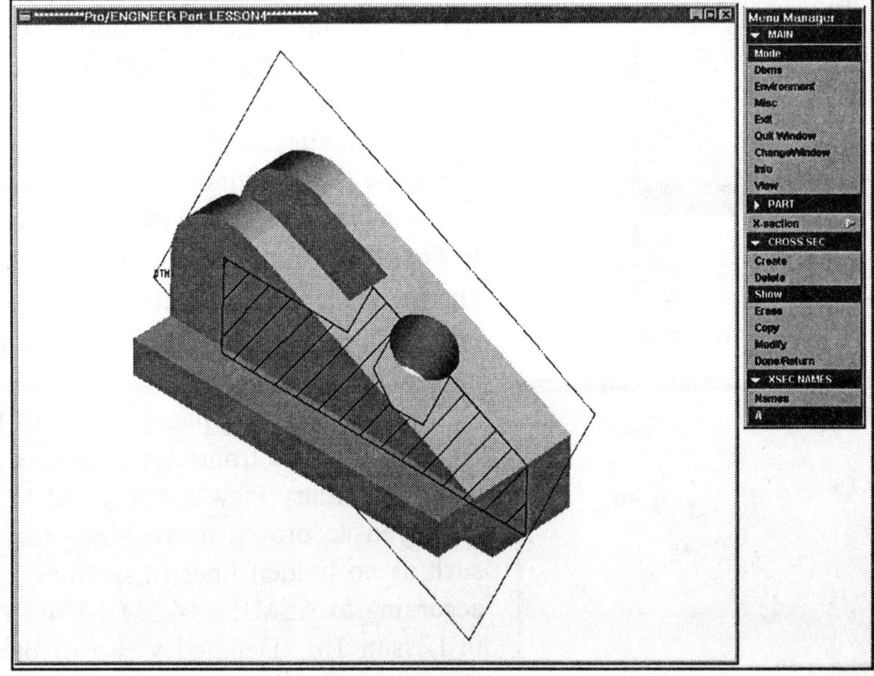

Figure 18.12
Section Created in Part Mode

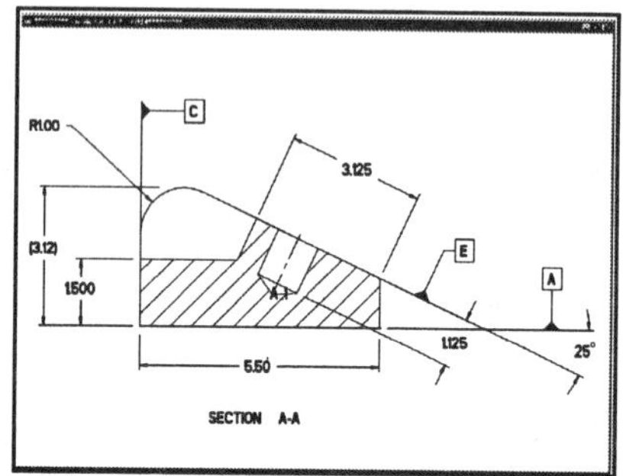

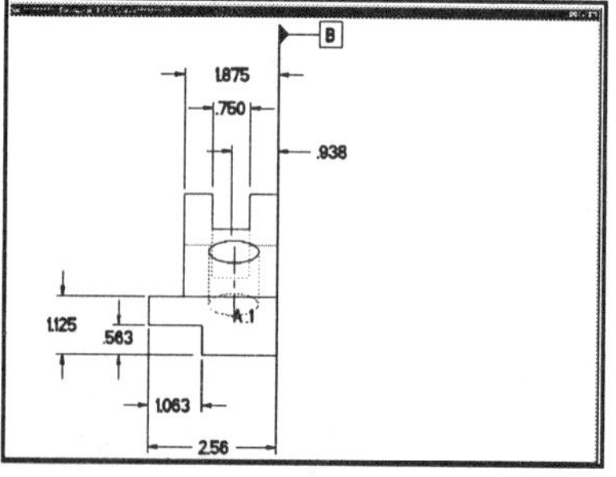

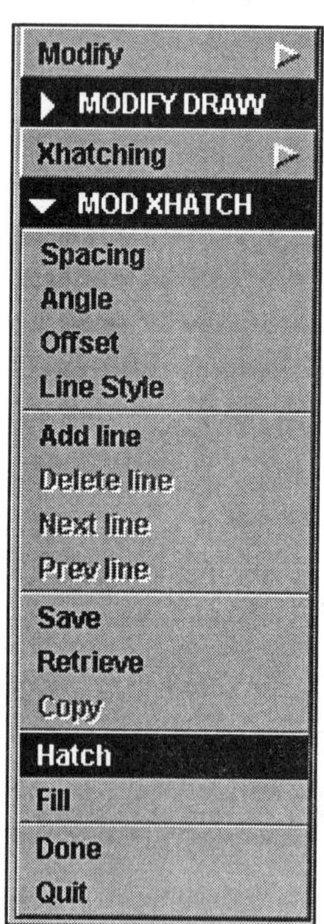

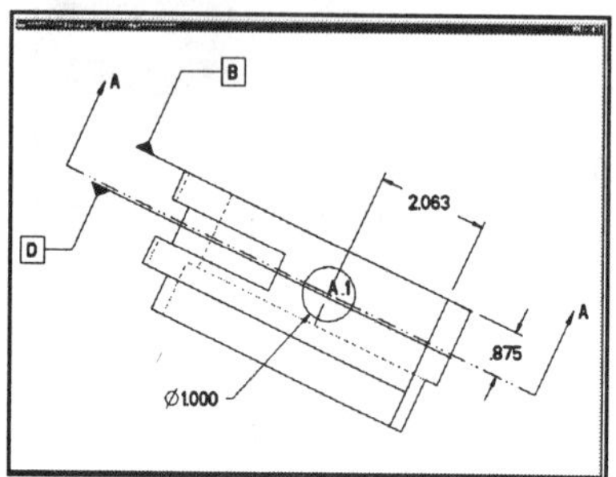

In Drawing mode, the **Planar** section option allows you to create a section that passes straight through a part without any "jogs" (steps) in the section line. **Planar** sections can be defined by retrieving a section, creating a datum plane, or selecting an existing datum plane to define the cut position.

When you want to create the section while in Drawing mode, you can select either a planar face or a **Datum Plane** as the "plane" the section cut passes through. Even though you can pick a planar face, this is generally not good practice. The section cut takes place *at* the plane you select. If you pick a face, this causes the section to be tangent to it. Not only is this poor drafting practice, but it also causes Pro/E to generate (though only in some cases) incomplete sections.

When you create a **Planar** section on the drawing, Pro/E first asks you to define the location of the center of the section view. It then displays the orthographic projection to which the section edges (created by the cut) and the crosshatching will be added. Choose **Create** ⇒ **Planar** ⇒ **Done**. Pro/E prompts you to enter a **name** for the section. This "name" is actually the section letter you wish to use. If you are creating your first section and you want it to have the letter **A** displayed at each arrow end and have the title **SECTION A-A**, you should enter **A** as the name of the section. It is a common mistake to type **AA**, thinking that you are establishing **SECTION A-A,** when in reality you are getting **SECTION AA-AA**. Type one letter only. You must then select a **Datum Plane** to define the section cutting plane. Pro/E then prompts you to pick a view in which the cutting plane is perpendicular to the screen. This is the view in which it will draw the section line and the view where the cut will actually take place. Pro/E displays the section line, with its arrows pointing in the direction it "thinks" you want to view the cut.

After the section has been created, use the right mouse button to get the pop-up menu item **Modify Item** ⇒ (pick the section view) ⇒ **Flip X-sec** (to flip the section identification arrows). The arrow direction does not affect the cross-sectioned area, which was defined when you indicated the location of the view. It does affect what you see in the "background," behind the cutting plane.

You can change the name of a section after it is created by going to **Part** mode and retrieving the part used to make the drawing. If you choose the option **X-section** ⇒ **Modify** ⇒ (pick the name to be changed) ⇒ **Name** ⇒ (enter a new name). The moment you change back to the drawing window, the section name will be updated.

In the following pages, you will be creating a detail drawing of the Anchor. The front view will be a full section. A right side view and an auxiliary view are required to detail the part. Views will be displayed according to visibility requirements per ANSI standards, such as no hidden lines in sections. The part is to be dimensioned according to ASME Y14.5M 1994. You will add the format created in Lesson 16. Detailed views of other parts will be introduced to show the wide variety of view capabilities of Pro/E's Drawing mode.

Using the Anchor model, create a drawing including a front (sectioned) view, a right side view, and an auxiliary view, as shown in Figure 18.13. The front view will be changed to a section view after the three views are established.

Dbms (**File**--PT/Modeler) ⇒
Save ⇒ **enter**
Purge ⇒ **enter** ⇒
Done-Return

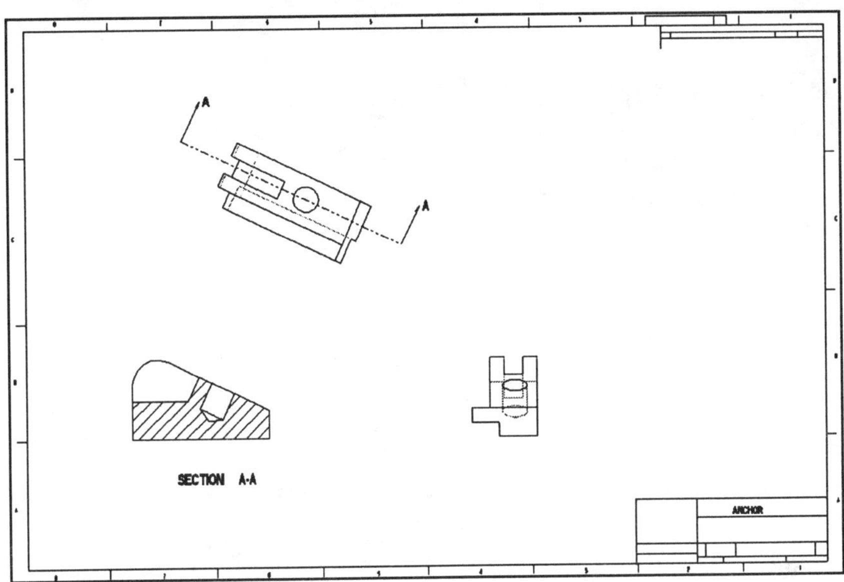

Figure 18.13
Front Section View, Right Side View, and Auxiliary View

HINT

Set Up ⇒ **Modify Val**	
draw_text_height	.25
default_font	filled
draw_arrow_style	filled
gtol_datums	STD_ASME
allow_3d_dimensions	YES

File ⇒ **Save** ⇒ **File** ⇒ **Exit**

Drawing ⇒ **Create** ⇒ (type **Anchor**) ⇒ **enter** ⇒ **Retr Format** ⇒ (?) ⇒ **enter** ⇒ [pick a previously created **D** size format (if none found ⇒ **Format Dir** ⇒ **d.frm** ⇒)] ⇒ **Set Up** ⇒ **Retrieve** ⇒ (pick your **.dtl** Drawing Set Up file from your directory list, and/or choose **Modify Val** to create a new **.dtl** file) ⇒ **Views** ⇒ (?) ⇒ **enter** ⇒ (locate **Anchor** part) ⇒ **Done** ⇒ (indicate the approximate center of the view, as shown in Fig. 18.14) ⇒ **Front** ⇒ (pick a front surface) ⇒ **Bottom** ⇒ (pick the bottom surface of the Anchor) ⇒ **Done/Return** (Fig. 18.15)

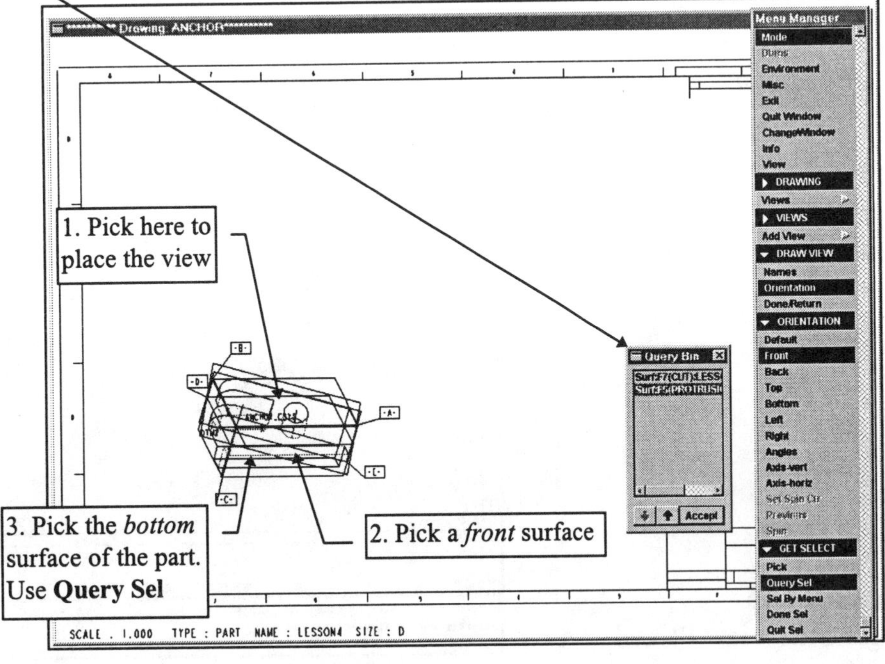

Figure 18.14
Orienting the Front View

HINT

Recall from Lesson 4 that Gtol datums were set (**Set Up ⇒ Geom Tol ⇒ Set datum**), and that assigning them to a layer is proper procedure.

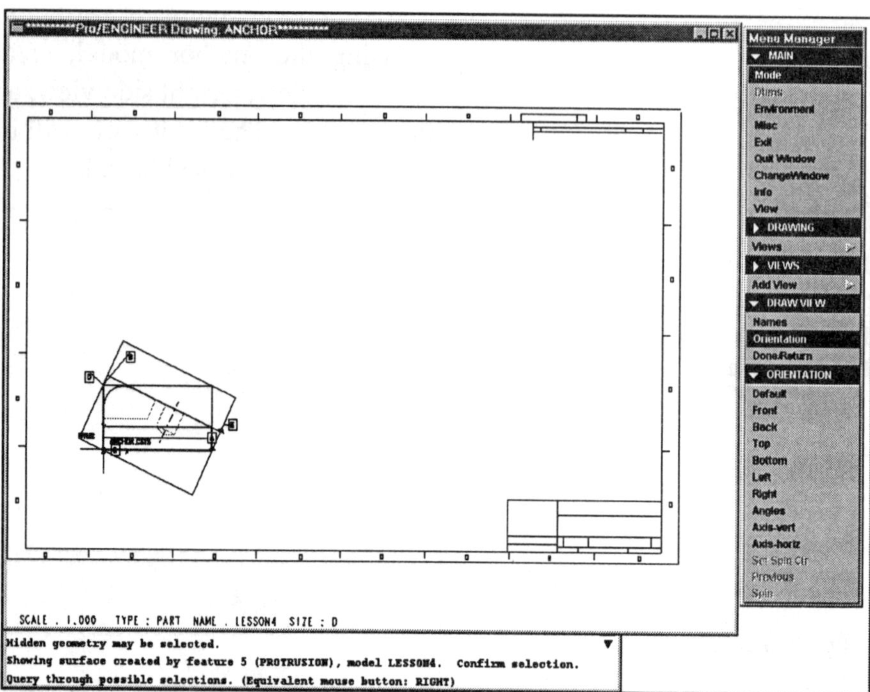

Figure 18.15
Front View Placed

Add the right side view, as shown in Figure 18.16. If you created and set geometric tolerance datums (basic) and set them on a layer in Part mode as recommended, you can now erase them from your drawing. Choose the following commands:

ENVIRONMENT

Environment ⇒

- ☐ **Disp DtmPln**
- ☐ **Disp Points**
- ☐ **Disp Axes**
- ☐ **Disp Csys**
- ☐ **Spin Center**
- ☐ **Grid Snap**
- **Hidden line**
- **No Disp Tan**

Done-Return

Add View ⇒ Done ⇒ (pick a position for the right side view as shown in Fig. 18.16) **⇒ Done/Return ⇒ Detail** (from DRAWING menu) **⇒ Show/Erase ⇒ Erase** (radio button) **⇒** (pick the geometric tolerance and datum plane radio buttons) **⇒ Erase All ⇒ Yes ⇒ Close ⇒ Done/Return**

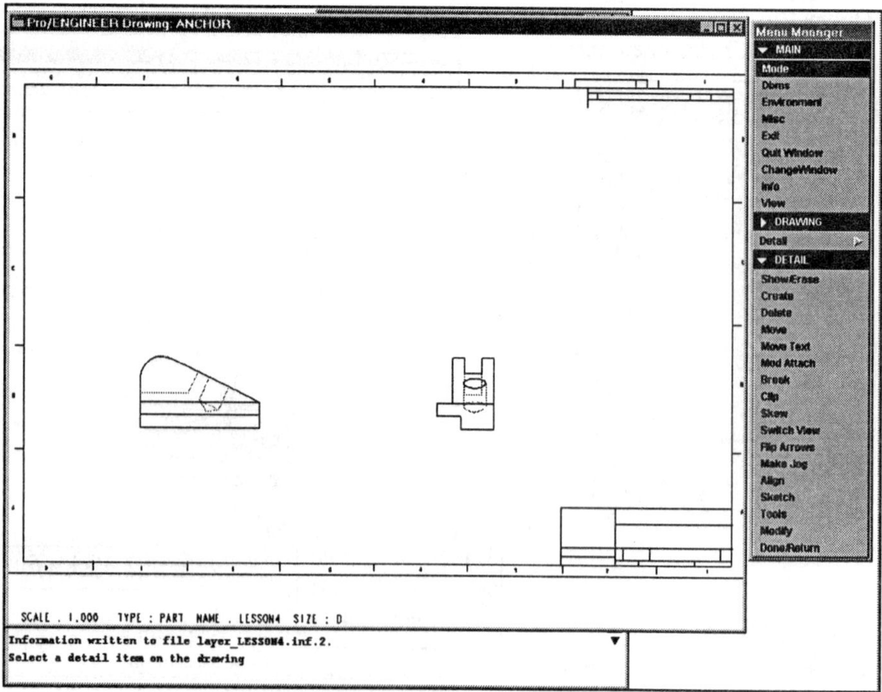

Figure 18.16
Right Side View Placed

Use the following commands to add the auxiliary view as shown in Figure 18.17:

Views (from DRAWING menu) ⇒ **Add View** ⇒ **Auxiliary** ⇒ **Done** ⇒ (pick a center point for view) ⇒ (select edge in main view)

HINT

Every time you add a new view, the set datums will appear in that view. If you don't want to see them, at this stage of the drawing, you must erase them.

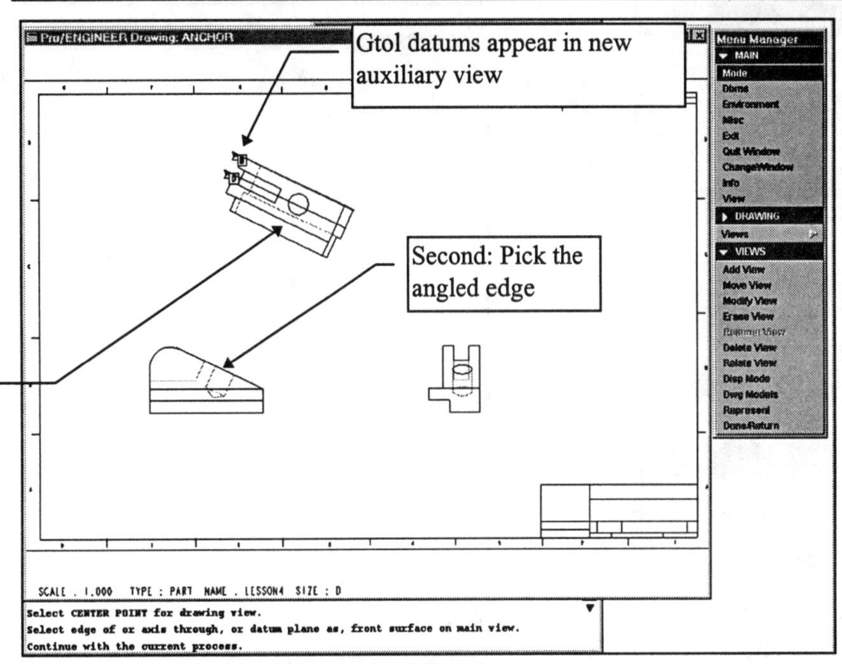

First: Pick here to locate the center of the auxiliary view

Figure 18.17
Auxiliary View

Now change the front view into a section view. The section **A** was created in Part mode. Use these commands (Fig. 18.18):

HINT

See hint on page 18-9.

Modify View ⇒ **View Type** ⇒ (pick the front view as the view to be modified) ⇒ **Section** ⇒ **Done** ⇒ **Full** (default) ⇒ **Total Xsec** (default) ⇒ **Done** ⇒ (pick section name from XSEC NAMES menu: **A)** ⇒ [pick a view for the arrows where the section cutting plane is *perpendicular* (Fig. 18.18)] ⇒ **Done Sel**

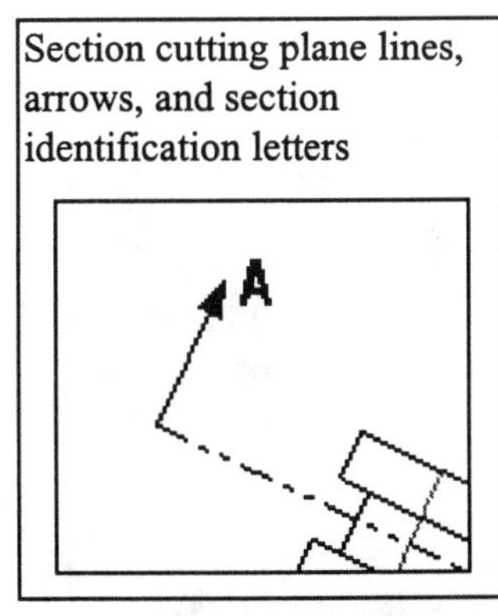

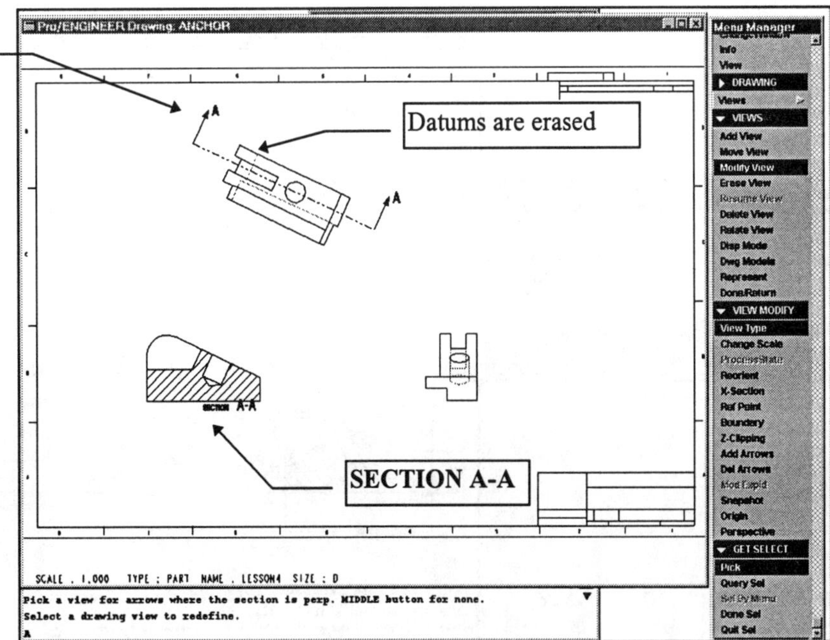

Figure 18.18
Front View Modified to be a Front Section View

Modify the **Disp Mode** of the views to remove all hidden lines:

Disp Mode ⇒ **View Disp** ⇒ (pick all three views) ⇒ **Done Sel** ⇒ **No Hidden** ⇒ **No Disp Tan** ⇒ **Done** (Fig. 18.19) ⇒ **Done Sel** ⇒ **Done/Return**

Figure 18.19
No Hidden, No Disp Tan

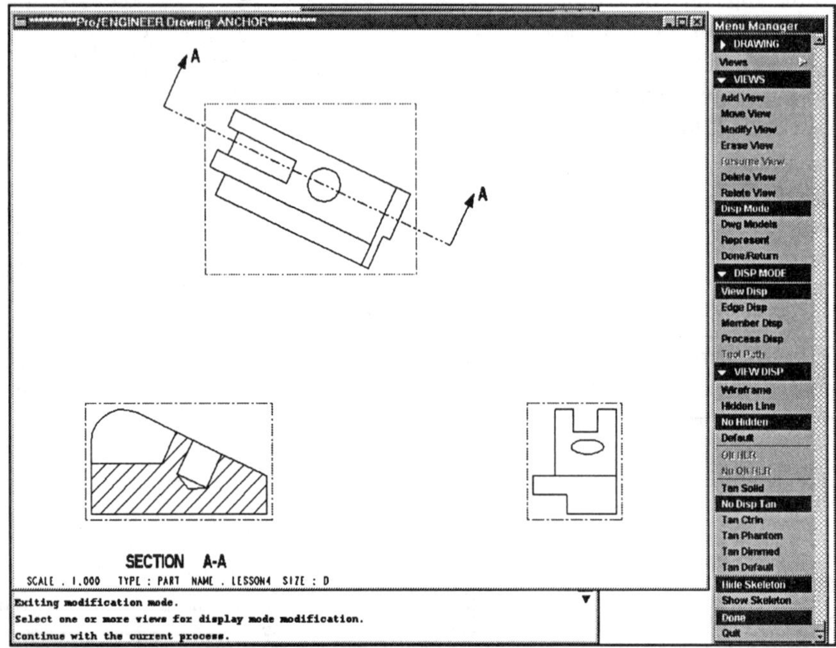

Show all dimensions, axis-centerlines, and tolerance datums by choosing:

Detail ⇒ **Show/Erase** ⇒ **Show** ⇒ pick the three radio buttons for *dimensions*, *datum plane*, and *axis* (Fig. 18.20) ⇒ **Show All** ⇒ **Yes**

HINT

Change the height of the section lettering to **.375**, either in the **.dtl** file (as the new default) or by choosing **Modify** ⇒ **Text** ⇒ **Text Style**

Use your *right* mouse button (**Modify Item**) to edit the dimensions, axes, and datum planes name positions until the drawing looks similar to Figures 18.21 and 18.22.

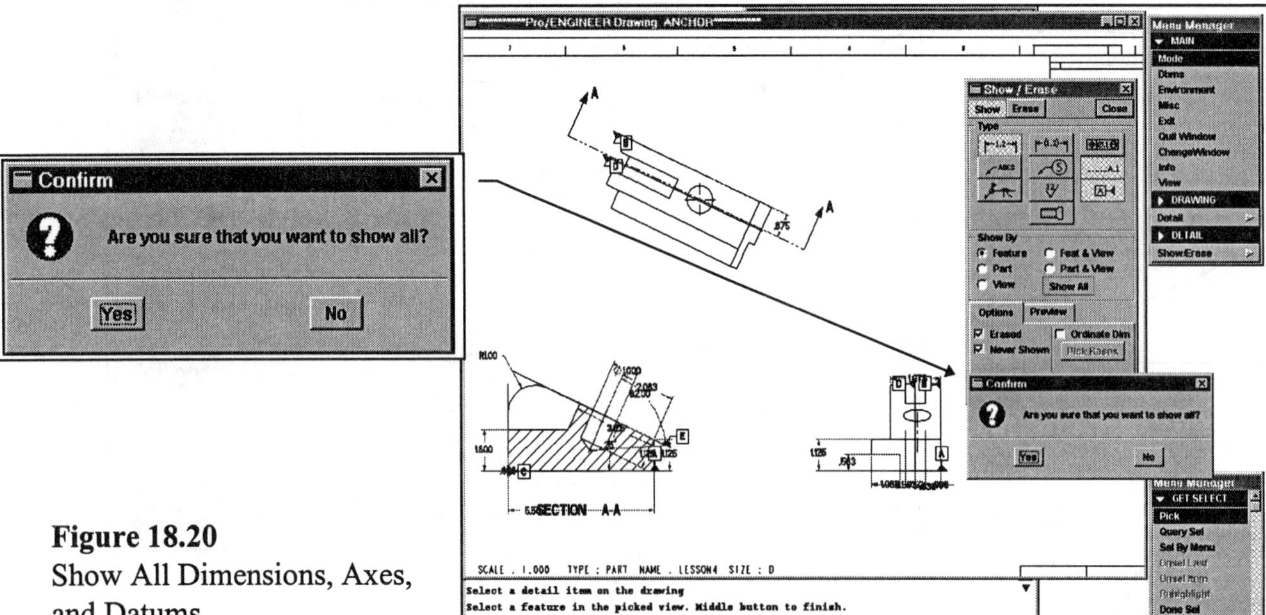

Figure 18.20
Show All Dimensions, Axes, and Datums

Dbms (File--PT/Modeler) ⇒
Save ⇒ enter
Purge ⇒ enter ⇒
Done-Return

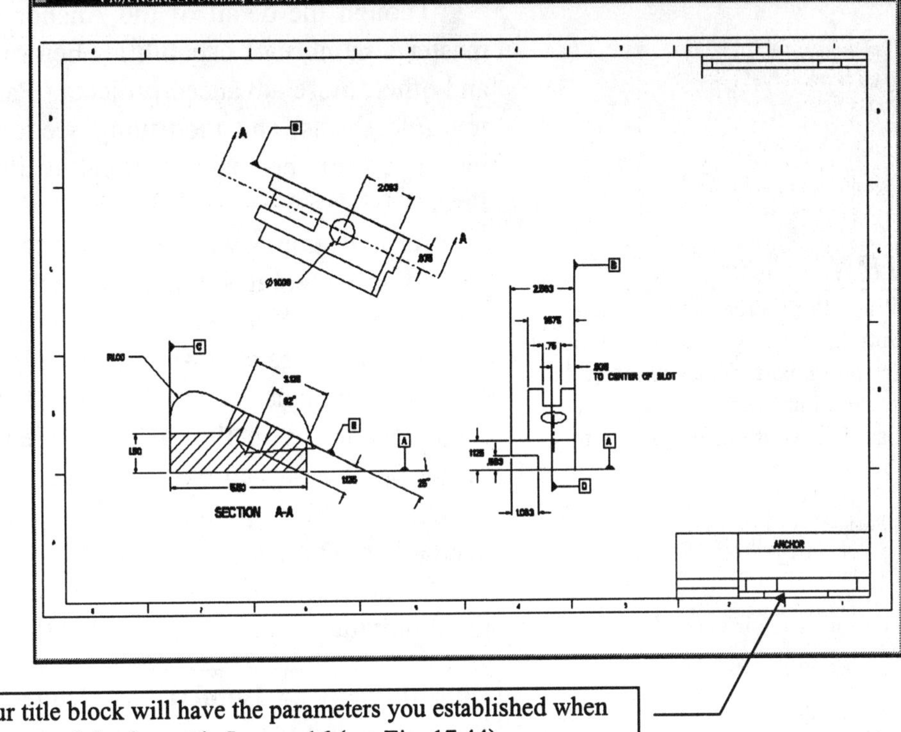

Figure 18.21
Completed Drawing

Your title block will have the parameters you established when you created the format in Lesson 16 (see Fig. 17.44)

Figure 18.22
Detail Drawing

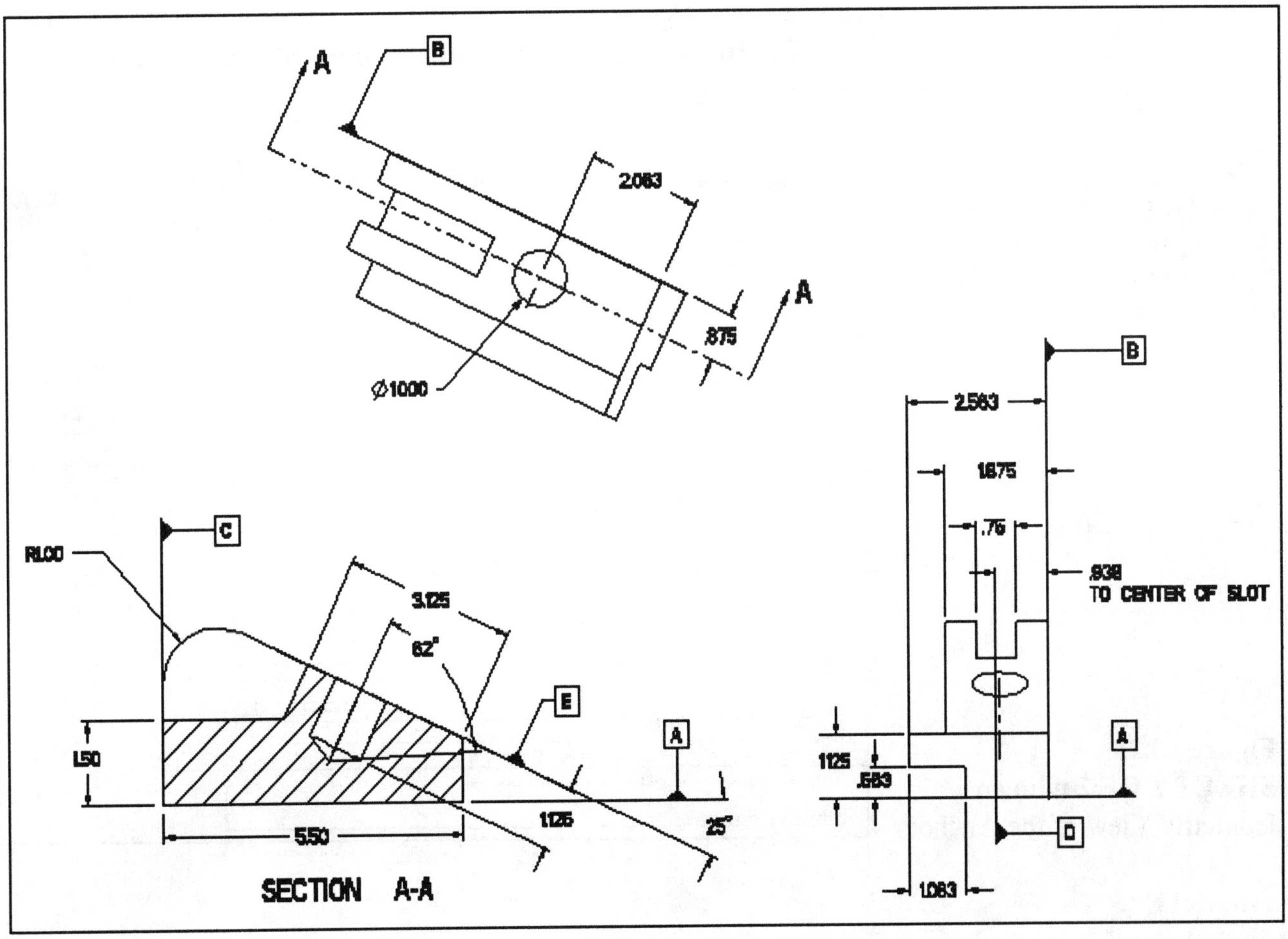

Though the detail of the Anchor is complete, you will need to master a number of capabilities before completing the lesson project and other, more advanced projects. Partial views, detail views, using multiple sheets, and modifying section lining (crosshatch lines) are just a few of the many options available in Drawing mode within Pro/E.

HINT

When Pro/E creates the section cutting plane line, it draws across the entire part. You can move the section line arrow to shorten it, based on your drafting standard.

As an example, when you want to break away a small portion of a part to see internal features in a local area, you can use the **Local** section option. You define the area of the **Local** breakout by drawing a spline. The area enclosed by the spline is then removed to the depth of a selected planar face or **Datum Plane**. You draw the spline in the same manner as in the **Detailed** view option. You must define a center point along an existing edge. The spline must be closed and contain the center point. This is just one of many detailing options available in Drawing mode. (Use Pro/HELP for more information)

Create a *second sheet* with an isometric view of the Anchor using the following commands (Fig. 18.23):

Environment ⇒ **Isometric** ⇒ **Done-Return** ⇒ **Sheets** ⇒ **Add** ⇒ **Format** ⇒ **Add/Replace** ⇒ **?** ⇒ **enter** ⇒ (pick the "C" size format you created in Lesson 16) ⇒ **Views** ⇒ **Add View** ⇒ **Scale** ⇒ **Done** ⇒ (pick a center point for the view) ⇒ (type **1.5** as the scale) ⇒ **enter** ⇒ **Done/Return** ⇒ **Done/Return** ⇒ **Sheets** ⇒ **Previous** (to return to **SHEET 1 OF 2**)

Dbms (File--PT/Modeler) ⇒
Save ⇒ **enter**
Purge ⇒ **enter** ⇒
Done-Return

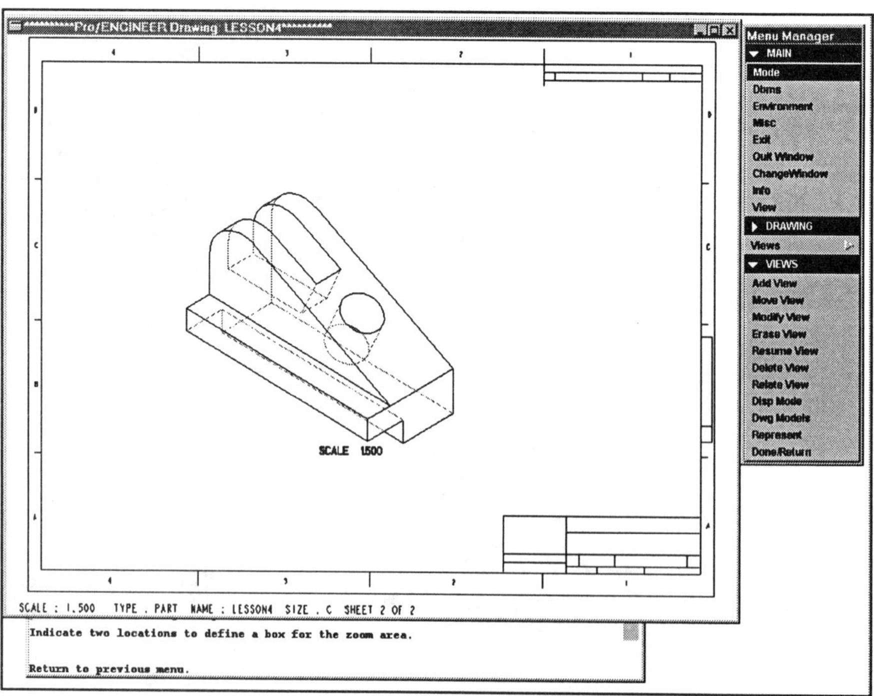

Figure 18.23
SHEET 2 OF 2 with an Isometric View of the Anchor

Create a detailed view with the following commands:

Views ⇒ Add View ⇒ Detailed ⇒ Done ⇒ (pick an open space on the drawing to position the new view, as in Fig. 18.24) ⇒ (type **1.5** for the view scale) ⇒ **enter** ⇒ (select the center point for the detail on an existing view, pick the top edge of the hole in the front section view) ⇒ (sketch a spline around the hole; use the left mouse button to sketch the spline and the middle button to end it) ⇒ (enter a name for the detail view: **A**) ⇒ **enter** ⇒ (pick **Circle** from the BOUNDARY TYPE menu) ⇒ (select the location of the note as shown in Fig. 18.25) ⇒ **Done/Return**

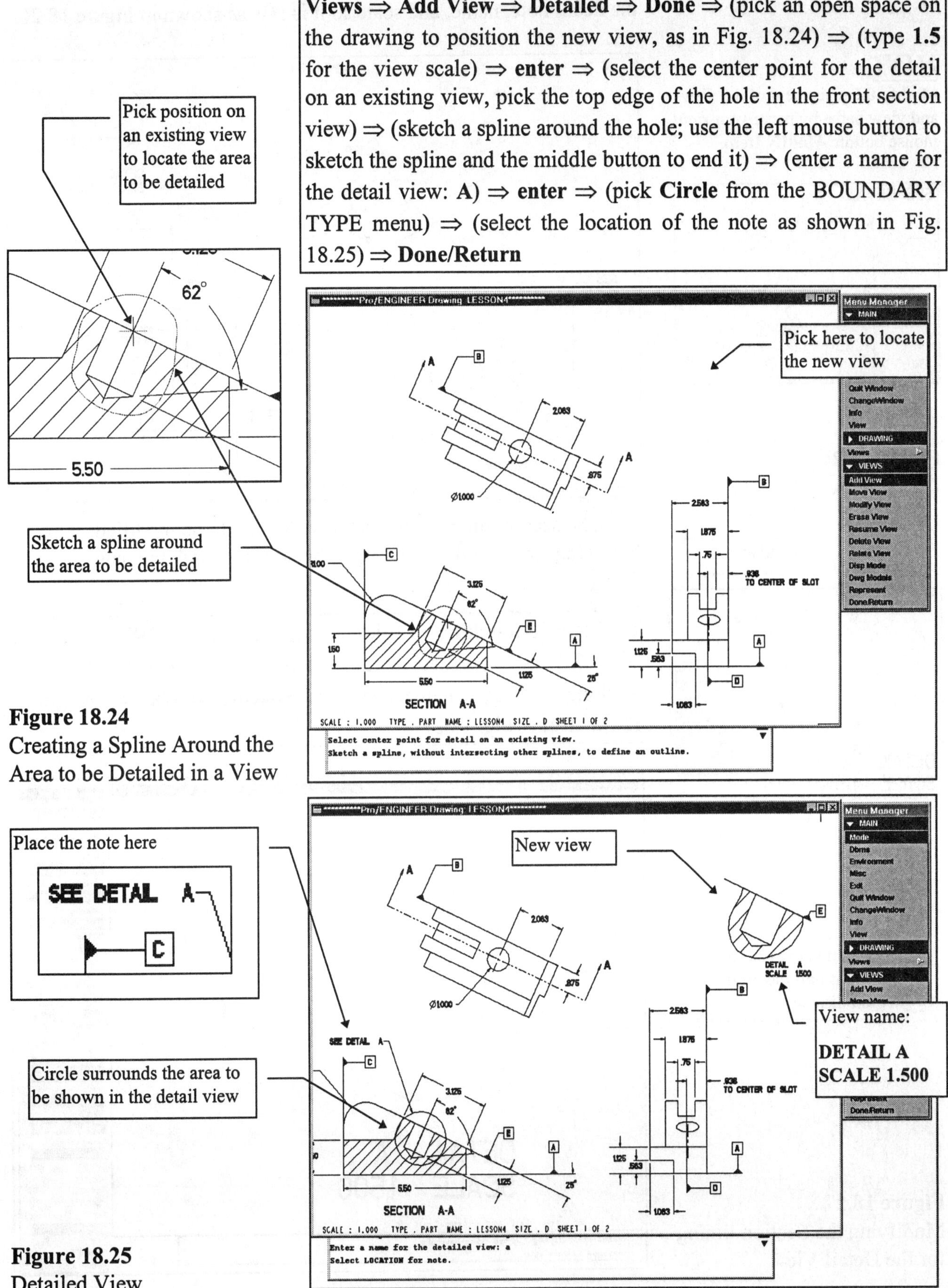

Figure 18.24
Creating a Spline Around the Area to be Detailed in a View

Figure 18.25
Detailed View

Add an axis to the detail of the hole, and modify the text height of the detail note, name, and scale so it is **.37**, as shown in Figure 18.26.

HINT

Modify the placement of the view and view name by using your *right* mouse button: **Modify Item.**

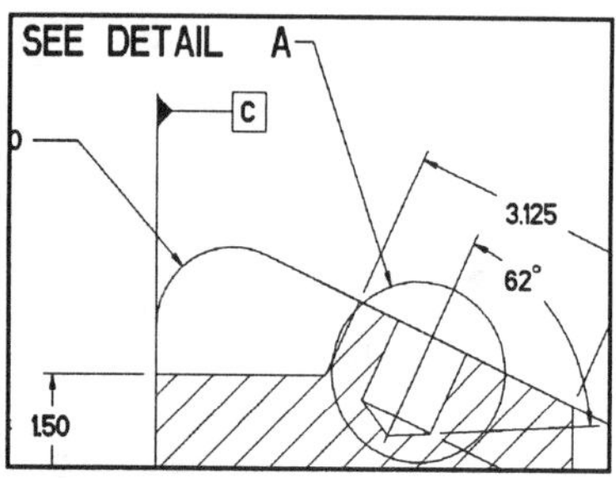

SEE DETAIL A
R1.00
C
3.125
62°
1.50
6.50
SECTION A-A
1.125
25°
A
B
E
2.063
.876
⌀1.000
DETAIL A
SCALE 1.500
2.563
1.876
.75
.938
TO CENTER OF SLOT
1.125
.563
1.063
D

Figure 18.26
DETAIL A

Add an axis

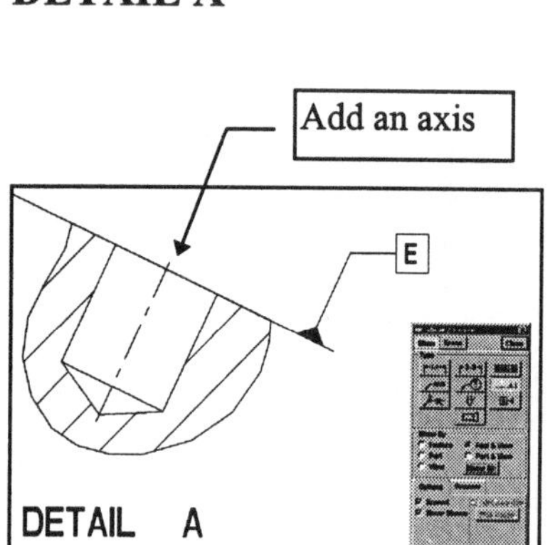

The section lining in the detail view should be modified. Use the following commands to change the spacing and the angle of the lining:

Modify ⇒ **Xhatching** ⇒ (pick the section lining in **DETAIL A)** ⇒ **Done Sel** ⇒ **Det Indep** (breaks the relationship to the parent view hatching, making it independent) ⇒ **Spacing** ⇒ **Half** ⇒ **Angle** ⇒ **120** ⇒ **Done** (Fig. 18.27)

Section lining is **120°** and spacing is half the original default distance

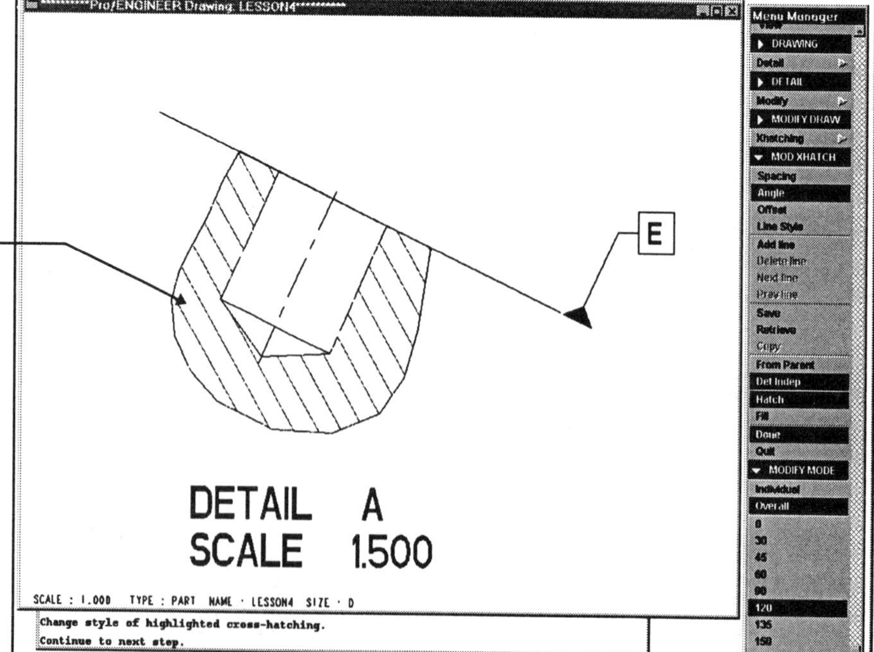

Figure 18.27
Modifying the Section Lining for the Detail View

Switch the hole depth dimension from the front section view to the detail view, as shown in Figure 18.28. The detail is now complete (Fig. 18.29).

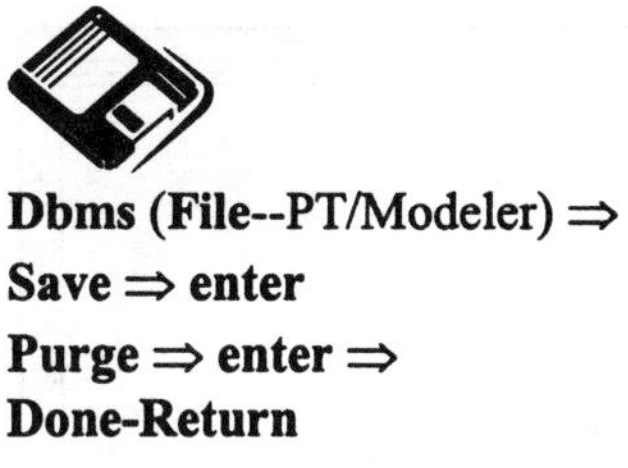

Dbms (File--PT/Modeler) ⇒
Save ⇒ enter
Purge ⇒ enter ⇒
Done-Return

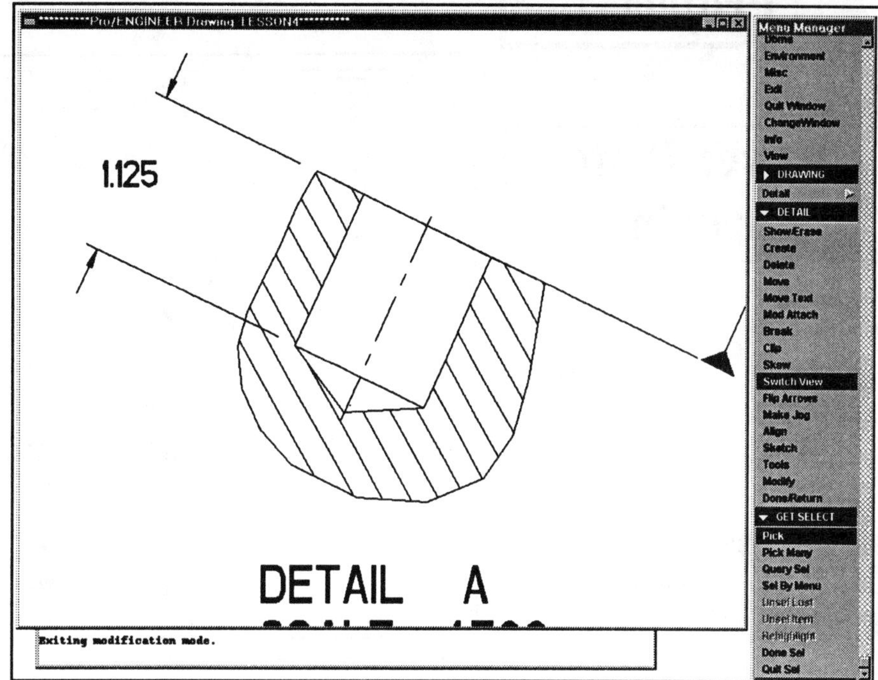

Figure 18.28
Hole Depth Dimension Shown in **DETAIL A**

Figure 18.29
Completed Detail Drawing (note that hidden lines are shown in the two nonsectioned views)

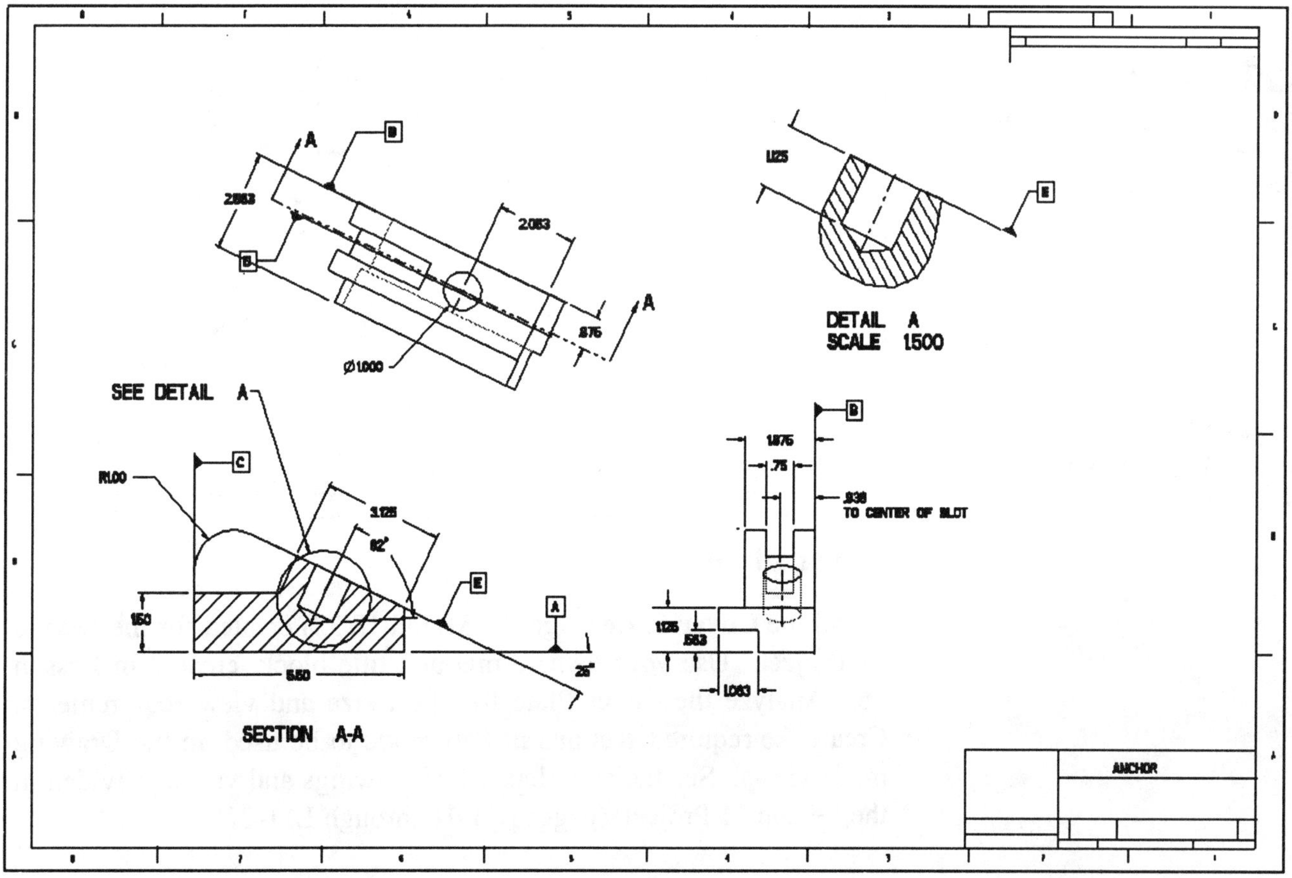

Lesson 18 Project

Cover Plate Drawing

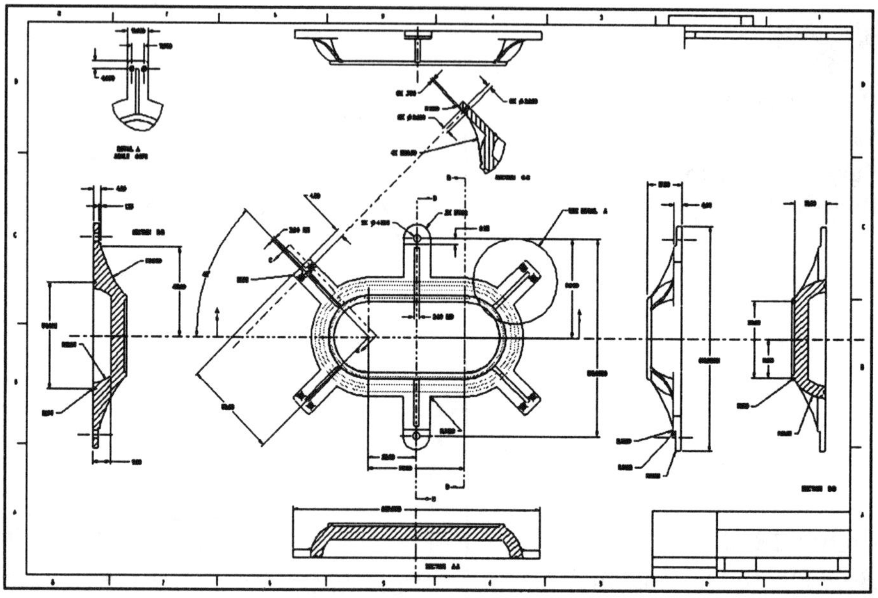

Figure 18.30
Cover Plate Drawing, Sheet One

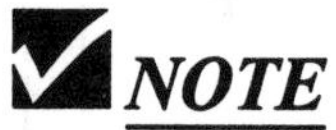

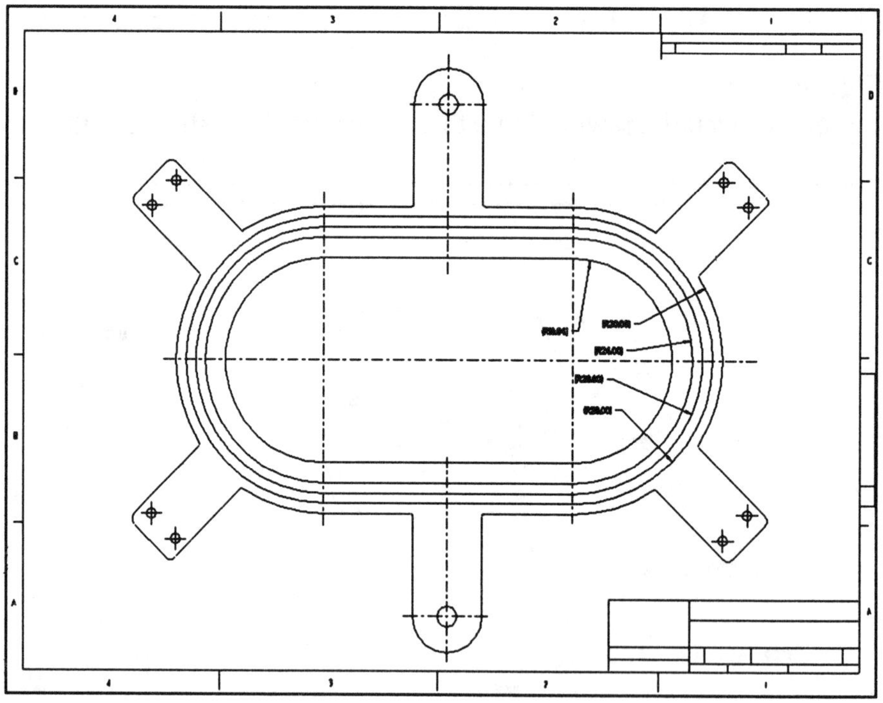

NOTE

You may detail any of the parts created in Lessons 1-13 at this time.

Figure 18.31
Cover Plate Drawing, Sheet Two

Cover Plate

Detail the **Cover Plate** (Figs. 18.31 and 18.32) created for the Lesson 11 Project. Use your own format and title block, created in Lesson 16. Analyze the Cover Plate for sheet size and view requirements. Create the required sections in Part mode to be used on the Drawing mode views. See the complete set of drawings and views provided in the Lesson 11 Project (pages L11-18 through L11-22).

Lesson 19

Assembly Drawings and BOM

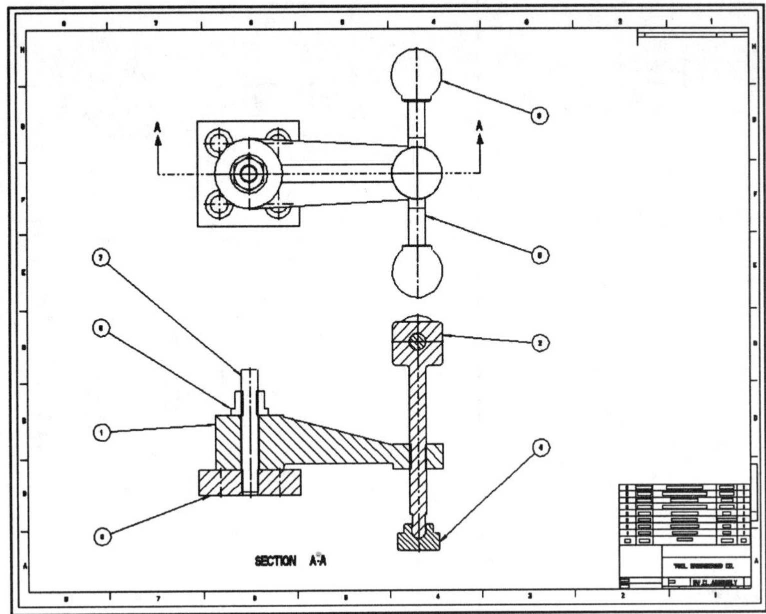

Figure 19.1
Swing Clamp Assembly Drawing

OBJECTIVES

1. **Create an assembly drawing**
2. **Generate a parts list from a bill of materials (BOM)**
3. **Balloon an assembly drawing**
4. **Create a section assembly view and change component visibility**
5. **Add parameters to parts**
6. **Create a table to generate a parts list automatically**

EGD REFERENCE
Engineering Graphics and Design with Graphical Analysis *or* **Fundamentals of Engineering Graphics and Design**
by L. Lamit and K. Kitto
Read Chapters 11, 17, 23
See pages 358-368, 662-671, and 810-846

COAch™ for Pro/ENGINEER

If you have **COAch for Pro/ENGINEER** on your system, go to SEARCH and do the Segments shown in Figures 19.4, 19.5, and 19.8.

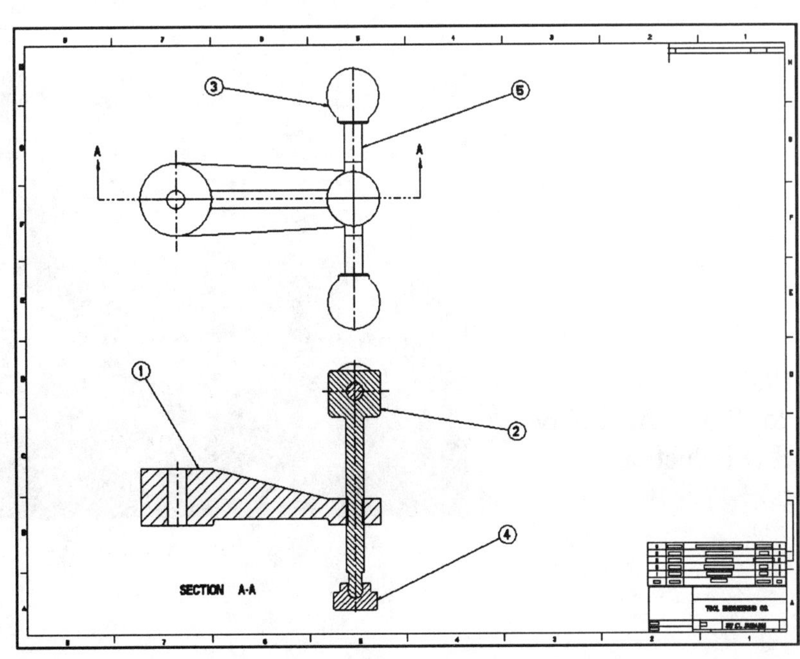

Figure 19.2
Swing Clamp Subassembly Drawing

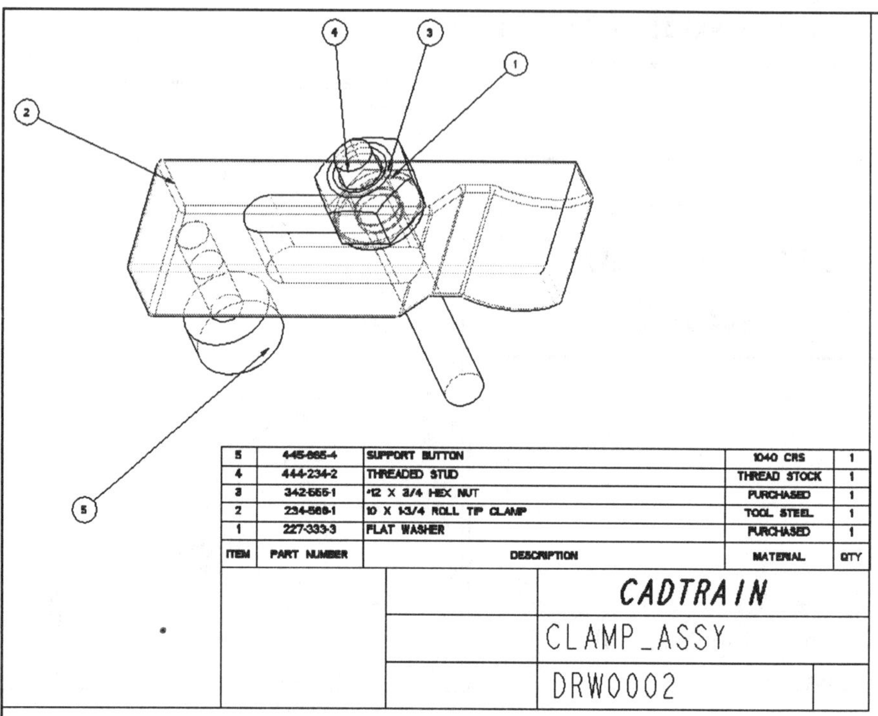

Figure 19.3
Strap Clamp with Parts List from CADTRAIN

ASSEMBLY DRAWINGS AND BOM

Pro/E incorporates a great deal of functionality into drawings of assemblies (Figs. 19.1 and 19.2). You can assign parameters to parts in the assembly that can be displayed on a *parts list* in an assembly drawing (Fig. 19.3). Pro/E can also generate the item balloons for each component (Fig. 19.4).

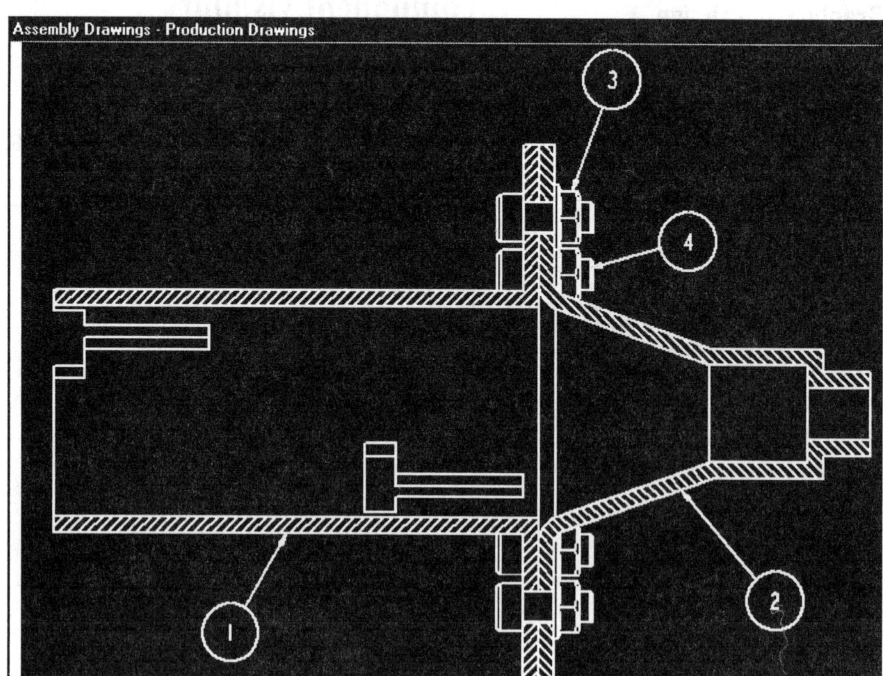

Figure 19.4
COAch for Pro/E, Assembly Drawings (Production Drawings)

Figure 19.5
COAch for Pro/E, Assembly Drawings (Adding Parts List Data)

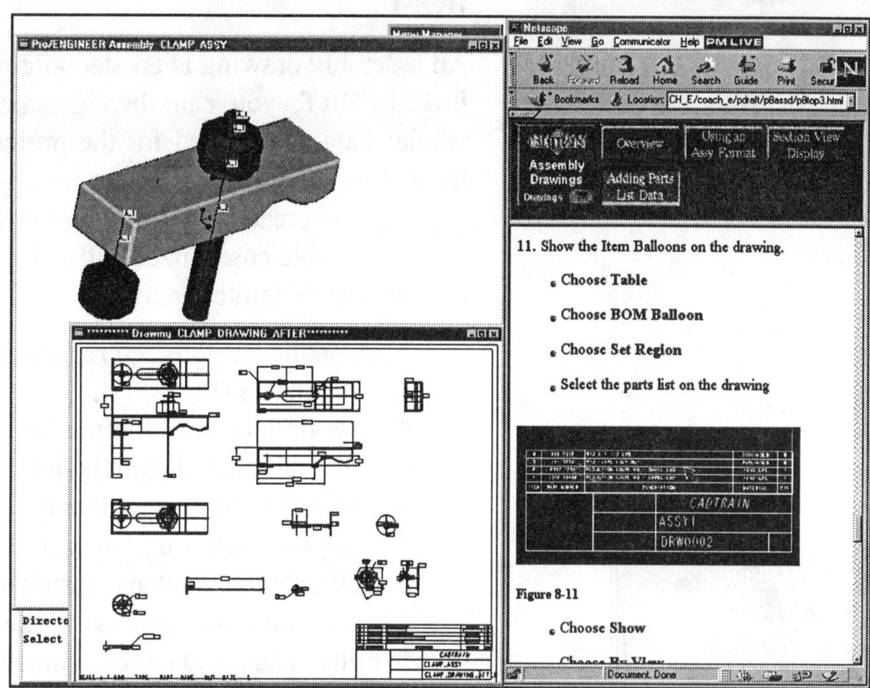

In addition, a variety of specialized capabilities allow you to alter the manner in which individual components are displayed in views and in sections (Fig. 19.5). The **format** for an assembly is usually (slightly) different from the format used for detail drawings. The most significant difference is the presence of a **parts list**.

The assembly format provided has been adapted from CADTRAIN's COAch for Pro/ENGINEER. As part of this lesson, you will create a set of assembly formats and place your standard parts list on them.

A parts list is actually a *Drawing Table object* that is formatted to represent a bill of materials on a drawing. By defining *parameters* in the parts in your assembly that agree with the specific format of the parts list, you make it possible for Pro/E to add pertinent data to the parts list automatically as components are added to the assembly. After the parts list and parameters have been added, Pro/E can balloon the assembly automatically (Fig. 19.6).

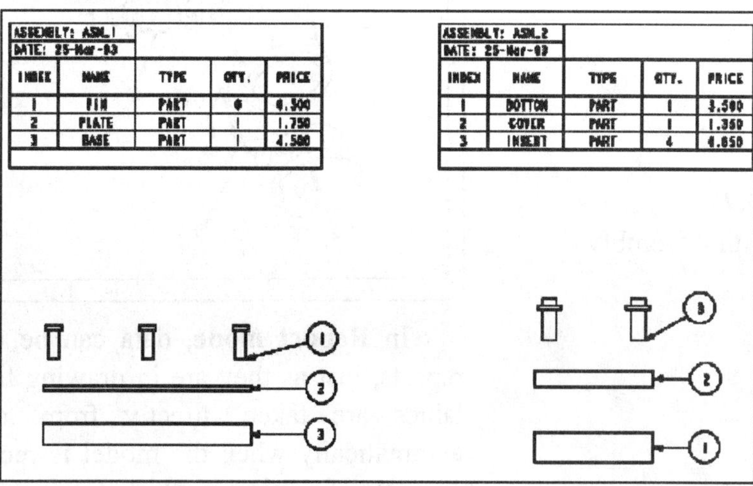

ASSEMBLY: ASM_1
DATE: 25-Mar-93

INDEX	NAME	TYPE	QTY.	PRICE
1	PIN	PART	6	0.300
2	PLATE	PART	1	1.750
3	BASE	PART	1	4.500

ASSEMBLY: ASM_2
DATE: 25-Mar-93

INDEX	NAME	TYPE	QTY.	PRICE
1	BOTTOM	PART	1	3.500
2	COVER	PART	1	1.350
3	INSERT	PART	4	0.050

Figure 19.6
Online Documentation, BOM Balloons

BOM

An assembly drawing is created after the assembly is complete. With Pro/REPORT, you can then generate a bill of materials or other tabular data as required for the project. **Pro/REPORT** introduces a formatting environment where text, graphics, tables, and data can be combined to create a dynamic report (Fig. 19.7). Specific tools enable you to generate customized **bills of materials** (BOMs), family tables, and other associative reports:

* Dynamic, customized reports with drawing views and graphics can be created (Fig. 19.7).
* User-defined or predefined model data can be listed on reports, drawing tables, or layout tables. These reported data can be sorted by any requested data-type display.
* Regions in drawing tables, report tables, and layout tables can be defined to expand and shrink automatically with the amount of model information you have asked to have displayed.
* Filters can be added to eliminate the display of specific types of data from reports, drawing tables, or layout tables.
* Recursive or top-level assembly data can be searched for display.
* Duplicate occurrences of model data can be listed individually or as a group in a report, drawing table, or layout table.
* Assembly component balloons can be linked directly to a customized BOM and automatically updated when assembly modifications are made.

BOM Balloon
Pagination
Save/Retrieve
Done/Return
▼ BOM BALLOONS
Set Region
Clear Region
Change Type
Set Param
Show
Alt Symbol
Merge
Detach
Redistribute
Split

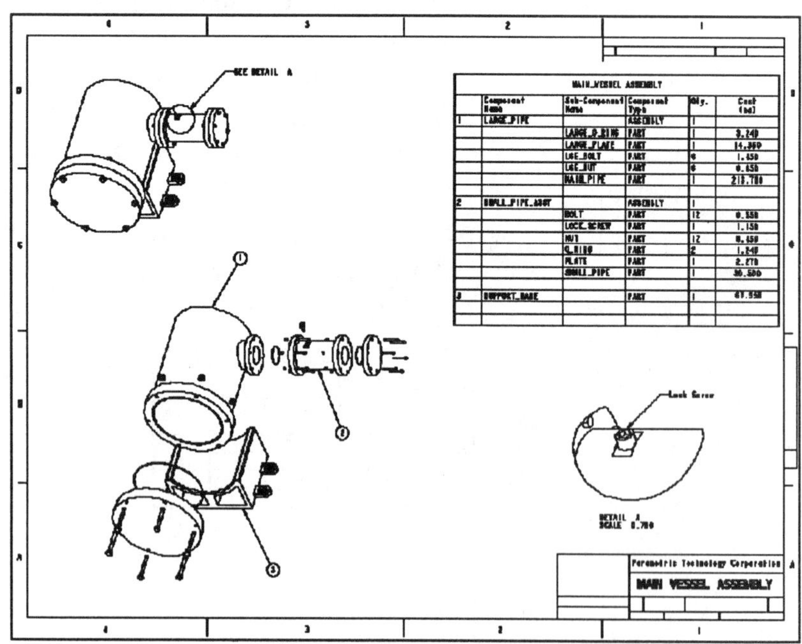

Figure 19.7
Report With Assembly Views

In **Report mode**, data can be displayed in a tabular form on reports, just as they are in drawing tables. The data reported on the tables are taken directly from a selected model and update automatically when the model is modified or changed. A common example of a report is a bill of materials report or a generic part table.

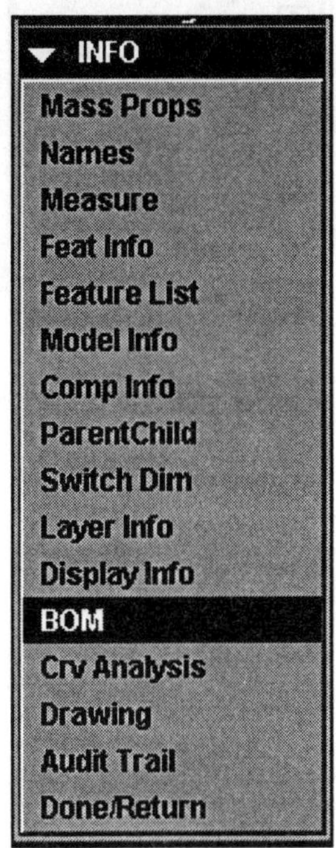

Figure 19.8
COAch for Pro/E, Parts List on Assembly Drawings

Including a Bill of Materials in a Drawing

If you do not have Pro/REPORT (Fig. 19.8) and want to add a BOM, create a BOM file in Assembly mode using the **Info** option in the MAIN menu. Choose **BOM** from the INFO menu, and add the BOM file to the drawing as a note entered from a file. To format or arrange the information in the BOM, you must use a text editor.

◆ Select the parts list on the drawing

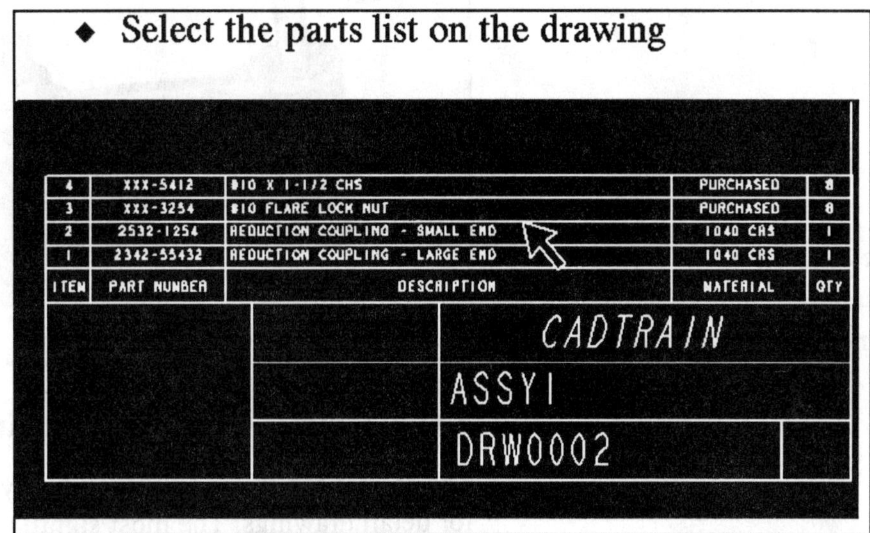

A BOM that is added to a drawing as a note is not connected with the BOM file that was used to create the note. If the composition of the assembly changes, you must create a new BOM file and add it to the drawing as a new note. You can fully edit the BOM displayed on drawings as a note without affecting the original BOM file.

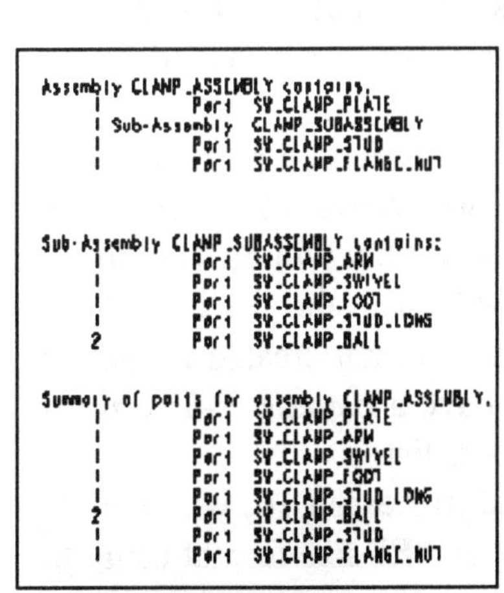

```
Assembly CLAMP_ASSEMBLY contains:
    1            Part  SW_CLAMP_PLATE
    1 Sub-Assembly  CLAMP_SUBASSEMBLY
    1            Part  SW_CLAMP_STUD
    1            Part  SW_CLAMP_FLANGE_NUT

Sub-Assembly CLAMP_SUBASSEMBLY contains:
    1            Part  SW_CLAMP_ARM
    1            Part  SW_CLAMP_SWIVEL
    1            Part  SW_CLAMP_FOOT
    1            Part  SW_CLAMP_STUD_LONG
    2            Part  SW_CLAMP_BALL

Summary of parts for assembly CLAMP_ASSEMBLY:
    1            Part  SW_CLAMP_PLATE
    1            Part  SW_CLAMP_ARM
    1            Part  SW_CLAMP_SWIVEL
    1            Part  SW_CLAMP_FOOT
    1            Part  SW_CLAMP_STUD_LONG
    2            Part  SW_CLAMP_BALL
    1            Part  SW_CLAMP_STUD
    1            Part  SW_CLAMP_FLANGE_NUT
```

To add a BOM to a drawing as a note:

1. Choose **File** from the NOTE TYPES menu. When adding a BOM to a drawing as a note, justify the note using **Default** or **Left**. If you use **Center** or **Right**, the BOM may be formatted incorrectly on the drawing.
2. Choose **Make Note** when you have finished choosing options.
3. Pick a location for the note (BOM) to appear.
4. **Enter NAME** and type the file name, including the **.bom** full extension.
5. The BOM is displayed on the drawing.

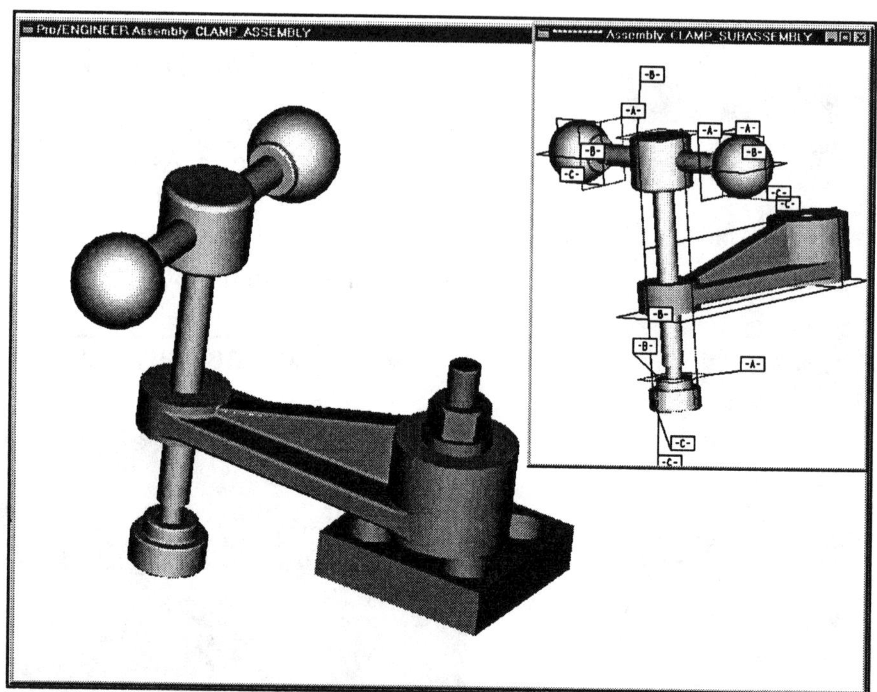

Figure 19.9
Swing Clamp Subassembly and Swing Clamp Assembly

Swing Clamp Assembly Drawing

The format for an assembly is usually different from the format used for detail drawings. The most significant difference is the presence of a parts list. We will create a standard "**E**" size format and place a standard parts list on it. You should create a set of assembly drawing formats on "**B**," "**C**," and "**D**" size sheets at your convenience.

A **parts list** is actually a *Drawing Table object* that is formatted to represent a bill of materials (BOM) on a drawing. By defining parameters for the parts in your assembly that agree with the specific format of the parts list, you make it possible for Pro/E to add pertinent data to the parts list as components are added to the assembly.

After you create an "**E**" size format sheet with a parts list table, you will create a new drawing with two views using your new assembly format. The Swing Clamp subassembly will be used for the first drawing. The second drawing will use the Swing Clamp assembly. Both drawings use the "**E**" size format created in the first section of this lesson. The format will have a parameter-driven title block (as in Lesson 16) and an integral parts list.

Use a format file name that will identify the format as an assembly format with a parts list, such as **ASM_PTL_E**.

Using steps similar to those outlined in Lesson 16, where a "**C**" size format was created and saved, create an "**E**" size format using the following commands:

Mode ⇒ **Format** ⇒ **Search/Retr** ⇒ **Format Dir** ⇒ **e.frm** ⇒ **Dbms** ⇒ **Save As** ⇒ **enter** ⇒ (type a unique name for your format: ***your_format_name***) ⇒ **enter** ⇒ **Quit Window** ⇒ **Format** ⇒ **Search/Retr** ⇒ (pick the new format: ***your_format_name*.frm**)

Figure 19.10 shows the standard "**E**" size format available from Pro/E using the format directory.

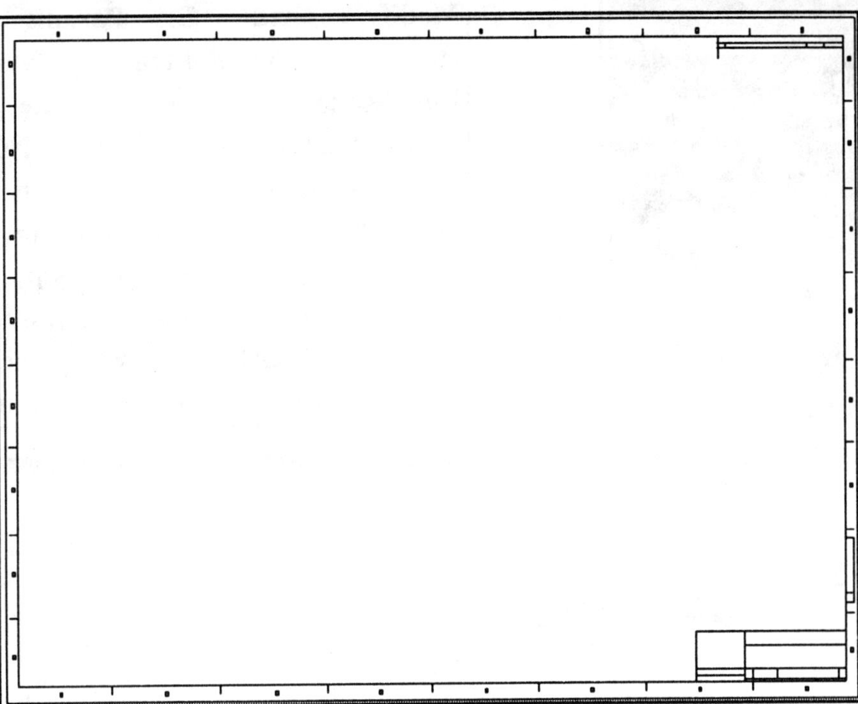

Figure 19.10
Standard "**E**" Size Format

Modify your **.dtl** file to have filled arrowheads and filled default font, as shown in Figure 19.11.

Set Up ⇒ **Modify Val** ⇒ (edit the file as required) ⇒ **File** ⇒ **Save** ⇒ **File** ⇒ **Exit** ⇒ **Quit** ⇒ **View** ⇒ **Repaint** ⇒ **Done-Return**

NOTE

Set the following in your **.dtl** file:
Set Up ⇒ **Modify Val**

default_font	**filled**
draw_arrow_style	**filled**
drawing_text_height	**.25**
draw_arrow_length	**.25**
draw_arrow_width	**.08**

File ⇒ **Save** ⇒ **File** ⇒ **Exit**

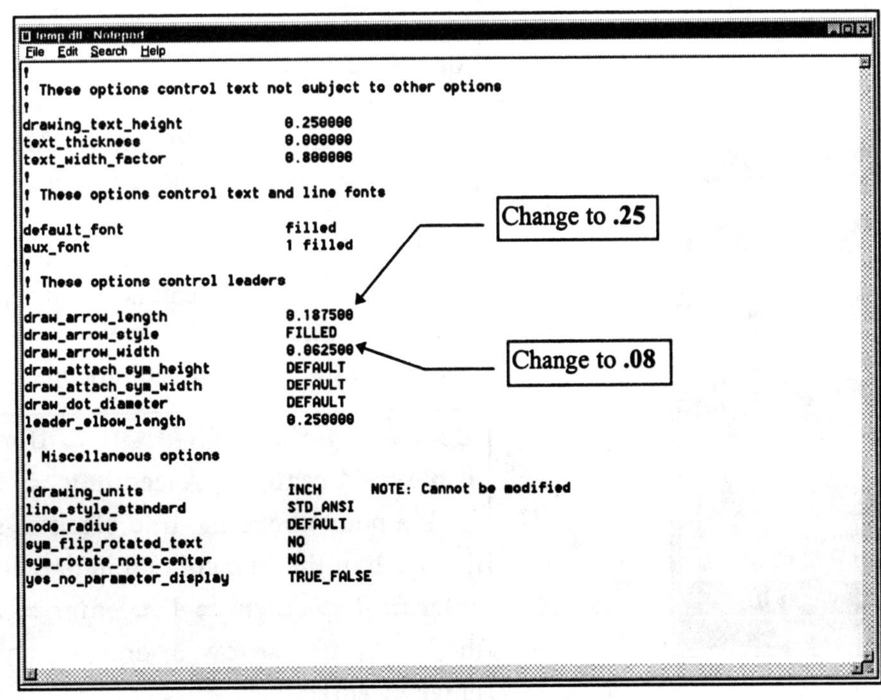

```
!
! These options control text not subject to other options
!
drawing_text_height          0.250000
text_thickness               0.000000
text_width_factor            0.800000
!
! These options control text and line fonts
!
default_font                 filled
aux_font                     1 filled
!
! These options control leaders
!
draw_arrow_length            0.187500
draw_arrow_style             FILLED
draw_arrow_width             0.062500
draw_attach_sym_height       DEFAULT
draw_attach_sym_width        DEFAULT
draw_dot_diameter            DEFAULT
leader_elbow_length          0.250000
!
! Miscellaneous options
!
!drawing_units               INCH       NOTE: Cannot be modified
line_style_standard          STD_ANSI
node_radius                  DEFAULT
sym_flip_rotated_text        NO
sym_rotate_note_center       NO
yes_no_parameter_display     TRUE_FALSE
```


Figure 19.11
.dtl Format File

DETAIL
Create
DETAIL ITEM
Note
GET POINT
Pick Pnt
Vertex
On Entity
Rel Coords
Abs Coords
Done
Quit

Zoom into the title block only and fill in the titles and parameters required to display the proper information, choose the following commands (you may wish to have **Grid Snap** on):

> **Modify** ⇒ **Grid** ⇒ **Grid Params** ⇒ **X&Y Spacing** ⇒ (type **.2**) ⇒ **enter** ⇒ **Done/Return** ⇒ **Grid On** ⇒ **Done/Return** ⇒ **Done/Return** ⇒ **Detail** ⇒ **Create** ⇒ **Note** ⇒ **Make Note** ⇒ (pick point for note) ⇒ (type **TOOL ENGINEERING CO.**) ⇒ **enter** ⇒ **enter** ⇒ **Make Note** ⇒ [create notes for **DRAWN** and **ISSUED**, place them in the title block (Fig. 19.12)]
>
> Make parametric notes: **Make Note** ⇒ (pick point for note) ⇒ (type **&dwg_name**) ⇒ **enter** ⇒ **enter** ⇒ [create parameter notes for **&scale** and **SHEET ¤t_sheet OF &total_sheets**, place them in the title block (Fig. 19.12)] ⇒ **Done/Return** ⇒ **Done/Return**

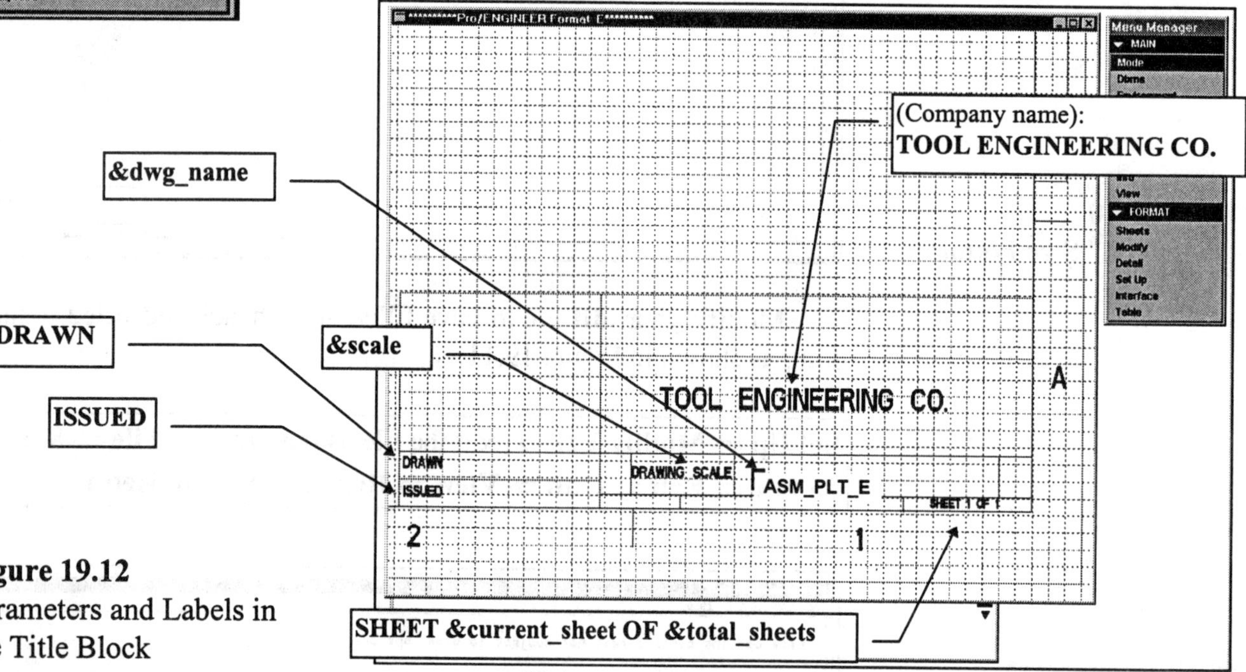

Figure 19.12
Parameters and Labels in the Title Block

Turn off the **Grid Snap**. **Modify** the text height and the placement of the notes so that they are placed similarly to those in Figure 19.12.

The parts list table can now be created and saved with this drawing format. You can add and replace formats and still keep the table associated with the drawing. Start the parts list by creating a table using the following commands:

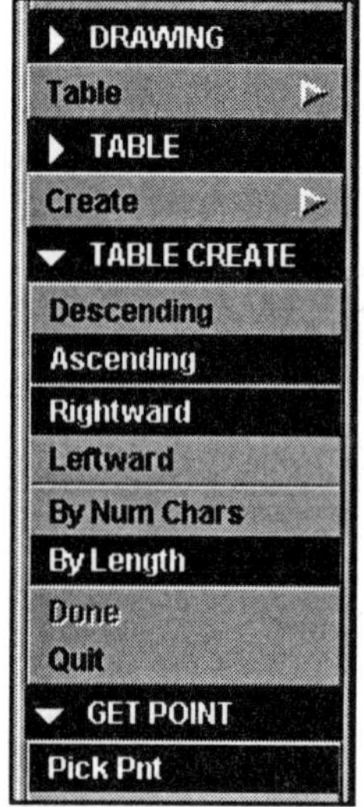

> **Modify** ⇒ **Grid** ⇒ **Grid Off** ⇒ **Done/Return** ⇒ **Done/Return** ⇒ **Table** ⇒ **Create** ⇒ **Ascending** ⇒ **Rightward** ⇒ **By Length** ⇒ (pick a point above the title block as shown in Fig. 19.13) ⇒ (enter the width of the first column in drawing units) ⇒ **1** ⇒ **enter** ⇒ **1** ⇒ **enter** ⇒ **4** ⇒ **enter** ⇒ **1** ⇒ **enter** ⇒ **.75** ⇒ **enter** ⇒ **enter** ⇒ (enter the height of first row in drawing units) ⇒ **.5** ⇒ **enter** ⇒ **.375** ⇒ **enter** ⇒ **enter**

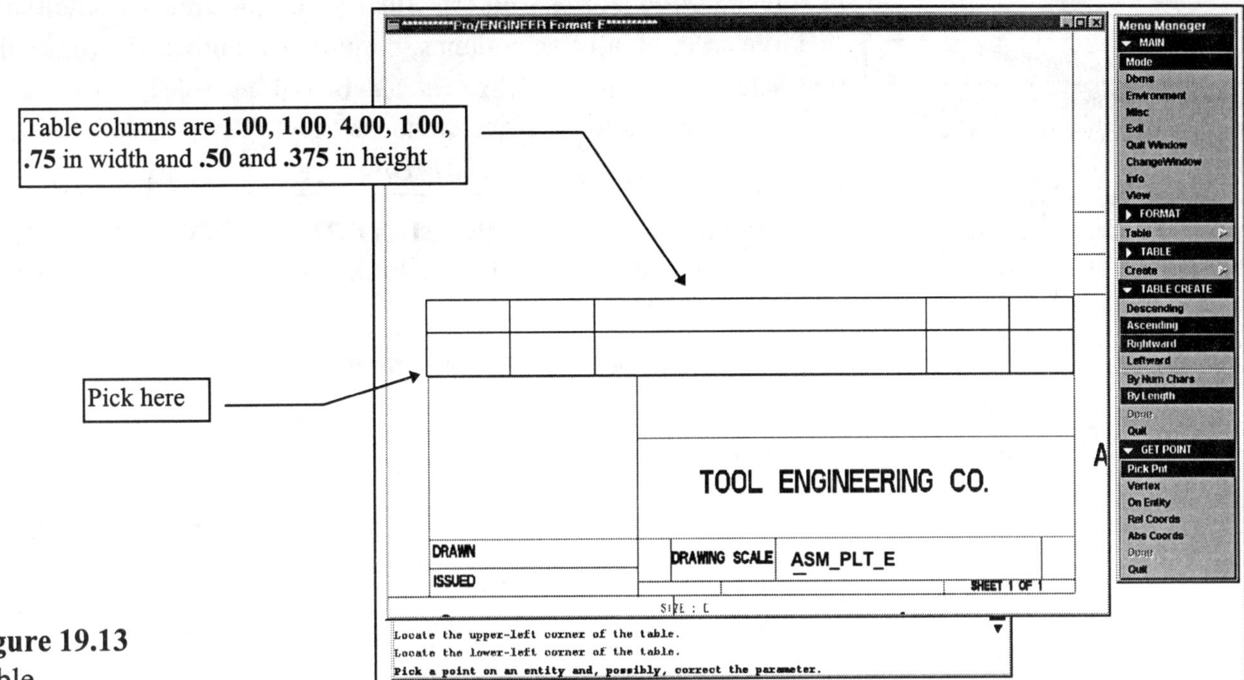

Figure 19.13
Table

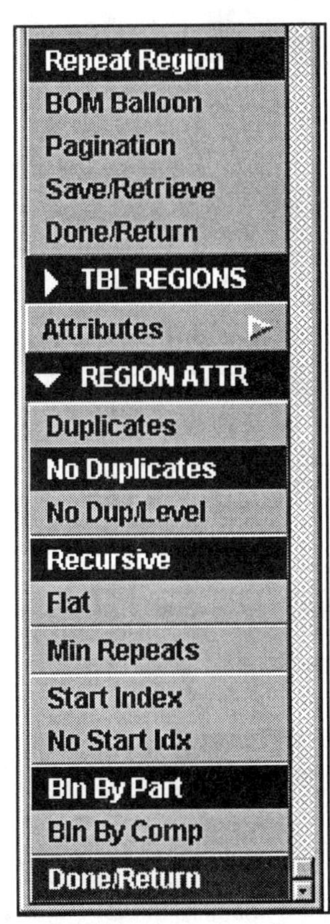

Next we will add a **Repeat Region** (***see PT/Modeler note on the next page***) to the table by continuing with the commands:

Repeat Region ⇒ **Add** ⇒ (locate the corners of the region as shown in Fig. 19.14) ⇒ **Attributes** (select the Repeat Region) ⇒ **No Duplicates** ⇒ **Recursive** ⇒ **Done/Return** ⇒ **Done/Return**

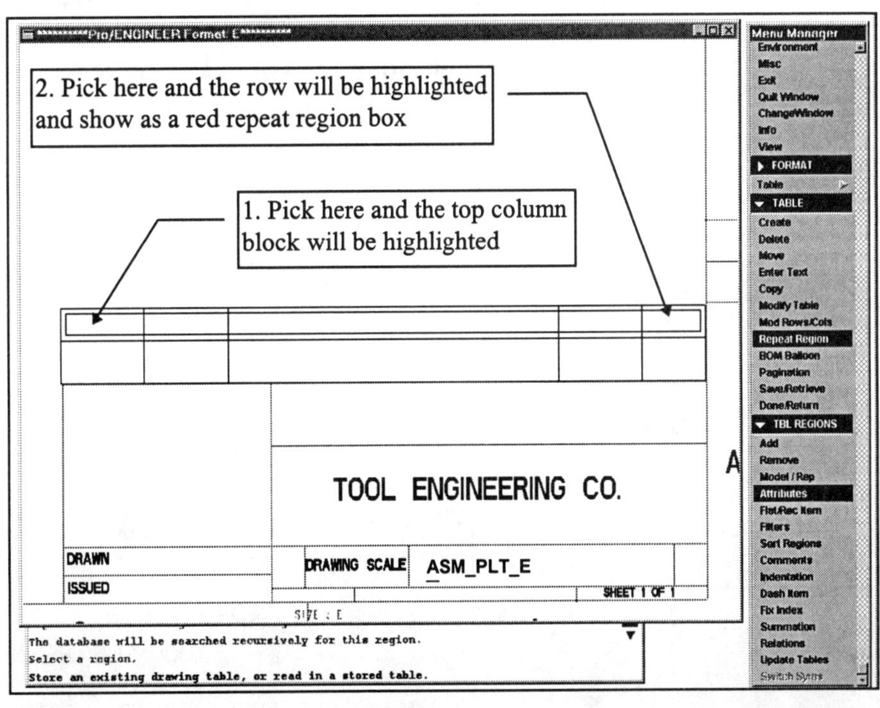

Figure 19.14
Repeat Region

The table must have parameters set in each appropriate block. The column headings should be inserted first, using plain text. Use the following commands:

PT/Modeler™

Repeat Region not supported. Page L19-5 describes how to include a text file note. Also, online documentation included with PT/Modeler describes formatting a BOM report.

Table ⇒ Mod Rows/Cols ⇒ Justify ⇒ Column ⇒ Center ⇒ Middle ⇒ (pick all five columns, they will be outlined in red as they are selected) **⇒ Enter Text ⇒ Keyboard ⇒** (pick the table cell where the text is to be placed, choose the **4.00** width column in the lower row) ⇒ (type **DESCRIPTION**) ⇒ **enter** ⇒ **enter**

Continue adding the titles **MATERIAL**, **QTY**, **ITEM**, and **PT NUM,** as shown in Figure 19.15. From the FORMAT menu, **Modify** the text height to **.125**.

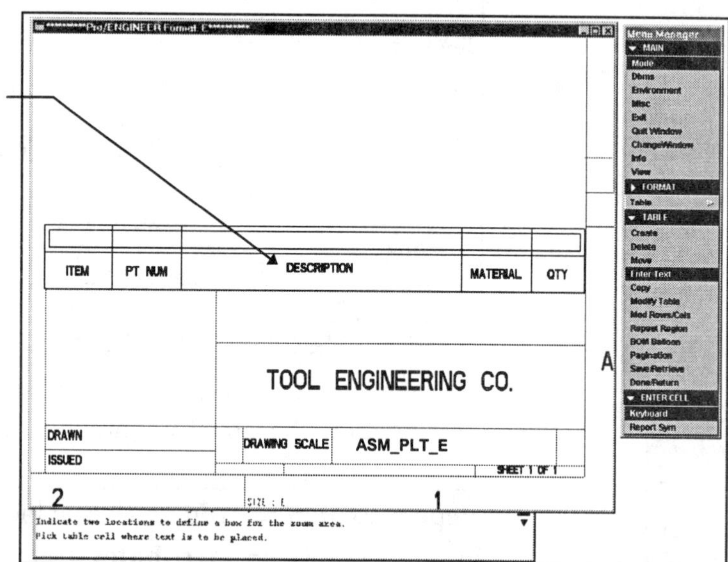

Figure 19.15
Entering Plain Text Headings for a Parts List

The **Repeat Regions** now need to have some of their headings correspond to the parameters created in Part mode for each component model. The **ITEM** and quantity (**QTY**) columns will have the **rpt.index** and **rpt.qty** parameters. Use the following commands from the TABLE menu (Fig. 19.16):

Table ⇒ Enter Text ⇒ Report Sym ⇒ (pick the first table cell of the Repeat Region) **⇒ rpt... ⇒ index**

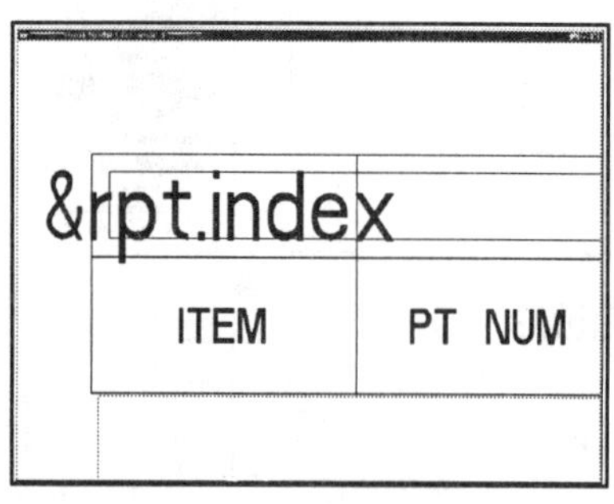

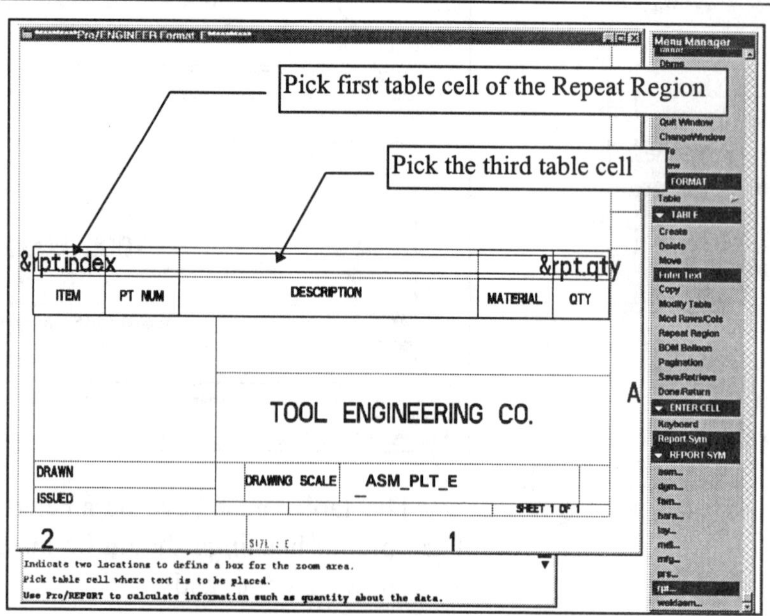

Figure 19.16
Entering Report Symbols

Enter Text ⇒ **Report Sym** ⇒ (pick the fifth table cell of the **Repeat Region**) ⇒ **rpt...** ⇒ **qty**

Now create the parametric **User Defined** text [pick the third table cell where the next text is to be placed (Fig. 19.16)] ⇒ **asm...** ⇒ **mbr...** ⇒ **User Defined** ⇒ (enter symbol text, type **DSC**) ⇒ **enter** ⇒ [pick the fourth table cell (Fig. 19.17)] ⇒ **asm...** ⇒ **mbr...** ⇒ **User Defined** ⇒ (enter symbol text, type **MAT**) ⇒ **enter** ⇒ (pick the second table cell) ⇒ **asm...** ⇒ **mbr...** ⇒ **User Defined** ⇒ (enter symbol text, type **PRTNO**) ⇒ **enter** (Fig. 19.18)

HINT

You will need to change the text height of the Report Symbols in the table cells of the **Repeat Region** to **.125**.

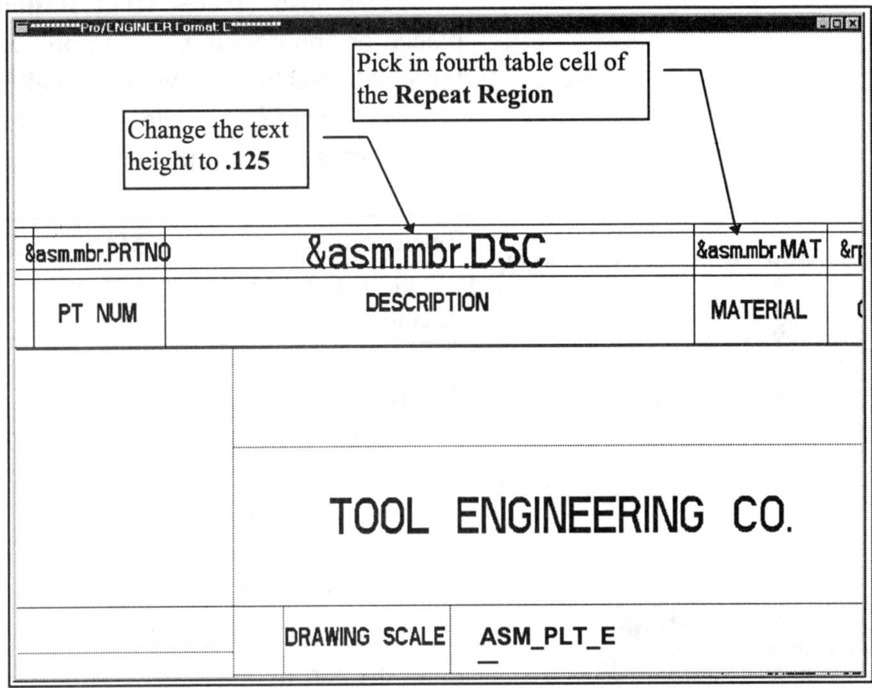

Figure 19.17
Entering Repeat Region Parameters

Dbms (File--PT/Modeler) ⇒ **Save** ⇒ **enter**
Purge ⇒ **enter** ⇒ **Done-Return**

Figure 19.18
Completed Repeat Region Parameters

Adding Parts List Data

When you save your standard assembly format, the Drawing Table that represents your standard parts list is now included. You must be aware of the titles of the parameters under which these data are stored so that you can add them properly to your parts.

As you add components to the assembly, Pro/E reads the parameters from them and updates the parts list. You can also see the same effect by adding these parameters after the drawing has been created.

Pro/E also creates **Item Balloons** on the first view that was placed on the drawing. To improve their appearance, you can move these balloons to other views and alter the locations where they attach.

In Part mode, add the three user defined parameters to each of the components used in the assembly using the following commands:

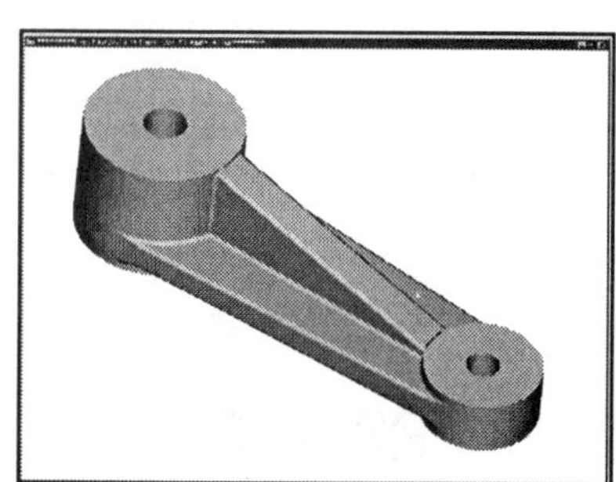

Retrieve the clamp arm: **Mode** ⇒ **Part** ⇒ **Search/Retr** ⇒ [choose **sw_clamp_arm.prt** (or the name you gave the part) from the directory list]

Add the parts list data:

Choose **Relations**

Choose **Add Param**

Choose **String**

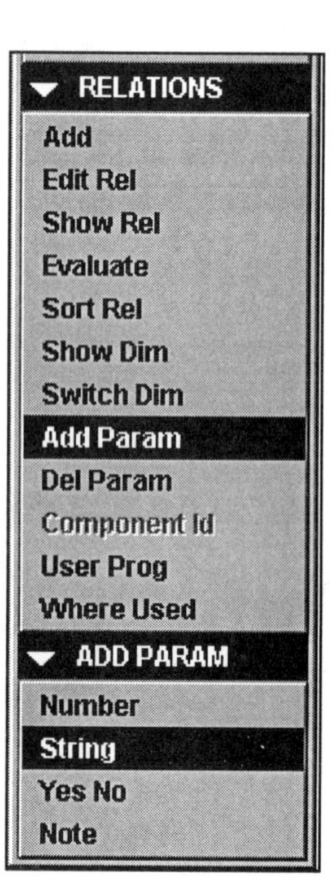

Enter the *exact* title of the parameter that you previously established in the assembly table parts list:

Type **PRTNO** ⇒ **enter**

Enter the part number:

Type **SW101-5AR** ⇒ **enter**

Choose **String**

Enter the *exact* title of the parameter:

Type **DSC** ⇒ **enter**

Enter the component description:

Type **SWING CLAMP ARM** ⇒ **enter**

Choose **String**

Enter the *exact* title of the parameter:

Type **MAT** ⇒ **enter**

Enter the material:

Type **STEEL** ⇒ **enter**

Save and quit the window:

Dbms ⇒ **Save** ⇒ **enter** ⇒ **Purge** ⇒ **enter** ⇒

Done-Return ⇒ **Quit Window**

Retrieve the clamp swivel: **Mode** ⇒ **Part** ⇒ **Search/Retr** ⇒ [choose **sw_clamp_swivel.prt** (or the name you gave the part)]

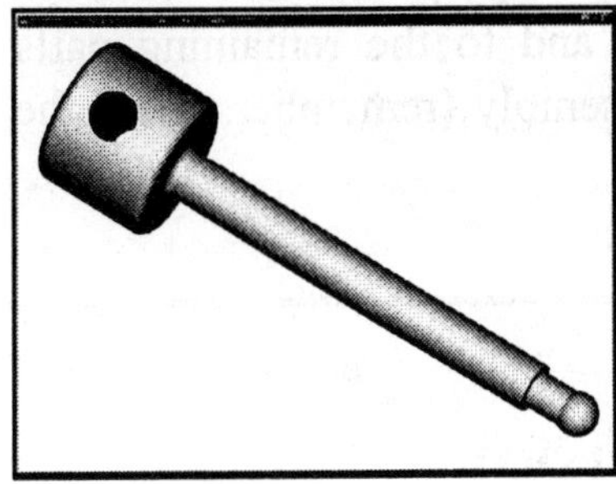

Add the parts list data:

Choose **Relations**

Choose **Add Param**

Choose **String**

Enter the *exact* title of the parameter:

Type **PRTNO ⇒ enter**

Enter the part number:

Type **SW101-6SW ⇒ enter**

Choose **String**

Enter the *exact* title of the parameter:

Type **DSC ⇒ enter**

Enter the component description:

Type **SWING CLAMP SWIVEL ⇒ enter**

Choose **String**

Enter the *exact* title of the parameter:

Type **MAT ⇒ enter**

Enter the material:

Type **STEEL ⇒ enter**

Save and quit the window:

Dbms ⇒ Save ⇒ enter ⇒ Purge ⇒ enter ⇒ Done-Return ⇒ Quit Window

Retrieve the part clamp ball: **Mode ⇒ Part ⇒ Search/Retr ⇒** [choose **sw_clamp_ball.prt** (or the name you gave the part)]

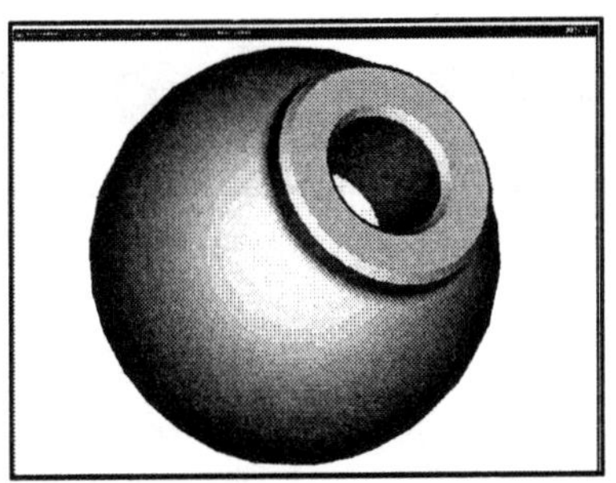

Add the parts list data:

Choose **Relations**

Choose **Add Param**

Choose **String**

Enter the *exact* title of the parameter:

Type **PRTNO ⇒ enter**

Enter the part number:

Type **SW101-7BA ⇒ enter**

Choose **String**

Enter the *exact* title of the parameter:

Type **DSC ⇒ enter**

Enter the component description:

Type **SWING CLAMP BALL ⇒ enter**

Choose **String**

Enter the *exact* title of the parameter:

Type **MAT ⇒ enter**

Enter the material:

Type **BLACK PLASTIC ⇒ enter**

Save and quit the window:

Dbms ⇒ Save ⇒ enter ⇒ Purge ⇒ enter ⇒ Done-Return ⇒ Quit Window

Use the following information to add parameters both to purchased components (standard parts) and to the remaining parts required for the subassembly and the assembly (remember to use the *exact* parameter names).

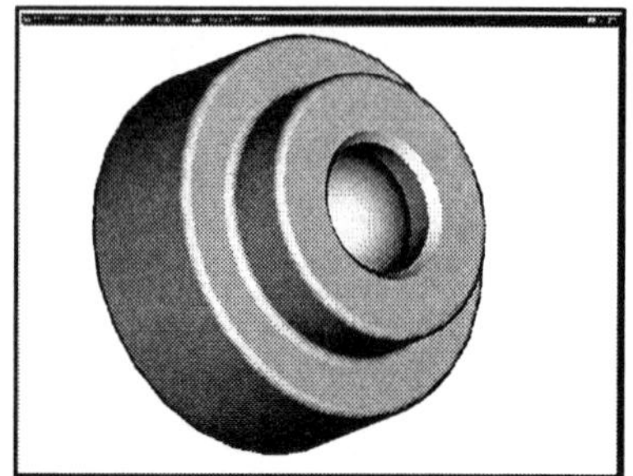

Component	**sw_clamp_foot**
Part Number	**SW101-8FT**
Description	**SWING CLAMP FOOT**
Material	**NYLON**

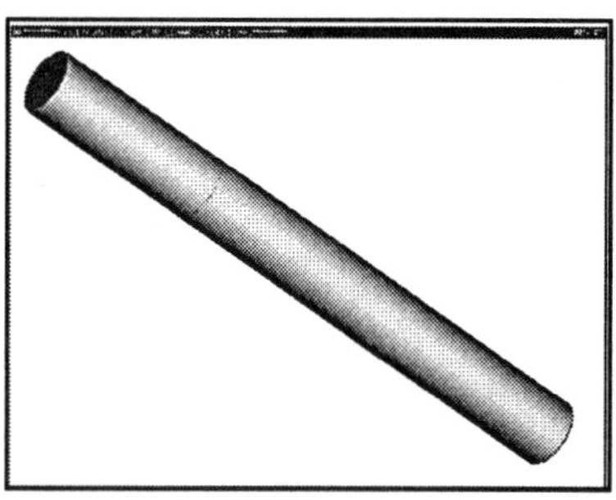

Component	**sw_clamp_stud_long**	**(STUD5)**
Part Number	**SW101-9STL**	
Description	**.500-13 X 5.00 DOUBLE END STUD**	
Material	**PURCHASED**	

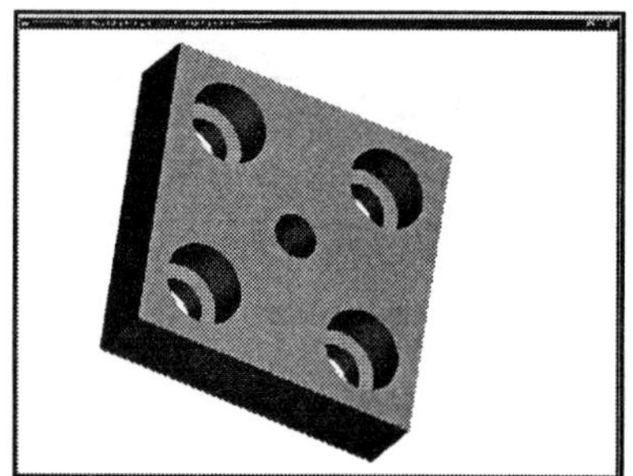

Component	**sw_clamp_plate**
Part Number	**SW100-20PL**
Description	**SWING CLAMP PLATE**
Material	**1020 CRS**

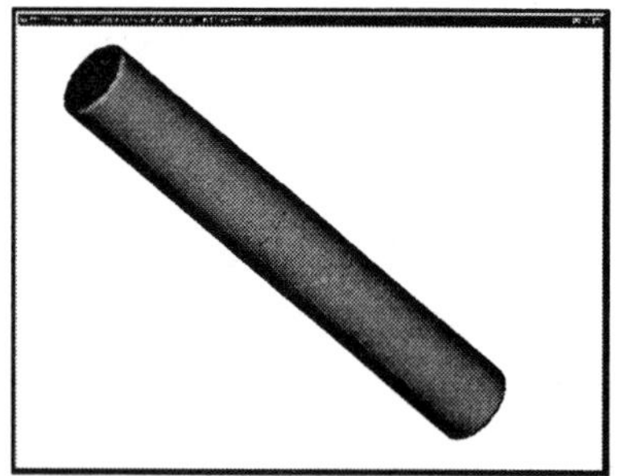

Component	**sw_clamp_stud**	**(STUD35)**
Part Number	**SW100-21ST**	
Description	**.500-13 X 3.50 DOUBLE END STUD**	
Material	**PURCHASED**	

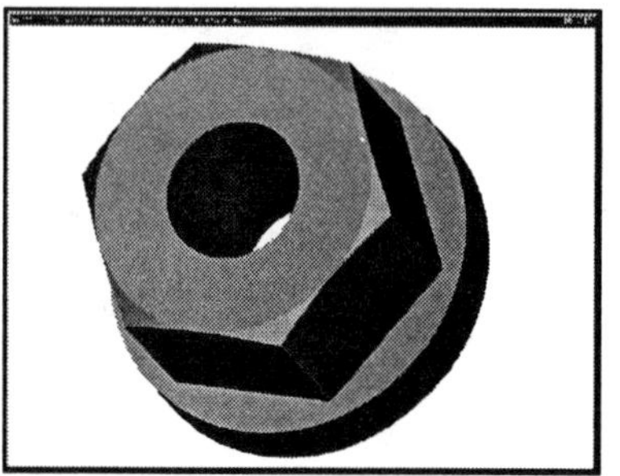

Component	**sw_clamp_flange_nut**	**(FLNGNUT)**
Part Number	**SW100-22FLN**	
Description	**.500-13 HEX FLANGE NUT**	
Material	**PURCHASED**	

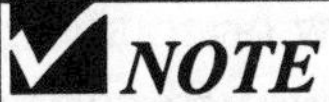

Parameters can be added, deleted, and modified in Part mode, Drawing mode, or Assembly mode. In Drawing mode and Assembly mode, choose **Set Up ⇒ Parameters**. You can also add a *parameters column* to the Model Tree and edit the parameter value by highlighting it in the tree and typing a new value.

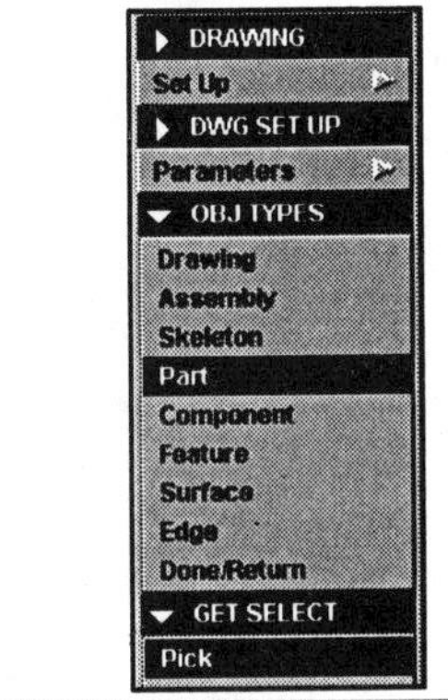

As you add the required relations to each component, you can see the parameters and check to see if they were input correctly by choosing **Show Rel** from the RELATIONS menu, as shown in Figure 19.19.

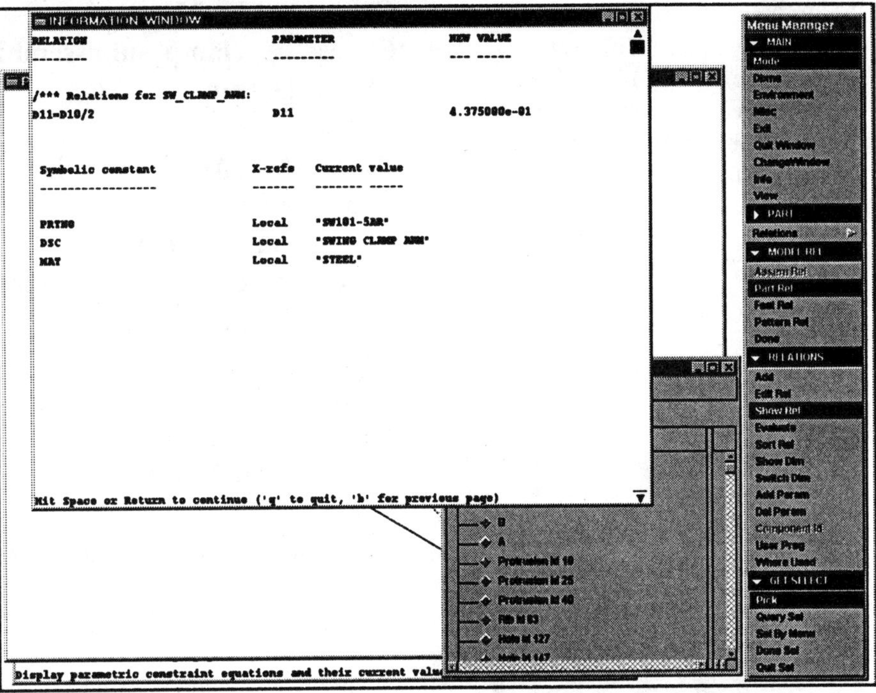

Figure 19.19
Relations for Clamp Arm

The parameters (and their values) have been established in each part. The assembly format with related parameters in a parts list table has been created and saved in the *read only* format directory. You can now create a drawing of the assembly, where the parts list will be generated automatically.

Create a new drawing with the following commands (Fig. 19.20):

Mode ⇒ Drawing ⇒ Create ⇒ (type **sw_cl_subasm**) **⇒ enter ⇒ Retr Format ⇒ ? ⇒ enter ⇒** (select the name of the assembly format produced in the previous segment)

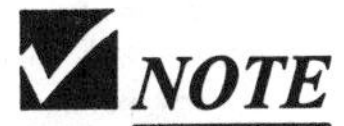

Set the following in your *Drawing* **.dtl** file:

Set Up ⇒ Modify Val ⇒

default_font	**filled**
draw_arrow_style	**filled**
drawing_text_height	**.50**
draw_arrow_length	**.50**
draw_arrow_width	**.17**
dim_leader_length	**1.00**
max_balloon_radius	**.50**
min_balloon_radius	**.50**

File ⇒ Save ⇒ File ⇒ Exit

To get the drawing to appear with the correct style, you must modify the values of the **.dtl** Drawing Set Up file using **Set Up ⇒ Modify Val.** Use the **Save** option (just under **Modify Val)** to save the settings to a file name so you can recall and use the file on another drawing.

Add the top view as the first view. Choose the following commands:

Views ⇒ ? ⇒ [pick **sw_clamp_subassembly** (or the name you gave the model)] **⇒ Done ⇒** (indicate the location for the center of the top view)

You will now use **datum planes** to establish the view orientation. *You must select assembly datums!* This would be very difficult to accomplish by picking on the screen, so you will use **Sel By Menu**:

Choose **Top**
Choose **Sel By Menu**
Choose **sw_clamp_subassembly** from the list
Choose **Datum**
Choose **Name**
Choose **ADTM2** from the list
Choose **Front**
Choose **Sel By Menu**
Choose **sw_clamp_subassembly** from the list
Choose **Datum**
Choose **Name**
Choose **ADTM3** from the list
Choose **Done/Return**
Choose **Done/Return**

HINT

Orient the view as shown in Figure 19.20. Your selections will be different if you have different datum plane names. If you do not get the correct view orientation, **Delete View** and **Add View** again (or pick Default and redo the orientation).

Modify the scale of the drawing so it is **2.00**:

Choose **Modify**

Select the drawing scale value **(1.00)** that is just below the drawing:

Type **2.00** ⇒ **enter**

Notice that the title block and parts list were filled in as the view was created.

5	SWK01-9STL	.500-13 X 5.00 DOUBLE END STUD	PURCHASED	1
4	SWK01-6FT	SWING CLAMP FOOT	NYLON	1
3	SWK01-7BA	SWING CLAMP BALL	BLACK PLASTIC	2
2	SWK01-6SW	SWING CLAMP SWIVEL	STEEL	1
1	SWK01-5AR	SWING CLAMP ARM	STEEL	1
ITEM	PT NUM	DESCRIPTION	MATERIAL	QTY

TOOL ENGINEERING CO.
DRAWN
ISSUED
2.000 SW.CL.SUBASM
SHEET 1 OF 1
2 1

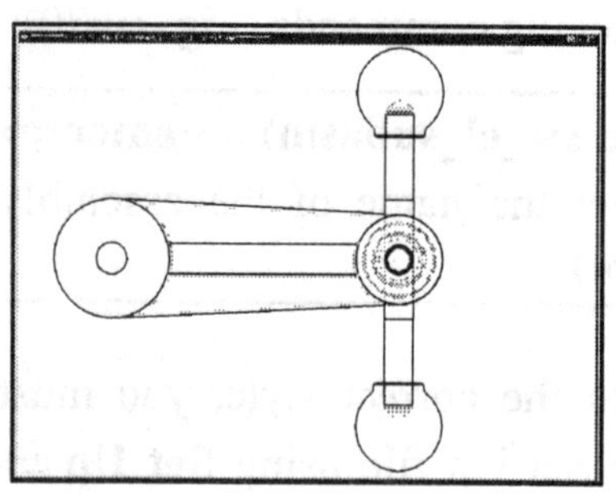

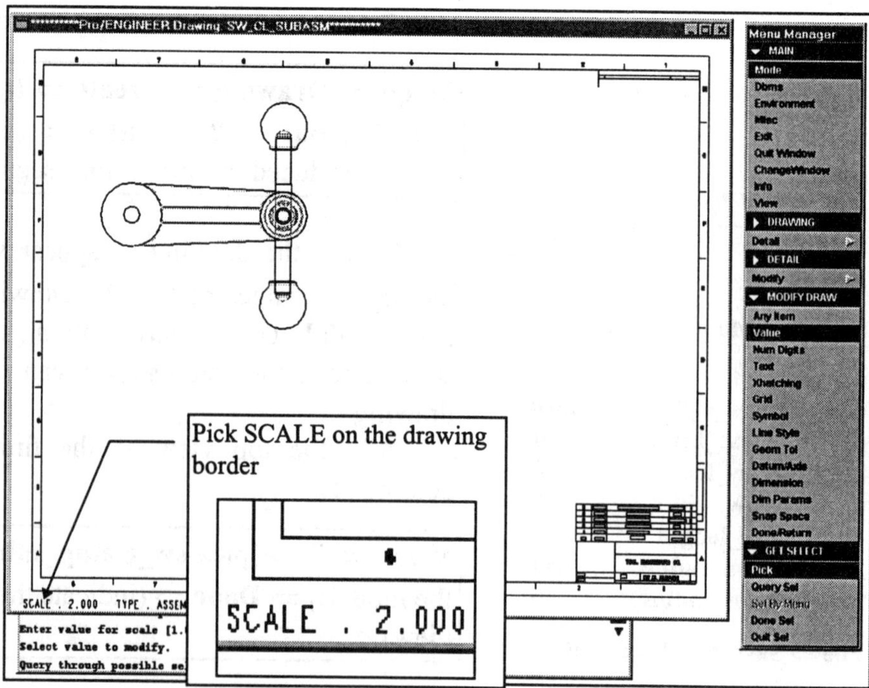

Figure 19.20
Swing Clamp Subassembly Drawing with Top View

HINT

If you haven't already done so, turn off the datums, points, coordinate system, spin center, axes, and grid snap in the ENVIRONMENT menu.

To erase the set datums use the following commands:

Detail ⇒ Show/Erase ⇒ Erase ⇒ (pick the datum radio button) **⇒ Erase All ⇒ Yes ⇒ Close ⇒ Done/Return**

You will need to do this after each view is placed on the drawing or after all views have been established.

The only other view required to show the assembly is the front view. We will be making a front section view using the following commands (Fig. 19.21):

Choose **Views**
Choose **Section**
Choose **Done**
Choose **Done** (since **Full** is the default)

Indicate the center of the view below the top view:

Choose **Create**
Choose **Done** (since **Planar** is the default)
Type **A ⇒ enter**
Choose **Sel By Menu**
Choose **sw_clamp_subassembly** from the list
Choose **Datum**
Choose **Name**
Choose **ADTM2** from the list

Now select any location in the *top* view to define it as the view where the section line, cutting plane line, arrows, and section identification lettering will be placed:

Choose **Done/Return**

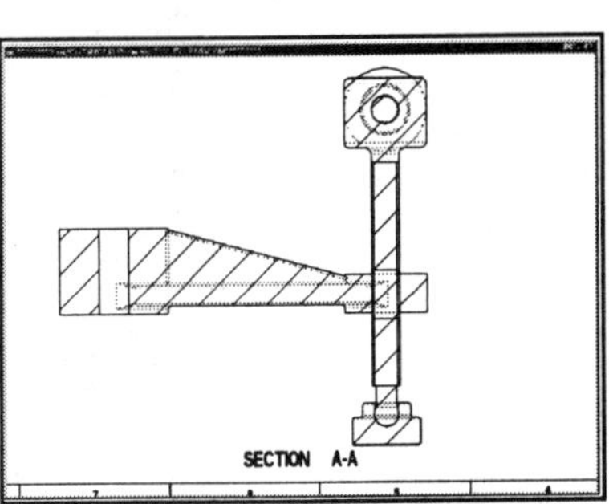

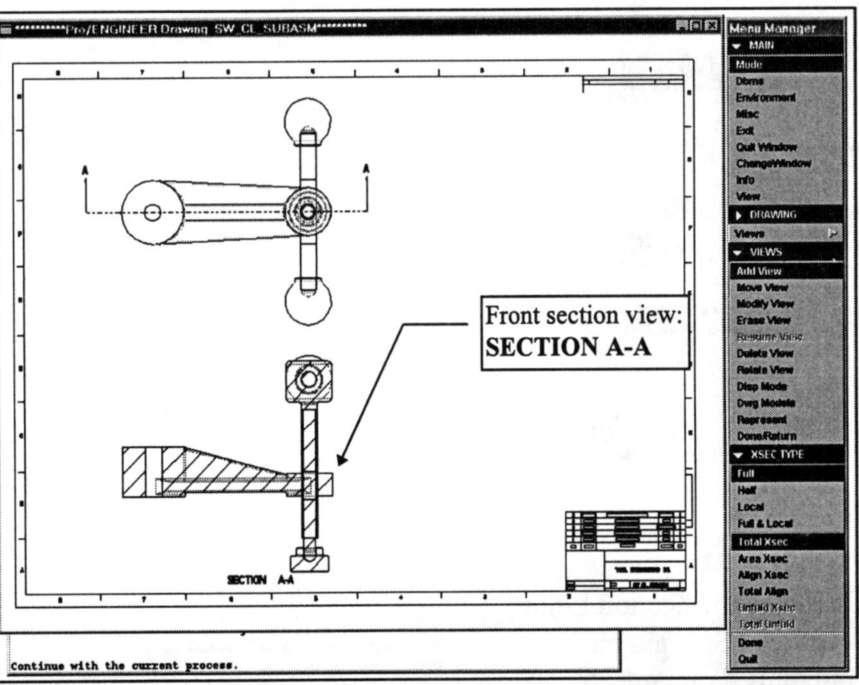

Figure 19.21
Swing Clamp Subassembly Drawing with Top View and Front Section View

Pro/E allows you to alter the display of the section view so an assembly makes more sense and to comply with industry standard practices.

Most companies require that the crosshatching on parts in section views of assemblies be "clocked" such that parts that meet do not use the same section lining (crosshatching) spacing and angle (Fig. 19.22). This makes the separation between parts more distinct. Clean up the section view to comply with industry practices. Change the spacing and the angle on the **Swivel** component (Fig. 19.22):

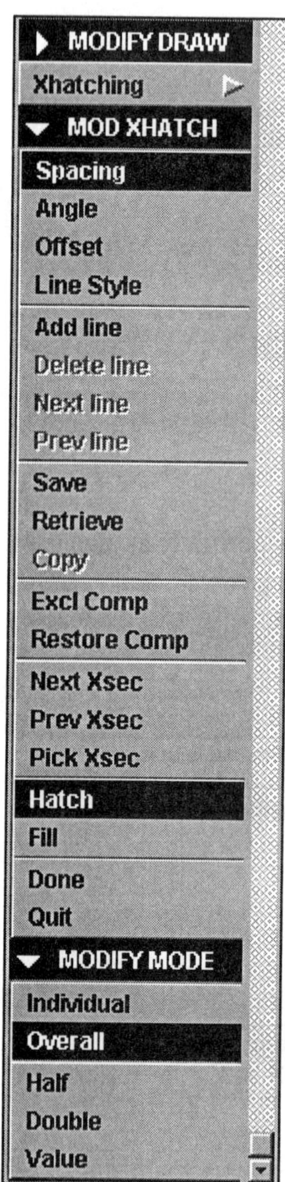

Choose **Detail**
Choose **Modfy**
Choose **Xhatching**

Select the crosshatching in the section view:

Choose **Done Sel**

Pro/E selects all the crosshatching in the view as a single object. You can cycle through the portion that lies on each component using the **Next Xsec**, **Prev Xsec**, and **Pick Xsec** options:

Choose **Next Xsec** until the Swivel (**sw_clamp_swivel**) is selected
Choose **Spacing**
Choose **Half** twice
Choose **Angle**
Choose **135**
Choose **Done**

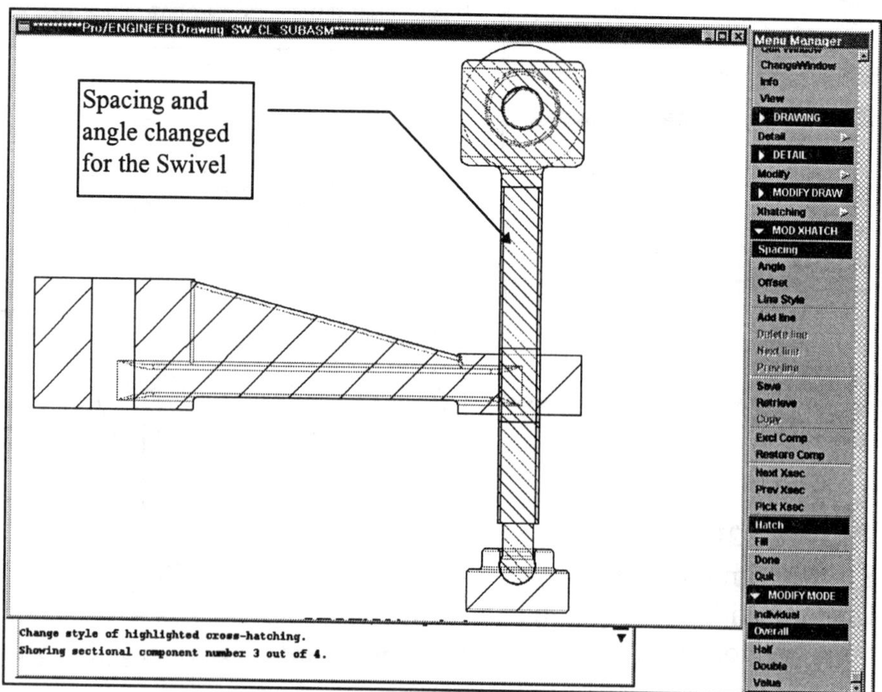

Figure 19.22
Changing the Section Lining Angle and Spacing on Assembly Components

Change the spacing on each component so it is similar to that shown in Figure 19.23. While you have the DETAIL menu up, erase the cosmetic threads from both views (Fig. 19.24).

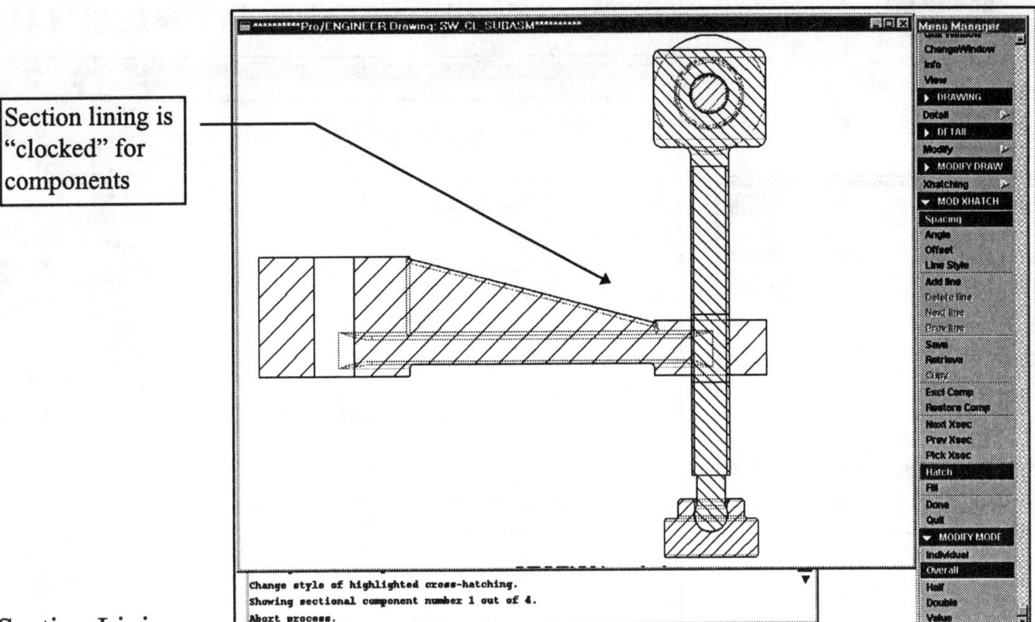

Figure 19.23
Changing the Section Lining

Since hidden lines are usually not shown on a section view, change the display of both views to remove hidden lines and tangent edges:

Choose DRAWING
Choose **Views**
Choose **Disp Mode**
Choose **View Disp**

Select both views:

Choose **Done Sel**
Choose **No Hidden**
Choose **No Disp Tan**
Choose **Done**

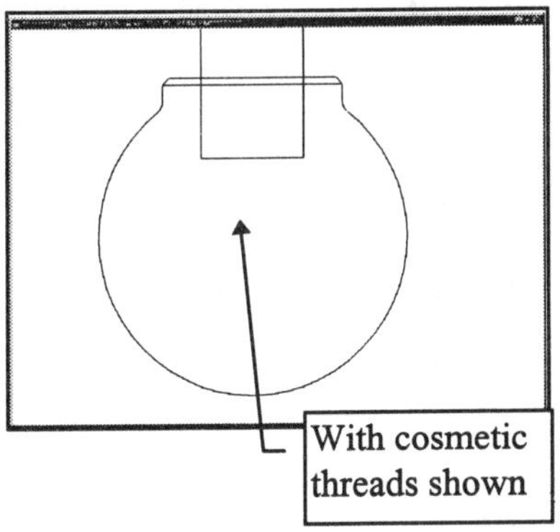

With cosmetic threads shown

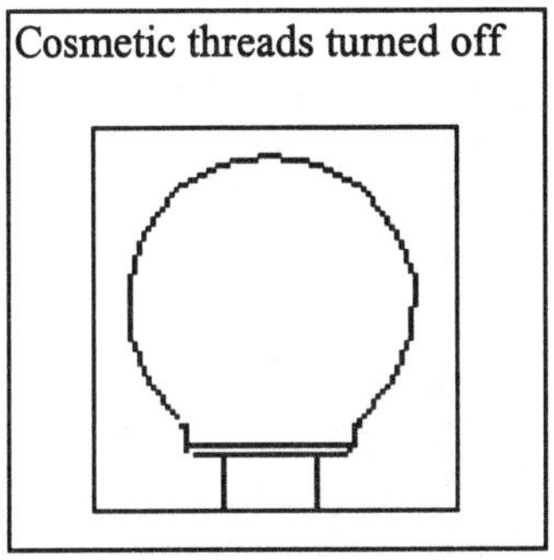

Cosmetic threads turned off

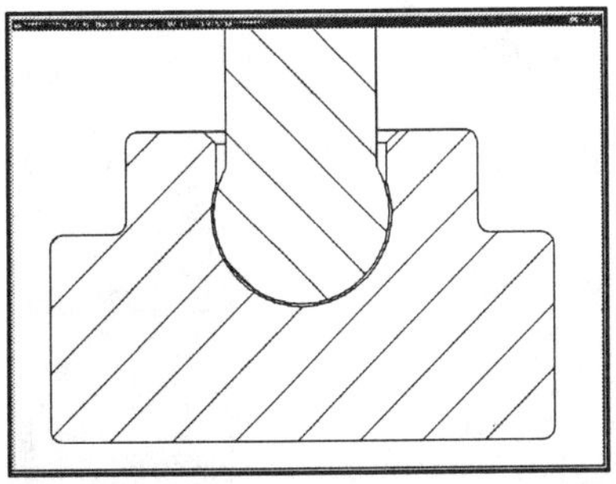

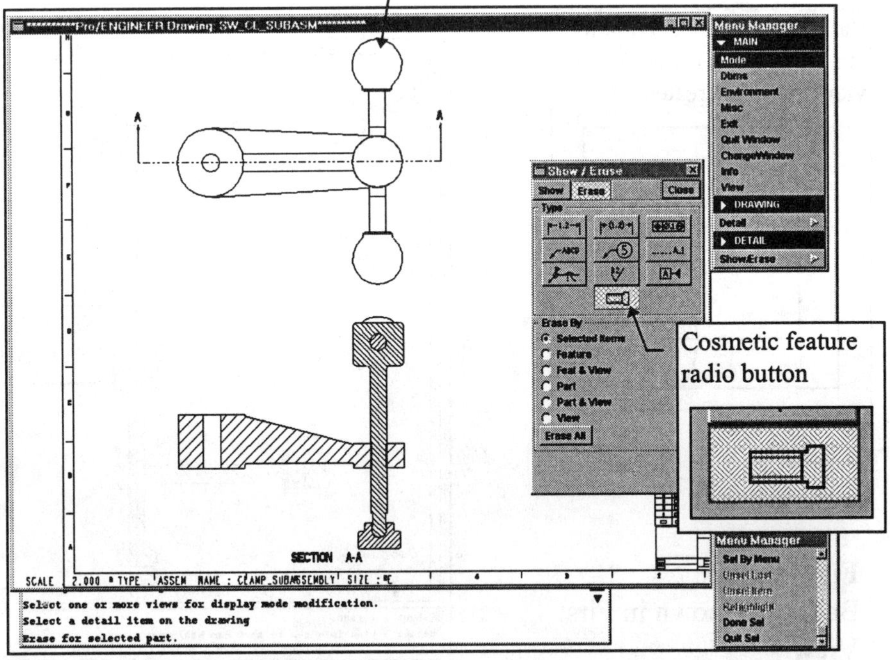

Cosmetic feature radio button

Figure 19.24
Cosmetic Threads, Hidden Lines, and Tangent Edges Removed from Views

To complete the views, show the axes (Fig. 19.25).

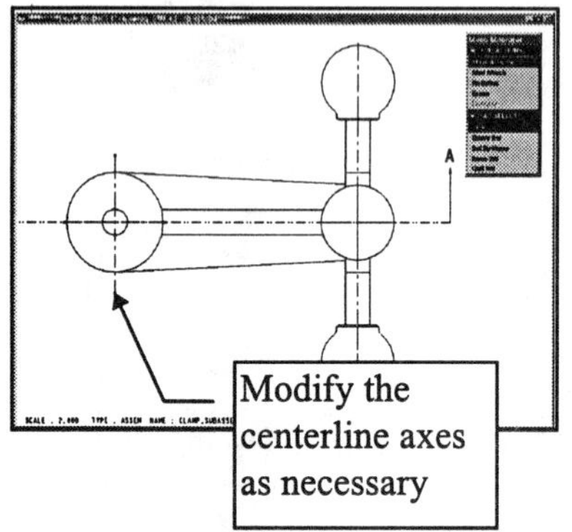

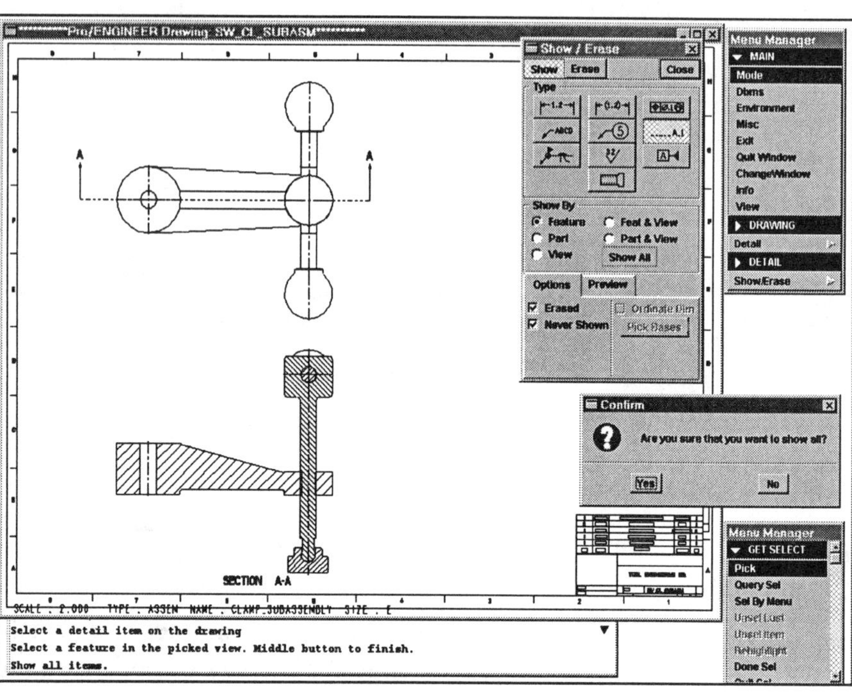

Figure 19.25
Centerline Axes Shown on Assembly Drawing

To finish the assembly drawing, you must display the balloons for each component. Show the **Item Balloons** on the drawing using the following commands (Fig. 19.26):

Choose **Table**
Choose **BOM Balloon**
Choose **Set Region**

Select the parts list on the drawing:

Choose **Show**
Choose **Show All**
Choose **Done/Return**

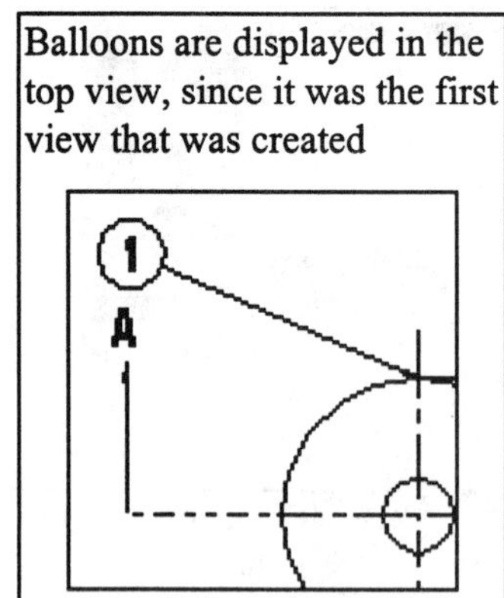

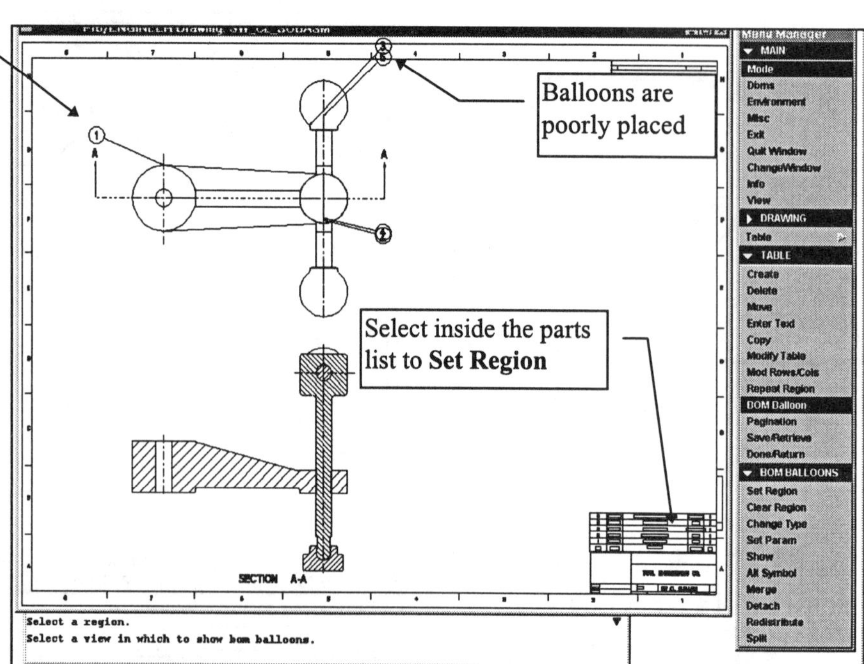

Figure 19.26
Balloons Shown in First View

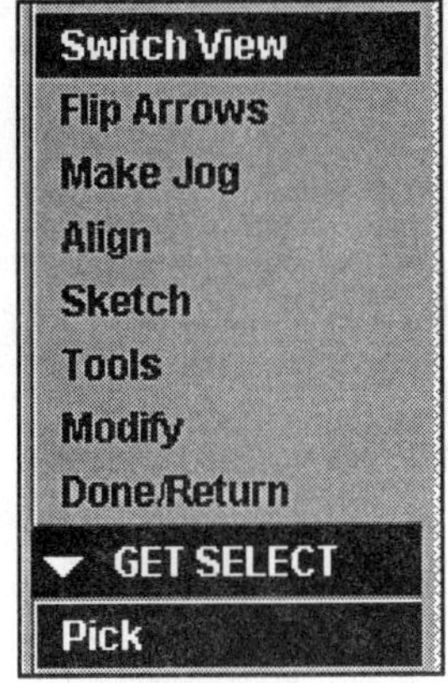

Switch the display for the Arm, Swivel, and Foot balloons to the front section view (Fig. 19.27). This is similar to switching the view of a dimension:

Choose **Detail ⇒ Switch View**
Select a balloon ⇒
Select a balloon ⇒
Select a balloon ⇒ **Done Sel**
Select the section view
Choose **View ⇒ Repaint ⇒ Done/Return**

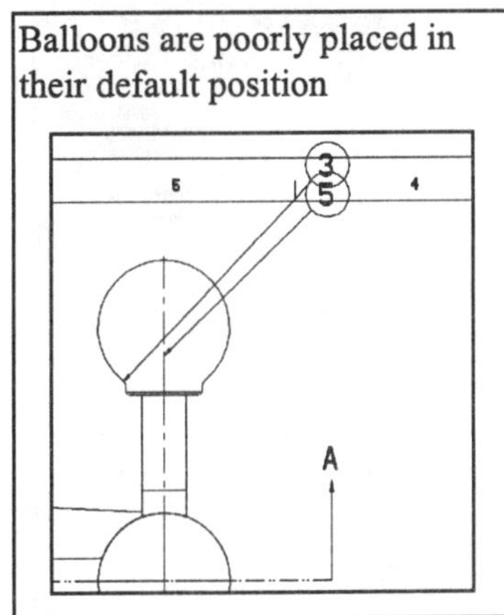

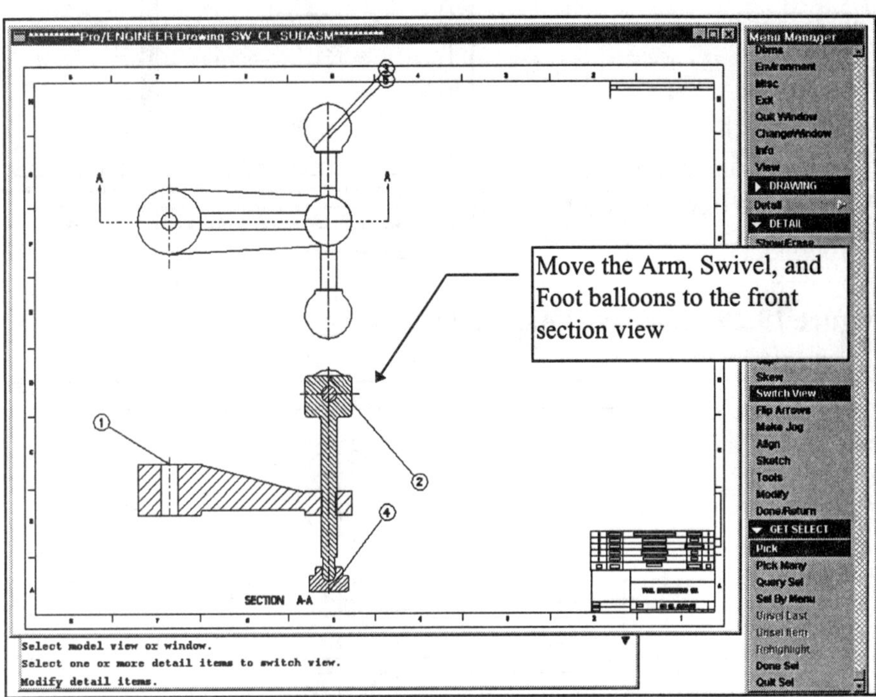

Figure 19.27
Switch Three Balloons to the Front View

Use the **Detail ⇒ Move** option to move the balloons away from each other (a bit). Move the attachment of the balloons (Figs. 19.28 and 19.29):

Choose **Mod Attach**

Select balloon item **1**:

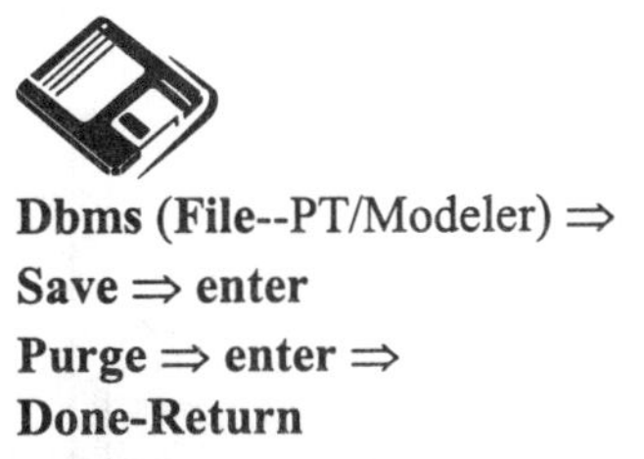

Dbms (**File**--PT/Modeler) ⇒
Save ⇒ enter
Purge ⇒ enter ⇒
Done-Return

Choose **Change Ref**

Change Ref allows you to pick a new edge to place the arrowhead on. Select the edge of the section as the new attachment:

Choose **Done Sel**
Choose **Done/Return**

Repeat the process until all the balloons are reattached as shown in Figure 19.29.

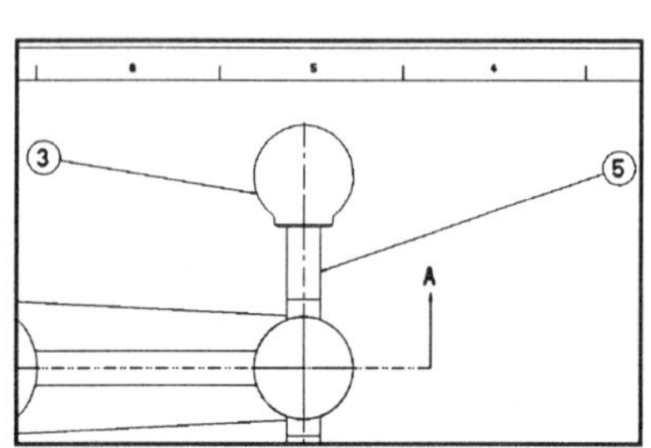

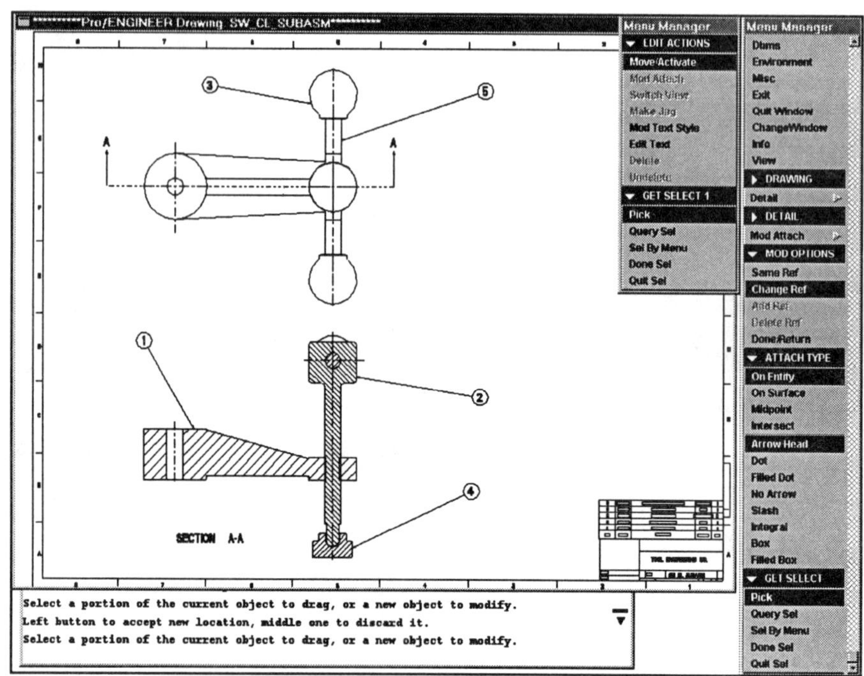

Figure 19.28
New Positions and Modified Attachment Locations for Balloons

Figure 19.29 Completed Assembly Drawing

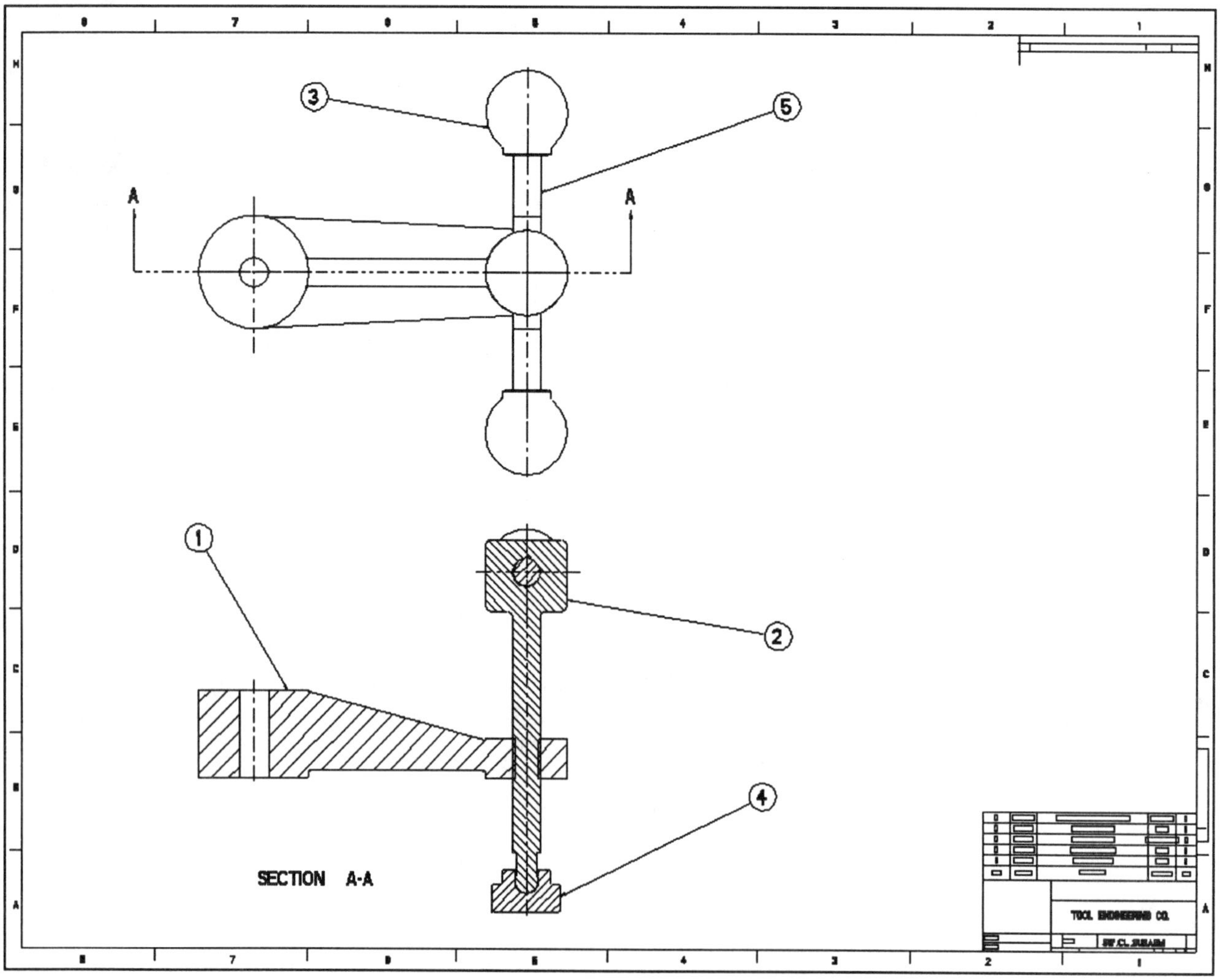

The next drawing we will create is the assembly drawing for the complete Swing Clamp. This assembly is composed of the subassembly in the previous drawing, the **Plate**, the short **Stud**, and the **Flange Nut**. The drawing will use the same format created for the subassembly. Formats are read-only files that can be used as many times as you want. Create the drawing using the following commands:

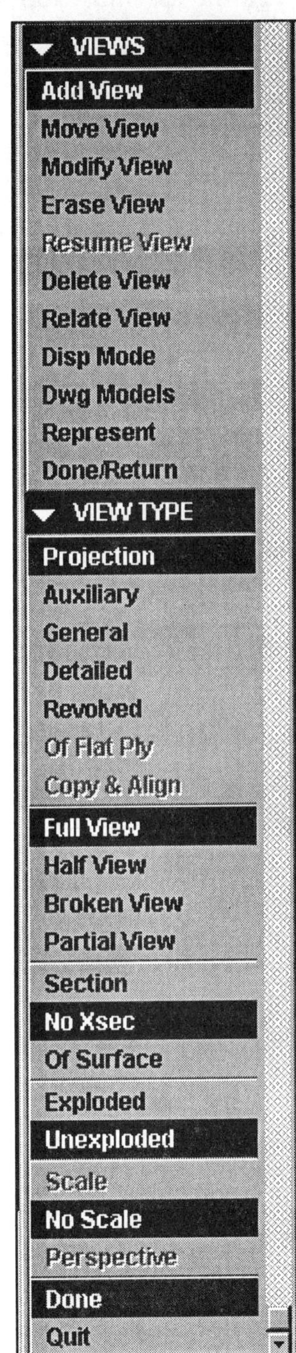

Mode ⇒ **Drawing** ⇒ **Create** ⇒ (type **sw_cl_assembly**) ⇒ **enter** ⇒ **Retr Format** ⇒ **?** ⇒ (select the name of the "E" size assembly format you created in this lesson)

To get the drawing to appear with the correct style, you must modify the values of the **.dtl** Drawing Set Up file using **Set Up** ⇒ **Modify Val** and modify the values as before. Another way would be to retrieve the previously used **.dtl** file if you have used the **Save** command just under **Modify Val**.

Now associate the complete assembly model to this Assembly format, and add two views. Add the top view as the first view. Use the following commands:

Views ⇒ **?** ⇒ [pick **sw_cl_assembly** (or the name you gave the assembly model)] ⇒ **Done** ⇒ (indicate the location of the center of the top view)

Orient the top view correctly as shown in Figure 19.30, and then add a front section view as was done for the subassembly.

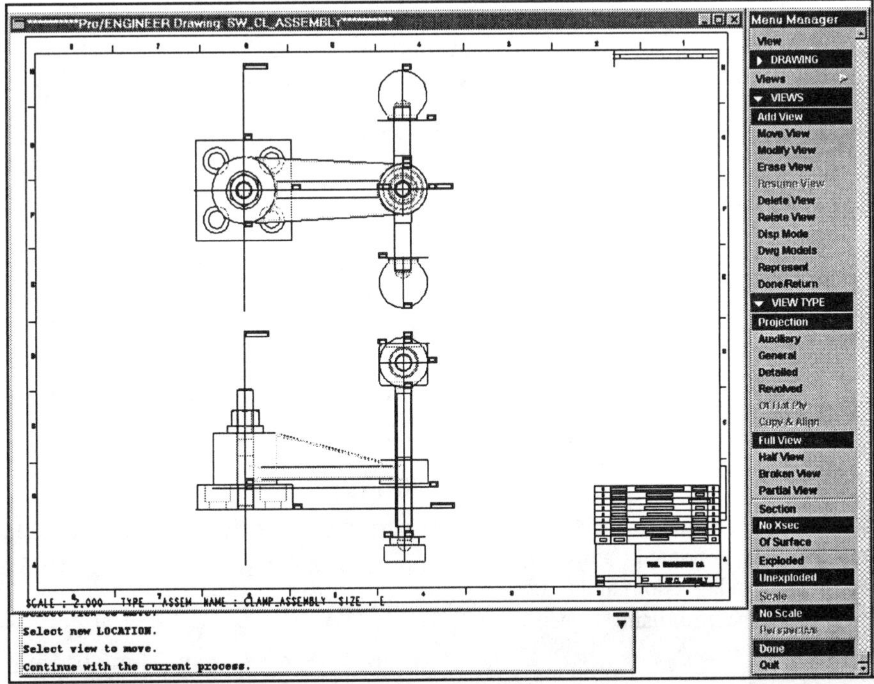

Figure 19.30
Assembly Drawing with Two Views

If you added the front view without choosing **Section** as one of the options as done in Figure 19.30, you must modify the view using the following commands (Fig. 19.31):

Views ⇒ **Modify View** ⇒ **View Type** ⇒ (pick the front view) ⇒ **Section** ⇒ **Done** ⇒ **Done** ⇒ **Create** ⇒ **Done** ⇒ (type **A**) ⇒ **enter** ⇒ [pick a datum plane that passes laterally through the assembly top view (Fig. 19.31)] ⇒ (pick the top view for the cutting plane line)

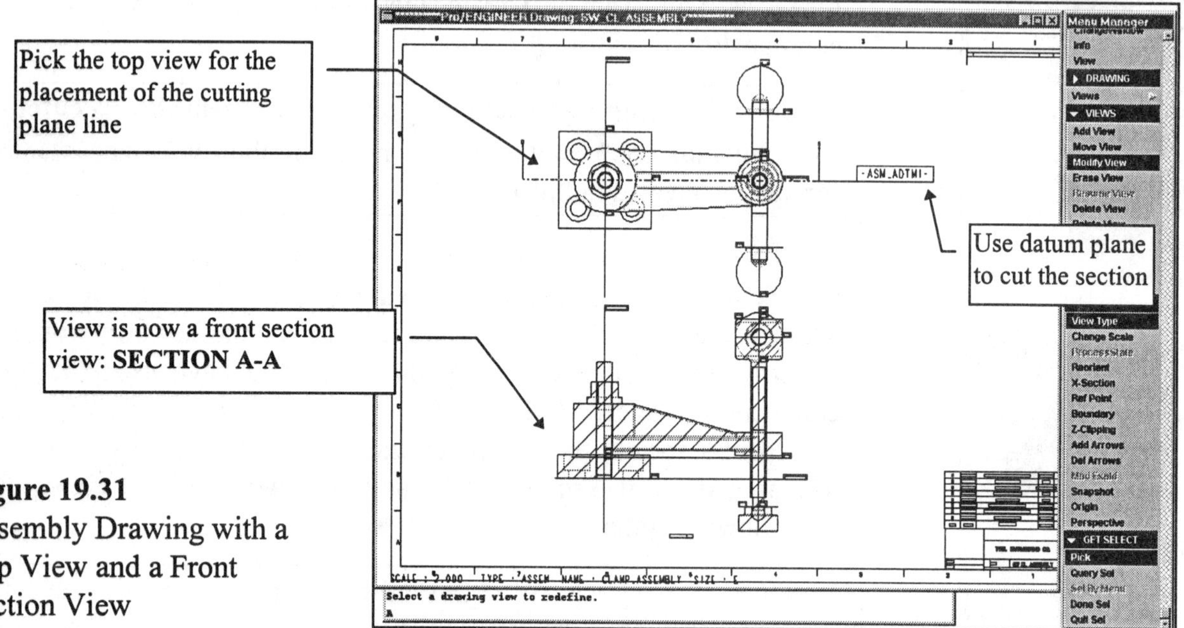

Figure 19.31
Assembly Drawing with a Top View and a Front Section View

Clean up the drawing by removing the cosmetic threads, datum axes, hidden lines, and display of the tangent edges as you did for the subassembly. Show the centerline axes and modify them so they are visually correct. Change the *scale* to **2.00.**

Most companies (and as per drafting standards) require that purchased round items, such as nuts, bolts, studs, springs, and die pins be excluded from sectioning even when the cutting plane passes through them.

NOTE
Modify the text height of the section identification if it is too small.

Remove the section lining (crosshatching) from the short **Stud (3.50** length) and the **Flange Nut** in the front section view:

Choose **Detail**
Choose **Modify**
Choose **Xhatching**

Select the crosshatching in the section view:
Choose **Done Sel**

Pro/E selects all the crosshatching in the view as a single object. You can cycle through the portion that lies on each component using the **Next Xsec**, **Prev Xsec**, and **Pick Xsec** options. Choose the next commands (Fig. 19.32):

Choose **Excl Comp** (this eliminates Xsec of flange nut)
Choose **Next Xsec** (the short stud happens to be next)
Choose **Excl Comp** (this eliminates Xsec of short stud)

*(See **HINT**)*

HINT
If you wish to eliminate the section lines from the long stud (end view), use the following commands after completing the command sequence to the right:

Choose **Next Xsec**
(the long stud happens to be next)
Choose **Excl Comp**
(this eliminates Xsec of long stud)

Or you can use **Query Sel** to pick the components:

Choose **Pick Xsec**
Choose **Query Sel**
Select the short **Stud** ⇒ **Accept**
Choose **Excl Comp**

and then the **Flange Nut**:

Choose **Pick Xsec**
Choose **Query Sel**
Select the **Flange Nut** ⇒ **Accept**
Choose **Excl Comp**

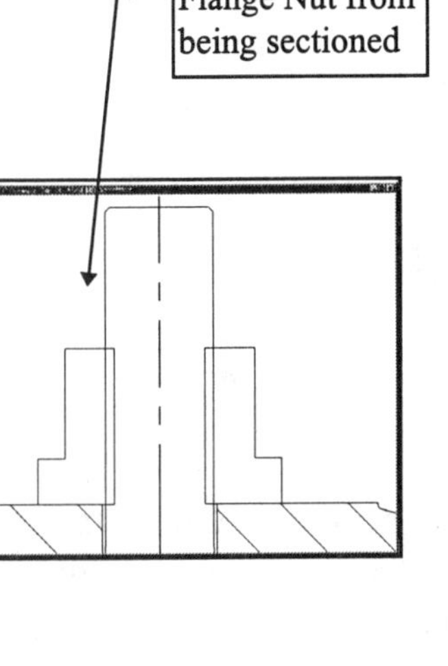

Choose **Done** ⇒ **Done Sel** ⇒ **Done/Return** ⇒ **Done/Return**

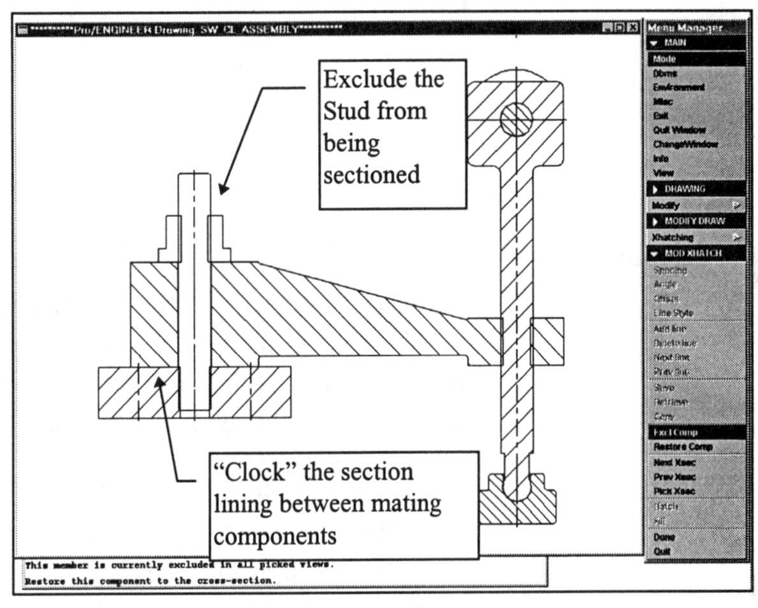

Figure 19.32
Assembly Drawing with Stud and Flange Nut Section Lining Excluded from the Views

The next step in completing the assembly drawing is to show the balloons. Choose the following commands (Fig. 19.33):

Table ⇒ **BOM Balloon** ⇒ **Set Region** ⇒ (select inside the parts list on the drawing) ⇒ **Show** ⇒ **Show All** ⇒ **Done/Return**

Balloons will show in the first view created

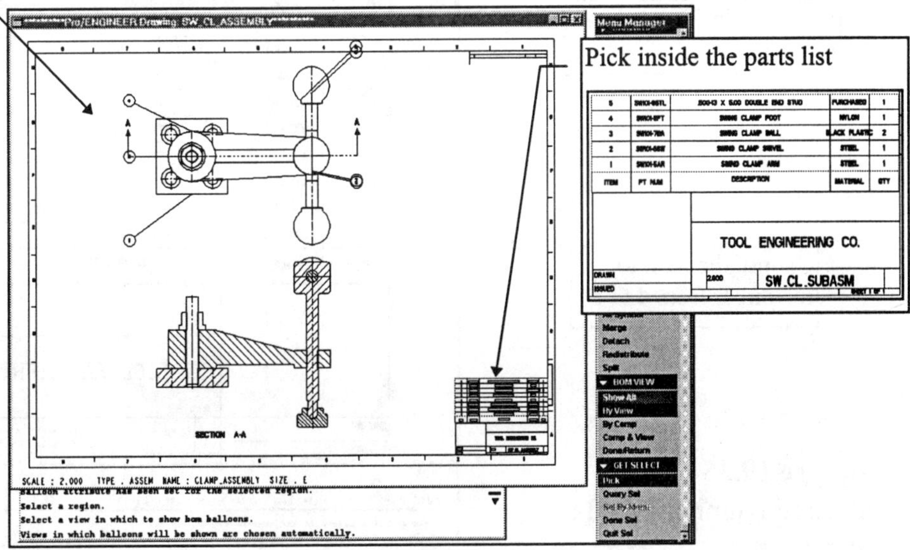

Figure 19.33
Showing the Balloons

Switch some of the balloons to the front view and reposition the others to make a more clear and balanced ballooning scheme (Fig. 19.34). You can line up the balloons by turning on **Grid Snap** and then moving the balloons to new positions.

Dbms (File--PT/Modeler) ⇒
Save ⇒ enter
Purge ⇒ enter ⇒
Done-Return

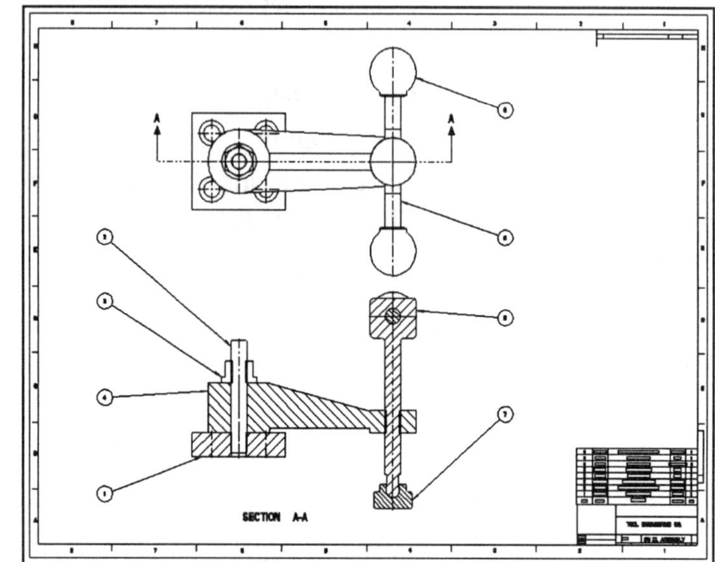

Figure 19.34
Showing the Balloons

The numbering of the components on the assembly may need to be different from the default setting. To change the ballooning you must use **Fix Index** from the TABLE menu. Use the following commands to change the numbering scheme (Fig. 19.35):

Table ⇒ Repeat Region ⇒ Fix Index ⇒ (select the parts list region) ⇒ (select a record in the repeat region; select the *Arm*, which is defaulted to item **4**) ⇒ (type **1**) ⇒ **enter** ⇒ (continue selecting and changing the numbering to that of the *subassembly* sequence: make the *Plate* **6**, the short *Stud* **7**, and the *Flange Nut* **8**) ⇒ **Done**

When you are finished, the parts list will sequence itself (Fig. 19.36) and the ballooning will update automatically (Fig. 19.37).

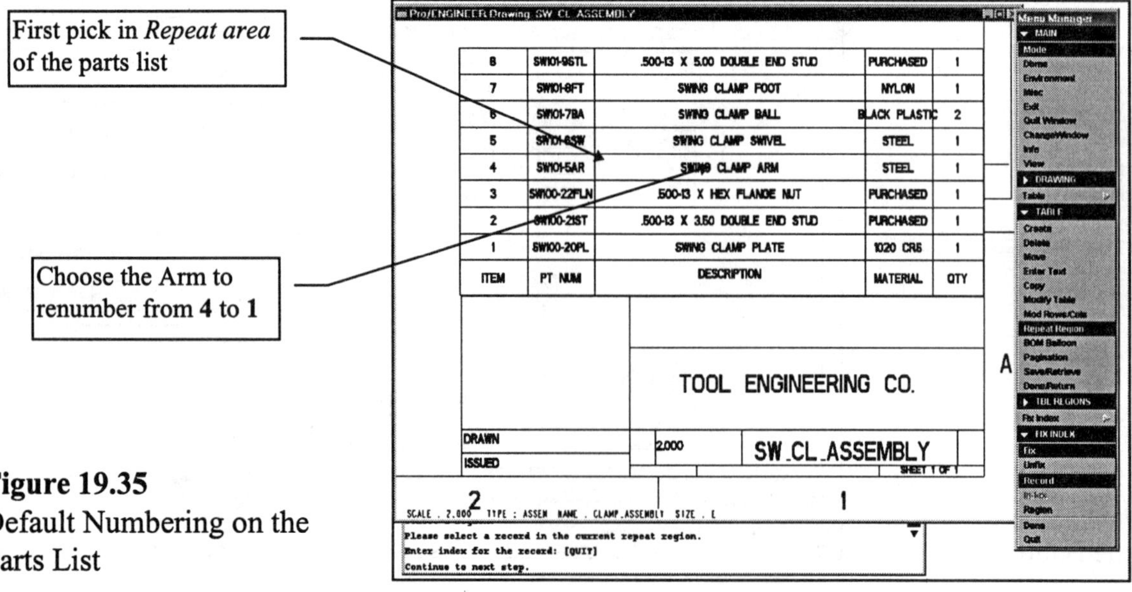

8	SW101-9STL	.500-13 X 5.00 DOUBLE END STUD	PURCHASED	1
7	SW101-8FT	SWING CLAMP FOOT	NYLON	1
6	SW101-7BA	SWING CLAMP BALL	BLACK PLASTIC	2
5	SW101-6SW	SWING CLAMP SWIVEL	STEEL	1
4	SW101-5AR	SWING CLAMP ARM	STEEL	1
3	SW100-22FLN	.500-13 X HEX FLANGE NUT	PURCHASED	1
2	SW100-21ST	.500-13 X 3.50 DOUBLE END STUD	PURCHASED	1
1	SW100-20PL	SWING CLAMP PLATE	1020 CRS	1
ITEM	PT NUM	DESCRIPTION	MATERIAL	QTY

Figure 19.35
Default Numbering on the Parts List

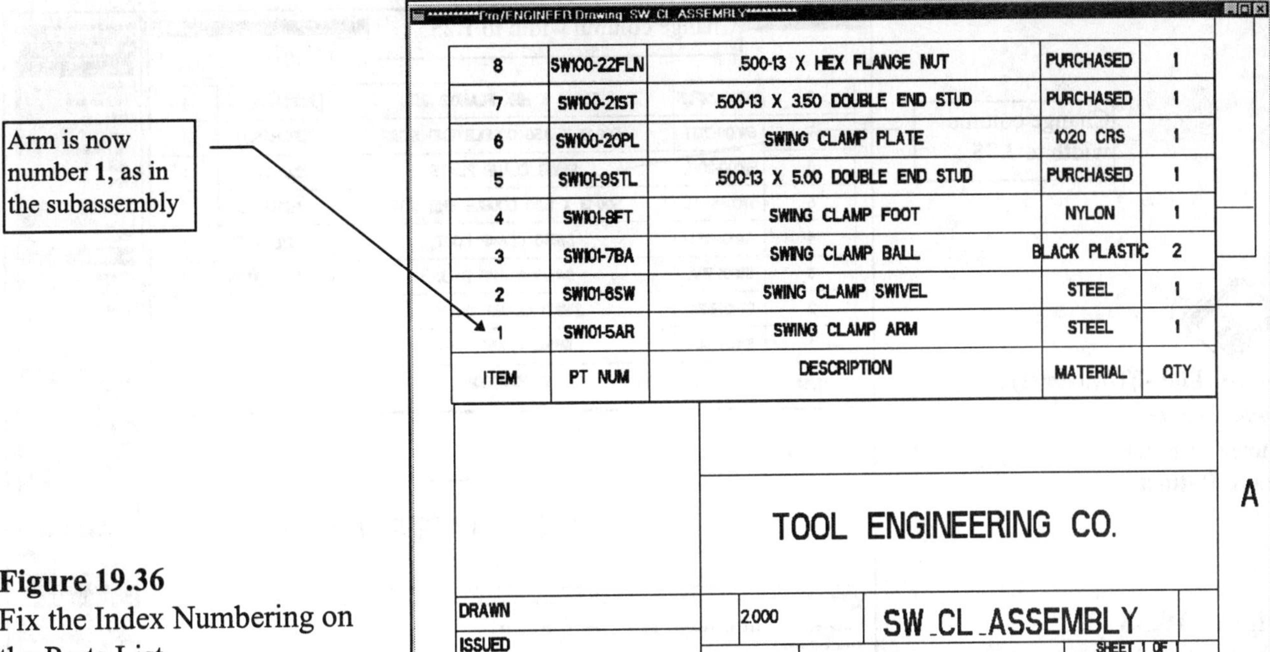

ITEM	PT NUM	DESCRIPTION	MATERIAL	QTY
8	SW100-22FLN	.500-13 X HEX FLANGE NUT	PURCHASED	1
7	SW100-21ST	.500-13 X 3.50 DOUBLE END STUD	PURCHASED	1
6	SW100-20PL	SWING CLAMP PLATE	1020 CRS	1
5	SW101-9STL	.500-13 X 5.00 DOUBLE END STUD	PURCHASED	1
4	SW101-8FT	SWING CLAMP FOOT	NYLON	1
3	SW101-7BA	SWING CLAMP BALL	BLACK PLASTIC	2
2	SW101-6SW	SWING CLAMP SWIVEL	STEEL	1
1	SW101-5AR	SWING CLAMP ARM	STEEL	1

TOOL ENGINEERING CO.

DRAWN

ISSUED

2.000

SW_CL_ASSEMBLY

SHEET 1 OF 1

A

Figure 19.36
Fix the Index Numbering on the Parts List

Dbms (File--PT/Modeler) ⇒ Save ⇒ enter
Purge ⇒ enter ⇒ Done-Return

Figure 19.37
Renumbered Balloons

NYLON	1
BLACK PLASTIC	2
STEEL	1

To change the numbering after it has been fixed, you must **Unfix** first and then **Fix Index** again.

Figure 19.36 shows that the MATERIAL cell column is not wide enough to accommodate the length of the Black Plastic material parameter. You can change the size of the column cell using **Mod Rows/Cols** from the TABLE menu. Use the following commands (Fig. 19.38):

Mod Rows/Cols ⇒ Change Size ⇒ Column ⇒ By Length ⇒ (pick the MATERIAL column) ⇒ (type **1.25**) ⇒ **enter** ⇒ (pick the DESCRIPTION column) ⇒ (type **3.75**) ⇒ **enter** (Fig. 19.39)

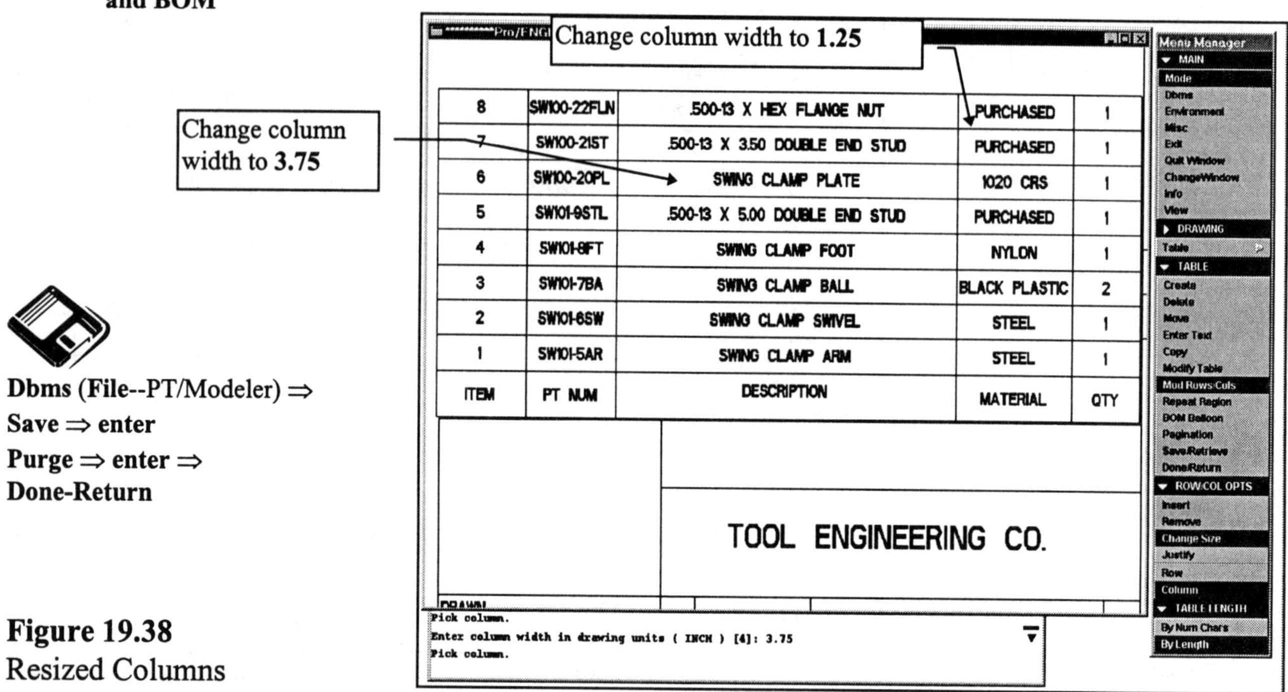

Dbms (File--PT/Modeler) ⇒ Save ⇒ enter
Purge ⇒ enter ⇒ Done-Return

Change column width to **1.25**
Change column width to **3.75**
8 SW100-22FLN .500-13 X HEX FLANGE NUT PURCHASED 1
7 SW100-21ST .500-13 X 3.50 DOUBLE END STUD PURCHASED 1
6 SW100-20PL SWING CLAMP PLATE 1020 CRS 1
5 SW101-9STL .500-13 X 5.00 DOUBLE END STUD PURCHASED 1
4 SW101-8FT SWING CLAMP FOOT NYLON 1
3 SW101-7BA SWING CLAMP BALL BLACK PLASTIC 2
2 SW101-6SW SWING CLAMP SWIVEL STEEL 1
1 SW101-5AR SWING CLAMP ARM STEEL 1
ITEM PT NUM DESCRIPTION MATERIAL QTY
TOOL ENGINEERING CO.
Pick column.
Enter column width in drawing units (INCH) [4]: 3.75
Pick column.
Menu Manager
MAIN
Mode
Dbms
Environment
Misc
Exit
Quit Window
ChangeWindow
Info
View
DRAWING
Table
TABLE
Create
Delete
Move
Enter Text
Copy
Modify Table
Mod Rows/Cols
Repeat Region
BOM Balloon
Pagination
Save/Retrieve
Done/Return
ROW/COL OPTS
Insert
Remove
Change Size
Justify
Row
Column
TABLE LENGTH
By Num Chars
By Length

Figure 19.38
Resized Columns

Figure 19.39 Assembly Drawing

Pro/ENGINEER Drawing: SW_CL_ASSEMBLY
A
A
SECTION A-A

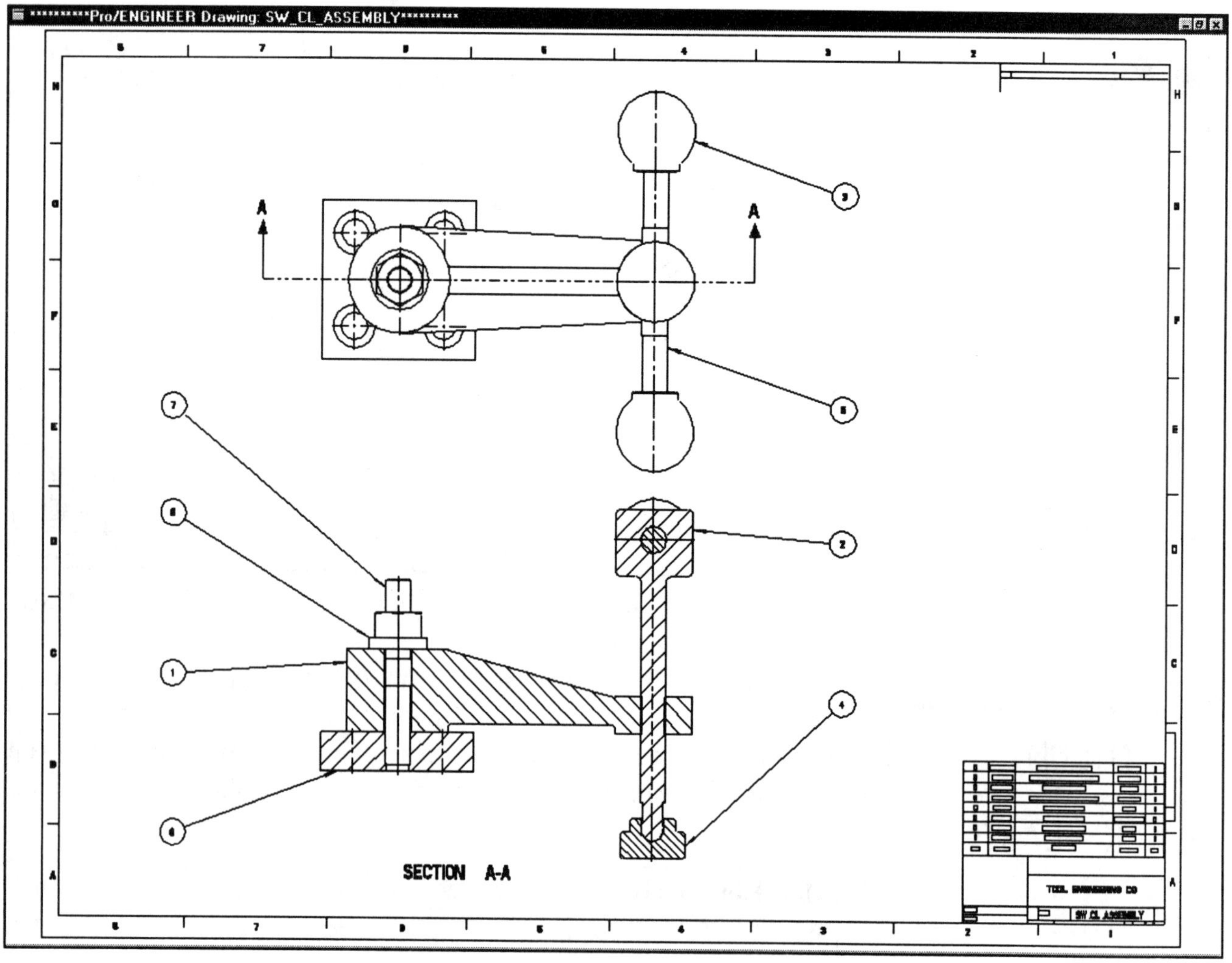

Lesson 19 Project

Coupling Assembly Drawing

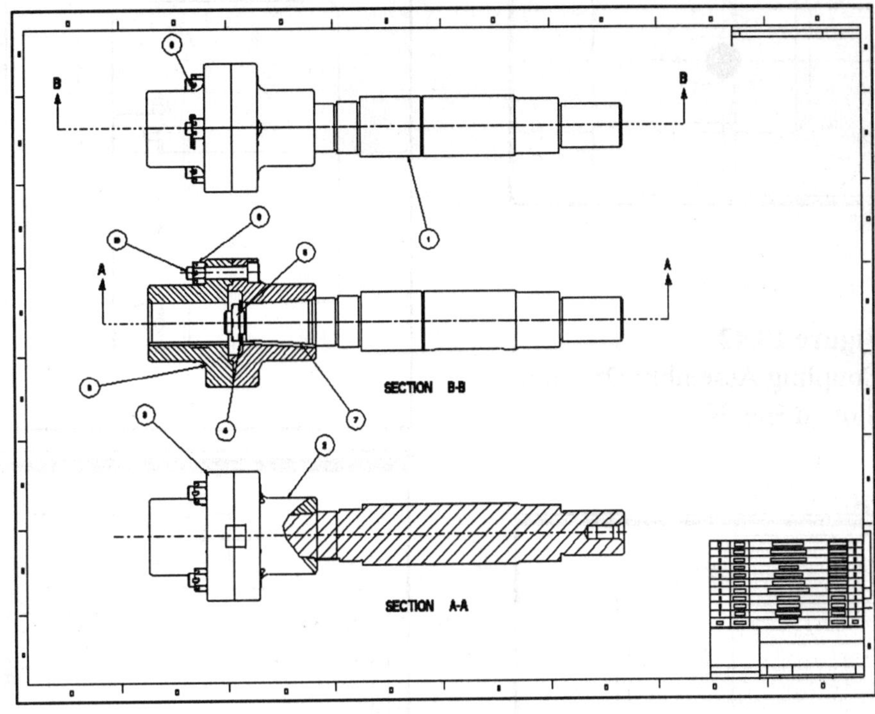

Figure 19.40
Coupling Assembly Drawing

Coupling Assembly Drawing

Create an assembly drawing of the Coupling Assembly using the format made in this lesson. Use Figure 19.40 through 19.44 as examples for this lesson project. The ballooned assembly drawing will have three views and a parts list.

Assign parameters to parts in the assembly so they can be displayed on a parts list in the assembly drawing and generate item balloons for each component.

☑ ***EGD REFERENCE***
Engineering Graphics and Design with Graphical Analysis *or* **Fundamentals of Engineering Graphics and Design**
by L. Lamit and K. Kitto
Read Chapter 23
See pages 865-866

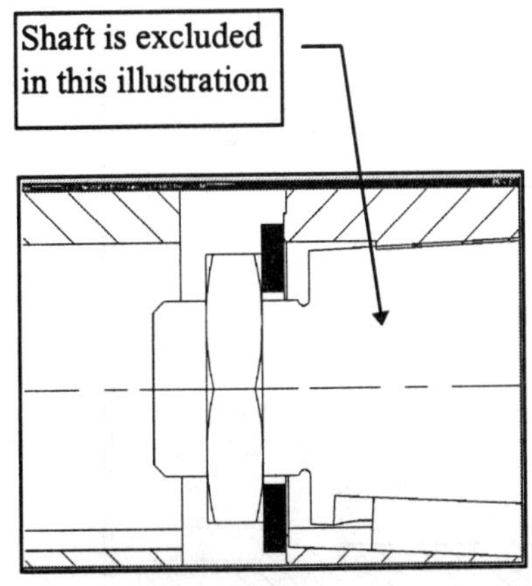

10	110-2CS	SOC HD CAP SCREW	PURCHASED	3
9	109-2SN	HEX SLOT NUT 16 X 2	PURCHASED	3
8	108-2CP	COTTER PIN .150 X 1.25	PURCHASED	3
7	107-2KY	KEY 14 X 61	PURCHASED	1
6	106-2DW	DOWEL 12OD X 70	PURCHASED	2
5	105-2HN	HEX NUT M30 X 3.5	PURCHASED	1
4	104-2WA	WASHER 33ID X 50OD X 4	PURCHASED	1
3	103-2CP2	COUPLING TWO	1040 CRS	1
2	102-2CP1	COUPLING ONE	1040 CRS	1
1	101-2SH	COUPLING SHAFT	1020 CRS	1
ITEM	PT NUM	DESCRIPTION	MATERIAL	QTY

Figure 19.41
Coupling Assembly Drawing Parts List

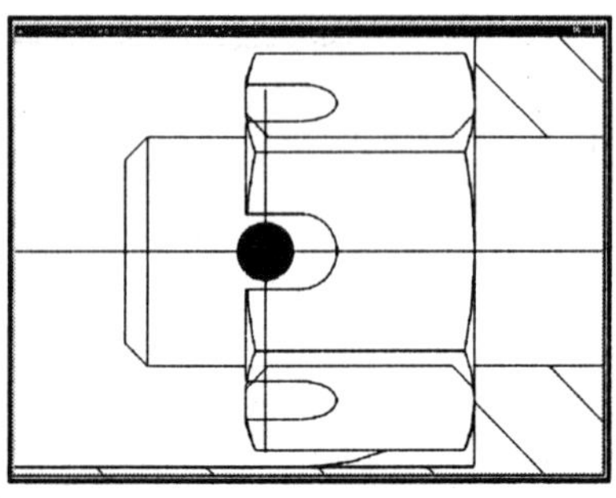

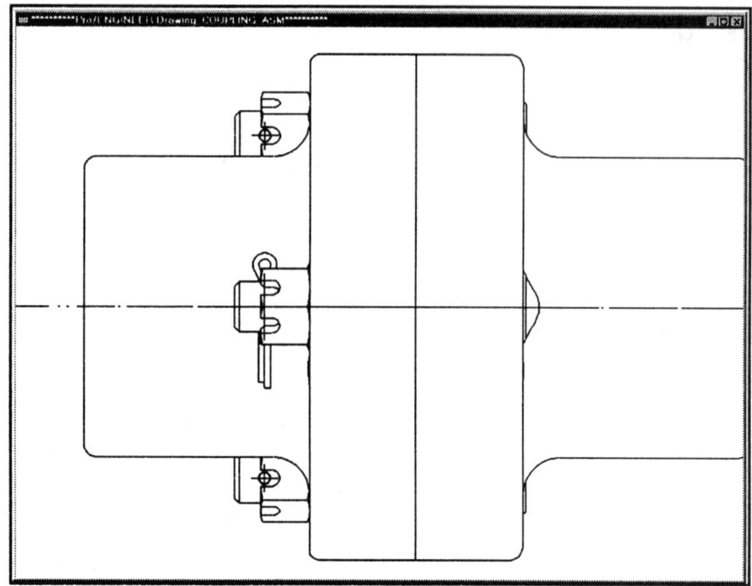

Figure 19.42
Coupling Assembly Drawing, Slotted Hex Nut

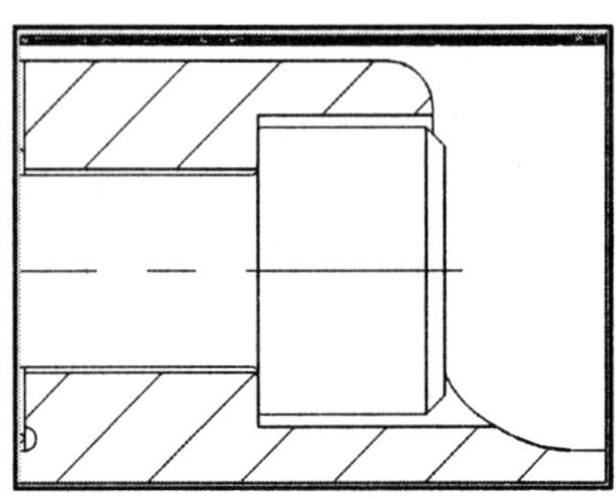

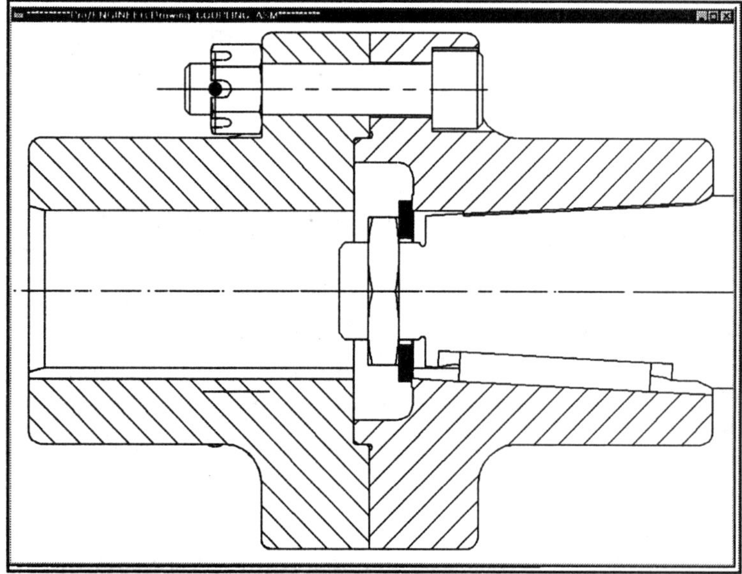

Figure 19.43
Coupling Assembly Drawing, Section Close-up

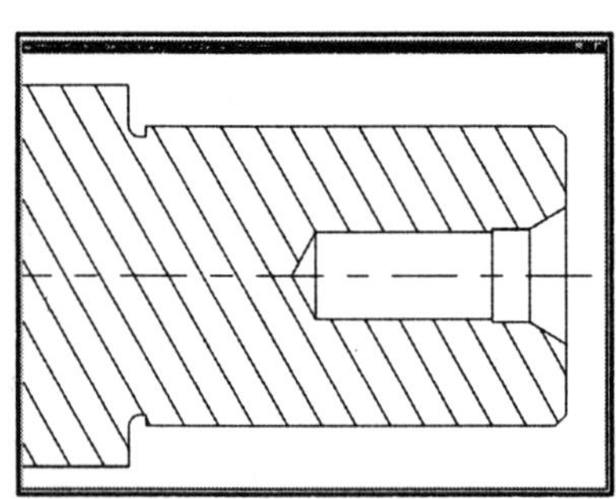

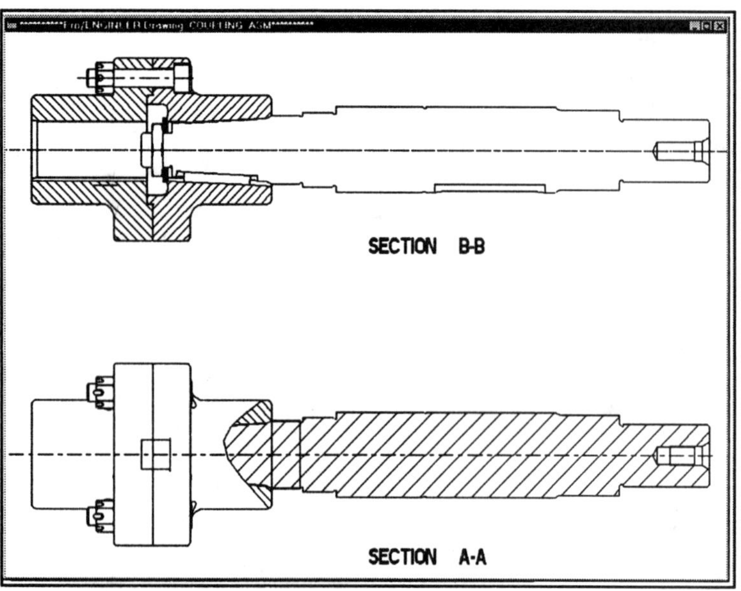

Figure 19.44
Coupling Assembly Drawing, **SECTION A-A** and **SECTION B-B**

Lesson 20

Exploded Assembly Drawings

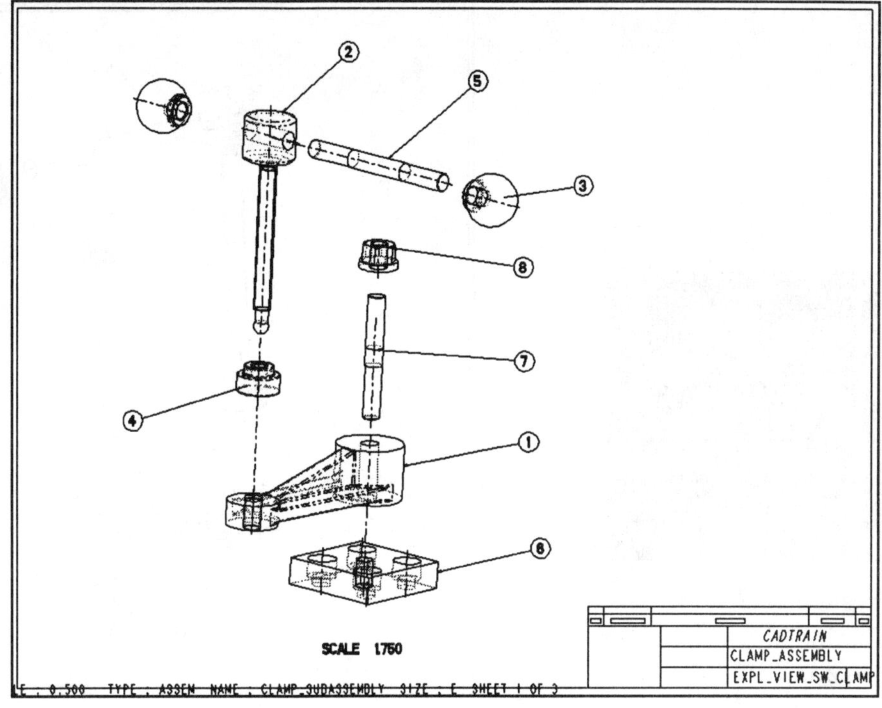

Figure 20.1
Sheet 1 Exploded Swing Clamp Drawing with Standard Format and Balloons and **Sheet 2** Exploded Trimetric View

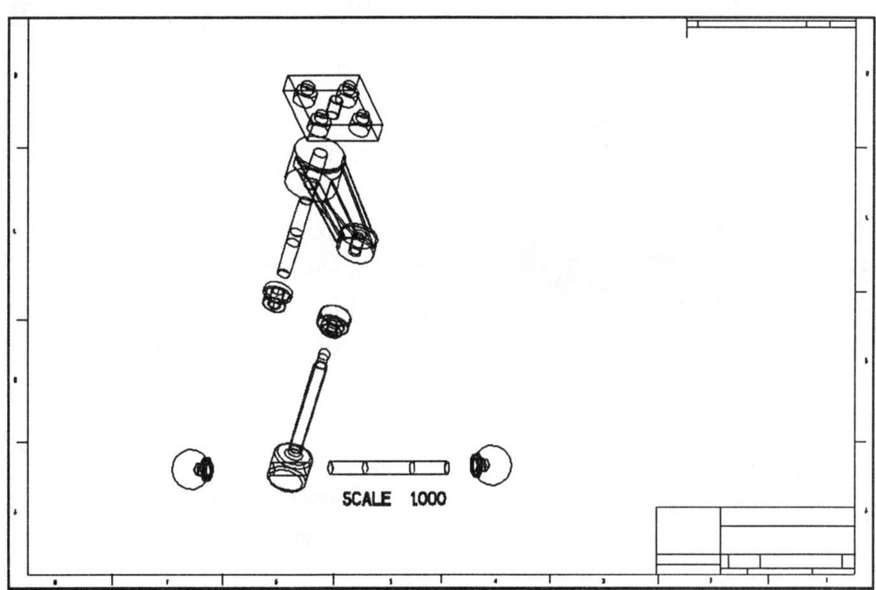

Figure 20.1 (continued)
Sheet 2 Exploded Trimetric View

☑ *EGD REFERENCE*

Engineering Graphics and Design with Graphical Analysis *or* **Fundamentals of Engineering Graphics and Design**
by L. Lamit and K. Kitto
Read Chapters 13, 23
See pages 447, 823, 836-837, 841

COAch™ for Pro/ENGINEER

If you have **COAch for Pro/ENGINEER** on your system, go to Assemblies, Exploded Assemblies.

OBJECTIVES

1. **Create drawings with exploded views**
2. **Use multiple sheets**
3. **Make assembly drawing sheets with multiple models**
4. **Create balloons on exploded assemblies**

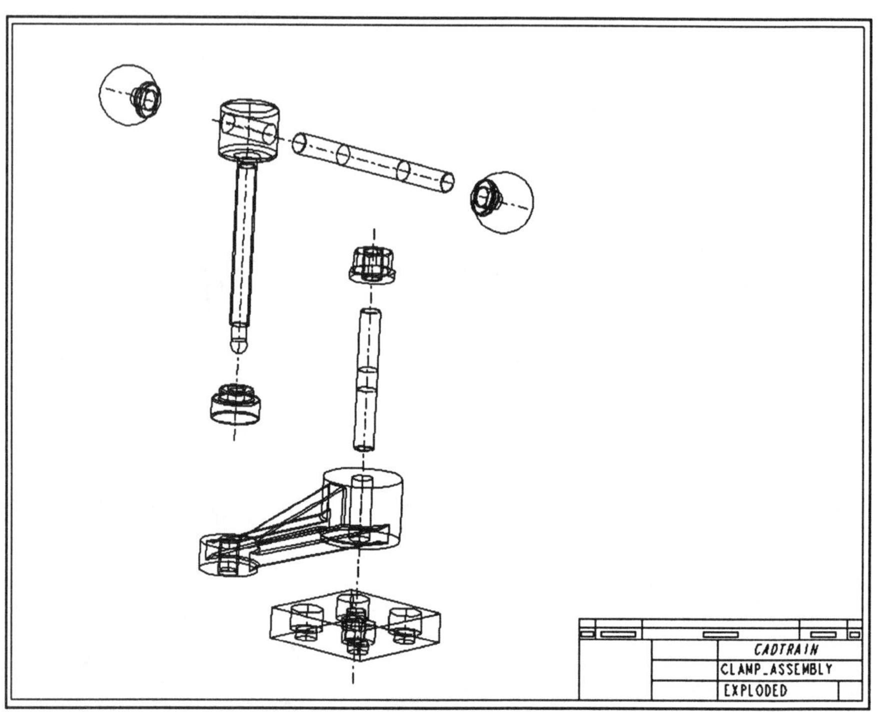

Figure 20.2
Exploded Swing Clamp Drawing

EXPLODED ASSEMBLY DRAWINGS

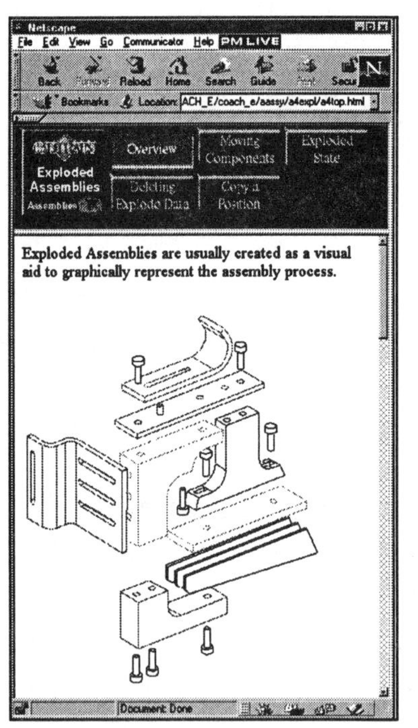

As we explained in Lesson 15, you can create an exploded view of an assembly. Exploding an assembly affects only the display of the assembly; it does not alter actual distances between components. *Exploded states* (the optional Pro/PROCESS_ASM license is required) are created and saved to allow a clear visualization and understanding of the positional relationships of all the components in an assembly. For each explode state, you can toggle the explode status of components, change the explode locations of components, and create explode offset lines. You can define multiple explode states for each assembly, then explode the assembly using any of these explode states at any time. You can also set an exploded state for each drawing view of an assembly.

The *exploded views* created and saved with the model in Lesson 15 will be used in this lesson. The **Swing Clamp Assembly** shown in Figures 20.1 and 20.2 is used for the lesson model, and the **Coupling Assembly** is used for the Lesson 20 Project.

If none of the exploded views you have created is exactly the exploded position you wish to use in this lesson, bring up the model (assembly), create new exploded views (Fig. 20.3), and save them for use in this lesson and lesson project.

You are required to create a complete documentation package for the two assemblies. *A documentation package contains all models and drawings required to manufacture the parts and assemble the components.* Your instructor may change the requirements, but, in general, create and plot the following:

Part Models for all swing clamp components
Detail Drawings for each nonstandard component, for example the Clamp Arm, Clamp Swivel, Clamp Foot, and Clamp Ball
Detail Drawings for standard components that have been altered; show only the dimensions required to alter the component, for example the socket head cap screw in the coupling assembly
Assembly Drawing using standard orthographic ballooned views
Exploded Assembly Drawing of the ballooned assembly
Exploded Subassembly Drawing of the ballooned subassembly

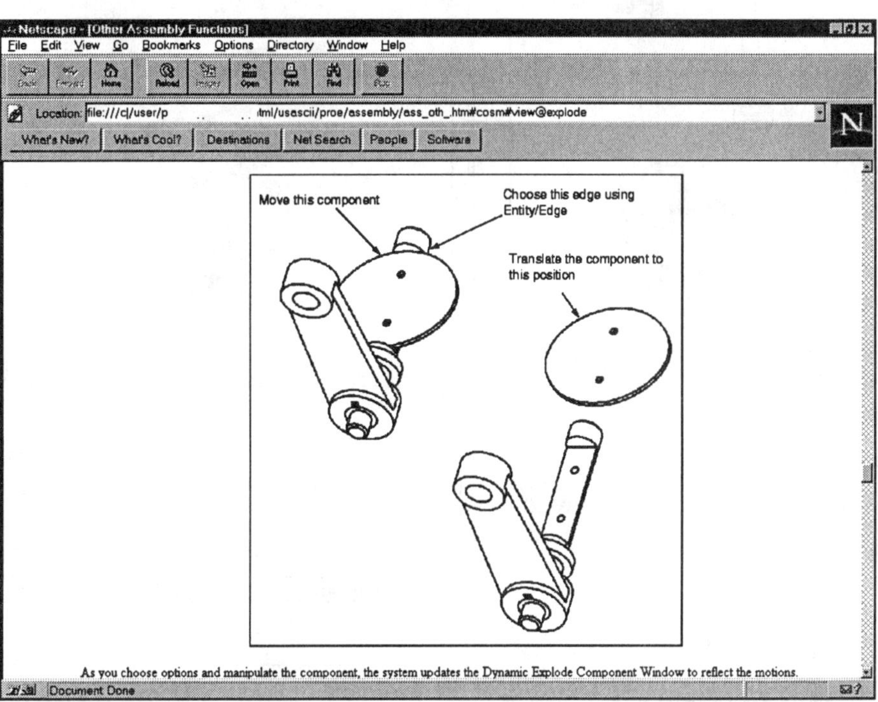

Figure 20.3
Online Documentation, Exploded Assemblies

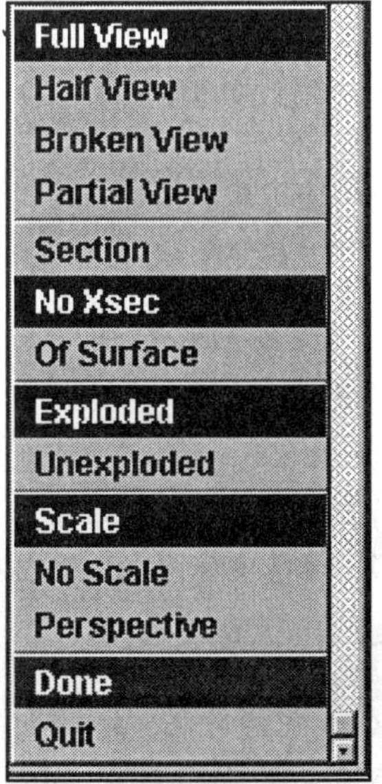

Exploded Swing Clamp Assembly Drawings

The process required to place an exploded view on a drawing is similar to that provided in Lesson 19 for adding assembly orthographic views to a drawing. Use the following commands:

Mode ⇒ **Drawing** ⇒ **Create** ⇒ (type **EXPL_VIEW_SW_CLAMP**) ⇒ **enter** ⇒ **Retr Format** ⇒ **?** ⇒ **enter** ⇒ **Format Dir** ⇒ **d.frm** ⇒ **Views** ⇒ **?** ⇒ **enter** ⇒ (pick the Swing Clamp assembly from the directory) ⇒ **Exploded** ⇒ **Scale** ⇒ **Done** ⇒ (pick the center of the drawing, as in Fig. 20.4) ⇒ (type **.75** as the scale) ⇒ **enter** ⇒ **Names** ⇒ (pick a saved view name from the list) ⇒ **Done/Return** ⇒ **Done/Return** ⇒ **Views** ⇒ **Modify View** ⇒ **Change Scale** ⇒ (pick the view) ⇒ (type **1** as the new scale) ⇒ **enter** ⇒ **Done/Return**

The drawing will have only one view. If you want to add a second sheet with a different exploded view orientation (Fig. 20.5), use the following commands:

Sheets ⇒ **Add** (sheet **2** has been added to the drawing) ⇒ **Views** ⇒ **Exploded** ⇒ **Scale** ⇒ **Done** ⇒ (pick the center of the new drawing sheet) ⇒ (type **1** as the scale) ⇒ **enter** ⇒ **Done/Return** (to keep the trimetric default orientation) ⇒ **Done/Return**

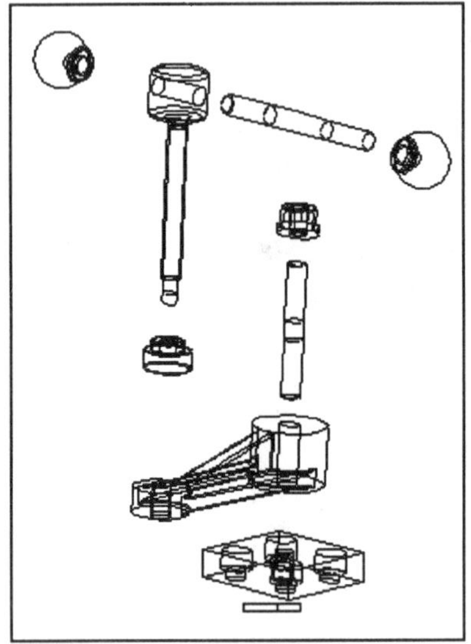

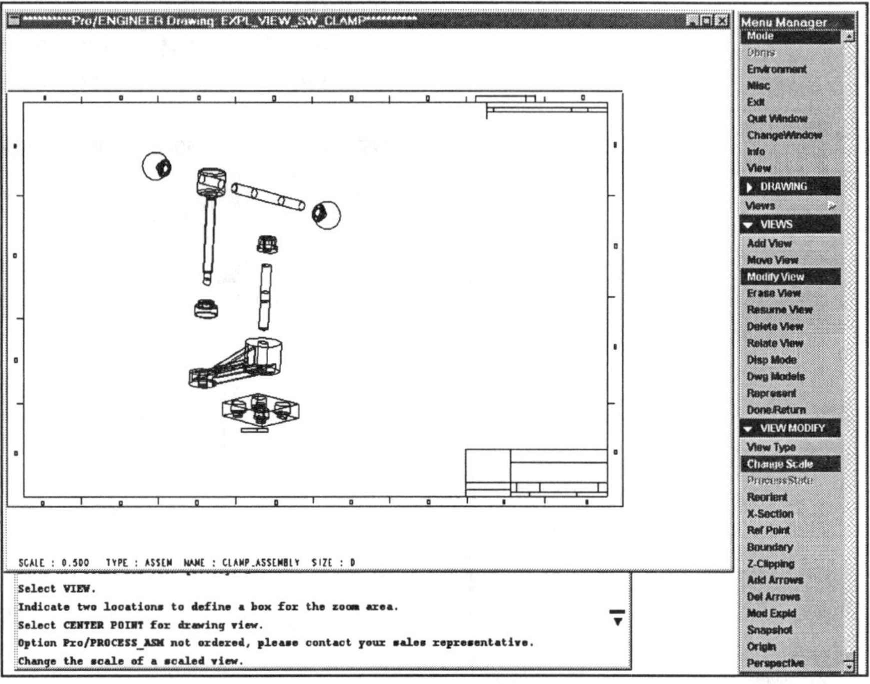

Figure 20.4
Adding an Exploded View to a Drawing

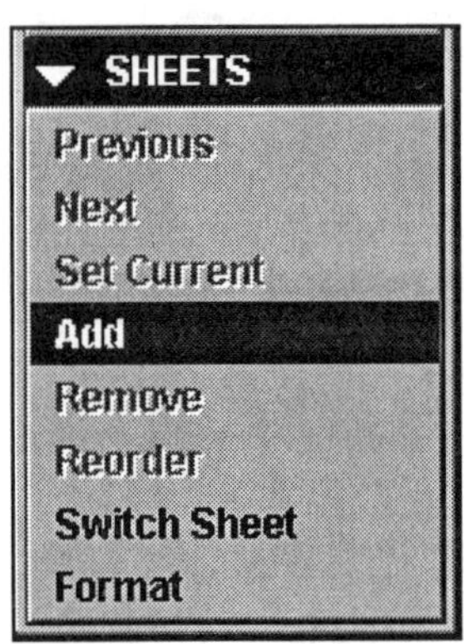

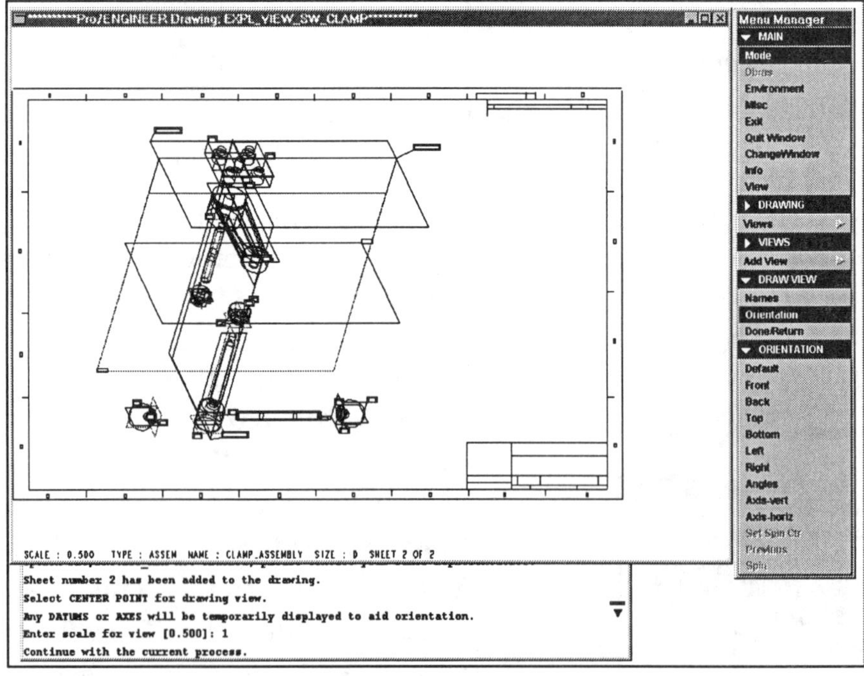

Figure 20.5
Adding a Second Sheet and a New View

Next, add another sheet and a different model. Use the **Swing Clamp subassembly** (Fig. 20.6). Modify the scale of the drawing after the view is placed. Use the following commands:

PT/Modeler™

Sheets ⇒ **Add** ⇒ **Format** ⇒ **Add/Replace** ⇒ (type **C**) ⇒ **enter** ⇒ **Dwg Models** ⇒ **Add Model** ⇒ **?** ⇒ **enter** ⇒ (pick the Swing Clamp subassembly from the directory list) ⇒ **Views** ⇒ **Exploded** ⇒ **Scale** ⇒ **Done** ⇒ (pick the center of the new drawing sheet) ⇒ **enter** (to accept default scale) ⇒ **Names** ⇒ (pick a saved view name ⇒ **Done/Return** ⇒ **Done/Return** ⇒ **Views** ⇒ **Modify View** ⇒ **Change Scale** ⇒ (select view) ⇒ (type **1.25**) ⇒ **enter** ⇒ **Done/Return**

Sheets ⇒ **Add** (sheet number **3** has been added to the drawing) ⇒ **Format** ⇒ **Add/Replace** ⇒ (type **C**) ⇒ **enter** ⇒ **Views** ⇒ **Dwg Models** ⇒ **Add Model** ⇒ **?** ⇒ **enter** ⇒ (pick the Swing Clamp subassembly from the directory list) ⇒ **Done/Return** ⇒ **Views** ⇒ **Exploded** ⇒ **Scale** ⇒ **Done** ⇒ (pick the center of the new drawing sheet) ⇒ (type **.75** as the scale) ⇒ **enter** ⇒ **Names** ⇒ (pick a saved view name from the list) ⇒ **Done/Return** ⇒ **Done/Return** ⇒ **Views** ⇒ **Modify View** ⇒ **Change Scale** ⇒ (pick the view on the window) ⇒ (type **1.25** as the scale) ⇒ **enter** ⇒ **Done/Return**

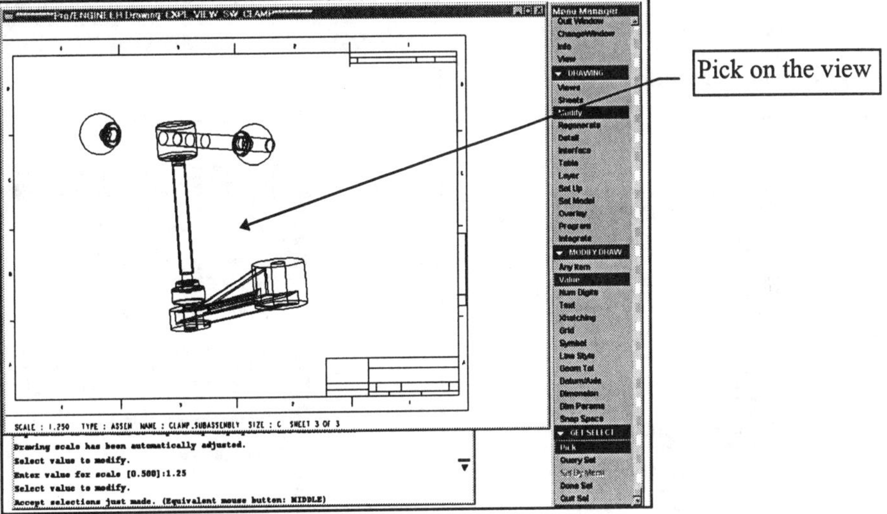

Figure 20.6
Sheet 3 with Exploded View of Subassembly

Show all the axes for the components on the first sheet. To switch between sheets, choose **Sheets** ⇒ **Next**. Use the following commands:

Detail ⇒ **Show/Erase** ⇒ **Show** (radio button) ⇒ **A_1** (radio button) ⇒ **Show All** (radio button) ⇒ **Yes** (radio button) ⇒ **Close** (radio button) ⇒ (press and hold down the right mouse button anywhere on the drawing, then **Modify Item)** ⇒ (pick an axis to be modified) ⇒ (modify each centerline (axis) to stretch between components that are in line, as in Fig. 20.7) ⇒ **Done/Return**

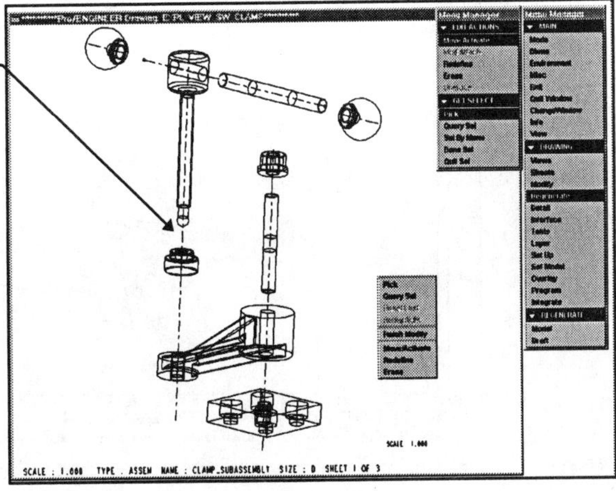

Figure 20.7
Showing and Modifying Axes

To add balloons to one of the sheets not using a parametric title block, you need to create each balloon separately. Before creating balloons, change the **.dtl** file so it has **1.00** diameter balloons and **.500** height for lettering (Fig. 20.8). Create balloons for the components on the first sheet. The balloons added must correspond to the component balloons on the assembly drawing completed in Lesson 19. Use the following commands:

Detail ⇒ Create ⇒ Balloon ⇒ Leader ⇒ Make Note ⇒ (pick the Swing Clamp Arm) **⇒ Done Sel ⇒ Done ⇒** (pick the location for the note) ⇒ (type **1**) ⇒ **enter ⇒ enter ⇒ Make Note ⇒** (continue until all eight components are ballooned, as in Fig. 20.9) ⇒ **Done/Return**

NOTE

To change the defaults on balloons:

Drawing ⇒ Setup ⇒ Modify Val ⇒

drawing_text_height .50
max_balloon_radius .50
min_balloon_radius .50

File ⇒ Save ⇒ File ⇒ Exit

Modify the defaults for balloon sizes

INCH
STD_ANSI
0.500000
0.500000
DEFAULT

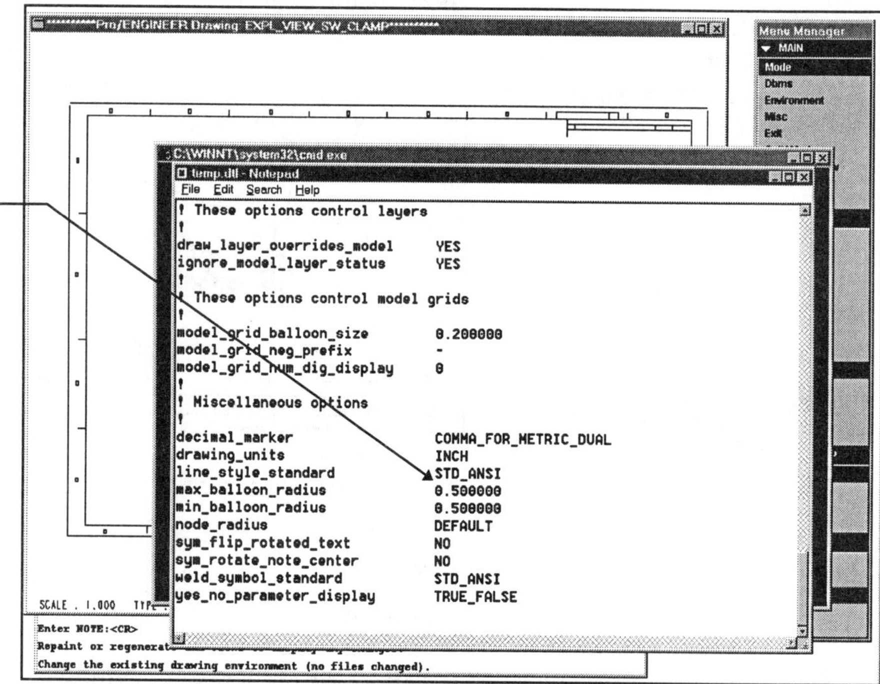

Figure 20.8
Modify Val

Dbms (File--PT/Modeler) ⇒ Save ⇒ enter
Purge ⇒ enter ⇒ Done-Return

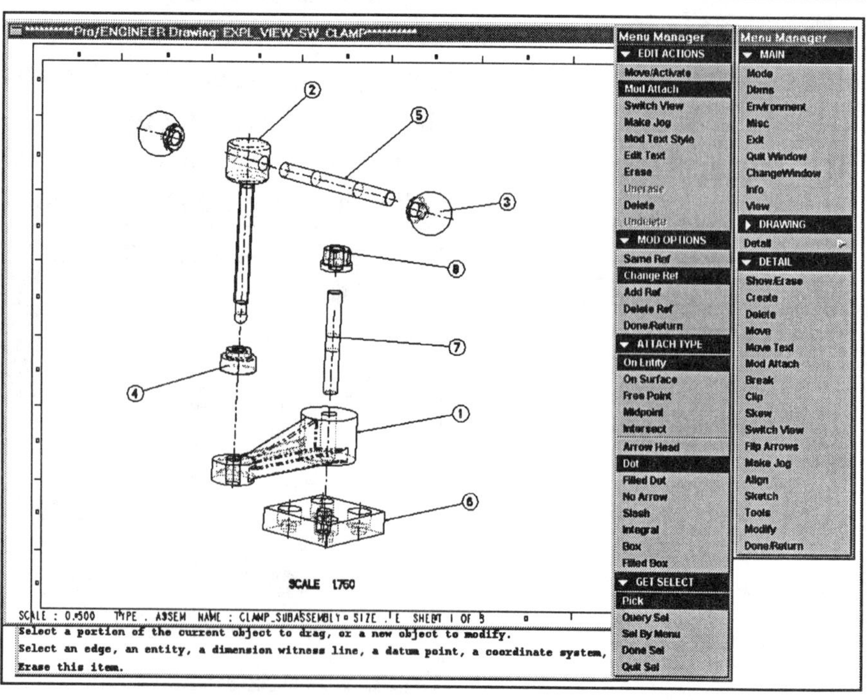

Figure 20.9
Ballooned Drawing

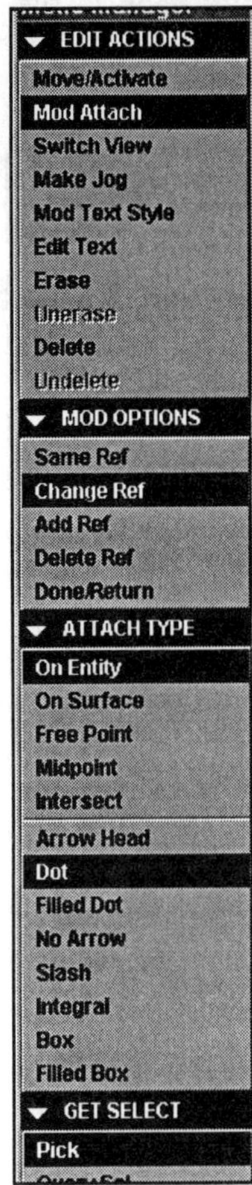

Figure 20.10
Modifying Balloons

After the ballooning is complete, move the balloons and their attachment points to clean up the drawing. Use the following commands to modify the attachment point from edge to surface and change the arrow to a dot (Fig. 20.10):

Modify Item (right mouse button) ⇒ (pick balloon **3**) ⇒ **Mod Attach** ⇒ **On Surface** ⇒ **Dot** ⇒ (pick a place on the ball's surface) ⇒ **Done/Return** ⇒ **Done Sel**

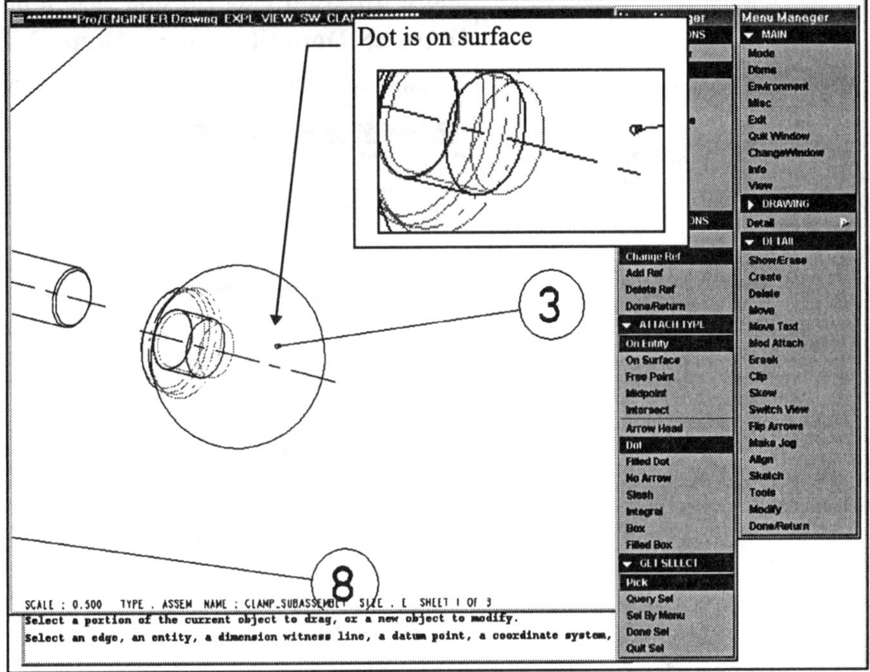

Complete the drawing by filling in the title block or switching to the parametric format you created in Lesson 19 (Fig. 20.11).

NOTE

Change the view display to show as **Hidden Line** and **Tan Phantom**.

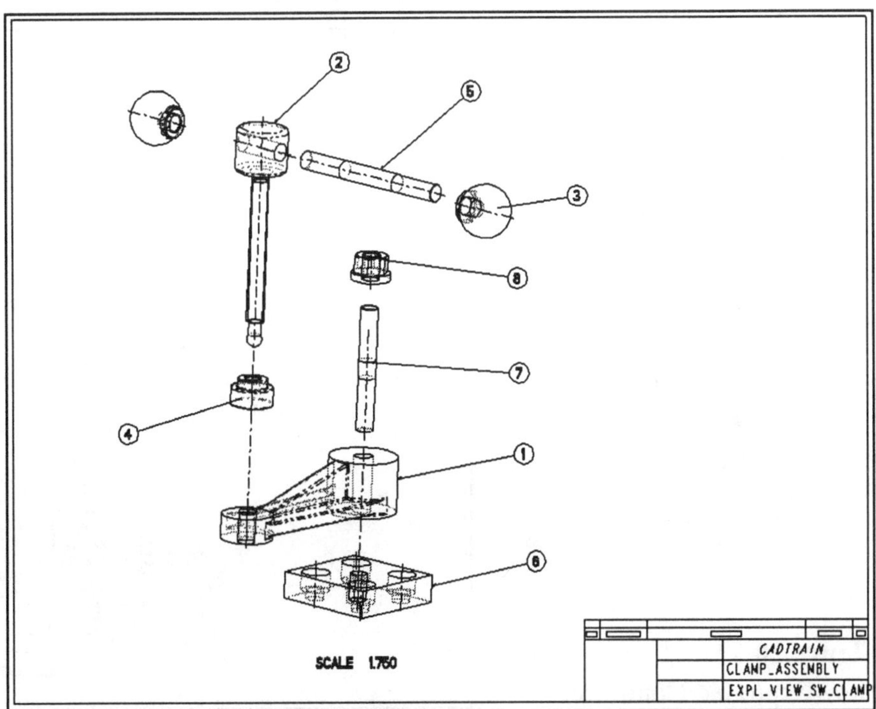

Figure 20.11
Ballooned Drawing

You can modify the exploded drawing without going back to the Assembly mode and editing the model (Figs. 20.12 and 20.13).

Use the following commands to modify the drawing:

PT/Modeler™
Views ⇒ **Modify View** ⇒ **Mod Expld** ⇒ (pick view) ⇒ **Move** ⇒ (move components) ⇒ **Done** ⇒ **Done/Return**

Views ⇒ **Modify View** ⇒ **Mod Expld** ⇒ (pick the view) ⇒ **Redefine** ⇒ **Position** ⇒ **Entity/Edge** ⇒ (select a *vertical* axis) ⇒ (pick the Arm) ⇒ (slide the Arm to a new position) ⇒ [Continue moving the components until they appear as in Fig. 20.13. To move the Balls and long Stud you will need to choose **Entity/Edge** again, and a different axis (in this case a *horizontal* axis).] ⇒ **Done Sel** ⇒ **Done** ⇒ **Done/Return** ⇒ **Done/Return** ⇒ **Done/Return** ⇒ (clean up balloon placement) ⇒ (change scale to **1.75**)

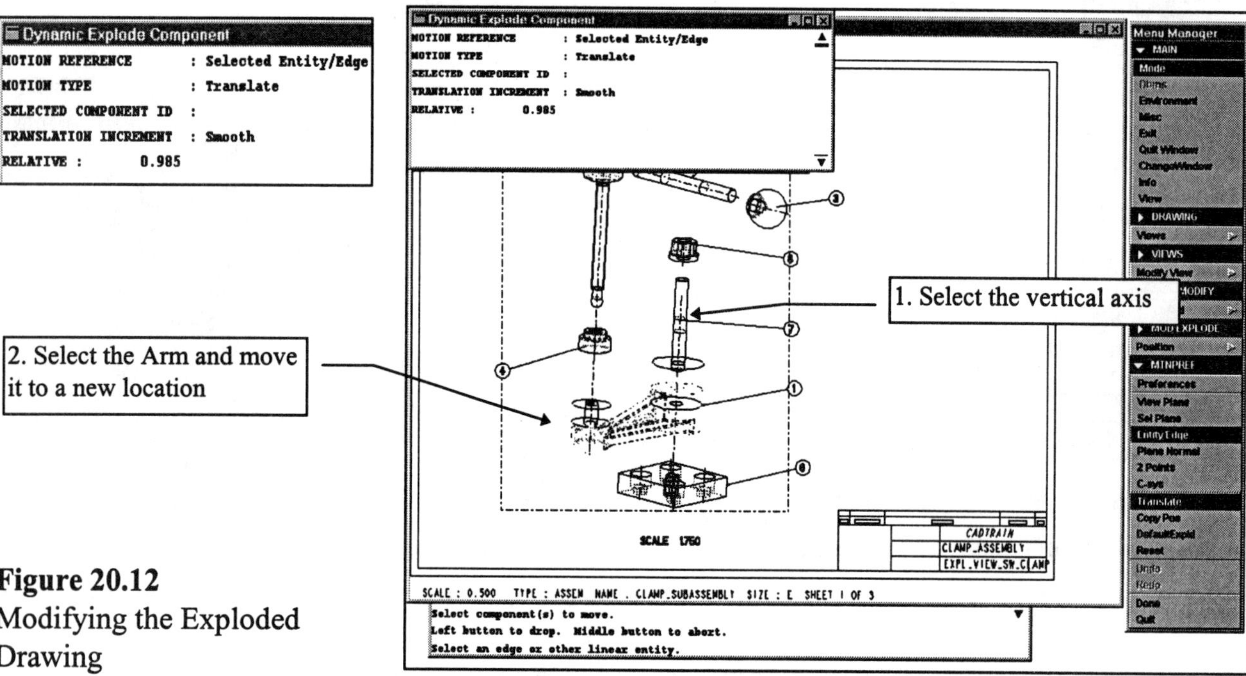

Figure 20.12
Modifying the Exploded Drawing

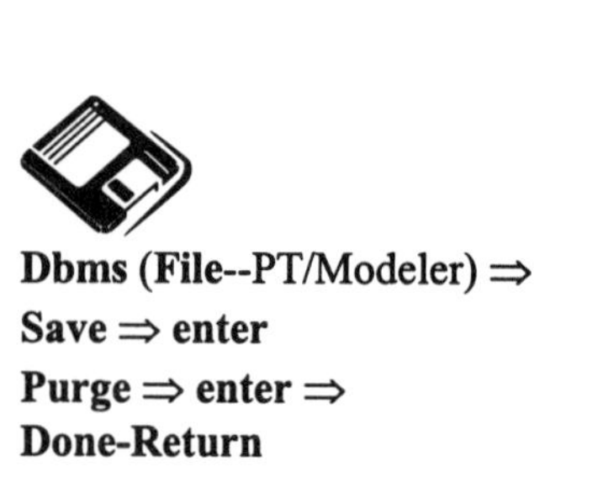

Dbms (**File**--PT/Modeler) ⇒
Save ⇒ **enter**
Purge ⇒ **enter** ⇒
Done-Return

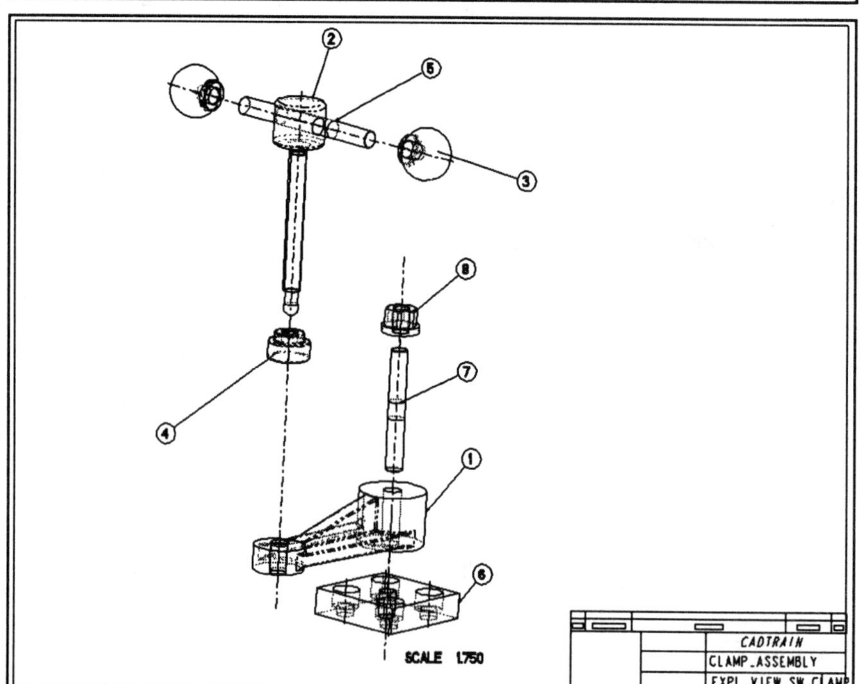

Figure 20.13
New Exploded Condition

Lesson 20 Project

Exploded Coupling Assembly Drawing

Exploded Coupling Assembly Drawing

Create a complete documentation package for the Coupling Assembly (Figs. 20.14 through 20.16). A documentation package includes all the models and drawings required to manufacture the parts and assemble the components. Some of the items listed here have been created in other lessons. Create or extract existing models and drawings and plot the following:

Part Models for all coupling assembly components
Detail Drawings for each nonstandard component, for example the Coupling Shaft
Detail Drawings for standard components that have been altered; show only the dimensions required to alter the component, for example the socket head cap screw in the coupling assembly
Assembly Drawing and **Parts List (BOM)** using standard orthographic ballooned views
Exploded Assembly Drawing of the ballooned assembly
Exploded Subassembly Drawing of the ballooned subassembly

Figure 20.14
Exploded Coupling Assembly Drawing

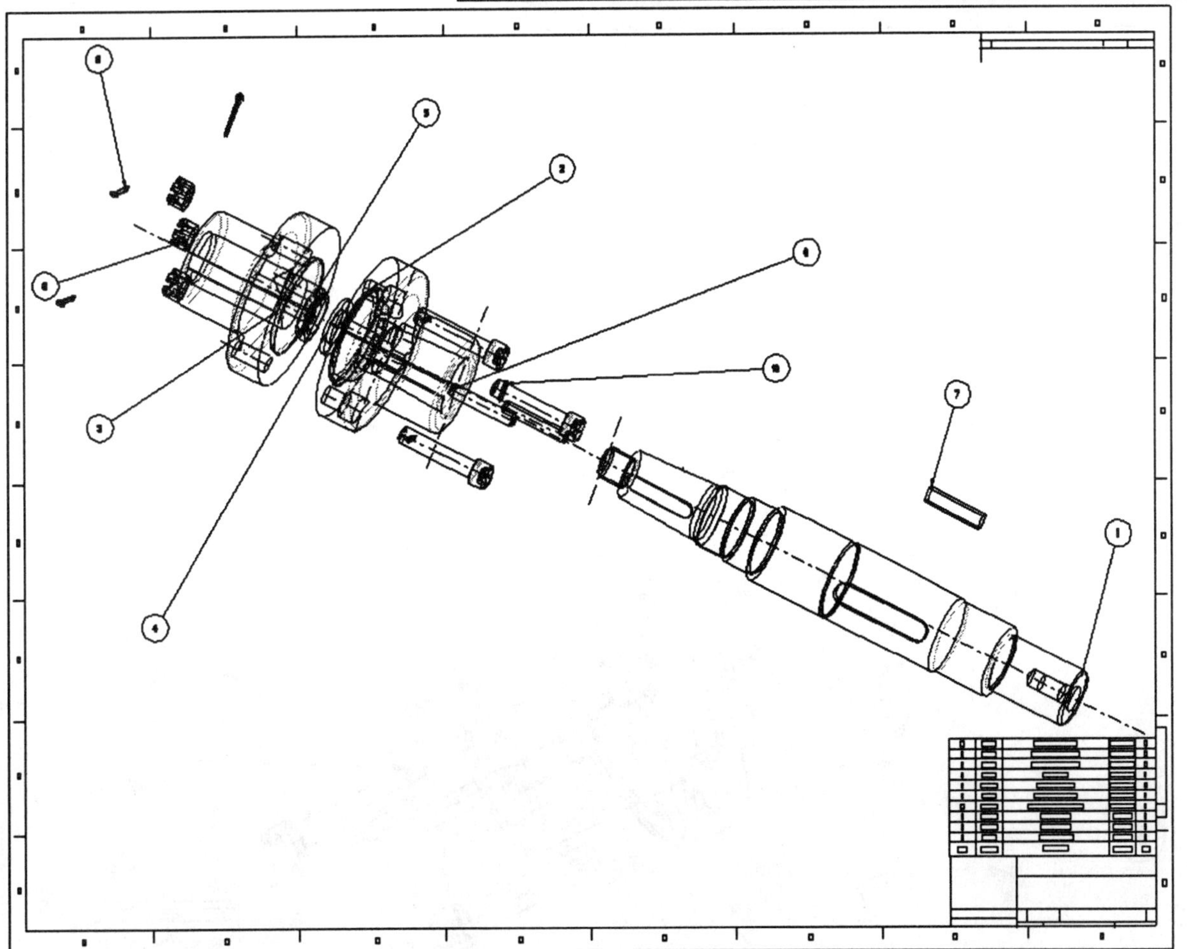

Figure 20.15 Coupling Assembly BOM

10	110-2CS	SOC HD CAP SCREW	PURCHASED	3
9	109-2SN	HEX SLOT NUT 16 X 2	PURCHASED	3
8	108-2CP	COTTER PIN .150 X 1.25	PURCHASED	3
7	107-2KY	KEY 14 X 61	PURCHASED	1
6	106-2DW	DOWEL 120D X 70	PURCHASED	2
5	105-2HN	HEX NUT M30 X 3.5	PURCHASED	1
4	104-2WA	WASHER 33ID X 500D X 4	PURCHASED	1
3	103-2CP2	COUPLING TWO	1040 CRS	1
2	102-2CP1	COUPLING ONE	1040 CRS	1
1	101-2SH	COUPLING SHAFT	1020 CRS	1
ITEM	PT NUM	DESCRIPTION	MATERIAL	QTY

Figure 20.16 Coupling Assembly, Close-up

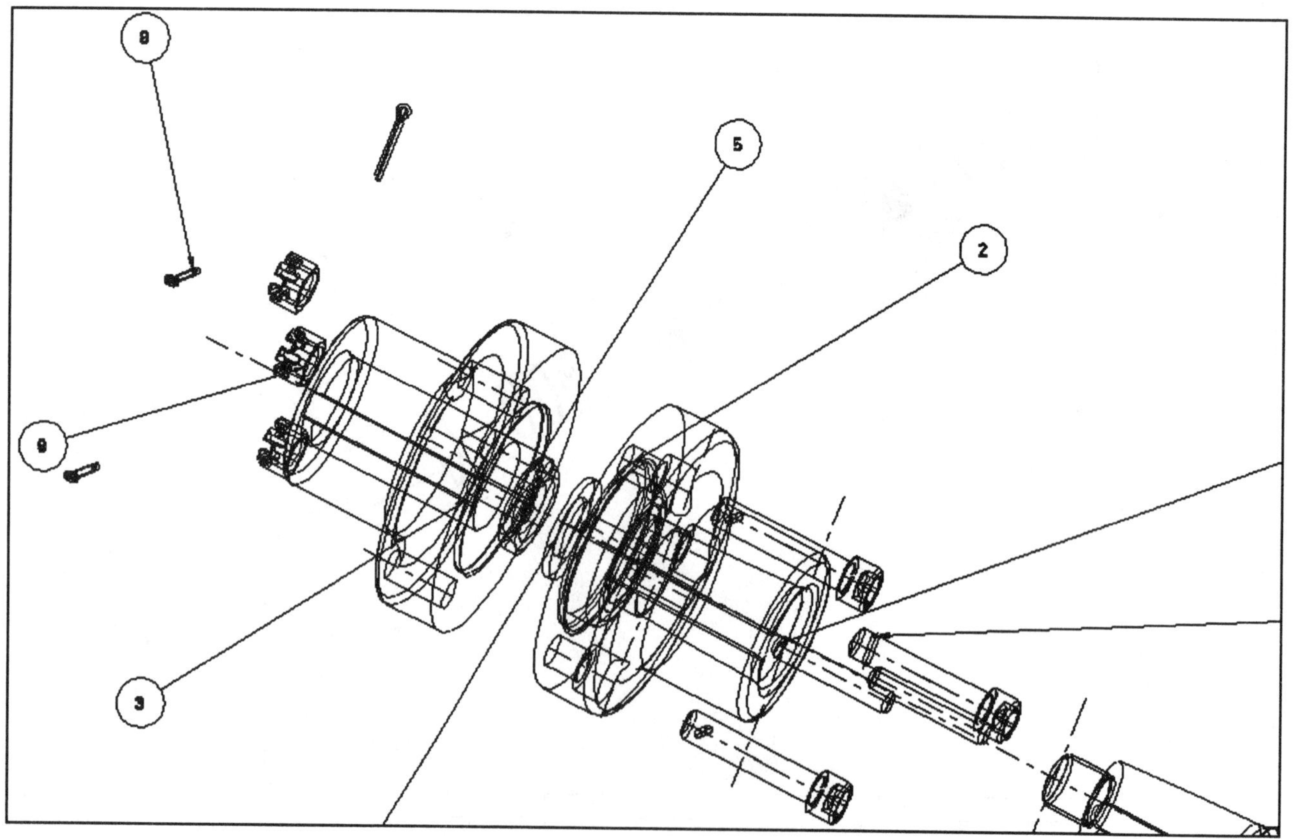

Part Four

Advanced Capabilities

Lesson 21 User Defined Features and Family Tables

ADVANCED CAPABILITIES

User Defined Features and Family Tables are covered in this part. With each new change of software from PTC and subsequent revision of the text, we hope to add one or more lessons.

In the future, lessons on **Top-Down Assembly and Part Design, Pro/MANUFACTURING and Fixture Design, Pro/SHEETMETAL**, and **Pro/SURFACE** are planned.

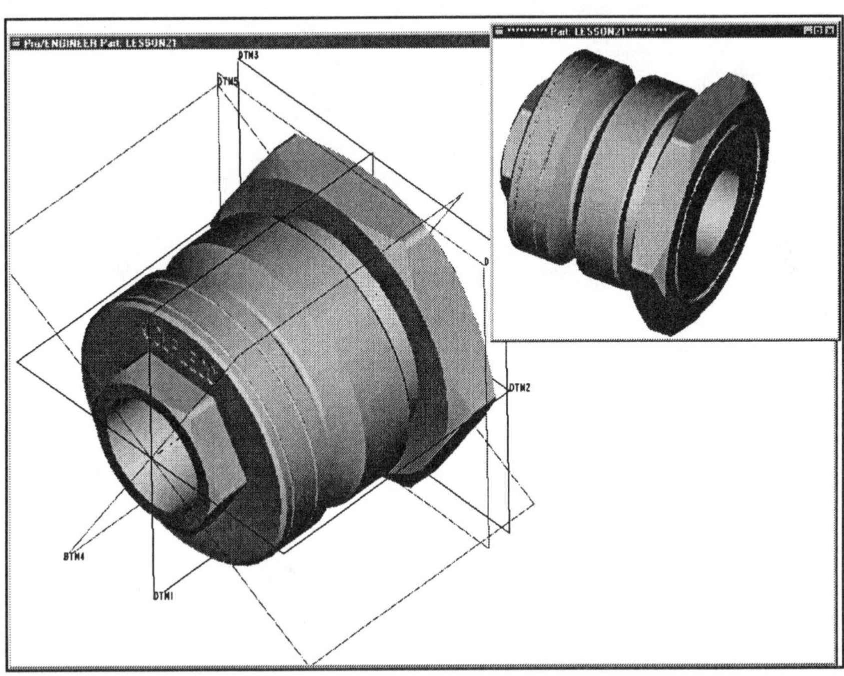

Coupling

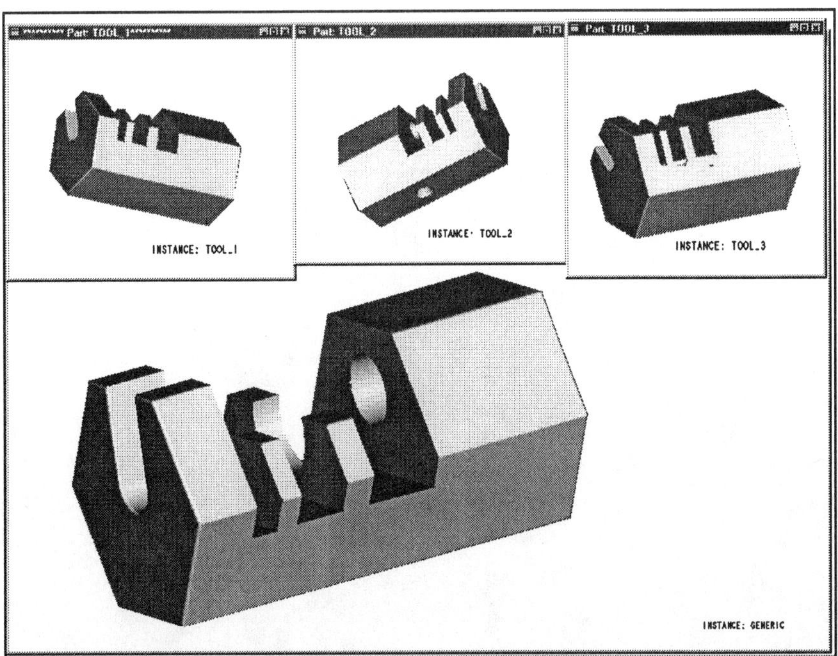

Tool Body

Lesson 21

User-Defined Features and Family Tables

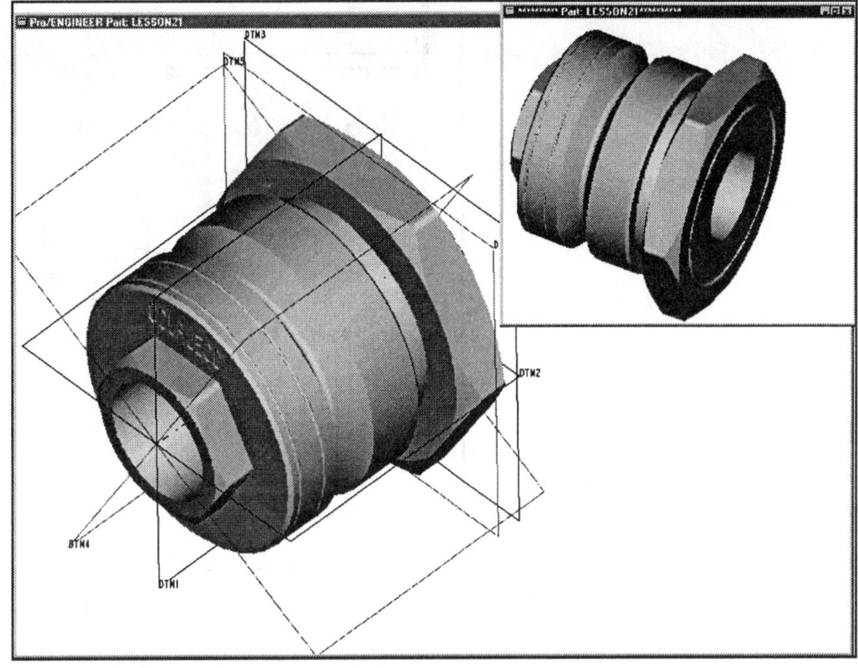

Figure 21.1
Coupling Fitting

OBJECTIVES

1. **Understand User-Defined Features (UDF)**

2. **Comprehend Family Tables, Generic Parts, and Instances**

3. **Create a Family Table for a part**

4. **Use User-Defined Features from Pro/LIBRARY**

PT/Modeler™

PT/Modeler (the Student Version) does not support UDFs at this time.

EGD REFERENCE

Engineering Graphics and Design with Graphical Analysis *or* **Fundamentals of Engineering Graphics and Design**
by L. Lamit and K. Kitto
Read Chapter 17
See pages 675-678

COAch™ for Pro/ENGINEER

If you have **COAch for Pro/ENGINEER** on your system, go to SEARCH and do the Segments shown in Figures 21.6 through 21.11.

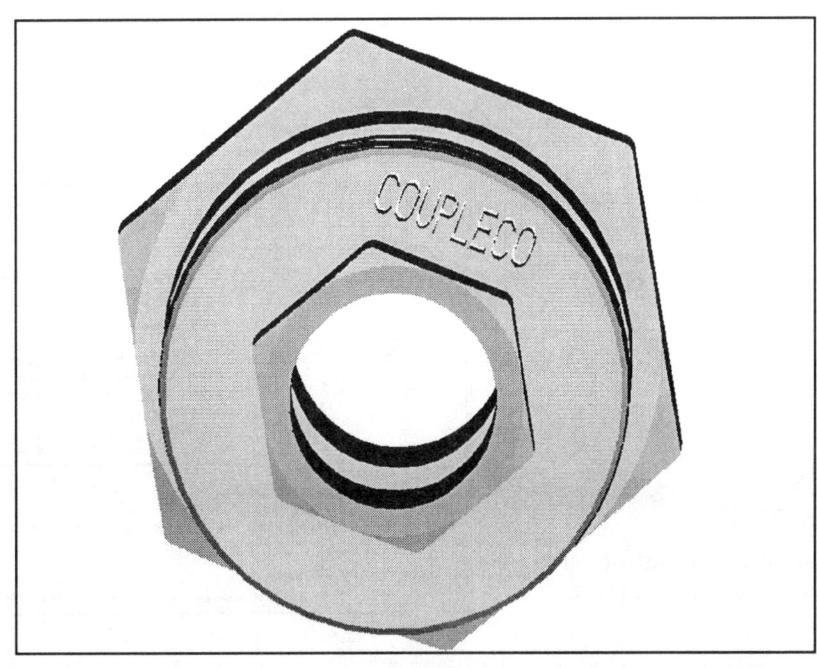

Figure 21.2
COUPLECO Coupling Fitting

INFORMATION WINDOW

! NAME	d1	d2	d4	d3	d5	F0056 ROUND
! GENERIC	5.0000	4.0000	2.2500	1.0000	0.3000	Y
BOLT1	5.0000	4.0000	2.2500	1.0000	0.3000	Y
BOLT2	4.0000	3.2500	1.8750	0.8750	*	N

Dimension in family table

Feature "ROUND" in family table

Generic bolt

BOLT2 does not have feature "ROUND"

Bolt instance "BOLT2"

Figure 21.3
Online Documentation, Family Table and Library Part

USER-DEFINED FEATURES AND FAMILY TABLES

User-Defined Features (UDFs) allow you to store a set of standard features, parts, or assemblies in a library for later use (Figs. 21.1 and 21.2). UDFs can then be added to existing parts (if they are features) or to existing assemblies (if they are parts or sub-assemblies).

A **Family Table** (Fig. 21.3) is a powerful tool for capturing engineering intelligence and promoting data reuse. Family Tables provide an automated and simple way for engineers to multiply a model into logical instances that reference the generic. This process delivers many models, which capture all the intelligence of the original, update as the generic updates, and use minimal disk space. Family Tables can be created within UDFs (Fig. 21.4).

Generic group name: RAD_HOLES

Name	d8 CHAMF_DIST	d1 HOLE_DIAM
GENERIC	0.220	1.000
QUARTR_IN	0.150	0.250
THREE-EIGH	0.300	0.375
ONE_HALF	0.220	0.500

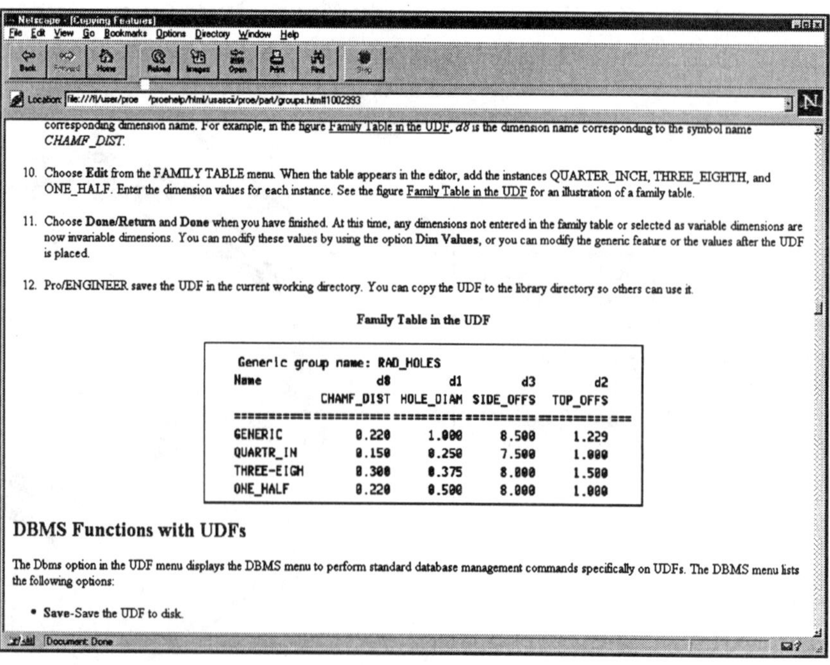

Netscape - [Copying Features]

File Edit View Go Bookmarks Options Directory Window Help

Location: file:///I/user/proe /proehelp/html/usascii/proe/part/groups.htm#1002993

corresponding dimension name. For example, in the figure Family Table in the UDF, *d8* is the dimension name corresponding to the symbol name *CHAMF_DIST*.

10. Choose **Edit** from the FAMILY TABLE menu. When the table appears in the editor, add the instances QUARTER_INCH, THREE_EIGHTH, and ONE_HALF. Enter the dimension values for each instance. See the figure Family Table in the UDF for an illustration of a family table.

11. Choose **Done/Return** and **Done** when you have finished. At this time, any dimensions not entered in the family table or selected as variable dimensions are now invariable dimensions. You can modify these values by using the option **Dim Values**, or you can modify the generic feature or the values after the UDF is placed.

12. Pro/ENGINEER saves the UDF in the current working directory. You can copy the UDF to the library directory so others can use it.

Family Table in the UDF

Generic group name: RAD_HOLES

Name	d8 CHAMF_DIST	d1 HOLE_DIAM	d3 SIDE_OFFS	d2 TOP_OFFS
GENERIC	0.220	1.000	8.500	1.229
QUARTR_IN	0.150	0.250	7.500	1.000
THREE-EIGH	0.300	0.375	8.000	1.500
ONE_HALF	0.220	0.500	8.000	1.000

DBMS Functions with UDFs

The Dbms option in the UDF menu displays the DBMS menu to perform standard database management commands specifically on UDFs. The DBMS menu lists the following options:

- **Save**-Save the UDF to disk.

Document: Done

Figure 21.4
Online Documentation, Family Table in the UDF

Figure 21.5
Online Documentation, User-Defined Feature (Pro/LIBRARY Countersunk Hole/Feature)

User-Defined Features

When you create a **UDF**, you must define the placement references for its use. When you place a UDF in your part, Pro/E prompts you to use these references in the same manner as Pro/ENGINEER does when you place a **Cut** or a **Slot** or a **Hole**. A variety of UDF features are available in Pro/LIBRARY, such as the countersink shown in Figure 21.5.

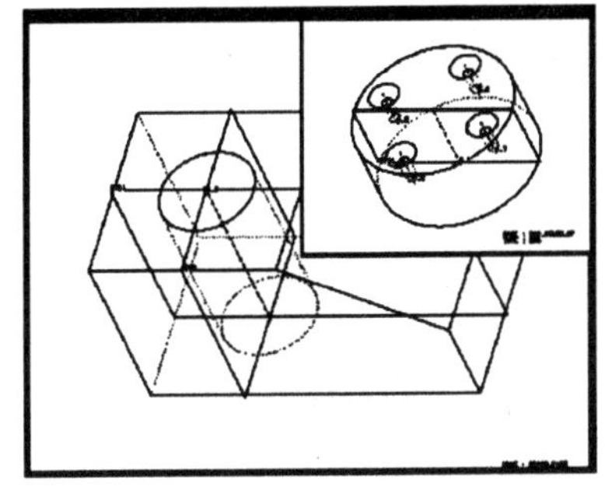

It is always advisable to plan out the references in a UDF before you create it, so that, when placed, the prompts and interaction mimic Pro/ENGINEER as much as possible.

If you create a UDF that represents a uniquely shaped feature (Figs. 21.6 and 21.7), define it in such a way that when it is placed, the user must select a placement plane and two reference edges to dimension from, then enter the distances of these dimensions and, if applicable, the size of the feature.

If you want the ability to place a UDF in several different ways, you are required to create a separate UDF for each method you want to use. In addition, you will have to create a separate model for each, since the manner in which you build the original defines the references and therefore controls how it can be placed.

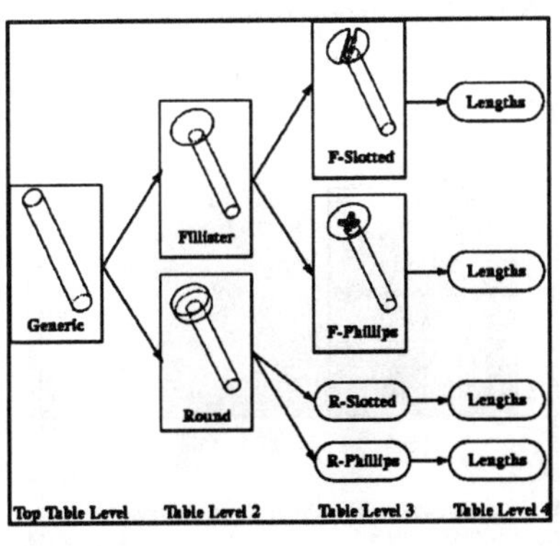

For example, if you want to be able to place a hole using either **Linear** or **Coaxial** references, you must create a hole using these methods, then create a UDF of each hole.

When you create a UDF, Pro/E creates a file, in the current directory, with the extension **.gph**. This is the UDF itself. If you also tell Pro/E to save a reference part, Pro/E also creates a part file with the text **_gp** added to the end of its name.

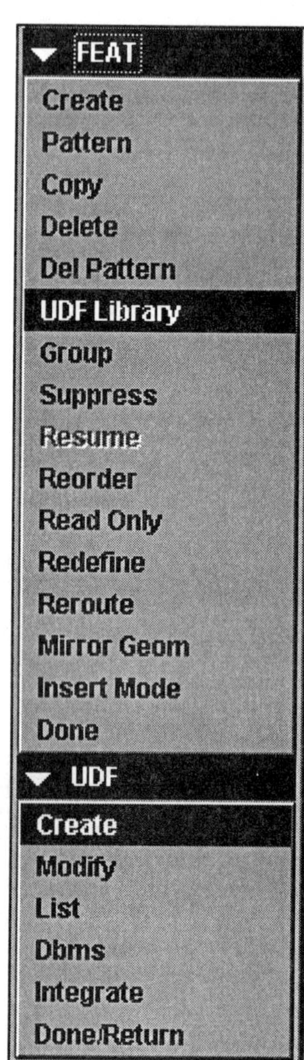

Creating a UDF

1. Choose **UDF Library** from the FEAT menu, then choose **Create** from the UDF menu. The menu lists the following options:

   Create Add a new UDF to the UDF library.
   Modify Modify an existing UDF. If there is a reference part, Pro/E displays the UDF in a separate part window.
   List List all the UDF files in the current directory.
   Dbms Perform database management functions for the current UDF.
   Integrate Resolve the differences between the source and the target UDFs.

2. Enter a name for the UDF.
3. Choose one of the options in the UDF OPTIONS menu, followed by **Done**:

   Stand Alone Pro/ENGINEER copies all the required information to the UDF. Respond to the prompt to include a reference part.
   Subordinate Pro/ENGINEER copies the information from the original part at run time.

4. Pro/E displays the UDF feature creation dialog box, which lists the required and optional elements.
5. Choose the **Features** element and **Define** from the dialog box.

When you create a UDF, it is not kept in RAM, but rather stored in your current directory. Only when you use it in a model (part or assembly) does it appear in RAM.

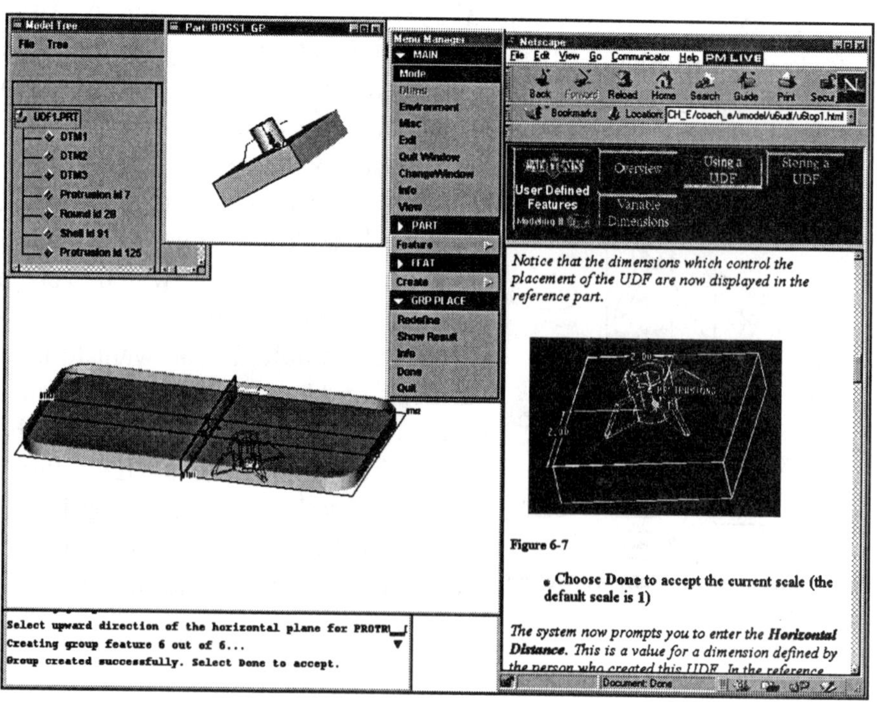

Figure 21.6
COAch for Pro/ENGINEER, Super Users (User-Defined Features)

6. Pro/ENGINEER displays the UDF FEATS menu, which lists the following options:

Add Add a feature to the UDF.
Remove Remove a feature from the UDF.
Show Highlight all the features in the UDF.
Info List all the features in the UDF.

7. Choose **Add** from the UDF FEATS menu.
8. Using the SELECT FEAT menu, select the features to add to the UDF. Note that you can include any feature except base features and shells. When you have finished, choose **Done** from the SELECT FEAT menu and **Done/Return** from the UDF FEATS menu.
9. Enter the prompts for the references used by the selected features. Pro/E highlights each reference and asks you to enter the prompt.

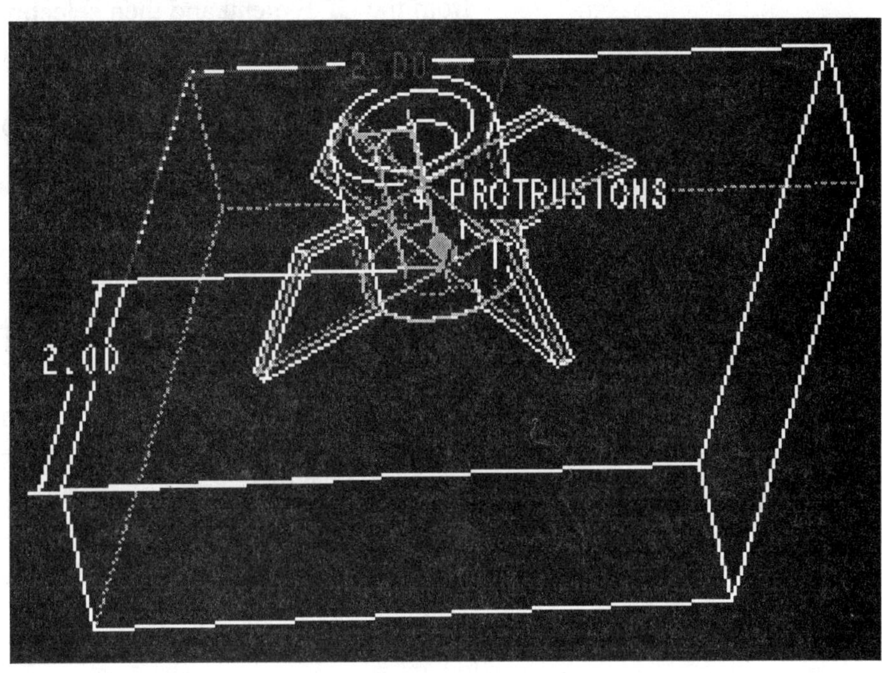

Figure 21.7
COAch for Pro/ENGINEER, Super Users (UDF)

When you specify a prompt for a placement reference that is used by more than one feature in the UDF, Pro/E lets you specify either single or multiple prompts for this reference. Choose the desired option from the PROMPTS menu and then select **Done**:

Single Specify a single prompt for the reference used in several features. When the UDF (Fig. 21.8) is placed, the prompt appears only once, but the reference you select for this prompt applies to all features in the group that use the same reference.
Multiple Specify an individual prompt for each feature that uses this reference. If you select **Multiple**, Pro/E highlights each feature that uses this reference, so you can enter a different prompt for each of them.

10. After you have entered all the prompts, Pro/E displays the MOD PRMPT and SET PROMPT menus so you can change any prompt as follows:

Use **Next** and **Previous** from the MOD PRMPT menu to select the prompt you want to change, and enter the new prompt instead.

To change a single prompt (specified for the placement reference used in several features) into multiple prompts, find a prompt that you want to change, choose **Multiple**, and enter an individual prompt for each feature, as prompted by Pro/E.

11. If you are satisfied with the prompts, choose **Done/Return** from the SET PROMPT menu.
12. The required elements in the UDF dialog box have been defined. You can complete the creation of the UDF by choosing **Done/Return** from the UDF menu and then selecting the **OK** button in the UDF dialog box or you can define other optional elements.

To define variable elements in the UDF:

1. Select the **Var Elements** element in the dialog box and click on the **Define** button.
2. Select a feature that belongs to the UDF for which you want to specify variable elements.
3. Pro/E displays the SEL ELEMENT menu, which lists the elements of the selected feature. Place a check mark in front of the elements you want to define as variable, then choose **Done**.
4. When you finish selecting variable elements, choose **Done Sel** from the GET SEL menu.

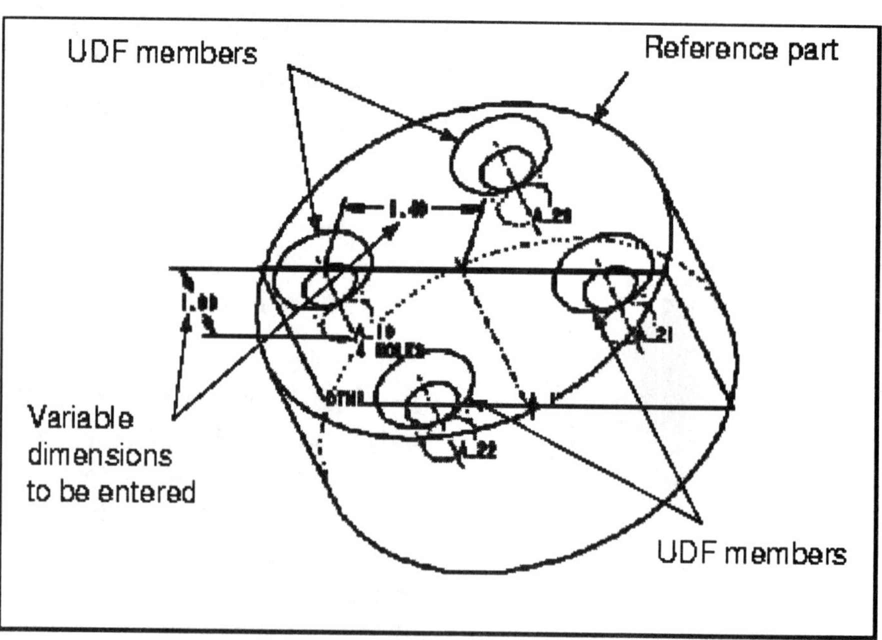

Figure 21.8
Online Documentation, UDF Patterns and Placement

Family Tables

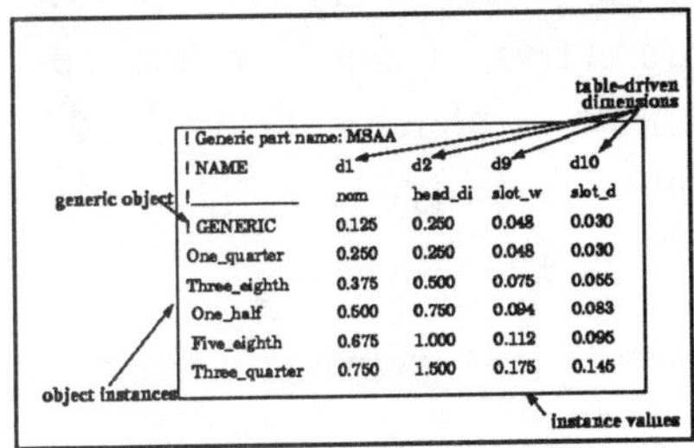

! Generic part name: MSAA				
! NAME	d1	d2	d9	d10
!__________	nom	head_di	slot_w	slot_d
! GENERIC	0.125	0.250	0.048	0.030
One_quarter	0.250	0.250	0.048	0.030
Three_eighth	0.375	0.500	0.075	0.055
One_half	0.500	0.750	0.094	0.083
Five_eighth	0.675	1.000	0.112	0.095
Three_quarter	0.750	1.500	0.175	0.145

Family Tables are effective for two main reasons: they provide a beneficial tool and they are easy to use. You need to understand the functionality of Family Tables and you must understand when a Family Table is required and what circumstances should promote their use.

Family Tables are used any time a part or assembly (Fig. 21.9) has several unique iterations developed from the original model. The iterations must be considered separate parts, not just iterations of the same model.

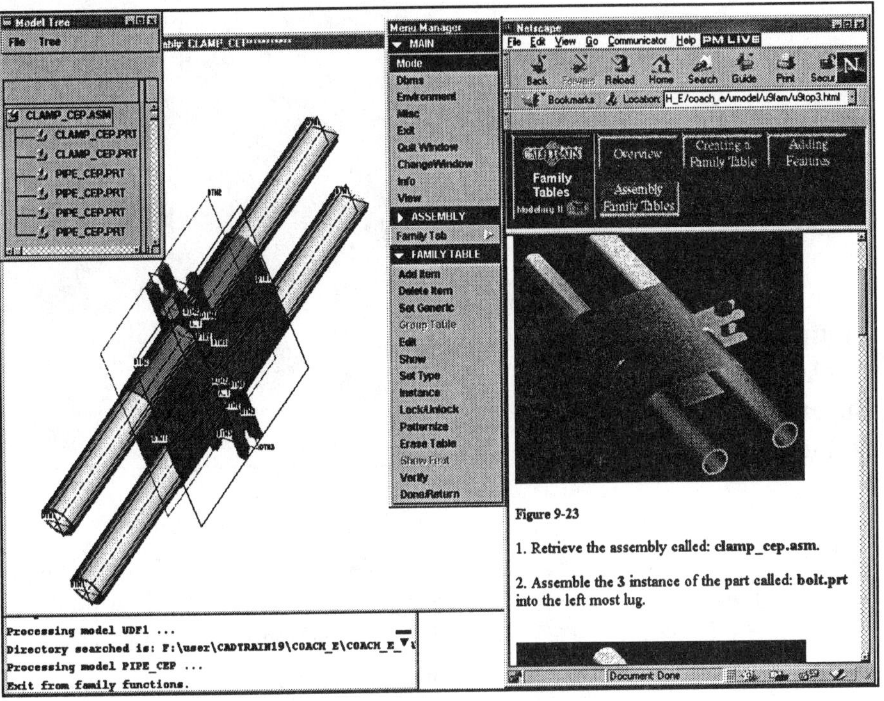

Figure 21.9
COAch for Pro/ENGINEER, Super Users (Family Tables)

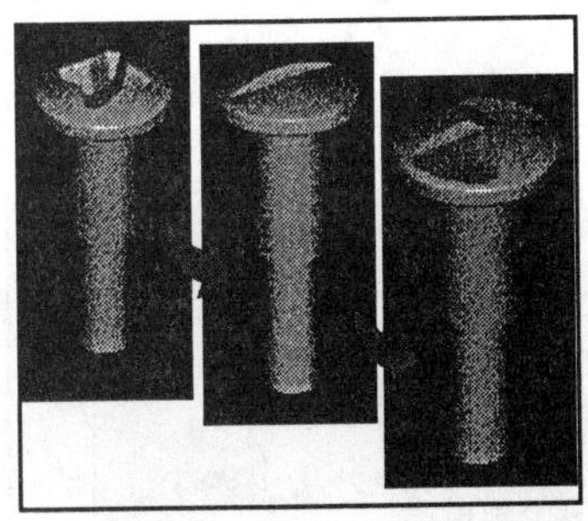

To determine whether a model is a candidate for a Family Table, establish whether the original and the variation would ever have to co-exist at the same time (both in the same assembly, both shown in the same drawing, both with independent Bill of Materials) and whether they should be tied together (most of the same dimensions, features, and parameters). If so, they are a candidate for the creation of a Family Table. Otherwise, the model may be a candidate for copying to an independent model or for a **Pro/PROGRAM**.

When the decision is made to create a Family Table, the layout of the table should be considered next. The three main considerations are *what is in the table, how deep the table should be,* and *how to organize the table.*

The *first consideration*, ***what is in the table,*** is defined by how the instances vary from the original. A table (Fig. 21.10) can include dimensions, features, components, parameters, groups, reference models, pattern tables, and system parameters (under Other). Each one should be considered as a variation possibility, although most tables vary in their dimensions, parameters, and features.

```
! FAMILY TABLE VALUES"
! "
! 1)This family table is for inspection ONLY; any changes
!   made wil NOT be saved."
! 2)Rows beginning with '@' are family table comments.
! 3)Rows beginning with '!' and empty rows are ignored."
! 4)Rows beginning with '$' contain locked instances.
! 5)'*' is used for the default value.
! 6)Generic names of features if appear are enclosed in ""
! 7)Feature identifications are their internal ids."
! "
! Generic part name: CAP-FAMILY"
! Name          d150   p1  F1730  SOURCE
!-----------------------------------------
! GENERIC       75.000 8   Y      generic
```

Figure 21.10
COAch for Pro/ENGINEER, Super Users (Family Table Values)

The *second consideration* is ***how deep the table should be.*** The model can have a single-level (e.g., different-sized Phillips flat-head screws) or multiple-level table (e.g., screws to flat-head screws to Phillips-head screws to individual sizes). This is dependent on a trade-off between complexity (a single-level table may be unwieldy if it becomes too large) and ease of use (a multiple-level table can make instance retrieval more cumbersome).

The *third consideration* is ***how to organize the table.*** Pro/ENGINEER will prompt users for information to identify an instance by the order in which it appears in the table (Fig. 21.11). Therefore, if color is the most important differentiating aspect of the instance, it should be the first item in the table.

```
! "
! Generic part name: CAP-FAMILY"
! Name          d150   p1  F1730  SOURCE
!-----------------------------------------
! GENERIC       75.000 8   Y      generic
  INSTANCE1     60.000 6   Y      ACME
  INSTANCE2     50.000 4   Y      UNIVERSAL
```

Figure 21.11
COAch for Pro/ENGINEER, Super Users (Family Table Generic Part)

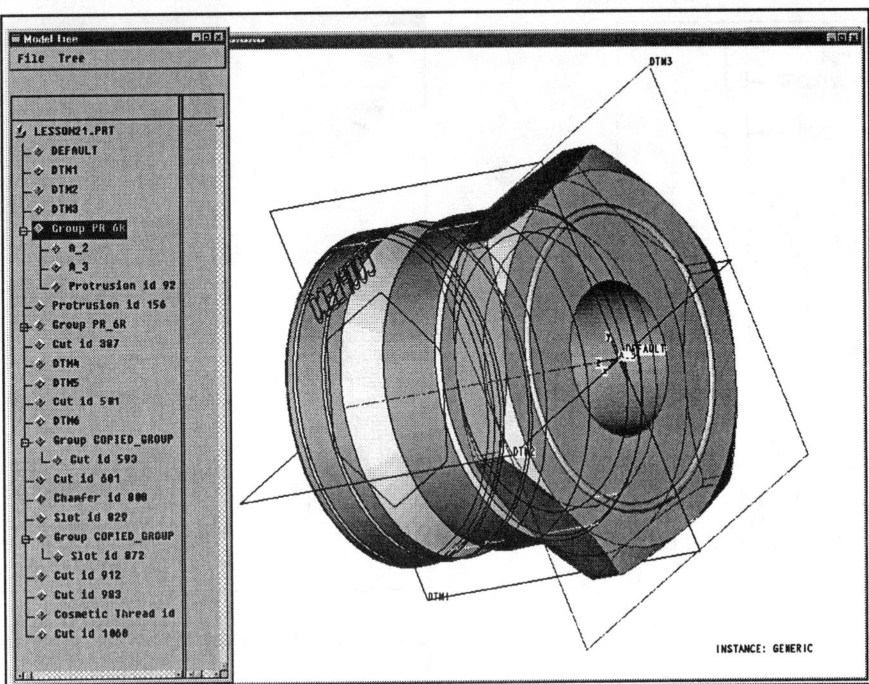

Figure 21.12
Coupling Fitting Showing Datum Planes, Coordinate System, and Model Tree

Coupling Fitting

The **Coupling Fitting** (Figs. 21.12 through 21.25) is designed so that three versions are available to the user. A Family Table is used to establish three instances of the coupling. The generic version is the smallest and the three instances vary in size and features. The coupling uses two hexagonal features that are taken from Pro/LIBRARY. If your system does not have access to Pro/LIBRARY, create a hexagonal UDF and store it in your own User-Defined feature library according to the information provided in the first part of this lesson.

NOTE

Set up the **COUPLING FITTING**:

- Material = Stainless Steel
- Units = Inches
- Default Datum Planes
- Default Coordinate System
- Layers = **DATUM_LAYER**

CONFIG.PRO

sketcher_dec_places	4
default_dec_places	4

File Edit View Format Help

FAMILY TABLE VALUES

1) This family table is for inspection ONLY; any changes made will NOT be saved.
2) Rows beginning with '@' are family table comments.
3) Rows beginning with '!' and empty rows are ignored.
4) Rows beginning with '$' contain locked instances.
5) '*' is used for the default value.
6) Generic names of features if appear are enclosed in [].
7) Feature identifications are their internal ids.

Generic part name: COUPLING-FITTING

Name	d30	d46	F1068 [CUT]	F829 [SLOT]	F872 [SLOT]
GENERIC	2.06250	0.31300	Y	Y	Y
CPLA	3.00000	0.50000	N	N	N
CPLB	3.25000	0.62500	N	Y	N
CPLC	3.50000	0.75000	Y	N	Y

Figure 21.13
Coupling Fitting Family Table

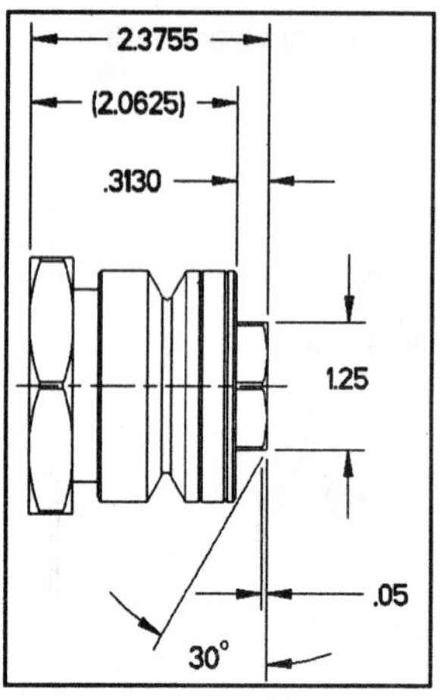

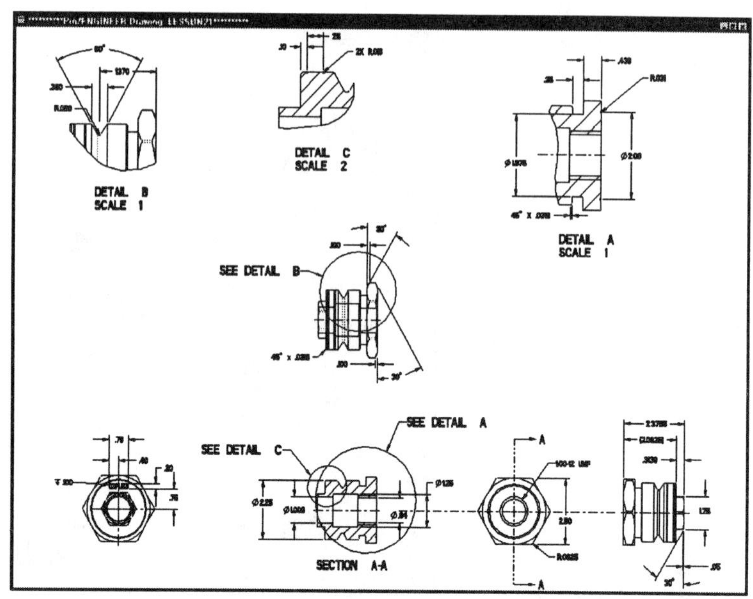

Figure 21.14
Coupling Fitting, Detail Drawing

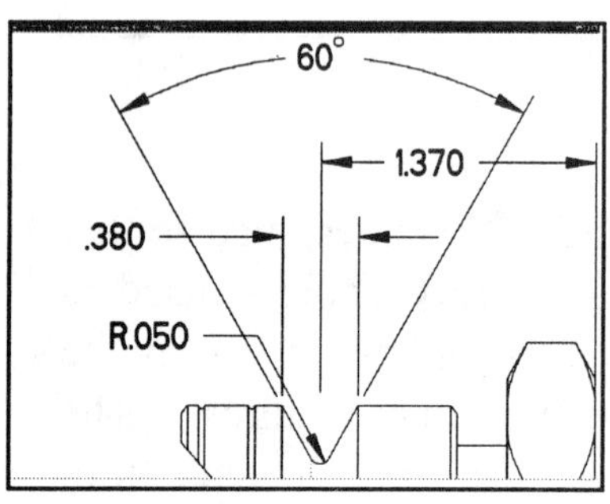

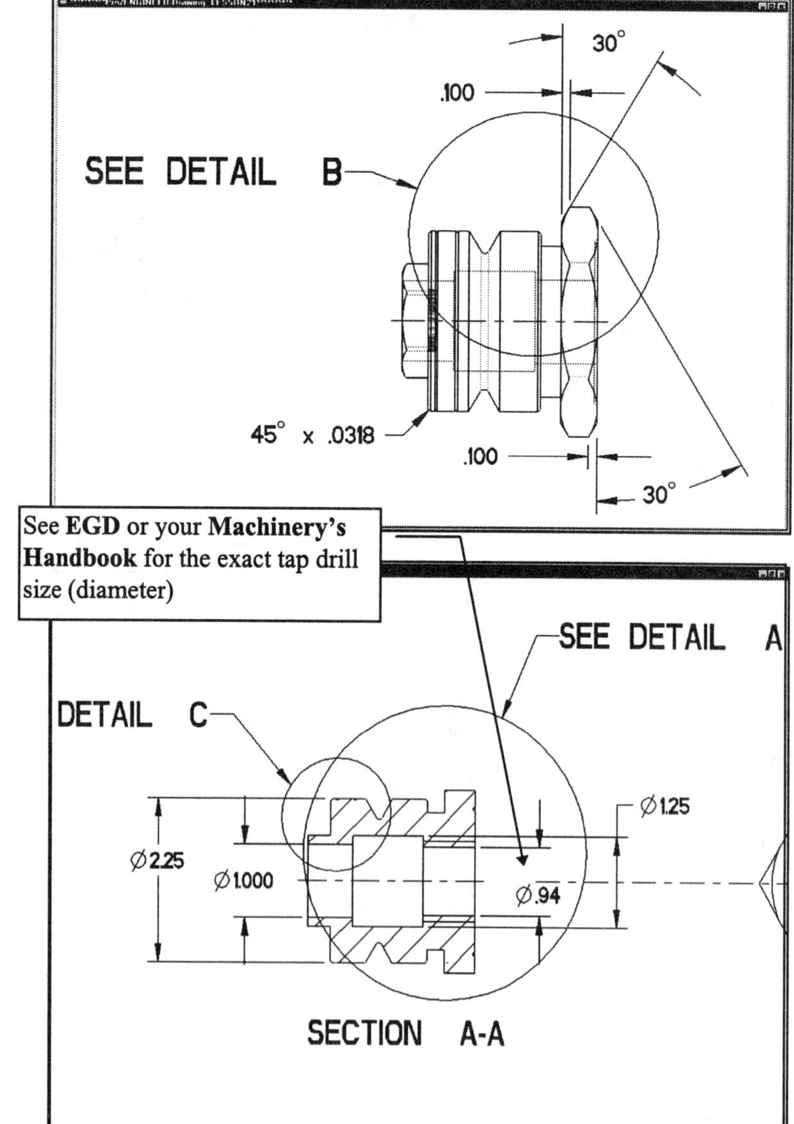

Figure 21.15
Coupling Fitting, Top View

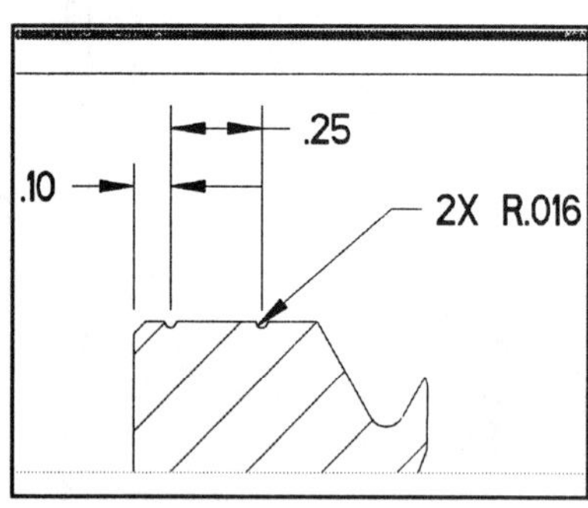

Figure 21.16
Coupling Fitting, **SECTION A-A**

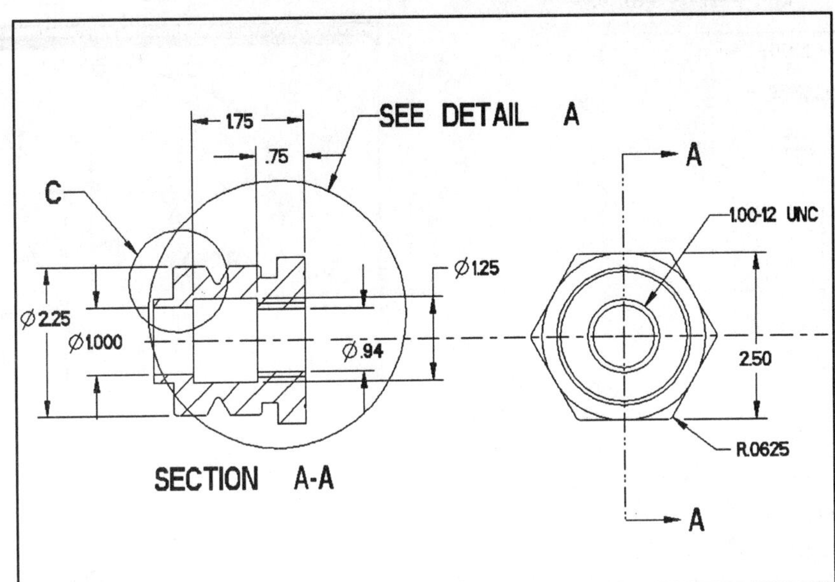

Figure 21.17
Coupling Fitting, Front View and Right Side View

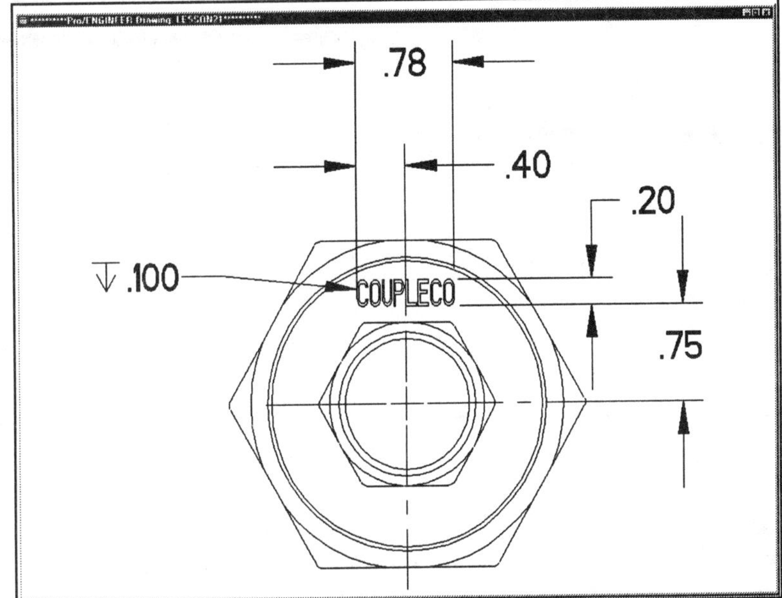

Figure 21.18
Coupling Fitting, Engraved Company Name **(COUPLECO)**

PT/Modeler™

PT/Modeler does not support extruded text cuts at this time.

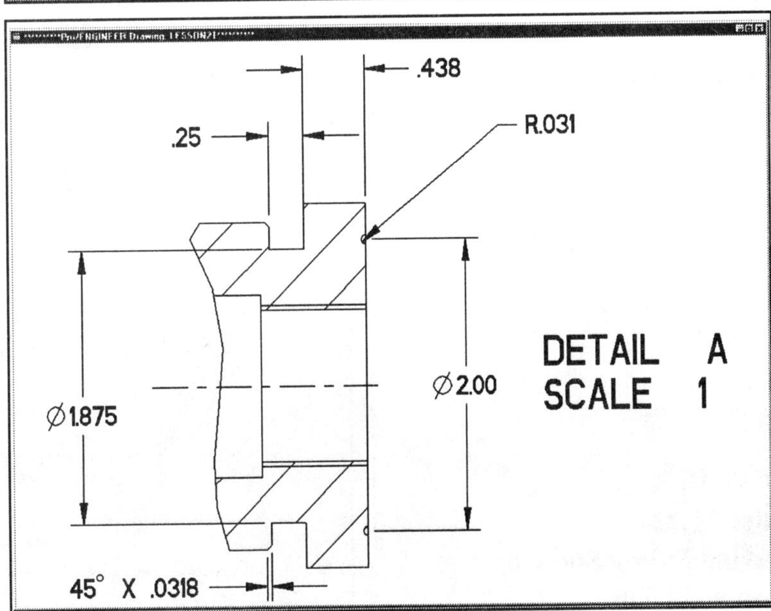

Figure 21.19
Coupling Fitting, **DETAIL A**

60°
1.370
.380
R.050
DETAIL B
SCALE 1

Figure 21.20
Coupling Fitting, **DETAIL B**

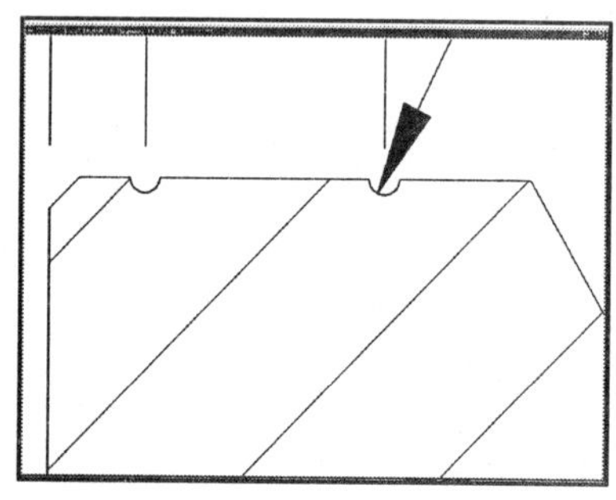

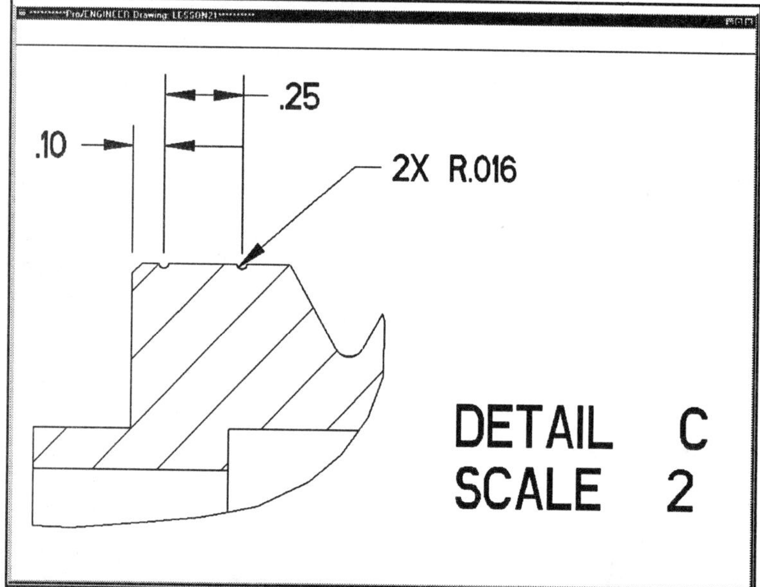

Figure 21.21
Coupling Fitting, **DETAIL C**

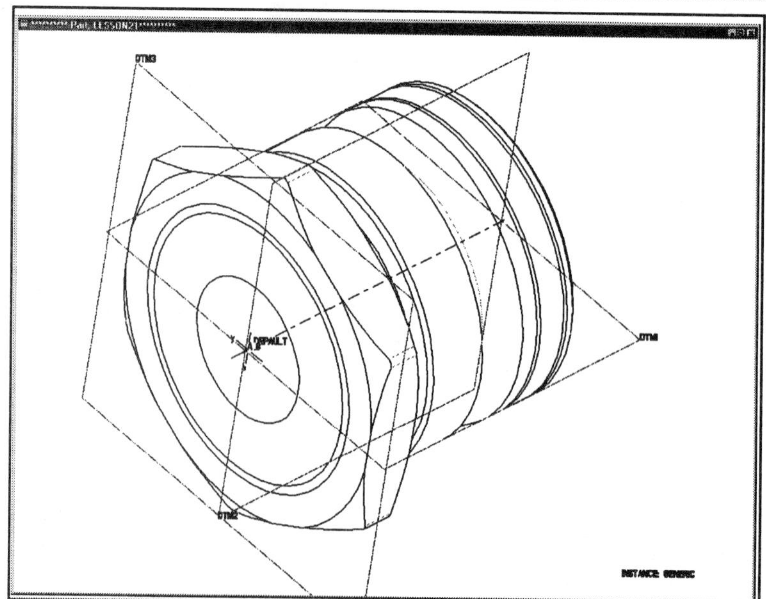

Figure 21.22
Coupling Fitting Showing Hexagonal UDF

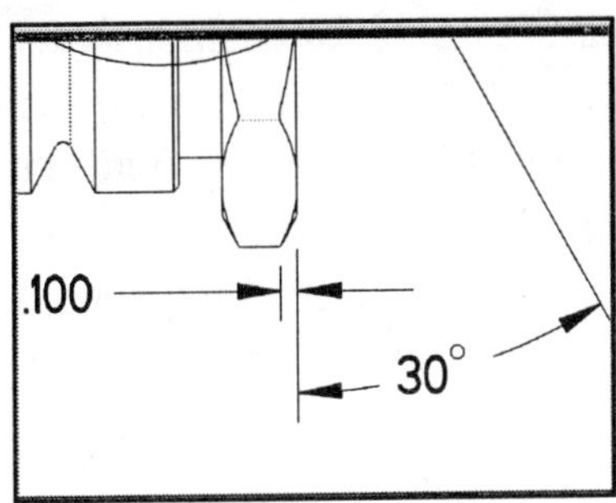

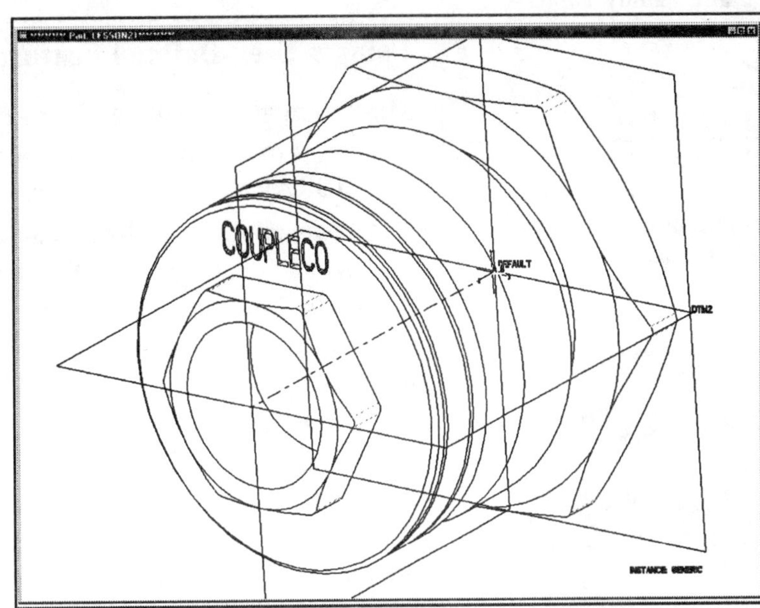

Figure 21.23
Coupling Fitting, Engraved End

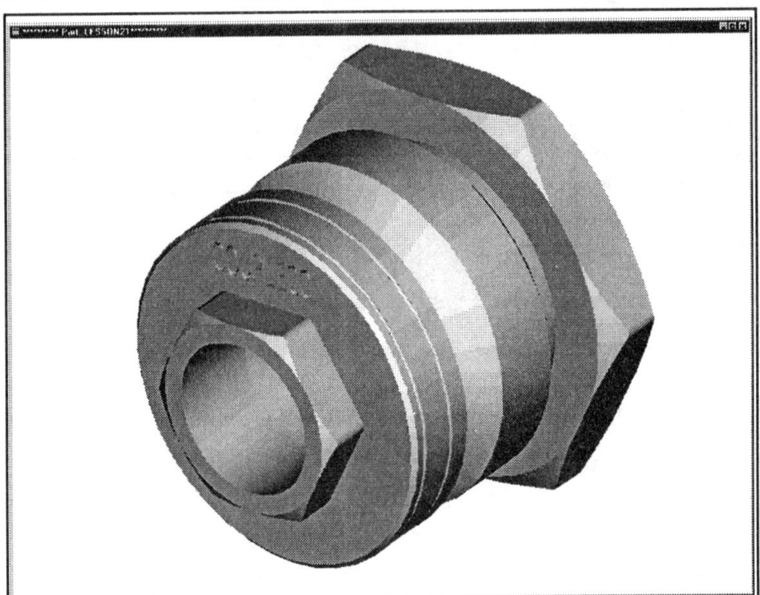

Figure 21.24
Coupling Fitting, Shaded Model (Generic)

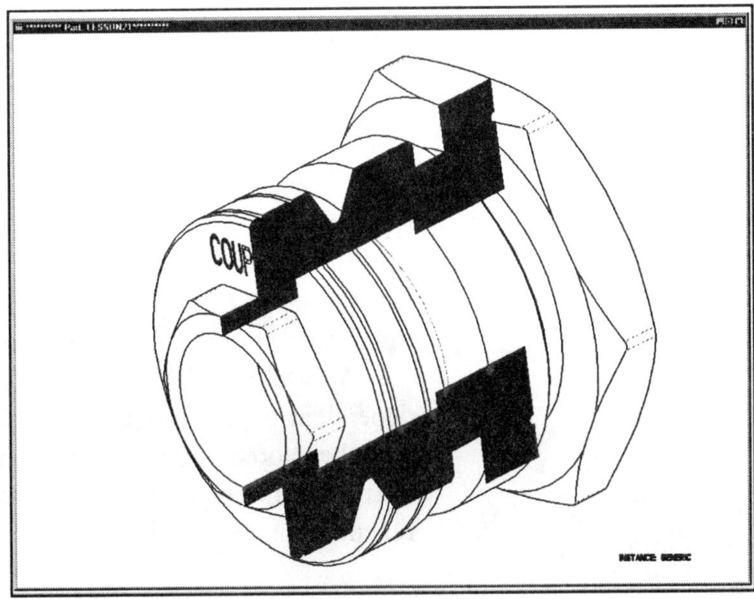

Figure 21.25
Coupling Fitting Showing Full Section

Using a User-Defined Feature

PT/Modeler™

Since PT/Modeler does not support UDFs, create the hexagonal features for this project using extruded protrusions.

Most companies create a library of standard features for each designer to use. Pro/ENGINEER has a wide range of features available in Pro/LIBRARY. You can access these features by choosing **Feature ⇒ Create ⇒ User-Defined ⇒ Search/Retr.**

When you choose a **Feature**, Pro/E asks you if you want to display a ***Reference Part***. If you respond **yes**, Pro/E opens a new window to show you what the UDF looks like. This can be extremely helpful if you have not used a particular UDF before. In general, it is good practice always to use the reference part. Another practice worth mentioning is the use of online documentation to see the UDF library features that are available and how to insert and use the features. Choose the following commands to see the online documentation for UDFs (Fig. 21.26):

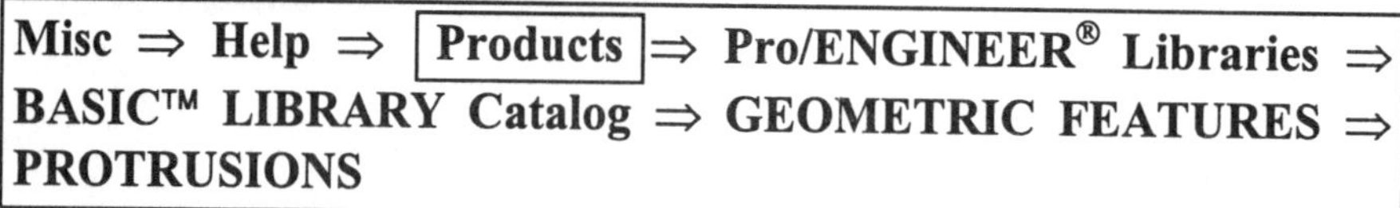
Misc ⇒ Help ⇒ Products ⇒ Pro/ENGINEER® Libraries ⇒ BASIC™ LIBRARY Catalog ⇒ GEOMETRIC FEATURES ⇒ PROTRUSIONS

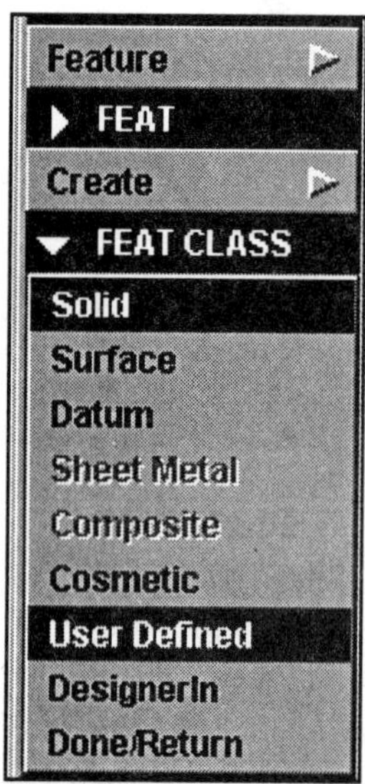

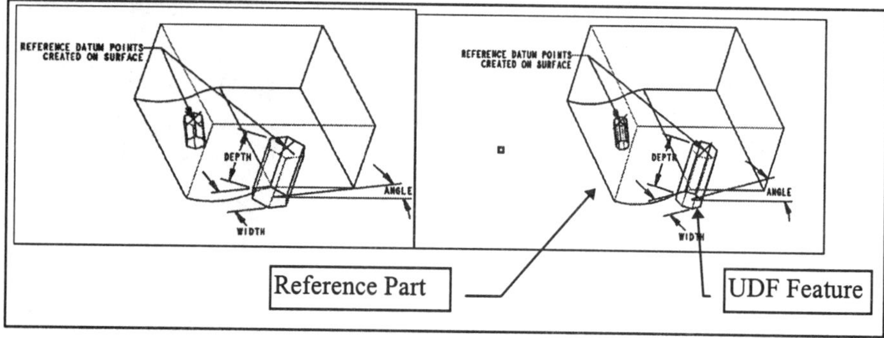

Figure 21.26
Pro/HELP For UDF Features Library Showing Hexagonal and Octagonal Protrusions Available as UDFs

Most UDFs are created on a **base feature** that is not actually part of the UDF. This feature is usually a flat rectangular solid. Even though this feature is displayed in the reference part, *it is not added to your part when you place the feature.*

The UDF is always stored when it is added to a part, and must be placed using the same references used to create the original. You must plan your construction method and references before you create the original model you use to store the UDF. When you place the UDF, Pro/E prompts you through the defining of the references. If the UDF was stored with variable dimensions (e.g., height, width, diameter), Pro/E also prompts you to enter those values.

After all the data are defined, the UDF is placed. If the UDF is a dependent one, a reference is established between the placed object and the stored UDF. Then whenever the original library version changes, the object in your part will also change (at the next regeneration or retrieval of your part).

The Coupling Fitting part is modeled using a default coordinate system and default datum planes. Leave the **Environment** setting on **Trimetric** during the creation of this part (Fig. 21.27). Since the reference part and the UDF are displayed in **Trimetric,** it is easier to answer the prompts for placing and sizing the UDF.

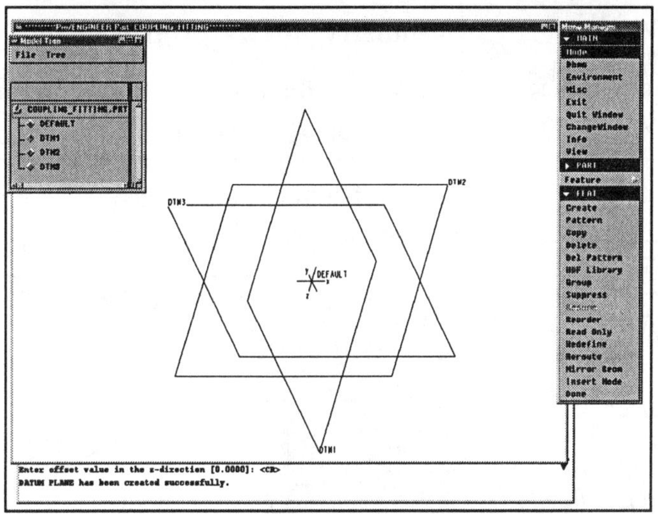

Figure 21.27
Start the Part with Default Coordinate System and Datum Planes in Trimetric

Create a new feature using a Pro/Library **UDF**. The first part feature is a hexagonal protrusion. Online documentation states: "When you use a protrusion in your part, you must have already defined a *datum point* on the surface of the part for the center of the group, and a *reference plane* for the direction." Create a datum point on **DTM3** using the following commands (Fig. 21.28):

COAch™ for Pro/ENGINEER

If your system has **COAch™** or your text comes with a **CADTRAIN** disk, review Super Users, User-Defined Features before completing this lesson.

Feature ⇒ **Create** ⇒ **Datum** ⇒ **Point** ⇒ **On Surface** ⇒ (pick **DTM3)** ⇒ (then pick somewhere *on* **DTM3)** ⇒ **Done Sel** ⇒ (*Pro/E responds:* **Select two placement PLANES/EDGES for the dimensions.** Select **DTM1** and **DTM2)** ⇒ (type the distance from each reference) ⇒ **0** ⇒ **enter** ⇒ **0** ⇒ **enter** ⇒ **Done (PNT0** is located at the intersection of the three datum planes) ⇒ **Done**

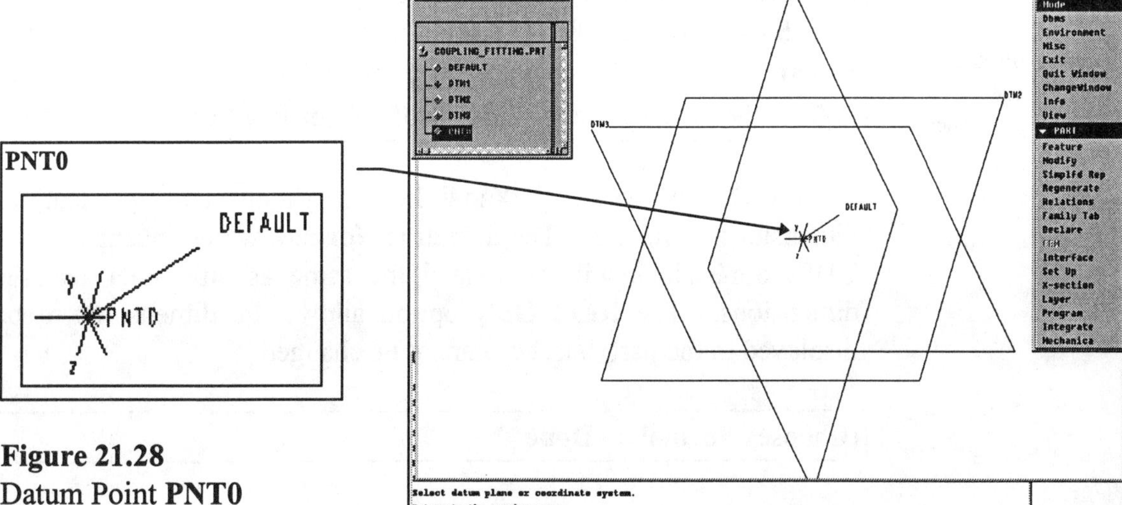

Figure 21.28
Datum Point **PNT0**

The first protrusion is created using a Pro/LIBRARY UDF feature, choose the following commands:

Feature ⇒ **Create** ⇒ **User-Defined** ⇒ **Search/Retr** ⇒ (filter through your directories to **Pro/Library** ⇒ **/objlib** ⇒ **/featurelib** ⇒ **/geometry_udf** ⇒ **/protrusion_udf** ⇒ **ugph)**

(Type **y** as the response to whether or not you want to see a reference part) ⇒ **enter** ⇒ **Done** (to accept the default of **Independent**)

NOTE
Your directory will probably be different. See your system manager for the required path to Pro/LIBRARY UDFs.

Notice that the dimensions that control the placement of the UDF are now displayed in the reference part, as shown in Figure 21.29.

Placement Dimensions

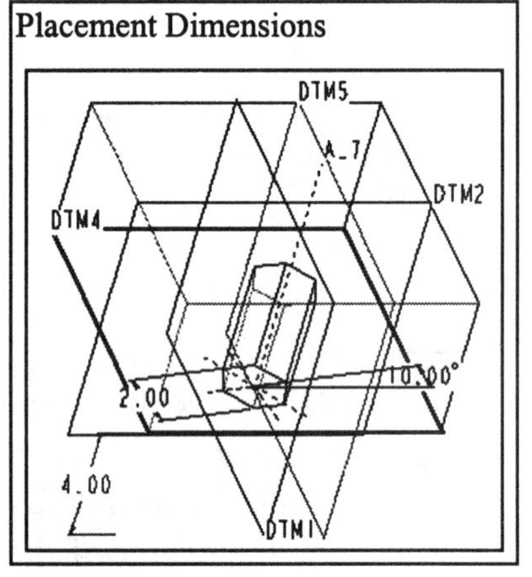

Figure 21.29
UDF with Reference Part Is Displayed with Placement Dimensions

Same Dims ⇒ **Done** (to accept the current scale--the default scale is 1) ⇒ (*Pro/E prompts with:* **Enter depth [4.0000]:** *type* **.438**) (**.438** is the thickness of the larger of the two hexagonal shapes) ⇒ **enter** ⇒ (*Pro/E prompts with:* **Enter angle from reference [10.0000]:** *type* **0**) ⇒ **enter** ⇒ (*Pro/E prompts with:* **Enter width [2.0000]:** *type* **2.50**) (**2.50** is the distance across the flats of the hexagon) ⇒ **enter**

HINT
As each prompt comes up, the dimensions in question are highlighted in the UDF reference part window

Pro/E now wants to know how it should add the features dimensions to the part. The default is **Normal,** which means that the UDFs dimensions will be treated the same as any other features dimensions. The **Read Only** option allows the dimensions to be displayed in the part, but they cannot be changed.

(Choose) **Normal** ⇒ **Done**

Pro/E now begins to display the placement references defined for this UDF and prompts you to select the appropriate surfaces on your part. The reference being asked for is highlighted in the reference part (Fig. 21.30).

(*Pro/E prompts with:* **Select point to locate center of feature.**) Select the datum point **PNT0** ⇒ (*Pro/E prompts with:* **Select plane to specify orientation.**) Select **DTM2,** which corresponds to the highlighted plane in the UDF window.

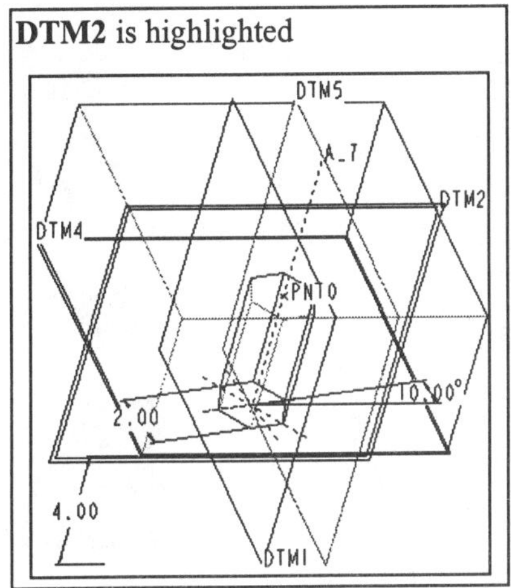

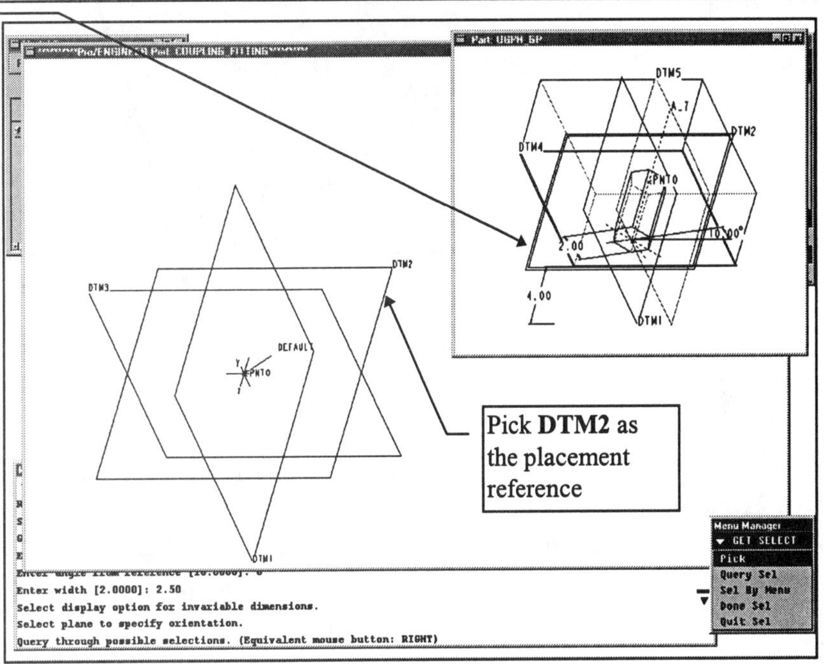

Figure 21.30
Picking the Reference Plane for UDF Orientation

Pro/E now displays an arrow to show you the direction in which it assumes that the positive axis points (Fig. 21.31). This direction affects not only the orientation of the feature, but the direction in which the distances you entered are measured.

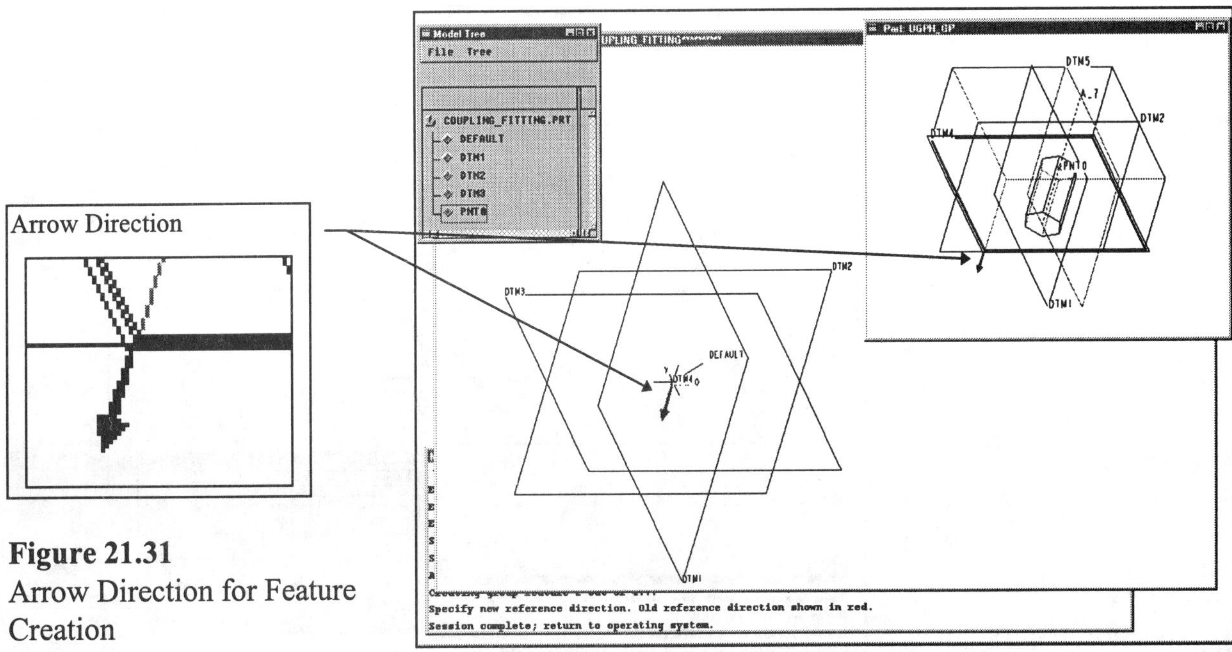

Figure 21.31
Arrow Direction for Feature Creation

Choose **Okay** ⇒ (*Pro/E prompts with:* **Specify new reference direction. Old reference direction shown in red.**) ⇒ **Okay** ⇒ **Done** ⇒ [*Pro/E responds with:* **Group created successfully.** (Fig. 21.31)]

Dbms (**File**--PT/Modeler) ⇒
Save ⇒ **enter**
Purge ⇒ **enter** ⇒
Done-Return

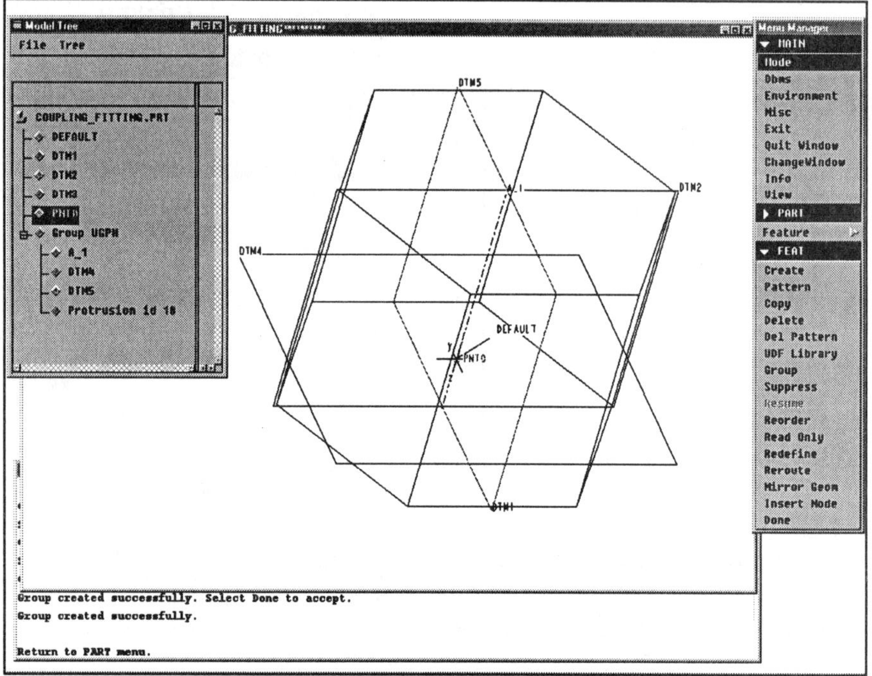

Figure 21.32
Group Created Successfully

To complete the feature, **Redefine** the UDF and change the depth to **Blind** (Fig. 21.33):

NOTE

When creating the **Tool Body** in the **Lesson Project**, use the following commands instead of what is shown to the right:

Redefine ⇒ (pick the UDF) ⇒ (Select **Depth** from the dialog box) ⇒ **Define** ⇒ **UpTo Surface** (pick **DTM4**) ⇒ **Done** ⇒ **OK**

Redefine ⇒ (pick the UDF from the screen) ⇒ (Select **Depth** from the dialog box) ⇒ **Define** ⇒ **Blind** ⇒ **Done** ⇒ (*Pro/E prompts with:* **Enter depth [1.4337]:**) *type* **.438** ⇒ **enter** ⇒ **OK** (Fig. 21.34)

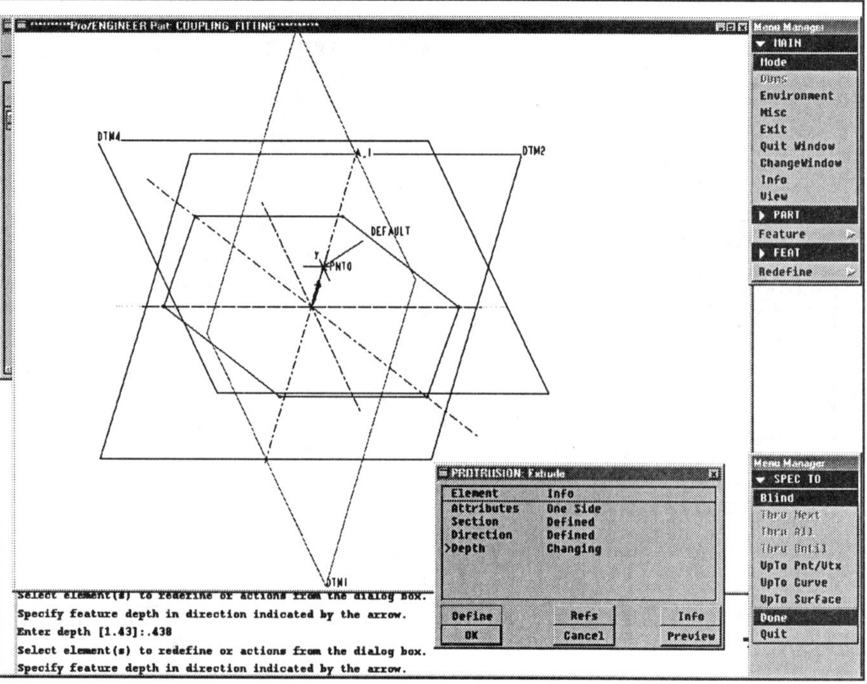

Figure 21.33
Redefining the UDF Depth

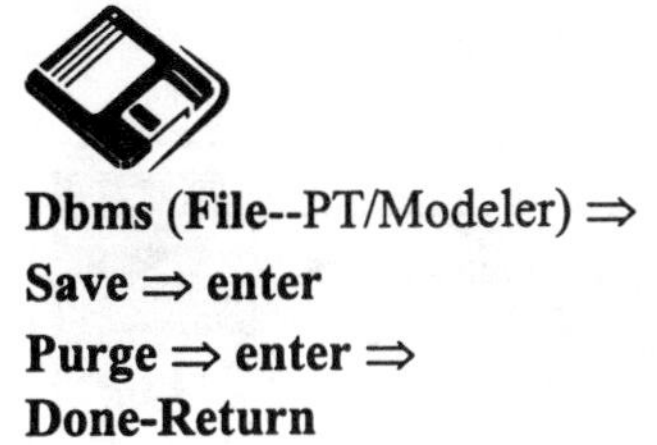

Dbms (File--PT/Modeler) ⇒
Save ⇒ enter
Purge ⇒ enter ⇒
Done-Return

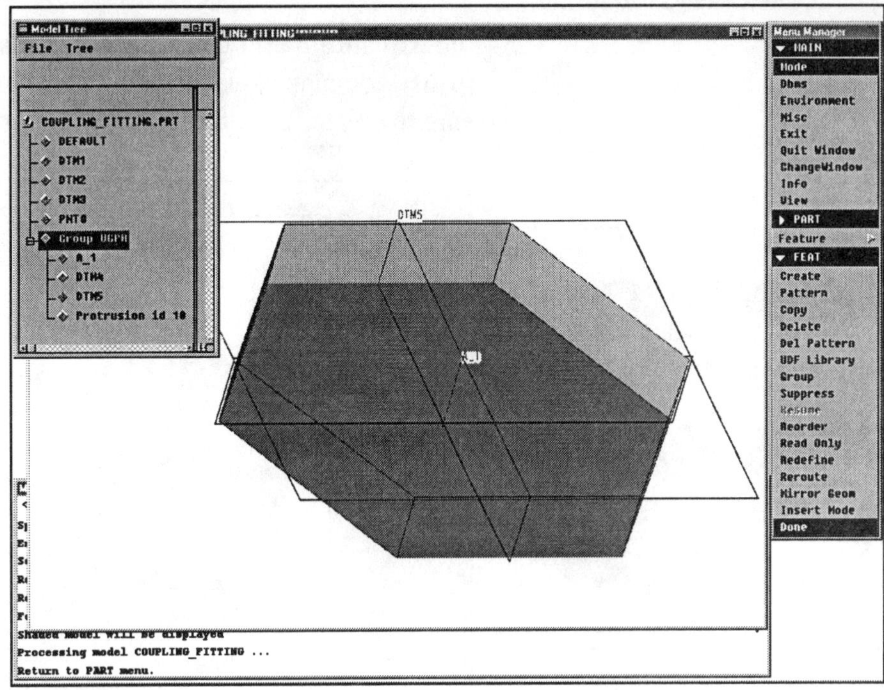

Figure 21.34
Completed UDF Feature

Seems like a lot of effort just to get a pre-made hexagonal protrusion. Actually, creating a hexagonal protrusion from scratch is even more effort! Remember that in general, UDFs can be created for almost any part or feature that is used frequently. This can be a great time saver in industry.

The next feature will be a revolved protrusion representing the body of the fitting (Fig. 21.35). You may include the **V** cut in this feature or add it later as a neck feature. It may be better design intent to make each cut separately so as to allow changes to the features individually and when suppressing in a Family Table. Sketch the revolved features on **DTM1**.

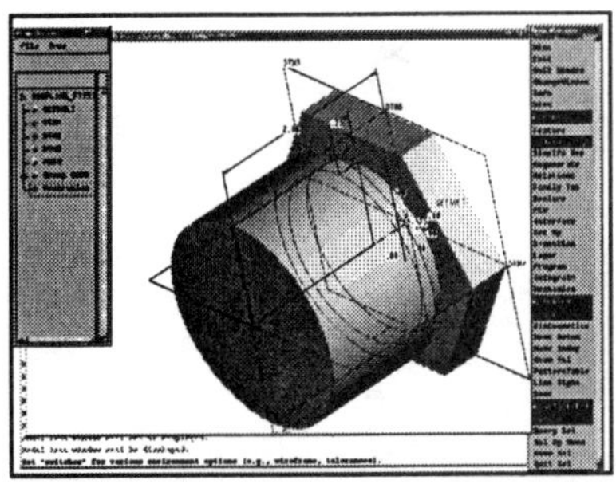

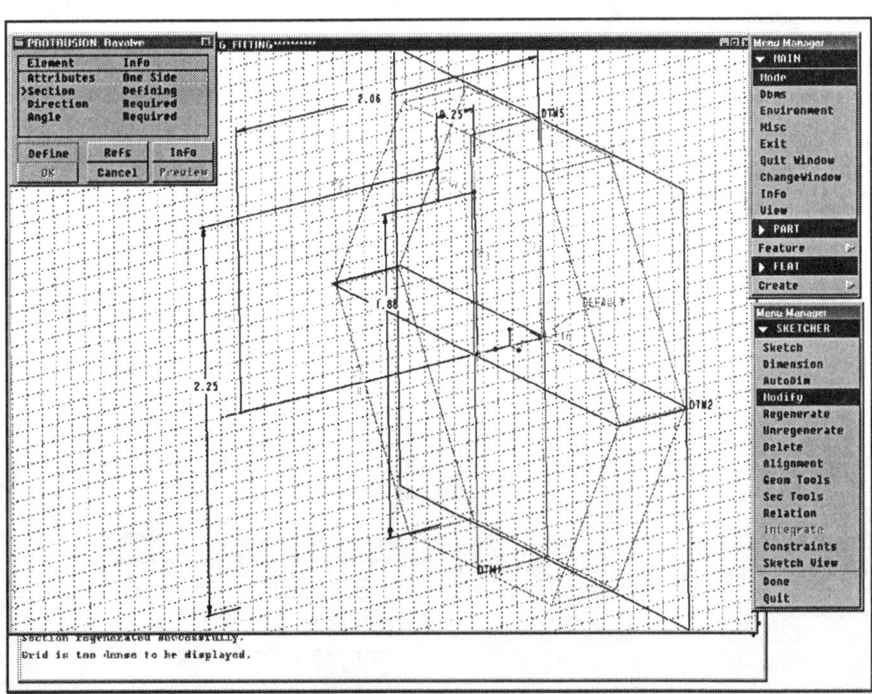

Figure 21.35
Section Sketch for Revolved Protrusion

Create a datum point on the planar end of the revolved protrusion to use for placement of a second hexagonal UDF (Fig. 21.36). Be sure to create the datum point as described previously.

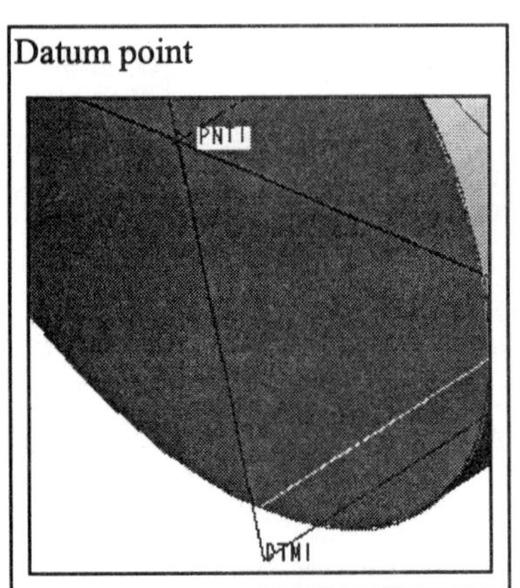

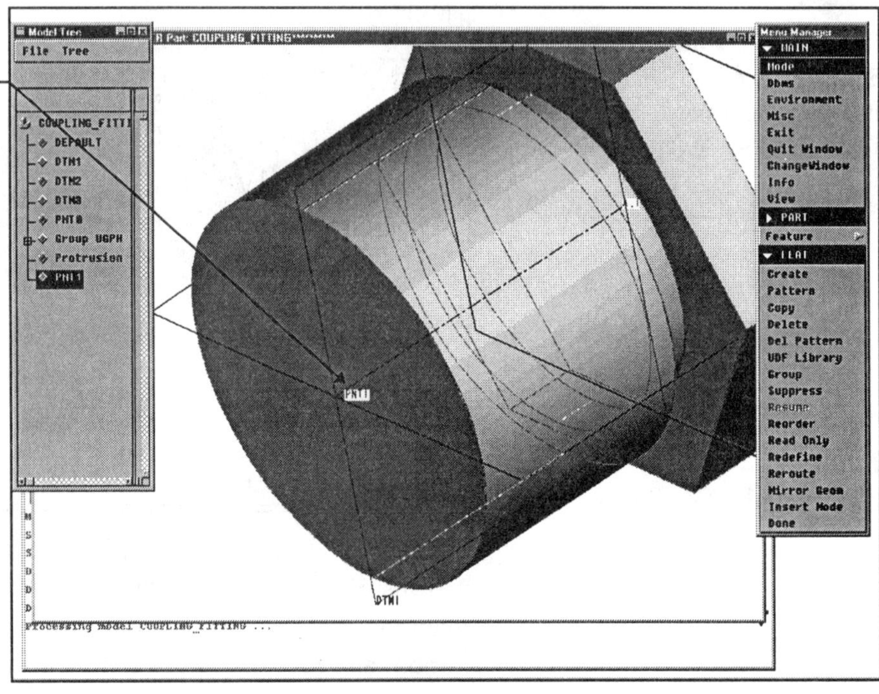

Figure 21.36
Datum Point Created on End of Revolved Protrusion

Use the same steps that were required to complete the first UDF. You should not need to redefine the UDF, since its depth of **.313** will establish the proper size of the feature (Fig. 21.37).

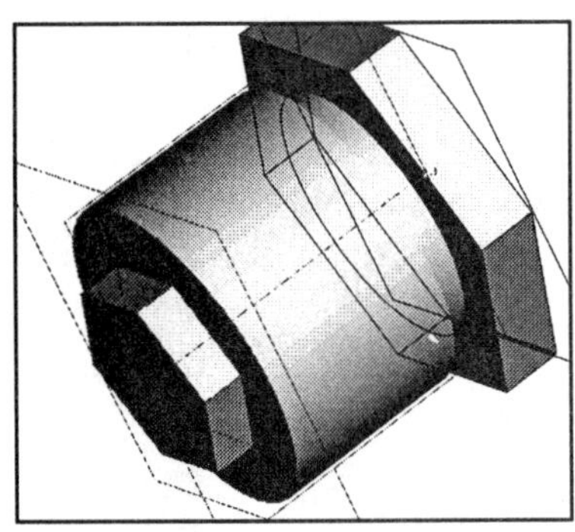

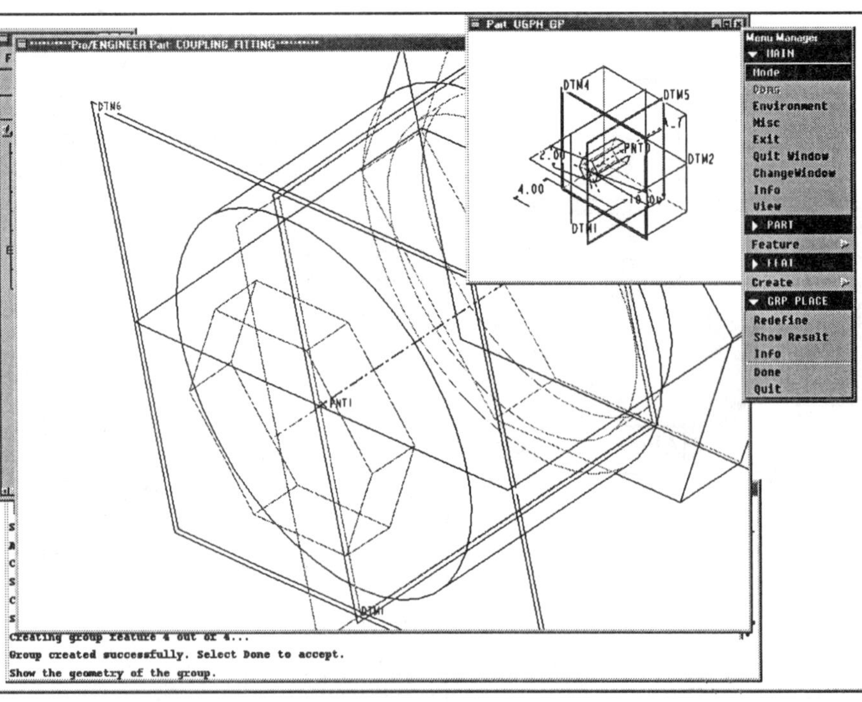

Figure 21.37
Second UDF Added

Complete the external features using the dimensions provided at the beginning of this lesson (Fig. 21.38). Add the internal cuts last.

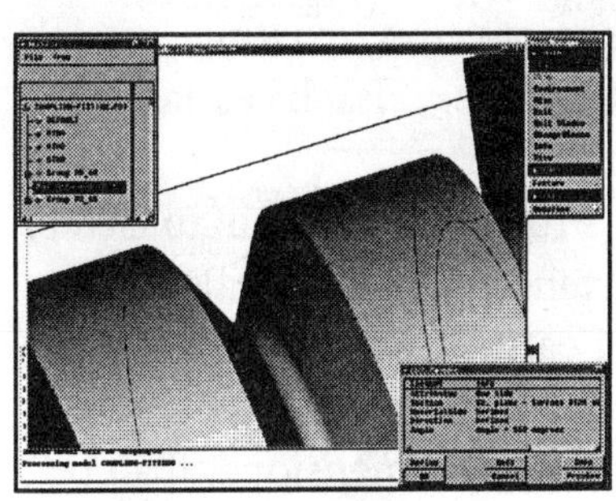

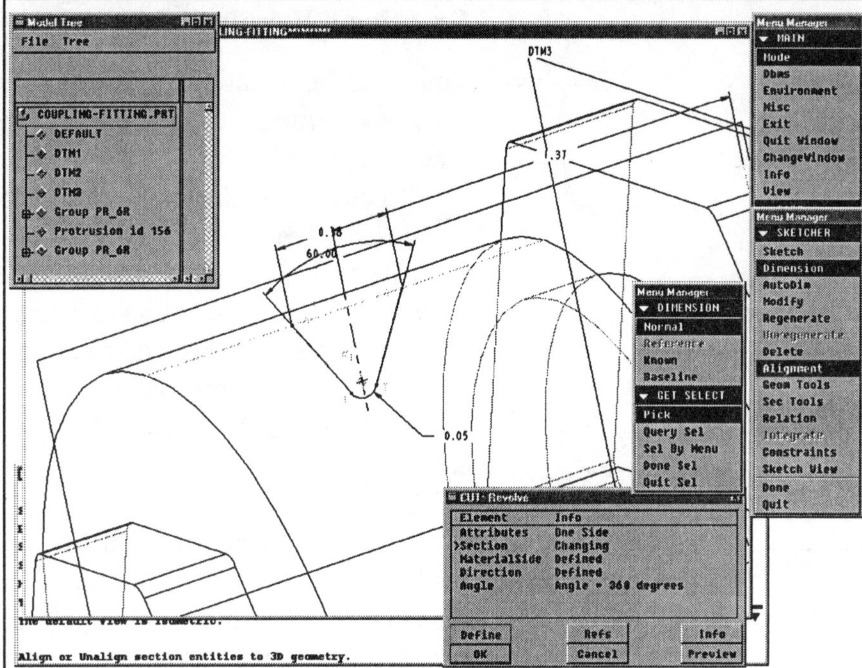

Figure 21.38
V Cut

The internal feature should be a revolved cut (Fig. 21.39) with both ends of the feature aligned to the outside of the part. If you align the *revolved cut* in this manner, the cut will remain *through* the part regardless of the changes different instances will have when a Family Table is created (Fig. 21.39). Do not use a *sketched hole,* since it would require a *relation* to establish the hole as *through* the part at all times.

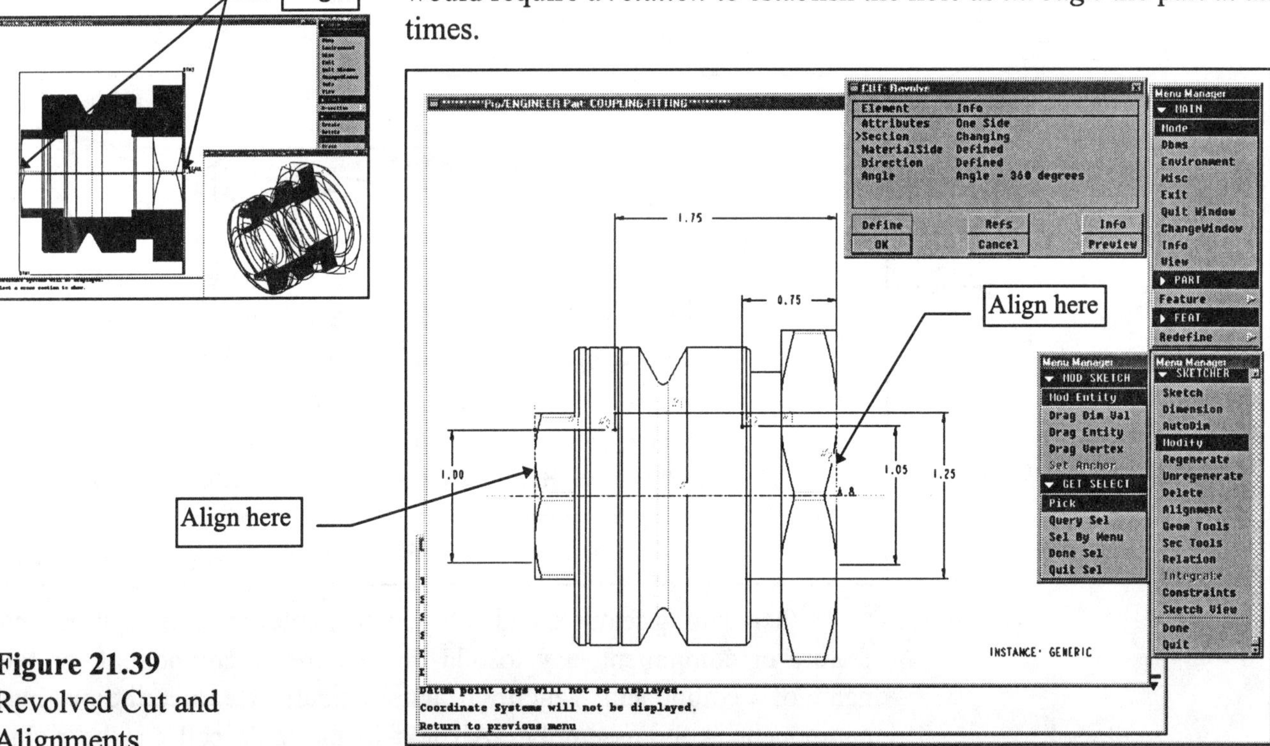

Figure 21.39
Revolved Cut and Alignments

Complete the generic part by adding cosmetic threads and cutting the **COUPLECO** company name into the part as shown in the figures provided.

Creating a Family Table

Dbms (**File**--PT/Modeler) ⇒
Save ⇒ **enter**
Purge ⇒ **enter** ⇒
Done-Return

You will be creating a Family Table from the existing model, Coupling Fitting. The model is the **Generic**. Each variation is referred to as an **Instance**.

When you create a Family Table, Pro/E allows you to *select dimensions,* which can vary between instances. You can also *select features* to add to the Family Table. Features can vary by being suppressed or resumed in an instance.

When you are finished selecting items (e.g., dimensions, features, parameters), the Family Table is automatically generated.

Edit the Family Table. You add instances by typing a new name and entering the values for the variables that belong to the Family Table.

COAch™ for Pro/ENGINEER

If your system has **COAch™** or your text comes with a **CADTRAIN** disk, review Super Users, Family Table

When you **Exit** the Family Table, Pro/E evaluates the Family Table entries and creates instances for each new name. The values in the table for each instance name are used to vary the dimension and parameter names that head each column in the Family Table.

Any line that starts with an exclamation mark (!) is treated as a comment line and cannot be altered (Fig. 21.40). Additional comments can be added if preceded by an @, but they will automatically be repositioned at the top of the table, before the Generic information.

Geometric tolerances, pattern dimensions, and dimensional tolerance values can be added as dimensions.

An * in the table will use the default value of the generic.

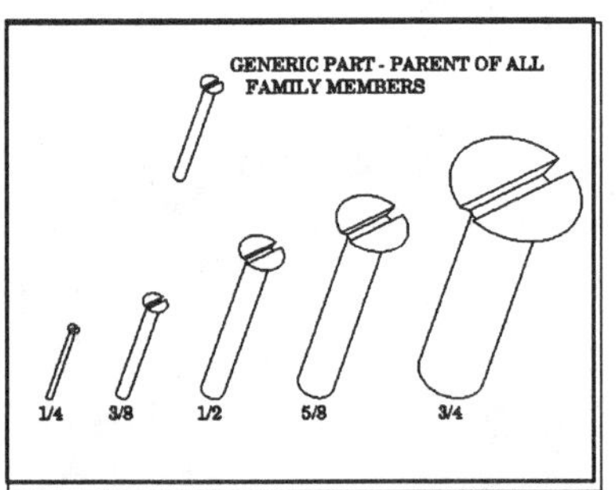

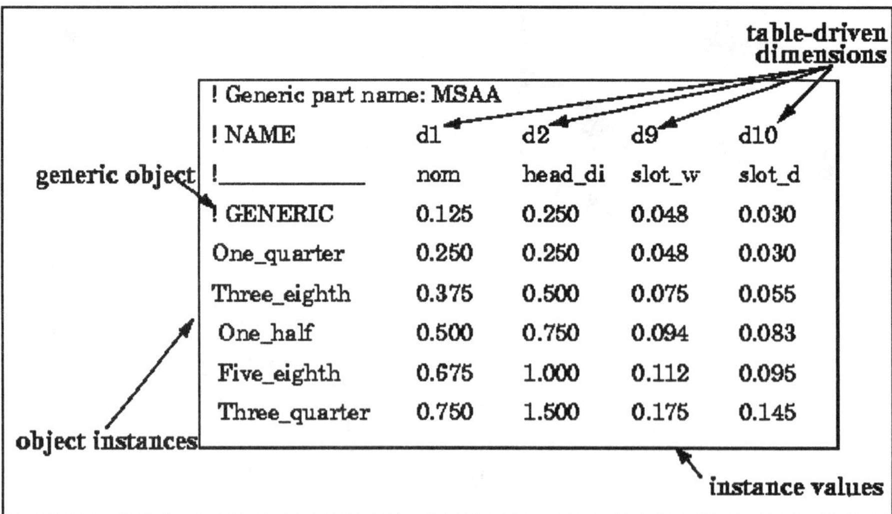

! Generic part name: MSAA				
! NAME	d1	d2	d9	d10
!__________	nom	head_di	slot_w	slot_d
! GENERIC	0.125	0.250	0.048	0.030
One_quarter	0.250	0.250	0.048	0.030
Three_eighth	0.375	0.500	0.075	0.055
One_half	0.500	0.750	0.094	0.083
Five_eighth	0.675	1.000	0.112	0.095
Three_quarter	0.750	1.500	0.175	0.145

Figure 21.40
Family Table

When adding features and components, enter an **n** to suppress the feature or component, a **y** to add the feature or component, or the name of a component, which will automatically **Replace** the original component in the assembly. All text in the table cell can be right-justified by preceding the text with a double quotation mark (").

For example, if you wanted to remove a bolt, the line would look like this:

INSTANCE BOLT
124828 "N

If you wanted a different bolt in the assembly, the line would look like this:

INSTANCE BOLT
124829 "Hex-Bolt-2163

Each instance must have a unique name. Consider each instance a separate model, as in the Family Table shown in Figure 21.41, where the bolts are named **BOLT1** and **BOLT2**.

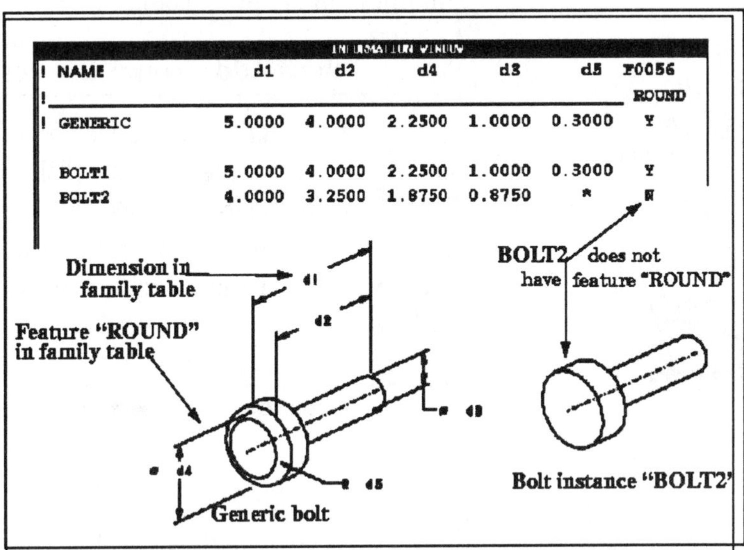

! NAME	d1	d2	d4	d3	d5	F0056 ROUND
! GENERIC	5.0000	4.0000	2.2500	1.0000	0.3000	Y
BOLT1	5.0000	4.0000	2.2500	1.0000	0.3000	Y
BOLT2	4.0000	3.2500	1.8750	0.8750	*	N

Figure 21.41
Family Table for a Bolt from Pro/HELP

If the variation can be described by a pattern, the **Patternize** option can automatically proliferate the table. Tables can have multiple levels, as shown in Figure 21.42 (taken from Pro/HELP).

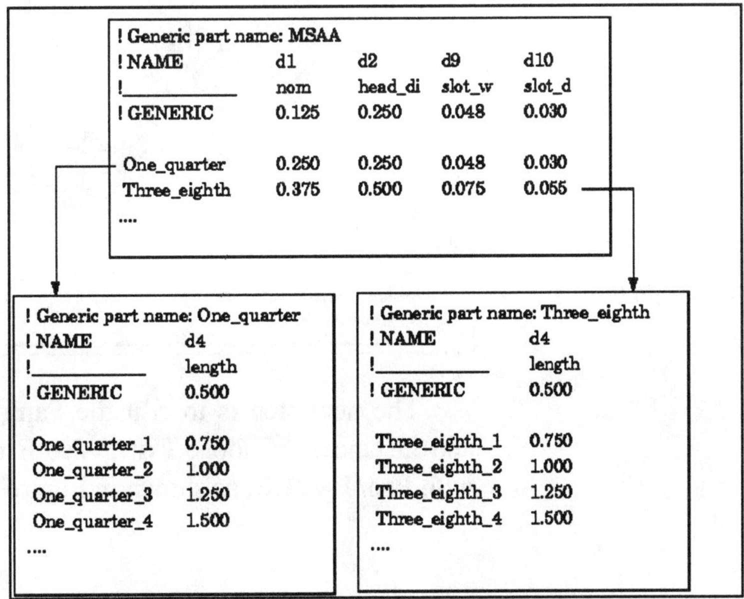

! Generic part name: MSAA

! NAME	d1	d2	d9	d10
!	nom	head_di	slot_w	slot_d
! GENERIC	0.125	0.250	0.048	0.030
One_quarter	0.250	0.250	0.048	0.030
Three_eighth	0.375	0.500	0.075	0.055
....				

! Generic part name: One_quarter

! NAME	d4
!	length
! GENERIC	0.500
One_quarter_1	0.750
One_quarter_2	1.000
One_quarter_3	1.250
One_quarter_4	1.500
....	

! Generic part name: Three_eighth

! NAME	d4
!	length
! GENERIC	0.500
Three_eighth_1	0.750
Three_eighth_2	1.000
Three_eighth_3	1.250
Three_eighth_4	1.500
....	

Figure 21.42
Family Table with Multiple Levels

When a generic is stored, a file called *directoryname*.idx is created. This file is automatically recreated and is stored with the generic, so the **.idx** file can be deleted. It is used in retrieving instances from disk directly. If that functionality is not required, the file can be deleted.

When the Family Table is edited, a file called *modelname*.ptd is created. Unless users want to edit the table outside of Pro/ENGINEER, the **.ptd** file can be deleted.

You can "lock" an instance by using the **Lock** option to prevent it from being modified.

Create three instances of the Coupling Fitting. The instances will have varying dimensions for the length of the body and the length of the small hexagonal feature. As you select a dimension, Pro/E adds a column in the Family Table. Use the following commands:

▶ PART
Family Tab
▼ FAMILY TABL
Add Item
Delete Item
Set Generic
Group Table
Edit
Show
Set Type
Instance
Lock/Unlock
Patternize
Erase Table
Show Feat
Verify
Done/Return
▼ ITEM TYPE
Dimension

Family Tab ⇒ Add Item ⇒ Dimension

Select the large cylindrical feature to display the dimensions as shown in Figure 21.43. Next, pick on the small hexagonal feature to display its dimensions. This will keep the display uncluttered. Pick on the **2.0625** dimension first and then the **.313** dimension. Choose **Done Sel** to end the selection of varying dimensions on the Family Table.

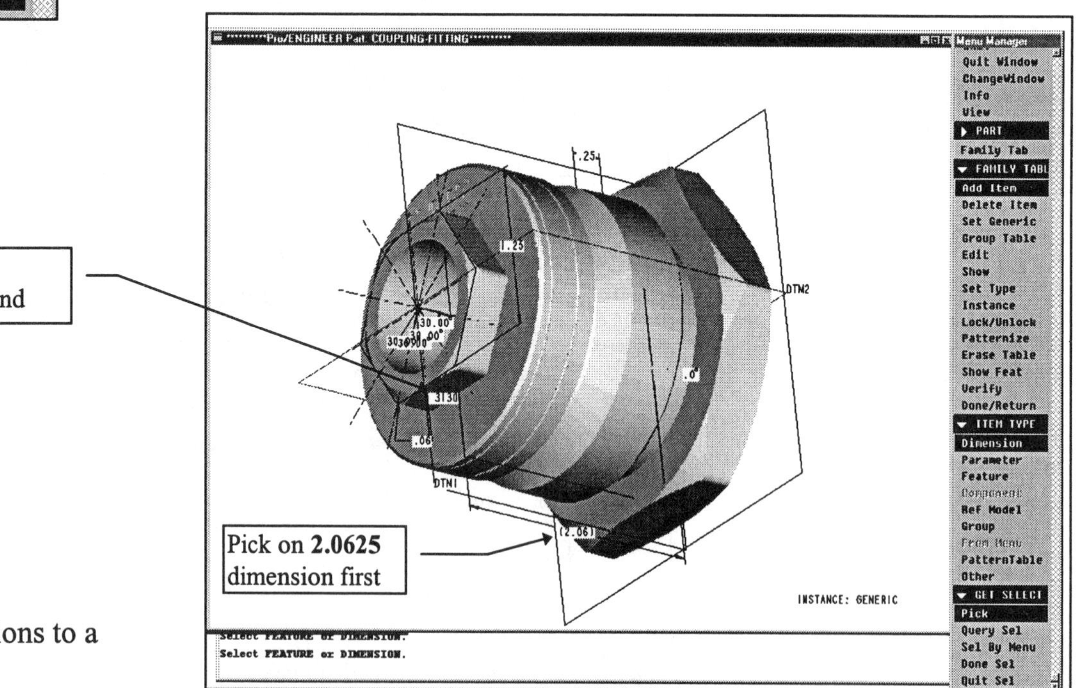

Figure 21.43
Adding Dimensions to a Family Table

The next step is to edit the Family Table to see the entries and add instances. Choose **Edit**. Pro/E now displays the Family Table with **Pro/TABLE,** as shown in Figure 21.44.

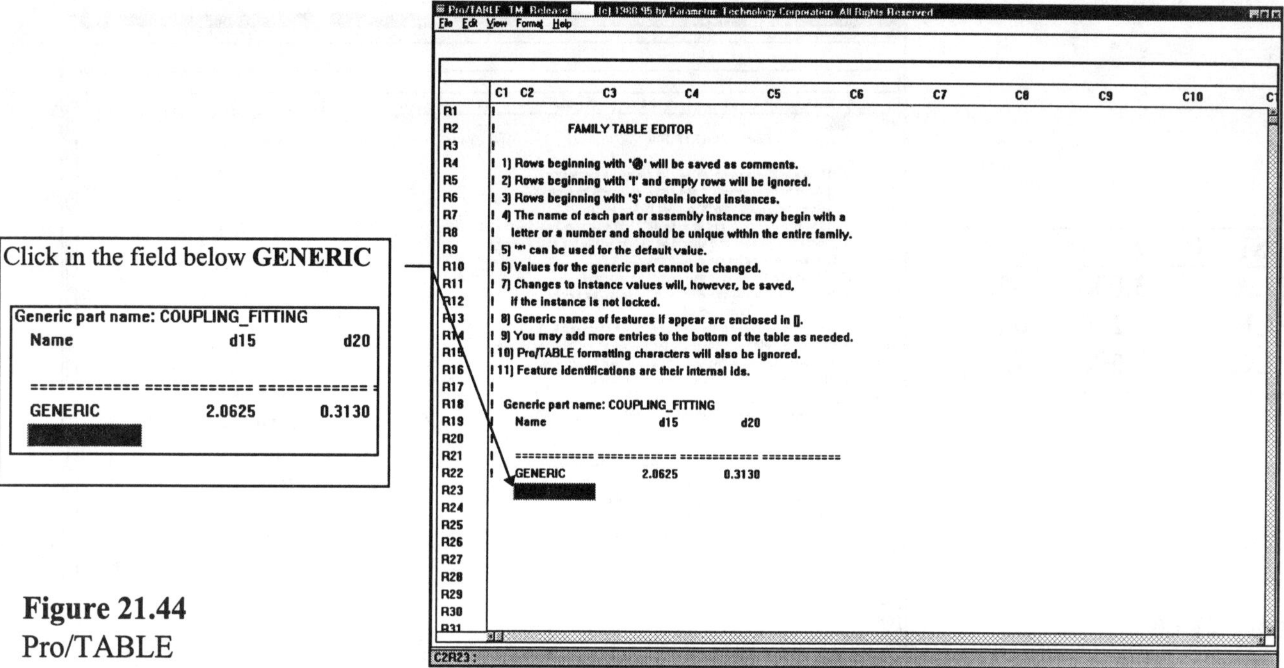

Figure 21.44
Pro/TABLE

Click in the field just below the text **GENERIC** (Fig. 21.44). Type **CPLA** as the name of the first instance. Click in the field below **CPLA** and type **CPLB** as the second instance. Add one more instance as shown in Figure 21.45.

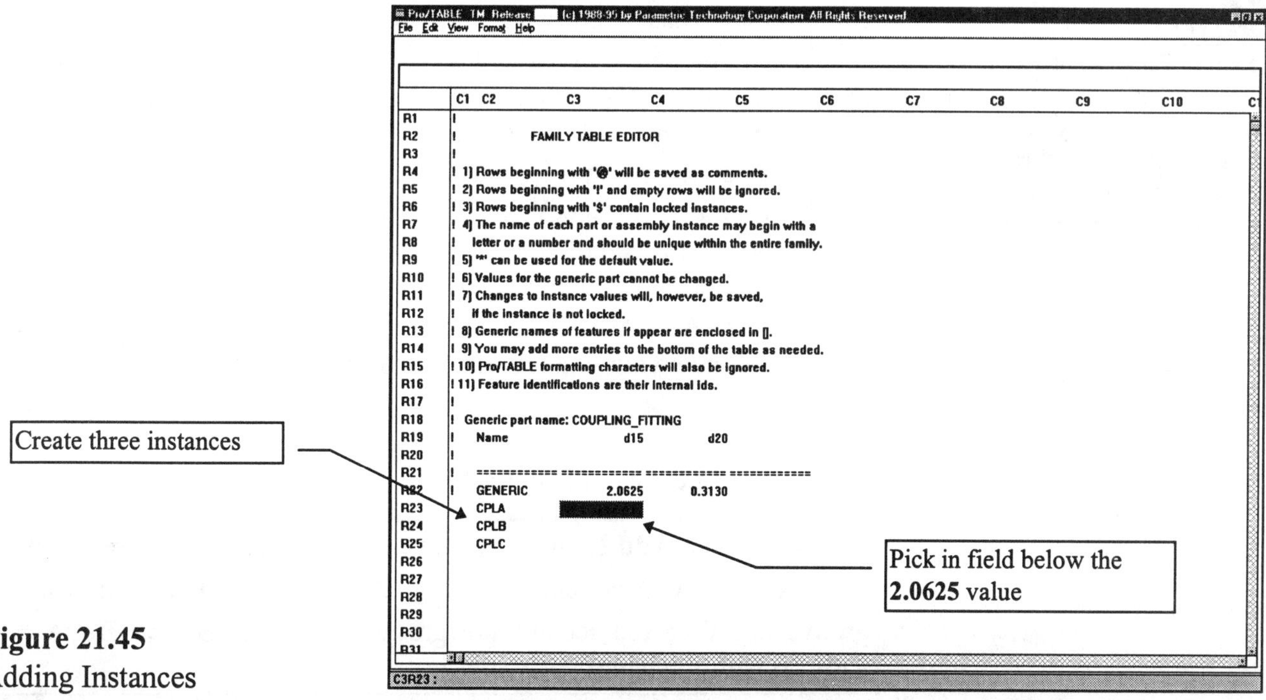

Figure 21.45
Adding Instances

To change the value for the first instance, click in the field just below the text **2.0625** (Fig. 21.45). Type **3.00** as the width of this instance. Click in the field just below the **.313** value and type **.500**. This completes the dimensional values for instance **CPLA**. Continue filling in the table as shown in Figure 21.46. Exit **Pro/TABLE** when you are finished.

GENERIC	2.0625	.313
CPLA	3.000	.500
CPLB	3.250	.625
CPLC	3.500	.750

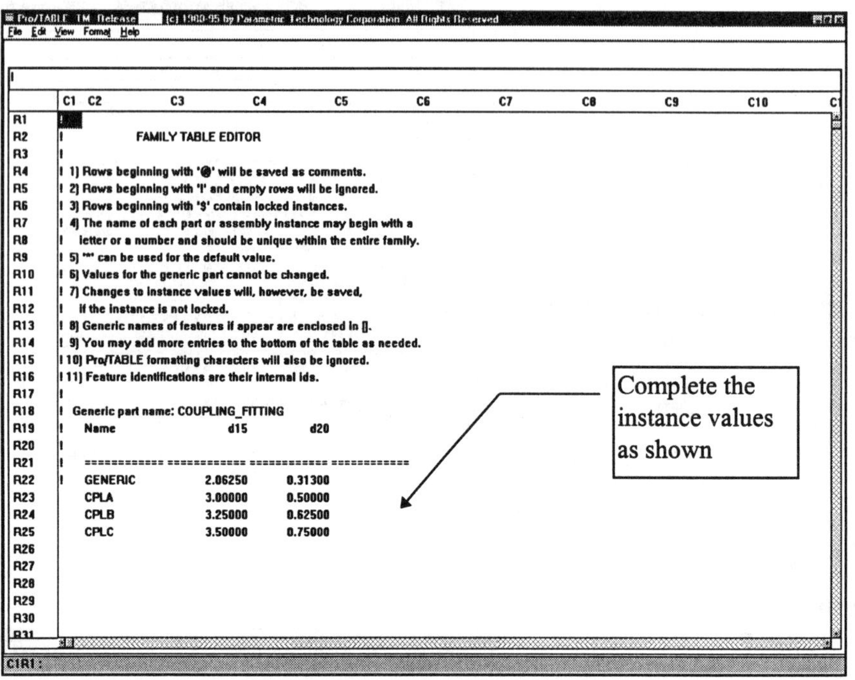

Figure 21.46
Changing Values

Features in Family Tables

You can also place *features* in Family Tables to control their presence in the design instance. Features in a table can be either present or suppressed. This allows you to create multiple versions of a design from the same basic part (using the same generic model).

When you add features to a table, they are also listed in columns to the right of the **Generic** name. As with dimensions, each feature has its own column. If the feature is displayed (and they all can be in the **Generic**), the column for that feature contains a **Y** for Yes.

NOTE
A Family Table is not about whether a feature is displayed, as with the layer display function, rather, it controls whether a feature is present (shown) or not for a given design instance.

You create instances in the same manner as you did for dimensions. In each column for the instance, you must enter a **Y**, an **N**, or an asterisk (*). The **N** causes Pro/E to suppress that feature in the instance. An asterisk (*) causes Pro/E to use the **Generic** value or feature. Since your choice is normally a **Y**, the feature will be displayed.

Add three different features to the Family Table:

Family Tab ⇒ **Add Item** ⇒ **Feature** ⇒ (pick the cut that created the text **COUPLECO** as shown in Fig. 21.47) ⇒ (pick the first slot-groove and then the second slot-groove) ⇒ **Done Sel** ⇒ **Done** ⇒ **Edit** (to **Edit** the table)

Since **Y** is the entry in all the columns in the **GENERIC**, it is also the default value for any other instance in the Family Table. Therefore, you could enter an asterisk (*) instead of a **Y** if you wanted. An implication of this would be that if anyone replaced the **GENERIC** later on, the asterisk would cause that change to be reflected in any instances with asterisks.

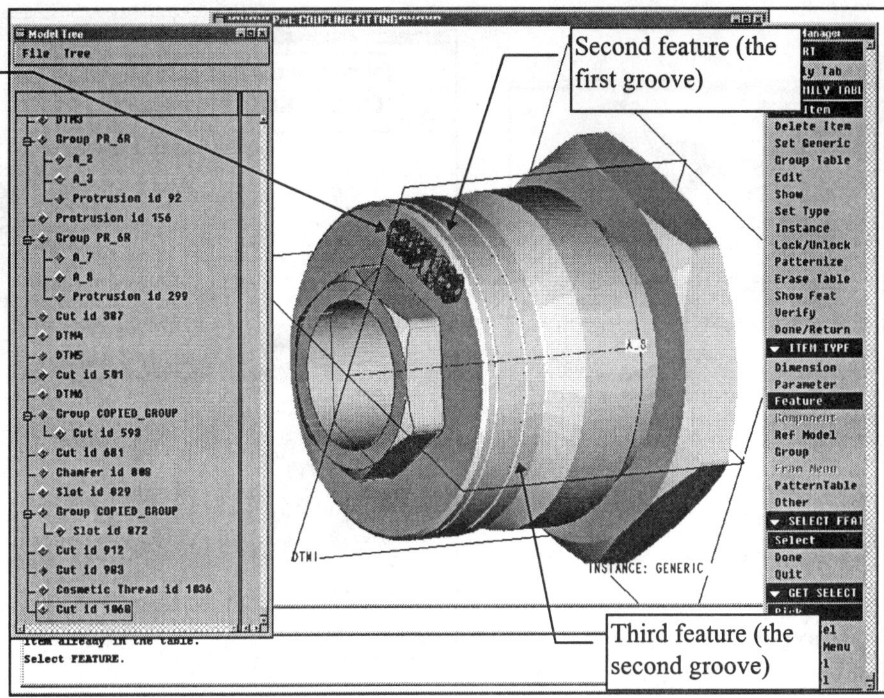

Figure 21.47
Selecting Features for Family Table

Complete the Family Table as shown in Figure 21.48. Pick in the field below each feature listing and type a **Y** or an **N** depending on the required value.

GENERIC	CUT	SLOT	SLOT
CPLA	**N**	**N**	**N**
CPLB	**N**	**Y**	**N**
CPLC	**Y**	**N**	**Y**

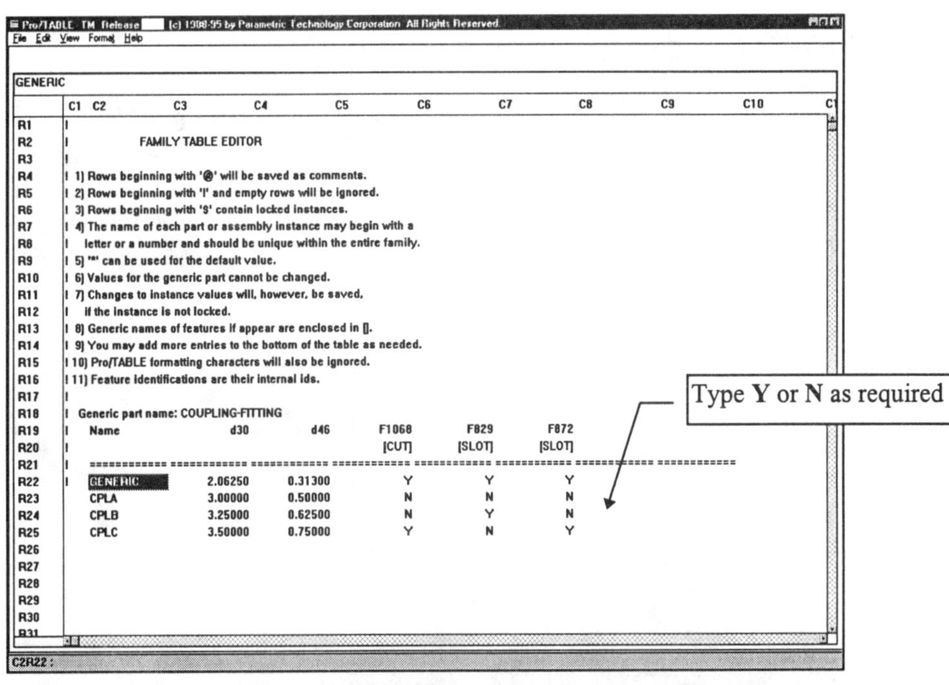

Figure 21.48
Editing the Features of Each Instance

Exit **Pro/TABLE**. Retrieve the **CPLA** instance you defined in the Family Table. Notice that only the features that have an entry of **Y** appear on the instance (Fig. 21.49).

Instance ⇒ CPLA ⇒ Change Window (pick in GENERIC part window) **⇒ Family Tab ⇒ Instance ⇒ CPLB** (Fig. 21.50).

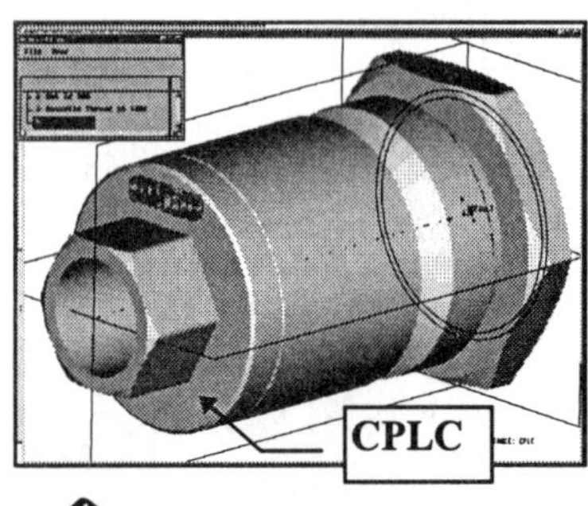

Dbms (**File**--PT/Modeler) ⇒
Save ⇒ **enter**
Purge ⇒ **enter** ⇒
Done-Return

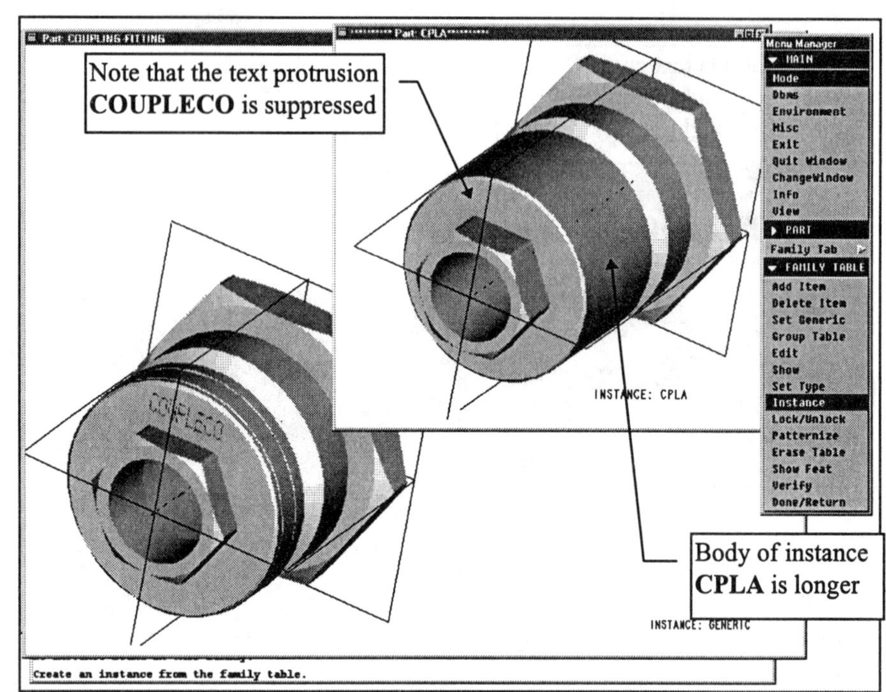

Figure 21.49
Displaying Instance **CPLA**

You now have three *instances* and a *generic* of the Coupling Fitting. When you open this part in **Part** mode, or add it to an assembly or drawing, you will be prompted to select the **Generic** or select from the **INSTANCES** list.

Figure 21.50 Displaying Instance **CPLB**

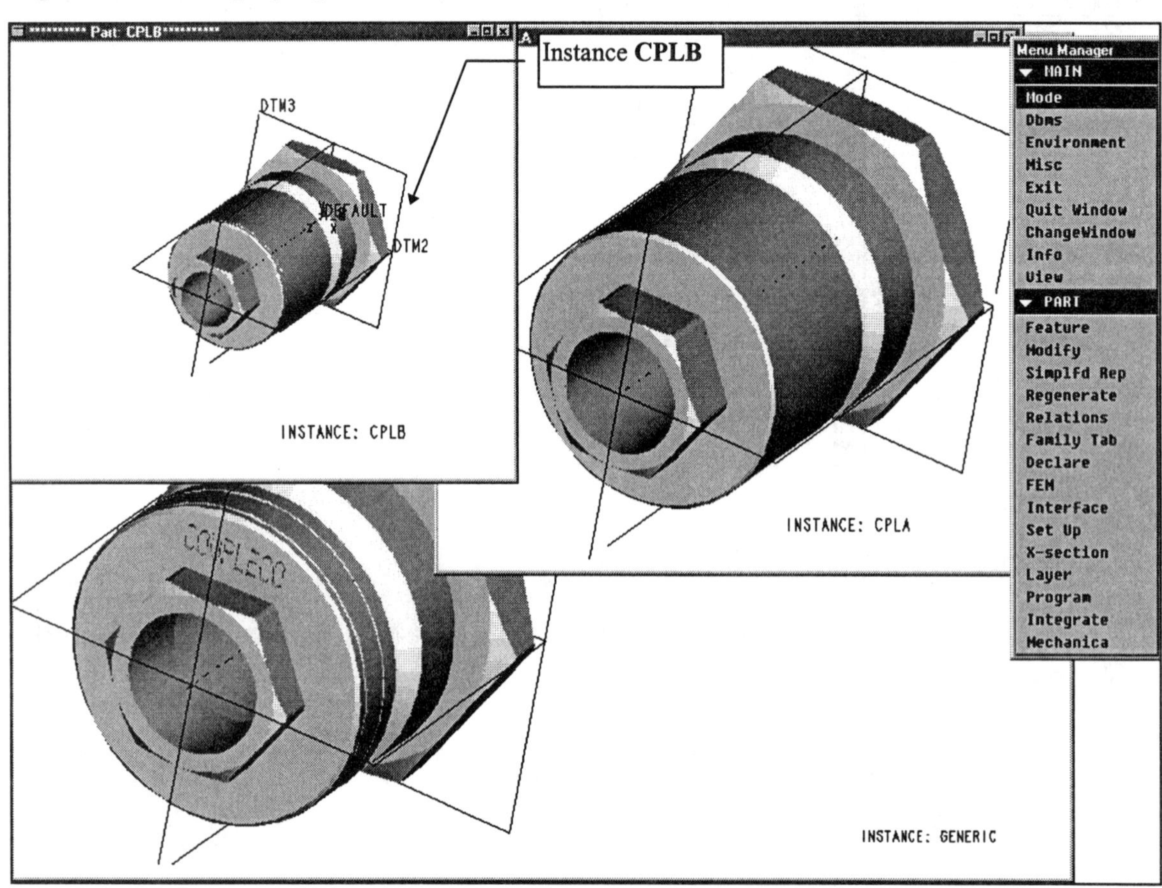

Lesson 21 Project

Tool Body

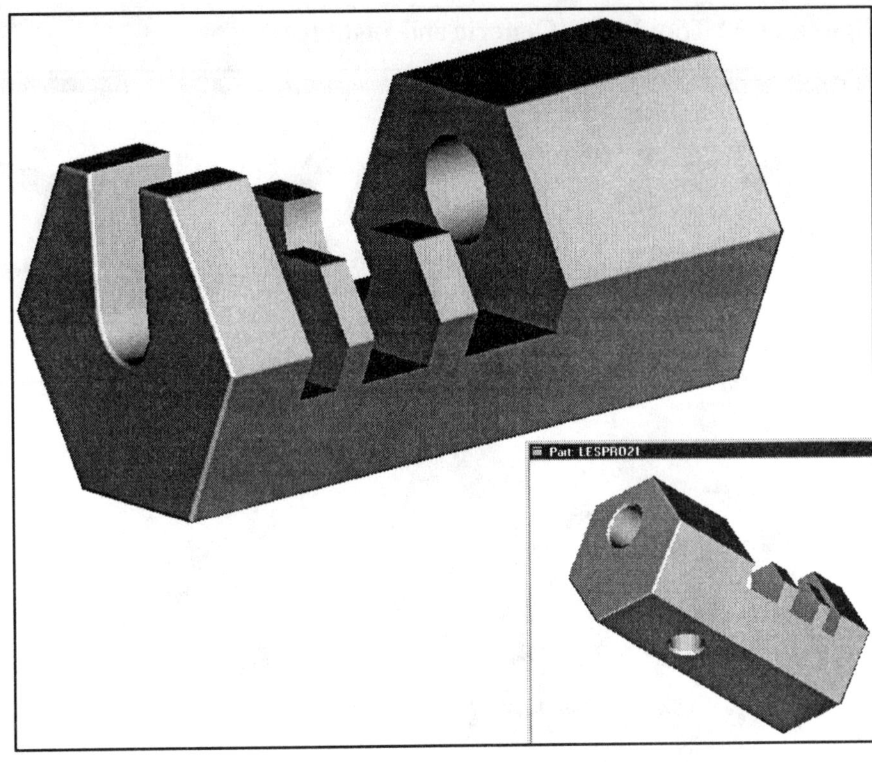

Figure 21.51
Tool Body **(Generic)**

Tool Body

EGD REFERENCE
Engineering Graphics and Design with Graphical Analysis *or* **Fundamentals of Engineering Graphics and Design**
by L. Lamit and K. Kitto
Read Chapter 21
See page 864

This advanced **lesson project** is a **Tool Body** which uses commands similar to those for creating the **Coupling Fitting**. Create the part shown in Figures 21.51 through 21.55. Use the **DIPS** in Appendix D to plan out the feature creation sequence and the parent-child relationships for the part. The Tool Body is made of *steel.*

Use a hexagonal UDF from your library for the first protrusion. If you do not have Pro/LIBRARY installed on your system, create a hexagonal UDF and save it in your own library. Create a Family Table with three instances. Add the following dimensions and features to the Family Table:

NOTE
When creating the **Tool Body** in the **Lesson Project**, use the following when redefining the UDF:

Redefine ⇒ (pick the UDF) ⇒ (Select **Depth** from the dialog box) ⇒ **Define** ⇒ **UpTo Surface** (pick **DTM4)** ⇒ **Done** ⇒ **OK**

	Length	Hexagon	Hole (.1875)
Generic	**1.500**	**.750**	**Y**
(Instances)			
Tool_1	**1.625**	**.875**	**N**
Tool_2	**1.750**	**1.000**	**Y**
Tool_3	**1.875**	**1.125**	**N**

Figure 21.52 Tool Body (**Generic** and **Instances**)

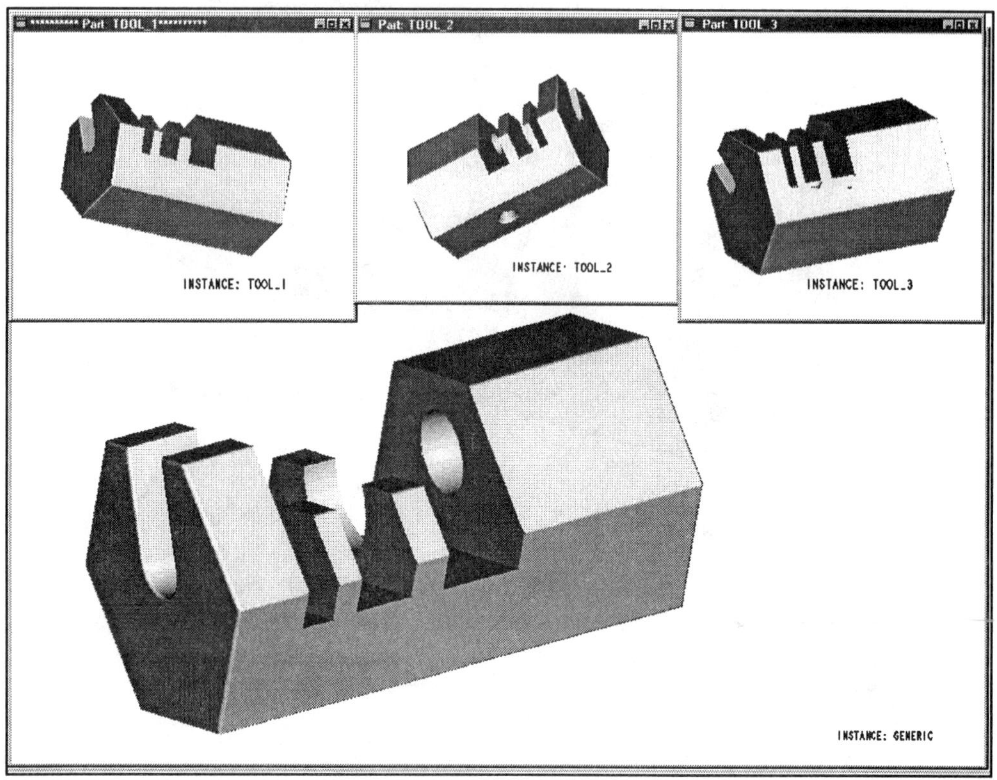

Figure 21.53 Tool Body Family Table

	C1	C2	C3	C4	C5	C6
R1	!					
R2	!	FAMILY TABLE VALUES				
R3	!					
R4	!	1) This family table is for inspection ONLY; any changes made will				
R5	!	NOT be saved.				
R6	!	2) Rows beginning with '@' are family table comments.				
R7	!	3) Rows beginning with '!' and empty rows are ignored.				
R8	!	4) Rows beginning with '$' contain locked instances.				
R9	!	5) '*' is used for the default value.				
R10	!	6) Generic names of features if appear are enclosed in [].				
R11	!	7) Feature identifications are their internal ids.				
R12	!					
R13	!	Generic part name: LESPRO21				
R14	!	Name	d2	d11	F306	
R15	!				[HOLE]	
R16	!	===========	===========	===========	===========	=====
R17	!	GENERIC	1.50000	0.75000	Y	
R18		TOOL_1	1.62500	0.87500	N	
R19		TOOL_2	1.75000	1.00000	Y	
R20		TOOL_3	1.87500	1.12500	N	
R21						

Figure 21.54 Tool Body (**Generic**) Detail Drawing

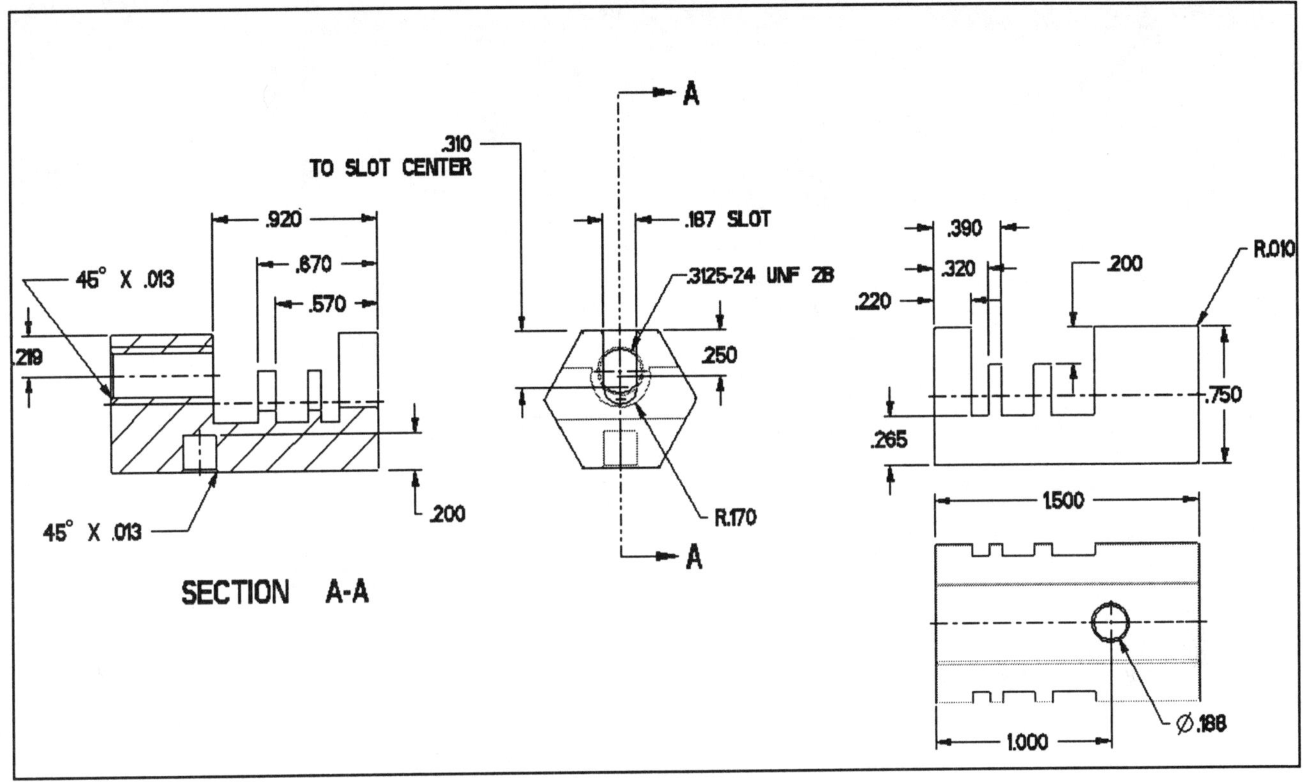

Figure 21.55 Tool Body Detail Drawing (**SECTION A-A**)

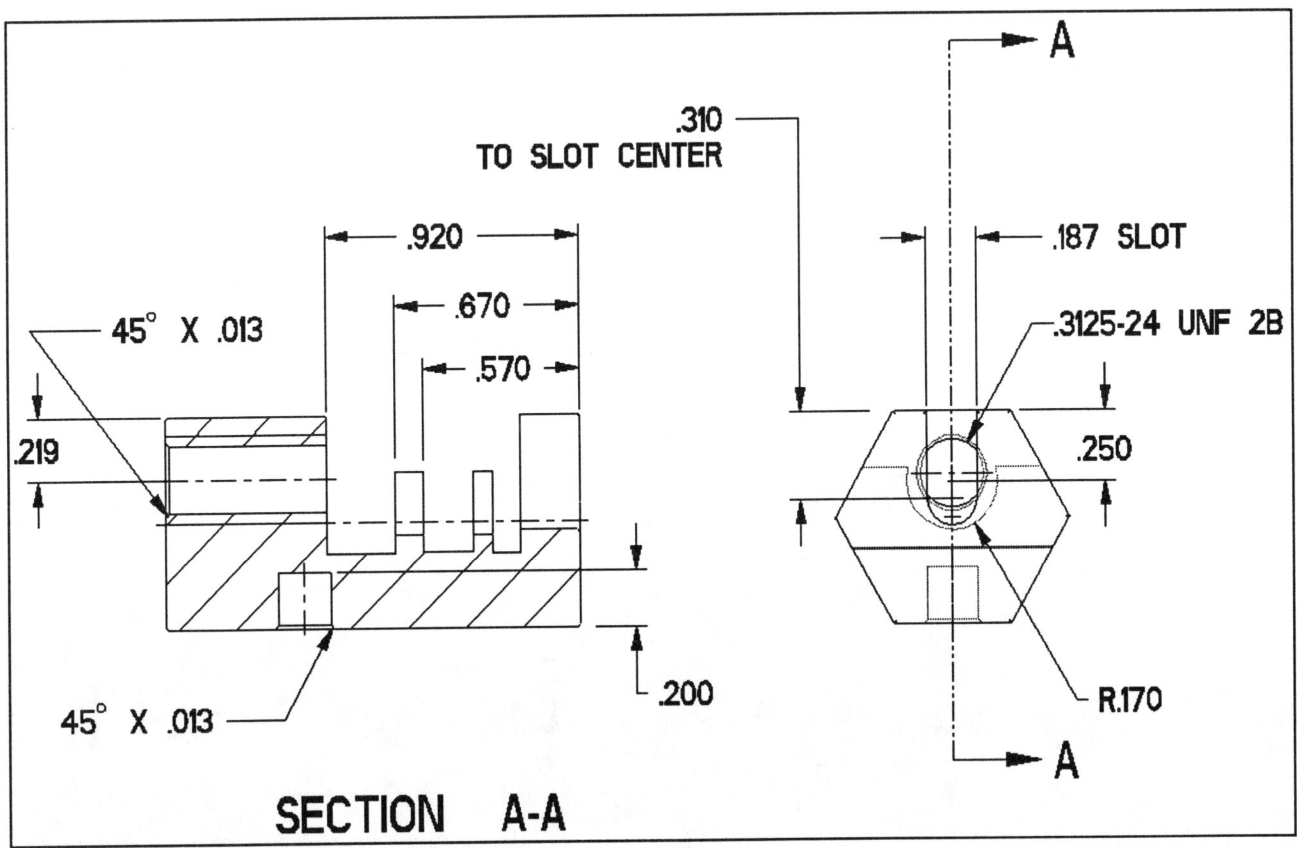

Appendixes

Appendix A Advanced Projects
Appendix B *Config.pro* Files and Mapkeys
Appendix C Glossary
Appendix D Design Intent Planning Sheets (DIPS)

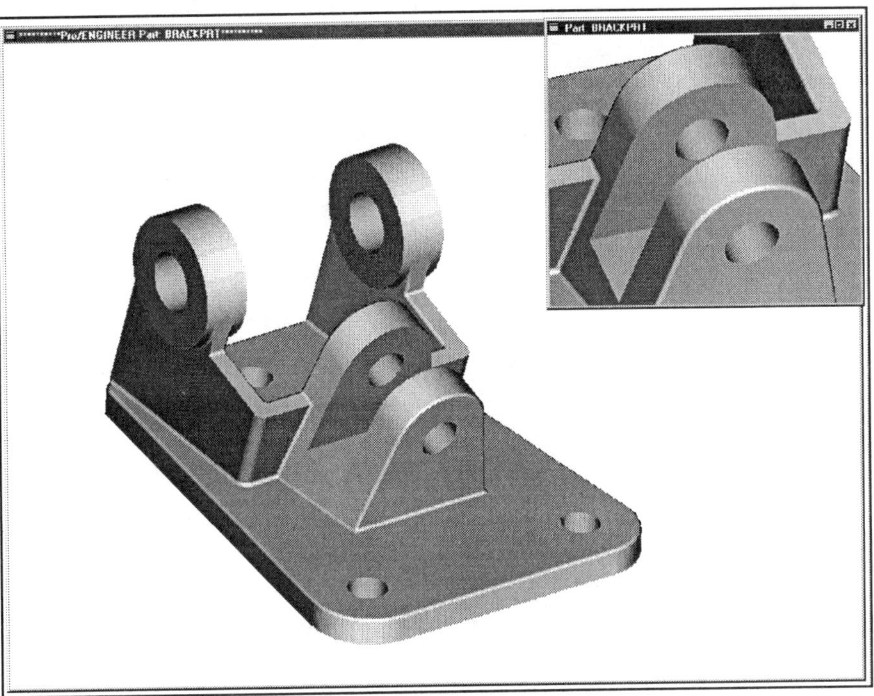

Advanced Project

Pro/TABLE TM Release 19.0 (c) 1988-95 by Parametric Technology Corporation All Rights Reserved.
File Edit View Format Help

mapkey

!ALL VALUES ARE FOR BASIC PRO/ENGINEER IN 20 LESSONS	
SKETCHER_ANIMATED_MODIFY	Yes
SKETCHER_DISP_VERTICES	Yes
START_APPMGR	Yes
TREE_PLACEMENT	RIGHT
mapkey	$F3 #VIEW;#REPAINT;#DONE-RETURN;
mapkey	%F4 #VIEW;#DEFAULT;#DONE-RETURN
mapkey(continued)	#VIEW;#DEFAULT;#DONE-RETURN;
mapkey	$f4 #ENVIRONMENT;#TRIMETRIC;#DONE-RETURN;#VIE\
SKETCHER_DISP_CONSTRAINTS	Yes
RESERVED_MENU_SPACE	1
INDEPENDENT_MENUS	Yes
FONTS_SIZE	LARGE
COMPANY_NAME	Lamit and Associates
DEFAULT_DEC_PLACES	2
ACCURACY_LOWER_BOUND	
HIGHLIGHT_NEW_DIMS	Yes
GRID_SNAP	Yes
MENUITEM_FONT	4
ACCURACY_LOWER_BOUND	
MODEL_TREE_START	Yes
PARENTHESIZE_REF_DIM	Yes
PROMPT_ON_EXIT	Yes
SKETCHER_DEC_PLACES	2
SKETCHER_DISP_CONSTRAINTS	Yes
SPIN_WITH_SILHOUETTES	Yes
TANGENT_EDGE_DISPLAY	DIMMED
TOL_MODE	NOMINAL

Config.pro File

Appendix A

Advanced Projects

Bracket
Shield
Air Foil
Connecting Rod
Cover

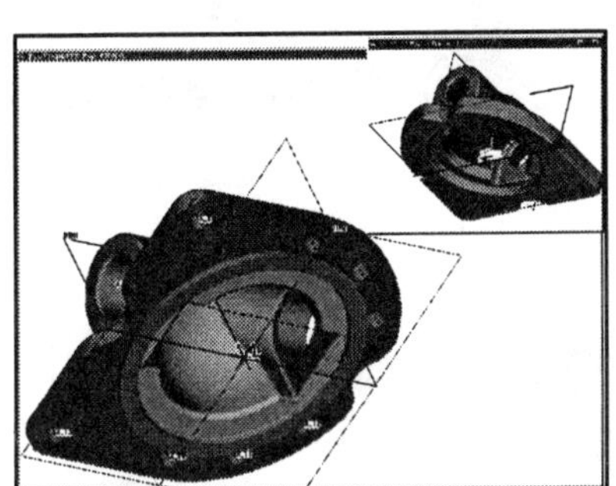

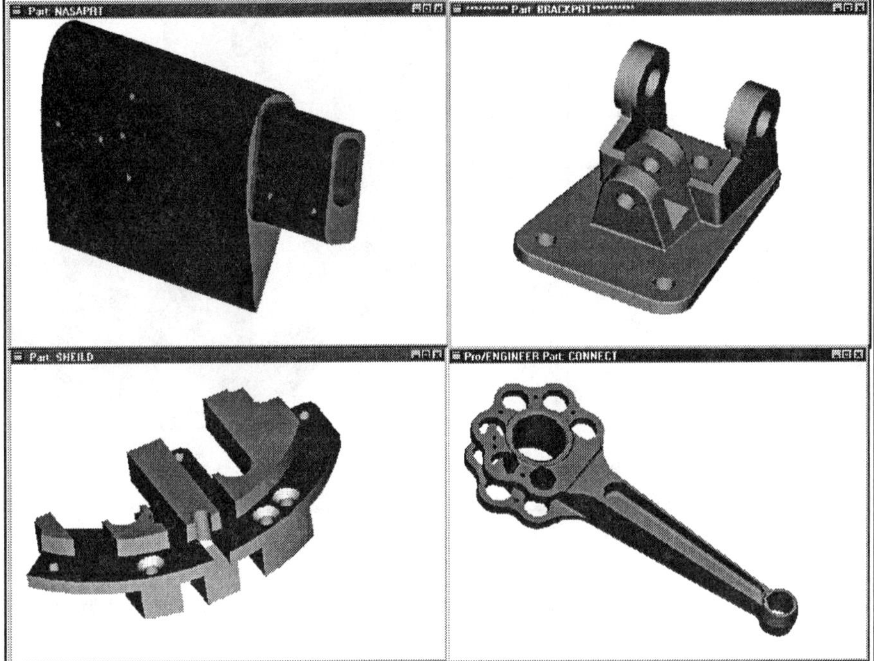

Figure A.1
Advanced Parts

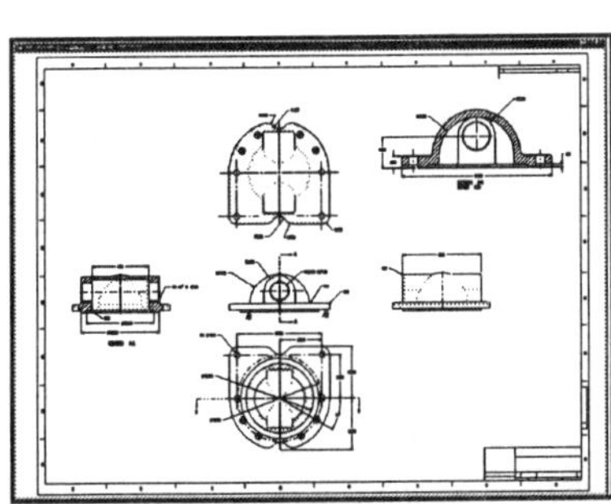

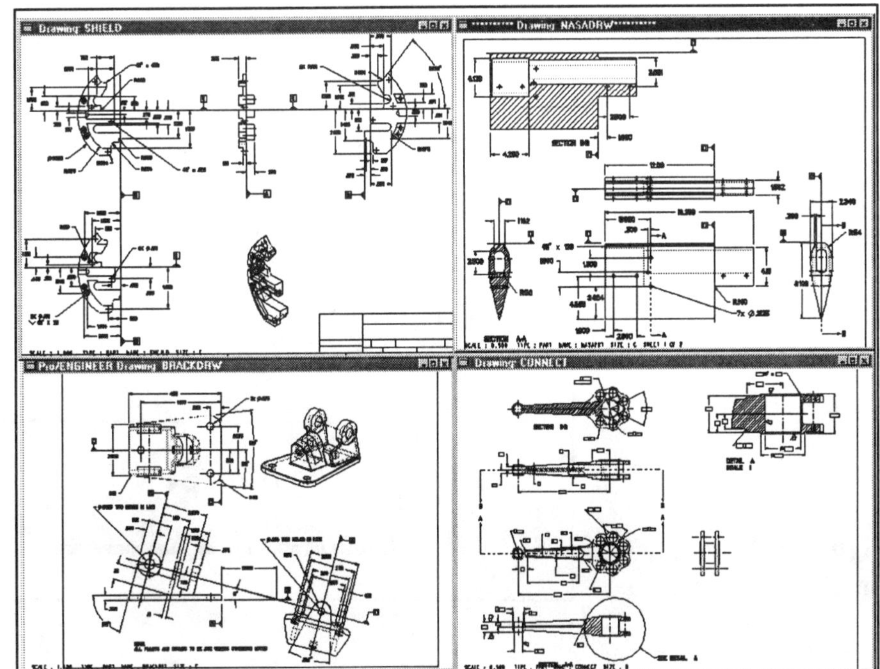

Figure A.2
Advanced Part Drawings

Bracket

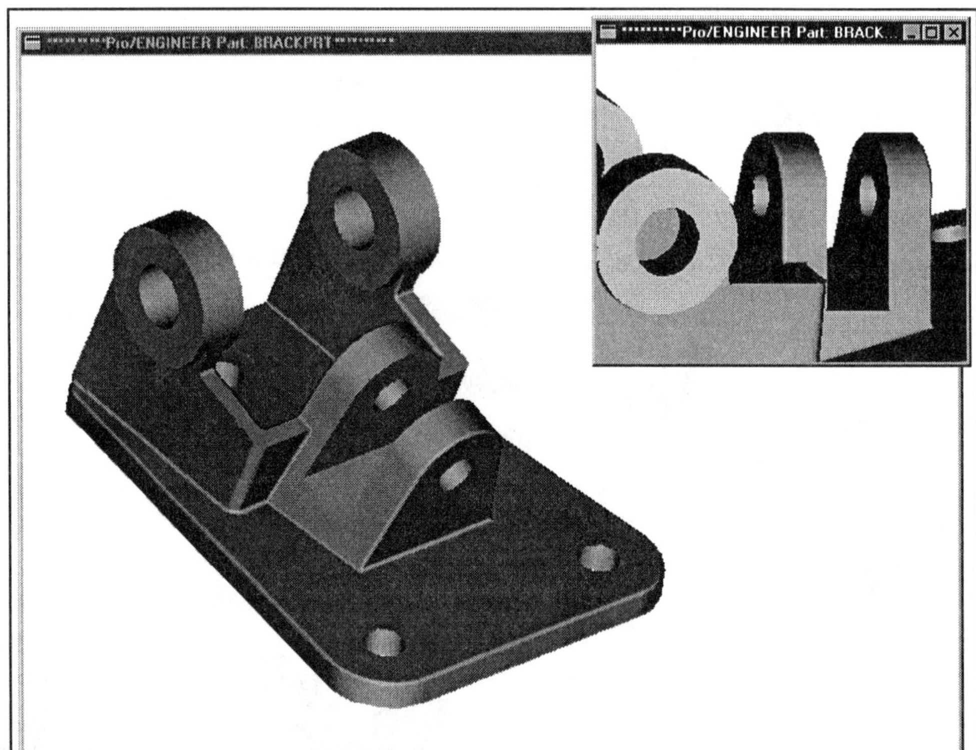

Figure A.3
Bracket Part Model

EGD REFERENCE

Engineering Graphics and Design with Graphical Analysis *or* **Fundamentals of Engineering Graphics and Design**
by L. Lamit and K. Kitto
See page 411
Problem 12.13

NOTE:
ALL FILLETS AND ROUNDS TO BE .062 UNLESS OTHERWISE NOTED

Figure A.4
Bracket Drawing

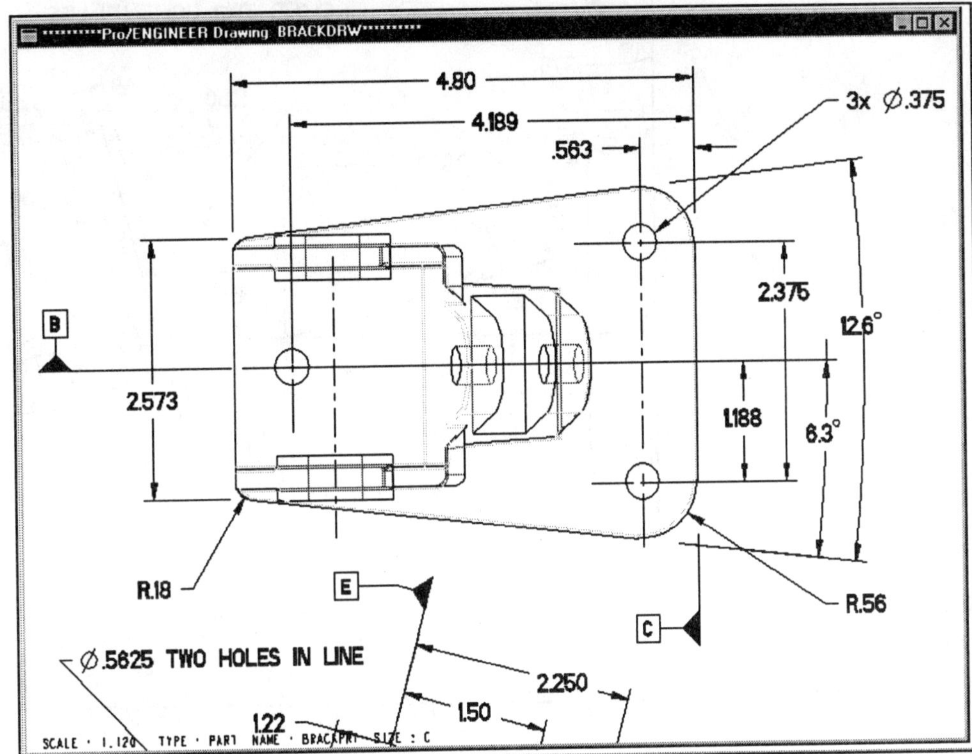

Figure A.5
Bracket Drawing,
Top View

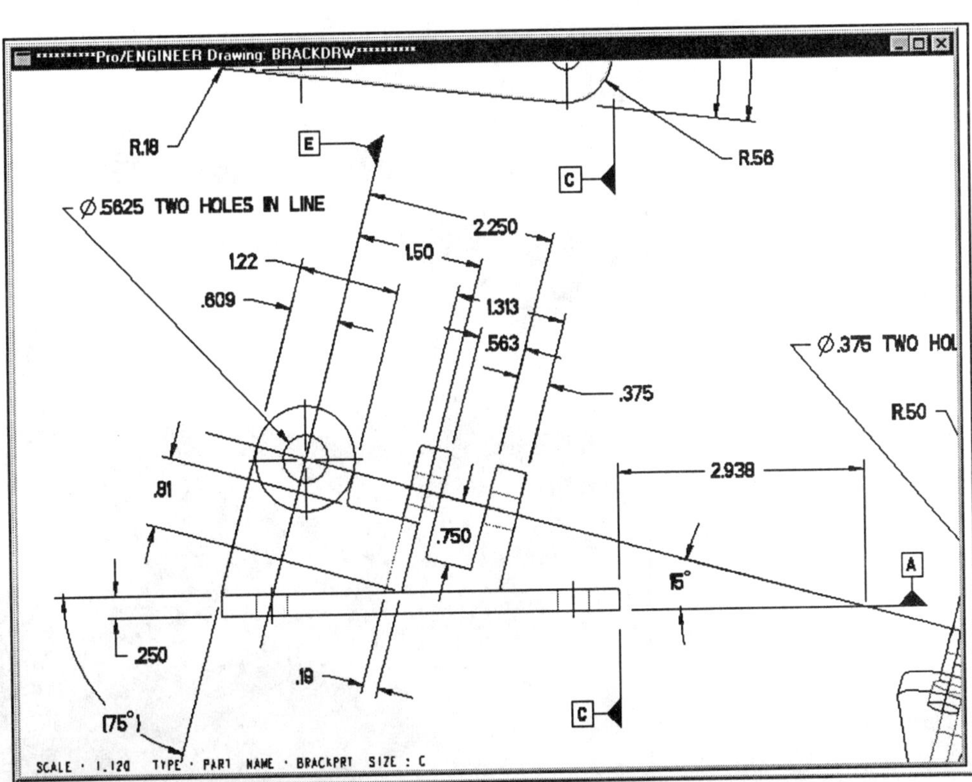

Figure A.6
Bracket Drawing,
Front View

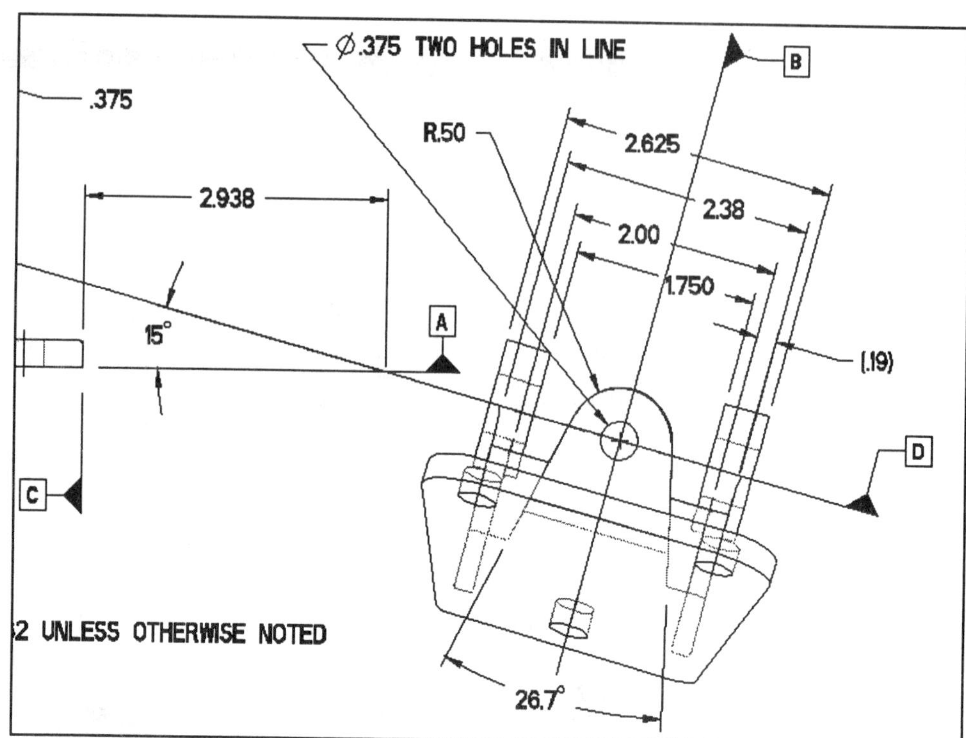

Figure A.7
Bracket Drawing, Auxiliary View

Shield

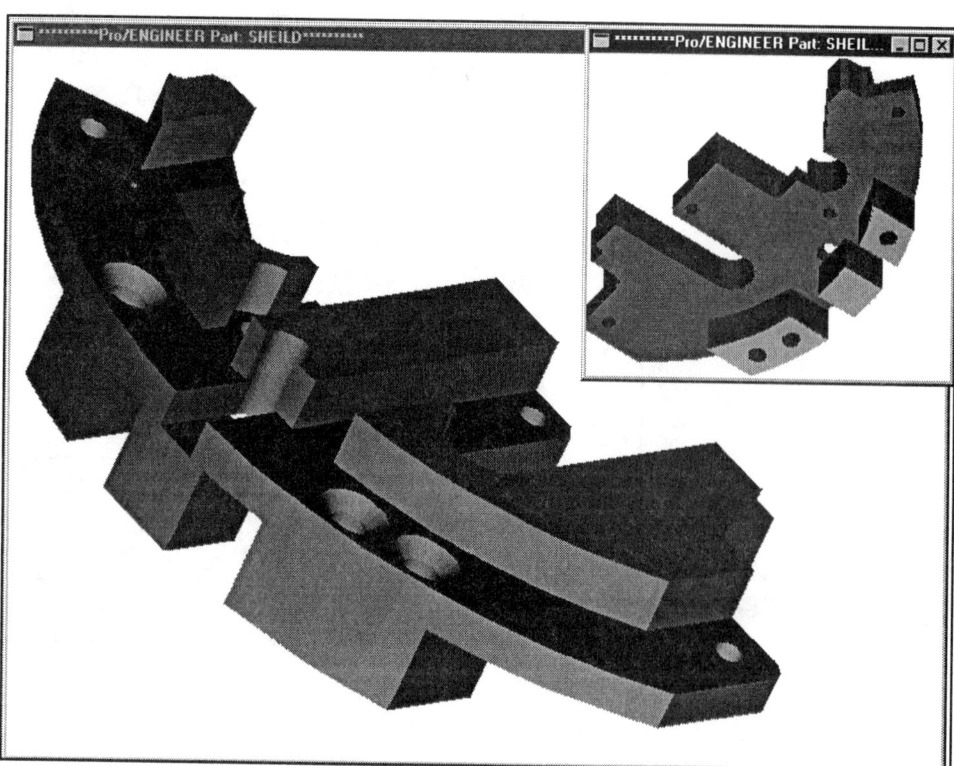

Figure A.8
Shield Part Model

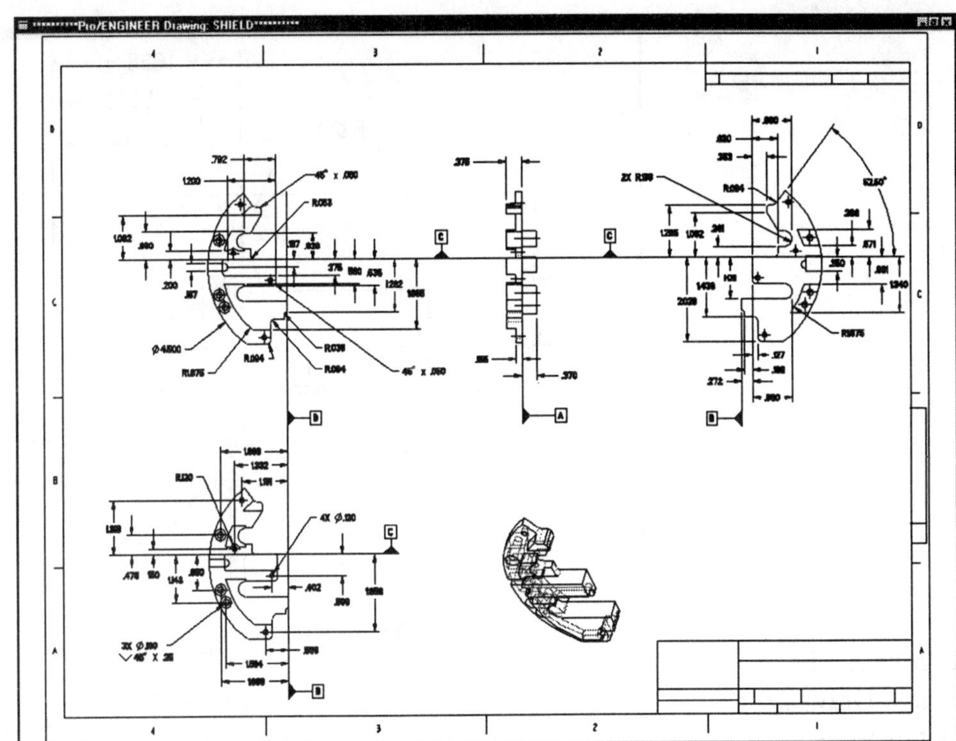

Figure A.9
Shield Drawing

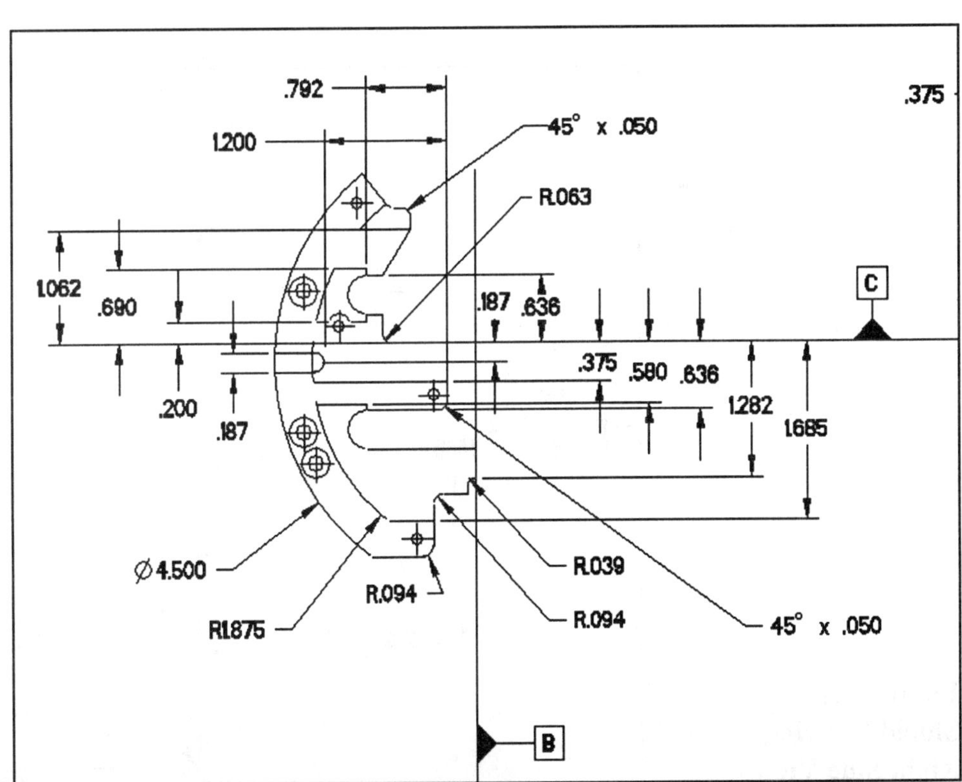

Figure A.10
Shield Drawing,
Front View

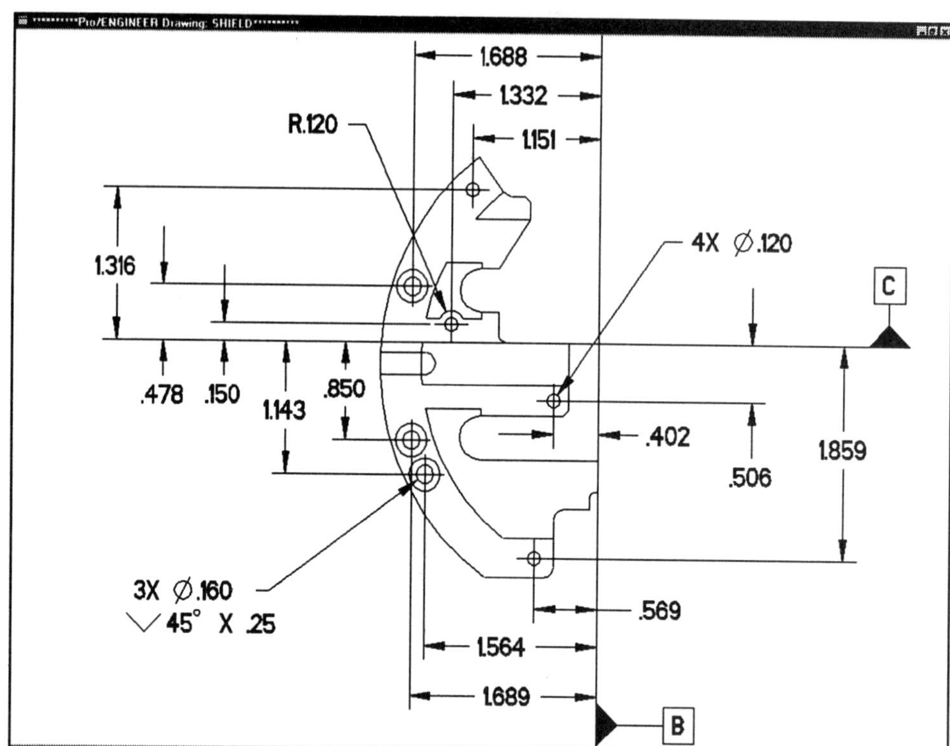

Figure A.11
Shield Drawing,
Detail View

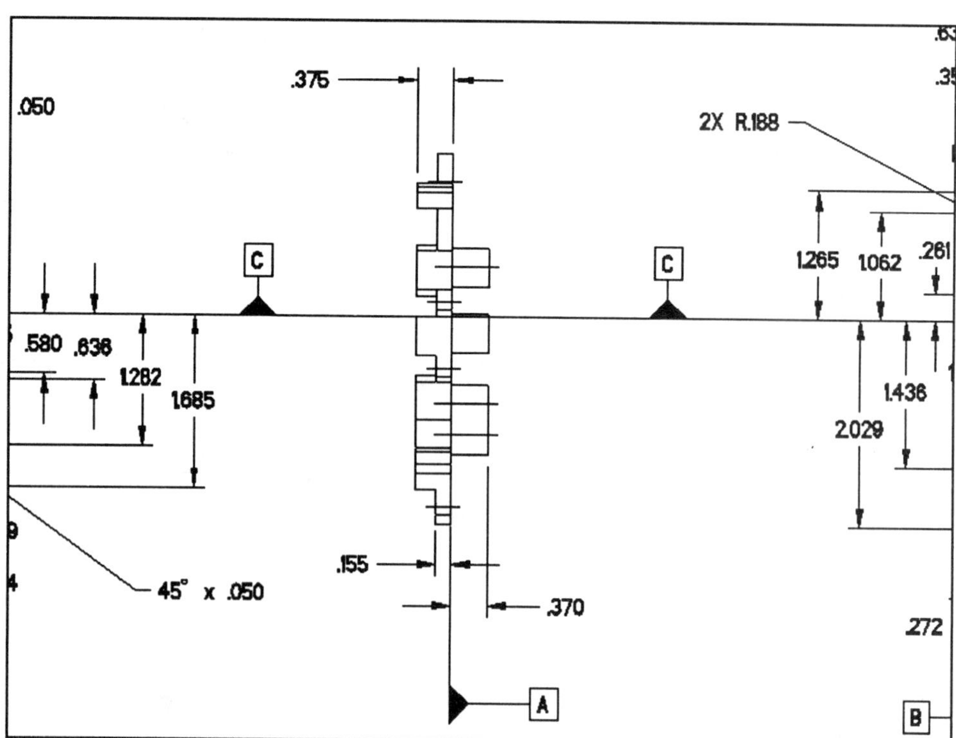

Figure A.12
Shield Drawing,
Right Side View

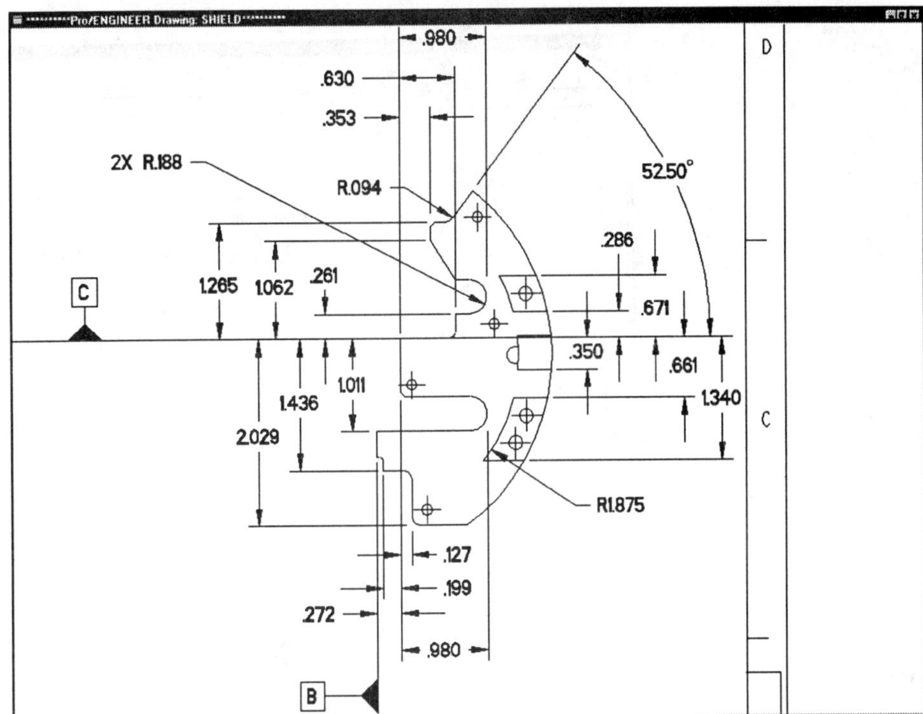

Figure A.13
Shield Drawing,
Back View

Airfoil

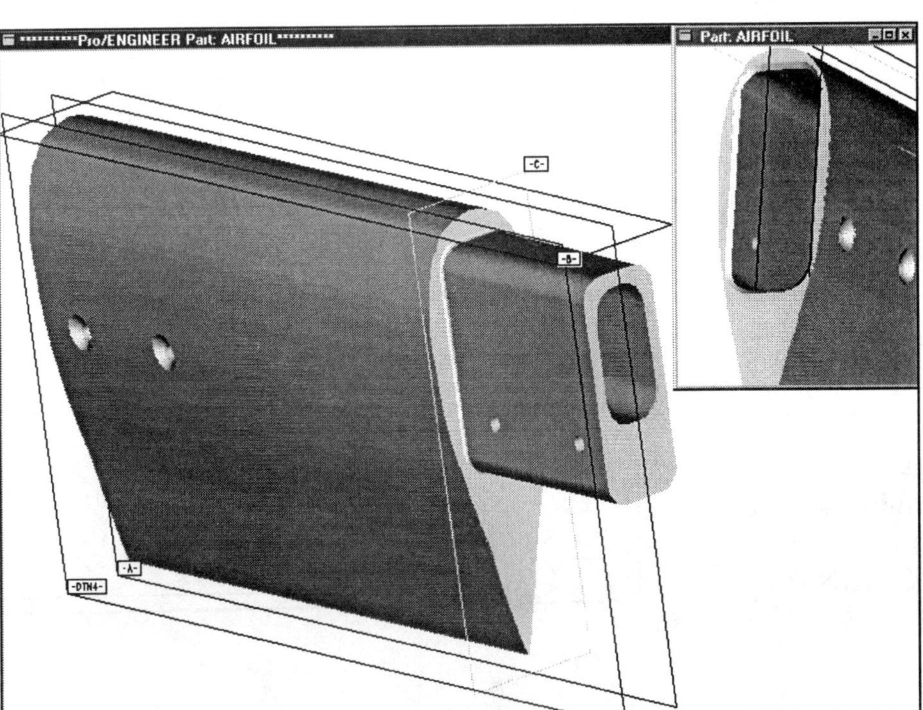

Figure A.14
Airfoil Part Model

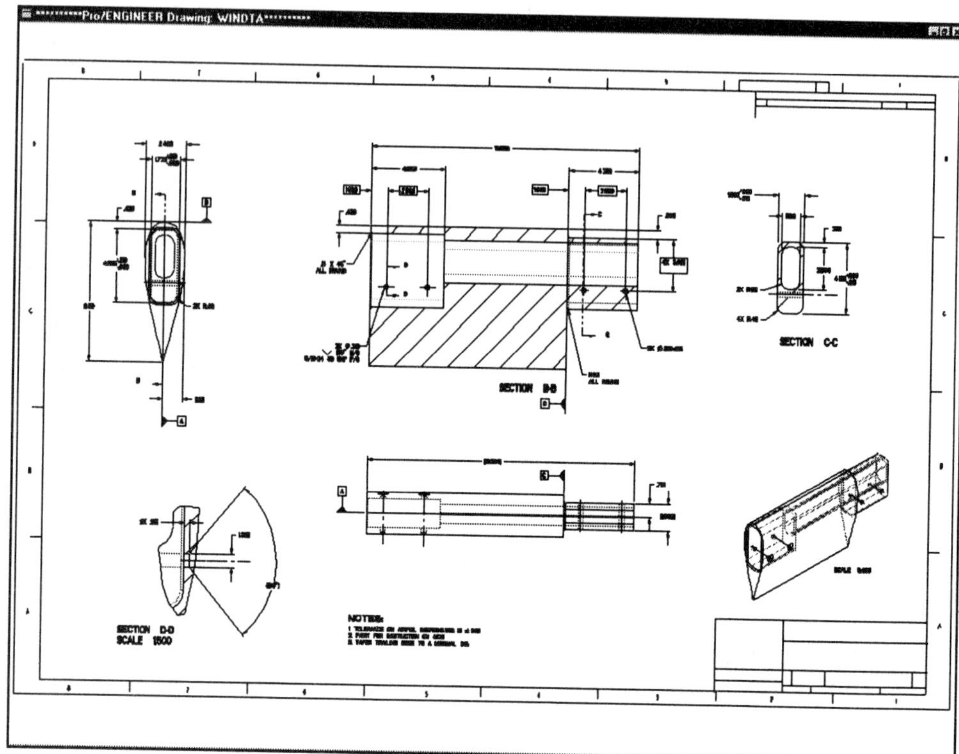

Figure A.15
Airfoil Drawing, Sheet One

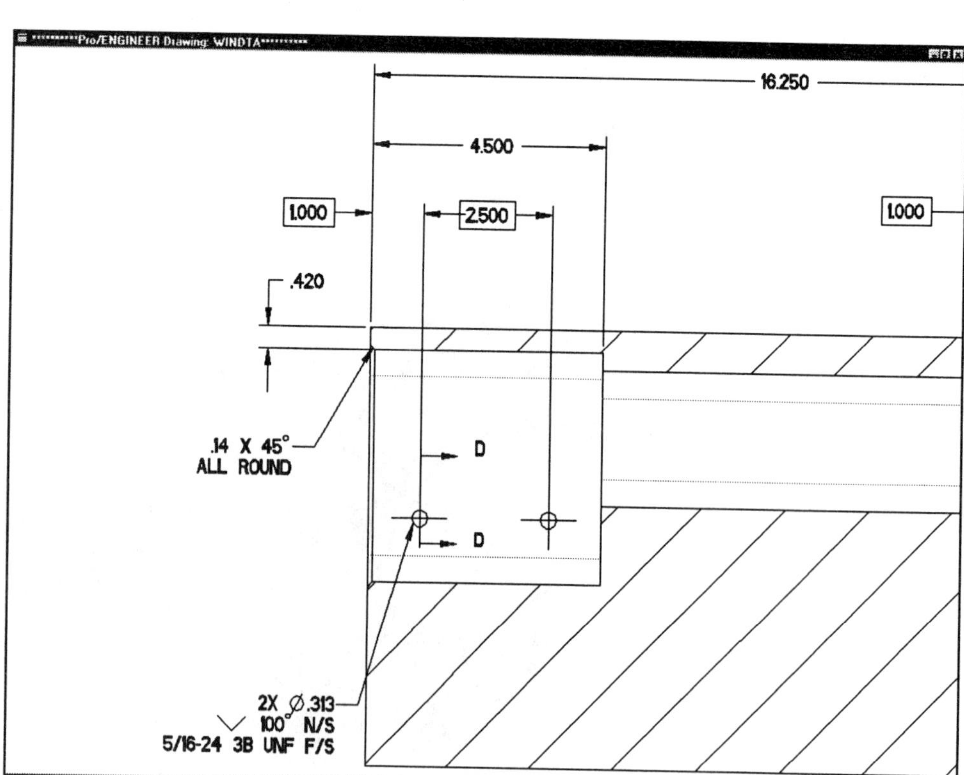

Figure A.16
Airfoil Drawing, Close-up, **SECTION B-B**

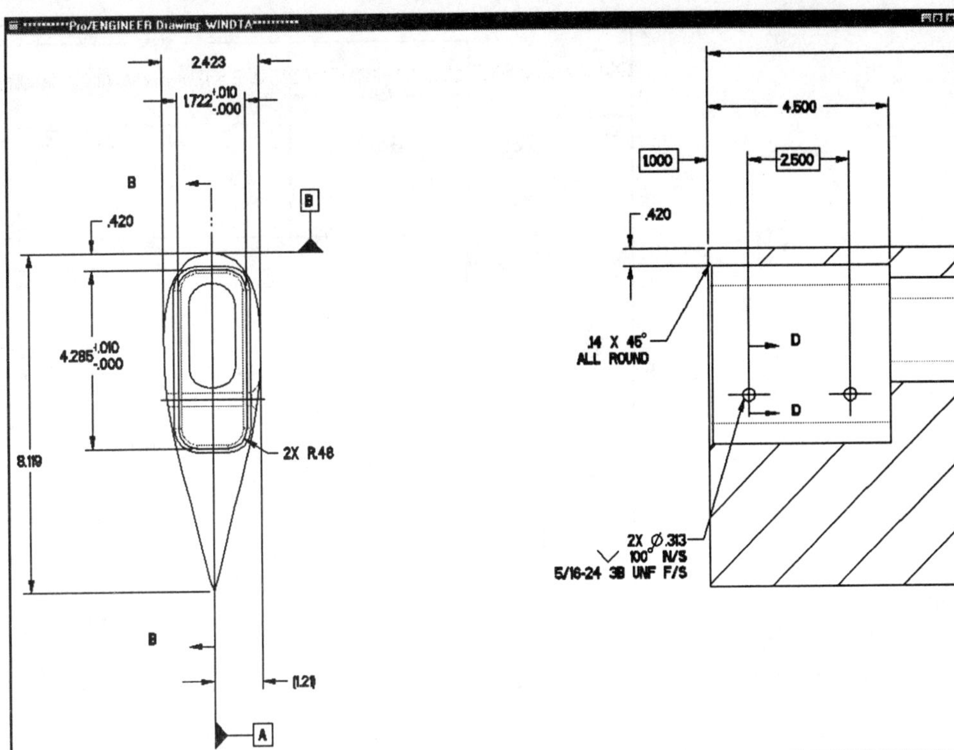

Figure A.17
Airfoil Drawing,
Left Side View

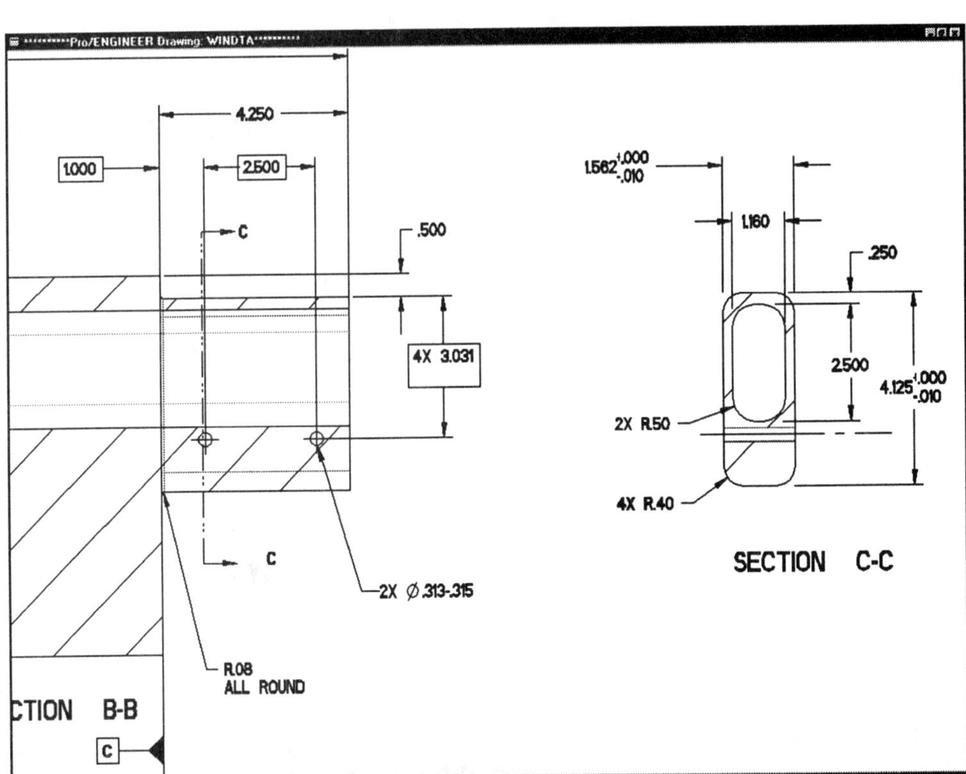

Figure A.18
Airfoil Drawing,
SECTION C-C

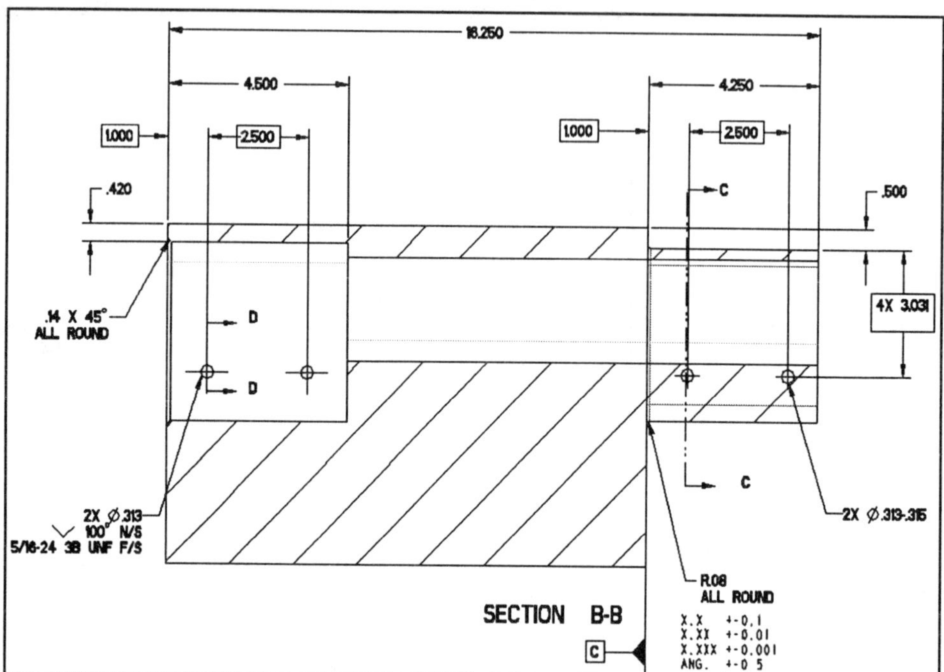

Figure A.19
Airfoil Drawing,
SECTION B-B

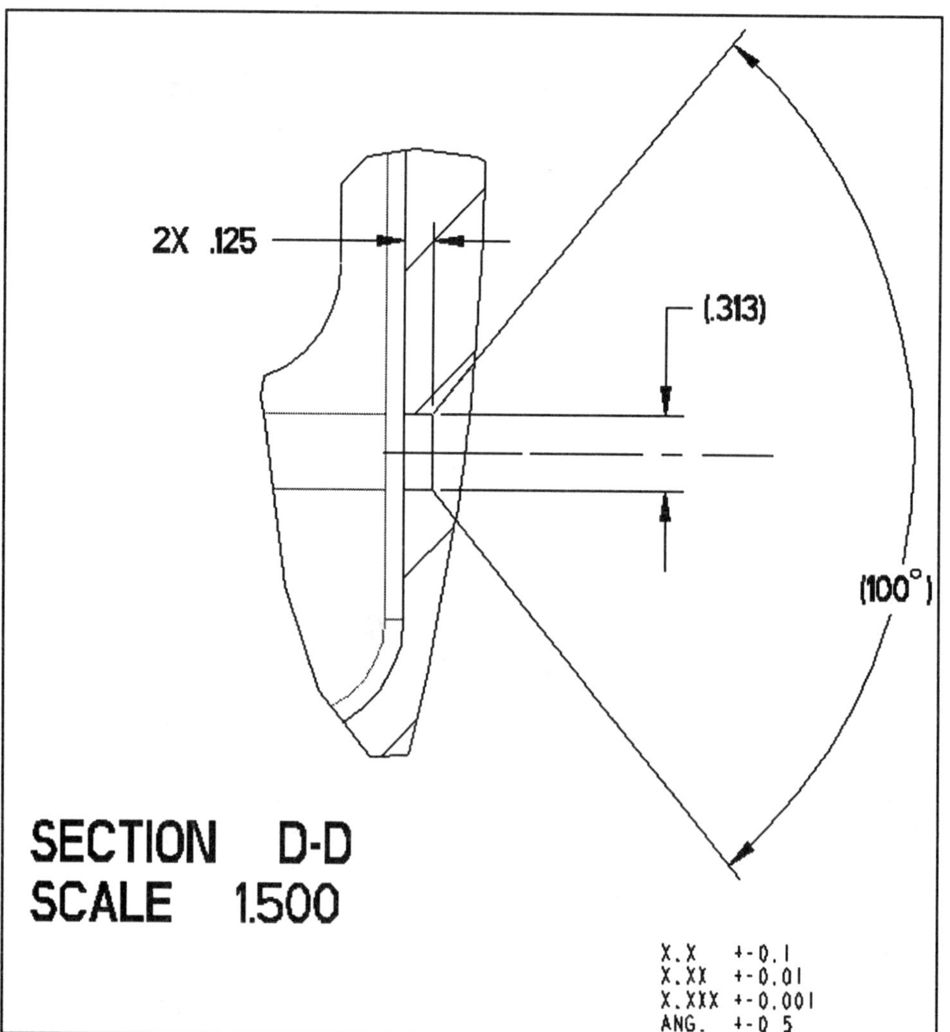

Figure A.20
Airfoil Drawing,
SECTION D-D

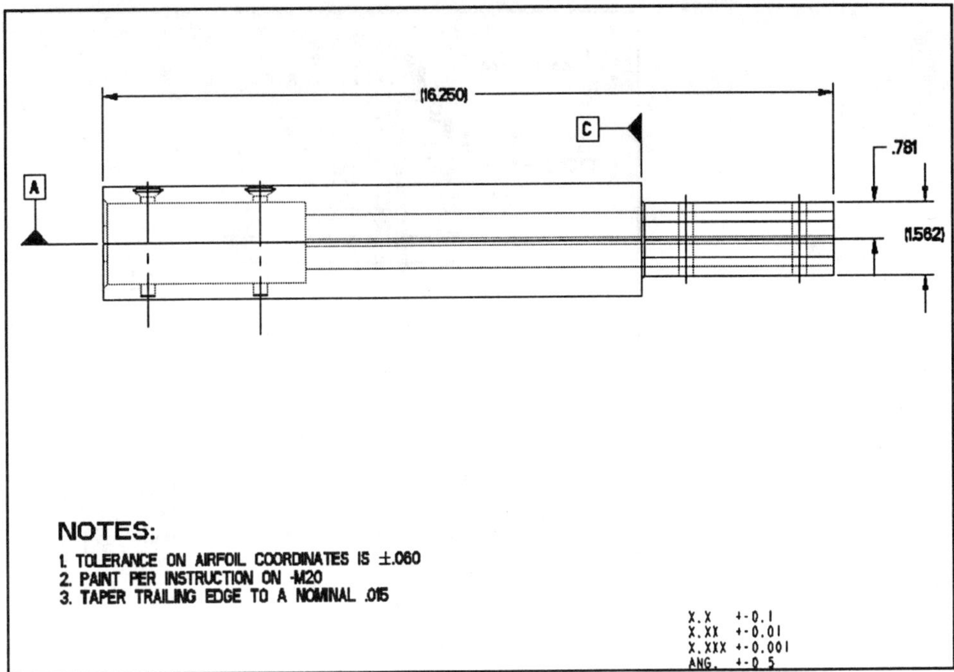

Figure A.21
Airfoil Drawing, Bottom View

X down	+Y to right	-Y to left
.000	.000	.000
.008	.098	.098
.032	.239	.239
.099	.393	.393
.191	.542	.542
.354	.711	.711
.510	.832	.832
.832	.969	.969
1.159	1.062	1.062
1.480	1.122	1.122
1.803	1.155	1.155
2.450	1.168	1.168
3.097	1.133	1.133
3.743	1.063	1.063
4.067	1.012	1.012
4.390	.949	.949
5.035	.801	.801
5.684	.640	.640
6.333	.477	.477
6.815	.357	.357
7.138	.278	.278
7.460	.201	.201
7.785	.144	.144
8.109	.017	.017
8.119	.000	.000

Figure A.22
Airfoil Drawing, Sheet Three Table

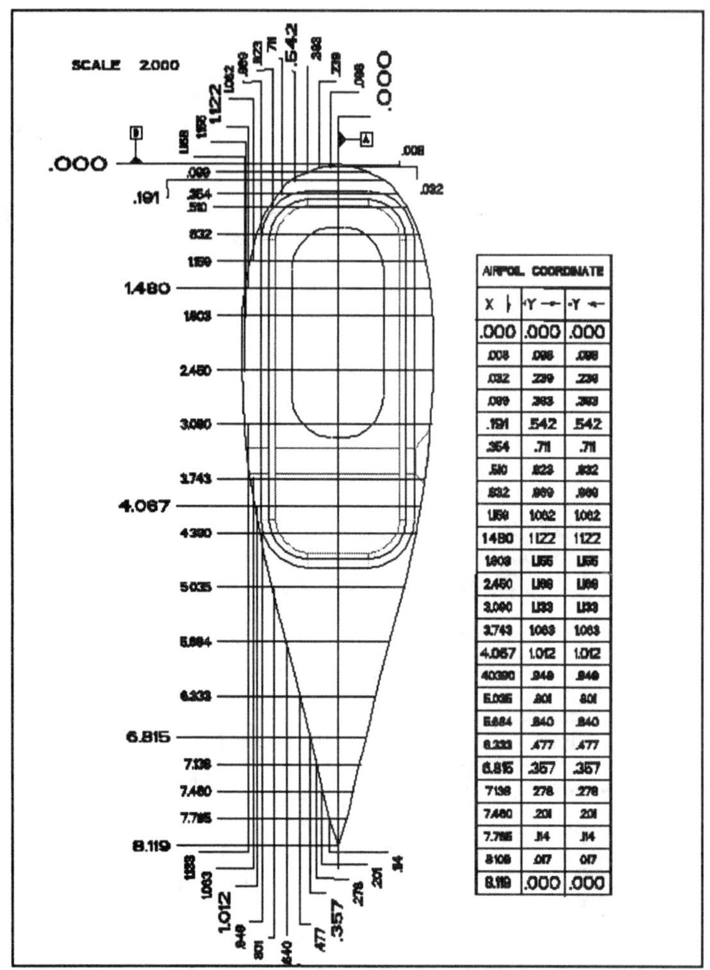

AIRFOIL COORDINATE		
X	+Y	-Y
.000	.000	.000
.008	.098	.098
.032	.239	.239
.099	.393	.393
.191	.542	.542
.354	.711	.711
.510	.823	.832
.832	.969	.969
1.159	1.062	1.062
1.480	1.122	1.122
1.803	1.155	1.155
2.450	1.168	1.168
3.090	1.133	1.133
3.743	1.063	1.063
4.067	1.012	1.012
4.0390	.948	.948
5.035	.801	.801
5.684	.640	.640
6.333	.477	.477
6.815	.357	.357
7.138	.278	.278
7.460	.201	.201
7.795	.114	.114
8.108	.017	.017
8.119	.000	.000

Figure A.23
Airfoil Drawing, Table and Section

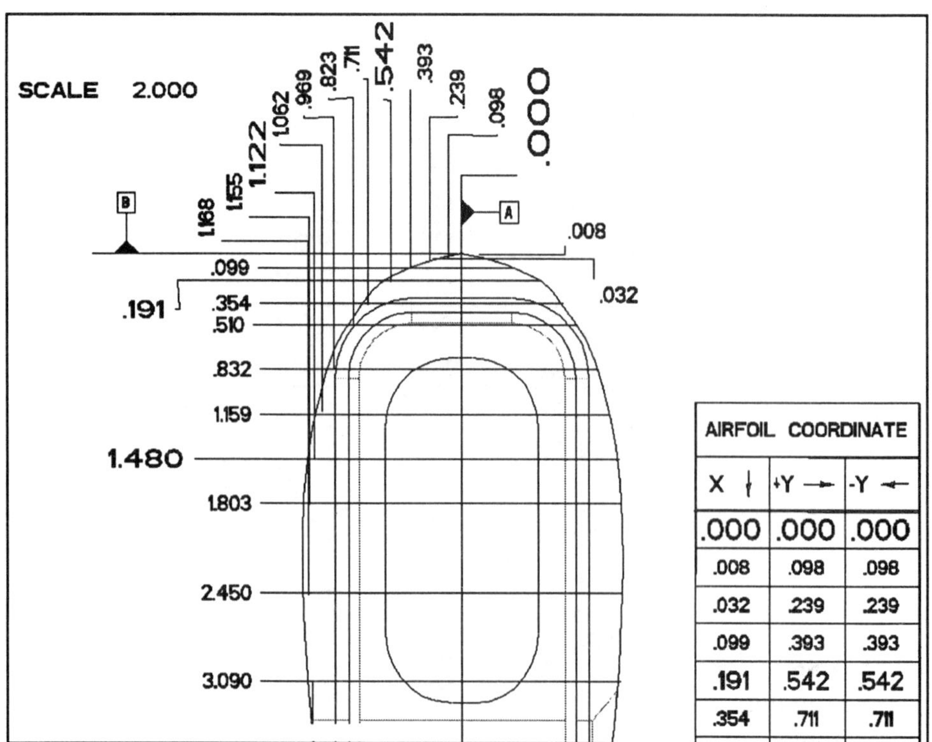

AIRFOIL COORDINATE		
X	+Y	-Y
.000	.000	.000
.008	.098	.098
.032	.239	.239
.099	.393	.393
.191	.542	.542
.354	.711	.711

Figure A.24
Airfoil Drawing, Upper Table

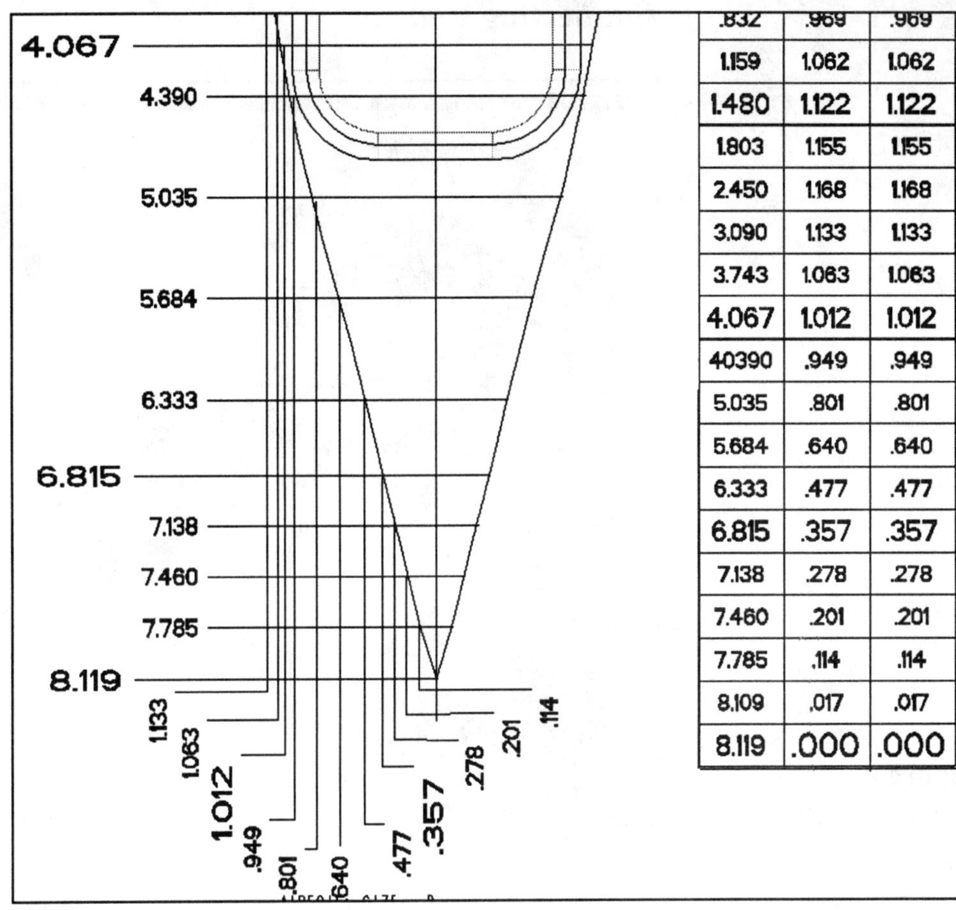

.832	.969	.969
1.159	1.062	1.062
1.480	1.122	1.122
1.803	1.155	1.155
2.450	1.168	1.168
3.090	1.133	1.133
3.743	1.063	1.063
4.067	1.012	1.012
40390	.949	.949
5.035	.801	.801
5.684	.640	.640
6.333	.477	.477
6.815	.357	.357
7.138	.278	.278
7.460	.201	.201
7.785	.114	.114
8.109	.017	.017
8.119	.000	.000

Figure A.25
Airfoil Drawing,
Lower Table

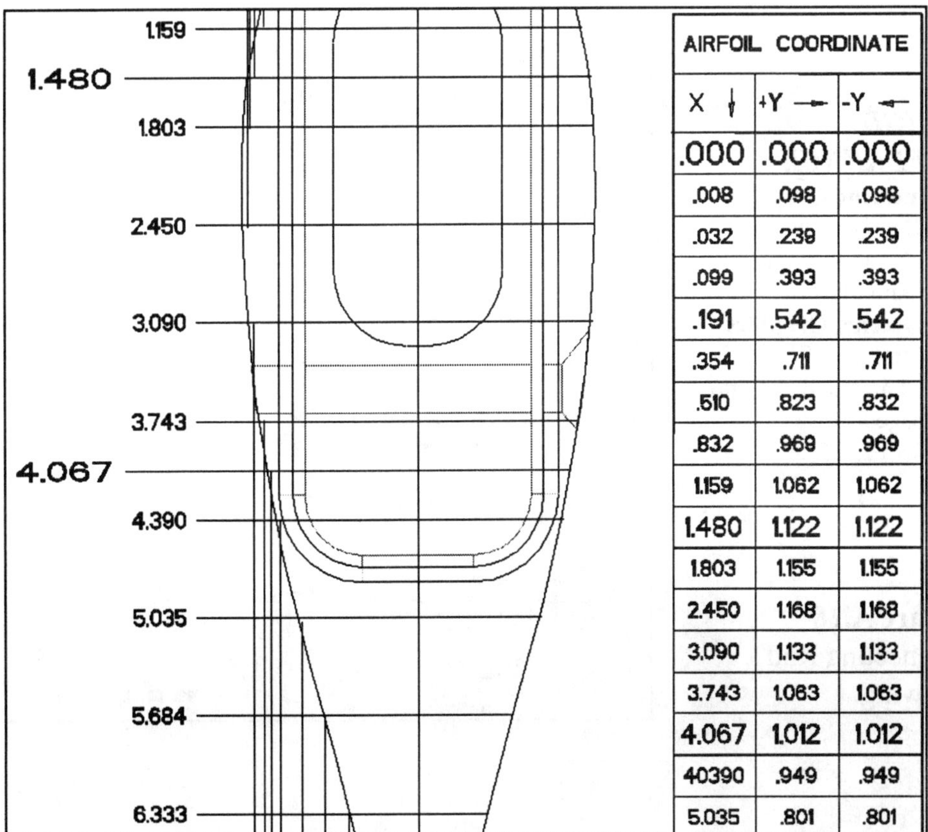

AIRFOIL COORDINATE		
X ↓	+Y →	-Y ←
.000	.000	.000
.008	.098	.098
.032	.239	.239
.099	.393	.393
.191	.542	.542
.354	.711	.711
.510	.823	.832
.832	.969	.969
1.159	1.062	1.062
1.480	1.122	1.122
1.803	1.155	1.155
2.450	1.168	1.168
3.090	1.133	1.133
3.743	1.063	1.063
4.067	1.012	1.012
40390	.949	.949
5.035	.801	.801

Figure A.26
Airfoil Drawing,
Middle Table

Connecting Rod

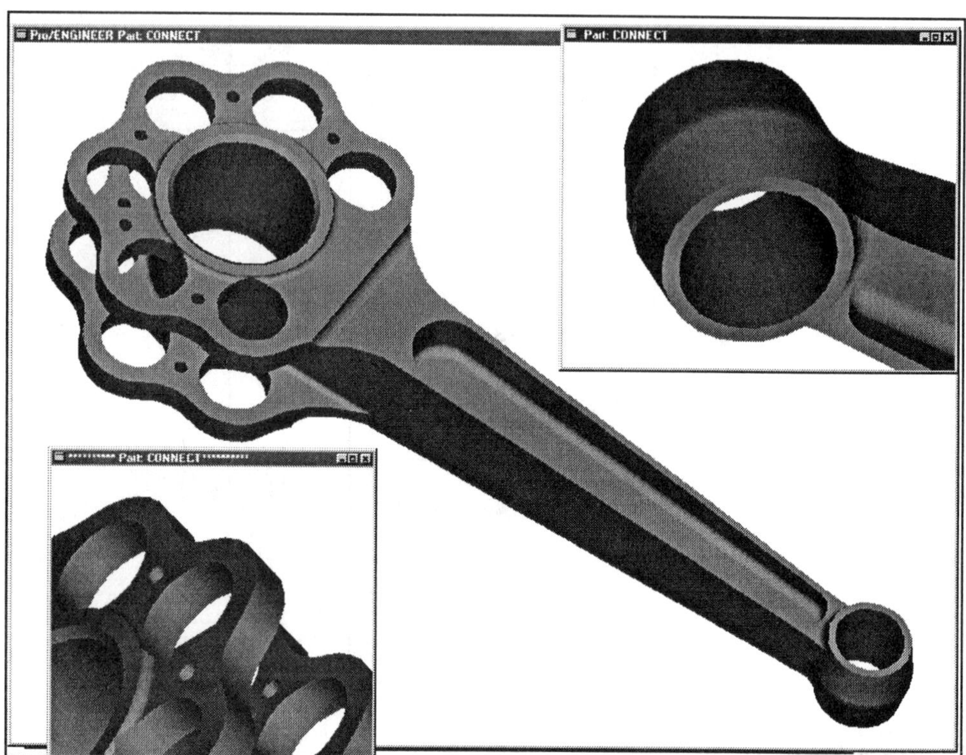

Figure A.27
Connecting Rod

EGD REFERENCE
Engineering Graphics and Design with Graphical Analysis *or* **Fundamentals of Engineering Graphics and Design**
by L. Lamit and K. Kitto
See page 857
Problem 23.13

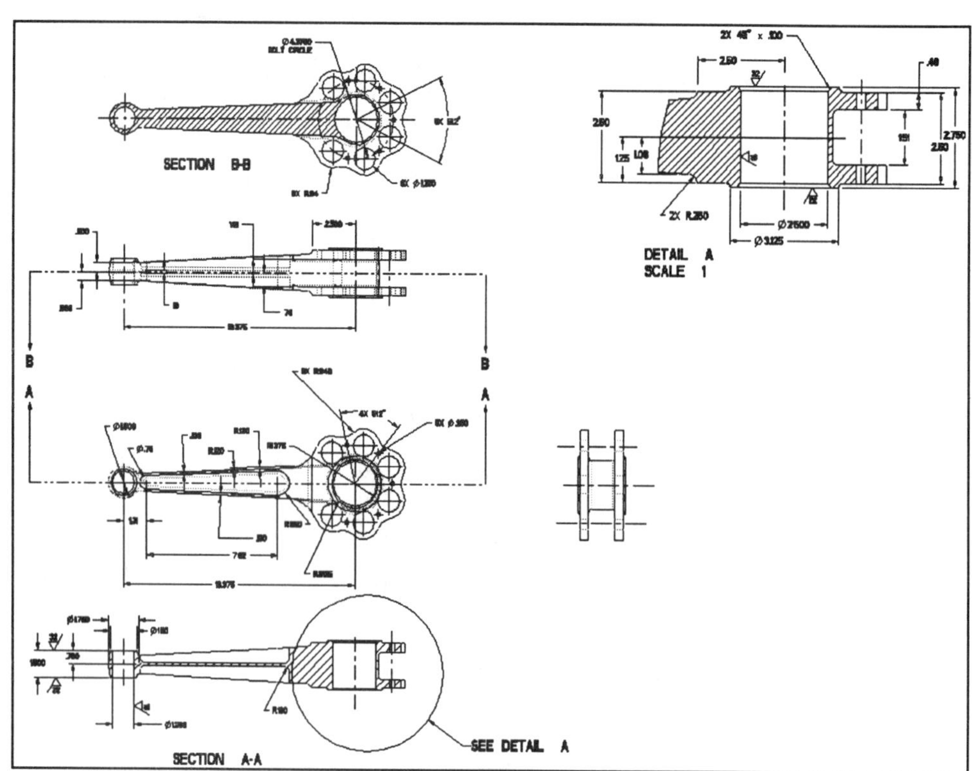

Figure A.28
Connecting Rod Drawing

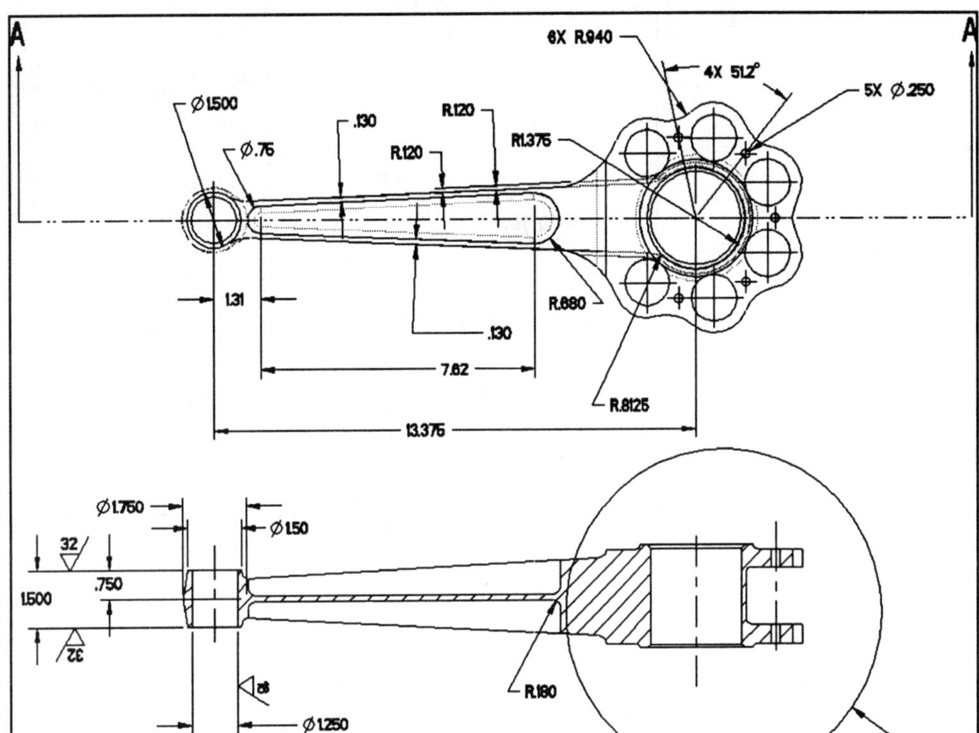

Figure A.29 Connecting Rod Drawing, Front and Top Views

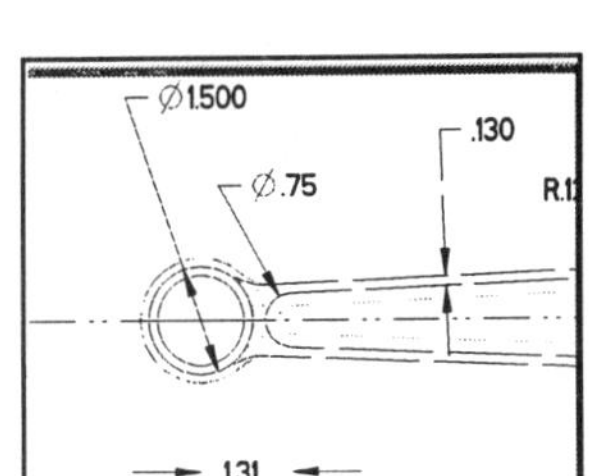

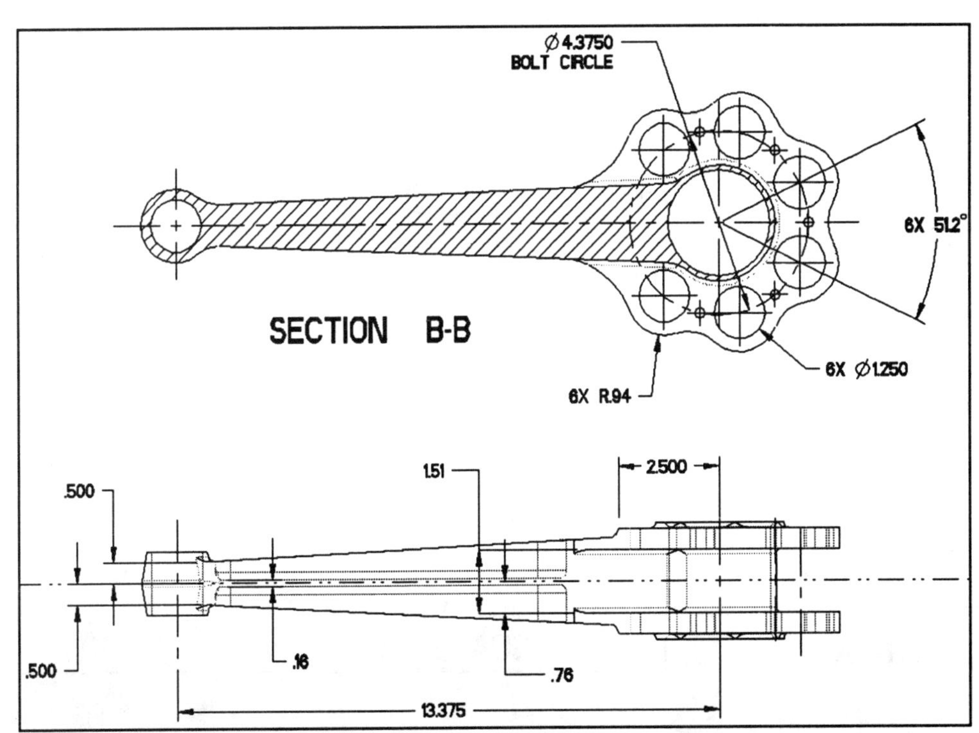

Figure A.30 Connecting Rod Drawing, Back and **SECTION B-B** Views

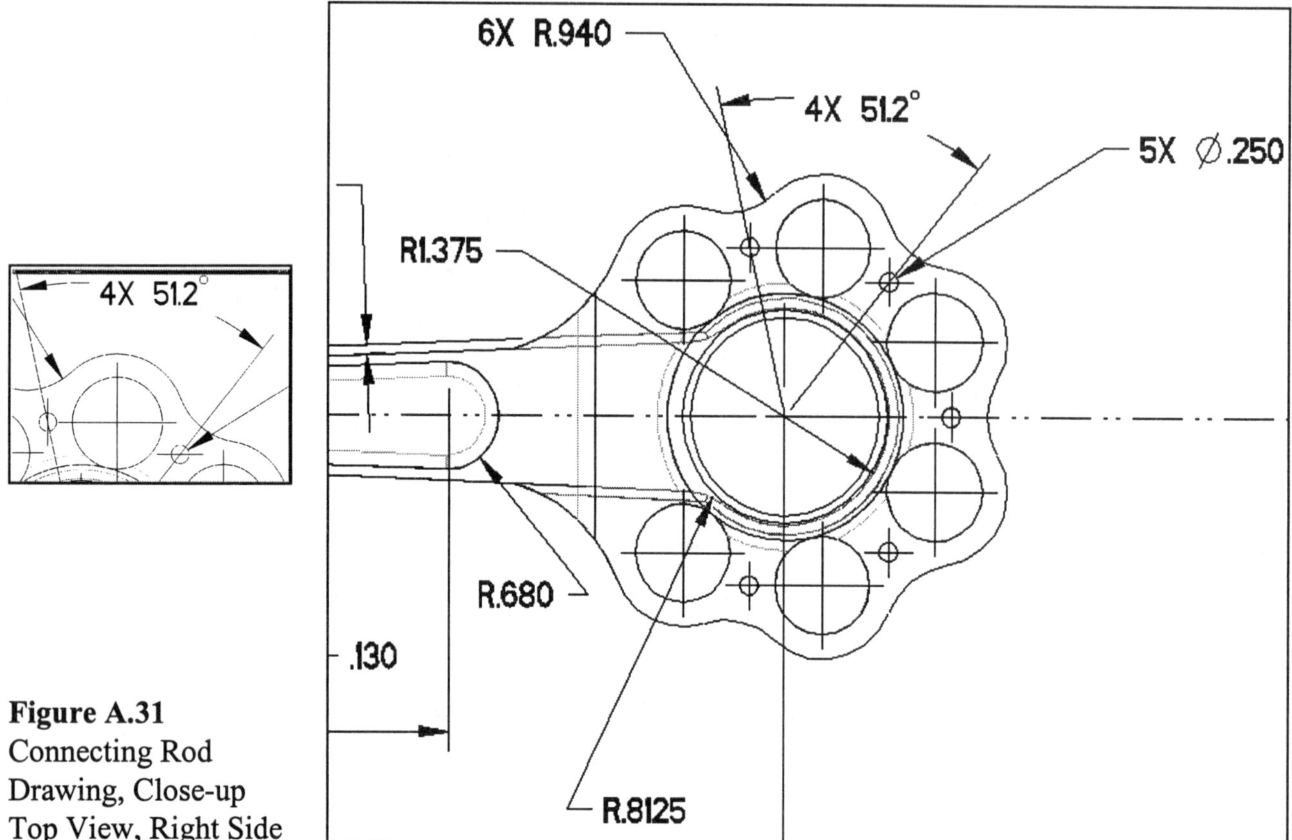

Figure A.31
Connecting Rod Drawing, Close-up Top View, Right Side

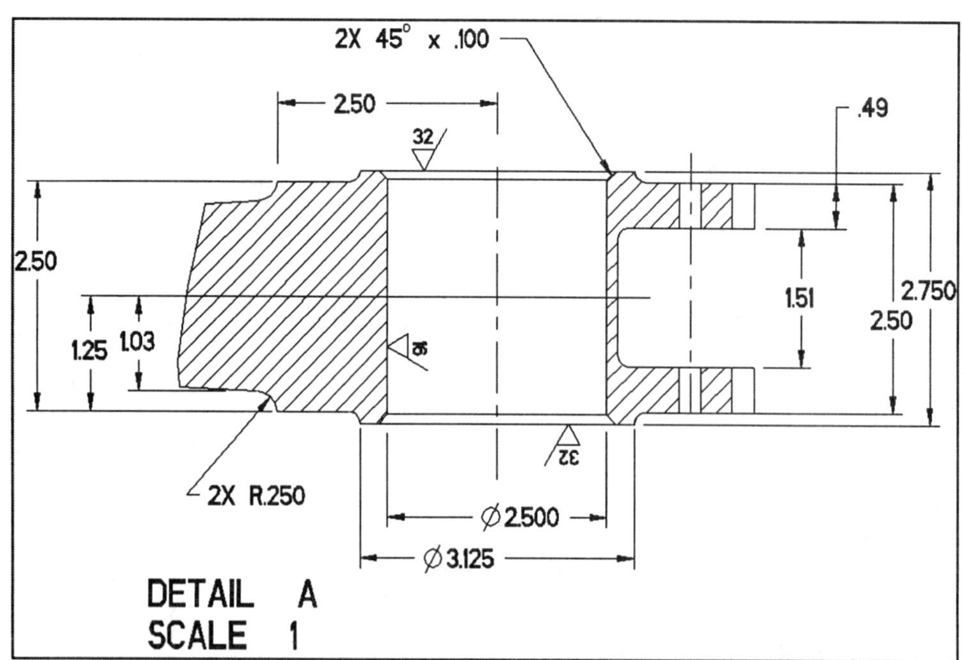

Figure A.32
Connecting Rod Drawing,
DETAIL A

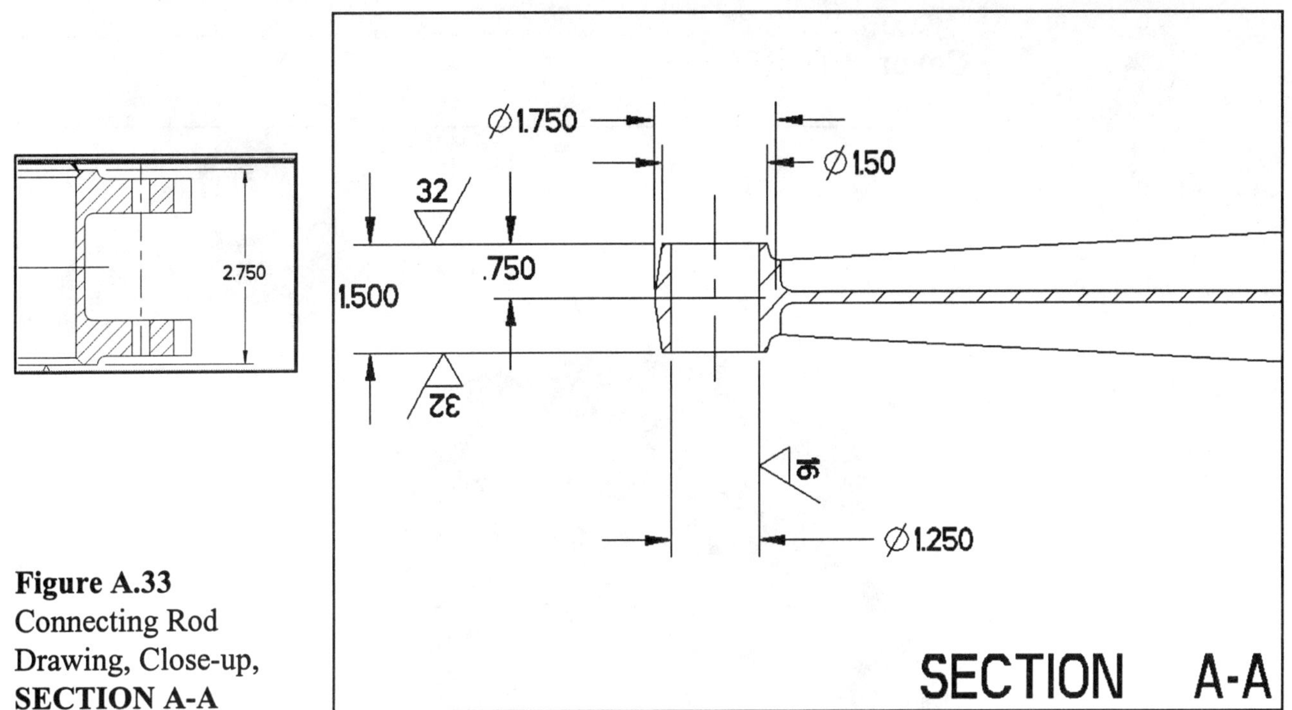

Figure A.33 Connecting Rod Drawing, Close-up, **SECTION A-A**

Figure A.34 Connecting Rod Drawing, Close-up, Top View

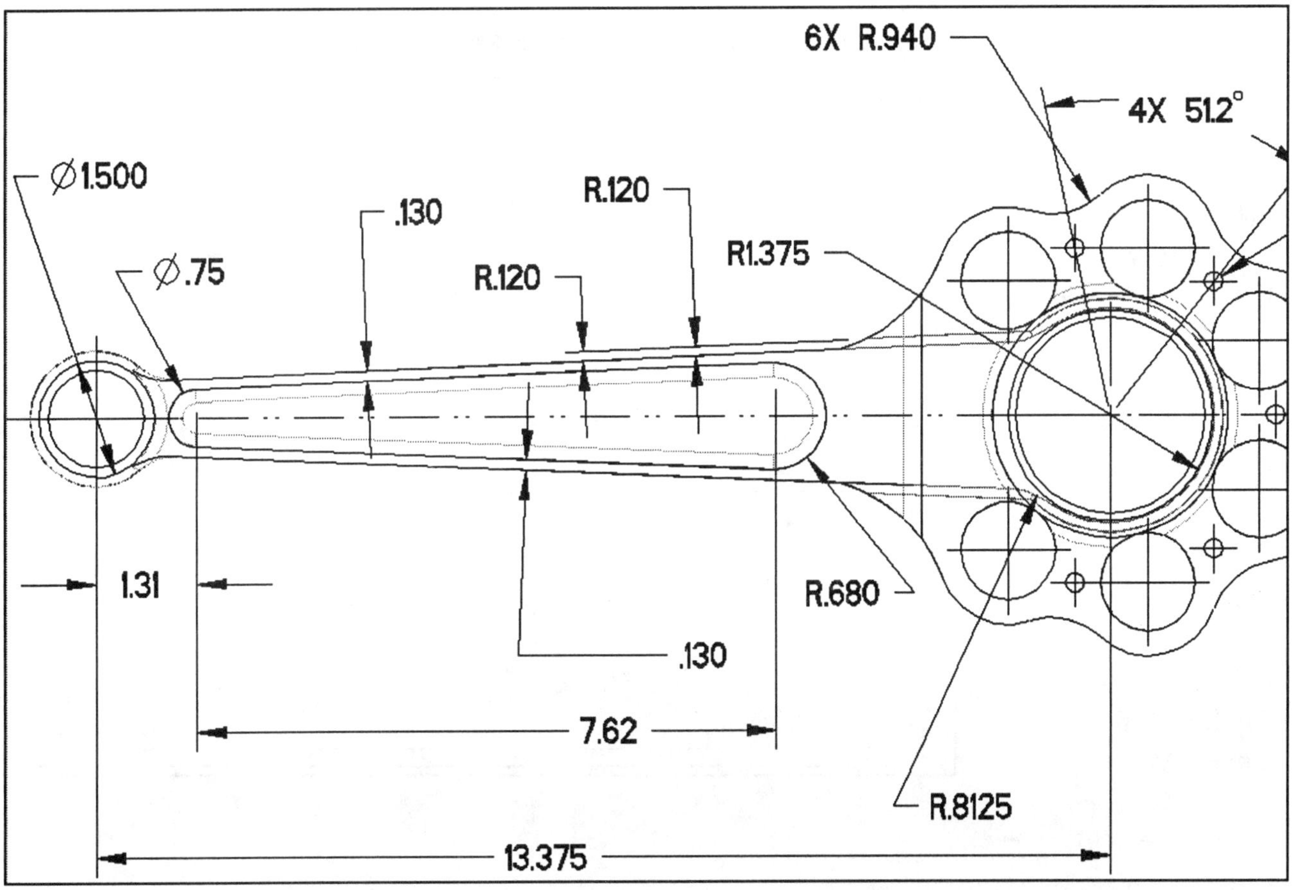

Cover

NOTE

Model the casting, save it under a new name, and continue modeling the machine part. When it is complete, you will have two separate parts, a casting (**workpiece**) and a machined part (**design part**), that can be used to create the *manufacturing model* for use in Pro/MANUFACTURING

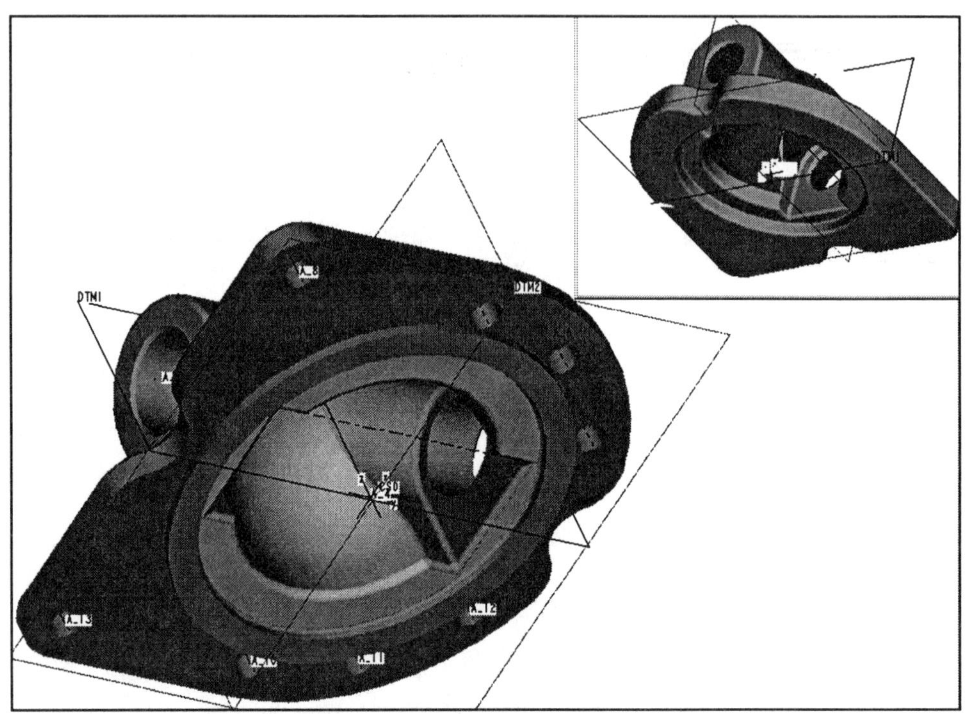

Figure A.35
Cover Machined Part and Casting (inset)

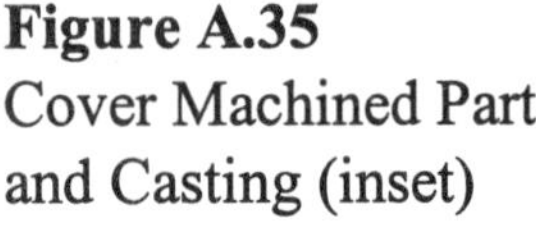

EGD REFERENCE

Engineering Graphics and Design with Graphical Analysis *or* **Fundamentals of Engineering Graphics and Design**
by L. Lamit and K. Kitto
See page 338
Problem 10.33

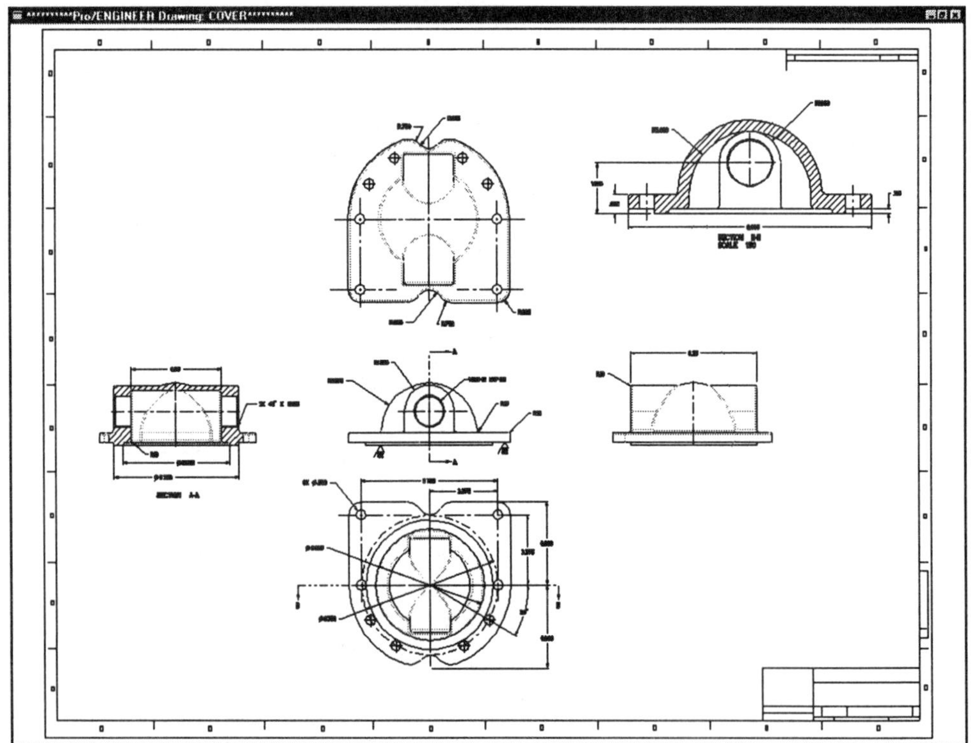

Figure A.36
Cover Drawing

Figure A.37
Cover *Casting* Dimensions

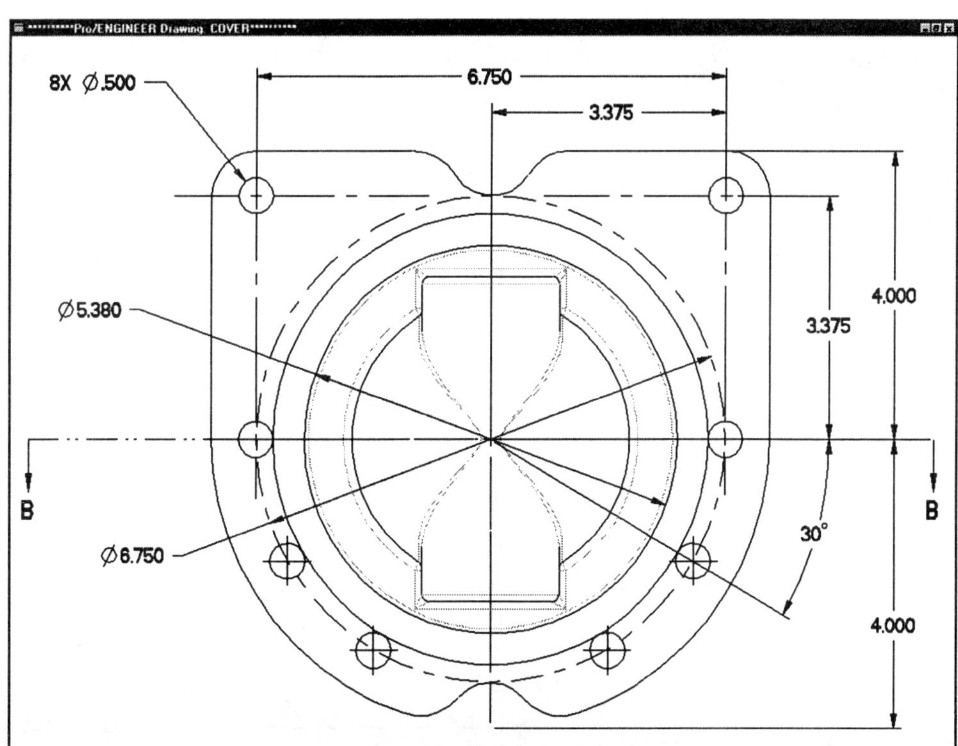

Figure A.38
Cover Drawing, Bottom View

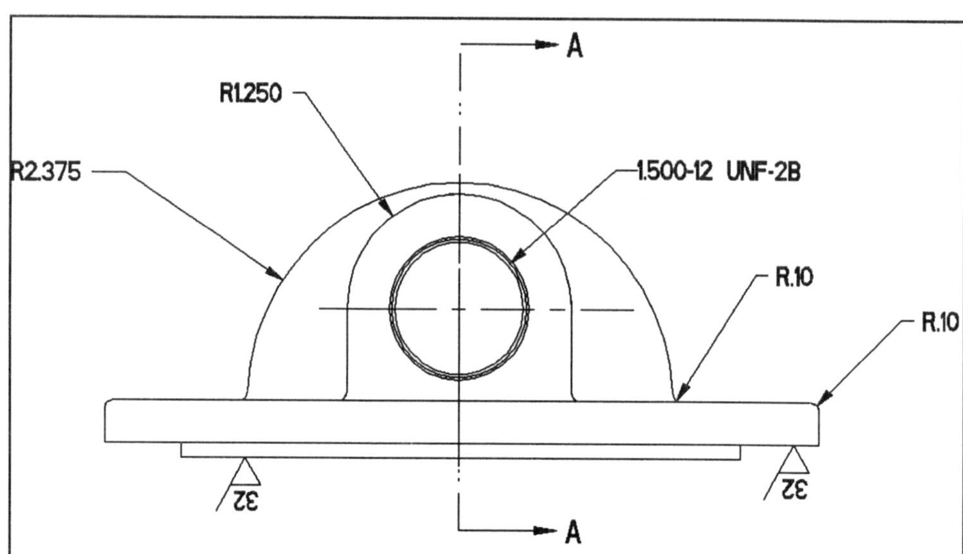

Figure A.39
Cover Drawing,
Front View

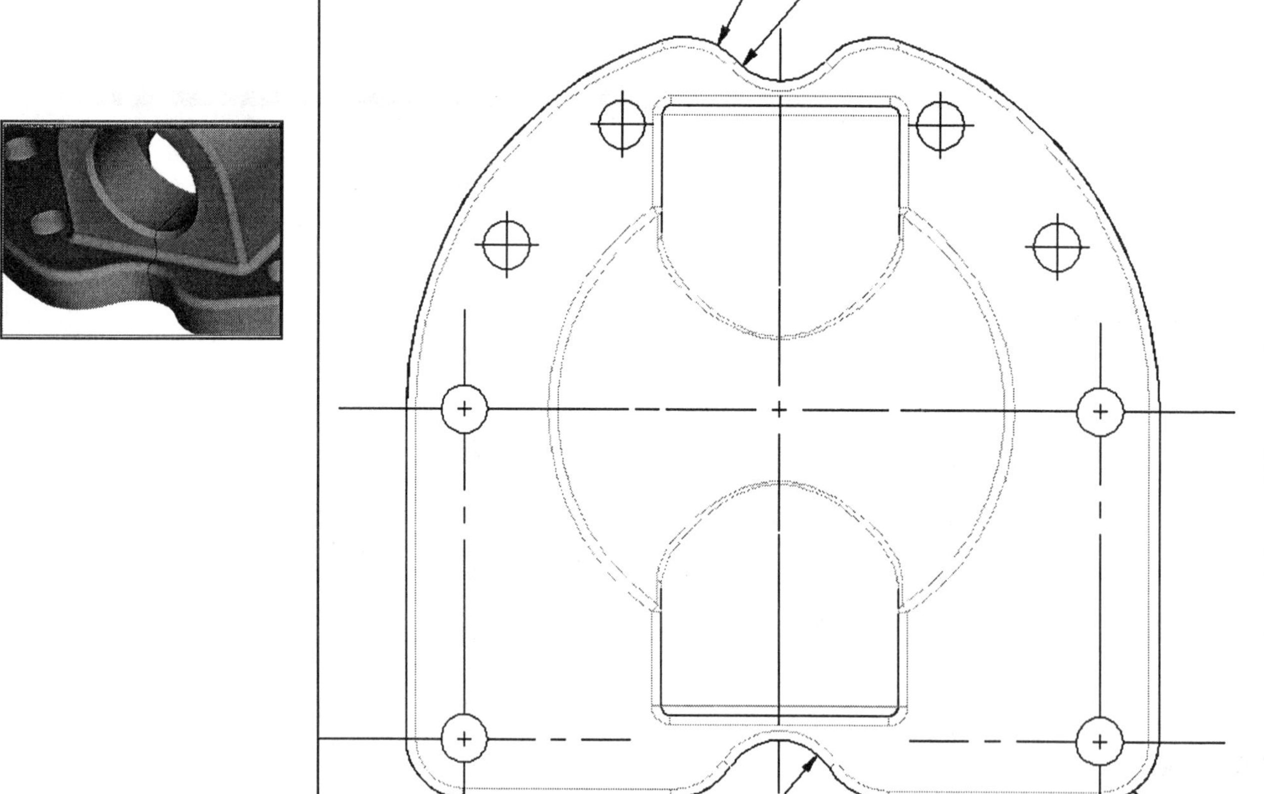

Figure A.40
Cover Drawing,
Top View

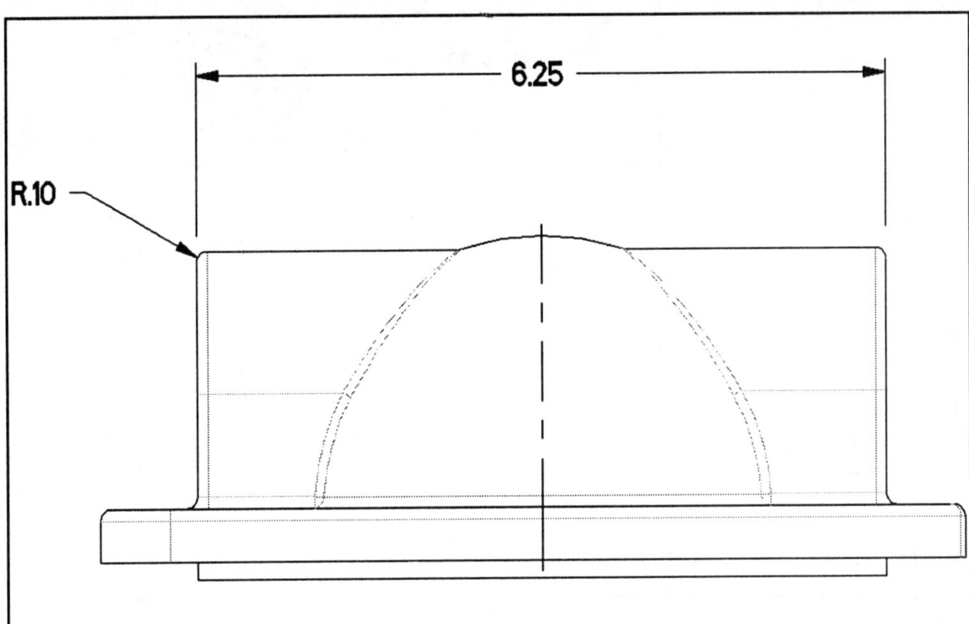

Figure A.41
Cover Drawing,
Right Side View

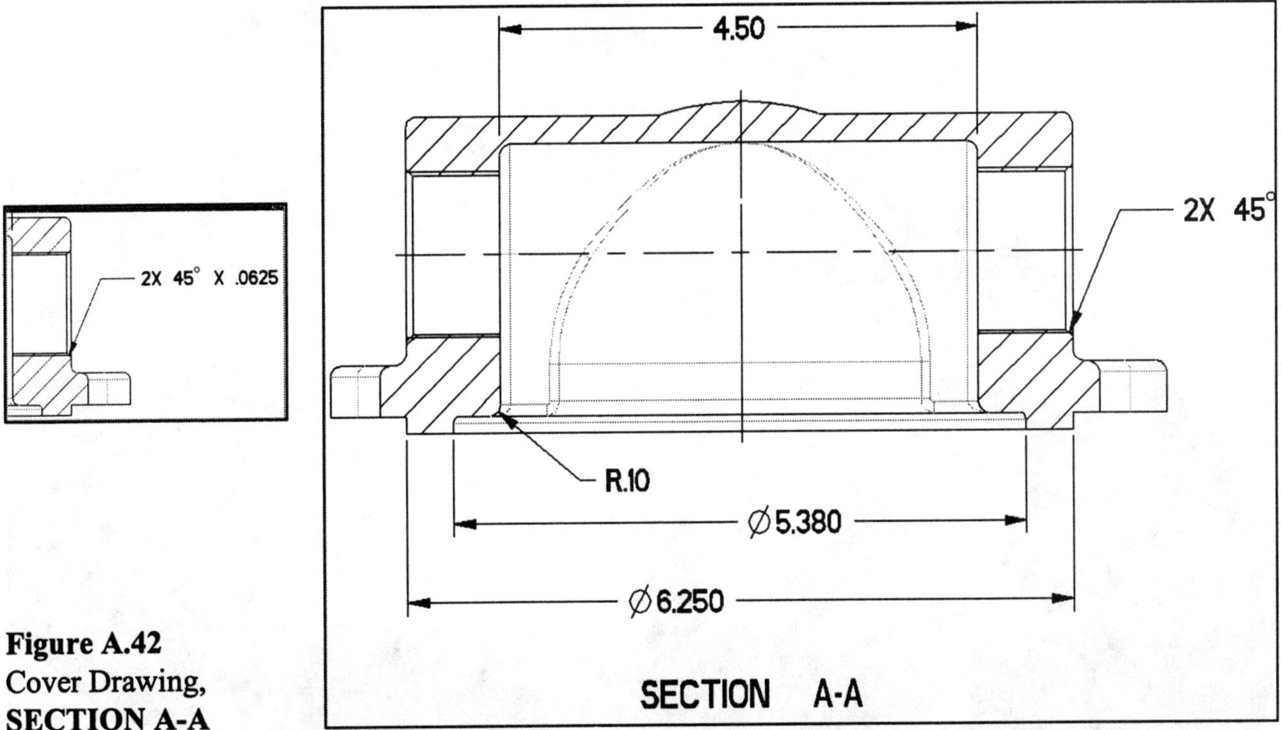

Figure A.42
Cover Drawing,
SECTION A-A

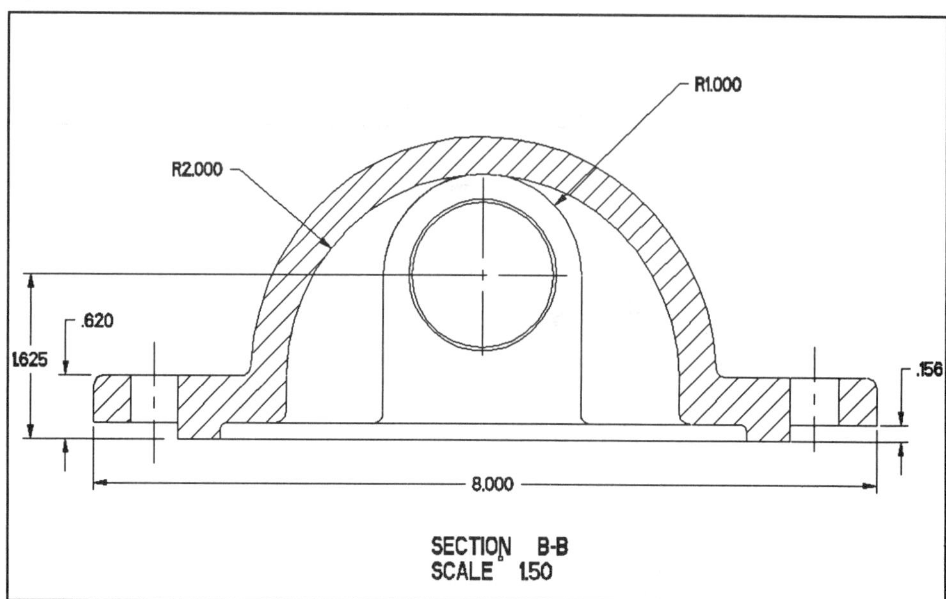

Figure A.43
Cover Drawing,
SECTION B-B

Figure A.44 Cover, Showing Sections

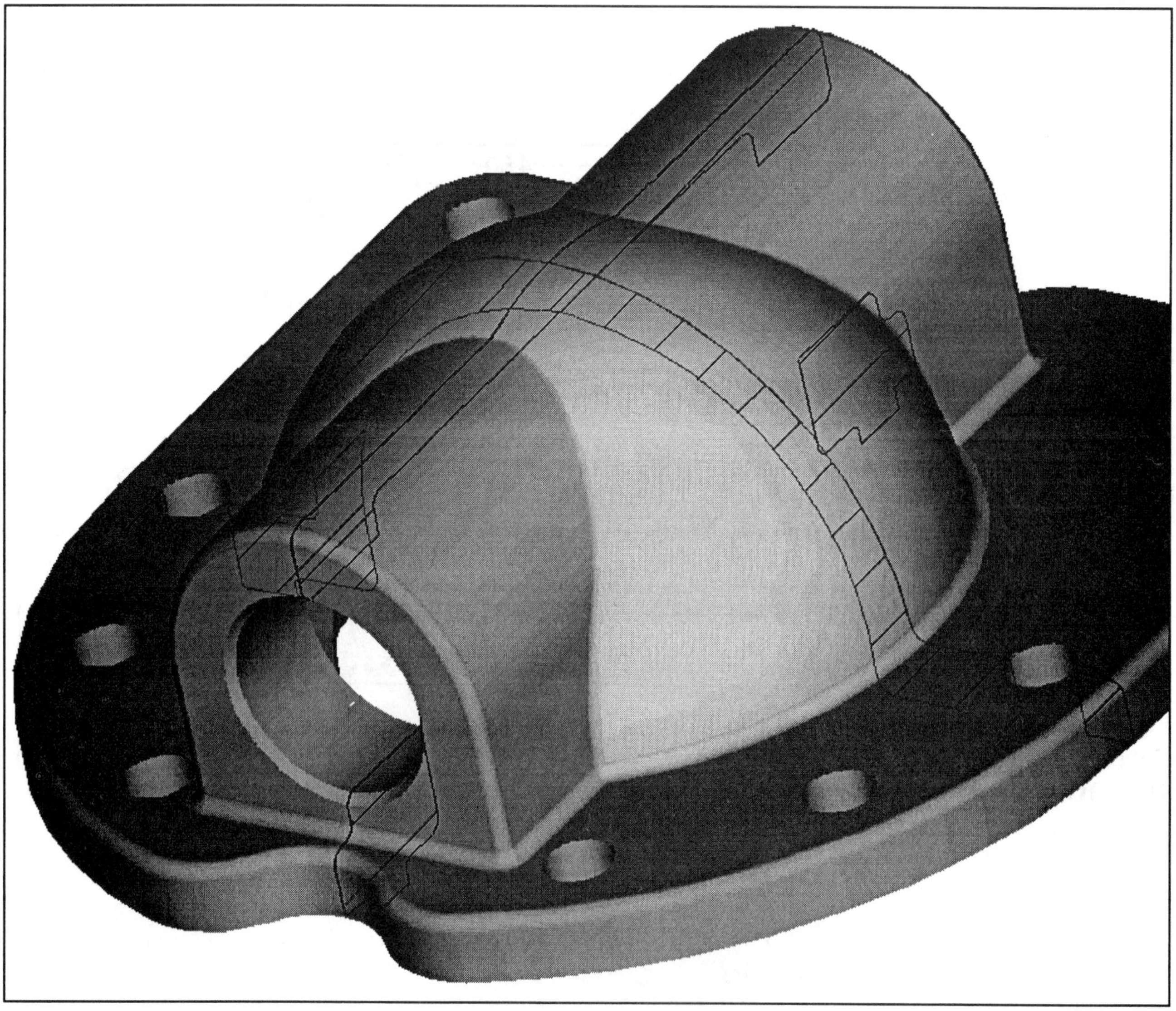

Appendix B

Config.pro Files and Mapkeys

mapkey
mapkey(continued)
mapkey
SKETCHER_DISP_CONSTRAINTS
RESERVED_MENU_SPACE
INDEPENDENT_MENUS
FONTS_SIZE
COMPANY_NAME
DEFAULT_DEC_PLACES
ACCURACY_LOWER_BOUND
HIGHLIGHT_NEW_DIMS
GRID_SNAP
MENUITEM_FONT
ACCURACY_LOWER_BOUND
MODEL_TREE_START
PARENTHESIZE_REF_DIM
PROMPT_ON_EXIT
SKETCHER_DEC_PLACES
SKETCHER_DISP_CONSTRAINTS
SPIN_WITH_SILHOUETTES
TANGENT_EDGE_DISPLAY
TOL_MODE
TREE_PLACEMENT
SPIN_CENTER_DISPLAY
SPIN_WITH_SILHOUETTES
DISPLAY
set_menu_width
mapkey
mapkey

Config.pro Files

The following is an example of various entries in a config.pro file.

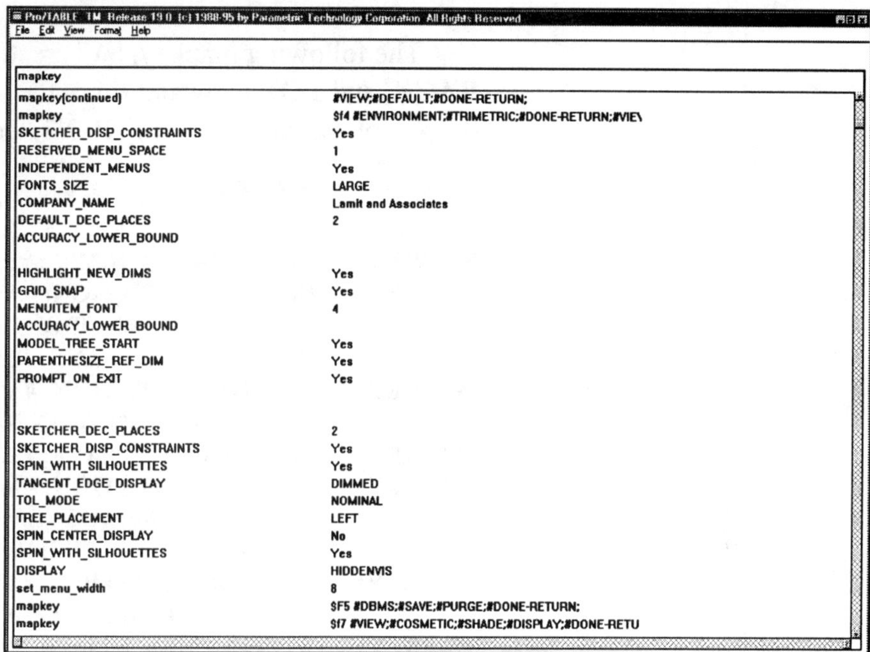

Pro/TABLE TM Release 19.0 (c) 1988-95 by Parametric Technology Corporation All Rights Reserved

File Edit View Format Help

mapkey	
mapkey(continued)	#VIEW;#DEFAULT;#DONE-RETURN;
mapkey	$f4 #ENVIRONMENT;#TRIMETRIC;#DONE-RETURN;#VIE\
SKETCHER_DISP_CONSTRAINTS	Yes
RESERVED_MENU_SPACE	1
INDEPENDENT_MENUS	Yes
FONTS_SIZE	LARGE
COMPANY_NAME	Lamit and Associates
DEFAULT_DEC_PLACES	2
ACCURACY_LOWER_BOUND	
HIGHLIGHT_NEW_DIMS	Yes
GRID_SNAP	Yes
MENUITEM_FONT	4
ACCURACY_LOWER_BOUND	
MODEL_TREE_START	Yes
PARENTHESIZE_REF_DIM	Yes
PROMPT_ON_EXIT	Yes
SKETCHER_DEC_PLACES	2
SKETCHER_DISP_CONSTRAINTS	Yes
SPIN_WITH_SILHOUETTES	Yes
TANGENT_EDGE_DISPLAY	DIMMED
TOL_MODE	NOMINAL
TREE_PLACEMENT	LEFT
SPIN_CENTER_DISPLAY	No
SPIN_WITH_SILHOUETTES	Yes
DISPLAY	HIDDENVIS
set_menu_width	8
mapkey	$F5 #DBMS;#SAVE;#PURGE;#DONE-RETURN;
mapkey	$f7 #VIEW;#COSMETIC;#SHADE;#DISPLAY;#DONE-RETU

Mapkeys

Mapkey entries can be added to a *config.pro* file at any time in the design process. These keyboard macros help streamline the design process. Feel free to experiment with your own macro creation. If you find yourself using a set of commands for a project over and over, it may be to your advantage to create a mapkey macro that will automate the process. The following listing provides a few suggestions for mapkey creation from Parametric Technology Corporation:

MAPKEY DR #DONE-RETURN
MAPKEY DT #ENVIRONMENT; #DISP DTMPLN; %DR
MAPKEY RP #VIEW; #REPAINT; %DR
MAPKEY SC #SKETCH; #LINE; #CENTERLINE
MAPKEY SA #SKETCH; #ALIGNMENT
MAPKEY SD #VIEW; #COSMETIC; #SHADE;
#DISPLAY; %DR
MAPKEY SN #VIEW; #ORIENTATION; #SPIN
MAPKEY SW #RELATIONS; #SWITCH DIM
MAPKEY TI #GEOM TOOLS; #INTERSECT
MAPKEY TT #GEOM TOOLS; #TRIM
MAPKEY TB #GEOM TOOLS; #TRIM; #BOUND
MAPKEY VT #VIEW; #NAMES; #RETRIEVE; #TOP; %DR
MAPKEY VR #VIEW; #NAMES; #RETRIEVE; #RIGHT; %DR
MAPKEY VF #VIEW; #NAMES; #RETRIEVE; #FRONT; %DR
MAPKEY VB #VIEW; #NAMES; #RETRIEVE; #BACK; %DR
MAPKEY VD #VIEW; #ORIENTATION; #DEFAULT; %DR

If the **MAPKEY** input is too long for one line, add a backslash

```
MAPKEY    VL  #VIEW; #NAMES; #RETRIEVE; #LEFT; %DR
MAPKEY    ZO  #VIEW; #PAN/ZOOM; #ZOOM OUT
MAPKEY    ZP  #VIEW; #PAN/ZOOM; #PAN
MAPKEY    ZR  #VIEW; #PAN/ZOOM; #RESET
```

The following mapkey, **SV**, creates a **FRONT**, **RIGHT**, **LEFT**, **BACK**, and **TOP** view and saves the views for later retrieval. Orient the part or assembly to the **FRONT** view before typing **SV**.

```
MAPKEY    SV   %S1V; %S2V; %S3V; %S4V:\
               %S5V; %S6V; %S7V; %S8V; %S9V
MAPKEY    S1V  #VIEW; #NAMES; #SAVE; FRONT
MAPKEY    S2V  #ORIENTATION; #ANGLES;\
               #VERT; -90 ; #DONE/ACCEPT
MAPKEY    S3V  #NAMES; #SAVE; RIGHT
MAPKEY    S4V  #ORIENTATION; #ANGLES;\
               #VERT; 180; #DONE/ACCEPT
MAPKEY    S5V  #NAMES; #SAVE; LEFT
MAPKEY    S6V  #ORIENTATION; #ANGLES;\
               #VERT; 90; #DONE/ACCEPT
MAPKEY    S7V  #NAMES; #SAVE; BACK
MAPKEY    S8V  #ORIENTATION; #ANGLES; #VERT;\
               180; #HORIZ; 90; #DONE/ACCEPT
MAPKEY    S9V  #NAMES; #SAVE; TOP
```

The following mapkey, **DD**, will create default datum planes and name them **TOP**, **FRONT**, and **SIDE**. **DD** should be typed immediately after typing in the name of a new part or new assembly. Use **DP** to create default datum planes (without renaming them).

```
MAPKEY    DP   #FEATURE; #CREATE; #DATUM; #PLANE;\
               #DEFAULT; #DONE
MAPKEY    DD   %DP; %D2D; %D3D; %D4D; #DONE
MAPKEY    D1D  #FEATURE; #CREATE; #DATUM; #PLANE
MAPKEY    D2D  #SET UP; #NAME;\
               #SEL BY MENU; #NAME; #DTM3; FRONT
MAPKEY    D3D  #SEL BY MENU; #NAME; #DTM2; TOP
MAPKEY    D4D  #SEL BY MENU; #NAME; #DTM1; SIDE
```

Appendix C

Glossary

Active Window The window to which the displayed menus apply. To change the active window, choose **Change Window** from the MAIN menu and click on the desired window.

Align Aligns a sketched entity with a part or assembly edge and is used as an assumption when you are solving the section.

Attributes (feature) The various characteristics of a feature. For example, a blind hole will have the following attributes: *position,* how the hole is placed (linearly, radially, coaxially); *section,* the cross section that defines the shape of the blind hole; *intersection,* simple or complex.

Attributes (user) The attributes that are added by a user to supply a description of the object beyond the geometric definition. For example, stock number, price, and cost per unit are all user attributes.

Base Feature The very first feature created for a part.

Base Member The very first component of an assembly.

Child An item, such as a view, part, or feature, that is dependent on another item for its existence. *See also* **Parent**

Collinear Two or more objects that occur along the same axis.

Configuration File A special text file that contains default settings for many Pro/ENGINEER functions. Default environment, units, files, directories, and so on are set when Pro/ENGINEER reads this file when it is started. A configuration file can reside in the startup directory to set the values for your working session only, or it can reside in the *load point directory* to set values for all users running a given version of Pro/ENGINEER. Also known as the *config.pro* file.

Dependent Parameter A parameter (dimension or user-defined) in a relation that is defined as a function of other parameters and values. A relation has one and only one dependent parameter, and it *always* appears on the left side of the equal sign.

Dimmed Option An option that appears grayed out when a menu is displayed. You cannot choose menu items when they are dimmed.

Entry Menu A menu that appears when you first enter a mode of Pro/ENGINEER. This menu allows you to create, retrieve, list, or import (in Part mode) an object.

Flip Arrow An arrow that appears on surfaces and edges of features to specify in which direction an operation should grow or project. If the arrow is pointing in the proper direction, choose **Okay**. If not, choose **Flip** and the arrow will be flipped **180°**. Then choose **Okay**.

Independent Parameter A parameter (dimension, value, relation) in a relation that is used to specify the value of the dependent parameter. Independent parameters *always* appear on the right side of the equal sign.

Information Window A Pro/ENGINEER window that displays information: object lists, mass properties, BOMs, and so on.

Load Directory The directory where Pro/ENGINEER is loaded.

Macro Keys The keyboard function keys or key sequences for which you predefine a menu option or sequence of menu options. This allows you to pick these keys to perform frequently used menu sequences.

Main View The first view added to a drawing.

Main Window The large primary window (main graphics window) created by Pro/ENGINEER.

Menu A list of options presented by Pro/ENGINEER that you select using the mouse or predefined macro keys.

Mode An environment in which Pro/ENGINEER allows you to perform closely related functions (e.g., Drawing, Sketcher).

Model A part or an assembly.

Object A Pro/ENGINEER object can be a drawing, a part, an assembly, a layout, and so on.

Parent An item that has other items dependent on it for their existence. For example, the base feature will have all other features dependent upon it. If a parent is deleted, all dependent items (**children**) will be deleted.

Part Type A part can be either standard or nonstandard (standard off-the-shelf component from the library). The part type affects the number of times a part can be assembled.

Pattern A method of feature creation in which one construction feature is used to create several related features.

Pick To select an item onscreen by clicking the left mouse button.

Placement Plane The plane on which a construction feature is located for placement.

Regenerate A menu option that recreates a model, incorporating any changes that have been made since the last time the model was stored or regenerated.

Startup Directory The directory from which you started Pro/ENGINEER.

System Editor The text set of editing functions available within your operating system.

Toggle A menu option that lets you switch between two settings, for example **Flip Arrow.**

Trail File A record of all the menu picks, screen picks, keyboard entries, and so on that occur during a Pro/ENGINEER session. The trail file can be run to recreate a work session.

Working Directory The working directory is the directory where you are using **Dbms** commands to store and retrieve your Pro/ENGINEER files. It may be in your startup directory or the directory you changed to using the **MISC** menu command **Change Dir**.

Work Session The period between your starting and stopping Pro/ENGINEER.

View A particular display of the model. View parameters include view orientation matrix, center, and scale.

Zoom Magnify or reduce your view of an object currently on the screen.

Appendix D

Design Intent Planning Sheets (DIPS)

This appendix provides a variety of sketching formats for planning your design. The ***design intent*** of a feature, a part, or an assembly (or even a drawing) should be established before any work is started with Pro/ENGINEER. Here we have provided a number of different formats in which to sketch and plan your feature, part, assembly, or drawing. Copy the sheets so that you have a number available for each lesson in the text.

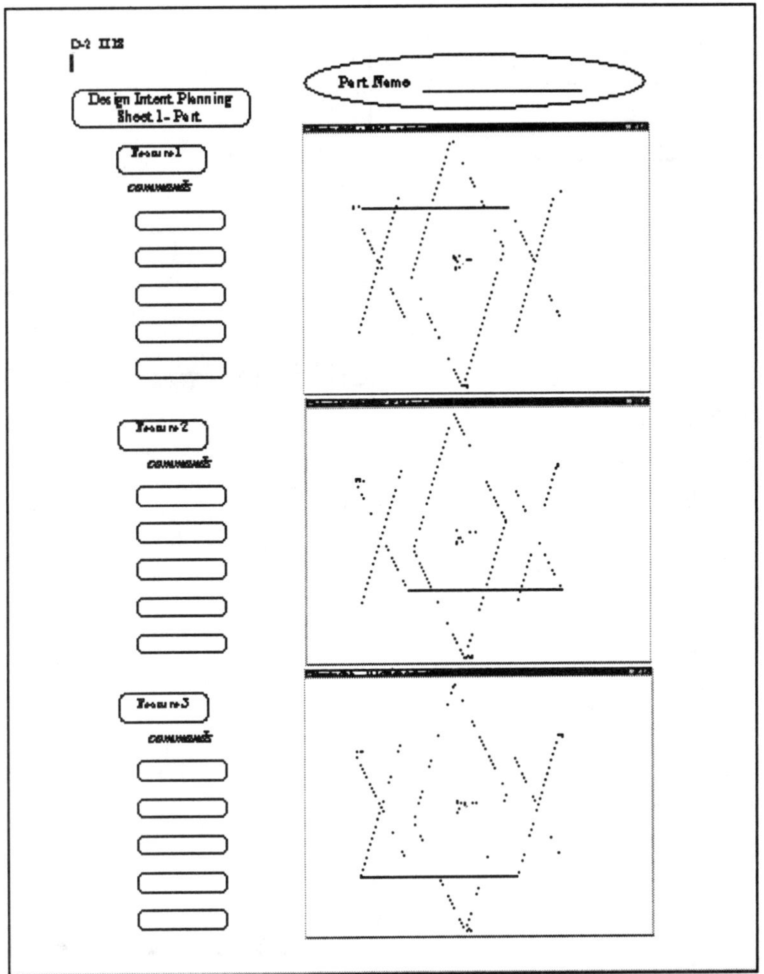

This appendix contains the following sheets:

DIPS1 Part Planning Sheet (trimetric)

DIPS2 Part Planning Sheet (pictorial with axes)

DIPS3 Assembly Planning Sheet (trimetric)

DIPS4 Assembly Planning Sheet (pictorial with axes)

DIPS5 Drawing Planning Sheet (no format)

DIPS6 Drawing Planning Sheet (format)

DIPS7 Feature/Sketch Planning Sheet (three sketches)

DIPS8 Feature/Sketch Planning Sheet (two sketches)

Part Name ______________________

Design Intent Planning Sheet 1 for Parts

Feature 1

commands

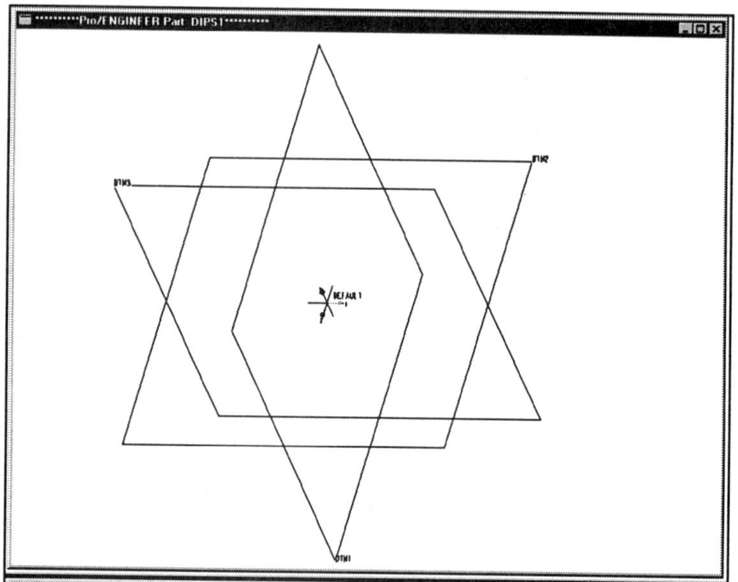

Feature 2

commands

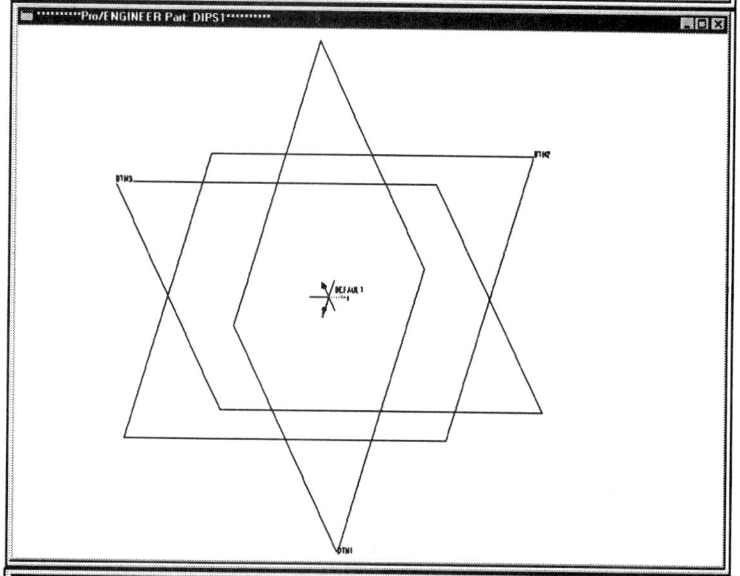

Feature 3

commands

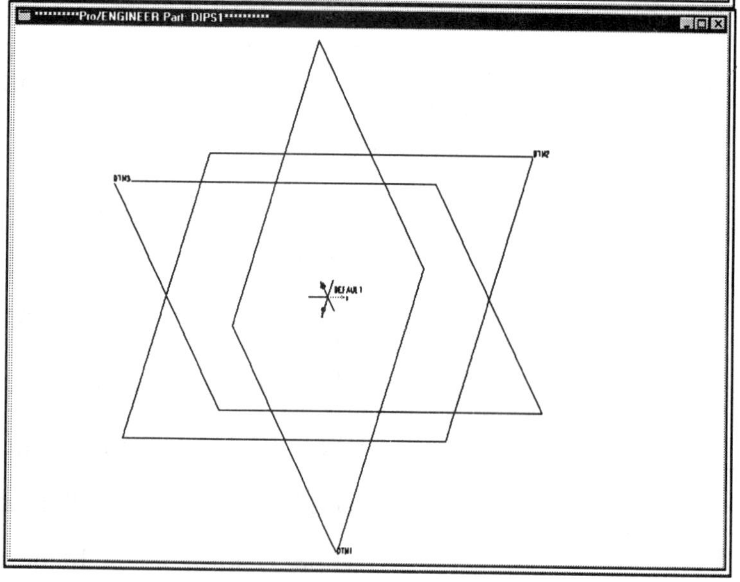

Part Name ______________

Design Intent Planning Sheet 2 for Parts

Feature 1

commands

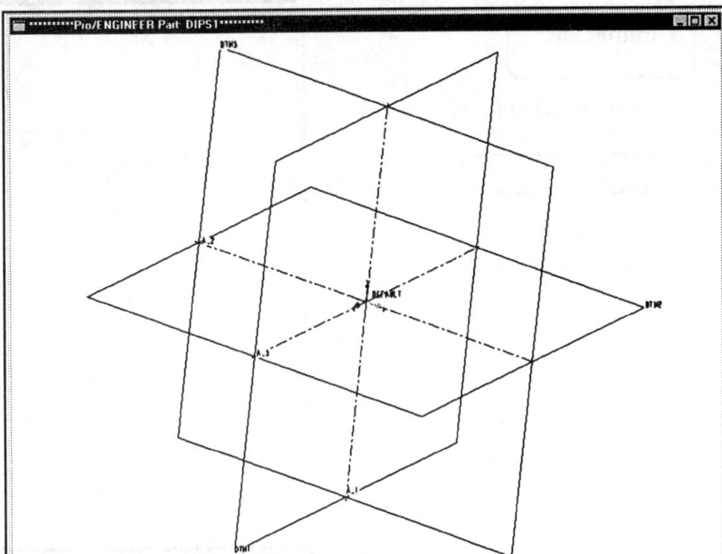

Feature 2

commands

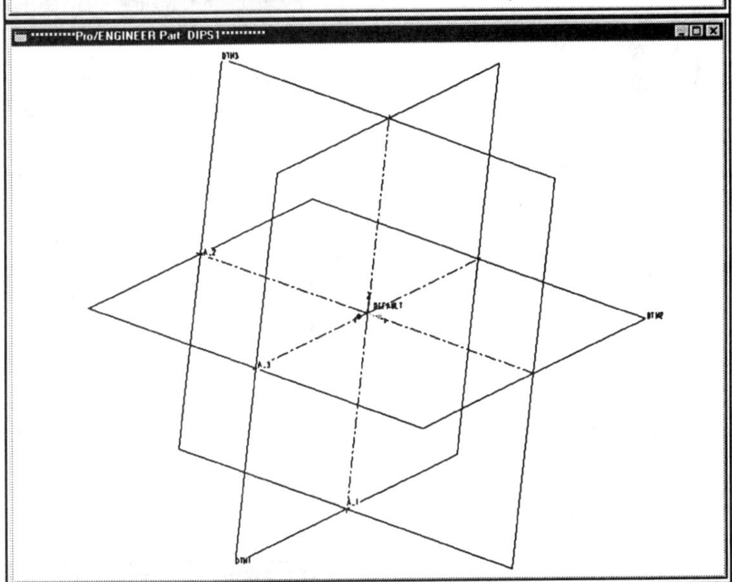

Feature 3

commands

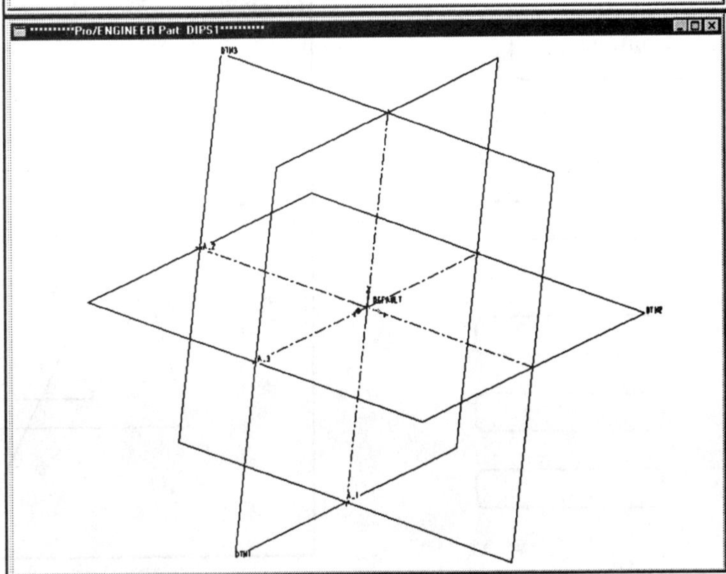

Design Intent Planning Sheet 3 for Assemblies

Assembly Name ______________________

Component

constraints

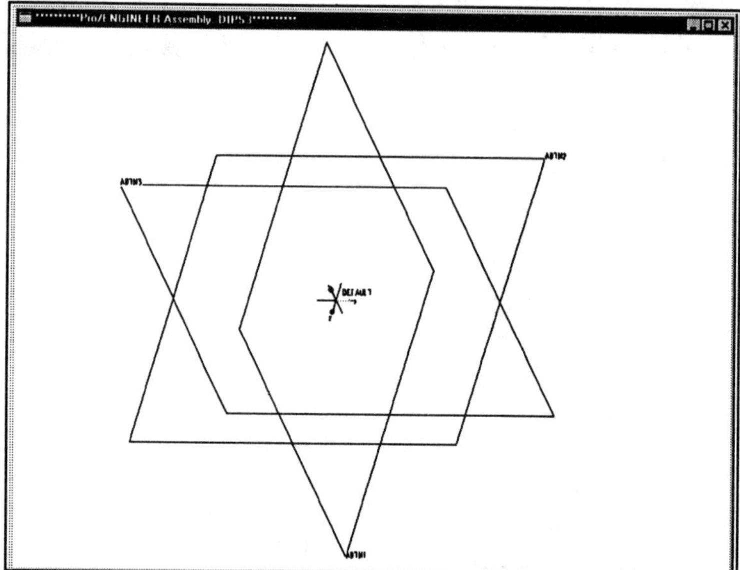

Component

constraints

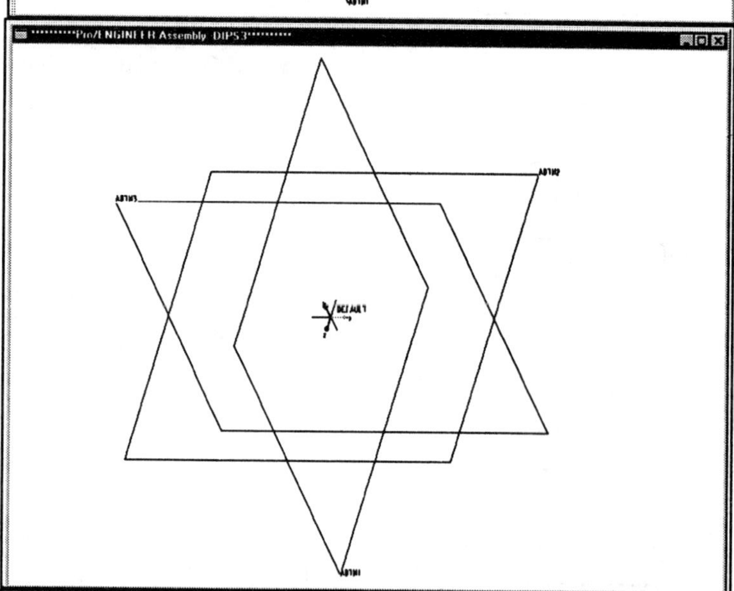

Component

constraints

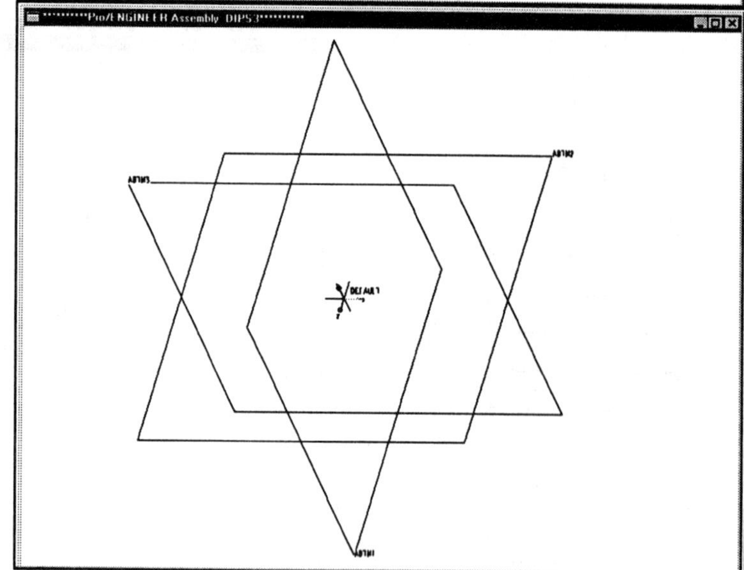

Design Intent Planning Sheet 4 for Assemblies

Assembly Name ______________________

Component

constraints

Component

constraints

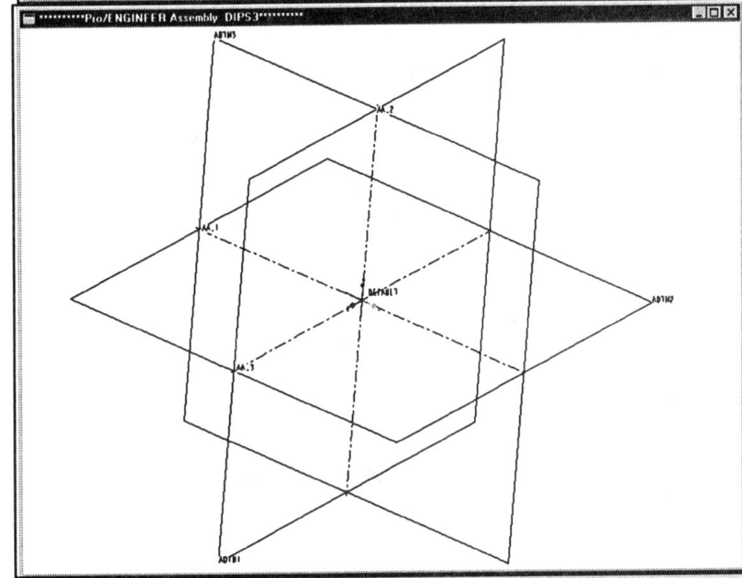

Component

constraints

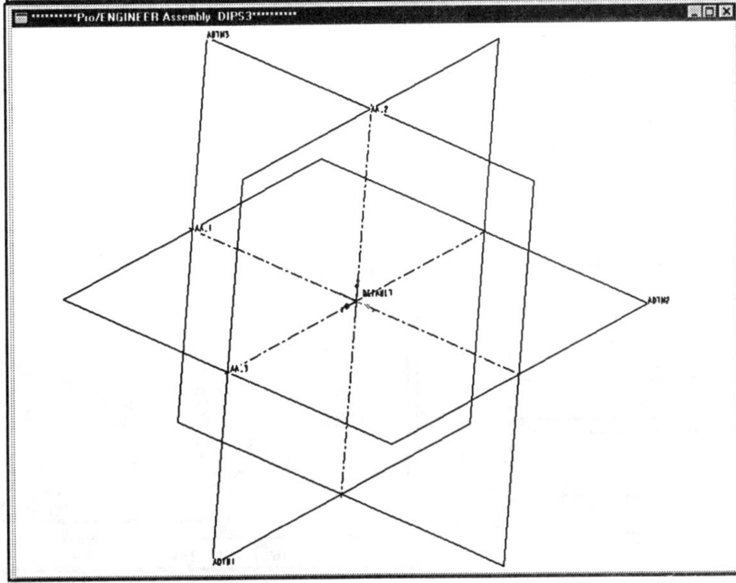

Design Intent Planning Sheet 5 for Drawings

Drawing Name ____________________

Part/Assembly

Specifications

Pro/ENGINEER Drawing: DIPS5

TYPE · DRAFT NAME · NONE SIZE · C

NOTES:

Pro/ENGINEER Drawing: DIPS5

TYPE · DRAFT NAME · NONE SIZE · C

Design Intent Planning Sheet 6 for Drawings

Drawing Name ______________

Part/Assembly

Specifications

Pro/ENGINEER Drawing: DIPS5

SCALE : 1.000 TYPE : ASSEM NAME : DIPS3 SIZE : C

NOTES:

Pro/ENGINEER Drawing: DIPS5

SCALE : 1.000 TYPE : ASSEM NAME : DIPS3 SIZE : C

Design Intent Planning Sheet 7 for Features/Sketches

Part Name ______________________

Sketch/Section 1

"*********Pro/ENGINEER Part: DIPS7*********"

Sketch/Section 2

"*********Pro/ENGINEER Part: DIPS7*********"

Sketch/Section 3

"*********Pro/ENGINEER Part: DIPS7*********"

Part Name ______________________

Design Intent Planning Sheet 8 for Features/Sketches

Sketch/Section 1

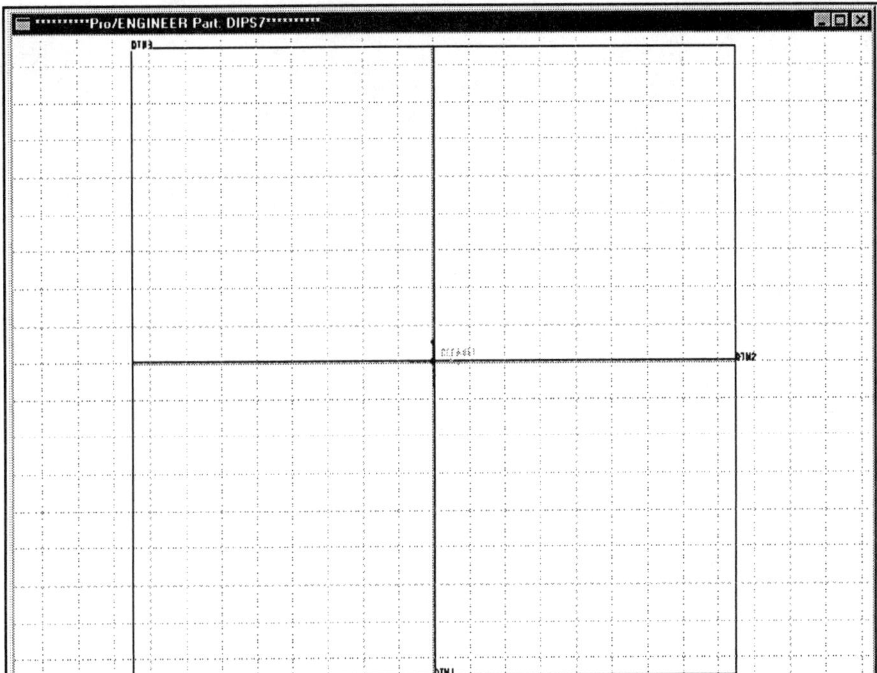

NOTES:

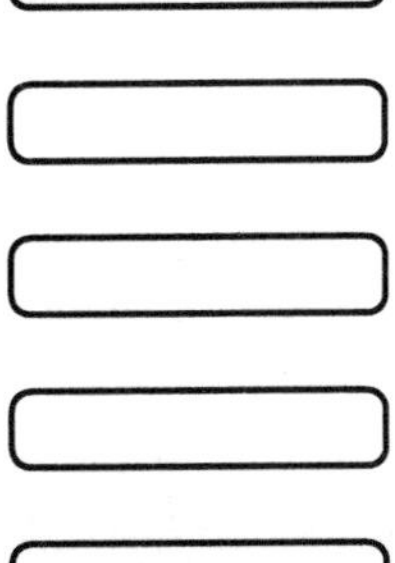

Sketch/Section 2

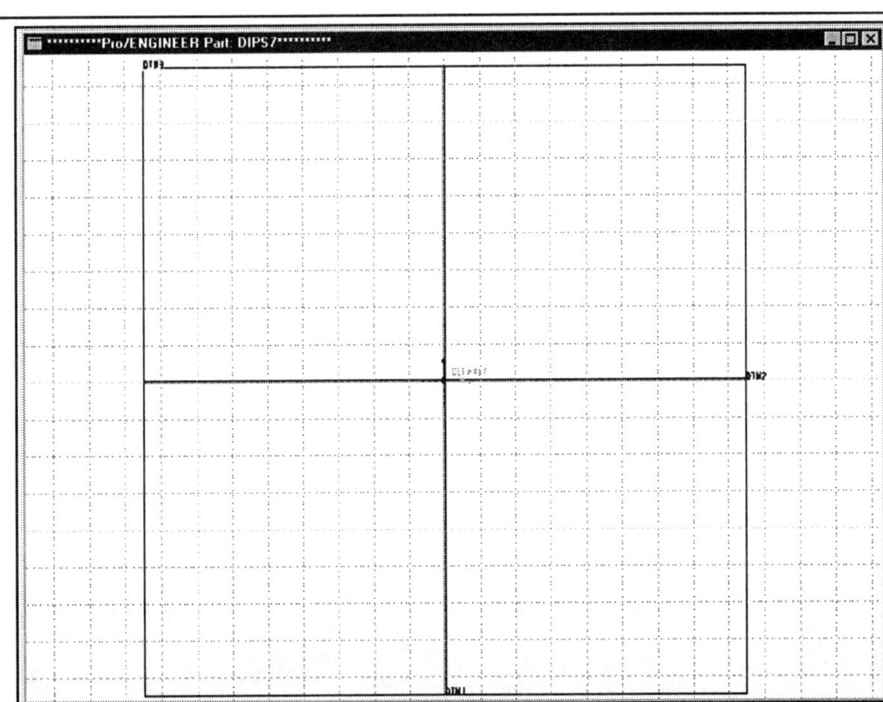

Index

E

F

S

T